W9-BKI-875

McGraw-Hill
Encyclopedia of
Astronomy

Sybil P. Parker
EDITOR IN CHIEF

McGraw-Hill Book Company

New York St. Louis San Francisco

Auckland Bogotá Guatemala Hamburg
Johannesburg Lisbon London Madrid Mexico Montreal
New Delhi Panama Paris San Juan São Paulo
Singapore Sydney Tokyo Toronto

Frontispiece:
Andromeda Nebula, M31 or NGC 224, the spiral galaxy which is nearest the Milky Way system. Photograph made with 48-in. (1.2-m) Schmidt telescope on Mount Palomar. (*Hale Observatories*)

1 2 3 4 5 6 7 8 9 0 KPKP 8 9 8 7 6 5 4 3

ISBN 0-07-045251-2

Library of Congress Cataloging in Publication data

McGraw-Hill encyclopedia of science & technology.
Selections.
McGraw-Hill encyclopedia of astronomy.

"All of the material in this volume has been published previously in the McGraw-Hill encyclopedia of science & technology, fifth edition"—T.p. verso.
Includes bibliographies and index.
1. Astronomy—Dictionaries. I. Parker, Sybil P.
II. Title.
QB14.M3725 1983 523′.003′21 82-21683
ISBN 0-07-045251-2

Editorial Staff

Sybil P. Parker, Editor in Chief

Jonathan Weil, Editor
Betty Richman, Editor

Edward J. Fox, Art director

Ann D. Bonardi, Art production supervisor
Cynthia M. Kelly, Art/traffic

Joe Faulk, Editing manager

Ruth L. Williams, Editing supervisor
Dorick Byard, Editing supervisor

Patricia W. Albers, Senior editing assistant
Judith Alberts, Editing assistant
Barbara Begg, Editing assistant

Consulting Editor

Prof. George O. Abell
Department of Astronomy
University of California, Los Angeles

Preface

Astronomy is the most ancient science. The constellations were recognized in Mesopotamia as early as 3000 B.C.; the Egyptians were using calendars based on solar events by 2000 B.C.; and Ptolemy provided a geometrical description of the motions of celestial bodies in the 2d century. By the 15th century we had entered the age of modern astronomy, and the 17th century witnessed major contributions from Sir Isaac Newton (interpreted the motions of the planets); Galileo (invented the telescope); C. Huygens (discovered a satellite of Saturn); and G. D. Cassini (observed Saturn's rings). While there were many noteworthy contributions to dynamical and observational astronomy during the 18th and 19th centuries, it was not until well into the 20th century—1957—that we entered the Space Age with the launch of *Sputnik 1*.

The science of astronomy is concerned with the study of radiation received from all parts of the universe. Such observation was limited to the visual portion of the electromagnetic spectrum during the greater part of human history. This limitation was eliminated by development of sophisticated instruments with sensitivities extending into other portions of the spectrum. The development of the photographic plate in the last half of the 19th century made possible the study of astrophysics, one of the most active branches of present-day astronomy. Over the past few decades, data have been accumulated and discoveries made at an unprecedented rate, and there are few signs of slowing down in the near future.

Today, astronomers are exploring the universe with satellites, space probes, rockets, and high-altitude balloons, as well as optical, radio, infrared, ultraviolet, x-ray, gamma-ray, cosmic-ray, and neutrino telescopes. The discovery of quasars, pulsars, complex interstellar molecules, cosmic microwave radiation, radio galaxies, supernovae, and peculiar objects, such as SS443, opened new vistas in observational astronomy. Similarly, theoretical astronomy and astrophysics have undergone revolutionary changes as scientists now focus on problems of cosmology and cosmogony—the expanding universe, stellar structure and evolution, the composition and age of the solar system, the interpretation of black holes—trying to answer the most fundamental questions about the nature of the universe.

The *Encyclopedia of Astronomy* provides authoritative and up-to-date coverage of this dynamic field of science. This is a reference work *of*, not *about*, astronomy. As such, the articles are arranged alphabetically, rather than topically, and have been prepared by more than one hundred of the world's leading scientists, each writing about his or her area of specialization. The Encyclopedia includes entries on the theoretical, observational, and experimental aspects of astronomy as well as on the instruments used to obtain the data.

The 230 articles were taken from the *McGraw-Hill Encyclopedia of Science and Technology* (5th ed., 1982). In addition to some 400 photographs charts, graphs, and drawings, there are 13 pages in full color. Most articles include a bibliography, and cross references are used extensively to guide the reader to related topics. There is also a detailed index for rapid access to the information.

Professor George O. Abell, who served as Consulting Editor in astronomy on the parent project mentioned above, has contributed invaluable guidance in creating this specialized volume. We thank him for his interest, and thank too the many Contributors whose expertise is reflected in the articles.

This Encyclopedia fills a need for a truly comprehensive and scholarly treatment of astronomy. It will be an indispensable tool for scientists, students, librarians, hobbyists, and all others who need accurate, detailed information in a readily accessible format.

Sybil P. Parker
EDITOR IN CHIEF

McGraw-Hill
Encyclopedia of
Astronomy

AZ

Aberration

The apparent change in direction of a source of light caused by an observer's component of motion perpendicular to the impinging rays.

To visualize the effect, first imagine a stationary telescope (illustration *a*) aimed at a luminous source such as a star, with photons traveling concentrically down the tube to an image at the center of the focal plane. Next give the telescope a component of motion perpendicular to the incoming rays (illustration *b*). Photons passing the objective require a finite time to travel the length of the tube. During this time the telescope has moved a short distance, causing the photons to reach a spot on the focal plane displaced from the former image position. To return the image to the center, the telescope must be tilted in the direction of motion by an amount sufficient to ensure that the photons once again come concentrically down the tube in its frame of reference (illustration *c*). The necessary tilt angle α is given by $\tan \alpha = vc$, where v is the component of velocity perpendicular to the incoming light and c is the velocity of light. (An analogy illustrating aberration is the experience that, in order for the feet to remain dry while walking through vertically falling rain, it is necessary to tilt an umbrella substantially forward.)

Aberration of light was discovered by the English astronomer James Bradley in 1725 while searching unsuccessfully for parallax of nearby stars, using the Earth's orbital diameter as baseline. He found instead the necessity to compensate continuously for the Earth's velocity in its elliptical orbit, an effect about a hundred times greater than the parallax of typical nearby stars. This discovery provided the first direct physical confirmation of the Copernican theory.

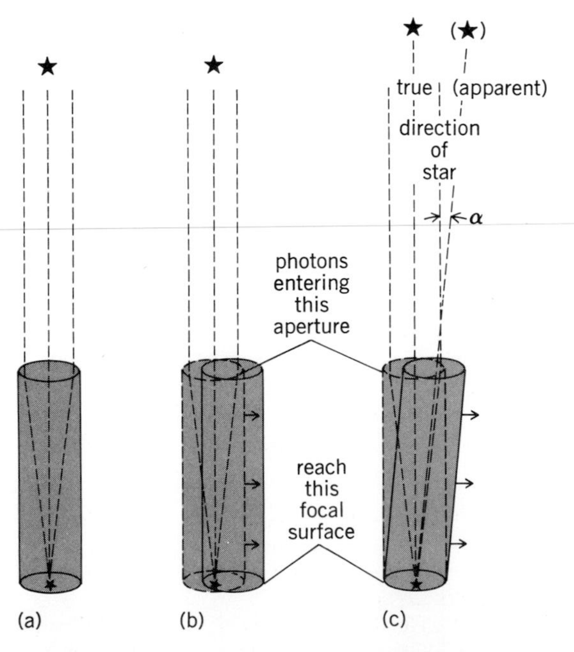

Demonstration of aberration. *(a)* Fixed telescope; photons form image at center of focal plane. *(b)* Moving telescope; image is displaced from center. *(c)* Tilted moving telescope is required to restore image to center.

Annual aberration causes stars to appear to describe ellipses whose major axes are the same length, the semimajor axis (usually expressed in arc-seconds) being the constant of aberration $\alpha = 20''.496$. The minor axis of the ellipse depends on the star's ecliptic latitude β, and is given by $2\alpha \sin \beta$. The small diurnal aberration arising from the Earth's axial rotation depends on geographic latitude ϕ, and is given by $0''.31 \cos \phi$.

A second important application of aberration has been its clear-cut demonstration that, as is axiomatic to special relativity, light reaching the Earth has a velocity unaffected by the relative motion of the source toward or away from the Earth. This is shown by the fact that the aberration effect is the same for all celestial objects, including some quasars with apparent recessional velocities approaching the speed of light *See* EARTH ROTATION AND ORBITAL MOTION; PARALLAX.

[HARLAN J. SMITH]

Active galactic nuclei

Cores of galaxies in which a substantial amount of energy is produced by explosive or violent events. Galactic activity involves a rich diversity of phenomena, including the production of unusual quantities of radiation at almost all wavelengths in the electromagnetic spectrum, the emission of jets of high-speed electrons, and the variability of the light from galactic nuclei on very short time scales. Astrophysicists believe that supermassive objects, possibly giant black holes, located at the cores of active galaxies are responsible for all these phenomena.

Types of active galaxies. It is possible to recognize an active galaxy in a variety of ways. Radio galaxies have a total radio power which exceeds the stellar power of a large spiral galaxy (10^{35} watts or 10^{42} ergs/s). In visible light, active galaxies show up as galaxies with brilliant, starlike nuclei (N galaxies) or as galaxies with peculiar spectra (Seyfert galaxies). The most powerful active galaxies are the BL Lac objects. At first glance, the BL Lacs look like stars, but their spectra are usually featureless, unlike stellar spectra. Deep photographs demonstrate that these objects are surrounded by a faint glow, which turns out to be the image of a galaxy much fainter than the overpowering nucleus. *See* RADIO SOURCES.

Still more energetic than the BL Lac objects are the quasars—hyperactive relatives of the active galaxies. In photographs, quasars, like BL Lacs, resemble stars. Unlike BL Lacs, their spectra show emission lines, concentrations of photons at particular wavelengths, which come from low-density gases illuminated by high-energy radiation. Quasars are the most distant discrete objects in the observable universe. The similarity between many of the phenomena seen in active galactic nuclei and those seen in quasars suggests a close relation between these two types of objects. *See* QUASARS.

High-speed electrons. When high-speed electrons, moving with nearly the speed of light, encounter magnetic fields they spiral around the field lines and emit a form of radiation known as synchrotron radiation. Depending on the energy of the electrons and the strength of the magnetic field, synchrotron radiation can appear at virtually any wavelength in the ultraviolet-to-radio range. Astrophysicists can recognize synchrotron radiation because its intensity increases with decreasing photon energy, even in the radio part of the spectrum, unlike the radio emission from hot gases.

In addition, synchrotron radiation is polarized, whereas most radiation emitted from cosmic objects is unpolarized. Because the spiraling motion of the electrons around the magnetic field is directional, the electric field of the radiation that is emitted is also directional, producing polarized radiation. The radio and optical radiation from active galactic nuclei is polarized by as much as 20–30%, indicating that the magnetic field lines in these synchrotron sources are well aligned rather than tangled.

Jets. Radio astronomers use pairs of telescopes, operated as interferometers, to map the structure of radio sources, including active galaxies. The clouds of radio-emitting gas often line up to form a jet of magnetized plasma. Some active galaxies show very tiny jets, only a light-year (1 light-year = 9.46×10^{12} km) or so long, which are aligned with enormous jets that stretch over hundreds of thousands or millions of light-years (Fig. 1). *See* RADIO TELESCOPE.

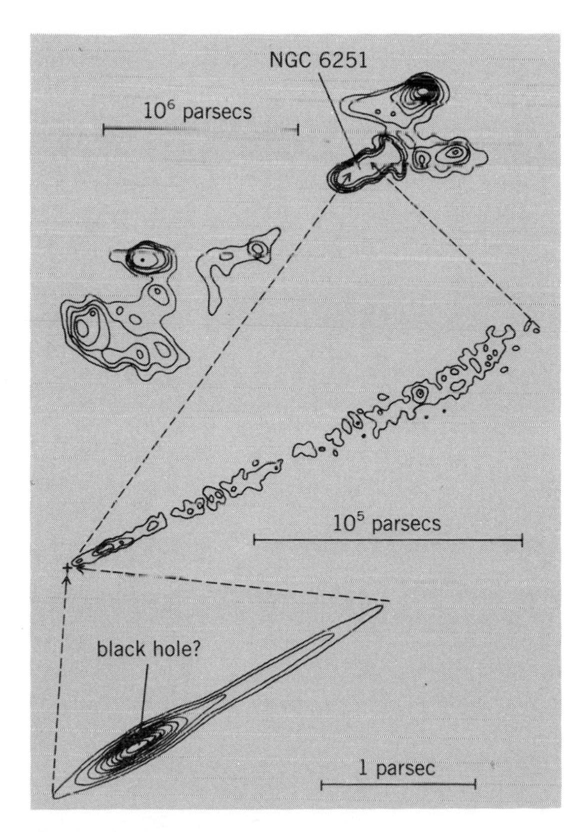

Fig. 1. Contour map showing the distribution of radio emission in the radio galaxy NGC 6251, sometimes called the Blowtorch Galaxy. The upper two contour maps were produced by an interferometer at Cambridge, England. The lower map is produced from Very Long Baseline observations. 1 parsec = 3.1×10^{13} km. *(After A. C. S. Readhead, M. H. Cohen, and R. D. Blandford, A jet in the nucleus of NGC 6251, Nature, 272:131–134, 1978)*

The rest of the synchrotron radiation from active galaxies comes from two types of roughly spherical clouds. (1) Huge clouds of plasma, millions of light-years across (the largest single objects in the universe), are often symmetrically placed on either side of an active radio galaxy. (2) Tiny clouds, in some cases no larger than the solar system (a few thousandths of a light-year across), can be found in the nearest active galaxies, galaxies which are quite feeble compared to the powerful BL Lacs. It is puzzling to see the same cosmic phenomenon occurring in active galaxies that vary in length by a factor of 10^9—especially when the alignment between small jets and large jets indicates a connection between the two.

Measurements of the positions of the smallest radio-emitting clouds made over periods of years by pairs of radio telescopes at opposite ends of the Earth have produced a startling discovery: in many objects, these small clouds are flying apart at speeds that seem to exceed the speed of light. At first glance, this discovery might seem to invalidate the special theory of relativity, or suggest that the galaxies are closer than had been believed. The true explanation is rather subtle. If a plasma jet is emitted in a direction almost toward an observer located on the Earth, the jet nearly catches up with the photons that it emits, and an optical illusion is produced that makes the end of the jet seem to move faster than light. The moving jet must have very high velocities, close to the speed of light, for this interpretation to work. *See* RELATIVITY.

The variability of some of the most active galactic nuclei also indicates their unusual properties. The brightness of some objects has been observed to change by a factor of two or so in less than a day. Because light has a finite speed and that emitted from the nearest and farthest extent of the source cannot reach the Earth simultaneously, such a source must be less than a light-day (2.6×10^{10} km) across. Direct measurements of such small sizes have been made in the nearer objects.

Infrared radiation. When a galactic core produces tremendous quantities of synchrotron radiation, this radiation interacts with the gas and dust surrounding the core and causes it to emit at other wavelengths, most particularly in the infrared range. In many objects, most of the total power is emitted as infrared radiation. In one extreme case, 10^{40} watts, 10^4 times the power emitted by a typical large spiral galaxy pours out in this part of the electromagnetic spectrum.

The infrared radiation could simply be synchrotron radiation. Because the infrared emission far exceeds that produced by a smooth connection of the optical and radio spectra, a separately produced jet of energetic electrons would be necessary. But another model has the infrared radiation coming from dust that is heated by the synchrotron radiation at other wavelengths. Both of these sources seem to operate in at least some cases. *See* INFRARED ASTRONOMY.

X-rays. The launch of the *Einstein* satellite in the late 1970s allowed astronomers to measure x-radiation from many active galaxies. It now seems that many, if not most, active galaxies are x-ray sources as well as radio and optical sources.

The x-rays are probably emitted when high-energy electrons collide with low-energy photons and give the photons a tremendous boost in energy through the Compton effect. *See* SATELLITE ASTRONOMY; X-RAY ASTRONOMY.

Emission lines. Some of the synchrotron and Compton radiation is emitted at such high energies that it can easily ionize atoms that are common in the interstellar gas, such as hydrogen, helium, carbon, and oxygen. These ionized atoms then recombine with electrons, forming neutral atoms again. If the newly formed atoms are in excited states, they then tumble down energy ladders, emitting photons with energies equal to the differences between energy levels in those atoms. The result of this ionization and recombination process is the conversion of high-energy radiation from the core of the active galaxy to optical and ultraviolet photons emitted at specific wavelengths. An emission-line spectrum is thus produced, as it is in any low-density gas that is illuminated by high-energy radiation.

The Seyfert galaxies were first recognized as being active because of their emission-line spectra. The breadth of these emission lines indicates that the filaments of gas that produce them are moving fast. Filaments moving away from the observer emit red-shifted photons, and filaments moving toward the observer emit blue-shifted photons, creating a broad line.

Galactic center. The Milky Way Galaxy is not, strictly speaking, an active galaxy, but there are many phenomena at the galactic core that are similar to those occurring in active galaxies. At the very center is a tiny core of radio-emitting plasma, a core which would fit inside the orbit of Saturn. The inner 3 light-years of the galaxy contain a spiral pattern of radio emission. Also near the center is a very small source of 511-keV gamma radiation produced when electrons and positrons collide and annihilate. Infrared observations show an emission line from ionized neon coming from the central region, but surprisingly this emission line indicates that the core has an axis of rotation that is perpendicular to the axis of the Galaxy as a whole. A larger radio source produces 10^{30} watts of radio power, which is large compared to the solar luminosity of 4×10^{26} watts but small compared to active galaxies.

Analysis of the motion of objects in the galactic core indicates that mass is heavily concentrated in the central few light-years, with 10^6 solar masses within 3 light-years of the center. In one other active galaxy, Messier 87, there is also some equivocal evidence for a very dense core. Here, the nucleus emits a concentration of light, different in color from the light of the surrounding stars, indicating rapid motion. But these data can be interpreted in different ways. In addition, Messier 87 produces a jet of optical radiation and x-radiation as well as a jet in the radio part of the spectrum. *See* GALAXY.

Interpretation. Astrophysicists struggle to provide a reasonably complete, coherent, and logical explanation of all these phenomena. Most, if not all, aspects of activity in galactic nuclei, ranging from the low power of the Milky Way to the very high luminosity of the BL Lac objects, seem to stem from the same underlying cause. Some central engine squirts out jets of high-speed electrons. The different phenomena observed in active galaxies then arise from the different ways of feeding the engine that powers the jet and from the varying ways in which the jet interacts with the environment. This engine can beam the jet in the same direction for millions of years, and can create huge clouds of radio-emitting plasma as well as tiny radio sources.

Study of a number of objects in the Milky Way Galaxy, far smaller than active galactic nuclei, can help provide clues to the nature of the engine. The Crab Nebula, the remnant of the supernova that exploded in A.D. 1054, has a synchrotron source which is surrounded by filaments producing emis-

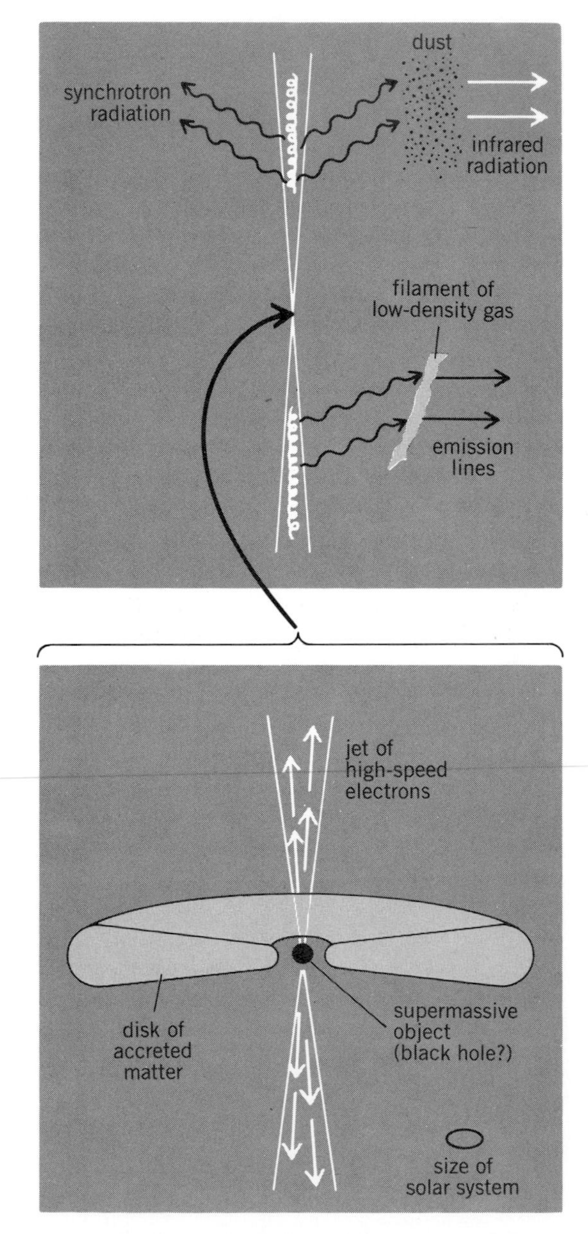

Fig. 2. Schematic illustration of an active galactic nucleus.

sion lines. A more significant analog is the object SS 433, a double-star system containing a neutron star or black hole which ejects twin jets that travel at one-quarter of the speed of light. The energy for SS 433 is believed to be provided by material falling into the gravitational field of a compact object. *See* CRAB NEBULA; SS 433.

The working model favored by most astrophysicists is that a very massive black hole, of 10^6 to 10^8 solar masses (1 solar mass $= 2 \times 10^{30}$ kg), is the central engine in a typical active galactic nucleus. Gas comes from somewhere in the environment, falls toward the black hole, whirls around it, and forms a nozzle near the poles of rotation of the accretion disk of swirling, compressed material. Jets of high-speed electrons squirt out through the nozzle at nearly the speed of light. These jets contain the electrons that produce the synchrotron radiation, and when the synchrotron radiation interacts with the surrounding materials, emission lines and large quantities of infrared radiation can be produced if low-density gas and dust are there (Fig. 2). Millions of light-years from the galactic nucleus, this jet is finally stopped by the surrounding intergalactic medium and produces the huge radio-emitting lobes of gas found on opposite sides of radio galaxies. *See* BLACK HOLE.

Identifying these ideas as a "working model" means that there is some favorable evidence, but that the concept is primarily regarded as a good subject for further observational tests, rather than a model which is accepted as definitive. The evidence does show that there is some kind of super massive object in the cores of active galaxies, but it might well be a supermassive star, sometimes called a magnetoid or spinar, rather than a black hole. *See* SUPERMASSIVE STAR.

[HARRY L. SHIPMAN]

Bibliography: R. D. Blandford, M. C. Begelman, and M. J. Rees, Cosmic jets, *Sci. Amer.*, 246:124–142, 1982; T. R. Geballe, The central parsec of the Galaxy, *Sci. Amer.* 241:60–85, 1979; C. Hazard and S. Mitton (eds.), *Active Galactic Nuclei*, 1979; G. K. Miley, The structure of extended radio sources, *Annu. Rev. Astron. Astrophys.*, 18:165–219, 1980; H. L. Shipman, *Black Holes, Quasars, and the Universe*, 2d ed., 1980.

Albedo

A term used to describe the reflecting properties of surfaces. White surfaces have albedos close to 1; black surfaces have albedos close to 0.

Two types of albedos are in common use: the bond albedo (A_b) determines the energy balance of a planet and is defined as the fraction of the total incident solar energy that the planet reflects back to space. The "normal albedo" of a surface, more properly called the normal reflectance (r_n), is a measure of the relative brightness of the surface when viewed and illuminated vertically. Such measurements are referred to as a "perfectly white Lambert surface"—a surface which absorbs no light, and scatters the incident energy isotropically—usually approximated by magnesium oxide (MgO), magnesium carbonate ($MgCO_3$), or some other bright powder. *See* PLANET.

Bond albedos for solar system objects range from 0.76 for cloud-shrouded Venus to values as low as 0.01–0.02 for some asteroids and satellites (Table 1). The value for Earth is 0.33. The bond albedo is defined over all wavelengths, and its value therefore depends on the spectrum of the incident radiation. For objects in the outer solar system, the values of A_b in Table 1 are estimates derived indirectly, since for these bodies it is impossible to measure the scattered radiation in all directions from Earth.

Normal reflectances of some common materials are listed in Table 2. The normal reflectances of many materials are strongly wavelength-dependent, a fact which is commonly used in planetary science to infer the composition of surfaces remotely. While the bond albedo cannot exceed unity, the normal reflectance of a surface can if the material is more backscattering at opposition than the reference surface.

In the case of solar system objects, the geometric albedo (p) is commonly defined as the ratio of incident sunlight reflected in the backscattering direction (zero phase angle or opposition) by the object, to that which would be reflected by a circular disk of the same size but covered with a perfectly white Lambert surface. Objects like the Moon, covered with dark, highly textured surfaces, show uniformly bright disks at opposition (no limb darkening); for these, $p = r_n$. At the other extreme, a planet covered with a Lambert-like visible surface (frost or bright cloud) will be limb-darkened as the cosine of the incidence angle at opposition, and $p = \frac{2}{3} r_n$.

For solar system objects, the ratio A_b/p is called the phase integral and is denoted by q. It corresponds to the scattering diagram of the object. Values range from near 1.5 for cloud-covered objects (1.3 for Venus) to 0.2 for the very dark and rugged satellites of Mars: Phobos and Deimos. The value

Table 1. Bond albedos and visual geometric albedos for solar system objects

Object	Bond albedo, A_b	Geometric albedo, p_v	Object	Bond albedo, A_b	Geometric albedo, p_v
Mercury	0.1	0.12	Moon	0.12	0.11
Venus	0.76	0.59	Io (J1)	0.56	0.63
Earth	0.33	0.37	Europa (J2)	0.58	0.68
Mars	0.25	0.15	Ganymede (J3)	0.38	0.43
Jupiter	0.42	0.45	Callisto (J4)	0.13	0.17
Saturn	0.4	0.46	Titan (S6)	0.30	0.16
Uranus	0.4	0.48	Ceres	0.03	0.05
Neptune	0.4	0.50	Vesta	0.12	0.23
Pluto	?	0.4			

Table 2. Normal reflectances of materials*

Material	Albedo	Material	Albedo
Lampblack	0.02	Granite	0.35
Charcoal	0.04	Olivine	0.40
Carbonaceous meteorites	0.05	Quartz	0.54
Volcanic cinders	0.06	Pumice	0.57
Basalt	0.10	Snow	0.70
Iron meteorites	0.18	Sulfur	0.85
Chondritic meteorites	0.29	Magnesium oxide	1.00

*Powders; for wavelengths near 0.5 micrometer.

for the Moon is about 0.6.

It is possible to define other types of albedos based on the geometry and nature (diffuse or collimated) of the incident beam and scattered beams. For such bidirectional albedos or reflectances, the angles of incidence and scattering as well as the angle between these two directions (phase angle) must be specified. Finally, in determining the energy balance of surfaces, one is often interested in the analog for a flat surface of the bond albedo, that is, the fraction of the incident energy that the surface reflects. For a plane parallel beam incident at an angle i, this quantity, $A(i)$, is generally close to the value A_b, but can be significantly larger than A_b for some surfaces, for large values of i.

[JOSEPH VEVERKA]

Aldebaran

Alpha Tauri, a red giant star of spectral type K5, temperature 3600 K, and luminosity about 400 times that of the Sun. Aldebaran is at a distance of 20 parsecs (6.2×10^{14} km); it is large enough to show a barely detectable disk, 0″.23 of arc, with the Mount Wilson stellar interferometer. Its size has been remeasured by two new techniques, speckle interferometry and lunar occultation, giving accordant results. Both depend on high-speed electronic recording of the light received, and reconstruction of the stellar image even though it is blurred to 40 times the true diameter by effects of the Earth's atmosphere. *See* ASTRONOMICAL PHOTOGRAPHY; STAR. [JESSE L. GREENSTEIN]

Almanac

A book that contains astronomical or meteorological data arranged according to the days, weeks, and months of a given year, and often includes a miscellany of other information. This article is restricted to astronomical and navigational almanacs.

The *Astronomical Almanac* is published annually by the U.S. Naval Observatory and the Royal Greenwich Observatory. It contains ephemerides, which are tabulations at regular time intervals of the positions of the Sun, Moon, planets, satellites, and some minor planets. It also contains mean places of stars, quasars, pulsars, galaxies, and radio sources and the time for certain astronomical phenomena such as eclipses, conjunctions, occultations, sunrise, sunset, twilight, moonrise, and moonset. This volume contains the fundamental astronomical data needed by astronomers, geodesists, navigators, surveyors, and space scientists. *See* ASTRONOMICAL COORDINATE SYSTEMS; EPHEMERIS.

The *Nautical Almanac* provides the data necessary for celestial navigation at sea. The annual volume contains an hourly tabulation of the positions of the Sun, Moon, and navigational planets. It also contains the positions of the navigational stars and times of sunrise, sunset, moonrise, and moonset. *See* CELESTIAL NAVIGATION.

The *Air Almanac*, published semiannually, contains the data necessary for celestial navigation in the air, but it can also be used at sea. The data are tabulated at 10-minute intervals, and sky diagrams are provided for rapid star identification.

The annual *Almanac for Computers* provides power series and Chebyshev coefficients, which permit the calculations of the data contained in the *Astronomical*, *Nautical*, and *Air Almanacs* for any specific time. A hand calculator, or small computer, is necessary for the computation, but interpolation is avoided. Algorithms are provided for astronomical and navigational computations.

The *Astronomical Phenomena*, published annually, contains the times for certain astronomical phenomena such as eclipses, conjunctions, occultations, sunrise, sunset, moonrise, moonset, moon phases, religious holidays, seasons, and visibility of planets. The *Planetary and Lunar Coordinates* contains the positions of the Sun, Moon, and planets in various coordinate systems at reduced accuracy for an extended period of time. This new publication replaces the Planetary Coordinates series; the first volume was published for 1980–1984, and the second will be for 1984–2000.

For surveyors, the Bureau of Land Management annually publishes the *Ephemeris*, and the Royal Greenwich Observatory prints the *Star Almanac.* The *Apparent Places of Fundamental Stars* is prepared annually by the Astronomisches Rechen Institut of Heidelberg, Germany.

The *Astronomical*, *Nautical*, and *Air Almanacs*, *Astronomical Phenomena*, and *Planetary and Lunar Coordinates* are cooperative publications of the U.S. Naval Observatory and the Royal Greenwich Observatory, and are available from the U.S. Government Printing Office and Her Majesty's Stationery Office. These publications are used in many countries, and similar references are published by the Spanish, Chinese, Soviet, Japanese, German, French, Indian, Argentinian, Brazilian, Danish, Greek, Indonesian, Italian, Korean, Mexican, Norwegian, Peruvian, Philippine, and Swedish governments.

The data from the publications are available in machine-readable form and will probably be available by direct computer access in the near future, but there appears to be a continuing requirement for the printed volumes.

[P. K. SEIDELMANN]

Alpha Centauri

The brightest star in the southern constellation Centaurus (visible only south of geographic latitude +30°). Alpha Centauri is one of a triple system, and is Earth's nearest neighbor in space, at a distance of 1.33 parsecs.

The visual binary of 80-year period contains

main-sequence stars of types G2 and K0, apparent magnitudes 0.1 and 1.4, respectively, separated by about 24 astronomical units, that is, by about the diameter of the solar system. The brighter of the visual binary is almost a twin of the Sun in luminosity, temperature, and mass. There is a faint companion, Proxima Centauri, of eleventh magnitude, spectral class M, which has the same measured distance and a proper motion which is nearly parallel in space with that of the brighter stars. The separation of 2°.2 in the sky corresponds to about 1.6×10^{12} km, 10,000 astronomical units. In spite of this separation, the triple system is probably gravitationally stable and the three stars probably had a common origin. *See* STAR.

[JESSE L. GREENSTEIN]

Andromeda Galaxy

The spiral galaxy of type Sb nearest to the Milky Way system. This galaxy is a member of a small cluster of galaxies known as the local group. This group contains also the Milky Way system, the Triangulum Nebula (M 33), the Large and Small Magellanic Clouds, NGC 6822, and several faint dwarf elliptical galaxies.

The Andromeda Galaxy M 31 or NGC 224, is particularly important because it is close enough for its stellar and other content to be studied in great detail. Observations show that the Great Galaxy M 31 contains bright blue stars of absolute magnitude $M_B \approx -9.5$, cepheid variable stars, bright supergiant red stars, clusters of stars (both globular and open), planetary nebulae, normal novae, gas, and dust.

The approximate distance to M 31 is known from the apparent luminosities of the cepheid variable stars. The period-luminosity relation for cepheids, combined with the apparent brightness of these stars; gives a distance of about 2,500,000 light-years. The accuracy of this value depends on the calibration of the period-luminosity relation for cepheids, which now appears to be well known. *See* CEPHEIDS; VARIABLE STAR.

The major axis of the photographic image of M 31 is 200 minutes of arc to an isophotal level of 25 magnitudes per square second of arc. This corresponds to a linear diameter of 150,000 light-years if the distance of 2,500,000 light-years is adopted.

The nebula is rotating about its center with a period of about 2×10^8 years. The evidence comes from radial velocities of selected objects in its disk. This rotational period, combined with the diameter, gives the mass of M 31 as 10^{11} solar masses. The total number of stars is greater than 2×10^{11}. The luminosity in the wavelength range from $\lambda = 3900$ A to $\lambda = 5000$ A is $M_B = -20.3$, which corresponds to 2×10^{11} equivalent suns. All studies show that M 31 is typical of other regular spiral galaxies of the Sb class. *See* GALAXY, EXTERNAL.

[ALLAN SANDAGE]

Antares

Alpha Scorpii, a bright, red supergiant of spectral type M1, temperature about 3000 K. Antares is variable in light and radial velocity; like other low-temperature stars of low density, it has an interferometrically measured diameter of 0".040, which at a distance of 100 parsecs corresponds to 450 times that of the Sun.

Antares has a close blue companion of spectral type B5, at a distance of 300 AU (astronomical units). The hot star is involved in nebulosity with an unusual spectrum. The emission lines are of forbidden Fe II and of Si II, without trace of hydrogen, which is normally strong in all other emission nebulae. It is possible that solid interstellar or circumstellar matter is being evaporated near the hot star. This type of circumstellar nebula has been found in many red supergiants, not in the normal spectral region but in the very-far-infrared. Apparently the strong stellar winds, or mass loss, from the red supergiants result in emission-line fluorescence of matter in the immediate surroundings at a rate so large that the nebula then becomes visible. *See* STAR.

[JESSE L. GREENSTEIN]

Aphelion

In astronomy, that point at one extremity of the major axis of an elliptical orbit about the Sun where the orbiting body is farthest from the Sun. The Earth passes aphelion on or near July 4, referred to as the time of aphelion passage. Because the orbit of the Earth is nearly a circle (eccentricity 0.017), the Earth is then only some 2.5×10^6 km farther from the Sun than at its mean distance of 149.6×10^6 km. *See* CELESTIAL MECHANICS; ORBITAL MOTION.

[RAYNOR L. DUNCOMBE]

Apogee

The position most distant from Earth in the orbit of a satellite, as in the orbit of the Moon or of an artificial satellite. The Moon at apogee is $5\frac{1}{2}$% further from Earth than at its mean distance; that is, its orbital eccentricity is 0.055. *See* APHELION; MOON; PERIGEE. [GERARD P. KUIPER]

Apsides

In astronomy, the two points in an elliptical orbit that are closest to, and farthest from, the primary body about which the secondary revolves. In the orbit of a planet or comet about the Sun, the apsides are, respectively, perihelion and aphelion. In the orbit of the Moon, the apsides are called perigee and apogee, while in the orbit of a satellite of Jupiter, these points are referred to as perijove and apojove. The major axis of an elliptic orbit is referred to as the line of apsides. *See* CELESTIAL MECHANICS; ORBITAL MOTION.

[RAYNOR L. DUNCOMBE]

Aquarius

The Water Bearer, in astronomy, a large zodiacal constellation visible in both summer and autumn. Aquarius is the eleventh sign of the zodiac. To ancients, the constellation resembled a man pouring a stream of water from a jar. Four stars, η, ζ, π, and γ, arranged like a Y form the head of the Water Bearer (see illustration). The stream of water flows into the Fish's Mouth (Fomalhaut) in the constellation Pisces Austrinus (Southern Fish). Fomalhaut, bright and solitary in this part of the sky, is one of the relatively important navigational stars. From earliest times this

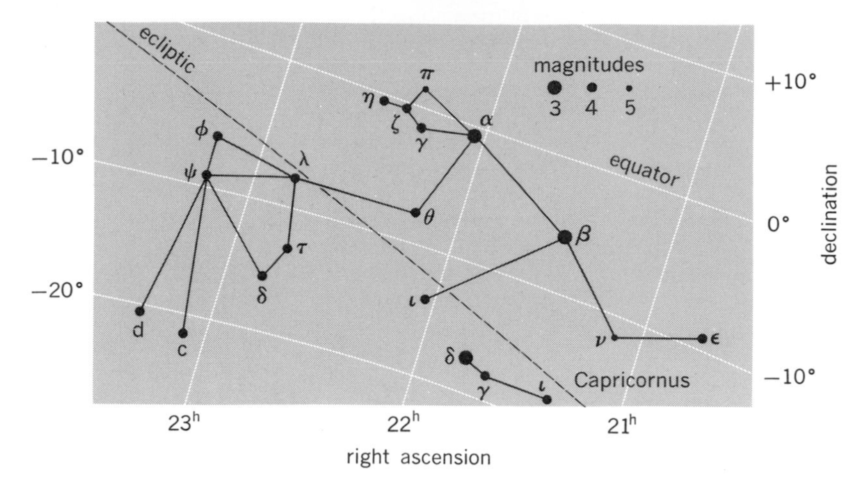

Line pattern of the constellation Aquarius. The grid lines represent the coordinates of the sky. The apparent brightness, or magnitude, of the stars is shown by the sizes of the dots, which are graded by appropriate numbers as indicated.

constellation has been associated with water, probably because the Sun is seen in Aquarius during the rainy season of February. *See* CONSTELLATION.

[CHING-SUNG YU]

Arcturus

Alpha Boötis, a red giant star of spectral type K2. Arcturus is one of the apparently brightest stars in the sky, of apparent magnitude 0.2. Because it is 11 parsecs (3.4×10^{14} km) distant, its absolute magnitude is −0.04, approximately 100 times brighter than the Sun, with a radius 20 times as large.

An interesting feature of Arcturus is its large space motion with respect to the Sun; it belongs to the high-velocity, or population II, group. The radial velocity is +90 km/s, and the observed proper motion in the plane of the sky corresponds to 120 km/s. Its galactocentric orbit is quite elliptical, unlike the Sun's, which is nearly circular.

W. W. Morgan recognized in Arcturus the spectroscopic features that have since become generally associated with population II. In particular, the bands of CN are weak compared to normal K giants of this luminosity class. Certain metallic lines indicate a high luminosity, the weakness of CN a lower luminosity. This contradiction is explained by an abnormally high value of the hydrogen abundance as compared with the metals.

Because of its brightness, high-resolution spectra were obtained by R. F. Griffin, using the Mount Wilson 100-in. (2.54-m) coudé spectrograph. They were compiled in 1968 in the form of a detailed spectrophotometric atlas, from 360.0 to 882.5 nm. Comparison between the spectrum of this red giant and of the Sun, a yellow main-sequence star, is now possible. A stellar spectrum is available for the first time at resolution comparable to that of the Sun, which was mapped in a similar fashion in 1940. *See* ASTRONOMICAL SPECTROSCOPY; STAR.

[JESSE L. GREENSTEIN]

Bibliography: R. F. Griffin, *A Photometric Atlas of the Spectrum of Arcturus*, 1968; M. Minnaert, G. R. W. Mulders, and J. Houtgast, *Photometric Atlas of the Solar Spectrum*, 1940.

Areal velocity

The rate at which a line that joins a fixed point and a moving particle sweeps out a surface area is called the areal velocity with respect to the fixed point.

In polar coordinates, where ϕ is the central angle and ρ is the distance between the fixed point and the moving particle, the areal velocity dA/dt equals $(1/2)\rho^2\, d\phi/dt$. In elliptical motion, if the origin is at one focus, $\rho^2\, d\phi/dt$ is a constant. In astronomy, Kepler's law of areas expresses this characteristic. *See* CELESTIAL MECHANICS.

[RAYNOR L. DUNCOMBE]

Aries

The Ram, in astronomy, a zodiacal and autumnal constellation. In Aries, there are a few stars brighter than fourth magnitude, three of which, α, β, and γ, form an obtuse triangle (see illustration). The bright star, α-Hamel (the Ram), an eye of the Ram, is a navigational star. Among the 12 zodiacal constellations, Aries was considered as the first, because about 2000 years ago when the zodiacal constellations were organized, the Sun was in Aries where it crossed the Equator at vernal equinox. Today, because of the precession of the equinoxes, this reference point has moved into the constellation Pisces. However, Aries remains the first sign of the zodiac. *See* CONSTELLATION.

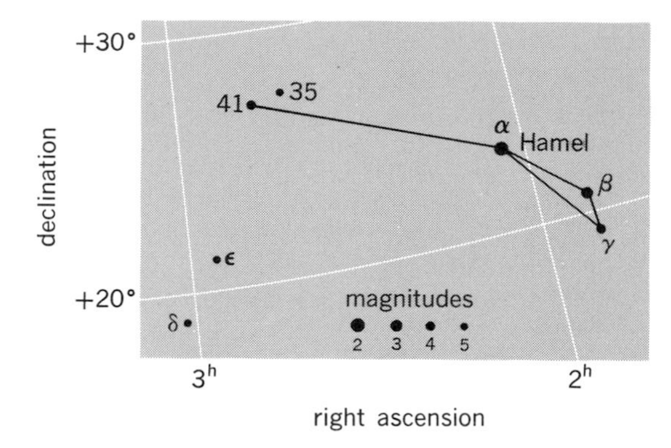

Line pattern of constellation Aries. Grid lines represent the coordinates of the sky. Apparent brightness, or magnitudes, of stars are shown by sizes of the dots, which are graded by appropriate numbers as indicated.

[CHING-SUNG YU]

Asteroid

One of the many thousands of small planets (minor planets) revolving around the Sun, mainly between the orbits of Mars and Jupiter. The presence of a gap in J. A. Bode's empirical law of planetary spacings motivated a search for the missing planet. The Italian astronomer G. Piazzi discovered Ceres on Jan. 1, 1801. Three other small planets were discovered in the next few years, leading H. Olbers to suggest that they were all fragments of a disrupted planet. Visual and, later, photographic searches for additional planets have continued to the present day. Newly discovered asteroids are assigned

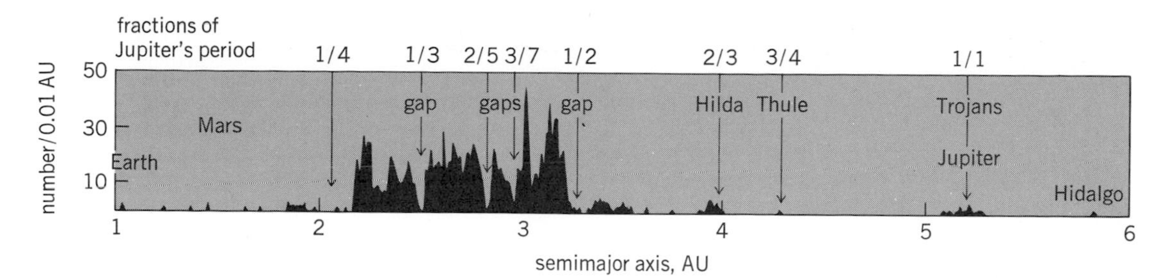

Fig. 1. Distribution of numbered asteroids, with distance from the Sun, between the Earth and Jupiter. Major fractions of asteroid orbital periods to Jupiter's period are shown. The arrows point to associated clusters or Kirkwood gaps.

a catalog number and name (such as 433 Eros) only after they are observed often enough to compute an accurate orbit. The Harvard-Smithsonian Center for Astrophysics maintains a complete file of all measurements of asteroid positions. The Institute for Theoretical Astonomy in Leningrad publishes an annual ephemeris of predicted asteroid positions for the over 2200 cataloged asteroids. *See* PLANET.

Orbits. The vast majority of asteroids have semimajor axes (mean distances to the Sun; symbolized a) between 2.2 and 3.2 astronomical units (1 AU = distance from Earth to the Sun = 149.6×10^6 km). However, several small asteroids orbit between Earth and Mars, and two groups, the Trojan asteroids orbit at Jupiter's distance from the Sun. The furthest asteroid discovered so far is 2060 Chiron, which is located between Saturn and Uranus at a = 13.6 AU. The innermost has a = 0.83 AU. *See* TROJAN ASTEROIDS.

Most asteroid orbits are more elliptical and inclined to the plane of the ecliptic than the orbits of major planets. Eccentricities (e) average about 0.15, and inclinations (i) about 10°; occasionally they exceed 0.5 and 30°, approaching the characteristics of short-period comet orbits. A number of small asteroids (Amor objects) cross, but do not intersect, the orbit of Mars, and a few even cross the Earth's orbit (Apollo objects).

Asteroids are not uniformly distributed in a, e, and i. Figure 1 shows the vacant lanes in the main asteroid belt, known as Kirkwood gaps, which occur at distances where the periods of revolution would be a simple fraction (such as ⅓ or ⅖) of Jupiter's period of 11.86 years. Asteroids in these gaps may have been destroyed by catastrophic collisions that they suffered more often than other asteroids because of enhanced speeds due to perturbations by Jupiter. Most asteroids originally beyond the ½ resonance were directly ejected from such orbits by nearby Jupiter early in the history of the solar system, except those grouped near the stable ⅔ (Hilda group) and ¾ (Thule) resonances, which survive because of the lack of nearby asteroids with which to collide.

Clusterings of asteroids with similar a, e, and i are known as Hirayama families, named for the Japanese astronomer K. Hirayama who discovered some of the major ones in 1918. About 40% of the numbered asteroids are members of over 100 families recently tabulated by J. Williams. They are probably composed of fragments of catastrophic collisions between asteroids.

Shapes, spins, and satellites. When the brightnesses of most asteroids are measured, they vary in a periodic manner, occasionally by more than one magnitude (a factor of 2.5), but more commonly 0.3 magnitude or less. The light curves usually are double-peaked (Fig. 2), characteristic of an irregularly shaped body spinning in space rather than the more complex curves that would result from albedo differences (spots) on a spherical body. Light curves have been measured for over 300 asteroids. The rotation periods are typically about 8 to 11 h, but range from 2½ to about 80 h. The fact that asteroids rotate with periods similar to the major planets suggests the spins may have a primordial origin. But there is some evidence that at least the smaller asteroids' spins may result from collisions they have suffered. Such collisions also produce the irregular shapes of asteroids. The larger asteroids, especially those composed of weak materials, cannot maintain highly nonspherical shapes against the compression of their own gravitational fields, which is why only smaller asteroids and some of the larger ones probably made of a strong material like iron have large light-curve amplitudes. Asteroids could not rotate much faster than 2½ h; otherwise the centripetal acceleration at their equator would exceed their gravity and they would fly apart.

Several kinds of observations have suggested

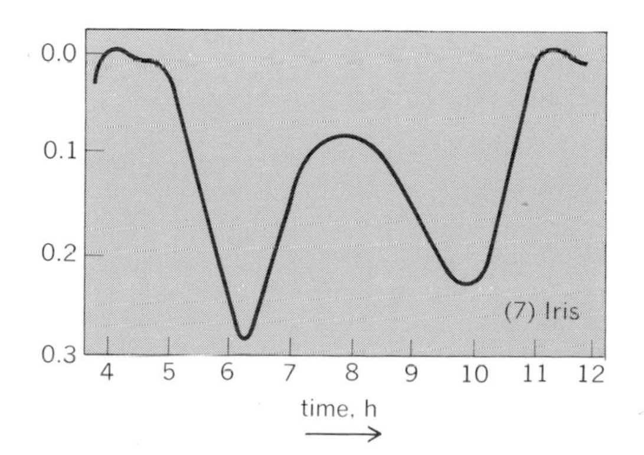

Fig. 2. Variation with time in the brightness of 7 Iris due to its irregular shape and 7.1-h rotation. Ordinate is in tenths of a stellar magnitude. (*After G. Kuiper.*)

Noteworthy asteroids*

Size rank	Number and name	Spectral type	Diameter, km	Spin period, h	Orbital elements *a*, AU	*e*	*i*
			Asteroids larger than 200-km diameter				
1	1 Ceres	C	1025	9.1	2.768	0.08	10°6
2	4 Vesta	Achondrite	555	5.3	2.362	0.09	7.1
3	2 Pallas	Unusual	538	7.9	2.773	0.23	34.8
4	10 Hygiea	C	443	18.	3.138	0.12	3.8
5	704 Interamnia	Unusual	338	8.7	3.060	0.15	17.3
6	511 Davida	C	335	5.2	3.181	0.17	15.9
7	65 Cybele	C	311	6.6	3.428	0.11	3.6
8	52 Europa	C	291	11.3	3.095	0.11	7.5
9	451 Patientia	C	281	20?	3.065	0.07	15.2
10	31 Euphrosyne	C	270	5.5	3.148	0.23	26.3
11	15 Eunomia	S	261	6.1	2.642	0.19	11.8
12	324 Bamberga	C	256	8?	2.685	0.34	11.2
13	107 Camilla	C	252	4.6	3.487	0.07	10.0
14	87 Sylvia	C?	251?	5.2	3.483	0.09	10.9
15	45 Eugenia	Unusual	250	5.7	2.720	0.08	6.6
16	3 Juno	S	249	7.2	2.671	0.25	13.0
17	16 Psyche	Unusual	249	4.3	2.922	0.14	3.1
18	24 Themis	C	249	8.4	3.129	0.13	0.8
19	13 Egeria	C	245	7.0	2.576	0.09	16.5
20	216 Kleopatra	C?	236?	5.4	2.790	0.25	13.2
21	624 Hektor	Unusual	300 × 150	6.9	5.153	0.03	18.3
22	165 Loreley	C	228	–	3.140	0.07	11.2
23	19 Fortuna	C	226	7.5	2.442	0.16	1.6
24	7 Iris	S	222	7.1	2.386	0.23	5.5
25	532 Herculina	S	217	9.4	2.774	0.17	16.3
26	702 Alauda	C?	217	–	3.195	0.03	20.5
27	88 Thisbe	C	214	6.0	2.769	0.16	5.2
28	250 Bettina	C?	211?	–	3.140	0.14	12.9
29	423 Diotima	C	209	–	3.069	0.03	11.2
30	121 Hermione	C	209	9?	3.460	0.14	7.6
31	747 Winchester	C	208	8.	3.004	0.34	18.2
32	6 Hebe	S	206	7.3	2.424	0.20	14.8
33	41 Daphne	C	204	6.0	2.767	0.27	15.8
34	386 Siegena	C	203	–	2.894	0.17	20.3
35	375 Ursula	C	200	–	3.133	0.10	15.9
			Other interesting asteroids				
	44 Nysa	Aubrite?	68	6.4	2.423	0.15	3.7
	349 Dembowska	Achondrite?	145	4.7	2.926	0.09	8.3
	433 Eros	Chon./S?	36 × 12	5.3	1.458	0.22	10.8
	1566 Icarus	Chon.?	2	2.3	1.078	0.83	22.9

*Most numbers are from the Tucson Revised Index of Asteroid Data.

that at least a few asteroids have "minor satellites" orbiting about them. The observations are controversial and need to be checked in the future.

Sizes and masses. Until the 1970s, asteroid sizes could be estimated only from their apparent brightnesses in the sky. But asteroids are faint, starlike objects; only Vesta is bright enough to be faintly seen with the unaided eye, and no asteroid shows a disk, even in the largest telescopes, large enough to measure accurately. It was impossible to tell, for instance, if 324 Bamberga is a relatively bright-colored body about 70 km in diameter or (as is now known) a very dark body 256 km in diameter. Two new techniques have been used in the 1970s to measure the diameters and albedos of over 130 asteroids: the polarimetric technique, based on an empirical correlation between the albedos of materials and how they polarize light; and the radiometric technique, which measures diameters by comparing the brightness of reflected visible sunlight from an asteroid with the brightness of the asteroid's emitted thermal radiation in the infrared. The new techniques have been checked in a few cases by timings of the disappearances of stars when asteroids chance to pass between the Earth and a star. *See* ALBEDO.

It is now known that asteroids are much darker, hence larger, than had been assumed before. There are about 35 asteroids larger than 200 km in diameter (see table); about 75% of them are soot-black (geometric albedos of 3–5%). Asteroids are much more numerous at smaller sizes, generally following a size distribution characteristic of fragmentation processes, as one would expect if the asteroids were smashing into each other. Indeed, there are so many large asteroids confined in the volume of the asteroid belt that collisions sufficient to destroy all but the larger asteroids occur every few billion years, and much more often for smaller ones. Thus the asteroids are mainly collisional fragments, grinding themselves down to dust.

Masses of asteroids are not easily measured

since they are so small as to exert a negligible effect on the orbits of other planets. But a few large asteroids regularly pass very close to some other asteroids and affect their orbits measurably. Measurements of these effects have yielded the masses of the three largest ones: Ceres (1.2×10^{24} g), Pallas (2.2×10^{23} g), and Vesta (2.7×10^{23} g). From the diameters and masses together, densities of 2.3, 2.6 and 3.3 g/cm^3 are implied, respectively. Vesta's density is similar to that of ordinary rocks, but the other two seem as underdense as carbonaceous chondritic meteorites, which contain a large fraction of volatiles. The total mass of all asteroids is only three times that of Ceres alone, or about 5% that of the Moon. *See* CERES.

Surface compositions. Spectra of sunlight reflected from asteroids have shapes, including absorption bands, characteristic of different rock-forming minerals (Fig. 3). Combined with the albedo data from polarimetry and radiometry, the spectra of surfaces of over 270 asteroids show that about 90% fall into two compositional groups: the C type, having very dark, neutral-colored surfaces of probable carbonaceous composition like the primitive carbonaceous chondritic meteorites; and the S-type, having moderate-albedo, reddish surfaces containing pyroxene and olivine silicates, probably mixed with metallic iron, similar to stony-iron meteorites or ordinary chondrites. About 75% of the asteroids are of the carbonaceous or C type (such as Ceres and Bamberga), and even more in the outer parts of the belt. Water has been detected in infrared spectra of several C-type asteroids. About 15% of asteroids are of the silicaceous or stony-iron or S type (such as Juno and Nausikaa), which is relatively common in the inner belt.

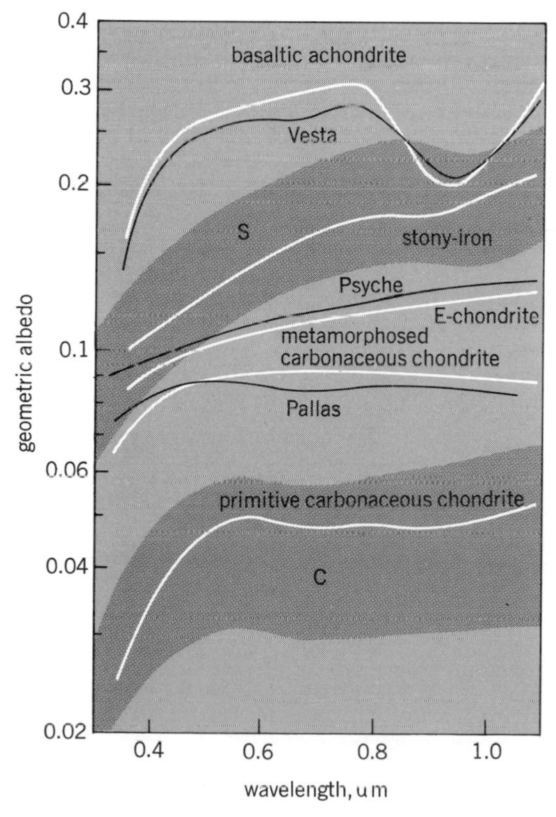

Fig. 3. Visible and near-infrared reflection spectra of the major S and C compositional classes of asteroids, plus several unusual ones. Laboratory spectra for some similar meteorites are shown with gray lines.

Only a few asteroids have spectra clearly indicating a composition like ordinary chondritic meteorites, the most common meteorites to fall on Earth; these include some small Earth-approaching bodies like Toro, and one small main-belt asteroid, Gryphia. Vesta is unique in having a surface like the basaltic achondritic meteorites. A few other meteorite types may be sparsely represented in the asteroid belt, but others have yet to be found at all. Since the two known celestial-mechanical modes for getting asteroid fragments from the belt to Earth are rather selective, it is not surprising that meteorite collections are not numerically representative of the asteroids. But it does seem likely that many meteorites are asteroid chips, although others may come from comets. *See* METEORITE.

Surface conditions. Surfaces of belt asteroids are cratered repeatedly. Larger asteroids may be fragmented and pulverized to great depths. But most excavated ejecta escape the weak gravity fields of smaller asteroids, so thick deposits of fragmental soil (regoliths) do not develop. Radar echoes from Eros, an elongated (36×12 km) Mars-crosser, show that it lacks a thick regolith on its probably crater-scarred surface. But polarimetry of asteroids suggests that even the smallest are dusty. Subsolar surface temperatures mostly range from 210 to 260 K. Asteroids lack even thin atmospheres because of their low gravity. *See* EROS.

Origin and evolution. Olbers's idea that the asteroids are fragments of an exploded planet is no longer attractive. It is hard to imagine what could produce an explosion sufficient to disperse a full-sized planet against its own gravity. Moreover, meteorites show evidence of having formed in bodies the size of present-day asteroids. Since current cosmogonical models for the origin of planets involve accretion from myriads of asteroidlike planetesimals, it is likely that asteroids are a remnant of the planetesimals that failed to accrete into a planet between Mars and Jupiter. Probably bombardment of the asteroid zone by large planetesimals scattered from massive, nearby Jupiter increased the relative veolcities of asteroids to the present value of 5 km/s so that asteroids fragment rather than accrete when they meet each other. Instead of forming a planet, the asteroids have been smashing each other to bits, and those seen now are mostly fragments, except for the "lucky few" that have escaped catastrophic collisions so far.

Evidently some asteroids of primitive, nonvolatile solar composition were heated within the first few hundred million years after the origin of the solar system, perhaps by the solar wind or extinct radionuclides, and they melted. Iron sank to their centers, forming strong stony-iron cores while basaltic lavas floated to their surfaces, producing minor planets like Vesta. The numerous collisions fragmented away the outer crusts and mantles of most of these bodies, with Vesta a prime, intact exception. The mantle materials of partially broken-up bodies may be exhibited on the surfaces of such unusual asteroids as Dembowska and Nysa. While the unmelted, weak, C-type asteroids may have been depleted by a large factor by the collisions, most of the strong stony-iron cores of the melted proto-asteroids have survived, perhaps as the S-type asteroids observed today. The asteroids still collide and fragment, occasionally spraying the inner solar system with chips that produce craters or fall as meteorites.

This picture of the evolution of the asteroids is incomplete and tentative. As new data are collected all the time, research continues on alternative interpretations. A spacecraft mission to several asteroids could reveal many further clues the asteroids contain about the formative period of solar system history. But if the pace of planetary exploration continues to slow down during the 1980s, it is unlikely such a mission can be flown until after the turn of the century.

[CLARK R. CHAPMAN]

Bibliography: C. R. Chapman, The nature of asteroids, *Sci. Amer.*, 232:24–33, Jan. 1975; C. R. Chapman, J. G. Williams, and W. K. Hartmann, The asteroids, *Annu. Rev. Astron. Astrophys.*, 16:33–75, 1978; T. Gehrels (ed.), *Asteroids*, 1979.

Astrometry

Measurement of space-time relations of celestial objects, primarily as observed on the celestial sphere, that is, perpendicular to the line of sight.

Visual star location. Over the past two centuries the positions of celestial objects have been measured visually with small, stable telescopes designed to measure meridian altitudes with high precision. These transit or meridian-circle telescopes and their variations, in combination with sidereal clocks, are the instrumental backbone of classical knowledge of the positions of celestial objects on the celestial sphere.

The positions are in the form of the two coordinates, right ascension and declination, referred to the celestial equator and vernal equinox. Results of these observations made at different observatories are compiled in catalogs of star positions. Difference in positions obtained at different epochs may be analyzed in terms of the precession of equator and ecliptic, and the proper motions of the stars. The latter, in turn, may be analyzed for solar motion and galactic rotation.

Faint-star photography. While the positions of Sun, Moon, and planets are thus easily obtained, only stars brighter than about the tenth magnitude can be observed in this fashion, because of the limited size of the instruments. Photographs, derived from standardized small telescopes in different locations, have yielded data known as the Astrographic Catalog or Carte du Ciel, which includes stars down to the thirteenth magnitude. The positions of stars down to the eighteenth magnitude and fainter are obtained photographically by specially designed powerful telescopes of short focal ratio. By referring the stars to distant galaxies, a close approach is obtained to absolute positions. The photographic method has become increasingly important and now has perhaps reached its limit of accuracy. Among the most important results of the photographic method is the wealth of newly discovered faint stars of large proper motions, all of which have proved to be comparatively nearby.

Interferometric methods provide limited astrometric data. Image-tube techniques, as well as photoelectric scanning, are becoming significant for binary stars.

Long-focus photography. A special branch of astrometry is long-focus photographic astrometry. Accurate stellar positions referred to faint, distant reference stars are measured by means of refractors of focal lengths of 8 m or more, commonly of large apertures of 40 cm or more; that is, with focal ratios up to *f*/20. Instrumental and atmospheric troubles caused by differences in colors of stars are minimized through proper choice of plates and filters; sharp photographs are thus obtained. Quartz reflectors are now extending long-focus photography to fainter stars. Originally, long-focus photographic astrometry was developed for the specific problem of stellar parallax, and as a result a great body of information on the distances of stars now exists. The method has been extended to other astrometric problems requiring high precision, such as the orbital motion of double stars, including the discovery and measurement of unseen companions from the perturbations on visible primary stars. The accuracy appears to be limited by the properties of the optical field of the telescope, of the photographic plate, and of the measuring machine. It is difficult to reach an accuracy higher than about 0.01 second of arc; fortunately many of these astrometric phenomena exceed this limit. *See* PARALLAX.

A special field of long-focus astrometry relates to positions of the components of double stars, measured relative to each other by visual, photographic, and interferometric methods. *See* CELESTIAL SPHERE.

[PETER VAN DE KAMP]

Radio astrometry. The field of radio astrometry is achieving very precise positions (0.005 arc-second) of celestial radio sources. The technique utilizes two radio telescopes configured as an interferometer. One method employs two radio telescopes, situated a few kilometers apart, linked by cables to a receiving system that instantly measures the difference in phase between the radio signals arriving at the telescope. Another technique, very-long-baseline interferometry (VLBI), utilizes a very accurate timing system, usually atomic clocks, to record the signals received at the end of a baseline several thousand kilometers apart. With the utilization of various types of high-speed electronic computers, a vast amount of radio data may be recorded, stored, and processed.

The National Radio Observatory has placed into operation a very large array (VLA) in New Mexico. The antenna configuration is Y-shaped, with arms 13, 13, and 11.8 mi (21, 21, and 19 km) long, making it the world's largest radio telescope. *See* RADIO TELESCOPE.

[BENNY L. KLOCK]

Bibliography: H. K. Eichhorn, *Astronomy of Star Positions*, 1974; W. A. Hiltner (ed.), *Stars and Stellar Systems*, vol 2, *Astronomical Techniques*, 1962; *New Problems in Astrometry*, IAU Symposium no. 61, 1974; K. A. Strand, Astrometry, *Smithson. Contrib. Astrophys.*, 1(1):21–24, 1956; The VLA takes shape, *Sky Telesc.*, 52(5): 320–322, November 1976; P. van de Kamp, Elements of long-focus photographic astrometry, *Photogramm. Eng.*, 22(2):314–325, 1956; P. van de Kamp, Long-focus photographic astrometry, *Pop. Astron.*, 59:65, 129, 176, 243, 1951; P. van de Kamp, *Principles of Astrometry*, 1967.

Astronomical coordinate systems

Systems of spherical coordinates serving to locate astronomical objects on the celestial sphere, which is the sphere of indeterminate radius, with its center at the observer, on the inside surface of which astronomical objects may be imagined to be projected. The celestial sphere has no physical existence; it is a concept convenient for specifying directions in space. The actual distances of most celestial objects are known with low precision and are not immediately perceptible; specification of the distance of an object would not assist in identifying it, and so for such purposes as aiming a telescope all objects are considered to be at the same unspecified distance.

Poles and coordinates. The system of astronomical spherical coordinates are all analogous to the system of latitudes and longitudes on the Earth, the most important difference being that the Earth is viewed from the outside and the celestial sphere is viewed from the inside. Every such system contains two diametrically opposite poles, a north pole and a south pole, the north pole being the one that is visible from the North Pole of the Earth. There is a primary circle, which is a great circle of the sphere, so placed that each pole is 90° from it. The latitudinal coordinate of an object is its angular distance measured northward or southward from the primary circle. Small circles parallel to the primary circle are called parallels. The longitudinal coordinate of an object is the angular distance measured along the primary circle from a specified origin to the point of intersection with a great circle passing through the object and the two poles; the one of the two intersections that is nearest the object is used. The great circles which pass through the poles and are perpendicular to the primary circle are sometimes called secondary circles.

In the terrestrial system the poles are the poles of rotation of the Earth, the primary circle is the Equator, the latitudinal coordinate is latitude, the parallels are parallels of latitude, and longitudinal coordinate is longitude, the origin of longitudes is the meridian of Greenwich, and the secondary circles are meridians of longitude.

Equatorial system. The astronomical system most used for locating objects is the equatorial system (Fig. 1). It is the geometrical extension of the terrestrial system. The poles are the intersections of the Earth's axis of rotation with the celestial sphere, and the primary circle is called the celestial equator, or simply equator. The latitudinal coordinate is called declination, and the parallels are parallels of declination. The longitudinal coordinate is called right ascension. The origin is the vernal equinox, which is the point where the Sun crosses the equator about Mar. 21. Right ascensions are measured eastward (in the direction of the Sun's apparent motion among the stars) from 0 to 360° and are commonly expressed in hours, minutes, and seconds, 360° being equal to 24 hours. In navigational practice, a coordinate measured in the same fashion as right ascension but in the opposite direction is used; it is named sidereal hour angle.

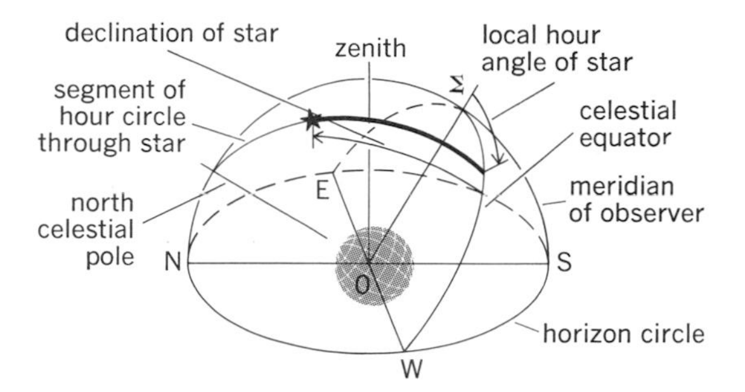

Fig. 1. Equatorial astronomical coordinate system.

The equatorial system is nearly but not quite fixed on the sphere. The poles are moving slowly, with the result that the right ascensions of equatorial stars increase about 46 seconds of arc annually.

Alternative systems. For some purposes it is convenient to modify the equatorial system by using an origin for the longitudinal coordinate that is fixed with respect to the observer. The origin is then the intersection of the celestial equator with the observer's meridian, and the corresponding coordinate is called the hour angle. It is commonly measured westward, so that the hour angle of any object increases by about an hour in an hour of time. The secondary circles, of which the meridian is one, are called hour circles.

Ecliptic system. Another system much used for theoretical studies of planetary motion is the ecliptic system. The primary circle is the ecliptic which, except for very small deviations, is the path traced out by the Sun on the celestial sphere during a tropical year. The latitudinal coordinate is called celestial latitude, and the longitudinal coordinate is called celestial longitude. The origin of longitudes is the vernal equinox, longitudes being measured eastward from 0 to 360°. The poles are called poles of the ecliptic.

Galactic system. A system used for studies of stellar motions is the galactic system. The primary circle is the plane of the galaxy and is called the galactic equator. The coordinates are called galactic latitude and galactic longitude. Until 1958 the origin of longitudes was the intersection of the galactic equator with the celestial equator in 18 hours 40 minutes of right ascension.

In 1958 a new standard system of galactic coordinates was adopted. The north galactic pole is at right ascension 12 hours 49 minutes and north declination 27.4 degrees. The origin of galactic longitude is at position angle 123 degrees with respect to the pole of the equator at 1950.0; thus, in the

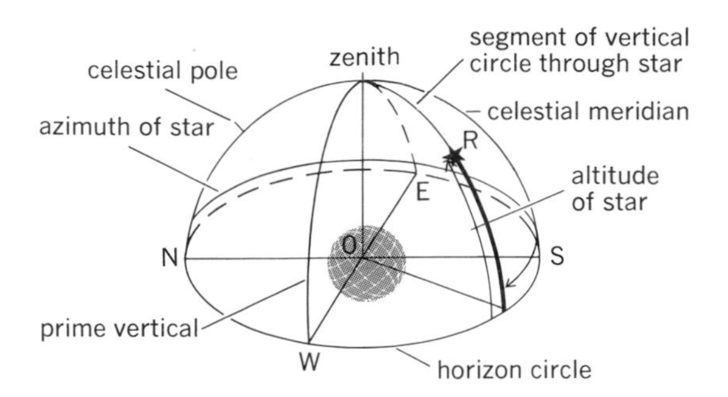

Fig. 2. Horizon system of astronomical coordinates.

new system the galactic longitude of a star is approximately 32.31 degrees greater than in the old system.

Horizon system. The horizon system is fixed relative to the observer (Fig. 2). The poles are the zenith, which is the point directly overhead, and the nadir. The primary circle is the horizon. The latitudinal coordinate is called altitude if it is measured upward from the horizon, and zenith distance if it is measured downward from the zenith; the parallels of equal altitude are called almucantars. The longitudinal coordinate is called azimuth. It may be measured in several ways, the commonest origin being the north point of the horizon. Two of the secondary circles have special names: The one passing through the north and south points of the horizon is called the meridian, and the one through the east and west points is called the prime vertical. The altitudes and azimuths of celestial objects are continuously changing. The system owes its importance to the ease and the precision with which altitudes can be measured. *See* AZIMUTH.

Transformations between systems. Coordinates of an object in one system can be transformed into coordinates of the same object in another system. An example is the calculation of the right ascension and declination of a planet if its celestial longitude and latitude are known. Such transformations are effected by the principles of spherical trigonometry; they form an important part of the subject of spherical astronomy. The most common transformation, which must be made for every observation taken in celestial navigation, is from the horizon system to the equatorial system. It involves the solution of the astronomical triangle, or ZPS triangle, which is a spherical triangle having vertices at the zenith, the pole, and the star. Much labor and ingenuity have been expended in facilitating the solution of the astronomical triangle.

[GERALD M. CLEMENCE]

Bibliography: V. V. Podobed, *Fundamental Astrometry: Determination of Stellar Coordinates*, 1965; Defense Mapping Agency, Hydrographic-Topographic Center, *Sight Reduction Tables for Air Navigation*, SRPUB 249, vol. 1, 1977, vol. 2, 1978, vol. 3, 1975.

Astronomical observatory

A building or group of buildings which house optical telescopes or the electronics required for radio telescopes or space telescopes, and the astronomers studying the observations.

Types. There are over 300 active astronomical observatories in at least 43 countries around the world, over 50 of them in the United States (see table). This omits hundreds of amateur telescopes, five or six Earth-orbiting space observatories controlled by radio and computers operated by astronomers on the ground, and at least one high-flying airplane equipped with telescopes. The optical telescopes used at these observatories vary widely in size, from the huge 200-in. (5.1-m) mirror reflector and 48-in. (1.2-m) Schmidt at Palomar Mountain in California, used mostly to study faint stars, nebulae, and galaxies, to the 10-in (0.25-m) lens telescopes at Zurich, Switzerland, used to study sunspots. Radio telescopes vary from the 1000-ft (305-m) bowl at Arecibo, Puerto Rico, through the 300-ft (91-m) dish at the National Radio Astronomy Observatory in West Virginia, to arrays of much smaller dishes in the United States and 25 other countries. The telescopes in orbit outside the Earth's atmosphere are smaller, and designed for special purposes, although the Space Telescope, to be put in orbit by NASA in 1984, has an aperture of 95 in. (2.4 m).

Affiliation. Most of these observatories are linked with universities or research institutes, some of them far from the telescope, which is best located on a remote mountain far from city lights. For instance, the Palomar Mountain Observatory is linked with the California Institute of Technology in Pasadena, some 125 mi (200 km) away. Universities provided the rapid increase in numbers of observatories in western Europe and the United States during the 19th century, when small lens telescopes were used by the faculty to teach astronomy students. In 1830 Harvard College Observatory was founded in Cambridge, MA, and has increased in size over the years to become the Center for Astrophysics, with over 200 employees. Earlier, some governments founded observatories for measuring time and helping navigators. For instance, the Royal Greenwich Observatory was founded in 1675 near London, England, and in 1830 the U.S. Naval Observatory was founded in Washington, DC. Even earlier, monarchs often supported small groups of astronomers or individuals such as Tycho Brahe, who had his own observatory built by the king's money at Uraniborg on a Danish island, before telescopes were invented. Tycho made many accurate measurements of the positions of planets by using a 19-ft (5.8-m) quadrant, a quarter circle marked from 0 to 90°. Earlier observatories in Samarkand, north of Afganistan in central Asia, and Peking, China, had elaborate pillars and bowls where the astronomers walked around to sight the stars and measure their positions. Stonehenge, in southern England, is thought to have been used in the same way as an observatory over 3000 years ago.

Equipment. The typical optical observatory has a telescope in a building with a hemispherical, rotatable dome in which there is a slot up one side with a cover that can be pulled aside from within, so that the telescope can "see out." Starting about 1890, astronomers took photographs, using the telescope as a large camera, so the observatories have been equipped with darkrooms where photographic plates can be developed, plate files where

Major astronomical observatories, by country*

Date founded, name, and place	Apertures and types of telescopes	Observing programs
Argentina		
1870, Córdoba (Bosque Alegre)	61-in. (1.5-m) mirror	Spectra, faint objects
1882, La Plata	33-in. (84-cm) mirror	Spectra, colors
	17-in. (43-cm) lens	Positions
	100-ft (30-m) radio dish	Interstellar gas
1964, Yale-Columbia Austral (El Leoncito)	20-in. (51-cm) lenses (2)	Positions, colors
Australia		
1924, Mt. Stromlo (Canberra)	74-, 50-in. (1.9-, 1.3-m) mirrors	Faint objects
	40-, 30-in. (1-, 0.8-m) mirrors	Spectra, colors
	26-in. (66-cm) lens	Positions
	20-in. (51-cm) Schmidt	Spectra, colors
1952, CSIRO (Parks)	210-ft (64-m) radio dish	Interstellar gas
1964, Sydney Univ. (Narrabri)	260-in. (6.6-m) mirrors (2)	Star diameters
	500-ft (150-m) radio array	Radio sources
1975, Anglo-Australian (Siding Springs)	150-in. (3.8-m) mirror	Faint objects, spectra
	48-in. (1.2-m) Schmidt	Faint objects
	40-in. (1-m) mirror	Colors, spectra
Canada		
1917, Dominion Astrophysical (Victoria)	72-, 48-in. (1.8-, 1.2-m) mirrors	Spectra, colors
1932, David Dunlap (Toronto)	74-, 24-in. (1.9-, 0.6-m) mirrors	Spectra, colors
1959, Penticton (B.C.)	84-ft (26-m) radio dish	Interstellar gas
	4000-ft (1220-m) radio reflectors	Interstellar gas
	5-ft (1.5-m) radio dish	Sun
	280-ft (85-m) radio reflector	Cosmic-ray effects
1960, Algonquin (Ontario)	150-ft (46-m) radio dish	Interstellar gas
Chile		
1962, Inter-American (Cerro Tololo)	150-in. (3.8-m) mirror	Faint objects
	60-, 36-in. (1.5-, 0.9-m) mirrors	Faint objects, spectra, planets
	24-in. (61-cm) Schmidt	Colors
1964, European Southern (La Silla)	140-in. (3.6-m) mirror	Faint objects
	60-, 40-in. (1.5-, 1-m) mirrors	Spectra, colors
	40-in. (1-m) Schmidt	Faint objects, colors
1968 Carnegie (Las Campanes)	100-in. (2.5-m) mirror	Faint objects
China		
(600 B.C.) 1958, Peking	24-in. (61-cm) Schmidt	Colors
	16-in. (41-cm) lenses (2)	Positions
(Sha-ho)	24-in. (61-cm) mirror	Sun
(Mi-yun)	35-ft (9-m) radio dishes (20)	Radio sources
(Hsing-lung)	78-in. (2-m) mirror	Faint objects, spectra
	24-in. (61-cm) Schmidt	Colors, spectra
	16-in. (40-cm) lens	Variable stars
1872, Shanghai	20-ft (6-m) radio dish	Time, positions
	16-in. (40-cm) lenses (2)	Positions
1934, Purple Mountain (Nanking)	24-in. (60-cm) Schmidt	Satellites, positions
	16-in. (40-cm) lens	Variable stars
	12-in. (30-cm) lens	Sun
1972, Yunnan (Kunming)	39-in. (1-m) mirror	Satellites, positions
	12-in. (30-cm) lens	Sun
Czechoslovakia		
1928, Ondrejov	78-in. (2.0-m) mirror	Spectra, faint objects
	25-in. (64-cm) mirror	Colors
	10-in. (25-cm) lens	Sun
1953, Skalnate Pleso	24-in. (61-cm) mirror	Colors
	12-in. (30-cm) lens	Positions
	8-in. (20-cm) lens	Sun
Denmark		
1963, Copenhagen (Brorfelde)	21-in. (53-cm) Schmidt	Spectra
	20-, 16-in. (50-, 40-cm) mirrors	Colors
Egypt		
1905, Helwan (Cairo)	30-in. (76-cm) mirror	Spectra, colors
(Kottamia)	74-in. (1.9-m) mirror	Faint objects, spectra

*For brevity, this list omits many active observatories. Moreover, new telescopes will be added in future years.

Major astronomical observatories, by country (cont.)

Date founded, name, and place	Apertures and types of telescopes	Observing programs
France		
1670, Paris (Meudon)	39-, 24-in. (1.0-, 0.6-m) mirrors	Spectra
	33-in. (84-cm) lens	Planets
	16-in. (40-cm) lenses (2)	Sun
1887, Nice	30-, 16-in. (75-, 40-cm) lenses	Double stars, positions
1930, Pic du Midi	41-in. (1-m) mirror	Spectra, Moon, planets
(Pyrenean Mountains)	24-in. (60-cm) lens	Planets
	16-in. (40-cm) lens	Sun
1938, Haute-Provence (St. Michele)	76-, 60-in. (1.9-, 1.5-m) mirrors	Spectra, faint objects
	24-in. (60-cm) Schmidt	Colors
1955, Univ. of Paris	650-ft (200-m) radio reflector	Interstellar gas
1975, CERGA (Grasse)	98-in. (2.5-m) Schmidt	Positions
	59-, 39-in. (1.5-, 1.0-m) mirrors	Moon and satellite ranging
	39-in. (1.0-m) mirrors (4)	Optical and infrared interferometry
Germany		
1705, Berlin (Babelsberg)	28-, 21-in. (71-, 53-cm) mirrors	Colors
	26-in. (65-cm) lens	Positions
1823, Hamburg	40-, 24-in. (100-, 60-cm) mirrors	Faint objects
(Bergedorf)	24-in. (60-cm) lenses (2)	Positions
	8-in. (20-cm) lens	Sun
	32-in. (80-cm) Schmidt	Spectra, colors
1836, Bonn (Hoher List)	42-, 14-in. (110-, 35-cm) mirrors	Colors, spectra
	13-in. (33-cm) Schmidt	Colors, spectra
	14-in. (35-cm) lens	Positions
	82-ft (25-m) radio dish	Interstellar gas
1960, Schwarzschild (Tautenburg)	54-in. (1.4-m) Schmidt	Faint objects
Great Britain		
1675, Royal Greenwich	36-, 30-in. (91-, 76-cm) mirrors	Colors
(Herstmonceux, 1948)	28-, 26-in. (71-, 66-cm) lenses	Positions
	6-, 4-in. (15-, 10-cm) lenses	Sun
(Canary Islands)	98-in. (2.5-m) mirror	Faint objects, spectra
1818, Edinburgh Royal	36-, 20-in. (91-, 51-cm) mirrors	Spectra, colors
	24-, 16-in. (60-, 40-cm) Schmidts	Positions, colors, spectra
1820, Cambridge	36-in. (91-cm) mirror	Spectra
	25-, 12-in. (63-, 30-cm) lenses	Sun
	17-in. (43-cm) Schmidt	Colors
1945, Mullard (Cambridge)	1-mi (1.6-km) radio array	Radio sources
	1500-ft (460-m) arrays (2)	Radio sources, Sun, planets
1949, Jodrell Bank	250-, 125-ft (76-, 38-m) dishes	Planets, interstellar gas
	218-ft (66-m) radio bowl	Radio sources
	120-ft (37-m) radio reflector	Radio sources
	25-, 10-ft (7.6-, 3.3-m) dishes	Radio sky survey
Italy		
1578, Vatican	25-in. (63-cm) Schmidt	Spectra, colors
(Castel Gandolfo)	24-in. (61-cm) mirror	Spectra
	16-in. (40-cm) lenses (2)	Positions
1880, Arcetri Royal	15-, 12-in. (38-, 30-cm) lenses	Sun
1940, Padua (Asiago)	72-, 48-in. (1.8-, 1.2-m) mirrors	Faint objects
	26-, 16-in. (65-, 40-cm) Schmidts	Faint objects, spectra
Japan		
1920, Tokyo (Mitaka)	36-, 12-in. (90-, 30-cm) mirrors	Colors
	26-in. (65-cm) lens	Spectra
	18-, 8-in. (45-, 20-cm) lenses	Sun
	20-in. (50-cm) Schmidt	Positions
	78-ft (24-m) radio bowl	Radio sources
	33-, 20-ft (10-, 6-m) dishes	Sun, polarization
1960, Okayama	74-, 36-in. (1.9-, 0.9-m) mirrors	Faint objects
	26-in. (65-cm) lens	Sun
Mexico		
1942, Tonantzintla	40-in. (1-m) mirror	Faint objects
	26-in. (65-cm) Schmidt	Colors, spectra
1978, Baja California	79-in. (2-m) mirror	Faint objects
(San Pedro Martir)		
Netherlands		
1955, Univ. of Leiden	82-ft (25-m) radio dishes (2)	Interstellar gas
(Dwingeloo)	33-, 25-ft (10-, 7.5-m) dishes	Interstellar gas
(Westerbork)	4700-ft (1.4-km) radio array	Radio sources

Major astronomical observatories, by country (cont.)

Date founded, name, and place	Apertures and types of telescopes	Observing programs
South Africa		
1820, Cape of Good Hope	40-, 30-in. (100-, 76-cm) mirrors	Colors
Royal	24-, 8-in. (61-, 20-cm) lenses	Positions, Sun
1928, Boyden (Mazelspoort)	60-in. (1.5-m) mirror	Spectra, colors
	32-in. (81-cm) Schmidt	Colors, positions
(1772) Radcliffe (Sutherland, 1978)	74-in. (1.9-m) mirror	Spectra, colors
Soviet Union		
1839, Pulkovo (Leningrad)	28-, 20-in. (70-, 50-cm) mirrors	Planets, spectra
	26-in. (65-cm) lens	Positions
	27-in. (69-cm) Maksutov	Faint objects, colors
	20-, 10-in. (50-, 25-cm) lenses	Sun
	12-in. (30-cm) transit	Positions
	394-ft (120-m) radio reflector	Sun, planets, radio sources
	52-, 39-ft (16-, 12-m) dishes	Radio sources, Sun, planets
	350-ft (105-m) radio array	Polarization
1948, Crimean Astrophysical	104-in. (2.5-m) mirror	Faint objects
	48-, 20-in. (1.2-, 0.5-m) mirrors	Planets, Sun
	25-in. (65-cm) Maksutov	Faint objects
	16-, 5-in. (40-, 13-cm) lenses	Sun
	72-ft (22-m) radio dish	Planets, interstellar gas
	23-, 13-ft (7-, 4-m) radio dishes	Sun
1952, Burakan (Armenia)	40-, 21-in. (1-, 0.5-m) Schmidts	Faint objects
	20-in. (50-cm) mirror	Faint objects
1976, Zelenchukskaya	236-in. (6-m) mirror (under reconstruction)	Faint objects, spectra
	1950-ft (600-m) radio array	Radio sources
Sweden		
1730, Uppsala	40-in. (1-m) Schmidt	Faint objects
1748, Stockholm	40-in. (1-m) mirror	Spectra, colors
(Saltsjöbaden)	24-, 16-in. (60-, 40-cm) lenses	Positions, colors
	20-in (50-cm) Schmidt	Spectra, colors
Switzerland		
1855, Zurich (Arosa)	10-in. (25-cm) lenses (2)	Sun
1895, Basel (Metzerlen)	24-in. (60-cm) mirror	Colors, faint objects
	16-in. (40-cm) Schmidt	Colors
1961, Geneva (Jungfraujoch)	30-in. (76-cm) mirror	Spectra, colors
United States (East and Midwest)		
1830, U.S. Naval (DC)	26-, 15-in. (65-, 38-cm) lenses	Planets, positions
	6-in. (15-cm) transit	Positions
	8-, 26-in. (20-, 65-cm) zenith tubes	Time
1830, Harvard (MA)	61-, 24-in. (1.5-, 0.6-m) mirrors	Faint objects, spectra
	15-in. (38 cm) lens	Colors
	Patrol cameras	Variable stars
1830, Yale (CT)	40-, 20-in. (1-, 0.5-m) mirrors	Positions, colors
	15-in. (38-cm) lens	Positions
	200-ft (60-m) radio array	Planets, Sun, polarization
1840, Van Vleck	24-in. (61-cm) mirror	Spectra, colors
(Wesleyan Univ., CT)	20-in. (51-cm) lens	Positions
1855, Univ. of Michigan (Ann Arbor)	38 in. (96-cm) mirror	Spectra
1866, Princeton (NJ)	36-in. (91-cm) mirror	Spectra
1878, Univ. of Wisconsin	36-in. (91-cm) mirror	Spectra, colors
(Madison)	16-in. (40-cm) lens	Colors
1883, Leander McCormick	32-in. (81-cm) mirror	Colors, spectra
(Univ. of Virginia)	26-in. (66-cm) lens	Positions
1897, Yerkes	41-, 24-in. (105-, 61-cm) mirrors	Spectra, colors, polarization
(Williams Bay, WI)	40-in. (1-m) lens	Spectra, positions
1917, Sproul (Swarthmore, PA)	24-in. (61-cm) lens	Positions
1954, Carnegie (DC)	120-, 97-, 60-ft (37-, 30-,18-m) radio dishes	Interstellar gas, sources
	650-ft (200-m) reflector	Radio sources, Sun
	3000-, 1500-ft (900-, 450-m) arrays	Radio sources, Sun
1956, Haystack Hill (MA)	120-, 84-ft (36-, 25-m) radio dishes	Planets, radio sources
1958, National Radio Astronomy	300-, 140-ft (91-, 43-m) dishes	Interstellar gas, planets
(West Virginia)	85-, 40-ft (26-, 12-m) dishes	Radio sources, planets
	15-ft (4.6-m) radio horn	Radio flux standard

Major astronomical observatories, by country (cont.)

Date founded, name, and place	Apertures and types of telescopes	Observing programs
United States (East and Midwest) (cont.)		
1962, Univ. of Illinois	600-ft (180-m) radio bowl	Radio sources, Moon
(Danville)	28-ft (8.5-m) radio dish	Sun
1963, Arecibo (Puerto Rico)	1000-ft (300-m) radio bowl	Planets, interstellar gas
United States (West)		
1875, Lick	120-, 36-in. (3-, 0.9-m) mirrors	Faint objects, spectra
(Mt. Hamilton, CA)	36-, 20-in. (91-, 51-cm) lenses	Positions
1894, Lowell	42-, 24-in. (1-, 0.6-m) mirrors	Faint objects, spectra
(Flagstaff, AZ)	24-, 13-in. (61-, 33-cm) lenses	Planets, positions
1904, Hale	100-, 60-in. (2.5-, 1.5-m) mirrors	Spectra, faint objects
(Mt. Wilson, CA)	12-in. (30-cm) mirrors (2)	Sun
1916, Steward (Kitt Peak)	85-in. (2.2-m) mirror	Faint objects
1937, McDonald	107-in. (2.7-m) mirror	Faint objects, planets
(Mt. Locke, TX)	82-, 36-in. (2.1-, 0.9-m) mirrors	Faint objects, spectra
	16-ft (5-m) radio dish	Interstellar gas
(Marfa, TX)	2-mi (3.3-km) radio array	Radio source survey
1940, High Altitude	16-, 10-in. (41-, 25-cm) mirrors	Sun
(Climax, CO)		
1948, Hale	200-, 60-in. (5.1-, 1.5-m) mirrors	Faint objects, spectra
(Palomar Mt., CA)	48-, 18-in. (120-, 46-cm) Schmidts	Faint objects, colors
1949, Sacramento Peak (NM)	16-, 12-in. (40-, 30-cm) mirrors	Sun
1955, U.S. Naval	61-, 40-in. (1.5-, 1-m) mirrors	Positions, colors
(Flagstaff, AZ)	23-in. (58-cm) lens	Colors
	12-in. (30-cm) mirror transit	Positions
1955, Univ. of Colorado (Boulder)	80-ft (25-m) radio dishes (2)	Sun, planets, radio sources
1958, Jet Propulsion Lab	210-ft (65-m) radio dish	Satellites, planets
(Goldstone, CA)	85-ft (26-m) radio dishes (2)	Planets, satellites
1959, Kitt Peak National	150-, 84-in. (3.8-, 2.1-m) mirrors	Faint objects, spectra
(Tucson, AZ)	36-, 16-in. (91-, 40-cm) mirrors	Colors, polarization
	60-, 36-in. (1.5-, 0.9-m) mirrors	Sun
1959, Stanford Univ. (CA)	150-ft (46-m) radio dish	Moon, planets, source sizes
	60-ft (18-m) radio dishes (3)	Radio sources, moon
	1150-ft (350-m) radio array	Planets, Sun, Moon
1961, Perkins (Flagstaff, AZ)	69-in. (1.7-m) mirror	Spectra, colors
1967, Mt. Hopkins (AZ)	6 × 72 in. (0.15 × 1.8 m) multiple mirror	Faint objects, positions, spectra
	60-, 12-in. (1.5-, 0.3-m) mirrors	Spectra, planets
	20-in. (51-cm) Schmidt	Positions
	34-ft (10-m) Cerenkov	Gamma-ray effects
1967, Hawaii (Mauna Kea)	84-, 60-in. (2.1-, 1.5-m) mirrors	Faint objects
Univ. of Michigan (Mauna Kea)	48-in. (1.2-m) mirrors (2)	Spectra, colors
1979, NASA-Hawaii (Mauna Kea)	120-in. (3-m) mirror	Infrared observations
1979, United Kingdom (Mauna Kea)	150-in. (3.8-m) mirror	Infrared observations
1979, France-Canada-Hawaii (Mauna Kea)	140-in. (3.6-m) mirror	Faint objects, infrared

they are stored, and measuring engines which are used to measure star images on the plates accurately. After 1910 or so, spectrographs were attached to the telescopes to spread the light of a star into a spectrum from red to blue. Photographs of these spectra (spectrograms) are measured on a one-dimensional measuring engine—a microscope mounted on a long, accurate screw.

Electronics. Since the mid-1930s, electronic detectors have been used on optical telescopes to measure starlight accurately. The electronics in large observatories have become increasingly complex. In the Multiple Mirror Telescope Observatory, opened in 1979 near Tucson, AZ, the astronomer need not peer through an eyepiece at the telescope, but can sit in a heated room watching a television screen and controlling the telescope from a computer terminal. Since radio and space telescopes are all electronic, modern observatories require electronic technicians and computers. There are also many electric motors required to rotate the dome of an optical observatory and to point the telescope. Large optical telescopes are housed in domes 60 ft (18 m) or more in diameter (although the Multiple Mirror Telescope is housed in a rectangular shed that rotates as a whole). Radio telescopes are mostly out in the open; space telescopes are in orbit many miles above the Earth's surface. The International Ultraviolet Explorer (IUE) is in geosynchronous orbit about 22,000 mi (35,500 km) from the Earth's center, but the NASA Space Telescope will be in a much lower orbit, circling the Earth every 90 min about 270 mi (430 km) above the surface. These two space telescopes are linked by radio to observatory control rooms in the Goddard Space Flight Center near Washington, DC. Other satellite telescopes have observatory control rooms at Princeton University and the Massachusetts Institute of Technology. *See* ASTRONOMICAL PHOTOGRAPHY; ASTRONOMICAL SPECTROSCOPY; OPTICAL TELESCOPE; RADIO ASTRONOMY; RADIO TELESCOPE; SCHMIDT CAMERA; TELESCOPE.

[THORNTON PAGE]

Bibliography: G. P. Kuiper and B. Middlehurst, *Telescopes*, 1960; T. Page, *Observatories of the World*, Smithsonian Astrophysical Observatory, 1970; T. Page and L. Page, *Telescopes*, 1966.

Astronomical photography

The application of the photographic process to astronomy. Over the past century, photographic observations have made profound contributions to astronomy and astrophysics; and for much of this period, photographs were the best light detectors available for a wide range of applications. Since the mid-1970s, a variety of image detectors utilizing photocathodes or silicon diode arrays have begun to live up to their anticipated potential, offering detective performance several times better than the best photographic materials. However, it is clear that many astronomers will continue to use photographic techniques for some time to come and for a number of reasons, including low initial cost, relative ease of use, and tradition. In particular, for applications where coverage of a wide field is required, photography is still the detector of choice, because emulsions can be manufactured in much larger formats than electronic detectors. Moreover, the emulsion itself is an effective and efficient storage device, requiring much less space than the equivalent digital image on a magnetic tape and presenting the information in a direct visual format. Indeed, many of the most important advances in astronomy and astrophysics have resulted from visual inspection and evaluation of photographs. The emphasis has been shifting from visual evaluation to pure machine measurements, partly because precision microphotometers are more readily available and partly because of advances in computer processing of digital images. Large photographic plates are still the best detectors for survey work, especially when used with Schmidt telescopes for direct imagery (Figs. 1 and 2) or with thin objective prisms for deep spectroscopic surveys. In modern astronomy the other chief uses of photography are photometry of stars and galaxies at the prime focus of large reflectors, astrometry with long-focal-length telescopes, and spectroscopy. For some of these applications, photography is often used as the readout for image intensifiers. *See* ASTROMETRY; ASTRONOMICAL SPECTROSCOPY; SCHMIDT CAMERA.

Fig. 2. A 48-in. (1.22-m) Schmidt plate of a field in the Milky Way showing bright and dark nebulosities. (*Hale Observatories*)

Fig. 1. The 48-in. (1.22-m) Schmidt telescope, Palomar Mountain.

Detective performance. Since the invention of the daguerreotype in 1839, dramatic improvements have been made in the sensitivity of photography (Fig. 3). Major revolutions occurred around 1851 and 1875 with the introduction of the wet plate and dry plate. In the 20th century, progress has largely been achieved by a long series of improvements in silver halide emulsions. Under optimum conditions modern emulsions can deliver a detective quantum efficiency of about 3% of the

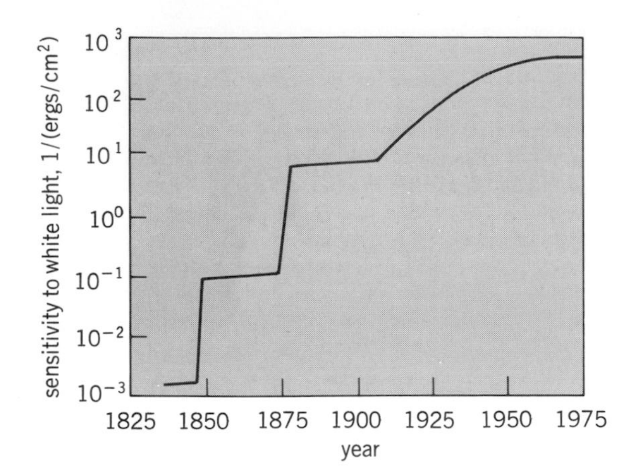

Fig. 3. History of photographic sensitivity. The vertical scale is the reciprocal of the energy, expressed in ergs/ cm^2, that is required to create an image. 1 erg/cm^2 = 10^{-3} J/m^2. *(Adapted from A. G. Smith and A. A. Hoag, Advances in astronomical photography at low light levels, Annu. Rev. Astron. Astrophys., 17:43–71, 1979)*

ideal, and in principle it should be possible to improve photographic performance by factors of 2 and more if some of the degrading mechanisms, such as recombination losses, could be eliminated.

Photographic photometry can be defined as any procedure that allows one to determine the amount of light that exposes each picture element (pixel). In general, the blackening (density) that results on a photograph depends on the details of the chemical processing as well as on the exposure. For photometry of spectra or diffuse images it is customary to calibrate the scale of relative intensity versus density on each photograph, typically by exposing an unused portion of the same photograph with a series of spots with known relative intensities. For stellar photometry it is better to have a sequence of stars with known magnitudes on the same photograph. The ratio of exposing signal to the uncertainty with which the signal was inferred (using photographic photometry) is often called signal-to-noise. For most photometric applications signal-to-noise is the most important characteristic of a photograph, because it offers a quantitative evaluation of the accuracy to expect from a measurement. *See* PHOTOMETRY.

Nearly all astronomical photography is done with special spectroscopic plates. These materials come in several classes of grain sizes. The ultimate accuracy with which photographic photometry can be carried out is set by the uncertainties due to graininess. In general, this uncertainty depends strongly on the density as well as the grain size of the emulsion, with optimum signal-to-noise for small pixel sizes occurring near density 1. For a pixel with area 1000 μm^2 the optimum signal-to-noise ranges approximately from 15 to 52 for various types of commonly used plates. The signal-to-noise varies with the square root of pixel area, [just as it should (Fig. 4)]. However, when the linear scale of a pixel begins to exceed 1 mm or so, large-scale nonuniformities can limit the signal-to-noise to about 100, or an accuracy of about 1%. For large plates very careful agitation is needed during development in order to keep the large-scale nonuniformities below 5 or 10% on an equivalent exposure basis.

Spectroscopic plates are available in several spectral sensitizations (Fig. 5). Types with the suffix -O are blue-sensitive, -J extends well into the green, -E and -F are panchromatic with good response out to about 700 nm, and -N is the best available in the near infrared out to 900 nm. The absolute values of the various curves in Fig. 5 are not reliable because of uncertainties in processing and emulsion type and batch, but the variation of sensitivity with wavelength is more accurate.

For long exposure times there is a serious breakdown in the reciprocity law that would predict that the photographic sensitivity should depend only on the product of the exposing intensity I and the exposure time t. The loss of sensitivity at low light levels and long exposure times is often called low-intensity reciprocity failure (LIRF). Fortunately, emulsion makers have discovered methods for reducing reciprocity failure. Spectroscopic plates that have been treated specially for low-intensity reciprocity failure usually are distinguished by an "a" in their designation. For example, the two plates compared in Fig. 6 have similar sensitivities for short exposure times, but the plate designated IIa-O is nearly 10 times less degraded than II-O for 1-hour exposures. However, even the specially treated plates still have substantial reciprocity failure for long exposures. During the 1970s there were major advances in understanding the sources of low-intensity reciprocity failure and in developing various treatments for its reduction. Some of the techniques are described below.

Hypersensitization techniques. Most hypersensitization techniques involve preexposure treatments such as evacuation, baking, bathing, or hypering with hydrogen gas. Many of the procedures work best if the exposures are made in a controlled atmosphere by using special cassettes. Since at least five different mechanisms are involved, it is often beneficial to combine treatments. However, since so many different combinations are possible, astronomers often disagree on what is the best recipe. Several of the most successful protocols are outlined below.

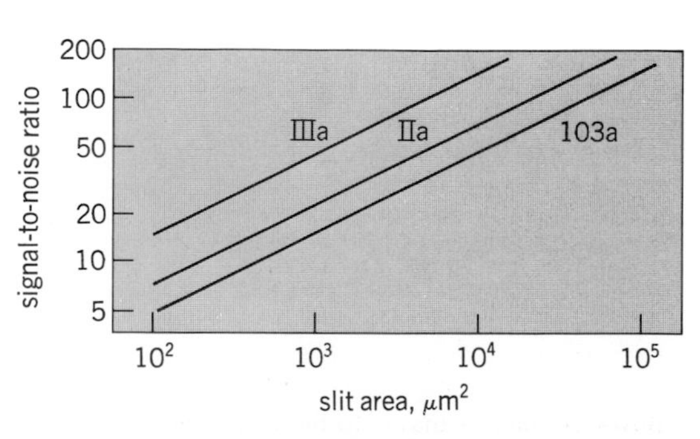

Fig. 4. Signal-to-noise as a function of microphotometer slit area for various photographic plates. *(From D. W. Latham, Signal-to-noise, in R. M. West and J. L. Heudier, eds., Modern Techniques in Astronomical Photography, European Southern Observatory, Geneva, pp. 141–152, 1978)*

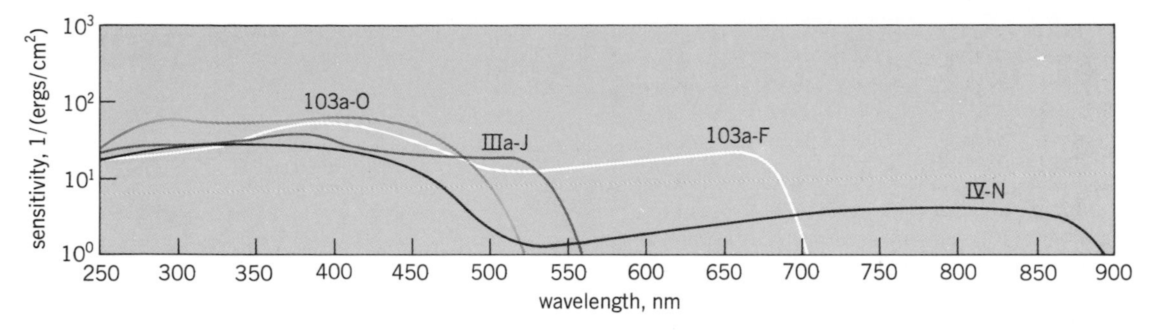

Fig. 5. Spectral sensitivity curves for various photographic plates. The vertical scale is the reciprocal of the exposure, expressed in ergs/cm², required to produce density of 0.6 above fog. The IV-N curve is with hypersensitization. 1 erg/cm² = 10^{-3} J/m². (*Adapted from Kodak Plates and Films for Scientific Photography, Kodak Publ. P-315, Eastman Kodak Co., 1973*)

Chemical treatments. Infrared plates are so slow that they are rarely used without some sort of preexposure chemical treatment. Simply bathing the plates in water and then drying before exposure can routinely gain a factor of 10 in speed, while factors of 100 are more common if the plates are exposed wet or in an oxygen-free 100% relative-humidity atmosphere. Another treatment that has been used for decades is bathing in ammonia. This can also provide gains as large as a factor of 100, but it is trickier to get good uniformity and low background fog. A treatment of bathing in dilute solutions of silver nitrate has had widespread success, and this is probably the best procedure for infrared plates. Bathing increases the Ag^+ concentration in the emulsion and may also increase the effectiveness of the dye sensitization. However, bathing is rarely used for plates other than the infrared, because more effective procedures than bathing are available for the other materials.

Removal of oxygen and water. In 1969 research showed conclusively that oxygen, especially in the presence of water, is a powerful desensitizer, with the effect being most detrimental at low light levels and long exposures. Although the exact mechanisms are not fully understood, careful removal of oxygen and water before and during exposure can all but eliminate low-intensity reciprocity failure for several types of astronomical plates. Two strategies for removal of oxygen and water are in common use: evacuation with a mechanical pump and trap for several hours is effective, and flushing with dry nitrogen for extended periods has also been used. Typical speed gains are factors of 2 to 4 for long exposures.

Baking. In order to provide a reasonable shelf life, the chemical sensitization of astronomical emulsions is not pushed to the optimum during manufacture. However, users can finish off the "ripening" process shortly before use by baking in vacuum or dry nitrogen. This technique is especially effective if oxygen and water have first been removed, and storage in a dry atmosphere often allows the baked plates to be useful days or weeks after treatment. The gains in speed from baking are typically factors of 2 to 5.

Hydrogen. In 1974 research showed that substantial speed gains could be achieved by soaking plates in hydrogen gas. The mechanism involved is probably reduction sensitization, and therefore can be used in conjunction with removal of oxygen and water and with baking to get the full benefit of all three techniques. A widely followed practice is to do the baking in a dilute mixture of hydrogen in nitrogen, known commercially as forming gas. This avoids the risks of a hydrogen explosion. The background fog comes up very quickly with excessive hydrogen treatment, and speed gains of more than a factor of 2 beyond that achieved with drying and baking are rarely practical. Moreover, drying and baking can actually improve the detective quantum efficiency, while hydrogen treatment only increases the speed.

Preflashing. Occasionally a better picture will result if a slight amount of uniform preexposure is impressed on the plate, but the signal-to-noise can only be degraded if the density of the preflash exposure exceeds about 0.2. In effect, the preflash moves faint features from a domain of low contrast near the photographic threshold into a domain of higher contrast. Thus the improvement is often one of perception, not of photographic detection.

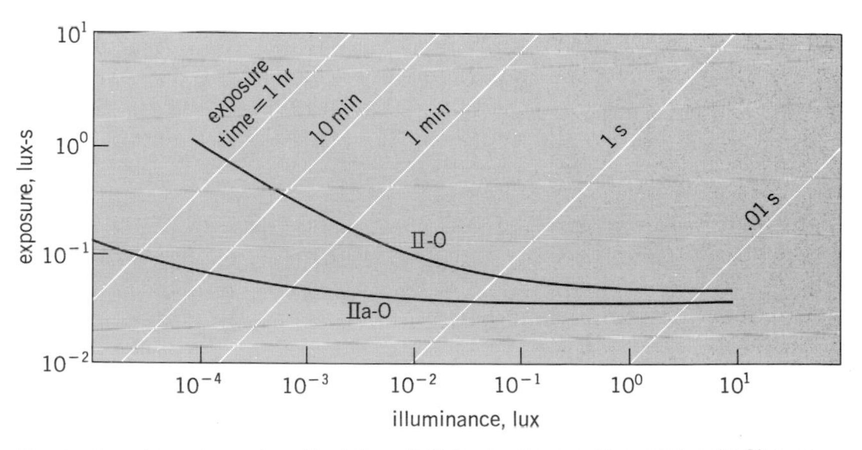

Fig. 6. Low-intensity reciprocity failure (LIRF) of a "normal" emulsion (II-O) compared with a similar material (IIa-O) that has been treated during manufacture to reduce reciprocity failure. Curves show exposure required to produce density 0.6 above fog; the rise of each curve at left thus indicates decreasing sensitivity at long exposures. (*Adapted from Kodak Films and Plates for Scientific Photography, Kodak Publ. P-315, Eastman Kodak Co., 1973*)

Push development. Extended development times can improve the speed and detective quantum efficiency of some materials. However, the newer fine-grained monodisperse emulsions show little or no gain from longer development.

Cooling. If emulsions are cooled during exposure, some of the low-intensity reciprocity failure can be eliminated. However, since removal of oxygen and water is a more effective way to reduce low-intensity reciprocity failure, cooling is rarely used with astronomical emulsions. It is occasionally justified when special materials with extreme low-intensity reciprocity failure, such as color film, are being used.

Intensified photography. Even when they are optimally hypersensitized, astronomical plates have a detective quantum efficiency of only about 2 or 3%. Photocathodes, on the other hand, can have quantum efficiencies as high as 25% or more. Even the image tubes produced in large numbers for military night-vision applications routinely achieve quantum efficiencies in excess of 10%. A stack of image tubes is often used at the focus of a spectrograph or direct camera, with the image from the final phosphor screen transferred optically onto a film or plate for recording a long exposure. The disadvantages of intensified photography are the distortions, nonuniformities, blemishes, and ion flashes that plague many tubes. Moreover, the tubes must be cooled to reduce the dark current for long exposures. Perhaps the biggest disadvantage for survey work is that the largest photocathodes are rarely 100 mm in diameter, and the military tubes are more often 25 mm than 40 mm.

Astronomers have also developed special image tubes which use a nuclear track emulsion to record directly the electrons produced at the photocathode. Several observatories have used these electronographic cameras successfully for both spectroscopy and direct photography. *See* IMAGE TUBE.

Electronic imaging devices. Enormous effort has gone into the development of detector systems which give out the image as an electrical signal which can be digitized either for immediate computer processing or magnetic storage for later analysis. These systems divide roughly into two classes. In the first class, a photocathode device such as an image intensifier chain is used to detect the incoming light, while an electronic device such as a television tube or diode array is used to read out the intensifier image. In the second class, silicon diode arrays are illuminated directly. The advantage of the intensified detectors is that they can detect individual photons and thus can be used at the very lowest light levels. The silicon arrays have the highest quantum efficiency, reaching nearly 75% in some devices, but the readout noise limits the operation to an integrating mode where a few tens of photons must be detected before looking at the picture. Size is also a problem; in 1980 the biggest chips were only 18 mm on a diagonal, but this is likely to improve.

Image processing. While the development of new electronic imaging hardware has improved the sensitivity of the big telescopes dramatically, perhaps the greatest advances in astronomical imagery, both photographic and electronic, have come in the area of image processing. Not only are modern high-precision microphotometers commonly available for digitizing photographic images, but elaborate computer schemes for processing and analyzing the images have been developed at several centers. Although it is still not common practice, it is feasible to extract nearly all of the information stored on a large direct photograph. For example, several groups have developed procedures for automatic computer identification and classification of the thousands of images on a single plate taken with a 4-m telescope or large Schmidt telescope. Other groups have software systems which process digitized spectra completely, identifying lines, solving for wavelengths and intensities, and determining radial velocities with the use of cross correlations. In some cases, these systems have been set up with electronic detectors and powerful displays right at the telescope so that much of the data analysis can go on in real time as the detector is integrating.

[DAVID W. LATHAM]

Bibliography: A. G. Smith and A. A. Hoag, Advances in astronomical photography at low light levels, *Annu. Rev. Astron. Astrophys.*, 17:43–71, 1979; R. M. West and J. L. Heudier (eds.), *Modern Techniques in Astronomical Photography*, European Southern Observatory, Geneva, 1978.

Astronomical spectroscopy

The use of spectroscopy as a technique for obtaining observational data on the compositions, physical conditions, and velocities of astronomical objects. Astronomical applications of optical spectroscopy from ground-based observatories cover the electromagnetic spectrum from the infrared through the visible to the near-ultraviolet. The advent of space-age techniques—high-altitude balloons, rockets, and satellites—has extended spectroscopic observations through the far-infrared to the domain of radio astronomy, and through the far-ultraviolet to the region of x-rays and gamma rays. Work in these longest and shortest wavelengths requires techniques other than those discussed here. *See* GAMMA-RAY ASTRONOMY; RADIO ASTRONOMY; X-RAY ASTRONOMY.

In optical spectroscopy a spectrograph is attached to a telescope, which serves as a light collector. If the telescope is of the reflector type, one distinguishes between prime-focus, newtonian, Cassegrain, and coudé spectrographs, depending on the focus at which they are used. The coudé focus is preferred for high-dispersion work. The image of the celestial body being studied is focused on the spectrograph slit by the telescope objective. The light from the slit passes through the collimator, which is a lens or mirror. The parallel rays of light are dispersed by glass or quartz prisms or, as in most modern spectrographs, by a diffraction grating. The grating is usually "blazed" by shaping its grooves in such a fashion that most of the dispersed light is cast into a chosen order of the spectrum. *See* OPTICAL TELESCOPE.

Dispersion. The linear dispersion provided by stellar spectrographs ranges from about 100 nm/mm to about 0.01 nm/mm, the higher dispersions usually being provided by stationary spectrographs at the coudé focus, while the other foci are employed for lower dispersions. The optical system of the coudé spectrograph of the 100-in (2.5-m) Mount Wilson telescope is shown in Fig. 1. An increasingly popular alternative arrangement to a coudé system for obtaining high dispersion is the echelle spectrograph, which achieves comparable results in a more compact instrument and can be mounted at one of the other foci. In an echelle, short-wavelength segments of successive high-order grating spectra, which would otherwise lie on top of each other, are separated by a prism, with dispersion at right angles to that of the grating, so that a large range of the spectrum is obtained in parallel strips on a rectangular area, rather than being spread out linearly, as in the coudé arrangement.

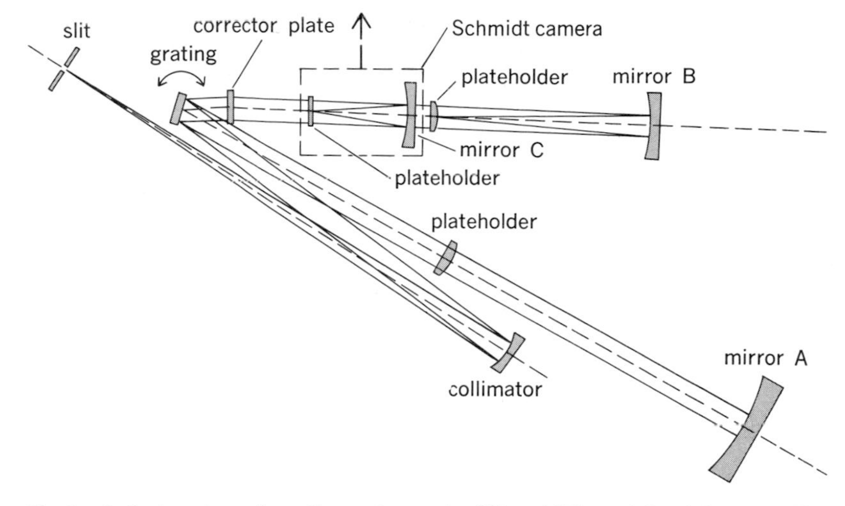

Fig. 1. Optical system of coudé spectrograph of Mount Wilson 2.5-m telescope. The three mirrors A, B, and C, with apertures of 91, 56, and 56 cm and focal lengths f of 290, 185, and 81 cm, are the camera objectives which can be chosen to photograph the spectrum; the camera with the focal length of 81 cm must be removed if the one with the focal length of 185 cm is to be used.

Recording methods. In the simplest form of an astronomical spectrograph, a prism is placed in front of the objective of the telescope; spectra of all stars in the field of the telescope are photographed in one exposure. Objective-prism spectrograms are used primarily for spectral and luminosity classification.

Until the 1970s, spectra were recorded photographically (Fig. 2). When high photographic speeds are required, Schmidt cameras are often used, as in Fig. 1. If only a small field (2–3 cm) in the focal plane suffices, an image tube can be used. The thus intensified image can be registered on a photographic plate or scanned electronically (Fig. 3) as in the Robinson-Wampler image tube scanner. Several image tubes are sometimes used in tandem, as in S. Shechtman's device at the coudé spectrograph. Spectral resolution tends to be degraded, but this problem is largely overcome in the image photon counting system devised by A. Boksenberg. Such devices employ high-speed computers so the scanner output can be displayed, processed, and stored. With large telescopes it is thus possible to record spectra of dim stars, remote galaxies, and faint quasars that could not have been attempted previously. *See* IMAGE TUBE.

Fourier transform spectroscopy. Fourier transform spectroscopy, used particularly in infrared work, employs a concept entirely different from the spectrographs described. Instead of being dispersed in a spectrograph, the light of a wide band of wavelengths is passed through a Michelson interferometer with variable spacing of its two apertures. The resulting interferogram, which is an

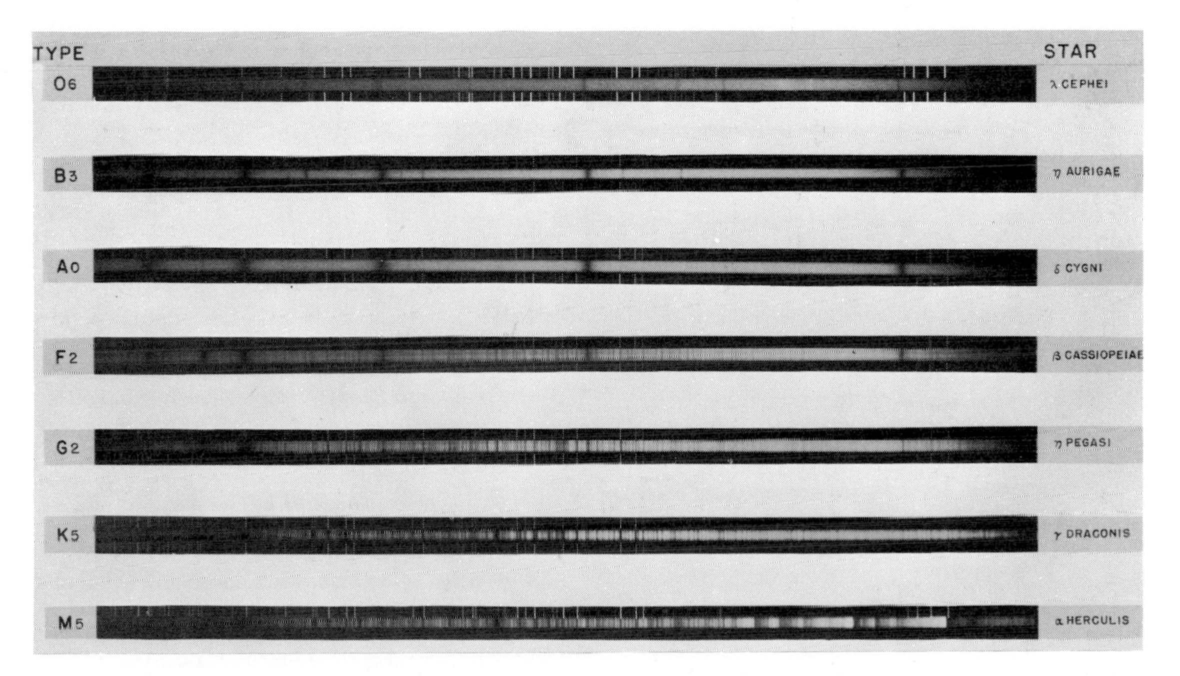

Fig. 2. Photographic recordings of principal types of stellar spectra. (*Hale Observatories*)

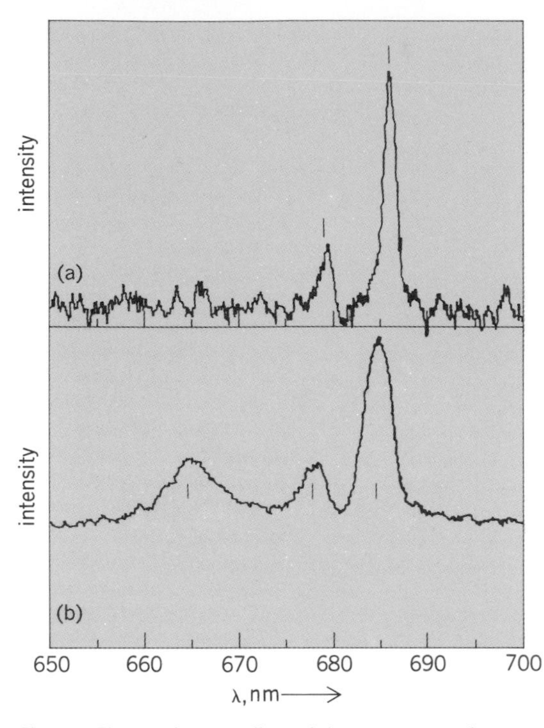

Fig. 3. Electronic recording of the spectrum of quasar 3C 48. The emission lines marked are produced in the hot gas of the quasar. (*a*) Nebulosity. (*b*) Nucleus. (*From E. J. Wampler et al., The nebulosity around 3C 48, Astrophys. J., 198:L50, 1975*)

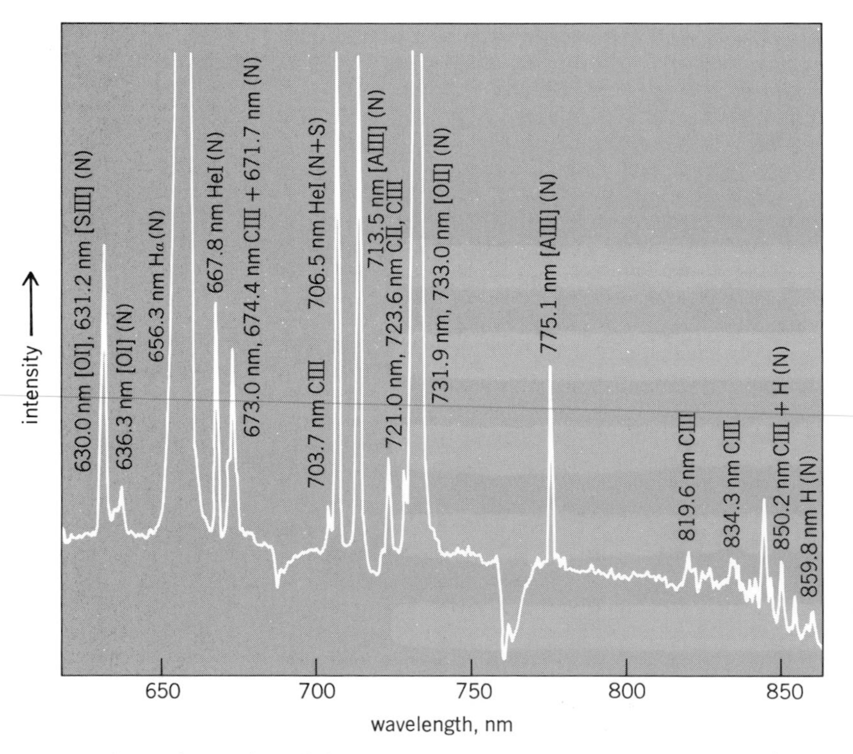

Fig. 4. Scan of a portion of the spectrum of the emission-line (Wolf-Rayet) star CD−30° 15469 surrounded by a small incandescent nebula; recorded by the Robinson-Wampler scanner at Lick Observatory. Lines of nebular origin are indicated by (N). The other lines come from the stellar envelope.

electronic record of the interference signal produced by the interferometer as the separation of the apertures is varied, is converted into a record of intensity versus wavelength by a high-speed computer.

Solar spectrum. The visual spectra of the Sun and most other stars consist of a continuous background ranging in color from deep red to violet, which is cut by numerous dark absorption lines (Fraunhofer lines), first mapped by J. von Fraunhofer in 1814. *See* SUN.

Planetary spectra. In the optical region the spectra of planets and their satellites are essentially reflected sunlight. When planets have atmospheres, absorption lines and bands due to molecules therein contained are also observed. From quantitative studies of these absorption bands, much information on chemical compositions is found. This spectroscopic method is so far the only one available for the planets beyond Saturn. *See* PLANET.

Emission-line spectra. Attenuated, incandescent, gaseous clouds in the Milky Way Galaxy and other galaxies produce bright line spectra, sometimes superposed on a continuous background. Extended stellar envelopes (Fig. 4). quasars (Fig. 4), and the outer envelopes of the Sun (chromosphere and corona) also produce emission-line spectra. *See* NEBULA; QUASARS.

Classification of stellar spectra. Although stellar spectra show a great variety of appearances, most can be classified in a two-parameter spectral class and luminosity system. Strengths and appearance of different spectral lines (usually on low-to moderate-dispersion spectrograms) supply classification criteria. Classification indices are usually temperature-sensitive lines; luminosity criteria are usually density-sensitive lines. Stars of high luminosity have lower atmospheric densities. The spectral sequence is described by letters O, B, A, F, G, K, M (ranging from hot, about 40,000 K, to cool, 3000 K) with decimal subdivisions I (very luminous) to V (main-sequence dwarf stars). The Sun's spectral class is G2V, which means it lies closer to GO than to KO and is a dwarf star. *See* STAR.

Spectrophotometry. For many purposes a plot of true intensity against wavelength is needed. Over a wide wavelength range (several hundred nanometers) the intensity distribution depends mostly on the temperature. With high spectral resolution one can measure the detailed profile of a line (Fig. 5), which supplies information on the abundance of the chemical element or compound involved, the temperature and density distribution in the stellar atmosphere, and large-scale mass motions and rotation. Such measurements are usually made with photoelectric scanners.

Radial velocity. Measurements of the wavelength displacement, due to the Doppler effect, between the spectral lines in a celestial object and a laboratory source (such as an iron arc or discharge tube) give the line-of-sight velocity of the source with respect to the observer. Radial velocity measurements enable establishment of orbits of close double stars (spectroscopic binaries, Fig. 6),

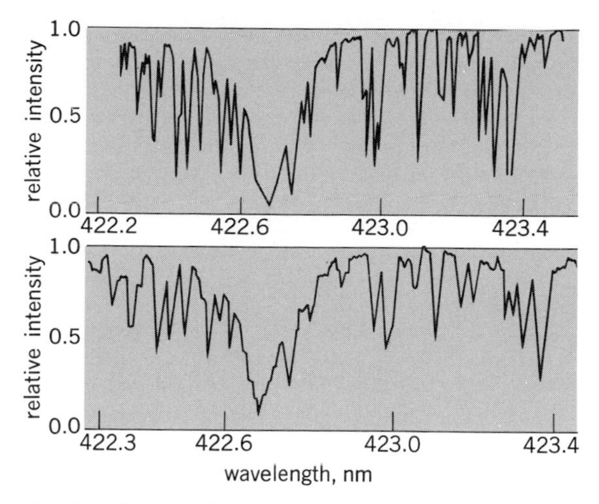

Fig. 5. Microdensitometer tracings of the solar spectrum as a function of wavelength. The same section of spectrum is shown; the upper tracing was made from a spectrogram with 0.03 nm/mm linear dispersion: the lower tracing was made from one with 0.28 nm/mm. (*Hale Observatories*)

and when combined with photometric data have yielded dimensions of the stars themselves. Also, such measurements have given the motion of the Sun with respect to nearby stars, the rotation of the Galaxy and other galaxies, and recession velocities of external galaxies (up to 200,000 km/s). Wavelength shifts of lines in spectra of some quasars imply velocities exceeding 90% of the speed of light. Radial velocity measurements revealed the remarkable nature of the object SS 433 with its jets moving at velocities of 50,000 km/s. *See* BINARY STAR; DOPPLER EFFECT; GALAXY, EXTERNAL.

[LAWRENCE H. ALLER]

Bibliography: G. O. Abell, *Exploration of the Universe*, 3d ed., 1975; L. H. Aller, *Atoms, Stars, and Nebulae*, 1971; D. F. Gray, *Observation and Analysis of Stellar Photospheres*, 1976; W. A. Hiltner (ed.), *Astronomical Techniques*, 1962; P. W. Merrill, *Space Chemistry*, 1963.

ASTRONOMICAL SPECTROSCOPY

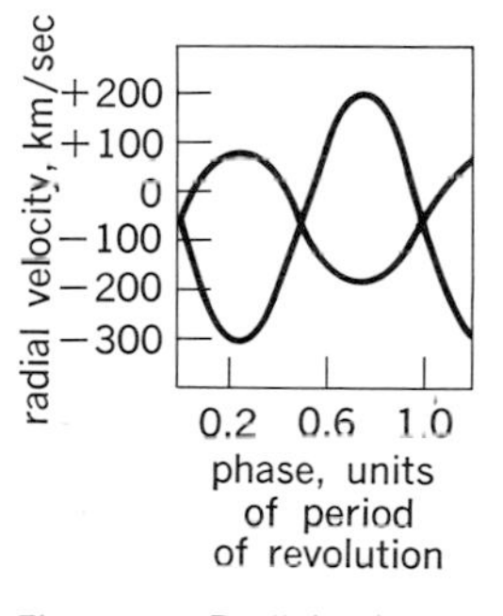

Fig. 6. Radial-velocity curve of a spectroscopic binary (W Ursae Majoris); the spectra of both components have been measured.

Astronomical unit

The basic unit of length in the solar system. The astronomical unit (AU) is used to a limited extent for interstellar distances as well through the definition of the parsec (1 pc = 206,265 AU). It is nearly equal to the mean distance a between the center of mass of the Sun and the center of mass of the Earth-Moon system ($a = 1.000\ 000\ 23$ AU), and for that reason it is often convenient to think of it as the mean distance between the Sun and Earth. *See* PARSEC.

The most accurate determination of the length of the astronomical unit in physical units, such as meters, is obtained from phase-modulated continuous-wave (CW) radio signals beamed to other planets. The round-trip travel times of the signals are determined by cross-correlating the returned signal from the planet with the transmitted signal, and as a result, planetary distances are measured directly.

Prior to the inception of planetary radar technology around 1961, determinations of the astronomical unit relied on parallax effects in the positions of planets and asteroids as viewed from distant points on the surface of the Earth, or on spectroscopic Doppler shifts in the light of stars near the plane of the ecliptic. These techniques were limited in accuracy to about 70,000 km. On the other hand, early radar determinations were accurate to 500 km, and by accumulating several years of planetary distance measurements, it was possible to refine this accuracy to 1 km, the limit of passive radar bounce technology. CW signals returned from the two Viking orbiters and two landers on the surface of Mars have made possible a determination of the astronomical unit to an accuracy of 100 m. *See*DOPPLER EFFECT; EARTH ROTATION AND ORBITAL MOTION; PARALLAX.

With an adopted value of 299,792.458 km/s for the speed of electromagnetic propagation in vacuum, the value of the astronomical unit from the Viking data is 149,597,870.7 km.

[JOHN D. ANDERSON]

Astronomy

The science concerned with the observation and interpretation of the radiation received in the vicinity of the Earth from the component parts of the universe. For the greater part of the history of humanity, the radiation was confined to the visual portion of the electromagnetic spectrum, and the eye was the sole receiver. Relatively crude direction-finding instruments aided the eye in following the motions of the celestial bodies on the sphere of the sky. The instruments made possible daily and seasonal timekeeping and elementary prediction of the positions of the Sun, Moon, and the five known planets, Mercury, Venus, Mars, Jupiter, and Saturn, even though these predictions were based on a geocentric theory of the universe. *See* CALENDAR; JUPITER; MERCURY; SATURN; TIME.

Fundamental astronomical constants. The invention of the telescope greatly enhanced the power of visual observation of celestial objects, revealing heretofore hidden details on the Sun, Moon, and planets and disclosing the existence of hordes of stars never before seen by the human eye. The increased light-gathering power of the telescope made possible greatly refined observations, which went hand in hand with growing philosophical and theoretical considerations of the nature of the solar system and the universe. These advances in turn led to the adoption of the heliocentric view of the solar system and the rise of celestial mechanics, which flowered under I. Newton and his followers and gave a detailed physical picture of the motion of celestial objects under gravitational forces. Comets, viewed with superstitious awe since the dawn of history, were shown to be tractable members of the solar system. With the advent of the telescope, early in the 17th century, the planets Uranus, Neptune, and Pluto were eventually discovered, and large numbers of tiny planets, the asteroids, were added to the Sun's family, as were the satellite systems of the major planets. Through increased precision made possible by larger and more specialized telescopes and auxiliary devices (cameras, micrometers, and the like), the determination of fundamental astronomical constants, such as precession, nutation, and aberration, and of precise star positions, as well as orbit computations and time determinations, became primary objectives of the astronomer. *See* ABERRATION; ASTEROID; CELESTIAL MECHANICS; COMET; CONSTELLATION; NEPTUNE; NUTATION; PLANET; PLUTO; SATELLITE; URANUS.

The telescope made possible the recognition of stars as distant suns through the measurement of their parallaxes and proper motions. Study of the physical nature of the stars became possible when the telescope was subsequently used in conjunction with auxiliary instruments such as the photometer and the spectrograph, designed to measure and analyze the radiation from the stars. The photometer allows quantitative measurement of stellar radiation in various portions of the electromagnetic spectrum, and the spectrograph makes possible detailed analysis of the spectra of stars, including measurement of the radial velocities of stars. *See* STAR; SUN; TELESCOPE.

Astrophysics. The introduction of the photographic plate was the significant step that made possible physical astronomy, or the application of modern physics to astronomy—astrophysics. Born only in the last half of the 19th century, astrophysics, the physical study of the components of the universe, has far outdistanced all other branches of astronomy. Among many other accomplishments, it has provided a convincing picture of the birth, life, and death of stars—stellar evolution—and the internal structure of stars. *See* ASTROPHYSICS.

Radio astronomy. The limitation of astronomical observations to the visual portion of the electromagnetic spectrum, imposed by the human eye, was eliminated by the extension of the spectral sensitivity of astronomical instruments and recording devices, including the photographic plate, to the infrared and ultraviolet portions of the spectrum. The most dramatic advance was the development of the radio telescope, which has made radio astronomy a major branch of astronomy. Radio telescopes, which are essentially large, highly directional radio antennas used in conjunction with extremely sensitive amplifiers and recording devices, have revealed not only the expected thermal radiation in the radio region of the spectrum but also the existence of nonthermal, or synchrotron, radiation, arising from highly accelerated relativistic electrons. The most prominent, as well as the most important, radio sources in the sky, radio galaxies, quasars, and supernova remnants, are nonthermal sources. The radio spectrum also contains "radio lines," both in emission and in absorption, arising from low-energy transitions in atoms and molecules. The most famous atomic radio line is the 21-cm line of neutral hydrogen. In the same manner, the use of radio telescopes has revealed the presence in space of over 30 relatively complex molecules. *See* INTERSTELLAR MATTER; QUASARS; RADIO ASTRONOMY; RADIO SOURCES; RADIO TELESCOPE; SUPERNOVA.

Observations above the atmosphere. The observation of the total electromagnetic spectrum of celestial sources came with the lifting of the opaque atmospheric veil by the use of high-altitude balloons, rockets, and craft orbiting well above the Earth's atmosphere. Space probes to the Moon and nearer planets placed astronomical equipment in their vicinity and even upon them. At this point, planetary and lunar astronomy suddenly made great leaps. Lunar features only feet and inches in size have been studied, and the visible surfaces of Mars and Venus have been brought under astronomical scrutiny to a degree that was considered to be utterly impossible even as late as the year 1950. *See* MARS; MOON; ROCKET ASTRONOMY; VENUS.

In radar astronomy, radio telescopes are used to transmit short bursts of energy toward a celestial target—of necessity in the solar system. The first glimpse of the surface of perpetually cloud-covered Venus was obtained in this way; improved radar maps of Venus have been made at Arecibo.

X-ray and gamma-ray astronomy. Rockets, high-altitude balloons, and satellites have spawned new branches of astronomy—x-ray and gamma-ray astronomy. The existence of sources of x-rays and gamma rays on the celestial sphere has been revealed in a manner reminiscent of the first detection of discrete radio sources. *See* CELESTIAL SPHERE; GAMMA-RAY ASTRONOMY; X-RAY ASTRONOMY; X-RAY STAR; X-RAY TELESCOPE.

Cosmology and cosmogony. The theoretical and philosophical aspects of astronomy, once concerned with the central position of the human race in a relatively tiny universe, are now concerned with the problems of cosmology and cosmogony. The march of astronomy has been one in which the place of the human race in the universe has been successively decentralized. Once humanity and the Earth were considered the center of the universe; then the Sun became the universal center toward which human beings turned; and in turn the Sun was deposed to an outlying region of the galaxy of about 200,000,000,000 stars to which it belongs. Now not only can the Milky Way Galaxy not be considered the center of the universe, but the concept of center must be entirely abandoned. Interpretation of the observed red shift of distant galaxies as evidence for the expansion of the universe is theoretically supported by relativistic cosmology, which presents a picture of the universe in which every galaxy appears to be a center from which all other galaxies are receding with speeds proportional to their distances from that given galaxy. This most surprising result could hardly have been imagined at the turn of the century. The march of astronomy shows little sign of slowing down. The discovery of quasars and of a host of interstellar molecules, pulsars, and cosmic infrared, x-ray, and gamma-ray sources represents a significant research challenge and, coupled with the large optical and radar telescopes, promises an auspicious future for astronomical research. *See* COSMOLOGY; GALAXY; GALAXY, EXTERNAL; INFRARED ASTRONOMY; PULSAR; RED SHIFT; RELATIVITY.

[J. ALLEN HYNEK]

Astrophysics

The application of modern physics to the problems of astronomy. Astrophysics embraces much of the activity of present-day astronomy; specifically excluded are the study of the motions of the planets and satellites (celestial mechanics), the measurement of positions and motions (astrometry), and usually also the structure and dynamics of the Galaxy and other galaxies, although such topics as density wave theory of spiral arms are sometimes considered to be part of astrophysics. *See* ASTRONOMY; CELESTIAL MECHANICS; GALAXY.

Instrumentation. From an operational point of view, astrophysics differs from terrestrial physics in that it is primarily an observational subject, although certain special problems lend themselves to experimental treatment. The basic problem is often the measurement of the quantity and quality, that is, the energy distribution and polarization, of the electromagnetic radiation of the Sun, stars, planets, and nebulae, and the study of their spectra. A variety of telescopes and detectors are employed. Detectors include the photographic plate with suitable filters, the photoelectric cell, lead sulfide and lead telluride cells, thermocouples, charge-coupled devices, reticons, image tube scanners (ITS), and image photon counting systems (IPCS). For many problems, the photographic plate is being replaced by electronic devices such as ITS and IPCS. The radio-frequency region requires antennas (large interferometer arrays and steerable dishes) and radio-frequency detectors appropriate to the problem at hand. *See* ASTRONOMICAL PHOTOGRAPHY; RADIO ASTRONOMY; TELESCOPE.

Spectrographs employ linear dispersions ranging from several millimeters per angstrom to several hundred angstroms per millimeter (1 angstrom = 0.1 nm). In solar work a host of special instruments are used to secure monochromatic photographs of the Sun, its faint outer envelopes, chromosphere, corona, and transient phenomena such as flares. Special instruments have been devised to measure the magnetic field of the Sun and to study radio-frequency radiation produced at times of high solar activity. Particle, x-ray, ultraviolet, and cosmic rays of solar origin are also observed. *See* ASTRONOMICAL SPECTROSCOPY.

Important advances have been made with the aid of telescopes equipped with x-ray and gamma-ray detectors flown in balloons (stratoscopes), rockets, and satellites. Among the most important of the last are the Copernicus and IUE satellites, which cover the ultraviolet wavelengths longer than the Lyman limit (91.2 nm). The Einstein satellite was equipped with a grazing incidence telescope that actually forms images in soft x-rays and produces high-resolution pictures of x-ray sources in the Galaxy and other galaxies. Some are optically identified. Solar x-rays and ultraviolet radiation can be monitored continuously, and it is possible to secure monochromatic pictures of the Sun in far-ultraviolet radiation. *See* GAMMA-RAY ASTRONOMY; X-RAY ASTRONOMY.

The field of experimental astrophysics involves measurement of shapes and strengths of spectral lines emitted under controlled conditions of temperature and pressure. Techniques involve the use of shock tubes, atomic beams, controlled arcs, beam foils, laser excitation of discrete levels, and so forth. Also involved are measurements of cross sections for atomic and molecular collisional excitation and ionization, low-energy nuclear transformations, and even synthesis of molecular fragments that may be found in interstellar space.

Divisions. Theoretical astrophysics constitutes a branch of theoretical physics and embraces many subjects that usually are considered the province of the latter. For purposes of discussion, astrophysics may be divided somewhat as follows, according to the types of celestial bodies that are involved.

Solar physics. Solar physics includes all phenomena connected with the Sun and overlaps with geophysics in the consideration of solar-terrestrial relationships, for example, the connection between solar activity and auroras, magnetic storms, and sudden ionospheric disturbances. Some aspects of solar physics are concerned with the quiet or undisturbed Sun, for example, the granulation observed in the white light photosphere, problems of the dark-line or Fraunhofer spectrum, the chromosphere, and the corona as observed at sunspot minimum. A study of the active Sun involves the 22.5-year magnetic cycle, sunspots and centers of activity, and flares, which often include ejection of streams of energetic particles, x-ray emission, and nonthermal radio-frequency emission. *See* SUN.

Solar system. The physics of the solar system includes the nature of planetary atmospheres and interiors and the chemical and physical constitution of comets, meteors, and small particles such as those that constitute the zodiacal cloud. In addition, much attention is paid to interplanetary plasmas (the solar wind) and magnetic fields, and to the magnetosphere of the Earth and Jupiter. The study of meteorites offers great opportunities in cosmochronology and cosmochemistry, since they supply essential information on the initial chemical composition and age of the solar system. The techniques involved are those of nuclear and microanalytical chemistry. *See* COMET; COSMOCHEMISTRY; MAGNETOSPHERE; METEOR; METEORITE; PLANETARY PHYSICS; SOLAR WIND.

Stellar atmospheres. The study of stellar atmospheres, including certain aspects of the solar atmosphere, constitutes an important and active field of astrophysics. Not only are apparently normal single stars such as the Sun considered, but also the envelopes of close binaries, variable stars, and stars of unusual chemical composition. *See* STAR.

Nebulae and interstellar matter. Much effort is devoted to gaseous nebulae and the interstellar medium. The former includes examples of both thermal excitation, as in Orion and planetary nebulae, and nonthermal excitation, as in the Crab Nebula. They may be excited by stellar radiation, by dissipation of shock waves, or by synchrotron radiation. The interstellar medium pervades the galactic plasma and extends beyond it as the galactic halo. It is very inhomogeneous, including cool (a few kelvins) clouds with solid grains and often complex molecules all shielded from ultraviolet starlight, relatively attentuated domains at temperatures of a few thousand kelvins, regions where hydrogen is ionized, and extremely hot (100,000-K) volumes presumable produced by supernova detonations. It is the milieu from which stars are being formed and to which they return their outer envelopes when they perish. *See* INTERSTELLAR MATTER; NEBULA.

Stellar structure and evolution. Stellar structure and evolution trace the development of a star from its condensation from the interstellar medium, through the initial gravitational contraction phase, to the state in which it converts hydrogen to helium in its core. It is then called a main-sequence star. When the hydrogen in the central region has all been transmuted into helium, the conversion takes place in a thin shell around the inert core and the star becomes a giant or supergiant. The evolution can now become very complex. Eventually, the outer stellar layers are lost to space, and the core normally settles down to become a dwarf. The escaping shell may form a planetary nebula. If the star is massive, a quiet demise may not occur—the star may explode as a supernova, and the remnant may become a neutron star or even a black hole. Evolution in close binaries may be very complicated as material from one star may be dumped upon the other, inducing outbursts, especially if the second star has evolved into a white dwarf. *See* BINARY STAR; BLACK HOLE; NOVA; STELLAR EVOLUTION; SUPERNOVA; WHITE DWARF STAR.

High-energy astrophysics. Radio astronomy data and observations secured from above the Earth's atmosphere have greatly stimulated the development of high-energy astrophysics, particularly the study of the acceleration of charged particles to high energies in supernova remnants, pulsars, and quasistellar sources. Much attention is paid to galaxies, both normal and exotic. Studies of normal galaxies are concerned with questions such as the relation between a galaxy's mass and angular momentum and its evolution, both dynamical and as reflected in the changing character of the stars that are involved and the chemical composition and physical state of the interstellar medium. Exotic galaxies such as strong radio emitters, Seyfert galaxies, and N galaxies usually contain active nuclei radiating large amounts of energy. *See* GALAXY, EXTERNAL; PULSAR; QUASARS.

General relativity. Verifications of general relativity are astronomical or astrophysical. In addition to classical tests such as deflection of light by the Sun or advance of the perihelion of Mercury, predictions of dissipation of energy by gravity waves have been verified by observations of a pulsar binary. General relativity has to be applied in interpretation of black holes, which are difficult to observe but are believed to occur in certain binary star systems and nuclei of galaxies. It is also used in constructing large-scale models of the physical universe, in cosmology, and in describing the first moments of the universe following the "big bang." *See* COSMOLOGY; RELATIVITY.

Relation to physics. The fields or problems of astrophysics may also be classified in terms of the underlying or supporting fields in physics. Classical mechanics is the basis of the mechanics of the solar and galactic systems; it supplies important data on the masses and density concentrations in the stars. Thermodynamics and statistical mechanics underlie the derivation of the basic equations of L. Boltzmann and M. N. Saha and the equations of molecular dissociation, which are necessary in the calculation of equilibrium conditions in stellar atmospheres and interiors. Geometrical optics and physical optics underlie both radio and optical astronomical instrumentation. Physical optics provides fundamental formulas for the dispersion and absorption of spectral lines and the scattering of light by small particles in space; it lies behind the theory of operation of all radio telescopes. Molecular and atomic structure provide the basis for an interpretation of spectra of the Sun, planets, comets, stars, and nebulae. Solid-state physics provides the background for various detection devices, for the investigation of the formation of grains in space, of meteorites, and of the internal structure of planets. Electricity and magnetism have become very important in astrophysics, particularly in connection with solar physics and synchrotron radiation. The origin and interpretation of the solar cycle, the shapes of gaseous nebulae, and the acceleration of high-energy particles in flares, the Crab Nebula, and other nonthermal sources all require techniques of magnetohydrodynamics. Nuclear physics furnishes information on problems of the generation of energy in stars, on stellar structure and evolution, and on the origin of elements. The importance of general relativity has been noted above.

It is now recognized that the origin of cosmic rays is an astrophysical problem concerned with the acceleration of charged particles to very high energies. Nearly every branch of physics has some application to astronomy. *See* COSMIC RAYS.

Unsolved problems. Some outstanding problems in theoretical and basic experimental astrophysics include the computation of basic atomic parameters such as f values, damping constants, target areas for the collisional excitation of various levels, and molecular dissociation energies, together with experimental checks whenever these can be obtained. Difficult problems are associated with stellar structure, even for normal stars during the quiet phases of their lives. The stability of compressible fluids in gravitational fields where there is energy loss by radiation as well as by convection currents will be an important area of study. The active regions of the Sun, where magnetic fields play an important role, as well as sunspots, the chromosphere, and the corona pose impressive challenges to both observer and theoretician.

The basic physics and chemistry of the interstellar medium are incompletely understood. It is not known how grains are formed, whether complex molecules are built up on the surfaces of grains, or how stars form out of the interstellar medium.

Although the broad principles of stellar structure and evolution seem well enough understood (except for rapid and catastrophic changes), failure to detect solar neutrinos demonstrates that there is an uncertainty either in stellar structure theory or in basic physics. The initial formation of a star is not well understood, and neither are the details of the transition from giant to white dwarf phase.

Among the most striking problems of all are those associated with high-energy astrophysics: the radio galaxies, with their giant lobes of nonthermal radiation; violent events in galactic nuclei; pulsars; and the quasistellar sources whose exact nature is not yet understood.

The most fundamental questions pertain to cosmology—the details of the big bang, and whether the universe will go on expanding forever or collapse upon itself. It is not yet known if the density of matter in the universe is sufficient to slow down and eventually stop the expansion. A better understanding of galaxies, clusters of galaxies, and the intergalactic medium is needed.

[LAWRENCE H. ALLER]

Bibliography: C. Allen, *Astrophysical Quantities*, 1973; L. Aller, *Atoms, Stars and Nebulae*, 1971; G. Burbidge (ed.), *Annual Reviews of Astronomy and Astrophysics*, vols. 1–16, 1963–1978; M. Harwit, *Astrophysical Concepts*, 1973; K. R. Lang, *Astrophysical Formulae*, 1978; B. Middlehurst and G. Kuiper (eds.), *Stars and Stellar Systems*, vols. 2, 4, 6–8, 1960–1978; Symposia of the International Astronomical Union, 1963–1979.

Atomic clock

An electronic clock whose frequency is supplied or governed by the natural resonance frequencies of atoms or molecules of suitable substances. These frequencies are generally in the microwave range from about 1400 to 40,000 MHz in the clocks presently used or under development. Some work is also being done in the optical and infrared frequency ranges. Atomic clocks are the most precise of all clocks; some of those presently available have errors sufficiently small that they would gain or lose less than 1 s in 30,000 years. The present internationally accepted unit of time, the atomic second, is based on an atomic clock and is defined as that time interval during which occur exactly 9,192,631, 770 cycles of the hyperfine resonance frequency of the ground state of the cesium-133 atom. *See* ATOMIC TIME.

Uses. Some uses that require the precision of these clocks are: (1) navigation systems such as loran C and Omega, in which position is determined by the relative times of arrival of radio pulses of electromagnetic energy emitted at precisely known times from several transmitters at different locations; (2) deep-space communications and Doppler navigation, both of which require extremely stable frequencies; (3) tests of relativity theory by comparison of rates of clocks carried on aircraft, rockets, or satellites with that of clocks on the ground. One test already performed involved cesium clocks flown in aircraft at a slow speed at about 10,000 m altitude for 15 hr. On return the airborne clocks gained about 45 nanoseconds with respect to the ground clocks, in agreement with predictions of relativity theory. The main portion of the time difference is due to gravitational potential difference, with a small contribution in the opposite direction from the velocity effect. *See* RELATIVITY.

Classification. The clocks presently in use or under development employ: (1) hyperfine structure (splitting of the spectral lines due to the nuclear magnetic moment) in the ground state of atoms such as cesium, rubidium, hydrogen, or thallium, or in ions such as singly ionized mercury, barium, or magnesium; (2) a molecular line such as the inversion line of ammonia; or (3) a narrow optical or infrared resonance in an atom, molecule, or ion. They may be further classed as passive or active. In a passive clock the atoms or molecules are irradiated with electromagnetic energy close in frequency to their resonance, and some method is provided to sense the difference in frequency and to control the source frequency so as to reduce the error toward zero. The source is usually a stable quartz oscillator at a frequency of a few megahertz followed by a frequency multiplier. Active clocks generate their own signal directly from the atoms or molecules, using the principle of oscillation by stimulated emission of radiation. The microwave frequency signal so generated is low in power, so an oscillator and frequency multiplier chain is usually synchronized to it. In both types some form of frequency synthesis (frequency multiplication by a rational fraction) is necessary to provide an output at a cardinal frequency, such as 1.0 MHz, suitable for clock use. Frequency synthesis into the infrared or optical range is very difficult and requires complex equipment.

Atomic-beam clocks. The atomic-beam magnetic resonance technique for passive clocks was first used by I. I. Rabi, J. R. Zacharias, S. Milman, and P. Kusch. In this method a beam of atoms traverses an evacuated space, in which they are irradiated with the electromagnetic energy and subsequently detected. Deflection magnets are arranged so that only those atoms that undergo a transition (caused by the excitation being at their resonance frequency) are deflected to the detector. The accuracy and stability (smallness of fluctuation) depend on the line width and the signal-to-noise ratio. The line width is inversely proportional to the time the atoms spend in the excitation region, and the signal-to-noise ratio is proportional to the square root of the number of atoms detected per unit time. A good clock will have a narrow line width and a high signal-to-noise ratio. Beam clocks generally use cesium-133 and have line widths of the order of 1 part in 10^7 to 1 part in 10^8 (Fig. 1). The number of atoms detected per second is 10^6 to 10^7. These clocks may be intercompared easily to 1 part in 10^{12} (Fig. 2). The use of thallium was proposed in 1957 by P. Kusch. It has the advantages over cesium of reduced sensitivity to magnetic fields and a higher frequency (21,300 MHz as compared to 9192 MHz) but is much harder to detect and deflect.

Rubidium-gas cells. R. H. Dicke showed that in an atmosphere consisting of an inert buffer gas

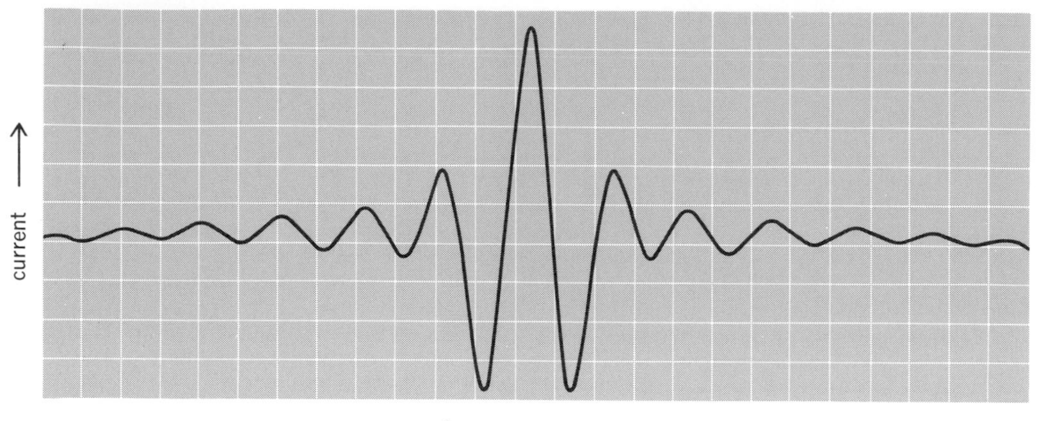

Fig. 1. Resonance line from a cesium-beam tube. The frequency width at half maximum is 250 Hz with the central frequency of 9192 MHz. This gives a line width of 2.7 parts in 10^8 and is very suitable for a good clock.

mixture of suitable pressure atoms could be confined by collisions with the buffer gas atoms which were not perturbing and thus could have a long interaction time with the exciting radiation, leading to narrow lines. This principle is used in passive and active rubidium-gas cell clocks. A small amount of rubidium and the buffer gas are placed in a glass cell inside a microwave cavity and kept at the proper temperature for optimum rubidium vapor pressure. Light from a rubidium lamp is passed first through a filter and then through the cell through windows in the cavity. In the passive clock under proper conditions the light transmitted through the cell is reduced in intensity, if the exciting radiation in the cavity is at the hyperfine resonance frequency (6834 MHz). This decrease in intensity is detected with a photocell and used to control the frequency of the exciting radiation, as described earlier. P. Davidovits and R. Novick were successful in making an active clock from the rubidium-gas cell in the following way. If the filter is properly designed, the proper buffer gas used, and the cavity losses low enough, then the light causes a population excess in the upper hyperfine level of the rubidium atoms and maser oscillator action takes place. In both rubidium clocks the collisions with the buffer gas atoms change the resonance frequency by a nonnegligible amount, and therefore the clocks must be calibrated initially and are subject to drifts in rate, as contrasted to the cesium- and thallium-beam devices.

Ammonia maser. The ammonia maser was first described by J. P. Gordon, H. Zieger, and C. H. Townes. One pair of energy levels in the ground state of the ammonia molecule is separated by 23,800 MHz (the inversion line). It is possible to focus ammonia molecules in the upper state into a cavity resonant at the inversion frequency by means of inhomogeneous electric fields. If the cavity losses are low enough and a large enough number of molecules per unit time enter, and if the cavity is long enough so that the interaction time is sufficient, maser oscillator action will occur. Unfortunately the interaction time is not very long, so that the line width is broad and the clock is not very accurate. The stability for very short times is good, however, as it depends on the power level and frequency of the maser, both relatively high.

Hydrogen maser. H. M. Goldenberg, D. Kleppner, and N. F. Ramsey constructed the first hydrogen maser. In this active clock hydrogen molecules are dissociated into a beam of hydrogen atoms. Those atoms in the upper hyperfine state are

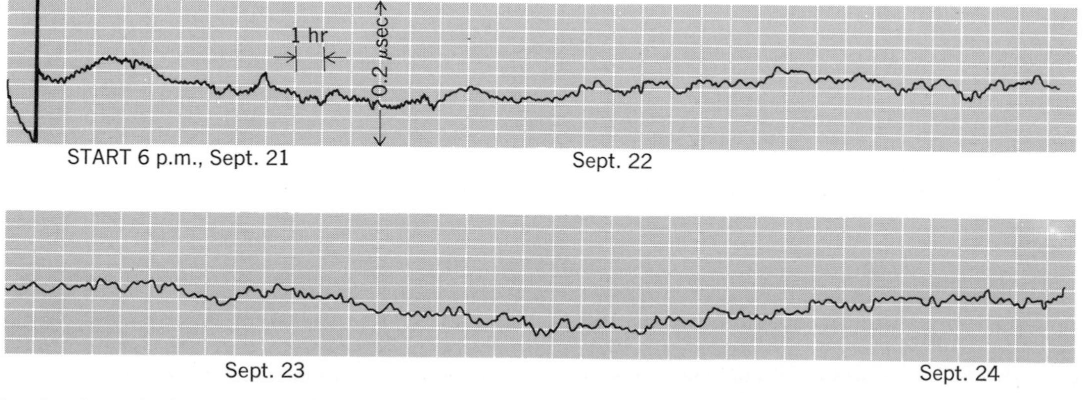

Fig. 2. Record of the time difference between two cesium-beam atomic clocks taken over a period of 3 days. The vertical scale is 2×10^{-7} sec full scale. The difference in rates is much less than 1 part in 10^{12}.

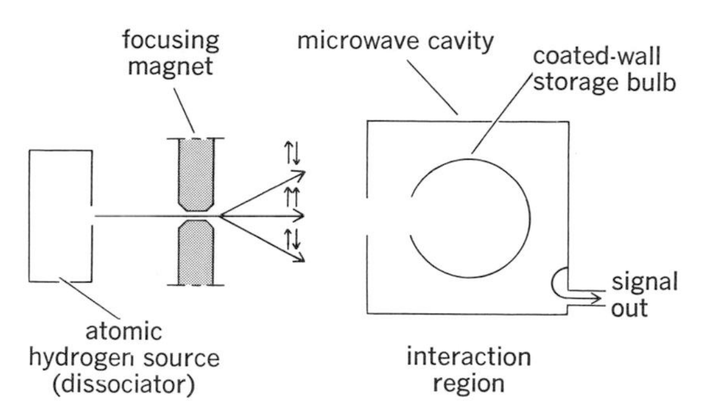

Fig. 3. Basic elements of the hydrogen maser. The hydrogen atoms in the upper hyperfine level are focused into the coated bulb, and the energy coupled out through a loop and coaxial cable.

then focused into a teflon-lined quartz bulb inside a cavity resonant at the hyperfine frequency of 1420 MHz (Fig. 3). The teflon coating allows the atoms to bounce off the walls without much perturbation. As a result of this and the small size of the entrance hole, the atoms can be confined in the bulb for a long time, resulting in interaction times with the exciting radiation of the order of a second. This produces a very narrow line width. If the cavity losses are sufficiently low, maser oscillations take place. The hydrogen maser oscillator is the most stable oscillator known. Since the atoms do make collisions with the wall, there is a slight pulling from the free atom frequency. This results in an uncertainty of only about 2 parts in 10^{12}, so that the hydrogen maser makes an excellent clock. By reducing the hydrogen flux below the threshold for oscillation and providing 1420 MHz excitation, the apparatus can also be used as a passive clock. This greatly reduces frequency pulling effects due to cavity tuning changes and thus reduces the requirements for precise cavity temperature control at the expense of losing the spectral purity of the maser oscillation.

Trapped ion standard. H. Dehmelt proposed trapping ions in a quadrupole radio-frequency trap to provide long life-times and consequently very narrow hyperfine resonance lines. Good preliminary results have been achieved. Work has also been done with a Penning trap.

Saturated absorption technique. R. Barger and J. Hall demonstrated a narrow saturation feature in the 3.39-μm absorption line of methane molecules at low pressure in an infrared cavity excited by laser radiation. The feature appears as a decrease in absorption of the laser light over a very narrow frequency range about the center of the methane absorption line. The standing wave in the cavity can be resolved into two equal-amplitude, oppositely directed running waves. Molecules with a longitudinal velocity component greater than about one-sixth wavelength per state lifetime will see the two running waves Doppler-shifted and clearly resolved in frequency. They can only interact strongly at their resonance frequency and consequently with only one of the running waves. Molecules with a longitudinal component smaller than this see both waves combined essentially coherently and consequently experience on the average an intensity twice that of the other molecules and, therefore, they are more saturated. The frequency width of the saturation feature grows with saturation level and has a minimum value of about $1/6T$, where T is the natural or collision-broadened life-time in seconds. Monitoring the intensity of the laser light transmitted through the methane-filled cavity allows one to slave the laser frequency to the center of the saturated absorption feature. The same technique has been applied to iodine molecules in the visible part of the spectrum. [LEONARD S. CUTLER]

Bibliography: R. E. Beehler, A historical review of atomic frequency standards, *Proc. IEEE*, June 1967; R. P. Feynman, R. B. Leighton, and M. Sands, *The Feynman Lectures on Physics*, 1965; H. Hellwig, K. M. Evenson, and D. J. Wineland, Frequency and physical measurement, *Phys. Today*, 31(12):23–30, December 1978; P. Kartaschoff, *Frequency and Time*, 1978; A. O. McCoubrey, Atomic frequency standards, *Proc. IEEE*, February 1966.

Atomic time

Time based on quantum transitions. According to the quantum theory, an atom or a molecule may exist in discrete states, each of which has a distinct energy level. Energy will be absorbed or radiated when the atom jumps between two states. The frequency of the electromagnetic radiation emitted or absorbed is related to the difference in energy: $f=(E_1-E_2)/h$. Here f is the frequency, E_1 is the initial energy, E_2 is the final energy, and h is Planck's constant. For a selected transition of a given atom, the frequency emitted will remain constant. In contrast, the frequency of oscillation of a pendulum or of a quartz crystal depends upon factors such as size, which may change. *See* QUARTZ CLOCK.

The transition first used to obtain very high precision, in 1955, is the one between the two hyperfine levels of cesium-133 in the ground state. The frequency 9,192,631,770 hertz (Hz) for cesium was determined in 1958. It was adopted in 1967 by the General Conference of Weights and Measures to define the unit of time, the second, in the International System of Units (SI).

A scale of International Atomic Time, or TAI (abbreviation in French), is formed from the operation of a number of cesium-beam atomic clocks, located at various observatories and laboratories about the world, by the Bureau International de l'Heure (International Time Bureau), or BIH, at Paris. Radio transmissions are used to transmit frequency and time. By definition, TAI equaled Universal Time (UT) at 0h 0m 0s on 1 January 1958.

Because of variations in the Earth's speed of rotation, TAI does not provide rotational time, which is needed for civil purposes, navigation, and geodesy. A time system was developed, called Universal Time (Coordinated), or UTC, which differs from TAI by an integral number of seconds. A single second, called a leap second, is added or subtracted from UTC so that UTC and rotational time (technically, denoted UT1) differ by less than 1 sec. The BIH announces when clocks should be changed. Clocks have been retarded once per year, generally, during recent years. In 1975 the General Conference of Weights and Measures recommended that international clock time should be based on UTC. Standard time, the time in common use, differs by only an integral number of hours from UTC. *See* EARTH ROTATION AND ORBITAL MOTION.

Various quantum devices, including masers (microwave region) and lasers (optical region), have been used to generate stable frequency of high precision, notably the hydrogen maser. The cesium-beam oscillator, however, was still the best control device for general use in 1976. Intercomparisons show that stabilities of about 1 part in 10^{12} to 10^{13} are attained for cesium. Relativistic effects become significant at these accuracies, and corrections will be needed for clocks at different locations for the gravitational red shift when further improvements are made. Atomic clocks carried on round-the-world flights in 1971 showed agreement with the predicted time dilation and red shift effects of relativity.

For possible differences between gravitational time and atomic time *see* ATOMIC TIME. *See also* ATOMIC CLOCK. [WILLIAM MARKOWITZ]

Azimuth

An angle measured in the plane of the observer's horizon from one of several arbitrary departure points; also the angle at the observer's zenith between the local meridian and the vertical circle passing through the object being observed. *See* ASTRONOMICAL COORDINATE SYSTEMS; CELESTIAL NAVIGATION.

In astronomical and surveying practice, azimuth has been referenced to each of the four cardinal points of the horizon: north, east, south, and west. In current navigation practice, however, azimuth is measured from the north point of the horizon eastward from 0 to 360°. The horizon angle measured from 0° at the north or south horizon point eastward or westward through 90 or 180° is connoted azimuth angle in contrast to azimuth.

[RAYNOR L. DUNCOMBE]

Bibliography: B. Dutton, *Navigation and Nautical Astronomy*, 12th ed., 1968; U.S. Navy Oceanographic Office, *American Practical Navigator*, H. O. Publ. no. 9, 1962.

Betelgeuse

Alpha Orionis, a very bright, cool, red supergiant star, spectral type M2, temperature near 3000 K. Typical of the low-temperature end of the supergiant sequence, Betelgeuse varies in light and velocity in periods of about 6 years. Because of high luminosity and low temperature, it has an enormous radius (about as big as the orbit of Mars), 300 times that of the Sun. The angular diameter was measured by the stellar interferometer on Mount Wilson and found to be variable, 0″.034–0″.047 of arc – the largest among the half-dozen stars with measurable disks. The star's mean density is only 5×10^{-7} g/cm³. Partly because of low surface gravity, the atmosphere is unstable and shows outward streaming of matter at low temperature and density.

This star has been observed to have a patchy, variable distribution of surface brightness, using the new method of speckle interferometry. Surrounding it is an envelope extending over 10^{15} m, in which emission lines of potassium occur. *See* ASTRONOMICAL PHOTOGRAPHY; STAR.

[JESSE L. GREENSTEIN]

Big bang theory

The theory that the universe began in a state of compression to infinite density and has been expanding since some particular instant that marked the origin of the universe. The big bang is the most generally accepted cosmological theory.

Two observations are at the base of observational big bang cosmology. First, the universe is expanding uniformly, with objects at greater distances receding at a greater velocity. Second, the Earth is bathed in an isotropic glow of radiation that has the characteristics expected from the remnant of a hot primeval fireball.

Cosmological theory in general and the big bang theory in particular are currently based on the theory of gravitation advanced by Albert Einstein in 1916 and known as the general theory of relativity. Though the predictions of the general theory have little effect in the limited sphere of the Earth, they dominate on as large a scale as the universe. *See* RELATIVITY.

Expansion of the universe. Only in the 1920s did it become clear that the "spiral nebulae," clouds of gas with arms spiraling outward from a core, were galaxies on the scale of the Milky Way Galaxy. This was established only in 1925, when Edwin Hubble reported that his observations of variable stars in several galaxies enabled the distance to these galaxies to be determined with some accuracy.

Starting in 1912, V. M. Slipher observed the spectra of several of these spiral nebulae, and discovered that they almost all had large red shifts. According to the Doppler effect, these large red shifts correspond to large velocities of recession from the Earth. Hubble estimated distances to the objects measured by Slipher, and in 1929 showed that there appeared to be a direct correlation between the distance to a galaxy and its velocity of recession. Hubble and M. L. Humason extended this work and by 1931 had clearly established the red shift–distance relation. *See* DOPPLER EFFECT; RED SHIFT.

The relation is known as Hubble's law: $v = H_0 d$, where v is the velocity of recession, d is the distance to the galaxy, and H_0 is a constant known as Hubble's constant. Determining Hubble's constant requires the independent measurement of the distances to galaxies; the red shift can easily be measured on spectra.

The data from which Hubble's law is derived may be plotted on the Hubble diagram (Fig. 1), in which the horizontal axis is the apparent magnitude, corrected for various instrumental and observational effects, and the vertical axis is the red shift $z = \Delta\lambda/\lambda$, where $\Delta\lambda$ is the shift in radiation of wavelength λ. The quantity cz (where $c \cong 3 \times 10^5$ km s^{-1} is the speed of light), also plotted on the vertical axis, is approximately equal to the recession velocity for velocities much smaller than c,

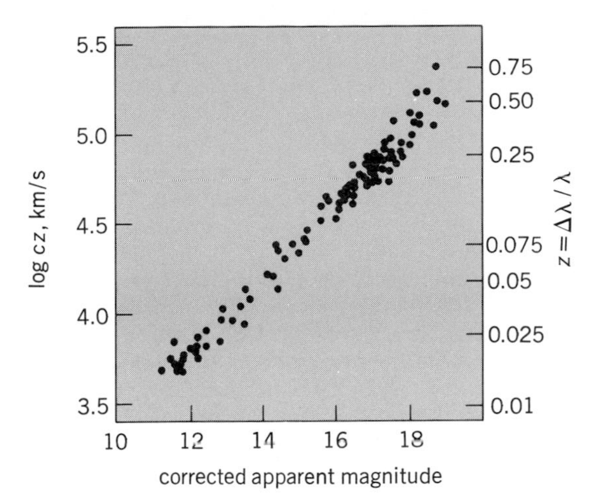

Fig. 1. Hubble diagram (*From J. Kristian, A. Sandage, and J. A. Westphal, The extension of the Hubble diagram, II. New redshifts and photometry of very distant galaxy clusters: First indication of a deviation of the Hubble diagram from a straight line, Astrophys. J., 221:383–394, 1978*)

and the distance to a galaxy of given luminosity is directly related to its apparent magnitude. Thus the Hubble law is expressed by the fact that the data points in Fig. 1 lie near a single straight line.

If velocity is expressed in kilometers per second, and distance is expressed in millions of parsecs (where 1 parsec is the distance from which the radius of the Earth's orbit would subtend 1 second of arc, and is equivalent to 3.26 light-years or 3.09×10^{13} km), then Hubble's constant H_0 is given in km s^{-1} Mpc^{-1}. The measurements of A. Sandage and G. Tammann with the 5-m Hale telescope on Palomar Mountain in California have given the standard value now in use for Hubble's constant, 50 km s^{-1} Mpc^{-1}. Other astronomers such as G. de Vaucouleurs have derived somewhat higher values of Hubble's constant, in the neighborhood of 75 to 100 km s^{-1} Mpc^{-1}. *See* GALAXY, EXTERNAL; HUBBLE CONSTANT.

Relativistic big bang theories. Tracing the expansion of the universe back in time, it is found that the universe was compressed to infinite density approximately 13 to 20×10^9 years ago. According to the leading theories of cosmology, at that time a "big bang" began the universe. It was the origin of time and space. From the time of the big bang, the universe has been expanding.

In 1917 Einstein found a solution to his own set of equations from his general theory of relativity that predicted the nature of the universe. His universe, though, was unstable: it could only be expanding or contracting. This seemed unsatisfactory at the time, for the expansion had not yet been discovered, so Einstein arbitrarily introduced a special term—the cosmological constant—into his equations to make the universe static. The need for the cosmological constant disappeared with Hubble's discovery of the expansion.

W. de Sitter worked out his own solution to Einstein's equations (with the cosmological constant) later in 1917. De Sitter's solutions, though, were valid only for a universe that did not contain any matter. This is not an impossible approximation to the current universe, however, because the cosmic density of matter is very low. Nonetheless, de Sitter's own work in 1930 found the density of the universe too high for his solution to be valid.

In 1922 A. Friedmann worked out solutions that are at the basis of the cosmological models that are now generally accepted. G. Lemaître in 1927 independently found similar solutions. Lemaître's solutions indicated that the original "cosmic egg" from which the universe was expanding was hot and dense. This is the origin of the current view that the universe began from a hot big bang. Friedmann's solution did not involve the cosmological constant and Lemaître's did, but otherwise the solutions were similar.

Early universe. Modern theoretical work has been able to trace the universe back to the first instants in time after the big bang. At the earliest instants, the universe may have been filled with exotic elementary particles of the types now being studied by physicists with large accelerators on Earth. Individual quarks may also have been present. By 1 μs after the big bang, the exotic particles and the quarks had been incorporated in other fundamental particles.

It is not definitely known why there is an apparent excess of matter over antimatter, though attempts in elementary particle physics to unify the electromagnetic, the weak, and the strong forces show promise in explaining the origin of the matter-antimatter asymmetry. The asymmetry seems to have arisen before the first millisecond after the big bang.

By 5 s after the big bang, the temperature had cooled to 10^9 K, and only electrons, positrons, neutrinos, antineutrinos, and photons were important. A few protons and neutrons were mixed in, and they grew relatively more important as the temperature continued to drop. The universe was so dense that photons traveled only a short way before being reabsorbed. By the time 4 min had gone by, elements had started to form.

After about a million years, when the universe cooled to 3000 K and the density dropped sufficiently, the protons and electrons suddenly combined to make hydrogen atoms, a process called recombination. Since hydrogen's spectrum absorbs preferentially at the wavelengths of sets of spectral lines rather than continuously across the spectrum, and since there were no longer free electrons to interact with photons, the universe became transparent at that instant. The average path traveled by a photon—its mean free path—became very large. The blackbody spectrum of the gas at the time of recombination was thus released and has been traveling through space ever since. As the universe expands, this spectrum retains its blackbody shape though its characteristic temperature drops.

Background radiation. Between 1963 and 1965 A. A. Penzias and R. W. Wilson, working with a well-calibrated horn antenna at the Bell Laboratories in Holmdel, NJ, discovered an isotropic source of radio noise whose strength was independent of time of day and of season, and whose intensity at the observing wavelength of 7 cm was equivalent

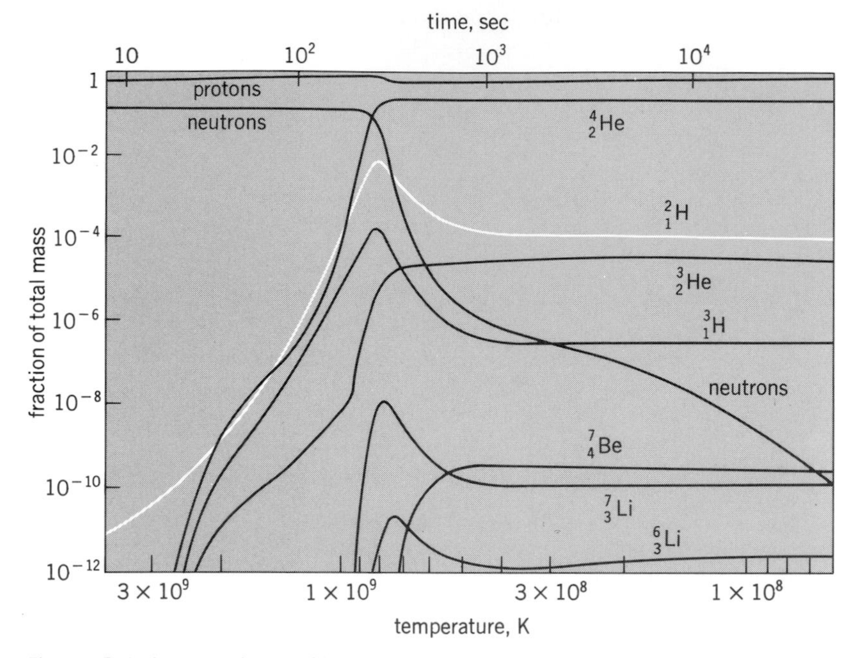

Fig. 2. Relative abundance of isotopes of the light elements in the first few minutes after the big bang according to the model of R. V. Wagoner, for a certain assumed density at some early time. *(From J. M. Pasachoff, Contemporary Astronomy, 2d ed., Saunders College/HRW, 1981)*

to that which would be emitted by a blackbody—an idealized radiating substance—at a temperature of about 3 K. P. J. E. Peebles in R. H. Dicke's group at Princeton had carried out calculations that indicated that the remnant radiation from the big bang might be detectable, and this led to the announcement of both the discovery of the radiation and its interpretation in terms of fossil radiation from the big bang. A prior theoretical prediction, made in the 1940s by G. Gamow, R. Alpher, and G. Hermann, was then recalled. Later other apparatus observed the radiation at other frequencies in the microwave region, but final confirmation of the blackbody nature of the radiation came only in 1975, when balloon observations succeeded in measuring the spectrum of the background radiation in a frequency range that included the peak of the blackbody curve and extended into the infrared.

The best current observations of the spectrum are balloon observations made by D. P. Woody and P. L. Richards. They are in close agreement with a blackbody curve at 2.94 K. *See* COSMIC MICROWAVE RADIATION.

Nucleosynthesis. As the early universe cooled, the temperatures became sufficiently low for element formation to begin. By about 100 s, deuterium (one proton plus one neutron) formed. When joined by another neutron to form tritium, the amalgam soon decayed to form an isotope of helium. Ordinary helium, with still another neutron, also resulted.

Some of the first work on nucleosynthesis was carried out in 1948 by Alpher and Gamow. The current formulation of nucleosynthesis in the big bang is that of R. V. Wagoner, W. A. Fowler, and F. Hoyle, updated in 1972 by Wagoner. The relative abundances of isotopes of the light elements in the first few minutes after the big bang, according to this model, are shown in Fig. 2. Time is shown on the top axis, and the corresponding temperature is shown on the bottom axis. The calculations shown are for a certain assumed density at some early time; if a different density were present, then the curves would be different.

The calculations show that within minutes the temperature drops to 10 K, too low for most nuclear reactions to continue. Most models give a resulting abundance of about 25% of the mass in the form of helium, regardless of the density of the universe. The helium abundance is hard to determine observationally. Current results, about 25% of the mass, are in rough agreement with the theoretical value.

The abundances of others of the light elements are more sensitive to parameters of matter soon after the big bang. Figure 3 shows abundances as a function of the present-day density of matter. From knowledge of the approximate rate of expansion of the universe, the density of matter soon after the big bang can be deduced from the current density, and abundances can then be calculated. In particular, the deuterium abundance is particularly sensitive to the cosmic density at the time of deuterium formation, because the rate at which deuterium is "cooked" into tritium increases rapidly with increasing density.

Big bang nucleosynthesis, although at first thought to be a method of forming all the elements, foundered for the heavy elements at mass numbers 5 and 8. Isotopes of these mass numbers are too unstable to form heavier elements quickly enough. The gap is bridged only in stars, through processes worked out by E. M. Burbidge, G. Burbidge, Fowler, and Hoyle in 1957. Thus the lightest elements were formed as a direct result of the big

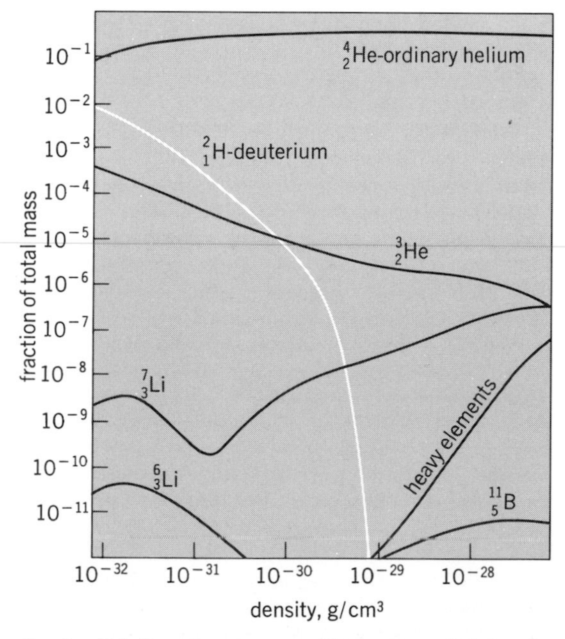

Fig. 3. Relative abundances of isotopes and elements as a function of present-day density of matter, according to the model of R. V. Wagoner. *(From J. M. Pasachoff, Contemporary Astronomy, 2d ed., Saunders College/HRW, 1981)*

bang (Fig. 3), while the heavier elements as well as additional quantities of most of the lighter elements were formed later in stars or supernovae. *See* ELEMENTS AND NUCLIDES, ORIGIN OF.

Open versus closed universe. The two basic possibilities for the future of the universe are that the universe will continue to expand forever, or that it will cease its expansion and begin to contract. It can be shown that the case where the universe will expand forever corresponds to an infinite universe. The term applied is the open universe. The case where the universe will begin to contract corresponds to a finite universe. The term applied is the closed universe.

Deceleration parameter. One basic way to test whether the universe is open or closed is to determine the rate at which the expansion of the universe is slowing, that is, the rate at which it is deviating from Hubble's law. The slowing is measured with a term called q_0, the deceleration parameter. The dividing line is marked by $q_0 = 1/2$, with smaller values corresponding to an open universe and larger values corresponding to a closed universe.

The most obvious way to measure q_0 is by looking at the most distant galaxies or clusters of galaxies, measuring their distances independently from Hubble's law, and plotting Hubble's law in a search for deviations. Deviations have been reported, for example, in the 1978 work by J. Kristian, Sandage, and J. A. Westphal. But all such observations are subject to the considerable uncertainty introduced by the fact that observations deep into space also see far back into time. The galaxies were surely different then than they are now, so evolutionary effects must be taken into account. It is not even known for certain whether galaxies were brighter or dimmer in the distant past, and thus the method may tell more about the evolution of galaxies than about cosmology.

A. Yahil, Sandage, and Tammann analyzed the deceleration parameter through studies of nearby galaxies. They were investigating the differential motions that have been generated since the beginning of the universe by the gravitational accelerations due to nearby fluctuations in the cosmic density. In particular, they studied motions in a complex of galaxies surrounding the Virgo Cluster of galaxies out to a radius of 1000 km s^{-1}, about 20 Mpc using a value of 50 for Hubble's constant. In 1979 they reported $q_0 = 0.06$, corresponding to an open universe.

Cosmic deuterium abundance. Because deuterium is a sensitive indicator of the cosmic density at the time of its formation soon after the big bang, and because deuterium can only be destroyed and not formed in stars, the study of the cosmic deuterium abundance is one of the best methods for determining the future of the universe. Basically, it involves assessing whether there is enough gravity in the universe to halt the expansion.

Counts can be made of all the stars, galaxies, quasars, interstellar matter, and so forth, and the sum gives a density much too low to close the universe. But this method does not assess the amount of invisible matter, which could be in the form of intergalactic gas, black holes, and so forth. Indeed, studies of the motions of galaxies in clusters of galaxies often indicate that much more mass is present inside galaxy clusters than is visible. The amount of this missing mass may be 50 times the amount of visible mass.

Assessing the density of the universe through studies of the deuterium abundance is independent of whether the matter is visible or invisible. Although deuterium is present even on Earth as a trace isotope, with an abundance of 1/6600 that of normal hydrogen in sea water, it is difficult to detect in interstellar space. Finally, in 1972, deuterium was first detected in the interstellar medium through radio astronomical investigations. A number of determinations have followed, including studies in the ultraviolet made with telescopes in space. Determinations of deuterium in the atmospheres of the planets have also been carried out. Though there are some discrepancies remaining between abundances determined in different locations and in different fashions, the deuterium observations seem to indicate clearly that the density of the universe is very low and hence that the universe is open.

X-ray and neutrino background. A diffuse background of x-rays had been observed over the decade of the 1970s. The first NASA High-Energy Astronomy Observatory, HEAO-1, measured its spectrum. These observations led to the suspicion that a lot of hot material, previously undiscovered, may have been present between the galaxies.

But the second High-Energy Astronomy Observatory, known as the Einstein Observatory, changed the picture in 1979. Einstein had the capability of forming x-ray images, and discovered that most or all of the x-ray background came from faint quasars, which appeared on long exposures of even fields that had been thought to be blank. Thus it does not appear that enough matter in the form of hot intergalactic gas is present to close the universe. *See* X-RAY ASTRONOMY.

Evidence suggesting that neutrinos may have a small rest mass could change this picture. Since there are about 100 neutrinos in each cubic centimeter of the universe, much mass would be in that form if the experimental values are confirmed.

Thus, several lines of evidence agree that the universe is open, including studies of differential velocities of nearby galaxies due to density perturbations, the cosmic abundance of deuterium, and the absence of hot intergalactic gas. The existence of neutrino mass is a piece of evidence favoring the opposite conclusion. *See* COSMOLOGY.

[JAY M. PASACHOFF]

Bibliography: J. M. Pasachoff, *Contemporary Astronomy*, 2d ed., 1981; A. A. Penzias,The origin of the elements, *Science*, 205:549–554, 1979; H. L. Shipman, *Black Holes, Quasars, and the Universe*, 2d ed., 1979; J. Silk, *The Big Bang*, 1979; S. Weinberg, *The First Three Minutes*, 1977; R. W. Wilson, The cosmic microwave background radiation, *Science*, 205:866–874, 1979.

Binary star

Two stars held together by their mutual gravitational attraction in a permanent (or at least long-term) association. The term binary system would actually be more appropriate. The two components of the binary system revolve in close elliptical (or circular) orbits around their common center of gravity, which carries them through space so that they also have a common proper motion. Usually treated together with binary stars are multiple systems, consisting of several stars.

Binary stars play an important role in astrophysics for three reasons:

1. They are a very common phenomenon. It is more likely for a given star to be a member of a binary or multiple system than to be single. Out of the 10 nearest stars visible to the naked eye, 6 are double or multiple objects and only 4 are single. Among the 4 is the Sun, which is not, strictly speaking, single because of its planetary system. If planetlike bodies are ignored, the 10 nearest naked-eye objects contain at least 18 individual stars, of which only 4 are single. These statistics are limited, but there is no doubt that binary and multiple stars are the rule rather than the exception; therefore their mere existence deserves attention.

2. Binary systems reveal much more information about stellar properties than do single stars. In particular, masses of stars can be measured only by their gravitational attraction on nearby bodies. The radii, luminosities, and effective temperatures of stars in binary systems can also be obtained, and the mutual relations of these quantities can then be studied. *See* HERTZSPRUNG-RUSSELL DIAGRAM; MASS-LUMINOSITY RELATION.

3. In binary stars where the components interact strongly, new phenomena and new types of stars occur which otherwise would not exist: novae, cataclysmic variables, and binary x-ray sources are examples. Binary nature also plays an important role in metallic (Am) stars, Wolf-Rayet stars, subgiants, and so forth.

Dimensions, orbits, and periods. The dimension of a binary system is determined by the separation of the component stars, that is, the distance between their centers. If the separation is denoted by A, then the period of the orbital motion, P, is given by the relation $P^2 = A^3/(M_1 + M_2)$. In this formula, component masses M_1 and M_2 are expressed in units of the mass of the Sun (2×10^{30} kg), and A is in astronomical units (1 AU $= 1.496 \times 10^8$ km); then the period is in years. According to circumstances, one can measure and observe the absolute motion of either component with respect to the center of gravity, or the relative motion of one component with respect to the other. For a circular orbit, the separation A is simply the radius of the relative orbit. Many orbits are ellipses; then the separation is understood to be the semimajor axis of the relative ellipse. For absolute orbits, referred to the common center of gravity, the period remains the same, the individual orbits are similar ellipses with the same eccentricities, and the semimajor axes are inversely proportional to masses.

Separations vary over an enormous range from one binary system to another. The upper limit is dictated by the general gravitational field of the neighboring stars, which in random encounters would disrupt a binary system in which the gravitational bond is too weak because of a large separation. This upper limit depends critically on the density of stars in space. In the broader vicinity of the Sun, it has been estimated to be between 5×10^3 and 4×10^4 AU. The lower limit on separations is given by the physical dimensions of the component stars. Two stars like the Sun will be in contact (at one point on the surface) when their separation is 1.8×10^6 km. Small compact stars like white degenerate dwarfs or neutron stars could be even closer together, if such pairs exist.

It is easier to measure periods rather than separations; it suffices to find the time interval after which the components return to the same relative positions. For a pair of Sunlike stars, the period corresponding to the upper limit of separations is $10^5 - 10^7$ years. Two Suns in contact would revolve about each other in one-third of a day. Very close pairs of dwarf stars in contact are very common (the W UMa stars). Ultrashort-period binaries are also known; one component is usually a degenerate dwarf. The recurrent nova WZ Sge has a period of 81.6 minutes. Pairs of white dwarfs, if they exist, could have periods shorter than 1 minute. *See* CELESTIAL MECHANICS.

Main categories. With the enormous range of separations and periods, it is not surprising that different observational techniques must be used. Systems of large separations are often observable directly through a telescope as two separate stars. They are then called visual binary stars. For a closer pair, the orbital velocities may be sufficiently large to be detectable by a periodic shift of spectral lines caused by the Doppler effect. These objects are called spectroscopic binaries. When the two stars are so close to each other that their dimensions are not negligible compared to their separation, they can eclipse each other, provided that the orbital plane is suitably oriented. These objects are then easily detected photometrically from the periodic changes of the light received from the system, and are called eclipsing binaries or photometric binary stars. *See* DOPPLER EFFECT.

The occurrence of two stars close together in the sky may be only the result of an accidental projection into the same direction of two stars at very different distances. Such objects are called optical double stars. The term double star includes both physical doubles (binaries) and optical pairs, but the distinction between the terms binary star and double star is often ignored.

If the separation of the components is so large as to make the period longer than about 10^3 years, the orbital motion will be practically imperceptible, but the binary system is nevertheless often recognizable as a common proper-motion pair. The nearest star to the Sun, Proxima Centauri, was discovered as late as 1915 just because of its common proper motion with the much brighter star α Centauri (itself a visual binary). Another famous common proper-motion pair is Mizar and Alcor in the "handle" of the Big Dipper, often used as an informal test of good sight.

Nature of component stars. It appears that two stars of almost any type can form a gravitating pair. Giants are seen combined with main-sequence dwarf stars. Often the difference in temperatures leads to a conspicuous contrast in colors, accentuated by the proximity of the stars as seen in the telescope. Nevertheless, there is a surprising number of pairs of nearly identical stars; this must be an effect related to the process by which binary stars originate. Naturally, faint components in dissimilar pairs are much more difficult to observe. Binary star statistics are very incomplete because of this selection effect.

History. The first visual binary, Mizar, was discovered by G. Riccioli (1650), shortly after the telescope had been introduced into astronomy. Such chance discoveries accumulated and led J. Mitchell (1767) to argue that so many pairs cannot be chance projections. W. Herschel (1803) was the first to detect orbital motion. F. Savary (1827) computed the first orbit, thereby demonstrating the validity of Newton's law of gravitation outside the planetary system.

At the same time when visual binaries were being recognized, J. Goodricke (1783) discovered periodic light changes in Algol and suggested that they might be due to periodic eclipses. Several other eclipsing binaries were subsequently discovered, but the proof of their duplicity had to wait until H. C. Vogel (1889) established spectroscopically the orbital motion in Algol.

Shortly after the introduction of spectroscopy into astronomy, attempts were made to use the Doppler effect for measuring radial velocities. The first spectroscopic binary to be detected was the primary component of Mizar (E. C. Pickering in 1889).

Micrometric observations of visual binary stars were brought to a high standard as early as 1821 by F. G. W. Struve. Soon adequate techniques for orbit computation were also developed. The same happened for the spectroscopic binaries at the turn of the century. H. N. Russell (1912) developed the first good method for interpreting light curves of eclipsing binaries. The precision of photometric observations greatly increased with the introduction of photoelectric photometers, pioneered by J. Stebbins as early as 1910, but not fully exploited before 1950.

Visual binaries. About 75,000 visual binary stars have been discovered and cataloged, but orbits are known for only about 700 of them. Each orbit determination requires many years of painstaking measurements. The relative position of the fainter component (the secondary) with respect to the primary is usually measured with a filar micrometer attached to a long-focus refractor. (This is one of the very few areas in astronomy where the human eye is still directly used in measurements.) As Fig. 1 shows, the orbit is plotted point by point by determining two polar coordinates, the angular separation and the position angle, as functions of time. This apparent orbit is the projection, onto the plane of the sky, of the actual orbital ellipse, which must be reconstructed mathematically.

The angular separation of the components is as a rule very small, and therefore the measurements are affected by serious observational errors. A telescope with an aperture of D centimeters can, under very good conditions, resolve a double star with an angular separation $a=12''/D$, but the atmosphere sets a limit for even the largest telescopes at about 0.″1. A reliable orbit requires

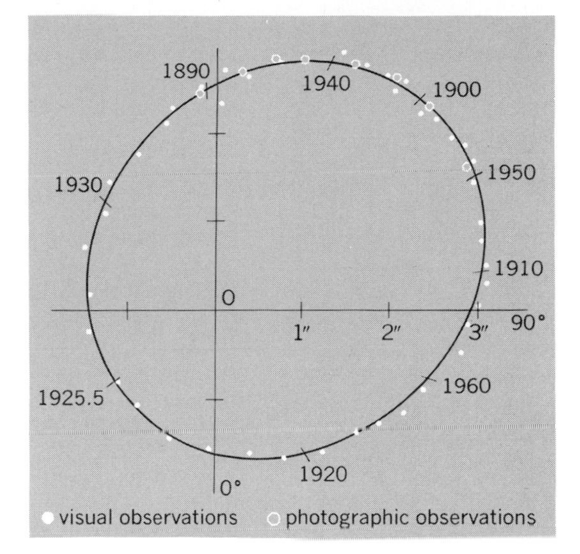

Fig. 1. Apparent orbit of visual binary Krüger 60.

separations of at least 0.″5. The relation between the angular separation a'' and the actual linear separation A in astronomical units is $a'' = A/d$, where d is the distance to the system in parsecs. Thus a binary at a distance of 20 parsecs must have a true separation at least 10 AU, or a period of 32 years for Sunlike stars. Larger separations, better for accurate measurements, quickly lead to long periods. Obviously also, visual binaries can be studied only among the nearby stars. These in turn are almost exclusively dwarf stars, smaller, less massive, and cooler than the Sun.

Within these limitations, visual binary stars furnish valuable fundamental data on stars. If the parallax of the primary star is reliably determined, the distance to the system is known, hence also the true linear separation of the components A. Knowing also the period P, one obtains the sum of the masses of the components, $(M_1 + M_2)$, from Kepler's third law formulated above. Individual masses can be obtained if the motion of each component is measured with respect to the center of mass of the system (that is, absolute orbits are determined). Good knowledge of the distance to the system is a prerequisite, but it helps to obtain luminosities as well, if the light flux from the components is measured photometrically. About 30 visual binary pairs have yielded reliable masses in this way. Their number is growing very slowly, since typical well-determined orbits correspond to periods between 20 and 350 years. The shortest-period orbit measured micrometrically has a period of 1.7 years, but this is a rarity.

Spectroscopic binaries. A binary component moving about the center of gravity of the system in a circular orbit with radius A_1 with a period P has an orbital speed $V_1 = 2A_1/P$. By measuring the periodic shifts of spectral lines caused by the Doppler effect, the radial component of this velocity (along the line of sight) can be determined.

The manner in which the radial velocity changes during the orbital cycle is shown in Fig. 2, which is drawn for the case when the observer is in the orbi-

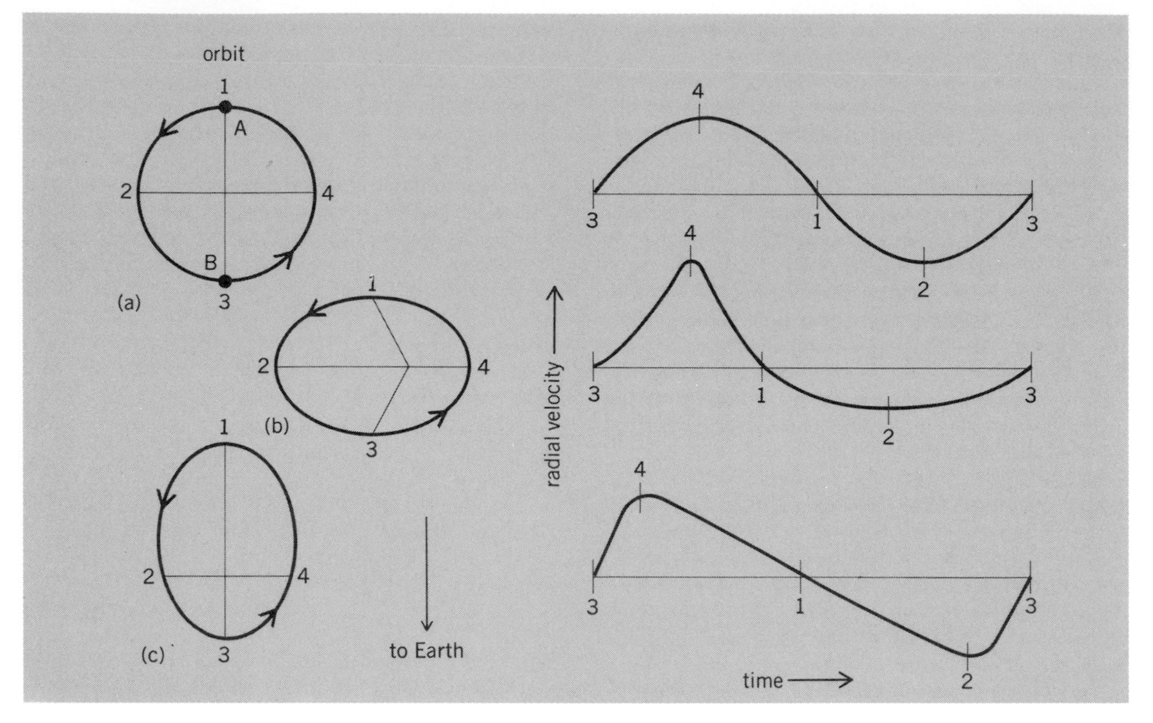

Fig. 2. Diagrams showing possible radial velocity curves (graphs of radial velocity versus time) of spectroscopic binaries. *(a)* Circular orbit. *(b)* Elliptical orbit with eccentricity $e = 0.5$ and major axis perpendicular to direction to Earth. *(c)* Elliptical orbit with $e = 0.5$ and major axis along direction to Earth.

tal plane of the system (here the plane of the paper). Figure 2*a* shows two identical stars A and B moving in circular orbits about their common center of gravity. When either star is in position 2, it is approaching the Earth with its full orbital speed. Since the Doppler shift in this case diminishes the wavelengths of the spectral lines, the radial velocity is taken as negative; thus position 2 corresponds to the bottom of the radial velocity curve (the graph of radial velocity versus time) plotted on the right. Half a period later, the same star will recede from the Earth at maximum speed: position 4 corresponds to the maximum point (4) on the radial velocity curve. When the stars are at points 1 and 3 (conjunctions), they move perpendicularly to the line of sight, and no radial velocity is observed for the components—except that the center of mass has its own radial velocity, which both components share. The radial velocity of only one component is plotted in Fig. 2. The other curve is its mirror image. Thus, at conjunctions, both stars have the same radial velocity, and their spectral lines coincide (Fig. 3*a*). At the quadratures, points 2 and 4, they are most widely split and accessible to observations (Fig. 3*b*).

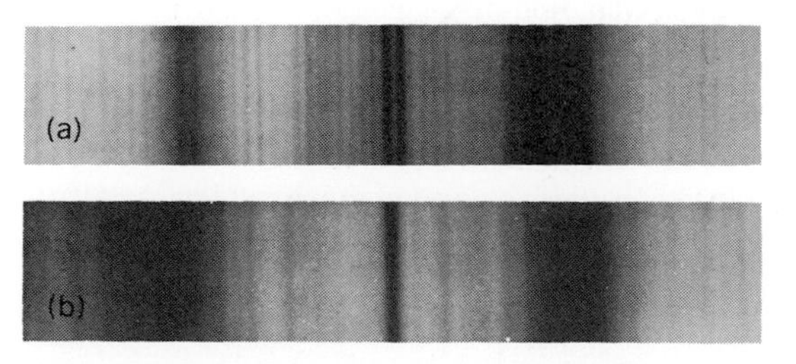

Fig. 3. Spectra of the close binary β Aurigae (period = 3.960 days) at: *(a)* conjunctions, with lines coincident; *(b)* quadratures, with the lines of the two components well separated.

If the orbits are ellipses, the radial velocity curves become skew-symmetrical. Also, the shape of the radial velocity curve depends on the orientation of the orbital ellipse with respect to the observer (Fig. 2*b* and *c*). In each case, the (unplotted) radial velocity curve of the other component is a mirror image of the curve shown. If the mass of the other component is smaller by the ratio M_B/M_A, the amplitude of its curve will be larger at each point by a factor M_A/M_B. Also, in general, the orbit will not be seen exactly edge-on as plotted here. Rather, the orbital plane will be tilted to the plane of the sky (which is perpendicular to the line of sight) at an angle i (the inclination). Inclination $i = 90°$ means that one is looking edge-on, and observes the maximum possible radial velocities. In the general case, one observes the velocities multiplied by a factor sin i. If one is looking at a binary orbit pole-on ($i = 0°$) no radial velocity variations at all are observed. If the two components greatly differ in spectral types but not in brightness, the two spectra may still be recognized; these pairs are called spectrum binaries.

The orbital velocities are larger when the two components are closer together (that is, for shorter periods). Also, larger component masses produce stronger gravitational fields and thus induce larger orbital velocities. Therefore spectroscopic binaries are more often detected among massive hot stars and among short-period systems; the probability of discovery does not directly depend on distance, in contrast to visual binaries.

About 2000 spectroscopic binaries are known, of which less than a half have known orbits. Most of them are single-spectrum binaries, in which the secondary star is so faint that its spectral lines cannot be detected or measured. More information is obtained from double-spectrum binaries, since the ratio of masses then follows from the ratio of the amplitudes of the radial velocity curves. Combined with the known period, these amplitudes also yield masses themselves, provided the inclination is known—but only eclipsing binaries can give this information. Fortunately spectroscopic binaries that are also eclipsing are not rare, and are the best source of data on stars. *See* ASTRONOMICAL SPECTROSCOPY.

Eclipsing binaries. At conjunctions (points 1 and 3 in Fig. 2), the two stars eclipse each other, since in the case represented by Fig. 2 they are exactly aligned on the line of sight. Binary systems with inclinations 90° are of course rare, but because of the finite size of the component stars, eclipses may occur at inclinations less than 90°, provided the separation between the components is sufficiently small.

The effect of the eclipses on the total light received from the binary system is seen from Fig. 4. Figure 4*a* shows the light changes in a typical system displaying alternate total and annular eclipses. The system consists of a smaller but hotter star and a larger but cooler secondary. Maximum light is obtained outside eclipses. Since it is assumed here that the components are spherical, there is no variation in the out-of-eclipse light (as indicated by the flat maxima of the light curve). When the secondary begins to eclipse the brighter primary, light drops rapidly during the phases of the partial eclipse, until the primary component is totally eclipsed. Then only the light of the cooler, fainter star is seen (corresponding to the flat bottom of the light curve). Half a period later, the secondary star is eclipsed, but the secondary minimum of light is less deep: firstly, the secondary contributes less light, and secondly, the primary star is smaller and causes only an annular eclipse of the primary. After another half period, the primary eclipse occurs again.

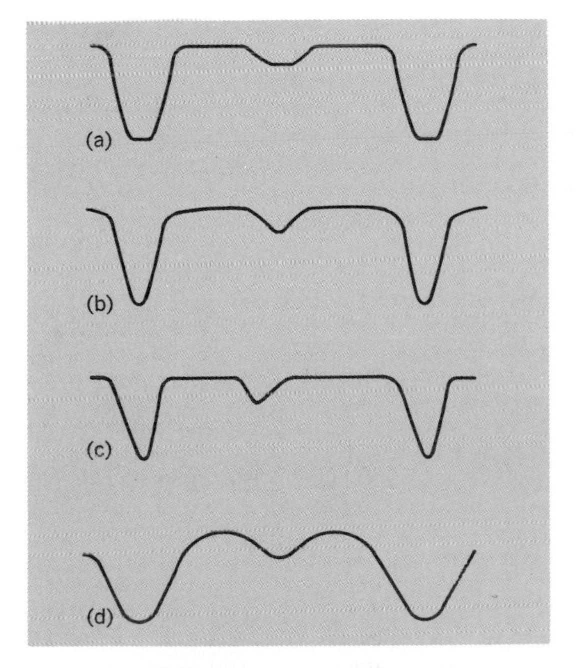

Fig. 4. Light variations from eclipsing binary systems. *(a)* Alternations of total and annular eclipses. *(b)* Partial eclipses; reflection effect present. *(c)* Eccentric orbit; partial eclipses. *(d)* Tidal deformation, showing convex light curve.

Figure 4*b* shows the light curve of a similar system seen at an angle significantly different from 90°. The eclipses are only partial, and their depths diminish as one proceeds to inclinations more and more different from 90°. Figure 4*c* shows again the same system, but with an eccentric orbit. Since the orbital velocity in an elliptical orbit is variable, the intervals between the primary and secondary eclipses are unequal, depending on the orientation of the orbital ellipse, as can be figured out by means of Fig. 2. Figure 4*b* and *d* indicates what kind of other complications may be encountered. When the two stars are close together, their mutual illumination invokes an additional light variation between the eclipses known as the reflection effect (Fig. 4*b*). If the proximity becomes such that the tides which the stars raise on each other appreciably deform their shapes, these tidally distorted ellipsoids will expose varying cross sections to the observer and produce an ellipticity effect (Fig. 4*d*). In systems of low inclination, all the observed light variation may be due to proximity effects only; such systems are known as ellipsoidal variables.

Uncomplicated eclipsing binaries are easy to discover even among faint stars. More than 1000 of them are fairly well known, although only about 200 have well-determined orbits. They yield the value of inclination needed to complete the determination of the spectroscopic elements and obtain masses; moreover, they yield radii from the relative durations of the eclipses. Most eclipsing binaries have periods of a few days, so that the accumulation of observational material proceeds much faster than with the visual binaries. Thus, eclipsing binaries are the most valuable source of data on stars. Such data represent normal single stars only if the presence of the other component does not alter significantly the structure and evolution of the star. In some binaries, this alternative does occur, and new interesting phenomena are observed. *See* ECLIPSING VARIABLE STARS.

Evolution. Stars spend most of their lifetime on the main sequence, generating energy by conversion of hydrogen into helium in their cores. When the hydrogen in the core is exhausted, the stars expand. Single stars encounter no problem in this expansion, but the gravitational field of a nearby companion sets a definite upper limit to the volume available for expansion to a binary star component; it also considerably distorts the shape of the expanded star. By a nomenclature introduced in 1967 by M. Plavec and B. Paczynski, systems in which evolution is substantially affected by the duplicity are called close binary systems. When the expanding component reaches the limiting surface, the evolution of both stars is completely changed. More massive stars evolve faster; therefore it is always the more massive component that reaches the limit first. *See* STELLAR EVOLUTION.

The sequence of shapes through which an expanding component passes is shown in Fig. 5. The curves drawn are cross sections of three-dimensional equipotential surfaces, for a system, with stars of masses M and M', viewed edge-on. Three forces are assumed to affect the structure of the expanding star's outer layers: its own gravitational attraction, the gravitational attraction of the other star, and centrifugal force due to the orbital evolution of the system. Axial rotation is assumed to be synchronized with orbital revolution, as indeed is mostly the case. This Roche model for a binary was introduced by G. Kuiper in 1941 and fully exploited by Z. Kopal in 1954. When a star is still small with respect to the dimensions of the system, it is little affected by the external force field, and is very nearly spherical. As it expands, its shape becomes that of a distorted ellipsoid more and more extended toward the companion, until the star reaches the critical surface marked with a heavy line in Fig. 5. The next equipotential surface already surrounds both stars. A star that has reached its critical surface will lose mass in the vicinity of the point L_1 (the first Lagrangian point) where the effective gravity is zero, and a gas stream will flow from there toward the other star. The stream either impacts directly on the companion or leads to the formation of an accretion disk surrounding it. In either case, the star gains (accretes) mass either directly from the stream or through the disk, in which viscosity makes some particles fall on the star. An as yet unknown part of the mass lost by the "loser" may, however, escape from the system. Observationally, gaseous streams and disks were detected by O. Struve about 1944.

Classification of close binaries. According to Kopal (1955), three main categories are distinguished: (1) In detached systems, both components are smaller than their respective critical lobes. Most unevolved binaries, with components on the main sequence, fall in this category; they are suitable for determining characteristics well representing ordinary single stars. (2) In semidetached systems, one component fills its critical Roche lobe. Many eclipsing binaries, often called Algols after their prototype, are of this type. (3) In contact systems, both components fill their critical lobes, or even overflow them and possess a common envelope. The W UMa systems, dwarf stars with periods less than 1 day, are of this type.

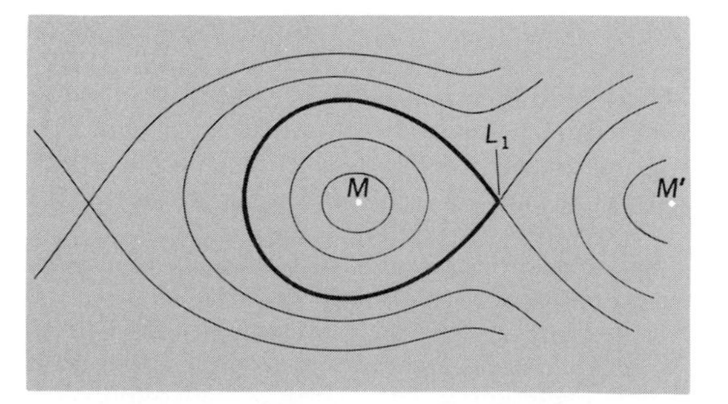

Fig. 5. Nature of equipotential surfaces. The Roche limit is marked by a heavy line.

Evolution with mass transfer. Since the more massive components expand first, it was surprising to find that in the semidetached systems such as Algol it is invariably the less massive star that is seen to fill the critical Roche lobe and lose mass. This Algol paradox was explained in 1967 by actual calculations of stellar evolution with mass loss. When the more massive star reaches its Roche lobe, it will begin to lose mass very rapidly, and in an astronomically short time it becomes the less massive component. Then the mass loss slows down considerably (so that about 10^{-8} solar mass per year is transfered), but the loser still fills its critical lobe. Mass loss stops when the loser either ignites helium (for more massive stars) or loses almost all its hydrogen-rich envelope. In both cases, it shrinks rapidly and becomes either a helium dwarf or a degenerate white dwarf. The rapid mass-transfer process (with rates of 10^{-5} solar mass per year or higher) is too short to be often observed. A few peculiar binaries such as β Lyrae are probably in the rapid stage. The peculiar phenomena seen in β Lyrae may be due to the reaction of the accreting star to the rapid mass influx. The accretion processes have been extensively studied but are still poorly understood.

Some stars, in particular supergiants, are known to lose mass through stellar wind even when they are smaller than their critical lobes; this phenomenon may be enhanced in binary systems or induced by irradiation.

X-ray binaries. Since 1970 x-ray satellites have discovered a number of strong sources of x-rays clearly associated with binary systems: the x-ray source undergoes eclipses or the optical counterpart is a binary, or both. Some sources emit x-rays in regular pulses (Cen X-3 with a period of 4.8 s, Her X-1 at 1.2 s), and the periodic modulation of this average pulsed period over the orbital period (which is 2.1 days for Cen X-3, 1.7 days for Her X-1), interpreted as Doppler shift, permits the orbital elements to be determined. Her X-1 was identified with a peculiar variable star, HZ Her, whose light variations are caused by one hemisphere being heated by x-rays. The source of the x-rays itself must be a compact star, since only such stars can pulse with so short a period; in most cases it is a neutron star, but a black hole is also a possible candidate. The optically visible components are as a rule supergiants, but in Her X-1 the visible star is a dwarf of about 2 solar masses.

The picture which has been developed from study of these systems is as follows. The optically visible star is unstable and loses mass, either by stellar wind or by Roche lobe overflow. The stream flows to the compact star and falls into a very deep "potential well," since the compact star is extremely small. Usually the accretion occurs through a viscous disk, but eventually always a large fraction of the potential energy of the particles is released as radiation (either abruptly in a series of shocks or more slowly in the viscous heating of the accretion disk). The hot gas reaches temperatures of order $50-500 \times 10^{6}$°C and radiates mostly in x-rays. Some of the x-ray binary sources are among the most powerful energy emitters in the Galaxy, radiating up to 10^{38} ergs/s or 10^{31} W (the Sun's total energy output is 4×10^{33} ergs/s or 4×10^{26} W). The regular rapid pulses of x-rays are believed to be generated by a rapid rotation of the magnetized neutron star. *See* PULSAR; X-RAY STAR.

Search for black holes in binary systems. The neutron stars found in binary x-ray sources are thought to be remnants of supernova explosions that in the past terminated the normal stellar lifetime of their parental stars. It is also believed that very massive stars leave black holes as the remnants after their supernova explosion, rather than neutron stars. The boundary between the two cases is determined by the mass of the remnant, and is at the present time very uncertain. Probably no neutron stars can exist with masses larger than about 3 solar masses. A black hole surrounded by an accretion disk in a mass-transferring binary system can produce x-rays equally well as a neutron star. Only the periodic pulses are not expected; rather, one should see rapid irregular intensity fluctuations. The x-ray binary source Cygnus X-1 does display such fluctuations. It is a binary with a 5.6 day period. The system does not eclipse, so the inclination is unknown, and the masses cannot be established reliably. The optically visible star is a hot supergiant, presumably with a mass larger than 20 solar masses. From its orbital motion one must conclude that the x-ray emitter can hardly be less massive than 6 solar masses. Consequently it appears to be a black hole. Other interpretations of the observations have been offered, but none is convincing. Cygnus X-1 is the best candidate for a black hole, but it is almost the only one. One or two other x-ray candidates exist, but the evidence in favor of a black hole is far less convincing.

Naturally, if the other component were rather small and stable against mass loss, no x-rays would come from a system containing a black hole. Since the x-ray evolutionary stage is short-lived, such quiescent systems containing black holes should be more frequent than the more conspicuous x-ray sources. However, the search for black holes in quiescent systsms is much more difficult. The proof must be a negative one: in the suspect system, there must be something massive but nonluminous and unexplained by other alternatives. Several peculiar binaries have been suggested to harbor a black hole (ϵ Aurigae, β Lyrae), but the consensus now goes in favor of alternative hypotheses. If black holes exist, they do not seem to play an important role among the known binary stars. On the other hand, it appears to be virtually impossible to detect black holes (as remnants of supernova explosions) other than in binary stars. The only thing that is left after the star has collapsed into a black hole is its gravitational field, which is always easiest to detect in binary systems. So the search for black holes in binary systems must continue if the question of the existence of these objects is to be resolved. *See* BLACK HOLE.

Cataclysmic variables. White degenerate dwarfs, the end products of evolution of stars of small masses, can hardly be strong x-ray sources because they are relatively large (about the size of the Earth), and therefore the free-fall velocity of the impacting material is not high enough to power x-ray sources stronger than about 10^{36} ergs/s (10^{29} W). Nevertheless, accreting white dwarfs do produce spectacular phenomena.

After several previous discoveries, a survey undertaken by R. P. Kraft (1964) showed that probably all novae are close binary systems. Novae are a special class (conspicuous because of a very high increase in light in an outburst) of cataclysmic variables. Other classes include recurrent novae (in which the eruptions are less powerful but repeatable over decades or centuries) and dwarf novae (U Gem stars, which display even smaller outbursts on time scales of months). All these objects appear to be binaries of a basically similar nature: a cooler object, usually a red dwarf (but occasionally, as in the recurrent nova T CrB, a giant), is unstable and transfers mass to a hotter and much more compact object, a degenerate white dwarf or a similar body. A viscous disk surrounding the accreting star plays an important role. Often it is so dense that a hot spot forms where the stream impacts. A hot spot or a hot region can naturally also form at the surface of the star proper. Variable rates of mass transfer, variable viscosity, and continual interaction between the star, the disk, and the stream then lead to instabilities and eruptive events. In the more violent cases, thermonuclear reactions at the surface of the accreting star are believed to be the source of the energy for a nova outburst. *See* NOVA; VARIABLE STAR.

Origin. Just what fraction of all stellar objects are binary stars is very difficult to determine because of the large number of selection effects in their discoveries. Estimates run between 30 and 60%. In any case, it is clear that the formation of binary stars must be a very common process related to stellar formation itself. Capture of one star by another requires the presence of a third body to carry away the excess of energy, and is therefore an inefficient process. Fission of a rapidly rotating star into two is hampered by the large density concentration in stars; perhaps it could occur at the very early stages of stellar evolution, and would then probably account for some binaries with rather unequal masses. According to L. Lucy, the bulk of binary systems are probably formed by successive fragmentations of protostars.

Triple and multiple systems. If fragmentation of the protostar is the main process producing binary systems, one should anticipate systems with more than two components. Indeed, it is estimated that about one-third of all binary stars are actually triple stars, and probably again one-third of all triple systems will eventually be found to be actually quadruple. Perhaps this chain continues. These multiple systems are as a rule hierarchically organized. A triple system consists typically of a close binary pair accompanied by a much more distant third component, which is as a rule less massive and fainter (Fig. 6*a*). A good example is the pair α Centauri accompanied by Proxima Centauri. Quadruple systems usually consist of two close pairs separated by a large distance (for example, ϵ Lyrae; Fig. 6*b*).

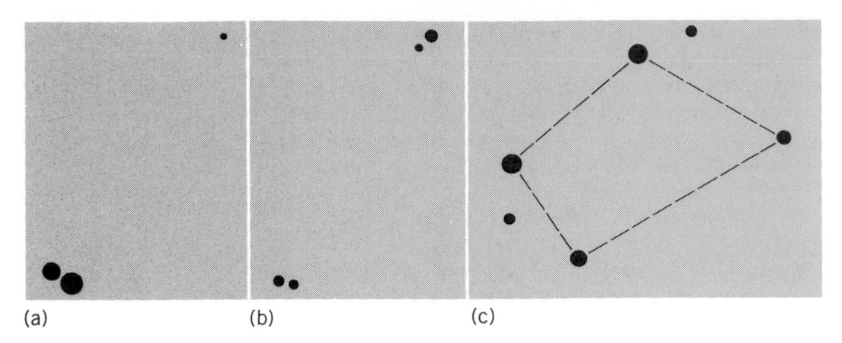

Fig. 6. Multiple stellar systems. *(a)* A hierarchical triple system (α Centauri). *(b)* A hierarchical quadruple system (ϵ Lyrae). *(c)* A system of the Trapezium type (θ' Orionis).

Numerical integrations of the equations of motion of systems with more than two components show that these systems are unstable if the components have comparable masses and are separated by comparable distances. In a relatively short time, one body (usually the least massive one) is ejected or pushed into a remote orbit, while the other two are locked in a close system—all in agreement with observations.

Seemingly contradicting these calculations are multiple systems of the trapezoid type (Fig. 6*c*), named after the famous Trapezium (θ Orionis) in the Orion Nebula. Trapezia are, however, always formed by very young luminous O and B stars, and are therefore young formations themselves, unstable on a longer time scale. Some may, however, outlive their member stars, which evolve extremely fast because of their large masses and high luminosities.

Astrometric binaries. Very careful measurements of the proper motions of the nearest stars disclosed that some of them show deviations from a rectilinear motion. Some of these deviations, although very minute, appeared periodic, and it was concluded that these stars (such as Barnard's Star or 70 Ophiuchi) are attended by companions of low mass. Systems discovered by these astrometric measurements are called astrometric binaries. The first white dwarf, Sirius B, was discovered in this way by F. W. Bessel in 1844, but subsequently observed directly by A. G. Clark in 1862. However, in the cases mentioned above, the companion masses turn out to be nearer to the mass of Jupiter than to the mass of a genuine star, and it must be assumed that these bodies may be nonluminous. They were considered as transitions between binary stars and planetary systems, and a similar origin for both types of systems was tentatively suggested. *See* ASTROMETRY.

However, more recently the discoveries of these planetlike bodies have been challenged because the periodicities may be spurious, and most of the observed deviations might apparently be explained as observational errors. The problem has not been decided definitely. It appears now that the origin of the binary stars (by fragmentation of a protostar, or by fission) is different from the origin of the planets, which are probably formed by condensations in the protoplanetary disk which surrounds an individual star after it has been fragmented out. *See* STAR.

[MIROSLAV PLAVEC]

Bibliography: A. H. Batten, *Binary and Multiple Systems of Stars*, 1973; W. Heintz, *Double Stars*, 1978; Z. Kopal, *Dynamics of Close Binary Systems, 1978;* J. Sahade and F. B. Wood, *Interacting Binary Stars*, 1978.

Black dwarf

A star that cannot generate thermonuclear energy. Not necessarily invisible, it may radiate its internal heat. None has ever been observed, but as-yet invisible companions of nearby faint stars have been suspected of having masses within the probable range. Isolated very low masses cannot form by gravitational contraction from interstellar gas and dust. Thermonuclear energy is released if the core temperatures (proportional to the mass of a self-gravitating gas sphere) exceed 2×10^6 K. Thus, black dwarf masses should be between 0.007 and 0.07 solar mass. During gravitational contraction, a black dwarf lies on a convective, Hayashi track for 10^7 to 10^8 years and is briefly luminous as it radiates gravitational energy. At its final radius, near 0.1 solar radius, the density rises and only the Fermi pressure of degenerate electrons can support the weight. A star of a given mass contracts to the unique radius proper to a degenerate hydrogen-rich white dwarf. Subsequent evolution is by loss of internal heat as core and surface cool. In this brown dwarf stage, after 5×10^9 years, a star of 0.04 solar mass would have 10^{-4} the luminosity of the Sun, a surface temperature of 800 K, visible only by infrared. *See* STELLAR EVOLUTION; WHITE DWARF STAR.

Jupiter, 0.00095 solar mass, has an infrared luminosity near 10^{-9} the Sun, from leakage of internal heat. Saturn possibly also radiates energy from phase changes and gravitational diffusion. Solids are strong enough to support a mass as large as that of Jupiter, a low-density planet. Appreciably larger masses must compress to the higher-density black dwarf stage. *See* JUPITER; SATURN.

[JESSE L. GREENSTEIN]

Black hole

One of the end points of gravitational collapse, in which the collapsing matter fades from view, leaving only a center of gravitational attraction behind. General relativity predicts that, if a star of more than about 3 solar masses has completely burned its nuclear fuel, it should collapse to a configuration known as a black hole. The resulting object is independent of the properties of the matter that produced it and can be completely described by stating its mass and spin. The most striking feature of this object is the existence of a surface, called the horizon, which completely encloses the collapsed matter. The horizon is an ideal one-way membrane; that is, particles and light can go inward through the surface, but none can go outward. As a result, the object is dark, that is, black, and hides from view a finite region of space (a hole). Arguments concerning the existence of black holes originally centered on the fact that there are many stars of over 3 solar masses and that there seemed to be no other outcome of collapse than the formation of a black hole. In 1971, however, some direct observational evidence was obtained for a black hole in the binary x-ray system

Cygnus X-1. *See* GRAVITATIONAL COLLAPSE; RELATIVITY.

Theory. Shortly after Albert Einstein formulated the general theory of relativity in 1916, the solution of the field equations corresponding to a nonrotating black hole was found. For many years this solution, called the Schwarzschild solution, was taken to represent only the gravitational attraction outside a spherical star. More than 20 years elapsed before it was shown that such a black hole could, and probably would, be formed in the gravitational collapse of a nonrotating star of sufficient mass. It was not until 1963 that the solution for a spinning black hole, the Kerr solution, was found. This was particularly important, since most stars are rotating, and the rotation rate is expected to increase when such stars collapse. However, to this date, no one has shown that collapsing rotating stars must form Kerr black holes, in the same manner as was done for nonrotating stars. Black hole solutions have also been found for the case in which the black holes have a charge, that is, an electrical as well as a gravitational influence. However, since matter on the large scale is electrically neutral, black holes with any significant charge are not expected in astronomy.

Uniqueness theorems about black holes make it likely that at least some Kerr black holes would be formed. These theorems show that only three numbers, mass, spin, and charge, completely specify a black hole. Further, any distortion of a black hole, such as is caused by a chunk of matter falling inside, is removed by a burst of radiation. Therefore, although the collapse of a rotating star would be quite complicated, it appears that the final system, the Kerr black hole, would be relatively simple and independent of the details of collapse.

The possible formation of black holes depends critically on what other end points of stellar evolution are possible. There can always be chunks of cold matter which are stable, but their mass must be considerably less than that of the Sun. For masses on the order of a solar mass, only two stable configurations are known for cold, evolved matter. The first, the white dwarf, is supported against gravitational collapse by the same quantum forces that keep atoms from collapsing. However, these forces cannot support a star which has a mass in excess of 1.2 solar masses, this limiting value being known as the Chandrasekhar limit. The second stable configuration, the neutron star, is supported against gravitational collapse by the same forces that keep the nucleus of an atom from collapsing. There is also a maximum mass for a neutron star, estimated to be between 1 and 3 solar masses, the uncertainty being due to the poor knowledge of nuclear forces at high densities. Both white dwarfs and neutron stars have been observed, the former for many years and the latter only recently in the studies of pulsars and binary x-ray sources. *See* PULSAR; WHITE DWARF STAR.

It would appear from the theory that if a collapsing star of over 3 solar masses does not eject matter, it has no choice but to become a black hole. Since there are many stars of over 3 solar masses, and more massive stars evolve more quickly, a considerable number of black holes should have formed. One major problem is that, since the black hole is dark, it is essentially unobservable. Fortunately, some black holes may be observable in the sense that the black hole maintains its gravitational influence on other matter, and thus it can make its presence known.

Structure. For a nonrotating black hole, the radius of the horizon (Schwarzschild radius) is determined entirely by the mass. Defining R so that the surface area of the spherical horizon is $4\pi R^2$, the equation relating R to the mass M is $R = 2GM/c^2$, where G is the constant of gravity and c is the speed of light. Classical general relativity would allow M to take on all possible values, but quantum effects suggest that the lowest possible value of M is about 10^{-8} kg. However, the lower-mass black holes may not be astronomically relevant, since collapsing stars with masses less than about a solar mass (2×10^{30} kg) would become white dwarfs or neutron stars. It is thought that low-mass black holes could exist only if they were created at the time of the origin of the universe.

An astronomical black hole of 5 solar masses would have a radius of about 20 km. This size is comparable to that expected for neutron stars. The density to which matter would have to be compressed in order to form such a black hole is comparable to that found in neutron stars or in the nuclei of atoms. Black holes of 1,000,000 to 100,000,000 solar masses may also exist in the centers of some galaxies, including the Milky Way. These could be formed either in the collapse of a supermassive star or in the coalescing of a large number of black holes of more modest mass. The density required to form these very massive black holes approaches that of ordinary terrestrial densities.

For nonrotating black holes, the horizon is also a surface of infinite red shift; that is, light emitted from just above the surface reaches a distant observer with a much lower frequency and energy than it had when it was emitted. As a result, an object which falls into a black hole appears to an observer to take an infinite time to reach the Schwarzschild radius, with the observed light coming from the object red-shifting quickly to darkness as the approach is made. The picture would be quite different for a person riding on the falling object. The "rider" would reach the Schwarzschild radius in a finite time, feeling nothing more unusual than gravitational tidal forces. However, once the person was inside, he would be trapped, since even light which moves outward cannot escape. The theory predicts that this individual, as well as everything else within the horizon, would be crushed to infinite density within a short time.

For rotating black holes, the surface of infinite red shift lies outside the horizon except at the poles, as illustrated in Fig. 1. The region between the two surfaces is called the ergosphere. This region is important because it contains particle trajectories which have negative energy relative to an outside observer. Roger Penrose showed that it is possible, though perhaps unlikely astronomically, to use these trajectories to recover even more than the rest mass energy of matter sent into a rotating black hole, the extra energy coming from the slowing down of the rotation of the black hole. Others have proposed that radiation incident on a

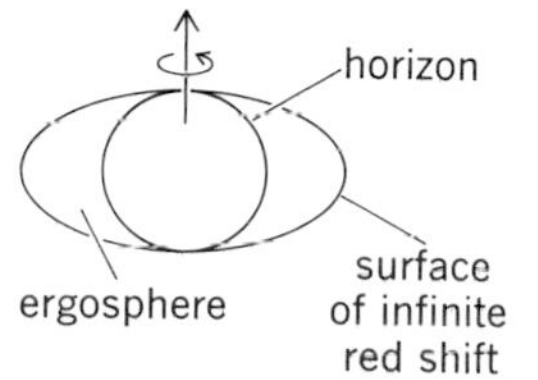

Fig. 1. The Kerr, or rotating, black hole. (*P. C. Peters, Black holes: New horizons in gravitational theory, Amer. Sci. 62(5):575–583, 1974*)

rotating black hole could be similarly amplified.

The black hole solutions of general relativity are completely stable. Once black holes form, they will remain forever, and subsequent processes, for example, the accumulation of matter, only increase their size. However, Steven Hawking showed that, if quantum effects are taken into account, black holes ought to radiate, with the smallest-mass black holes emitting the strongest radiation For astronomical black holes formed from collapse, this loss by radiation is insignificant compared to the gain by adsorption of radiation which exists throughout space.

Observation. Because black holes themselves are unobservable, their existence must be inferred from their effect on other matter. Thus a theoretical model must be constructed to explain the observational data and to show that the most reasonable explanation is that a black hole is responsible for the observed effects. Such is the case with the binary x-ray star system Cygnus X-1. There are a number of binary x-ray systems known. The model which best explains the data is one in which a fairly normal star is in mutual orbit about a very compact object. Because these two are so close, mass flows from the star onto an accreting disk about the compact object, as shown in Fig. 2. As the mass in the disk spirals inward, it

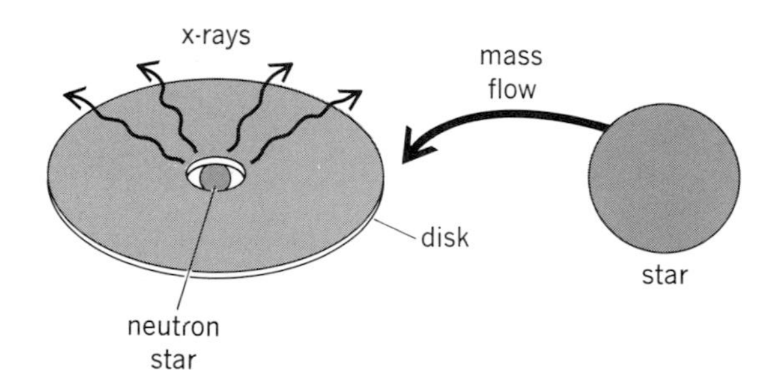

Fig. 2. Schematic picture of a binary x-ray source. (*P. C. Peters, Black holes: New horizons in gravitational theory, Amer. Sci. 62(5):575–583, 1974*)

heats up by frictional forces. Because the central body is so compact, the matter heats to a temperature at which thermal x-rays are produced. The only compact objects known that could accomplish this are neutron stars and black holes. The existence of very short time bursts of radiation also points to an object of small diameter, that is, compact. In some of these binary x-ray systems, there is also a regular pulsed component to the x-rays, indicating a rotating neutron star (by reasoning similar to that given for pulsars). In these systems, the compact object could not be a black hole because that would imply a more complicated structure than a black hole would allow. In other systems, however, there are only irregular pulsations or fluctuations; they are candidates for possible black holes. In one of these systems, the Cygnus X-1, a rough mass determination has been made, with the result that the compact object must be more massive than 5 solar masses. According to theory, there is no compact system with this mass other than a black hole. It is in this sense that it can finally be said that black holes have been observed. *See* BINARY STAR; X-RAY ASTRONOMY; X-RAY STAR. [PHILIP C. PETERS]

Bibliography: R. Penrose, *Sci. Amer.*, 226(5): 38–46, 1972; P. C. Peters, *Amer. Sci.*, 62(5):575–583, 1974; R. Ruffini and J. A. Wheeler, *Phys. Today*, 24(1):30–41, 1971; K. S. Thorne, *Sci. Amer.*, 217(5):88–98, 1967, and 231(6):32–43, 1974.

Boötes

The Bear Driver, in astronomy, a northern and summer constellation. Boötes is one of the earliest recorded constellations. Arcturus, an orange, first-magnitude navigational star, dominates the constellation. It has been known and admired for ages and is one of the few stars mentioned in the Bible. Five prominent stars in this constellation, β, γ, ρ, ϵ, and δ, form a pentagon, shaped much like an elongated kite, with Arcturus at the junction of the tail. Boötes is conventionally pictured as a driver of the Bear (Ursa Major) nearby. In a more recent version he is seen seated, smoking a pipe, his feet dangling and with one hand holding the leash of the Hunting Dogs (Canes Venatici). *See* CONSTELLATION. [CHING-SUNG YU]

Calendar

A list, usually in the form of a table, showing the correspondence between days of the week and days of the month; also, a list of special observances with their dates, especially those that fall on different dates in different years, as a church calendar; also, a set of rules that serves to attach to any day a specific number or name or combination of number and name. The calendars used at various times and places are numerous and diverse. Most represent attempts to divide the synodic month or the tropical year into numbered days. According to whether the emphasis is on month or year, calendars are called lunar or solar.

Gregorian calendar. The calendar used for civil purposes throughout the world, known in Western countries as the Gregorian calendar, was established by Pope Gregory XIII, who decreed that the day following Thursday, Oct. 4, 1582, should be Friday, Oct. 15, 1582, and that thereafter centennial years (1600, 1700, and so on) should be leap years only when divisible by 400 (1600, 2000, and so on), other years being leap years when divisible by four, as previously. The effect of the reform was to restore the vernal equinox (beginning of spring, which governs the observance of Easter) to Mar. 21, and to reduce the number of days in 400 calendar years by 3, making the average number of days in the calendar year 365.2425, instead of 365.25 as formerly, and bringing it into closer agreement with the tropical year, which has 365.24220 days. The new calendar was adopted by Great Britain in 1752, but did not become universal until well into the 20th century.

Although the Gregorian calendar is a solar calendar, lunar vestiges remain in it, as is shown by the division of the year into months. The week is independent of any astronomical phenomenon; it was established before the beginning of the Christian era, and the cyclical succession of the days of the week has not since been disturbed. The connection between the week and the ancient Jew-

Perpetual calendar*

Day of the week for any known date from the beginning of the Christian Era to the year 2400

Year				Julian calendar							Gregorian calendar				
			Century	0 700 1400	100 800 1500†	200 900	300 1000	400 1100	500 1200	600 1300	1500‡	1600 2000	1700 2100	1800 2200	1900 2300
0				. . .	ED	FE	GF	AG	BA	CB	. . .	BA	C	E	G
1	29	57	85	B	C	D	E	F	G	A	F	G	B	D	F
2	30	58	86	A	B	C	D	E	F	G	E	F	A	C	E
3	31	59	87	G	A	B	C	D	E	F	D	E	G	B	D
4	32	60	88	FE	GF	AG	BA	CB	DC	ED	CB	DC	FE	AG	CB
5	33	61	89	D	E	F	G	A	B	C	A	B	D	F	A
6	34	62	90	C	D	E	F	G	A	B	G	A	C	E	G
7	35	63	91	B	C	D	E	F	G	A	F	G	B	D	F
8	36	64	92	AG	BA	CB	DC	ED	FE	GF	ED	FE	AG	CB	ED
9	37	65	93	F	G	A	B	C	D	E	C	D	F	A	C
10	38	66	94	E	F	G	A	B	C	D	B	C	E	G	B
11	39	67	95	D	E	F	G	A	B	C	A	B	D	F	A
12	40	68	96	CB	DC	ED	FE	GF	AG	BA	GF	AG	CB	ED	GF
13	41	69	97	A	B	C	D	E	F	G	E	F	A	C	E
14	42	70	98	G	A	B	C	D	E	F	D	E	G	B	D
15	43	71	99	F	G	A	B	C	D	E	C	D	F	A	C
16	44	72		ED	FE	GF	AG	BA	CB	DC	. . .	CB	ED	GF	BA
17	45	73		C	D	E	F	G	A	B	. . .	A	C	E	G
18	46	74		B	C	D	E	F	G	A	. . .	G	B	D	F
19	47	75		A	B	C	D	E	F	G	. . .	F	A	C	E
20	48	76		GF	AG	BA	CB	DC	ED	FE	. . .	ED	GF	BA	DC
21	49	77		E	F	G	A	B	C	D	. . .	C	E	G	B
22	50	78		D	E	F	G	A	B	C	. . .	B	D	F	A
23	51	79		C	D	E	F	G	A	B	. . .	A	C	E	G
24	52	80		BA	CB	DC	ED	FE	GF	AG	. . .	GF	BA	DC	FE
25	53	81		G	A	B	C	D	E	F	. . .	E	G	B	D
26	54	82		F	G	A	B	C	D	E	C	D	F	A	C
27	55	83		E	F	G	A	B	C	D	B	C	E	G	B
28	56	84		DC	ED	FE	GF	AG	BA	CB	AG	BA	DC	FE	AG

Month					Dominical letter						
Jan., Oct.					A	B	C	D	E	F	G
Feb., Mar., Nov.					D	E	F	G	A	B	C
Apr., July					G	A	B	C	D	E	F
May					B	C	D	E	F	G	A
June					E	F	G	A	B	C	D
Aug.					C	D	E	F	G	A	B
Sept., Dec.					F	G	A	B	C	D	E
1	8	15	22	29	Sun.	Sat.	Fri.	Thurs.	Wed.	Tues.	Mon.
2	9	16	23	30	Mon.	Sun.	Sat.	Fri.	Thurs.	Wed.	Tues.
3	10	17	24	31	Tues.	Mon.	Sun.	Sat.	Fri.	Thurs.	Wed.
4	11	18	25		Wed.	Tues.	Mon.	Sun.	Sat.	Fri.	Thurs.
5	12	19	26		Thurs.	Wed.	Tues.	Mon.	Sun.	Sat.	Fri.
6	13	20	27		Fri.	Thurs.	Wed.	Tues.	Mon.	Sun.	Sat.
7	14	21	28		Sat.	Fri.	Thurs.	Wed.	Tues.	Mon.	Sun.

To find the calendar for any year of the Christian Era, first find the Dominical letter for the year in the upper section of the table. Two letters are given for leap years; the first is to be used for January and February, the second for the other months. In the lower section of the table, find the column in which the Dominical letter for the year is in the same line with the month for which the calendar is desired; this column gives the days of the week that are to be used with the month. For example, in the table of Dominical letters the letter for 1962 is G; in the line with July, this letter occurs in the first column; hence July 4, 1962, is Wednesday.

*From *Smithsonian Physical Tables*, 9th ed., Washington, D.C.
†On and before 1582, Oct. 4 only. ‡On and after 1582, Oct. 15 only.

ish cycle of seven days has not been established; for example, it is not known whether the modern Saturday is identical with the ancient seventh day.

The calendar generally used before the adoption of the Gregorian calendar is called the Julian calendar; their relationship is shown in the table. Dates recorded for civil and commercial purposes before Oct. 4, 1582, may be assumed to be in the Julian calendar, but dates recorded since then may require considerable care in interpretation.

Calendar reform. The Gregorian calendar is not perfectly adjusted to the tropical year, the error accumulating to 1 day after about 3300 years. Numerous proposals for calendar reform have been made with the object of reducing the error, and there have also been many proposals for altering the succession of days for the purpose of bringing greater regularity into the calendar. For example, one proposal would divide the year into 13 months of 4 weeks each, which would make 364 days. The 365th day, and the 366th in leap years, would have a special name and would not belong to any month or to any week. Thus, every month of every year would begin on Sunday and end with Saturday, and annual calendars would become unnecessary. The 13-month calendar has the disadvantage, serious from the standpoint of business, of not being divisible into quarters. Another proposal, which overcomes this difficulty, retains 12 months but would give them successively 31, 30, 30, 31, 30, 30, 31, 30, 30, 31, 30, 30 days, the extra day or days again being outside the system. No attempt to regularize the calendar can succeed if the succession of weekdays and the average length of the calendar year are both preserved. *See* TIME. [GERALD M. CLEMENCE]

Bibliography: E. Achelis, *Of Time and the Calendar*, 1955; H. J. Cowan, *Time and Its Measurement: From the Stone Age to the Nuclear Age*, 1958.

Cancer

The Crab, in astronomy, a winter constellation and the faintest of the zodiacal groups. Cancer, the fourth sign of the zodiac, is important because during early times it marked the northernmost limit of the ecliptic, when the zodiacal system was adopted. The Tropic of Cancer takes its name from this constellation. The four faint stars, α, β, δ, and ι, form a rough Y outline, which is suggestive of a crab (see illustration). In the center of Cancer is a hazy object. This is a magnificent cluster of faint stars called Praesepe (the Beehive) or the Manger. *See* CONSTELLATION. [CHING-SUNG YU]

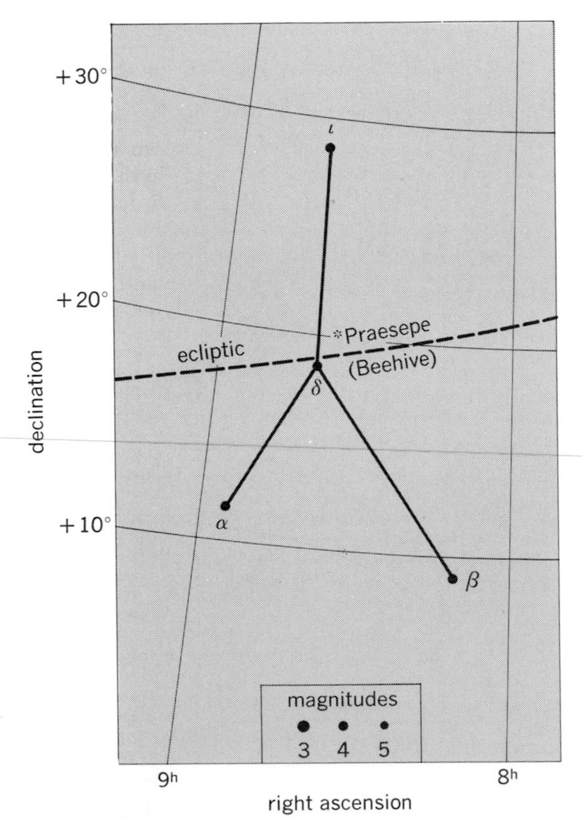

Line pattern of constellation Cancer. Grid lines represent the coordinates of the sky. Apparent brightness, or magnitudes, of stars are shown by sizes of dots, which are graded by appropriate numbers as indicated.

Capricornus

The Sea Goat, in astronomy, an inconspicuous zodiacal constellation in the southern sky lying between Aquarius and Sagittarius. Capricornus is the tenth sign of the zodiac. It has two third-magnitude stars, β and δ, the remainder being of fourth magnitude or fainter. The constellation has been described from the earliest times as a goat, or as a figure that is part goat with the tail of a fish (see illustration). Together with the neighboring zodiacal constellations Aquarius and Pisces, Capricornus forms the great heavenly sea; all the names are related to water. The Tropic of Capricorn originates from this constellation, which marked the southern limit of the ecliptic in ancient times. *See* CONSTELLATION. [CHING-SUNG YU]

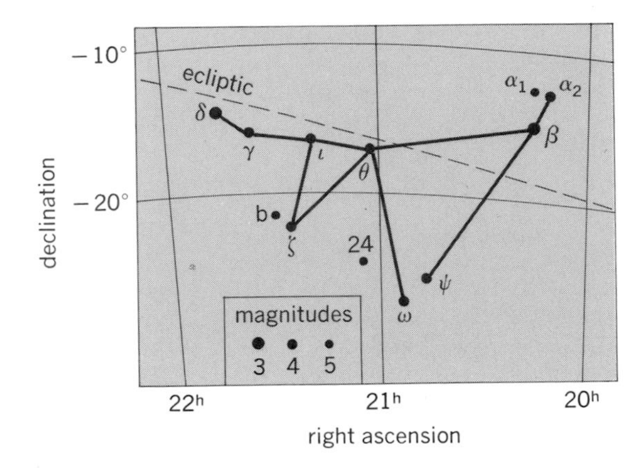

Line pattern of constellation Capricornus. Grid lines represent coordinates of the sky. Apparent brightness, or magnitudes, of stars are shown by size of dots, which are graded by appropriate numbers as indicated.

Carbon star

Any of a class of stars with an apparently high abundance ratio of carbon to hydrogen. In normal stars this ratio also prevails, oxygen is more abundant than carbon, and carbon and nitrogen are about equal. In carbon stars, however, carbon is more abundant than oxygen. The majority of the carbon-rich stars are low-temperature red giants of the C class, showing bands of C_2, CN, and CH (the older designation of this is R or N). A few hot carbon-rich stars are also known. An excess abundance of carbon is also found in the S (zirconium oxide) stars and the barium (heavy-element) stars. The C-class stars have surface temperatures in the range 5500–2500 K. Heavy absorption produced by bands ascribed to C_3 and SiC_2 exists in cooler objects. An unexpected peculiarity is the strength of the resonance lines of neutral lithium; this element is easily destroyed by nuclear reactions with hydrogen. Its synthesis by helium burning may occur in late stages of red giant evolution.

The degree of enhancement of carbon varies from star to star. In a related phenomenon found in globular clusters, variations in the C/N/O ratios are found. An outstanding peculiarity is the presence of the isotope ^{13}C in some of these stars, as shown by the presence of $^{12}C^{13}C$ and $^{13}C^{13}C$ isotope bands, with an abnormally high $^{13}C/^{12}C$ ratio, between 1 to 4 and 1 to 10. In some other C-class stars the $^{12}C^{13}C$ bands are absent, indicating a low $^{13}C/^{12}C$ ratio, possibly near the terrestrial value of 1 to 90. An important discovery was that nearly all red giants, not only carbon stars, have systematically higher values of the $^{13}C/^{12}C$ ratio than the Sun.

The difference between normal red giant spectra and those of the carbon-rich group arises apparently because of the strong binding of the compound CO. In stars for which there is more oxygen than carbon, at temperatures below 5000 K, almost all the carbon is bound in the form of CO and therefore unavailable to form further carbon compounds. In the carbon-rich stars the CO consumes all the oxygen, but leaves some carbon in free atomic form, available for carbon compounds. The importance of the C/O ratio lies in the theory of the origin of chemical elements, in which it is shown that carbon is easily synthesized at temperatures found in cores of evolving red giants. *See* ELEMENTS AND NUCLIDES, ORIGIN OF; STAR.

[JESSE L. GREENSTEIN]

Carbon-nitrogen-oxygen cycles

A group of nuclear reactions that involve capture of protons (nuclei of hydrogen, or ^{1}H, atoms) by carbon, nitrogen, and oxygen nuclei. These cycles are believed to be the source of energy in main-sequence stars which are more massive than the Sun. Completion of any one of the cycles results in consumption of four protons (4 ^{1}H), synthesis of one helium nucleus (^{4}He) and two neutrinos, and 26.73 MeV of energy. This energy E reflects the difference in mass m between the four protons and the helium nucleus and is equal to the mass difference times the square of the velocity of light c, as is known from Einstein's statement of mass-energy equivalence, $E = mc^2$. Because the nuclear fuel consumed in these processes is hydrogen, they are referred to as hydrogen burning by means of the carbon-nitrogen-oxygen (CNO) cycles.

Carbon-nitrogen cycle. The original carbon-nitrogen (CN) cycle was suggested independently by H. A. Bethe and C. F. von Weizsäcker in 1938 as the source of energy in stars. In the first reaction of the carbon-nitrogen cycle, a carbon nucleus of mass 12 captures a proton, forming a nitrogen nucleus of mass 13 and releasing a photon of energy 1.944 MeV. This may be written: $^{12}C + {}^{1}H \rightarrow {}^{13}N + \gamma$ or, in shorthand notation, $^{12}C(p,\gamma)^{13}N$. ^{13}N is an unstable nucleus that decays by emitting a positron (e^+) and a neutrino (ν). In equation form, $^{13}N \rightarrow {}^{13}C + e^+ + \nu$ or $^{13}N(e^+\nu)^{13}C$. The cycle continues through ^{14}N, ^{15}O, and ^{15}N by the reactions and decays shown in shorthand notation by

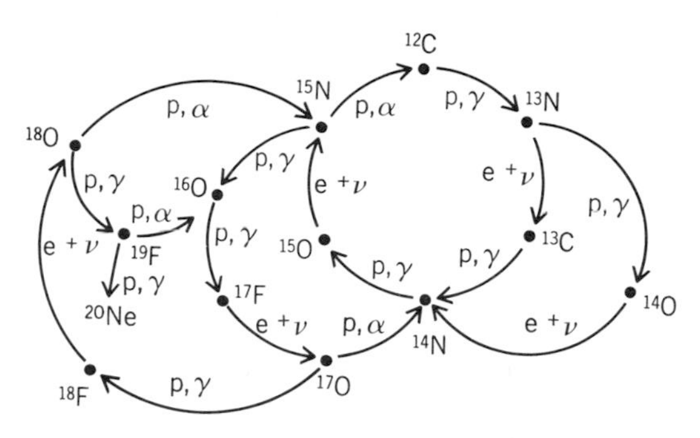

Carbon-nitrogen-oxygen cycles. (*From G. R. Caughlan, The CNO cycles, in J. Audouze, ed., CNO Isotopes in Astrophysics, D. Reidel, Dordrecht, pp. 121–131, 1977*)

$^{12}C(p,\gamma)^{13}N(e^+\nu)^{13}C(p,\gamma)^{14}N(p,\gamma)^{15}O(e^+\nu)^{15}N(p,\alpha)^{12}C$. These reactions form the central cycle in the illustration. The reaction $^{15}N(p,\alpha)^{12}C$ represents emission of an alpha particle (^{4}He nucleus) when ^{15}N captures a proton and the cycle returns to ^{12}C. Because of the cycling, the total number of carbon, nitrogen, and oxygen nuclei remains constant, so these nuclei act as catalysts in the production of a helium nucleus and two neutrinos with the release of energy. The positrons that are created annihilate with free electrons rapidly after creation, so the energy used in their creation is returned to the energy fund. Synthesis of some of the carbon, nitrogen, and oxygen nuclei is accomplished through hydrogen burning by means of the carbon-nitrogen-oxygen cycles.

CNO bicycle. Nuclear research in the 1950s led to the addition of a second cycle to the processes of hydrogen burning by the carbon-nitrogen-oxygen cycles. Laboratory research has shown that one in 880 proton captures by ^{15}N results in the formation of an oxygen nucleus of mass 16, leading to reactions which may cycle back to ^{14}N by $^{15}N(p,\gamma)$-$^{16}O(p,\gamma)^{17}F(e^+\nu)^{17}O(p,\alpha)^{14}N$. The pair of cycles shown above, forming the main cycle and the first cycle to the left of it in the illustration is called the carbon-nitrogen-oxygen bicycle.

Other CNO cycles. In the years of research since 1960, it has become apparent that many possible branches among the nuclei must be included in any analysis of hydrogen burning by carbon, nitrogen, and oxygen nuclei. For example, if the unstable nucleus ^{13}N manages to capture a proton before it decays in its mean lifetime of 862 seconds, a third cycle can occur through $^{13}N(p,\gamma)$-$^{14}O(e^+\nu)^{14}N$. This cycle is displayed in the illustration by the branch on the right-hand side of the main carbon-nitrogen cycle and is known as the "fast" or the "hot" carbon-nitrogen cycle. The other possible branches leading to additional cycles shown in the diagram are due to competition between (p,γ) and (p,α) reactions. The added reactions are $^{17}O(p,\gamma)^{18}F(e^+\nu)^{18}O(p,\alpha)^{15}N$ and $^{18}O(p,\gamma)^{19}F(p,\alpha)^{16}O$. The reaction $^{19}F(p,\gamma)^{20}Ne$ shown in the diagram leads out of the carbon, nitrogen, and oxygen nuclei and hence away from carbon-nitrogen-oxygen cycles.

There are two additional branches that may occur if the unstable fluorine nuclei (^{17}F and ^{18}F)

capture protons before they can decay. *See* ELEMENTS AND NUCLIDES, ORIGIN OF; PROTON-PROTON CHAIN; STELLAR EVOLUTION.

[GEORGEANNE R. CAUGHLAN]

Bibliography: H. A. Bethe, Energy production in stars, *Phys. Rev.*, 55:103, 434–456, 1939; E. M. Burbidge et al., Synthesis of elements in stars, *Rev. Mod. Phys.*, 29:547–650, 1957; G. R. Caughlan, The CNO cycles, in J. Audouze (ed.), *CNO Isotopes in Astrophysics*, pp. 121–131, 1977; W. A. Fowler, *Nuclear Astrophysics*, 1967; C. F. von Weiszäcker, Uber Elementwandlungen im Innern der Sterne, II, *Phys. Z.*, 39:633–646, 1938.

Cardinal points

The four intersections of the horizon with the meridian and with the prime vertical circle, or simply prime vertical, the intersections with the meridian being designated north and south, and the intersections with the prime vertical being designated east and west (see illustration). The cardinal points are 90° apart; they lie in a plane with each other and correspond to the cardinal regions of the heavens. The four intermediate points, northeast, southeast, northwest, and southwest, are the collateral points.

[FRANK H. ROCKETT]

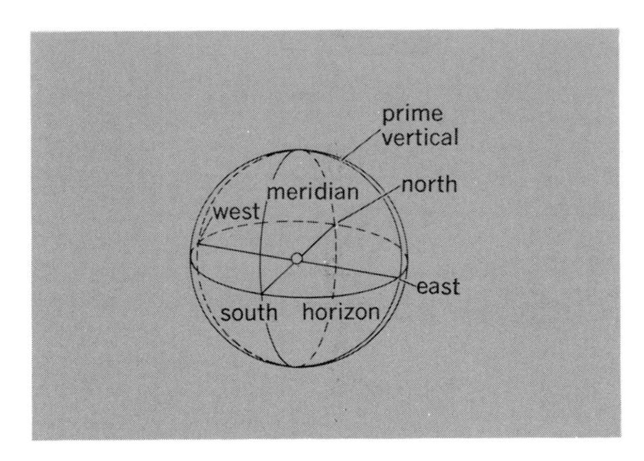

Cardinal points around the horizon.

Cassiopeia

In astronomy, a prominent northern circumpolar constellation as seen from the middle latitudes. The five main bright second- and third-magnitude stars of Cassiopeia form the rather distorted W or M by which the constellation is usually identified (see illustration). The W is also called Cassiopeia's Chair, because on its side the W suggests the shape of a chair. Most old sky maps show Cassiopeia as a queen seated upon a throne. Lying as it does in the Milky Way, the entire region of Cassiopeia is rich in star fields containing several beautiful star clusters. Cassiopeia and the Big Dipper lie across the opposite sides of the North Celestial Pole. *See* CONSTELLATION; URSA MAJOR.

[CHING-SUNG YU]

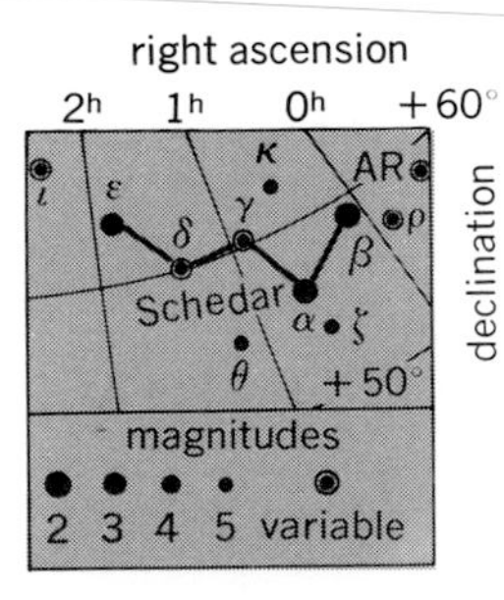

Line pattern of the constellation Cassiopeia. The grid lines represent the coordinates of the sky. Magnitudes of the stars are shown by the sizes of the "dots."

Celestial mechanics

The field of kinematics and dynamics applied to celestial objects. Originally concerned with bodies of the solar system—planets, satellites, asteroids, and comets—it has come to include the motions of double and multiple stars, stellar associations, open and globular star clusters, and the Milky Way system and other galaxies. The designation astromechanics is therefore also appropriate.

Newton's law of gravitation. The basic law of celestial mechanics, Isaac Newton's law of universal gravitation [Eq. (1)], states that the mu-

$$f=-G\frac{(M_1+M_2)}{r^2} \quad (1)$$

tual acceleration f of two particles is proportional to the sum of their masses M_1 and M_2, and inversely proportional to the square of their distance r from each other. (The constant of proportionality G is discussed below.) The negative sign indicates an attraction; in the case of orbital motion, a "falling around."

This law follows from a synthesis of Newton's laws of motion, partly foreseen by Galileo Galilei, and Kepler's laws of planetary motion.

Newton's laws of motion. These are as follows.

1. Law of inertia: Every particle persists in a state of rest or of uniform rectilinear motion, unless an external action causes a change in this motion, that is, an acceleration.

2. Definition of force: This acceleration is inversely proportional to the mass of the particle and may be attributed to a "force" to which it is proportional.

3. Action equals reaction: The mutual forces of any two particles on each other are equal in amount and opposite in direction along the straight line connecting the two particles.

Kepler's laws of planetary motion. Astromechanics may be said to have started with the heliocentric viewpoint proclaimed in 1543 by Nicolaus Copernicus. A quantitative understanding of the laws at work, however, awaited the observations of Tycho Brahe, made in Denmark over a period of 20 years, from which Johannes Kepler derived his three classical laws of planetary motion.

Kepler's first two laws (1609) are:

1. Law of ellipses: Each planet describes an elliptical orbit with the Sun at one focus.

2. Law of equal areas: The line joining planet and Sun sweeps over equal areas in equal times.

These first two laws are referred to as Keplerian motion (Fig. 1). They were followed a decade later by:

3. Kepler's third or harmonic law (1619): The cubes of the semimajor axes or "mean distances" a of the planetary orbits are in a constant ratio to the squares of their periods of revolution P.

In polar coordinates, r and θ, the first law is given by Eq. (2), where r is the instantaneous

$$r=\frac{p}{1+e\cos\theta} \quad (2)$$

radius vector joining planet and Sun, e the eccentricity of the orbits, and θ the true anomaly—the angle, measured at the Sun, between the planet and its point of closest approach (perihelion, periastron). The parameter p (half latus-rectum) is given by Eq. (3). The second law states, for any

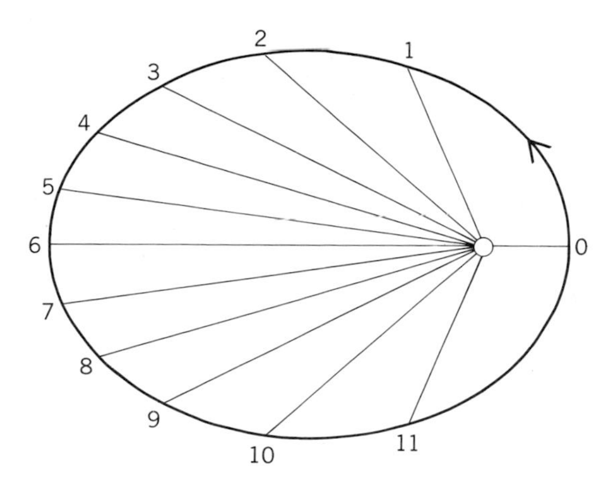

Fig. 1. Keplerian motion: elliptic motion and law of equal areas with respect to a focus. Orbital positions are marked for each time interval of one-twelfth of a period of revolution, beginning with perihelion. (*From P. van de Kamp, Elements of Astromechanics, W. H. Freeman, 1964*)

$$p = a(1-e^2) \tag{3}$$

one orbit, the constancy of the areal velocity A, given by Eq. (4), where t is the time. *See* ORBITAL MOTION; PLANET.

$$A = \frac{r^2 d\theta}{2dt} \tag{4}$$

Derivation of Newton's law. The acceleration of a planet may be expressed in terms of its components f_r and f_θ in polar coordinates. The component f_θ is related to the areal velocity A by Eq. (5). Kep-

$$f_\theta = \frac{2dA}{rdt} \tag{5}$$

ler's second law implies that $f_\theta = 0$; that is, there is only a central ("focal") acceleration along the radius vector joining the Sun and planet. The first law, Eq. (2), leads to Eq. (6) for the value

$$f_r = \frac{-4A^2}{p} \cdot \frac{1}{r^2} \tag{6}$$

of this acceleration, which is the famous inverse-square law of attraction for any one planetary orbit. The subscript may now be dropped. Since the areal velocity is given by Eq. (7), Eq. (6) can be transformed to Eq. (8).

$$A^2 = \pi^2 p \frac{a^3}{P^2} \tag{7}$$

$$f = -4\pi^2 \frac{a^3}{P^2} \cdot \frac{1}{r^2} \tag{8}$$

Kepler's third law, Eq. (9), leads to the gener-

$$\frac{a^3}{P^2} = C \tag{9}$$

alization of the inverse-square law given by Eq. (10),

$$f = -4\pi^2 C \frac{1}{r^2} \tag{10}$$

where C is the same constant for all planetary orbits. Obviously, this constant indicates the dominance of the Sun. But since the Sun is also subject to minute accelerations from the planets, a necessary consequence of Newton's third law of motion, a further generalization leads to Newton's law of gravitation, Eq. (11), where m and M are the

$$f = -G\frac{M+m}{r^2} \tag{11}$$

masses of planet and Sun, respectively, and G is Newton's constant of gravitation, whose value in cgs is given by Eq. (12).

$$G = 6.67 \times 10^{-8} \tag{12}$$

Introducing the concept of force F, Eq. (11), in inertial coordinates, may be written in the form of Eq. (13).

$$F = G\frac{Mm}{r^2} \tag{13}$$

It is easily seen that conversely Kepler's three laws follow from Newton's law of gravitation, with corrections for small mutual attractions or perturbations of the planets, and with a more precise expression for Kepler's third law, namely Eq. (14).

$$\frac{a^3}{P^2} = C\left(1+\frac{m}{M}\right) \tag{14}$$

Because of the dominance of the Sun's mass (1047 times that of Jupiter, 330,000 times that of Earth), Eq. (15) is valid to a high degree of approxi-

$$f = -G\frac{M}{r^2} \tag{15}$$

mation. But the solar-planetary situation is unique, and Newton generalized his law in the form of Eq. (1), which proves to be universally valid. Newton's law of gravitation is formulated for particles, but holds equally well for two objects possessing spherical symmetry, a situation which with great accuracy is the case for stars and planets.

Two-body problem. Of particular interest is the two-body problem, such as Sun and any one planet, the two components of a binary star, and so forth. Any secondary effects in the two-body situation may be considered as a perturbation of the pure two-body problem. Newton's law of gravitation implies that the orbital motion of either object with respect to the other lies in a conic section no longer limited to an ellipse; Eq. (2) holds, where the value of p determines the extent (Fig. 2), and that of e the shape, of the orbit: circle: $e = 0$; ellipse: $0 < e < 1$; parabola: $e = 1$; hyperbola: $e > 1$. By far the most common case is the elliptic orbit, of which the circle frequently is a good approximation.

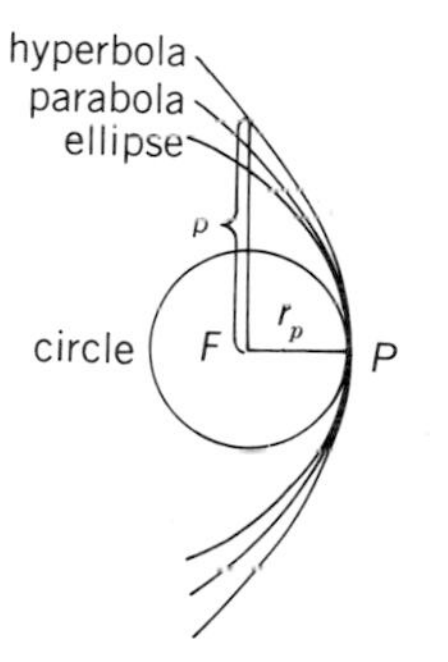

Fig. 2. Relation of parameter p to distance of closest approach r_p for circular, elliptic, parabolic, and hyperbolic orbits. P= point of closest approach. F= focus of conic section at which primary object is located. (*From P. van de Kamp, Elements of Astromechanics, W. H. Freeman, 1964*)

The energy relation in an orbit is expressed by the law of conservation of energy, Eq. (16), where

$$T + \phi = E \tag{16}$$

T is the kinetic energy, ϕ the potential energy, and E the total energy. For the two-body gravitational field this relation becomes Eq. (17), where V is the

$$\frac{V^2}{2} - \frac{G(M_1+M_2)}{r} = \frac{-G(M_1+M_2)}{2a} \tag{17}$$

relative velocity, and a the semimajor axis of the relative orbit. The total negative value simply results from the potential energy being chosen to be zero at infinity and hence negative at any finite distance.

Equation (17) may be written as Eq. (18), which

$$V^2 = G(M_1 + M_2)\left(\frac{2}{r} - \frac{1}{a}\right) \quad (18)$$

is of particular interest in the study of orbital velocity and leads to the important relation given by Eq. (19), where V_p is the parabolic velocity or

$$V_p = V_c\sqrt{2} \quad (19)$$

velocity of escape and V_c is the circular velocity. Equation (18) is called the equation of energy, integral of energy, or vis-viva integral. For a circular orbit Eqs. (20) hold. This is a special case of

$$T = -E \qquad \phi = 2E \quad (20)$$

the virial theorem, a statistical statement which holds for time averages of T and ϕ in a stable system, that is, a system which does not disintegrate with time.

Restricted three-body problem. While the two-body problem mathematically has been completely solved, no general solution exists for triple and multiple systems. However, there are special cases of theoretical and practical (observational) interest of the three-body problem. The "restricted" three-body problem for an object of negligible mass relative to two objects (of masses $M + m$) was first solved by J. L. Lagrange in 1772. While, in the two-body problem, conic sections are the natural trajectories, the orbits are more complicated in the restricted three-body problem. Figure-eight orbits are possible and desirable solutions, dictated by energy and stability requirements.

The restricted three-body problem is easily solved for the case of a relative circular orbit of M and m. Of interest are the gravitational equilibrium locations for an object of negligible mass. There are five such libration points designated $L_1 - L_5$ (Fig. 3); three are located along the line through M and m, two more are in the plane of the orbit at the vertices of two equilateral triangles, whose base is the segment Mm. The latter two points are stable, the former are unstable. Astronomical examples of the stable solution include the so-called Trojan asteroids, some of which precede and others follow Jupiter in orbit around the Sun. See TROJAN ASTEROIDS.

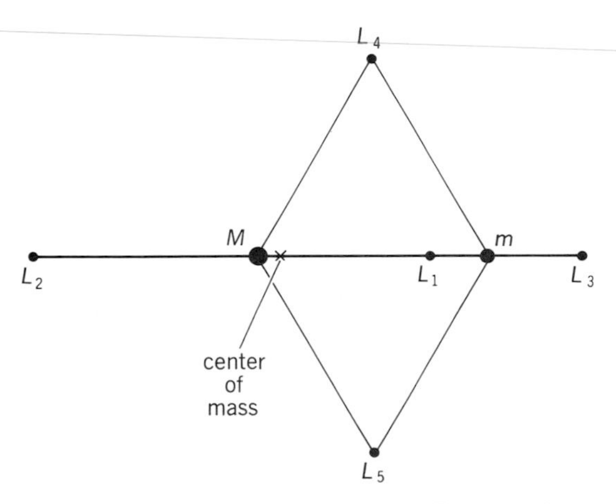

Fig. 3. The five libration points. (*From P. van de Kamp, Elements of Astromechanics, W. H. Freeman, 1964*)

Mass determinations. The space-time-mass relation embodied in Newton's law of gravitation is the basis of mass determination of binary situations, such as Sun-planet or planet-satellite systems, Combining Eqs. (1) and (8) one has Eq. (21),

$$M_1 + M_2 = \frac{4\pi^2}{G}\frac{a^3}{P^2} \quad (21)$$

in metric units. It is often more convenient and practical to introduce the astronomical units of space, time, and mass, that is, the mean Earth-Sun distance, the sidereal year, and the Sun's mass, which lead to the harmonic relation, Eq. (22), a

$$M_1 + M_2 = \frac{a^3}{P^2} \quad (22)$$

basic formula used in the determination of cosmic masses.

In the two-body problem one distinguishes between the relative orbit of one object (secondary) with respect to another (primary) and the orbit of each component around the center of mass. In the case of planetary motions around the Sun, or satellite orbit around planets, the center of mass is often within the surface of the primary and the mass of the secondary is often virtually negligible. In other cases, such as binary stars with components of comparable masses, the mass-ratio must be evaluated in order to divide the total mass over the two components. See BINARY STAR.

The virial theorem permits the determination of the masses of star clusters. The total kinetic energy may be obtained from the estimated numbers of stars in the cluster and their average velocities, measured spectroscopically. The resulting potential energy depends on the size and mass of the cluster and the internal distribution of stars within the cluster; hence the total mass of the cluster may be determined. See STAR CLUSTERS.

Understanding was reached about the kinematical and dynamical structure of the Milky Way galactic system in the 1920s, after a long accumulation of observational facts. The first step was the galactocentric viewpoint proposed in 1918 by Harlow Shapley. An important milestone was the theory of galactic rotation (Bertil Lindblad, 1925), observationally confirmed by Jan H. Oort in 1926. As a first rough approximation a strong concentration of matter (galactic nucleus) may be assumed, at a distance of some 24,000 light-years (2.3×10^{20} m) in the direction of the constellation Sagittarius. The Sun and neighboring stars fall around that center approximately in circular paths, with a speed of about 250 km/s. A simple calculation yields a mass of some 10^{11} solar masses for that distant nucleus. Analogous to the situation of the planets in the solar system, within the stellar neighborhood there proves to be a "shearing effect" in the sense of increasing velocity with decreasing distance to the galactic center. This, another example of Kepler's third law, and various other aspects of stellar motions have put understanding of the kinematical and dynamic structure of the galactic system on a firm footing. Analogous situations have been observed in galaxies beyond the Milky Way. *See* GALAXY; GALAXY, EXTERNAL.

Perturbations. An important aspect of celestial mechanics is the discovery of unseen objects, which reveal their existence through their gravitational effect on visible stars or planets. Classical examples include the discovery in 1846 of Neptune from its perturbing effect on Uranus, and in 1844 of the white dwarf companions of Sirius and of Procyon, through the perturbation in the proper motions of these stars. These discoveries were subsequently followed by visual detection. There are numerous other examples of detection of stellar companions, mostly resulting from precision photographic techniques with long-focus telescopes. While most companions of stars thus discovered have proved to be stars of low luminosity, there appear to be several objects of very low mass, below 6% of the Sun's mass, considered as the lower limit below which an object cannot be self-luminous. A new field is thus opened, that of substellar masses, even including planetary companions of stars other than the Sun. These studies are astrometric, that is, based on positions on the celestial sphere as contrasted with radial velocity observations in the line of sight, using the Doppler principle. Companions found by the latter method generally are objects of larger masses, solar or more. *See* PERTURBATION.

Tides, precession, Roche's limit. While the gravitational effect of Sun and Moon on the Earth is proportional to M/r^2 (mass/distance^2), the differential effect on opposite tides of the Earth is proportional to $2M/r^3$ (and to the diameter of the Earth). The comparative nearness of the Moon, 1/389 that of the Sun, results in a tidal effect on the oceans which averages 2.2 times that caused by the Sun. The same relative importance exists for the precession, a conical motion of the Earth's axis with a period of 25,800 years, due to the differential gravitational attraction of Moon and Sun for the Earth's equatorial bulge, and the fact that the Earth's equator is tilted 23°.5 with respect to the ecliptic (Earth's orbital plane). *See* PRECESSION OF EQUINOXES; TIDE.

A related problem is that of possible tidal disruption. For a spherical satellite of a planet of equal density, tidal disruption occurs within a distance (from the center of the planet) of 2.44 times the radius of the planet, assuming no structural cohesion of the satellite but only mutal gravitation between its mass elements. This critical distance is called Roche's limit; an example appears to be the ring systems of Saturn, Uranus, and Jupiter which are well within this limit.

Within the rings of Saturn there are empty spaces corresponding to periods of revolution which are simple fractions of the period of some of Saturn's inner satellites. The corresponding orbits would be "resonant," that is, would be gradually distorted and destroyed by the gravitational effects caused by some of the satellites. *See* SATURN.

Space flight. Assuming spherical symmetry for the Earth (mass M, radius R), surface gravity g is given by Eq. (23). The gravitational acceleration

$$g=-G\frac{M}{R^2} \qquad (23)$$

above the surface at distance r from center is given by Eq. (24). Hence the energy required to lift

$$f=-G\frac{M}{r^2}=-g\frac{R^2}{r^2} \qquad (24)$$

1 gram vertically against gravity from surface to infinity is given by Eq. (25), which may be provided

$$gR^2\int_R^\infty \frac{dr}{r^2}=gR \qquad (25)$$

by initial kinetic energy $V_p^2/2$, where V_p is the parabolic or escape velocity. Hence the escape velocity is given by Eq. (26). The circular velocity for an

$$V_p^2=2gR \qquad (26)$$

object skimming the surface of the Earth is given by Eq. (27). Both Eqs. (26) and (27) follow also from Eq. (18).

$$V_c^2=gR \qquad (27)$$

The numerical velocities are V = 11.19 km/s, and V_c = 7.91 km/s with a period of 84.4 min. For a circular orbit with radius r the period of revolution is given by Eq. (28), which is another version of

$$P^2=\frac{4\pi^2}{g}\frac{r^3}{R^2} \qquad (28)$$

Kepler's third law. For an object circling the Earth in 24 h, the orbital radius is 42,000 km. Since kinetic energy is proportional to the square of velocity, the energy required to escape from the Earth is double that required to skim the surface of the Earth.

The basic formulas of astromechanics determine such values as the time and initial velocity to reach the Moon or any other object. The initial "free" velocity is obtained as the end product of the reaction, or successive reactions, from the loss of mass (fuel) ejected at high exhaust velocity (Newton's third law of motion). For a rocket, starting from rest in a nongravitational field, one has Eq. (29), where V is the ultimate velocity resulting

$$V=v\log_e R \qquad (29)$$

from a constant exhaust velocity v, and $R = m_0/m_1$ is the mass ratio of the rocket; m_0 and m_1 are the initial mass and the payload of the rocket.

Escape from the Earth's gravitational field is governed by Eq. (30), where the mass-ratio R_n is

$$V\left(\frac{n+1}{n}\right)=v\log_e R_n \qquad (30)$$

now given by Eq. (31) for a vertical velocity V at

$$R_n=\exp\left[\frac{V}{v}\cdot\frac{n+1}{n}\right] \qquad (31)$$

engine cut-off and exhaust velocity v; n is the fractional acceleration of the rocket in terms of g, the gravitational acceleration at the Earth's surface.

In practice, multiple rocket assemblies are required to obtain velocities over 11.2 km/s, the escape velocity from the Earth.

Einstein's law of gravitation. Newton's law of gravitation describes very accurately the motions of celestial bodies. The notions of space, time, and mass are kept independent, while positions and motions are measured by the method and properties of euclidean geometry. A new and more nearly accurate statement is given by Albert Einstein. His description discards the euclidean background, but stresses that the distribution of mass in space and time is the "real" framework of the universe, which determines the choice of a fitting geometrical background that is, in general, not quite euclidean. Einstein's approach implies the equivalence of gravitational and any other acceleration. His view generalizes the law of inertia into free motion of any object as long as a proper geometric choice is made of the background to which the motion is referred. Galileo's and Newton's classical law of inertia simply follows for the special case of negligible gravitational effects. One thus may speak of Einstein's law of gravitation, motion, or inertia, which all are identical concepts when viewed in proper geometrical perspective; the concept of force becomes unnecessary.

The superiority of Einstein's law to Newton's is illustrated by certain cosmic effects, such as small changes in the orbit of Mercury. However, Newton's description is adequate for the vast majority of problems in celestial mechanics, and is widely employed. *See* GRAVITATION; RELATIVITY.

[PETER VAN DE KAMP]

Bibliography: G. Abell, *Exploration of the Universe*, 1975; D. Brouwer and G. M. Clemence, *Methods of Celestial Mechanics*, 1961; J. M. A. Danby, *Fundamentals of Celestial Mechanics*, 1962; P. van de Kamp, *Elements of Astromechanics*, 1964.

Celestial navigation

Navigation with the aid of celestial bodies, primarily for determination of position when landmarks are not available. In celestial navigation, position is not determined relative to the objects observed, as in navigation by piloting, but in relation to the points on the Earth having certain celestial bodies directly overhead.

Celestial bodies are also used for determination of horizontal direction on the Earth, and for regulating time, which is of primary importance in celestial navigation because of the changing positions of celestial bodies in the sky as the Earth rotates daily on its axis.

The navigator is concerned less with the actual motions of celestial bodies than with their apparent motions as viewed from the Earth. He pictures the heavens as a hollow celestial sphere of infinite radius, with the Earth at its center and the various celestial bodies on its inner surface. He visualizes this sphere as rotating on its axis once in about 23 hr, 56 min—one sidereal day. The stars are then back where they were when the period started.

Bodies closer to the Earth appear to change position at a different rate than do those at a distance. The Sun appears to make a complete revolution among the stars once a year, as the Earth makes one revolution in its orbit. The apparent motion is along a great circle called the ecliptic, which is inclined nearly 23.5° to the plane of the Equator of the Earth. All of the planets stay within 8° of the ecliptic, in a band called the zodiac. Within this band they appear to move among the stars. The Moon, too, stays within the zodiac as it revolves around the Earth—or more properly as the Earth and Moon revolve around their common center of mass—once each lunar month.

Body selection and identification. The navigator uses a limited number of celestial bodies—the Sun, Moon, 4 planets, and perhaps 20–30 stars. Although 173 stars are listed in the *Nautical Almanac*, and 57 of these are listed in the *Air Almanac* and on the daily pages of the *Nautical Almanac*, the majority of them are normally not used unless the navigator's favorite ones are unavailable. Some of the navigational stars are not seen at the latitudes traveled by many navigators.

With relatively few bodies in use and little change in star positions from one evening to the next, identification is seldom a problem. A tentative selection is often made in advance by means of some form of star finder. When set for the latitude of the observer and the time and date of observation, this device provides a graphical indication of the approximate altitude and azimuth of each star shown. The relative positions of other bodies can be plotted by the user. Star finders, tables, and star charts may be used to identify bodies observed before identification.

Celestial equator coordinate system. Several systems of coordinates are available to identify points on the celestial sphere. The celestial equator system of coordinates is an extension of the equatorial system commonly used on the Earth. The intersection of the plane of the terrestrial Equator, extended, with the celestial sphere is a great circle called the celestial equator. The Earth's axis, extended, intersects the celestial sphere at the north and south celestial poles. Small circles parallel to the celestial equator, similar to parallels of latitude on the Earth, are called parallels of declination. Each of these connects points of equal declination, the celestial coordinate similar to latitude on the Earth.

Great circles through the celestial poles, similar to meridians on the Earth, are called celestial meridians if they are considered to remain fixed in relation to terrestrial meridians, and hour circles if they are considered to remain fixed on the rotating celestial sphere.

Several different quantities on the celestial sphere are analogous to those of longitude on the Earth. Greenwich hour angle is measured westward from the Greenwich celestial meridian, through 360°. Local hour angle is similarly measured from the celestial meridian of the observer. Meridian angle is measured eastward and westward from the local meridian, through 180°. Sidereal hour angle is measured westward from the hour circle of the vernal equinox, the point at which the Sun crosses the celestial equator on its northward travel in spring, through 360°. Right ascension is measured eastward from the hour circle of the vernal equinox, usually in hours, minutes, and seconds, from 0 through 24 hr. The various relationships of the celestial equator system

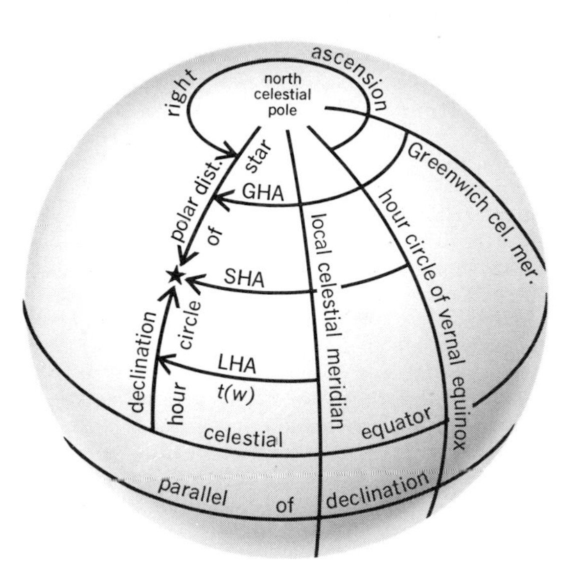

Fig. 1. The celestial equator system of coordinates.

are shown in Fig. 1, where GHA is Greenwich hour angle, SHA is sidereal hour angle, LHA is local hour angle, and t is meridian angle (shown here as westerly).

With the exception of right ascension, all of these quantities are customarily stated to a precision of one minute of arc by the air navigator, and to one-tenth of a minute of arc by the marine navigator. The celestial equator system is used in the almanacs for indicating positions of celestial bodies at various times.

Horizon system of coordinates. The navigator also uses the horizon system of coordinates, which is similar to the celestial equator system. The primary great circle is the horizon of the observer. The pole vertically overhead is the zenith, and the opposite pole is the nadir. Small circles parallel to the horizon are called parallels of altitude, each connecting all points having the same altitude. Angular distance downward from the zenith is called zenith distance. Great circles through the zenith and nadir are vertical circles. The prime vertical circle passes through the east and west points of the horizon. Azimuth is measured clockwise around the horizon, from 000° (it is generally expressed in three figures) at the north point, to 360°. Azimuth angle is measured eastward and westward to 180°, starting from north in the Northern Hemisphere, and from south in the Southern Hemisphere. Thus, it starts directly below the elevated pole, the celestial pole above the horizon. The various relationships of the horizon system are shown in Fig. 2. The navigator uses the horizon system because it offers the most practical references for the origin of his measurements.

Altitude and zenith distance are customarily stated by the air navigator to a precision of one minute of arc, and azimuth and azimuth angle to a precision of one degree; the marine navigator states them to one-tenth of a minute and one tenth of a degree, respectively.

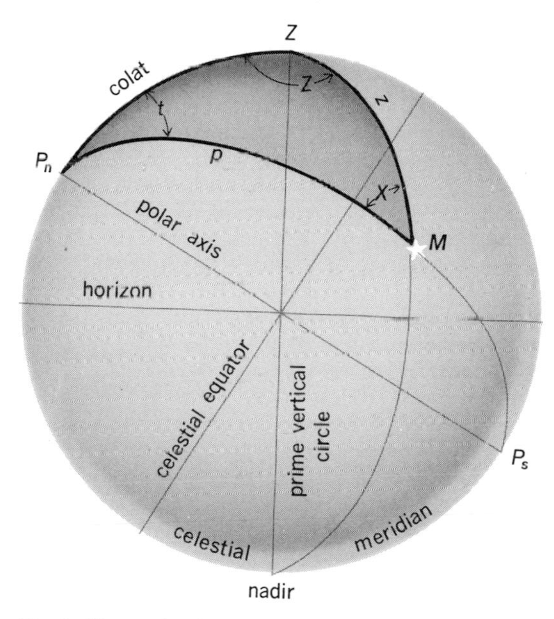

Fig. 3. The navigational triangle.

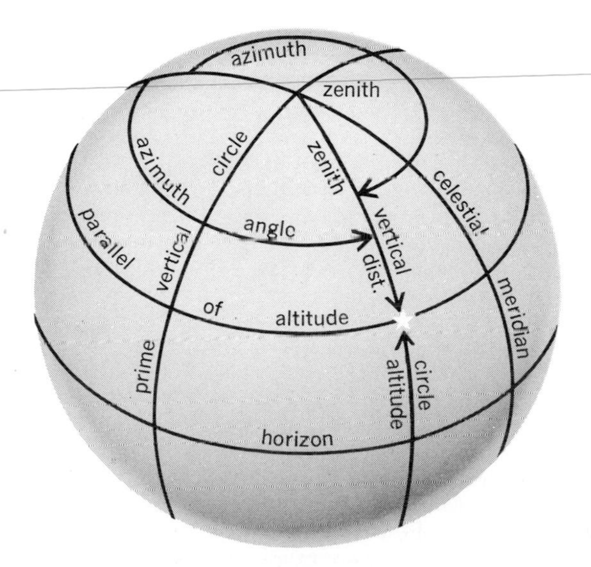

Fig. 2. The horizon system of coordinates.

Two similar systems based upon the ecliptic and the galactic equator are used by astronomers, but not by navigators.

Position determination. Position determination in celestial navigation is primarily a matter of converting one set of coordinates to the other. This is done by solution of a spherical triangle called the navigational triangle.

The concept of the spherical navigational triangle is graphically shown in Fig. 3, a diagram on the plane of the celestial meridian. The celestial meridian passes through the zenith of the observer, and is therefore a vertical circle of the horizon system. Elements of both systems are shown in Fig. 3, indicating that an approximate solution can be made graphically.

The vertices of the navigational triangle are the elevated pole (P_n), the zenith (Z), and the celestial body (M). The angles at the vertices are, respectively, the meridian angle (t), the azimuth angle (Z), and the parallactic angle (X). The sides of the triangle are the codeclination of the zenith or the colatitude (colat) of the observer, the coaltitude or zenith distance (z) of the body, and the codeclination or polar distance (p) of the body.

A navigational triangle is solved, usually by computation, and compared with an observed altitude to obtain a line of position by a procedure known as sight reduction.

Observed altitude. To establish a celestial line of position, the navigator observes the altitude of a celestial body, noting the time of observation. Observation is made by a sextant, so named because early instruments had an arc of one-sixth of a circle. By means of the double reflecting principle, the altitude of the body is double the amount of arc used. Similar instruments were called octants, quintants, and quadrants, depending upon the length of the arc. Today, all such instruments, regardless of length of the arc, are generally called sextants.

The marine sextant uses the visible horizon as the horizontal reference. An air sextant has an artificial, built-in horizontal reference based upon a bubble or occasionally a pendulum or gyroscope. The sextant altitude, however measured, is subject to certain errors, for which corrections are applied.

When a marine sextant is used, observations can be made only when both the horizon and one or more celestial bodies are visible. This requirement generally eliminates the period between the end of evening twilight and the beginning of morning twilight. The navigational stars and planets are therefore usually observable only during twilight. Air navigation is not subject to this limitation, star and planet observations being available all night.

In selecting bodies for observation, the navigator considers difference in azimuth, magnitude (brightness) of the body, altitude (avoiding both extremes), and sometimes other factors. If he is particularly concerned about his speed, he selects a body nearly ahead or astern to provide a speed line. A body near the ship's beam provides a course line. One north or south provides a latitude line, while one east or west provides a longitude line. One perpendicular to a shoreline provides an indication of distance offshore.

Many navigators prefer to observe three bodies differing in azimuth by about 120°, or four bodies differing by 90°, and preferably at about the same altitude. In this way any constant error in the altitudes is eliminated.

Sight reduction. The process of deriving from an observation the information needed for establishing a line of position is called sight reduction. A great variety of methods has been devised. That now in general use is called the Marcq St.-Hilaire method, after the French naval officer who proposed it in 1875, and is described as follows.

Having obtained the body's corrected sextant altitude, called the observed altitude, the navigator uses the almanac to obtain the Greenwich hour angle and declination of the body. He converts the former to local hour angle or meridian angle for an assumed position in the vicinity of his actual position. With these two quantities and the latitude of his assumed position, the navigator solves the navigational triangle for altitude and azimuth.

Each altitude of a given celestial body at any one instant defines a circle having the geographical position of the body (the point on the Earth at which the body is momentarily vertically overhead) as the center, and the zenith distance as the radius. The difference between the computed altitude for the assumed position and the observed altitude at the actual position of the observer, called the altitude difference or altitude intercept, is the difference in radii of the circles of equal altitude through the two positions.

This difference is measured along an azimuth line through the assumed position, each minute of altitude difference being considered one nautical mile. A perpendicular through the point so located is considered a part of the circle of equal altitude through the position of the observer. This is the line of position sought. The intersection of two or more nonparallel lines of position adjusted to a common time defines a fix (or running fix if the elapsed time between observations is more than a few minutes) locating the position of the observer at the time of observation. Although two lines of position are sufficient for a fix, most navigators prefer to observe three or more to provide a check and to decrease somewhat the probable error of the fix.

Sometimes, however, the navigational triangle is solved in reverse, starting with altitude, assumed latitude, and declination, and solving for meridian angle. This is then compared with Greenwich hour angle at the time of observation to determine the longitude at which the line of position crosses the assumed latitude. An American, Capt. Charles H. Sumner, used this method in his discovery in 1837 of the celestial line of position. Others used one point, found by this so-called time sight method, and the azimuth of the celestial body to establish a line of position.

Special short-cut methods have been used for a body observed on or near the celestial meridian and for Polaris. Since simple methods of sight reduction have become widely used, the popularity of such special methods has decreased.

A large number of methods has been devised for solution of the navigational triangle. The most widely accepted have been mathematical, but a great many graphical and mechanical solutions have been proposed. Many of these methods are discussed in U.S. Defense Mapping Agency Hydrographic Topographic Center Publication no. 9, *American Practical Navigator*, originally by N. Bowditch. Most American navigators and many of other nationalities now use one of two methods, both published by the U.S. Defense Mapping Agency Hydrographic Topographic Center. Publication no. 229, *Sight Reduction Tables for Marine Navigation*, is intended for use with the *Nautical Almanac* for marine navigation. Publication no. 249, *Sight Reduction Tables for Air Navigation*, is intended for use with the *Air Almanac* for air navigation. Editions of both of these sets of tables are also published in certain other countries.

With the emergence of electronic computers and hand-held calculators, sight reduction has been performed increasingly with limited use or elimination of tables. The extent to which a computer replaces tables depends upon the capability of the available computer and the preference of the navigator. The principal use of a nonprogrammable hand-held calculator is to perform simple arithmetical calculations such as addition, subtraction, and interpolation, with respect to both the almanac and sight reduction tables. If a programmable calculator with stored programs is available, the complete sight reduction process can be performed without the aid of tables, and if suitable data are available, even the ephemerides of celestial bodies can be computed, eliminating the need for an almanac in the usual form. Since the edition for 1977, an *Almanac for Computers*, providing the essential data, has been published by the U.S. Naval Observatory. In the most sophisticated systems, a central computer combines sight reduction data with outputs of electronic positioning systems, perhaps including those associated with navigational satellites, and dead-reckoning data to provide continuous readout of the current most probable position.

Time relationships. Time is repeatedly mentioned as an important element of a celestial observation because the Earth rotates at the approximate rate of 1 minute of arc each 4 sec of time. An error of 1 sec in the timing of an observation might introduce an error in the line of position of as much as one-quarter of a mile. Time directly affects longitude determination, but not latitude. The long search for a method of ascertaining longitude at sea was finally solved two centuries ago by the invention of the marine chronometer, a timepiece with a nearly steady rate.

Several different kinds of time are used by the navigator. His timepiece, which keeps watch time, usually has a small watch error. When this is applied to watch time, the result is usually zone time. This is familiar to most people as standard time (such as Pacific Standard Time) or, when clocks are set an hour ahead, as daylight saving time (such as Central Daylight Saving Time). At sea the zones may be set by each vessel or aircraft, but they are generally 15° wide, centered on the meridians exactly divisible by 15°.

When zone time is increased or decreased by 1 hr per 15° longitude (the amount of the zone description, for example, +7 for Mountain Standard Time), Greenwich mean time is obtained. This is the time used in the almanacs.

Local mean time differs from zone time by the difference in longitude between the meridian of the observer and the zone meridian, at the rate of 4 min of time for each degree of longitude, the meridian to the eastward having the later time. Local mean time is used in tables indicating time of sunrise, sunset, moonrise, moonset, and beginning and ending of twilight.

All forms of mean time are based upon apparent motions of a fictitious mean sun which provides an essentially uniform time. Apparent time, based upon the apparent (visible) Sun, may differ from mean time by a maximum of nearly 16½ min. Apparent time, plus or minus 12 hr, indicates the actual position of the Sun with respect to the celestial meridian. At local apparent noon the Sun is on the celestial meridian, and the local apparent time is 1200. (Navigators customarily state time in four digits without punctuation, from 0000 at the start of a day to 2400 at the end of a day.)

Sidereal time, based upon motion of the stars, is used (indirectly in many cases) with a star finder or a star chart.

The custom of setting navigational timepieces to Greenwich mean time is growing, particularly among air navigators. This time is used almost invariably in polar regions.

Time signals are broadcast from a number of stations throughout the world to permit checking of standard timepieces. Marine chronometers are not reset by the user, an accurate record being kept of chronometer time, chronometer error, and chronometer rate.

Day's work. A typical day's work of a marine navigator at sea, when using celestial navigation, is as follows:

1. Plot of dead reckoning.
2. Morning twilight observations for a fix.
3. Report of 0800 position to the commanding officer.
4. Morning Sun line and compass check.
5. Winding of chronometers and determination of error and rate.
6. Noon Sun line, advance of morning Sun line for running fix, and report of 1200 position to the commanding officer.
7. Afternoon Sun line and compass check.
8. Determination of time of sunset and preparation of a list of bodies available for observation during evening twilight.
9. Evening twilight observations for a fix.
10. Report of 2000 position to the commanding officer.
11. Determination of time of beginning of morning twilight, time of sunrise, and preparation of a list of bodies available for observation during morning twilight.
12. Time of moonrise and moonset.

Electronic applications, however, are gradually changing the pattern of celestial navigation in at least three important respects: (1) by providing noncelestial position information at sea, (2) by providing devices for automatic observation and sight reduction, and (3) by extending use of celestial navigation to all weather conditions at all times of day or night, by the use of electronic star trackers and of radio astronomy. *See* RADIO ASTRONOMY; TIME. [ALTON B MOODY]

Bibliography: E. S. Maloney, *Dutton's Navigation and Piloting*, 1978; U.S. Defense Mapping Agency Hydrographic Topographic Center, *Air Navigation* (Publ. no. 216, 1966), *American Practical Navigation* (Publ. no. 9, 1977), *Sight Reduction Tables for Air Navigation* (3 vols., Publ. no. 249), *Sight Reduction Tables for Marine Navigation* (6 vols., Publ. no. 229); U.S. Naval Observatory, *Air Almanac*, *Almanac for Computers*, and *Nautical Almanac*.

Celestial sphere

A hypothetical sphere of infinite radius with its center at the observer. Because all stars are so far from an Earth-bound observer that they can be located solely by their direction, it is convenient to consider them as located on the inside of a spherical surface centered on the observer. Because the

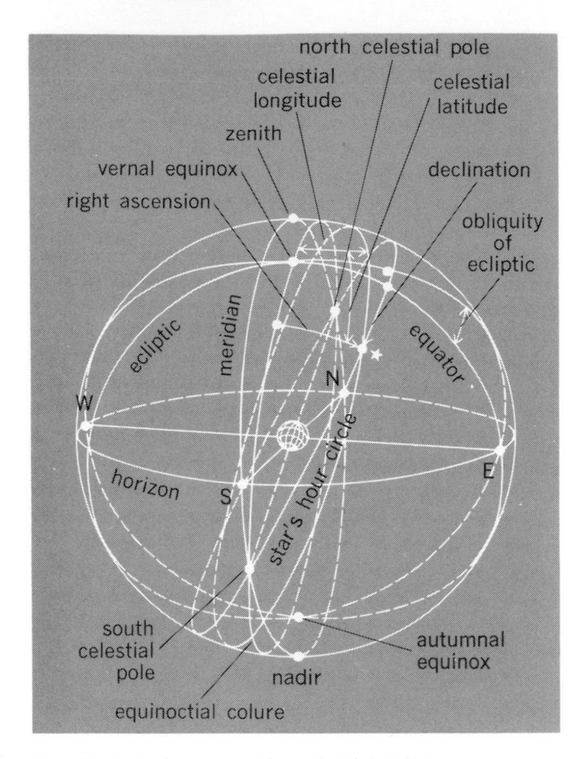

Coordinates of a star on the celestial sphere.

radius of Earth is negligible, it is frequently convenient to consider Earth, rather than the terrestrial observer, centered in the celestial sphere. Stars are designated by their location on this sphere (see illustration), and planets and other heavenly bodies are located against this background. *See* ASTRONOMICAL COORDINATE SYSTEMS.

[FRANK H. ROCKETT]

Centaurus

The Centaur, in astronomy, one of the most magnificent of the southern constellations. Two first-magnitude navigational stars, Alpha and Beta Centauri, mark the right and left front feet, respectively, of the centaur. The former is Rigil Kentaurus, or simply Rigil Kent, and the latter Hadar. Rigil Kent is the third brightest star in the whole sky, fainter than only Sirius and Canopus. The line joining Rigil Kent and Hadar points to the constellation Crux (Southern Cross). Thus they are also called the Southern Pointers, in contradistinction to the northern pointers of the Big Dipper. *See* CONSTELLATION; URSA MAJOR.

[CHING-SUNG YU]

Central force

A force whose line of action is always directed toward a fixed point. The central force may attract or repel. The point toward or from which the force acts is called the center of force. If the central force attracts a material particle, the path of the particle is a curve concave toward the center of force; if the central force repels the particle, its orbit is convex to the center of force. Undisturbed orbital motion under the influence of a central force satisfies Kepler's law of areas. *See* CELESTIAL MECHANICS.

[RAYNOR L. DUNCOMBE]

Cepheids

A class of brightness-variable stars whose prototype is the star Delta Cephei in the constellation Cepheus. While both bluer and redder stars also vary in their intrinsic light, the properties of these β Cephei, ZZ Ceti, RV Tauri, and Mira variables are much less understood than the yellow-color Cepheids. These yellow stars are known to be pulsating in radius by as much as 10% or more. Their light variations are due to their changing surface area and, more importantly, their changing surface temperature. Larger yellow stars are intrinsically brighter because they have more surface area, and they have larger pulsation periods because they have a larger radius. *See* STAR; VARIABLE STAR.

The interest in these stars is twofold: If their intrinsic brightnesses can be inferred from their pulsation period, the brightnesses can be used as indicators of their distance from the Earth. The observed period and a calibrated period-luminosity relation is used to give an intrinsic brightness. The observed distance-dependent apparent brightness then gives the actual distance. The second, and more current, interest in Cepheids is that their pulsation properties reveal their masses and internal structure, which help in understanding how stars age. Thus, Cepheids and the related classes of yellow pulsating stars have been extremely useful in mapping the scale of the universe and in probing the details of stellar interiors.

Types of cepheids. Classical Cepheids are high-luminosity variables with masses between 3 and 18 times that of the Sun. Their pulsation periods range from 2 to over 100 days. They are nearing the end of their thermonuclear energy production lives, having exhausted most of their hydrogen fuel by converting it to helium in their central regions. In most of these variable stars, helium at the very center is being further converted to carbon and oxygen by even higher-temperature thermonuclear burning.

The luminosity of more slowly evolving stars with lower mass, radius, and pulsation period is still due to hydrogen converting to helium. Those stars with masses between 1 and 3 solar masses and periods between about 0.1 and 1 day are in the related class of stars called Delta Scuti variables. Both the Cepheids and the Delta Scuti variables have original hydrogen, helium, and heavier-element compositions similar to that of the Sun, and all are called population I stars.

As stars of about 1 solar mass age, they become Delta Scuti variables and later evolve again to pulsation conditions after some loss of mass. They then are called RR Lyrae variables and are seen mostly in the very old globular star clusters, with their most massive stars already evolved to their death as white dwarfs or other compact objects. These RR Lyrae variables are also seen in the general galactic field. The name population II has been given to these very old stars which have masses like the Sun but ages two or more times longer. Since they were born before some of the elements such as carbon, nitrogen, oxygen, and iron were made in earlier generations of stars, they are very deficient in these elements. As these RR

Lyrae variables age and grow even larger in radius, converting their central helium to carbon and oxygen, they become population II Cepheids with periods about the same as the classical, massive, and younger population I Cepheids. *See* ELEMENTS AND NUCLIDES, ORIGIN OF.

Thus the population I stars become variables when their surface temperatures are 7000–8000 K as Delta Scuti variables, and, for more massive and luminous stars, when their surface temperatures are 5000–7000 K as Cepheids. The population II variable-star history with mass much like that of the Sun is as a Delta Scuti variable (7000–8000 K), an RR Lyrae variable (6500–7500 K), and finally as a population II Cepheid (5000–7000 K). The radii of both population types range from just over the solar radius for the shortest-period Delta Scuti variable to over 100 times that value for the longest-period Cepheids. The mean luminosity of these variables is from less than 10 to over 100,000 times that of the Sun.

Observation of pulsations. The pulsations of the Cepheids and the related classes can be detected not only by their varying luminosity but also by their outward and inward motions as detected by the periodic Doppler shifts of their spectral lines. The light range of the Cepheid variation is about one magnitude, and the peak-to-peak velocity range is between 10 and 70 km/s. These variations, which are mirror images of one another with maximum light at minimum radial velocity, are frequently smooth. The interpretation of this relative variation is that the surface temperature is largest when the star is passing through its mean position in its outward expansion. However, for the Cepheids between 5- and 15-day periods, bumps occur at a phase that is late in declining light for a period of 5 days, at the peak of the luminosity for a period of 10 days, and at the very start of rising light at a period of 15 days. All these variable stars pulsate in radial modes, that is, they retain their spherical shapes even as they pulsate.

Period-luminosity-color relation. The use of the Cepheids as distance indicators requires calibration of the period-luminosity-color (P-L-C) relation. Observations in 1912 of the Cepheids in the Small Magellanic Cloud showed that, for these stars all at the same distance from the Earth, the mean luminosity increased with period. This period-luminosity relationship can also be derived now from stellar evolution and pulsation theories. A further small effect is also known from observation and theory, that is, the period depends also on color or, physically, on the surface temperature. Two stars with the same mean luminosity can have two different mean radii, that is, different temperatures and colors, and therefore have two different periods. This P-L-C relation is calibrated from stars in nearby star clusters or associations with distances known by other considerations. Then the intrinsic mean luminosity and color are directly related to an observed period. Care must be taken that the color is properly corrected for the absorption of light by interstellar matter. Since Cepheids are seen in several other nearby galaxies, they serve to measure these galaxy distances with fair accuracy.

Structure and evolution. Cepheids have also given insight into stellar evolution theory. These yellow variables pulsate because of an instability in the near-surface layers which periodically inhibits the flow of radiation to the surface. The detailed properties of hydrogen and helium as these elements become ionized give the instability, which grows until internal radiative damping exactly balances the ionization driving. Many linear (assuming an infinitesimal amplitude) and nonlinear (allowing for an observable amplitude) calculations have clarified the pulsation details in the several longest-period modes with only a few internal nodes.

There is a definite surface temperature, which is only slightly hotter or cooler for lesser or greater luminosity, above which the Cepheids are not observed to pulsate. The theoretical explanation for this cutoff is that when the ionization mechanisms get too shallow they do not involve much mass and do not give enough driving to overcome the deep damping. This observed blue edge of the pulsation instability strip in the $\log L - \log T_e$ (where L is the luminosity and T_e is the electron temperature), or Hertzsprung-Russell, diagram can be used to determine a minimum helium content, because the ionization of the element helium causes most of the driving.

An observed red edge of the instability strip seems to be caused by the occurrence of convection in the surface layers. While the detailed pulsation damping mechanisms of convection are not so well understood, it is clear that there is a red (or cool) limit to the ionization instability mechanism, even though redder, often irregular variables exist.

The light curve bumps for 5–15-day Cepheids and the ratio of two periods seen simultaneously in about 11 short-period Cepheids indicate through pulsation calculations that masses are considerably less than those given by evolution theory and some other aspects of pulsation theory. The resolution of this mass anomaly is an important problem in astrophysics, and it may lead to new ideas about the internal structure of highly evolved stars.

The yellow pulsating stars of all classes mentioned here can reveal their mean radii, which can be compared to those predicted from stellar evolution theory. At two different phases of the pulsation cycle, when the surface temperatures are equal, the differing luminosity of the stars is due only to the differing surface areas. Use of simultaneous observations of the light and radial velocity variations can give the actual motion of the stellar surface during this time between equal-temperature phases. Analysis then gives the mean (or Wesselink) radius, which even for the anomalous-mass Cepheids agrees with evolution theory. Further, the use of the theoretical relation between the period and the mean density of the star gives a mass consistent with evolution theory. *See* STELLAR EVOLUTION.

[ARTHUR N. COX]

Bibliography: S. A. Becker, L. Iben, and R. S. Tuggle, On the frequency-period distribution of Cepheid variables in galaxies in the Local Group, *Astrophys. J.*, 218:633–653, 1977; M. Breger, Delta Scuti and related stars, *Pub. Astron. Soc.*

Pacific, 91:5–26, 1979; A. N. Cox, Cepheid masses from observations and pulsation theory, *Astrophys. J.*, 229:212–222, 1979; R. P. Kraft, The absolute magnitudes of classical Cepheids, in K. A. Strand (ed.), *Stars and Stellar Systems*, vol. 3, pp. 421–447, 1963.

Ceres

The first asteroid discovered. It was found serendipitously by G. Piazzi of Palermo, Sicily, on Jan. 1, 1801. Although it is located at the heliocentric distance predicted by the Titius-Bode law for the "missing" planet between Mars and Jupiter, additional asteroids were not located thus, and so the significance, if any, of this "law" is unknown.

With a diameter of approximately 1000 km, Ceres is the largest asteroid but not the brightest since it reflects only 6% of the visual light it receives. (Vesta, though only half as large as Ceres, is the brightest asteroid since it is closer to the Sun, 2.4 versus 2.8 astronomical units, or 3.6 versus 4.2×10^8 km, and reflects 25% of its incident light.) Ceres' mass of 1.17×10^{21}kg (that is, 2×10^{-4} that of the Earth) contains approximately 40% of the asteroid belt's total mass. Its relatively low density (approximately 2.3 g/cm^3) may, at least in part, be due to the fact that its surface material, and hence possibly its interior, contains water of hydration.

Ceres is a nearly spherical object, displays no large-scale color or brightness variations over its surface, and has a rotation period of 9^h03^m. *See* ASTEROID; ASTRONOMICAL UNIT; PLANET.

[EDWARD F. TEDESCO]

Bibliography: J. Degewij et al., Albedo and color contrasts on asteroid surfaces, *Icarus*, vol. 40, 1980; L. Lebofsky, Asteroid 1 Ceres: Evidence for water of hydration, *Mon. Not. Roy. Astron. Soc.*, 182:17p–21p, 1978; J. Schubart and D. L. Matson, Masses and densities of asteroids, in T. Gehrels (ed.), *Asteroids*, 1980.

Chromosphere

A transparent tenuous layer of gas in the atmosphere of the Sun resting on the photosphere. The chromosphere is between 3000 and 5000 km thick. It is a thermal buffer zone between the low temperature at the top of the photosphere (4400°K) and the high temperature of the corona (1 or 2×10^6°K). The structure of the chromosphere consists of a homogeneous atmosphere, in which are embedded forests of hairy spikes known as spicules. The largest spicules extend 10,000 km up into the corona above the homogeneous layer. They are believed to carry the energy that maintains the high coronal temperature. *See* SUN.

[JOHN W. EVANS]

Chronograph

A device for recording the epoch of an event. In its older form the astronomical chronograph includes a drum rotated at constant speed and a pen actuated by an electromagnet. A spiral line is traced on paper on the drum. Signals from a clock produce a set of time marks and signals made by an observer produce other marks, so that the epoch of the events observed can be determined.

Modern chronographs are of the digital printing type. In one form rotating type wheels carry numbers. In another form the time is accumulated electronically, and when a signal is given the output is printed by keys. *See* TIME-INTERVAL MEASUREMENT.

[WILLIAM MARKOWITZ]

Chronometer

A large, strongly built watch especially designed for precise timekeeping on ships at sea. The name is sometimes loosely applied to any fine watch.

The features that distinguish a chronometer from a watch are (1) a heavy balance wheel, the axis of which is kept always vertical by mounting the entire instrument within two concentric rings, so pivoted as to permit the chronometer to remain undisturbed despite considerable tilting of the box containing it, as the ship rolls and pitches; (2) a balance spring wound in cylindrical shape, instead of a nearly flat helix; (3) a special escapement; and (4) a fusee, by means of which the power of the mainspring is made to work through a lever arm of continuously changing length, being shortest when the spring is tightly wound and longest when it has run down, thus regulating the transmitted power so that it is approximately constant at all times.

Oceangoing ships during a voyage formerly relied completely on chronometers keeping Greenwich mean time as a means for determining longitude. The broadcasting of radio time signals that became widespread in the decade 1920–1930 has made Greenwich mean time available to mariners at almost any time of day, and chronometers are no longer indispensable for determining longitude at sea.

[GERALD M. CLEMENCE]

Coal Sack

An area in one of the brighter regions of the Southern Milky Way which to the naked eye appears entirely devoid of stars and hence dark with respect to the surrounding Milky Way region. Telescopic and spectroscopic observations reveal that the Coal Sack is a cloud of small, solid particles approximately 120 parsecs (400 light-years) from Earth and 10–15 parsecs in diameter. The cloud not only absorbs about two-thirds of the visible light passing through it but also strongly reddens the transmitted light. The Coal Sack is located just to the southeast of the Southern Cross. *See* INTERSTELLAR MATTER.

[WILLIAM LILLER]

Comet

One of the major types of objects that move in closed orbits around the Sun. Compared to the orbits of planets and asteroids, comet orbits are more eccentric and have a much greater range of inclinations to the ecliptic (the plane of the Earth's orbit). Physically, a comet is a small, solid body which is approximately a kilometer in diameter, contains a high fraction of icy substances, and shows a complex morphology, often including the production of a tail, as it approaches the Sun. *See* ASTEROID; PLANET.

About 10 comets are discovered or rediscovered each year. On the average, one per year is a bright comet, visible to the unaided eye and generating much interest among the public as well as among comet workers.

Astronomers consider comets to be worthy of detailed study for several reasons: (1) They are intrinsically interesting, involving a large range of physical and chemical processes. (2) They are valuable tools for probing the solar wind. (3) They are considered to be remnants of the solar system's original material and, hence, prime objects to be studied for clues about the nature of the solar system in the distant past.

Appearance. Comets are nebulous in appearance, and the tail is usually the most visually striking feature. This tail can in some cases stretch along a substantial arc in the sky. An example is given in Fig. 1, which shows Comet West dominating the eastern sky on March 9, 1976, with tails some 30° in length. Some fainter comets, however, have little or no tail.

The coma or head of a comet is seen as the ball of light from which the tail or tails emanate. Within the coma, and not yet seen, is the nucleus, the origin of the material in the tail and coma.

Discovery and designation. Comets are discovered by both amateur and professional astronomers. The fainter ones are often discovered by professionals on wide-field photographic plates taken for other purposes. The Czech astronomer L. Kohoutek discovered his famous comet in 1973 while taking plates for asteroid studies. Amateurs usually carry out systematic searches of the sky using wide-field binoculars or telescopes. Comet Ikeya-Seki was discovered in 1965 by two Japanese amateurs. Discoveries are communicated to the Bureau for Astronomical Telegrams, Smithsonian Astrophysical Observatory, Cambridge, MA, and are then announced by International Astronomical Union telegrams.

Normally, comets are named after their discoverers, and up to three independent codiscoverers are allowed. An example is Comet Kobayashi-Berger-Milon. The need for such a rule is easily understood when one realizes that some bright comets have been discovered almost simultaneously by dozens of individuals. Comets are numbered in two ways. The first comet discovered in 1980 would be 1980a, the second 1980b, and so on. After the orbits have been calculated, they are assigned a roman numeral in order of perihelion (point of closest approach to the Sun) passage. The third comet to pass perihelion in 1980 would be 1980 III. Halley's Comet at its last appearance was first comet 1909c and then comet 1910 II. Halley's Comet is also designated P/Halley, the P indicating a periodic comet.

Occasionally, a comet has been named after the

Fig. 1. Comet West as photographed on March 9, 1976, showing the general appearance of a bright comet. The left-hand, fan-shaped structure emanating from the head or coma is the dust tail, while the single straight structure to the right is the plasma tail. (*S. M. Larson, Lunar and Planetary Laboratory, University of Arizona*)

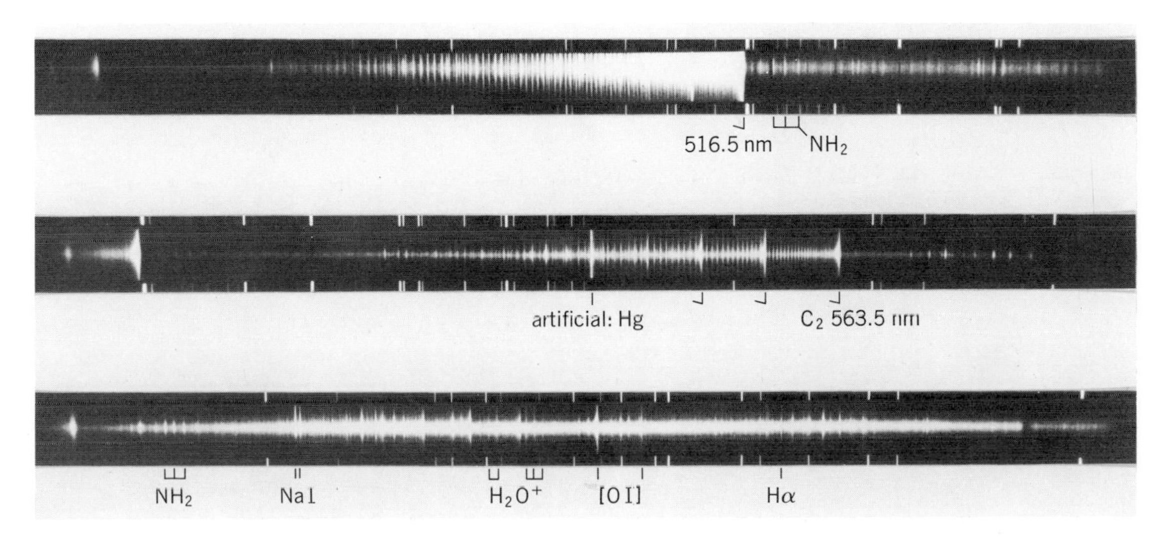

Fig. 2. High-dispersion spectrograms of Comet Kohoutek on January 9 and 11, 1974. (*Lick Observatory, University of California*)

person who computed its orbit. Examples are Halley's Comet and Encke's Comet.

Oort's cloud and comet evolution. A major step in understanding the origin of comets was taken in 1950 by J. Oort. He developed the idea that comets are in effect "stored" in a spherical cloud a great distance from the Sun. Occasionally gravitational perturbations by passing stars send comets into the inner solar system, where their phenomena are observed. This process, of course, leads to the eventual destruction of the comet, because passages near the Sun cause a loss of material which is not recovered by the comet. Very rough estimates indicate that a comet loses 1% of its mass at each perihelion passage. Thus, a comet with radius of 1 km would lose a layer approximately 3 m thick on each passage. The origin and evolution of comets are discussed in more detail below.

Observations. Observations of comets run the gamut of modern observing techniques. These include photographs in visual and ultraviolet wavelengths, photometry (accurate brightness measurements) in visual and infrared wavelengths, spectral scans in many wavelength regions, radio observations, and observations in extreme ultraviolet wavelengths from rockets and orbiting spacecraft above the Earth's atmosphere. Observations of Comet Kohoutek were carried out by the Skylab astronauts in 1973 and 1974. A sample spectrum in the visual wavelength range with some constituents identified is shown in Fig. 2. *See* ASTRONOMICAL SPECTROSCOPY.

Orbits. The first closed comet orbit to be calculated was Edmond Halley's elliptical orbit for the comet of 1680. This work indicated that comet orbits were ellipses with the Sun at one focus. In subsequent work, Halley noticed the striking similarity of the orbits of what were thought to be three different comets observed in 1531, 1607, and 1682. In 1705 he concluded that these were the same comet with a period of 75 or 76 years, and predicted its return in 1758. This comet is the one bearing Halley's name. It made an appearance in 1910 (Fig. 3), and is next due in 1985 and 1986. *See* HALLEY'S COMET.

The second comet to have its return successfully predicted was named after J. F. Encke. This comet has the shortest known period, 3.3 years.

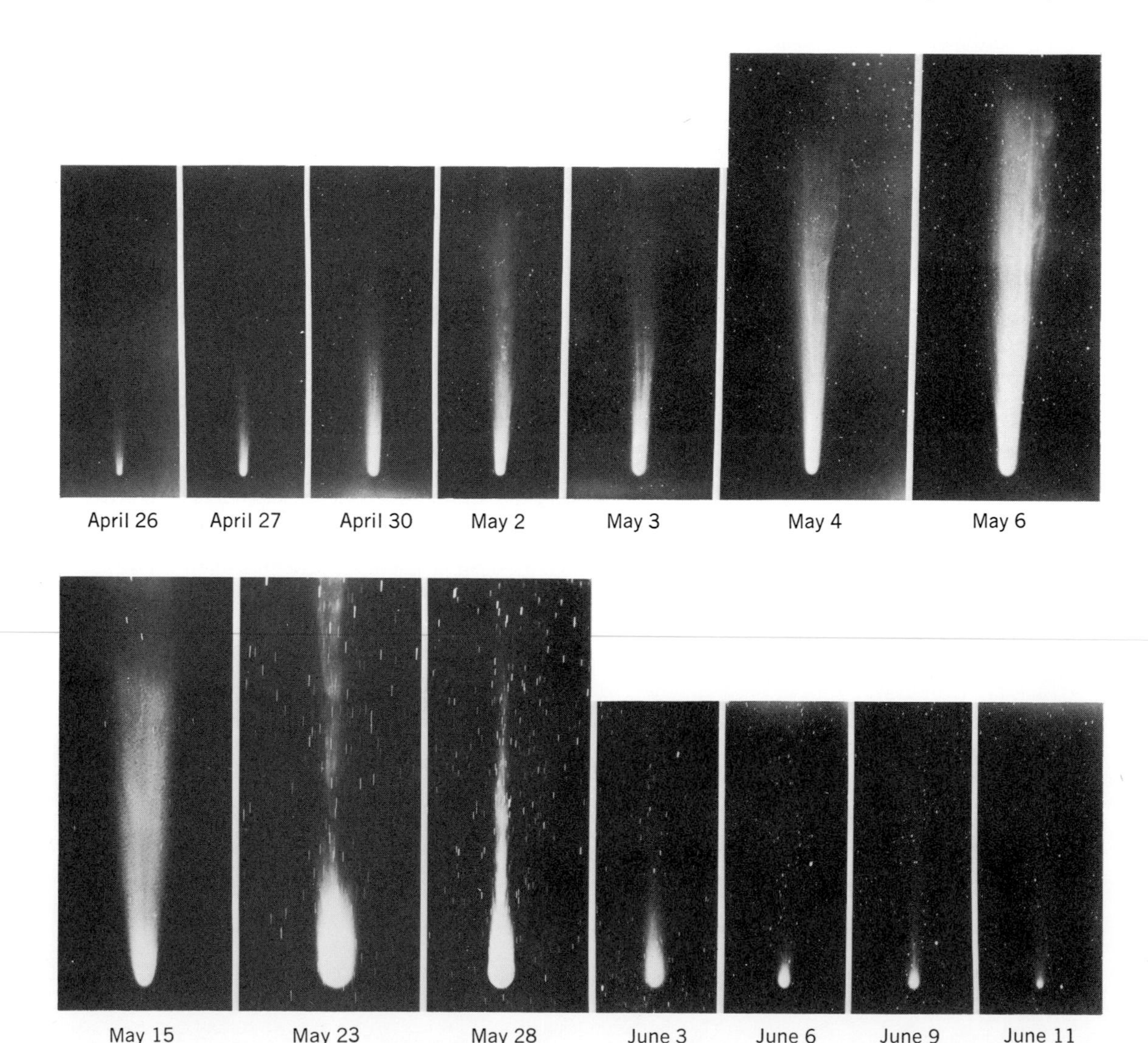

Fig. 3. Sequence of photographs showing the changing appearance of Halley's Comet in 1910. *(Hale Observatories)*

At its 1838 return, Encke's comet showed another common property of comets, that is, a steadily changing period. This phenomenon is now known to result from the so-called nongravitational forces which must be explained by any successful comet model. *See* ENCKE'S COMET.

Six paratemers are necessary to describe completely the orbit of a comet. They specify the orientation of the orbital plane in space, the orientation of the orbit in this plane, the size and shape of the ellipse, and the position of the comet along the orbit. In principle, three observations of position on the celestial sphere are sufficient, because each observation consists of independent measures of two coordinates. In practice, definitive orbits are derived from many observations, often hundreds.

While comet orbits are represented by ellipses to a good approximation, there are departures caused by the nongravitational forces and by the gravitational perturbations of the planets. When orbital parameters are listed for a comet, they refer to an ellipse which exactly matches the comet's position and motion at a specific time. Such an osculating orbit, as it is called, forms the starting point for studies of orbital evolution and for accurate predictions of the time and location of a comet's appearance in the sky. *See* CELESTIAL MECHANICS.

Orbital parameters have been determined for over 600 individual comets. Over 500 are classified as long-period comets, that is, comets with orbital periods greater than 200 years. The orbital planes of the long-period comets have approximately random inclinations with respect to the ecliptic. This means that there are as many comets with direct orbits (revolving around the Sun in the same sense as the planets) as with retrograde orbits (revolving in the sense opposed to the planets' motion). Careful examination of the original (that is, inbound) orbits of the long-period comets shows none that are hyperbolas; that is, no interstellar comets have yet been observed. This fact strongly implies that the cloud of comets is gravitationally bound to the Sun and therefore is a part of the solar system.

The over 100 short-period comets are mostly in direct orbits with inclinations of less than 30°. The distribution of periods shows a peak between 7 and 8 years, and the majority have an aphelion (point of greatest distance from the Sun) near the orbit of Jupiter.

Composition. The results of many spectroscopic studies on the composition of comets are summarized in the table, where the atoms, molecules, ions, and classes of substances that have been observed are listed. Fairly complex molecules

Atomic and molecular species observed in comets*

Coma (neutrals)	Tail (ions)
H, OH, O, S	CO^+, CO_2^+, H_2O^+, OH^+
C, C_2, C_3, CH, CN, CO, CS	CH^+, N_2^+, Ca^+, C^+, CN^+
HCN, CH_3CN, NH	
NH_2, Na, Fe, K	
Ca, V, Cr, Mn, Co, Ni, Cu	

*Silicates observed in coma and dust tail.

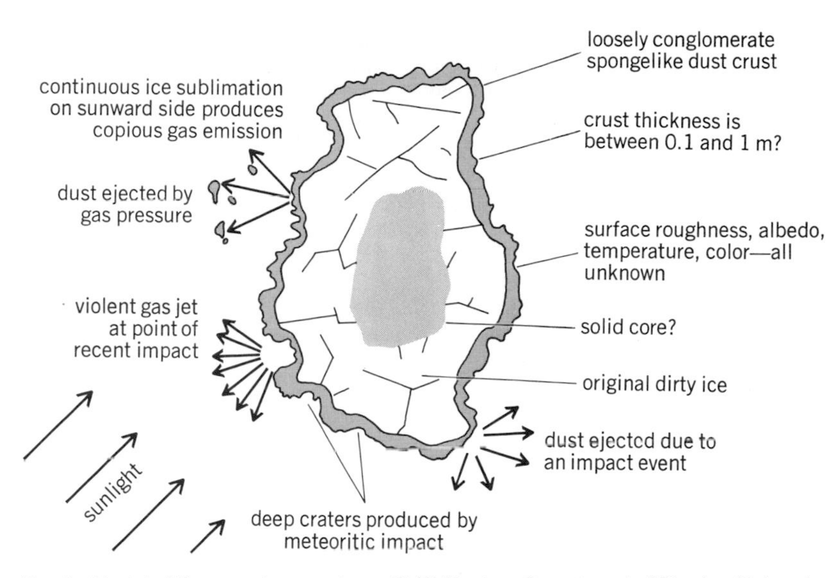

Fig. 4. Model of the cometary nucleus. *(D. W. Hughes, Department of Physics, University of Sheffield)*

such as hydrogen cyanide (HCN) and methyl cyanide (CH_3CN) are present. Water, which is considered the principal constituent of the nucleus (for observational and theoretical reasons mentioned later), has not been observed directly.

Structure. The main components of a comet are the nucleus, coma, hydrogen cloud, and tails.

Nucleus. Strong indirect evidence points to the existence of a central nuclear body or nucleus from which all cometary material, both gas and dust, originates. Unfortunately, no photograph of a nucleus exists.

In the early 1950s, F. L. Whipple proposed an icy conglomerate model of the cometary nucleus. In this model, the nucleus is not a cloud of particles (the "sandbank" model), but a single mass of ice with embedded dust particles, commonly called a "dirty snowball." Such a nucleus could supply an adequate amount of gas to explain cometary phenomena and last through many apparitions because only a thin surface layer would be eroded away (by sublimation) during each passage near the Sun.

A schematic drawing of a cometary nucleus, embodying the best available information, is given in Fig. 4. It is probably somewhat irregular, with radius ranging from a few hundred meters to 10 km. Masses would range from 10^{11} to 10^{16} kg, with roughly equal parts of ices and dust. The average density would be about 2 g/cm^3. Nuclei of comets sometime split into two or more pieces; Comet West (1976 VI) is an example.

Coma. The coma is observed as an essentially spherical cloud of gas and dust surrounding the nucleus. The principal gaseous constituents are the neutral molecules listed in the table. The coma can extend as far as 10^5 to 10^6 km from the nucleus, and the material is flowing away from the nucleus at a typical speed of 0.5 km/s. As the gas flows away from the nucleus, the dust particles are dragged along. For comets at heliocentric distances greater than 2.5 to 3 astronomical units

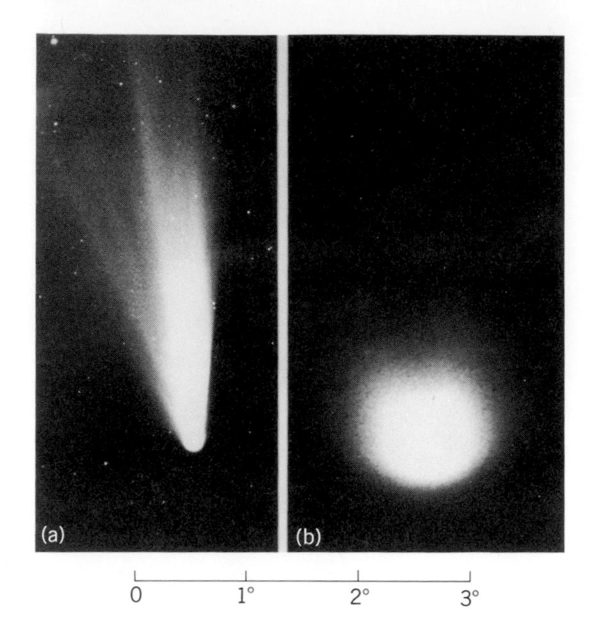

Fig. 5. Comparison of photographs of Comet West, obtained from a rocket on March 5, 1976, and printed to same scale. *(a)* Visual photograph *(P. D. Feldman, Johns Hopkins University)*. *(b)* Ultraviolet photograph showing hydrogen cloud *(C. B. Opal and G. R. Carruthers, Naval Research Laboratory)*.

(AU), the coma is not normally visible and is presumed not to be present.

Hydrogen cloud. In 1970, observations of Comet Tago-Sato-Kosaka (1969g) and Comet Bennett (1969i) from orbiting spacecraft showed that these comets were surrounded by a giant hydrogen cloud extending to 10^7 km, or a size larger than the Sun. The observations were made in the resonance line of atomic hydrogen at 121.6 nm. Hydrogen clouds have since been observed for three other comets. A photograph of Comet West's hydrogen cloud is given in Fig. 5, along with a visual photograph to the same scale for comparison. Bright comets (such as Bennett) have a hydrogen production rate by sublimation from the nucleus at heliocentric distance of 1 AU in the range 3 to 8×10^{29} atoms/s. The size of the hydrogen cloud depends on the velocity of the outflowing hydrogen atoms, and 8 km/s has been derived. This velocity would arise (from energy balance considerations) if most of the hydrogen in the cloud were produced by the photodissociation of the hydroxyl radical (OH).

Tails. Photographs of bright comets generally show two distinct types of tails (Fig. 1): the dust tails and the plasma tails. They can exist separately or together in the same comet. In a color photograph, the dust tails appear yellow because the light is reflected sunlight, and the plasma tails appear blue from emission due to ionized carbon monoxide, CO^+.

Studies of features in plasma tails led L. Biermann in 1951 to postulate the existence of a continuous outflow of ionized material from the Sun, which he called the solar corpuscular radiation, now called the solar wind. The interaction of the solar wind and its magnetic field, as suggested by H. Alfvén in 1957, plays an important role in cometary physics and in the formation of plasma tails. Understanding of comets will not be complete until this interaction is understood. *See* SOLAR WIND.

The plasma tails are generally straight and have lengths in the range 10^7 to 10^8 km. The Great Comet of 1843 had a plasma tail extending over 2 AU in length. The plasma in these tails is composed of electrons and molecular ions. The dominant visible ion is CO^+, and the other ions known to be present are listed in the table. The zone of production for the molecular plasma appears to be in the coma near the sunward side of the nucleus. The material in the plasma tails is concentrated into thin bundles or streamers. This structure is strong evidence for a magnetic field threading the plasma tail. Additional structure is found in the tail in the form of knots and kinks. These features appear to move along the tail away from the head at speeds which range from 10 km/s to 200 km/s. Plasma tails are generally not observed beyond heliocentric distances of 1.5 to 2 AU; an exception is Comet Humason, which showed a spectacular, disturbed plasma tail well beyond the normal distances.

The dust tails are usually curved and have lengths in the range 10^6 to 10^7 km. Normally, the dust tails are relatively homogeneous; an exception, Comet West, is shown in Fig. 1. Observations indicate that the dust particles are typically 1 μm in diameter and are probably silicate in composition. Occasionally dust tails are seen which appear to point in the sunward direction, the so-called antitails. Examples are Comet Arend-Roland in 1957 and Comet Kohoutek in 1973 and 1974. These are not truly sunward appendages but are the result of projection effects.

The structure of the constituent parts of comets—nucleus, coma, hydrogen cloud, and tails—is summarized schematically in Figs. 6 and 7.

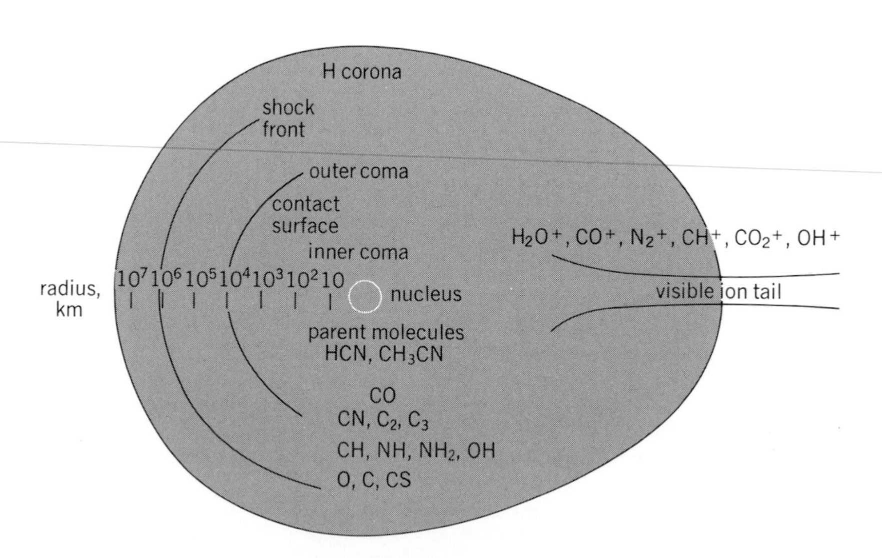

Fig. 6. Gaseous features of a comet diagrammed on a logarithmic scale. *(Report of the Comet Science Working Group, NASA–Jet Propulsion Laboratory, NASA Tech. Mem. 80542, 1979)*

Modern theory. The goal of modern comet theory is to explain the facts about comets outlined above. The theory appears to be in reasonably good shape and, moreover, can be tested directly

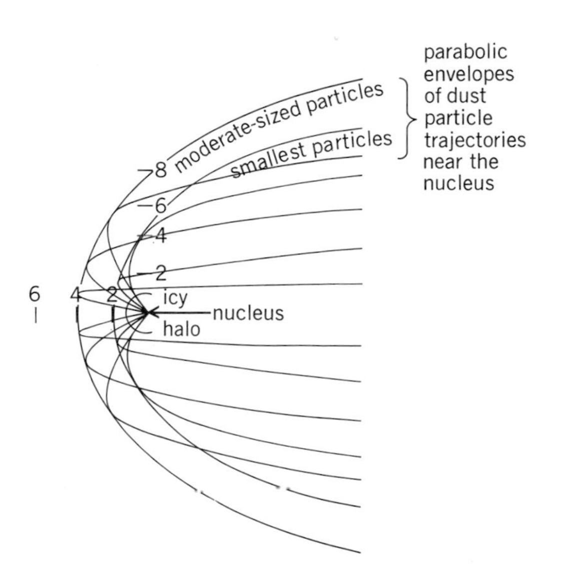

Fig. 7. Dust features of a comet, diagrammed on a linear scale, unit 10,000 km. The icy halo, a cloud of icy particles which may surround the nucleus, is also shown. *(Report of the Comet Science Working Group, NASA–Jet Propulsion Laboratory, NASA Tech. Mem. 80542, 1979)*

when a spacecraft is sent to a comet.

Sublimation from nucleus. The cornerstone of the current best ideas is F. L. Whipple's icy conglomerate model of the nucleus as further developed by A. Delsemme. As a comet approaches the Sun on its orbit, sunlight supplies radiant energy to the surface of the nucleus. The energy received heats the nucleus when it is far from the Sun. As the comet continues toward the Sun, the temperature of the surface layers increases to a value, determined by the thermodynamic properties of the ice, where sublimation (passage from the solid state directly to the gaseous state) occurs. In this situation, most of the incident energy goes to the sublimation of ices.

The onset of activity in comets at 2.5 to 3 AU is entirely consistent with water ice as the dominant ice constituent of the nucleus. Even though a definitive observation of H_2O has not been made, substances likely having H_2O as the parent molecule have been observed; hence, the existence of water is most probable. But a problem still exists. Most other possible constituents are predicted from theory to begin sublimation much farther from the Sun than ice, but, for example, molecular emissions from CN are visible essentially at the onset of activity.

The accepted solution to the problem is to assume that comets are made up of the type of ice called a clathrate hydrate. Very simply, ice in the crystalline form has cavities which are formed by the bonds that hold the crystal together and into which other substances can be trapped. Thus, the release of most minor constituents is controlled by the thermodynamic properties of water ice and not by the properties of the minor constituent. If all the available cavities are filled, the minor constituents can amount to 17% of the icy lattice material by number of atoms. If more minor constituents are present in the nucleus, they would be vaporized well beyond 3 AU on the comet's first approach to the Sun. This phenomenon could explain comets found to be anomolously bright at large distances from the Sun and the dimming of some comets after their first perihelion passage.

Origin of nongravitational forces. The nongravitational forces have a ready explanation if the cometary nucleus is rotating. The result of radiant energy producing sublimation of ices in an ice-dust mixture is to leave a crust of dusty material. Thus, there is a time lag while the heat traverses the dust layer between maximum solar energy received and maximum loss through sublimation. If the nucleus were not rotating, the maximum mass loss would be directly toward the Sun. For a rotating nucleus, the mass loss occurs away from the sunward direction, toward the afternoon side of the comet. The analogous situation on Earth (that is, the time lag between cause and effect) produces the warmest time of day in the afternoon, not at noon. The mass loss under these circumstances produces a force on the nucleus via the rocket effect (reaction force), and this force can accelerate or retard the motion of the comet in its orbit, as shown schematically in Fig. 8. Detailed studies of the nongravitational forces in comets show that they are entirely consistent with water ice as the controlling substance.

Formation of coma. The sublimated gases, mostly neutral molecules, flow away from the nucleus, dragging some of the dust particles with them to form the coma. Close to the nucleus, the densities are high enough that chemical reactions can occur between molecular species. Thus, the molecules observed spectroscopically far from the nucleus may not be the same as the initial composition.

Formation of dust tails. The dust particles carried away from the nucleus by the flow of coma gases are blown in the antisolar direction by the Sun's radiation pressure to form the dust tails. The general theory has been developed by M. L. Finson and R. F. Probstein, and good agreement can be obtained with the observed shapes and sizes of the tails if the emission of dust from the nucleus has a peak before perihelion. The larger particles liberated from the nucleus can orbit the Sun and reflect sunlight to produce the zodiacal light.

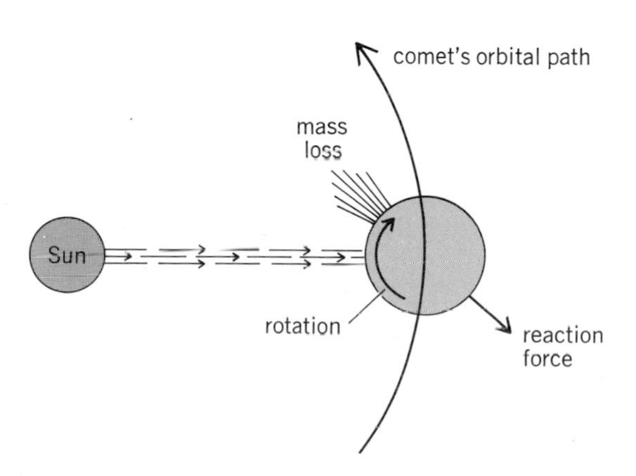

Fig. 8. Mechanism by which mass loss can produce a reaction force on the nucleus by the rocket effect.

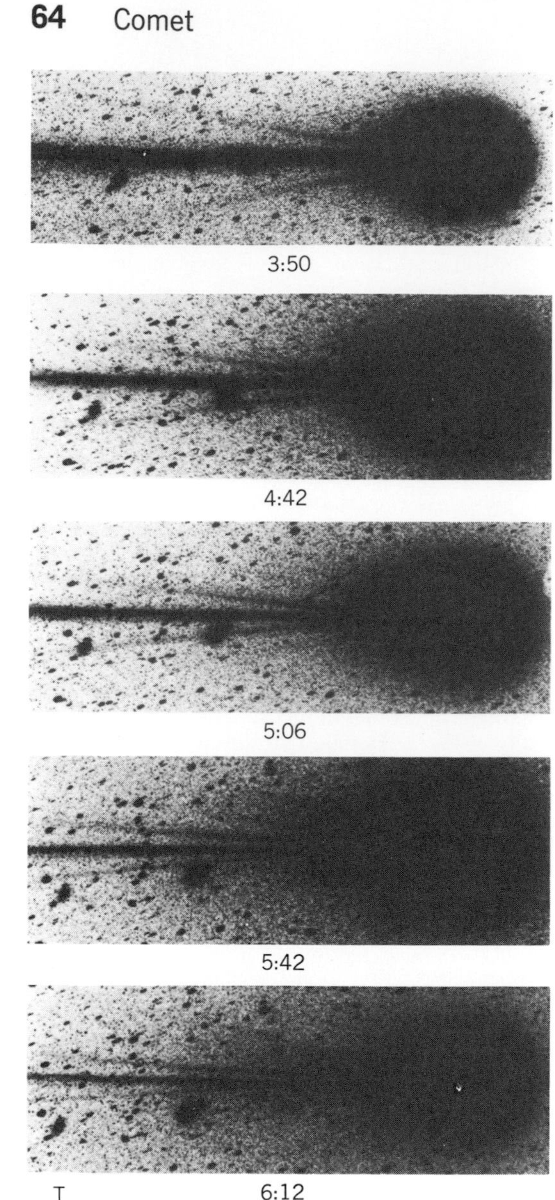

Fig. 9. Sequence of photographs of Comet Kobayashi-Berger-Milon on July 31, 1975, showing the capture of magnetic field from the solar wind. The dominant pair of tail streamers visible on either side of the main tail lengthen and turn toward the tail axis in this sequence. (*Joint Observatory for Cometary Research, operated by NASA–Goddard Space Flight Center and New Mexico Institute of Mining and Technology*)

Formation of plasma tails. The gas flowing away from the nucleus has a more involved fate. Under normal circumstances, when a comet's heliocentric distance is about 1.5 to 2 AU, significant ionization of the coma molecules occurs (probably by solar radiation), and this triggers a reaction with the solar wind. At the Earth, the solar wind, a fully ionized proton-electron gas, flows away from the Sun at 400–500 km/s and has an imbedded magnetic field. Because of the magnetic field, the ionized cometary molecules cause the solar-wind field lines to slow down in the vicinity of the comet while proceeding at the full solar-wind speed away from the comet. This situation causes the field lines and the trapped plasma to wrap around the nucleus like a folding umbrella, to form the plasma tail. The folding can be seen and photographed because of the emission from the trapped ions (such as CO^+) which serve as tracers of the field lines. A photographic sequence showing this phenomenon is given in Fig. 9. Thus, while the ionized molecules are indeed swept in the antisolar direction by the solar wind, the plasma tail should be thought of as a part of the comet attached to the near-nuclear region by the magnetic field captured from the solar wind.

Exceptions occur at times when the polarity of the solar-wind magnetic field changes. This disrupts the magnetic connection to the near-nuclear region and literally causes the old plasma tail to disconnect while the new tail is forming (Fig. 10).

Fate of comets. When the process of sublimation has been carried out over an extensive period of time, as would be the case for the short-period comets, the ices will be exhausted and the inactive or "dead" comet should consist of dust particles and larger-sized rocky material. These remnants are dispersed along the comet's orbit by perturbations, and are the particles responsible for produc-

Fig. 10. Comet Morehouse. (*a*) Beginning of separation of tail from comet's head on September 30, 1908. (*b*) Tail widely separated from the head on October 1, 1908. (*Yerkes Observatory, University of Chicago*)

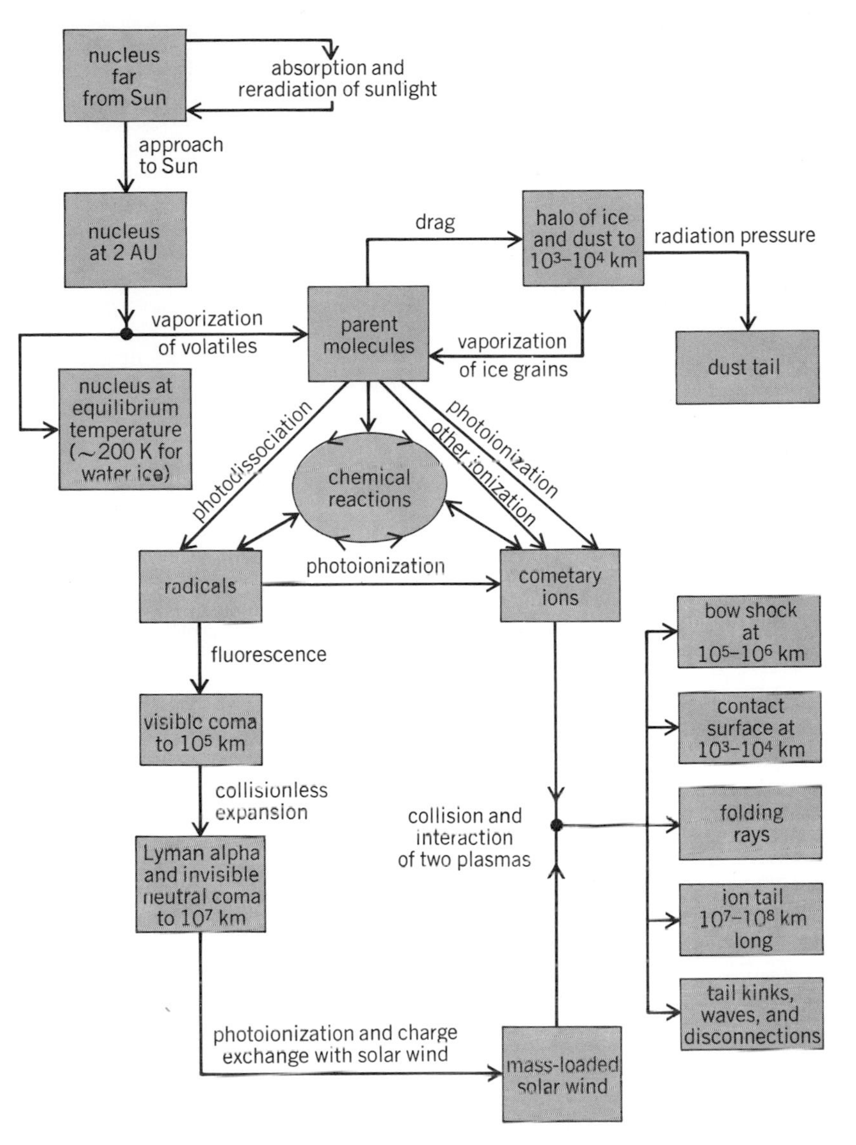

Fig. 11. Physical and chemical processes involved in the interaction of a comet with sunlight and the solar wind. *(Report of the Comet Science Working Group, NASA–Jet Propulsion Laboratory, NASA Tech. Mem. 80542, 1979)*

ing meteors when they enter the Earth's atmosphere. Small particles, very probably of cometary origin, have been collected by high-flying aircraft. *See* METEOR.

A detailed summary of the physical processes in comets is given in Fig. 11.

Origin. While not definitely proved, it seems likely that the population of short-period comets can be produced from the population of the long-period comets—those in the Oort Cloud—by their gravitational interaction with Jupiter. Thus, the following discussion concentrates on the origin of the long-period comets.

The evidence from statistics of cometary orbits indicates an essentially spherical cloud of comets with dimensions in the range 10^4 to 10^5 AU. Gravitational perturbations from passing stars have several effects on the cloud. They limit its size and tend to make the orbits random (as observed). Most importantly, the perturbations continually send new comets from the cloud into the inner solar system, where they are observed.

The current consensus on the origin of comets holds that they condensed from the solar nebula at the same time as the formation of the Sun and planets. In other words, although the details are sketchy, comets are probably a natural by-product of the solar system's origin. They may be remnants of the formation process, and their material may be little altered from the era of condensation to the present time. *See* SOLAR SYSTEM.

Some scenarios for formation and evolution assign additional roles to the comets. They may have been a major source of the atmospheres of the terrestrial planets and might even have provided the organic molecules necessary for the evolution of life on Earth. While some of these ideas are speculative, they help show why comets are objects of keen interest and active study, and why the National Aeronautics and Space Administration will probably launch a deep-space probe to comets in the near future.

Space missions to comets. Interest in sending a spacecraft to a comet has been heightened by the expected appearance of Halley's Comet in 1985 and 1986. This large, active comet displays the entire range of cometary phenomena. Plans have been approved by Japan and by the European Space Agency to send spacecraft to the vicinity of Halley's Comet. Similar plans have also been considered by the United States and the Soviet Union. All of these missions would execute a flyby of the comet, but would not rendezvous with it. Experiments on board the spacecraft would photograph the nucleus and carry out a large number of measurements along the trajectory of the spacecraft. Rendezvous missions to comets would probably be carried out in the late 1980s. *See* SPACE PROBE.

Comets and the public. There is no scientific evidence for the belief that comets are harbingers, omens, or actual producers of evil, disasters, or natural calamities. Some facts about comets can be distorted to provide an apparent basis for some fears. There are toxic gases in comets, but the amount is very small and, moreover, the cometary gas would not penetrate the Earth's atmosphere. The only real problem would be an impact of a comet on the Earth's surface. The odds of an impact are very small, and only one plausible site on the Earth's surface is thought to have witnessed a cometary event. The event took place in central Siberia on June 30, 1908, and general devastation occurred for a radius of approximately 35 km. The most spectacular result was a large number of trees blown down with their trunks pointing radially away from the explosion. Studies of the evidence indicate that the explosion took place in the air and that the body probably was a large, inactive chunk of Encke's Comet. The chances of such a body striking the Earth have been estimated as on the average once every 2000 years. Moreover, only 10^{-5} of the Earth's surface was affected.

The opportunities for much of the public to view bright comets are becoming increasingly rare due to the combination of atmospheric pollution and bright city lights. In January 1974 there was general public disappointment over the poor visibility of Comet Kohoutek, particularly as viewed from cities in the populous northeastern United States, although it was a conspicuous object to astronomers at mountaintop observatories in the southwestern United States.

[JOHN C. BRANDT]

Bibliography: J. C. Brandt and R. D. Chapman, *Introduction to Comets*, 1981; J. A. Hynek (ed.), *Astrophysics*, 1951; M. Neugebauer et al., (eds.), *Space Missions to Comets*, NASA CP-2089, 1980; F. L. Whipple, Background of modern comet theory, *Nature*, 263:15–19, 1976.

Constellation

A name used in astronomy originally to designate any one of the star groups that were imagined to form configurations in the sky. Now the term constellation also refers to any one of the definite areas of the sky. The division of the heavens into constellations is the oldest part of astronomy. From great antiquity bright stars in the same region of the sky with conspicuous geometric patterns have been arranged into artificial groupings to resemble familiar objects, animals, and personages. This was done to facilitate distinguishing and recognizing the stars. *See* STAR.

Like cities on the surface of Earth, which can be located by their coordinates (longitude and latitude), stars on the celestial sphere also possess coordinates (right ascension and declination), which define their positions accurately in the sky. However, it is easier to find a city on the map by knowing its relationship to a coastline, a river, or a

Table 1. Traditional constellations

Ancient name	Modern name
Andromeda	Andromeda
Aquarius*	Water Bearer
Aquila	Eagle
Ara	Altar
Argo†	Ship
Aries*	Ram
Auriga	Charioteer
Boötes*	Bear Driver
Cancer*	Crab
Canis Major	Greater Dog
Canis Minor	Lesser Dog
Capricornus*	Sea Goat
Cassiopeia*	Cassiopeia
Centaurus*	Centaur
Cepheus	Cepheus
Cetus	Whale
Corona Australis	Southern Crown
Corona Borealis	Northern Crown
Corvus	Crow
Crater	Cup
Cygnus*	Swan
Delphinus	Dolphin
Draco	Dragon
Equuleus	Little Horse
Eridanus	River Po
Gemini*	Twins
Hercules	Hercules
Hydra	Snake
Leo*	Lion
Lepus	Hare
Libra*	Balance
Lupus	Wolf
Lyra*	Lyre
Ophiuchus	Serpent Carrier
Orion*	Warrior
Pegasus*	Winged Horse
Perseus*	Perseus
Pisces*	Fishes
Pisces Austrinus	Southern Fish
Sagitta	Arrow
Sagittarius*	Archer
Scorpius*	Scorpion
Serpens	Serpent
Taurus*	Bull
Triangulum	Triangle
Ursa Major*	Great Bear
Ursa Minor*	Little Bear
Virgo*	Maiden

*See separate article.

†The large Ptolemy constellation Argo (Ship) is now divided into four smaller ones, namely, Carina (Keel), Puppis (Stern), Pyxis (Compass), and Vela (Sail).

Table 2. Bayer's constellations of the southern hemisphere

Latin name	English equivalent
Apus	Bird of Paradise
Camelopardalis	Giraffe
Canes Venatici	Hunting Dogs
Chamaeleon	Chameleon
Columba	Dove
Dorado	Sword Fish
Grus	Crane
Hydrus	Water Snake
Indus	Indian
Lacerta	Lizard
Leo Minor	Lesser Lion
Lynx	Lynx
Monoceros	Unicorn
Musca	Fly
Pavo	Peacock
Phoenix	Phoenix
Scutum	Shield
Sextans	Sextant
Triangulum Australe	Southern Triangle
Tucana	Toucan
Volan	Flying Fish
Vulpecula	Little Fox

mountain. Similarly, a star is located faster by a recognition of the constellation in which it is situated. The constellations are the landmarks, and the bright stars the guide posts. Constellations are useful in helping astronomers, navigators, and surveyors to find the stars they need. In this space age constellations are indispensable to moon watchers for tracking artificial satellites and to astronauts navigating their spacecraft.

The system of constellations in use today was begun thousands of years ago by Chaldean and Egyptian astronomers and was further developed by the Greeks and Romans. Selecting conspicuous groups of stars, they fitted around them the imagined forms and figures from their mythology and wove about them legend and folklore. Some constellations vividly represent their designations, while others require a considerable stretch of the imagination to visualize any resemblance.

People from different parts of the world naturally arranged the stars differently. Even prominent configurations were represented by different objects. Thus the Oriental nations, such as China, Japan, Korea, and India, and the early American Indians have their own systems of sky maps, which are, with few exceptions, distinct from the Greco-Semitic patterns used today.

The ancients originated most of the constellations visible in northern latitudes. Ptolemy, the Alexandrian astronomer, recorded 48 of them 2000 years ago, as listed in Table 1; they are still used.

The ancients living in northern latitudes were unable to see the southern sky; hence it was not mapped. In 1603 J. Bayer charted the southern regions and named 13 new southern constellations, and in 1690 Hevelius added nine more, as presented in Table 2. Finally in 1763 N. L. de Lacaille completed the list by adding 14 names to new constellations, given in Table 3.

These constellations, together with Coma Berenices (Berenice's Hair) whose origin is uncertain although it is sometimes credited to Tycho Brahe, make up a total of 88 groups, now recognized and universally accepted (see illustration).

Of the constellations in the sky, those along the zodiac are of special significance. The zodiac is an imaginary belt or zone of the sky, 18° wide and centered on the ecliptic (the path of the Sun). Within this zone move the Sun, Moon, and planets. The stars in the zodiac were grouped into constellations, most of which have the figures of animals; hence the name zodiac, or animal circle.

The zodiac is divided equally into 12 sections, called signs, in each of which the Sun is situated for 1 month of the year. Each sign, 30° in length, is named from a constellation with which the sign originally coincided. These 12 zodiacal constellations are—beginning from the vernal equinox, the first point of Aries, in order westward—Aries (Ram), Taurus (Bull), Gemini (Twins), Cancer (Crab), Leo (Lion), Virgo (Maiden), Libra (Balance), Scorpius (Scorpion), Sagittarius (Archer), Capricornus (Sea Goat), Aquarius (Water Bearer), and Pisces (Fishes).

When the zodiacal constellations first came into use over 2000 years ago, the vernal equinox was at Aries. Because of the precession of the equinoxes, each of the zodiacal signs has since moved westward 30° along the ecliptic. Thus all the signs are now displaced into the neighboring constellation. For instance, the sign Aries is now in the constellation Pisces. Unsuccessful attempts have been made to change the names of the constellations or to introduce new ones, such as to replace the old pagan nomenclature with Christian names by substituting the names of the 12 Apostles for the 12 zodiacal constellations.

Some of the old constellations have been given new significance. For instance, Cygnus today is better known as the Northern Cross. The Big Dipper is the most familiar part of Ursa Major. Delphinus is sometimes called Job's Coffin. Also some familiar designations bear no relation to the constellations' names, such as the Great Square in Pegasus, the Sickle in Leo, and the Milk Dipper in Sagittarius. Constellation boundaries were irregular. When only bright stars were considered, there was no ambiguity in locating them. But when fain-

Table 3. Lacaille's additional southern constellations

Latin name	English equivalent
Antlia	Air Pump
Caelum	Chisel
Circinus	Compasses
Crux*	Cross
Fornax	Furnace
Horologium	Clock
Mensa	Table
Microscopium	Microscope
Norma	Rule
Octans	Octant
Pictor	Easel
Recticulum	Net
Sculptor	Workshop
Telescopium	Telescope

*See separate article.

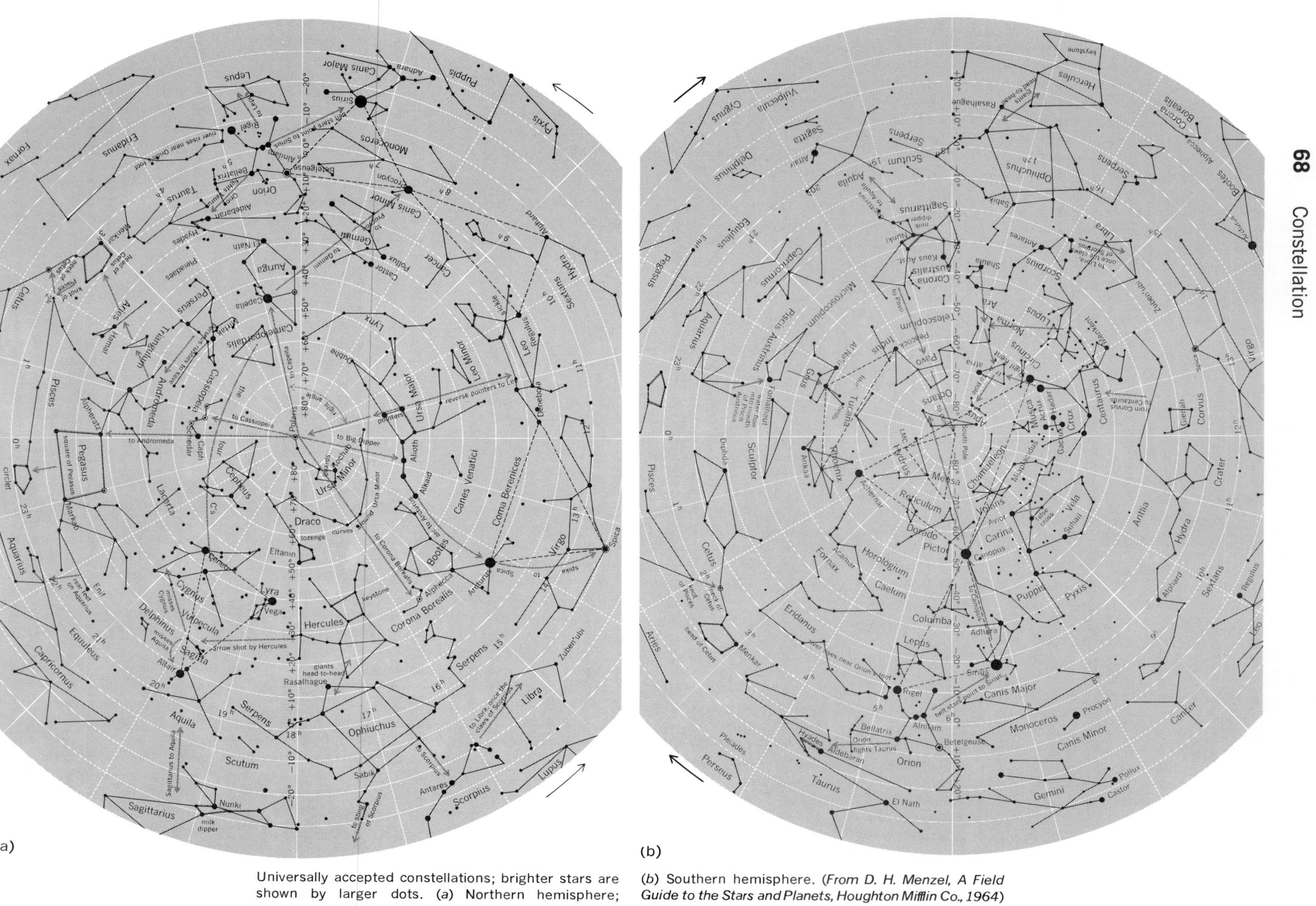

Universally accepted constellations; brighter stars are shown by larger dots. (*a*) Northern hemisphere; (*b*) Southern hemisphere. (*From D. H. Menzel, A Field Guide to the Stars and Planets, Houghton Mifflin Co., 1964*)

ter and fainter stars were considered and new and more constellations were added, the sky areas became so crowded that uncertainty and confusion naturally resulted at the boundaries. The situation became even more hopeless when different sky cartographers would draw the constellation figures and boundaries according to their own whims and fancies. To remedy the confusion and to secure a general agreement on the boundaries, the International Astronomical Union in 1928 instituted a much-needed reform by remapping all the constellations. Replacing the irregular and uncertain outlines, the new boundaries now run only from north to south and from east to west, though they zigzag considerably to contain as many stars as possible in their former constellations. [CHING-SUNG YU]

Bibliography: R. K. Marshall and I. M. Levitt, *Star Maps for Beginners*, 1980; D. H. Menzel, *A Field Guide to the Stars and Planets*, 1964; S. Mitton (ed.), *Star Atlas*, 1979; P. Moore, *The New Guide to the Stars*, 1976; W. T. Olcott and R. N. Mayall, *Field Book of the Skies*, 1954; H. A. Rey, *Find the Constellations*, rev. ed., 1976; H. Vehrenberg and D. Blank, *Handbook of the Constellations*, 1977.

Coronagraph

A specialized astronomical telescope used for observation of the corona of the uneclipsed Sun. This observation is difficult because the brightness of the corona is only 10^{-6} that of the Sun. The stray light surrounding the solar image in a conventional telescope is about 100 times brighter, and its glare drowns the corona completely. The coronagraph is designed to minimize this glare by artificially eclipsing the solar image itself with an occulting disk. Further, it suppresses the diffracted light from the edge of the primary objective and the doubly reflected light from its center by forming an image of the objective on an annular screen which occults these areas, as illustrated. In addition to such matters of design, success depends on an objective which produces the least possible scattered light. It must be made of glass free of all defects, polished to the highest perfection, and scrupulously cleaned. When the atmosphere above the coronagraph is free of light-scattering dust and smoke and the sky is blue right up to the edge of the Sun, the corona can be observed through an appropriate filter or with a spectrograph. *See* SOLAR CORONA.

[JOHN W. EVANS]

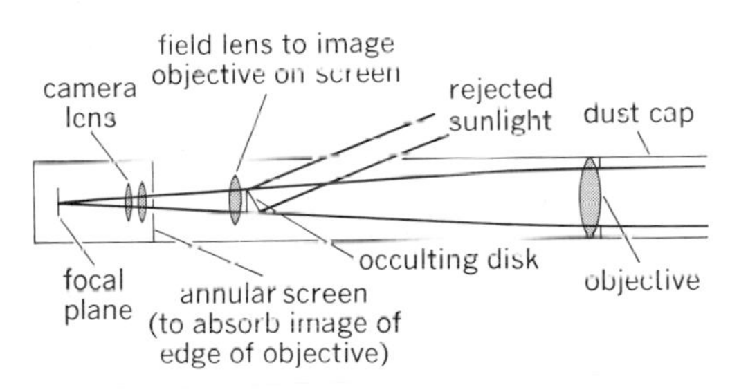

The coronagraph of B. Lyot.

Cosmic electrodynamics

The science concerned with electromagnetic phenomena in ionized media encountered in interstellar space, in stars, and above the atmosphere. Because these ionized materials are excellent electrical conductors, they are strongly linked to magnetic-field lines; they can travel freely along but not across the field lines. Statistically this linkage tends to equalize the energies in the magnetic field and in the turbulent motion of the ionized material. Phenomena treated under cosmic electro-dynamics include acceleration of charged particles to cosmic-ray energies, both in the galaxy and on the surface of the Sun; collisions between galaxies; the correlation of magnetic fields with galactic structure; sunspots and prominences; magnetic storms; aurora; and Van Allen radiation belts.

Alfven waves—transverse waves which travel along the magnetic-field lines in a manner similar to the way waves travel along a stretched string—are often important.

[ROLF K. M. LANDSHOFF]

Bibliography: H. Alfvén and C. G. Falthammer, *Cosmical Electrodynamics*, 2d ed., 1963; R. L. Boyd, *Space Physics: The Study of Plasmas in Space*, 1975; S. Chapman, *Solar Plasma: Geomagnetism and Aurora*, 1964; J. W. Dungey, *Cosmic Electrodynamics*, 1958; J. H. Piddington, *Cosmic Electrodynamics*, 2d ed., 1978.

Cosmic microwave radiation

A nearly uniform flux of microwave radiation that is believed to permeate all of space. The discovery that "empty" space is filled with microwave radiation has had a profound impact on understanding the nature and history of the universe. Although the discovery of the radiation was made in 1965 by A. Penzias and R. W. Wilson, it was only in the late 1970s that its origin as the remnant fireball from the "big bang" became nearly universally accepted by astrophysicists. The correct prediction of the entire spectrum by P. J. E. Peebles, R. H. Dicke, R. G. Roll, and D. T. Wilkinson, after the observation of the intensity at only one frequency, was one of the great triumphs of cosmological theory.

Origin of the radiation. Cosmic microwave radiation is the strongest evidence available for a hot, compressed early universe. In the theory of the big bang, as originally detailed by G. Gamow, H. Alpher, and R. Hermann, the universe began with an explosion about 15×10^9 years ago. This was not an explosion of matter into empty space, but an explosion of space itself. In the big bang theory the entire early universe is assumed to be filled with dense, hot, glowing matter. There is no region of space which is free of matter or radiation. The explosion of space increases the volume of the matter and radiation, and thus also reduces the density and the temperature. The initial temperature is so high that, even after 500,000 years, the temperature of the universe is still several thousand kelvins, the temperature of the surface of the present-day Sun. The matter of the universe is in the form of a plasma of photons, electrons, protons, and alpha particles (the nuclei of helium).

About this time the expansion of the universe causes the temperature to drop to the point that the electrons and protons combine to form neutral hydrogen. The universe, which was previously opaque to visible light, suddenly becomes transparent, thus decoupling matter and radiation. This happens because the visible photons are much less strongly absorbed by neutral hydrogen than by the plasma. From that time until now the cosmic blackbody photons have been traveling virtually unscattered, carrying information about the nature of the universe at the time of the decoupling (Fig. 1).

To an observer moving with the plasma, the photons have a blackbody spectrum with a characteristic temperature of a few thousand kelvins. (A blackbody spectrum is the characteristic emission from a perfectly absorbing object heated to the characteristic temperature. The orange glow emitted from a heated pan is approximately blackbody, as is the light emitted from the filament of a light bulb or from the surface of the Sun.) Although the glow from the plasma is in the visible region, due to the recessional velocity of the plasma from the Earth, the radiation is red-shifted from the visible by a factor on the order of 1000 into the microwave region, with a characteristic temperature of 3 K. When one measures the radiation, one is really "looking at" the shell of matter which last scattered the radiation. *See* RED SHIFT.

This microwave radiation is coming from the most distant region of space ever observed, and was emitted earlier in time than any other cosmological signal. The radiation was originally termed cosmic background radiation by Penzias and Wilson, because they foresaw that it would cause a background interference with satellite communications, but the term has taken on a new meaning: the radiating shell of matter forms the spatial background in front of which all other astrophysical objects, such as quasars, lie. Until methods are devised to detect the neutrinos or gravity waves that were emitted earlier, there will be no direct means of viewing beyond this background.

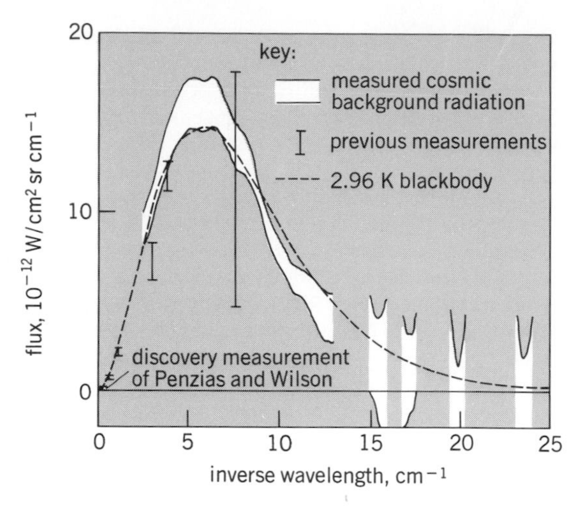

Fig. 2. Measured spectrum of cosmic background radiation. One-standard-deviation error limits are shown for the measured cosmic background and for previous measurements. Frequency resolution of the measurements = 1 cm^{-1}.

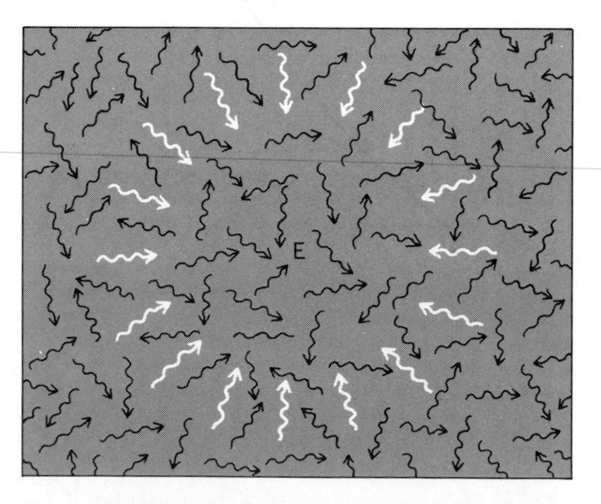

Fig. 1. Origin of cosmic background radiation. Arrows represent photons at a time when the universe became transparent, about a half million years after the big bang. Those photons which reached Earth in 1978 are indicated by the white arrows; the future position of Earth is indicated by letter E.

Spectrum measurements. After the discovery of the radiation, the initial work was concerned primarily with the measurement of the radiation's "color," that is, its spectrum of intensity at different frequencies. Much of this work was done by the group led by Dicke, and within a few years the low-frequency (long-wavelength) spectrum was known. But there were alternate theories that could explain the shape of the low-frequency end of the spectrum, and most astrophysicists wanted to observe the predicted peak intensity at millimeter wavelengths and the drop off at shorter wavelengths (the Wien tail) before they were willing to accept the big bang as the origin of the radiation. Measurements in this region were very difficult because the atmosphere is not transparent at millimeter wavelengths. Worse than that, the atmosphere is an emitter of millimeter radiation, with an intensity much greater than that expected from the cosmic signal. Finally in the late 1970s, in a series of experiments flown aboard a balloon gondola above 99.9% of the Earth's atmosphere, P. L. Richards and D. Woody observed the full spectrum and verified that the shape was essentially as predicted in the big bang theory (Fig. 2).

The striking confirmation of the characteristic blackbody shape is the most important result of the experiment of Richards and Woody. However, they saw a small (but believed to be statistically significant) departure from that shape: the observed spectrum is slightly high in the region of 3 to 9 cm^{-1} and low from 9 to 12 cm^{-1} (cm^{-1} is a unit of frequency commonly used in radio astronomy, and is equal to 30×10^9 Hz, the frequency corresponding to an electromagnetic wave of wavelength 1 cm). Nobody expected the assumptions of the big bang theory (such as complete uniformity of matter distribution) to be exact, and deviations may be clues which will help to develop the full picture.

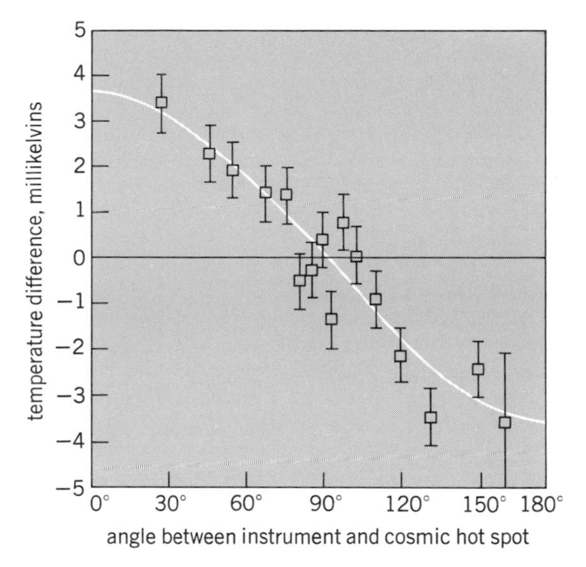

Fig. 3. Anisotropy of cosmic background radiation. Points and bars represent measured data; the curve gives the cosine fit.

Isotropy measurements. In describing their discovery, Penzias and Wilson said that the intensity of radiation in different directions was uniform to better than 10%. Within a few years measurements were made that showed that the radiation was isotropic (that is, equal in intensity in all directions) to about 0.1%. Although the simple big bang theory predicts total isotropy, there were good reasons to expect some departure from total uniformity due to effects which are not included in that theory. A nonuniformity (or anisotropy) could be caused by nonuniformity in the explosion or by a slow rotation of the entire universe. Other possible causes include the existence of very-long-wavelength gravitational radiation and a nonuniform distribution of matter at the time of the decoupling. Since the universe is quite lumpy now, with most of the matter distributed in clusters of galaxies, stars, and planets, one might expect to see some evidence of lumps at the time of decoupling.

Anisotropy has now been observed by B. Corey and Wilkinson and by M. Gorenstein, P. Lubin, R. Muller, and G. Smoot. The latter group's data (Fig. 3) were taken with a radiometer flown aboard the NASA-Ames Earth Survey Aircraft (U-2) in a series of flights in the Northern and Southern hemispheres. Anisotropy, expressed as a temperature difference in millikelvins from the average value, is plotted in Fig. 3 against the angle between the instrument axis and the observed direction of maximum temperature (the cosmic hot spot). The smooth "cosine" behavior of the data suggests that its cause is not cosmological but local: the motion of the Earth relative to the radiation. If the Earth is moving relative to the shell of matter which emitted the radiation, the red shift is not going to be the same in all directions. Radiation coming from the direction in which the earth is moving will be somewhat more intense, and radiation coming from the opposite direction will be weaker, due to the Doppler effect. (The effect is similar to that seen in a moving car in a rainstorm: more rain hits the front window than the rear one. In this case instead of raindrops one has cosmic photons.) The intensity of radiation due to motion should be proportional to the cosine of the angle between the direction of motion and the direction of observation, and this angular dependence is exactly what is observed in the data. *See* DOPPLER EFFECT.

From the maximum amplitude of the anisotropy, which is about 10^{-3} of the intensity of the 3-K signal, one can deduce that the velocity of the Earth is 10^{-3} of the speed of light, or about 300 km/s. The direction of motion is the same as the direction of maximum intensity, given in astronomical coordinates as right ascension = 11 hours, declination = 10 degrees toward the constellation Leo. But the Earth is in the part of the Milky Way Galaxy which is rotating backward, away from Leo, so the net motion of the entire Milky Way Galaxy must be much larger. Taking into account the rotational velocity of the solar system around the center of the Milky Way, one can calculate that the entire Galaxy must be moving at about 500 km/s relative to the microwave radiation. Since it is known that nearby galaxies are approximately at rest with respect to the Milky Way Galaxy, it follows that a huge region of space, including the Milky Way, the Andromeda Galaxy, and other nearby galaxies are all moving together at this very high velocity. (It was because all nearby galaxies are moving with the Milky Way that the motion was not detected before.) The reason for this motion is not definitely understood, but it is suspected that it may represent a gravitational acceleration of the local group

Fig. 4. Planned COBE (Cosmic Background Explorer). Apertures at center are for the 3.3- to 0.33-mm spectrum radiometer and for the diffuse infrared photometer; the four pairs of microwave horns are for the isotropy radiometers. Radiometers are shielded from the Sun and Earth by the large conical shade. (*NASA*)

of galaxies toward the center of a large supercluster of galaxies (the Local Supercluster). *See* GALAXY; GALAXY, EXTERNAL.

Except for the cosine component, there is no other statistically significant deviation from isotropy. There is no evidence that the universe began to clump at the time of decoupling. The uniformity of the radiation implies that if the universe is rotating, the rotation must be very small, less than one-billionth of a second of arc per century. The uniformity of the radiation also implies that the universe is not full of intense long-wave gravitational waves.

Future experiments. The slight departure from the blackbody spectrum and the lack of clumping observed in the isotropy measurements are worthy of future study. Theory suggests that a more sensitive experiment should detect anisotropy in addition to the part due to the motion of the Galaxy. But the sensitivity that can be achieved from the Earth's surface has now been reached, and future experiments must be done from a satellite. NASA has undertaken planning of just such an experiment, called COBE (Cosmic Background Explorer). This satellite (Fig. 4) is to make a very sensitive measurement of the spectrum for every direction in the sky, and in addition should make a detailed map of the anisotropy with a sensitivity much greater than previously possible. In addition to studying the possible deviations from the blackbody spectrum, the anisotropy measurements may show for the first time the primordial clumping of the early universe. *See* COSMOLOGY.

[RICHARD A. MULLER]

Bibliography: R. A. Muller, The cosmic background radiation and the New Aether Drift, *Sci. Amer.*, 238(5):64–74, 1978; P. J. E. Peebles, *Physical Cosmology*, 1971; S. Weinberg, *The First Three Minutes*, 1977; D. Woody and P. L. Richards, Spectrum of the cosmic background radiation, *Phys. Rev. Lett.*, 42:925–929, 1979.

Cosmic rays

Electrons and the nuclei of atoms—largely hydrogen—that impinge upon Earth from all directions of space with nearly the speed of light. These nuclei with relativistic speeds are often referred to as primary cosmic rays, to distinguish them from the cascade of secondary particles generated by their impact against air nuclei at the top of the terrestrial atmosphere. The secondary particles shower down through the atmosphere and are found all the way to the ground and below.

Cosmic rays are studied for a variety of reasons, not the least of which is a general curiosity over the process by which nature can produce such energetic nuclei. Apart from this, the primary cosmic rays provide the only direct sample of matter from outside the solar system. Measurement of their composition can aid in understanding which aspects of the matter making up the solar system are typical of the Galaxy as a whole and which may be so atypical as to yield specific clues to the origin of the solar system. Cosmic rays are electrically charged; hence they are deflected by the magnetic fields which are thought to exist throughout the Galaxy, and may be used as probes to determine the nature of these fields far from Earth. Outside the solar system the energy contained in the cosmic rays is comparable to that of the magnetic field, so the cosmic rays probably play a major role in determining the structure of the field. Collisions between the cosmic rays and the nuclei of the atoms in the tenuous gas which permeates the Galaxy change the cosmic-ray composition in a measurable way and produce gamma rays which can be detected at Earth, giving information on the distribution of this gas.

This modern understanding of cosmic rays has arisen by a process of discovery which at many times produced seemingly contradictory results, the ultimate resolution of which led to fundamental discoveries in other fields of physics, most notably high-energy particle physics. At the turn of the century several different types of radiation were being studied, and the different properties of each were being determined with precision. One result of many precise experiments was that an unknown source of radiation existed with properties that were difficult to characterize. In 1912 Viktor Hess made a definitive series of balloon flights which showed that this background radiation increased with altitude in a dramatic fashion. Far more penetrating then any other known at that time, this radiation had many other unusual properties and became known as cosmic radiation, because it clearly did not originate from the Earth or from any of the known properties of the Earth's atmosphere.

Unlike the properties of alpha-, beta-, gamma-, and x-radiation, the properties of cosmic radiation are not of any one type of particle, but are due to the interactions of a whole series of unstable particles, none of which was known at that time. The initial identification of the positron, the muon, the π-meson or pion, and certain of the K-mesons and hyperons were made from studies of cosmic rays, which are still the only source of particles with energies of 400 GeV and higher.

Thus the term cosmic ray does not refer to a particular type of energetic particle, but to any energetic particle being considered in its astrophysical context.

Cosmic-ray detection. Cosmic rays are usually detected by instruments which classify each incident particle as to type, energy, and in some cases time and direction of arrival. A convenient unit for measuring cosmic-ray energy is the electronvolt (eV), which is the energy gained by a unit charge (such as an electron) accelerating freely across a potential of 1 volt. One electronvolt equals about 1.6×10^{-19} joule. For nuclei it is usual to express the energy in terms of electronvolt/nucleon, since as a function of this variable the relative abundances of the different elements are nearly constant. Two nuclei with the same energy/nucleon have the same velocity.

Flux. The intensity of cosmic radiation is generally expressed as a flux by dividing the average number seen per second by the effective size or "geometry factor" of the measuring instrument. Calculation of the geometry factor requires knowledge of both the sensitive area (in square centimeters) and the angular acceptance (in steradians) of the detector, as the arrival directions of the cosmic rays are randomly distributed to within 1% in most cases. A flat detector of any shape but with area of 1 cm^2 has a geometry factor of π cm^2-sr if it is sensitive to cosmic rays entering from one side only. The total flux of cosmic rays in the vicinity of the Earth but outside the atmosphere is about 0.3 nuclei/cm^2-s-sr. Thus a quarter dollar, with a surface area of 4.5 cm^2, lying flat on the surface of the Moon will be struck by $0.3 \times 4.5 \times 3.14 = 4.2$ cosmic rays per second.

Energy spectrum. The flux of cosmic rays varies as a function of energy. This type of function is called an energy spectrum, and may refer to all cosmic rays or to only a selected element or group of elements. Since cosmic rays are continuously distributed in energy, it is meaningless to attempt to specify the flux at any one exact energy. Normally one speaks of an integral spectrum, in which the function gives the total flux of particles with energy greater than the specified energy (in particle/cm^2-s-sr), or a differential spectrum, in which the function provides the flux of particles in some energy interval (typically 1 MeV/nucleon-wide) centered on the specified energy (in particles/cm^2-s-sr-MeV/nucleon). The basic problem of cosmic-ray research is to measure the spectra of the different components of cosmic radiation and to deduce from them and other observations the nature of the cosmic-ray sources and the details of where the particles travel on their way to Earth and what they encounter on their journey.

Types of detectors. All cosmic-ray detectors are sensitive only to moving electrical charges. Neutral cosmic rays (neutrons, gamma rays, and neutrinos) are studied by observing the charged particles produced in the collision of the neutral primary with some type of target. At low energies the ionization of the matter through which they pass is the principal means of detection. Such detectors include cloud chambers, ion chambers, spark chambers, Geiger counters, proportional counters, scintillation counters, solid-state detectors, photographic emulsions, and chemical etching of certain mineral crystals or plastics in which ionization damage is revealed. The amount of ionization produced by a particle is given by the square of its charge multiplied by a universal function of its velocity. A single measurement of the ionization produced by a particle is therefore usually not sufficient both to identify the particle and to determine its energy. However, since the ionization itself represents a significant energy loss to a low-energy particle, it is possible to design systems of detectors which trace the rate at which the particle slows down and thus to obtain unique identification and energy measurement.

At energies above about 500 MeV/nucleon, almost all cosmic rays will suffer a catastrophic nuclear interaction before they slow appreciably. Some measurements are made using massive calorimeters which are designed to trap all of the energy from the cascade of particles which results from such an interaction. More commonly an ionization measurement is combined with measurements of physical effects which vary in a different way with mass, charge, and energy. Cerenkov detectors and the deflection of the particles in the field of large superconducting magnets or the magnetic field of the Earth itself provide the best means of studying energies up to a few hundred GeV/nucleon. Detectors employing the phenomenon of x-ray transition radiation promise to be useful for measuring composition at energies up to a few thousand GeV/nucleon. Transition radiation detectors have already been used to study electrons having energies of 10–200 GeV which, because of their lower rest mass, are already much more relativistic than protons of the same energies.

Above about 10^{12} eV, direct detection of individual particles is no longer possible since they are so rare. Such particles are studied by observing the large showers of secondaries they produce in Earth's atmosphere. These showers are detected either by counting the particles which survive to strike ground-level detectors or by looking at the flashes of light the showers produce in the atmosphere with special telescopes and photomultiplier tubes. It is not possible to directly determine what kind of particle produces any given shower. However, because of the extreme energies involved, which can be measured with fair accuracy and have been seen as high as 10^{20} eV (16 J), most of the collision products travel in the same direction as the primary and at essentially the speed of light. This center of intense activity has typical dimensions of only a few meters, allowing it to be tracked (with sensitive instruments) like a miniature meteor across the sky before it hits the Earth at a well-defined location. In addition to allowing determination of the direction from which each particle came, the development of many such showers through the atmosphere may be studied statistically to gain an idea of whether the primaries are protons or heavier nuclei. Basically the idea behind these studies is that a heavy nucleus, in which the energy is initially shared among several neutrons and protons, will cause a shower that starts higher in the atmosphere and develops more regularly than a shower which has the same total energy but is caused by a single proton.

Atmospheric cosmic rays. The primary cosmic-ray particles coming into the top of the terrestrial atmosphere make inelastic collisions with nuclei in the atmosphere. The collision cross section is essentially the geometrical cross section of the nucleus, of the order of 10^{-26} cm^2. The mean free path for primary penetration into the atmosphere is given in Table 1. (Division by the atmospheric density in g/cm^3 gives the value of the mean free path in centimeters.)

Table 1. Mean free paths for primary cosmic rays in the atmosphere

Charge of primary nucleus	Mean free path in air, g/cm^2
$Z=1$	60
$Z=2$	44
$3 \leq Z \leq 5$	32
$6 \leq Z \leq 9$	27
$10 \leq Z \leq 29$	21

When a high-energy nucleus collides with the nucleus of an atom in the Earth's atmosphere, a number of things usually occur. Rapid deceleration of the incoming nucleus leads to production of pions with positive, negative, or neutral charge; this meson production is closely analogous to the generation of x-rays, or bremsstrahlung, produced when a fast electron is deflected by impact with the atoms in a metal target. The mesons, like the bremsstrahlung, come off from the impact in a narrow cone in the forward direction. Anywhere from 0 to 30 or more pions may be produced, depending upon the energy of the incident nucleus. The ratio of neutral to charged pions is ~0.75.

A few protons and neutrons (in about equal proportions) may be knocked out with energies a significant fraction of that of the incoming nucleus. They are called knock-on protons and neutrons.

A nucleus struck by a proton or neutron of the nucleonic component with an energy greater than approximately 300 MeV may have its internal forces momentarily disrupted so that some of its nucleons are free to leave with their original nuclear kinetic energies of about 10 MeV. The nucleons freed in this fashion appear as protons, deuterons, tritons, alpha-particles, and even somewhat heavier clumps, radiating outward from the struck nucleus. In photographic emulsions the result is a number of short prongs radiating from the point of collision, and for this reason is called a nuclear star.

All these protons, neutrons, and pions generated by collision of the primary cosmic-ray nuclei with the nuclei of air atoms are the first stage in the development of the secondary cosmic-ray particles observed inside the atmosphere. Since several secondary particles are produced by each collision, the total number of energetic particles of cosmic-ray origin will increase with depth, even while the primary density is decreasing. Since electric charge must be conserved and the primaries are positively charged, the positive particles outnumber the negative particles in the secondary radiation by a factor of about 1.2. This factor is called the positive excess.

Electromagnetic cascade. The uncharged π^0-mesons decay into two γ-rays with a life of $\sim 8 \times 10^{-17}$ s. The decay is so rapid that π^0-mesons are not directly observed among the secondary particles in the atmosphere. The two γ-rays, which together have the rest energy of the π^0, about 140 MeV, plus the π^0 kinetic energy, each produce a positron-electron pair. Upon passing sufficiently close to the nucleus of an air atom deeper in the atmosphere, the electrons and positrons convert their energy into bremsstrahlung. The bremsstrahlung in turn creates new positron-electron pairs, and so on. This cascade process continues until the energy of the initial π^0 has been dispersed into a shower of positrons, electrons, and photons with insufficient individual energies ($\leqq 1$ MeV) to continue the pair production. The shower, then being unable to reproduce its numbers, is dissipated by ionization of the air atoms. The electrons and photons of such showers are referred to as the soft component of the atmospheric (secondary) cosmic rays, reaching a maximum intensity at an atmospheric depth of 150–200 g/cm² and then declining by a factor of about 10^2 down to sea level.

Muons. The $\pi^{\pm}$-mesons produced by the primary collisions have a life about 2.5×10^{-8} s before they decay into muons:

$$\pi^{\pm} \rightarrow \mu^{\pm} + \text{neutrino}$$

With a life of this order a $\pi^{\pm}$ possessing enough energy (> 10 GeV) to experience significant relativistic time dilatation may exist long enough to interact with the nuclei of the air atoms. The cross section for $\pi^{\pm}$ nuclear interactions is approximately the geometrical cross section of the nucleus, and the result of such an interaction is essentially the same as for the primary cosmic-ray protons. Most low-energy $\pi^{\pm}$ decay into muons before they have time to undergo nuclear interactions.

The muons will not interact with nuclei, and are too massive (207 electron masses) to produce bremsstrahlung. They can lose energy only by the comparatively feeble process of ionizing an occasional air atom as they progress downward through the atmosphere. Because of this ability to penetrate matter, they are called the hard component. Their life is 2×10^{-6} s before they decay into an electron or positron and two neutrinos. Even with the time dilatation of their high energy, only 5% of the muons reach the ground. Their interaction with matter is so weak that they penetrate deep into the ground, where they are the only charged particles of cosmic-ray origin to be found. At a depth equivalent of 300 m of water the muon intensity has decreased from that at ground level only by a factor of 20; at an equivalent depth of 1400 m it has decreased by a factor of 10^3.

Nucleonic component. The high-energy nucleons—the knock-on protons and neutrons—produced by the primary-particle collisions and a few pion collisions proceed on down into the atmosphere. They produce nuclear interactions of the same kind as the primary nuclei, though of course with diminished energies. This cascade process constitutes the nucleonic component of the secondary cosmic rays.

When the nucleon energy falls below about 100 MeV, stars and further knock-ons can no longer be produced. At the same time the protons are rapidly disappearing from the cascade because their ionization losses in the air slow them down before they can make a nuclear interaction. The neutrons are already dominant at 3500 m, about 300 g/cm² above sea level, where they outnumber the protons four to one. Thus the final stages in the lower atmosphere are given over almost entirely to neutrons in a sequence of low-energy interactions which convert them to thermal neutrons (neutrons of kinetic energy of about 0.025 eV) in a path of about 90 g/cm². These thermal neutrons are readily detected in boron trifluoride (BF_3) counters. The nucleonic component increases in intensity down to a depth of about 120 g/cm², and thereafter declines in intensity, with a mean absorption length of about 200 g/cm².

The various cascades of secondary particles in the atmosphere are shown schematically in Fig. 1. Note that about 48% of the initial primary cosmic-ray energy goes into charged pions, 25% into neutral pions, 7% into the nucleonic component, and 20% into stars. The nucleonic component is produced principally by the lower-energy (~5 GeV) primaries. Higher-energy primaries put their energy more into meson production. Hence in the lower atmosphere, a Geiger counter responds mainly to the higher-energy primaries (~15 GeV) because it counts the muons and electrons, whereas a BF_3 counter detecting thermal neutrons responds more to the low-energy primaries.

Neutrinos. Instruments have been developed which detect the flux of neutrinos produced in this atmospheric cascade. No primary neutrinos have been detected, but this is not surprising given the present level of sensitivity. One of the primary goals of such experiments is to detect bursts of neutrinos from supernova explosions in an attempt to answer questions about what is happening deep inside them. Measurement of the flux of solar neutrinos, which is really quite a different problem, has begun to cause fundamental changes in thought about the physics of the Sun. The most ambitious of the proposals to increase the detection sensitivity of these experiments is the Deep Underwater Muon and Neutrino Detector (DUMAND) project. As presently conceived, this would involve placing 10^5 photomultipliers in a three-dimensional lattice throughout a region containing 10^9 metric tons of water located some 5 km below the surface of the sea. *See* SOLAR NEUTRINOS.

Relation to particle physics. Investigations of cosmic rays continue to make fundamental contributions to particle physics. Neutrino detectors have set the best limit yet (about 10^{30} years) on the lifetime of the proton, whose value has been the object of considerable theoretical attention. These systems are ideal to study the neutrino and specifically to search for oscillations, or spontaneous conversion of one type of neutrino into another.

Geomagnetic effects. The magnetic field of Earth is described approximately as that of a magnetic dipole of strength 8.1×10^{25} cgs (centimeter-gram-second) units located near the geometric center of Earth. Near the Equator the field intensity is 0.3 gauss (1 gauss = 10^{-4} tesla), falling off in space as the inverse cube of the distance to the Earth's center. In a magnetic field which does not vary in time, the path of a particle is determined entirely by its rigidity, or momentum per unit charge; the velocity simply determines how fast the particle will move along this path. Momentum is usually expressed in units of eV/c, where c is the velocity of light, because at high energies, energy and momentum are then numerically almost equal. By definition, momentum and rigidity are numerically equal for singly charged particles. The unit so defined is normally called the volt, but should not be confused with the standard and nonequivalent unit of the same name. Table 2 gives examples of these units as applied to different particles with rigidity of 1 GV. This corresponds to an orbital radius in a typical interplanetary (10^{-5} gauss) magnetic field of approximately 10 times the distance from the Earth to the Moon.

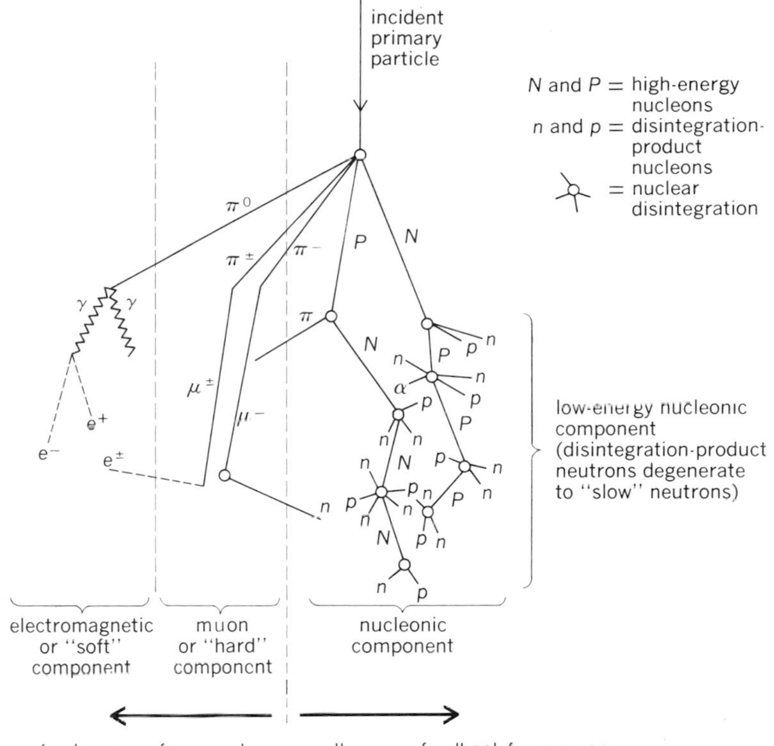

Fig. 1. Cascade of secondary cosmic-ray particles in the terrestrial atmosphere.

The minimum rigidity of a particle able to reach the top of the atmosphere at a particular geomagnetic latitude is called the geomagnetic cutoff rigidity at that latitude, and its calculation is a complex numerical problem. Fortunately, for an observer near the ground, obliquely arriving secondary particles, produced by the oblique primaries, are so heavily attenuated by their longer path to the ground that it is usually sufficient to consider only the geomagnetic cutoff for vertically incident primaries, which is given in Table 3. Around the Equator, where a particle must come in perpendicular to the geomagnetic lines of force to reach

Table 2. Properties of particles when all have a rigidity of 1 GV

Particle	Charge	Nucleons	Kinetic energy, MeV	Kinetic energy, MeV/nucleon	Momentum, MeV/c
Proton	1	1	430	430	1000
^{3}He	2	3	640	213	2000
^{4}He	2	4	500	125	2000
^{16}O	8	16	2000	125	8000

Table 3. Geomagnetic cutoff

Geomagnetic lat.	Vertical cutoff, GV
0°	15
±20°	11.5
±40°	5
±60°	1
±70°	0.2
±90°	0

Earth, particles with rigidity less than 10 GV are entirely excluded, though at higher latitudes where entry can be made more nearly along the lines of force, lower energies can reach Earth. Thus, the cosmic-ray intensity is a minimum at the Equator, and increases to its full value at either pole—this is the cosmic-ray latitude effect. Even deep in the atmosphere the variation with latitude is easily detected with BF_3 counters, as shown in Fig. 2. North of 45° the effect is slight because the additional primaries admitted are so low in energy that they produce few secondaries.

Accurate calculations of the geomagnetic cutoff must consider the deviations of the true field from that of a perfect dipole and the change with time of these deviations. Additionally the distortion of the field by the pressure of the solar wind must often be accounted for, particularly at high latitude. Such corrections vary rapidly with time because of sudden bursts of solar activity and because of the rotation of the Earth. Areas with cutoffs of 400 MV during the day have no cutoff at all during the night. This day-night effect is confined to particles with energies so low that neither they nor their secondaries reach the ground, and is thus observed only on high-altitude balloons or satellites.

Since the geomagnetic field is directed from south to north above the surface of Earth, the incoming cosmic-ray nuclei are deflected toward the east. Hence an observer finds some 20% more particles incident from the west. This is known as the east-west effect.

Solar modulation. Figure 3 presents portions of the proton and alpha-particle spectra observed near the Earth but outside of the magnetosphere in 1973. Below 20 GeV/nucleon the cosmic-ray intensity varies markedly with time. S. Forbush was the first to show that the cosmic-ray intensity was low during the years of high solar activity and sunspot number, which follow an 11-year cycle. This effect is clearly seen in the data of Fig. 2 and has been extensively studied with ground-based and spacecraft instruments. While this so-called solar modulation is now understood in general terms, it has not been calculated in detail, in large part because of the lack of any direct measurements of conditions in the solar system out of the ecliptic plane, to which all present spacecraft are confined because of limitations on the power of rockets.

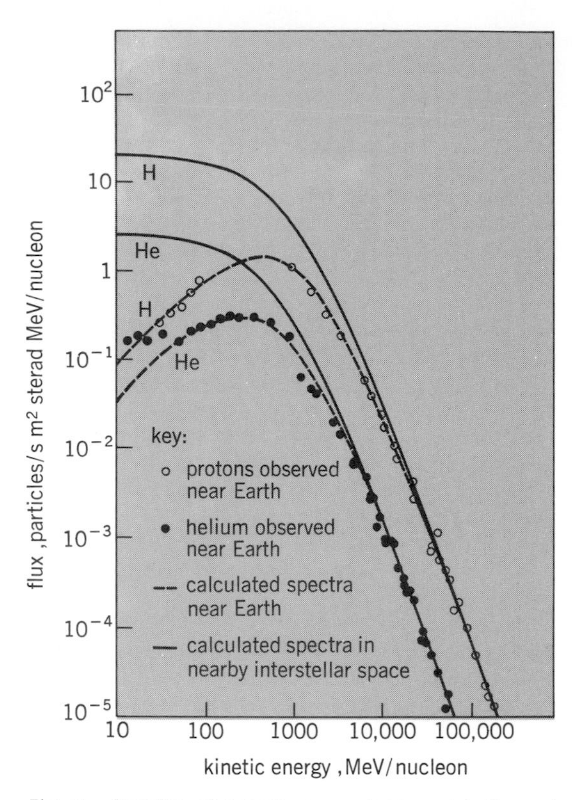

Fig. 3. Spectra of cosmic-ray protons and helium at Earth and in nearby interstellar space, showing the effect of solar modulation. Observations were made in 1973, when the Sun was quiet. (*Courtesy of Dr. M. Garcia-Munoz, University of Chicago*)

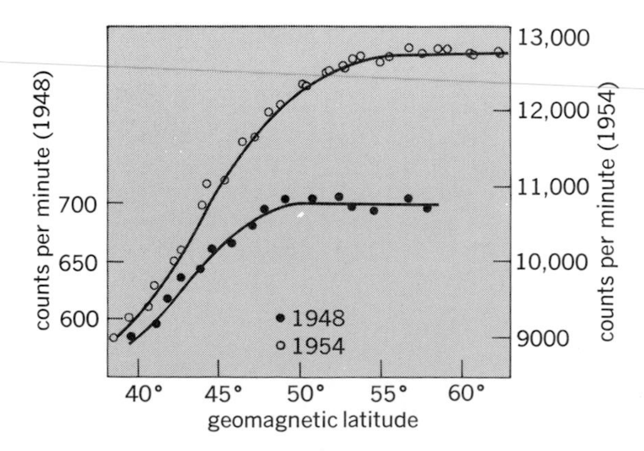

Fig. 2. Latitude variation of the neutron component of cosmic rays in 80°W longitude and at a height corresponding to an atmospheric pressure of 22.5 cm of mercury in 1948, when the Sun was active, and 1954, when the Sun was deep in a sunspot minimum.

There is indirect evidence that the solar system is not at all spherically symmetric and that conditions near the ecliptic plane are quite special. The International Solar Polar Mission, scheduled for a 1985 space shuttle launch, should provide the first good look at conditions out of the ecliptic plane.

The primary cause of solar modulation is the solar wind, a highly ionized gas (plasma) which boils off the solar corona and propagates radially from the Sun at a velocity of about 400 km/s. The wind is mostly hydrogen, with typical density of 5 protons/cm³. This density is too low for collisions with cosmic rays to be important. Rather, the high conductivity of the medium traps part of the solar magnetic field and carries it outward. The rotation of the Sun and the radial motion of the plasma combine to create the observed archimedean spiral pattern of the average interplanetary magnetic field. Turbulence in the solar wind creates fluctuations in the field which often locally obscure the average direction and intensity. This complex system of magnetic irregularities propagating outward from the Sun deflects and sweeps the low-rigidity cosmic rays out of the solar system.

In addition to the bulk sweeping action, another effect of great importance occurs in the solar wind, adiabatic deceleration. Because the wind is blowing out, only those particles which chance to move upstream fast enough are able to reach Earth. However, because of the expansion of the wind, particles interacting with it lose energy. Thus, particles observed at Earth with 10 MeV/nucleon energy actually started out with several hundred MeV/nucleon in nearby interstellar space, and those with only 100–200 MeV/nucleon initial energy probably never reach Earth at all. This is particularly unfortunate because at these lower energies the variation with energy of nuclear reaction probabilities would allow much more detailed investigation of cosmic-ray history. Changes in the modulation with solar activity are caused by the changes in the pattern of magnetic irregularities rather than by changes in the wind velocity, which are quite small.

There are several phenomenological classes of cosmic-ray variation besides the 11-year variation which are associated with the short-term variations of the solar wind and generally affect only low-energy particles.

A sudden outburst at the Sun, at the time of a large flare, yields a cosmic-ray decrease in space which extends to very high energies, 50 GeV or so. The magnetic fields carried in the blast wave from the flare sweep back the cosmic rays, causing a decrease in their intensity. This is termed the Forbush-type decrease, where the primary intensity around the world may drop in an irregular way as much as 20% in 15 h or 8% in 3 h, slowly recovering in the days or weeks that follow. Often, but not always, the Forbush decrease and geomagnetic storms accompany each other. Striking geographical variation is to be seen in the sharper fluctuations during the onset of a Forbush decrease.

The region in space where solar modulation is important is probably a sphere with a radius of 30–50 AU, although this is not at all certain. (1 AU, or astronomical unit, is the mean Earth-Sun separation, 1.49×10^8 km.) The *Pioneer 11* spacecraft did not find abrupt changes in intensity indicative of crossing any type of boundary out to a distance of 20 AU, although the fluxes detected clearly increased as it traveled outward. A distinct change in interplanetary phenomena seems to take place in the 10–15-AU range to a quieter, more azimuthally uniform, and less temporally varying situation. *See* SOLAR MAGNETIC FIELD; SOLAR WIND; SUN.

Composition of cosmic rays. Nuclei ranging from protons to lead have been identified in the cosmic radiation. The relative abundances of the elements up to nickel are shown in Fig. 4, together with the best estimate of the "universal abundances" obtained by combining measurements of solar spectra, lunar and terrestrial rocks, meteorites, and so forth. Most obvious is the similarity between the two distributions. However, a systematic deviation is quickly apparent: the elements lithium-boron and scandium-manganese

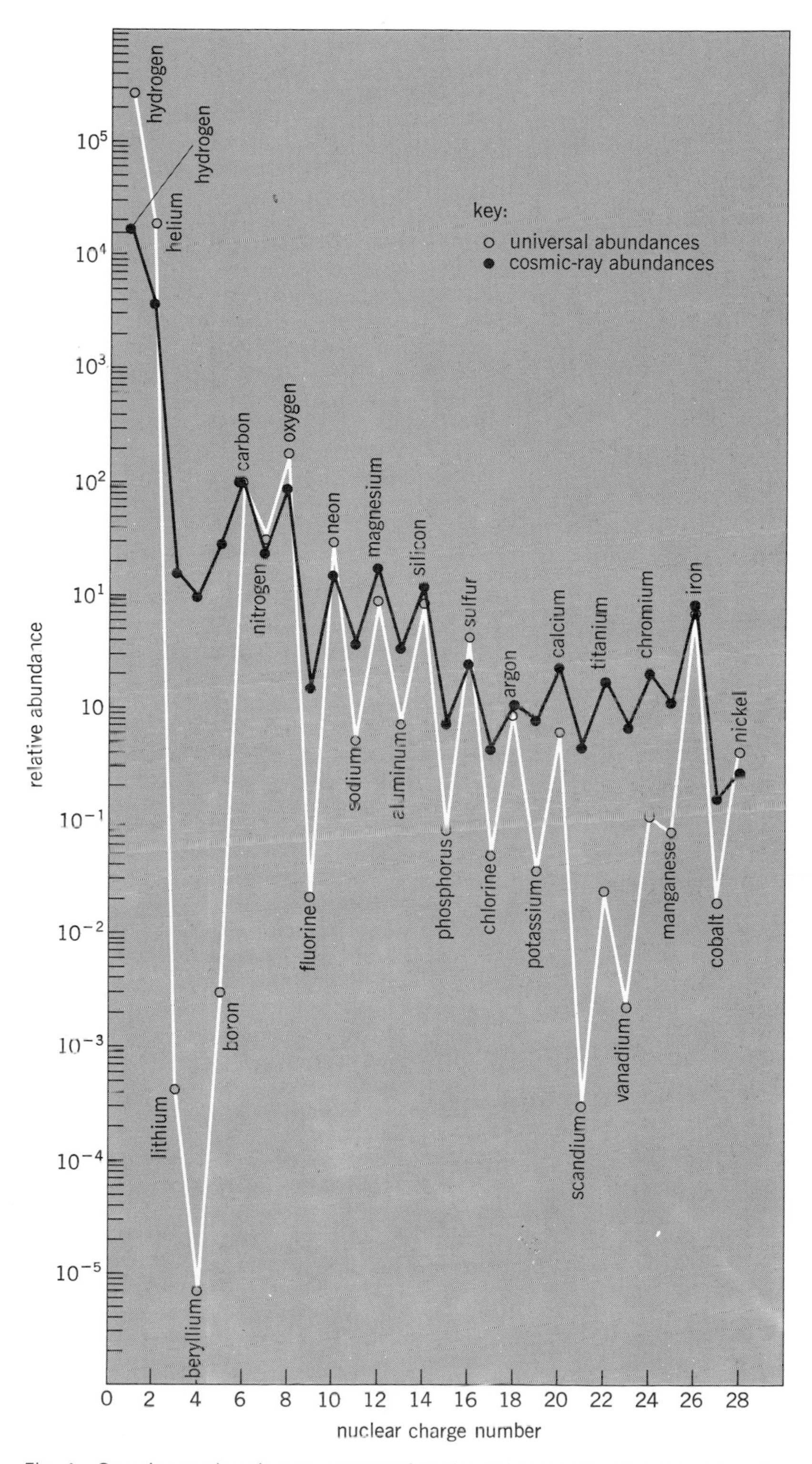

Fig. 4. Cosmic-ray abundances compared to the universal abundances of the elements. Carbon is set arbitrarily to an abundance of 100 in both cases. (*Courtesy of Dr. M. Garcia-Munoz, University of Chicago*)

as well as most of the odd-charged nuclei are vastly overabundant in the cosmic radiation. This effect has a simple explanation: the cosmic rays travel great distances in the galaxy and occasionally collide with atoms of interstellar gas—mostly hydrogen and helium—and fragment. This fragmentation, or spallation as it is called, produces lighter nuclei from heavier ones but does not change the energy/nucleon very much. Thus the energy spectra of the secondaries are similar to those of the primaries.

Calculations involving reaction probabilities determined by nuclear physicists show that the overabundances of the secondary elements can be explained by assuming that cosmic rays pass through an average of about $5/cm^2$ of material on their way to Earth. Although an average path length can be obtained, it is not possible to fit the data by saying that all particles of a given energy have exactly the same path length; furthermore, results indicate that higher-energy particles traverse less matter in reaching the solar system, although their original composition seems energy independent. *See* ELEMENTS, COSMIC ABUNDANCE OF.

When spallation has been corrected for, differences between cosmic-ray abundances and solar-system or universal abundances still remain. The most important question is whether these differences are due to the cosmic rays having come from a special kind of material (such as would be produced in a supernova explosion), or simply to the fact that some atoms might be more easily accelerated than others. It is possible to rank almost all of the overabundances by considering the first ionization potential of the atom and the rigidity of the resulting ion, although this calculation gives no way of predicting the magnitude of the enhancement expected. It is also observed that the relative abundances of particles accelerated in solar flares are far from constant from one flare to the next.

Isotopes. The possibility of such preferential acceleration is one of the reasons why much cosmic-ray study is concentrated on determining the isotopic composition of each element, as this is much less likely to be changed by acceleration. It is apparent that the low-energy helium data in Fig. 3 do not fit the calculated values. Since it is known that this low-energy helium is nearly all ^{4}He, whereas the higher-energy helium contains 10% ^{3}He, one can be fairly certain that the deviation is due to a local source of energetic ^{4}He within the solar system rather than a lack of understanding of the process of solar modulation. Similarly, a low-energy enhancement of nitrogen is pure ^{14}N, whereas the higher-energy nitrogen is almost 50% ^{15}N.

High-accuracy measurements of isotopic abundance ratios became available only in the late 1970s, but already one puzzle has emerged. The ratio of ^{22}Ne to ^{20}Ne in the cosmic-ray sources is estimated to be 0.37, while the accepted solar system value for this number is 0.12, which agrees well with the abundances measured in solar-flare particles. However, another direct sample of solar material—the solar wind—has a ratio of 0.08, indicating clearly that the isotopic composition of energetic particles need not reflect that of their source. Conclusions drawn from the observed difference in the solar and cosmic-ray values must be viewed as somewhat tentative until the cause of the variation in the solar material is well understood.

Electron abundance. Cosmic-ray electron measurements pose other problems of interpretation, partly because electrons are nearly 2000 times lighter than protons, the next lightest cosmic-ray component. Protons with kinetic energy above 1 GeV are about 100 times as numerous as electrons above the same energy, with the relative number of electrons decreasing slowly at higher energies. But it takes about 2000 GeV to give a proton the same velocity as a 1-GeV electron. Viewed in this way electrons are several thousand times more abundant than protons. Electrical neutrality of the Galaxy is maintained by lower-energy ions which are more numerous than cosmic rays although they do not carry much energy.) It is thus quite possible that cosmic electrons have a different source entirely from the nuclei. It is generally accepted that there must be direct acceleration of electrons, because calculations show that more positrons than negatrons should be produced in collisions of cosmic-ray nuclei with interstellar gas. Measurements show, however, that only 10% of the electrons are positrons. As the number of positrons seen agrees with the calculated secondary production, added confidence is gained in the result that there is indeed an excess of negatrons.

Electrons are light enough to emit a significant amount of synchrotron radiation as they are deflected by the 10^{-6}-gauss galactic magnetic field. Measurement of this radiation by radio telescopes provides sufficient data for an approximate calculation of the average energy spectrum of electrons in interstellar space. Comparison of spectra of electrons and positrons measured at Earth with those calculated to exist in interstellar space provides the most direct measurement of the absolute amount of solar modulation. *See* RADIO ASTRONOMY.

Properties of energy spectrum. At energies above 10^{10} eV, the energy spectra of almost all cosmic rays are approximated over many decades by functions in which the flux decreases as the energy raised to some negative, nonintegral power referred to as the spectral index. Such a power-law relationship is of course a straight line when plotted using logarithmic axes. A steep (that is, more rapidly falling with increasing energy) spectrum thus has a higher spectral index than a flat spectrum. The straight-line regions of the spectra in Fig. 3 correspond to a variation of flux with a spectral index of −2.7. A spectral index of −2.7 provides a good fit with the data up to 10^{15} eV total energy. Between 10^{15} and 10^{19} eV a steeper spectrum, with an index around −3.0, seems to be well established. Above 10^{19} eV the spectrum surprisingly flattens once more, returning to an index of about −2.7. The spectral index above 10^{20} eV has not been determined, because particles are so rare that they are almost never seen, even in detectors which cover several square kilometers and operate for many years. At such high energies, the individual particles are not identified, and changes in the measured-energy spectrum could be the result of composition changes. However, the evidence available indicates that the composition is essentially unchanged.

Age. Another important result which can be derived from detailed knowledge of cosmic-ray isotopic composition is the "age" of cosmic radiation. Certain isotopes are radioactive, such as ^{10}Be with a half-life of 1.6×10^6 years. Since Be is produced entirely by spallation, study of the relative abundance of ^{10}Be to the other Be isotopes, particularly as a function of energy to utilize the relativistic increase in this lifetime, will yield a number related to the average time since the last nuclear collision. Measurements show that ^{10}Be is nearly absent at low energies and yield an estimate of the age of the cosmic rays of approximately 10^7 years. An implication of this result is that the cosmic rays propagate in a region in space which has an average density of 0.1–0.2 atoms/cm^3. This is consistent with some astronomical observations of the immediate solar neighborhood.

Very-high-energy particles cannot travel long distances in the 2.7-K blackbody-radiation field which permeates the universe. Electrons of 15 GeV energy lose a good portion of their energy in 10^8 years by colliding with photons via the (inverse) Compton process, yet electrons are observed to energies of 100 GeV and over. A similar loss mechanism becomes effective at approximately 10^{20} eV for protons. These observations are of course not conclusive, but a safe statement is that a cosmic-ray age of 10^7 years is consistent with all currently available data. *See* COSMIC MICROWAVE RADIATION.

Several attempts have been made to measure the constancy of the cosmic-ray flux in time. Variations in ^{14}C production, deduced from apparent deviations of the archeological carbon-dating scale from that derived from studies of tree rings, cover a period of about 10^3 years. Radioactive ^{10}Be in deep-sea sediments allows studies over 10^6 years, whereas etching of tracks left by cosmic rays in lunar minerals covers a period of 10^9 years. None of these methods has ever indicated a variation of more than a factor of 2 in average intensity. There are big differences in these time scales, and the apparent constancy of the flux could be due to averaging over variations which fall in the gaps as far as the time scales are concerned. Nevertheless, the simplest picture seems to be that the cosmic rays are constant in time at an intensity level which is due to a long-term balance between continuous production and escape from the Galaxy, with an average residence time of 10^7 years.

Origin. Although study of cosmic rays has yielded valuable insight into the structure, operation, and history of the universe, their origin has not been determined. The problem is not so much to devise processes which might produce cosmic rays, but to decide which of many possible processes do in fact produce them.

In general, analysis of the problem of cosmic-ray origin is broken into two major parts: origin in the sense of where the sources are located, whatever they are, and origin in the sense of how the particles are accelerated to such high energies. Of course these questions can never be separated completely.

Location of sources. It is thought that cosmic rays are produced by mechanisms operating within galaxies and are confined almost entirely to the galaxy of their production, trapped by the galactic magnetic field. The intensity in intergalactic space would only be a few percent of the typical galactic intensity, and would be the result of a slow leakage of the galactic particles out of the magnetic trap. It has not been possible to say much about where the cosmic rays come from by observing their arrival directions at Earth. At lower energies (up to 10^{15} eV) the anisotropies which have been observed can all be traced to the effects of the solar wind and interplanetary magnetic field. The magnetic field of the Galaxy seems to be completely effective in scrambling the arrival directions of these particles.

Between 10^{15} and 10^{19} eV a smoothly rising anisotropy is measured, ranging from 0.1 to 10%, but the direction of the maximum intensity varies in a nonsystematic way with energy. At these energies, particles have a radius of curvature which is not negligible compared to galactic dimensions, and thus their arrival direction could be related to where they came from but in a complex way.

Above 10^{19} eV the radius of curvature in the galactic magnetic field becomes comparable to or larger than galactic dimensions, making containment of such particles in the Galaxy impossible. Only about 50 events greater than 10^{19} eV have been detected, but the directions from which they have come are plainly nonrandom. A clear minimum appears in the direction of the disk of the Galaxy and a clear maximum near the north galactic pole, which is also the direction of the majority of the galaxies in the so-called local supercluster. The average distance of these galaxies is such that these particles in fact would be able to propagate to the Earth from them without losing their energy to photon collisions. This clearly defined anisotropy begins at the same energy at which the energy spectrum changes—further indication that these particles may have a different source from the low-energy particles. An alternative explanation, that particles are absorbed by material in the galactic plane, cannot at present be ruled out because no large detectors presently view the sky in the region of the south galactic pole. Thus it is not known whether the presence of the cluster of galaxies in the direction of the maximum flux is only coincidental. *See* SUPERCLUSTERS.

Direct detection of cosmic rays propagating in distant regions of the Galaxy is possible by observing the electromagnetic radiation produced as they interact with other constituents of the Galaxy. Measurement of the average electron spectrum using radio telescopes has already been mentioned. Proton intensities are mapped by studying the arrival directions of gamma rays (at about 50 MeV) produced as they collide with the interstellar gas. Unfortunately, the amount of radiation in these processes depends upon both the cosmic-ray flux and the magnetic-field intensity or density of interstellar gas. Areas where cosmic rays are known to exist can be pointed out because the radiation is observed. But where no radiation is seen, it is not known whether its absence results from lack of cosmic rays or lack of anything for them to interact with. In particular, very little radiation is seen from outside the Galaxy, but there is also very little gas or magnetic field there. There is therefore no direct evidence either for or against galactic containment.

A major difficulty with the concept of cosmic radiation filling the universe is the large amount of energy needed to maintain the observed intensity in the face of an expanding universe—probably more energy than is observed to be emitted in all other forms put together. *See* COSMOLOGY.

Confinement mechanisms. Three possible models of cosmic-ray confinement are under investigation. All assume that cosmic rays are produced in sources, discrete or extended, scattered randomly through the galactic disk. Most popular is the "leaky box" model, which proposes that the particles diffuse about in the magnetic field for a few million years until they chance to get close to the edge of the Galaxy and escape. This is a phenomenological model in that no mechanism is given by which either the confinement time or the escape probability as a function of energy can be calculated from independent observations of the galactic structure. Its virtue is that good fits to the observed abundances of spallation products are obtained by using only a few adjustable parameters. Variations of the model which mainly postulate boxes within boxes—ranging from little boxes surrounding sources to a giant box or static halo surrounding the whole Galaxy—can be used to explain variations from the simple predictions. However, all attempts to calculate the details of the process have failed by many orders of magnitude, predicting ages which are either far older or far younger than the observed age.

A newer model is that of the dynamical halo. Like the earlier static-halo model, it is assumed that cosmic rays propagate not only in the galactic disk but also throughout a larger region of space, possibly corresponding to the halo or roughly spherical but sparse distribution of material which typically surrounds a galaxy. This model is based on the observation that the energy density of the material which is supposed to be contained by the galactic magnetic field is comparable to that of the field itself. This can result in an unstable situation in which large quantities of galactic material stream out in a galactic wind similar in some respects to the solar wind. In this case the outward flow is a natural part of the theory, and calculations have predicted reasonable flow rates. In distinction to the solar wind, in which the cosmic rays contribute almost nothing to the total energy density, they may provide the dominant energy source in driving the galactic wind.

A third model assumes that there is almost no escape; that is, cosmic rays disappear by breaking up into protons which then lose energy by repeated collision with other protons. To accept this picture, one must consider the apparent 5-g/cm^2 mean path length to be caused by a fortuitous combination of old distant sources and one or two close young ones. Basically, the objections to this model stem from the tendency of scientists to accept a simple theory over a more complex (in the sense of having many free parameters) or specific theory when both explain the data. *See* GALAXY.

Acceleration mechanisms. Although the energies attained by cosmic-ray particles are extremely high by laboratory standards, their generation can probably be understood in terms of known astronomical objects and laws of physics. Even on Earth, ordinary thunderstorms generate potentials of billions of volts, which would accelerate particles to respectable cosmic-ray energies (a few GeV) if the atmosphere were less dense. Consequently, there are many theories of how the acceleration could take place, and it is quite possible that more than one type of source exists. Two major classes of theories may be identified—extended-acceleration regions and compact-acceleration regions.

Extended-acceleration regions. Acceleration in extended regions (in fact the Galaxy as a whole) was first proposed by E. Fermi, who showed that charged particles could gain energy from repeated deflection by magnetic fields carried by the large clouds of gas which are known to be moving randomly about the Galaxy. Many other models based on such statistical acceleration have since been proposed, the most recent of which postulates that particles bounce off shock waves traveling in the interstellar medium. Such shocks, supposed to be generated by supernova explosions, undoubtedly exist to some degree but have an unknown distribution in space and strength, leaving several free parameters which may be adjusted to fit the data.

Compact-acceleration regions. The basic theory in the compact-acceleration class is that particles are accelerated directly in the supernova explosions themselves. One reason for the popularity of this theory is that the energy generated by supernovas is of the same order of magnitude as that required to maintain the cosmic-ray intensity in the leaky box model.

However, present observations indicate that the acceleration could not take place in the initial explosion. Cosmic rays have a composition which is similar to that of ordinary matter and is different from the presumed composition of the matter which is involved in a supernova explosion. At least some mixing with the interstellar medium must take place. Another problem with an explosive origin is an effect which occurs when many fast particles try to move through the interstellar gas in the same direction: the particles interact with the gas through a magnetic field which they generate themselves, dragging the gas along and rapidly losing most of their energy. In more plausible theories of supernova acceleration, the particles are accelerated gradually by energy stored up in the remnant by the explosion or provided by the intense magnetic field of the rapidly rotating neutron star or pulsar which is formed in the explosion.

Such acceleration of high-energy particles is clearly observed in the Crab Nebula, the remnant of a supernova observed by Chinese astronomers in A.D. 1054. This nebula is populated by high-energy electrons which radiate a measureable amount of their energy as they spiral about in the magnetic field of the nebula. So much energy is released that the electrons would lose most of their energy in a century if it were not being continuously replenished. Pulses of gamma rays also show that bursts of high-energy particles are being produced by the neutron star—the gamma rays coming out when the particles strike nuclei in the atmosphere of the neutron star. Particles of cosmic-ray energy are certainly produced in this object, but it is not known whether they escape from the trapping magnetic fields in the nebula and join the freely propagating cosmic-ray population. *See* CRAB NEBULA; NEUTRON STAR; PULSAR; SUPERNOVA; X-RAY ASTRONOMY.

Acceleration in the solar system. The study of energetic particle acceleration in the solar system is valuable in itself, and can give insight ino the processes which produce galactic cosmic rays. Large solar flares, about one a year, produce particles with energies in the GeV range, which can be detected through their secondaries even at the surface of the Earth. It is not known if such high-energy particles are produced at the flare site itself or are accelerated by bouncing off the shock fronts which propagate from the flare site outward through the solar wind. Nuclei and electrons up to 100 MeV are regularly generated in smaller flares. X-ray, γ-ray, optical, and radio mapping of these flares are used to study the details of the acceleration process. By relating the arrival times and energies of these particles at detectors throughout the solar system to the observations of their production, the structure of the solar and interplanetary magnetic fields may be studied in detail.

In addition to the Sun, acceleration of charged particles has been observed in the vicinity of the Earth, Mercury, Jupiter, and Saturn—those planets which have significant magnetic fields. Again, the details of the acceleration mechanism are not understood, but certainly involve both the rotation of the magnetic fields and their interaction with the solar wind. Jupiter is such an intense source of electrons below 30 MeV that it dominates other sources at the Earth when the two planets lie along the same interplanetary magnetic field line of force. Although the origin of the enhanced flux of ^{4}He has not been identified with certainty, it may be generated by the interaction of the solar wind with interstellar gas in the regions of the outer solar system where the wind is dying out and can no longer flow smoothly. *See* JUPITER; MERCURY; PLANETARY PHYSICS; SATURN.

Direct observation of conditions throughout most of the solar system will be possible in the next few decades, and with it should come a basic understanding of the production and propagation of energetic particles locally. This understanding will perhaps form the basis of an understanding of the problem of galactic cosmic rays, which will remain for a very long time the only direct sample of material from the objects of the universe outside the solar system. [PAUL EVENSON]

Bibliography: V. L. Ginzburg and S. I. Syrovatskii, *Origin of Cosmic Rays*, 1964; S. Hayakawa, *Cosmic Ray Physics*, 1969; A. M. Hillas, *Cosmic Rays*, 1972; F. B. McDonald and C. E. Fichtel (eds.), *High Energy Particles and Quanta in Astrophysics*, 1974; J. L. Osborne and A. W. Wolfendale (eds.), *Origin of Cosmic Rays: Proceedings of the NATO Advanced Study Institute*, 1975; M. A. Pomerantz, *Cosmic Rays*, 1971; G. D. Rochester and A. W. Wolfendale (eds.), A discussion on the origin of the cosmic radiation, *Phil. Trans. Roy. Soc. London*, ser. A, 277:317–501, 1974; B. B. Rossi, *Cosmic Rays*, 1964.

Cosmochemistry

The science of the chemistry of the universe (excluding the Earth), that is, the abundance and distribution of elements, chemical compounds, and minerals; chemical processes, particularly in the formation of cosmic bodies; isotopic variations; radioactive transformations; and nuclear reactions, including those by which the elementary constituents were formed. Cosmochemistry is often functionally considered as a branch or extension of the older and more developed science of geochemistry. It is also closely related to astronomy.

TECHNIQUES AND SOURCES OF INFORMATION

Many techniques of geochemistry can be applied to extraterrestrial samples which can be obtained for analysis in the laboratory. These include chemical analyses by a variety of methods, including electron-microprobe examination of very small areas of polished surfaces and neutron-activation analysis for trace elements; petrographic-microscope and x-ray–diffraction characterization of minerals and rocks; studies of magnetic properties and remnant magnetism; determination of relative abundances of isotopes of the elements; radiometric age determinations; and measurement of cosmic ray-induced radioactivities.

The most important of the available samples are meteorites, which are generally considered to be fragments of minor planets and possibly outgassed comets, and lunar samples which have been brought to Earth by United States and Soviet spacecraft. Remote in-place chemical studies have been made on the surfaces of the Moon, Venus and Mars, and in the atmospheres of Venus, and Mars. *See* METEORITE.

For most of the bodies and objects of the universe, chemical information comes from an analysis and theoretical interpretation of the radiations received from them on Earth and by satellites above the atmosphere. Spectral analysis of the light (visible, ultraviolet, infrared) received from stars and luminous nebulae gives information on their elementary contents and on the few compounds which are stable in their exposed regions. Radio astronomy provides information on the interstellar medium, including a considerable number of molecular constituents, mainly in cool dust clouds. Very-high-temperature regions of stars, nebulae, and galaxies are studied in the short-wavelength ultraviolet and x-ray spectral regions by instruments outside the atmosphere, principally in artificial satellites. *See* ASTRONOMICAL SPECTROSCOPY; RADIO ASTRONOMY; SATELLITE ASTRONOMY; ULTRAVIOLET ASTRONOMY; X-RAY ASTRONOMY.

Nonluminous bodies of the solar system can be

investigated by their reflection spectra, that is, albedo as a function of wavelength. This yields information about atmospheres, clouds, and visible surfaces. Meteor spectra give information about these small fragments of asteroids and comets. Observations of the solar wind and energetic solar particles give chemical information about the Sun's outer layers, especially the corona. Cosmic rays bring similar information about more distant regions, although the curvature of their paths obscures the directions of their sources. *See* ALBEDO; COSMIC RAYS; METEOR; SOLAR WIND.

THE SOLAR SYSTEM

The elementary composition of the Sun, which accounts for nearly 99.9% of the mass of the solar system, is essentially that of the system as a whole and of the presumed precursor "solar nebula." Lines of over 60 elements have been identified in the solar spectrum, and it is probable that all of the 81 stable and 2 long-lived elements are present. Like most other stars, the Sun is believed to contain about 77% hydrogen and 21% helium by mass. The relative abundances of carbon, nitrogen, oxygen, sulfur, and the noble gases are taken to be similar to those in other population I stars and luminous nebulae. The relative abundances of the nonvolatile elements are assumed to be the same as in the type I carbonaceous chondrite meteorites, which are believed to have condensed from the solar nebula with little fractionation. The illustration summarizes one set of results. *See* SOLAR SYSTEM; SUN.

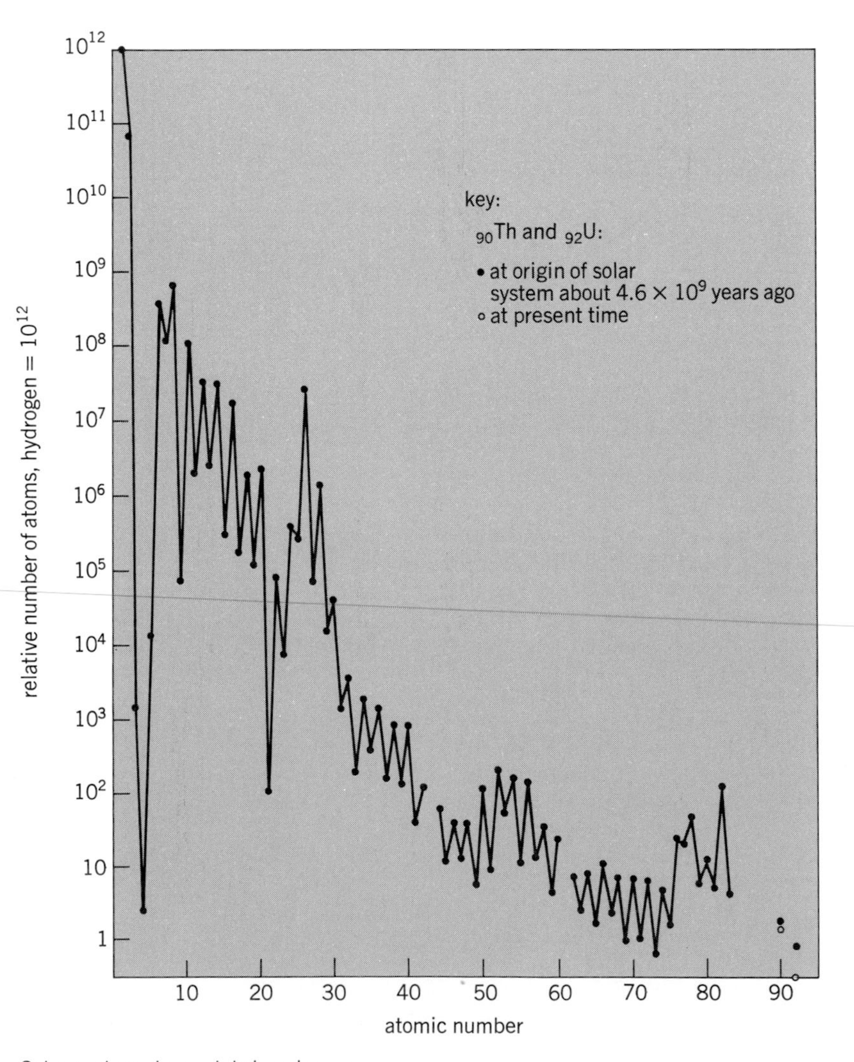

Solar system elemental abundances.

Planets. The terrestrial planets (Mercury, Venus, Earth, Mars), their satellites, and the asteroids consist largely of solids which have been separated from volatile elements. The more massive Jovian planets (Jupiter, Saturn, Uranus, Neptune) have retained large amounts of hydrogen, helium, other noble gases, and volatile compounds such as methane (CH_4), ammonia (NH_3), and water (H_2O). Jupiter probably has an overall composition similar to that of the Sun, with the proportion of light elements decreasing for the less massive planets. For a detailed comparison of the compositions of the planets *see* PLANETARY PHYSICS. For discussions of the compositions of the individual planets *see* JUPITER; MARS; MERCURY; NEPTUNE; PLUTO; SATURN; URANUS; VENUS. For discussions of satellites see the articles on their planets; for the Earth's Moon *see* MOON.

Asteroids. Reflection spectroscopy of asteroids indicates considerable variety in surface compositions, many corresponding to those of known types of meteorites. The most abundant have low albedos (about 2–5%) and spectra resembling carbonaceous chondrites. Next in frequency are more reflective bodies (about 8–21%) with spectra resembling stony meteorites, mostly siliceous with embedded metal particles. A smaller group has a similar albedo range (about 8–19%) and spectra indicating a predominant nickel-iron-metal composition. A very few have quite high albedos (30–40%), some with spectra resembling enstatite (low-iron) chondrites or achondrites. *See* ASTEROID.

Comets. The fluorescent emission spectra of the nearly straight type I tails of luminous comets show the presence of atoms, molecules, radicals, and ions of hydrogen, carbon, nitrogen, oxygen, and sulfur. In the comas of bright comets metallic elements are also seen. Huge atomic-hydrogen clouds have been observed in the antisolar directions. Shower meteors, which are outgassed fragments of comets, show spectral lines of common metallic elements. These observations are consistent with the icy-conglomerate model, according to which comets were formed as masses of ices of water (H_2O), methane (CH_4), ammonia (NH_3), and carbon dioxide (CO_2), in which are embedded organic compounds and metallic, siliceous, and carbonaceous particles. When expelled, the particles form curved type II reflecting tails. *See* COMET; INTERPLANETARY MATTER.

THE STELLAR UNIVERSE

Most stars in the Milky Way and other galaxies can be grouped into populations having various distinguishing characteristics, but which are most fundamentally related to their ages or times of formation. Within each population most main-sequence stars and many of the evolved stars exhibit similar surface chemical compositions. Contents of heavier elements, such as carbon, nitrogen, oxygen, neon, and metallic elements, increase with decreasing population age. Regardless of age, the hydrogen/helium ratio seems to be roughly constant at about 0.1 on an atomic basis, and the relative amounts of the heavy elements to each other are roughly constant.

These observations can be explained by two sets of circumstances: (1) The interstellar medium, from which stars are formed, has gradually increased its content of heavy elements (thus also dust), rapidly in the early history of the Galaxy and slowly recently, but the amount of helium resulting from hydrogen fusion and escaping from stars is small relative to the primordial amount. (2) All but the least luminous main-sequence stars and some moderately evolved stars have nonconvecting (radiative-energy-transporting) zones somewhere in their interiors, so the products of nuclear transformations in their cores are never mixed into their surface layers; and low-luminosity stars have as yet transformed very little of their hydrogen into helium in their lifetimes. Therefore, all of these stars have preserved in their atmospheres the elementary composition of the media from which they were born.

Some main-sequence and slightly evolved stars are peculiar in the sense of having distinctly nonstandard contents of hydrogen, helium, or some heavy elements. While some of these are regarded as showing surface contaminations or induced nuclear transformations in the stellar atmospheres, many highly evolved stars have spectra showing nonstandard elementary contents which can be attributed to the presence in their surfaces of products of energy-generating thermonuclear transformations in their interiors. Study of such stars helps to understand the locales and mechanisms of energy generation and nucleosynthesis. *See* ELEMENTS AND NUCLIDES, ORIGIN OF; STAR; STELLAR EVOLUTION.

Interstellar matter. The several percent of the mass of the Milky Way in the form of gas and dust is concentrated in the spiral arms. In the vicinity of hot ultraviolet-emitting stars the dust is vaporized and the gas is excited to thermal and fluorescent emissions, forming ionized-hydrogen (H II) luminous nebulae, and spectral analysis shows elementary compositions generally similar to the young stars which have been formed from this matter. In cooler regions most of the heavy elements are condensed into solid grains (oxides, silicates, metal, graphite) forming about 1% of the interstellar mass. The hydrogen is present mostly as neutral atoms (H I regions), and is detected by its characteristic 21-cm radio emission. Absorption lines of neutral metallic atoms (Ca, K, Na) and a few molecules (CH, CN, H_2) are superposed on the continuous spectra of distant stars, particularly in the ultraviolet region. The abundances of heavy elements relative to H are lower than in population I stars, corresponding to the expectation that the heavy elements are concentrated in the solid grains. Radio surveys of OH and CO show that these molecules are widely distributed in the disk and central regions of the Galaxy.

When the gas and dust are concentrated into dense clouds, and particularly into dark globules, the opacity to starlight and transparency to infrared and microwave radiation allow the interiors to cool to very low temperatures. Then most of the hydrogen is present as H_2, and many other di- and polyatomic molecules are present. These are evidently formed by heterogeneous catalysis at grain surfaces and are shielded from destructive ultraviolet starlight. About three dozen molecules have been detected in such regions by their characteristic radio spectra, mostly in emission. *See* INTERSTELLAR MATTER.

Cosmic rays. Galactic cosmic rays (those from outside the solar system) are highly energetic atomic nuclei, whose relative abundances reflect the composition of their source regions, modified by mass discrimination in the acceleration process and by spallation reactions in space.

Galactic composition. The mass fraction of heavy elements is enhanced in the central regions of the Milky Way and of similar spiral galaxies such as the Andromeda Galaxy (Messier 31). Most galaxies seem to have chemical compositions similar to the Milky Way, the variation depending chiefly on the different stellar population distributions. Even the Seyfert galaxy nuclei, quasars, and other enormously luminous objects show emission and absorption lines of familiar elements, although excitation conditions are different and much of their luminosity is in the form of synchrotron radiation, which is not element-specific. *See* GALAXY; GALAXY, EXTERNAL; QUASARS.

COSMOCHRONOLOGY

The radioactive dating methods based on primary natural radionuclides which are useful in geochemistry are also applicable to cosmochemical samples in terrestrial laboratories. Methods based on short lived extinct natural radionuclides are applicable to times very early in the solar system and thus are important in cosmochemistry, though not in geochemistry. Methods based on cosmic-ray-induced nuclides are especially useful for meteorites and lunar-surface materials, which have not been shielded by an atmosphere.

A number of meteoritic uranium, thorium-lead (U, Th-Pb), rubidium-strontium (Rb-Sr), and samarium-neodymium (Sm-Nd) ages have been obtained which are close to 4.56×10^9 years, and no reliable ages are older. This is taken as the age of solid objects in the inner solar system. A few younger dates indicate subsequent disturbances due to igneous fractionation, metamorphism, or impact-induced shock, which "reset the clocks."

Short-lived extinct radionuclides. Excess ^{129}Xe from the decay of now-extinct ^{129}I has been observed in a number of meteorites, especially chondrites. This indicates that they were formed as cool solids no later than about 10 half-lives of ^{129}I (15.7×10^6 years) following its last prior nucleosynthesis in a supernova. Xe isotopes attributable to spontaneous fission of ^{244}Pu (half-life 81×10^6 years) are also widely observed, especially in achondrites. Aluminum-correlated ^{26}Mg excesses have been found in high-temperature condensates occurring as inclusions in carbonaceous chondrites. There is little doubt that ^{26}Al is the parent, but because of its short half-life (7.2×10^5 years) it is questionable whether it was present in the solar nebula; some suggest that those inclusions were local condensations of the expanding and cooling ejecta of the supernova which produced the ^{26}Al.

Nucleosynthesis of solar-nebula material. The nuclides ^{129}I, ^{244}Pu, ^{235}U, ^{238}U, and ^{232}Th all result from the *r*-process of nucleosynthesis and were contributed by many supernovae to the interstellar medium from which the solar nebula formed. Analysis of the relative abundances of these radionuclides in meteorites at the time of their formation indicates that nucleosynthesis began about $8-15 \times 10^9$ years ago and terminated about $1-2 \times 10^8$ years before meteorite formation, with several percent of the *r*-products coming from the last contributing supernova. The ^{26}Al observations are interpreted by some as indicating a supernova contribution much later.

Age of bodies in solar system. Measurements of stable and radioactive cosmogenic nuclides have shown that most iron meteorites have existed as small bodies in space for less than 10^9 years, and most stones, which are more fragile, for less than 20×10^6 years. Dating of lunar soil, which is a mixture of many rock fragments and volatilization condensates, indicates an age of the Moon as a body of about $4.4-4.6 \times 10^9$ years, about the same as that of the meteorites. However, intact igneous lunar rocks are generally younger.

Isotopes. Stable-isotope studies of meteoritic elements show anomalies, mostly in minor mineral components, indicating nucleosynthetic heterogeneity of the source elements of the solar system. The noble gases show strong effects of chemical fractionation within the solar system.

CHEMISTRY IN COSMIC EVOLUTION

According to current orthodoxy, the matter of the universe was created in an expanding "primordial fireball" about $10-20 \times 10^9$ years ago, mostly in the form of ^{1}H and ^{4}He in close to the present proportions, about 10:1 on an atomic basis. Small amounts of ^{2}H, ^{3}He, and ^{7}Li were also formed then. *See* BIG BANG THEORY; COSMOLOGY.

Successive condensations in the expanding universe produced galaxies, star clusters, and individual stars. The more massive early stars generated heavy elements in their interiors and ejected them into the interstellar medium, mainly through supernova eruptions. Later-formed stars, formed before the Galaxy had collapsed to a disk, thus inherited small amounts of heavy elements. Subsequent star formation occurred mainly in the disk, with a rather rapid buildup of heavy elements, reflected in the initial compositions of successive generations of stars. A variety of nucleosynthetic mechanisms were involved in different locales.

Solar system formation. About 4.7×10^9 years ago a gas-dust cloud in a spiral arm of the Milky Way collapsed gravitationally to form the solar nebula. The collapse may have been initiated by a pressure wave from a nearby supernova, some of whose newly formed stable and radioactive nuclides may have been incorporated into the nebula. In the center of the nebula, accelerated contraction produced the Sun, initially much more luminous than today, while the outer parts collapsed into a disk. The inner regions at least were heated by gravitational energy and solar radiation to temperatures sufficient to vaporize the dust grains. As the Sun decreased in luminosity and became a stable main-sequence star, the nebula cooled while maintaining a distribution of temperature decreasing with increasing distance from the Sun. As the temperature at a particular radial distance decreased, a series of solids condensed from the gas.

When the temperature had dropped and perhaps stabilized at a value characteristic of the solar distance, grains of equilibrium composition at that temperature adhered and accreted into small particles and larger bodies called planetesimals. The increased transparency allowed solar radiation pressure in the inner regions to drive away the remaining gases. The planetesimals coalesced further to form planets. In the outer regions, where enormous amounts of solids and gases were present, gravitational attraction became important, and uncondensed gases and unaccreted grains were also swept up by the planets. In an intermediate zone conditions were such that a single large planet did not form, but instead a number of minor planets, the progenitors of the asteroids, resulted.

The decreasing densities and other compositional trends of the terrestrial planets and the larger satellites of the Jovian planets are generally consistent with this equilibrium-condensation model. However, for the minor planets represented by the chondrite meteorites there is evidence based mainly on trace-element analyses that the temperature was still decreasing during the accretion, so that a layered structure resulted.

Planetary evolution. After and even during accretion, planetary evolution occurred. Most chondrites show evidence that each was at one time a part of the regolith of a bombarded planetary body, and that it was later subjected to higher temperatures and pressures in the interior which resulted in lithification and varying degrees of metamorphism. The major heat sources were probably radioactive-disintegration and gravitational energy. Small bodies do not generate appreciable gravitational energy and do lose heat rapidly, so evolutionary changes are minor except for those produced by impacts. Larger bodies generate more gravitational energy on accretion and retain heat better, so their interiors rise to higher temperatures. If melting occurs, mass redistribution releases still more gravitational energy, especially if an iron or troilite core is formed. Melting, gravitational redistribution, recrystallization, and outgassing produce stratified interiors and crusts, hydrospheres and atmospheres, volcanism and other geologic phenomena, and igneous rocks, both plutonic and extrusive. *See* PLANETARY PHYSICS.

[TRUMAN P. KOHMAN]

Bibliography: D. Flanagan et al. (eds.), *The Solar System: A Scientific American Book*, 1975; O. Gingerich (ed.), *New Frontiers in Astronomy: Readings from Scientific American*, 1975; W. O. Mulligan (ed.), *Proceedings of the Robert A. Welch Foundation Conferences on Chemical Research, XXI: Cosmochemistry*, 1978; V. Trimble, The origin and abundances of the chemical elements, *Rev. Mod. Phys.*, 47:877–976, 1975.

Cosmology

The study of the large-scale structure and the evolution of the universe, and including, in its modern connotation, that part of cosmogony that deals with the origin of the universe and of the chemical elements. *See* UNIVERSE.

Structure of the universe. Beginning about 1924, E. P. Hubble developed ways of estimating distances to remote galaxies, and also studied their distribution in space. From counts of galaxy images on photographs of 1283 sample regions of the sky, Hubble concluded that on the large scale the distribution of galaxies is homogeneous and isotropic. Hubble's finding was the first observational evidence for the cosmological principle, which states that at any instant of cosmic time the universe is, on the large scale, the same as seen by all hypothetical observers; that is, it is the same everywhere (is isotropic and homogeneous). *See* GALAXY, EXTERNAL.

Galaxy clusters. On the small scale, however, Hubble found a tendency for galaxies to cluster. A few dozen individual clusters of galaxies were known when Hubble carried out his survey, and he selected the sample fields to avoid these known clusters. There were, however, too many fields with too few galaxies, and too many fields with too many for their distribution to be completely random in space. Hubble suggested that all galaxies either are members of, or were formed in, groups and clusters. The Milky Way Galaxy, in fact, is a member of a system of about two dozen galaxies, most of them much smaller than the Milky Way, spread over a region of space with a linear diameter of about 3×10^6 light-years (1 LY $= 9.46 \times 10^{12}$ km); the system is called the Local Group.

The general tendency of galaxies to cluster was confirmed by the analysis of the galaxy distribution in two post–World War II photographic surveys of the sky: the Lick Astrographic survey with the 20-in. (50-cm) astrographic camera, and the Palomar Sky Survey with the 48-in. (1.2-m) telescope. The distribution of galaxy images on the Lick photographs was found to be compatible with a statistical model of complete clustering. Occasional clusters, the great clusters, contain thousands of member galaxies each, and have radii of 5×10^6 to 2×10^7 LY.

Superclusters. G. O. Abell cataloged 2712 of these great clusters, and defined a homogeneous statistical sample of about 1800. On the large scale, the clusters in the statistical sample confirm Hubble's finding of homogeneity and isotropy—further observational evidence for the cosmological principle. On the small scale, however, even the clusters are clumped into larger aggregates, and Abell identified 17 obvious superclusters. Subsequently, statistical studies of catalogs of galaxies and clusters of galaxies confirmed that the spatial distribution of galaxies is correlated over scales that are large compared to sizes of individual clusters.

The Local Group appears to be a member of such a larger aggregate, the Local Supercluster, whose center may lie in the general direction of the nearest fairly rich cluster of galaxies, the Virgo cluster, some $30-50 \times 10^6$ LY distant. Several other individual superclusters have been studied in some detail, and evidence on their structures is rapidly accumulating. Typical superclusters are $1-3 \times 10^8$ LY across and contain one or two great clusters, although some contain 10 or more. They also contain dozens of lesser clusters, and hundreds of groups like the Local Group; they probably contain individual galaxies as well, and possibly intercluster gas, although at present there is firm evidence for intergalactic gas only in certain rich clusters; this gas is at a temperature of 10^8 K and is detected by its thermal emission of x-rays. Some investigators are of the opinion that all of the matter of the universe is in superclusters, and there is growing evidence (by no means yet conclusive) that this is so. *See* SUPERCLUSTERS; X-RAY ASTRONOMY.

Large-scale homogeneity. Present observations suggest that superclusters are the end of the hierarchy. One way to sample the distribution of matter to various depths is to count galaxies to successively fainter limits. If counts are extended to a limit of faintness such that the line of sight from the Earth encounters only one or two superclusters, the presence of those aggregates is clearly apparent in the irregular way that the counts increase with faintness. But if the counts are continued to very faint limits, they always smooth out, consistent with a large-scale homogeneous distribution of matter, washing out all traces of superclustering. In 1977 G. Rainey made such counts in three widely separated directions in the sky and found that he could definitely rule out general clustering of matter on scales of 10^9 LY or more. He also found a remarkably high degree of isotropy in those different directions. Other similar studies, some to much fainter limits, are in agreement with Rainey's and further strengthen observational support for the cosmological principle.

In summary, the structure of the universe is observed to be very lumpy on scales of up to a few hundred million light-years, but on larger scales appears to be the same everywhere. The cosmological principle, therefore, is a starting assumption for nearly all theories of cosmology, and for all that are widely considered today. There is, of course, no guarantee that the observed large-scale homogeneity extends throughout the entire universe, but without that assumption there is no basis for formulating a general cosmological model.

Gravitation in the universe. The present-day density of matter in the universe exceeds the mass equivalent of the energy of radiation by at least a factor of 10^2; thus the universe today is dominated by matter. Moreover, bulk matter is electrically neutral to a very high degree. Hence the only known long-range force that can be important in influencing the dynamical evolution of the universe is gravitation.

In the 17th century Newton considered the effects of gravitation on the universe at large, and concluded that in an infinite universe the potential due to gravitation would everywhere be the same, and hence an infinite universe could be stable against gravitational collapse (except locally, where irregularities in the density could cause local collapses that might become stars).

On the other hand, the new theory of gravitation

embodied in Einstein's general theory of relativity leads to field equations that, in their simplest form, do not permit a static stable universe even if it is infinite. In general relativity the field equations describe the curvature of space-time by matter; it is in curved space-time that all objects, material bodies as well as photons of light, move along unaccelerated paths (geodesics). Imbued with the idea that the universe should be static, Einstein, in his effort to apply the field equations to cosmology, modified them with the introduction of a term now called the cosmological constant. The cosmological constant, if positive, implies a repulsive force that is greater in proportion to distance. By choosing just the correct value for the constant, the force can be made to balance gravitation on the large scale, and permit a static universe. *See* GRAVITATION; RELATIVITY.

The cosmological constant does not violate the assumptions on which the field equations are derived, for it enters as a constant of integration, and in the most general mathematicial case it should be present. But there is no analog for such a repulsion in newtonian theory, and there is no observational evidence for its existence; just as one usually evaluates a constant of integration from the boundary conditions of the physical problem involved, so one would normally assign to the cosmological constant the value zero. Its introduction here with just the correct value to provide a static universe seemed an ad hoc procedure adopted for the sole purpose of saving the theory.

Expanding universe. In the 1920s it occurred to others that the universe need not necessarily be static, and that the field equations in their simplest form (zero cosmological constant) can apply perfectly well to a uniformly expanding or contracting universe. In particular, the Soviet mathematician Alexandre Friedmann in 1922, and independently the Belgian cosmologist Georges Lemaître in 1927, proposed general relativistic models of an expanding universe. Indeed, Lemaître suggested observations that could test the hypothesis.

These were observations begun about 1912 by V. M. Slipher of the spectra of the "nebulae" that were later found to be galaxies. Slipher found the features in the spectra of the objects to be displaced from their normal wavelengths by the Doppler effect, indicating that the nebulae are moving away from the Earth at high speeds. *See* DOPPLER EFFECT; RED SHIFT.

Now, if the cosmological principle is applied to an expanding universe, the universe must expand uniformly; otherwise, irregularities would be generated that would make the universe inhomogeneous. In a uniform expansion, the separations of all pairs of objects must increase by the same factor in the same time, which means that widely separated objects, having further to move apart, must separate at a greater speed than objects close together, which have less far to move to increase their separations by the same relative amount. In other words, every observer, everywhere in an expanding universe, will see every other object moving away from him at a speed that is proportional to its distance; it is a necessary and unique condition for a uniform expansion. Of course, since all observers see the same effect, the fact that objects are moving away from a given observer at speeds proportional to their distances from him gives him no information about his location in the universe, nor of any "center." In an infinite expanding universe, the concept of "center" has no meaning.

By 1929 Hubble had found approximate distances to more than 40 of the galaxies whose radial velocities (speeds of recession) had been measured by Slipher. Hubble's analysis showed there to be a significant correlation between velocity and distance, with velocities ranging up to nearly 2000 km/s. Meanwhile, M. L. Humason had begun observing the spectra of still more distant objects with the 100-in. (2.54-m) telescope, and in 1931 Hubble and Humason jointly published a comparison of the distances and radial velocities of galaxies receding at speeds up to 15 times those of the most remote galaxies observed by Slipher. The velocities and distances were in direct proportion; the expansion of the universe had been firmly established. The observed relation between the velocities and distances of galaxies is now called the Hubble law.

As noted above, the cosmological principle applies only to the universe on the large scale. Similarly, the uniform expansion applies only to the large-scale universe—not to individual galaxies and clusters of galaxies. Like the Earth, solar system, and stars, galaxies are bound tightly together with their own mutual gravitation; they are gravitationally stable and do not expand with the universe—a point often misunderstood. At least the rich clusters of galaxies also appear to be gravitationally bound, and gravitation must play some role in the dynamics of superclusters, although just how much is not yet known.

The mutual gravitational attractions of the member galaxies in a cluster accelerate them to high speeds (often 10^3 km/s or more) in their more or less random paths back and forth through the cluster. When the radial velocity of a galaxy in a group or cluster is observed, therefore, the velocity of recession of the cluster itself caused by the expansion of the universe always has superimposed on it the line-of-sight component of the galaxy's velocity due to the local gravitational accelerations by other cluster members. In a few nearby clusters the speeds of individual galaxies within them are great enough that a few galaxies are approaching the Earth, even though the clusters to which they belong are receding.

Cosmological models. The Hubble law alleviates the need for the positive cosmological constant introduced by Einstein to allow a static universe. Nevertheless, theoreticians have carefully considered the much larger range of cosmological models that are possible with both positive and negative values of the cosmological constant, and it cannot be guaranteed that future observations will never require such nonzero values. The cosmological constant is not, however, needed at present, and the only models for the universe that receive the wide attention of both theoreticians and observers are those based on general relativity with zero cosmological constant. These models will now be described qualitatively.

Because the cosmological principle requires

that the universe expand uniformly, its evolution can be described uniquely with a simple change in scale. Thus the actual distance $r(t)$ between two remote objects in space at time t can be expressed as the product of a suitably normalized time-varying scale factor, usually denoted as $R(t)$, and the fixed distance u between those objects at some arbitrarily chosen time; thus, Eqs. (1) hold.

$$r(t) = R(t)u \qquad (1a)$$

$$u = r(t)/R(t) \qquad (1b)$$

The radial velocity V_r of a galaxy is simply the rate of change of $r(t)$. Since u is constant, Eq. (2) holds.

$$V_r = \frac{dr(t)}{dt} = \frac{dR(t)}{dt}u = \frac{1}{R(t)}\frac{dR(t)}{dt}r(t) \qquad (2)$$

According to the Hubble law, Eq. (3) is valid,

$$V_r = Hr(t) \qquad (3)$$

where the constant of proportionality, H, is called the Hubble constant. Note that the Hubble law does not imply that a given galaxy speeds up as it recedes to greater distance; rather at any given time the galaxies that happen to be more remote must recede faster. *See* HUBBLE CONSTANT.

Comparison of Eqs. (2) and (3) shows that Eq. (4)

$$H = \frac{1}{R(t)}\frac{dR(t)}{dt} \qquad (4)$$

holds, so the Hubble "constant" is not really constant over time, but has a value that changes as the universe evolves and the scale $R(t)$ changes with time. Even if all galaxies maintained their present speeds, H would diminish as the scale grows. In fact, however, the mutual gravitation of the objects in the universe must slow its expansion, so that galaxies, as they recede from each other, are gradually reducing their relative speeds. A major goal of observational cosmology is to determine how strongly the gravitation of the universe decelerates its expansion.

Closed, open, and flat universes. In general relativity, photons and material objects flow unaccelerated along geodesics in space-time, and that space-time is curved by the presence of matter. In a random distribution of massive objects the curvature can be very complex, but for the universe at large, adoption of the cosmological principle results in an enormous simplification. Except for small-scale local effects, the curvature is the same everywhere. If the mean universal density of matter and energy is high enough, the curvature is positive, in which case the universe is unbounded but finite, for information can never be received from beyond a particular volume of space. Such a universe is said to be closed, and the geometry of space-time is elliptical (in which the volume v of a sphere of radius r is greater than $4\pi r^3/3$). If the universal density is below a critical value, however, the curvature is negative, the universe is open and infinite, and the geometry of space-time is hyperbolic (in which v is less than $4\pi r^3/3$). The dividing line between these two classes of space-time, requiring the critical density that does not quite close the universe, is flat space-time, in which the universe is infinite and the geometry is euclidean.

A closed universe is closed not only to radiation but to matter, for it implies that the geodesics of particles of matter and of photons are closed; such a universe will eventually cease expanding and start to contract. The contraction will continue until all of the matter and radiation approaches a singularity at the origin of a universal black hole—the "big crunch," as J. A. Wheeler described it. It has been speculated that such a universe may undergo another "big bang" to begin a new cycle, and thenceforth oscillate between successive expansions and contractions, each contraction followed by a new big bang. There is, however, no physical theory to account for future big bangs and subsequent expansions, so the "oscillating" universe is at present no more than a speculative extension of the closed universe.

Alternatively, the open universe will expand forever into greater and greater infinities; neither radiation nor material objects will ever come together again. Closed and open universes are analogous to rockets fired from the surface of the Earth at respectively less than and greater than the velocity of escape. A rocket launched with greater than the escape velocity is always slowed by the Earth's gravitation, but it has enough energy that it never stops; similarly, in the open universe the mean density of matter and radiation is too low to provide enough gravitation to ever stop the expansion.

The flat universe has the critical density to barely stop the expansion, analogous to a rocket launched with precisely the escape velocity. In this case the scale $R(t)$ increases as $t^{2/3}$. It is mathematically equivalent to a model by Einstein later modified by the astronomer W. de Sitter, and so is also called the Einstein–de Sitter universe. Figure 1 shows the scale of the universe R plotted against time in various cosmological models.

Age of universe. All general relativistic models of the universe with zero cosmological constant have a unique beginning, usually called the big bang, and a finite age. The maximum possible age for any of these models is that of the open universe with so low a mean density that gravitation can be ignored, so that the galaxies continue to recede at constant speed. In this case, since the time it takes a given galaxy to reach its present distance is that

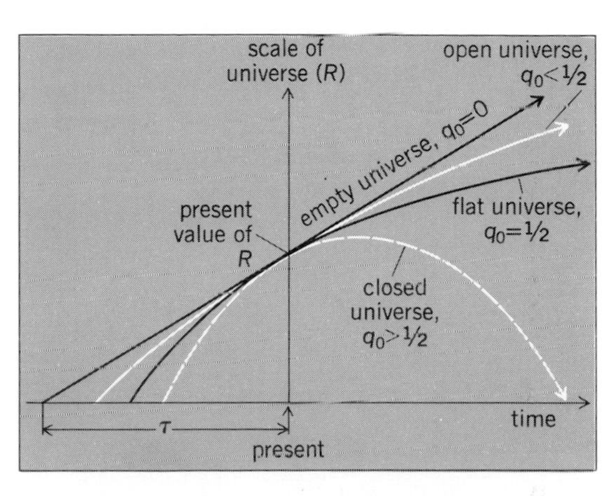

Fig. 1. Scale of the universe R plotted against time in various cosmological models; τ is the maximum possible age for universes with cosmological constant equal to 0.

distance divided by its speed, the time since that galaxy began receding with zero distance, or the age of the expansion, is simply the reciprocal of the Hubble constant. With the range of current estimates for the Hubble constant, that maximum possible age lies in the range $1-2 \times 10^{10}$ years. The age of the Einstein – de Sitter universe is just two-thirds of the reciprocal of the Hubble constant; open universes have greater ages than the Einstein – de Sitter universe, and closed universes have smaller ages.

Selecting between cosmological models. The future of the universe, that is, whether it is open or closed, depends on how rapidly the expansion is being slowed by gravitation. A measure of this slowing is called the deceleration parameter, denoted q_0. (The formal definition of q_0 is $-\ddot{R}_0 R_0/\dot{R}_0^2$, where the dots indicate time derivatives and the zero subscripts the present-day value of the scale, R.) The value of q_0 is proportional to the mean density of the universe, and so is always positive; that is, any density produces gravitation that results in a deceleration. For the Einstein – de Sitter universe, q_0 has the value ½; it is less than ½ for the open and greater than ½ for the closed universes. It is possible to find q_0 directly from observations, but those observations are difficult to make and the present-day evaluations of q_0 are highly uncertain. Some of the procedures will now be discussed.

Precise form of the Hubble law. As explained above, any uniform expansion results in a linear relation between velocity and distance (so long as the velocities are small compared to that of light; at very great distances the velocities become very large, and special relativity requires that they asymptotically approach the speed of light). Instantaneous velocities and distances are not observed, however, because of the finite speed of light. The velocity observed for a galaxy 10^8 LY away is the speed of recession it had 10^8 years ago when the light, by which it is now observed, was emitted. Both the distance and speed of recession of the galaxy have subsequently changed, and how much they have changed depends in a predictable way on q_0—that is, how rapidly the expansion is slowing. A plot of the Hubble law therefore shows deviations from linearity for objects at large distances (Fig. 2), and comparison of the observed deviations from those predicted for various values of q_0 can, in principle, determine q_0. Velocities of galaxies can be measured with sufficient precision by measuring the Doppler shifts of their spectral lines, but relative distances cannot be determined accurately enough to make the test. The main problem is that distances of galaxies must be estimated from their apparent brightnesses and their assumed intrinsic luminosities. But galaxian luminosities are expected to change with time because of the evolution of their member stars, and the evolutionary corrections that must be applied are not yet well determined. Estimates of those corrections lead to an interpretation of the observed Hubble law that is compatible with an open universe, but a closed universe cannot be ruled out. Angular sizes of galaxies or cluster of galaxies, again in principle, can also be used as distance indicators when proper allowance is taken of photometric and special relativistic effects, but the present precision of measurements is not good enough to improve the discrimination between open and closed models.

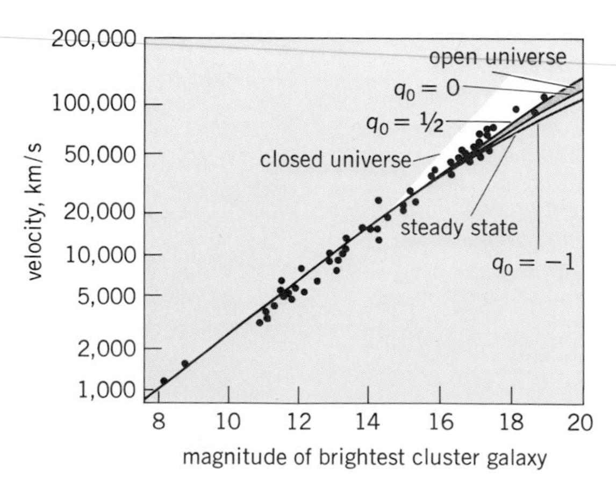

Fig. 2. Hubble law curves for various cosmological models. Data points indicate observed values.

Counts of galaxies and other objects. In a static homogeneous universe the numbers of objects should increase as the cube of the distance to which they are counted, but in an expanding universe the density decreases with time. Thus when one looks out to great distance, one also looks back in time to a denser universe; the counts should increase at a greater rate than with the cube of the distance, and how much greater depends on q_0. The difficulty in applying this test lies in the problem of knowing the distances of the counted objects, particularly because their intrinsic luminosites can evolve in time. The quasars, for example, have very large radial velocities indicating enormous distances; quasars must therefore be objects that existed in the remote past but no longer today. *See* QUASARS.

Deuterium-hydrogen ratio. In the first few minutes of the big bang, temperatures were high enough to support nuclear fusion. Calculations show that deuterium was formed, but would have fused almost completely into helium unless the density of the matter were relatively low; for appreciable deuterium to survive the big bang, the density then would have to have been low enough that the present-day density would be far below the critical value to close the universe. Observations of the present ratio of deuterium to hydrogen indicate a value in the range 10^{-5} to 10^{-4}, high enough to imply an open universe. This conclusion of an open universe must be regarded with caution, however, for it is not certain that the observed deuterium is primordial. *See* BIG BANG THEORY.

Mean universal density. Several other tests for determining q_0 from observations have been attempted, but the one with the most significant (although not definitive) result is that of the mean density of matter in the universe. The critical density to close the universe is of the order 10^{-30} to 10^{-29} g/cm³ (depending on the value of the Hubble constant). Fairly reliable estimates of the density corresponding to visible matter in the universe lead to values smaller than the critical density by one to two orders of magnitude. These estimates of the mass density include any invisible gas or dust within clusters of galaxies, whose total masses can be found from the measured speeds of their member galaxies, and probably within superclusters as well. To make a high enough density to close the universe, there would have to exist invisible matter more or less uniformly permeating the universe with an average density at least ten to a hundred times that of all observed matter. Any intergalactic gas, and even primordial black holes, would be expected to follow the clumpy distribution of matter in clusters and superclusters, and hence would already be counted in the estimate of the density. The standard model of the big bang predicts an average numerical density of neutrinos in the universe of the order $10^2/cm^3$, but if these neutrinos have zero rest mass they contribute negligibly, and if they do have rest mass they can be important only if they cooled to nonrelativistic velocities very early in the big bang, in which case they would also probably clump with clusters and superclusters. Whereas all of these arguments are model-dependent, and are subject to possible large systematic errors of observation, the estimate of the present mean universal density nevertheless provides the best evidence for the future evolution of the universe, and strongly suggests that it is open.

Early universe. In 1931 Lemaître proposed that the universe had an explosive beginning with the fission of a primeval atom, which ultimately disintegrated to produce the atoms that make up the universe today. In 1948 G. Gamow, R. A. Alpher, and R. C. Herman proposed that the present elements of the universe were formed by fusion in the big bang. Modern calculations show that elements heavier than helium cannot be produced in material with the combination of temperature and density expected in the big bang, and the current view is that the heavy elements are produced in the cores of stars advanced in their evolution and in supernovae explosions. But otherwise the fusion hypothesis of Gamow and his associates is now accepted. The best current model, called the standard model of the big bang, is that of R. A. Wagoner, W. A. Fowler, and F. Hoyle. *See* ELEMENTS AND NUCLIDES, ORIGIN OF.

At an age of 1 s, the temperature was about 10^{10} K, and matter and radiation were in equilibrium. In the early universe the temperature varied with time as $t^{-1/2}$, and after a few seconds pair production and annihilation ceased, and the present-day neutrinos were released. After a few minutes fusion ceased, and the only nuclei to survive were those of hydrogen (about 75% by mass), helium, and a trace of deuterium. After a million years, when the temperature had dropped to about 3000 K and the density of protons to $10^3/cm^3$, protons and electrons combined to make neutral hydrogen, and the universe became transparent to visible radiation. It is presumed that stars and galaxies formed during the next 10^9 years, but there is not yet a satisfactory theory for galaxy formation. It is not even known whether stars, galaxies, or clusters formed first, although observations of differences between the morphological forms of galaxies inside and outside rich clusters provide circumstantial evidence that the present rich clusters are reflections of primordial condensations.

Whereas the formation and early evolution of matter in the universe is very uncertain, and is the subject of considerable theoretical controversy, the general scenario of the evolution of the universe from a hot, dense state has strong support from three very different kinds of observations:

1. The present expansion of the universe, verified by Hubble and Humason in 1931.

2. The present relative abundance of hydrogen and helium. The standard model of the big bang predicts that the ratio by mass should be 3:1, independent of the value of q_0. Within observational uncertainty this is just the ratio observed today, except for small local enhancements of helium that are understood in terms of fusion in stellar interiors.

3. Microwave radiation arriving isotropically from space, and with the spectrum characteristic of that from a black body of temperature 2.7 K. This cosmic background radiation is described next.

Cosmic background radiation. When hydrogen became neutral at a temperature of about 3000 K, the opacity of matter dropped suddenly, and matter and radiation were effectively decoupled. At that point the radiation was in the form of visible light characteristic of that emitted by a 3000-K black body, but to observe the radiation today one must look into the past to that time of decoupling. The only parts of the universe from which radiation emitted $1-2 \times 10^{10}$ years ago can still be observed are those that lie at a distance of $1-2 \times 10^{10}$ LY, and because of the expansion of the universe those parts are receding at a speed that is within two parts in a million of that of light. Consequently, that visible light is Doppler-shifted to radio radiation like that from a black body a thousand times cooler.

Cosmic background radiation was predicted independently by a number of workers from before 1950; it was discovered accidentally in 1965 by A. A. Penzias and R. W. Wilson. Penzias and Wilson were attempting to use a large horn antenna, designed for satellite communication, for the absolute calibration of galactic radio sources when they encountered the unexpected background noise. Many subsequent observations at many wavelengths, from high-altitude balloons and a U-2 reconnaissance aircraft (converted by NASA for space research), as well as from the ground, have confirmed the black body spectrum of the background radiation.

It is important to understand that there is no "site" for the big bang; it was everywhere. The Earth is in the universe now and has always been surrounded by it; consequently the background radiation comes from all directions in space, and, of course, gives no indication of any center or preferred point of observation. The background radiation is also the present "horizon" of the observable universe. To look farther into space, it would be necessary to look farther back into time, and hence into the opaque fireball which vision cannot penetrate. The background radio radiation is from that last radiating surface before the universe became transparent, and that surface serves as an opaque curtain completely surrounding the Earth. That horizon recedes in time, however, for in the remote future it would be possible to look a farther distance into the past, and thus farther into space.

Steady-state universe. The microwave background is the best evidence that the universe has evolved from a hot dense state, and is generally regarded as specifically ruling out an alternative cosmology advanced by Hoyle, H. Bondi, and T. Gold in 1948. Their steady-state universe was based on the perfect cosmological principle, namely that the universe is the same (on the large scale) not only everywhere, but for all time. As the universe expands, new matter would have to be created to maintain a constant density, and at just the right rate to maintain the same ratio of young and old galaxies. The steady-state universe was eternal, with no specific origin, and could show no hot opaque radiating surface in the past.

Possible deviations from blackbody spectrum. High-altitude balloon observations of the cosmic background radiation revealed small discrepancies in the spectrum of the microwave background radiation from that of a blackbody; the spectrum seems to lie slightly above that of the blackbody spectrum at the peak wavelength (about 1 mm) and slightly below it at shorter wavelengths. The discrepancies appear to be significant, but the corrections for atmospheric and instrumental effects are difficult, and the observations are in need of confirmation. These deviations from a blackbody, if real, are not yet understood, but in any case are not expected to alter the principal interpretation of the cosmic background radiation.

Anisotropy. On the small scale (a few minutes of arc) the background radiation is exceedingly isotropic—to better than a few parts in 10,000—which places a corresponding upper limit on the temperature variations in the emitting matter. Density fluctuations in that material, which must certainly have been present in order to provide gravitational instabilities required to condense galaxies and clusters, would produce temperature variations close to this limit, so it is anticipated that small-scale inhomogeneities will be found when observations of higher sensitivity are possible.

A large-scale anisotropy in the background radiation has, however, been observed independently by a group headed by D. T. Wilkinson, who have sent a horn antenna to an altitude of 27 km in a balloon, and by a group headed by G. Smoot, who have flown a similar antenna on a U-2 aircraft at an altitude of 20 km. The anisotropy is compatible with a dipole of amplitude 0.003 K superimposed on an isotropic background, and is generally interpreted as being due to a peculiar motion of the Sun with respect to the background itself of about 300 km/s. The radiation from the direction in the sky toward which the Sun approaches is Doppler-shifted to slightly shorter wavelengths, relative to the average, and appears as a very slightly hotter blackbody, and that from the opposite direction in the sky is shifted the opposite way, and mimics a slightly cooler blackbody. When account is taken of the rotation of the Galaxy and of its motion in the Local Group, it is found that the Local Group appears to be moving at about 500 km/s with respect to the background toward a direction about 40° south of the Virgo cluster. One interpretation is that the Local Supercluster has enough gravitation to slow its expansion somewhat with respect to that of the universe as a whole, so that the Local Group is moving away from the center of the supercluster with a speed a few hundred kilometers per second less than it would if there were no such gravitational drag and the Local Group were moving with the general expansion of the universe. *See* COSMIC MICROWAVE RADIATION.

[GEORGE O. ABELL]

Bibliography: P. J. E. Peebles, *Physical Cosmology*, 1971; D. W. Sciama, *Modern Cosmology*, 1971; J. Silk, *The Big Bang*, 1980; S. Weinberg, *The First Three Minutes*, 1977; S. Weinberg, *Gravitation and Cosmology*, 1972.

Crab Nebula

The Crab Nebula in Taurus (see illustration) is the most remarkable known gaseous nebula. Observed optically, it consists of an amorphous mass that radiates a continuous spectrum and is involved in a mesh of delicate filaments which radiate a bright line spectrum characteristic of typical gaseous nebulae. *See* NEBULA.

This nebula has been identified as the expanding remnant of a supernova that appeared in A.D. 1054, reached an apparent magnitude of −5, corresponding to an absolute magnitude of about −18 (roughly 1.6×10^9 times as bright as the Sun). From a comparison of the angular expansion rate of the elliptical nebular shell with the expansion velocity in kilometers per second, measured by the Doppler shift, distance estimates ranging from 1300 to 2000 parsecs (1 parsec $= 3.1 \times 10^{13}$ km) have been obtained. If a distance of 1800 parsecs is adopted, the nebular size is 3.1×2.1 parsecs. The total mass of the nebula is comparable to that of the Sun. Presumably it comprises only material from the star not yet mixed with the interstellar medium. *See* DOPPLER EFFECT; SUPERNOVA.

The Crab Nebula also radiates strongly in the radio range, and the infrared, ultraviolet, x-ray, and gamma-ray spectral regions. Furthermore, the

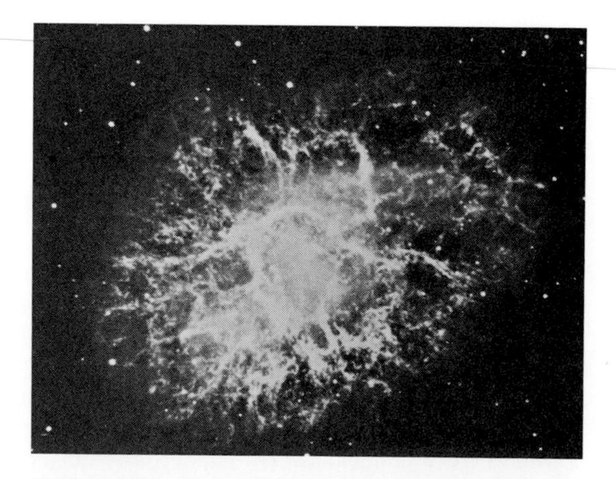

Crab Nebula, in the constellation Taurus, emitter of strong radio waves and of x-rays.

radiation of the central amorphous mass is strongly polarized, even in the x-ray region—indicating that the source of the radiation must be the synchrotron emission of electrons accelerated in a field on the order of 10^{-8} tesla. Energy must be supplied to these electrons at a rate equivalent to about 30,000 times the solar power output.

The source of energy for the Crab Nebula appears to be a remarkable, rapidly spinning object known as a pulsar, or neutron star, which is presumed to be the residue of the supernova that created the nebula. Its period of variability (which is associated with its rotation period) is 0.033 s. The period increases uniformly with time except for sudden spin-ups or glytches, when the rotation suddenly speeds up. These glytches are interpreted as readjustments in a crust of a neutron star, that is, as "star-quakes." Energy output variations are observed in the radio, infrared, optical, and x-ray range. The total amount of electromagnetic energy emitted by the star is about 10^{36} ergs/s (10^{29} W), but this is only a small fraction of the amount supplied to high-energy particles. The complex pulses observed both optically and at radio frequencies can be interpreted by an oblique magnetized rotator, but the mechanisms responsible for accelerations of particles to high energies are not fully understood. The ultimate source of the energy of the Crab Nebula is the rotational energy of the neutron star; in the period 1054–1975 the pulsar lost about 10^{40} kilojoules, mostly in particle energy. At present, the pulsar is losing about 10^{28} kilowatts, but the rate of energy loss may have been much higher at an earlier epoch. The Crab Nebula may be regarded as the Rosetta Stone of high-energy astrophysics. *See* PULSAR.

One might anticipate that a search of the sky would find some object roughly similar to the Crab Nebula. Supernova (SN) remnants (for example, the Veil Nebula in Cygnus) abound in the Milky Way Galaxy, in the Magellanic Clouds, and in the Triangulum Spiral, M33, but in many instances what is observed is material thrown off by the supernova mixed with the neighboring interstellar medium. There is no trace of the residual star. Only the Vela supernova pulsar has been found optically, a twenty-sixth-magnitude star detected with the Anglo-Australian telescope. It is the faintest star ever observed. Remote pulsars cannot be detected, but it would seem that in some instances the star simply shattered completely or the residue disappeared as a black hole.

[LAWRENCE H. ALLER]

Bibliography: K. M. V. Apparao, *Astrophys. Space Sci.*, 25:3–116, 1973; *Crab Nebula, 46th International Astronomical Union Symposium*, Manchester, 1971.

Crux

The Southern Cross, in astronomy, the most celebrated of the constellations of the far south. The four principal bright stars of the group α, γ, β, and δ form the figure of a cross, giving the constellation its name, in contradistinction to the Northern Cross of Cygnus in the north, which is larger and more distinct, having a bright star to mark its center (see illustration). The brightest star in

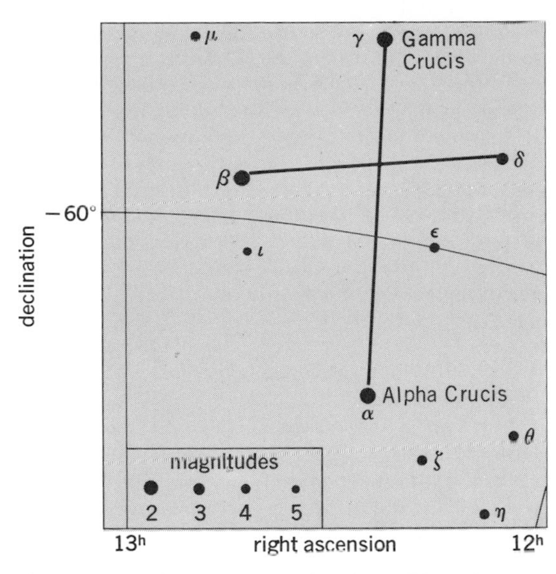

Line pattern of the constellation Crux. The grid lines represent the coordinates of the sky. The apparent brightness, or magnitude, of the stars is shown by the sizes of dots, graded by appropriate numbers as indicated.

Crux is Alpha Crucis at the foot of the cross, which has received the artificial name Acrux. This is a navigational star. The star at the top of the cross is Gamma Crucis, sometimes called Gacrux. The line joining these two stars points approximately toward the South Celestial Pole. *See* CONSTELLATION; CYGNUS.

[CHING-SUNG YU]

Cygnus

The Swan, in astronomy, is a conspicuous northern summer constellation. The five major stars of the group, α, γ, β, ε, and δ, are arranged in the

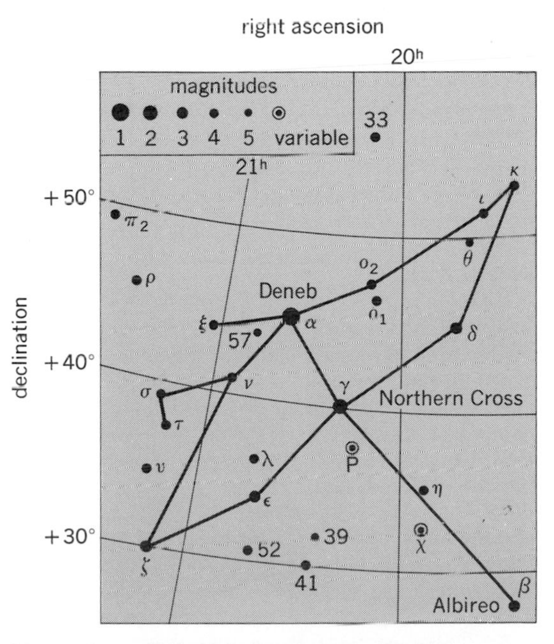

Line pattern of constellation Cygnus. Grid lines represent coordinates of sky. Apparent brightness, or magnitudes, of stars is shown by sizes of dots, which are graded by appropriate numbers as indicated.

form of a cross (see illustration). Hence Cygnus is often called the Northern Cross, to distinguish it from the Southern Cross of the constellation Crux. The constellation is represented by a swan with widespread wings flying southward. The bright star Deneb, signifying tail, is the tail of the swan; it lies at the head of the cross. Albireo, a beautiful double star of contrasting orange and blue colors, is the head of the swan. The whole constellation lies in, and parallel to, the path of the Milky Way. It contains several splendid star fields. *See* CONSTELLATION; CRUX.

[CHING-SUNG YU]

Day

A unit of time equal to the period of rotation of Earth. Different sorts of day are distinguished, according to how the period of rotation is reckoned with respect to one or another direction in space.

Solar day. The apparent solar day is the interval between any two successive meridian transits of the Sun. It varies through the year, reaching about 24 hr 30 sec of ordinary clock time in December and about 23 hr 59 min 39 sec in September.

The mean solar day is the interval between any two successive meridian transits of an imagined point in the sky that moves along the celestial equator with a uniform motion equal to the average rate of motion of the Sun along the ecliptic. Ordinary clocks are regulated to advance 24 hr during a mean solar day.

Sidereal day. The sidereal day is the interval between any two successive meridian transits of the vernal equinox. Similarly, as for the solar day, a distinction is made between the apparent sidereal day and the mean sidereal day which, however, differ at most by a small fraction of a second. A mean sidereal day comprises 23 hr 56 min 4.09054 sec of a mean solar day.

The period of rotation of Earth with respect to a fixed direction in space is 0.0084 sec longer than a sidereal day. No special name has been given to this kind of day, and although of theoretical interest, it is not used in practice.

Variations in duration. The mean solar day, the sidereal day, and the day mentioned in the preceding paragraph all vary together in consequence of variations in the speed of rotation of Earth, which are of three sorts: seasonal, irregular, and secular. The seasonal variations are probably caused, at least in part, by the action of winds and tides; the effect is to make the day about 0.001 sec longer in March than in July, and is nearly repetitive from year to year. The irregular variations are probably the result of interactions between motions in the core of Earth and the outer layers; the effect is to cause more or less abrupt changes of several thousandths of a second in the length of the day, which persist for some years. The secular variation is the result of tidal friction, mainly in shallow seas, which causes the duration of the day to increase about 0.001 sec in a century. *See* EARTH ROTATION AND ORBITAL MOTION; TIME.

[GERALD M. CLEMENCE]

Bibliography: H. Jeffreys, *The Earth: Its Origin, History and Physical Constitution*, 6th ed., 1976; D. D. McCarthy and J. D. H. Pilkington (eds.), *Time and the Earth's Rotation*, 1979.

DOPPLER EFFECT

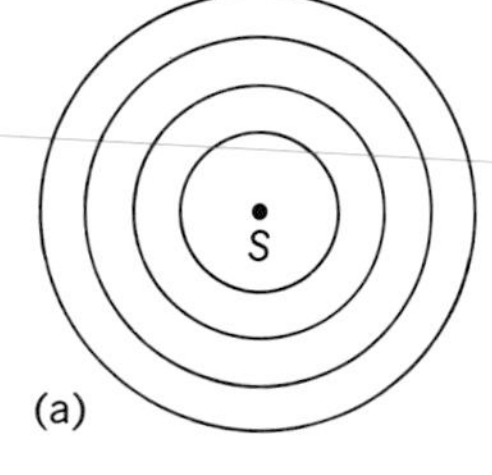

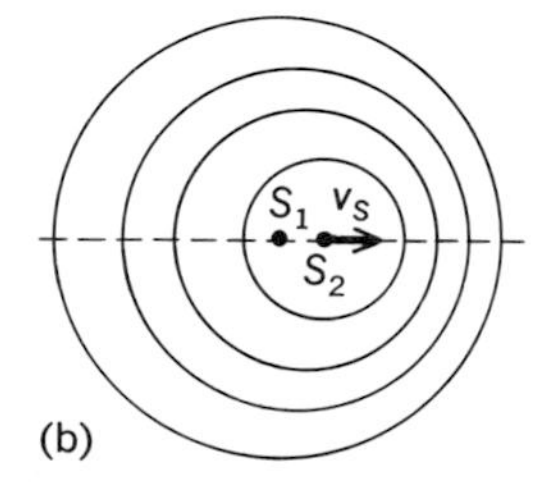

Fig. 1. Spherical waves from a point source. (*a*) At rest; (*b*) in motion.

Doppler effect

A change in the observed frequency of sound, light, or other waves, caused by motion of the source or of the observer. A familiar example for sound waves is the increase (decrease) in pitch of a train whistle as the train approaches (passes). The optical phenomenon is shown in the altered frequencies of spectral lines in the light emitted from a moving star. If the star and the Earth are moving closer to one another, more light pulses are received in a given time interval, and the color emitted from the star appears to be shifted toward the violet end of the spectrum. When the distance between the Earth and the star is increasing, the observed light is shifted toward the red end of the spectrum. The color shifts of remote galaxies are taken as evidence that the universe is expanding. *See* RED SHIFT.

In astronomy, color differences between the approaching sides and receding sides have been used to compute the rotation of the Sun and planets.

Acoustical Doppler effect. Acoustical observations of a moving source emitting sound at a constant frequency make its pitch appear greater when the source is approaching the listener, and smaller when the source-to-listener distance is increasing. The effect is based on the fact that the listener perceives as frequency the number of sound waves arriving per second.

The acoustical Doppler effect deals with cases of relative motion between the listener and the source, and includes the effects of motion of the medium itself relative to both the source and the listener. The wave velocity u of the sound in the medium is a property of the medium and its value is referred to that medium. The wavelength λ, frequency f, and velocity u are related in wave propagation by the equation $u = f\lambda$.

A distinction needs to be drawn between the case in which the source moves relative to the listener fixed in the medium, and the case in which the listener moves with respect to the source fixed in the medium.

In the first case, if the source moves toward the fixed observer with a velocity v_S, waves emitted with a frequency f_S appear to have their wavelength shortened in the ratio $(u - v_S)/u$, because of a crowding of the waves (Fig. 1) which, however, still arrive at the listener with a velocity u.

In the second case, if the listener moves toward the fixed source, the waves appear to arrive with a velocity $(u + v_L)$. The wavelength of the sound in the medium is unchanged in this case. In addition, the wavelength is equal to that measured when both the listener and the source are fixed in the medium.

Consider now the effect of the velocity of the medium relative to the listener and the source. If v_M is the component of this velocity taken positive in the direction from the listener to the source, and if v_L and v_S are the velocity components along the line joining the listener to the source and are now taken to be positive in the direction from the listener to the source, then the general equation relating the observed frequency f_L and the source frequency f_S is given by Eq. (1).

$$\frac{f_L}{u+v_L-v_M}=\frac{f_S}{u+v_S-v_M} \qquad (1)$$

Optical Doppler effect. This phenomenon seems at first to be analogous to the acoustical Doppler effect, but the causes, detailed effects, and explanation of the optical Doppler effect are fundamentally different and result from the relativistic behavior of light. *See* RELATIVITY.

Differences between the two effects. Three fundamental differences exist between the acoustical and the optical Doppler effects.

1. The optical frequency change does not depend upon whether it is the source or the observer that is moving with respect to the other, whereas the acoustical frequency change is different in the two cases.

2. No effect is observable in the acoustical case when the source, or the observer, moves at right angles to the line connecting the source and the observer. An optical frequency change is observable under such conditions.

3. The motion of the medium through which the waves are propagated does not affect the observed optical frequency, whereas it does affect the observed acoustical frequency.

Light source in motion. The mathematical expressions of the observable effects involving light and other electromagnetic waves are arrived at by noting that the propagation of a given plane wave must be described by the same law in the source frame and the observer frame, according to the relativistic principle of equivalence. Accordingly, the equation of propagation of the plane wave written for the source frame of coordinates is transformed to the observer frame of coordinates with the help of the well-known Lorentz transformations, and the relevant factors on the two sides of the resulting equation, identifying the descriptions in the two frames, are identified. The result is expressed in Eqs. (2) and (3). These equations relate the frequency f_O and angle θ_O measured in the observer frame to the frequency f_S and angle θ_S that would be measured in the source frame, under the conditions in which the source frame is measured (in the observer frame) to move with a velocity v relative to the observer frame; c is the velocity of light in free space.

$$f_O=\frac{f_S\sqrt{1-(v^2/c^2)}}{1-(v/c)\cos\theta_O} \qquad (2)$$

$$\cos\theta_O=\frac{\cos\theta_S+(v/c)}{1+(v/c)\cos\theta_S} \qquad (3)$$

Examination of the frequency relation shows that it incorporates two factors: a purely relativistic direction-independent factor, $f_O \sim f_S \cdot \sqrt{1-(v^2/c^2)}$, according to which the observed frequency will be smaller than the source frequency regardless of the apparent direction of motion of the source (transverse Doppler effect); and a direction-dependent factor, $f_O \sim f_S/[1-(v/c)\cos\theta_O]$, showing a further dependence on the direction of relative motion. Like the acoustical Doppler effect, this factor is understandable on the basis of classical arguments.

The part involving the direction of relative velocity (radial Doppler effect) can be derived by counting as the observed frequency f_O the number of waves arriving in a time interval dt_O, corresponding to the difference in the times of arrival of a first wave and of a last wave traveling with a velocity c toward the observer. The waves are emitted at a frequency f_S by a source traveling with a velocity v at an angle θ_O with respect to the observer. During a given time interval dt_S the source emits a total of $f_S\, dt_S$ waves. The relativistic velocity-dependent part is then included by noting that the source frequency will appear to be $f_S\sqrt{1-(v^2/c^2)}$ according to the special theory of relativity.

H. E. Ives and G. R. Stilwell (1938), skeptical as to the conclusions of the special theory of relativity, set out to verify the velocity-dependent part of the frequency shift (transverse Doppler effect) observed at zero angle ($\theta_O=0$). They measured the wavelengths of the H_β line in the direction of motion ($\theta_O=90°$) of hydrogen canal rays at 18,000 volts (Fig. 2) and in the opposite direction (θ_O =

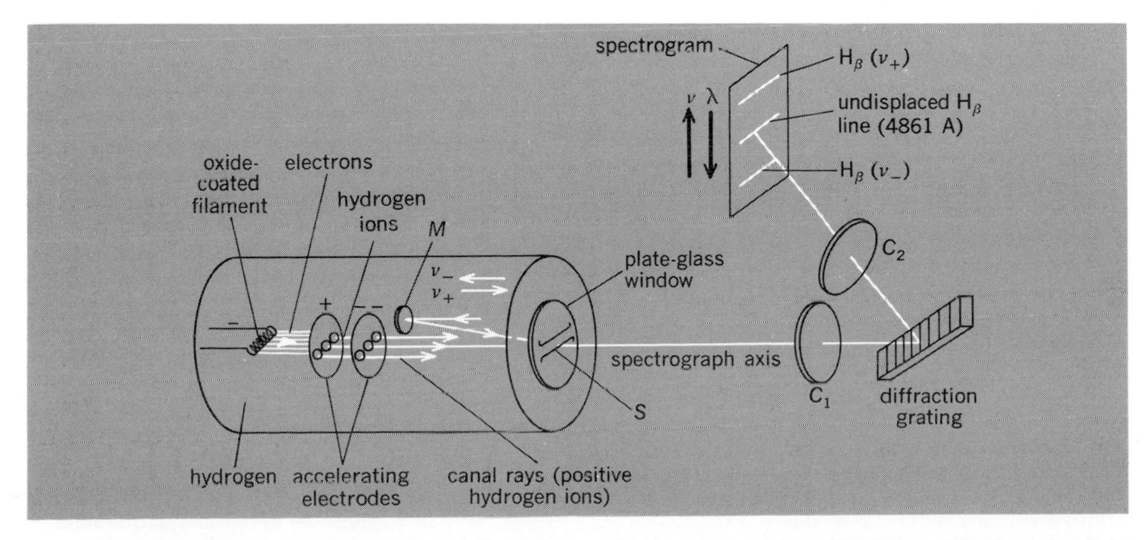

Fig. 2. Ives and Stilwell experiment. C_1 and C_2, collimating and camera lenses; *S*, spectrograph entrance slit; *M*, concave mirror focused on slit, observing light from hydrogen ions moving from slit. (*G. W. Stroke*)

−90°) for which the frequencies are respectively $f_{O1}=f_S\sqrt{1-\beta^2}/1-\beta)$ and $f_{O2}=f_S\sqrt{1-\beta^2}/(1+\beta)$, where $\beta=v/c$. They determined the average $f_O=(f_{O1}+f_{O2})/2=f_S/(\sqrt{1-\beta^2})$. This result was found to be in accord with the theoretical value $f_O=f_S/\sqrt{1-\beta^2}$, thus providing a direct proof of the "dilatation of time," according to which the observer thinks that the source period T_S is $T_O/\sqrt{1-\beta^2}$ and therefore greater than the observer period T_O.

[GEORGE W. STROKE]

Bibliography: A. Einstein et al., *The Principle of Relativity*, 1923; T. P. Gill, *The Doppler Effect*, 1965; W. C. Michels, Phase shifts and the Doppler effect, *Amer. J. Phys.*, 24(2):51–53, 1956; R. Resnick and D. Halliday, *Physics*, 3d ed., 1978; A. J. W. Sommerfeld, *Lectures on Theoretical Physics*, vol. 4, 1954; F. W. Sears et al., *University Physics*, 5th ed., 1976.

Dwarf star

The most common type of star in the Galaxy, also known as a main-sequence star. The Sun is a typical dwarf, with surface temperature of 5750 K, radius of 6.96×10^5 km, mass of 2×10^{30} kg, and luminosity of 4×10^{33} ergs/s (that is, 4×10^{23} kW). These numbers are fairly typical as to radius and mass. However, in luminosity, main-sequence stars occur over a range from O and B stars, up to more than 10^4 times brighter than the Sun, with surface temperatures of 35,000–15,000 K, down to the faintest M dwarfs, 10^{-4} as bright as the Sun, with a temperature of only 2500 K. Dwarfs form a single-parameter family in which the significant variable is mass; members range in mass from 20 Suns down to less than 0.1 Sun. Significant variations in composition occur, distinguishing the oldest dwarfs from those more recently formed. Thus, abundance ratios of metals to hydrogen are lower by factors of up to 1000 in some of the fastest-moving stars of the halo population. But all main-sequence stars have the same helium-to-hydrogen ratio, reflecting the primordial helium production in the big bang. The M or red dwarfs are the most common stars in space, and because of their low luminosity have the longest life. Almost all other types of stars have evolved from main-sequence stars. *See* STAR.

[JESSE L. GREENSTEIN]

Dynamical time

A form of time which appears in the equations of motion and in ephemerides of bodies moving in accordance with Newton's laws of dynamics and gravitation, as modified by relativity. The Earth's orbital motion about the Sun, for example, is a function of dynamical time, but not its rotational motion, which is affected by nongravitational forces, including tidal friction.

Current observational accuracies require several forms of dynamical time because of relativistic differences for coordinate systems in relative motion and for different gravitational potentials. The equations of motion of a deep-space body, for example, are best referred to a heliocentric or a barycentric system, while coordinates and clock times are measured in a geocentric system.

Secular and irregular variations in rotational time, for example, mean solar, with respect to dynamical time were long suspected but not definitely confirmed until 1939. By 1955 quartz crystal clocks had confirmed periodic, seasonal variations. A time scale free of the above variations was needed.

Astronomers had used dynamical time under names such as newtonian and gravitational. In 1958 the International Astronomical Union (IAU) adopted a specific definition based on the orbital motion of the Earth given by Simon Newcomb's *Tables of the Sun* of 1895 and designated it ephemeris time (ET). An observed position of the Sun relative to the stars (opposite to the Earth's direction from the Sun) was to be entered in the Sun's ephemeris and ET taken out. The time in the *Tables* had been called mean solar, but it was now recognized as being ET. In practice, short intervals of ET were obtained more accurately by observing the Moon and using the lunar ephemeris. The Moon is not a satisfactory source of ET over centuries, however, because an uncertain tidal couple affects its motion. *See* CELESTIAL MECHANICS; EPHEMERIS; ORBITAL MOTION.

From a joint experiment of 1955–1958, the frequency of the cesium-beam atomic clock constructed in 1955 by the National Physical Laboratory in Britain was found to be 9,192,631,770 cycles per second of ET as determined by photographic observations of the Moon and stars made at the U.S. Naval Observatory, Washington, DC. In 1967 the General Conference of Weights and Measures defined the International System (SI) unit of time, the second, as the duration of this number of periods of cesium radiation. The roles of dynamical and atomic times were thereby interchanged. The second of SI became the standard. *See* ATOMIC TIME.

In 1976 the IAU defined timelike arguments for use in dynamical astronomy, and in 1979 designated them as follows: Terrestrial Dynamical Time (TDT) is the scale for apparent geocentric ephemerides, and Barycentric Dynamical Time (abbreviated TDB in French) is the scale for equations of motion referred to the barycenter (center of mass) of the solar system. TDT and TDB will supersede ET in national ephemerides as of January 1, 1984.

TDT is defined as TAI + 32.184 s for 1977.0, where TAI is International Atomic Time. TAI, and therefore TDT, can be obtained immediately with very high precision from clocks and radio signals. The constant makes TDT continuous with ET at 1977.0. TDT is thus a highly precise atomic time scale since 1955, derived from and used in dynamical astronomy. For earlier dates, dynamical scales of lower precision can be formed, continuous with TDT in time and rate at 1955, but dependent upon the combination of solar, planetary, and lunar observations used.

TDB has various forms, which depend upon the particular theory of relativity used. However, TDT and any TDB differ only by periodic terms. The largest is an annual term with coefficient 1.6 ms, due to the annual change in potential with respect to the Sun, caused by the Earth's eccentric orbit.

A question of much interest is whether dynamical and atomic times are physically equivalent. If observed solar and planetary longitudes since 1955, referred to TDT, show only accordant deviations linear with time, then they are; the equations of motion would be changed, but not the definition of TDT. However, if there are other deviations, for example, quadratic, then the times are not equivalent. *See* EARTH ROTATION AND ORBITAL MOTION; TIME. [WILLIAM MARKOWITZ]

Bibliography: *Explanatory Supplement to the Astronomical Ephemeris and the American Ephemeris and Nautical Almanac*, H. M. Stationery Office, 1961; E. W. Woolard and G. M. Clemence, *Spherical Astronomy*, 1966.

Earth

The third planet, of nine, from the Sun. The Sun is an average star about two-thirds of the way out from the center of the Milky Way Galaxy, a typical spiral galaxy. The Earth is unique, so far as is known, in having life, although statistics strongly suggest that many similar planets exist, and some probably also have life. The earth is orbited by the Moon and many artificial satellites.

The Earth's nearest neighbors in space, other than the Moon, are the planets Venus, which is about 108×10^6 km (67×10^6 mi) from the Sun, and Mars, 227×10^6 km (141×10^6 mi) from the Sun. Earth is 150×10^6 km (93×10^6 mi) from the Sun. Certain asteroids move in eccentric orbits around the Sun and occasionally come closer to Earth.

Motions. There are two types of Earth motions, orbital and rotational.

Orbit. The Earth's orbit around the Sun is an ellipse, with the Sun at one focus. Perihelion, or the closest approach to the Sun, occurs about January 3, and aphelion, farthest from the Sun, about July 4. Because of the Earth's orbital precession, these dates will be reversed in about 10,500 years. At perihelion the Earth is about 147×10^6 km (91.5×10^6 mi) from the Sun, and at aphelion 152×10^6 km (94.5×10^6 mi), giving an average distance of 150×10^6 km (93×10^6 mi). The Earth's velocity varies systematically, being greatest at perihelion and least at aphelion. The average velocity is 107,000 km/h (66,600 mi/h) or 29.6 km/s (18.5 mi/s). This changing velocity affects measurements of solar time. *See* CELESTIAL MECHANICS.

The Earth's orbital period is the year, whose length is determined by the average distance between Earth and Sun. The calendar year or tropical year is defined as the time between two successive crossings of the celestial equator by the Sun at the time of the vernal equinox. It is exactly 365 days 5 h 48 min 46 s. The fact that the year is not a whole number of days has affected the development of calendars. *See* CALENDAR.

As the Earth moves in its orbit around the Sun, its north spin axis, or geographic pole, points in the direction of the star Polaris, making it the North Star or polestar. One obvious result is that different parts of the Earth receive differing amounts of sunlight; this is the primary cause of seasons.

The Earth's axis has not always pointed in the direction of Polaris. Because of the gravitational attraction of the Moon and Sun on the Earth's equatorial bulge, the direction in which the Earth's axis points moves very slowly. This motion is called precession, and a complete cycle requires about 25,800 years. Superimposed on the precession is a very much smaller fluctuation, called nutation, that results from the movement of the Sun twice a year and the Moon twice a month across the celestial equator. *See* PRECESSION OF EQUINOXES.

Rotation. The Earth's period of rotation is the day, and the day is used to define the second, which is the basic unit of time. The length of the solar day varies because the Earth's orbital velocity changes predictably with distance from the Sun, and less regularly owing to tidal friction, changes in the Earth's core, and seasonal atmospheric circulation. For this reason, the second was defined as 1/86,400 of a mean solar day averaged over a year. To avoid the obvious problems in determining the length of the mean solar day, the second is now defined atomically. *See* DAY; TIME.

The rotation speed is maximum in late July and early August, and minimum in April; the difference in the length of the day is about 0.0012 s. Since about 1900 the Earth's rotation has been slowing at a rate of about 1.7 s per year. In the geologic past the Earth's rotational period was much faster. Daily, monthly, tidal, and annual growth rings on fossil marine organisms reveal that about 350,000,000 years ago (Middle Devonian Period) the year had 400–410 days, and 280,000,000 years ago (Pennsylvanian Period) the year had 390 days. These observations are very close to the calculated values of the effect of tidal friction. The slowing of the Earth's rotation is caused by tidal friction between the sea floor and the ocean water. The Moon is the main cause of tides, and the total rotational energy in the Earth-Moon system is conserved, so that the energy lost by the Earth is gained by the Moon. This causes the Moon to move farther from Earth, and this in turn lengthens the period of the Moon's revolution. *See* EARTH ROTATION AND ORBITAL MOTION; TIDE.

Satellites. The Moon is Earth's only natural satellite, although a number of artificial satellites have been put into orbit about the Earth. The Moon's mass is 1/81.3 of Earth's, and its average distance from Earth is 383,403 km (238,247 mi). The center of mass of the Earth-Moon system is within the Earth, about 4645 km (2886 mi) from the Earth's center, and it is about this point that the Earth-Moon system revolves.

The Moon's period of revolution and rotation is 27 days 7 hr 43 min 11.5 s. The orbital plane of the Earth-Moon system is inclined to the orbital plane of the Earth-Sun system at an angle of 5°8′33″. At times the Sun, Earth, and Moon are in a line, and an eclipse of the Sun or the Moon occurs. *See* ECLIPSE; MOON.

Size, shape, mass, and density. The Earth is not quite a sphere, but has an equatorial bulge caused by its rotation. Its shape therefore is an oblate spheroid. The equatorial radius (semimajor axis) is 6,378.160 km, and the polar radius (semiminor axis) is 6,356.775 km.

The Earth's mass is 5.975×10^{27} g, giving it an average density of 5.53 g/cm^3. This figure is about twice the density of the common rocks that form

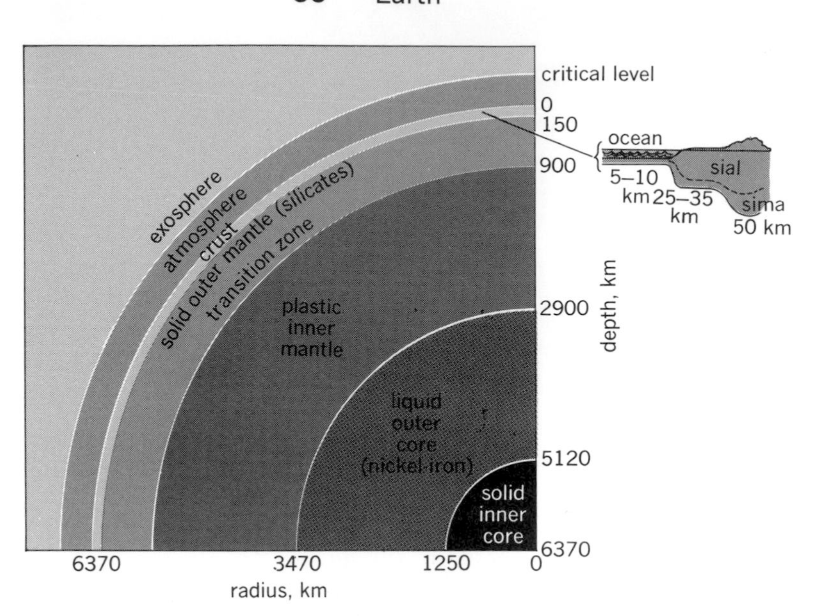

Fig. 1. The principal layers of the Earth.

the Earth's surface and strongly suggests that the interior is denser than the surface. Seismic studies have confirmed that the Earth's interior has a layered structure (Fig. 1).

Internal structure. The deepest layer is the core, which is divided into a solid inner core and a liquid outer core. The core is believed to be composed of nickel-iron and probably lighter elements such as sulfur. Electric currents moving in this liquid conducting core are believed to be the origin of Earth's magnetic field. The layer between the crust, or outermost rock layer, and the core is called the mantle. The mantle is believed to be composed of silicate minerals, probably the rocks dunite or eclogite. The composition of the core and mantle are obtained from analogy to meteorites, which are believed to be similar in structure and composition to Earth, and from theoretical calculations. Near the top of the mantle, the temperature and pressure are such that the rocks are partially melted, giving the low-seismic-velocity zone. The horizon that separates the mantle from the crust is the Mohorovičić seismic discontinuity, a narrow zone of pronounced change in the velocity of seismic waves believed to be caused by a change in rock type.

The crust is the thinnest of the rock layers, and is composed of 30–60 km of granitic rocks under the continents and 5 km of basaltic rocks under the oceans. The continents are much older than the oceans, and deformed rocks as old as 3,900,000,000 years indicate an eventful history. No ocean-floor rocks older than about 180,000,000 years have yet been found.

Crustal structures and surface features. The diverse elevations and shapes of the Earth's surface, as well as the deformed and metamorphosed underlying rocks of various ages that are exposed, record a complex history (Fig. 2). On the continents, mountain belts are the most dramatic features. They range in elevation from Mount Everest, 8848 m (29,030 ft), in the Himalaya Mountains to older, rounded, deeply eroded ranges that barely rise above the surrounding plains. Granitic and metamorphic rocks are generally exposed in the cores of mountain ranges. The overlying rocks are generally sedimentary rocks, mainly of shallow marine origin, that have been deformed. The deformation is the result of compression that causes folding and faulting, and may be accompanied by intrusion and metamorphism. Mountains generally are formed by several phases of deformation spread over several tens of millions of years. The regions deformed in this manner are generally places where shallow-water marine sedimentary rocks accumulated in great thicknesses. Some topographic mountains are formed by volcanoes, and some are the result of erosion of uplifted undeformed rocks, but most mountain ranges are formed of deformed rocks.

Much of the surface of the continents is covered by a thin veneer of sedimentary rocks. Where the underlying rocks of the plains and hills that make up most of the continents are exposed, the rocks and their structures are similar to those found in most mountain ranges. This leads to the theory that continents are formed by the deeply eroded remnants of earlier mountain ranges.

The topographic features underlying the oceans are similarly diverse and reveal more evidence of a dynamic Earth. The continental shelf, an area covered by shallow water, generally less than 150 m (500 ft) deep, surrounds the continents at most places. Such areas are generally underlain by continental, granitic rocks, and are submerged parts of the continents.

Continental slopes are the transition between the continental shelf and the ocean floors. Their tops are generally less than 150 m below sea level, and they slope down to about 4400 m. They are narrow, steep features, with slopes generally between 2 and 6°, but some are up to 45°. They are generally underlain by thick accumulations of sedimentary rocks.

Submarine trenches and their associated volcanic island arcs are another, very different type of border between continent and ocean. The lowest elevations on Earth's surface are found in the

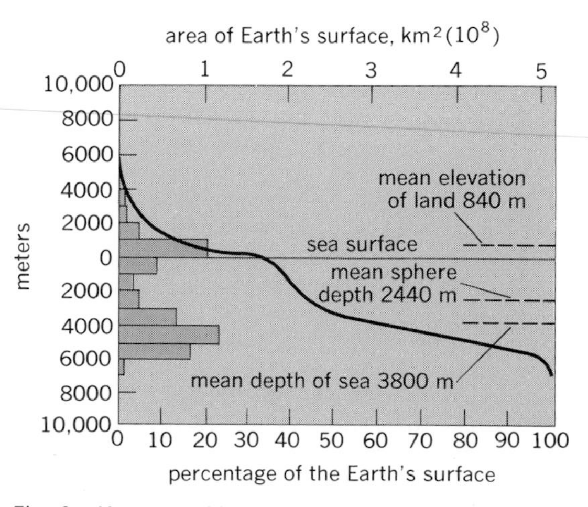

Fig. 2. Hypsographic curve of land surface elevation. (*After E. Kossinna, Die Tiefen des Weltmeeres, Veröff. Inst. Meereskunde Univ. Berlin, Neue Folge A, Geogr.-naturwiss., 9:1–70, 1921*)

submarine trenches. The deepest place on Earth is the Marianas Trench, 11,022 m (36,152 ft). The continental side of such trenches is generally an area of active volcanoes that protrude above the ocean, forming islands. In some cases a shallow sea exists between the volcanic islands and the continent. Much of the western border of the Pacific Ocean is of this type.

The ocean floors are the most widespread surface feature of Earth. Beneath an average of 4.4 km (2.75 mi) of sea water are about 2.3 km (1.4 mi) of sedimentary rocks with some intercalated basalt, and below that is the oceanic crust, consisting of 4.8 km (3 mi) of basaltic rocks. Interrupting the ocean floor at many places are submarine mountains formed by basalt volcanoes. Some of these volcanoes are very large and form oceanic islands such as the Hawaiian Islands. At other places vertical movements of the ocean floor are revealed by sunken islands whose flat tops were formed by wave erosion (guyots) and coral atolls.

The ocean floors rise gradually to the mid-ocean ridges, a more or less continuous feature through all the oceans with some branches and offsets. The ridges range between 480 and 4800 km (300 to 3000 mi) wide and are much more rugged and irregular than the ocean floors. They rise about 3000 m (10,000 ft) above their base on average. Lines of parallel volcanoes and steep scarps mark the mid-ocean ridges. The central parts of the generally more or less symmetrical ridges are the most active volcanically and seismically.

Surface processes. The source of the energy that deforms the surface rocks is probably crystallization of the liquid core and the decay of radioisotopes within the whole Earth. The energy released at the bottom of the mantle may cause convection in the mantle, which in turn deforms the crust, but the actual processes are probably much more complex.

It is clear that at places under the oceans new oceanic (basaltic) crust is created at the mid-ocean ridges, and this newly formed thin crustal plate moves away from the ridges (Fig. 3). The crustal plates formed in this way may carry continents on them, and are believed to be the mechanism of continental drift. Paleomagnetic data from the continents indicate that the continents have moved relative to each other. The crustal plates are consumed at the trench–volcanic island arc areas. As well as the ridge and the trench, a third type of plate boundary is where two plates pass each other at a transform fault. Moving plates may collide in several ways (Fig. 4). Such collisions may account for the deformed rocks found in the crust.

The evidence for continental drift in the geologic past includes matching of rock types, ages, fossils, climates, and structures (mountain ranges), as well as the paleomagnetic data. Evidence showing or suggesting present movements consists of shallow earthquakes along mid-ocean ridges and the transform faults that offset them; deep earthquakes associated with deep-sea trench–volcanic island arc areas; direct measurement of movement; volcanic activity at mid-ocean ridges; and volcanic activity at trench-island arc areas.

Atmosphere. The Earth's temperature and gravitation are such that an atmosphere is present. The major constituents are nitrogen and oxygen. The atmosphere, especially oxygen, and the presence of water, both at the surface and in the atmosphere, make life possible. Precipitation, mainly rain, results in running water such as streams and rivers on the continents. Running water is the main cause of erosion of the continents, and most of the landscapes are eroded by water, although some are eroded by wind or ice (glaciers).

The atmosphere shields the Earth from most meteorites. Meteors are caused by frictional heat when such high-velocity objects move through the atmosphere. A few of these objects pass through the atmosphere and strike the Earth, causing craters. Other bodies, such as the Moon, are not protected by an atmosphere and so are struck by more meteorites. The craters formed on the Moon are

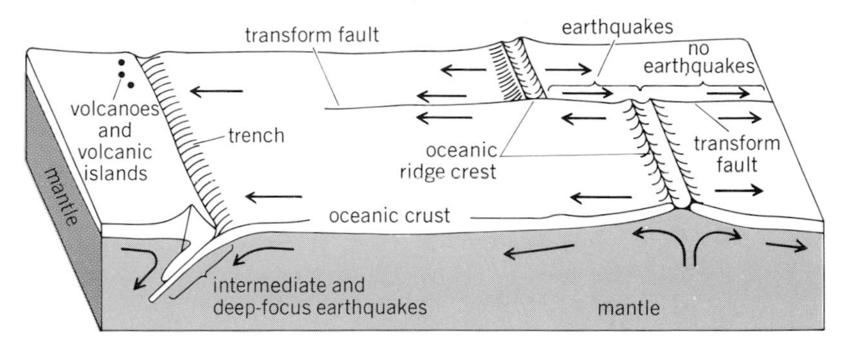

Fig. 3. Sea-floor spreading. *(From G. Gross, Oceanography, 3d ed., Charles E. Merrill, 1976)*

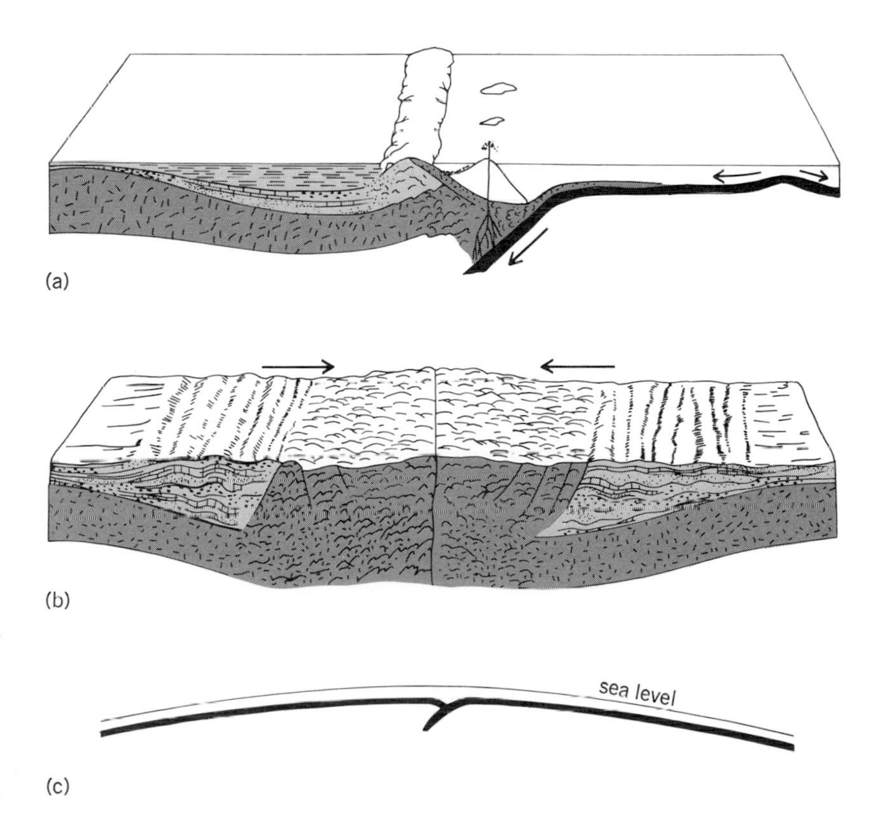

Fig. 4. Plate collisions: (*a*) continent-ocean plate collision; (*b*) continent-continent plate collision; (c) ocean-ocean plate collision. *(From R. J. Foster, Physical Geology, 3d ed., Charles E. Merrill, 1979)*

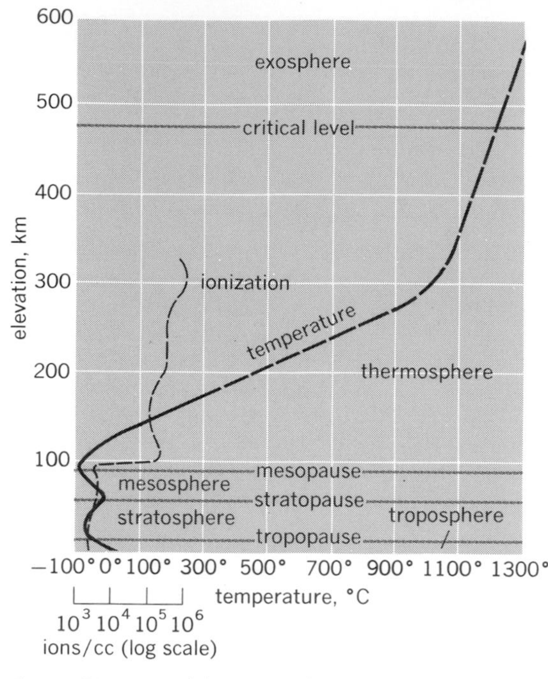

Fig. 5. Structure of the atmosphere.

preserved, because without an atmosphere almost no erosion occurs. *See* METEORITE.

A thin ozone layer in the atmosphere also shields the Earth from lethal ultraviolet radiation from the Sun. The structure and composition of the atmosphere is shown in Fig. 5.

Age. The Earth with the rest of the solar system is believed to have formed about 4,500,000,000 years ago. This age is determined by dating radioactive isotopes in meteorites. Meteorites are believed to be fragments produced by collisions among small bodies formed by the same process that created the solar system. Theoretical studies of the Sun and other studies of radioactive isotopes also suggest a similar age. *See* COSMOLOGY; PLANET; SOLAR SYSTEM.

[ROBERT J. FOSTER]

Bibliography: A. Holmes, *Holmes' Principles of Physical Geology*, 3d ed. rev. by D. L. Holmes, 1978; A. N. Strahler, *The Earth Sciences*, 2d ed., 1971.

Earth rotation and orbital motion

The rotation of the Earth about its axis is demonstrated by the classical Foucault pendulum experiment. Its revolution in its orbit around the Sun is shown by the annual parallactic displacement of relatively nearby stars against the background of more distant stars. However, because the Earth is not truly a rigid symmetric body and because it interacts with other members of the solar system gravitationally, these motions vary with time. *See* FOUCAULT PENDULUM; PARALLAX.

ROTATION OF THE EARTH

Until recent times the rotation of the Earth has served as the basis for timekeeping. The assumption was made that the rotational speed of the Earth was essentially constant and repeatable, and that the length of the day which resulted from this constant rotational speed was naturally useful as a measure of the passage of time. Astronomical observations, however, have shown that the speed with which the Earth is rotating is not constant with time. It appears that the variations in rotational speed may be classified into three types: secular, irregular, and periodic. The secular variation of the rotational speed refers to the apparently linear increase in the length of the day due chiefly to tidal friction. This effect causes a slowing of the Earth's rotational speed resulting in a lengthening of the day by about 0ˢ.0005 to 0ˢ.0035 per century.

The irregular changes in speed appear to be the result of random accelerations, but may be correlated with physical processes occurring on or within the Earth. These irregular changes cause the length of the day to vary by as much as 0ˢ.01 over the past 200 years. Periodic variations are associated with periodically repeatable physical processes affecting the Earth. Tides raised in the solid Earth by the Moon and the Sun produce periodic variations in the length of the day of the order of 0ˢ.0005 with periods of 1 year, 1/2 year, 27.55 days, and 13.66 days. Seasonal changes in global weather patterns occurring with approximately annual and semiannual periods also cause variations in the length of the day of this order. *See* TIDE.

Knowledge of the rotational speed of the Earth is required for observers on the Earth who find it necessary to know the orientation of the Earth in an inertial reference frame. This includes navigators, astronomers, and geodesists. The rotational speed of the Earth remains essentially unpredictable in nature due to the incompletely understood irregular variations. Because of this, astronomical observations continue to be made regularly with increasing accuracy, and the resulting data are the subject of continuing research in the field.

Observations of rotational speed. Astronomical observations of stars, radio sources, the Moon, and artificial Earth satellites are used to determine a time scale which is based strictly on the rotation of the Earth within an inertial reference system defined by the positions and motions of the celestial objects. This Universal Time scale (UT1) is compared with time scales known to be more uniform in nature such as that determined from the motion of solar system objects such as the Earth about the Sun (Ephemeris Time) or by atomic clocks (International Atomic Time). Variations in the differences among these types of time scales may be used to determine variations in the rotational speed of the Earth. Astronomical observations of time are made routinely by a number of observatories located around the world for this purpose. *See* ATOMIC CLOCK; ATOMIC TIME; TIME.

The times and locations of past eclipse observations have also been analyzed to provide information on the length of the day. If the Earth has changed its speed of rotation since ancient times, the path of an eclipse which occurred 2000 years ago would be displaced in longitude with respect to the path that would have occurred if the rotational speed had remained constant. Ancient records of eclipses, while not made with great accuracy, are valuable for the fact that they were made long ago.

Comparison of very old observations of the longitude of the Sun with current theories of the motion of the Sun based on a uniform time scale has also been used to estimate the increase in the length of the day since ancient times. *See* ECLIPSE.

Careful analyses of this information reveal the three types of variations in the speed of rotation. These ancient observational data form the basis for estimates of the secular deceleration in the speed of rotation. The more recent information, having been obtained with higher accuracy and more regularity, has shown the apparently random changes in the acceleration causing the irregular variations in the length of the day. These data have also been used to detect the periodic variations in the length of the day.

Causes of variations. The conservation of angular momentum of the Earth requires that changes in the Earth's moments of inertia produce changes in the speed with which the Earth is rotating. The moments of inertia are dependent on the distribution of mass on and within the Earth. This includes the mass contained in the atmosphere and in the oceans. As mass is redistributed, the moments of inertia change, producing the subsequent changes in the rotational speed of the Earth. An example of such an effect is that of the spinning skater who is able to change rotational speed by redistributing mass through the changing of the positions of the arms.

The crust and the interior of the Earth are not strictly rigid. Tides generated by the Sun and the Moon change the shape of the Earth. The Earth is thought to have a fluid core with material possibly in motion. Coupling between such a core and the mantle would change the speed of rotation of the mantle from which the observations are made. The Earth may undergo changes in the moments of inertia as landmasses move or rise and fall. All of these motions cause changes in the Earth's moments of inertia and may cause changes in the rotational speed. Seasonal variations are caused by an exchange of momentum between winds and the crust of the Earth.

Another known cause of variations in the rotational speed is tidal friction. The Moon raises tides in the ocean. Friction carries the maximum tide ahead of the line joining the center of the Earth and Moon (Fig. 1). The resulting couple diminishes the speed of rotation of the Earth, and this reacts on the Moon to increase its orbital momentum. The sum of the angular momentum of the Earth and the orbital momentum of the Moon remains constant. This produces an increase in the size of the orbit of the Moon and a reduction of its angular speed about the Earth. Tidal friction should be distinguished from actual changes in the moments of inertia of the Earth brought about by the tides.

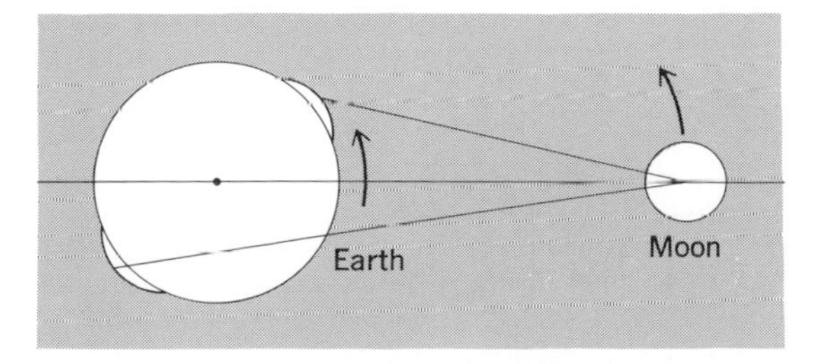

Fig. 1. Couple produced by tidal friction.

The effect of tidal friction is to increase both the distance of the Moon and the length of the lunar month measured with a uniform time scale. Because of the change in the lunar month, the Moon is observed to have an orbital deceleration in terms of a uniform time scale. The proportional change in the distance, however, is greater than in the length of the month. Hence, in terms of Universal Time, the Moon appears to have a secular orbital acceleration, an effect discovered by E. Halley in 1693 from a study of ancient eclipses. In 1853 J. C. Adams found that gravitational theory could account for only about half of the acceleration found by Halley. It has become clear since then that the rotational speed is also decreased because of tidal friction.

REVOLUTION ABOUT THE SUN

The motion of the Earth about the Sun is seen as an apparent annual motion of the Sun along the ecliptic. That the effect is caused by the motion of the Earth and not that of the Sun is proved by the annual parallactic displacement of nearby stars and by the aberration of light, causing an apparent annual displacement of all stars on the celestial sphere. *See* ABERRATION.

Orbit of the Earth. A large number of astronomical observations of the positions of the Sun and other solar system objects have been made and are being made continuously. This information is required to determine the nature of the motion of the Earth about the Sun. Observations are analyzed using the mathematical methods of celestial mechanics to provide improved estimates of the motions of the solar system objects in the future and to describe the past motions of the objects. The description of the apparent motion of the Sun in the sky provides the determination of the orbit of the Earth.

Period of revolution. The true period of the revolution of the Earth around the Sun is determined by the time interval between successive returns of the Sun to the direction of the same star. This interval is the sidereal year of 365 days 6 h 9 min 9.5 s of mean solar time or 365.25636 mean solar days. The period between successive returns to the moving vernal equinox is known as the tropical year of 365 days 5 h 48 min 46.0 s or 365.24220 days. The length of the tropical year is regarded as the length of the year in common usage for calendars. The period of time between successive passages at perihelion (the closest approach of the Earth to the Sun) is called the anomalistic year of 365 days 6 h 13 min 53.0 s or 365.25964 days. The length of each of these years depends on observationally determined astronomical quantities. Improvements in the determination of these quantities will result in slight changes in the numerical values. The lengths of the years listed above are given for the year 1900. These values vary slowly as a consequence of the long period perturbations of the Earth's orbit by other planets. *See* CALENDAR; PERTURBATION.

Mean radius of orbit. The mean distance from the Earth to the Sun, or the semimajor axis of the Earth's orbit, was the original definition of the astronomical unit (AU) of distance in the solar system. Its absolute value fixes the scale of the solar system and the whole universe in terms of terrestrial standards of length. The distance between the Earth and the Sun can be determined by a variety of methods; the results are usually expressed in terms of solar parallax. The Sun's mean equatorial horizontal parallax p is the angle subtended by the equatorial radius r of the Earth at the mean distance a of the Sun. Mathematically, $a = r/\sin p = 206\,265\, r/p$, if p is in seconds of arc. The equatorial radius of the Earth is 6 378 137 m. Geometrical, gravitational, and physical methods have been used at various times to measure the solar parallax.

Geometrical methods. The geometrical methods involve the direct measurement by optical triangulation of the parallax of a nearby planet (Mars or Venus) or other solar system object such as the minor planet Eros at its closest approach to the Earth. Because the relative distances in the solar system are accurately known in terms of the astronomical unit, the absolute measurement of one very accurate distance can be used to calibrate the scale of the system (Fig. 2).

Gravitational methods. The gravitational methods involve the determination of the ratio of the mass m of the Earth to the mass M of the Sun from the perturbations in the motion of a minor planet (such as Eros) caused by the Earth. The method also depends essentially on an accurate determination of the acceleration of gravity g and the length of the sidereal year T in seconds of mean solar time. The distance a is then found from the application of Kepler's third law, $4\pi^2 a^3/T^2 = gr^2(1 + M/m)$. Because g and T are known with high accuracy, a determination of M/m gives a directly in terms of terrestrial standards of length. Although it is customary to express this result as a parallax angle, the Earth's radius is not needed to compute a, but it is necessary to make allowances for the mass of the Moon relative to the mass of the Earth.

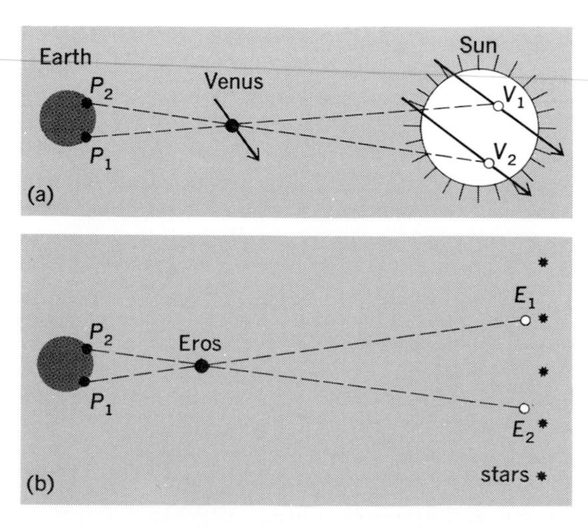

Fig. 2. Determination of solar parallax. (*a*) Observation of Venus in transit. (*b*) Observation of Eros in opposition.

Physical methods. One physical method depends on a determination of the ratio of the mean orbital velocity, $V = 2\pi a/T$, to the accurately known velocity of light. This ratio can be derived either from the annual variation of the radial velocities of ecliptic stars (or occasionally planets) determined by observations of the Doppler shift of spectral lines, or with less accuracy, from the constant of aberration. *See* DOPPLER EFFECT.

A far more precise physical method relies on measurement of the travel time of radar signals reflected from objects in the solar system. The distance of the planet is the ratio of half of the round trip travel time of the radar signal to the velocity of the radar waves. Since the distance of the planet is known precisely in astronomical units from the orbital ephemerides, measuring the actual distance in meters at some given time permits the astronomical unit to be calibrated in terms of terrestrial units of length. Also the Doppler shift of radar reflections gives the radial velocity in $m \cdot s^{-1}$, which, on comparison with the known value in $AU \cdot s^{-1}$, gives the astronomical unit in meters. The tracking of artificial space probes to solar system objects has also been used in similar ways to estimate the scale of the solar system. The currently adopted value of the astronomical unit is $149\,600 \times 10^6$ m. The value to be used in ephemerides after 1984 is $149\,597\,870 \times 10^3$ m. *See* ASTRONOMICAL UNIT.

Eccentricity of orbit. The eccentricity of the Earth's orbit can be determined by the variations of the apparent diameter of the Sun's disk. Determinations of higher accuracy are based on the variable speed of the Sun's apparent motion along the ecliptic and the laws of elliptic motion. The nonuniformity of the Sun's motion manifests itself in the equation of time, which is the difference between solar time determined from the actual observation of the Sun, and mean solar time which is based on the motion of a fictitious point having a uniform motion close to the average of the Sun's motion. The difference in the two types of time arises in part from the obliquity of the ecliptic and in part from the eccentricity of the Earth's orbit. The adopted value of the eccentricity is 0.01675. The number varies slowly with time due to perturbations from the planets. The Earth is at perihelion on January 2. It reaches its greatest distance from the Sun on July 2. *See* EQUATION OF TIME.

Seasons. The fact that the Equator of the Earth is inclined in space by about 23.5° to the orbital plane of the Earth (the ecliptic) causes the Northern Hemisphere to be exposed to the more direct rays of the Sun during part of the Earth's revolution around the Sun. The Southern Hemisphere receives the more direct rays 6 months, or a half revolution, later. This effect causes the seasons.

OTHER MOTIONS

In addition to the rotation of the Earth and its orbital motion about the Sun, the Earth experiences various small motions about its center of mass. Precession and nutation are examples, and these are caused by the gravitational attraction of the Sun and Moon on the nonspherical Earth. Because the Earth is ellipsoidal in shape, the gravitational attraction of these bodies produces a couple acting on the equatorial bulge, changing the orientation of the Earth about its center of mass. This motion can be predicted with accuracy, based on the knowledge of the shape of the Earth and of the motion of the Earth around the Sun and that of the Moon about the Earth. Precession causes the axis of angular momentum of the Earth to describe a 23.5° cone in space with a period of approximately 26,000 years. Nutation causes the axis of angular momentum to "nod" slightly in space as it executes the precessional motion. The main period of the nutational motion is 18.6 years. Other periodic motions due to the gravitational attraction of the Sun and the Moon are also included in nutation. *See* NUTATION; PRECESSION OF EQUINOXES.

Because the axis of symmetry of the Earth is not aligned precisely with the axis of rotation, the Earth also executes a motion about its center of mass known as polar motion. This motion, caused by geophysical and meteorological effects on and within the Earth, is not predictable with accuracy, and must be observed continuously to provide the most precise information on the orientation of the Earth. Polar motion is characterized mainly by an approximately 435-day and a 365-day periodic circular motion of the axis of rotation on the surface of the Earth. The radius of the circular motion is of the order of 5 m, but this may vary considerably. *See* EARTH; PLANET.

[DENNIS D. MC CARTHY]

Bibliography: D. D. McCarthy and J. D. H. Pilkington (eds.), *Time and the Earth's Rotation*, 1979; W. H. Munk and G. J. F. MacDonald, *The Rotation of the Earth*, 1975; E. W. Woolard and G. M. Clemence, *Spherical Astronomy*, 1966; S. P. Wyatt, *Principles of Astronomy*, 1971.

Eclipse

The darkening of one celestial body by the shadow of another. A solar eclipse occurs when the Moon passes directly between Earth and the Sun, hiding at least a part of the Sun from view. The Moon is eclipsed when it goes into Earth's shadow. This article deals only with these two kinds of eclipses, but there are other similar phenomena. An occultation occurs when the Moon or a planet moves in front of a star. When either of the inner planets, Mercury and Venus, passes in front of the Sun, it appears as a small black dot on the Sun's disk. If the two components of a double star revolve around each other in a plane which is nearly edgewise to Earth, they eclipse each other, and the system is called an eclipsing binary. Finally, each of the four galilean satellites of Jupiter can be eclipsed by entering the shadow of that planet. *See* ECLIPSING VARIABLE STARS; JUPITER; OCCULTATION; TRANSIT.

Seasons. The Moon's orbit around Earth is inclined 5° to the plane of Earth's orbit around the Sun. The latter is called the ecliptic, because eclipses can occur only when the Moon is on or near it. Figure 1 shows four positions of the Moon's orbit along a part of Earth's orbit. Half of the Moon's orbit is above or north of the ecliptic, while the other half is below or south of it. When Earth is at A, B, or C, the Moon's shadow at new moon is

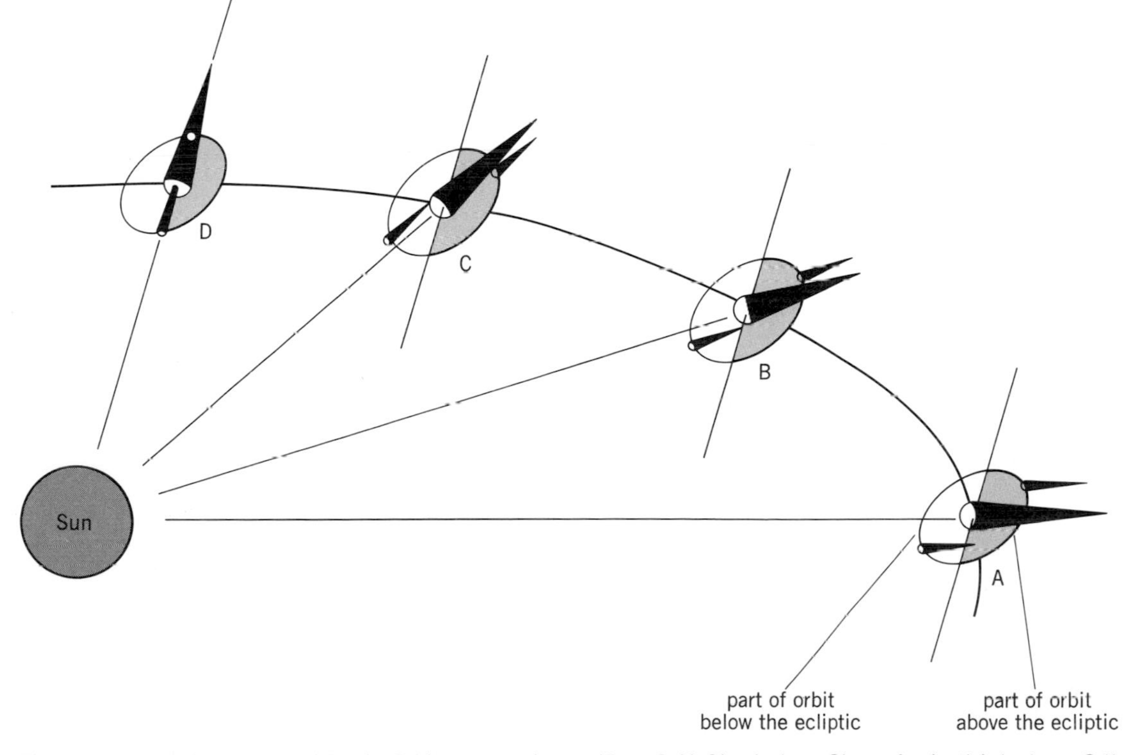

Fig. 1. Orbits of the Moon and Earth. Orbits are not in the same plane, so that eclipses can occur only at two opposite seasons of the year about 6 months apart. (*From C. H. Cleminshaw, Sizes of celestial shadows, Griffith Observer, 23(10):134–138, 1959*)

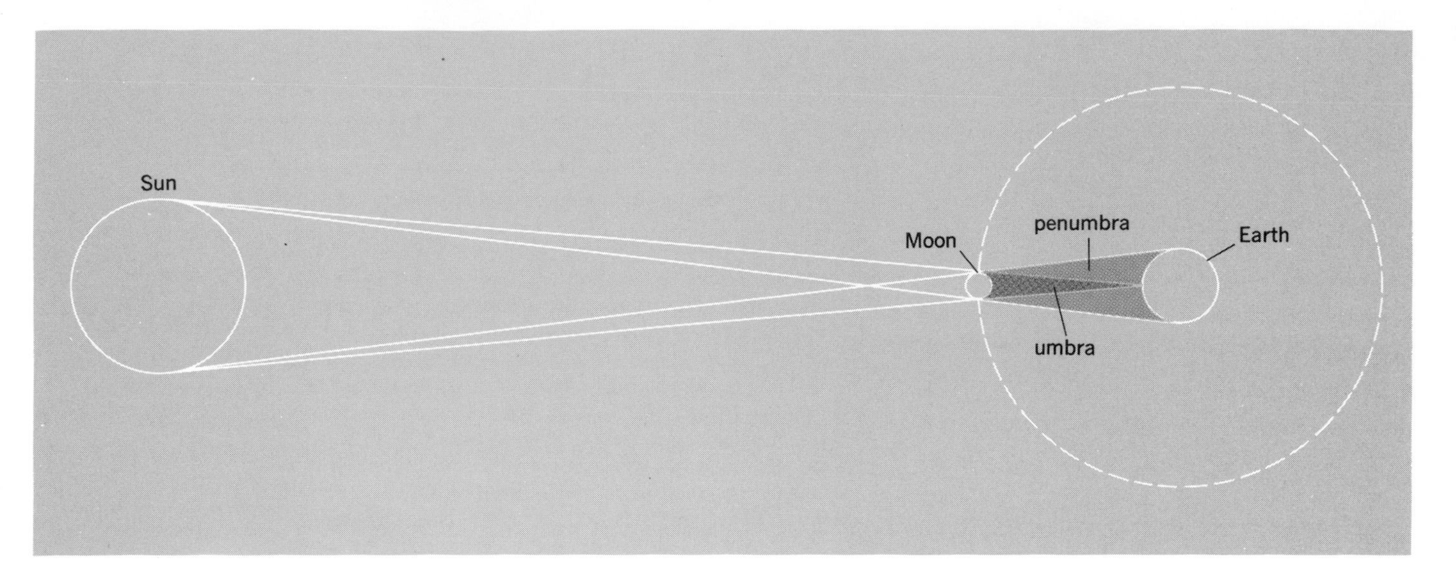

Fig. 2. Diagram of a solar eclipse.

below Earth and there can be no solar eclipse. Also, the Moon when full passes above Earth's shadow and no lunar eclipse can occur.

The two points where the Moon's orbit crosses the ecliptic are called nodes. The line connecting them and passing through Earth's center is the line of nodes. When Earth is at D, this line points to the Sun, so that the Moon's shadow strikes Earth, producing a solar eclipse. About 2 weeks before or after that event, the Moon can go through Earth's shadow, producing a lunar eclipse. This interval during which the Earth-Sun line is approximately along the line of nodes is called an eclipse season. Another one occurs about 6 months later.

The Moon's orbit is not fixed in position, being disturbed by the gravitational pull of the Sun. This causes the line of nodes to move in a direction opposite to that of Earth's orbital motion. This effect is called the regression of the nodes. It makes the eclipse seasons occur earlier each year by about 20 days, resulting in an interval of 346.62 days, which is called the eclipse year. At least one solar eclipse must occur during an eclipse season, making a minimum of two solar eclipses in a year. If one occurs early in January, a second can occur in the same month, with a lunar eclipse between them. A similar group of three eclipses may happen in the middle of the year. A fifth solar eclipse may occur in December, making a total of seven eclipses, the greatest number in one year. This number can also be reached by four eclipses of the Sun and three of the Moon. *See* MOON.

Solar eclipses. A total eclipse of the Sun is one of the greatest spectacles of nature and results from one of the most fortunate coincidences of nature. In round numbers, the Sun's diameter is 400 times that of the Moon, and the Sun's distance from Earth is also 400 times that of the Moon. Both subtend nearly the same angle of about 0.5°, so that the Moon's disk can appear just large enough to cover the Sun's disk. For no other planet is such a circumstance true. As seen from any other planet, each of its satellites appears either much smaller or much larger than the Sun.

Since the Moon and Earth are smaller than the Sun, the shadow of each is a cone having its point directed away from the Sun. All sunlight is geometrically excluded within this region, which is called the umbra. The Moon's umbra is formed by tangents to the Sun and Moon, which intersect in Fig. 2 on or near Earth's surface. Surrounding it is the larger inverted cone of the penumbra, formed by tangents intersecting between the Sun and Moon. A total eclipse can be observed within the umbra, and a partial eclipse within the penumbra.

The length of the Moon's umbra is very nearly 1/400 of its distance from the Sun. At new moon it averages about 374,000 km (232,000 mi), and it varies about 6400 km (4000 mi) either way. Since the Moon's distance ranges from about 357,000 km (221,000 mi) to 407,000 km (253,000 mi), its umbra is sometimes long enough to reach Earth, but more often falls short. In the first case an observer located on the line of centers of the Sun and Moon can see a total solar eclipse. In the second case the Moon appears smaller than the Sun and the observer sees an annulus or ring of sunlight around the Moon. Hence this is called an annular eclipse.

In the most favorable case of a total eclipse, the diameter of the circular umbra where Earth's surface cuts it squarely is 269 km (167 mi). This occurs near the Equator. The path of totality traced out by the umbra is so narrow compared with Earth's diameter that the average interval between total eclipses at any one place on Earth is 360 years.

Since the Moon moves eastward in its orbit at about 3400 km (2100 mi) per hour, its umbra sweeps eastward at the same speed. However, Earth rotates eastward at the Equator at very nearly one-half that speed. Therefore, an observer at the Equator would be passed by the umbra with a speed of about 1700 km (1050 mi) per hour. At higher latitudes, where the rotational surface velocity is less, and near sunrise and sunset, the speed of the umbra is greater. Under the most favorable conditions a total solar eclipse can last at a given point near the Equator for about 7½ min.

Principal total solar eclipses, 1980–2000

Date	Duration of totality, min	Where visible
1980, Feb. 16	4.1	Central Africa, India
1981, July 31	2.0	Central Asia
1983, June 11	5.2	Indonesia
1984, Nov. 22	2.0	New Guinea
1988, Mar. 18	3.8	Indonesia, Philippines
1990, July 22	2.5	Finland, northeastern Asia
1991, July 11	6.9	Hawaii, Central America, Brazil
1992, June 30	5.3	South Atlantic Ocean
1994, Nov. 3	4.4	South America
1995, Oct. 24	2.2	Southern Asia
1997, Mar. 9	2.8	Northeastern Asia
1998, Feb. 26	4.1	Colombia
1999, Aug. 11	2.4	Central Europe, southern Asia

The maximum possible duration of an annular eclipse is nearly 12½ min.

Saros cycle. From the beginning of the 12th century to the present time, the three total solar eclipses with the longest duration of totality occurred on June 8, 1937 (7^m04^s), June 20, 1955 (7^m08^s), and June 30, 1973 (7^m04^s). The interval, called the saros, from one of these eclipses to the next is 6585.32 days or 18 years, 11⅓ days (10⅓ days when there are 5 leap years). It consists of 223 synodic months (from new moon to new moon) and is very nearly equal to 19 eclipse years (6585.78 days). Also, 239 anomalistic months (from perigee to perigee) amount to 6585.54 days. This means that if at one eclipse the Moon is at perigee (closest to Earth and making a long eclipse), it will be almost exactly at perigee again when a saros has passed. The effect of the one-third of a day in the period is to shift the region of the following eclipse 120° west in longitude. After three periods the path of totality is nearly the same again. There is also a small shift in latitude. Thus the next eclipse in this saros cycle occurs on July 11, 1991, lasting 7 min and crossing Hawaii, Central America, and Brazil, about 10° north of the eclipse on June 8, 1937. The pairs of eclipses of 1980 and 1998 and of 1981 and 1999 listed in the table are examples of two other saros cycles.

Appearance. Since the Moon moves its own diameter in about 1 h, the complete phenomenon of a total solar eclipse lasts about 2 h: 1 h each in covering and uncovering the Sun and not more than a very few minutes for the total phase. Staring at the Sun during the partial phases with unprotected eyes can cause severe eye damage. The safest method of viewing a solar eclipse is to use pinhole projection. Make a tiny hole in a piece of cardboard and hold it so that sunlight passes through the hole and onto a second piece of cardboard, where the Sun's image appears. The same principle applies to the shadows cast by foliage. The light shining through the small openings among the leaves forms images of the Sun, which are usually circles on the ground, but become crescents during the partial phases of the eclipse.

About 15 min before totality the sky darkens noticeably and the remaining light takes on a strange quality. This is due to the fact that it comes from the edge of the Sun's disk, and hence from the upper layers of the Sun's atmosphere. Animals grow disturbed, birds go to roost, and the temperature falls. Shadow bands can be seen on white surfaces a few minutes before totality. They appear as rapidly moving dark and light waves, and are visible only when the Sun's crescent is extremely narrow. At the last moment it breaks up into "Baily's beads," which are due to sunlight shining through valleys on the edge of the Moon's disk.

With the arrival of totality, the pearly white corona appears (Fig. 3). It is the Sun's outer atmosphere, extending in all directions for millions of miles. It is bright at its inner edge and fades out in irregular streamers. Flamelike prominences are often visible to the unaided eye and usually with binoculars. They are protuberances from the red chromosphere, which is the layer of gases just above the Sun's visible surface. The brighter stars and planets are usually visible to the unaided eye. Totality ends as suddenly as it began, and the events of the partial eclipse are repeated in reverse order. Annular eclipses are more frequent than total eclipses, but they are not as spectacular. The corona and prominences are not visible, and the sky does not become dark enough for any stars or planets to appear. *See* SUN.

Lunar eclipses. Although lunar eclipses do not occur as often as solar eclipses, they are seen by many more people. They occur during full moon, which is opposite the Sun and can be observed at any one instant by all of the night half of Earth. Also, since a total lunar eclipse can last for more than an hour, it can be seen from a little more than half of Earth's surface. On the other hand, solar eclipses are visible even in the partial phases from limited areas of Earth.

Fig. 3. Solar corona of June 30, 1973, photographed from a jet aircraft flying over Africa at an altitude of 37,800 ft (11.5 km). This picture is a composite of a 0.5-s exposure with no filter and a 1-s exposure with a radial-gradient neutral-density filter. (*Los Alamos Scientific Laboratory, University of California*)

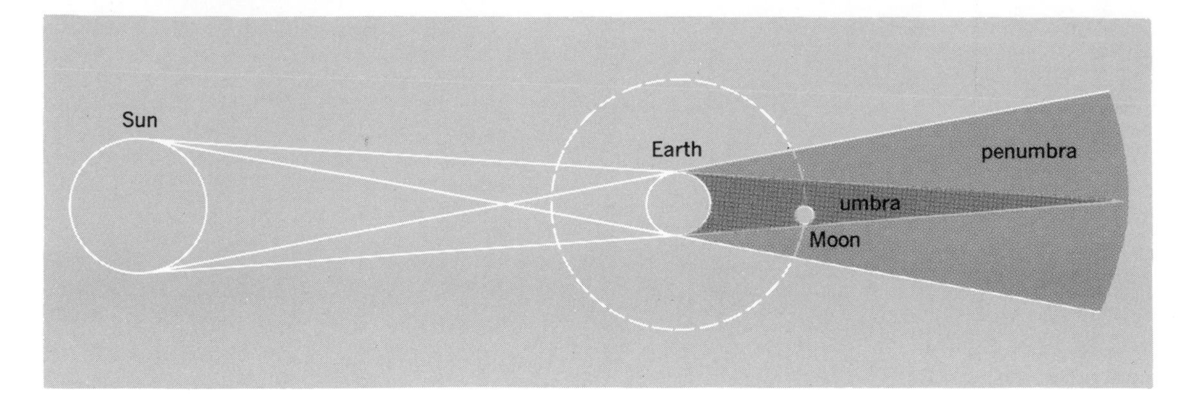

Fig. 4. Diagram of a lunar eclipse.

When the Moon passes through the center of Earth's shadow, it takes about 5 h 40 min. The first hour is spent in the penumbra (Fig. 4). No darkening is noticeable until near the end of that period, because all of the Moon's side facing Earth is still receiving some direct sunlight. Then as the Moon enters the umbra, the edge of the Moon appears nearly black by contrast with the bright parts of the lunar surface. The second hour is required for all of the Moon to get into the umbra.

The diameter of the umbra where the Moon crosses it is about 2⅔ times the Moon's diameter. Since the Moon moves approximately its own diameter every hour, it can remain totally eclipsed for 1 h 40 min. The next hour is spent in moving out of the umbra, and the final hour in passing through the penumbra.

While the Moon is going into the umbra and going out of it, the edge of the umbra is always a part of a circle. As was realized by some of the ancient Greeks, this is a proof that Earth is a sphere, since a sphere is the only object casting a circular shadow, no matter what direction the light comes from (Fig. 5).

If Earth had no atmosphere, the Moon would disappear from view while in the umbra. However, the Earth's atmosphere acts like a lens and bends the sunlight into the umbra. The longer waves of red light penetrate the atmosphere better than the shorter waves of blue light, which are scattered to form the blue of the sky. An observer on the Moon would see Earth's disk surrounded by a thin ring of bright sunset colors. This accounts for the usual reddish color of the totally eclipsed Moon. However, if there happened to be many clouds around Earth's rim, nearly all light would be cut off and the Moon would appear very dark, and on rare occasions might even disappear from view.

[CLARENCE H. CLEMINSHAW]

Bibliography: G. O. Abell, *Exploration of the Universe*, 3d ed., 1975; J. Meeus, C. C. Grosjean, and W. Vanderleen, *Canon of Solar Eclipses*, 1966; S. A. Mitchell, *Eclipses of the Sun*, 5th ed., 1951; T. Oppolzer, *Canon der Finsternisse (Canon of Eclipses)*, 1962.

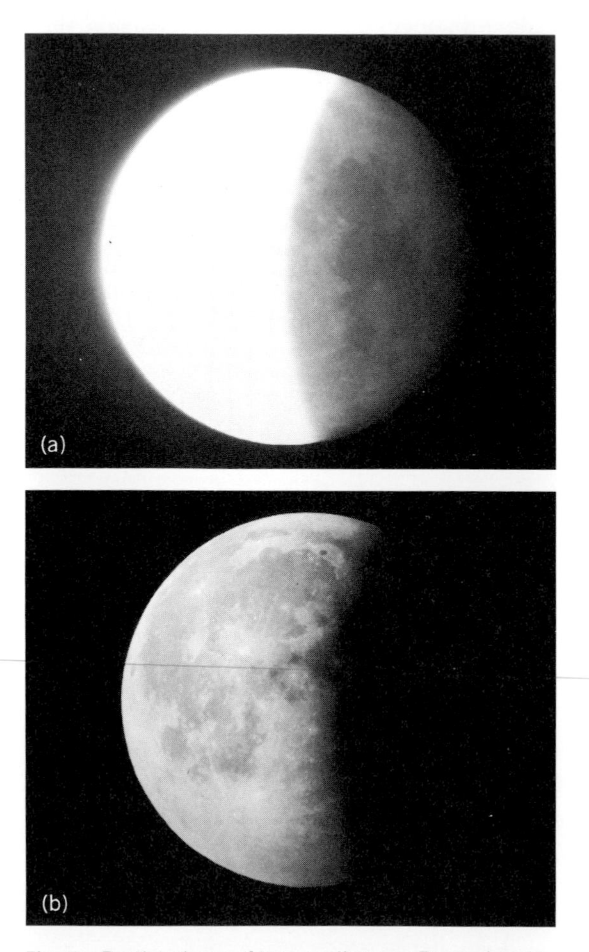

Fig. 5. Partial phase of lunar eclipse on December 18, 1945. (*a*) An exposure of 25 s was used to show the eclipsed part of the Moon, resulting in an overexposure of the part receiving direct sunlight. (*b*) An exposure of only 0.5 s reveals details of the lunar surface in direct sunlight, but is much too short to show any of the eclipsed part. The Moon was moving out of Earth's shadow, and this picture was taken 5 min after *a*. (*Paul Roques, Griffith Observatory*)

Eclipsing variable stars

Double star systems in which the two components are too close to be seen separately but which reveal their duplicity by periodic changes in brightness as each star successively passes between the other and the Earth, that is, eclipses the other. Studies of the light changes and the radial velocity changes of each component permit the computation of the radii, masses, and densities of the components—important quantities that cannot be measured directly in single stars. In addition, these close double stars are useful in studies of mass loss and of stellar evolution. Since eclipsing stars are variable in light, they are included in general variable star catalogs under the same system of nomenclature. *See* BINARY STAR; VARIABLE STAR.

Periods. The periods of light variation range from less than 3 hr for very close systems to over 27 years for the peculiar system Epsilon Aurigae. However, the majority of the periods lie between 0.5 and 10 days. In many cases the periods are not constant but change with time. In a few cases the variation is caused by a slow change in the orientation of the major axis of an elliptical orbit; in such cases the rate of change combined with other quantities gives information concerning the manner in which the density of the star increases from the outer layers to the center. In most cases, however, the changes are unpredictable and are probably connected with ejections of matter from one of the stars.

Velocity curves. The radial velocity (velocity of approach or recession) of each star can be determined at any time by the displacement of the spectral lines. A plot of velocities against time over one period of orbital revolution is known as a velocity curve. The maximum radial velocity of approach or recession depends on the true orbital velocity of the star and the "inclination," or the amount by which the plane of the orbit is tilted relative to the line of sight to the Earth. (Technically, the inclination is the angle between a perpendicular to the orbit plane and the line of sight; when the inclination is 90°, the eclipses are central.) If the inclination is known, the orbital velocity of each star can be calculated from the radial velocity. The orbital velocity multiplied by the period will give the circumference of the orbit and from this the radius of each star about the center of mass. From the size of the orbit and the period, by using the law of gravitation, the mass of each star can be calculated in terms of the Sun's mass. However, the inclination cannot be determined from the velocity curve alone; the quantities finally determined are $m_1 \sin^3 i$, $m_2 \sin^3 i$, and $a \sin i$, where i is the inclination, m_1 and m_2 the stellar masses, and a the radius of a circular orbit or half the major axis of an elliptical one.

Light curve. The light curve shows the changes in brightness of the system throughout one orbital revolution. The manner in which the light changes during the eclipse of each star by the other depends very strongly on three factors: the size of each star relative to the radius of the orbit (r_1/a and r_2/a) and the inclination. The determination of precisely what relative sizes and inclination give a computed curve which will approximate satisfactorily the observations is one of the more difficult problems of modern astronomy. However, by use of a complex set of tables computed by John E. Merrill, or another set using a different approach computed by Masatoshi Kitamura, the problem can be solved in many cases. Numerical values for these three quantities are thus found. Several alternate methods have been developed which make use of the capabilities of electronic computers. These take into account simultaneously the various interaction effects such as distortion of the shape of the components in the earlier methods. These had to be removed from the light curve before a solution could be carried out. *See* ECLIPSE; LIGHT CURVES.

Absolute dimensions. The masses, radii, and densities of the stars, usually expressed in comparison with those of the Sun, are called the absolute dimensions of the system. Combining the results from the light and velocity curves yields this fundamental information which cannot be obtained from either approach alone. The inclination, determined from the shape of the light curve, can be substituted in the quantities $m_1 \sin^3 i$, $m_2 \sin^3 i$, and $a \sin i$ to find m_1, m_2, and a. Thus the masses of the stars in terms of the Sun's mass and the size of the orbit in terms of the Sun's radius can be found. Since the radii of the stars in fractions of the size of the orbit have been determined from the light curve, the sizes of the stars relative to the size of the Sun (either in miles or kilometers) can now be computed. Figure 1 shows the relative sizes and separation of the stars in a typical eclipsing system.

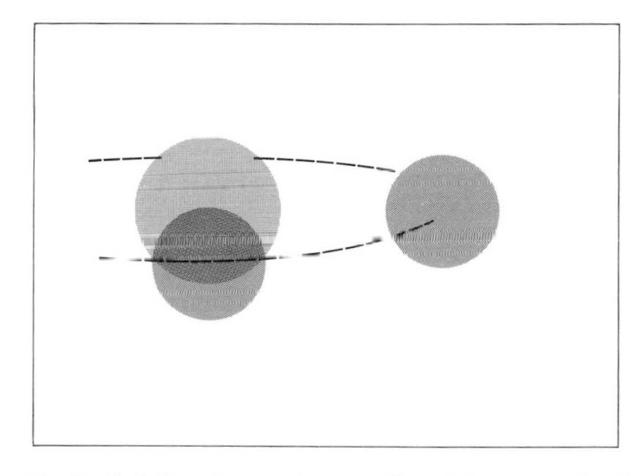

Fig. 1. Relative sizes and separation of components of R Canis Majoris. The smaller star represents the cooler component. It is shown in two positions: at the middle of an eclipse and when separation would be greatest as viewed from Earth. Even at the greatest separation, all such systems appear as single stars. (*Computed by Robert H. Koch, Steward Observatory, University of Arizona*)

Complications. Before the light curve can be "solved" to give the above quantities, corrections must be made for other factors which influence the light changes. This is done by studying the light between eclipses where, were it not for these complications, the brightness of the system would not change. One of these effects is called ellipticity. The tidal attraction of each star for the other has caused distortions until the stars in extreme cases resemble footballs more than baseballs in shape. The technical term is prolate ellipsoids, although when the stars differ in size and mass they will also differ in shape. Further, the radiation of each star falling on the side of the other nearest it will cause this side to be brighter than the side turned away. The difference will be most marked for the cooler component where the effect of the intense radiation from the nearby hotter star is most strongly evident.

There are other effects, some very poorly understood, which cause light changes between eclipses, and all of these must be carefully studied before the analysis of the eclipse begins.

Evolutionary changes. Studies of single stars indicate that, when the hydrogen in the center of the star has been converted into helium, the star undergoes a relatively rapid expansion in size. The presence of a nearby companion complicates the picture considerably, but is does seem clear that much of the mass of the star must be lost to the system or possibly transferred to the other component. Studies of this mass loss and its effect on the evolution of close double stars form one of the exciting branches of modern astronomy.

The mass-losing star eventually becomes a collapsed object—a white dwarf, a neutron star, or a black hole. Each of these types has been identified in at least one binary system. Then when the secondary, originally less massive star begins its expansion, matter from it is transferred to the collapsed object, often with dramatic results. *See* BLACK HOLE; WHITE DWARF STAR.

It is now believed that all explosive variables (novae, recurrent novae, and so forth), with the exception of supernovae, are members of close binary systems. At least some of the x-ray sources are close binaries in this state, although the detection (by instruments aboard satellites) of x-radiation from Algol and other systems which are not yet in this state indicates other physical mechanisms may also be responsible. Some systems show intermittent bursts of radiation at radio frequencies. Evidence indicates the presence of clouds of circumstellar material.

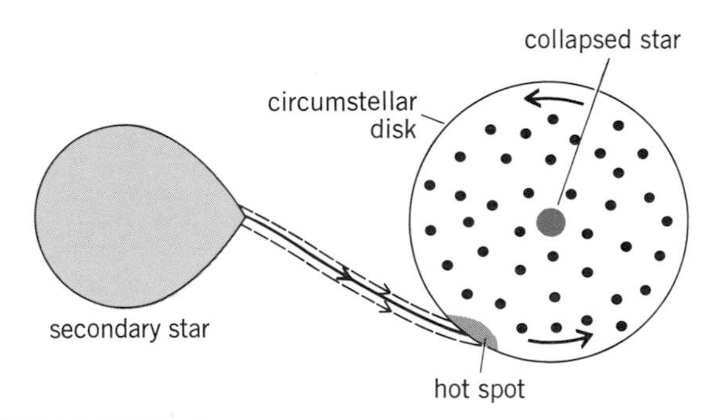

Fig. 2. Model generally accepted for certain types of eclipsing binary stars, in particular those known as dwarf novae. (*From W. W. Richardson, University of Florida*)

In a few of the eruptive variables, particularly those known as dwarf novae, rapid scintillation is found, presumably in each system from a hot spot where the transferring mass collides violently with a circumstellar disk of relatively low-density material revolving around the collapsed star (Fig. 2); the scintillation stops periodically when the spot is eclipsed by the other component. Instruments on satellites have extended observations to the far ultraviolet, as well as the x-ray, regions of the spectrum. Thus, in addition to the classical reasons for studying eclipsing variable stars, observation of them leads into many branches of astrophysics. *See* NOVA; STELLAR EVOLUTION; X-RAY ASTRONOMY. [FRANK BRADSHAW WOOD]

Bibliography: A. H. Batten, *Binary and Multiple Systems of Stars*, 1973; J. Sahade and F. B. Wood, *Interacting Binary Stars*, 1978.

Ecliptic

The path in the sky traced by the Sun in its apparent annual journey as Earth revolves around it. The ecliptic is a great circle on the celestial sphere, inclined about 23°.5 to the celestial equator, the angle of inclination being called the obliquity of the ecliptic. *See* ASTRONOMICAL COORDINATE SYSTEMS. [GERALD M. CLEMENCE]

Einstein shift

A shift toward longer wavelengths of spectral lines emitted by atoms in strong gravitational fields. One of three famous predictions of the general theory of relativity, this shift results from the slowing down of all periodic processes in a gravitational field. The amount of the shift is proportional to the difference in gravitational potential between the source and the receiver. For starlight received at the Earth the shift is proportional to the mass of the star divided by its radius. In the solar spectrum the shift amounts to about 0.001 nm at a wavelength of 500 nm. In the spectra of white dwarfs, whose ratio of mass to radius is about 30 times that of the Sun, the shift is about 0.03 nm, which can easily be measured if it can be separated from the Doppler effect. This was done by W. S. Adams for the companion of Sirius, a white dwarf whose true velocity relative to the Earth can be deduced from the observed Doppler effect in the spectrum of Sirius. The measured shift agreed with the prediction based on Einstein's theory and on independent determinations of the mass and radius of Sirius B. A more accurate measurement was carried out in 1954 by D. M. Popper, who measured the Einstein shift in the spectrum of the white dwarf 40 Eridani B. Similar measurements, all confirming Einstein's theory, have since been carried out for other white dwarfs. Attempts to demonstrate the Einstein shift in the solar spectrum have thus far proved inconclusive, because it is difficult to distinguish the Einstein shift from so-called pressure shifts resulting from perturbations of the emitting atoms by neighboring atoms. *See* WHITE DWARF STAR.

Attempts have also been made to deduce stellar masses from measurements of the Einstein shift, but the difficulty of allowing properly for the Doppler effect and for pressure shifts renders these determinations very uncertain. *See* RELATIVITY.

Some scientists have suggested that red shifts observed in the spectra of quasars are Einstein shifts, resulting from strong gravitational fields at the surfaces of these objects, rather than Doppler shifts resulting from their recession. But so far no plausible model embodying this suggestion has been put forward. *See* QUASARS; RED SHIFT.

[DAVID LAYZER]

Elements, cosmic abundance of

The abundance of the elements in surface rocks of the Earth, in the Earth as a whole, in meteorites, in our solar system, in our galaxies, or in the total universe corresponds to the average relative amounts of the chemical elements present, or, in other words, to the average chemical composition of the respective object. Element abundances are given in numbers of atoms of one element relative to a certain number of atoms of a reference element. Silicon is commonly taken as the reference element in the study of the composition of the Earth and the meteorites, and the data are given in atoms per 10^6 atoms of silicon. The results of astronomical determinations of the composition of the Sun and of the stars are often expressed in atoms per 10^{10} atoms of hydrogen. Ordinary chemical analyses, including advanced techniques for trace element studies (such as neutron activation or isotope dilution), are used for determination of the composition of rocks and meteorites. The composition of the Sun and of stars can be derived by quantitative spectral analysis. On the surface of the Earth, the most abundant elements are oxygen, silicon, magnesium, calcium, aluminum, and iron. In the universe as a whole, hydrogen and helium constitute more than 95% of the total matter. *See* ASTRONOMICAL SPECTROSCOPY.

Abundances in the Sun and stars. The possibility of a quantitative spectral analysis of the Sun, stars, and planetary nebulae is based on the fact that the intensity of the absorption lines in the spectrum, the Fraunhofer lines, depends on the concentration of the atoms causing the absorption. In order to calculate the relation of line intensity with atomic concentration, a number of physical properties of the absorbing atoms, as well as the thermodynamic state of the absorbing stellar matter, have to be known in detail. Furthermore, a knowledge of the depth of the layer in which the absorption occurs, the thermal velocity of the absorbing atoms, their macroscopic turbulent motion, and other characteristics is necessary before the exact functional dependence of line intensity and atomic concentration can be calculated.

The first abundance data by spectral analysis were obtained by C. H. Payne-Gaposchkin in 1925 and by H. N. Russell in 1928. Relatively few stars have been analyzed, and the data from spectral analyses are far from complete. However, the data have shown that the chemical composition of the universe is remarkably uniform, although systematic variations in the composition of stars seem to exist, depending on age and position in the galaxies. *See* SOLAR RADIATION; SUN.

Abundances in the Earth and meteorites. The abundance of the elements in terrestrial rocks was first investigated by F. W. Clarke and H. S. Washington during the last decade of the 19th century. These investigators compared numerous rock analyses, gave average figures for the occurrence of each element in the various types of terrestrial rocks, and hoped that some sort of regularity might become apparent. They expected that the chemical composition of the terrestrial rocks would reflect some fundamental quantity connected with the relative amounts in which the elements occur in nature in general. Since then, it has become obvious that meteorites are better objects for the study of a primeval abundance distribution of the elements. The composition of the Earth corresponds to the nonvolatile part of a primeval cloud from which the planets originated. The meteorites have formed from the same cloud, but they have undergone less chemical fractionation than any material on Earth. Meteoritic matter shows, in general, separation into three chemical phases—metal, sulfide, and silicate—in a ratio of about 10.6:1:100, respectively. The elements that concentrate in the metal phase are called siderophile, those in the sulfide phase, chalcophile, and those in the silicate phase, lithophile elements. A large fraction of meteorites, the chondrites, contain all three phases in relatively constant proportions. It is generally believed that the chondrites contain the nonvolatile components of the primeval solar matter in essentially unchanged proportions, because it seems improbable that chemically similar elements were separated from each other under conditions that did not lead to an effective separation of the three main phases from each other. The giant planets (Jupiter, Saturn, Uranus, Neptune) have retained to a large degree volatile substances, including hydrogen and helium, and elements such as carbon, nitrogen, and oxygen in the form of methane, ammonia, and water, respectively. *See* METEORITE.

Nuclear abundances. Most elements are composed of more than one isotope. The isotopic composition of the elements is practically the same in all terrestrial material and in meteorites. Small variations can be observed only in light elements as a consequence of small differences in the chemical properties owing to the difference in mass. Variations also occur if an isotope is produced by radioactive decay. From the isotopic composition of an element and its cosmic abundance, the nuclear abundances of its isotopes can be calculated. A number of empirical rules exist for the abund-

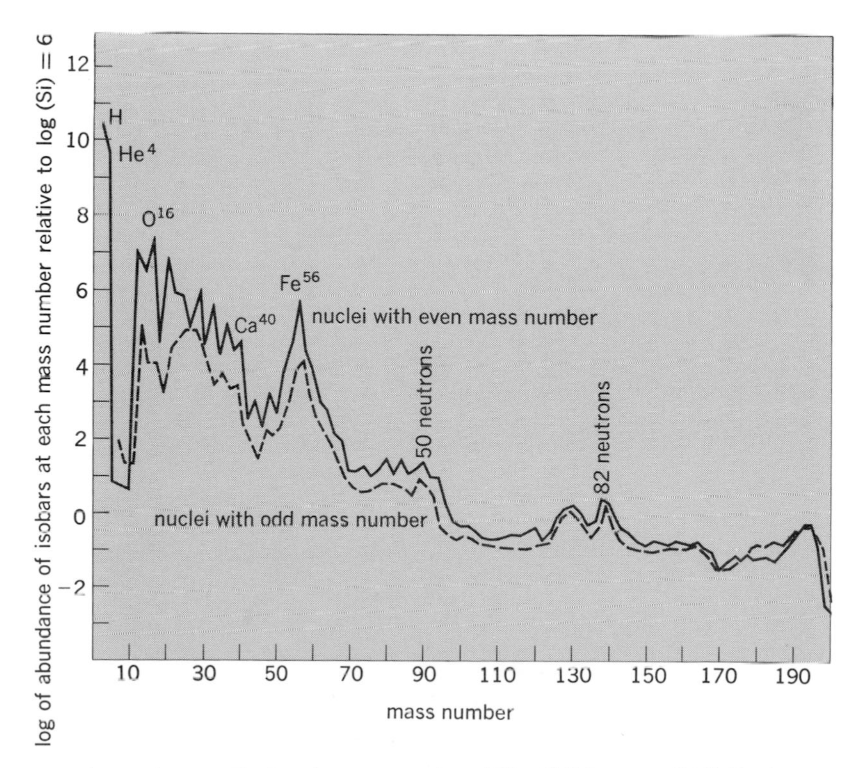

Plot of abundances of isobars by mass number. (*After H. E. Suess and H. C. Urey*)

Table of abundance values of elements

Element	Parts per million by weight			Atoms per 10^6 atoms of silicon	
	Meteorites*				
	Metal	Sulfide	Silicate	Astronomical values†	Cosmic abundances‡
1 H			630?	3.2×10^{10}	3.2×10^{10}
2 He				4.1×10^{9}	4.1×10^{9}
3 Li			3	0.23	100
4 Be			1	1.0	20
5 B			3		24
6 C	1100		400	11×10^{6}	11×10^{6}
7 N			1	3×10^{6}	3×10^{6}
8 O			4×10^{5}	3.1×10^{6}	3.1×10^{6}
9 F			40		1600
10 Ne				1.7×10^{7}	8.6×10^{6}
11 Na			7.8×10^{3}	6.2×10^{4}	4.4×10^{4}
12 Mg	320		1.6×10^{5}	4.8×10^{6}	9.1×10^{5}
13 Al	40		1.7×10^{4}	5.0×10^{4}	9.5×10^{4}
14 Si	40		2.1×10^{5}	1×10^{6}	1×10^{6}
15 P	2200	3100	1600	9×10^{3}	1×10^{4}
16 S	360	3.4×10^{5}	1.8×10^{5}	43×10^{4}	37×10^{4}
17 Cl			900	3×10^{5}	9×10^{3}
18 Ar				1×10^{5}	1.5×10^{5}
19 K			2000	2800	3200
20 Ca	500		2×10^{4}	74,000	49,000
21 Sc			5.8	49	28
22 Ti	100		1000	2900	2400
23 V	6		90	330	220
24 Cr	240	1200	3500	4900	7800
25 Mn	300	460	3000	6200	6900
26 Fe	9.1×10^{5}	6.1×10^{5}	1.6×10^{5}	1.8×10^{5}	6.0×10^{5}
27 Co	6300	100	200	1700	1800
28 Ni	8.6×10^{4}	1000	1400	2.9×10^{4}	2.7×10^{4}
29 Cu	310	4200	1.6	170	212
30 Zn	110	1500	3.4	1000	490
31 Ga	50	0.5	0.5	4.5	11.4
32 Ge	190	600	10	60	50
33 As	360	1000	20		4
34 Se	3	100	13		68
35 Br	1		25		13
36 Kr					51
37 Rb			4.5	5.2	6.5
38 Sr			26	19.5	19
39 Y			6.5	30	9
40 Zr	8		100	4.5	55
41 Nb	0.2		0.5	3	1.0
42 Mo	16	11	2.5	2.8	2.4
43 Tc					
44 Ru	10	4.2		8.8	1.5
45 Rh	4	1		11.5	0.2
46 Pd	3.7	4.5		0.3	0.7
47 Ag	3.3	21		0.01	0.3
48 Cd	8	30	1.6	1.8	0.9
49 In	1	0.8	0.2	0.2	0.1
50 Sn	80	15	3	1.0	1.3
51 Sb	2	8	0.1		0.25
52 Te		17			4.7
53 I	0.6		1.3		0.8
54 Xe					4.0
55 Cs			0.1		0.5
56 Ba			9.0	5.6	3.7
57 La			2.2		2.0
58 Ce			2.5		2.3
59 Pr			1.0		0.4
60 Nd			3.7		1.4
61 Pm					
62 Sm			1.3		0.7
63 Eu			0.3		0.19
64 Gd			2.0		0.68
65 Tb			0.6		0.10
66 Dy			2.5		0.56
67 Ho			0.7		0.12
68 Er			2.1		0.32

Table of abundance values of elements (cont.)

Element	Parts per million by weight			Atoms per 10^6 atoms of silicon	
	Meteorites*				
	Metal	Sulfide	Silicate	Astronomical values†	Cosmic abundances‡
69 Tm			0.4		0.03
70 Yb			2.0	1 ?	0.22
71 Lu			0.7		0.05
72 Hf			1		0.44
73 Ta	0.1		0.4		0.07
74 W	8.2		18 ?		0.5
75 Re	0.85				0.14
76 Os	7.6	10			1.0
77 Ir	3	0.5			0.82
78 Pt	19	30	0.1		1.63
79 Au	1.8	0.5	0		0.15
80 Hg		0.2 ?	0.01 ?	26 ?	0.02 ?
81 Tl		0.3 ?	0.15 ?		0.1 ?
82 Pb	60 ?	20 ?	2 ?	15 ?	0.1 ?
83 Bi	0.5 ?	2			0.01 ?
90 Th	0.04		2		0.03
92 U	0.01		0.4		0.018

*K. Rankama and G. Sahama. †L. H. Aller. ‡H. E. Suess and H. C. Urey.

ances of nuclear species. The most important one is Harkins' rule, which states that isotopes with an odd mass number are less abundant than their even-mass-numbered neighbors. In certain ranges it is possible to modify, within limits of error of the analytical data, the values for the abundances of the elements in such a way that the abundance values of the individual nuclear species, as a function of their mass number, form regular smooth lines for the odd-mass-numbered species, and at the same time, for the sum of the abundances of isobars at even mass numbers. Irregularities occur where the number of neutrons or protons reaches a so-called magic number, connected with a nuclear shell closure. Modified abundance values by H. E. Suess and H. C. Urey are compared with the best empirical data on stellar spectra by L. H. Aller and on meteorites in the accompanying table.

The nuclear abundances as a function of mass number are shown graphically in the illustration. At even mass numbers, where more than one isobar exists, the sum of the abundances is given. The upper line refers to those nuclei with even mass number, the lower broken line to those with odd mass number.

Nuclear abundance values show a clear correlation with certain nuclear properties, and can be assumed to represent in good approximation the original yield distribution of the thermonuclear processes that lead to the formation of the elements. The empirical abundance values can therefore serve as the basis for theoretical considerations about the origin of matter and of the universe, and have led to the following conclusion: No simple, single mechanism exists by which the elements in their observed isotopic composition can have formed. The matter of the cosmos appears to be a mixture of material that formed under different conditions by different types of nuclear processes. *See* ELEMENTS AND NUCLIDES, ORIGIN OF.

[HANS E. SUESS]

Bibliography: L. H. Aller, *The Abundance of the Elements*, 1961; S. Fluegge (ed.), *Handbuch der Physik*, vol. 51, 1958; V. M. Goldschmidt, *Geochemistry*, 1954; H. E. Suess and H. C. Urey, The abundance of the elements, *Rev. Mod. Phys.*, 28: 53–74, 1956.

Elements and nuclides, origin of

Theories of the origin of the elements involve synthesis with nucleons (neutrons and protons), the elementary building blocks of the nucleus. The nuclear theory, comprising nine distinct processes, is commonly called nucleosynthesis. Any acceptable theory of nucleosynthesis must lead to an understanding of the cosmic abundance of the elements observed in the solar system, stars, and the interstellar medium. The curve of these abundances is shown in Fig. 1. Hydrogen and helium comprise about 98% of the total element content, and there is an almost exponential decrease with increasing nuclear mass number A. The processes of nucleosynthesis described in this article attempt to understand these facts and the peaks that occur at certain mass regions. *See* ELEMENTS, COSMIC ABUNDANCE OF.

Observations of the expanding universe and of the 3-K background radiation lend credence to the belief that the universe originated in a primordial event known as the "big bang" some $1-2 \times 10^{10}$ years ago. Absence of stable mass 5 and mass 8 nuclei preclude the possibility of synthesizing the major portion of the nuclei of masses greater than 4 in those first few minutes of the universe, when density and temperature were sufficiently high to support the necessary nuclear reactions to synthesize elements. *See* COSMOLOGY.

The principal source of energy in stars is certainly nuclear reactions which release energy by fusion of lighter nuclei to form more massive nuclei. Since the masses of the products are less than those of the constituent parts, the energy released is $E = mc^2$, where m is the mass difference and c is

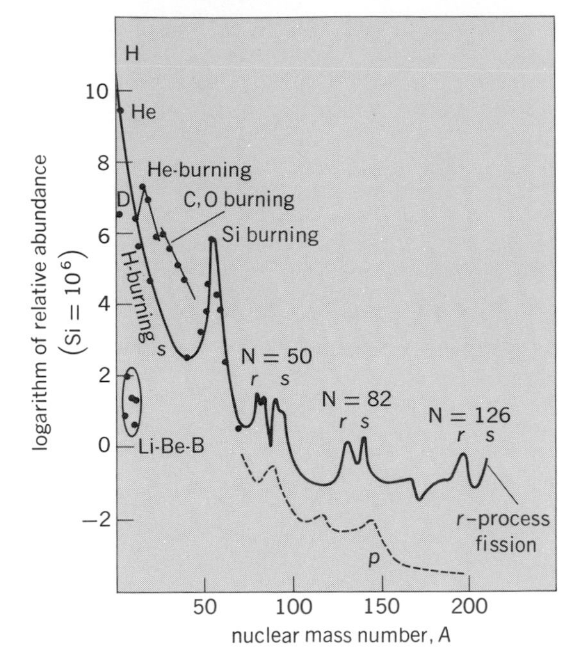

Fig. 1. Schematic diagram of cosmic abundance as a function of nuclear mass number *A*, based on data of H. E. Suess and H. C. Urey. Predominant nucleosynthetic processes are indicated. *(Adapted from E. M. Burbidge et al., Synthesis of elements in stars, Rev. Mod. Phys., 29: 547–650, 1957)*

the velocity of light. In 1957 E. M. Burbidge, G. R. Burbidge, W. A. Fowler, and F. Hoyle laid the foundations for astrophysical research into nucleosynthesis, stellar energy generation, stellar structure, and stellar evolution.

There is considerable evidence that nucleosynthesis is going on in stars, and has been doing so for billions of years. Observations make it possible to determine the relative ages of some stars, and it is found that the oldest stars show a ratio of iron and heavier elements to hydrogen that is 100–1000 times smaller than in the Sun. This is understood on the basis of nuclear reactions occurring in previous-generation stars that evolved to the point of exploding as supernovae thus enriching the interstellar medium with the massive nuclei which the stars had synthesized during their lifetimes. Later-generation stars were then formed from the interstellar gas and dust that is still primarily hydrogen and helium, but with small amounts of the more massive nuclei added. In certain of the carbon stars, the ratio of ^{12}C to ^{13}C is as low as 4, much less than the nearly universal ratio of 90 for those isotopes of carbon. This is believed to reflect the synthesis of ^{13}C in the central regions of those stars. Some heavy-metal stars show lines of technetium in their spectra. That technetium must have been produced in those stars themselves or have accreted there from relatively recent supernovae in their neighborhood. No technetium is found naturally on Earth since there is no stable nucleus of technetium, and the stars showing technetium spectra are obviously older than its less than a million years half-life.

Hydrogen burning. The first of the nine processes of nucleosynthesis converts hydrogen nuclei into helium. In stars of 1.2 or less solar masses, hydrogen burning proceeds by the proton-proton chain. Such reactions occurring during the big bang are believed to be responsible for the helium and deuterium, and possibly some of the ^{7}Li observed today. In more massive stars where temperatures are equal to or greater than 2×10^7 K, hydrogen burning is accomplished through proton captures by carbon, nitrogen, and oxygen nuclei, in the carbon-nitrogen-oxygen (CNO) cycles, to form ^{4}He. Hydrogen burning is responsible for the synthesis of helium, but much of the helium produced in stars may be consumed in later stages of nucleosynthesis in those stars. The observed abundances of some carbon, nitrogen, oxygen, and fluorine nuclei are attributed to hydrogen burning in the carbon-nitrogen-oxygen cycles. The cycles are believed to be responsible for the anomalous abundances of ^{13}C observed in some carbon stars. *See* CARBON-NITROGEN-OXYGEN CYCLES; PROTON-PROTON CHAIN.

Helium burning. When the hydrogen fuel is exhausted in the central region of the star, the core contracts and its temperature and density increase. When the core density is some 10^4 g cm^{-3} and its temperature is of the order 10^8 K, helium becomes the fuel for further energy generation and nucleosynthesis. The basic reaction in this thermonuclear phase is the three-alpha process in which three ^{4}He nuclei (three alpha particles) fuse to form a carbon nucleus of mass 12 (^{12}C). Capture of an alpha particle by ^{12}C forms oxygen-16. This reaction can be represented by the expression $^{12}C + {}^{4}He \rightarrow {}^{16}O + \gamma$, where γ represents energy in the form of electromagnetic radiation. It is convenient to express the reaction in shorthand notation as $^{12}C(\alpha,\gamma)^{16}O$. Other reactions that are usually included in helium burning are $^{16}O(\alpha,\gamma)^{20}Ne$, $^{20}Ne(\alpha,\gamma)^{24}Mg$, $^{14}N(\alpha,\gamma)^{18}F$, and $^{18}O(\alpha,\gamma)^{22}Ne$. The fluorine-18 that is produced when ^{14}N captures an alpha particle is unstable and decays by emitting a positron (e^+) and a neutrino (v) to form oxygen-18 (that is, $^{18}F(e^+v)^{18}O$). Because there is likely to be ^{13}C in the core of the star if hydrogen burning proceeded by the carbon-nitrogen-oxygen cycles, one should also include the neutron-producing reaction $^{13}C(\alpha,n)^{16}O$ with the helium-burning reactions. Helium burning is probably responsible for much of the ^{12}C observed in the cosmic abundances, although in more massive stars the later stages of nucleosynthesis will consume the ^{12}C produced in their interiors.

Carbon burning. Upon exhaustion of the helium supply, if the star has an initial mass of at least 7.5 solar masses, gravitational contraction of the core can lead to a temperature of about 5×10^8 K, where it becomes possible for two ^{12}C nuclei to overcome their high mutual Coulomb-potential barrier and fuse to form ^{20}Ne, ^{23}Na, and ^{24}Mg through the reactions $^{12}C(^{12}C,\alpha)^{20}Ne$, $^{12}C(^{12}C,p)^{23}Na$, and $^{12}C(^{12}C,\gamma)^{24}Mg$. Carbon burning can produce a number of nuclei with masses less than or equal to 28 through proton and alpha-particle captures by nuclei that are present.

Oxygen burning. Carbon burning is immediately followed by a short-lived stage, sometimes referred to as neon burning, in which ^{20}Ne is photodisintegrated via the reaction $^{20}Ne(\gamma,\alpha)^{16}O$. The eventual result is that most of the carbon from helium burning becomes oxygen, which supplements the original oxygen formed in helium burning. This is all followed by the fusion of oxygen nuclei at significantly higher temperatures. Temperatures of about 10^9 K are probably required to overcome the high Coulomb barrier. Fusion reactions of oxygen are: $^{16}O(^{16}O,\alpha)^{28}Si$, $^{16}O(^{16}O,p)^{31}P$, and $^{16}O(^{16}O,\gamma)$ ^{32}S. Nuclei of masses up to $A = 40$ may be produced in this phase through proton, neutron, and alpha-particle captures by nuclei from oxygen fusions.

Silicon burning. This process commences when the temperature reaches 3×10^9 K. In this phase, photodisintegration of ^{28}Si and other intermediate-mass nuclei in the neighborhood of $A = 28$ produces copious supplies of protons, alpha particles, and neutrons. Capture of these by other intermediate-mass nuclei synthesizes nuclei up to a mass A of about 60, and results in the buildup of the iron-group peak near $A = 56$ (Fig. 1).

Synthesis of nuclei by charged-particle reactions beyond mass A ~ 60 is difficult because of increasing Coulomb barriers and the fact that the maximum binding energy per nucleon occurs in that mass range. Additional nucleosynthetic processes must be invoked to understand the cosmic abundances of the more massive nuclei.

The s-process. Because neutrons are neutral particles, their capture is not affected by the Coulomb barrier that inhibits charged-particle capture by the massive nuclei. If the number of neutrons per seed nucleus is small, so that time intervals between neutron captures are long compared to the β-decay lifetimes of unstable nuclei that are formed, the *s*-process (slow process) takes place. The seed nuclei are probably primarily those in the iron peak, but neutron processes probably alter some of the lighter-mass nuclei as well. In this slow process, after neutron capture a nucleus that is unstable because of excess neutrons decays by electron and antineutrino emission to a stable isobar which may capture another neutron, leading to a chain of captures and decays to synthesize nuclei of masses up to about $A = 209$, when alpha decay becomes a deterrent to further synthesis by the *s*-process. The *s*-process is believed to occur in red giant stars. The reactions $^{13}C(\alpha,n)^{16}O$, $^{21}Ne(\alpha,n)$ ^{24}Mg, and $^{22}Ne(\alpha,n)^{25}Mg$ are probable sources of neutrons during phases when both hydrogen burning and helium burning are occurring in concentric shells around the hydrogen- and helium-exhausted cores of the red giant stars. The technetium observed in some heavy-metal stars is probably synthesized by the s-process. Since the cross sections for neutron capture by nuclei with magic numbers of neutrons ($N = 50$, 82, and 126) are small, the *s*-process builds high abundances of these nuclei and produces the peaks at $A = 90$, 138, and 208, labeled *s* in the atomic abundance curve in Fig. 1. The *s*-process path among nuclei is represented by the single line in Fig. 2, and lies near the beta stability line through the nuclei. Capture of a neutron increases the number of neutrons *(N)* in the nucleus by one, beta decay increases the number of protons *(Z)* by one, and the path follows a stepwise curve near the line of beta stability in the *N-Z* plane, since time between neutron captures allows the unstable nuclei to decay.

The r-process. This process occurs when a large neutron flux allows rapid captures of neutrons. Up to 100 captures can occur in 1–100 s. Only in the massive explosions of supernovae are such large neutron fluxes conceivable. With such rapid successive captures, it is possible to proceed through unstable nuclei before they can either β-

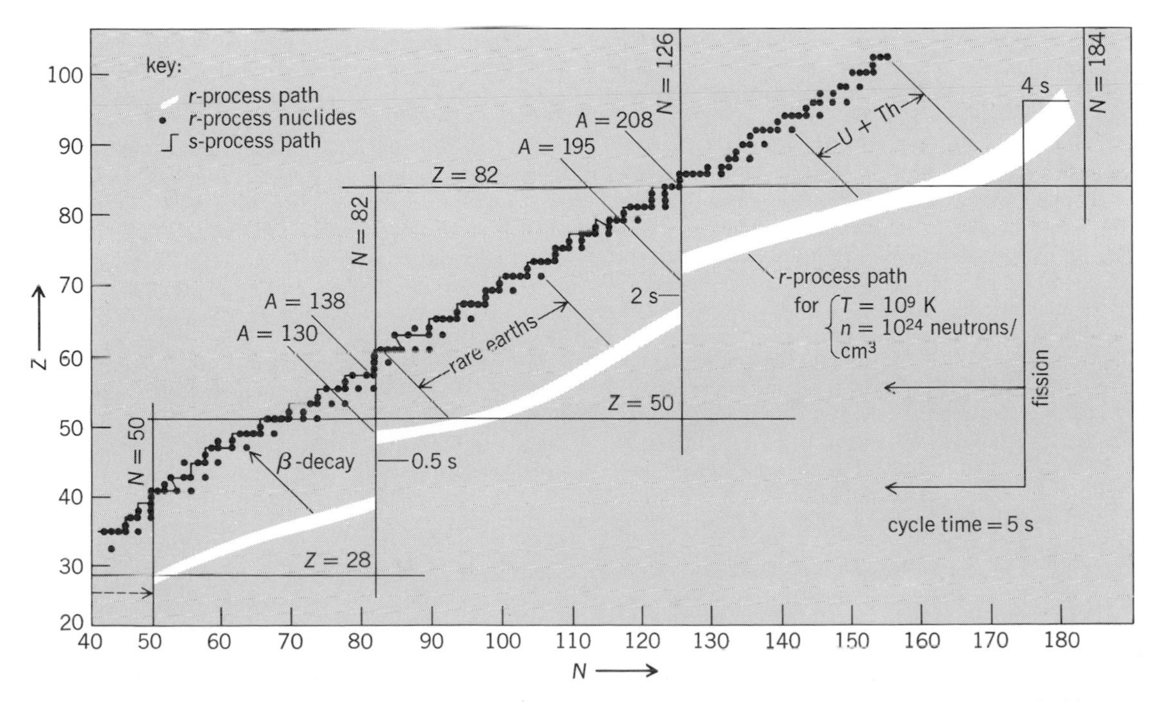

Fig. 2. Neutron-capture paths among the nuclei plotted in the *N-Z* plane, showing nuclei identified by the number of neutrons *(N)* and number of protons *(Z)* in the nucleus. (*From P. A. Seeger, W. A. Fowler, and D. D. Clayton, Nucleosynthesis of heavy elements by neutron capture, Astrophys. J., Suppl., 11(97):121–166, 1965*)

or α-decay. In this manner it is possible to synthesize nuclei all the way into the transuranic elements. The *r*-process path is shown by a band in Fig. 2. The path depends on the temperature and the neutron density at the site of the explosion. The path in Fig. 2 is based on a reasonable assumption of a temperature $T = 10^9$ K and neutron density $n = 10^{24}$ neutrons cm^{-3}. The path lies to the right of the line of beta stability and represents the progenitors of the nuclei that result from the *r*-process. The circles in Fig. 2 show the *r*-process nuclei that result after the unstable progenitor nuclei β-decay under these conditions. Diagonal lines indicate the β-decay lines along isobars to the region of stability. The times shown are intervals to reach locations along the *r*-process path for these conditions. *See* SUPERNOVA.

At $Z = 94$, the transuranic elements are unstable to neutron-induced fission, and *r*-process synthesis of more massive nuclei ceases. The resultant fission leads to production of intermediate-mass nuclei which may then act as seed nuclei to introduce a cycling effect. The peaks at $A = 80$, 130, and 195 labeled *r* in Fig. 1 are due to the *r*-process effects at magic numbers of neutrons ($N = 50$, 82, and 126) and the subsequent decays of unstable nuclei.

The p-process. This process produces proton-rich heavier elements probably by modifying a small fraction of the *s*-process and *r*-process nuclei through (p,γ), (p,n), and (γ,n) reactions. This would be expected to occur in the red giant stars and in phases of explosive nucleosynthesis. Nuclei produced in this process are indicated by the curve marked *p* in Fig. 1.

The l-process. The lithium, beryllium, and boron found in the cosmic abundances cannot have survived processing in stellar interiors because they are readily destroyed by proton bombardment. Although some ^{7}Li may have originated in the big bang, that primordial event is probably not responsible for all the ^{7}Li in existence. Spallation of more abundant nuclei such as carbon, nitrogen, and oxygen by protons can account for the low-abundance nuclides ^{6}Li, ^{9}Be, ^{10}B, and ^{11}B and for some ^{7}Li. This breaking up of the more massive nuclei to form the light nuclei is called the *l*-process. The most probable site for the *l*-process is in the interstellar medium, where high-energy galactic cosmic rays bombard the interstellar gas. The surfaces of magnetic stars have also been suggested as possible sites for the *l*-process, but it is difficult to show that sufficient energy is available there to produce the observed abundances. The nuclei produced in the *l*-process are indicated by Li-Be-B in Fig. 1.

Conclusions. Although stars spend the majority of their lifetimes in hydrogen-burning phases and most of the remaining time in burning helium, the vast majority of the elements are synthesized in the relatively brief time spans of later stages of evolution and the processes described. Figure 1 serves as a very brief summary of those processes and the mass ranges of elements for which they are believed to be responsible.

Nucleosynthesis remains a developing field of research. There are many unanswered questions regarding which processes are mainly responsible for certain elements. In Fig. 1 the schematic diagram is a modification of one that was developed by Burbidge and colleagues in 1957. The C-, O-, and Si-burning processes replace the phases originally called α-process (because of the number of α-particles and $A = 2Z$ nuclei involved) and the *e*-process (because of the quasiequilibrium extant during silicon burning). Research has shown that the terms C-, O-, and Si-burning processes are more realistic. Although many of the nuclear reaction rates for intermediate-mass nuclei ($A = 20-60$) are still theoretical estimates, nuclear physicists are continuing to improve knowledge of the rates at which nuclear reactions occur under conditions believed to be present at the sites of stellar nucleosynthesis. Astrophysicists are increasing the knowledge of the probable thermodynamic conditions prevailing in stars. As a result, further modifications in the theory of the processes of nucleosynthesis may occur in the future. Clarification of some of the issues that are cloudy today can be expected, but new puzzles will undoubtedly arise. *See* STELLAR EVOLUTION.

[GEORGEANNE R. CAUGHLAN]

Bibliography: B. J. Allen, J. H. Gibbons, and R. L. Macklin, Nucleosynthesis and neutron-capture reactions, *Advan. Nucl. Phys.*, 4:205–259, 1971; C. A. Barnes, Nucleosynthesis by charged particles, *Advan. Nucl. Phys.*, 4:133–204, 1971; E. M. Burbidge et al., Synthesis of elements in stars, *Rev. Mod. Phys.*, 29:547–650, 1957; W. A. Fowler, *Nuclear Astrophysics*, 1967; W. A. Fowler and W. E. Stephens, Resource letter OE-1 on origin of elements, *Amer. J. Phys.*, 36:289–302, 1968; P. A. Seeger, W. A. Fowler, and D. D. Clayton, Nucleosynthesis of heavy elements by neutron capture, *Astrophys. J., Suppl.*, 11(97):121–166, 1965; V. Trimble, The origin and abundance of the chemical elements, *Rev. Mod. Phys.*, 47:877–976, 1975; G. Wallerstein, Observational evidence concerning nucleosynthesis, in W. D. Arnett et al. (eds.), *Nucleosynthesis*, pp. 29–47, 1968.

Encke's Comet

A member of the solar system whose return in 1822 was predicted by J. F. Encke, who had computed an elliptical orbit for a faint comet observed earlier. Encke further showed that the same object had been under observation in 1805, 1795, and 1786 and that it represented a new type of object, the short-period comet. The period of 3.3 years still remains the shortest known for any comet.

At its brightest Encke's Comet is ordinarily slightly fainter than the naked-eye limit. A perihelion passage took place on Apr. 28, 1974, and subsequent perihelion passages have occurred or will occur at intervals of approximately 3.30 years. The other five orbital elements and the perihelion distance q are listed as follows: q, .339; a, 2.22 AU; e, .847; i, 12°; ω, 185°; and ☊, 335°. *See* CELESTIAL MECHANICS.

Encke's Comet is responsible for two meteor showers: the extended Taurid shower of October and November and a daytime shower the β-Taurids, discovered by radar observations. *See* METEOR.

The comet is remarkable in one other respect.

The orbit is systematically becoming smaller and more nearly circular. The change is almost entirely in the decreasing value of aphelion. The period has decreased by over 2 days in the past century. The acceleration has been explained by F. L. Whipple as resulting from the jet action of gases evaporating from the comet when it is near the Sun. The loss of 0.2% of its mass per revolution is sufficient to cause the effect.

The comet is about two magnitudes fainter today than it was 100 years ago. By interpreting this change as a result of decreasing comet size, Whipple and D. H. Douglas-Hamilton have deduced that the object will disappear by the year 2000. *See* COMET. [RICHARD E. MC CROSKY]

Bibliography: G. S. Hawkins, *Meteors, Comets, and Meteorites*, 1964; F. L. Whipple, A comet model, I: Acceleration of Comet Encke, *Astrophys. J.*, 111:375–94, 1950.

Ephemeris

A table of data, especially astronomical data, that depend on the time, usually arranged with values of the time in the left-hand column. A lunar ephemeris, for example, may give the right ascension and declination of the Moon for every hour of a particular year. *The American Ephemeris and Nautical Almanac* is an annual volume published by the Nautical Almanac Office, U.S. Naval Observatory, with ephemerides of the Sun, Moon, planets, and satellites, and other astronomical data.

[GERALD M. CLEMENCE]

Equation of time

The quantity added to mean solar time to obtain apparent solar time. (Formerly, when apparent solar time was in common use, the opposite convention was used.)

Apparent solar time equals the hour angle of the true Sun (angle west of the meridian) plus 12 hr. The Sun moves apparently with respect to the stars, in the ecliptic inclined $23\frac{1}{2}°$ to the Equator, and not at a uniform angular rate because of the Earth's elliptic orbit. In consequence, apparent solar time has annual variations. Variations in speed of rotation of the Earth, which are very small, are not considered here. *See* TIME.

The introduction of fairly accurate clocks and watches about 1800 made it necessary to introduce the term mean solar time, defined as 12 hr plus the hour angle of the mean sun. The mean sun is a fictitious body that moves in the celestial equator at the same average speed as the true Sun and has the same average position.

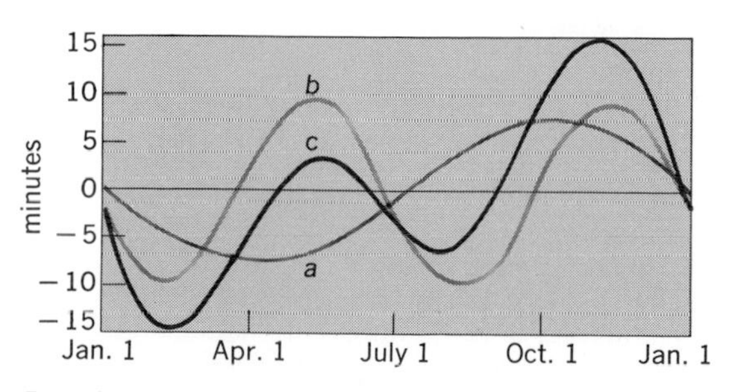

Equation of time *c* is the sum of elliptic motion effect *a* and inclination effect *b*.

Some values of the equation of time, in minutes, are: January 1,−3; February 12, −14; April 14, 0; May 14, +4; June 14, 0; July 25, −6; August 30, 0; November 2, +16; and December 24, 0. The dates for a given value may change by a few days from year to year. The illustration shows the equation of time and the two effects which cause it.

To obtain mean solar time from a sundial, which indicates apparent solar time, the equation of time is subtracted from apparent solar time, and a correction is applied for longitude; 4 min are added for every degree east of the central meridian of the time zone. *See* SUNDIAL.

[WILLIAM MARKOWITZ]

Equinox

The date of the year and the point on the celestial sphere at which the Sun's rays at noon are 90° above the horizon at the Equator, or at an angle of 90° with the Earth's axis, and neither North nor South Pole is inclined toward the Sun. This phenomenon occurs on two days of the year, approximately March 21 and September 23. In the Northern Hemisphere the event in March is referred to as the vernal equinox, and that in September as the autumnal equinox. As conditions are the same on the Earth at both the vernal and autumnal equinoxes, no distinction need be made here between the two. *See* ECLIPTIC.

With the Sun's rays at 90° to the Earth's axis, the rays illuminate half the globe, from pole to pole, thus bisecting all parallels. Therefore the Sun appears at all places on the Earth and gives 12 hr of sunlight and 12 hr of darkness, not considering twilight. In other words, at all places the Sun at the equinox appears in the east at 6 A.M., local Sun time, and sets at 6 P.M., in the west; thus the term equinox, meaning equal night.

The situation at the two poles is slightly different; there on this day the Sun rises on the eastern horizon and makes one circuit of the horizon in 24 hr. On the day after the vernal equinox, the Sun rises higher and remains above the horizon on the North Pole, but drops below the horizon at the South Pole. The reverse is true following the autumnal equinox, when the Sun remains above the horizon at the South Pole. The vernal equinox is considered the first day of spring in the Northern Hemisphere, and the autumnal equinox the first day of fall.

At the time of the equinox the altitude of the noon Sun at any place is the same as the colatitude (90° minus the latitude) of the place. Thus at the Equator it is 90°, at the poles 0°, and at 45° north or south it is 45°. *See* ASTRONOMICAL COORDINATE SYSTEMS; PRECESSION OF EQUINOXES.

[VAN H. ENGLISH]

Eros

A minor planet discovered by G. Witt at Berlin in August 1898, at a time when there were proposals to discontinue further attention to the rapidly growing number of minor planet discoveries brought about by the introduction of photography.

But the orbit of Eros was unusual, as the planet passed within 2.3×10^7 km of the Earth, and provided an excellent independent method of finding the distance to the Sun. *See* PARALLAX.

Eros was well observed in 1901 and 1931. It revolves around the Sun in 643.2 days in an ellipse with 0.223 eccentricity, and rotates on its own axis in 5 hr 16 min. On Jan. 24, 1975, it occulted the star Kappa Geminorum, and it was observed in Connecticut. The observations did not yield conclusive results, but Eros has a maximum diameter of about 32 km and is probably a very elongated object. Radar observations of the planets now offer a better means for finding the solar parallax. *See* ASTEROID; PLANET. [PAUL HERGET]

Bibliography: G. O. Abell, *Exploration of the Universe*, 3d ed., 1975; C. R. Chapman, J. G. Williams, and W. K. Hartmann, The asteroids, *Annu. Rev. Astron. Astrophys.*, 16:33–75, 1978; T. Gehrels (ed.), *Asteroids*, 1979.

Fluorescence astronomy

The study of astronomical objects in which certain elements absorb photons of energy greater than their ionization energy, thereby causing the elements to fluoresce, or to emit luminescence. The conversion of photons of high energy to lower energy occurs in a variety of astronomical objects. Examples are the astrophysically most abundant elements, such as hydrogen, 13.54 eV; helium, 24.48 eV; and ionized helium 54.17 eV. The commonest physical environment is a low-density gas cloud near or surrounding a hot star. *See* STAR.

Mechanism. The star's photons are absorbed by photoionization, with a cross section of about 10^{-18} cm^2, from the ground states. The free electrons and ions are thermalized by collision to an electron temperature which is lower than that of the star's radiation distribution or color temperature. They then recombine at a rate set by the electron density and temperature. Only a fraction of the recombinations occur to the ground states, and those captured on high, excited levels are followed by cascades as the electrons fall through intermediate levels, or to the ground state. Thus, one or several photons may be emitted, with energies less than that of the original photoionizing quantum. In a sense the gas becomes an ultraviolet photon counter. The spectrum emitted by transitions between higher levels consists of many lines which leak out of the gas cloud. Only the photons produced by transitions to and from the ground state are likely to be absorbed again. The particular case of hydrogen is of special simplicity. The figure shows a photon (1) absorbed from the ground state, $n=1$, into a free state with electron kinetic energy equal to the photon energy minus 13.54 eV. After thermalization the electron may be captured (2), for example, in $n=4$, emitting a photon in an infrared continuum; it drops to $n=3$, emitting the first line of the Paschen series (Paschen alpha), then from $n=3$ to $n=2$, emitting (3) Balmer alpha (656.3 nm), and then from $n=2$ to $n=1$, emitting (4) Lyman alpha (121.6 nm). From the ground one cannot observe Lyman alpha, but can easily observe fluorescent Balmer alpha, and it is by this fluorescent red light that hot gas clouds are most easily photographed.

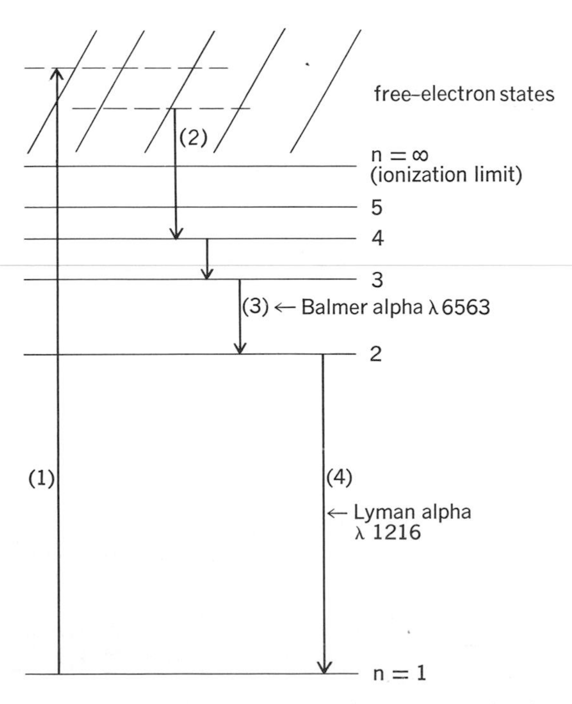

Energy-level diagram for the hydrogen atom showing fluorescent conversion of ultraviolet light into longer wavelength line and continuum radiations.

Indirect consequences of such gaseous fluorescence occur through the transfer of electron kinetic energy, by inelastic collision, into excitation energy of other atoms or ions which have low-energy excited levels. Then downward transitions produce spectral lines from transitions which may be normally forbidden. These, for example, occur in singly and doubly ionized oxygen, producing forbidden lines characteristic of gaseous nebulae, at wavelengths of 372.6 and 372.9 nm [O II] and wavelengths of 495.9 and 500.7 nm [O III].

Solid-state fluorescence. The lunar surface has been subject to long-term solar-wind and cosmic-ray bombardment. Its crystalline structure has suffered radiation damage, and its minerals may fluoresce under both photon and particle impact. [JESSE L. GREENSTEIN]

Foucault pendulum

A pendulum or swinging weight, supported by a long wire, by which J. B. L. Foucault demonstrated in 1851 the rotation of Earth on its axis. Foucault used a 28-kg iron ball suspended on about a 60-m wire in the Pantheon in Paris. The upper support of the wire restrains the wire only in the vertical direction. The bob is set swinging along a meridian in pure translation (no lateral or circular motion). In the Northern Hemisphere the plane of swing appears to turn clockwise; in the Southern Hemisphere it appears to turn counterclockwise, the rate being 15 degrees times the sine of the local latitude per sidereal hour. Thus, at the Equator the plane of swing is carried around by Earth and the pendulum shows no apparent rotation; at either pole the plane of swing remains fixed in space while Earth completes one rotation each sidereal day. *See* DAY.

[FRANK H. ROCKETT]

Bibliography: G. O. Abell, *Exploration of the Universe*, 3d ed., 1975; C. Kittel, W. D. Knight, and M. A. Ruderman, *Mechanics*, Berkeley Physics Course, vol. 1, 2d ed., 1973; O. Struve, B. Lynds, and H. Pillans, *Elementary Astronomy*, 1959.

Fundamental stars

The term applied to a relatively small number of the brighter stars, distributed as uniformly as possible over the entire sky. For metrical purposes the sky is considered as the inner surface of a sphere; astrometry consists in measuring the spherical coordinates of the stars, analogously to measuring the latitudes and longitudes of points on the surface of the Earth. The fundamental stars are the celestial analog of the so-called first-order triangulation points on the Earth. Thus the coordinates of the fundamental stars are first measured, using elaborate special techniques designed to measure large angular distances; the coordinates of the remaining stars are then inferred by other, less elaborate techniques. *See* ASTROMETRY; ASTRONOMICAL COORDINATE SYSTEMS.

Since the coordinates of the stars are continuously changing as a result of the motion of the Earth and, to a lesser degree, the motions of the stars themselves, it is equally necessary to measure the changes as to measure the coordinates at a definite specified time. Thus the determination of the coordinates is a continuing process, with no definite termination.

From time to time the available knowledge of the coordinates of the fundamental stars, and of their changes, is compiled and published in a so-called fundamental catalog. By international agreement it is used by astronomers throughout the world. It contains the coordinates and their variations of 1535 stars for the years 1950 and 1975; from this information the coordinates at any other time may be derived.

[GERALD M. CLEMENCE]

Bibliography: E. W. Fricke and A. Kopff (eds.), *Fourth Fundamental Catalogue (FK4)*, Veroffentlichungen des Astronomisches Rechen-Instituts, Heidelberg, 1963.

Galaxy

The large aggregation of stars and interstellar gas and dust of which the Sun is a member. Except for three nearby external galaxies, which appear as hazy patches of light, all objects visible to the unaided eye are members of the Galaxy. The Milky Way is the faint band of light which encircles the sky and results from the combined light of many stars. Running through the Milky Way band is a dark lane of interstellar material which obscures the stars behind it (Fig. 1). The Milky Way provides a view of the Galaxy seen from within and off-center. The astronomer must uncover the structure of the entire Galaxy from this limited inside view.

Shape and content. Studies of other galaxies reveal three main types: elliptical, spiral, and irregular. Elliptical galaxies are ellipsoidal (flattened) in shape and contain little or no interstellar material. Spiral galaxies have flattened spiral shapes and contain moderate amounts of interstel-

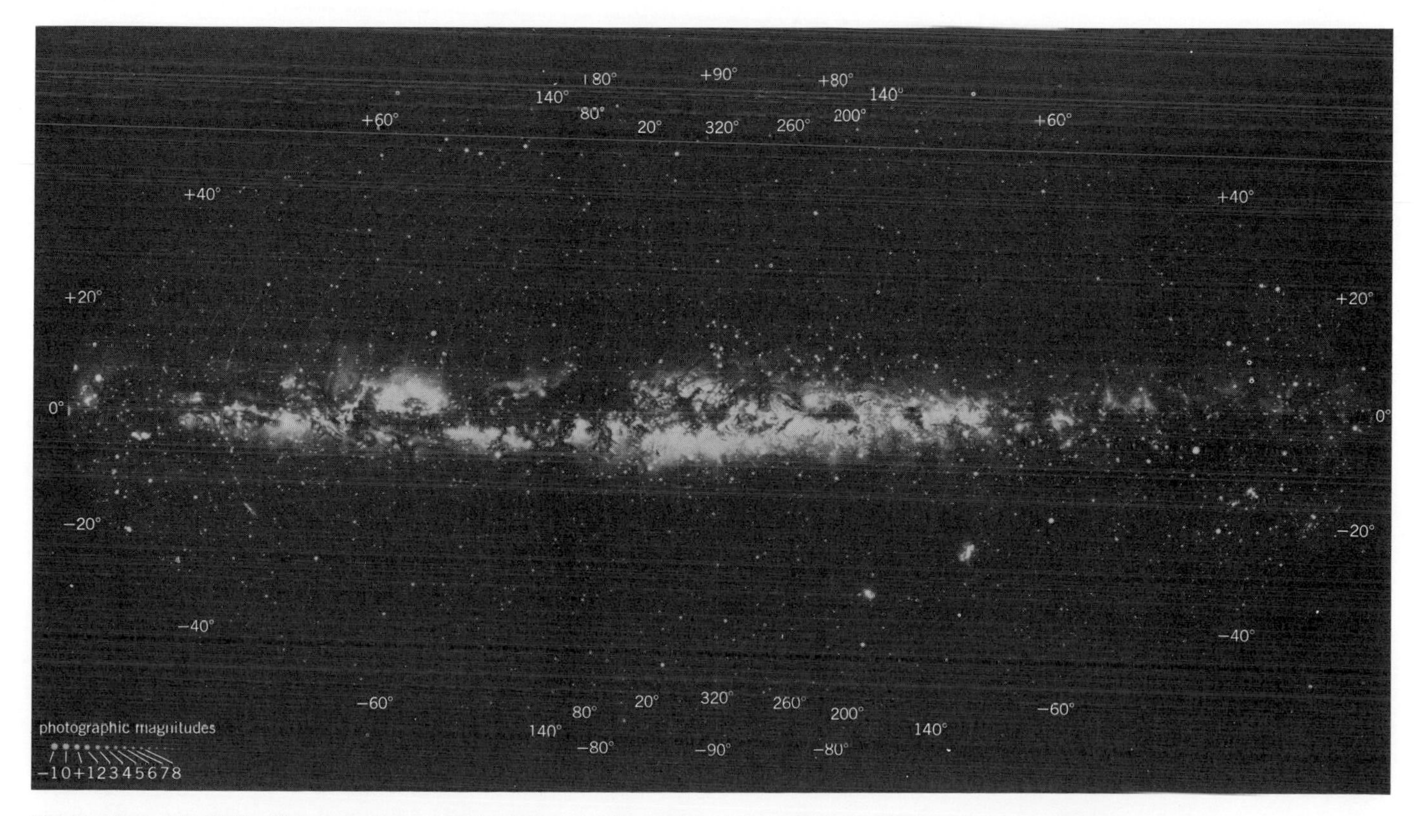

Fig. 1. Map of the Milky Way, based on a variety of photographic sources. The star clouds of Sagittarius, marking the direction toward the center of the Galaxy, are at the center of the map. Galactic coordinates are shown. (*Courtesy of M. and T. Kesküla, Lund Observatory, Sweden*)

Fig. 2. NGC 4565, a spiral galaxy in Coma Berenices, seen edge-on. (*Hale Observatories*)

lar material concentrated in a plane, while irregular galaxies are amorphous, with a great deal of interstellar matter. *See* GALAXY, EXTERNAL.

Counts of stars of different brightnesses, or apparent magnitudes, show that in the vicinity of the Sun the Galaxy is quite flat. This result and the observation of interstellar material near the Sun indicate that the Galaxy is probably a spiral. Figures 2 and 3 show approximately how the Galaxy would appear from outside, as viewed edge-on and nearly face-on. The central bulge, or nucleus, is in the direction of the constellation Sagittarius. There the stars in the Milky Way are closest together (Fig. 1).

Stars of different types occupy different regions of the Galaxy. The youngest and brightest stars of population I, together with interstellar material, are found in the spiral arms in an extremely flat distribution, the galactic disk. Older population I stars form a slightly less flattened subsystem, while the oldest stars, population II, occupy a spherical distribution, or halo, centered about the galactic nucleus. Population II stars are also found in the nucleus. *See* STAR.

In addition to individual stars, the Galaxy contains many star clusters. These are groups of stars which remain together as a result of gravitational attraction and which move through the Galaxy as a unit. Open clusters contain population I stars and are concentrated in the disk of the Galaxy near the spiral arms. Globular clusters are compact, symmetric groups of population II stars and are scattered throughout the halo. *See* STAR CLUSTERS.

The stars in the Sun's vicinity represent a fair sampling of different types of stars. Near the Sun less than 10% of the mass is in interstellar matter. *See* INTERSTELLAR MATTER.

Size and subdivisions. Until early in the 20th century, it was believed that the Galaxy was a huge flattened stellar system with a diameter of about 10,000 parsecs (1 parsec $= 3 \times 10^{13}$ km $= 3.26$ light-years) with the Sun near its center. However, studies of globular clusters by Harlow Shapley revealed that the Sun lies far from the galactic center. The distribution of globular clusters and the magnitudes of RR Lyrae variable stars indicate that the Sun is about 8500 parsecs from the galactic center. Interstellar material dims light passing through it and makes a distance determination uncertain if the amount of material is not well known.

Several subdivisions of the Galaxy are recognized (Fig. 4). The Sun is a fairly mediocre star in the central plane of the Galaxy, the plane of the Milky Way. The nuclear bulge of the Galaxy extends to about 5000 parsecs from the galactic center in the galactic plane. The thickness of the nuclear bulge is about 3000 parsecs. The total mass of the central section of the Galaxy is of the order of 1.4×10^{11} solar masses. The interstellar gas and cosmic dust and the young stars of population I are found only within the inner curve of Fig. 4, hence close to the central plane. The inner section is enveloped by a thin halo probably of ellipsoidal shape. Its maximum radius is 20,000 parsecs. Its mass is probably close to another 1.4×10^{11} solar masses, contained mostly in vast numbers of population II stars. The RR Lyrae variable stars are the most conspicuous population II stars, especially those found in the outlying globular clusters.

The halo is enveloped in a vast corona of unknown composition but with a very great mass. This corona stretches to at least 60,000 parsecs from the center, more likely to 100,000 parsecs. Its estimated mass is in the range between 1 and 2×10^{12} solar masses, far exceeding the mass of the core region and the conventional halo. The star clouds of Magellan, 7 dwarf galaxies (population II), and at least 11 globular star clusters are recognized components of the corona, but they represent only the tip of the iceberg as far as total population and mass are concerned. The Galaxy is comparable in mass and size to the nearest major neighbor, the Andromeda Spiral Galaxy. *See* ANDROMEDA GALAXY; MAGELLANIC CLOUDS.

The outer boundary of the nuclear bulge in the galactic plane at 5000 parsecs from the center is marked by a ring of giant gaseous nebulae. In addition to interstellar gas and cosmic dust (with a considerable number of young stars and star clusters), there are vast numbers of old population II stars within the bulge. Inside there is a ringlike or spirallike, expanding gaseous feature that appears as though it had been expelled from the center 3×10^7 years ago. Infrared and radio data indicate the presence of another smoke ring at 300 parsecs from the center, composed of high-temperature gas. The ring contains a mixture of giant molecular clouds and some very hot gaseous nebulae with associated clusters of hot and young blue-white stars. There are vast numbers of old and dying stars within 10 parsecs or so of the very center of the Galaxy, and radio evidence suggests that there exists a core object with a mass of 5×10^6 solar masses, possibly a massive black hole. *See* BLACK HOLE; SUPERMASSIVE STARS.

Structure. Young, bright stars and emission nebulae lie in the spiral arms, and local spiral structure can be deduced from the directions and distances of these objects from the Sun. After correction for dimming by interstellar material, the results indicate that the Sun lies near the inner edge of a spiral arm, called the Orion arm because of the presence of the Orion Nebula. Another arm, the Perseus arm, lies about 2000 parsecs beyond the Sun. A third arm, the Sagittarius-Carina arm, is about 2000 parsecs closer to the galactic center than the Sun.

Since radio waves are not absorbed by the interstellar medium, spiral structure far from the Sun is mapped by radio telescopes, which detect the continuous radiation from the interstellar medium and the 21-cm line of neutral hydrogen. The density of gas is greater in a spiral arm than between arms, and the amount of radiation detected is high when the telescope is directed tangent to a spiral arm. Distances can be determined from motions of the gas, and an approximate picture of the overall spiral structure results. *See* INTERSTELLAR MATTER; RADIO ASTRONOMY; RADIO TELESCOPE.

The Galaxy has a weak magnetic field whose lines of force run parallel to the spiral arms. The field causes interstellar particles to align along the lines of force, and light passing through the material is polarized. The average strength of the field is estimated to be 10^{-5} gauss.

Fig. 3. M81, a spiral galaxy in Ursa Major, seen nearly face-on, which resembles the Milky Way Galaxy. (*Hale Observatories*)

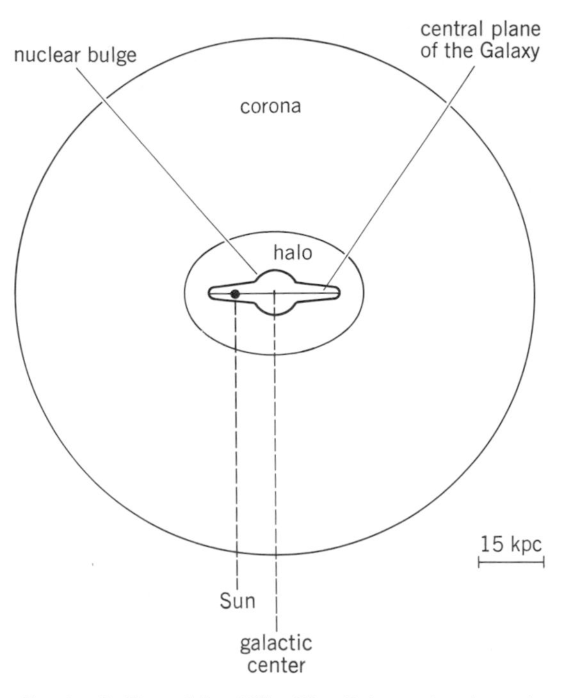

Fig. 4. Outline of the Milky Way Galaxy, showing subdivisions. (*From B. J. Bok, The Milky Way Galaxy, Sci. Amer., vol. 244, copyright 1981 by Scientific American, Inc.; used with permission*)

Structure of the halo is obtained by extending star counts of population II stars above and below the galactic disk. The RR Lyrae variables all have nearly the same intrinsic luminosity. This luminosity and the mean apparent magnitudes of RR Lyrae variables give distances to these stars, and the structure of the halo can be mapped.

Dynamics. Observations of other spiral galaxies show that they rotate about an axis through their centers and perpendicular to their disks. Rotation is also apparent from their highly flattened shapes. The Galaxy, as a spiral, must also rotate, and the Sun and other stars must move in orbits about the galactic center.

The motion of a star relative to the Sun along the line of sight causes a shift in the lines in the star's spectrum. Such motion is radial velocity, and the shift in the spectral lines is known as the Doppler shift. The motion of a star across the line of sight shows up as a change in the direction of the star over a long period of time and is called proper motion. The radial velocity, proper motion, and distance of a star together give the space motion of the star relative to the Sun. Space motions can be determined only in the Sun's vicinity, where most stars move in similar orbits about the nucleus. To learn about the Sun's galactic orbit, distant objects which do not participate in the local motion must be observed. *See* DOPPLER EFFECT.

The globular clusters, for example, are not confined to a flat region and do not rotate with the galactic disk. The motion of the Sun relative to the globular cluster system can be deduced from the radial velocities of the clusters. Radial velocities of nearby groups of galaxies also provide information on the Sun's motion in the Galaxy. Results of these analyses show that the Sun is moving with a speed of about 220–250 km/sec in a nearly circular orbit close to the galactic disk. The period of the Sun's motion is about 200,000,000 years.

Motions of most stars near the Sun are similar to the Sun's motion. However, some stars have high space motions and large motions perpendicular to the galactic disk. These are population II stars and occupy a less flattened region; hence their large perpendicular motions are not surprising.

The 21-cm radiation from neutral hydrogen provides information on the large-scale galactic rotation. When the line of sight is tangent to a spiral arm, the intensity detected by the radio telescope is large and the radial velocity is a maximum. The distance from the center of the Galaxy to the hydrogen cloud being observed depends on the Sun's distance from the center and the angle between the directions of the center and the cloud, known as the galactic longitude. The Sun's motion is subtracted from the observed radial velocity, and circular motions are assumed for both the Sun and the cloud. For each hydrogen cloud observed in this fashion the distance of the cloud from the galactic center R and the circular velocity of the cloud about the center V are found. From these pairs of numbers R and V a graph can be plotted called the rotation curve.

Such a curve is shown in Fig. 5. The distance from the Sun to the galactic center, R_0, is assumed to be 10,000 parsecs (10 kiloparsecs) instead of the preferred value of 8500 parsecs. The inner part of the curve has been known since the 1950s, but the section beyond R_0 dates from about 1980. It is based mostly on data obtained from radio observations of the 21-cm line of neutral atomic hydrogen, plus some optical measurements of radial velocities for gaseous nebulae. Most workers on the motions of the Galaxy had expected that the curve would show a steep downward trend beyond R_0. It does exactly the opposite and meanders upward to 300 km/s at a distance of 20,000 parsecs from the center of the Galaxy. This is conclusive evidence

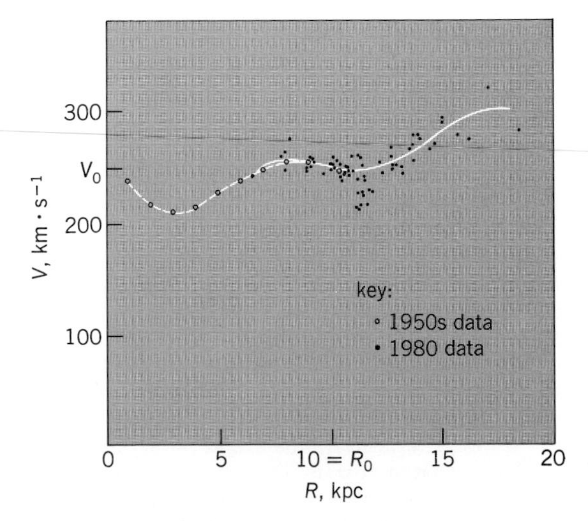

Fig. 5. Rotation curve of the Galaxy. (*From L. Blitz, International Astronomical Union, Star Clusters, Symp. vol. no. 85, 1980*)

for the presence of much mass beyond the limits of the conventional halo shown in Fig. 4, hence for the corona. The evidence from radial velocities of very distant globular clusters indicates that the rotational velocity of 300 km/s persists to at least 60,000 parsecs from the galactic center, which can only mean that there is much matter in the outer corona of the Galaxy.

Evolution. The evolution of the Galaxy must be reconstructed from knowledge of present observed motions and spatial distributions of different types of stars. Initially the Galaxy was a large, essentially spherical mass of gas. This sphere contracted and stars began forming from the gas. The gas continued to contract, and star formation increased as the density of the gas increased. It contracted rapidly to a disk with a central concentration. Stars which were formed before the collapse today have large motions perpendicular to the disk and are spatially distributed in a spherical region—the halo. These stars are old and are members of population II. Stars which formed after the collapse now have small perpendicular motions and are population I stars. Stars born during the contraction now have moderate perpendicular motions and are intermediate population II objects. Star formation is still going on in the spiral arms, where there is interstellar material, but the rate may be slower than in the early days of the disk since the interstellar material is being used up. However, the steady replenishment of interstellar gas through supernova explosions and the like should be taken into account. See STELLAR EVOLUTION; SUN.

[B. J. BOK]

Bibliography: B. J. Bok, The Milky Way Galaxy, *Sci. Amer.*, vol. 244, 1981; B. J. Bok and P. F. Bok, *The Milky Way*, 5th ed., 1981; T. R. Geballe, The central parsec of the Galaxy, *Sci. Amer.*, 241(1):60–70, July 1979; International Astronomical Union, *The Large-Scale Characteristics of the Galaxy*, Symp. vol. no. 84, 1979, *Star Clusters*, Symp. vol. no. 85, 1980, *Radio Molecules*, Symp. vol. no. 87, 1980.

Galaxy, external

One of the large, self-gravitating aggregates of stars, gas, and dust which contain nearly all of the observed matter in the universe. Typical large galaxies have symmetric and regular forms, are about 50,000 light-years (1 ly = 9.46×10^{15}m) in diameter, and are roughly 5×10^{10} times more luminous than the Sun. The stars and other material within a galaxy move through it, often in regular rotation, with periods of a few hundred million years. The characteristic mass associated with a large galaxy is a few times 10^{12} solar masses (1 solar mass $= 2 \times 10^{30}$ kg). Galaxies often occur in associations containing from two to many thousands of individual galaxies and ranging in size from a few hundred thousand to many millions of light-years. Considerable progress has occurred in the study of galaxy nuclei, formation, evolution, and interactions. The nearest observed galaxies are about 75,000 light-years away; the farthest, almost 10^{10}. Galaxies are the landmarks by which cosmologists survey the large-scale structure of the universe.

Composition. The hundreds of billions of stars making up a galaxy are not generally individually observable with current telescope technology because they are too faint and distant. Only the brightest stars in the nearest galaxies can be observed directly with large telescopes. Such stars are of three types: very young, blue, and massive hydrogen-burning (that is, their energy comes from the nuclear fusion of hydrogen into helium) main-sequence stars; very old, red, helium-burning giant stars near the end of their life cycle; and exploding (nova and supernova) or violently variable stars. Although such stars may contribute most of the total visible light from a galaxy, they are few in number; most of the stars and most of a galaxy's mass are in the form of much fainter and lower individual-mass hydrogen-burning main-sequence stars and in the faint burned-out remnants of dead stars. The faint stars in galaxies can be studied only indirectly through the properties of their combined light. See NOVA; STELLAR EVOLUTION; SUPERNOVA.

Two general types of stellar populations are distinguished: One type (Population I) is characterized by the presence of young stars and by ongoing star formation. It is usually associated with the presence of gas. The second type (Population II) shows an absence of gas and young stars as well as other indications that star formation ceased long ago. The Sun is a Population I star. See STAR.

Galaxies contain gas (mostly un-ionized hydrogen) in amounts varying from essentially zero up to a considerable fraction of their total mass. Dust in galaxies, although small in mass (typically 1% of the gas mass), is often dramatic in appearance because it efficiently obscures the starlight. See INTERSTELLAR MATTER.

Form and size. Galaxies generally display strikingly regular forms. The most common form is a disk with a central bulge. The disk is typically 100,000 light-years in diameter and only about 1000 light-years thick. Its appearance is characterized by radially decreasing brightness with a superposed spiral or bar pattern, or both (Figs. 1 and 2). The central bulge may vary in size from hundreds to many thousands of light-years and may be spherical or barlike in shape. Such galaxies are classified as spirals (S) or barred spirals (SB) and subclassified a, b, or c (for example, Sa, SBc) to distinguish increasingly open spiral structure and small bulge size. The disks of these galaxies are dominated by Population I stars, while their bulges contain mainly Population II stars. The Milky Way Galaxy is an Sb type. See GALAXY.

Another common type of galaxy is a featureless ellipsoid with radially decreasing brightness. These galaxies are classified as ellipticals (E) and subclassified according to their axial ratios by a number from 0 (E0 = round; Fig. 3) to 7 (E7 = 3-to-1 axial ratio). They may vary in size from thousands to several hundred thousand light-years. They are most commonly found in clusters of galaxies and rarely contain gas or dust. The brightest galaxies are usually ellipticals. They are dominated by Population II stars.

Other, more rare forms of galaxies include a class called S0 which has a disk superimposed on

Fig. 1. The great spiral galaxy in Andromeda (M31, NGC 224) and its two small elliptical companions (NGC 205 and 221), photographed with the 48-in. Schmidt telescope. (*Hale Observatories*)

an otherwise elliptical type of light distribution, and an irregular (Irr) class composed of galaxies with chaotic forms (Fig. 2). The latter class contains no more than a few percent of all galaxies. Irregulars are generally of low total luminosity and rich in gas, dust, and Population I stars.

Internal motions. The motions of the stars and gas within galaxies are of two types, random and rotational. All of the motions are the result of the gravitational interactions of the stars with each other. The galaxies are supported against gravitational collapse by these motions in the same sense in which the planets of the solar system are kept from falling into the Sun by their orbital motions. The random motions are complex and result in highly eccentric and irregular orbits for the individual stars. The rotational motions correspond to ordered and systematic circular orbits. Usually, the inner regions of a galaxy undergo solid body rotation (velocity proportional to radius), while the outer regions rotate differentially (velocity constant). The total velocity of the material within galaxies varies from one to several hundred kilometers per second. Typical orbital periods for stars are several hundred million years.

Fig. 2. The "Whirlpool" galaxy (NGC 5194), type Sc, and a companion irregular satellite (NGC 5195).

The distribution of kinetic energy into random and rotational motions varies with the galaxy type. The disk of a spiral galaxy may have only about 1% of its total kinetic energy in random motions, while an E galaxy may have essentially all of it there.

A few peculiar galaxies have ionized gas flying out of their nuclei (that is, their very central regions) with velocities ranging up to tens of thousands of kilometers per second. The detailed origin of these motions is unknown but is thought to be associated with violent explosions in the galactic nuclei.

Luminosities. The number of galaxies with total luminosities L is roughly proportional to L^{-y}, where y is between 1 and 1.5 for luminosities less than about 3×10^{10} solar luminosities. The number is roughly exponentially cut off for higher luminosities. The brightest observed galaxies are fainter than 2×10^{11} solar luminosities; the faintest, brighter than about 10^6. Galaxies with old stellar populations (Population II) give off more of their total luminosity in the red than galaxies with young

Fig. 3. Photograph of the E0 galaxy M87 (NGC 4486) in the Virgo cluster constellation. This galaxy is a source of radio emission, and it has a nucleus which is active. (*Hale Observatories*)

stellar populations (Population I). The Milky Way Galaxy's total luminosity is roughly 10^{10} times greater than the Sun's.

This distribution of luminosities is such that while there are very many faint galaxies, they do not contribute a large fraction of the total light given off by galaxies. For instance, if $y = 1$, galaxies brighter than 4×10^9 solar luminosities are responsible for almost 90% of all the light from galaxies. This is important because only the brighter galaxies, visible for great distances through space, can be observed easily and in great numbers.

It appears that the masses of galaxies are roughly proportional to their luminosities. If true, this would mean that the distribution of galaxy masses is described by a relation similar to that for their luminosities (that is, a power law with an exponential cutoff). It would also mean that the bright galaxies contain most of the mass.

Clustering. Although galaxies are scattered through space in all directions for as far as they can be observed, their distribution is not uniform or random. Most galaxies are found in associations (Fig. 4) containing from two to hundreds of individual galaxies (with a median number of about five). The clustering of galaxies may be described by a covariance function which gives the excess probability (above random) of finding two galaxies with a separation r. The observed covariance function is roughly proportional to $r^{-1.8}$. P. J. E. Peebles and collaborators, who pioneered in the measurement of the covariance function, have claimed that this result indicates that galaxy clustering arose from the gravitational growth of random density fluctuations in the early universe.

Like the stars within a galaxy, the galaxies within a cluster move about under the influence of their mutual gravitational attraction. The motions are generally random and show little evidence of rotation. Typical velocities range from about a hundred up to several thousand kilometers per second.

For reasons which are unknown, there is a tendency for E and S0 galaxies to be concentrated in large clusters more strongly than spirals.

The clustering of galaxies may be a hierarchial process. This would mean that the clusters themselves are clustered to form superclusters, the superclusters are in turn clustered to form yet higher order clusters, and so on. Although this is the most popular theory of galaxy clustering and is supported by the existence of several observed superclusters, it is far from an established fact. The search for a clear understanding of galaxy clustering is still underway. *See* SUPERCLUSTERS.

Masses. There are two methods for measuring the masses of galaxies: Either the velocities of stars within a galaxy or of galaxies within a cluster are measured. Then, given the size of the system, it is possible to deduce the amount of mass required to gravitationally generate the observed

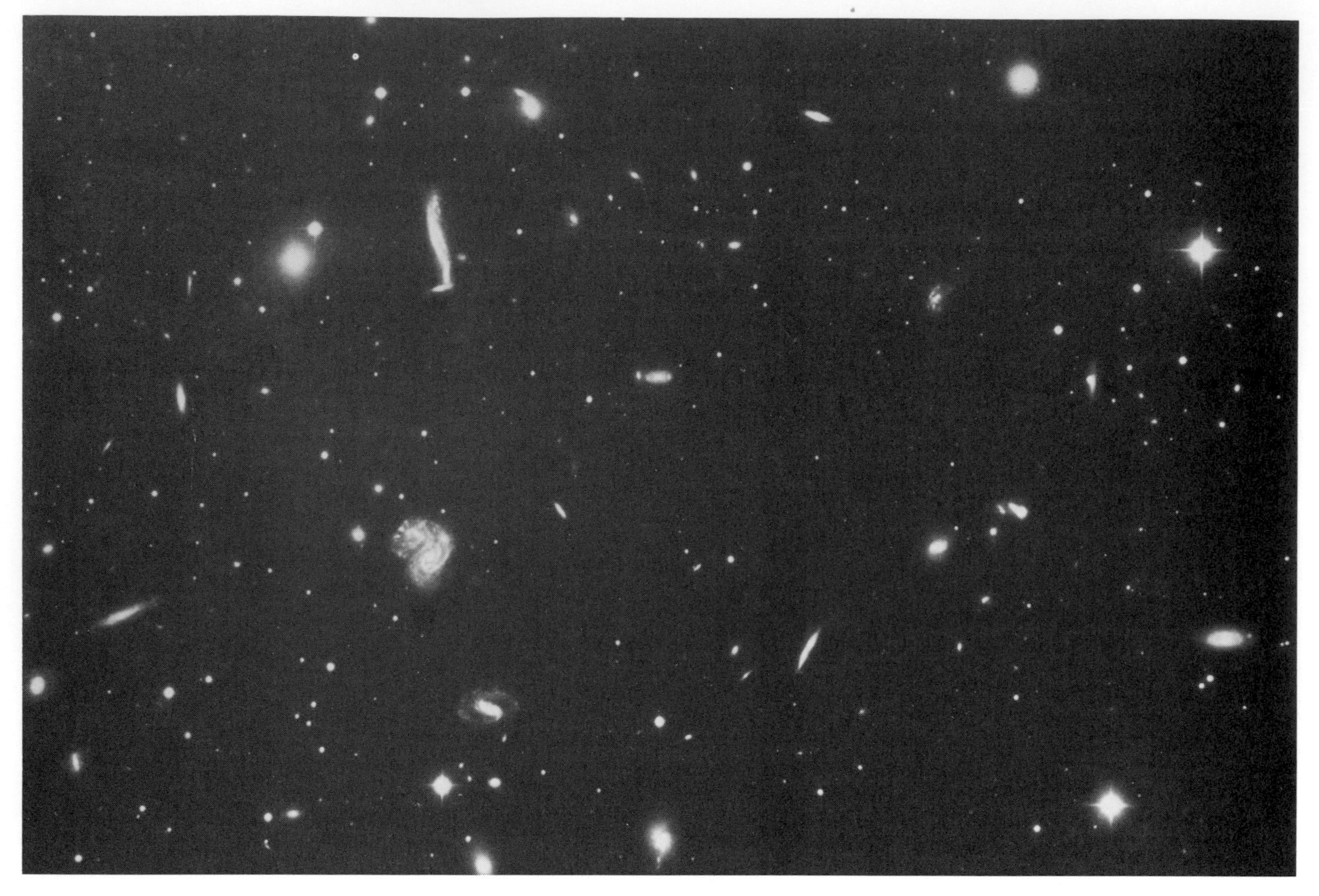

Fig. 4. Photograph illustrating the clustering characteristic of galaxies. This association of galaxies in the constellation Hercules includes many different types of galaxies. (*Hale Observatories*)

velocities. When applied to stars within individual galaxies, the method gives typical masses near 2×10^{11} solar masses. But when applied to galaxies within clusters, typical masses per galaxy come out in the range of 3×10^{12} solar masses. This order-of-magnitude disagreement has led J. P. Ostriker and others to suggest that galaxies may be enclosed in extensive halos containing most of the galaxy's mass but giving off little or no light. Although not directly observable, such halos would explain the mass discrepancy and help to explain certain other properties of galaxy rotation velocities. If correct, the massive halo hypothesis means that the systems normally referred to as galaxies are no more than the bright central regions of much larger and more massive aggregates of dark matter. Another possible explanation of the mass discrepancy is that clusters contain large amounts of unobserved material in some unknown form. Whatever its ultimate solution, the mass problem is one of the most important puzzles in the study of galaxies.

Interactions. As galaxies move about within clusters, they will occasionally pass very near one another or even collide directly. The effect this has on the individual galaxies depends critically on the details of the encounter. Possible outcomes include loss of material from the galaxies' outer regions, transfer of material from one galaxy to another, merger of the two galaxies, modification of the galaxies' forms by tidal perturbations, and loss of gas and dust due to collisional heating. Alar and Juri Toomre have explained the 10 or so most peculiar observed forms of galaxies as tidal perturbations during grazing collisions.

Nuclei. In the very central regions (sizes at least as small as a light-year) of galaxies, violent and apparently explosive behavior is often observed. This activity is manifested in many ways, including the high-velocity outflow of gas, strong nonthermal radio emission (implying relativistic particles and magnetic fields), intense and often polarized and highly variable radiation at infrared, optical, ultraviolet, and x-ray wavelengths, and ejection of jets of relativistic material. In the most extreme cases (called Seyfert and N galaxies) the energy in the nuclear activity surpasses that in the rest of the galaxy combined. The ultimate source of this activity, whether in some extreme stellar population or in some previously unobserved type of object (for example, black holes) remains completely unknown. There are many astronomers who suspect that active galactic nuclei are related to the even more extreme, but in many ways similar, phenomenon of quasistellar objects. *See* BLACK HOLE; QUASARS.

Distances. The nearest galaxies to the Milky Way Galaxy are the Large and Small Magellanic Clouds, two small irregulars lying about 75,000 light-years away. They are visible to Southern Hemisphere residents as two faint patches of light in the constellations of Norma and Toucan. The nearest bright galaxy is the Andromeda Nebula, M31 (Fig. 1), visible to the naked eye in the constellation of the same name. It is at a distance of almost 2×10^6 light-years. These galaxies and a handful of fainter ones (all within a few million light-years) make up the Local Group, of which the Milky Way Galaxy is a member. *See* ANDROMEDA GALAXY; MAGELLANIC CLOUDS.

Within a distance of about 6×10^7 light-years, there are hundreds of galaxies which have large total luminosities but which are too far away to be seen with the unaided eye. Many of these can be seen or photographed with a small telescope. Some of the most spectacular of these are contained in Messier's famous list (M51, M81, M82, M87, and so on). Some are members of the nearest significant cluster of galaxies in the constellation Virgo. *See* MESSIER CATALOG.

The Coma cluster is the best-observed large cluster of galaxies. Its distance of about 3×10^8 light-years requires large (usually professional) telescopes. The most distant galaxies detectable with the largest telescopes are nearly 10^{10} light-years away.

Cosmology. Galaxies are closely associated with cosmology because they are the beacons by which the universe is mapped. In 1929 Edwin Hubble discovered that the universe is expanding by showing that galaxies are receding from the Milky Way Galaxy at a speed v proportional to their distance d ($v = H_0 d$ where H_0 is called the Hubble constant). The Hubble relation in turn led to the conclusion that the universe was formed in a Big Bang about 10^{10} years ago. *See* BIG BANG THEORY; HUBBLE CONSTANT.

Studies of very distant galaxies can, in principle, reveal whether or not the mutual gravitational attraction is sufficient to stop and then reverse the Hubble expansion. Despite intense efforts by a number of astronomers, this goal has not been clearly achieved. The best available results seem to indicate that the deceleration is not sufficient.

The same problem can be attacked by attempting to estimate the density of the universe from a study of relatively nearby galaxies. If the density exceeds a certain critical value (2×10^{-29} g cm^{-3}), the universe will eventually stop expanding and recollapse. The best available data indicate that the density of galaxies is a few tenths of the critical density. Thus, barring large amounts of unobserved mass, the universe will continue to expand forever.

[EDWIN L. TURNER; JOYCE B. TURNER]

Evolution. How these diverse objects originated and evolved into their present form is a topic of intense speculation among astrophysicists.

Evidence from structural properties. Some clues can be discerned in certain structural properties of galaxies. One of the most regular forms is the smooth and round light distribution of an elliptical galaxy. Astronomers calculate that an inhomogeneous distribution of stars will take many billions of years to relax into such a uniform pattern. In fact, the most plausible explanation is that the stars formed out of a collapsing gas cloud. The rapidly changing gravitational pull experienced by different stars as the collapse proceeds has been shown by means of sophisticated computer experiments to rearrange the stars into the observed shape of an elliptical galaxy.

The highly flattened disks of spiral galaxies must have formed during a similar collapse, but it is believed that most star formation did not occur until the rotating gas cloud had already flattened into a pancakelike shape. Had the stars formed at an earlier stage of the collapse, their rapid motions would have led to the formation of an elliptical galaxy. However, if the cloud stays gaseous until it flattens, much of the kinetic energy of the collapse is radiated away by gas atoms. Subsequent star formation is found to maintain a highly flattened, disklike shape, characteristic of spiral galaxies.

The flattening occurs in part because of the centrifugal forces in the rotating cloud. A confirmation of this picture has come from the discovery that elliptical galaxies rotate much less rapidly than spiral galaxies. This raises the question of the origin of the rotation itself. A natural explanation seems to lie in the action of the gravitational torques exerted by neighboring galaxies (or rather, protogalaxies: galaxies in the process of formation). Much as the Moon and Sun cause tides in the Earth's ocean, so can a massive neighbor induce a protogalaxy to begin to rotate.

Evidence from composition. Another aspect of galaxies that has evolutionary significance is their composition, and in particular, the actual distribution of heavy elements. The amount of heavy-element enrichment can be inferred by measuring the color of the starlight, blue stars characteristically being metal-poor and red stars being metal-rich. Galaxies are found to be significantly bluer in their outermost regions and redder toward their central nuclei. Such gradients in color can have developed only during the galaxy formation stage, since most of the starlight from the galaxy displays this trend. The explanation seems to be that the galaxy formed out of a collapsing gas cloud that formed stars in a piecemeal fashion. As stars formed, they evolved, underwent nuclear reactions, produced heavy elements, and eventually shed enriched material (some stars even exploding as supernovae). Successive generations of stars formed out of the debris of earlier stars, and in this way the stellar content of the galaxy systematically became enriched. The greatest enrichment would naturally occur toward the center of the galaxy, where the gaseous stellar debris tended to collect. *See* ELEMENTS AND NUCLIDES, ORIGIN OF.

In this manner, theories of galactic evolution have been constructed that explain in outline many of the observed characteristics of galaxies. Detailed models of protogalaxies have even been developed, but because of the very limited knowledge of the fundamental process of star formation, quantitative explanations for most properties of the galaxies have not yet been found. *See* STELLAR EVOLUTION.

Origin. An outstanding and unresolved issue concerns the origin of the primordial gas clouds out of which the galaxies evolved. As scientists attempt to look back into the early stages of the universe, the view becomes increasingly obscure. The cosmic background radiation yields a glimpse of the universe prior to the epoch of galaxy formation. This radiation has a blackbody spectrum that corresponds to a temperature of only 3 kelvins above absolute zero. It is the remnant of the primeval fireball radiation, created in the early stages of the big bang, when the universe was less than a year old. The universe is now completely transparent to the cosmic background radiation. However, at an epoch corresponding to about 10^9 years after the big bang, the radiation was sufficiently hot that matter was ionized, and was also sufficiently dense to render the universe completely opaque to the radiation. To look out to greater and greater distances is also to look back in time. The cosmic background radiation originates in the most distant regions of the universe that can be "seen" with optical or radio telescopes. To observe the background radiation now (some 10^{10} years later) is to see back to this early epoch, known as the decoupling epoch: at earlier times, matter and radiation were intimately linked, and subsequently the radiation propagates freely to the present time. (However, it is possible that after galaxies have begun to form, there may be sufficient ionizing radiation produced to reionize the intergalactic medium and cause additional scattering of the cosmic background radiation: this could happen at epochs as late as 10^9 years after the big bang.) *See* COSMIC MICROWAVE RADIATION.

Theory of formation from fluctuations. The cosmic background radiation is found by radio astronomers to be very uniform. This indicates that the matter distribution in the early universe also was uniform, to at least 1 part in 10^3. However, it cannot have been completely without any structure, otherwise galaxies could not have formed. This apparent paradox is resolved by the assumption that small inhomogeneities or fluctuations in the matter distribution were present in the early universe. The mutual action of gravity exerted between these infinitesimal fluctuations results in their gradual enhancement. Eventually, great gas clouds develop that will collapse to form galaxies. Thus the seeds from which galaxies grew by the action of gravitational instability are infinitesimal density fluctuations in the very early universe. The required amplitude for these primordial seed fluctuations must be of the order of 1 part in 10^3.

Numerical simulations of galaxy clustering have enabled the spectrum of fluctuation length-scales and amplitudes to be inferred. It seems likely that these density inhomogeneities were distributed in a random fashion, not unlike that of noise or turbulence. However, the source of the fluctuations remains a mystery. One possibility is that the fluctuations developed from random statistical fluctuations. If the initial stages of the big bang were matter-dominated, rather than being predominantly radiation, this mechanism yields a promising source of fluctuations. Unfortunately, the theory of the form of nuclear matter at the incredibly high densities that are attained in the big bang does not allow a definite choice between a cold big bang (one in which the initial explosion is matter-dominated) or a hot big bang (one in which it is radiation-dominated). Even if the initial big bang was hot, many cosmologists believe that the big bang was itself not completely uniform, but already contained at the initial instant of expansion the nascent seeds of future galaxies.

Theory of chaotic early universe. According to an alternative viewpoint, the early universe was extremely irregular and chaotic. In this case, cosmologists appeal to physical processes to remove the inhomogeneities in a chaotic universe rather than to create mild fluctuations in a uniform universe, on the grounds that a highly irregular initial state is more likely than the very specific state of the idealized uniform big bang cosmology. This program has not yet met with success, in part because highly inhomogeneous cosmological models are not readily amenable to study. It seems inevitable, however, that primordial black holes form in a chaotic universe, and these could be the nuclei around which galaxies eventually are accreted. There remains the important constraint on this hypothesis of accounting for the isotropy of the cosmic background radiation, for the processes that have yielded structure on galactic scales have evidently conspired to leave the matter and radiation distribution exceedingly uniform on much larger scales. Cosmologists conclude that both approaches, that of the relatively smooth early universe and that of primordial chaos, are confronted with difficulties, and they cannot yet choose between these alternatives.

Isothermal and adiabatic fluctuations. The most notable success of studies of fluctuations in the early universe has been the realization that fluctuations can be categorized into distinct varieties. Of particular importance for galaxy formation are density fluctuations that are found to generally be a combination of two basic types: isothermal and adiabatic. Primordial isothermal fluctuations consist of variations in the matter density, without any corresponding enhancement in the radiation density. Consequently, in the radiation-dominated early phase of the big bang, isothermal fluctuations neither grow nor decay, as the uniform radiation field prevents any motion. Once the universe becomes transparent, the matter fluctuations respond freely to gravity and grow if above a certain critical size. This critical scale (the Jeans length) represents the balance point between attractive gravitational and expansive pressure forces. The smallest isothermal fluctuations that can become enhanced and form gas clouds contain about 10^6 solar masses. On the other hand, primordial adiabatic fluctuations are analogous to a compression of both matter and radiation. The diffusive tendency of the radiation tends to smooth out the smaller adiabatic fluctuations. This process remains effective until the decoupling epoch, and only adiabatic fluctuations which contain upward of 10^{12} solar masses can survive to eventually recollapse into gas clouds and galaxies.

Many cosmologists believe that the emergence of these two mass scales by means of a cosmic filter from a random spectrum of primordial fluctuations should be seriously examined: The minimum scale of isothermal fluctuations corresponds to the masses of globular star clusters, known to contain the oldest stars and speculated to be the building blocks of the luminous regions of galaxies. The minimum scale of adiabatic fluctuations corresponds to that of massive galaxies; indeed it is possible that all luminous galaxies have associated massive halos that contain 10^{12} solar masses or more of dark matter of unknown nature. Conceivably, both types of fluctuations may have been present in the early universe. Discovery of small fluctuations in the cosmic microwave background radiation, required at some level by all theories of galaxy formation, should eventually allow a choice between the alternative possibilities. *See* COSMOLOGY; STAR CLUSTERS.

[JOSEPH SILK]

Bibliography: T. Ferris, *Galaxies*, 1980; J. R. Gott, III, et al., Will the universe expand forever?, *Sci. Amer.*, 243(3):62–79, 1976; W. J. Kaufmann, III, *Galaxies and Quasars*, 1979; A. Sandage, *The Hubble Atlas of Galaxies*, 1961; A. Sandage, M. Sandage, and J. Kristian (eds.), *Stars and Stellar Systems: Galaxies and the Universe*, 1976; R. J. Tayler, *Galaxies: Structure and Evolution*, 1979.

Gamma-ray astronomy

The study of gamma rays of cosmic origin. The field may be divided into the study of high-energy gamma rays and that of soft gamma-ray bursts. High-energy gamma-ray astronomy originated about 1965, but the most relevant results have been obtained only since the mid-1970s. Gamma-ray bursts were first observed in 1967, and their discovery announced in 1973. The so-called "high-energy" range for gamma rays is at the far end of the electromagnetic spectrum: the photons observed have an energy of more than about 30 MeV, up to a few gigaelectronvolts. The gamma-ray bursts have a typical energy 100 times lower, and do not seem at the present time to be related to the high-energy gamma rays.

HIGH-ENERGY GAMMA RAYS

The operating principles and basic limitations of gamma-ray telescopes will be discussed, and the results of observations with these instruments will then be described.

Gamma-ray telescopes. Because, in the high-energy gamma-ray range, the photons (simply called gamma rays in the following) are completely absorbed by the atmosphere, it is necessary to place the gamma-ray detectors aboard high-altitude balloons, or better still, artificial satellites. Most of the pioneering work on extraterrestrial gamma rays was done with balloons, but although technical constraints are far more severe, only satellites are used now. Most of the results given below were obtained by the American *SAS 2* satellite launched in November 1972, and the European *COS-B* satellite launched in August 1975. *See* SATELLITE ASTRONOMY.

The heart of a gamma-ray telescope is a spark chamber, similar to chambers used with particle accelerators, but much smaller (Fig. 1). The incoming gamma-rays are absorbed in the plates and converted into electron-positron pairs. Since the gamma rays are invisible, it is the analysis of the electron and positron tracks, visualized by sparks, that gives the directional and energetic information about the detected photons. The location of the sparks is obtained through a digitized wire array. The spark chamber is triggered by the passage of electrons through various electronic devices combined with photomultiplier tubes at the bottom of the spark chamber. Incoming charged cosmic rays do not trigger the spark chamber because of an anticoincidence system.

Telescope limitations. The information obtained by gamma-ray telescopes is limited by various factors, although these limitations should become less restrictive with the development of more advanced detectors. These limitations are both instrumental and astronomical in nature.

Instrumental limitations. The main instrumental limitations are in angular resolution and energy resolution.

1. Angular resolution. The scattering of the electrons and positrons as they cross the successive spark chamber plates results in a severe loss of directional information. The accuracy with which the original direction of a single photon is reconstructed is about 5° at 100 MeV, somewhat better at higher energies, but worse at lower energies. An actual observation, however, involves many such photons, and statistical considerations make it possible to lower this figure: typical positional error boxes on the sky have a 1° radius for *COS-B*. In other words, the actual image of a point in the sky is a blurred spot of about 5° radius (this is the angu-

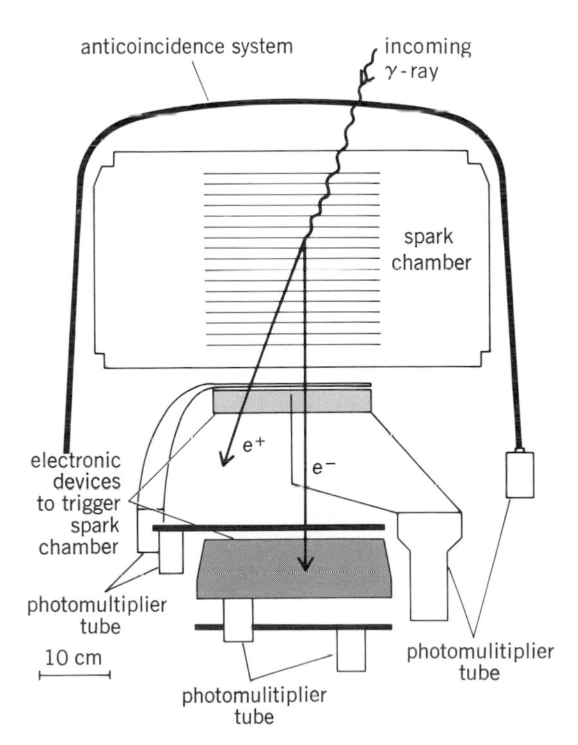

Fig. 1. Gamma-ray telescope aboard the *COS-B* satellite. (*From L. Scarsi et al., The COS-B experiment and mission, in Proceedings of the 12th ESLAB Symposium, Frascati, Italy, May 24–27, 1977, European Space Agency, SP-124, pp. 3–11, 1978*)

lar resolution), but its position can be determined with a 1° accuracy, sometimes better. This is a crucial factor when interpreting the data. Nevertheless, because the field of view of a gamma-ray telescope is about 20° in radius, one is able to obtain images of the sky, even if of a relatively poor quality, analogous to the images of an out-of-focus photographic lens.

2. Energy resolution. The measurement of the energy of the incoming gamma rays is not very precise. At present, it involves an uncertainty of about 50–100%. It is not impossible, in principle, to obtain satisfactory continuous spectra, but limitations, of an astronomical nature, which are discussed below, do make it impossible to obtain such spectra in small regions of the sky, except in a few cases.

Astronomical limitations. The astronomical limitations of gamma-ray telescopes result from the weak intrinsic fluxes of gamma rays and from background problems. The fluxes of extraterrestrial gamma rays are on the order of 10^{-2} per square meter per second; as a consequence, a typical gamma-ray observation requires long exposures, of about 1 month duration, in order to get a significant image. The main background sources originate outside the instrument: cosmic rays interact with it to give secondary gamma rays, and in some regions the sky itself is relatively so bright that it becomes difficult to resolve structures or localized sources of gamma rays. Even when this is possible, the image contrast is in most cases so low that one cannot extract a significant spectral information on these sources.

Gamma-ray sky. At wavelengths visible to the naked eye, the strong concentration of stars and dust which makes up the Milky Way is conspicuous and provides an edge-on view of the disk of the Galaxy. The gas contained in the Galaxy is invisible at other than radio or soft x-ray wavelengths. Not all the sky has been investigated by *SAS 2* or *COS-B*; however, the overall view obtained with their out-of-focus gamma-ray "lenses" (Fig. 2) is basically similar to the naked-eye sky, but also shows very significant differences.

The gamma-ray image of the Milky Way appears as a bright lane of diffuse emission, much like the optical image, although thinner (2° wide). It spans roughly the same angular extent, from −60° to +60° in galactic longitude (that is, with respect to the galactic center, located in the constellation Sagittarius). Two bright spots, in the constellations Cygnus and Vela, are also visible. There is a fainter emission in other parts of the galactic disk.

The gamma-ray sky is almost devoid of "stars," except along the galactic plane, where there is a strong concentration of gamma-ray sources (that is, "stars" of 1° radius). This is remarkable, as it is very different from what is seen at other wavelengths, even in x-rays.

Diffuse emission. The galactic diffuse emission is thought to be mostly the result of the interaction of cosmic rays (electrons and protons) with the interstellar gas throughout the Galaxy. Relativistic electrons (of energies above 30 MeV) radiate gamma rays via the bremsstrahlung process (braking radiation) when they cross the electric fields of nuclei; fast protons (of energies above 1 GeV) indirectly generate gamma rays when they collide with protons at rest, because in the course of the reaction elementary particles called π^0 (neutral

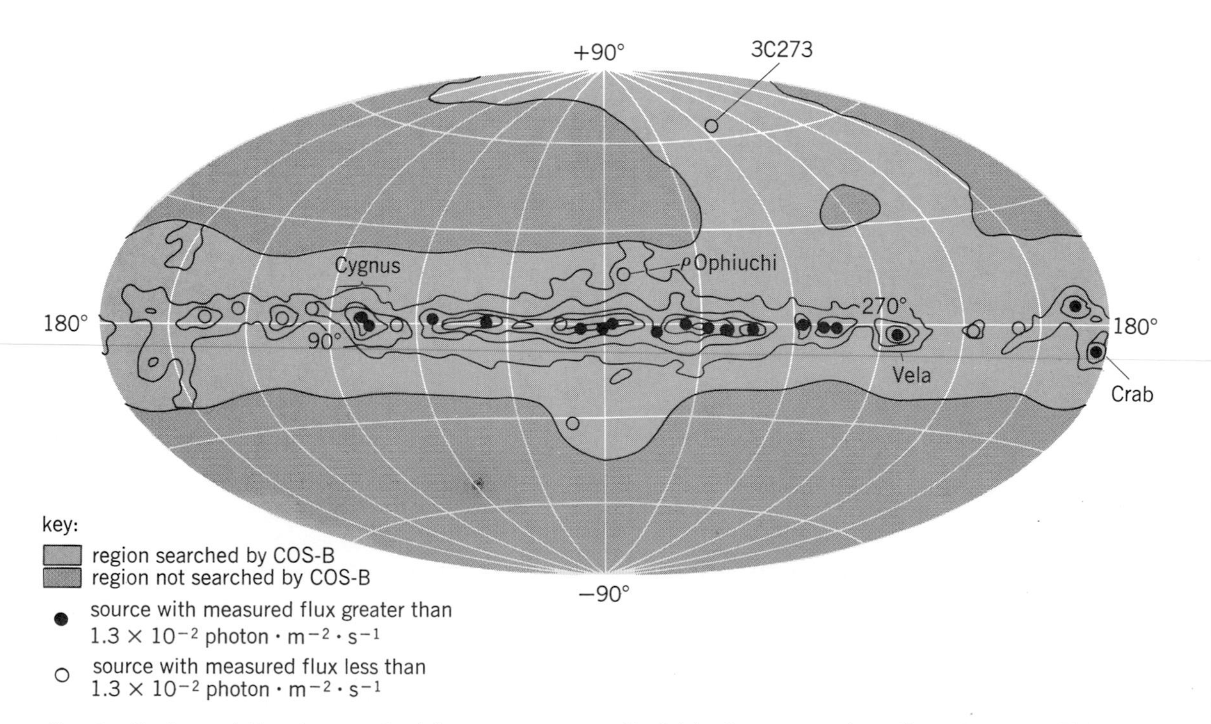

Fig. 2. Regions of the sky searched for gamma-ray sources by *COS-B* and sources detected above 100 MeV as of June 1979. Each source is located within a 1° radius error box. The diffuse gamma-ray emission of the galactic disk is shown as a schematic contour map. There is a factor of 2 in intensity between each contour, starting from a background of 4×10^{-3} photon · s^{-1} · sr^{-1} recorded by the telescope.

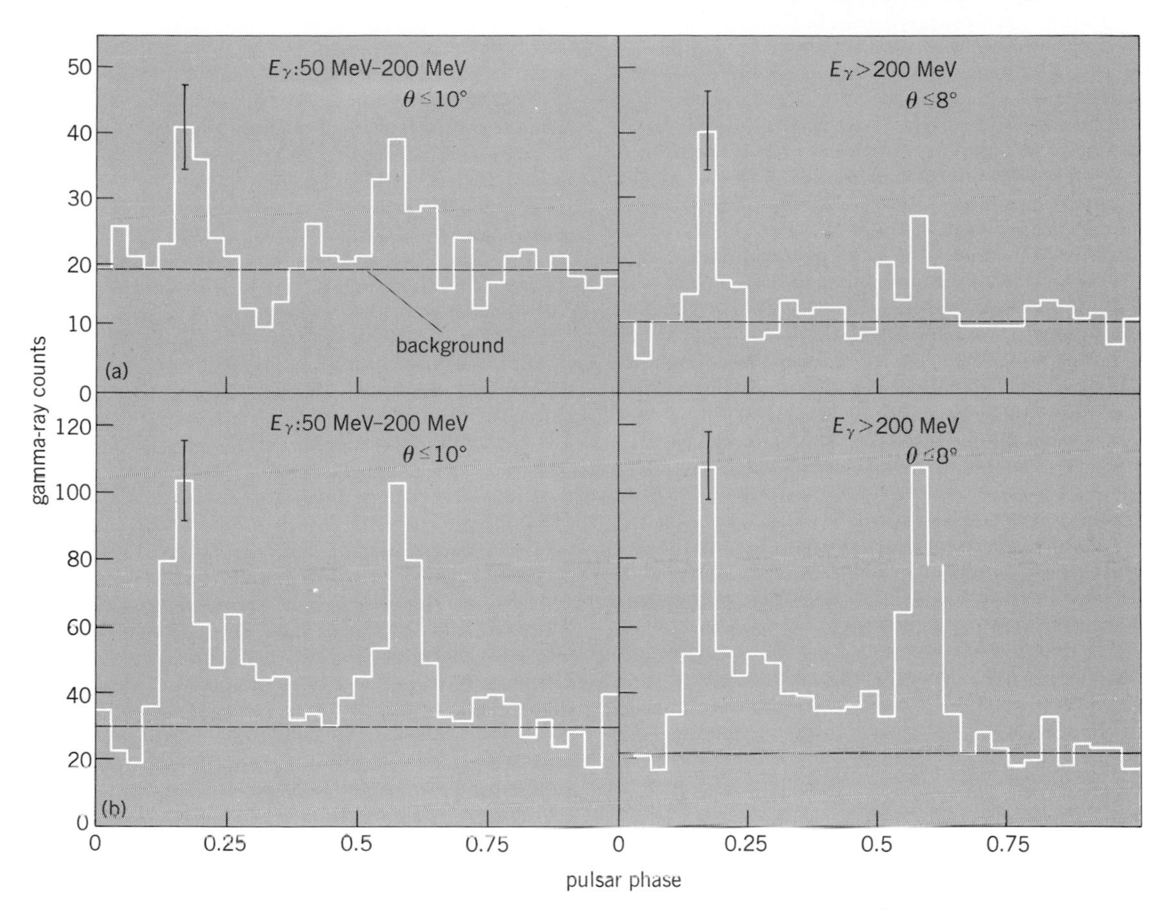

Fig. 3. Light curves of the (*a*) Crab and (*b*) Vela pulsars, from the *COS-B* data. Each light curve shows the number of photons above a given energy E_γ, coming within an angle θ to the axis of the gamma-ray telescope, and collected here as a result of several months of pointing toward the source. (*From K. Bennett et al., COS-B observations of pulsed gamma-ray emission from PSR 0531+21 and PSR 0833−45, Astron. Astrophys., 61:279–284, 1977*)

pions) are created, which decay almost instantaneously in two gamma rays. Bremsstrahlung and π^0 decay contribute about equally to the diffuse gamma-ray emission from the galactic plane.

Given the distribution of the interstellar gas in the Galaxy, one can in principle use gamma rays as a tracer of cosmic rays and deduce their density at any point. This is quite central to the problem of the origin of cosmic rays in general, and to other problems such as the structure of the Galaxy. The distribution of the interstellar gas is still controversial, and little new quantitative information on cosmic rays can actually be deduced from diffuse gamma rays. There is nevertheless evidence that some areas are devoid of cosmic rays. *See* COSMIC RAYS.

Also contributing to the diffuse galactic emission are unresolved sources, but their overall distribution is essentially unknown. *See* GALAXY.

Localized sources. Resolved, localized sources of gamma rays numbered 25 in 1980, and new ones are certain to be discovered. Most of them have been isolated by *COS-B* but the first ones were discovered by *SAS 2*. Any astronomical object with an angular extent up to 1° can be in principle a possible counterpart for one of these sources.

The only certain identifications to date are those of the Crab and Vela pulsars, because the photon arrival-time analysis gives a light curve (Fig. 3) in phase with that observed at other wavelengths. These light curves show conspicuously two peaks, separated by a fraction 0.4 of the pulsar period (33 ms for the Crab, 89 ms for Vela), whereas these two pulsars are markedly different at other wavelengths. The Vela pulsar is the brightest gamma-ray source in the sky. The gamma-ray emission from pulsars, which carries the bulk of the energy released by them, is not understood at present. *See* PULSAR.

Probable identifications, based on positional coincidences in low-sky-background regions, include one of the nearest quasars, 3C273, and a nearby dense interstellar cloud, near the star ρ Ophiuchi. In the quasar, the first extragalactic gamma-ray source ever discovered, the energy released in gamma rays is nearly as great as at all other wavelengths combined. This poses severe, unsolved physical problems. On the other hand, the flux coming from the ρ Ophiuchi cloud implies a cosmic-ray density at least five times higher than in the solar neighborhood: for some reason, this cloud, or objects nearby, appears to be a cosmic-ray source. No such source has ever been isolated so far. Suggested physical mechanisms include cosmic-ray acceleration by supersonic winds from very young stars, associated with the cloud.

Out of the remaining sources, possible identifications have been proposed for nearly a half, in terms of supernova remnants interacting with massive dense interstellar clouds or with the very young stars, traced by "OB associations," to which they are thought to give birth; for instance, the Cygnus region. Cosmic-ray densities would be very high (10–100 times the solar neighborhood value). Along these lines, the birthplace of stars could also be a major birthplace of cosmic rays. Possible physical processes for the generation of cosmic rays include particle acceleration to cosmic-ray energies by supernova shock waves or by supersonic stellar winds (like in the ρ Ophiuchi cloud). These ideas should be taken as tentative at the present time. *See* SUPERNOVA.

No identifications have been proposed yet for the remaining sources; in some cases, a possible counterpart does not even exist, since nothing is visible at other wave lengths.

[T. MONTMERLE]

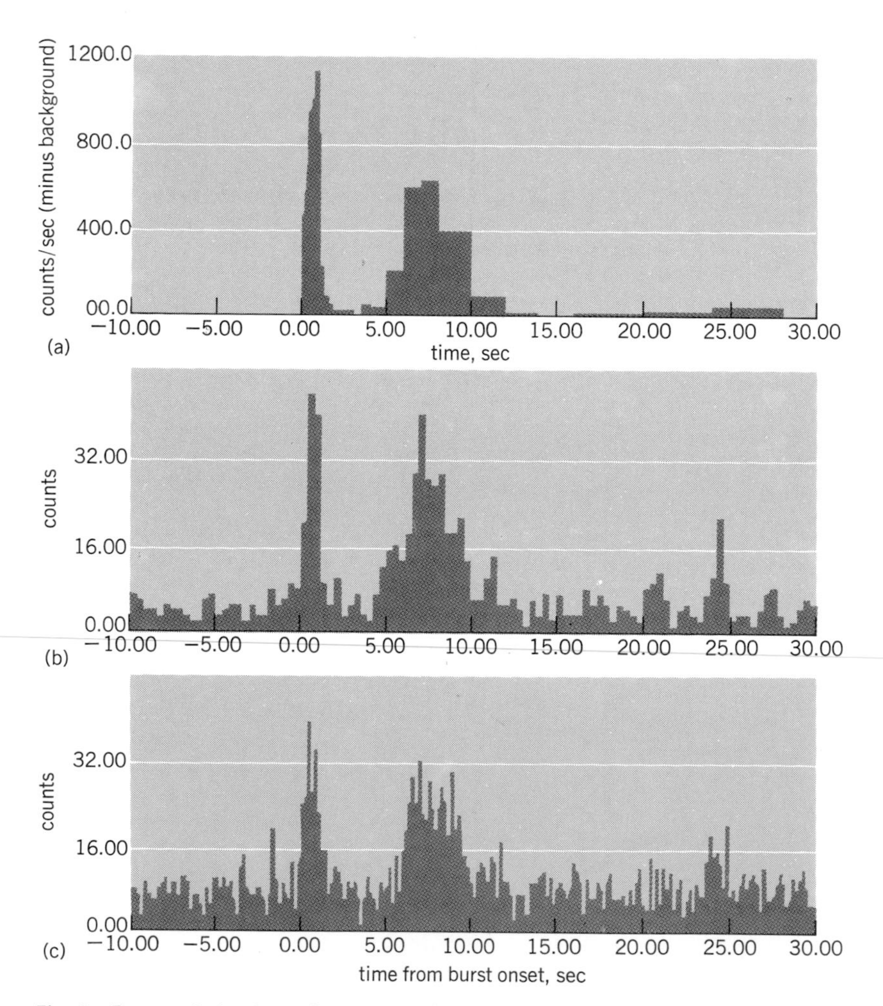

Fig. 4. Temporal structure of gamma-ray burst 71-5 as revealed by detectors aboard satellites. *(a) Vela 6B* satellite. *(b) Uhuru* satellite FWHM (full width at half maximum) 5° collimated detector. *(c) Uhuru* satellite ½° FWHM collimated detector. *(From D. Koch et al., Gamma-ray bursts seen by Uhuru, in I. Strong, ed., Proceedings of the Conference on Transient Cosmic Gamma Ray and X-Ray Sources, Los Alamos Lab. Rep. LA-5505-C, 1974)*

GAMMA-RAY BURSTS

In 1973 scientists of Los Alamos, NM, announced the detection by Vela satellites of brief, intense bursts of cosmic gamma rays. These bursts are not obviously associated with any known class of astronomical objects. Their discovery has again underlined the tendency in astronomy for exploration of new time scales in new regions of the electromagnetic spectrum to reveal entirely unexpected phenomena, and the important role of serendipity in such discoveries.

Discovery. The Vela satellites were developed to monitor the Nuclear Test Ban Treaty. In 1967, with an improved detection and analysis system, project scientists sought to demonstrate that no natural sources of gamma rays existed that might be confused with nuclear tests. However, examination of satellite data for July 2, 1967, clearly showed that a gamma-ray burst characteristic of nuclear debris had been detected. Subsequent analysis of the event and of similar later events showed, however, that they did not come from any object in the solar system.

The omnidirectional monitoring capability of the Vela satellites and their nearly continuous coverage in time, together with the possibility of coincident detections by more than one satellite, were vital to the discovery and verification of events that occur only four or five times a year at a level exceeding the Vela detector thresholds. Subsequent observations by many other satellites, among them the *IMP 6*, *IMP 7*, and *Uhuru*, have confirmed the bursts and have provided valuable additional information on the temporal and spectral characteristics of the bursts.

Character of the bursts. The soft gamma-ray bursts typically last 1–10 sec; however, some are as short as 100–300 msec, and the longest burst observed so far had a duration of about 80 sec. Most of the bursts exhibit multiple (nonperiodic) pulses with significant structure occurring on time scales of 100 msec. Some even show variations on a time scale as short as the 16-msec resolution of the Vela detectors. Such short time scales imply a source size less than 10^9 cm (10^7 m) or a highly relativistic emission process. Figure 4 shows the temporal structure of the event 71-5 (the fifth event in 1971 cataloged).

Because the individual Vela satellites lack directional sensitivity, directional information has come primarily from time-of-flight measurements between the satellites (typically about 0.8 sec). All the Vela satellites, however, lie in a single orbital plane approximately perpendicular to the galactic plane. Consequently, detection by a pair of satellites provides a circle of position on the celestial sphere, but detection by three or more satellites reduces this circle only to a pair of directions that are mirror-symmetric with respect to the orbital plane and that lie at approximately the same galactic latitude. The error boxes for the two possible directions are now as large as $10° \times 10°$ owing to uncertainties in the time-of-flight measurements.

A concentration of sources close to the galactic equator would provide unambiguous evidence that the soft gamma-ray-burst sources lie within the Galaxy. The distribution of possible directions of observed sources, however, is consistent with a uniform distribution in both galactic latitude and galactic longitude. This suggests that the sources either lie rather nearby, with a characteristic distance of about 100 parsecs or less (1 parsec equals approximately 3 light-years, or 3×10^{16} m) or are extragalactic, with distances greater than about 10 Mparsecs.

The soft gamma-ray bursts have typical fluxes of approximately 10^{-5} erg cm^{-2} sec^{-1} (10^{-8} joule m^{-2} sec^{-1}) and integrated fluxes of about 10^{-4} erg cm^{-2} (10^{-7} joule m^{-2}). An estimate for the total energy of a burst can be derived from these flux values by using the two characteristic distances mentioned above. This energy is approximately 10^{38} ergs (10^{31} joules) for a source distance of 100 parsecs, or about 10^{48} ergs (10^{41} joules) if the distance is 10 Mparsecs.

Information on the spectral characteristics of the soft gamma ray bursts has been gleaned from data provided by satellites which, of course, were not designed to study the bursts. Almost all the data are contaminated by the effects of secondary particles that the gamma rays produce through interactions with the matter in the satellite; consequently there are substantial uncertainties in the spectral data. However, there is evidence indicating that the gamma rays that make up a burst may have a characteristic energy of about 150 keV. The multiple pulses observed in most bursts are believed to have relatively hard spectra and are superimposed on a softer background that grows and decays in intensity on a longer time scale. At least in some bursts, the energy contained in x-rays may be comparable to that observed in soft gamma rays.

Models. If the sources of the gamma-ray bursts are assumed to lie outside the Galaxy, general arguments show that it is difficult to achieve the required luminosity of 10^{48} erg s^{-1} (10^{41} J s^{-1}) within the constraints on source temperature (equivalent to an energy of about 150 keV) and size (less than 10^9 cm, or 10^7 m), imposed by observations. Partly for this reason, Type II supernovae in massive stars (with masses greater than about 30 times that of the Sun) have been the only extragalactic source model elaborated in detail. *See* SUPERNOVA.

The duration and the fast-time-scale structure of the gamma-ray bursts lie in the range that is characteristic of the dynamical time scales (10^{-4} to 1 s) associated with degenerate dwarfs, neutron stars, and black holes. There are believed to be a large number of such compact objects within the 100-parsec distance already mentioned as characteristic if the gamma-ray-burst sources lie within the Galaxy. These considerations have led theorists to propose models involving such compact objects; several of these models invoke neutron star starquakes. *See* BLACK HOLE; GRAVITATIONAL COLLAPSE; WHITE DWARF STAR.

A characteristic distance of 100 parsecs also implies a source luminosity of about 10^{37} erg s^{-1} (10^{30} J s^{-1}) which is typical of the luminosities encountered earlier in the compact x-ray sources. This correspondence has provided the impetus for models in which accretion onto a compact object is the source of the bursts. Proposed ideas include sudden changes in the rate of mass transfer in close binaries containing a compact object (due, for example, to flaring of the companion star or to a sudden change in the flow pattern of accreting matter), the impact of comets onto neutron stars, and nuclear explosions on degenerate dwarfs. *See* BINARY STAR; X-RAY ASTRONOMY; X-RAY STAR.

The short duration and the extremely fast-time-scale structure of the bursts have motivated a number of models involving intense flares on stars ranging from more or less normal stars and flare stars to degenerate dwarfs and neutron stars. More exotic models have included the collision of chunks of antimatter with normal stars, and relativistic iron grains entering the solar system.

Despite a plethora of proposed models, the limited observational data make it difficult to judge if any of these approach the true explanation. Thus the intense gamma-ray bursts remain an enigma.

[D. Q. LAMB]

Bibliography: K. Bennett et al., *COS-B* observations of pulsed gamma-ray emission from PSR 0531 + 21 and PSR 0833 − 45, *Astron. Astrophys.*, 61:279–284, 1977; M. Cassé and J. Paul, Local gamma-rays and cosmic-ray acceleration by supersonic stellar winds, *Astrophys. J.*, 237:236–243, 1980; R. H. Hartman et al., Galactic plane gamma-radiation *Astrophys. J.*, 230:597–606, 1979; R. Klebesadel, I. Strong, and R. Olson, Observations of gamma-ray bursts of cosmic origin, *Astrophys. J.*, 182:L85–L88, 1973; H. A. Mayer-Hasselwander et al., *COS-B* observations of the Milky Way in high-energy gamma rays, in Proceedings of the 9th Texas Symposium, Munich, Germany, Dec. 15–19, 1978, *Ann. N.Y. Acad. Sci.*, 336:211–222, 1980; T. Montmerle, On gamma-ray sources, supernova remnants, OB associations, and the origin of cosmic rays, *Astrophys. J.*, 231:95–110, 1979; *Proceedings of the Conference on Transient Cosmic Gamma-Ray and X-Ray Sources*, Los Alamos Lab. Rep. LA-5505-C, 1974; M. Ruderman, Theories of gamma-ray bursts, in Proceedings of the 7th Texas Symposium on Relativistic Astrophysics, Dallas, Dec. 16–20, 1974, *Ann. N.Y. Acad. Sci.*, 262: 164–180, 1975; L. Scarsi et al., The *COS-B* experiment and mission, in *Proceedings of the 12th ESLAB Symposium, Frascati, Italy, May 24–27, 1977*, European Space Agency, SP-124, pp. 3–11, 1978; F. W. Stecker, Observations of galactic gamma-rays and their implications for galactic structure studies, *Astrophys. J.*, 212:60–70, 1977; I. Strong, R. Klebesadel, and W. Evans, Observations of gamma-ray bursts, in Proceedings of the 7th Texas Symposium on Relativistic Astrophysics, Dallas, Dec. 16–20, 1974, *Ann. N.Y. Acad. Sci.*, 262:145–158, 1975; R. D. Wills et al., High-energy gamma-ray sources observed by *COS-B*, in Proceedings of the COSPAR Symposium on Non-Solar Gamma-Rays, Bangalore, India, May 30–June 1, 1979, *Advances in Space Exploration*, 7: 43–47, 1980.

Gemini

The Twins, in astronomy, is a winter zodiacal constellation. Gemini is the third sign of the zodiac. It is conspicuous, containing first-, second-, and third-magnitude stars. These stars, α, β, γ, and μ, form a rough quadrilateral figure (see illustration).

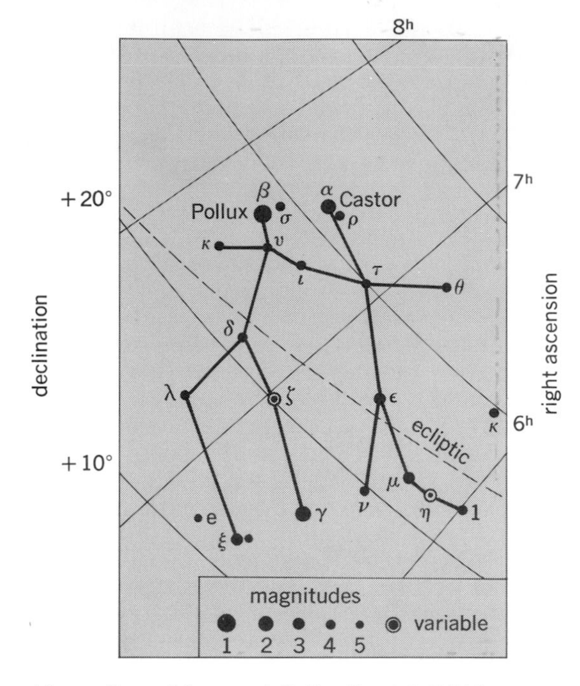

Line pattern of the constellation Gemini. Grid lines represent coordinates of the sky. Apparent brightness, or magnitudes, of stars is shown by sizes of dots graded by appropriate numbers as indicated.

The constellation is pictured as the figures of the twin heroes Castor and Pollux, with the two brightest stars of the same names representing the heroes' heads. Pollux, slightly brighter than Castor, is a navigational star. The Sun is in this constellation at the time of the summer solstice. The southwest corner of Gemini, being in the Milky Way, contains star fields which are among the finest in the sky. *See* CONSTELLATION.

[CHING-SUNG YU]

Giant star

A member of one of the families into which stars are divided in the Hertzsprung-Russell diagram; the term giant refers to true brightness rather than to size. Although giant stars are frequent among the brighter, naked-eye stars, such as Arcturus and Capella, they are comparatively rare in space. They have luminosities approximately 40–200 times that of the Sun, and occur most commonly in a range of temperature from 2500 to 7000 K; a few hotter giants are known. The common red giant stars have radii about 10 times that of the Sun, masses from 1.5 to 4 times that of the Sun, and mean densities about 10^{-3} g/cm^3. They represent a comparatively late stage in stellar evolution. Some cooler red giants vary periodically in brightness. Many are losing mass due to instabilities of various types, such as "stellar winds." Infrared surveys have revealed the existence of still cooler giants, some with temperatures as low as 1000 K. Water vapor and CO have been found in circumstellar clouds. *See* HERTZSPRUNG-RUSSELL DIAGRAM; STAR; STELLAR EVOLUTION.

[JESSE GREENSTEIN]

Globule

A roundish blob of cosmic dust (that is, a roundish dark nebula) seen projected against the starlit sky. The diameter of a typical nearby globule is generally about 3–5 minutes of arc. Globules look like "holes in the heavens" and are made detectable by the fact that the cosmic dust inside is so thick that the stars on the far side of the globule cannot be seen. There are about 200 globules within 500 parsecs (1 parsec = 3.1×10^{13} m) of the Sun; the Milky Way Galaxy is estimated to contain 25,000 globules.

The dust content of a globule can be estimated from counts of stars in the region of the globule, compared to the counted number for a region of about the same angular diameter near, but well outside, the globule. Since the globule cannot be penetrated in normal blue or visible light by photography, it must be recorded in the near infrared, and preferably in the far infrared, at wavelengths from 1 to 1000 micrometers. *See* INFRARED ASTRONOMY.

There are various ways of estimating the distance to the globule. From the star counts an approximate dust mass can be calculated. Since it seems likely that for every gram of cosmic dust there are about 150 grams of molecular hydrogen (H_2), the true mass of the globule can be estimated. For the smallest globules, those with diameters of about half a parsec, the total mass is thus estimated at 20 solar masses. The larger globules have diameters up to 1, or a little over 1, parsec and total masses up to 80 solar masses.

Radio astronomy has contributed much to the study of globules. In the radio spectrum there are two closely spaced radio spectral lines attributable to the carbon monoxide molecule, one produced by the common ^{12}CO molecule, the other by the rare isotope ^{13}CO. From the study of these features the probable carbon content of the globule and its internal temperature can be derived. Since the ratio between the CO mass and the H_2 mass is known fairly well, the total mass of the globule can once more be estimated. The radio values for the masses agree well with the masses derived from star counts at optical wavelengths. The most common temperatures inside globules are only 5–9 K. *See* RADIO ASTRONOMY.

Theoretical calculations show that globules should collapse into protostars in times of the order of a half million years. There will probably be considerable mass loss in the process of collapse. The result of the collapse will probably be a star (or binary pair of stars) with masses of 1 to a few solar masses. It seems probable that a solar system accompanying the star or binary will be formed as a by-product of the collapse. Eight varieties of organic molecules have been found in globules by radio astronomers. Hence, conditions for the start of life in the young system seem to be present.

It appears that about one in every five to eight stars in the Milky Way Galaxy is born from a globule. *See* INTERSTELLAR MATTER; STELLAR EVOLUTION.

[BART J. BOK]

Gravitation

The mutual attraction between all masses and particles of matter in the universe. In a sense this is one of the best-known physical phenomena. During the 18th and 19th centuries gravitational astronomy, based on Newton's laws, attracted many of the leading mathematicians and was brought to such a pitch that it seemed that only extra numerical refinements would be needed in order to account in detail for the motions of all celestial bodies. In the present century, however, Albert Einstein shattered this complacency, and the subject is currently in a healthy state of flux.

Until the 17th century, the sole recognized evidence of this phenomenon was the gravitational attraction at the surface of the Earth. Only vague speculation existed that some force emanating from the Sun kept the planets in their orbits. Such a view was expressed by Johannes Kepler (1571–1630), the author of the laws of planetary motion. But a proper formulation for such a force had to wait until Isaac Newton (1643–1727) founded Newtonian mechanics, with his three laws of motion, and discovered, in calculus, the necessary mathematical tool. *See* CELESTIAL MECHANICS; PLANET.

Newton's law of gravitation. Newton's law of universal gravitation states that every two particles of matter in the universe attract each other with a force that acts in the line joining them, the intensity of which varies as the product of their masses and inversely as the square of the distance between them. Put into symbols, the gravitational force F exerted between two particles with masses m_1 and m_2 separated by a distance d is given by Eq. (1), where G is called the constant of gravitation.

$$F = Gm_1m_2/d^2 \quad (1)$$

A force varying with the inverse square power of the distance from the Sun had been already suggested—notably by Robert Hooke (1635–1703) but also by other contemporaries of Newton, such as Edmund Halley (1656–1742) and Christopher Wren (1632–1723)—but this had only been applied to circular planetary motion. The credit for accounting for, and partially correcting, Kepler's laws and for setting gravitational astronomy on a proper mathematical basis is wholly Newton's.

Newton's theory was first published in *The Principia* in 1686. According to Newton, it was formulated in principle in 1666 when the problem of elliptic motion in the inverse-square force field was solved. But publication was delayed in part because of the difficulty of proceeding from the "particles" of the law to extended bodies such as the Earth. This difficulty was overcome when Newton established that, under his law, bodies having spherically symmetrical distribution of mass attract each other as if all their mass were concentrated at their respective centers.

Newton verified that the gravitational force between the Earth and the Moon, necessary to maintain the Moon in its orbit, and the gravitational attraction on the surface of the Earth were related by an inverse square law of force. Let E be the mass of the Earth, assumed to be spherically symmetrical with radius R. Then the force exerted by the Earth on a small mass m near the Earth's surface is given by Eq. (2), and the acceleration of gravity on the Earth's surface, g, by Eq. (3).

$$F = GEm/R^2 \quad (2)$$

$$g = GE/R^2 \quad (3)$$

Let a be the mean distance of the Moon from the Earth, M the Moon's mass, and P the Moon's sidereal period of revolution around the Earth. If the motions in the Earth-Moon system are considered to be unaffected by external forces (principally those caused by the Sun's attraction), Kepler's third law applied to this system is given by Eq. (4).

$$4\pi^2a^3/P^2 = G(E+M) \quad (4)$$

Equations (3) and (4), on elimination of G, give Eq. (5).

$$g = 4\pi^2 \frac{E}{E+M} \frac{a^2}{R^2} \frac{a}{P^2} \quad (5)$$

Now the Moon's mean distance from the Earth is $a = 60.27R = 3.84 \times 10^{10}$ cm and the sidereal period of revolution is $P = 27.32$ days $= 2.361 \times 10^6$ sec. These data give, with $E/M = 81.35$, $g = 977$ cm sec^{-2}, which is close to the observed value. *See* EARTH.

This calculation corresponds in essence to that made by Newton in 1666. At that time the ratio a/R was known to be about 60, but the Moon's distance in miles was not well known because the Earth's radius R was erroneously taken to correspond to 60 mi per degree of latitude instead of 69 mi. As a consequence, the first test was unsatisfactory. But the discordance was removed in 1671 when J. Picard's measurement of an arc of meridian in France provided a reliable value for the Earth's radius.

Gravitational constant. Equation (3) shows that the measurement of the acceleration due to gravity at the surface of the Earth is equivalent to finding the product G and the mass of the Earth. Determining the gravitational constant by a suitable experiment is therefore equivalent to "weighing the Earth."

In 1774 N. Maskelyne determined G by measuring the deflection of the vertical by the attraction of a mountain. This method is much inferior to the laboratory method in which the gravitational force between known masses is measured. In the torsion balance two small spheres, each of mass m, are connected by a light rod, suspended in the middle by a thin wire. The deflection caused by bringing two large spheres each of mass M near the small ones on opposite sides of the rod is measured, and the force is evaluated by observing the period of oscillation of the rod under the influence of the torsion of the wire (see illustration.) This is known as the Cavendish experiment, in honor of H. Cavendish who achieved the first reliable results by this method in 1797–1798. More recent determinations using various refinements yield the results: constant of gravitation $G = 6.67 \times 10^{-11}$ mks units; mass of Earth $= 5.98 \times 10^{27}$g. The uncertainty of these results is probably about one-half unit of the last place given.

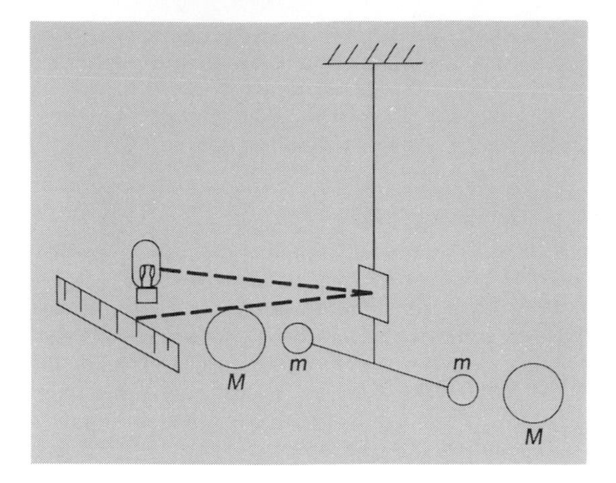

The torsion balance.

In Newtonian gravitation G is an absolute constant, independent of time, place, and the chemical composition of the masses involved. Partial confirmation of this was provided before Newton's time by the experiment attributed to Galileo when different weights released simultaneously from the top of the tower of Pisa reached the ground at the same time. Newton found further confirmation, experimenting with pendulums made out of different materials. Early in this century, R. Eötvös found that different materials fall with the same acceleration to within 1 part in 10^7. The accuracy of this figure was extended by R. H. Dicke to 1 part in 10^{11}, using aluminum and gold, and by V. B. Braginskii and V. I. Panov to 0.9×10^{-12} with a confidence of 95%, using aluminum and platinum.

With the discovery of antimatter, there was speculation that matter and antimatter would exert a mutual gravitational repulsion. But experimental results indicate that they attract one another according to the same laws as apply to matter of the same kind.

Mass and weight. A cosmology with changing physical "constants" was first proposed in 1937 by P. A. M. Dirac. Field theories applying this principle have since been proposed by P. Jordan and D. W. Sciama and, in 1961, by C. Brans and R. H. Dicke. In these theories G is diminishing; for instance Brans and Dicke suggest a change of about 2×10^{-11} per year. This would have profound effects on phenomena ranging from the evolution of the universe to the evolution of the Earth. For instance, stars evolve faster if G is greater, so that stellar evolutionary ages computed with constant G at its present value would be too great. The Earth, compressed by gravitation, would expand, having a profound effect on surface features. Planetary orbits would gradually be increasing in size. About 3×10^9 years ago the Sun would have been hotter than it is now, and the Earth and its orbit would have been smaller, so that the temperature on the Earth's surface might have approached the boiling point of water; this would be important for the origin of life on the Earth. Astronomical observations of the planets over the past few hundred years are not accurate enough for the predicted change to be detected; but T. C. Van Flandern has reported that observations of the motion of the Moon are consistent with the predicted change in G. He suggests a change of the order of 1 part in 10^{-10} per year. From radar ranging experiments I. I. Shapiro gives a limit of no greater than 4×10^{-10}. However, using cosmological arguments, J. D. Barrow has argued that the figure can be no greater than 1.5×10^{-12}.

In the equations of motion of Newtonian mechanics, the mass of body appears as inertial mass as a factor of the acceleration, and as gravitational mass in the expression of the gravitational force. The equality of these masses is confirmed by the Eotvos experiment. It justifies the assumption that the motion of a particle in a gravitational field does not depend on its physical composition. In Newton's theory the equality can be said to be a coincidence, but not in Einstein's theory, where inertia and gravitation are unified.

While mass in Newtonian mechanics is an intrinsic property of a body, its weight depends on certain forces acting on it. For example, the weight of a body on the Earth depends on the gravitational attraction of the Earth on the body and also on the centrifugal forces due to the Earth's rotation. The body would have lower weight on the Moon, even though its mass would remain the same.

Gravity. This should not be confused with the term gravitation. Gravity is the older term, meaning the quality of having weight, and so came to be applied to the tendency of downward motion on the Earth. Gravity or the force of gravity are today used to describe the intensity of gravitational forces, usually on the surface of the Earth or another celestial body. So gravitation refers to a universal phenomenon, while gravity refers to its local manifestation.

A rotating planet is oblate (or flattened at the poles) to a degree depending on the ratio of the centrifugal to the gravitational forces on its surface and on the distribution of mass in its interior. The variation of gravity on the surface of the Earth depends on these factors and is further complicated by irregular features such as oceans, continents, and mountains. It is investigated by gravity surveys and also through the analysis of the motion of artificial satellites. Because of the irregularities, no mathematical formula has been found that satisfactorily represents the gravitational field of the Earth, even though formulas involving hundreds of terms are used. The problem of representing the gravitational field of the Moon is even harder because the surface irregularities are proportionately much larger.

In describing gravity on the surface of the Earth, a smoothed out theoretical model is used, to which are added gravity anomalies, produced in the main by the surface irregularities.

Gravity waves are waves in the oceans or atmosphere of the Earth whose motion is dynamically governed by the Earth's gravitational field. They should not be confused with gravitational waves, which are discussed below.

Gravitational potential energy. This describes the energy that a body has by virtue of its position in a gravitational field. If two particles with masses m_1 and m_2 are a distance r apart and if this distance is slightly increased to $r + \Delta r$, then the work done against the gravitational attraction is $Gm_1m_2\Delta r/r^2$. If the distance is increased by a finite amount, say from r_1 to r_2, the work done is given by Eq. (6). If $r_2 \to \infty$, Eq. (7) holds.

$$W_{r_1,r_2} = Gm_1m_2 \int_{r_1}^{r_2} dr/r^2 = Gm_1m_2(1/r_1 - 1/r_2) \quad (6)$$

$$W_{r_1,\infty} = Gm_1m_2/r_1 \quad (7)$$

If one particle is kept fixed and the other brought to a distance r from a very great distance (infinity), then the work done is given by Eq. (8).

$$-U = -Gm_1m_2/r \quad (8)$$

This is called the gravitational potential energy; it is (arbitrarily) put to zero for infinite separation between the particles. Similarly, for a system of n particles with masses $m_1, m_2, \ldots, m_n$ and mutual distance r_{ij} between m_i and m_j, the gravitational potential energy $-U$ is the work done to assemble the system from infinite separation (or the negative of the work done to bring about an infinite separation), as shown in Eq. (9).

$$-U = -G \sum_{i<j} m_i m_j / r_{ij} \quad (9)$$

A closely related quantity is gravitational potential. The gravitational potential of a particle of mass m is given by Eq. (10), where r is distance

$$V = -Gm/r \quad (10)$$

measured from the mass. The gravitational force exerted on another mass M is M times the gradient of V. If the first body is extended or irregular, the formula for V may be extremely complicated, but the latter relation still applies.

A good illustration of gravitational potential energy occurs in the motion of an artificial satellite in a nearly circular orbit around the Earth which is affected by atmospheric drag. Because of the frictional drag the total energy of the satellite in its orbit is reduced, but the satellite actually moves faster. The explanation for this is that it moves closer to the Earth and loses more in gravitational potential energy than it gains in kinetic energy.

Similarly, in its early evolution a star contracts, with the gravitational potential energy being transformed partly into radiation, so that it shines, and partly into kinetic energy of the atoms, so that the star heats up until it is hot enough for thermonuclear reactions to start. *See* STELLAR EVOLUTION.

Another related phenomenon is that of speed of escape. A projectile launched from the surface of the Earth with speed less than the speed of escape will return to the surface of the Earth; but it will not return if its initial speed is greater (atmospheric drag is neglected). For a spherical body with mass M and radius R, the speed of escape from its surface is given by Eq. (11). For the Earth V_e is 11.2

$$V_e = (2MG/R)^{1/2} \quad (11)$$

km/sec; for the Moon it is 2.4 km/sec, which explains why the Moon cannot retain an atmosphere such as the Earth's. By analogy, a black hole can be considered a body for which the speed of escape from the surface is greater than the speed of light, so that light cannot escape. P. S. Laplace speculated along these lines; but the analogy is not really exact since Newtonian mechanics is not valid. The question as to whether the universe will continue to expand can be considered in the same way. If the density of matter in the universe is great enough, then expansion will eventually cease and the universe will start to contract. At present the density cannot be found with sufficient accuracy to decide the question. *See* BLACK HOLE; COSMOLOGY.

Application of Newton's law. In modern times Eq. (5), in a modified form with appropriate refinements to allow for the Earth's oblateness and for external forces acting on the Earth-Moon system, has been used to compute the distance to the Moon. The results have only been superseded in accuracy by radar measurements and observations of corner reflectors placed on the lunar surface.

Newton's theory passed a much more stringent test than the one described above when he was able to account for the principal departures from Kepler's laws in the motion of the Moon. Such departures are called perturbations. One of the most notable triumphs of the theory occurred when the observed perturbations in the motion of the planet Uranus enabled J. C. Adams in 1845 and U. J. Leverrier in 1846 independently to predict the existence and calculate the position of a hitherto unobserved planet, later called Neptune. When yet another planet, Pluto, was discovered in 1930, its position and orbit were strikingly similar to predictions based on the method used to discover Neptune. But the discovery of Pluto must be ascribed to the perseverance of the observing astronomers; it is not massive enough to have revealed itself through the perturbations of Uranus and Neptune. *See* PERTURBATION.

F. W. Bessel observed nonuniform proper motions of Sirius and Procyon and inferred that each was gravitationally deflected by an unseen companion. It was only after his death that these bodies were telescopically observed, and they both later proved to be white dwarfs. More recently P. van de Kamp has accumulated evidence for the existence of some planetary masses around stars. The discovery of black holes (which will never be directly observed) hinges in part on a visible star showing evidence for having a companion of sufficiently high mass (so that its gravitational collapse can never be arrested).

Newton's theory supplies the link between the observed motion of celestial bodies and certain physical properties, such as mass and sometimes shape. Knowledge of stellar masses depends basically on the application of the theory to binary-star systems. Analysis of the motions of artificial satellites placed in orbit around the Earth has revealed refined information about the gravitational field of the Earth and of the Earth's atmosphere. Similarly, satellites placed in orbit around the Moon have yielded information about its gravitational field, and other space vehicles have yielded the best information to date on the masses and gravitational fields of other planets. *See* SPACE PROBE.

Newtonian gravitation has been applied without apparent difficulty to the motion in distant star systems. But over very great distance (or over very small distances, when gravitation is swamped by other forces) it has not been confirmed or disproved.

Accuracy of Newtonian gravitation. Newton was the first to doubt the accuracy of his law when he was unable to account fully for the motion of the perigee in the motion of the Moon. In this case he eventually found that the discrepancy was largely removed if the solution of the equations were more accurately developed. Further difficulties to do with the motion of the Moon were noted in the 19th century, but these were eventually resolved when it was found that there were appreciable fluctuations in the rate of rotation of the Earth, so that it was the system of timekeeping and not the gravitational theory that was at fault. *See* ORBITAL MOTION.

A more serious discrepancy was discovered by Leverrier in the orbit of Mercury. Because of the action of the other planets, the perihelion of Mercury's orbit advances. But allowance for all known gravitational effects still left an observed motion of about 43 seconds of arc per century unaccounted for by Newton's theory. Attempts to account for this by adding an unknown planet or by drag with an interplanetary medium were unsatisfactory, and a very small change was suggested in the exponent of the inverse square of force. This particular discordance was accounted for by Albert Einstein's general theory of relativity in 1916, but the final word on the subject has yet to be said. *See* RELATIVITY.

Gravitational lens. Light is deflected when it passes through a gravitational field, and an analogy can be made to the refraction of light passing through a lens. It has been suggested that a galaxy situated between an observer and a more distant source might have a focusing effect, and that this might account for some of the observed properties of quasi-stellar objects. The multiple images of one quasar (QSO 0957 + 567 A,B) are almost certainly caused by the light from a single body passing through a gravitational lens. *See* QUASARS.

Testing of gravitational theories. One of the greatest difficulties in investigating gravitational theories is the weakness of the gravitational coupling of matter. For instance, the gravitational interaction between a proton and an electron is weaker by a factor of about 5×10^{-40} than the electrostatic interaction. (If gravitation alone bound the hydrogen atom, then the radius of the first Bohr orbit would be 10^{13} light-years, or about 1000 times the radius of the Hubble universe!) The contrast between the accuracy achieved in the laboratory when measuring G and the accuracy required if the possible inconstancy of G is to be investigated shows that gravitation is not a laboratory subject. But the astronomical universe provides a wealth of situations for investigating gravitation. Their main drawback is that they must be taken as they occur: with the exception of some experiments included in the space program, the situations cannot be controlled, modified, or repeated for experimental convenience.

In the solar system there are planetary orbits around the Sun and satellite orbits around the planets. Gravitational effects on electromagnetic radiation as well as those on orbiting bodies can be observed and tested against theory. There are stars that rotate, oscillate, explode, and collapse, including white dwarfs, neutron stars (observed as pulsars), and probably black holes, some with enormous masses. There are binary systems, some with orbital periods as low as 1000 sec. There are star clusters, galaxies (some of them exploding), clusters of galaxies, and finally there is the universe itself. Of special interest is the "binary pulsar" PSR 1913+16. The observed star is a pulsar, and it has been speculated that the unseen companion is also a highly condensed object. The period is almost 8h, the orbital eccentricity is 0.6, and the motion of periastron is 4.2° per year. *See* PULSAR; SOLAR SYSTEM.

In the solar system, increased observational accuracy is needed to detect small departures from Newtonian gravitational theory. In other situations, these departures may be very great, but observable effects are attenuated due to great distances. Of greatest interest are systems where there is a high concentration of matter and rapid motion, involving rotation, revolution, or collapse.

Relativistic theories. Before Newton, detailed descriptions were available of the motions of celestial bodies—not just Kepler's laws but also empirical formulas capable of representing with fair accuracy, for their times, the motion of the Moon. Newton replaced description by theory, but in spite of his success and the absence of a reasonable alternative, the theory was heavily criticized, not least with regard to its requirement of "action at a distance" (that is, through a vacuum). Newton himself considered this to be "an absurdity," and he recognized the weaknesses in postulating in his system of mechanics the existence of preferred reference systems (that is, inertial reference systems) and an absolute time. Newton's theory is a superb mathematical one that represents the observed phenomena with remarkable accuracy.

Einstein showed in his special theory of relativity that these postulates were physically unacceptable. In his general theory of relativity he incorporated gravitational phenomena in such a way that there was no longer any preferred reference system. He treated the phenomenon of gravitation as a consequence of the geometric properties of space-time, this geometry being affected by the presence of gravitating matter. The acceleration of a body is determined by the local geometry, and the Newtonian concepts of gravitational force and action at a distance are abandoned. Mathematically the theory is far more complicated than Newton's. Instead of the single potential described above, Einstein worked with 10 quantities that are members of a tensor.

Principle of equivalence. An important step in Einstein's reasoning is his "principle of equivalence": that a uniformly accelerated reference system imitates completely the behavior of a uniform gravitational field. Imagine, for instance, a scientist in a space capsule infinitely far out in empty space so that the gravitational force on the capsule is negligible. Everything would be weightless; bodies would not fall; and a pendulum clock would not work. But now imagine the capsule to be accelerated by some agency at the uniform rate of 981 cm/sec^2. Everything in the capsule would then behave as if the capsule were stationary on its launching pad on the surface of the Earth and therefore subject to the Earth's gravitational field. But after its original launching, when the capsule is in free flight under the action of gravitational forces exerted by the various bodies in the solar system, its contents will behave as if it were in the complete isolation suggested above. Note that this principle requires that all bodies fall in a gravitational field with precisely the same acceleration and that this is confirmed by the Eotvos experiment mentioned earlier. Also, if matter and antimatter were to repel one another, it would be a violation of the principle.

Einstein's theory requires that experiments should have the same results irrespective of the location or time. This has been said to amount to the "strong" principle of equivalence.

Classical tests. The ordinary differential equations of motion of Newtonian gravitation are replaced in general relativity by a nonlinear system of partial differential equations for which general solutions are not known. Apart from a few special cases, knowledge of solutions comes from methods of approximation. For instance, in the solar system, speeds are low so that the quantity v/c (v is the orbital speed and c is the speed of light) will be small (about 10^{-4} for the Earth). The equations and solutions are expanded in powers of this quantity; for instance, the relativistic correction for the motion of the perihelion of Mercury's orbit is adequately found by considering no terms smaller than $(v/c)^2$. This is called the post-Newtonian approximation. (Another approach is the weak-field approximation.)

Einstein's theory has appeared to pass three famous tests. First, it accounted for the full motion of the perihelion of the orbit of Mercury. (Mercury is the most suitable planet, because it is the fastest-moving of the major planets and has a high eccentricity, so that its perihelion is relatively easily studied.) Second, the prediction that light passing a massive body would be deflected has been confirmed with an accuracy of about 5%. Third, Einstein's theory predicted that clocks would run more slowly in strong gravitational fields compared to weak ones; interpreting atoms as clocks, spectral lines would be shifted to the red in a gravitational field. This, again, has been confirmed with moderate accuracy.

I. I. Shapiro has confirmed predictions of the theory in an experiment in which radar waves were bounced off Mercury; the theory predicts a delay of about 2×10^{-4} sec in the arrival time of a radar echo when Mercury is on the far side of the Sun and close to the solar limb. Tests, similar in principle, have been conducted using observations of the Mariner space vehicles, the accuracy of confirmation being in the region of 4%. E. Fomalont and R. Stramek observed the deflection of microwave radiation passing close to the Sun, using radio interferometry with a baseline of 35 km. The amount of bending they found is 1.015 ± 0.011 times the amount predicted by general relativity. In another test, the precession of a gyroscope in orbit around the Earth is to be studied for evidence of the so-called geodetic precession. The motion of a perihelion is suitable for study since its effects continue to accumulate. Other periodic (noncumulative) orbital effects have until recently been too small to observe. But the current revolution in observational techniques and accuracy has changed the situation; post-Newtonian terms are now routinely included in many calculations of the orbits of planets and space vehicles, and comparison with observations will furnish tests of the theory.

The observation and analysis of gravitation waves, discussed below, will constitute further tests.

Mach's principle. One of the most penetrating critiques of mechanics is due to E. Mach, toward the end of the 19th century. Some of his ideas can be traced back to Bishop G. Berkeley early in the 18th century. Out of Mach's work there has arisen what is known as Mach's principle; this is philosophical in nature and cannot be stated in precise terms. The idea is that the motion of a particle is only meaningful when referred to the rest of the matter in the universe. Geometrical and inertial properties are meaningless for an empty space, and the motion of a particle in such space is devoid of physical significance. Thus the behavior of a test particle should be determined by the total matter distribution in the universe and should not appear as an intrinsic property of an absolute space. If this is so, then the quantitative aspects of physical laws (that is, the various constants involved) should be dependent on position.

Brans-Dicke theory. The field theory developed in 1961 by Brans and Dicke is perhaps the best-known theory in conformity with Mach's principle. For instance, the expansion of the universe causes G to diminish in time. In this theory the gravitational field is described by a tensor and a scalar, the equations of motion being the same as those in general relativity. The addition of a scalar field leads to the appearance of an arbitrary constant, whose value is not known exactly. The Brans-Dicke theory predicts that the relativistic motion of the perihelion of Mercury's orbit is reduced compared with Einstein's value, and also that the light deflection should be less. With regard to the orbit of Mercury, Dicke pointed out that if the Sun were oblate, this might account for some of the motion of the perihelion. In 1967 he announced that measurements showed a solar oblateness of about 5 parts in 100,000 (or a difference in the polar and equatorial radii of about 34 km). His observations and discussion are still subject to some controversy. The difference between the theory and that of general relativity can be parameterized by the number ω, where $\gamma = (1+\omega)/(2+\omega)$; for general relativity $\gamma = 1$. Dicke has proposed $\omega \sim 7.5$; but the results of Fomalont and Stramek indicate a value of ω greater than 23, for which the predictions of the two theories would not be greatly different.

There are, of course, many other theories not mentioned here.

Supergravity. This is the term applied to a highly mathematical theory of gravitation forming part of a unified field theory in which all types of forces are included.

Gravitational waves. The existence of gravitational waves, or gravitational "radiation," was predicted by Einstein shortly after he formulated his general theory of relativity. They are now a feature of any relativity theory. Gravitational waves are "ripples in the curvature of space-time." In other words, they are propagating gravitational fields, or propagating patterns of strain, traveling at the speed of light. They carry energy and can exert forces on matter in their path, producing, for instance, very small vibrations in elastic bodies. The gravitational wave is produced by change in the distribution of some matter. It is not produced by a rotating sphere, but would result from a rotating body not having symmetry about its axis of rotation: a pulsar, perhaps. In spite of the relatively weak interaction between gravitational radiation and matter, the measurement of this radiation is now technically possible and may already have been achieved. This is due to the work of Joseph Weber, whose original and pioneering work has led to a very exciting situation in science. The present situation contains some uncertainties; but gravitational-wave astronomy has been added to other branches of astronomy, and a new window is opening to the universe.

A classical problem, solved by Einstein, concerns the gravitational radiation from a rod spinning about a perpendicular axis through its center. If the rod has moment of inertia about the axis of spin I ($I=Md^2/3$, where M is the mass of the rod in kilograms and $2d$ its length in meters) and angular velocity ω, the power of the radiation in watts (1 W $=10^7$ ergs/sec) is given by Eq. (12), where G

$$P=\frac{32GI^2\omega^6}{5c^5}=1.73\times10^{-52}I^2\omega^6 \qquad (12)$$

is the constant of gravitation in mks units and c is the speed of light in meters per second. A calculation using a steel rod of mass 4.9×10^5 kg (490 metric tons), length 20 m, and angular velocity $\omega=28$ rad/sec, limited by the balance between centrifugal force and tensile strength, gives 2.2×10^{-29} W. So the problem of the generation and detection of gravitational waves in the laboratory is at present somewhat academic.

In electromagnetic theory, electric-dipole radiation is dominant. The gravitational analog of the electric dipole is the mass dipole moment whose time rate of change is the total momentum of the system; since this is constant, there is no gravitational dipole radiation; the principal power is in quadrupole radiation. The radiation has fairly elaborate polarization properties.

Binary systems. Consider a binary star system having period P hr and masses m_1, m_2, where the relative orbit is circular. If $M=m_1+m_2$ and $\mu=m_1m_2/M$, the power output by gravitational radiation is given by Eq. (13), where $M_\odot$ is the mass of

$$P_B=\left(\frac{\mu}{M_\odot}\right)^2\left(\frac{M}{M_\odot}\right)^{4/3}P^{-10/3}\,3.0\times10^{26}\text{ W} \qquad (13)$$

the Sun. For the orbit of the Earth around the Sun, P_B is about 200 W. The gravitation radiation extracts energy from the system. If a binary system has a relative elliptic orbit, then most of the energy is extracted at the closest point of approach and the orbit approaches a circle; then the orbit will gradually shrink, with the bodies colliding after a "spiral time" given by Eq. (14), where a_0 is the

$$\tau_0=\frac{5c^5}{256G^3}\frac{a_0^{\,4}}{\mu M^2} \qquad (14)$$

initial radius of the relative orbit. Under this mechanism the Earth would have fallen toward the Sun less than 1 cm in the life of the solar system!

Clearly one must look outside the solar system for promising sources. Ordinary binaries are not helpful. Sirius and its companion, with a spiral time of 7×10^{21} years, radiate at 10^8 W, the flux received at the Earth being 10^{-31} W. The closer the members of the system are to each other, the more promising they are; some eclipsing binaries can generate power that would be observed on the Earth at about 10^{-20} W, and have spiral times of the order of 10^{10} years. The shortest periods known are for close pairs consisting of a white dwarf and a main sequence star; here spiral times can be as low as 10^9 years, and the predicted flux at the Earth for the most promising candidate, ι Boo, is 18×10^{-18} W. For these binaries, gravitational radiation appears to play an important part in their physical characteristics. Matter flows toward the white dwarf from the companion star, causing flickering and occasional nova outbursts. The stars are very close, and it seems that the contraction of the orbit, caused by gravitational radiation, plays a crucial part in instigating the flow of matter. Closer binaries can at present only be generated by hypothesis. Two neutron stars, having solar masses and with 10^4-km separation, radiate at 3×10^{34} W and have a spiral time of 3 years. With such a system the formulas given above show that the evolution becomes increasingly rapid and the power input increases, so that the final stages of collapse constitute a burst of radiation. The period of the "binary pulsar" has been observed with high precision over several years. J. H. Taylor, L. A. Fowler, and P. M. McCulloch have reported that the period is diminishing at the annual rate of 1 part in 10^9, or 3 parts in 10^{12} per revolution. This is consistent with the prediction of general relativity, and encourages belief in the reality of gravitational radiation, although this particular radiation would not be directly observable.

Pulsars. The most rapidly rotating single objects that have been observed are pulsars. These are neutron stars rotating with periods mostly less than 1 sec. From their irregular light curves it is reasonable to suppose that they do not possess symmetry about the axis of rotation. Suppose that they are assumed homogeneous and the equatorial section is an ellipse with axes a and b, and that the ellipticity of the equator is $\epsilon=(a-b)/a$. If the star rotates with angular velocity ω, then the power radiated is given by Eq. (15), where I is the moment

$$P_R=\frac{32G\omega^6I^2\epsilon^2}{5c^5} \qquad (15)$$

of inertia about the axis of rotation. A promising

candidate here is the pulsar in the Crab Nebula, remnant of a supernova; the period of rotation is 0.033 sec; the moment of inertia is likely to be of the order of 4×10^{37} kg m^2, and the power output can be estimated by writing Eq. (15) as Eq. (16),

$$P_R=\left(\frac{I}{4\times10^{37}\text{ kg m}^2}\right)^2\left(\frac{P}{0.033\text{ sec}}\right)^{-6}\left(\frac{\epsilon}{10^{-3}}\right)^2 10^{31}\text{ W} \quad (16)$$

where P is the period. Clearly it is important to estimate ϵ. The periods of rotation are known to be increasing, and this puts an upper limit on ϵ. It is estimated that the flux received from the Crab pulsar would be less than 3×10^{-20} W. Some pulsars occasionally show sudden changes or glitches in their rotational period. These could be due to starquakes (neutron stars have solid surfaces) and might lead to strong bursts of gravitational radiation. See PULSAR.

Explosive events. The gravitational collapse involved in a supernova explosion might produce the strongest radiation that can be observed. The processes involved can only be tentatively estimated, and unfortunately supernova occur about once every 100 years in the Galaxy. But their radiation, probably in short bursts, could be sufficiently powerful for them to be observed from other galaxies. It is possible that stellar collapse takes place without the display of a supernova, so estimates of frequency may be much too low. See SUPERNOVA.

Many galaxies show evidence of explosive activity. For quasars, gravitational radiation at 10^{38} W has been suggested, and for explosions in galactic centers, 10^{30} W; but these estimates are not at all definitive.

As matter falls into a black hole, it will release a burst of gravitational radiation; the energy released is proportional to the square of the mass captured and inversely proportional to the mass of the black hole; the time of the outburst is proportional to the mass of the black hole. It has been suggested that there might be a large black hole at the center of the glaxy; if its mass were 10^8 solar masses and it captured a star of 1 solar mass, a burst of energy 10^{37} W might be produced. If the black hole were rotating or the infalling star somehow had a speed greater than that acquired from falling from infinity, then the energy could be greater.

Nature of radiation. The radiation discussed could be continuous or in bursts. The radiation would have a spectrum that might be discrete, as in the case of rotation or orbital revolution, where the fundamental frequency is 2ω (there will also be harmonics), or broadband in the case of explosive events. The longest wavelengths suggested are from the primordial history of the universe, when they could be greater than the size of a galaxy; the shortest, in supernovae and stellar collapse.

Dirac worked out a quantum theory for this radiation; the graviton is a theoretically deduced particle postulated as the quantum of the gravitational field.

Detection of gravitational waves. When a gravitational wave interacts with a system of particles, the particles wiggle slightly; in the case of a solid body, strains are set up in the body; what is actually measured is a sort of tidal effect. Most of the detectors currently under consideration involve strains in solid bodies and they involve the principle of resonance; that is, they react much more strongly to radiation of a given frequency than to other frequencies. When radiation of the correct frequency impinges on the detector it oscillates, as if ringing; radiation at other frequencies is essentially ignored.

Weber's experiment. Weber's detectors principally consist of cylinders suspended in vacuum. They are typically of aluminum, 66 cm in diameter and 153 cm long, weighing 1.5×10^3 kg and resonant at 1661 Hz. They are directional, being most strongly sensitive to radiation traveling perpendicular to the axis of the cylinder. The strains in the cylinder are converted into measurable voltages by quartz strain gages: piezoelectric crystals bonded around the girth of the cylinder. Strains of the order of 1 part in 10^{17} can be detected. There will be continued background thermal "noise," a random effect due to the thermal motion of individual molecules. Since the gravitational signals are of the same order of magnitude as this noise, it is necessary to have two indpendent receivers, and to look for coincidences in the signals received.

In the principal experiments, Weber analysed signals received 1000 km apart. Since 1969 Weber has reported coincidences at the average rate of about three times a day; typical displacements are around 5×10^{-17} m. He has looked for possible correlations between his coincidences and solar flares, electric storms, surges in the interstate power grid, network television broadcasting, seismic events, and cosmic rays, with no success, and so concludes that the observations are consistent with the detection of gravitational radiation.

Interpretation of Weber's results. In 1970 Weber reported that the coincident pulses were strongly correlated with sidereal time in a manner consistent with the antenna pattern if the gravitational radiation were coming from the center of the Galaxy. This leads to an astrophysically satisfying (in principle) source for the radiation and also a distance for the source: the galactic center. It therefore makes possible the calculation of the energies at the source that would produce the flux observed at Earth. Immediately problems arise. Each burst is what would be expected from a supernova or stellar collapse; if so, these events are more than 1000 times more frequent that had been expected. By using the formula $E=mc^2$, the mass loss that would be the equivalent to this energy can be calculated, and it appears that the Galaxy loses at least 500 solar masses each year; since Weber cannot be observing all events, perhaps this figure should be 10 times as great! But this is not acceptable; the Galaxy would be used up in a time of about one-hundredth of its known age.

The calculation of total energy uses the assumption that the energy is beamed equally in all directions and over a broad band. The band must be at least wide enough to include 1580 Hz, since Weber has observed coincidences at this frequency also. Many suggestions have been made to ease the situation: (1) Gravitational waves have been emitted at this rate for only a small fraction of the age of the Galaxy; this would imply that there is something special about the present age, and that there is something even more special about gravitational waves, since no other observed phenomenon shows the same preference for this era. (2) The source might be much nearer to Earth than the center of the Galaxy. (3) The radiation is strongly focused; into the galactic plane, for instance. (4) There may be some mechanism by which the radiation is magnified between the source and the Earth. (5) There is also the possibility that Weber's events are not caused by gravitational waves. Theories for the origin of the radiation include the possible presence of a very massive black hole at the galactic center, perhaps even disk-shaped, and the possibility of synchrotron-type gravitation radiation.

In another experiment Weber used a disk-shaped antenna. This was designed to search for scalar gravitation radiation. The result was negative, but the disk reacted in a way that was consistent with a source of tensor gravitational waves at the galactic center. So Weber claims similar results from two quite different antennas.

Other experiments. No other experimenter has reproduced Weber's results, although many have tried. A second generation of instruments is in operation or is being devised; greater sensitivity has been sought by making the detector more massive, cooler, or purer (that is, using large crystals such as silicon or sapphire that naturally have less noise). In another approach a laser beam is split into two parts along freely suspended perpendicular arms, each beam being reflected many times by mirrors at the ends; if the length of one arm changes relative to the other, this will be shown through interference when the beams are recombined. It has been suggested that the very accurate tracking of spacecraft could be exploited to detect the effect of gravitational waves on Doppler shifts. *See* DOPPLER EFFECT.

[J. M. A. DANBY]

Bibliography: R. H. Dicke, *Gravitation and the Universe*, 1970; G. Gamow, *Gravity*, 1962; J. L. Logan, Gravitational waves: A progress report, *Phys. Today*, 26:44–52, March 1973. C. W. Misner, K. S. Thorne, and J. A. Wheeler. *Gravitation*, 1973; R. Pinheiro, Gravity wave astronomy, *Astronomy*, 7(6):6–14, 1979; W. H. Press, and K. S. Thorne, Gravitational wave astronomy, *Annu. Rev. Astron. Astrophys.*, 10:335–374, 1972; T. J. Sejnowski, Sources of gravity waves, *Phys. Today*, 27:40–48, January 1974; K. S. Thorne, The future of gravitational wave astronomy, *Mercury*, 7(3):58–61, 1979; S. Weinberg, The forces of nature, *Amer. Sci.*, 65:171–176, 1977; D. T. Whiteside, *J. Hist. Astron.*, 1:5–19, 1970; E. T. Whittaker, *From Euclid to Eddington*: *A Study of Conception of the External World*, 1949; E. T. Whittaker, *A History of the Theories of the Aether and Electricity*, 1954.

Gravitational collapse

The rapid implosion of a star or other astronomical body from its initial, normal size to a size hundreds or thousands of times smaller. Gravitational collapse has been proposed as a mechanism, considerably more efficient than thermonuclear reactions, for converting the mass of an astronomical object into explosive energy. However, general relativistic phenomena may often prevent the energy released by collapse from escaping into the surrounding space.

Energy release. Since 1963 astrophysicists have devoted considerable effort to understanding the mechanisms behind such explosive astrophysical phenomena as supernovae, quasars, the origin of strong radio sources, and explosions in the nuclei of galaxies. Because the explosive energy in these objects may be a sizable fraction of the rest mass, and because thermonuclear reactions, such as occur in the Sun and in hydrogen bombs, can convert no more than 0.9% of the rest mass to energy, a search has been made for other, more powerful mechanisms of energy release. *See* QUASARS; RADIO SOURCES; SUPERNOVA.

The most likely alternative mechanism seems to be gravitational collapse. If a star or other body collapses to a small fraction of its original size and then releases the final kinetic energy of collapse as radiation and high-energy particles, the fraction of its original mass converted to energy is approximately as shown in notation (1). Here $M/M_\odot$ is the

$$\frac{\text{Energy released}}{\text{Initial mass}} \approx 1 - \left(1 - \frac{M}{M_\odot}\frac{3.0\text{ km}}{R}\right)^{1/2} \approx \frac{M}{M_\odot}\frac{1.5\text{ km}}{R} \qquad (1)$$

final mass of the body in units of solar masses, and R is its final radius. If the final radius is near $3.0(M/M_\odot)$ km, the energy released can be near 100% of the rest mass. It will be shown below that the final radius can never be less than $3.0(M/M_\odot)$ km.

There was, as of 1975, no fully compelling reason to believe that gravitational collapse is important in quasars, in the origin of strong radio sources, or in galactic explosions. However, theoretical and observational studies provided convincing evidence that collapse is important in supernovae.

Supernova models. There is no generally accepted detailed theory of the supernova explosions of stars. However, idealized supernova models constructed on computers by Stirling Colgate, Richard White, and David Arnett suggest that the energy source for supernovae is probably gravitational collapse. One such model is shown in Fig. 1.

Each solid curve in Fig. 1 represents the radius as a function of time for a shell of matter inside which a certain fraction of the star's mass is contained. The curves are labeled by this "mass fraction." At time $t = 0$, the star in this model, which has already burned most of its nuclear fuel, is set into collapse by an endothermic nuclear reaction that removes its pressure-sustaining electrons. After 1.863 sec of collapse, the inner 50% of the star is collapsing with half the speed of light, and the central core has reached nuclear density (about 10^{14} g/cm^3). At this point, repulsive nuclear forces suddenly create a huge internal pressure, which halts the collapse and sends a shock wave propagating out through the core of the star. In the shock wave, the core's huge kinetic energy of collapse is converted to heat; then, by way of temperature-sensitive elementary-particle reactions, the heat is tranformed into neutrinos. The neutrinos diffuse out into the still-collapsing envelope, where they deposit much of the core's original kinetic energy of collapse and thereby create an exploding envelope shock wave. This second shock wave blows the envelope away from the collapsed core with nearly the speed of light and with temperatures of billions of degrees, and it converts the outer 0.01% of the envelope into cosmic rays. *See* COSMIC RAYS.

In 1968, long after the calculations for Fig. 1 were made, astronomers discovered that neutron stars do indeed exist in the centers of supernova explosions. The neutron stars make themselves known by radio waves which they emit and which arrive at Earth in pulses; hence the name "pulsar." This discovery proves that the gravitational collapse theory of supernovae is correct. *See* PULSAR.

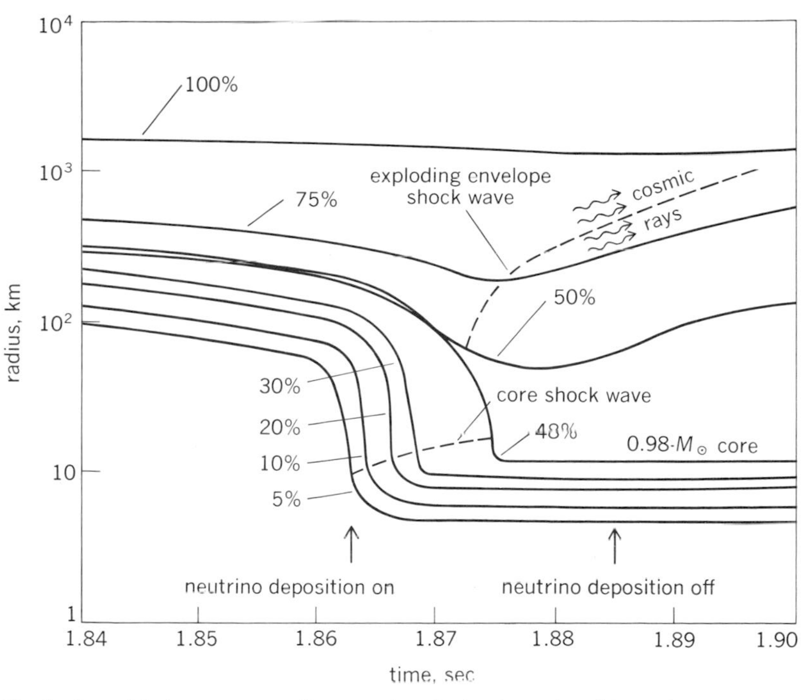

Fig. 1. A model of supernova collapse and explosion for a star of two solar masses, as calculated by Stirling Colgate and Richard White.

General relativistic collapse. The effectiveness of gravitational collapse as a mechanism for converting mass to explosive energy is severely limited by general relativistic phenomena. J. Robert Oppenheimer and his students showed in 1939 that any spherical star which collapses to a radius smaller than its Schwarzschild radius, Eq. (2), can

$$R_{\text{Sch}} = 3.0(M/M_\odot)\ \text{km} \tag{2}$$

no longer release its kinetic energy of collapse in any form to the outside universe; it must continue to collapse until gravitation forces have compressed it to infinite density (general relativistic singularity). The work of Oppenheimer and his students and of S. Chandrasekhar also revealed that any star exceeding about 1.5 solar masses (the critical value is uncertain by a factor 2) must eventually reduce its mass below this value—for example, by a supernova explosion—or face inevitable collapse past its Schwarzschild radius and into the singularity.

These long-established results on spherical collapse, together with the newer astrophysical interest in gravitational collapse, have stimulated extensive theoretical research since 1964 on the inevitability of collapse to a singularity and on the dynamics of such collapse. Perhaps the most important outcome of this research is a set of theorems due to R. Penrose and S. Hawking. These theorems generalize to situations without spherical symmetry the concept of being inside the Schwarzschild radius, and the consequent inevitability of continued collapse to a singularity. Other research has revealed that the "singularities" which are created by nonspherical collapse can have a wide variety of forms. Some of the forms do not entail infinite density, but they generally entail infinitely strong and disruptive gravitational forces.

Singularities in any physical theory are believed to signal a breakdown in the theory. Consequently, a vigorous effort is being made to determine whether quantum gravitational effects, which general relativity does not take into account but which should be important at densities above 10^{90} g/cm³, can prevent singularities from arising in gravitational collapse.

The collapse of a nonrotating, spherical star through its Schwarzschild radius and into a singularity is illustrated in Fig. 2. The gravity-generated curvature of space-time constrains the orbiting observer and the star's surface to move along curved paths in the diagram rather than along

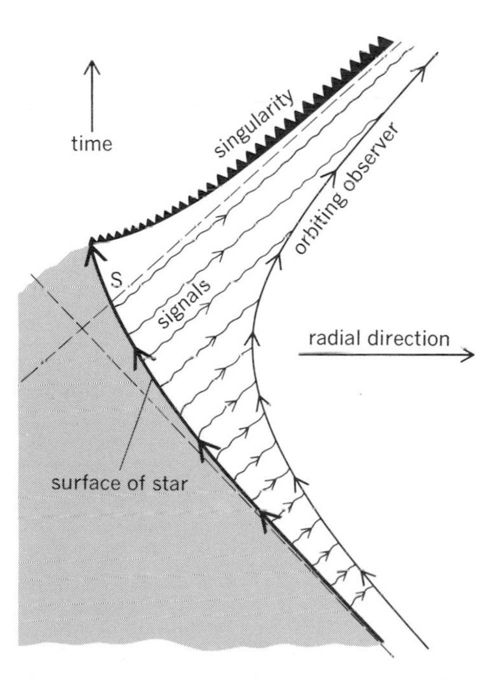

Fig. 2. Space-time diagram for the collapse of a star past its Schwarzschild radius and into a general relativistic singularity of infinite density.

straight lines. As the collapse proceeds, a man on the surface of the star sends radio signals (wavy 45° lines) describing the collapse to a friend who is orbiting the star. The signals, which are sent at equally spaced intervals, are received by the orbiting observer at more and more widely spaced intervals.

This shift, which is known as the Einstein red shift, results from a slowing down of all periodic processes in a gravitational field. A signal sent from just outside the Schwarzschild radius S is not received until an infinite time has elapsed, as seen by the observer's clock. Moreover, none of the signals, nor any energy released by the collapse, sent from inside the Schwarzschild radius ever reach the observer. They, like the star, get swallowed and destroyed in the singularity. *See* EINSTEIN SHIFT.

Consequently, as seen by the orbiting observer, the star never collapses beyond its Schwarzschild radius S. Rather, as the star nears the Schwarzschild radius, it appears to collapse more and more slowly, and light from it becomes more and more red-shifted. Eventually, the star appears to be a dark sphere hovering just outside the Schwarzschild radius. Because it is then invisible but can capture matter and light from outside, it is called a black hole.

By contrast, as seen by the man on the star's surface (Fig. 2), the star collapses freely past the Schwarzschild radius and into the singularity in a time of about $0.00001(M/M_{\odot})$ sec. At the singularity, infinitely strong gravitation forces destroy the man on the surface of the star by compressing him to infinite (or near infinite) density. *See* STELLAR EVOLUTION.

Evidence has suggested strongly that a black hole, produced millions of years ago by stellar collapse, is now in orbit around a massive, bright star in the constellation Cygnus. The hole seems to be pulling gas off its companion. As the gas falls inward toward the hole's Schwarzschild radius, it emits x-rays. Satellites above the Earth's atmosphere have discovered the x-rays, and scientists are studying them. The name of this possible black hole is Cygnus X-1. *See* BLACK HOLE.

[KIP S. THORNE]

Bibliography: P. G. Bergmann, *Riddle of Gravitation: From Newton to Einstein to Today's Exciting Theories*, 1968; C. W. Misner, K. S. Thorne, and J. A. Wheeler, *Gravitation*, 1973; K. S. Thorne, Gravitational collapse, *Sci. Amer.*, November, 1967; K. S. Thorne, The death of a star, *Sci. Year*, 1968; K. S. Thorne, The search for black holes, *Sci. Amer.*, 231(6):32–43, 1974.

Gravitational lens

A massive body producing distorted, magnified, or multiple images of more distant objects when its gravitational fields bend the paths of light rays. Lenses have been observed when the light from very distant quasars is affected by intervening galaxies and clusters of galaxies, producing several different images of the same quasar. Albert Einstein predicted the occurrence of this phenomenon in 1936, but the discovery of real gravitational lenses did not occur until 1979. Gravitational lenses, in addition to being intrinsically interesting, can reveal the intrinsic properties of galaxies, active galaxies, and quasars, and provide information on the overall geometry and scale of the universe.

Action of gravity. The lens phenomenon exists because gravity bends the paths of light rays, which is predicted by Einstein's general theory of relativity. Since photons, the carriers of light energy, have no mass, Newton's theory of gravity indicates that light would always travel in a straight line even if there were heavy, massive objects between the source and the observer. (Even if photons are given mass in Newton's theory, the predicted bending of light is different from the result in general relativity.) But in general relativity, gravity acts by producing curvature in space-time, and the paths of all objects, whether or not they have mass, are also curved if they pass near a massive body.

Numerous eclipse observations have confirmed Einstein's prediction with modest accuracies of 20–30%. Radio astronomers have measured changes in the positions of quasars that occur

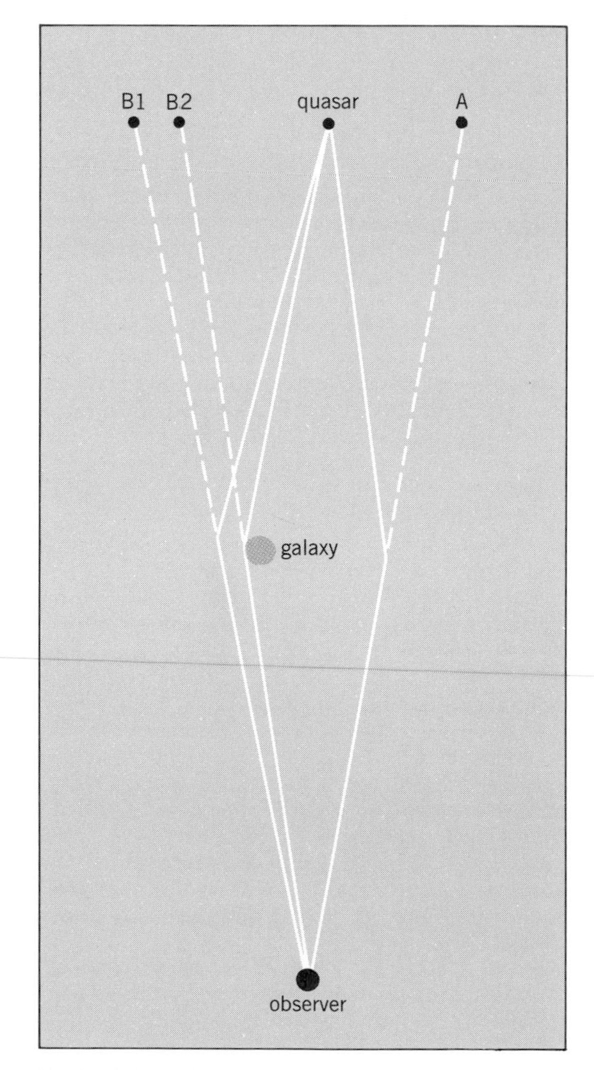

Fig. 1. Schematic illustration of a gravitational lens. The angles are exaggerated for clarity. Here, the lens action produces three images of a quasar (A, B1, and B2).

when the Sun passes near them in the sky. The precision of these experiments, which fit Einstein's predictions, is now at the level of tenths of a percent. *See* RELATIVITY.

A massive object acts as a gravitational lens when light rays from a distant quasar are bent around or through it and are focused to form an image, which can be seen or photographed by an astronomer on Earth, as shown in Fig. 1. Here, three different images of the quasar at A, B1, and B2 are seen, since light from the quasar can travel along three different curved paths and still reach the observer.

Sometimes a lens can amplify the total intensity of light in a quasar image, making it considerably brighter than the quasar would appear to be in the absence of a lens. If the galaxy and quasar are sufficiently well aligned, several images of the same quasar will appear, since light can travel on many different paths and still arrive at the detecting telescope. It has been shown that if there are multiple images of the same quasar, there must be an odd number of them, as long as the galaxy is big enough so that it does not act as a point mass.

Discovery of lenses. Astronomers treated gravitational lenses as curiosities for a long period of time. Space is so empty that the probability of two stars being aligned accurately enough is extremely small. But the discovery of quasars, hyperactive galaxies which are bright enough to be visible even though they are nearly 10^{10} light years (1 ly = 9.46×10^{12} km) away, made it possible to probe a much larger volume of space. Now, there was a reasonable chance that a galaxy might lie in the path of light traveling from a quasar to the Earth. *See* QUASARS.

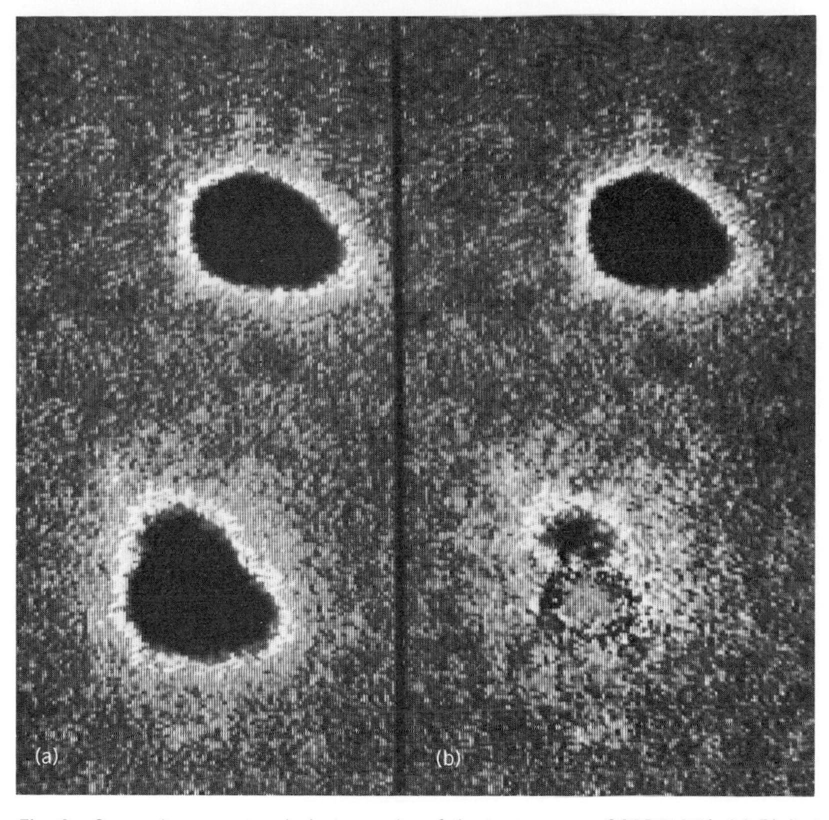

Fig. 2. Computer-processed photographs of the lens quasar Q0957+561. (*a*) Digital superposition of five 1-min exposures. (*b*) Same superposition with 0.7 of the northern (A) quasar image subtracted from the southern image, showing the galaxy just above the southern image. (*Institute of Astronomy, University of Hawaii*)

In March 1979, spectra were obtained of a pair of quasars located very close to each other, only 6 seconds of arc apart (the size of a dime at 600 meters away; Fig. 2*a*). The similarity of their spectra indicated that the twin quasars might be two images of the same object. More measurements of the radio, infrared, and ultraviolet radiation from the twin quasars eventually confirmed the gravitational lens interpretation. This object is known as Q0957+561 A,B, where the numbers serve to indicate the position of the object in the sky, and the letter Q indicates that the object is a quasar. The letters A and B show that there are two objects at that location.

In the succeeding 3 years two more gravitational lenses were discovered. A group of astronomers obtained measurements of a triple quasar, PG1115+080 A,B,C (PG = Palomar-Green; its unusual nature was first recognized in a survey of faint blue objects made at Palomar Observatory by R. Green; Fig. 3), and another double, designated 2345+007, was discovered in August 1981. The survey techniques which were responsible for the discovery of the second and third lenses have been responsible for the discovery of more than a thousand quasars, and there are only three good examples of the gravitational lens effect.

Double quasar. The best-studied gravitational lens consists of the first double quasar Q0957+561 and an unnamed galaxy which lies between it and the Earth. The first evidence that this was indeed a lens came from optical spectroscopy. The red shifts, the strength of the emission lines, and the position of the absorption lines in the two quasars were identical. The southern, or B, quasar was about 0.7 times as bright as the northern, or A, quasar. Since the red shifts, spectra, and line strengths of different quasars vary considerably, it would be a remarkable coincidence if two quasars with identical properties just happened to be very close to each other.

But coincidences do happen, and more observations were needed in order to confirm the lens interpretation. It was important to determine whether the two images really were identical twins and whether a reasonable type of galaxy could produce the bending needed to produce the observed positions of the images. Since lens galaxies are supposed to produce an odd number of images, the location of the third one was a mystery, and the possibility arose that the intervening object was a huge black hole instead of a galaxy. It was also important to detect the galaxy which was responsible for bending the light from the quasar and producing the multiple images.

Further evidence in support of the lens model comes from observations in the infrared and ultraviolet parts of the spectrum. If this object is a lens, then the intensity of the B image should always be a certain fixed percentage of the intensity of the A image, as long as the measure-

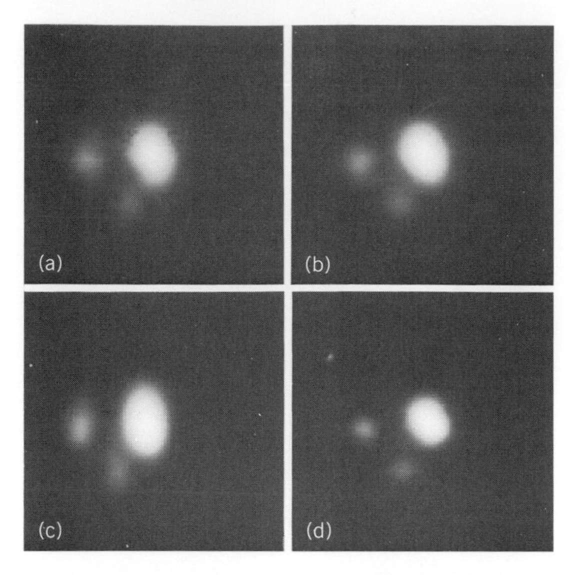

Fig. 3. Images of the triple quasar PG1115+08. (*a*) Blue filter. (*b*) *V* band (yellow) filter. (*c*) Unfiltered image. (*d*) Red filter. (*From E. K. Hege et al., Morphology of the triple QSO PG1115+08, Nature, 287:416–417, 1980*)

ments are taken at the same time. Allowing for possible long-term variations in the brightness of the quasar, this constraint indicates that the B image should be 70–80% as bright as the A image in all wavelengths. Infrared and ultraviolet observations have shown that this prediction is correct.

At first, radio observations indicated some possible problems with the lens interpretation. Maps made with the Very Large Array (VLA) showed a number of radio features apparently connected with the northern quasar, and no corresponding features associated with the southern quasar. These observations are shown in Fig. 4; in addition, the optical images A and B1 are also radio sources. The B image would be expected to have a jet and three blobs like those of the A image, but the VLA map did not show them. Another trouble with the lens interpretation was that there were two images rather than three. *See* RADIO TELESCOPE.

But there is a way out of these problems, a way which leaves little doubt that the double quasar is a lens. A galaxy in exactly the right place would produce triple images of the quasar and the inner part of the jet, and single images of the outer part of the jet and the radio blobs. The third image, B2, is much fainter than image A or B1, and may lie exactly where the galaxy does. The radio work did detect a blob of radio emission north of the southern quasar, B1. This feature may be the missing third image, or be a second image of the inner part of the jet, or be associated with the lens galaxy.

The discovery of the lensing galaxy in November 1979 demonstrated the correctness of the lens theory. A charge-coupled device (CCD) was used as a detector to obtain images of the double quasar. Independently, a photograph of the twins was obtained through a superbly steady atmosphere. Analysis of these two photographs indicated that there was a large galaxy immediately to the north of the southern quasar. The galaxy is shown in Fig. 2*b*, where 0.7 of the A image is subtracted from the B1 image. The missing third image (B2 in Fig. 4) is submerged by the light from the galaxy, possibly contributing to it in a small way.

The model shown in Fig. 4 is not absolutely final. It is also possible that the third, or B2, image is very close to the second, or B1, image and is impossible to separate from it. Furthermore, the galaxy is located in a larger cluster of galaxies which also deflects light and radio waves from the distant quasar. But these are matters of detail; it seems well confirmed that this quasar is an example of a gravitational lens.

The two other known examples of gravitational lenses have not been modeled as elaborately. The identical spectra of the multiple images show that these two other objects are definitely gravitational lenses. There are a few other examples of quasars which are close together in the sky, and when the spectra of these objects are obtained, the number of known gravitational lenses may increase.

Implications. The discovery of gravitational lenses affects understanding of the universe as a

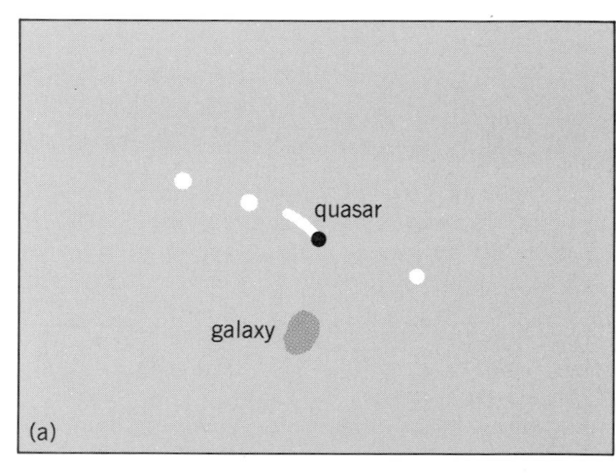

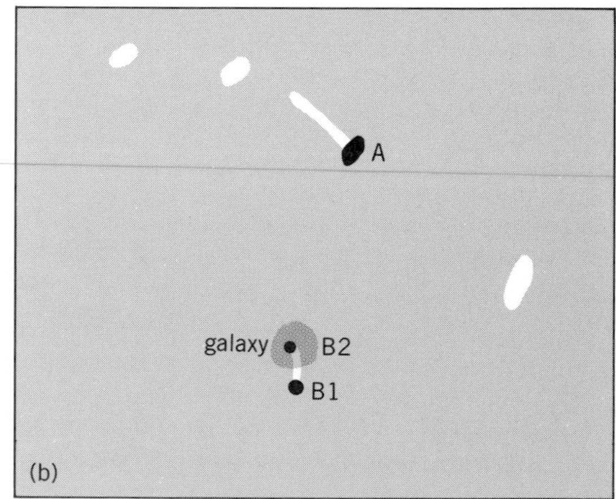

Fig. 4. Configuration of the double quasar Q0957+561. White areas represent areas of radio emission; dark gray and black areas represent optical images. (*a*) Configuration which would appear with the lens action of the galaxy turned off. (*b*) Configuration which is actually observed, with the lens action on.

whole. The very existence of this phenomenon demonstrates that at least three quasars are more distant than the galaxies which are focusing their light. Thus lenses provide one more way of showing that quasars are extremely distant objects, and that their red shifts are produced by the expansion of the universe, rather than by some kind of exotic new physics as some astrophysicists have argued.

Gravitational lenses can make distant quasars appear brighter than they would otherwise be, since their focusing action beams more light in the direction of the Earth. This effect could cloud the interpretation of the quasar statistics, which currently argues for a "quasar era" early in cosmic evolution. Further, lenses acting as quasar amplifiers might explain the high luminosity of some extreme quasars. But they cannot account for the high power of quasars considered as a class, since it is very unlikely that all of the 1500 known quasars are single-image lenses.

When the brightness of the multiple images in gravitational lenses has been monitored for a long time, lenses may reveal the cosmos in some more fundamental ways. If the energy-emitting regions of quasars are small (as is believed), then stars in the halos of galaxies could also act as lenses, changing the intensity of the light from quasar images when the stars pass between the Earth and the quasar. The lens phenomenon could be used to detect the existence of these stars, even if they were too faint to see optically. It is even possible, in principle, to detect Jupiter-sized objects in the halo of a galaxy acting as a gravitational lens, though other variations in the intensity of radiation from quasars might make it very difficult to unscramble the amplification produced by the low-mass objects from other fluctuations in the quasar brightness.

Variations in the brightness of multiple images of the same quasar could also determine the distance to the lensing galaxy and thus the scale of the universe. The distance to the quasar along the path that corresponds to one image differs from the distance along the path corresponding to the other image. Thus, if the quasar suddenly increased its power, one image would brighten first, before the other changed. The time lag could be measured, and the difference in path lengths thus determined, since light follows both paths at the same speed. The difference in path lengths is related to the distance to the quasar and to the lens galaxy. By measuring the distance to these very distant objects, a much better measurement of the overall cosmic distance scale would be obtained. *See* COSMOLOGY.

[HARRY L. SHIPMAN]

Bibliography: F. H. Chaffee, Jr., The discovery of a gravitational lens, *Sci. Amer.*, 243(5):70–78, November 1980; J. K. Lawrence, Gravitational lenses and the double quasar, *Mercury*, 9(3):66–72, May–June 1980; D. H. Roberts and B. F. Burke, Gravitational lenses: From Einstein to the double quasar, *Technol. Rev.* 83(4):68–78, February–March 1981; P. Young et al., Q0957+561: Detailed models of the gravitational lens effect, *Astrophys. J.*, 244:736–755, March 15, 1981.

Halley's Comet

The most famous of all comets, associated with many important events in history. Records of Halley's Comet appear at least as far back as 240 B.C., and they are found in the *Nuremberg Chronicle* (the apparition of A.D. 684) and the Bayeux Tapestry (the apparition of A.D. 1066). The comet's size, activity, and favorably placed orbit (with the perihelion roughly halfway between the Sun and the Earth's orbit) ensure its visibility to the naked eye at each apparition.

History. This comet was the first to have its return predicted, a feat accomplished by Edmond Halley in 1705. He computed the orbits of several comets with Isaac Newton's then new gravitational theory. The orbits of comets observed in 1531, 1607, and 1682 were remarkably similar. Halley assumed that the sightings were of a single comet and predicted its return in 1758–1759. The prediction was verified, and the comet was named in his honor. Halley's Comet was observed in 1835 by F. W. Bessel and in 1910 by E. E. Barnard and many others (see illustration).

Orbital properties. The comet's orbit is a very elongated ellipse (with an eccentricity of 0.967) which has a perihelion of 0.59 astronomical unit (AU = the Earth-Sun distance = 1.496×10^{11} m) and an aphelion of 35 AU, between the orbits of Neptune and Pluto. Halley's Comet was farthest from the Sun in 1948 and since then has been moving toward the Sun. The average period of revolution is 76 years, and the comet's motion is retrograde, that is, opposite the planets' motion. The present relative positions of the comet's orbit and the Earth's orbit mean that the comet can approach Earth as close as 0.15 AU at the ascending node and 0.05 AU at the descending node. *See* CELESTIAL MECHANICS.

Physical properties. Halley's Comet is thought to have a solid, icy nucleus roughly 2.5 km in radius. It displays the gamut of known cometary phenomena, including a long tail (see illustration). *See* COMET.

1986 apparition. The next perihelion passage of the comet is confidently expected on February 9, 1986. The best times for public viewing and for

Halley's Comet on May 13, 1910. The planet Venus is in the lower part of the photograph. The tail stretched approximately 45° on this date. (*Lowell Observatory*)

ground-based observation of the comet are during November 1985 and April 1986.

The full range of phenomena plus the known, predictable orbit makes this comet a prime target for the first space probe to a comet in 1985 or 1986.

Associated meteors. Halley's Comet is associated with two meteor showers, the η-Aquarids and the Orionids. The origin of these showers undoubtedly involves meteoroids from the comet's nucleus that were gravitationally perturbed into their current orbits, which intersect the Earth's orbit. *See* METEOR. [JOHN C. BRANDT]

Bibliography: N. T. Bobrovnikoff, Halley's Comet in its apparition of 1910–11, *Publications of Lick Observatory*, vol. 17, 1931; J. C. Brandt and R. D. Chapman, *Introduction to Comets*, 1981; M. Neugebauer et al., *Space Missions to Comets*, NASA CP-2089, 1979.

Hertzsprung-Russell diagram

A graphical representation of the absolute magnitude and the surface temperature of stars. It is named after E. Hertzsprung of Denmark and H. N. Russell, working at Princeton University. In 1911 Hertzsprung plotted the magnitudes versus the colors of the stars in several clusters. In 1913 Russell examined the stars in the solar neighborhood by plotting the absolute magnitude versus their spectral classes. Their plots were the prototypes of contemporary Hertzsprung-Russell (H–R) diagrams. The ordinate is commonly the absolute visual magnitude. In the case of a cluster of stars, all of which may be assumed to be at the same distance, the apparent magnitude may be used in place of the absolute magnitude. For comparison with theory, the logarithm of the luminosity, in units of the Sun's luminosity, is used. The abscissa is the spectral type or the color index, both of which are related to the star's surface temperature. For comparison with theory, the logarithm of the effective temperature is used, the effective temperature being closely related to the surface temperature. When the absolute or apparent magnitude is plotted against a color index, the H–R diagram is also called a color-magnitude diagram.

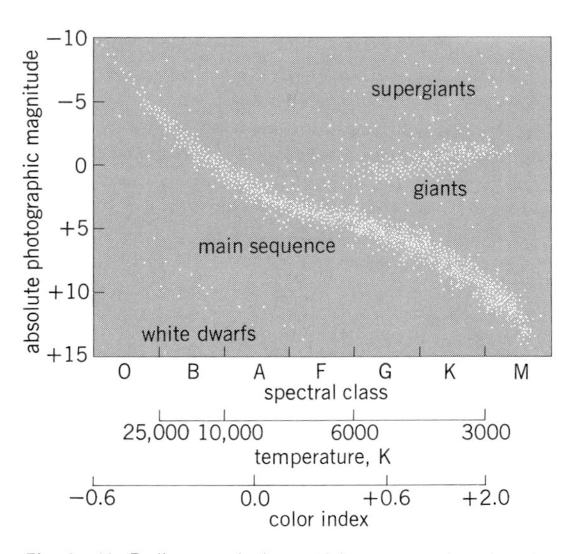

Fig. 1. H–R diagram (schematic) representing the distribution of stars of known distance. (*From G. O. Abell, Exploration of the Universe, 3d ed., Holt, Rinehart and Winston, 1975*)

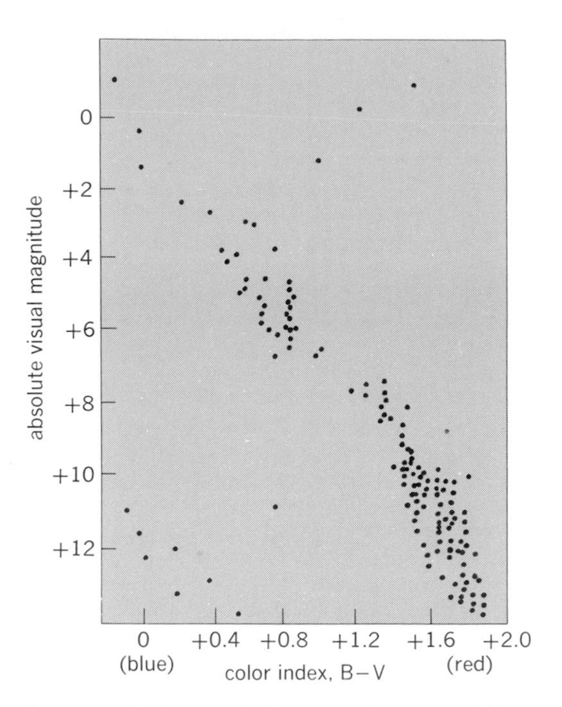

Fig. 2. H–R diagram limited to all the stars within 10 parsecs from the Sun. (*Adapted from H. L. Johnson and W. W. Morgan, Astrophys. J., 117:313, copyright © 1953 by University of Chicago*)

Distribution of stars. An H–R diagram of all stars with known absolute magnitudes and surface temperatures is shown in Fig. 1. By far the majority of the stars fall within a band running from the lower right to the upper left. These stars are called main-sequence stars. A few stars are found in the lower left part of the diagram; these are the white dwarfs. The group of stars in the middle right

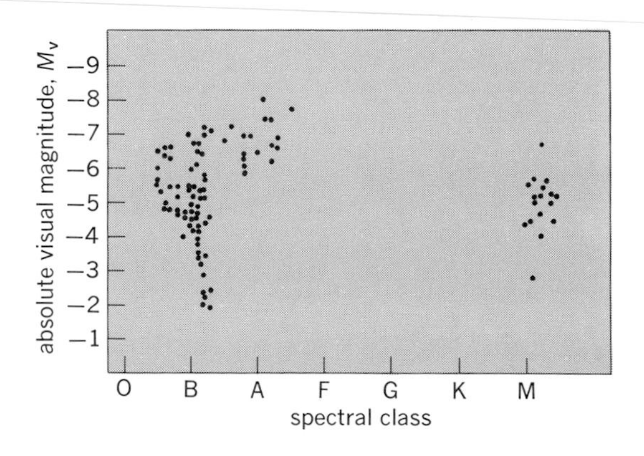

Fig. 3. H–R diagram of h and χ Persei. (*From R. T. Wildey, Astrophys. J., suppl. ser., 8:489, 1964*)

part of the diagram are the giants, and the stars along the upper right-hand part of the diagram are the supergiants. The region between the main sequence and the giants and supergiants, where few stars are found, is called the Hertzsprung gap. The distribution shown in Fig. 1 does not represent the true proportion of stars in a given volume of space. This is because the intrinsically faint stars, such as the white dwarfs and red main-sequence stars (red dwarfs), are underrepresented due to the difficulty in observing them. Figure 2 shows the stars within 10 parsecs (3.086×10^{14} km), the largest distance within which a nearly complete sample is available. *See* DWARF STAR; GIANT STAR; RED DWARF STAR; SUPERGIANT STAR; WHITE DWARF STAR.

The total range in absolute magnitude is about 27 magnitudes, from +18 to −9, or 10^{11} in brightness, and the total range of surface temperature is from about 2300 to about 50,000 K. The Sun is a G-type main sequence star with an intermediate absolute magnitude (+5) and an intermediate surface temperature (about 6000 K). In a given volume of space around the Sun, about 90% of the stars are main-sequence stars, about 10% are white dwarfs, and about 1% are giants or supergiants. Along the main sequence, from the lower right toward the upper left, the stars become less and less common.

Cluster H-R diagrams. The open double cluster, h and χ Persei (Fig. 3), shows a blue main sequence plus some supergiants. Each of the open clusters, the Pleiades (Fig. 4) and Praesepe (Fig. 5), shows a long main sequence with a few giants in Praesepe. The open cluster M67 (Fig. 6) shows a very short main sequence and a scattering of giants. A composite H−R diagram for the four open clusters is shown in Fig. 7. Globular clusters, such as M3 (Fig. 8), show a pattern similar to M67 but with important differences. There is a large number of stars extending up the right-hand side of the diagram; this is the giant branch. There is also a band of stars extending horizontally across the diagram from the middle to the giant branch; this is the horizontal branch. Although M67 shows a few stars in a similar region, they are not true horizontal-branch stars. The main sequences turn off at about the same color of +0.4 to +0.5 in $B-V$, but the giants extend up about 2 magnitudes higher in M3 than in M67.

Stellar evolution. The H−R diagram is useful for exhibiting the effects of stellar evolution. Stars evolve as a result of the nuclear reactions in their central regions; these reactions supply the energy which is radiated as light and heat from their surfaces. An example is the conversion of hydrogen into helium, which is occurring in the Sun and

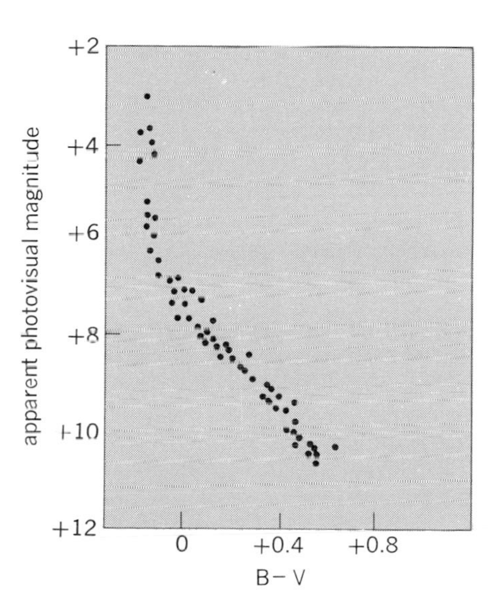

Fig. 4. H−R diagram of the Pleiades. (*Adapted from H. L. Johnson and W. W. Morgan, Astrophys. J., 117:313, copyright © 1953 by University of Chicago*)

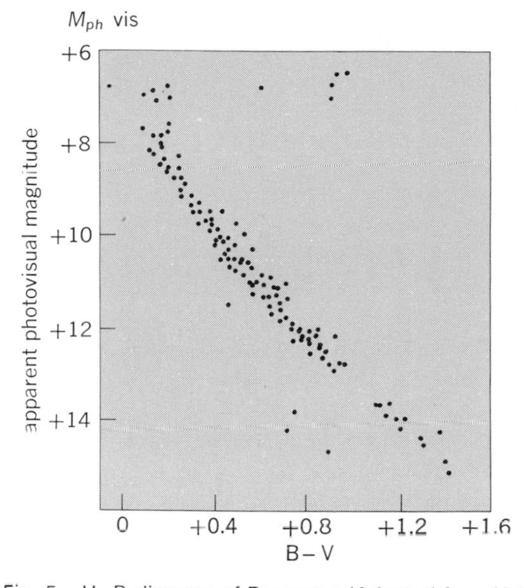

Fig. 5. H−R diagram of Praesepe (*Adapted from H. L. Johnson and W. W. Morgan, Astrophys. J., 117:313, copyright © 1953 by University of Chicago*)

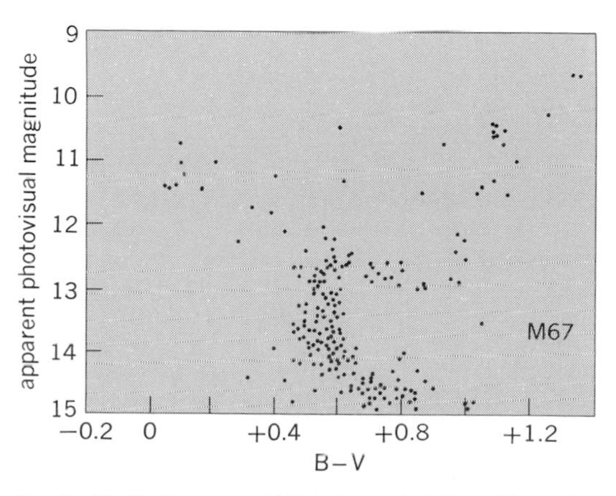

Fig. 6. H−R diagram of M67. (*Adapted from H. L. Johnson and W. W. Morgan, Astrophys. J., 117:313, copyright © 1953 by University of Chicago*)

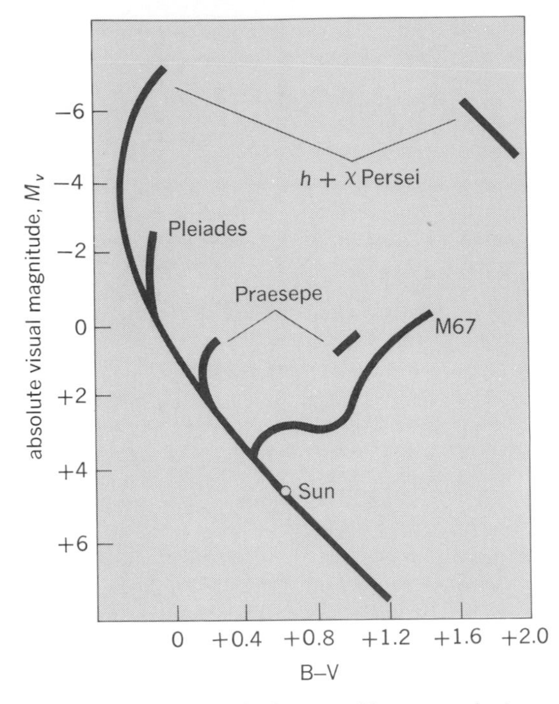

Fig. 7. A composite H-R diagram of four open clusters.

other main-sequence stars. As a result of its evolution, a star's position in the H−R diagram changes with time. The time scales for noticeable changes in the position of a star in the H−R diagram generally are on the order of millions to billions of years. The Sun has been in nearly the same position for about 5×10^9 years, but about 5×10^9 years from now it will begin to move into the giant region. Because stars evolve so slowly, the evolutionary characteristics of the H−R diagram are discovered by using computer calculations of model evolving stars.

Because of stellar evolution, the H−R diagram is dynamic, rather than static, with stars moving into and out of each region. Those regions in which there are few stars, such as the Hertzsprung gap, are regions in which the stars are unstable and pass through relatively rapidly. Those regions where there are many stars, such as the main sequence, are those in which the stars stay for relatively long periods of time.

Cluster ages. Along the main sequence, from the lower right toward the upper left, the masses of the stars are found to increase from about 1/10 of the Sun's to almost 100 times the Sun's. That is, the more massive stars are brighter and have higher surface temperatures than the less massive ones. They also evolve more quickly, so that the more massive stars spend less time on the main sequence than less massive ones. After a star has used up essentially all of the hydrogen in its central regions, it will begin to move toward the right-hand side of the diagram. Thus, for a cluster of stars, all of which were formed at the same time, the H−R diagram initially has a main sequence extending into the upper left-hand portion of the diagram. As time passes, the massive stars begin to leave the main sequence, so that the older the cluster the shorter its main sequence. In Fig. 7, then, the age sequence (youngest to oldest) is: *h* and χ Persei, the Pleiades, Praesepe, and M67. The cluster M67 has an age of a few billion years and is one of the oldest open clusters known. The globular clusters are much older, with ages estimated to lie in the range 10 to 15×10^9 years.

Effects of initial composition. The differences between the open clusters and the globular clusters can be explained by their initial composition. The open clusters are members of population I, to which the Sun also belongs. The globular clusters belong to population II. The population I stars formed from gas clouds in which the "metals" were roughly 1% of the mass of the gas. (In astrophysics the term metal includes all of the elements heavier than helium.) Population II stars formed from gas clouds in which the metals were from ten to several hundred times rarer than in the case of the Sun. The population II stars represent an extremely old generation of stars, whereas the population I stars represent newer generations of stars, none of which are nearly as old as population II stars but which include stars formed up to the present. The fact that the turnoff of the main sequence occurs at nearly the same color in M3 and M67, in spite of the much greater age of M3, is a consequence of the difference in initial composition. The brighter giants in M3 are also a consequence of the difference in initial composition of the two clusters.

The position of an individual star changes with time, but the distribution of stars changes much less rapidly, particularly in the case of population I stars. New population I stars are continually being formed, and old ones are dying. There are no new population II stars being formed, so the distribution of population II stars is changing slowly with time. *See* STAR; STAR CLUSTERS; STELLAR EVOLUTION. [DONALD S. HAYES]

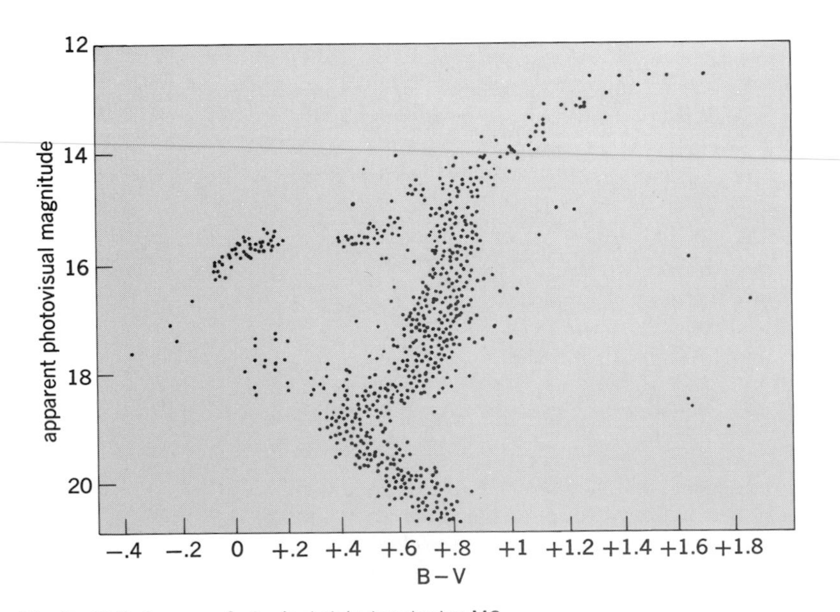

Fig. 8. H-R diagram of a typical globular cluster M3.

Bibliography: G. O. Abell, *Exploration of the Universe*, 3d ed., 1975; C. Payne-Gaposchkin,

Stars and Clusters, 1979; A. G. D. Philip and D. S. Hayes (eds.), *IAU Symposium 80: The H–R Diagram*, 1978; O. Struve and V. Zebergs, *Astronomy of the 20th Century*, 1962.

Horizon

The visible horizon is the apparent boundary line between sky and earth or sea. The astronomical horizon is the great circle of the celestial sphere 90° from the zenith and the nadir. *See* ASTRONOMICAL COORDINATE SYSTEMS.

[GERALD M. CLEMENCE]

Horology

Measurement of the time dimension. In practice, horology is: the search for a steady or repetitive action; and the design of an instrument to perform that action and to indicate (read out) a measure of the action. Until early in the 20th century, horology dealt with mechanical instruments, with effort distributed between improving accuracy and decreasing size of timepieces. Increasingly, however, electronic instruments provided means for meeting these objectives.

Early instruments. The rotation of the Earth on its axis provides a naturally repeating action that produces a directly accessible indication of the flow of time. A sundial increases the precision of readout in subdivisions of the diurnal unit. In modern times the photographic zenith tube is used to identify the orientation of the Earth against the background of the fixed stars relative to the Sun. *See* EARTH ROTATION AND ORBITAL MOTION; OPTICAL TELESCOPE; SUNDIAL.

Among the earliest instruments for indicating the flow of time was a burning candle. Although highly sensitive to its own structural variations and to such environmental influences as air currents, the candle transformed the measurement of time into the measurement of height, which could be done accurately. A somewhat related currently used action is radioactive decay of an isotope.

Mechanical clocks. Galileo first observed the isochronism of the pendulum, its period depending primarily on its length. By choice of configuration and of material, the length of a pendulum from its hinge point to its center of gravity was made relatively independent of temperature. Many mechanisms were devised to transfer energy from a falling weight to the pendulum in synchronism with its swing, to make up for frictional losses and drive the readout mechanism. Modern pendulum clocks can be accurate to 1 part in 10^8 during the course of a day.

The swinging of the pendulum about its pivotal support required that the clock be fixed vertically. To overcome this restriction, an alternative timing mechanism was developed in the form of the balance wheel and hairspring. The moment of inertia of the balance and the compliance of the spring jointly determine the frequency of oscillation. Again, by temperature compensation and by release of energy from a main spring through an escapement, the clock is isolated from its environment. In the shipboard chronometer, the gimbals isolate the movement from pitch and roll and a massive balance wheel decouples it from yaw. A fusee between main spring and great wheel, and a helical hairspring further improve accuracy. *See* CHRONOMETER.

Electronic clocks. The piezoelectric crystal replaced the assembly of balance wheel and hairspring with a naturally oscillatory structure. The piezoelectric property of the crystal couples its mechanical vibration into an electrical circuit. Its mechanical frequency is the consequence of its physical constants: density and elastic compliance; the crystal is precisely ground to produce the desired frequency, in the region of kilohertz to megahertz. The electronic circuit around the crystal replaces the escapement to drive the crystal in forced oscillation with an alternating voltage that the circuit develops from a direct current source such as a battery to sustain vibration and to power the readout display. Thus, time measurement becomes essentially frequency measurement. A quartz clock may drift as little as about 1 part in 10^{11} during a day. *See* QUARTZ CLOCK.

The next advance in accurate measurement of time came by replacement of dynamic mechanical oscillators with quantum energy transitions. In the ammonia-controlled clock, those molecules in a beam of ammonia that are in the higher of two energy states are isolated in a cavity. There they fall to the lower state, emitting electromagnetic radiation in the microwave region. This emission line in the molecular spectrum serves as the timing means to regulate an auxiliary oscillator. The inversion frequency of 2.387×10^{10} Hz is inherently stable; however, the timing accuracy achieved from the clock is masked by drifts and noise in the electronic pumping circuits that perform a function analogous to the escapement. *See* ATOMIC CLOCK.

Greater stability is obtained from the quantum emissions that accompany the transition in the precession axis of the outer unpaired electron of vaporized cesium-133. In the cesium-beam primary clock, a beam of cesium-133 atoms is passed through a magnetic field. The field deflects each atom in which the valence electron has the desired one of two possible spins into a cavity. There, microwave energy derived from an auxiliary crystal oscillator and frequency-multiplier chain aligns the electrons to the opposite spin. The cesium beam exits the cavity through a second magnetic field that directs the realigned electrons to a detector whose output slaves the auxiliary oscillator so that the energy fed to the cavity is at the precise frequency required to stimulate the quantum transition in spin direction. The auxiliary oscillator also feeds a counter and readout. The transition frequency is inherently invariant and highly independent of environmental influence. As a consequence, the cesium-beam oscillator is accurate to 1 part in 10^{12}, and the transition frequency is used to define the second.

In any sensitive system, mechanical or electronic, thermal fluctuations set a noise threshold to short-term stability. To combat this limitation, various electronic amplifiers and oscillators operate in cyrogenic enclosures. Also, at low temperature, hydrogen molecules exist predominantly at two energy states in which the spins of the two nuclei are respectively in opposition to and parallel to each other. Consequently operation of a hydrogen maser at low temperature acquires both the advantages of lower-noise electronics and more discrete energy states. A hydrogen maser clock with the cavity cooled by carbon tetrafluoride to −248°C provides a stability of 1 part in 10^{14}. *See* TIME.

[FRANK H. ROCKETT]

Bibliography: J. T. Fraser and N. Lawrence (eds.), *The Study of Time II*, 1975; J. L. Jesperson et al., Special issue on time and frequency, *Proc. IEEE*, (60)5:476–638, 1972; P. Kartaschoff, *Frequency and Time*, 1978.

Hubble constant

The rate at which the velocity of recession of the galaxies increases with distance. Edwin Hubble established the existence of this motion observationally in 1929. *See* RED SHIFT.

The value of the expansion rate is still not agreed upon. The two values which seem to have the widest acceptance are 15 km per second per million light-years (1.6×10^{-18} s^{-1}) and 29 km per second per million light-years (3.1×10^{-18} s^{-1}).

If the expansionary flow of the galaxies were completely regular, the best determination of the Hubble constant would be made from relatively nearby, bright objects. For such galaxies the distance could be determined from the difference between the known luminosity of various stellar components and their measured brightness, and the velocities of recession could be found from relatively high dispersion spectra. However, it has been suggested that perturbations result from the inhomogeneous distribution of galaxies in the Local Supercluster, and in particular that the gravitational attraction of the Ursa Major and Virgo clusters deforms the local velocity field so that determination of the Hubble constant from measures made on relatively nearby galaxies does not reflect the rate of expansion for the universe as a whole.

A. Sandage and G. A. Tammann used as distance indicators the period-luminosity law for Cepheids for the nearest objects, H II regions for those at intermediate distances, and the established properties of the brightest class of late-type spirals for those farthest away. They obtained a value of 15 km per second per million light-years. They found no significant difference in the Hubble constant (a) for objects with recessional velocities below 2000 km·s^{-1} and for those above 6000 km·s^{-1}; (b) for galaxies in the hemisphere toward the Virgo cluster and those in the opposite hemisphere; and (c) for various regions along the equator of the supergalaxy. *See* CEPHEIDS; INTERSTELLAR MATTER.

M. Aaronson, J. Huchra, and J. Mould first established that there is a close correlation between the infrared magnitudes of late-type galaxies and the separation of the two components of their 21-cm line of hydrogen. This correlation is to be expected since the separation increases with the rate of rotation, which itself increases with total mass, and the total luminosity in the infrared also increases with mass. Using this correlation and known recessional velocities for relatively nearby galaxies, they obtained a value of the Hubble constant in agreement with that of Sandage and Tammann. Next Aaronson and colleagues extended their work to galaxies beyond the Virgo cluster, obtaining 29 km per second per million light-years. Aaronson attributed the difference in the two results to the Milky Way Galaxy falling toward Virgo. *See* COSMIC MICROWAVE RADIATION.

If the expansion rate has been constant, then the age of universe is the reciprocal of the Hubble constant. If, as seems more probable, the expansion has decelerated due to the gravitational attraction of the galaxies, the reciprocal of the Hubble constant gives an upper limit to the age, approximately 19 and 10×10^9 years respectively for the two values quoted above. Attempts to determine the deceleration from observations of galaxies so distant they they are observed at a substantially earlier time, when they may have been receding more rapidly, are vitiated by the possible evolutionary changes in luminosity which may have occurred. W. A. Fowler gave the age of the universe as determined from the oldest globular clusters in the Galaxy as between 7 and 20×10^9 years, and from the abundances of the various radioactive element as between 10 and 15×10^9 years. It therefore appears that either value of the Hubble constant mentioned above is consistent with the other estimates of the age of the universe. *See* COSMOLOGY; GALAXY, EXTERNAL; STAR CLUSTERS.

[LOUIS C. GREEN]

Hyades

A group of stars known from early times, scattered through the V in the constellation of Taurus. An estimated 350 stars form the Hyades cluster with half of its mass in a sphere of diameter of 40 light-years (1 ly $= 9.46 \times 10^{12}$ km). Its brightest stars are yellow to red. Because of the proximity of this cluster, which permits intensive observations, its main sequence stars are used as luminosity standards for the zero-age main sequence, from which the ages and distances of other clusters may be determined. The age of the Hyades is 7×10^8 years. However, its precise distance is still controversial. Various geometric methods of distance measurement are sensitive to sizes of positional errors, and photometric methods involve assumptions of chemical composition. Reliable distance determinations fall between 140 and 150 light-years, but even a small error in the distance of the Hyades will magnify errors in distance determinations for more distant clusters.

The Hyades is also the most prominent moving cluster. Known as the Taurus moving cluster, the group has a velocity of 45 km/s toward a convergent point in the constellation of Orion. Similarities in age, chemical composition, and motions between members of the Hyades and Praesepe open clusters lead to a hypothesis of common origin. *See* CONSTELLATION; STAR; STAR CLUSTERS.

[HELEN S. HOGG]

Image tube

A photoelectric device for intensifying faint astronomical images. The photographic emulsion (plate or film), which is the time-honored method for recording astronomical images, has many excellent characteristics for the purpose. It has high resolution; reasonably high signal-to-noise ratio images; good dimensional stability and permanence; and wide spectral response. However, it lacks one characteristic of greatest importance to the astronomer: high quantum efficiency, or the ability to record a high percentage of the light quanta originally in the incident image. Even the most sensitive photographic emulsions are able to record only a few light photons in every thousand that the telescope collects and focuses into an image. All the rest, and the information they bring, are lost. The photographic process has two other drawbacks for astronomical applications. The sensitivity, already low, becomes lower during the very long exposures that astronomers frequently use. Ordinarily one thinks of an emulsion as responding to a given total amount of light representing the exposure regardless of whether it comes in the form of an intense burst of short duration or as a fainter image exposed for a correspondingly longer time. But when exposures exceed a few seconds, the intensity and exposure time are no longer inversely proportional, a phenomenon called reciprocity failure. The net effect of reciprocity failure is that the emulsion speed decreases drastically as exposures become very long. The other drawback of the photographic process is its nonlinearity. There is a linear relation between exposure and resulting emulsion density only over a relatively narrow range of exposures, and beyond this range for either higher or lower exposures the response is nonlinear.

Photoemissive process. The photoemissive process embodied in modern photomultiplier tubes is free from the drawbacks cited above. In these tubes the photosensitive surface, or photocathode, receives light and converts it into electrons that are emitted from the cathode into the vacuum of the tube where, by means of electric and magnetic fields, they can be collected for measurement.

The process of converting light photons to photoelectrons can be quite efficient, in some cases yielding as much as one electron for every four incident light quanta, though the average yield over a reasonably wide region of the spectrum would be more like one electron per 10–20 photons, an amount that is still impressive compared with photographic efficiencies. The photoemissive process also responds linearly in terms of photocurrent versus light intensity over an extremely wide range, and brightness ranges of more than a billion to one can be accommodated without difficulty; and equally important, the photoemissive process does not have reciprocity failure. It is no wonder, then, that the photomultiplier has become an essential astronomical research tool, for its characteristics are best just where the photographic emulsion is most deficient. Unfortunately, the photomultiplier can look at only one small part of the image at a time, whereas the photographic emulsion may receive light from millions of tiny picture elements simultaneously. In this respect the photographic emulsion is far superior, and photography and photoelectric photometry have continued to coexist as complementary methods for obtaining astronomical data.

The photoelectric image tube represents a device which in principle combines the best features of photography and photoelectric photometry. An image tube is conceptually like a vast array of microscopic phototubes independently and simultaneously receiving information from each picture element as the photographic emulsion does, but handling it with the efficiency and linearity of the photomultiplier. The potential advantages of such devices for astronomy are tremendous; they combine the high resolution and simultaneous reception from all elements of an image that characterize the photographic emulsion with the high quantum efficiency and wide, linear dynamic range available from photoemissive devices.

Types of tubes. Basically the image tube is very simple. It consists of a photoemissive surface inside a vacuum tube on which a light image from the telescope or its associated spectrograph falls. Electrons emitted from this surface represent an electronic replica of the original light image. This electron image is electrically accelerated and focused down the tube by means of electric or combined electric and magnetic fields (Fig. 1). Up to

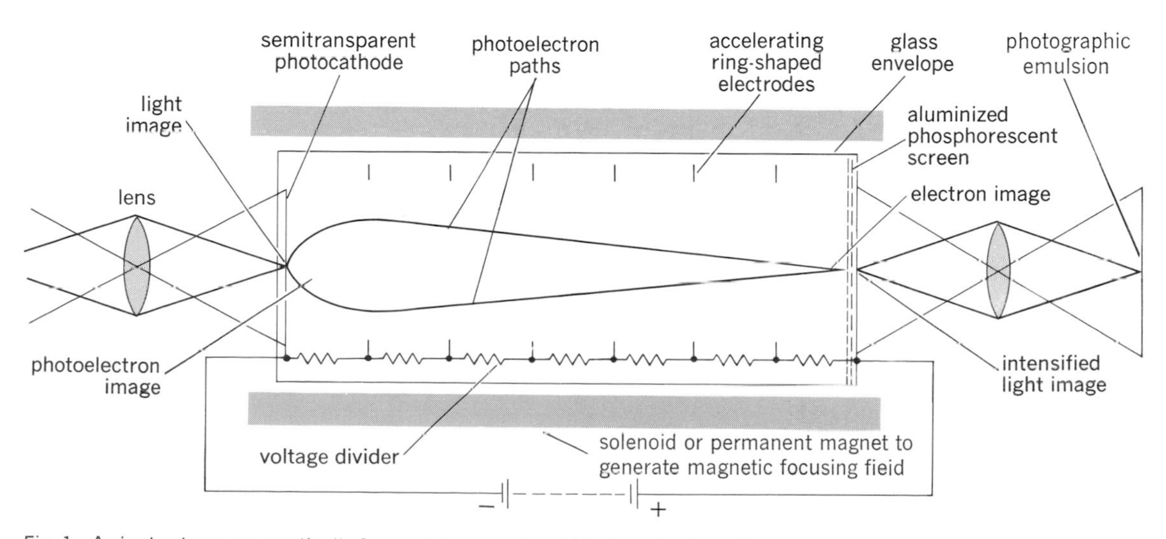

Fig. 1. A single stage, magnetically focused image tube. (*U.S. Army Engineer Development Corps*)

this point all image tubes are very much the same. The different types are distinguished largely by what happens to the electron image at its focal plane down the tube.

Electronographic tube. One way to handle the electron image is to record it directly upon film or plates just as is done in the electron microscope. Direct recording of electron images is called electronography, and image tubes using this method of recording are referred to as electronographic tubes. This recording technique can be very efficient if the electrons strike the photographic emulsion with energies corresponding to acceleration through voltages of 15–40 kV, and it is possible in this way to record nearly every electron in the electron image. Not only is recording efficiency high in electronography, but very-fine-grain, high-resolution emulsions can be used. These emulsions are generally too insensitive to light to be useful for direct astronomical photography, but some of them have sufficiently good response to fast electrons to be valuable for electronographic recording. Furthermore, some of the emulsions have a linear relationship between density and exposure to electrons, thus preserving the linearity of response of the photocathode. Electronographic cameras using high-resolution plates are capable of producing excellent results with resolutions close to 100 line pairs per millimeter. Unfortunately, the electronographic camera has a technical drawback that has not been fully surmounted. The photographic emulsion when placed in vacuum gives off gases such as water vapor that destroy the photocathode of the image tube in a short time. Various schemes have been successfully employed to prevent gas evolved from the emulsion from getting to the photosurface. However, because of the complexity of electronographic tubes, they are used only at a very few observatories which have the necessary staff and equipment to operate them successfully. *See* ASTRONOMICAL PHOTOGRAPHY.

Fiber-optically coupled multistage tube. Another type of image tube converts the electron image back into a light image by means of a phosphor screen placed at the focus of the electron image. Since the electrons are accelerated and gain energy by electric fields in the tube, the final light image can be appreciably brighter than the original, and this device functions as an image intensifier. The intensification process is usually repeated one or more times by coupling single-stage image tubes so the output of the first is imaged on the photocathode of the second, and so on.

It is inefficient to transfer the image between stages by means of lenses that collect only about 5% of the light from the phosphor screen. To overcome this loss, the individual stages can be fiber-optically coupled. In this case the output window of the first tube is a vacuum-tight fiber-optic faceplate, and the phosphor screen of the first tube is put directly upon it. Likewise, the photocathode of the second tube is deposited on an input fiber-optic faceplate, and the tubes are operated in series with their fiber-optic windows in optical contact. The image transfer efficiency is much higher with this arrangement, with values around 50% common. The electron multiplication (number of secondary electrons emitted per initial input photoelectron) is about 50–100 at each phosphor-photocathode sandwich. The final intensified image is transferred from the output phosphor screen to a recording detector either by a fiber-optic faceplate or by a fast transfer lens. Figure 2 shows a tube of this type.

Cascade image tube. Alternately, the phosphor of one stage may be deposited on a very thin transparent membrane of mica, glass, aluminum oxide, or similar material, with the photocathode of the following stage deposited on the other side of the membrane which is enclosed in the vacuum tube. Close proximity between phosphor and photocathode in this thin sandwich gives very good light transfer efficiency without serious loss of resolution, provided the membrane is thin enough. The

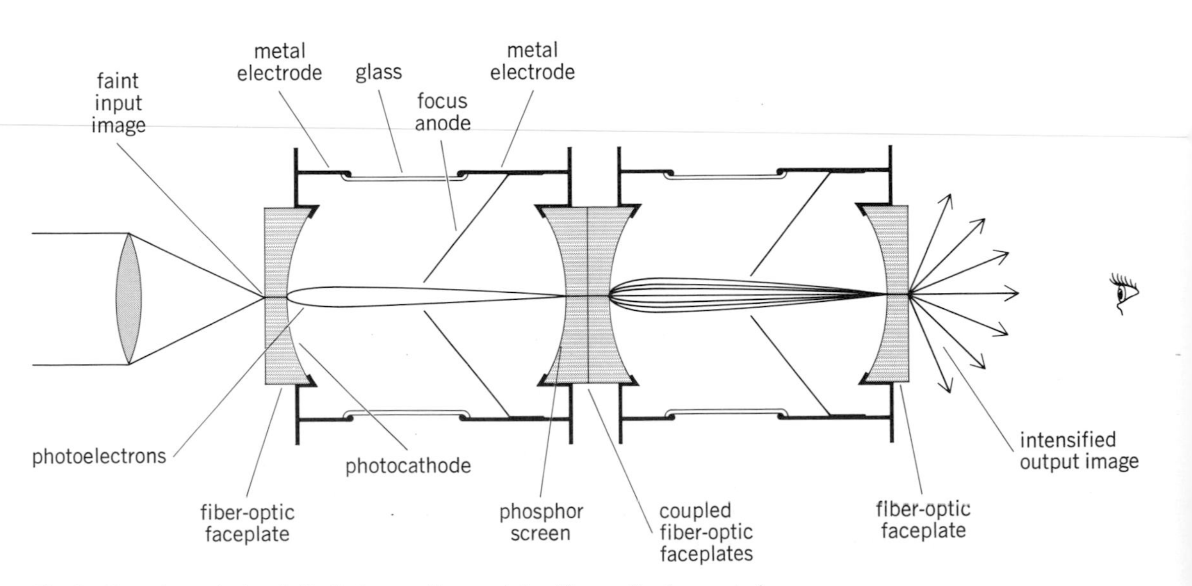

Fig. 2. Two-stage electrostatically focused image tube, fiber-optically coupled.

cascading of two or more stages may be accomplished in this manner to obtain high electron multiplication and high brightness gain.

TSEM-type tube. Another way to provide intensification is to focus the electrons on a succession of very thin membranes (dynodes) in the tube, each causing high-energy primary electrons to knock out several secondary electrons directly (without converting electrons into photons, and photons back into electrons). This process is known as transmission secondary emission multiplication (TSEM). Although TSEM dynodes are simpler than the phosphor-photocathode sandwiches, dynodes used in past tubes have had the disadvantage of producing, on an average, only five to seven secondary electrons for each primary electron. Thus, more stages of amplification are required to produce the same overall gain. Worse, one primary electron may produce 10 secondaries, whereas another primary electron may produce only one or two—or zero (indeed, zero is the most probable number). For this reason, high-gain multistage TSEM intensifiers show poor multiplication statistics, with photoelectron scintillations at the final output phosphor screen varying in brightness over a range of 100:1. Because of their inferior performance, TSEM tubes are no longer used in astronomy, but new dynode materials may alter this in the future.

Channel plate multiplier tube. Another means of obtaining electron multiplication within an image tube is to focus the electrons on a channel plate multiplier. The channel plate multiplier (Fig. 3) consists of an array of closely spaced tubes, each coated on the inside with a secondary-electron-emitting material, so that when a primary electron enters the channel, it collides with the side wall and releases several secondary electrons. The secondary electrons are then accelerated further down the channel by an electric field, and they, in turn, collide with the side wall and are multiplied in a similar fashion. The process is repeated several times in a given channel, so that by the time the secondary electrons exit the rear of the channel plate they have been multiplied some 1000–10,000 times. These electrons are then focused on an output phosphor screen.

The channel plate multiplier suffers from a type of poor multiplication statistics similar to that of the TSEM dynode, and for this reason the technique is not widely used in astronomical detectors. However, because of the much shorter overall length and lower voltage requirements of this device compared to the others, channel plates have been used where size or operating voltage are important considerations.

Compared with the electronographic tube, all other types of tubes described above suffer some loss of resolution, and resolutions obtained in practice generally fall in the range of 25–45 line pairs/mm.

Recording images. Except for the electronographic tube, the image tubes described above do not incorporate a means of recording their output image; they merely intensify the incoming light. The output light must be transferred by means of a fiber-optic faceplate or a transfer lens to another

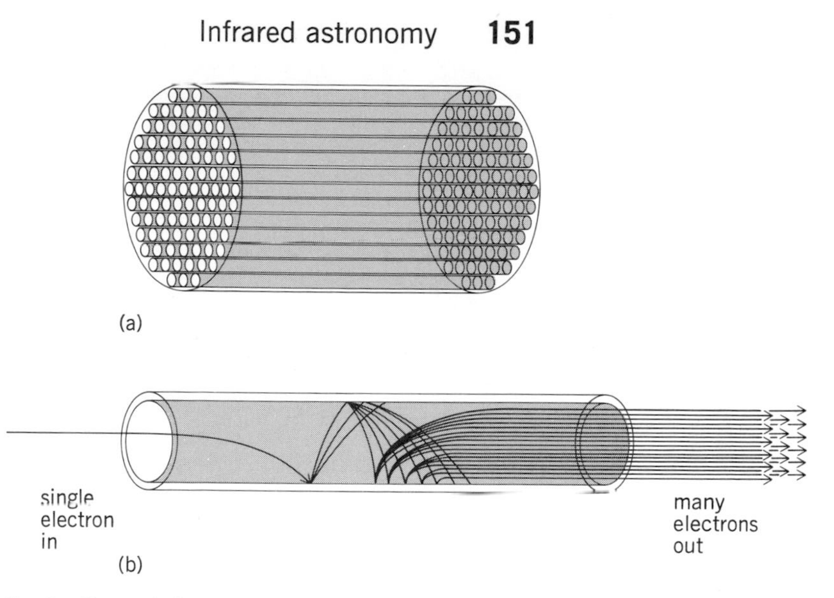

Fig. 3. Channel plate multiplier. (*a*) Array of tubes. (*b*) Details of a single channel

detector to make a permanent record of the image. The photographic emulsion is perhaps the most straightforward and convenient detector to use for this purpose. When a photographic emulsion is optimally combined with a high-quality image tube, the result is a photographic detector that has very nearly the same high quantum efficiency as the image tube photocathode and also has a large number of resolution elements. For detecting the very faintest images, multistage intensifiers can provide sufficient gain so that each photoelectron is recorded as a discrete clump of photographic grains. Then the overall photographic system behaves like a photon counter.

Steps have been made to record image tube outputs with various electronic detectors, thus eliminating the photographic emulsion altogether. Once the output is recorded electronically, the imagery data are much more readily available for quantitative study, including computer analysis. Image tube systems have been developed that use photodiode arrays, image dissector tubes, and various types of television camera tubes, all with promising success.

Several advances have been made in image tube technology since 1960, and the field of astronomy has greatly benefited from these advances. Image intensifiers are widely used at nearly every major observatory for the detection of faint astronomical sources. Knowledge about quasistellar objects, distant galaxies, star clusters, and many other astronomical objects has been considerably enhanced through the use of image tube detectors.

[RICHARD H. CROMWELL]

Bibliography: J. D. McGee et al. (eds.), *Advances in Electronics and Electron Physics*, vols. 16, 22, 28, 33, 40, and 52, 1962–1979.

Infrared astronomy

Astronomy in the 1- to 300-μm range of the electromagnetic spectrum. It has been rendered possible by technological advances, such as solid-state sensors, which have changed it from a difficult, laboratory-type effort to a relatively straightforward technique at conventional and specialized telescopes. This has culminated in the construction of a 120-in. (3.0-m) infrared telescope on the 13,900-ft-high (4,200-m) peak of Mauna Kea, Hawaii, above most of the water vapor. Still another successful technique has been to fly a battery of infrared instruments at the highest possible altitude, well above most of the water vapor; NASA's infrared airplane has provided over a thousand hours of such observation.

Technique. The energy maximum of a blackbody is located at a wavelength of 2900 $\mu m/T$, where T is its thermodynamic temperature in kelvins. The limitations of minimum detectable signal level are set by (1) thermal noise in the telescope and superthermal noise in the detectors, and (2) fluctuations of thermal noise in the Earth's atmosphere (which is near 300 K). The first can be reduced by proper shielding and operation at very low temperature (liquid hydrogen or helium), the second by selection of a high-altitude, dry location. Both noise sources can be most effectively monitored by a chopper, alternating between sky and sky-plus-object, followed by synchronous detection of the difference signal. The wavelength region is marked by absorption bands of H_2O, CO_2, and O_2. Windows at 10 and 20 μm permit some observations; at 1-mm wavelength, work is done with a solid-state detector fed by a waveguide. Above 1 mm, radio-astronomical techniques are used. Two precision telescopes of 33-ft (10-m) diameter, designed to work at and below 1 mm, have been completed (1980); four will be coupled as an interferometer, and still larger telescopes are being planned. Most observations have been in the 1- to 30-μm range; surveys originally confined to shorter wavelengths, have been extended to 20 μm.

The stellar magnitude scales are defined so that an A0 star has the same apparent magnitude at all wavelengths. The conventional photoelectric (UBV) scales and the infrared photometry of H. Johnson's group and G. Neugebauer's group are now calibrated. The table gives the flux for a star of zero apparent magnitude; at other magnitudes (m), multiply by $10^{-0.4\,m}$. PbS cells work to 3.5 μm, and several thousand stars have been observed. Various bolometers have flat response from 1 to 1000 μm, and stars are measured to 20 μm. A special 36-ft (11-m) moderate-resolution telescope has been built at Kitt Peak, AZ; it is used also to study molecular lines, essentially by radio techniques. One fundamental program of infrared astronomy is to establish the temperature scale for cool stars by measurements of these magnitudes over wide spectral regions. *See* STELLAR MAGNITUDE.

Calibration of the magnitude scale*

Region	λ_0, μm	Flux in units of 10^{-23} watt/m²/Hz
U	0.36	1.88
B	0.43	4.44
V	0.54	3.81
R	0.70	2.88
I	0.90	2.24
J	1.25	1.77
H	1.60	1.09
K	2.2	0.63
L	3.4	0.31
M	5.0	0.18
N	10.2	0.043
Q	22.0	(0.0102)

*Flux from a star of apparent magnitude 0.0.

Planetary temperature. Because of their low temperature (700–100 K), planets radiate largely in the infrared; their atmospheres may block this reradiation and, by the greenhouse effect, raise surface temperatures. On Venus, the high clouds (visual region) give 220 K, the surface gives about 700 K (infrared), and along the equator the temperature hardly changes with distance from the subsolar point. On the Moon, without atmosphere, the soil characteristics set the temperature after sunset; highland areas stay warmer, and local hot spots exist, dependent on conductivity. Rings have been detected around Jupiter and Uranus, and hot spots found on the satellite Io. *See* MOON; PLANETARY PHYSICS; VENUS.

Cool stars. A northern sky survey at 2.2 μm, with an automatic scanning 62-in. (157-cm) telescope, revealed very red and, probably, very cool stars. Some are nearly invisible at the V wavelength but extremely bright in the infrared. Colors suggest $T = 700$ K (red giants are seldom cooler than 1500 K). Some infrared objects are also highly reddened by interstellar dust. Those observable spectroscopically in normal wavelengths are very cool red giants. However, still cooler objects exist; for example, a large, cold nebula inside the very bright, hot Orion Nebula has $T = 70$ K, and a stellar object with $T = 700$ K was found near it. The interstellar radio frequency OH line, 1667 MHz, comes from certain small portions of gaseous emission nebulae. Surveys indicate that many OH sources are also infrared stars, like the one in Orion. It seems probable that these sources are groups of stars in the process of condensation. The surrounding dust traps the low-temperature stellar radiation. Protostar formation can thus be monitored, and infrared emission lines permit velocity mapping.

Stars (T Tauri type) in a later stage of formation are observable at normal wavelengths; infrared measures show them to be surprisingly red. Their total light reaches 100 times that observed below 0.7 μm. The increased luminosity and trapped infrared radiation suggest convective energy transport and plasma dissipation of magnetic fields and rotational energy. *See* STAR; STELLAR EVOLUTION.

Other sources. Infrared energy emission mechanisms are common and efficient in a wide variety of objects. Infrared excess energy from dust has been found in hot, gaseous planetary nebulae. Startlingly nonblackbody excesses are also detected in peculiar galaxies of the Seyfert type. These have small nuclei in which violent explosions have occurred. The infrared excess is enormous in the quasars, where it sometimes contains most of the radiated energy. The center of the Galaxy has been mapped as a complex of dense, rapidly moving gas clouds, emitting forbidden lines in the infrared. These clouds are so crowded and move so rapidly that they can only have short lifetimes, and must be formed repeatedly. This active nucleus may be linked to a giant black hole that may exist at the center of the Galaxy. *See* GALAXY; GALAXY, EXTERNAL; NEBULA; QUASARS.

[JESSE L. GREENSTEIN]

Interplanetary matter

Gas and small solid particles in the space between the planets in the solar system. Most of the interplanetary gas originates in the Sun, flowing outward in the solar wind. Prior to its direct detection by spacecraft, the existence of such a wind had been hypothesized as an explanation of why gaseous tails of comets were directed radially outward from the Sun. The widespread presence of microscopic particulate matter has been inferred from observations of the zodiacal light. Somewhat larger meteoroids, in their chance encounters with the Earth, are manifested as meteors and fireballs, and, for the most massive, as meteorites. The solid material originates in the asteroids and comets.

Interplanetary gas. The interplanetary gas is essentially the solar wind, which sweeps out and replenishes interplanetary space at an average rate of $10^{12} g \cdot s^{-1}$. Enhancements in the flow are associated with increased solar activity, particularly solar flares. On average, the contribution by the wasting of comets is much smaller, probably by several orders of magnitude, although locally the cometary gas can dominate for a time. Even in the gaseous tail of the comet, however, the sweeping action of the solar wind can already be seen, and ultimately all the cometary gas is cleared out of interplanetary space.

The solar wind, which originates in the solar corona, consists primarily of protons, reflecting the composition of the Sun. It is a supersonic flow, with a typical speed of 400 $km \cdot s^{-1}$. At 1 AU (1 astronomical unit $= 1.496 \times 10^8$ km, the mean distance from Earth to Sun), the particle density in this spherically symmetric wind has fallen (as r^{-2} where r is the distance from the Sun) to 5 cm^{-3}. Being highly ionized, this plasma carries out the solar magnetic field, which also falls as r^{-2} to about 8×10^{-5} gauss (8 nT) at 1 AU. Beyond about 5 AU (the size of the orbit of Jupiter), both the particle density and field strength have fallen below the values typical of the interstellar medium. The question therefore arises of the radial extent of the solar wind or, alternatively, of the penetration of interstellar gas into the solar system. An estimate of the size of the spherical interstellar bubble carved out by the solar wind can be obtained by balancing the ram pressure of the wind with the thermal pressure of the interstellar gas. The result is 100 AU, which exceeds the size of the orbit of Pluto (40 AU), but is much smaller than the solar system cometary cloud and the distances to the nearest stars (10^5 AU). The transition from wind to interstellar medium takes place through a shock. *See* INTERSTELLAR MATTER; SOLAR WIND.

The solar wind, or interplanetary gas, does not impact directly on the Earth's atmosphere, but is forced to flow around by the magnetic pressure of the terrestrial field. The size of the resulting magnetosphere in the solar direction, obtained by balance with the local ram pressure, is about 10 earth radii. (For comparison, the size of the exosphere is only 500 km or 0.1 earth radii.) The interaction zone just beyond this distance is a collisionless shock. In the antisolar direction the wake has a much greater extent. *See* MAGNETOSPHERE.

The magnetic field carried in the solar wind is responsible for excluding low-rigidity cosmic rays from the solar system. Significant solar modulation occurs for energies less than 20 GeV $nucleon^{-1}$. *See* COSMIC RAYS.

Interplanetary dust. The largest nonplanetary bodies in the solar system are the asteroids (or minor planets) and the comets. Collisions between asteroids and the breakup of short-period (captured) comets result in the smaller solid particles, or meteroids, to be discussed below. Because the orbits of asteroids and periodic comets lie near the mean plane of the solar system, the meteoric complex is a flattened system, unlike the spherically symmetric solar wind. Its limited radial extent, only a few astronomical units, also reflects the sources of dust. *See* ASTEROID; COMET.

Composition. The largest pieces (masses greater than 10^3 g) of rubble are potential survivors of a fiery passage through the Earth's atmosphere, to become meteorites. However, the observed meteorites are not necessarily representative of the interplanetary population. The most common belt asteroids, those of C type, are thought to create the largest number of interplanetary fragments, a few of which are ultimately seen on Earth as the rare carbonaceous chondrites. On the other hand, the most numerous meteorites, ordinary chondrites, seem to be fragments from a rarer type of asteroid, among which are some with favorable Earth-approaching orbits. Cometary debris is apparently so friable that even massive bodies break up in the atmosphere. *See* METEORITE.

There are much larger numbers of smaller meteoroids. Particles of mass on the order of 1 g produce visual meteors in the Earth's atmosphere, whereas meteoroids of about 10^{-5} g produce ionization trails which are detectable by radar. Micrometeorites (10^{-9} g) are detected by impact experiments on spacecraft and have been collected in the upper atmosphere. Scattering of solar radiation by particles of this size (10–30 μm) is responsible for the zodiacal light. Modeling of observations of the brightness and polarization distributions of this diffuse light as a function of elongation angle and heliocentric distance indicates an oblate zodiacal cloud, with an axial ratio of about 7, lying in the plane of the solar system. The particle density falls off as $r^{-1.5}$ out to a cutoff near 3.3 AU. *See* METEOR; MICROMETEORITE; ZODIACAL LIGHT.

Evolution of meteoric complex. There are a number of direct clues as to the sources of the meteoroids. For example, meteorites appear to originate in asteroids, whereas shower meteors are clearly associated with the dissipation of short-period comets. Comets have dust tails. The relative contributions of primary particles, which will ultimately produce the smaller-sized population through mutual collisions of fragments, are difficult to assess, however. The extent of the zodiacal cloud, which lies interior to the asteroid belts, is not necessarily conclusive support of an asteroidal origin, just as the lack of an enhancement in the density of micrometeoroids within the asteroid belts is not a firm rebuttal. Micrometeoroids are subject to nongravitational forces which soon smooth out all traces of their origins.

What does appear reasonable is that the steep power-law mass distribution observed down to 10^{-5} g is a steady state resulting from successive mutual collisions of primary particles. The flattening of the mass distribution function below 10^{-5} g suggests two conclusions. First, particles with masses around 10^{-5} g dominate the meteoric complex, whose integrated mass is $10^{19}-10^{20}$ g. Second, additional destructive processes appear to be depleting the distribution at smaller masses.

The indicated mechanism, which first becomes important in just this mass range, involves the Poynting-Robertson effect. These meteoroids lose a substantial amount of their orbital momentum to the solar photons which they intercept, and thus continuously spiral in toward the Sun. So dramatic is this effect that the zodiacal cloud would become devoid of small particles (masses less than 10^{-6} g) on a time scale of only 10^5 years, were it not for continuous replenishment by collisions between more massive meteoroids. To maintain the observed steady state, injection of primary particles at a rate near $10^7 g \cdot s^{-1}$ is required. This is close to the contribution estimated to be available from the destruction of comets. Over the age of the solar system, the accumulated mass requirement would be 10^{24} g, the mass of a large asteroid, so that an asteroidal origin is not ruled out either.

The evolution of the particles as they spiral into the inner solar system can be modeled. In addition to the destructive effects of mutual collisions, which become increasingly important, further degradation by sputtering and by sublimation must be considered, especially at heliocentric distances less than 1 AU. As the characteristic particle size decreases, the relative importance of radiation pressure force to gravity (the ratio β) grows, possibly leading to a timely increase in orbit size. No particle would survive an orbital decay to less than 0.1 AU. It is encouraging that steady-state size distributions and radial distributions predicted by models including these processes are at least qualitatively like those observed for micrometeoroids in the zodiacal cloud.

A most dramatic piece of evidence which ties in with this scheme is the discovery of submicrometer-sized meteoroids streaming out of the inner solar system into the interstellar medium at speeds of 100 $km \cdot s^{-1}$. It is hypothesized that these are fragments which have become so small that β exceeds unity and they experience a net outward force. They have been named the β-meteoroids.

[P. G. MARTIN]

Bibliography: J. A. M. McDonnell (ed.), *Cosmic Dust*, 1978; J. A. Wood, *The Solar System*, 1979.

Interstellar matter

The material between the stars, constituting several percent of the total mass of the Galaxy. Being the reservoir from which new stars are born in the Galaxy, interstellar matter is of fundamental importance in understanding both the processes leading to the formation of stars, including the solar system, and ultimately the origin of life in the universe. Among the many ways in which interstellar matter is detected, perhaps the most familiar are attractive photographs of bright patches of emission-line or reflection nebulosity. However, these nebulae furnish an incomplete view of the large-scale distribution of material, because they depend on the proximity of one or more bright stars for their illumination. Radio observations of hydrogen, the dominant form of interstellar matter, reveal a widespread distribution throughout the thin disk of the Galaxy, concentrating in the spiral arms. Mixed in with the gas are small solid particles, called dust grains, of characteristic radius 0.1 μm. Although by mass the grains constitute less than 1% of the material, they have a pronounced effect through the extinction of starlight. Striking examples of this obscuration are the dark rifts seen in the Milky Way. On average, the density of matter is only one hydrogen atom per cubic centimeter (equivalently 10^{-24} gm cm^{-3}), but because of the long path lengths over which the material is sampled, this tenuous medium is detectable. Radio and optical observations of other spiral galaxies show a similar distribution of interstellar matter in spiral arms in the galactic plane.

A hierarchy of interstellar clouds, concentrations of gas and dust, exists within the spiral arms. Many such clouds or cloud complexes are recorded photographically. However, the most dense, which contain interstellar molecules, are often totally obscured by the dust grains and so are detectable only through their infrared and radio emission. These molecular clouds contain the birthplaces of stars. *See* GALAXY; GALAXY, EXTERNAL.

Gas. Except in the vicinity of hot stars, the interstellar gas is cold, neutral, and virtually invisible. However, collisions between atoms lead to the production of the 21-cm radio emission line of atomic hydrogen. Because the Galaxy is quite transparent at 21 cm, surveys with large radio telescopes have produced a hydrogen map of the entire Galaxy. Different emission regions in the Galaxy along the same line of sight are moving with different velocities relative to Earth and so are distinguishable by their different Doppler shifts. Supplemental information is obtained from 21-cm absorption-line measurements when hydrogen is located in front of a strong source of radio emission. These radio studies show that the gas is concentrated in clouds within the spiral arms, with densities typically 30 times the average of one atom per cubic centimeter. The cold clouds (70 K) appear to be in near pressure equilibrium with a more tenuous, hotter phase (7000 K) of neutral hydrogen in which they are embedded. *See* RADIO ASTRONOMY.

Other species in the gas are detected by the absorption lines they produce in the spectra of stars, and so are observable only in more local regions of the Galaxy. Interstellar lines are distinguished from stellar atmospheric lines by their extreme narrowness and different Doppler shifts. High dispersion spectra show the lines are composed of several components possessing unique velocities which correspond to the individual clouds detected with the 21-cm line. Elements such as Ca, Na, and Fe are detected in optical spectra, but the more abundant species in the cold gas, such as H, C, N, and O, which produce lines only in the ultraviolet require observations from satellites outside the Earth's atmosphere.

A broad ultraviolet line of O VI (oxygen atoms from which five electrons have been removed) has also been discovered, quite unexpectedly because collisional ionization to such an extent requires a very hot gas (10^5-10^6 K). However, such a hot gas emits soft x-rays and has been independently confirmed as a contributor to the diffuse x-ray background radiation. This hot low-density component (0.01 cm^{-3}) is thought to permeate the galactic disk, perhaps as a network of tunnels created and maintained in the cooler H I gas by supernova remnants.

In the earliest stages of the Galaxy only H and He were present, but by nuclear burning processes in stellar interiors and supernova explosions the abundances of heavier elements have been built up through many cycles of star formation and mass loss from dying stars. The present relative abundances in interstellar matter are expected to be the same as seen in the atmospheres of the later-generation stars like the Sun. However, examination of the abundances obtained from interstellar lines reveals a considerable depletion of heavy atoms relative to hydrogen. The atoms missing in the interstellar gas can be accounted for by the interstellar matter seen in solid particle form. *See* ELEMENTS AND NUCLIDES, ORIGIN OF.

Molecules. Inside interstellar clouds, molecules form either on the surfaces of dust particles or by gas-phase reactions. In dense (greater than 10^3 cm^{-3}) clouds where molecules are effectively shielded from the ultraviolet radiation that would dissociate them, the abundances become appreciable. Hydrogen is converted almost completely to its molecular form, H_2, which has been detected in ultraviolet spectra along with the next most abundant molecule, CO. The high abundance of CO may be attributed to the high cosmic abundances of C and O and the great stability of CO.

Most molecules, with the notable exception of H_2, are discovered by their millimeter-line absorption or emission. Dark nebulae are found to contain H_2, CO, OH, and H_2CO. These clouds are very cold (20 K). The Orion molecular cloud, of which the Orion Nebula, an H II region, is a fragment, has a mass of at least 200 solar masses and is one of the chief regions for detection of a large number of more complex polyatomic organic molecules. The other region, the Sagittarius B2 cloud near the galactic center, has a mass exceeding 200,000 solar masses. Altogether, over three dozen species have been detected, many of which could be classed as prebiotic. It is interesting that the first stages of organic evolution can occur in such low-density environments. *See* ORION NEBULA.

The existence of molecular clouds throughout the galactic disk has been traced by using the 2.6-mm emission line of CO, in a manner analogous to mapping with the 21-cm line. It appears that more than half of interstellar hydrogen may be in molecular rather than atomic form, especially within the inner 5000 parsecs (1 parsec = 3.26 light-years = 3.09×10^{13} km) of the Galaxy.

Particles. The light of stars near the galactic disk is dimmed by dust grains which both absorb the radiation and scatter it away from the line of sight. The amount of extinction at optical wavelengths varies approximately as the reciprocal of the wavelength, resulting in a reddening of the color of a star, much as molecular scattering in the Earth's atmosphere reddens the Sun, especially near sunrise and sunset. The dependence of extinction on wavelength is much less steep than it is for molecular scattering, indicating the solid particles have radii about 0.1 μm. Satellite observations show a continued rise in the extinction at ultraviolet wavelengths which seems to require a component of smaller interstellar grains.

By comparison of the observed color of a star with that predicted from its spectral features, the degree of reddening or selective extinction can be fairly accurately determined, but the total extinction at any given wavelength is more difficult to measure. The best estimates suggest that extinction by dust over a pathlength of 1000 parsecs in the galactic plane will, on average, reduce a star's brightness by 80%. This requires the mass density of grains to be about 1% of the gas density. Since pure H or He grains cannot exist in the interstellar environment, a major fraction of the heavier elements must be in the solid particles. The number density of 0.1-μm grains would be about 2000 km^{-3}. Studies of reddening in conjunction with measurements of the 21-cm and ultraviolet Lyman-α lines of hydrogen show that dust and gas concentrations are well correlated.

The light of reddened stars is partially linearly polarized, typically by 1% but reaching 10% for the most obscured stars. The broad peak in polarization at yellow light, together with the correlation of the degree of polarization with reddening, suggests that the polarization and extinction are caused by the same dust grains. The grains must be both nonspherical and spinning about a preferred direction in space to produce polarization. The agent for the large-scale ordering required to explain the strong tendency of the planes of polarization of stars in some directions in the Milky Way to lie parallel to the galactic plane is believed to be the galactic magnetic field. Minute amounts (0.01%) of circular polarization have also been used to study the topology of the magnetic field. *See* COSMIC RAYS.

The possible types of grain material can be restricted through considerations of relative cosmic abundances, but detailed identification is difficult because of the paucity of spectral features in the extinction curve. Although ice has been detected by its 3.1-μm absorption band, it is not nearly as abundant as expected. Silicates are suggested by 10-μm absorption in front of strong infrared sources, and in the ultraviolet an absorption peak at 2200 A (220 nm) could be explained by a component of small graphite particles. A popular theory of grain formation begins with the production of small silicate particles in the extended atmospheres of red supergiant stars. While in the circumstellar region, these grains are warmed by the starlight, so that they are detectable by their thermal emission in the near infrared (10 μm). Radiation pressure ejects these particles into the interstellar gas, where they become much colder (15 K). A dielectric mantle is then built up by accretion of the most abundant elements, H, C, N, and O.

Because of the high concentration of interstellar material toward the galactic plane, it is extremely difficult to detect radiation with a wavelength less than 1 μm coming a large distance through the plane. Conversely, a line of sight to a distant object viewed out of the plane is much less obscured because the disk is so thin. The zone of avoidance, corresponding roughly to the area occupied by the Milky Way, is that region of the sky in which essentially no extragalactic object can be seen because of intervening dust. The dark rifts in the Milky Way result from the same obscuration. The component of starlight scattered rather than absorbed by the grains can be detected as a diffuse glow in the night sky near the Milky Way. However, it must be carefully separated from other contributions to the night sky brightness: the integrated effect of faint stars, zodiacal light from dust scattering within the solar system, and airglow (permanent aurora). *See* INTERPLANETARY MATTER.

Dark nebulae. A cloud of interstellar gas and dust can be photographed in silhouette if it appears against a rich star field. The largest and most dense clouds are most easily detected because of the large contrast produced in the apparent star density. A distant cloud is difficult to find because of many foreground stars. Many large dark nebulae or groups of nebulae can be seen in the Milky Way where the material is concentrated. The distance to a dark nebula can be estimated using the assumption that statistically all stars are of the same intrinsic brightness. When counts of stars within a small brightness range are made in the nebula and an adjacent clear region, the dimming effect of the cloud will appear as a sudden relative decrease in the density of stars fainter than a certain apparent brightness, which corresponds statistically to a certain distance. Alternatively, a lower limit to the distance is provided by the distance to the most distant unreddened stars in the same direction. One of the best-known and nearest dark nebulae is the Coal Sack, situated at a distance of 175 parsecs. Another example is the "Gulf of Mexico" area in the North America Nebula. *See* COAL SACK.

Fig. 1. Quadrant of the shell-shaped Rosette Nebula, showing dense globules of obscuring dust and gas silhouetted on the bright emission-line background of an H II region. Central hole may have been swept clear of gas by radiation pressure from central star (lower left) acting on the dust grains. Photographed in red light with the 48-in. (122 cm) Schmidt telescope of the Hale Observatories.

Fig. 2. NGC 6611 (M16), a complex H II region in which the exciting stars are members of a cluster. Note dark globules and elephant-trunk structures, and the bright rims where ionizing radiation is advancing into more dense neutral gas. Photographed in Hα+[N II] with the 200-in. (508 cm) telescope of the Hale Observatories. (*Hale Observatories*)

Obscuring clouds of all sizes can be seen against the bright H II regions described below. In many cases the H II regions and dark nebulae are part of the same cloud. The bay in the Orion Nebula is one such region, but perhaps even more familiar is the spectacular Horsehead Nebula. Even smaller condensations, called globules, are seen in the Rosette Nebula (Fig. 1) and NGC 6611 (M16; Fig. 2). The globules, which are almost completely opaque, have masses and sizes which suggest they might be the last fragments preceding the birth of stars. *See* NEBULA.

Bright nebulae. An interstellar cloud can also become visible as a bright nebula if illuminated by a nearby bright star. Whether or not an H II region or a reflection nebula results depends on the quantity of ionizing radiation available from the star. To be distinguished from H II regions, but often also called bright gaseous nebulae, are shells of gas that have been ejected from stars. Included in this latter category are planetary nebulae, nova shells and supernova remnants which have a bright emission-line spectrum similar to that of an H II region. *See* CRAB NEBULA; NEBULA.

H II regions. A star whose temperature exceeds

about 25,000 K emits sufficient ultraviolet radiation to completely ionize a large volume of the surrounding hydrogen. The ionized regions (Figs. 1 and 2), called H II regions, have a characteristic red hue resulting from fluorescence in which hydrogen, ionized by the ultraviolet radiation, recombines and emits the Hα line at 6563 A (656.3 nm). Optical emission lines from many other elements have been detected, including the well-known "nebulium" line of oxygen at 5007 A (500.7 nm).

An H II region can be extended with a relatively low surface brightness if the local density is low, as in the North America Nebula. However the best-known regions, such as the Orion Nebula, are in clouds that are quite dense (10^3 to 10^4 cm^{-3}) compared to the average; dense clouds use up the ionizing radiation in a region closer to the star and consequently are smaller with a higher surface brightness. Since the brightest stars are also the youngest, it is not surprising to find them still embedded in the dense regions from which they formed. Later in its evolution an H II region can develop a central hole if radiation pressure on the dust grains is sufficient to blow the dust and gas away from the star, as in the Rosette Nebula (Fig. 1). H II regions are also conspicuous sources of free–free radio emission characteristic of close electro proton encounters in the 10,000 K gas. Some H II regions are seen only as radio sources because their optical emission is obscured by dust grains. Radio recombination lines of H, He, and C, which result when the respective ions recombine to highly excited atoms, are also important.

Reflection nebulae. In the absence of sufficient ionizing flux, the cloud may still be seen by the light reflected from the dust particles in the cloud. The scattering is more efficient at short wavelengths, so that if the illuminating star is white the nebula appears blue. The absorption or emission lines of the illuminating star appear in the nebular spectrum as well. The Orion Nebula has an underlying reflection nebula arising from the dust in the gas cloud which produced the H II region. However, in this and other H II regions the emission-line radiation rather than the reflected light dominates the nebulosity.

Reflection nebulae are strongly polarized, by as much as 40%. Both the color and the polarization can be explained by dust grains similar to those which cause interstellar reddening. Some reflection nebulae, such as those in the Pleiades, result from a chance close passage of a cloud and an unrelated bright star, providing a unique look at interstellar cloud structure. Other reflection nebulae appear to be intimately related to stars in early or late stages of stellar evolution.

Star formation. Superluminous stars such as those exciting H II regions cannot be very old because of the tremendous rate at which they are exhausting their supply of hydrogen for nuclear burning; the most luminous are under 100,000 years old. With this clear evidence for recent star formation, observations have been directed toward discovering stars even closer to their time of formation. Compact H II regions, such as the Orion Nebula, appear to be the first fragments of much larger molecular clouds to have formed stars. Ultracompact H II regions, seen at radio wavelengths in molecular clouds but totally obscured optically, appear to be an earlier stage in which the protostellar core has just become highly luminous. These are often called cocoon stars. Even earlier still are the compact infrared sources in which hot dust grains in a protostellar cloud are being detected at wavelengths of 5–100 μm. These earliest phases are often associated with intense H_2O and OH molecular maser emission. Examples of all of these stages are often found in the same region of space. In addition to the Orion molecular cloud, many regions such as the W3 radio and infrared sources associated with the visible H II region IC 1795 have been studied extensively. *See* INFRARED ASTRONOMY; STELLAR EVOLUTION.

[P. G. MARTIN]

Bibliography: W. B. Burton, The morphology of hydrogen and of other tracers in the Galaxy, *Annu. Rev. Astron. Astrophys.*, 14:275–306, 1976; T. Gehrels (ed.), *Protostars and Planets*, 1978; P. G. Martin, *Cosmic Dust*, 1978; D. E. Osterbrock, *Astrophysics of Gaseous Nebulae*, 1975; L. Spitzer, Jr., *Physical Processes in the Interstellar Medium*, 1978.

Jupiter

The largest planet in the solar system, and the fifth in the order of distance from the Sun. It is visible to the naked eye, except for short periods when in near conjunction with the Sun. Usually it is the second brightest planet in the sky; only Mars at its maximum luminosity and Venus appear brighter. Jupiter is brighter than Sirius, the brightest star.

Planet and its orbit. The main orbital elements are a semimajor axis, or mean distance to the Sun, of 484 × 10^6 mi (778 × 10^6 km); an eccentricity of 0.048, causing the distance to the Sun to vary about 47 × 10^6 mi (75 × 10^6 km) between perihelion and aphelion; sidereal revolution period of 11.86 years; mean orbital velocity of 8.1 mi/s (13.1 km/s); and inclination of orbital plane to ecliptic of 1°.3. *See* PLANET.

The apparent equatorial diameter of its disk varies from about 47″ at mean opposition (50″ at perihelic opposition, 44″ at aphelic opposition) to 32″ at conjunction. The polar flattening due to its rapid rotation is considerable and is easily detected by visual inspection; the ellipticity is $(r_e - r_p)/r_e = 0.065$, where r_e is the equatorial radius and r_p is the polar radius. The equatorial diameter is about 88,700 mi (142,700 km), and the polar diameter is 82,800 mi (133,300 km). The volume is about 1317 (Earth = 1) with an uncertainty of several percent. The mass is about 318.4 (Earth = 1), and is accurately determined from the motion of the four major satellites. The mean density is 1.34 g/cm^3, a low value characteristic of the four giant planets; the corresponding value of the mean acceleration of gravity at the visible surface is about 26 m/s^2; however, because of the large radius and rapid rotation, the centrifugal force at the equator amounts to 2.25 m/s^2, reducing the effective acceleration of gravity to about 24 m/s^2.

Phases. As an exterior planet, Jupiter shows only gibbous phases. Because of the large size of Jupiter's orbit compared with that of the Earth, the

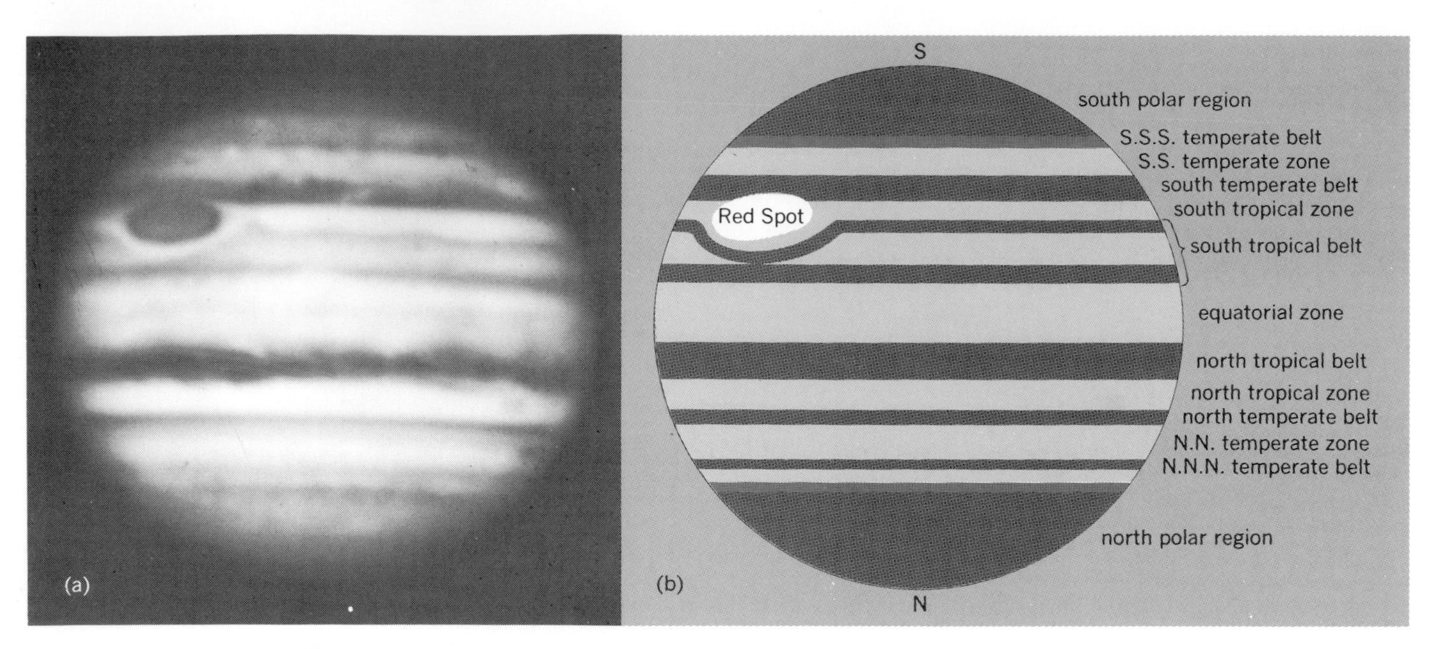

Fig. 1. Jupiter. (*a*) Telescopic appearance (*Hale Observatories*). (*b*) Principal bands.

maximum phase angle is only 12° at quadratures and the phase effect shows up only as a slightly increased darkening of the edge at the terminator. The apparent visual magnitude at mean opposition is −2.4, and the corresponding value of the reflectivity (geometrical albedo) is about 0.4; the physical albedo is 0.45, with some uncertainty due to the small range of phase angle observable. The high value of the albedo, characteristic of the four giant planets, indicates the presence of a dense, cloud-laden atmosphere. *See* ALBEDO.

Telescopic appearance. Through an optical telescope Jupiter appears as an elliptical disk, strongly darkened near the limb, and crossed by a series of bands parallel to the equator (Fig. 1). Even fairly small telescopes show a great deal of complex structure in the bands and disclose the rapid rotation of the planet. The period of rotation, determined from long series of observations of the transits of spots at the central meridian, is very short, about 9h55m, the shortest of the giant planets. The details observed, however, do not correspond to the solid body of a planet but to clouds in its atmosphere, and the rotation period varies markedly with latitude. The nomenclature and mean rotation periods of the main belts of the clouds are given in Table 1. The rotation period of any given zone is not exactly constant but suffers

Table 1. Mean latitudes and rotation periods of Jupiter's bands (after B. M. Peek)

Band	Region	Latitude	Period, 9h+	Remarks
North polar region		>+48°	55m42s	
N.N.N. temperate belt	Center	+43°	55m20s	Temporary
N.N. temperate belt	Center	+38°	55m42s	Temporary
	South edge	+35°	53m55s	
North temperate belt	North edge	+31°	56m05s	
	Center	+27°	53m17s	
	South edge	+23°	49m07s	
North tropical zone	Center	+18°	55m29s	
North tropical belt	Center	+13°	54m09s	Temporary
	South edge	+6°	50m24s	
Equatorial zone	Middle	0°	50m24s	
South tropical belt	North edge	−6°	50m26s	
	Center	−10°	51m21s	
	South edge	−19°	55m39s	
South tropical zone	Center	−23°	55m36s	
Great Red Spot	Center	−22°	55m38s	
South temperate belt	North edge	−27°	55m02s	Temporary
	Center	−29°	55m20s	
	South edge	−31°	55m07s	
S.S. temperate zone	Center	−38°	55m07s	
S.S.S. temperate zone	Center	−45°	55m30s	Temporary
South polar region		<−45°	55m30s	

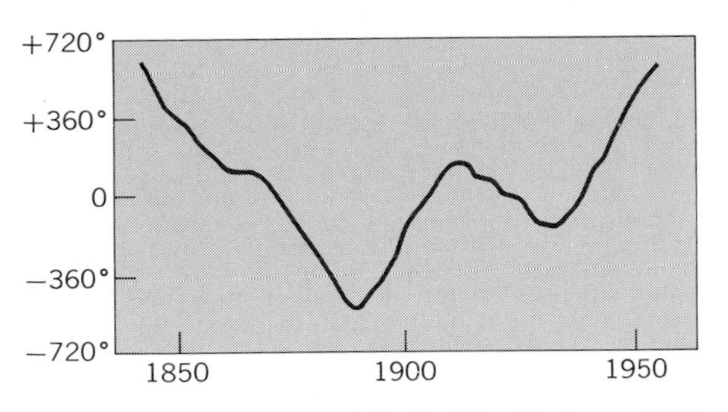

Fig. 2. Drift in longitude of the Red Spot between 1840 and 1955. (*After B. M. Peek*)

continual fluctuations about a mean value. Occasionally, short-lived atmospheric phenomena may depart more strongly from the mean rotation period of the zone in which they appear and thus drift rapidly with respect to other details in the zone. The rotation axis is inclined only 3° to the perpendicular to the orbital plane, so that seasonal effects are practically negligible.

Red Spot. Apart from the constantly changing details of the belts, some permanent or semipermanent markings have been observed to last for decades or even centuries, with some fluctuations in visibility. The most conspicuous and permanent marking is the great Red Spot, intermittently recorded since the middle of the 17th century and observed continually since 1878, when its striking reddish color attracted general attention. It was conspicuous and strongly colored again in 1879 1882, 1893–1894, 1903–1907, 1911, 1914, 1919–1920, 1926–1927, and especially in 1936–1937, 1961–1968, and 1973–1974; at other times it has been faint and only slightly colored, and occasionally only its outline or that of the bright "hollow" of the south temperate zone which surrounds it has remained visible.

The mean rotation period of the Red Spot between 1831 and 1955 was 9h55m37.5s, with a range of variation of about ±6 s (Fig. 2). The mean dimensions of the Red Spot are about 30,000 mi (48,000 km) in longitude and 10,000 mi (16,000 km) in latitude.

The *Pioneer 10* and *11* and *Voyager 1* and *2* spacecraft flybys have shown that the clouds making up the Red Spot rotate approximately every 6 days and that the spot itself consists of an area of uplifted clouds which are about 10 mi (16 km) higher than the surrounding atmospheric levels and about 10°C colder (Fig. 3). The vivid red coloration remains unexplained; however, red phosphorus (P_4) has been suggested as a possibility. Perhaps this coloring agent is plentiful in the interior and is brought to the surface by convection, or is a transformation product of phosphine gas (PH_3) into red phosphorus by the action of sunlight.

Other markings. Another remarkable, semipermanent marking, the South Tropical Disturbance, was intermittently observed on Jupiter between 1901 and 1935 and possibly also in 1940. This marking circulated in the same zone as the Red Spot, but with a shorter mean rotation period, about 9h55m16s, and periodically came into conjunction with it. A large number of temporary features of shorter duration have been observed, but their mechanisms of formation and laws of motions are not understood.

Atmosphere. Limits on atmospheric pressure at the top of the clouds can be set by observing pressure broadening of spectral lines; the pressure is probably a bit less than 3 atm (300 kPa). The atmosphere must be largely composed of hydrogen and helium, as is shown from the detection of weak spectral lines arising from molecular hydrogen.

The radiometrically determined temperature of the visible disk of Jupiter, about 130 ± 30 K, is in fairly good agreement with the value theoretically estimated from the assumption that the visible cloud layer is mainly composed of ice crystals of solidified ammonia (about 160 K). *Pioneer 10* and *11* flybys showed that Jupiter's zones and belts consist of gases at different altitudes and temperatures. The coloring agents of the bands are unknown; the only substances positively identified in Jupiter's atmosphere spectroscopically have been hydrogen, helium, ammonia, methane, and water, while the presence of hydrogen sulfide is inferred. These are all colorless, and other molecules have been proposed as coloring agents; among these are ammonium sulfide, ammonium hydrosulfide, free radicals, various organic compounds, and complex inorganic polymers. However, when *Voyager 2* observed regions deep in the Jovian atmosphere, it was found that transparent hydrogen can be seen due to the fact that solar light undergoes Rayleigh scattering (appearing blue in color) in much the same manner as the terrestrial blue sky.

The cloud tops of the near-white zones have been found to be the highest and coldest features, consisting probably of ammonia crystals. At lower

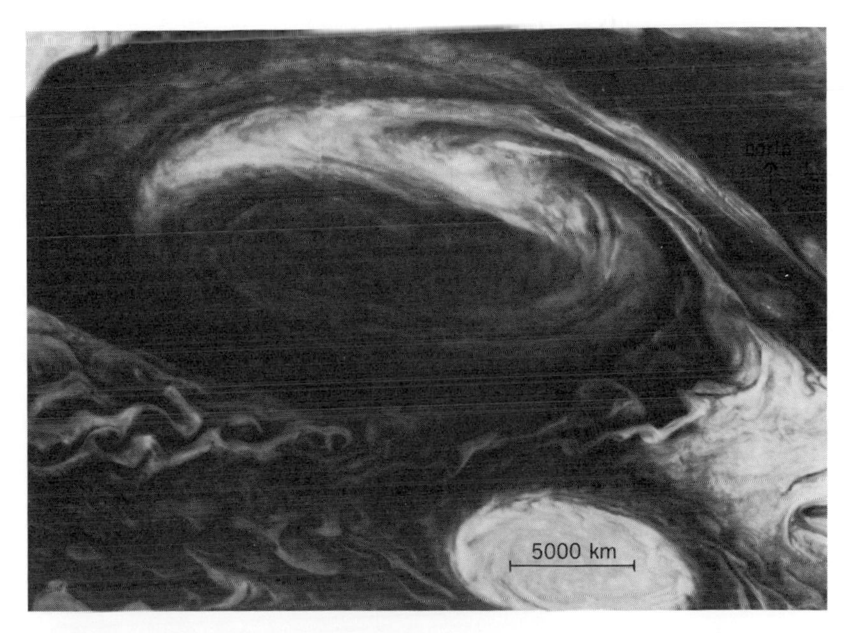

Fig. 3. Details of the Red Spot as seen by the *Voyager 1* flyby. Visible features include a white oval with a wake of counterrotating vortices, puffy features inside the Red Spot, and reverse-S spirals inside both the Red Spot and the oval. The large white feature extending over the northern part of the Red Spot was observed to revolve about the Red Spot center with a period of 6 days. (*NASA*)

altitudes, the melting point of ammonia is reached and the colored compounds are found. At the next lower level, it is likely that water is present, first as ice crystals, and then, at still lower levels, as droplets.

The Voyager probes showed that the Jovian clouds are affected by the same varieties of weather patterns that produce terrestrial weather, though on a much larger scale (Fig. 4). Measurements of the velocities of surface features show that the Jovian atmosphere consists of regular patterns of alternately east and west wind velocities (Fig. 5). While the major atmospheric features no doubt persist due to inertia, it is unknown what agent powers the Jovian weather system, though undoubtedly it comes from within the planet.

Interior composition and structure. Jupiter is primarily made up of liquid and metallic hydrogen. Its composition is similar to that of the Sun, namely 10 times as much hydrogen as helium (by numbers of atoms), and undoubtedly reflects the primordial composition of the solar system. This is not surprising since its strong gravitational forces would have captured and retained most of the primitive material.

Under the extreme pressures of the interior of the planet, the hydrogen would remain in a liquid metallic state, and this material is thought to make up the inner 60% of the radius of the planet. Above this layer it is speculated that there lies liquid hydrogen, with the pressure and temperature lowering sufficiently to become gaseous in nature at approximately 600 mi (1000 km) below the cloud tops. It seems likely that the core of Jupiter is made up of heavy iron-silicate molten material of several times the Earth's total mass and has a probable radius of a few thousand miles.

Jovian magnetosphere. Jupiter possesses the strongest magnetic field intrinsic to a planet in the solar system. This field rotates with the rotational period of the planet and contains an embedded plasma trapped in the field. At the distance of the satellite Io, the field revolves faster than the satellite, and so numerous collisions occur with the atmospheric gas of that body, resulting in the stripping away of $10^{28}-10^{29}$ ions (mainly atomic sulfur and atomic oxygen) per second. The energy involved slows the magnetic field, and so, beyond Io, the magnetic field no longer rotates synchronously with the planet. The ions removed from Io spiral around the magnetic lines of force, oscillating above and below the plane of Io's orbit. This ring of sulfur and oxygen ions is known as the Io plasma torus and emits strongly in the ultraviolet.

Fig. 4. Fine details of Jovian atmospheric features as photographed by *Voyager 1*. The dark halo surrounding the smaller bright spot to the right of large oval is a hot spot, a warmer region of Jupiter's atmosphere. *(NASA)*

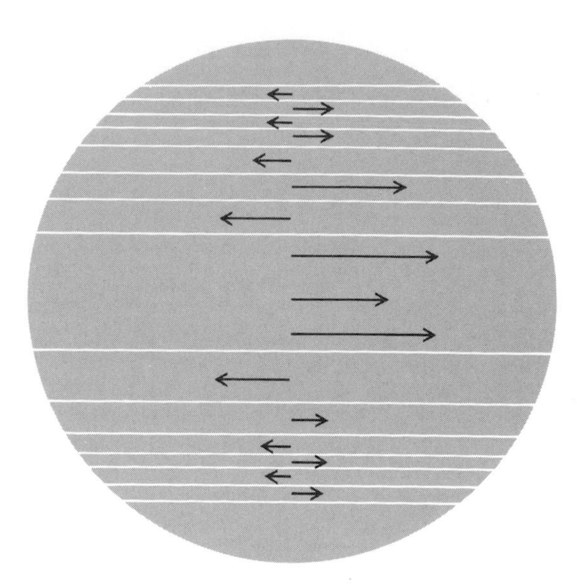

Fig. 5. Jovian circulation patterns as observed from cloud features photographed by the *Voyager 1* spacecraft. *(From Return to Jupiter, Astronomy, 7(9):6–24, 1979)*

The torus powers a series of belts of auroras (discovered by *Voyager 1*) which encircle both the Jovian poles. Eventually some of the particles move out into the extended magnetospheric disk which is approximately 2×10^6 mi (3.2×10^6 km) in radius, and eventually escape into interplanetary space through the magnetotail (Fig. 6). *See* MAGNETOSPHERE.

Under the influence of the solar wind, the Jovian magnetic field is severely disturbed. When solar wind particles encounter the magnetic field, they form a bow shock wave, some 2×10^6 mi (3.2×10^6 km) from the cloud surfaces. The solar wind particles are then deflected around the planet. *See* SOLAR WIND.

Except near the planet, the major component of the Jovian magnetic field is dipolar and opposite in direction from that of the Earth. At a distance of three Jovian radii, the field strength is 0.16 gauss (16 μT). Closer to the planet, the *Pioneer 11* probe has shown the field to be quadrupolar and octopolar. There the field strength varies from 3 to 14 gauss (0.3 to 1.4 mT).

Radio astronomy. Jupiter is known to produce three distinct types of radio emission. Thermal radiation from the high stratosphere is detectable at wavelengths below about 10 cm and indicates temperatures in the upper emitting layers of 100–130 K. Microwave nonthermal emission is observed in the band from about 3 to 70 cm, and is known to arise from synchrotron radiation from relativistic electrons in extended Jovian Van Allen belts (Fig. 7). Sporadic decametric radio noise storms are heard on wavelengths longer than 8 m; these seem consistent with gyrofrequency radiation produced by spiraling electrons, perhaps near the magnetic poles of Jupiter. The existence of both kinds of these nonthermal emissions suggests a Jovian polar surface magnetic field.

Some radio noise is also thought to originate in lightning storms such as those observed on the dark side of the planet by *Voyager 2*.

Periodicities in the radio-noise storms and rocking of the polarization plane of the microwave nonthermal emission led to the well-determined radio rotation period 9h55m29.7s. The difference between the radio and the various other observed Jovian rotation periods suggests that the core of Jupiter is rotating about 13 s faster than the mantle, that the atmosphere has vast wind currents with relative velocities up to 500 km/h, and that angular momentum may be significantly exchanged among these regions over periods of years. *See* RADIO ASTRONOMY.

Jovian ring. *Voyager 2* discovered a faint ring about the planet on its closest passage. The brightest part of this ring is approximately 4000 mi (6000 km) in diameter, but the ring itself may possibly extend all the way to the planetary surface. It is made up of particles of strong forward-scattering material that are thought to be quite small. This poses a problem since such a ring should be reasonably quickly destroyed by collisions and the gravitational attractions of the satellites, and not last longer than a few million years. Therefore, the ring material may perhaps be constantly replenished from some outside source.

Satellites. Jupiter has 15 known satellites of which the four largest, I Io, II Europa, III Ganymede, and IV Callisto, discovered by Galileo in 1610, are by far the most important (Table 2).

The four Galilean satellites are of fifth and sixth stellar magnitudes and would be visible to the naked eye if they were not so close to the much brighter parent planet. All the others are faint telescopic objects.

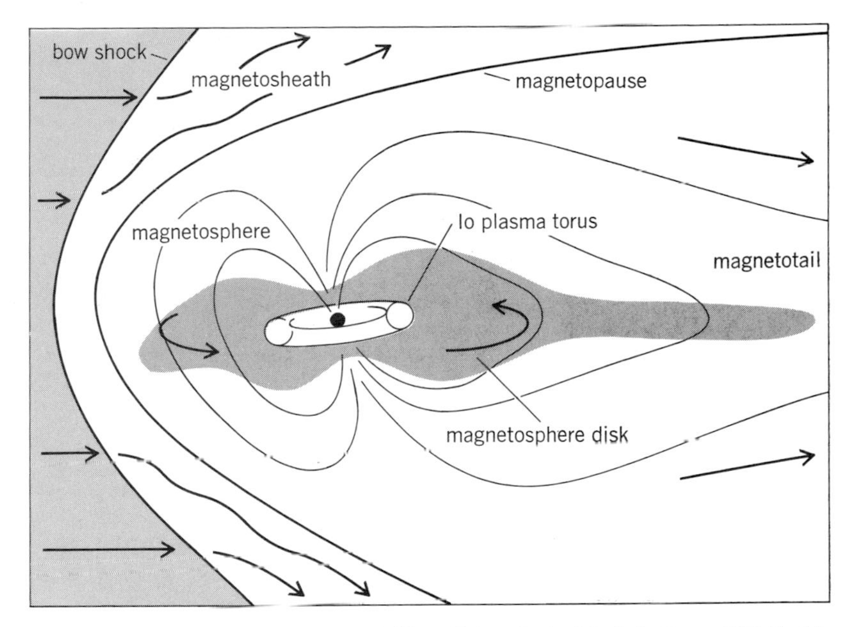

Fig. 6. Jovian magnetic environment. (*From Return to Jupiter, Astronomy, 7(9):6–24, 1979*)

The masses of the major satellites can be roughly estimated from their mutual perturbations; in terms of the mass of the Moon as a unit, the mass of III is about 2, of I about 1, of IV about 4/3, and of II about 2/3. The apparent diameters are on the order of 1–2″, and the corresponding linear diameters are listed in Table 2.

The planes of the orbits of the major satellites are inclined less than 0°.5 to the equatorial plane of Jupiter, so that with the occasional exception of IV, they are eclipsed in Jupiter's shadow at each revolution and also project their shadows on Jupiter and transit in front of its disk near conjunction.

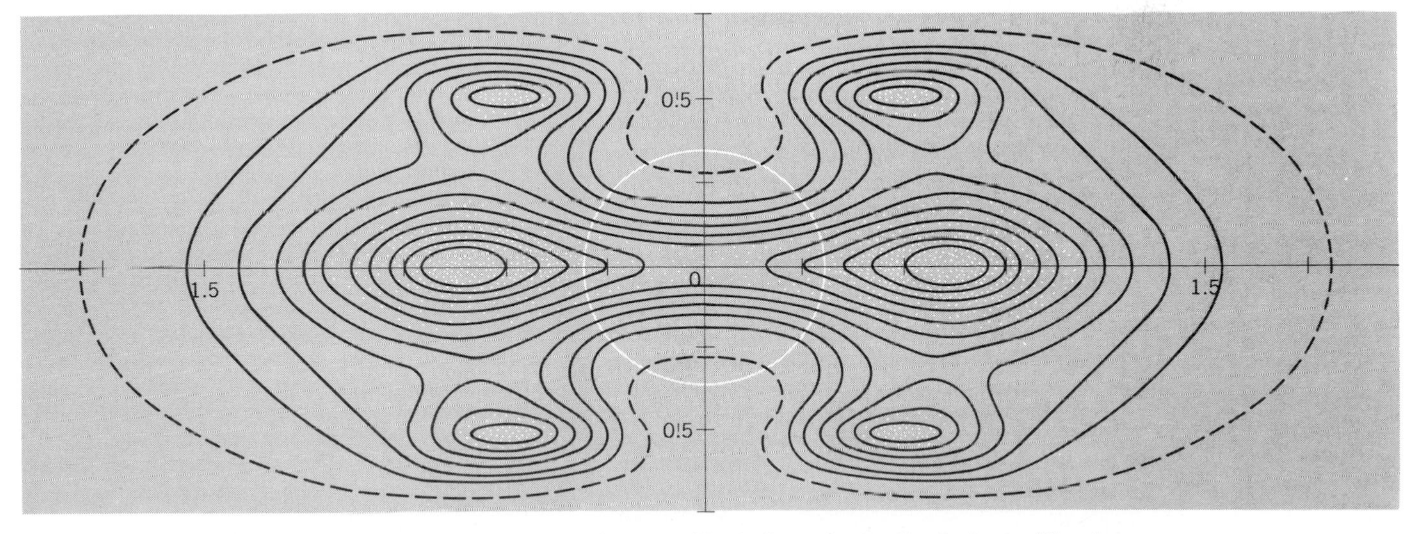

Fig. 7. Contours of radio emission from Jupiter originating from Van Allen belts, and extending far beyond the planet's disk.

Table 2. Satellites of Jupiter

Satellite	Mean distance from Jupiter, 10^3 mi*	Sidereal period, days	Diameter, mi*	Magnitude at mean opposition	Albedo (visual)	Mass (Moon = 1)
I Io	262	1.769	2015	5.5 (visual)	0.37	1.05
II Europa	417	3.551	1941	5.7 (visual)	0.39	0.65
III Ganymede	666	7.155	3276	5.1 (visual)	0.20	2.1
IV Callisto	1,170	16.689	2993	6.3 (visual)	0.03	1.2
V Amalthea	113	0.498	150	13.0 (photographic)		
VI Himalia	7,120	250.6	35	17.0 (photographic)		
VII Elara	7,290	259.6	20	18.0 (photographic)		
VIII Pasiphae	14,600	739.	35	17.0 (photographic)		
IX Sinope	14,700	758.	17	18.6 (photographic)		
X Lysithea	7,300	260.5	15	18.8 (photographic)		
XI Carme	14,000	700.	19	18.4 (photographic)		
XII Ananke	13,000	625.	14	18.9 (photographic)		
XIII Leda	7,000	260.	–	20 (photographic)		
XIV 1979J1	60	0.255	–	–		
XV 1979J2	159	0.676	50	–		

*1 mi = 1.609 km.

The eclipses, transits, and occultations of Jupiter's satellites led to the discovery of the finite velocity of propagation of light by O. Roemer in 1675. Satellites VIII, IX, XI, and XII have a retrograde motion. The very small outer satellites of Jupiter are probably captured asteroids; their orbits are subject to large perturbations by the Sun. Thus at least the more distant satellites probably form part of a fluctuating population gained and lost over very long time spans. *See* ECLIPSE; OCCULTATION; TRANSIT.

The close approaches of the Voyager spacecraft have shown that the four Galilean moons are quite different from each other (Fig. 8). I Io is perhaps the most unusual; its surface has been observed to include at least eight active volcanoes, lava flows, and volcanic fissures. Much of the surface is covered with the ash and dust of igneous material spewed from these volcanoes, averaging approximately 0.1 to 1 mm/year. It is likely that Io has a core of molten rock surrounded by silicate rock of a complexly layered nature. The surface is overlaid with a thick layer of solid sulfur through which the volcanoes periodically erupt. It is likely that Io's proximity to Jupiter has caused the volcanic activity by physically bending the surface over a range of several hundred feet during every revolution (due to Jupiter's tidal forces on Io), which heats the crust, melts the rocks, and powers the volcanoes.

Fig. 8. *Voyager 1* photographs of the Galilean satellites, shown to scale. *(a)* Io. *(b)* Europa. *(c)* Ganymede. *(d)* Callisto. *(NASA)*

II Europa and Io differ from the remaining two Galilean moons by being nearly twice as dense. Europa is primarily a rocky body which may be rich in silicates and lightweight water ices. Voyager showed a satellitewide system of cracks on the surface, running for thousands of miles, which were approximately 10–25 mi (16–40 km) in width. There are few elevations on the satellite, and the surface is remarkably free from craters.

III Ganymede and IV Callisto probably share the same structure: a liquid core and an icy crust. Ganymede is a mixture of heavily cratered dark and old regions and lighter grooved surfaces thought to be younger. The cracks and grooves are thought to be faults and breaks of the icy surface.

IV Callisto is perhaps the most heavily cratered body in the solar system, thereby showing that its surface is very old. Large body impacts are shown by the presence of immense concentric crater basins. Callisto's low density requires a composition of liquid water or slush with a thin, rigid ice crust. There also exists a class of blue-haloed craters which may reveal material from deeper within. *See* PLANETARY PHYSICS. [ELAINE M. HENDRY]

Bibliography: G. P. Kuiper and B. M. Middlehurst (eds.), *Planet and Satellites*, 1961; B. M. Peek, *The Planet Jupiter*, 1958; Return to Jupiter, *Astronomy*, 7(9):6–24, 1979; Reports of *Voyager 1* encounter with Jupiter, *Science*, 204:945–1008, 1979; Reports of *Voyager 2* encounter with Jupiter,

Science, 206:925–996, 1979; J. H. Wolfe, Jupiter, *Sci. Amer.*, 233(3):118–126, 1975.

Kepler's equation

The mathematical relationship between two different systems of angular measurement of the position of a body in an ellipse; specifically the relation between the mean anomaly M and eccentric anomaly θ', $M = \theta' - e \sin \theta'$, where e is the eccentricity of the ellipse. *See* PLANET.

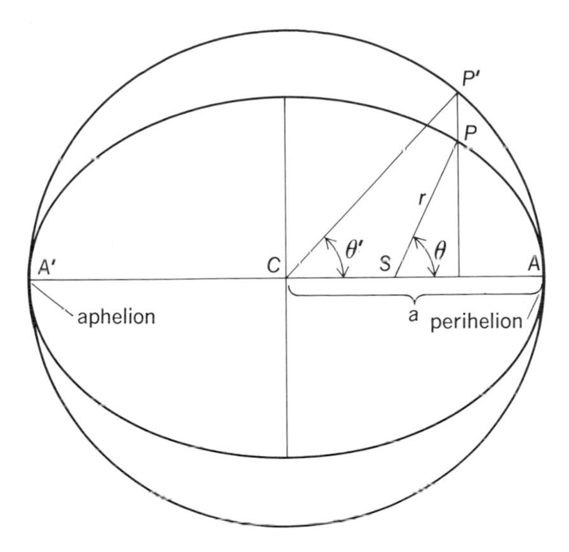

True and eccentric anomaly.

The true position of a planet P in an elliptical orbit can be represented by the angle θ (true anomaly) measured at the focus S between the line directed to the planet and the line directed to the perihelion A (see illustration). The radius vector r from the focus to the planet can be expressed by $r = a(1 - e^2)/(1 + e \cos \theta)$, where a is the semimajor axis. The radius vector may also be expressed in terms of the eccentric anomaly θ' by $r = a(1 - e \cos \theta')$, where θ' is the angle at the center of the ellipse measured from perihelion along the circumscribed circle of radius a to the point whose projection perpendicular to the major axis passes through the planet. The true anomaly θ may be expressed in terms of the eccentric anomaly θ' by $\tan (\theta/2) = [(1+e)/(1-e)]^{1/2} \tan (\theta'/2)$. In actual practice, however, it is more convenient to describe the angular position of a planet in an elliptical orbit at anytime t by means of its average angular velocity n (called mean motion) and the time T of last perihelion passage. This angle M (mean anomaly) is expressed by $M = n(t - T)$. Therefore, given the orbital elements a, e, n, and T, it is possible by means of Kepler's equation and the intermediary angle θ' to evaluate the true anomaly θ and the actual position of the planet in the orbit for any instant t.

Several practical methods for the iterative solution of this transcendental equation exist: (1) Starting with the value of M and denoting approximate values of θ' and $\sin \theta'$ by θ'_0 and $\sin M$, solve $\theta'_0 = M + e \sin M$. Denoting a second approximation to θ' by θ'_1, solve $\theta'_1 = M + e \sin \theta'_0$. Indicating the third approximation to θ' by θ'_2 solve $\theta'_2 = M + e \sin \theta'_1$, continuing the iteration until the required convergence of θ'_i is obtained. (2) Starting with M and an approximate value of θ' solve $M_0 = \theta'_0 - e \sin \theta'_0$, where M_0 corresponds to θ'_0. Set $M - M_0 = \Delta M_0$ and let $\Delta\theta'_0$ be a first correction to θ'_0, then $\Delta\theta'_0 = \Delta M_0/1 - e \cos \theta'_0$. Set $\theta'_1 = \Delta\theta'_0 + \Delta\theta'_0$, evaluate M_1 corresponding to θ'_1, form $M - M_1 = \Delta M_1$, and solve $\Delta\theta'_1 = \Delta M_1/1 - e \cos \theta'_1$. Then $\theta'_2 = \theta'_1 + \Delta\theta'_1$, and the process is repeated until the required convergence of $\Delta\theta'_i$ is reached.

[RAYNOR L. DUNCOMBE]

Bibliography: J. M. A. Danby, *Fundamentals of Celestial Mechanics*, 1962; G. R. Smith, A simple efficient starting value for the iterative solution of Kepler's equation, *Celestial Mech.*, 19(2):163–166, 1979.

Latitude and longitude

The latitude of a location specifies the angle between an imaginary line directed generally toward the center of the Earth and the Equator. The longitude measures the angle between the meridian (the plane defined by the Earth's axis and this local reference direction) and the plane of the Greenwich meridian (illustration *a*).

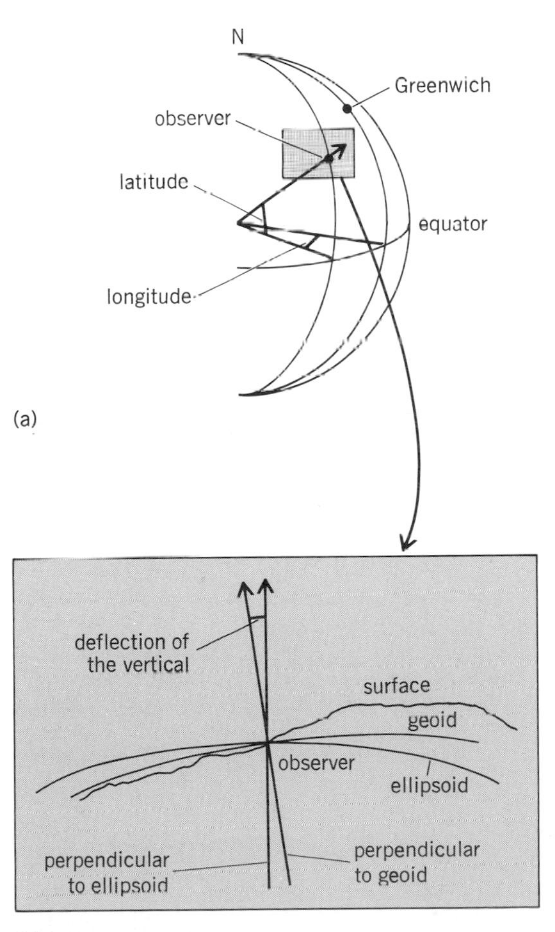

Latitude and longitude. (*a*) Angles formed by reference direction, defining latitude and longitude. (*b*) Detail, showing angle between reference directions in astronomical and geodetic coordinates, the deflection of the vertical.

Astronomical coordinates. Astronomical (or astronomic) latitude and longitude use the direction of gravity for the reference direction. This direction, known as the astronomical vertical, is perpendicular to the equipotential surface of the Earth's gravitational field at the location of the observer.

Geodetic coordinates. A particular geopotential surface approximating mean sea level in the open ocean is called the geoid. A mathematical surface in the form of an oblate ellipsoid may be constructed to approximate the geoid. The direction perpendicular to this reference ellipsoid at the observer's location is used as the reference direction in defining geodetic latitude and longitude. A geodetic datum is defined by its reference ellipsoid, the adopted coordinates of a reference station, and the azimuth of a reference line. Many geodetic datums are in use.

Geocentric coordinates. Geocentric latitude and longitude are defined by a reference direction which passes precisely through the center of mass of the Earth. These coordinates are determined mathematically from the geodetic latitude and longitude, assuming a fixed relationship between the center of the geodetic datum and the center of mass and knowing the mathematical shape of the ellipsoid.

Determination. Astronomical latitude is determined by observing the altitude on the meridian of a celestial object whose declination is known. Astronomical longitude is identical with the difference between Universal Time and local mean time. Local mean time is determined by observing the time of meridian transits of celestial objects with known positions. The direction of gravity is established with the aid of a spirit level, a liquid surface, or observation of the horizon corrected for the elevation of the observer. It varies locally, and is generally not directed toward the center of the Earth. *See* TIME.

Geodetic latitude and longitude are determined by referring precise measurements of distance and direction, made with geodetic instruments on the surface of the Earth, to the datum. Geodetic coordinates are used for location of sites, while astronomical latitude and longitude are used to determine the angular orientation of celestial objects with respect to observers on the Earth. The difference between astronomical and geodetic coordinates is called the deflection of the vertical (illustration *b*), and commonly amounts to some seconds of arc, occasionally reaching a minute of arc. *See* EARTH.

[DENNIS D. MC CARTHY]

Bibliography: I. I. Mueller, *Spherical and Practical Astronomy as Applied to Geodesy*, 1969.

Leo

The Lion, in astronomy, is a magnificent zodiacal constellation appearing during spring and early summer. It is the fifth sign of the zodiac. Leo is well defined and bears a close resemblance to the creature it represents. The head is outlined by a group of six stars, $\alpha,\eta,\gamma,\zeta,\mu$, and ϵ, called the Sickle. The first-magnitude star, Regulus (Little Ruler), forms the handle of the sickle. Three stars to the east, β,δ, and θ, forming a small triangle, constitute the Lion's haunches, with the bright star, Denebola (tail of the lion), a navigational star (see illustration). Associated with this constellation are the famous Leonids shower of meteors, which can be seen radiating from Leo in November of each year and appearing especially brilliant at intervals of about 33 years. *See* CONSTELLATION; METEOR.

[CHING-SUNG YU]

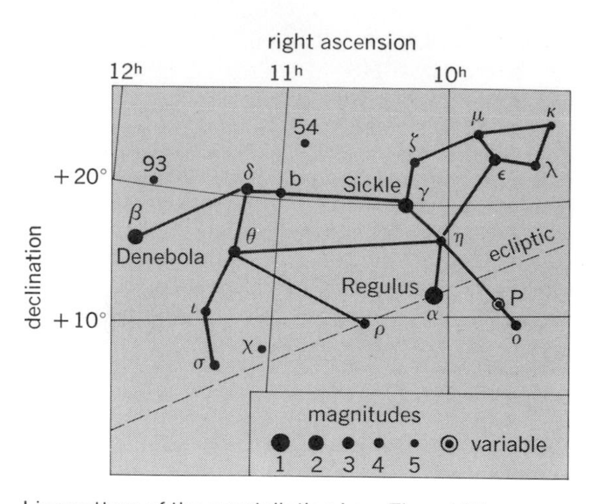

Line pattern of the constellation Leo. The grid lines represent the coordinates of the sky. Right ascension (E-W) in hours, and declination (N-S) in degrees, corresponding to the longitude and latitude of the Earth. The apparent brightness, or magnitudes, of the stars is shown by the sizes of the dots, which are graded by appropriate numbers as indicated.

Libra

The Balance, in astronomy, appearing during the spring. It is the seventh sign of the zodiac. The constellation consists of faint stars and is not conspicuous. It lies just west of the claws of Scorpius. The principal stars, α, β, γ, and σ, outline a four-sided figure resembling a balance with beam and

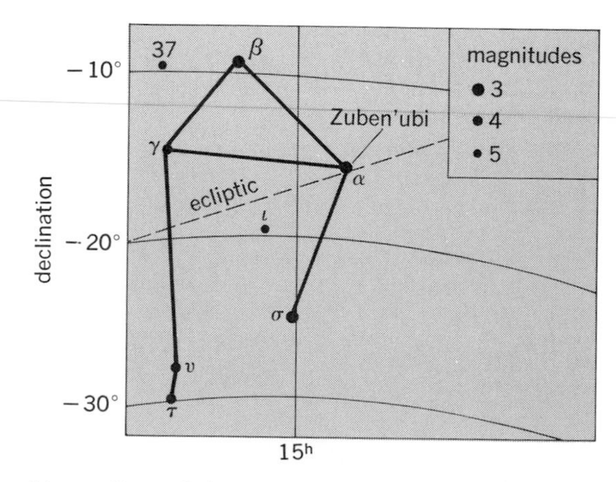

Line pattern of the constellation Libra. The grid lines represent the coordinates of the sky. The apparent brightness, or magnitude, of the stars is shown by the sizes of the dots, graded by appropriate numbers.

pans (see illustration). The balance might have been held originally in the hand of Virgo, a zodiacal constellation nearby, who was identified with the Goddess of Justice. Another possible reason for identifying the constellation with the balance is that 2000 years ago the Sun was in Libra at the then autumnal equinox, at which time days and nights are of equal length. *See* CONSTELLATION.

[CHING-SUNG YU]

Light curves

Graphical descriptions of the light changes of variable stars. If the light changes occur in a regular manner, that is, if they are periodic or nearly so, a plot can be made of the brightness versus the phase. The phase is the time interval between an observation and the instant when the star was at some easily determined portion of the light curve. For eclipsing variables, this "epoch" will be at the time of greatest light loss when the star is faintest; for intrinsic variables the phase is usually reckoned from the time of maximum light when the star is brightest. *See* ECLIPSING VARIABLE STARS.

Interpretation. Figures 1 and 2 are examples of light curves. In these the phases are expressed in fractions of the period, and the brightness changes are in magnitudes—the standard method in astronomy for describing the brightness of a star. As is customary, the brightness is plotted as difference of magnitude, Δm_y or $V - C_y$, between the variable and a nearby comparison star; ΔC is the difference in color. For other examples of light curves *see* VARIABLE STAR.

The magnitude scale is a logarithmic one. This means that large changes in brightness can be expressed with relatively few numbers. A change of brightness of 5 magnitudes is equivalent to a factor of 100 in the total luminosity output. That is to say that if one star is described as being 5 magnitudes brighter than another, it actually emits 100 times as much light. A star 10 magnitudes brighter than another would thus emit 100 × 100 or 10,000 times as much radiation. Such great changes of light occur in certain red variable stars. Even greater changes occur in novae and far greater yet in supernovae. *See* NOVA; SUPERNOVA.

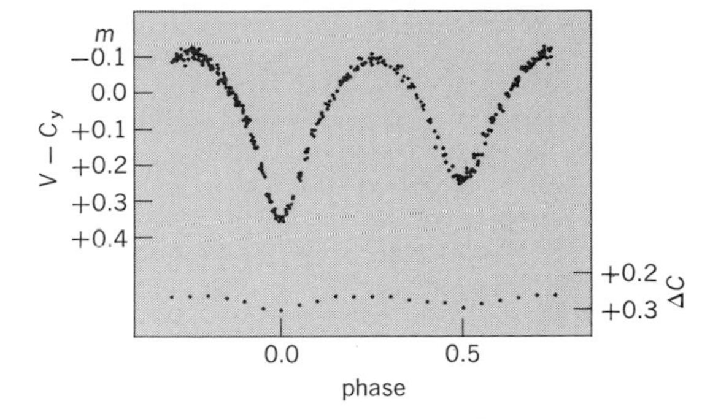

Fig. 2. Variable star XY Leonis. The top observations represent the light curve; the lower ones represent the color curve, showing the changes of color during the light variation. (*Based on photoelectric observations made by R. H. Koch at the Steward Observatory*)

Applications. Studies of light curves are helpful in many ways. Careful analysis of those of eclipsing stars combined with spectrographic data enables one to determine the sizes, masses, and densities of the stars themselves. When a star has been identified as a certain type of pulsating variable (Cepheid or RR Lyrae star) with the help of the shape of the light curve, the period tells the intrinsic brightness of the star, that is, its real output of light. Measures of its apparent brightness combined with this enables one to compute its distance and, hence, the distance to any association within which it lies. The distances to the nearer external galaxies are based on this kind of analysis of the variable stars found in them.

The precise appearance of a light curve will frequently depend on the color in which the observations are made. In eclipsing stars when the hotter star passes behind the cooler, a much greater fraction of the light is lost when observed in the shorter wavelengths (for example, ultraviolet or blue light) than when observed in the longer wavelengths (for example, yellow or red). The reverse is true when the cooler star is eclipsed by the hotter. In certain pulsating stars (Cepheid variables), the exact time when the light is a maximum depends on the color in which the observations are made. *See* CEPHEIDS.

The measures from which light curves are formed are nearly always expressed as the differences of magnitude between the variable star and a nearby star assumed to be of constant light. This is because some light is always lost in passage through the Earth's atmosphere. This extinction varies not only with the angular distance of the star above the horizon but also with atmospheric conditions. By observing the variable and comparison stars in rapid succession—or even simultaneously in some extremely sophisticated light-measuring instruments—the errors of measurement caused by extinction can be made a minimum.

When the light changes do not repeat in a regular manner, no phase can be computed and the brightness must simply be plotted against the date

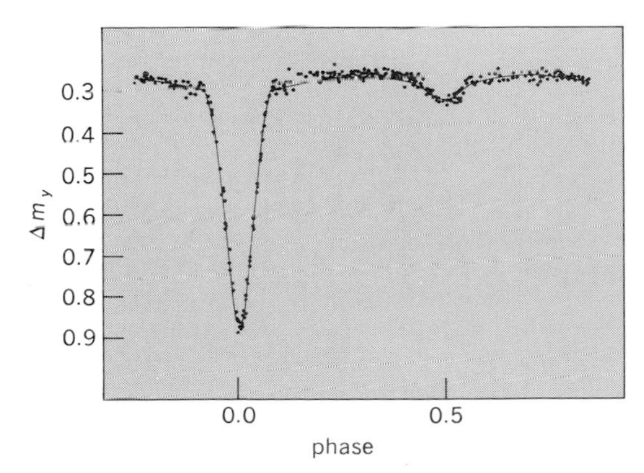

Fig. 1. Light curve of the variable star R Canis Majoris. Each dot represents one measure of brightness. The phase is expressed in fractions of the period of light variation. (*Based on photoelectric observations made by R. H. Koch at the Steward Observatory*)

and time of observation. This is necessary in the case of almost all red variable stars, as well as explosive variables, and certain stars which at irregular intervals lose a great deal of light only to regain it after an unpredictable interval.

Light curves can be obtained by visual, photographic, or photoelectric methods. Visual observations are now confined chiefly to amateurs, but they are useful in studies of long-period variables whose light changes are large; there are not enough professional astronomers to keep these stars under constant scrutiny. Photographic means are chiefly used for discovery and classification of variable stars, and for preliminary studies to determine which merit more precise observation. The most accurate studies of light curves are carried on by photoelectric techniques, in which the precision considerably exceeds that possible by visual or photographic methods. It is customary to use color filters to isolate certain regions of the spectrum; most studies are now made in two, three, or even more such color bands.

[FRANK B. WOOD]

Bibliography: W. Strohmeier, *Variable Stars*, 1972; H. Weaver, Photographic photometry, in S. Fluegge (ed.), *Encyclopedia of Physics*, vol. 54, 1962; A. E. Whitford, Photoelectric techniques, in S. Fluegge (ed.), *Encyclopedia of Physics*, vol. 54, 1962; F. B. Wood, *Photoelectric Astronomy for Amateurs*, 1963.

Light-year

A unit of measurement of astronomical distance. A light-year is the distance light travels in 1 sidereal year. One light-year is equivalent to 9.461×10^{12} km, or 5.879×10^{12} mi. Distances to some of the nearer celestial objects, measured in units of light time, are shown in the table.

Distances from the Earth to some celestial objects

Object	Distance from Earth (in light time)
Moon (mean)	1.3 sec
Sun (mean)	8.3 min
Mars (closest)	3.1 min
Jupiter (closest)	33 min
Pluto (closest)	5.3 hr
Nearest star (Proxima Centauri)	4.3 years
Andromeda Nebula (M31)	2,300,000 years

This unit, while useful for its graphic presentation of the enormous scale of stellar distances, is seldom used technically except in cosmology. *See* ASTRONOMICAL UNIT; PARALLAX; PARSEC.

[JESSE L. GREENSTEIN]

Local Group

The cluster of galaxies to which the Milky Way Galaxy belongs. The Galaxy is believed to be a normal spiral. The main body of its disk is at least 10^5 light-years (1 LY $= 9.46 \times 10^{12}$ km) in diameter, but an extended, roughly spherical halo of a sparse distribution of stars, and perhaps of unseen matter as well, is now thought to reach a diameter at least twice that. The Sun (with its planetary system) lies in the disk, either in or near a spiral arm, at a distance of about 3×10^4 LY from the galactic center, about which it revolves in a nearly circular orbit at a speed estimated to be 220–250 km/s. *See* GALAXY.

It has been known from the 1920s that other galaxies exist and that sometimes they are found in clusters. By the mid-1930s Mount Wilson Observatory astronomer Edwin Hubble's study of the large-scale distribution of galaxies in space led him to speculate that most or all galaxies are clustered, and that conclusion is now generally accepted. The great clusters, which contain thousands of member galaxies, are rare, but can be recognized to very large distances, and thousands are cataloged. Clusters of smaller population are far more numerous. Most common are systems of only a few dozen or less galaxies; these are usually called groups to distinguish them from the richer clusters. The Local Group is such an entity. *See* COSMOLOGY.

Types of galaxies. The total membership of the Local Group is about two dozen. A precise figure cannot be given, partly because a few outlying galaxies are considered members by some authorities and not so by others; moreover, new members, usually dwarf galaxies, are discovered from time to time. Those galaxies generally agreed to belong to the Local Group include three spirals: the Milky Way Galaxy; the Andromeda Galaxy, also known as Messier 31 (M31) and New General Catalog Number 224 (NGC 224); and the spiral in Triangulum–M33 or NGC 598. There are four irregular galaxies, including the nearest galaxian neighbors, the Large and Small Magellanic Clouds, and two others; irregular galaxies contain stars and interstellar matter (as do spirals) but lack axial symmetry and well-organized spiral arms. Four Local Group galaxies are relatively small ellipticals; among them are the two companions to M31 visible on most photographs of that object. Elliptical galaxies seem either spherical or elliptical in shape, are amorphous in appearance, and consist almost entirely of stars, having little or no interstellar matter. They range from giants, far larger than large spirals like the Galaxy, down to dwarfs of only a few million member stars. The Local Group does not happen to contain any giant ellipticals, but in addition to the four small ones already mentioned, it has at least 10 dwarf ellipticals. *See* ANDROMEDA GALAXY; GALAXY, EXTERNAL; MAGELLANIC CLOUDS.

Distances. The Local Group galaxies are the first ones whose distances were determined, and still serve as important calibrators for the extragalactic distance scale. The Large and Small Magellanic Clouds, not visible from northern latitudes, are easily visible with the unaided eye, and appear in the sky as detached pieces of the Milky Way. Although they are only 1.6×10^5 and 1.8×10^5 LY away, respectively, because the largest telescopes of the 1920s lay in the Northern Hemisphere, the Magellanic Clouds did not play the key role in establishing the true nature and distances of galaxies. That honor belongs to M31, in which Hubble recognized a familiar kind of variable star (a cepheid variable) in 1923. Subsequently, Hubble found additional cepheids in M31, M33, and NGC 6822, and by applying the inverse-square law of light propagation to the comparison of the apparent faintnesses and known intrinsic brightnesses of those stars, by the end of 1924 he had established the extragalactic nature of their parent galaxies. *See* CEPHEIDS.

In addition to the Magellanic Clouds, at least three dwarf ellipticals are within a half million light-years of Earth. The dwarf ellipticals contain only a few million stars, or at most a few tens of millions of stars, and even though they are believed to be the most common kind of galaxy, they can be seen only in the nearest other groups and clusters. M31, the nearest large spiral more or less like the Galaxy, is just over 2×10^6 LY away, and appears in the autumn sky as a faint patch of light a couple of times the size of the Moon. It and the Magellanic Clouds are the only external galaxies visible to the unaided eye. About half of the Local Group galaxies lie at a distance similar to that of M31. The entire Local Group covers a region roughly 3×10^6 LY across. The nearest cluster beyond the Local Group, a comparable group of galaxies centered on the large spiral M81, is about 10^7 LY away.

Properties of galaxies. The table describes gross properties of those galaxies generally agreed to be members of the Local Group. For most, the distances are uncertain by 10–20%, and those distances given in parentheses are estimates only. The tabulated diameters refer to the regions of the galaxies that show easily on photographs, but do not include the faint extended halos believed to surround at least the large spirals. The luminosities, in units of the Sun's rate of emission of radiant energy, are uncertain by at least 20%, and for some galaxies only estimates are given (in parentheses). Masses, in units of the Sun's mass, are computed from the law of gravitation and measures of the internal motions in or rotations of the galaxies, and, again, correspond only to those parts of the galaxies containing objects bright enough to measure; an unknown amount of additional mass may exist in the extended halos of

The Local Group of galaxies

Galaxy	Type	Right ascension (1950)		Declination (1950)		Visual magnitude	Distance 10^3 LY	Diameter, 10^3 LY	Solar luminosities	Radial velocity, km/s	Solar masses
		h	m	°	′						
Milky Way Galaxy	Sb	–	–	–	–	–	–	100	(2×10^{10})	–	2×10^{11}
Large Magellanic Cloud	Irr	5	26	−69		0.9	160	30	10^9	+276	2.5×10^{10}
Small Magellanic Cloud	Irr	0	51	73		2.5	180	25	3×10^8	+168	
Ursa Minor system	E4 (dwarf)	15	8.6	+67	11		220	3	(3×10^5)		
Sculptor system	E3 (dwarf)	0	58.9	−33	48	8.0	270	7	4×10^6		$(2-4 \times 10^6)$
Draco system	E2 (dwarf)	17	19.9	+57	56		330	4.5	(8×10^5)		
Carina system	E3 (dwarf)	6	41.1	−50	57		(550)	(4.8)	(8×10^5)		
Fornax system	E3 (dwarf)	2	38.9	−34	36	8.4	800	15	2×10^7	+39	(2×10^7)
Leo II system	E0 (dwarf)	11	12.4	+22	16		750	5.2	(8×10^5)		(1.1×10^6)
Leo I system	E4 (dwarf)	10	7.4	+12	24	12.0	900	5	10^6		
NGC 6822	Irr	19	43.8	−14	50	8.9	1500	9	7×10^7	−58	
NGC 147	E6	0	32.0	+48	23	9.73	1900	10	5×10^7	−168	
NGC 185	E2	0	37.8	+48	14	9.43	1900	8	7×10^7	−208	
NGC 205	E5	0	39.2	+41	35	8.17	2200	16	3×10^8	−240	
NGC 221 (M32)	E3	0	41.6	+40	46	8.16	2200	8	3×10^8	−216	
IC 1613	Irr	1	2.1	+1	51	9.61	2200	16	6×10^7	−235	
Andromeda Galaxy (NGC 224; M31)	Sb	0	41.6	+41	10	3.47	2200	130	3×10^{10}	−301	3×10^{11}
And I	E0 (dwarf)	0	44.6	+37	54	(14)	(2200)	(1.6)	(2×10^6)		
And II	E0 (dwarf)	1	15.2	+33	18	(14)	(2200)	(2.3)	(2×10^6)		
And III	E3 (dwarf)	0	34.2	+36	24	(14)	(2200)	(0.9)	(2×10^6)		
NGC 598 (M33)	Sc	1	32.7	+30	33	5.79	2300	60	3×10^9	−180	8×10^9

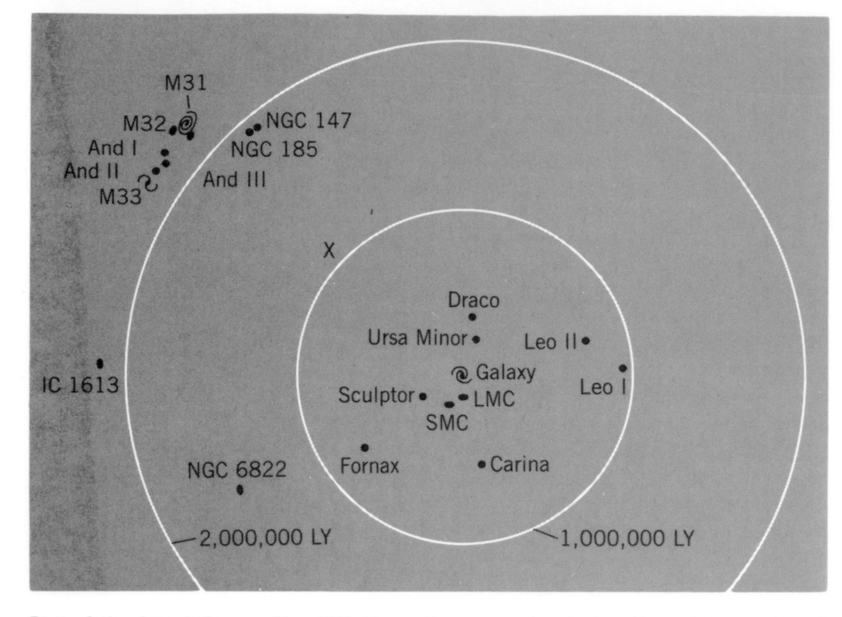

Plot of the Local Group. The "X" shows the approximate location of the center of mass of the group.

the large spirals. The mass estimates in parentheses are estimated from the luminosities. The radial (line-of-sight) velocities, determined from the Doppler shifts of the lines in the galaxian spectra, are with respect to the Sun, and have not been corrected for the Sun's orbital velocity to indicate what would be observed from the center of the Galaxy; positive velocities are of recession, negative ones of approach.

The illustration is a plot of the Local Group; the galaxies have been projected onto an arbitrary plane centered on the Galaxy, then their distances from the center of the plot have been increased so that these galaxies are shown at the correct relative distances from the Milky Way Galaxy. The "X" marks the approximate center of mass of the Group.

Dynamics of the systems. The Galaxy and M31 are estimated to make up more than 95% of the mass of the Local Group. Hence the center of mass of the group is roughly on the line between the centers of the Galaxy and M31, and at a distance that depends on the relative mass of the two galaxies. If it is assumed that the other Local Group galaxies that are not so near either the Galaxy or M31 as to be satellites of those systems are moving at random, a statistical analysis of their radial velocities yields a solution for the velocity of the Sun with respect to the center of mass of the Local Group. Hence the center of with knowledge of the Sun's orbital speed (220–250 km/s), makes it possible to find the velocity of the Galaxy with respect to the center of mass. That velocity, the observed radial velocity of M31, and the assumption that M31 and the Galaxy have equal and opposite momenta with respect to the center of mass, lead to the relative orbital velocity of the two: A 1982 analysis yielded the value 137 ± 20 km/s; this defines the size and shape of the relative orbit of M31 and the Galaxy. Finally, by assuming the time since the two were nearest together (presumably, the time since their formation), the total mass of the Local Group can be estimated. That value is highly uncertain, but probably lies in the range $3-6 \times 10^{12}$ solar masses.

[G. O. ABELL]

Bibliography: G. O. Abell, *Exploration of the Universe*, 4th ed., chap. 37 and 38, 1982; J. Einasto and D. Lynden-Bell, On the mass of the local group and motion of its barycentre, *Monthly Notices, Roy. Astron. Soc.*, 199:67–80, 1982. A. R. Sandage, *The Hubble Atlas of Galaxies*, 1961.

Lunar laser ranging

An optical technique for monitoring the topocentric distance from Earth to specific points on the surface of the Moon with extremely high accuracy. While it is conceptually similar to terrestrial laser geodometry and artificial satellite applications of laser ranging (also called lidar, optical radar, or laser radar), there are significant differences in the techniques required in making the observations. The applications encompass the geometry and dynamics of both Earth and Moon. The features peculiar to lunar laser ranging, as contrasted with other laser applications, are related to the Moon's distance and motion.

Technique. The basic measurement consists of the precise time at which a light pulse is emitted

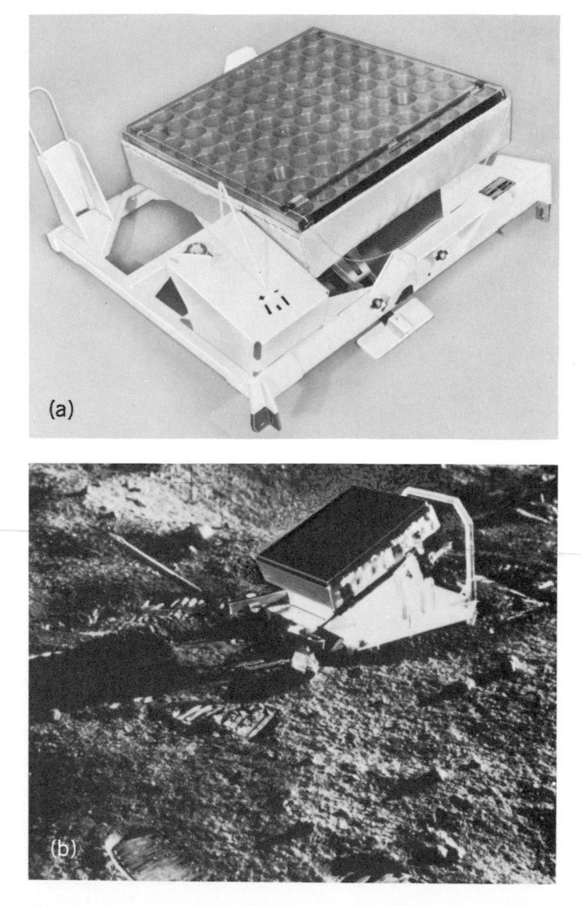

Fig. 1. Laser retroreflectors. (*a*) *Apollo II* array before the mission. (*b*) Array of retroreflectors after deployment on lunar surface.

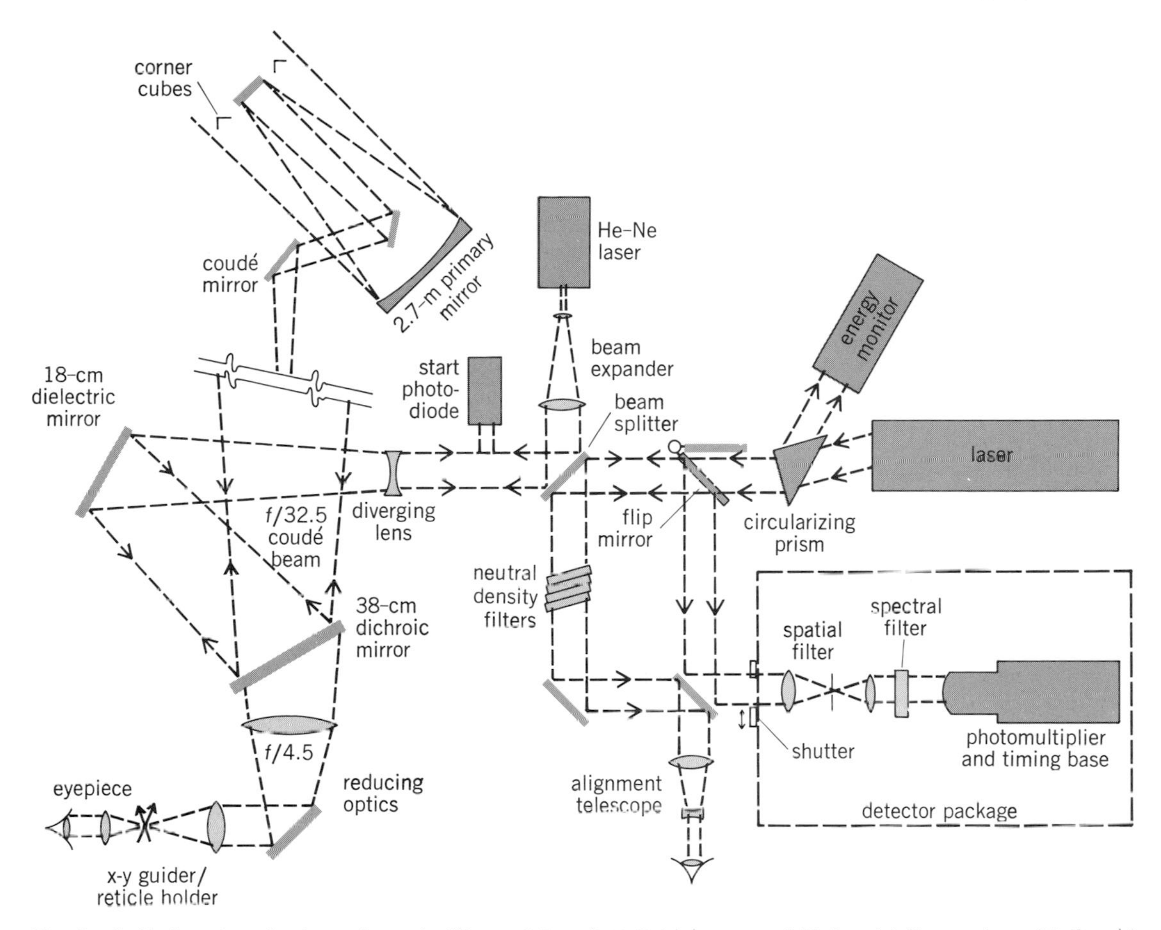

Fig. 2. Optical system for lunar laser facility on 2.7-m (107-in.) telescope of McDonald Observatory. (*McDonald Observatory*)

from the laser, together with the elapsed time required for the pulse to travel from the transmitter to a known point on the Moon and return to an earthbound detector. A precise definition of the lunar target is provided by using a retroreflector (Fig. 1*a*), an array of mirrors whose design ensures that the incoming light will be reflected through almost exactly 180°, so that it will return to the emitting source. Five such arrays have been placed on the Moon (Fig. 1*b*), in Mare Tranquilitatis (*Apollo 11*, 1969) and Mare Crisium (*Luna 17*, 1970) and at Fra Mauro (*Apollo 14*, 1971), Rima Hadley (*Apollo 15*, 1971), and Crater LeMonnier (*Luna 21*, 1973); the Lunakhod reflectors are of French design and construction.

The fundamental laser property that makes any form of lidar possible is the phase coherence of the light produced, which results in near-parallel beams, with little scattering. In practice, however, there are no truly parallel beams, and it is necessary to control and reduce the divergence of the outgoing beam after it leaves the laser, to ensure that an adequate number of photons strike the lunar mirror. This is done by projecting the laser beam through an astronomical telescope. This also provides a convenient pointing and guiding mechanism (Fig. 2). Such facilities operate at Fort Davis, TX, Orroral Valley (Australia), and Simeis (Soviet Union). Other facilities are under construction or testing, at Dodaira (Japan), Fort Davis, Grasse (France), Maui, HA, Simeis (Soviet Union), and Wettzell (West Germany). Temporary stations have operated at Bagnères (France), Boston, MA, San Jose, CA, and Tucson, AZ.

The collimating ability of a telescope depends on its size; the larger the mirror, the smaller the beam divergence. The practical limit is imposed by the atmosphere, whose turbulence defocuses the beam after it leaves the telescope. Consequently, under good conditions, the laser pulse illuminates at least a 2-km-diameter circle when it strikes the lunar surface. The reflectors are 1 m^2 or less, so despite their extremely high individual reflectivity, the global reflective efficiency of a lunar laser ranging system is quite low. The most successful of the present systems transmits a pulse of 10^{18} photons, of which only one comes back down the telescope some 2.5 s later. Because of this, lunar laser ranging is critically dependent on detection technology.

The low global quantum efficiency requires the dependence on detection of individual photons, placing an emphasis on high-efficiency optics and photomultipliers. Stray photons from other sources are filtered both spatially (with diaphragms) and by color; bandpasses as small as 0.12 nm are used. Despite these precautions, only about 10% of the photons detected are true signals. An observation consists of a cluster of such signal photons, which can be discriminated from the noise by statistical techniques with better than 99% certainty.

The critical aspect of lunar laser ranging is that the accuracy depends on the shortness of the pulse, while the signal level depends on the average power (number of photons per unit time). One cannot increase the energy or decrease the pulse length at will, because there are limits on the energy density that the laser crystal can withstand without material damage. Thus, there must be a careful match between telescope size, pulse length, and energy output. Improvements in laser technology since 1969, when lunar laser ranging operations began, are such that the 76-cm system under construction at Fort Davis is expected to replace the present 272-cm system with no loss in average signal.

Applications. In principle, lunar laser ranging observations can be used to study any of the phenomena causing relative motion between the reflectors and telescopes: surface coordinates of telescopes and reflectors, the lunar orbit, rotation of Earth and Moon, gravitational fields of Earth and Moon, gravitational theory, tides, precession and nutation of the Earth's axis, and tectonic motions.

The most significant results of these applications include: (1) A 50-fold improvement in knowledge of the orbital distance of the Moon. (2) A null result for the "Nordtvedt effect" in relativity theory, essentially providing a confirmation of the equivalence of gravitational and inertial mass for large bodies to about 7 parts in 10^{12}. This supports Einstein's general theory against competitors, specifically the Brans-Dicke-Jordan theory. (3) Determination of the effect of tidal friction on the lunar orbit, an acceleration of about 0.12 mrad (24 arc-sec) per century^2, corresponding to a value of $Q = 12$ for the terrestrial tidal-effective dissipation parameter. (4) Improved values for the parameters of the lunar gravitational field which, together with other data, imply a value $C/Mr^2 = 0.3905$ (where C = the greatest polar moment of inertia, M = the lunar mass, and r = the lunar radius) for the lunar moment of inertia parameter; this suggests a sizable metallic core.

Several of the potential applications require either improved accuracy or a wider distribution of observing sites. The uncertainty in a typical observation is 10 cm in equivalent distance, with some systems promising a reduction to 3 cm. See MOON.

[J. DERRAL MULHOLLAND]

Bibliography: J. D. Mulholland (ed.), *Scientific Applications of Lunar Laser Ranging*, 1977; J. D. Mulholland, Scientific achievements from ten years of lunar laser ranging. *Rev. Geophys. Space Phys.*, 18:549–564, August 1980; A. Orszag, *Mesure par laser de la distance terre-lune.*, Université de Paris-Sud (Orsay), 1972; E. C. Silverberg, Operation and performance of a lunar laser ranging station, *Appl. Opt.*, 13:565–574, 1974.

Lyra

The Lyre, in astronomy, a summer constellation, small but important. Lyra has a first-magnitude star, Vega, a navigational star and the most brilliant star in this part of the sky. Vega forms, with

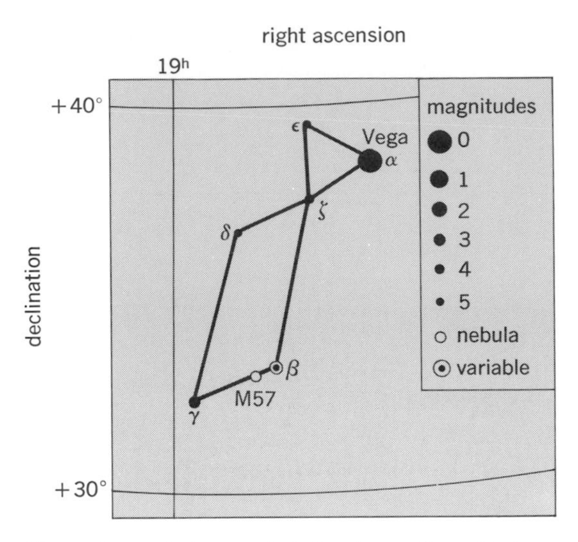

Line pattern of the constellation Lyra. The grid lines represent the coordinates of the sky. The apparent brightness, or magnitudes, of the stars is shown by the sizes of the dots, graded by appropriate numbers.

two faint stars ϵ and ζ to the east, an almost perfect equilateral triangle (see illustration). The southern one in turn forms, with three brighter stars, β, γ, and δ, to the south, an approximate parallelogram. The resulting overall figure resembles a tortoise more than a stringed musical instrument. However, according to legend, Mercury made the first lyre from a turtleshell by placing strings across it. Hence the two different representations are not incompatible. In this constellation lies the famous Ring Nebula. See CONSTELLATION; NEBULA.

[CHING-SUNG YU]

Magellanic Clouds

Of all easily observable external galaxies, the one nearest to the Milky Way galactic system. The distance of the Large Megallanic Cloud (LMC) is 55,000 parsecs (180,000 light-years or 1.70 × 10^{21} m); that of the Small Magellanic Cloud (SMC) is 63,000 parsecs (205,000 light-years or 1.94 × 10^{21} m); these distances are uncertain by about 10%. Because of their proximity, many types of galactic objects, such as bright main sequence stars, supergiants, classical cepheid variables, novae, and planetary nebulae can be studied. Since all these objects are at nearly the same distance from the observer, differences in apparent brightness correspond to differences in true brightness. Thus, the period-luminosity relation for classical cepheids was first established in the small Magellanic Cloud by H. Leavitt. Once the zero point of that relation (that is, the true brightness of at least one cepheid) was found, the distance of any stellar system containing classical cepheids could be established. Thus, the clouds served as a "Rosetta Stone" to astronomy. See CEPHEIDS; GALAXY, EXTERNAL; VARIABLE STAR.

Both Clouds are visible to the naked eye; the Large Magellanic Cloud has an angular diameter of about 7°; the Small Magellanic Cloud an approximate diameter of 3°. E. P. Hubble classified them as irregular galaxies, although there is now evi-

Chemical composition of Magellanic Clouds and Milky Way Galaxy, in numbers of atoms per 10^6 hydrogen atoms

Galaxy	He	N	O	Ne	S	Ar	Fe
Large Magellanic Cloud	87,000	9.8	270	57	10	2.3	15
Small Magellanic Cloud	83,000	2.9	120	25	3	0.8	4
Milky Way Galaxy*	100,000	87	690	105	16	5	35

*Galaxy values are taken mostly from solar data.

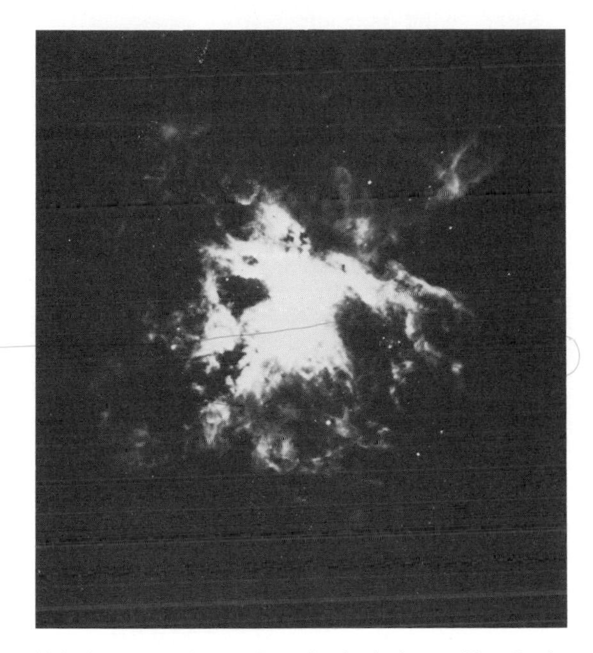

Nebula centered on 30 Doradus in the Large Magellanic Cloud. (*Photograph by S. Z. Czyzak and L. H. Aller at Cerro Tololo International Observatory*)

dence (including the nature of the internal motions) that supports the hypothesis that the Large Magellanic Cloud (at least) may be a barred spiral. The Large Magellanic Cloud is much more massive than the Small Magellanic Cloud; mass estimates of 5×10^9 and 4×10^8 solar masses have been given. Both Clouds contain populations of old stars and also numerous young stars and associations. Nebulosities are recorded strikingly on narrow-wavelength-range photographs centered on the Hα line. The Large Magellanic Cloud shows many wispy nebulosities many times larger than the Orion Nebula. Most outstanding is the complex centered on 30 Doradus (see illustration), an intricate, extremely filamentary gaseous nebula with graceful loops suggestive of several supernova remnants. Its angular diameter or 7′ corresponds to a linear diameter of 110 parsecs or 360 light-years (3.40×10^{18} m). It is larger than any optically observed HII region in the Galaxy or local group. The Small Magellanic Cloud nebulosities are less impressive, but here, as in the Large Magellanic Cloud, the presence of luminous bright blue stars in H II regions and of supergiants shows that a process of star formation is going on. *See* NEBULA; SUPERGIANT STAR.

Other evidence, such as chemical composition determinations of stars and nebulosities, shows that the rate of element building and therefore of star formation is slower in the Large Magellanic Cloud than in the Galaxy and slower yet in the Small Magellanic Cloud (see table). Element building in the Clouds and in the Galaxy differs, and although the same kinds of stars are found in each, chemical composition differences promise to throw significant light on the problem of stellar evolution. *See* ELEMENTS AND NUCLIDES, ORIGIN OF; STAR; STELLAR EVOLUTION.

[LAWRENCE H. ALLER]

Magnetosphere

A comet-shaped cavity or bubble around the Earth, carved in the solar wind. This cavity is formed because the Earth's magnetic field represents an obstacle to the solar wind, which is a supersonic flow of plasma blowing away from the Sun. As a result, the solar wind flows around the Earth, confining the Earth and its magnetic field into a long cylindrical cavity with a blunt nose (Fig. 1). Since the solar wind is a supersonic flow, it also forms a bow shock a few earth radii away from the front of the cavity. The boundary of the cavity is called the magnetopause. The region between the bow shock and the magnetopause is called the magnetosheath. The Earth is located about 10 earth radii from the blunt-nosed front of the magnetopause. The long cylindrical section of the cavity is called the magnetotail, which is on the order of a few thousand earth radii in length, extending approximately radially away from the Sun. *See* SOLAR WIND.

The concept of the magnetosphere was first formulated by S. Chapman and V. C. A. Ferraro in 1931, but the term magnetosphere was introduced in 1959 by T. Gold, who defined it as "the region above the ionosphere in which the magnetic field of the Earth has a dominant control over the motions of gas and fast charged particles." It was, however, only during the 1970s that the magneto-

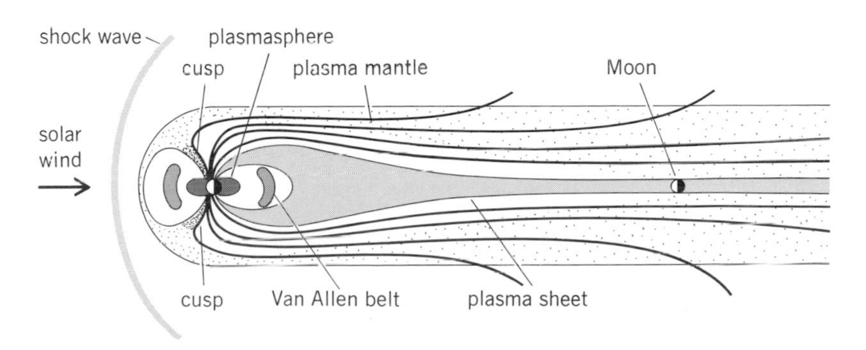

Fig. 1. Noon-midnight cross section of the magnetosphere and the distribution of various plasmas inside it.

sphere was extensively explored by a number of satellites carrying sophisticated instruments. This exploration was a part of the effort to understand that region of the Earth's environment above the level where most meteorological phenomena take place, namely, the space between the Sun and the stratosphere/mesosphere. The space which surrounds the magnetosphere is often referred to as geospace.

Magnetospheric plasma. The satellite observations have indicated that the cavity is not an empty one, but is filled with plasmas of different characteristics (Fig. 1). Just inside the magnetopause some solar wind plasma blows; this particular plasma region is called the plasma mantle. There are also two funnel-shaped regions (one in each hemisphere), extending from the front magnetopause to the polar ionosphere, which are also filled with solar wind plasma; these regions are called the cusp. The extensive tail region of the magnetosphere is divided into the northern and southern halves by a thin sheet of plasma, called the plasma sheet. The distant tail region (beyond the lunar distance) has been infrequently explored when it was traversed by space probes on their way to the planets. The inner magnetosphere is occupied by the Van Allen belts, the ring current belt, and the plasmasphere. All of them are doughnut shaped and surround the Earth. The ring current belt is colocated with the Van Allen belts and is so named because it carries a large amount of a westward-directed electric current around the Earth.

It is generally believed that the main part of the plasma sheet is of solar origin; the mantle plasma may be an important source of it. The plasmasphere is formed mainly by ionospheric plasma which diffuses outward along magnetic field lines. At times the ring current belt contains heavy ions of ionospheric origin, such as helium and oxygen ions, as well as protons of solar origin. The heavy ions are found also in the plasma sheet.

The Earth's dipolar magnetic field is considerably deformed by these plasmas and the electric currents generated by them. The dipolar field is compressed in the dayside part of the magnetosphere, mainly by the action of the impacting solar wind, whereas it is considerably stretched along the equatorial plane in the magnetotail (Fig. 1).

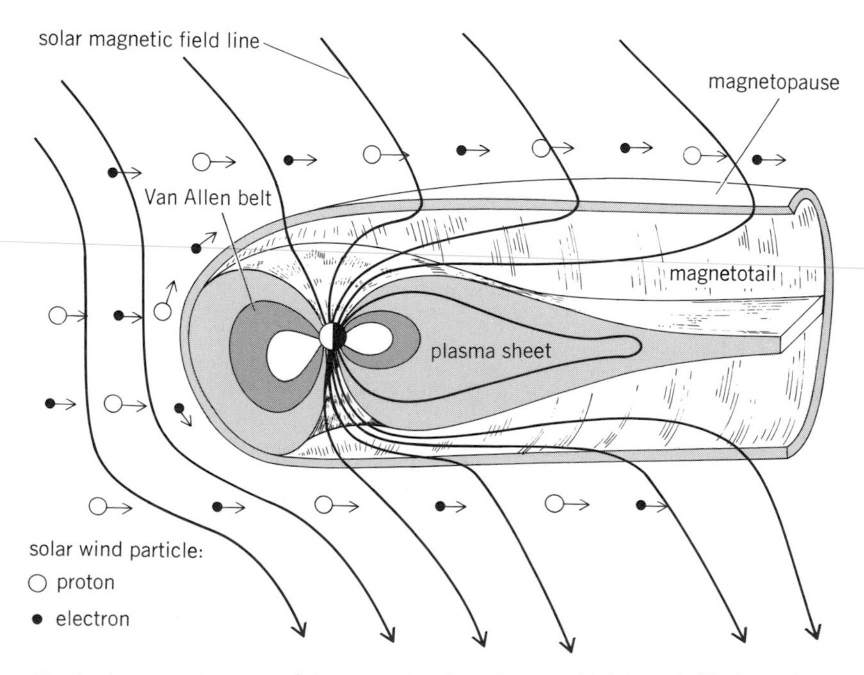

Fig. 2. Internal structure of the magnetosphere (noon-midnight cut). Motions of solar wind protons and electrons around the magnetopause and the linkage of the solar wind magnetic field lines and Earth's magnetic field lines are also shown.

Solar wind–magnetosphere generator. If there were no magnetic field in the solar wind, the Earth's magnetic field would simply be confined by the solar wind into a teardrop-shaped cavity (rather than a comet-shaped cavity), and the cavity would be almost empty. It is the magnetic field of the solar wind which introduces considerable complexity in the interaction between the solar wind and the magnetosphere. Some of the magnetic field lines originating from the polar region are almost always connected to the solar wind magnetic field lines across the magnetopause (Fig. 2). In fact, because the solar wind magnetic field lines originate from the Sun, it may be said that the Sun and the Earth are connected by magnetic field lines. More specifically, the connected field lines originate from an oval area in the polar region. This area is called the polar cap and is bounded by a belt of the aurora called the auroral oval. The magnetic field lines originating from the northern polar cap are stretched along the northern half of the magnetotail as a bundle of magnetic field lines and are connected to the solar wind magnetic field lines across the magnetopause. Similarly, the magnetic flux from the southern polar cap occupies the southern half of the magnetotail.

This connection between the solar wind magnetic field lines and the Earth's magnetic field lines provides the most important aspect of the solar wind–magnetosphere interaction. Since the solar wind (a fully ionized plasma and thus a conductor) must blow across the connected magnetic field lines as it flows along the magnetopause (Fig. 2), an electromotive force is generated. This force powers the solar wind–magnetosphere generator and thus provides practically all the electric power for various magnetospheric processes, such as auroral phenomena and magnetospheric disturbances, including geomagnetic storms. It is precisely in this way that the solar wind (a flow of magnetized plasma) couples its energy to the magnetosphere. Therefore, if there were no connection of the field lines, there would be no energy transfer from the solar wind to the magnetosphere, and thus there would be no auroral phenomena, which are one of the most spectacular wonders of nature.

Because the solar wind–magnetosphere generator is operated on the magnetopause, where the solar wind blows across the connected field lines, the whole cylindrical surface of the magnetotail can be considered to be the generator in which the current flows from the duskside magnetopause to the dawnside magnetopause along the cylindrical surface. Much of the current thus generated flows in the plasma sheet across the magnetotail; this part of the circuit may be called the cross-tail circuit (Fig. 3). Thus, the magnetotail may be viewed as two long solenoids, producing two bundles of

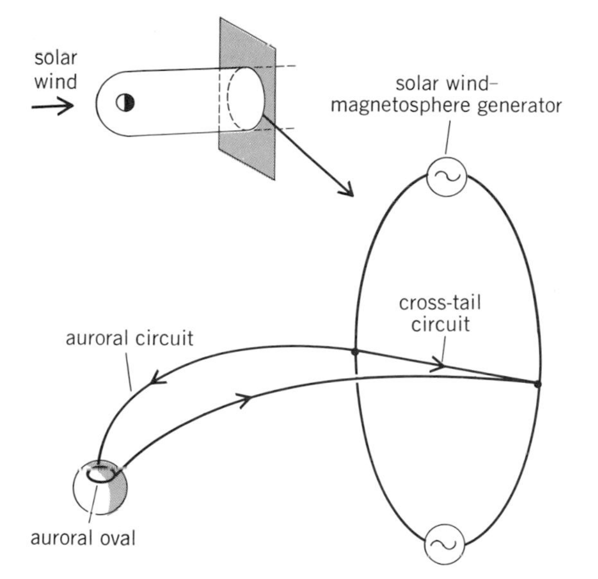

Fig. 3. The solar wind–magnetosphere generator located on the magnetopause (one cross section is shown) and its two main circuits: the cross-tail circuit and the auroral circuit.

magnetic field lines separated by the plasma sheet. As a result, the Earth's dipolar field is stretched along the magnetotail in the antisolar direction. The generator also powers a large-scale motion of plasmas in the magnetosphere. In the main part of the plasma sheet, the plasma flows toward the Earth. *See* SOLAR MAGNETIC FIELD.

Magnetosphere and auroral phenomena. Another important circuit that is connected to the generator consists of a current flow along magnetic field lines from the morning side of the magnetotail to the morning half of the auroral oval, a flow along the auroral ionosphere (the auroral oval), and a flow along magnetic field lines from the evening half of the auroral oval to the evening side of the magnetotail. This discharge circuit may be called the auroral circuit because it is responsible for causing acceleration of current-carrying charged particles (Fig. 3). When the accelerated particles collide with upper atmospheric particles, they excite or ionize them. The auroral lights are emitted by those excited or ionized particles when they return to a lower or the ground state.

The total power generated by the solar wind–magnetosphere generator depends on the magnitude and direction of the solar wind magnetic field, as well as on the speed of the solar wind. In particular, when the solar wind magnetic field is directed southward, the generated power becomes maximum. On the other hand, when the solar wind magnetic field is directed northward, the power becomes minimum. The power is also greater if the magnitude of the solar wind magnetic field B and the speed of the solar wind V are higher. The power P is given by the equation below, where θ is

$$P = VB^2 \sin^4 \left(\frac{\theta}{2}\right) l_0^2$$

the polar angle of the solar wind vector projected to a plane perpendicular to the Sun-Earth line and l_0 is a constant of about 7 earth radii.

When the solar wind magnetic field vector turns gradually from a northern to a southern direction in the vicinity of the magnetosphere, the power generated by the dynamo begins to increase. As a result, the cross-tail electric current in the plasma sheet, is increased. As the generated power is very high, the consumption of the power by the magnetosphere also becomes high. An indication of this high consumption rate is the phenomenon called the magnetospheric substorm which manifests itself in various magnetospheric and polar upper-atmospheric disturbances. As a result of this process, the earthward edge of the plasma sheet is suddenly pushed toward the Earth, and the plasma thus injected into the inner magnetosphere forms the ring current belt.

At the same time, the electric current along the auroral circuit is also intensified. As a result, the aurora becomes bright and active. The enhanced ionospheric portion of the auroral circuit is called the auroral electrojet. Intense magnetic field variations are experienced under and near the auroral electrojet. In this way, effects of the enhanced power of the solar wind–magnetosphere generator are manifested most conspicuously in the polar region. An intense auroral activity associated with this phenomenon is called the auroral substorm. The associated intense magnetic disturbances are called polar magnetic substorms. The ionosphere is also greatly disturbed in this period, and this phenomenon is called the ionospheric substorm. The ionospheric disturbances greatly affect high frequency and very-high-frequency radio communication in high latitudes. Even satellite microwave communication is at times disturbed by auroral activity.

Because the Earth is a good conductor, the auroral electrojets induce a potential difference on the Earth's surface. Therefore, if two points, along which the potential difference is present, are connected by a conductor, an electric current flows along it. It is for this reason that electric currents are induced in oil and gas pipelines and in power transmission lines in high latitudes during auroral activity. In the power lines there are severe power fluctuations when the electrojets are particularly intense.

Various solar activities are responsible for the time variations of the three solar wind quantities which determine the power of the generator. In particular, an intense explosion on the solar surface, a solar flare, causes a gusty solar wind and a large-scale deformation of the solar wind magnetic field. If this deformation happens to produce a large southward-directed magnetic field around the magnetosphere for an extended period (approximately 6 to 24 h), the generated power is considerably enhanced, and intense magnetospheric substorms occur. As a result, an unusually intense ring current grows, carrying a westward current surrounding the Earth and producing a southward-directed magnetic field around the Earth. Since the Earth's magnetic field is directed northward, this addition of the southward field is observed as a decrease of the Earth's magnetic field, one typical aspect of geomagnetic storms.

The Sun has also a few regions from which a high-speed stream is generated, particularly during the declining epoch of the 11-year sunspot cycle. Such a stream lasts for an extended period, from several months to more than 1 year. The solar wind magnetic field is markedly variable in the high-speed stream. As the Sun and the stream rotate with a period of about 27 days, the Earth is immersed in the stream once each period. As a result, the power of the solar wind–magnetosphere dynamo is increased and fluctuates considerably, causing magnetospheric substorms and auroral activity. *See* SUN; SUNSPOT.

Magnetosphere of planets. All other magnetic planets, such as Mercury, Jupiter, and Saturn, have magnetospheres which are similar in many respects to the magnetosphere of the Earth. The magnetosphere of Mercury is much smaller than that of the Earth. Since the solar wind pressure is higher at Mercury's than at the Earth's distance and since the magnetic field of Mercury is weaker than that of the Earth, the blunt-nosed front of the magnetosphere of Mercury is located at a distance of only 0.5 Mercury radii from its surface. On the other hand, because of a weak solar wind, a very strong magnetic field, and a fast rotation, the magnetosphere of Jupiter is gigantic and has an extensive current disk which lies along its equator. The distance to the blunt-noted front is about 100 Jupiter radii. The *Voyager 2* space probe found that the magnetosphere of Jupiter is filled with sulfuric ions, presumably injected from one of its satellites, Io. The space probe also traversed its gigantic magnetotail. The magnetosphere of Saturn is a little smaller than that of Jupiter, but extends well beyond its rings. *See* JUPITER; MERCURY; SATURN.

A great variety of plasma processes, plasma waves, and plasma instabilities are involved in all magnetospheric phenomena. Some of them occur commonly in a thermonuclear machine. The magnetosphere provides a natural laboratory in which these plasma behaviors can be studied without the so-called wall effects. Among a variety of astrophysical conditions, the magnetospheres of the Earth and the planets are the only regions in which plasma behaviors can be studied in natural conditions. Therefore, a study of the magnetosphere provides an important foundation upon which astrophysical plasma theories can be tested and constructed. It has also been speculated that many magnetic stars, pulsars, and even some galaxies have structures similar in some respects to the Earth's magnetosphere. In fact, since a magnetosphere results from an interaction of magnetized plasma flow and a magnetized celestial body, it is quite likely that it is of common occurrence in galaxies. *See* GALAXY; PULSAR.

[S.-I. AKASOFU]

Bibliography: S.-I. Akasofu, *Physics of Magnetospheric Substorms*, 1977; S.-I. Akasofu and S. Chapman, *Solar-Terrestrial Physics*, 1972.

Mars

The planet fourth in distance from the Sun. It is visible to the naked eye as a bright red star, except for short periods near its conjunctions with the Sun. Having a mean (synodic) period between oppositions of 780 days, Mars is in opposition with the Sun every other year. The main orbital elements are: semimajor axis (mean distance to the Sun) = 1.524 astronomical units = 228×10^6 km; eccentricity = 0.093, one of the largest of the main planets (this eccentricity causes the distance to the Sun to vary from 206×10^6 km at perihelion to 248×10^6 km at aphelion); sidereal period of revolution = 1.881 years = 686.98 days; mean orbital velocity = 24.1 km/s; and inclination of the orbital plane to the ecliptic = 1° 51′. *See* PLANET.

Elements of the globe. The mean diameter of the globe of Mars is 6762 km, or about 53% that of the Earth. The planet's volume is therefore about 15% of Earth's. The mass of Mars, about 11% of Earth's, has been well determined, first by the motions of its natural satellites, and more recently with greater precision by the motion of spacecraft orbiting about the planet. The mean density is about 3.94 g/cm^3, less than that of any other terrestrial planet. The surface gravity is about 38% of the Earth's, or more precisely 375 cm/s^2 at the equator and 380 cm/s^2 at the poles. The escape velocity is about 5.0 km/s. The rotation period is very accurately determined from several centuries of observation of the times of transits of surface markings at the central meridian of the disk; the sidereal period is 24h37m22.66s; the corresponding length of the mean solar day is 24h39m25.0s. It follows that there are about 646 Martian sidereal days and 645 mean Martian solar days in the planet's year. The inclination of the polar axis to a line perpendicular to the orbital plane is 25°.0, only slightly greater than that of the Earth, so the seasons are similar but longer and also more unequal because of the greater eccentricity of the orbit. In the northern hemisphere of the planet the spring lasts 199 (terrestrial) days; summer, 182 days; autumn, 146 days; and winter, 160 days. Further, because the passage at perihelion (which takes place at heliocentric longitude 357°), the cold season is longer and colder, and the warm season is shorter and hotter in this hemisphere. This asymmetry has a pronounced effect on the Martian climates and seasonal variations, especially in the polar regions.

Telescopic appearance. In a telescope, Mars usually appears as a bright reddish disk marked by complex, semipermanent dark regions and variable white polar caps. In general, surface albedo features are not visible on photographs taken in violet and ultraviolet light because of decreasing contrast between the light and dark regions at these shorter wavelengths. Scientists have long known of atmospheric phenomena, variable polar caps of ice and clouds, and large bright and dark regions on the surface with dimensions of several hundred kilometers or more. But a multitude of other small surface formations with characteristic scales less than 200 km manifest their existence by the optical illusion of "canals" and "oases." These ephemeral features, some of which vary with the Martian seasons, were long thought by some astronomers to be due to vegetation responding to the Martian seasons. As revealed by Mariner and Viking spacecraft, however, these features are now known to be a complex of smaller, dark splotchy areas whose contrast changes with the shifting Martian sands driven by seasonal winds.

The apparent diameter of the disk varies from a minimum of 3″.5 at conjunction to a maximum of 25″.1 at the most favorable perihelic opposition, when the distance to the Earth is only 56×10^6 km. At aphelic oppositions the apparent diameter may not exceed 13″.8. The most favorable oppositions with Mars (near its perihelion) occur every 15 years; those of the past century, during which astronomers closely scrutinized its surface, took place in 1877, 1892, 1909, 1924, 1939, 1956, and 1971. The next perihelic opposition will occur in 1986.

As a "superior" planet (orbiting beyond the Earth), Mars presents only gibbous phases when it is not in opposition to or conjunction with the Sun; the maximum phase angle (Sun-Mars-Earth angle) is 48°, when Mars is in quadrature. The apparent visual magnitude of Mars in conjunction averages −1.5, but it brightens to −1.2 at perihelic opposition (Mars is then brighter than any other planet except Venus).

The visual albedo of Mars is about 0.15, half again that of the Moon or Mercury, but less than half the values for the planets surrounded by dense atmospheres. The reflectivity of Mars increases rapidly from a low value of only 5% in the near ultraviolet to more than 30% in the near infrared. This variation accounts for the reddish color of Mars. At ultraviolet wavelengths shorter than 0.3 μm, observable only from high-altitude rockets or satellites, the albedo increases toward shorter wavelengths because of an increasing contribution of scattering in the atmosphere from molecules and aerosol particles. In the infrared, at wavelengths between 1 and 2 μm, the albedo curve of Mars shows only slight variations. These resemble the spectral reflectivity curve of iron oxides, especially geothite and hematite. *See* ALBEDO.

Surface. The Martian surface, which initially resembled the heavily cratered lunar uplands, has been modified extensively by the processes of volcanism, faulting, and the effects of wind and water. The northern hemisphere tends to be characterized by large volcanic lava plains, whereas the southern hemisphere appears to be older and more heavily cratered. Several large impact basins associated with the final accretion of the planet are easily seen even in Earth-based telescopes. The largest of these, Hellas, is approximately 2000 km across and at least 4 km deep. *See* MOON.

Volcanoes. Among the more impressive features on the Martian surface are several enormous shield volcanoes, larger than any mountains on Earth. The largest and perhaps the youngest is Olympus Mons, which is nearly 600 km across at its base and stands approximately 26 km above the surrounding terrain. Three other large shield volcanoes, Ascraeus Mons, Pavonis Mons, and Arsia Mons, lie along the nearby Tharsis Ridge. The Tharsis volcanoes are approximately 400 km across and 20 km high. These features may be compared with the largest volcanic constructs on Earth, Mauna Loa and Mauna Kea in Hawaii, which together are approximately 200 km across and stand 9 km above the ocean floor.

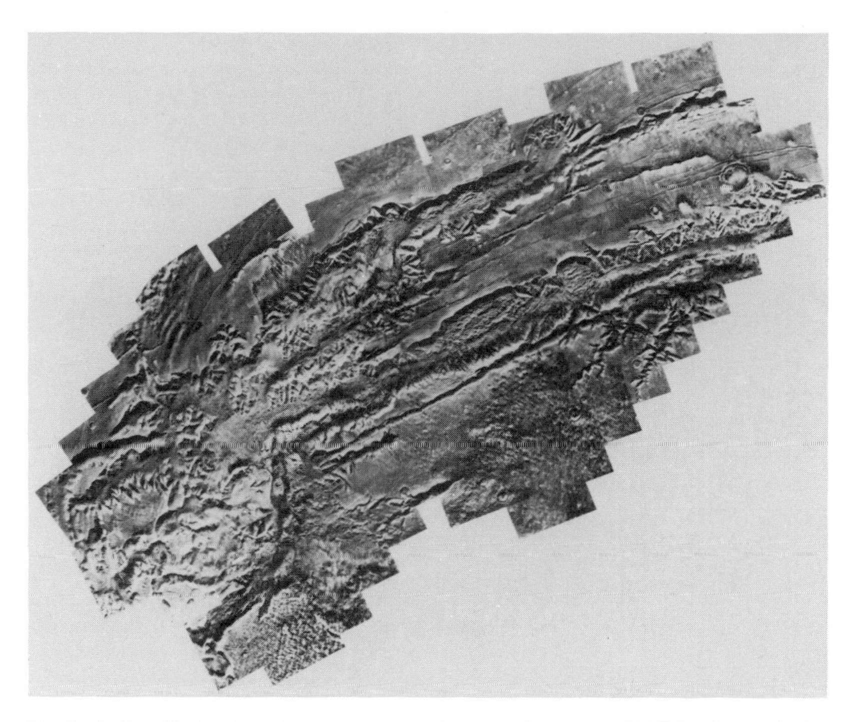

Fig. 1. Valles Marineris, a huge canyon system on Mars, over 5000 km long; photographed from *Viking Orbiter 1*. (*NASA*)

In shape and structure the Martian shield volcanoes bear a strong resemblance to their Hawaiian counterparts. At the summit of each shield is a complex of calderas, collapsed craterlike features that were once vents for lava. The central smooth-floored caldera of Arsia Mons is the largest, about 130 km in diameter. Studies of Viking data suggest this caldera may be the youngest surface feature on the planet, perhaps less than 300×10^6 years old. About 5000 km to the west of the Tharsis area is another volcanic field, dominated by the symmetrical shield volcano, Elysium Mons, approxi-

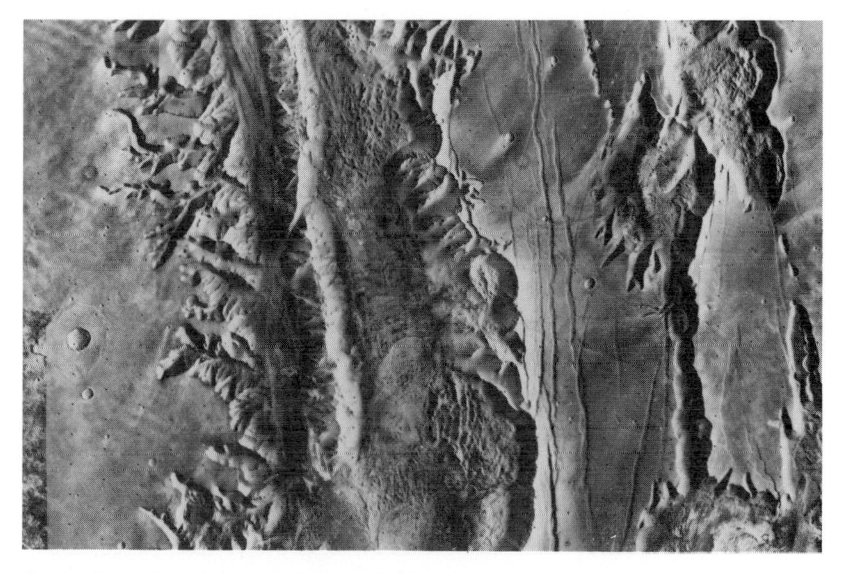

Fig. 2. Part of western end of Valles Marineris, showing canyon walls; photographed by *Viking Orbiter 1*. (*NASA*)

mately 225 km across. Many other volcanoes with dimensions of up to 100 km and in various states of preservation are found scattered over the Martian surface, mostly in the northern hemisphere. However, while no large, young volcanoes appear in the southern hemisphere, volcanic activity has, nonetheless, played a significant role there. Close examination of spacecraft images shows that nearly one-third of this hemisphere is covered by relatively smooth plains of volcanic origin.

A volcanic formation known as a patera (Latin for "saucer"), a shallow, complex crater with scalloped edges, occurs on Mars but has not been observed on any other planet. Alba Patera, northeast of Olympus Mons, is between 1500 and 2000 km across but rises only 3 to 5 km above the surrounding plains. Enormous quantities of highly fluid lava must have emerged from its central caldera, flowing to great distances along channels and tubes.

The Tharsis ridge, or bulge, is an uplifted portion of the surface that stands more than several kilometers above the mean elevation of the planet. Most of Mars's tectonic features are associated with Tharsis, and it affects approximately one-quarter of the entire surface. Its formation appears to have begun some 3 to 4×10^9 years ago, long before the volcanoes that now dot its crest and surroundings. A fact puzzling to scientists is that Tharsis is, in effect, top-heavy—its bulk should not be sitting so high on the crust, but rather sink in (much as an iceberg submerges most of its volume beneath the ocean's surface).

Canyon lands. Perhaps the most spectacular features on the Martian surface are the huge canyons located primarily in the equatorial regions. Valles Marineris, actually a system of canyons, extends for over 5000 km along the equatorial belt (Fig. 1). In places, the canyon complex is as much as 500 km wide and drops to more than 6 km below the surrounding surface. Dwarfing the Grand Canyon, Valles Marineris is comparable in size to the great East African Rift Valley. In general, the walls of the canyon are precipitous, have well-defined edges, and show evidence of slumping and landslide activity (Fig. 2). An intricate system of smaller canyons extending back from the main rim may have developed during melting and evaporation of subsurface ice, or it may have been caused by wind or water erosion. Valles Marineris's principal canyons appear to be linked with the formation of the Tharsis uplift. They may have formed by faulting along lines of tension within the crust. In essence, such canyons (called graben) are pairs of parallel faults with a sunken area between them. While erosion continues to eat away at the canyon walls, that is an unlikely origin for Valles Marineris and similar complexes. These enormous chasms have no outlets, and the only obvious way to remove debris is by wind; yet the material to be transported out of the canyons is so great as to cast serious doubt on the effectiveness of this mechanism operating by itself.

Channels. An astonishing class of features on the Martian surface is a widespread network of channels that bear a very strong resemblance to dry river beds (Fig. 3). Ranging in size from broad, sinuous features nearly 60 km wide, to small, narrow networks less than 100 m wide, the channels present the strong impression that they were created by water erosion. Early analysis assumed the apparently dendritic nature of some channels to be evidence of watershed rain collection and episodic flow. On closer inspection, however, many systems previously considered dendritic now appear to result from more gradual formation through groundwater sapping, not rainfall. The largest channels must have been formed by enormous torrents of water, presumably released in some catastrophic manner. Yet, liquid water on Mars, under conditions that prevail today, cannot exist, because the atmospheric pressure and surface temperature keep most available water trapped as subsurface ice or locked up in the permanent polar caps. As a way out of this enigma, it has been suggested that the long-term climate of Mars is variable, and that in the past the planet might have had conditions suitable for free running water over its surface. Long-term changes in the inclination of Mars's axis to its orbital plane and in the eccentricity of its orbit tend to support this supposition.

These channels do not coincide with the "channels" or "canals" which, in the past, were fre-

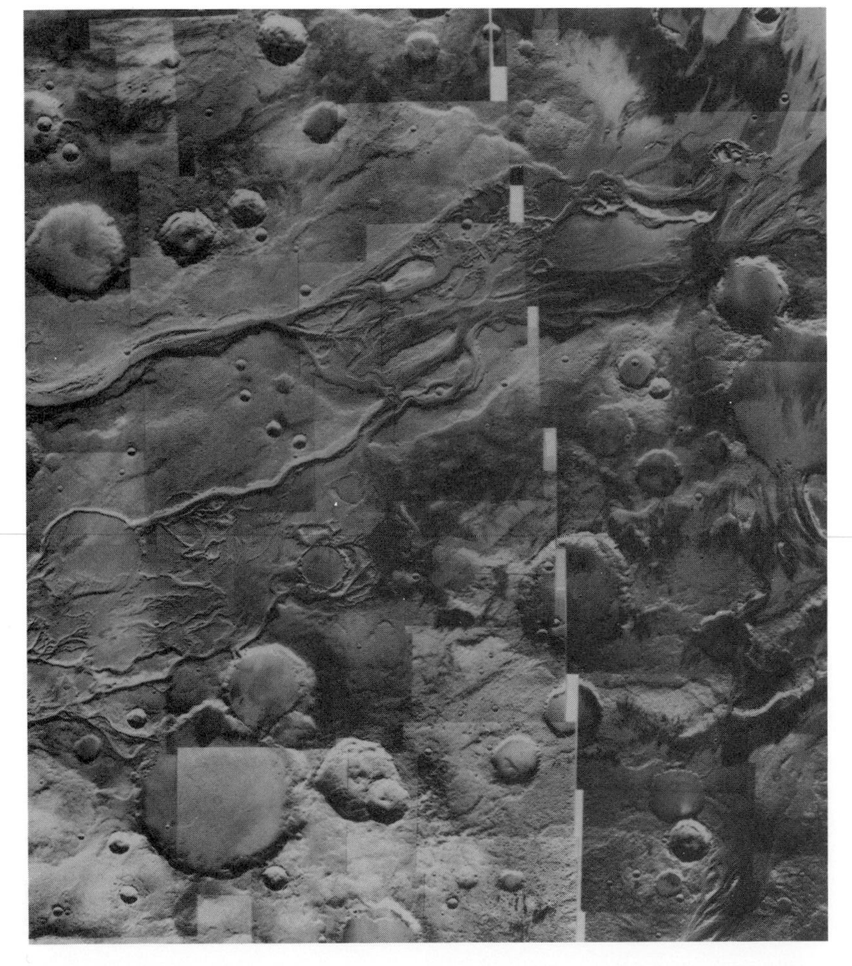

Fig. 3. Channel system in the Mangala Vallis region of Mars; photographed by *Viking Orbiter 1*. (*NASA*)

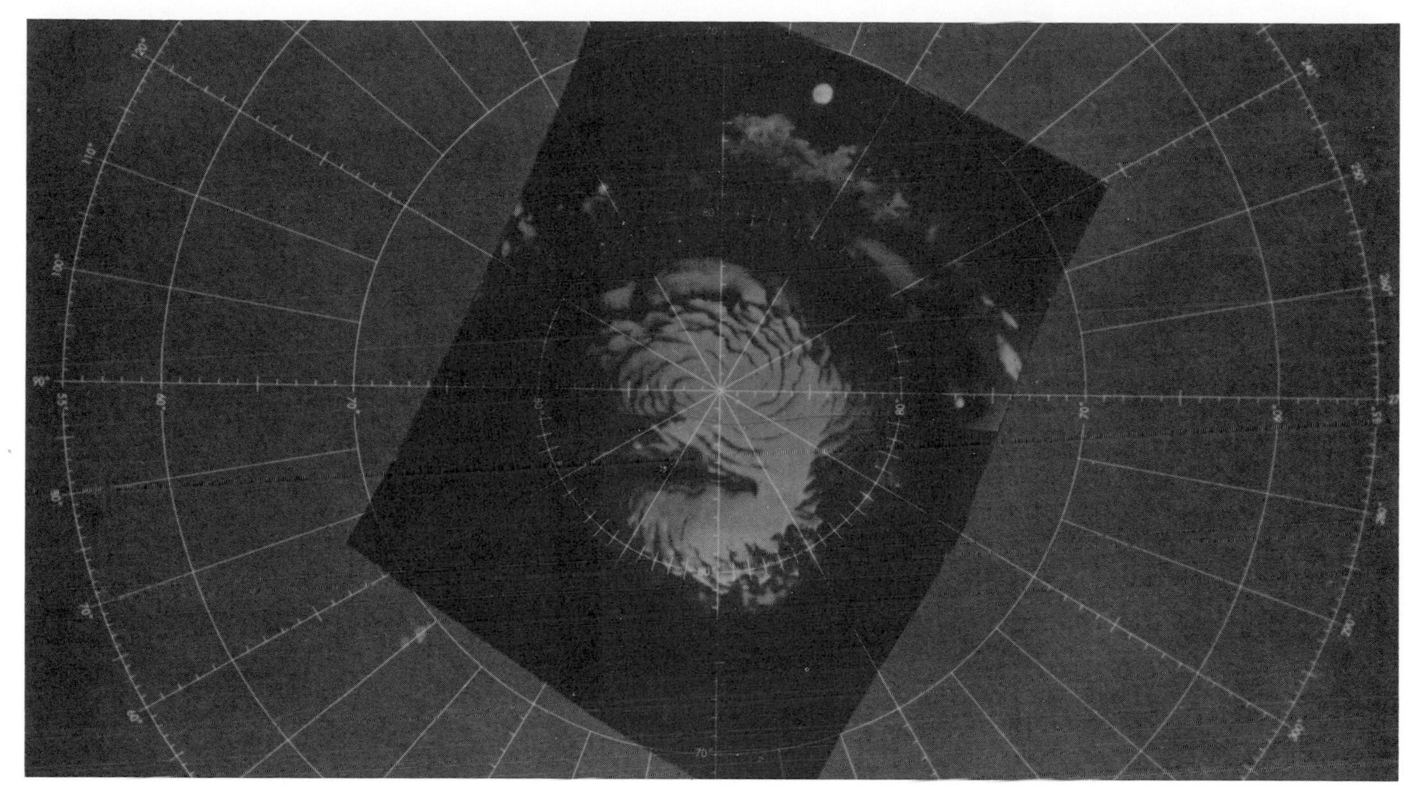

Fig. 4. North polar region of Mars, photographed by *Mariner 9*, showing polar cap nearing its minimum extent approximately two weeks after the summer solstice. Crater at upper right trapped and shielded frost from sun, leaving a large patch on its floor. (*NASA*)

quently reported by visual observers of Mars. Only Valles Marineris can be associated with a feature previously seen from Earth.

Polar caps. The seasonal cycle of growth and decay of the bright polar caps has long been taken as evidence of the presence of water on Mars. However, evidence from the Mariner spacecraft identified carbon dioxide as the principal constituent of the polar snow, with lesser amounts of water ice also present. Formed during autumn and winter by condensation and deposition of the icy mist covering the polar regions, the polar caps at the end of winter cover a vast area extending down to latitude 60° in the southern hemisphere and 70° in the northern hemisphere. The surfaces covered are, respectively, 1 and 4.5×10^6 km. The thickness of the cap is unknown but presumably decreases radially outward from the pole. During spring and summer the edges of the caps retreat, reaching minimum diameters of about a few hundred and a thousand kilometers, respectively. The residual caps (Fig. 4), which never completely disappear, are believed to be composed almost entirely of water ice. It is estimated that, if the total water component of the polar caps were distributed uniformly over the surface of Mars, it would produce a layer 10 m deep. A far greater amount of water may exist as subsurface ice in nonpolar regions.

The polar regions are characterized by layered terrain, consisting of narrow, evenly spaced strata from 20 to 50 m in thickness. A sequence com-

Fig. 5. Martian landscape, photographed by *Viking Lander 1*. Picture covers 100°, looking northeast at left, southeast at right. (*NASA*)

posed of more than 100 such layers has been measured in the south polar regions. The absence of craters in these layered terrains suggests that they represent either the youngest areas on Mars or, more likely, the most heavily eroded.

Atmosphere. The Martian atmosphere is very thin, with surface pressure (measured at the Viking landing sites) varying between 7 and 10 millibars (700 and 1000 Pa) and averaging about 8 mb (800 Pa), or less than 1% of the pressure at the Earth's surface. It is composed principally of carbon dioxide, but contains nitrogen, argon, oxygen, and a trace of water vapor totaling about 5%. If all the water remaining in the atmosphere of Mars were to condense onto the surface, it would form a film only 10 μm thick.

Clouds. Both carbon dioxide and ice form clouds in the Martian atmosphere. Certain water-ice condensations are associated with the largest volcanic peaks, while carbon dioxide clouds occur frequently in the polar regions. Often, clouds will appear in the hours just after sunrise, when frosts formed during the night sublime back into the atmosphere.

Dust storms. Localized dust storms appear quite frequently on Mars. Because of the extreme thinness of the Martian atmosphere, wind velocities greater than 40–50 m/s are needed to set surface dust grains in motion, although they can then be carried to high altitudes by winds of lesser velocity. The most abundant particles in a typical Martian dust storm are estimated to be about 10 μm in diameter, similar to finely powdered talc. The very smallest particles may be carried to heights as great as 50 km above the surface.

Some dust storms develop with such intensity that their total extent may be hemispheric or even global. Such was the case of the great dust storm of 1971, which began in September and completely enshrouded the planet by the time *Mariner 9* arrived in November. Major dust storms tend to begin at the end of spring in the southern hemisphere, when Mars is near perihelion. The initial dust-raising disturbance is probably related to temperature instabilities in the Martian atmosphere at a time when solar heating is at a maximum. Once the dust is in the air, increased absorption of solar energy causes the dust-laden air to rise, producing strong surface winds which rush in to replace the rising air mass. Thus the storm propagates, growing in intensity and extent. When the distribution of dust finally reaches global dimensions, the temperature differences over the planet are diminished, winds correspondingly subside, and dust begins to settle out of the atmosphere. The settling process may take weeks, or in the case of a severe storm, several months. From beginning to end, the 1971 storm lasted for approximately 5 months.

Fig. 6. Martian landscape, photographed by *Viking Lander 2*, looking northeast. (*NASA*)

Viking landing sites. In 1976 two American spacecraft—the Viking landers—successfully dropped onto the surface of Mars. *Lander 1* set down in a region known as Chryse Planitia (22°.27N, 47°.97W) on July 20, and *Lander 2* arrived at Utopia Planitia (47°.67N, 225°.74W) on September 3. From these locations, the spacecraft collected more than 1400 photographs and large quantities of other data concerning the character, composition, and organic content of the surface, and the composition and climatic patterns of the lower atmosphere. Both landers operated long after the completion of their basic 90-day missions, and one continued to collect and relay data through 1980.

Lander 1 sits in a moderately cratered, low-lying volcanic plain near the mouth of a large outflow channel. In Fig. 5, craters up to 600 m across near the horizon are probably the source of rocks littering the scene. These rocks range in size and color, and some bedrock may also be exposed; fine-grained material occasionally collects into large drifts. *Lander 2*, situated 1500 km farther north, is in a region of fractured plains about 200 km south of the large crater Mie; it probably sits atop a lobe of the crater's ejecta blanket. This site (Fig. 6) is very flat and sparsely cratered, subdivided by polygonal fractures and troughs crisscrossing the area. While no drifts are visible and rocks cover a higher proportion of the surface than at the other site, rocks surrounding *Lander 2* look remarkably similar to those near *Lander 1*, with numerous pits giving them a spongy appearance. During the course of a full Martian year, the landers recorded only slight variations in the brightness and color of the surrounding surface, which can be explained by the deposition and removal of a thin veneer of dust and, for *Lander 2*, frost.

Five experiments sampled the soil within reach of the 3.2-m mechanical arm aboard each spacecraft. The arms themselves were used to test the properties of rocks and soils, and these ranged in character from loose piles of drift material near *Lander 1* to rocks at both sites that were too hard to be chipped or scratched. Three biological investigations and a gas chromatograph/mass spectrometer attempted to settle the question of whether life existed on the planet. The latter instrument detected no organic compounds at all (within the limits of its sensitivity). One of the trio of biological experiments gave a positive result—the release of oxygen from a soil sample when humidified—but there is no consensus as to the source of this reaction. Some scientists have suggested inorganic superoxides or other catalysts as the substance involved. Most of the data concerning the composi-

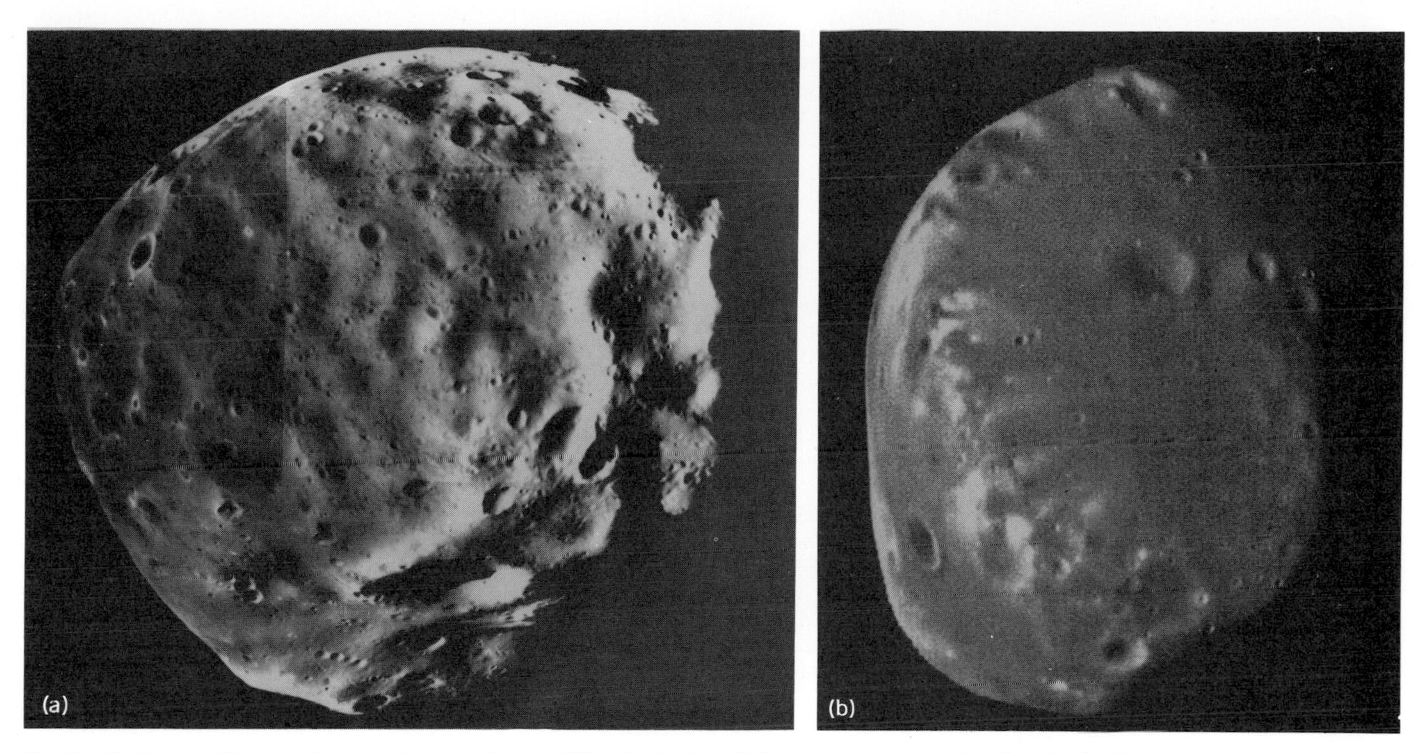

Fig. 7. Martian satellites. (*a*) Phobos, photographed by *Viking Orbiter 1*. (*b*) Deimos, photographed by *Viking Orbiter 2*. (*NASA*)

tion of the surface came from x-ray fluorescence spectrometers, which showed that small particles were remarkably similar at both sites, with silicon and iron accounting for about two-thirds of the samples' content. The sulfur content was unexpectedly high, 100 times greater than in the Earth's crust, while potassium was found to be present at one-fifth of Earth's proportion. Small particles also clung to magnets on each spacecraft, and these are probably composed of the mineral maghemite. In general, the surface material at both sites can be characterized as iron-rich clays.

The weather (atmospheric pressure, temperature, wind velocity, and direction) was monitored by both landers for more than a full Martian year. Temperatures ranged from a low of 150 K at *Lander 2* to a high above 250 K at *Lander 1*; on any given day, the diurnal variation at each site ranged between 35 and 50 K, and dust storms tended to moderate temperatures. In the Martian summer, as atmospheric pressure dropped toward its minimum (because of frost condensing onto the south polar cap), the weather became rather monotonous: diurnal variations were very consistent and winds averaged 1–2 m/s. But toward autumn, day-to-day variations increased (especially at *Lander 2*), and a regular sequence of cyclones and anticyclones were observed to pass eastward over the more northerly lander with a frequency of about once per week. Some of these moving systems were apparently associated with frontal systems, such as the advance of the so-called north polar hood.

Internal constitution. Accurate measurements of the planet's gravitational field made by three orbiting spacecraft have shown that Mars has a dense core and thus is differentiated. The crust, composed of lighter rocks, enriched in silicon and aluminum and deficient in magnesium, is believed to be about 50 km thick. Melting of the crust or upper mantle, over geologic time, has caused lighter materials to float to the surface, creating the observed lava plains and volcanic structures. This volcanism is believed to have continued episodically up to the present.

Satellites. Mars has two small satellites, Phobos and Deimos, discovered by A. Hall in 1877. The apparent visual magnitudes at mean opposition are 11.5 and 12.5, respectively. Although they are not intrinsically very faint, a large telescope is required to see them clearly against the bright glare surrounding the planet. The outer satellite, Deimos, moves at a mean distance from the planet's center of 23,500 km in a sidereal period of 30h18m, only slightly longer than the period of rotation of Mars. The orbital plane is inclined about 1° to the planet.

The inner satellite, Phobos, moves at a mean distance from the planet's center of 9350 km in a sidereal period of 7h39m, which is less than one-third of the mean Martian solar day. It is the only known satellite whose period of revolution is less than the period of revolution of the planet; it follows that, as seen from the surface of Mars, Phobos rises in the west and sets in the east twice daily, moving apparently in the opposite direction to all other celestial bodies, including Deimos. The inclination of the orbital plane to the equatorial plane of the planet is 1°.7. The mean distance of Phobos to the surface of Mars is only 6000 km, and the apparent diameter of Mars seen from Phobos is 42°; conversely, Phobos cannot be seen from regions of Mars within 21° of the poles. Seen overhead near the Martian equator, Phobos would appear less than half the size of the Moon viewed from the Earth, and Deimos would be a barely discernible disk.

The attraction of the equatorial bulge of Mars causes the lines of the apsides of the satellites' orbits to advance and the lines of the nodes to retrograde in a period of about 55 years for Deimos and 2.2 years for Phobos. Neither satellite is massive enough to be gravitationally contracted to a spherical shape (Fig. 7). Phobos measures only 20 × 23 × 28 km, and Deimos just 10 × 12 × 16 km. The very low surface gravity of these tiny satellites leads to an escape velocity of only 10 m/s. A person standing on Deimos or Phobos could quite easily throw rocks which would leave the satellite forever.

Both satellites are saturated with impact craters; the largest is Stickney, an 8-km-diameter crater located on Phobos. Had the impacting object which excavated Stickney been much larger, it probably would have shattered Phobos. In fact, it has been suggested that Deimos and Phobos may be the shattered and, perhaps, partially reconstituted remains, of an earlier, single, large satellite of Mars. Both satellites have the same low albedo, approximately 0.05, surfaces that are as dark as the very darkest regions of the Moon. This low reflectivity suggests that the Martian satellites may have had their origin in the asteroid belt. *See* ASTEROID. [J. KELLY BEATTY]

Bibliography: J. K. Beaty and B. O'Leary, *The New Solar System*, 1981; M. H. Carr, The volcanoes of Mars, *Sci. Amer.*, 234:33–43, 1976; C. W. Snyder, The planet Mars as seen at the end of the Viking mission. *J. Geophys. Res.*, 84:8487–8519, 1979; Viking Lander Imaging Team, *The Martian Landscape*, 1978; J. A. Wood, *The Solar System*. 1979.

Mass-luminosity relation

The relation, observed or predicted by theory, between the quantity of matter a star contains (its mass) and the amount of energy generated in its interior (its luminosity). Because of the great sensitivity of the rate of energy production in a stellar interior to the mass of the star, the mass-luminosity relation provides an important test of theories of stellar interiors. This predicted relation is a corollary of the more general principle that if a star is in equilibrium, that is, if it is neither expanding nor contracting, all its other properties depend only on its mass and the distribution of chemical elements within it. For a family of stars with different masses but with the same mixture of chemical elements uniformly distributed throughout the stellar volumes, there will be a unique mass-luminosity relation. Since most of the stars in the solar neighborhood have about the same chemical composition, the observed relation, obtained from binary stars for which masses and luminosities can be observationally evaluated, conforms reasonably well with theory. It was this agreement that showed the pioneering studies of stellar structure by A. S. Eddington in the 1920s to be on a sound basis. *See* BINARY STAR; STAR.

Because of the observational difficulties in obtaining masses of high precision as well as in evaluating moderate differences of chemical composition among stars, there is some controversy over the uniqueness of the mass-luminosity relation for all main-sequence stars. On the other hand, there are classes of stars not on the main sequence which depart markedly from the mass-luminosity relation for main-sequence stars. The theory of stellar evolution predicts that changes in interior chemical composition will result from conversion of hydrogen into helium and heavier elements in the process of nuclear energy generation, which is the source of stellar luminosity. Accompanying these changes in composition there should be changes in luminosity. Giants and supergiants are examples of such evolved stars.

Fragmentary observational evidence indicates that they are more luminous than main-sequence stars of the same mass, in agreement with theory. White dwarf stars, which are known to be extremely underluminous for their masses, are thought to be stars in which all sources of nuclear energy have been exhausted. If a star is a member of a close binary system, it may be prevented by the presence of its companion from evolving by expansion into a giant star. The evolutionary histories of stars in such systems are predicted to involve mass exchange between the components. The relation between masses and luminosities will depend upon the state of evolution of the individual system. In some cases, gross departures from the mass-luminosity relation for main-sequence stars are predicted. A star which has lost the bulk of its mass to its companion may radiate as much energy as a normal star. X-ray binaries are thought to be extreme examples of advanced evolution of close binary systems. A number of the predictions of the evolution of close binaries appear to be verified observationally. *See* GIANT STAR; STELLAR EVOLUTION; SUPERGIANT STAR; WHITE DWARF STAR; X-RAY STAR.

[DANIEL M. POPPER]

Bibliography: W. D. Heintz, *Double Stars*, 1978; E. Novotny, *Introduction to Stellar Atmospheres and Interiors*, 1973; J. Sahade and F. B. Wood, *Interacting Binary Stars*, 1978.

Mercury

The planet nearest to the Sun. It is visible to the naked eye shortly after sunset or shortly before sunrise when it is near its greatest angular distance from the Sun, which is never more than 28°. The diameter, measured with high accuracy by radar, is 4878 ± 2 km. The mass, determined from the gravitational perturbations produced by Mercury on the motions of the other planets and of space probes, is 3.301×10^{26} g, or 0.055 times the mass of the Earth.

Mercury has the highest density (5.44 g cm^{-3}) of any planet except Earth, indicating that iron and nickel are its main chemical constituents. The surface of the planet is composed of silicate rock, heavily cratered from more than 4×10^9 years of exposure to impacting meteoritic debris. There is no atmosphere, so that the surface retains the record of nearly the entire past geological history.

Orbit and rotation. The average distance of Mercury from the Sun is 57.9×10^6 km. Since the eccentricity of the orbit is large (0.206), the distance from the Sun varies from 45.9×10^6 km at perihelion to 45.9×10^6 km at perihelion to 69.8×10^6 km at aphelion. The inclination of the orbital plane to the ecliptic (the plane of the Earth's orbit) is 7°; both the eccentricity and the inclination are greater for Mercury than for any other planet except Pluto. The sidereal period of revolution is 87.969 Earth days. The period of rotation on its axis (the sidereal day) is exactly two-thirds of the sidereal year, or 58.646 days, as was first revealed by radar studies in 1965. This peculiar two-thirds synchronism results in a solar day (the interval from one sunrise to the next) that is twice the sidereal year and three times the sidereal day in length.

The rotation period of Mercury was locked into its present value as a result of the large tidal force produced by the relatively nearby Sun. Because of the high eccentricity of its orbit, Mercury's rotation for several days near each perihelion passage is such as to keep approximately the same face toward the Sun, resulting in the present stable ratio of axial and orbital periods. *See* TIDE.

The motion of Mercury, which is more influenced than the other planets by the gravitational field of the Sun, played an important role in the development of the general theory of relativity. When all known forces that act on the planet are taken into account, the observed motion of the perihelion point of the orbit exceeds by 43 arc-seconds per century, the motion calculated on the basis of the purely Newtonian theory of gravitation. This discrepancy, discovered in the 19th century, remained unexplained until 1915, when Einstein's general theory of relativity predicted a perihelion motion in exact agreement with observation. *See* RELATIVITY.

Telescopic observation. The apparent diameter of Mercury varies from 5 arc-seconds at superior conjunction to 13 arc-seconds when closest to the Earth. Because of the proximity of the planet to the Sun, observations must be carried out during daylight. Being an inferior planet, Mercury exhibits phases, going through a complete cycle in about 4 months. Under excellent observation conditions, faint surface markings can be seen with moderately large telescopes. The albedo (reflectivity) is 10% at visual wavelengths, very similar to that of the Moon, and the color is nearly neutral except for a decrease in reflectivity in the ultraviolet. The reflectivity varies with viewing angle, being substantially higher near full phase, in a manner quantitatively very similar to the Moon's reflectivity. Thus, in spite of the difficulty of discerning surface detail, ground-based telescopic studies before the space age strongly suggested that the surface of Mercury might be similar to that of the Moon. Spectroscopic studies also showed that the planet has no appreciable atmosphere, nor was one expected in view of the high surface temperatures and low escape velocity. *See* ALBEDO; MOON.

Magnetic field. Understanding of Mercury was greatly advanced in 1974 and 1975 when the United States spacecraft *Mariner 10* made three passes close to the planet. On the first and last of these encounters, *Mariner 10* passed within 1000 km of the surface and directly investigated Mercury's near environment. An intrinsic magnetic field with a dipole moment equal to about 0.004 that of the Earth was discovered. The magnetic field is of sufficient strength to hold off the solar wind, creating a bow shock, and to accelerate charged parti-

Fig. 1. Mercury at quarter phase as seen by the approaching *Mariner 10* spacecraft in March 1974. (*Jet Propulsion Laboratory*)

cles from the solar wind in the vicinity of the planet. The magnetic field may originate in remanent magnetism of the crustal rocks that is preserved from an earlier epoch when the surface was hotter and the solar magnetic field stronger, but it is more likely due to a self-exciting dynamo in which the field is generated in a liquid iron core, drawing its energy from the planet's rotation. The details of the dynamo mechanism are poorly understood, however, for both Mercury and the Earth. *See* MAGNETOSPHERE.

Temperature. Although it is closest to the Sun, Mercury is not the hottest planet; that distinction belongs to Venus, with its atmospheric greenhouse effect. On Mercury, the maximum surface temperature varies with distance from the Sun, at perihelion reaching 700 K at the equator. At night, however, the unshielded surface cools rapidly to below 110 K, producing the greatest diurnal variations in temperature in the solar system. A meter below the surface, the temperature hardly varies from an average value of about 350 K. High-resolution infrared temperature measurements from the *Mariner 10* spacecraft demonstrated that the crust is a good thermal insulator, indicating that it consists of a porous soil or rock powder like the lunar regolith, rather than of bare rock. This is to be expected on a planet with no atmosphere whose surface is shattered and stirred by meteorite infall. *See* VENUS.

Fig. 3. A close view of Mercury from *Mariner 10* showing a heavily cratered region. Running vertically down the left is Discovery Scarp, a major compressional feature about 500 km long and up to 3 km in height. (*Jet Propulsion Laboratory*)

Fig. 2. The 1300-km Caloris Basin, the largest impact feature on the observed part of Mercury, as seen from *Mariner 10* shortly after its flyby of Mercury in March 1974. (*Jet Propulsion Laboratory*)

Geology. *Mariner 10* obtained photographic coverage of nearly half of Mercury, with 20% of the surface imaged at a resolution of 1 km or better (Figs. 1–3). Information on the surface of Mercury is thus about equivalent in resolution and coverage to that available for the Moon before the space age. Mercury is much like the Moon geologically. Its basaltic crust is apparently like the crusts of Earth, Moon, and Mars, and its composition indicates past melting and geochemical differentiation. The dominant geological structures are impact craters, which have been formed by infalling meteoritic debris and reach a saturated crater density over much of the surface. At the resolution of Fig. 1 (about 4 km), the planet is nearly indistinguishable from the Moon. It is probable, but cannot be demonstrated, that the large, greatly degraded craters are at least 4×10^9 years old. The largest impact feature on the observed part of Mercury is the Caloris Basin (Fig. 2), about 1300 km in diameter, which has a raised mountain rim (the Montes Caloris) and a lava-flooded interior marked by concentric fissures and an intricate system of cracks and ridges caused by cooling and subsidence. The basin is surrounded by relatively flat plains, apparently also flooded by ancient lava, extending all the way across Fig. 2. About 20 other basins with diameters greater than 200 km are known.

The face of Mercury shows evidence of internal (endogenic) geologic processes as well as the scars of impacts. Much of the surface appears to be flooded with lava, in the same manner as the lunar maria, probably as the result of extensive past volcanism. However, no volcanic mountains (such as those on Mars or Earth) have been identified. There are also many great compressional scarps that cut across other topographic features and may have been caused by shrinkage of the planet as its interior cooled (Fig. 3). Mercury is the only planet except Earth with such widespread compressional, as opposed to tensional, tectonic features. *See* EARTH; MARS.

Past history. Mercury was presumably formed from planetesimals in the cooling protosolar nebula at the same time as the other planets, about 4.5 × 10^9 years ago. Being closer to the Sun and therefore hotter, it did not accrete volatile substances and therefore has a relatively great abundance of iron and other metals. Since the iron is now concentrated in the core and the crust has a silicate composition, while the surface bears very ancient impact scars, it seems that differentiation must have occurred early and the crust must have cooled before the end of the period of planetary accretion. Since that time, the crust has been molded by both impacts and endogenic processes, while the liquid iron core has remained little changed except for a gradual cooling. A similar early history may have been experienced by the other terrestrial planets, but their more extensive geologic activity has since blurred the record. *See* PLANET; PLANETARY PHYSICS.

[DAVID MORRISON]

Bibliography: W. K. Hartmann, *Moons and Planets*, 1972; B. Murray, Mercury, *Sci. Amer.*, 233(3):58–68, 1975; Various papers reporting *Mariner 10* results, *Science*, 185:141–180, July 12, 1974, and *J. Geophys. Res.*, 80:2341–2514, June 10, 1975.

Meridian

That half of a great circle on Earth that passes through points having the same longitude and terminates at the North and South poles. The meridian from which longitudes are measured is the one passing through Greenwich, near London, England; it is called the prime meridian.

The celestial meridian is a great circle on the celestial sphere passing through the two celestial poles and the observer's zenith. Two branches of it are distinguished, each extending from pole to pole, the upper branch containing the zenith and the lower the nadir. The meridian passage, or culmination, of a celestial object is its crossing over the celestial meridian, upper and lower culminations referring to the upper and lower branches. Stars nearer to the celestial pole than a distance equal to the latitude of the observer are above the horizon at both upper and lower culminations, and are called circumpolar stars. *See* ASTRONOMICAL COORDINATE SYSTEMS.

[GERALD M. CLEMENCE]

Messier catalog

An early listing of nebulae and star clusters. Charles Messier (1730–1817) was primarily interested in discovering comets, so he compiled a list of objects that might be confused with comets in small telescopes. His first catalog was published in the Royal Academy of Sciences of France in 1771; later lists appeared in the *Connaissance des Temps*. Some of the later objects were discovered by Pierre Mechain. The compilation (see table) contains both open and globular star clusters, diffuse galactic nebulae, so-called planetary nebulae, and external galaxies. A star cluster in Scorpio,

The Messier catalog of nebulae and star clusters*

Messier number	NGC or (IC)	Right ascension (1980)	Declination (1980)	Apparent visual magnitude	Description
1	1952	5h 33.3m	+22°01′	8.4	Crab Nebula in Taurus; remains of SN 1054
2	7089	21h 32.4m	−0°54′	6.4	Globular cluster in Aquarius
3	5272	13h 41.2m	+28°29′	6.3	Globular cluster in Canes Venatici
4	6121	16h 22.4m	−26°28′	6.5	Globular cluster in Scorpio
5	5904	15h 17.5m	+2°10′	6.1	Globular cluster in Serpens
6	6405	17h 38.8m	−32°11′	5.5	Open cluster in Scorpio
7	6475	17h 52.7m	−34°48′	3.3	Open cluster in Scorpio
8	6523	18h 02.4m	−24°23′	5.1	Lagoon Nebula in Sagittarius
9	6333	17h 18.1m	−18°30′	8.0	Globular cluster in Ophiuchus
10	6254	16h 56.1m	−4°05′	6.7	Globular cluster in Ophiuchus
11	6705	18h 50.0m	−6°18′	6.8	Open cluster in Scutum Sobieskii
12	6218	16h 46.3m	−1°55′	6.6	Globular cluster in Ophiuchus
13	6205	16h 41.0m	+36°30′	5.9	Globular cluster in Hercules
14	6402	17h 36.6m	−3°14′	8.0	Globular cluster in Ophiuchus
15	7078	21h 28.9m	+12°05′	6.4	Globular cluster in Pegasus
16	6611	18h 17.8m	−13°47′	6.6	Open cluster with nebulosity in Serpens
17	6618	18h 19.6m	−16°11′	7.5	Swan or Omega Nebula in Sagittarius
18	6613	18h 18.7m	−17°08′	7.2	Open cluster in Sagittarius
19	6273	17h 01.4m	−26°14′	6.9	Globular cluster in Ophiuchus
20	6514	18h 01.2m	−23°02′	8.5	Trifid Nebula in Sagittarius

*From G. O. Abell, *Drama of the Universe*, Holt, Rinehart and Winston, 1978.

The Messier catalog of nebulae and star clusters *(cont.)*

Messier number	NGC or (IC)	Right ascension (1980)	Declination (1980)	Apparent visual magnitude	Description
21	6531	18h 03.4m	−22°30′	6.5	Open cluster in Sagittarius
22	6656	18h 35.2m	−23°56′	5.6	Globular cluster in Sagittarius
23	6494	17h 55.8m	−19°00′	5.9	Open cluster in Sagittarius
24	6603	18h 17.3m	−18°26′	4.6	Open cluster in Sagittarius
25	(4725)	18h 30.5m	−19°16′	6.2	Open cluster in Sagittarius
26	6694	18h 44.1m	−9°25′	9.3	Open cluster in Scutum Sobieskii
27	6853	19h 58.8m	+22°40′	8.2	"Dumbbell" planetary nebula in Vulpecula
28	6626	18h 23.2m	−24°52′	7.6	Globular cluster in Sagittarius
29	6913	20h 23.3m	+38°27′	8.0	Open cluster in Cygnus
30	7099	21h 39.2m	−23°16′	7.7	Globular cluster in Capricornus
31	224	0h 41.6m	+41°10′	3.5	Andromeda Galaxy
32	221	0h 41.6m	+40°46′	8.2	Elliptical galaxy; companion to M31
33	598	1h 32.7m	+30°33′	5.8	Spiral galaxy in Triangulum
34	1039	2h 40.7m	+42°43′	5.8	Open cluster in Perseus
35	2168	6h 07.5m	+24°21′	5.6	Open cluster in Gemini
36	1960	5h 35.0m	+34°05′	6.5	Open cluster in Auriga
37	2099	5h 51.1m	+32°33′	6.2	Open cluster in Auriga
38	1912	5h 27.3m	+35°48′	7.0	Open cluster in Auriga
39	7092	21h 31.5m	+48°21′	5.3	Open cluster in Cygnus
40		12h 21m	+59°		Close double star in Ursa Major
41	2287	6h 46.2m	−20°43′	5.0	Loose open cluster in Canis Major
42	1976	5h 34.4m	−5°24′	4	Orion Nebula
43	1982	5h 34.6m	−5°18′	9	Northeast portion of Orion Nebula
44	2632	8h 39m	+20°04′	3.9	Praesepe; open cluster in Cancer
45		3h 46.3m	+24°03′	1.6	The Pleiades; open cluster in Taurus
46	2437	7h 40.9m	−14°46′	6.6	Open cluster in Puppis
47	2422	7h 35.7m	−14°26′	5	Loose group of stars in Puppis
48	2548	8h 12.8m	−5°44′	6	"Cluster of very small stars"; identifiable
49	4472	12h 28.8m	+8°06′	8.5	Elliptical galaxy in Virgo
50	2323	7h 02.0m	−8°19′	6.3	Loose open cluster in Monoceros
51	5194	13h 29.1m	+47°18′	8.4	Whirlpool spiral galaxy in Canes Venatici
52	7654	23h 23.3m	+61°30′	8.2	Loose open cluster in Cassiopeia
53	5024	13h 12.0m	+18°16′	7.8	Globular cluster in Coma Berenices
54	6715	18h 53.8m	−30°30′	7.8	Globular cluster in Sagittarius
55	6809	19h 38.7m	−30°59′	6.2	Globular cluster in Sagittarius
56	6779	19h 15.8m	+30°08′	8.7	Globular cluster in Lyra
57	6720	18h 52.8m	+33°00′	9.0	Ring Nebula; planetary nebula in Lyra
58	4579	12h 36.7m	+11°55′	9.9	Spiral galaxy in Virgo
59	4621	12h 41.0m	+11°46′	10.0	Spiral galaxy in Virgo
60	4649	12h 42.6m	+11°40′	9.0	Elliptical galaxy in Virgo
61	4303	12h 20.8m	+4°35′	9.6	Spiral galaxy in Virgo
62	6266	16h 59.9m	−30°05′	6.6	Globular cluster in Scorpio
63	5055	13h 14.8m	+42°07′	8.9	Spiral galaxy in Canes Venatici
64	4826	12h 55.7m	+21°39′	8.5	Spiral galaxy in Coma Berenices
65	3623	11h 17.9m	+13°12′	9.4	Spiral galaxy in Leo

denoted as M7, and the Praesepe cluster, M44, have been known since classical antiquity, while the Pleiades, M45, have been known since prehistoric times. Messier's catalog included the supernova remnant, the Crab Nebula, M1, the Andromeda spiral galaxy, M31, the Triangulum spiral galaxy, M33, the giant elliptical galaxy in Virgo, M87, the globular clusters, M13 and M15, the Great Orion Nebula, M42, and the Ring Nebula, M57 or NGC 6720.

The New General Catalog (NGC) and the supplementary Index Catalogs (IC) published by J. L. E. Dreyer at the end of the 19th century were far more extensive than that of Messier. Consequently the NGC numbers are generally used for all except a few objects whose Messier numbers had become firmly established. Some Messier objects have not been identified. For example, M102 is probably a repeat observation of the great spiral galaxy M101, although Dreyer suggested it could be identified with a nearby, much fainter galaxy, NGC 5928. *See* GALAXY, EXTERNAL; NEBULA; STAR CLUSTERS.

[LAWRENCE H. ALLER]

The Messier catalog of nebulae and star clusters *(cont.)*

Messier number	NGC or (IC)	Right ascension (1980)	Declination (1980)	Apparent visual magnitude	Description
66	3627	11h 19.2m	+13°06′	9.0	Spiral galaxy in Leo; companion to M65
67	2682	8h 50.0m	+11°53′	6.1	Open cluster in Cancer
68	4590	12h 38.4m	−26°39′	8.2	Globular cluster in Hydra
69	6637	18h 30.1m	−32°23′	8.0	Globular cluster in Sagittarius
70	6681	18h 42.0m	−32°18′	8.1	Globular cluster in Sagittarius
71	6838	19h 52.8m	+18°44′	7.6	Globular cluster in Sagitta
72	6981	20h 52.3m	−12°38′	9.3	Globular cluster in Aquarius
73	6994	20h 57.8m	−12°43′	9.1	Open cluster in Aquarius
74	628	1h 35.6m	−15°41′	9.3	Spiral galaxy in Pisces
75	6864	20h 04.9m	−21°59′	8.6	Globular cluster in Sagittarius
76	650	1h 41.0m	+51°28′	11.4	Planetary nebula in Perseus
77	1068	2h 41.6m	−0°04′	8.9	Spiral galaxy in Cetus
78	2068	5h 45.7m	0°03′	8.3	Small emission nebula in Orion
79	1904	5h 23.3m	−24°32′	7.5	Globular cluster in Lepus
80	6093	16h 15.8m	−22°56′	7.5	Globular cluster in Scorpio
81	3031	9h 54.2m	+69°09′	7.0	Spiral galaxy in Ursa Major
82	3034	9h 54.4m	+69°47′	8.4	Irregular galaxy in Ursa Major
83	5236	13h 35.4m	−29°31′	7.6	Spiral galaxy in Hydra
84	4374	12h 24.1m	+13°00′	9.4	Elliptical galaxy in Virgo
85	4382	12h 24.3m	+18°18′	9.3	Elliptical galaxy in Coma Berenices
86	4406	12h 25.1m	+13°03′	9.2	Elliptical galaxy in Virgo
87	4486	12h 29.7m	+12°30′	8.7	Elliptical galaxy in Virgo
88	4501	12h 30.9m	+14°32′	9.5	Spiral galaxy in Coma Berenices
89	4552	12h 34.6m	+12°40′	10.3	Elliptical galaxy in Virgo
90	4569	12h 35.8m	+13°46′	9.6	Spiral galaxy in Virgo
91	Omitted				
92	6341	17h 16.5m	+43°10′	6.4	Globular cluster in Hercules
93	2447	7h 43.7m	−23°49′	6.5	Open cluster in Puppis
94	4736	12h 50.0m	+41°14′	8.3	Spiral galaxy in Canes Venatici
95	3351	10h 42.9m	+11°49′	9.8	Barred spiral galaxy in Leo
96	3368	10h 45.7m	+11°56′	9.3	Spiral galaxy in Leo
97	3587	11h 13.7m	+55°07′	11.1	Owl Nebula; planetary nebula in Ursa Major
98	4192	12h 12.7m	+15°01′	10.2	Spiral galaxy in Coma Berenices
99	4254	12h 17.8m	+14°32′	9.9	Spiral galaxy in Coma Berenices
100	4321	12h 21.9m	+15°56′	9.4	Spiral galaxy in Coma Berenices
101	5457	14h 02.5m	+54°27′	7.9	Spiral galaxy in Ursa Major
102	5866(?)	15h 05.9m	+55°50′	10.5	Spiral galaxy (identification as M102 in doubt)
103	581	1h 31.9m	+60°35′	6.9	Open cluster in Cassiopeia
104†	4594	12h 39.0m	−11°31′	8.3	Spiral galaxy in Virgo
105†	3379	10h 46.8m	+12°51′	9.7	Elliptical galaxy in Leo
106†	4258	12h 18.0m	+47°25′	8.4	Spiral galaxy in Canes Venatici
107†	6171	16h 31.4m	−13°01′	9.2	Globular cluster in Ophiuchus
108†	3556	11h 10.5m	+55°47′	10.5	Spiral galaxy in Ursa Major
109†	3992	11h 56.6m	+53°29′	10.0	Spiral galaxy in Ursa Major
110†	205	0h 39.2m	+41°35′	9.4	Elliptical galaxy (companion to M31)

†Not in Messier's original (1781) list; added later by others.

Bibliography: K. G. Jones, *Messier Nebulae and Star Clusters*, 1968; J. H. Mallas, and E. Kreimer, *Messier Album*, 1978.

Meteor

Observable phenomenon resulting from entry of a particle of matter, called a meteoroid, into the Earth's atmosphere at a speed between 11 and 72 km/sec. The lower limit of speed represents the velocity attained by free fall starting a great distance from the Earth. The upper limit is the combined velocity of the Earth's motion about the Sun and the velocity of a particle (at the Earth's distance) in a parabolic orbit about the Sun. This velocity is realized when the Earth and meteoroid meet head-on. There is no evidence of an appreciable number of meteoroids in hyperbolic orbits, which might lead to velocities greater than 72 km/sec. Meteoric material, therefore, originates in the solar system and is attributed to asteroidal and cometary debris.

The mass of individual particles extends from

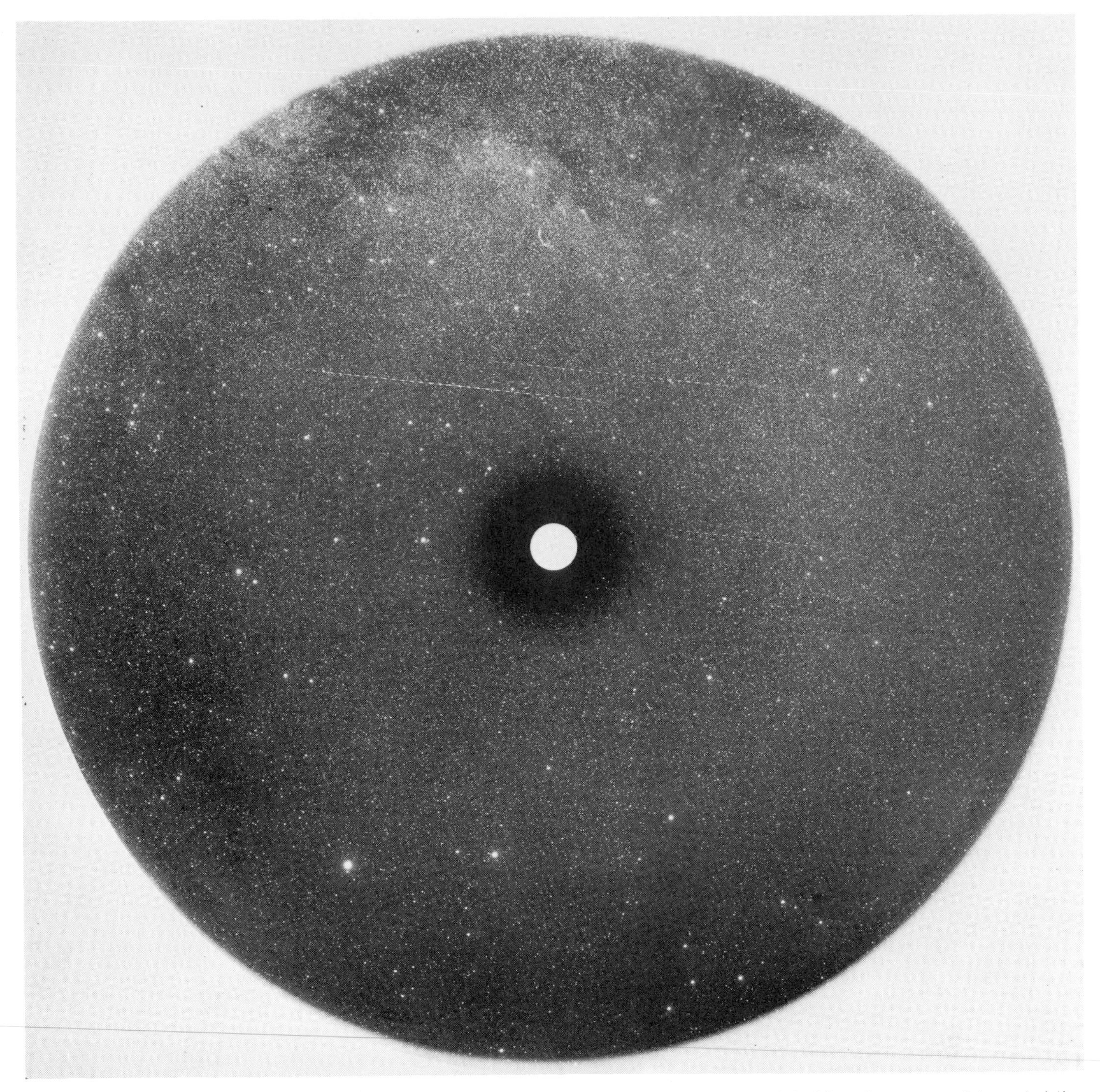

Photographic trails of three meteors of the Perseid shower, extended back to the left, converge toward a common radiant. A rotating focal-plane shutter in the Baker Super-Schmidt meteor camera interrupted the trails each 1/60 sec. Hub of shutter caused unexposed center. (*Harvard University Meteor Project*)

micron-size dust specks to complete asteroids with weights in the thousands of tons. These extreme cases represent special fields of study. *See* METEORITE.

Meteoroids become visible at altitudes of from 80 to 120 km, depending on their velocities. At these altitudes the energy transferred to meteoroids by collisions with atmospheric atoms and molecules becomes sufficient to vaporize the solids. Further collisional interactions of the vaporization products with the air produce luminosity and ionization. The detectable portions of trajectories, during which the meteoroids are luminous, are called trails. These phenomena may be observed by visual, photographic, telescopic, or radar techniques.

Trajectories and radiants. When still some distance from the Earth, meteors travel in elliptical or parabolic orbits about the Sun. As they approach the Earth, their orbits are perturbed by the Earth's gravitational field until the orbit is best described as a hyperbola about the Earth's center. However, the arc of the hyperbola for the few tens of kilometers of trail may almost always be described as a straight line. Only in cases of extremely long and slow meteors is curvature of the trail detectable by the most accurate photographic and measuring procedures. *See* CELESTIAL MECHANICS.

This straight-line trajectory is a great circle as seen projected on the celestial sphere. If the same meteor is observed from several different locations, the extension, backward along the trail, of the great circle observed at each station intersects the same point on the celestial sphere. This is the meteor's radiant point and is determined solely by the direction of motion of the meteoroid. In the same way, if a number of meteors traveling parallel to one another are observed from a single location, the backward extensions of their trails intersect at the common radiant point. Meteors with parallel motion and identical velocities must have identical orbits and, very likely, identical origins. The groups of meteors displaying these characteristics are called meteor showers (see illustration).

Showers and comets. On the average, a visual observer may expect to see about five meteors per hour. At specific times of the year, however, this rate may increase greatly when the Earth's orbit intersects the orbit of a stream of meteors. Each year on Aug. 12 in the early morning an observer in the Northern Hemisphere can see as many as 20 meteors per hour. Most of these have a common radiant point in the constellation of Perseus at right ascension (RA) 46° and declination (Dec) +58°. These meteors, named Perseids after the constellation from which they appear to emanate, have nearly identical velocities of 60 km/sec before they are retarded by the atmosphere.

Again on Dec. 13 at a few hours past midnight a comparable number of meteors may be seen with a radiant at RA 113° and Dec +32° in Gemini. These are Geminids; their velocity is 36.5 km/sec (see table).

The times and dates listed are for the maximum rate. The showers may be seen at a lesser rate on several days preceding and following the date of maximum, and at other hours of the night when the radiant is above the horizon. The width of the stream, in the direction of the Earth's orbit, must be several tens of millions of miles for the meteors to occur over several days. Also, because the showers repeat regularly each year and because the

Meteor shower and cometary associations

Shower	Date of max., universal time	Extreme limits	Radiant, degrees RA	Radiant, degrees Dec	Velocity, km/sec	Strength*	Parent comet
Quadrantids	Jan. 3	Jan. 1 4	230	+48	43	M	
Coma Berenicieds	Jan. 17	Jan. 13 23	186	+20	65	W	1913I
Virginids	Mar. 13	Mar. 5 21	183		31	W	
Lyrids	Apr. 21	Apr. 20 23	270		48	M	1861I
η-Aquarids	May 4	May 2 6	336	+0	64	M	Halley
Daytime Arietids	June 8	May 29 June 18	44	+23	39	S	
Daytime ζ-Perseids	June 9	June 1 16	62	+23	29	S	
Daytime β-Taurids	June 30	June 24 6	86	+19	32	S	Encke
δ-Aquarids	July 30	July 14 Aug. 19	339	−11	43	M	
ι-Aquarids		July 16 Aug. 25	331	−5	31	W	
α-Capricornids	Aug. 1	July 17 Aug. 21	309	−10	26	W	1948n
Perseids	Aug. 12	July 29 Aug. 17	46	+58	60	S	1862III
Draconids (Giacobinids)	Oct. 10	Oct. 10	264	+54	23	Periodic	Giacobini-Zinner
Orionids	Oct. 22	Oct. 18 26	94	+16	66	S	Halley
Leo Minorids	Oct. 24	Oct. 22 24	162	+37	65	W	1739
Taurids	Nov. 1	Sept. 15 Dec. 15	51	+17	30	M	Encke
μ-Pegasids	Nov. 11	Nov. 11	335	+21	16	W	1819IV
Leonids	Nov. 17	Nov. 14 20	152	+22	72	M	1866I
Geminids	Dec. 14	Dec. 7 15	113	+32	36	S	

*Shower strength (W, weak; M, medium; S, strong) is a relative measure of the number of meteors that may be observed visually, except for those daylight showers which are observed by radar.

orbital period of the meteors is neither exactly 1 year nor even commensurable with the Earth's period, the meteoric material must be distributed rather uniformly all about the stream's orbit.

There are sometimes condensations of meteoric material within an orbit. This was amply demonstrated in 1833 and 1866 when it was reported that the Leonids, ordinarily a shower of moderate activity, fell as thickly as snowflakes. Estimations of the maximum hourly rate in 1833 were 35,000. The average separation between particles in the stream at this time was about 20 km. The display of the Giacobinids on Oct. 10, 1946, yielded visual counts of 1000/hr at maximum activity. In most years this shower is nonexistent.

In 1866 G. V. Schiaparelli showed that the orbit of the Perseids corresponded with that of the Comet 1862III. Shortly after that, three other comet-meteor stream associations were made, and the hypothesis that meteoric material represents debris from existing comets was firmly established. At present one can give 10 cases where an association assuredly exists (see table). Nearly half the known meteor streams are included. The presumption is strong that the remaining showers also arose from comets which have either completely disintegrated or have been perturbed by planets into orbits too distant from the Sun for observation.

Meteoric material is expelled from the comet along with gaseous material. The velocity of departure from the comet nucleus is small enough to maintain a near identity of comet and meteoroid orbits, but large enough to assure that, in time, individual particles will be distributed rather completely about the comet's orbit. Effects of solar radiation are also important in perturbing the smallest particles into different orbits. *See* COMET; ZODIACAL LIGHT.

Occasional large observed rates for Leonids and Giacobinids represent examples of collisions with meteoroid streams that have not had sufficient time to disperse. The Giacobinids, debris from the Giacobini-Zinner Comet, are important only if the Earth crosses the comet's orbit at about the time the comet is near that position of its orbit. The 1946 shower occurred 15 days after the comet made its closest approach to the Earth's orbit. In 1939, although the approach to the Earth's orbit was closer, the distance from the comet to the Earth was considerably greater (136 days late) and no shower was observed.

Observations. Meteors are observed to determine their brightness, velocities, and trajectories in the atmosphere and, through these latter two, their orbits about the Sun. To determine the trajectory, one must introduce a distance in the measures, either with a range measurement by radar or by triangulation between two stations observing the same meteor.

Visual observations. The apparent angular velocity of meteors ranges upward to tens of degrees per second. The eye is an exceedingly effective mechanism for detecting fast-moving objects; thus visual observations of meteors have been made extensively, notwithstanding their low accuracy. A skilled observer can plot the observed great circle motion of a meteor with an error of about 1°. Two observers at different stations can often determine the radiant with an error of only a few degrees, a sufficient accuracy for some types of analysis. However, the not surprising inabilities to estimate with accuracy the total angular path of the meteor and the time to traverse this path limit the value of visual observations in the determination of velocities.

Perhaps the most impressive results of visual observations are those by completely unskilled observers who, by chance, sight extremely bright meteors (commonly called fireballs). Many such observations of one fireball are often reported and enable an investigator familiar with the common errors of observations to estimate a space trajectory with sufficient accuracy to initiate a successful search for an unconsumed portion, the meteorite, that reached the ground.

Only with the best photographic techniques is it possible to record meteors fainter than can be seen by the naked eye. A few such photographs exist. Some telescopic data on fainter meteors have been obtained by a few excellent observers. However, the small field of view of a telescope limits the number of observations, and the high apparent speed with which the meteor traverses the field reduces the accuracy of the positional measures.

Photographic observations. Because neither time of appearance nor location of a meteor can be predicted, meteor photography is carried out on a patrol basis; that is, cameras are continuously open, and success is achieved when a sufficiently bright meteor chances to appear in the field. Usually two similar cameras, separated by a baseline of 20–80 km, are operated simultaneously. These are directed at the same region in the atmosphere at a height of about 90 km. Both cameras photograph the same meteor, and the two photographs supply the necessary trajectory data. To determine velocities, a time scale is introduced in the meteor trail. The most common device is a rotating shutter which alternately exposes and occults the objective or the focal plane at known intervals. Interruption rates from 20 to 60 per second are desirable.

Cameras suitable for meteor photography possess (1) a wide field of view to patrol a large region of the sky, (2) a large aperture to gather as much light as possible at any instant during the meteor's flight, and (3) a short focal length to decrease the angular rate of the image and, consequently, to concentrate the light on a small area of the photographic emulsion.

The most successful instrument in these respects is the Baker Super-Schmidt meteor camera. Its focal length is 8 in., and its effective aperture is approximately 10 in. With fast blue-sensitive emulsions, slow meteors of fifth magnitude and fast meteors of third magnitude can be photographed. The field of view of the camera is 57°. Approximately four meteors are photographed during each hour of exposure.

Radar observations. Radar observations of meteors depend on the reflection, or backscatter, of the radar pulse from the ionized column of gases formed as the meteor passes through the upper atmosphere. The echo reaches a pronounced maximum when the meteor trail is at right angles to the radar beam. The distance to this point is determined by the time lapse between the transmitted and received pulses. If a continuous-wave transmission is used, the pattern of the reflected signal shows oscillations as a result of the diffraction produced by the growing line source. A measurement of the rate of oscillation, together with a measurement of radar range, is sufficient to determine the velocity.

Radiants of meteor showers are determined by a series of observations with the radar directed toward the horizon. When the detection rate reaches a maximum, the meteor trajectories are at right angles to the beam. The center of the beam then represents the pole of the great circle containing the radiant; the radiant is 90° from the beam. Observations at differing azimuths yield two great circles, the intersection of which must be the radiant.

Spectroscopic observations. The first photographic observation of a meteor spectrum was made in 1897; since then only a few hundred more have been obtained. Bright meteors are required; a single camera equipped with a transmission grating or prism may acquire less than 10 meteor spectra per year.

Spectra show that the luminosity arises primarily from emission lines of neutral atoms and once-ionized atoms. The predominant lines of ionized calcium and much of the light produced by iron atoms account for the blueness of all meteors. The color index of photographic meteors, on the average, is a -1.8 magnitude.

The only molecular bands detected in meteor spectra arise from the first and second positive groups of atmospheric nitrogen. A comparison of elements responsible for the spectra and the elements known to occur in meteorites proves that most photographic meteors are stone rather than iron. Only a few spectra exist that may be attributed to iron.

The highest observed excitation potentials (upper states) in meteor spectra are due to hydrogen and ionized silicon. These are between 12 and 13 ev. The degree of excitation and ionization increases as the meteor progresses into the atmosphere. High-velocity meteors show greater excitation and ionization than those of low velocity. Many bright meteors display sudden and short-lived increases in brightness caused by the detachment of small droplets or particles from the meteoroid. In these bursts the degree of excitation always decreases.

Observations from spacecraft. The relative speed between a spacecraft and a meteoroid can exceed 10 km/sec. A chance collision with even a small meteoric body, of the order of 10^{-2} g, can produce intolerable damage. To determine the protection required by a spacecraft and its contents, a number of United States satellites have been instrumented to determine the penetration rate in metal sheets a few thousandths of an inch thick. The metal is punctured by meteoroids of mass about 10^{-9} g and larger, and each square meter of surface undergoes several hundred impacts each year by meteoroids of mass greater than 10^{-10} g.

While the early satellite experiments did little more than supply counts of meteoroids, one can foresee future experiments that will provide information on velocities and orbits and, most important, composition of the meteoric material.

Masses, magnitudes, and numbers. The brightness of a meteor is generally described in terms of its apparent stellar magnitude at maximum light. The number of visual meteors increases with decreasing brightness, although the observed rate of meteors near the visual limit is low because the sensitivity of the eye is not uniform over the field of vision. It is estimated that an experienced observer sees fewer than 10% of the meteors in the field between the fifth and zero magnitude and essentially all meteors brighter than zero magnitude.

Visual and photographic observations show that the number of meteors between magnitudes M and $M+1$ is about 3.5 times the number between magnitudes $M-1$ and M. About 10^8 meteors of the fifth and fainter magnitudes strike the Earth daily.

Although some uncertainty is still present in the mass determination of meteors, an order of magnitude estimate may be given. A 30-g meteroid with velocity of 30 km/sec will produce a zero-magnitude meteor.

The total mass of meteors, including those fainter than the visual limit, may be obtained if one cares to risk an extrapolation of the number-magnitude law to faint meteors. Such an estimate gives an incoming mass of several thousand tons per day. Future data from satellites on their rate of impact with small particles should improve this estimate.

Wakes and trains. Meteors are primarily a moving point source of light; that is, most of the luminosity occurs only within a small region surrounding the moving meteoroid. However, there is some additional light outside this region. One example is the wake, a short appendage behind the meteor proper. The wake appears on a photograph as luminosity which partially fills in the breaks of the trail introduced by the rotating shutter. Spectra of the wake show the material to be compatible with the meteor itself, although with the lower excitation expected of a smaller particle.

The wake may be caused either by meteoric gases or by small fragments which have been retarded with respect to the parent body. Existing observations are insufficient to assign the responsibility to one mechanism or the other. Quite probably both play a role.

The second special case of meteor luminosity is the meteor train. This is a relatively long-persisting luminosity which may be seen for some seconds, or more rarely, some minutes after the meteoroid has passed. It is a streak of light that marks the path of the meteor. The phenomenon is most often produced by high-velocity meteors.

Spectra of meteor trains are exceedingly rare; most information concerning them is derived from spectrographic observations with a camera that is equipped with a rotating shutter. If the brightness of a given spectral line is nearly constant in both the exposed portion of the meteor trail and the portion occulted by the shutter, it must result from luminosity that persisted through many shutter rotations.

P. M. Millman first obtained examples of such persistent luminosity; the lines were attributed to magnesium, calcium, iron, and, probably, sodium.

More recently a line at 5577 A was photographed in a number of train spectra. This radiation arises from a metastable state of the oxygen atom and is also responsible for green auroras. The mechanism of train production, other than by oxygen, is not yet understood. The half-life of the auroral oxygen line is about 1 sec and this may solve the problem for the short-duration trains. There is some evidence that this type of train is produced more readily when the 5577-A radiation from the sky as a whole is pronounced.

Trains persisting for some seconds or more frequently depart from linearity as the upper atmospheric winds blow them about. Time-lapse photography of trains is employed and to determine the winds at altitudes of between 80 and 100 km.

Nature of meteoric matter. Information on the composition and structure of meteoric material is derived directly from the analysis of meteorites or indirectly from observations of meteor phenomena in the atmosphere. The first source, in which the investigator may actually handle the body he studies, is unique in the field of astronomy, and would appear to reduce the problem to that of finding a sufficient number of meteorites for statistical study. However, in addition to stony and iron meteorites, at least one other class of meteoric material exists. It is represented in essentially all of the ordinary visual meteors.

Apart from the elements they contain, visual meteors may bear no similarity to meteorites. There is good evidence that most meteroids fragment as they pass through the atmosphere, and some evidence that nearly complete disruption of certain meteoroids occurs whenever the dynamic pressures on the meteoroids exceed 0.5 psi. The general fragmentation may come about by molten droplets blowing off a substantial meteoroid, but the destruction of an entire meteoroid by so small a pressure, and not by thermal effects, is indicative of extreme fragility.

Other observations indicate that meteoroids may have an overall density much less than that of the stone or iron meteorites (meteoroids as they finally reach the ground). The density of meteoroids may possibly be less than that of water. It is presumed that the low density and extreme fragility are due to porosity of the meteoric material.

On the basis of F. L. Whipple's icy-conglomerate model of comets, one would expect porosity which results from the evaporation of the volatile ices from between the solid structure.

Artificial meteors. A new approach to meteor physics was initiated on Oct. 16, 1957, when three artificial meteors were flown to an altitude of 80 km in a rocket over New Mexico. The meteors were formed by high-velocity ejection of the liners of shaped charges, an explosive device designed to direct the energy of the explosion along a specific line.

Two of the charges were successfully fired, and one yielded some information. The fraction of a gram of micron-sized particles of aluminum ejected by this charge reached a velocity of 15 km/sec and was clearly recorded on photographic films as a trail about 1000 m long. By the end of their visible trail the particles were essentially completely vaporized and the remains brought to a stop by atmospheric drag. More sophisticated artificial iron meteoroids of known mass have been produced by accelerating the shaped charge with rockets fired back into the atmosphere. The velocity of the meteoroid, which is the sum of the shaped charge velocity and that of the rockets, can exceed 16 km/sec. Observations of these events by the standard photographic techniques have given an important calibration of the mass of natural meteors. [RICHARD E. MC CROSKY]

Bibliography: J. A. McDonnell (ed.), *Cosmic Dust*, 1978; J. A. Wood, *The Solar System*, 1979.

Meteorite

A natural, nonterrestrial object which survives passage through the Earth's atmosphere and arrives as a solid at the Earth's surface. As the meteorite passes through the atmosphere, it may be viewed as a falling star, with all of the optical brilliance and auditory effects characteristic of meteors. However, only an extremely small fraction of the meteors survive their atmospheric entry. Hence, meteorites represent rare events, but meteors can be seen on almost any clear night if moonlight or the lights of a city do not overilluminate the nighttime sky. *See* METEOR.

Meteorites consist mostly of metal and silicate material. They can be classified as irons, stony irons, or stones, depending on the relative proportions of the iron and the silicates. There are only about 2500 meteorites known. These do not include the tiny micrometeorites that have been found in oceanic sediments and in high-altitude dust. Micrometeorites include both interplanetary dust and some of the numerous bits of debris that are generated when larger objects pass through the atmosphere.

Meteorites represent samples of the primitive interplanetary rubble that orbits the Sun but did not aggregrate into a planetary-sized object. This rubble consists of bodies that range from hundreds of kilometers in diameter to tiny particles of interplanetary dust. Collisions between these bodies are frequent, particularly in the thickly populated region between the orbits of Mars and Jupiter. Meteoroids are generally believed to be fragments that have been thrown into Earth-crossing orbits by these collisions. Such fragments move in orbits that also have a relatively high probability of colliding with one of the other planets or their satellites. Analyses of the meteorites that strike the Earth indicate that most were fragmented about 5×10^6 years before their arrival.

The larger bodies that are located between the orbits of Mars and Jupiter are called asteroids. More than 1600 asteroids are known, and they appear to have the same range of densities and chemical compositions that are observed in meteorites. *See* ASTEROID.

Comets are another possible source of meteorites, but these bodies consist primarily of very volatile, low-density material that would not survive passage through the Earth's atmosphere. Cometary debris is probably the source of many meteors, including the more spectacular meteor showers, but there is no compelling evidence that comets are the source of any known meteorites.

See COMET; SOLAR SYSTEM.

It is estimated that the Earth acquires over 100,000 metric tons of meteoritic material each year, but only a small fraction of this tonnage consists of meteorites weighing more than a few grams. Although it is estimated that the Earth's upper atmosphere is penetrated each decade by thousands of 100-ton meteoroids, more than a hundred 1000-ton meteoroids, about ten 15,000-ton meteoroids, and one 100,000-ton meteoroid, most of these large bodies are destroyed in the atmosphere. Thus, the Earth's yearly influx of meteorite material is primarily in the form of particles of dust-sized grains.

In all major countries there are laboratories which devote a large part of their efforts to the investigation of meteorites. Scientifically, the single most important aspect of meteorites is their extreme antiquity. Over 20 years ago, Clair Patterson showed that meteorites are 4.55×10^9 years old. Many have not undergone the secondary reprocessing experienced by terrestrial and lunar material. Meteorites, therefore, provide the only available record of events that occurred during that period of time.

More recent studies have shown that many of the minerals which occur in meteorites were formed almost immediately after the isotopes of the chemical elements were produced. The time period between the end of element synthesis and the condensation of meteorite minerals was so short that several short-lived radioactive isotopes were trapped as the minerals formed. Even the isotopes of individual elements were not well mixed. In one mineral, an element may be enriched in a particular isotope. In another mineral, that same isotope may be underabundant.

In the 1950s astrophysicists predicted that different isotopes of the same element would be produced in different regions and at different stages during a star's evolution. There is little doubt that meteorites contain material that can best be described as condensed stellar debris, that is, stardust. The heterogeneous distributions of elements and their isotopes in meteorites appear to confirm the general astrophysical model of stellar synthesis of the chemical elements.

After the following review of other properties of meteorites, a brief summary will be given of the information which meteorite analyses yield for the early history of the solar system, the formation of the Earth, the origin of life, and the origin of the chemical elements.

Historical evidence. A falling meteoroid may be accompanied by a fireball and luminous trail that are visible over great distances, by sounds like thunder, and by explosions if the meteoroid disintegrates in flight. It is no surprise that the falling of a stone from heaven, heralded by such an awesome display, was once viewed as a message from the gods. The arrival of an iron meteorite, particularly in a region where iron smelting was unknown, or a stone with surfaces blackened from its encounter with the atmosphere could hardly fail to excite the imaginations of people.

Recorded history contains many references to the falling of stones from the sky, but the records are often too vague or too embellished by legend to be considered as evidence of a meteorite fall. Ancient Chinese, Greek, and Roman writings contain several accounts of such events, and there is a Biblical story of a battle in which humans were killed by stones from heaven.

The locations where meteorites have been found provide evidence that some meteorites were cult objects. The most nickel-rich meteorite known (62% Ni) was found in an Indian burial mound in Oktibbeha County, MS. The Casa Grande meteorite from Chihuahua, Mexico, was found wrapped in linen cloth and buried in the middle of a room in a temple. This iron meteorite was a new kind of substance to the Montezuma Indians, and they obviously regarded it as a special object. The Black Rock of Mecca, which is built into the Kaaba at Mecca, is also regarded as a probable meteorite. This stone is an object of great respect for many Muslims.

There is also ample evidence that iron meteorites played an important historical role in the making of tools. Analyses on an iron-bladed copper ax from an ancient tomb of the Shang Dynasty (circa 1400 B.C.) indicate that this was forged from an iron meteorite. From about 400 B.C. to A.D. 300, the populace in some areas used meteorites to make metallic coins.

The 127-kg stone which fell at Ensisheim, France, in 1492, represents the oldest witnessed meteorite fall that has been preserved. The main body of this meteorite is still kept in the town hall of Ensisheim.

With the development of modern science, the stories of stones falling from the sky were in disrepute among the enlightened. President Thomas Jefferson was openly critical of the suggestion that the stone which fell at Weston, CT, in 1807 was extraterrestrial. When a stone fell at Luce, France, in 1768, Father Barchley reported the event to the Royal Academy of Science in Paris, and the Academy appointed a commission to investigate this fall. A. L. Lavoisier was on that commission. After some study, the Academy reported that the Luce stone was a terrestrial rock. Another stone fell in Barbotan, France, on July 24, 1790; another at Siena, Italy, on June 16, 1794; and another at Wold Cottage, England, on December 13, 1795. Finally, a shower which consisted of thousands of stones fell at L'Aigle, France, on April 26, 1803. This event changed the thinking of the Royal Academy, and the new commission that was appointed reported that these stones did indeed come from the sky.

Meteorite falls. The path of a meteoroid, before its encounter with the Earth, is controlled by the gravitational forces of the Sun. If the meteoroid closely approaches the Earth, it will be attracted by the Earth's gravitational field and accelerated toward an impact. The object must pass through the atmosphere before reaching the Earth's surface, and this encounter tends to destroy all but the very large and the very small. It also selectively destroys structurally weak objects.

The atmosphere thus acts as a filter. This filtering action does not, however, act in the same manner on micrometeorites. The collection of micrometeorites from the stratosphere and from deep-sea sediments seems to confirm the suspicion that the "macro" meteorites, which survive atmospheric entry, are strongly biased in favor of those that are structurally strong. Micrometeorites, such as the one shown in Fig. 1, are highly enriched in materials whose composition is similar to that of carbonaceous chondrites, although such volatile-rich meteorites amount to less than 5% of the objects that have been classified as meteorites.

Most meteoroids are stony masses that lack the durability to survive passage through the atmosphere. These objects usually have internal fractures and tend to break along preexisting planes of weakness during their fall. Their external surfaces, prior to their entry into the atmosphere, are angular and rough, but after entering the atmosphere these surface features are ablated. Thus, recovered meteorites generally have rounded forms.

Meteoroids enter the atmosphere with speeds of about 10 to 40 km per second. Although initially cold, their surfaces are heated as the objects pass through the air. A region of compressed, ionized air collects at the front surface and causes surface melting, aerodynamic shaping, a luminous fireball glow, and often explosive fragmentation. The meteroid may first appear as a fireball at a height of about 130 km. As the meteoroid is slowed by denser air, the fireball becomes smaller and usually disappears at a height of about 20 km.

A meteoroid usually tumbles in flight so that all sides are exposed to the disruptive forces generated by highly compressed gases in front of it. A few meteorites maintain one orientation in flight. These may escape fragmentation and be shaped by ablation into a blunt nose cone. However, most stone meteoroids break and scatter pieces along the line of flight. Usually the smaller pieces do not travel as far along the flight path as the larger ones. Iron meteoroids are less subject to break up in flight, probably because of their greater structural strength.

The heated zone of these falling bodies is thin, so the heat is quickly lost. The surfaces of meteoroids are covered by a paper-thin layer of molten material. This layer is thinner on the leading face and on the forward parts of the side faces, but on the trailing surface a thicker layer of the fusion crust accumulates (Fig. 2). On stone meteorites, the fusion crust of the forward face is thin and contains small bubbles, because this surface is under considerable air pressure. On the rear face, more of the slag accumulates and the gas bubbles are much larger, because immediately behind a falling object there is an area of low air pressure, almost a vacuum.

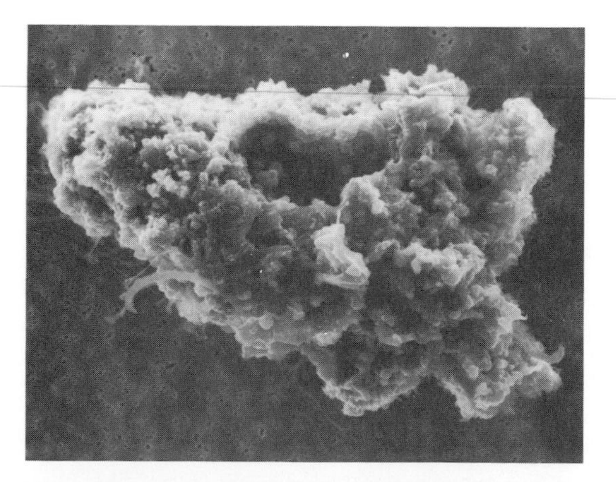

Fig. 1. Micrometeorite collected in the stratosphere by filters mounted on a high-altitude aircraft. The proportion of volatile-rich meteorites is much higher in these micrometeorites than in the "macro" meteorites. *(Courtesy of Dr. D. E. Brownlee)*

Fig. 2. Amherst (No. 2), Neb., meteorite showing dome-shaped forward face with flight markings which radiate from center. Deflection of flow lines around pits indicates pits complicated airflow over this leading face.

When a metallic body falls through the air, the general conditions are essentially the same as for a stony mass, but enough heat penetrates the metal to modify the internal structure. Thus, when an iron meteorite is cut, polished, and etched, the depth of the thermally treated metal can be seen.

The fusion crust on stony meteorites generally is black, because the most common minerals in them contain considerable iron. However, there are a few stony meteorites in which the minerals are essentially free of iron, so on these the crust will have just a tinge of brownish color.

Although a falling meteoroid may appear to be hot, the object itself is not hot, but the air surrounding it is. Two interesting situations illustrate the low temperature of meteorites when they impact the Earth. On July 14, 1860, at Dharmasala, India, and on July 14, 1917, at Colby, WI, stones fell, and both were quickly coated with frost. The humidity at those places was high, and the objects that fell were cold enough to condense the moisture from the air and form frosty coats.

Recovery. One of the most important developments for the study of meteorites was the 1969 discovery of large numbers of meteorites in the region of the Yamato Mountains, Antarctica. These meteorites were found lying on "blue ice" areas where the snow cover had been removed by the wind. Over 1000 specimens have been collected by subsequent expeditions, and a joint American-Japanese program has been established for future expeditions. The United States collection is under the supervision of the Polar Programs Division of the National Science Foundation, and the specimens are being stored at the Johnson Space Center in Houston.

The discovery of the Antarctic meteorites is important because the specimens have not suffered the same collection bias of most meteorite finds. It is also significant that the meteorites have been preserved in a cold, relatively clean environment since impact.

To recover newly fallen meteorites, the searcher should get reports from witnesses, preferably from those who have seen the fireball from different directions. The searcher should evaluate a number of accounts to establish the terminal point of the fireball. Reports from people who saw the fireball from a moving car or who were in an open space lacking reference objects while viewing the sky generally will provide the least reliable information. Such people rarely can relocate the exact spot where they sighted the fireball; consequently, their directions are likely to be inaccurate. When a number of accounts indicate that the end point is confined to a limited area, the searcher can go to that place and check roofs for punctures and ask occupants if they heard sounds indicating that something had fallen on their roofs. Local people and the weather bureau can be asked if there were noticeable winds at the time of the fireball, because as a relatively small body approaches the end of its flight, it loses most of its momentum and a crosswind can blow it off course. A field search requires numerous traverses across the estimated end point of the fireball. The searcher must look for broken limbs on trees, small patches of freshly disturbed ground, and any object that resembles a meteorite.

Witnessed falls of meteorites depend on the presence of an observer and must therefore be related to the population density. However, most of the known meteorites are classified as "finds," indicating that they were discovered without any witnesses to the fall. One very important factor in the discovery of a meteorite is the use that is made of the land. A disproportionately large fraction of the meteorites are found in areas where the land is cultivated. There are two factors which might explain this observation: the number of rocks per unit area of cultivated land is small, and the plow is an efficient instrument for locating meteorites.

For meteorites recovered from witnessed falls, there are about 28 stony meteorites for each iron meteorite, but the proportion of iron meteorites is much higher among meteorite finds. In the United States there are almost twice as many iron meteorites as stony meteorites among meteorite finds. This is usually attributed to the more rapid weathering of stony meteorites and to the relative ease with which an iron meteorite is recognized as being different from local rocks.

Craters. Air resistance virtually stops a falling body of moderate size at a height of about 20 km from the Earth's surface, but massive bodies may impact its surface at high speeds. It has been estimated that the terrestrial influx rate of 100-metric ton meteoroids averages one per day. Bodies of this size are not appreciably slowed by air resistance, but most meteoroids are stone, and these tend to fragment before descending to an altitude of 20 km.

The impact of a large meteorite, one weighing more than a few hundred tons, releases so much energy that the meteorite explodes. This produces a depression called a meteorite crater. There was little interest in terrestrial meteorite craters prior to the observation that impact craters are responsible for many of the surface features of the Moon and Mars. The results of the respective space probes have aroused interest in the role that meteorite impacts play in shaping the surface features of the Earth. There are now 13 terrestrial craters whose meteoritic origins have been proved by the recovery of meteoritic fragments. One of these is shown in Fig. 3. There are over 80 probable meteorite craters in which shock features of the rocks suggest a definite impact origin.

A meteorite which forms a crater is never buried in the crater. The greater part of the impacting mass, if it is not atomized, is thrown out of the crater almost immediately by the rebound of the impacted rock. The rejected fragments generally are hurled radially outward from the rim of the crater

Fig. 3. Canyon Diablo, Arizona, meteorite crater. The rocks immediately surrounding this impact crater are turned upward. Meteorites from this crater were found on the plains around the depression.

along straight lines. These lines are called rays. Rim rocks surrounding an impact crater are upturned, and frequently parts of the rim are flipped over. In time, a crater rim may be eroded away; consequently, old craters may not have prominent rims. An object which hits the Earth does not have to be the size of the crater that it forms, because much of the country rock around the impact point is broken to bits and thrown from the crater.

Composition. Meteorites contain the same elements that are found in terrestrial rocks, and these elements combine to form many of the same minerals that are found in terrestrial rocks. The chemical compositions of some stony meteorites, called achondrites, are very similar to those of basic igneous rocks.

The minerals in some meteorites, such as the achondrites, represent the products of chemical differentiation and recrystallization. However, the rocks and minerals in other meteorites are aggregates of grains that formed in radically different environments. A meteorite of one classification may contain large fragments of material which fit the classification scheme of another class of meteorites. Some stony meteorites contain compacted, fine-grain material that earlier resided as soil on the surface of an asteroid-sized body. Some iron meteorites contain silicate inclusions that are clearly chondritic in composition. These inclusions and their metallic host were produced under different conditions and then mechanically mixed together, but the chemical elements in the two phases never equilibrated. Even on a microscopic scale, many meteorites appear to be a happenstance collection of unrelated particles, and analyses have shown that the same element may have different isotopic compositions in different grains.

The number of minerals found in meteorites has grown since 1885, when Gustav Tschermark listed 16 different meteoritic minerals. Today, with the help of modern instruments, over 105 minerals have been identified, and 25 of these occur only in meteorites (Table 1). Many of the minerals listed there are trace constituents. The common minerals of meteorites are olivine, pyroxenes, and metallic nickel-iron. Iron, silicon, magnesium, and oxygen are the major elements of these minerals, and these four elements account for over 90% of the atoms in practically all meteorites.

Types. There are three major classes of meteorites: stony, intermediate, and iron.

I. Stony meteorites.

A. Chondrites, the most abundant of the stony meteorites, are characterized by small, spherical silicate particles, which are referred to as chondrules (Fig. 4). Chondrules are glassy and lack volatiles. In chondrites, the chondrules are bonded by a fine-grained, volatile-rich matrix. Chondrules formed as molten droplets in space, either by direct condensation from the primitive nebula or by secondary melting of matrix material. They escaped recrystallization, but by some mechanical process were transported and later accumulated before being bonded together. Thus, chondrites were probably produced by the original process that formed planetary solids from the solar nebula. Consequently, this type of meteorite is important scientifically.

There are two groups of chondrites: equilibrated and unequilibrated. Equilibrated chondrites are the most abundant, and in these, the silicates within the chondrules and in the interstitial matrix are essentially the same. Their chondrules are usually small and so well bonded that their outlines are not well defined. The major silicates of the unequilibrated group have variable compositions, and the aggregates are so friable that they tend to crumble. The chondrules of this group occur in well-defined, rounded forms, but the matrix is fine-grained (an example is the Chainpur, India, chondrite). There are five types of chondrites:

1. *Enstatite chondrites* are usually nearly pure $MgSiO_3$. Metal inclusions may be abundant and usually are low in nickel. These meteorites are highly reduced and often contain silicon.
2. *Olivine-bronzite chondrites* are more abundant than the enstatite chondrites and contain about equal amounts of olivine and bronzite.
3. *Olivine-hypersthene chondrites* are the most abundant chondritic meteorites and generally contain more olivine than pyroxene. This hypersthene contains 12–20% iron, hence these meteorites have a darker color than the bronzite chondrites. Their metal grains usually contain 7–12% nickel.
4. *Olivine-pigeonites* differ from the olivine-hypersthene type by having pigeonite rather than hypersthene. Olivine is the predominant mineral. The metal is usually rich in nickel; and troilite, an iron sulfide common in other types of chondrites, may be replaced by pentlandite.

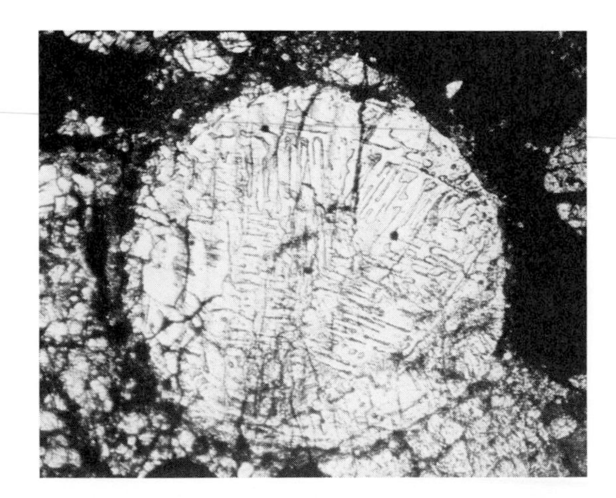

Fig. 4. Ioka, Utah, meteorite: a cross section through an olivine chondrule in transmitted light. Lighter bands are olivine and the darker are glass. Dark area outside this chondrule is due to troilite or nickel-iron.

Table 1. Minerals found in meteorites

Mineral	Formula or symbol
Native elements, metallic compounds	
*Awaruite	$Ni_{13}Fe$
Carbon	C
Chaoite	C
*Lonsdalite	C
Copper	Cu
Diamond	C
Gold	Au
Sulfur	S
*Kamacite	α-(Fe,Ni)
*Taenite	γ-(Fe,Ni)
Carbides, phosphides, nitrides, silicides	
Cohenite	$(Fe,Ni)_3C$
*Carlsbergite	CrN
*Haxonite	$Fe_{23}C$
Moissanite	γ-SiC
*Osbornite	TiN
*Perryite	$(Ni,Fe)_5(Si,P)_2$
*Rhabdite	$(Ni,Fe)_3P$
Schreibersite	$(Fe,Ni)_3P$
*Senoite	Si_2N_2O
Sulfides	
Alabandite	(Mn,Fe)S
Bornite	Cu_5FeS_4
Bravoite	$(Fe,Ni)S_2$
*Brezinaite	Cr_3S_4
Chalcopyrite	$CuFeS_2$
Chalcopyrrhoite	$CuFe_4S_5$
Cubanite	$CuFe_2S_3$
*Daubreelite	$FeCr_2S_4$
Djerfisherite	$K_3CuFe_{12}S_{14}$
*Gentnerite	$Cu_8Fe_3Cr_{11}S_{18}$
Heazelwoodite	Ni_3S_2
Mackinawite	FeS
*Niningerite	(Mg,Fe)S
*Oldhamite	CaS
Pentlandite	$(FeNi)_9S_8$
Pyrrhotite	$Fe_{1-x}S$
Sphalerite	(Zn,Fe)S
Troilite	FeS
Vallerite	$4(Fe,Cu)S \cdot 3(Mg,Al)(OH)_2$
Violarite	Ni_2FeS_4
Chlorides	
Lawrencite	$(Fe,Ni)Cl_2$
Oxides	
Chromite	$FeCr_2O_4$
Christobalite	SiO_2
Hematite	Fe_2O_3
Ilmenite	$FeTiO_3$
Maghemite	α-Fe_2O_3
Perovskite	$CaTiO_3$
Quartz	SiO_2
Rutile	TiO_2
Spinel	$MgAl_2O_4$
Tridymite	SiO_2
Wüstite	FeO
Carbonates, phosphates, sulfates	
Apatite	$Ca(PO_4)_3Cl$
Bloedite	$Na_2Mg(SiO_4)_24H_2O$
Breunnerite	$(Mg,Fe)CO_3$
*Brianite	$Na_2MgCa(PO_2)$
Calcite	$CaCO_3$
Cassidyite	$Ca_2(Ni,Mg)(PO_4)_2 \cdot 2H_2O$
Collinsite	$Ca_2(Mg,Fe)(PO_4)_2 \cdot 2H_2O$
Dolomite	$CaMg(CO_3)_2$
Epsomite	$MgSO_4 \cdot 7H_2O$
Farringtonite	$Mg_3(PO_4)_2$
Graftonite	$(FeMn)_3(PO_4)_2$
Gypsum	$CaSO_4 \cdot 2H_2O$
Lipscomite	$(Fe^{+2},Mn)Fe_2^{+3}(PO_4)_2(OH)_2$
Magnesite	$(Fe,Mg)CO_3$
Merrillite	$Ca_9(Mg,Fe)H(PO_4)_7$
*Panethite	$(CaNa)_2(MgFe)_2(PO_4)_2$
Siderite	$FeCO_3$
*Stanfieldsite	$Ca_4(MgFe)_5(PO_4)_6$
Whitlockite	$Ca_9(Mg,Fe)H(PO_4)_7$
Organic minerals, oxalates	
Whewellite	$CaC_2O_4 \cdot H_2O$
Silicates	
Andradite	$Ca_3Al_2Si_3O_2$
Augite	$(Ca,Na)(Mg,Fe,Al,Ti)$-$(SiAl)_2O_6$
Clinopyroxene	$(CaMgFe)SiO_3$
Chlorite	$(Ni,Mn,Cr,Li,Mg,Fe,Al)_6(AlSi)_4O_{10}(OH)_8$
Cordiorite	$Mg_2Al_4Si_5O_{15}$
Diopside	$(Ca,Mg)(SiO_3)_2$
Enstatite	$Mg_2Si_2O_6$
Gehlenite	$Ca_2Al_2SiO_7$
Grossularite	$Ca_3Al_2Si_3O_{12}$
Hypersthene	$(Mg,Fe)_2Si_2O_6$
*Krinovite	$NaMg_2CrSi_3O_{10}$
*Majorite	$Mg_3(Mg,Si)Si_3O_{12}$
Melilite	$Ca_2(Mg,Al)(Si,Al)_2O_7$
*Merrihueite	$(K,Na)_2(Fe^{+2},Mg)_5Si_{12}O_{30}$
Monticellite	$Ca(Mg,Fe)SiO_4$
Nepheline	$NaAlSiO_4$
Olivine	$(Mg,Fe)_2SiO_4$
Plagioclase	$(Ca,Na)(Al,Si)_4O_8$
Potash feldspar	$(K,Na)AlSi_3O_8$
Richterite	$Na_2CaMg_5Si_8O_{22}F_2$
Rhonite	$CaMg_2TiAl_2SiO_{10}$
*Roedderite	$(K,Na)_2Mg_5Si_{12}O_{30}$
*Ringwoodite	$(Mg,Fe)_2SiO_4$
Sodalite	$Na_8Al_6Si_6O_{24}Cl_2$
Wollastonite	$CaSiO_3$
*Ureyite	$NaCrSi_2O_6$
*Yagiite	$(K,Na)_2(Mg,Al)_5(SiAl)_{12}O_{30}$
Hydrous oxides	
Akaganeite	FeO(OH)
Garnierite	$(NiMg)_3Si_2O_5(OH)_4$
Goethite	γ-FeO(OH)
Lepedrosite	FeO(OH)
Limonite	$FeO(OH)H_2O$
Opal	$SiO_2 \cdot nH_2O$
Reevesite	$Ni_6Fe_2(OH)_{16}(CO_3) \cdot 4H_2O$

SOURCE: From B. Mason, Mineralogy of meteorites, *Meteoritics*, 7:309–326, 1972.
*Found only in meteorites.

5. *Carbonaceous chondrites* have dark colors as a result of their carbon content. Unlike other meteorites, carbonaceous chondrites contain a considerable amount of water. The carbon is present in complex compounds. There are three types. Type I is strongly magnetic, contains sulfates, and has a carbon content of about 3.5%. Type II is weakly magnetic, and most of its sulfur is present as free sulfur; this type contains about 2.5% carbon. Type III contains the lowest percentage of water, has the highest density, and often contains pigeonite (examples are the Allende, Mexico, and Isna, Egypt, chondrites).

B. Achondrites are coarsely crystalline meteorites, which are very similar to terrestrial rocks, such as basalts and dunites (an example is the Moore County, NC, achondrite). They usually contain less olivine than the chondrites, and their feldspar is generally more calcic. There are two groups of achondrites: calcium-poor and calcium-rich. Of the former, there are four types, and of the latter, five types.

1. *Calcium-poor achondrites.*
   a. *Aubrites* usually contain almost pure $MgSiO_2$ and have minor quantities of accessory minerals.
   b. *Diogenites* contain hypersthene and modest amounts of nickel-iron and other accessory minerals.
   c. *Chassignite*, a type with only one example, consists almost entirely of olivine.
   d. *Ureilites* are a type of achondrite that contains graphite and occasionally small diamonds. The metal is low in nickel.
2. *Calcium-rich achondrites.*
   a. *Nakhlites* contain diopside and olivine.
   b. *Angrites* are augite-bearing achondrites.
   c. *Eucrites* are achondrites composed of plagioclase and pigeonite.
   d. *Howardites* are achondrites composed of hypersthene and plagioclase.
   e. *Whitleynite* has only one example, which is composed of enstatite and dark angular chondritic inclusions.

II. Intermediate meteorites (stony and iron meteorites).

A. *Pallasites* are the most abundant of this class and consist of olivine in a nickel-iron matrix.
B. *Siderophyres* consist of bronzite and tridymite in nickel-iron.
C. *Lodranites* consist of bronzite and olivine in nickel-iron.
D. *Mesosiderites* have about equal amounts of silicates and nickel-iron.
E. *Sorotites* are similar to the pallasites, with troilite replacing olivine.

III. Iron meteorites (see Table 2).

A. *Hexahedrites* are the simplest type of meteorite. They consist of one phase of nickel-iron, kamacite (alpha iron). These irons often contain accessory minerals and usually display Neumann lines (Fig. 5). The nickel content ranges 5–6.5%.
B. *Octahedrites* have nickel values of about 6.5–16%. These are often grouped into coarse, medium, and fine octahedrites, depending upon the widths of the kamacite lamellae. No sharp distinction exists between the hexahedrites and octahedrites or between the other three groups. Octahedrites contain two distinct phases of nickel-iron: kamacite and taenite.
C. *Ataxites* have structures that are usually fine-grained and are unlike either of the other groups of iron meteorites. Some of the ataxites have nickel contents that are similar to those of hexahedrites; others have nickel contents higher than the octahedrites. Ataxites with a high nickel percentage usually contain a considerable number of schreibersite inclusions (Fig. 6).

The structures used in classifying iron meteorites become apparent when sections are cut through the meteorites and the cut surfaces are highly polished. The details of these structures are best seen after the polished surfaces are etched

METEORITE

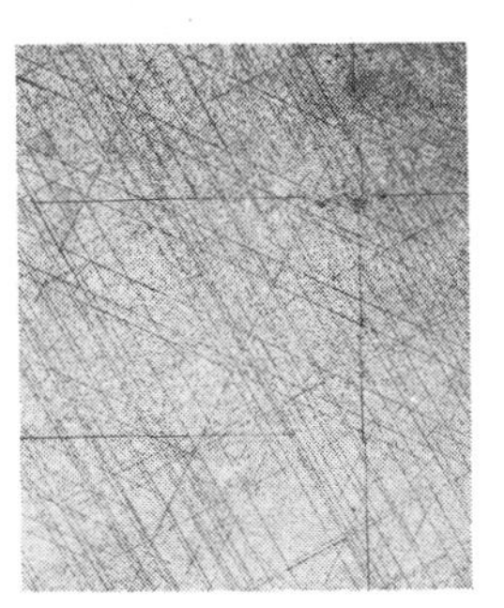

Fig. 5. Etched surface of Edmonton, Canada, meteorite. Parallel Neumann lines of kamacite are found in all hexahedrites. Two prominent lines intersecting at right angles are cleavage directions and parallel a cube face.

Table 2. Classification of iron meteorites

Class	Common symbol	Kamacite band widths, mm	Remarks
Hexahedrites	H	–	Neumann lines abundant, cubic cleavage
Coarse octahedrites	Ogg	> 3.3	Large grains of kamacite, taenite rare, usually no well-developed octahedral pattern
Coarse octahedrites	Og	1.3–3.3	–
Medium octahedrites	Om	0.5–1.3	–
Fine octahedrites	Of	< 0.2–0.5	–
Finest octahedrites	Off	< 0.2	Well-developed octahedral structure, narrow kamacite bands
Ataxites	D	–	Fine structure, spindles of kamacite phosphide inclusions
Anomalous	Anomalous	–	All meteorites which require special descriptions

with nital, a reagent that attacks the two nickel-iron phases differently. The patterns that develop depend upon the amount of nickel that is present and on the rate at which the metal is cooled. When the nickel content is below about 6%, the predominating nickel-iron phase is kamacite, alpha iron; however, some high-nickel phase taenite, gamma iron, may form at the edges of the kamacite grains. As the nickel increases, more taenite forms, and the widths of the kamacite lamellae decrease.

For many years it was believed that iron meteorites formed the core of a planetary-sized object by melting and differentiating in a manner similar to that which produced the Earth's core. Analyses now suggest, however, that some of the iron meteorites formed directly from the primitive nebula and not by the fractionation of metal from silicate.

Record of antiquity. The Earth and Moon formed about the same time as the meteorites, 4.55×10^9 years ago. Terrestrial and lunar rocks have been altered by subsequent metamorphic processes; none of these rocks is as old as typical meteorites. Many chondrites escaped the pressure, heat, and chemical effects that occur on planetary-sized objects. These contain grains that probably formed directly from the primitive nebula.

Other meteorites have been extensively recrystallized; however, this recrystallization generally occurred during the first 10^8 years of the meteorites' existence. Most recrystallized meteorites are at least 4.4×10^9 years old, far older than any known terrestrial or lunar rocks.

Meteorites thus contain the best available record of events that occurred in the early history of the solar system. Studies of these objects provide information on several fundamental questions. The meteoritic record of events which are pertinent to four of these questions follows.

Formation of solar system. Trends in the relative abundances of the chemical elements in meteorites demonstrate that these bodies, the Sun, and the planets were produced from a common source. Analyses of meteorites indicate that the parent source for some of the material was a gaseous nebula that was extremely hot.

Many carbonaceous chondrites contain inclusions of highly refractory minerals embedded in a matrix of fine-grained, volatile-rich material. The structures and chemical compositions of the inclusions are those that one would expect to find in material that condensed directly from the primitive nebula at temperatures above 1500 K (1200°C or 2200°F). The fine-grained, volatile-rich material probably formed at much lower temperatures, in a cooler region of the nebula.

The formation of meteoritic minerals occurred very soon after the chemical elements were formed. The stable decay products of two extinct radioisotopes, ^{129}I and ^{244}Pu, have been observed in many meteorites. These two isotopes have half-lives of 1.6×10^7 and 8.2×10^7 years, respectively. They were produced by the violent nuclear reactions that occur when a massive star explodes as a supernova. The amounts of these two isotopes that were trapped in meteorites indicate that no more than 2×10^8 years lapsed between the supernova explosion and the formation of the meteorites.

The decay product of ^{129}I has also been found inside the Earth, and the decay product of ^{244}Pu has been found in lunar samples. These findings suggest that solids of the Earth, the Moon, and the meteorites formed at the same time. Other long-lived radioisotopes, such as ^{235}U, ^{238}U, and ^{232}Th, were also produced in the supernova explosion. These isotopes still exist, but only about 1% of the initial supply of the shortest-lived one, ^{235}U, remains today.

Some meteorites also trapped radioactive ^{107}Pd (half-life = 6.5×10^6 years) and ^{26}Al (half-life = 7.3×10^4 years). If these were also produced by a supernova explosion, then meteoritic solids must have formed within about 2×10^6 years of that event.

In summary, the solar system formed from a hot, gaseous nebula very soon after a supernova explosion.

Formation of Earth. For many years it was thought that the solar system condensed from a homogeneous nebula and that the Earth accreted as a homogeneous body. Studies on meteorites have shown, however, that both concepts are incorrect. The concept of homogeneous accretion of the Earth requires a later period of at least partial melting and differentiation to produce the Earth's core. According to this view, metal-silicate partitioning also formed iron meteorites as the core of a sizable body, which later fragmented. This view was first challenged by geochemists, who noted that the terrestrial distribution of other elements and the high fractional abundance of iron in the four inner planets could best be explained if the iron cores formed first and then acted as nuclei on which silicates accreted. The geochemists suggested that the cores of the terrestrial planets initially formed in a central iron-rich region of the primitive nebula.

Analyses of iron meteorites have shown that the silicate inclusions in some of these have the same mineralogy and chemical composition as chondrites. This finding demonstrates that there has been no chemical partitioning of elements between the metal and silicate fractions. It has also been found that the silicate and troilite inclusions in these meteorites formed as early as the chondrites. This leaves essentially no time for differentiation in a large body. These results suggest that the metal and the inclusions formed in separate, chemically distinct regions of the nebula, and the inclusions in some iron meteorites were produced by mechanical mixing of the material which formed in different regions of a chemically heterogeneous nebula.

Primordial helium was found in meteorites over 2 decades ago. This helium is similar to that in the Sun; it is not the monoisotopic ^{4}He that is produced by the α-decay of heavy elements. Primordial helium has been found in gases leaking from the Earth's mantle. In view of the ease with which helium diffuses through solids, particularly at elevated temperatures, it seems unlikely that the Earth's interior would retain helium if it transferred iron to the core during a period of partial melting.

The gaseous decay product of extinct ^{129}I, ^{129}Xe, has been observed in meteorites and in samples from the Earth's mantle. The presence in the Earth of this gaseous decay product from a rela-

METEORITE

Fig. 6. Freda, N. Dak., meteorite. Etched surface is typical of ataxites. Inclusions are kamacite and schreibersite in acicular groundmass of kamacite and taenite. Light areas around the inclusions are taenite.

tively short-lived radioisotope also suggests that the Earth was never hot enough to form an iron core.

Finally, it should be noted that there is evidence that terrestrial kimberlites, the source of diamonds produced by high pressures inside the Earth, are derived from primitive material. This conclusion is based on the fact that the abundance of ^{143}Nd, the decay product of ^{147}Sm, in kimberlites of different ages is in excellent agreement with the abundance of ^{143}Nd that would be produced by the partial melting of chondritic material at the times indicated by the kimberlite ages.

In summary, the internal structure of the Earth was probably formed by heterogeneous accretion, with the iron core forming first. Its thermal history has been sufficiently mild to permit the retention of primordial helium, radiogenic ^{129}Xe, and many of the characteristics of chondrites in the mantle. The outer portions of the Earth may have accreted from volatile-rich material, such as that found in carbonaceous chondrites.

Origin of life. No reliable evidence of extraterrestrial life has been found in meteorites, lunar samples, or Martian soils. However, some carbonaceous chondrites contain the chemical components that are most important for life (carbon, hydrogen, nitrogen, and oxygen) and the conditions necessary to form the molecular precursors of life. For example, the Murchison carbonaceous chondrite, which fell on September 28, 1969, was found to contain a large number of amino acids. These amino acids are optically inactive (racemic) and are unlike terrestrial material that is of biologic origin.

Extraterrestrial amino acids have also been found in a few other carbonaceous chondrites, including two of the meteorites discovered in Antarctica. The origin of life may be related to the influx of such bodies, but it now appears that the evolution of life in the solar system was restricted to the planet Earth.

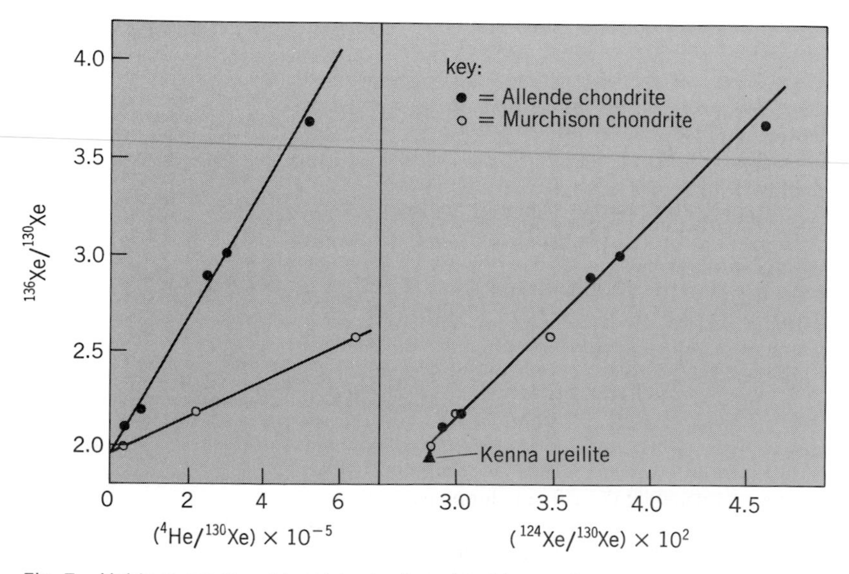

Fig. 7. Noble gases in acid-resistant minerals of two carbonaceous chondrites, Allende and Murchison.

Origin of chemical elements. As noted earlier, the solar system probably formed from a hot, gaseous nebula very soon after a supernova explosion. In addition to providing information on the time when the synthesis of the elements ceased, recent analyses have shown that the meteorites retain a record of the nuclear processes that synthesized individual isotopes of the elements.

Different isotopes of the same element may be produced in different regions or at different evolutionary stages of a star. Analyses of the isotopes, which make up many of the elements in meteorites, demonstrate that meteoritic grains formed even before the isotopes produced by the different stellar nucleosynthesis reactions became mixed. Strange isotopic compositions that were produced by stellar nuclear reactions are commonly called nucleogenetic isotopic anomalies. The first clear evidence that meteorites had trapped an element of unusual isotopic composition was noted in 1972. In the period since then, nucleogenetic anomalies have been recognized in the isotopes of about 20 elements, which include oxygen, mercury, xenon, krypton, uranium, barium, samarium, calcium, and tellurium.

Large differences in the isotopic composition of an element are even found in adjoining minerals of the same meteorite. This is illustrated for the nucleogenetic component of ^{16}O in adjacent minerals of melilite, pyroxene, and spinel from the Allende meteorite. The large inclusions are about 1.5 cm in diameter. The ^{16}O content of oxygen in the different minerals of that inclusion varies by up to 4%.

Stellar debris is expected to be both chemically and isotopically heterogeneous. As a star evolves, all of the light elements are destroyed in its interior, where pressures and temperatures become so intense that nuclei of light elements initially there are fused together to make nuclei of heavier elements. Thus, the major chemical components of a highly evolved, massive star exist in layers with elements such as iron and silicate toward the center, and elements such as hydrogen and helium toward the outer portions. Carbon and oxygen would exist between these extremes. *See* STELLAR EVOLUTION.

Small amounts of heavy elements, such as iron and silicon, may exist in the outer regions of a star, but none of the light elements, such as hydrogen and helium, can survive at the elevated temperatures of the star's interior. Consequently, the debris produced by the terminal supernova explosion of an evolved star will be chemically, as well as isotopically, heterogeneous. Variations in the abundances of the chemical elements are therefore expected to accompany the nucleogenetic isotopic anomalies that have been observed in meteorites.

When the solids condensed from the primitive nebula, most of the elements were partitioned into different minerals, depending on their chemical properties and the condensation temperatures of their compounds. This redistribution of elements during condensation tended to erase evidence of earlier chemical heterogeneities in the primitive nebula, except among elements with very similar chemical properties. The rare-earth elements can be used to detect differences in the amount of oxygen present during condensation, because these 14 elements have almost identical chemical properties but their oxides have different volatilities. Thus, members of the rare-earth elements which condensed from the oxygen-rich zone would form oxides and be trapped with a distinctive elemental abundance pattern. Discovered variations in the relative abundances of the rare-earth elements in individual meteorite grains follow the pattern expected if some condensed from such an oxygen-rich zone.

Another group of elements, the five noble gases, have similar physical properties and are essentially inert to chemical reactions. It is perhaps not surprising that the distribution of these elements in meteorites reveals evidence of the elemental and isotopic heterogeneities expected in supernova debris. Xenon is the heaviest stable noble gas, and it is composed of nine isotopes. The lightest, ^{124}Xe, and the heaviest, ^{136}Xe, are enriched by almost a factor of 2 in the carbon-rich material that remains after other minerals in a bulk carbonaceous chondrite are dissolved by etching in a mixture of acids. These two xenon isotopes may be produced in the outer regions of a supernova, which is the explosive stage that occurs shortly after a star reaches the advanced evolutionary stage.

Analyses of the noble gases in acid-etched residues of carbonaceous chondrites reveal high concentrations of helium associated only with that xenon which is enriched in ^{124}Xe and ^{136}Xe. The analyses also show that there is no helium present when the xenon is not enriched in ^{124}Xe and ^{136}Xe. The total absence of helium is expected in material from a stellar interior, and the close association of helium with ^{124}Xe and ^{136}Xe is expected if these two isotopes had come from outer stellar layers.

The correlations of helium with excess ^{136}Xe in acid-resistant minerals of the Allende and Murchison carbonaceous chondrites are shown on the left side of Fig. 7. Differences in the slopes of the two lines suggest that the carbonaceous material of the Murchison chondrite was more efficient in trapping helium than was that of the Allende chondrite, but in both cases the analyses indicate that there is no helium unless the xenon is enriched in ^{136}Xe.

The correlation between enrichments of ^{124}Xe and ^{136}Xe in these meteorites is shown in the right section of Fig 7. The common correlation between these two isotopes in minerals from Murchison and Allende suggests that the carbon-rich material of both meteorites sampled the same isotopically anomalous xenon.

The abundances of ^{136}Xe and ^{124}Xe in the xenon from a bulk sample of an achondrite, the Kenna ureilite, are also shown in Fig. 7. It should be noted that the correlation of helium with anomalous xenon, as shown on the left section of Fig. 7, indicates that there would be no helium associated with xenon which has the isotopic composition observed in Kenna. No trapped helium was found in Kenna, although the sample contained high concentrations of xenon and other heavy noble gases.

Analyses of the noble gases in the carbon-rich residues of the Lancé, Kainsaz, and Ornans chondrites show the same trends as those seen in the Allende and Murchison chondrites. Helium is associated with xenon enriched in ^{124}Xe and ^{136}Xe, and there is no helium when these isotopes of xenon are not enriched.

Thus, the isotopic and elemental abundances of noble gases in meteorites and the large number of nucleogenetic anomalies that have been observed in other elements of meteorites demonstrate that the chemical elements of the solar system initially contained many of the heterogeneities expected in supernova debris. Details of the mechanism that generated each anomaly are still a matter of debate, but the meteoritic record of early events leaves little doubt that the origin of the elements and the birth of the solar system were intimately related, perhaps by the violent death of a massive star. *See* ELEMENTS AND NUCLIDES, ORIGIN OF.

[OLIVER K. MANUEL]

Micrometeorite

A meteorite less than about 0.1 mm in diameter. The terms meteoroid and meteorite are sometimes mistakenly used interchangeably. Strictly speaking, a meteorite is an interplanetary body of any size capable of surviving entry into the atmosphere of the Earth and reaching its surface, whereas the term meteoroid refers to the same bodies in space before they encounter the atmosphere. If the body is larger than about 1 cm in diameter, some fraction may be expected to survive the atmospheric encounter. Micrometeorites can also be expected to survive, since their size permits them to radiate the tremendous heat energy generated by friction with the atmosphere before it can cause vaporization. These smaller bodies are usually called micrometeoroids while still in space. The Earth collects micrometeorites at a rate of about 10 tons (9 metric tons) per day. Those larger than micrometeorite size are destroyed during atmospheric entry and produce the visible phenomena known as meteors. Studies have shown that micrometeoroids yield important clues to the nature, origin, and evolution of the solar system and contribute significantly to the physical and chemical nature of the Earth's upper atmosphere.

Measurement techniques. Establishing the physical nature and chemical composition of the small-particle interplanetary complex has been a persistent problem. Measurement techniques have included visual, photographic, and video observation of meteors in the upper atmosphere. For particles too small to be detected by visible radiation, measurements have been made by bouncing radar signals off the electron trails left during atmospheric entry, by physical collection of such particles in places where contamination by terrestrial particles is minimized, and by direct measurement in space.

Collections. Microparticle collections are more or less routinely taken from deep-sea sediments, from remote arctic ice cores, from high-altitude balloons and rockets designed specially for this purpose, and from manned Earth-orbiting satellites. The problem with such collections has been to distinguish the meteoric material from terrestrial contaminants which, in the case of high-altitude collections, may be carried up with the collecting vehicle. Coring collections, both deep-sea and arctic, have focused on magnetic spherules. With the exception of certain limited welding operations, human activity does not normally produce these tiny iron-laden spheres. The natural magnetic spherules are believed to result from the ablation of larger meteorites and so give little clue to the nature of the smaller bodies. Further, it is known from the large finds that the iron variety probably constitutes less than 10% of all meteorites which strike the Earth. Balloon and aircraft collections have returned micrometeorites to the laboratory in what is believed to be their pristine condition. Because these particles radiate the tremendous frictional heat as fast as they receive it, little melting is believed to have occurred. Further, the atmosphere acts as a cushion, thus preventing the impact destruction common to all micrometeorite studies performed above the atmosphere. The illustration is an electron micrograph of one such particle. Composition analyses indicate that these particles have about the same elemental abundance as the large chondritic meteorites, which leads scientists to believe that they are true micrometeoroids as they existed in space.

Beta meteoroids. Direct measurements have been performed from high-altitude rockets, Earth-orbiting space vehicles, and deep-space probes. The lunar rocks returned by the Apollo missions have been examined for microcraters formed by meteoroid impacts over the long history of the lunar surface. Experiments with spaceborne detectors capable of measuring particles less than 0.1 μm in diameter have given new insight into the history of these tiny particles in the solar system. These detectors measure the ionization produced when the particle strikes a target. Data have been collected from two such instruments flown on the *Pioneer 8* and *9* spacecraft circling the Sun in an Earthlike orbit. The name which is being applied to these extremely small particles is beta meteoroids. Beta (β) is the symbol which is commonly used for the ratio of solar light pressure to gravitational attraction. For these very small particles, β is greater than 1. This means that the pressure of sunlight on the particle exceeds the gravitational pull of the Sun, and the particle is forced out of the solar system. These particles are believed to result from collisions between meteoroids passing close to the Sun. From the *Pioneer 8* and *9* measurements, it has been established that the Earth collects about 10^{-4} of these beta meteoroids per square meter per second as they are being "blown" outward from the Sun at velocities of the order of 100 km/s. These velocities are far in excess of the solar escape velocity which, at 1 astronomical unit (the distance at which the Earth orbits the Sun), is only 41 km/s.

Asteroid belt. Meteoroid detectors were also flown on the *Pioneer 10* and *11* missions which passed through the asteroid belt and encountered Jupiter in close flybys. The penetration detectors on those missions established that the micrometeoroid concentration in the asteroid belt is no greater than elsewhere in the inner solar system. A new meteoroid detector flown on those missions measured particles larger than 0.1 mm diameter by reflected sunlight. These detectors showed a modest increase of these larger particles in the asteroid belt, but still no substantial hazard to spacecraft passing through. These light-measuring detectors further showed that the particles are spun up by solar radiation to rotational rates of the order of thousands of revolutions per second, and that as they rotate they reflect sunlight in tiny glints like a beacon. The spinning up and glinting could only occur if the particles were highly irregular, which, from the illustration, one can see is the case.

Sources. Because of the identified associations of meteor streams with comet orbits, it was long believed that the predominant source of the meteoric complex was cometary. Several studies and experiments have cast doubt upon this widely held belief. For one, estimates from comets indicate that they are capable of producing only about 10% of the particulate material in the solar system. Further, *Pioneer 10* measurements of the zodiacal light (a feature of the night sky which is now known to be caused by sunlight reflected from interplanetary micrometeoroids) have shown that practically all of the reflecting particles are inside of the asteroid belt. This leads one to believe that at least this material may be made up largely of debris coming from these belts. On the other hand, the *Pioneer 10* penetration detectors, which measured particles of about 10 μm diameter, did not show any discontinuity in passing through the asteroid belt, and seemed to indicate about the same particulate concentration out to Jupiter and beyond. Thus, the small particles in the inner solar system may be fed by two different sources, one cometary and the other asteroidal. Collisions between the two inside the asteroid belt may give rise to many more smaller particles that spiral inward

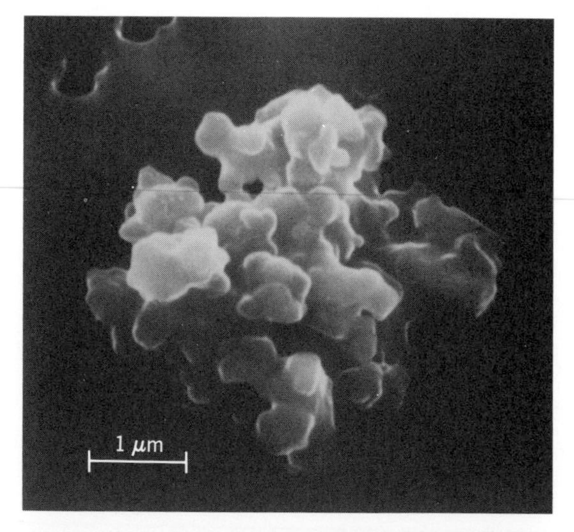

Electron micrograph of a micrometeorite recovered by high-altitude aircraft sampling. (*From D. E. Brownlee et al., The physical properties of interplanetary grains, in B. Donn et al., eds., The Study of Comets, NASA SP Ser., 1975*)

toward the Sun. Some of these are finally drawn into the Sun, while others further collide or partially evaporate near the Sun to produce the β meteoroids which complete their life cycle by being blown out of the solar system. In this very small size regime (β becomes greater than 1 for particles smaller than about 1 μm diameter), the Sun contributes particles to the interstellar medium. These β meteoroids may be similar to the interstellar grains known to populate the universe. More sensitive detectors are being developed to investigate how far into the solar system interstellar grains from other stars may be penetrating. Other instruments capable of measuring the composition of micrometeoroids and interstellar grains are already in use. It is expected that these new measurement techniques will provide greater insight into the physical and chemical nature of micrometeorites. See METEORITE.

[ROBERT K. SOBERMAN]

Bibliography: D. E. Brownlee et al., The physical properties of interplanetary grains, in B. Donn et al. (eds.), *The Study of Comets*, NASA SP Ser., 1975; R. K. Soberman, The terrestrial influx of small meteoritic particles, *Rev. Geophys. Space Phys.*, 9(2):239–258, 1971; R. K. Soberman, S. L. Neste, and K. Lichtenfeld, *Optical measurement of interplanetary particulates from Pioneer 10, J. Geophys. Res.*, 79(25):3685–3694, 1974; H. A. Zook and O. E. Berg, A source for hyperbolic cosmic dust particles, *Planet. Space Sci.*, 23(1):183–203, 1975.

Midnight sun

A phenomenon observed in the polar zones of the Earth near the time of the summer solstice, when the Sun remains visible above the horizon at midnight and reaches its minimum altitude without setting. The midnight sun is a consequence of the inclination of the rotation axis of the Earth, by which the Earth presents in turn each pole to the Sun for 6 months. The length of the period of uninterrupted daylight decreases as one goes away from the poles and, in principle, would vanish at the Arctic and Antarctic circles (latitude $\pm 66°33'$). However, because atmospheric refraction at the horizon raises the Sun's image by 34′, the midnight sun can be seen for a few days around the summer solstice from all points north of +66° or south of −66° latitude. See EARTH ROTATION AND ORBITAL MOTION.

[GERARD DE VAUCOULEURS]

Mira

The first star recognized to be a periodic variable. Called "The Wonderful" by the ancients, Mira was discovered by Fabricius in 1596 and has a period of 332 ±9 days. It is the prototype of long-period variables, and like most of its class, changes considerably from epoch to epoch. At times the maximum is as bright as second magnitude, and sometimes it is barely fifth magnitude.

Mira (*o* Ceti) has a faint companion discovered by R. G. Aitken after A. H. Joy predicted its presence from a study of the spectrum. The companion at times is about as bright as Mira is at minimum; the companion has a dwarf class B spectrum, with peculiar bright lines. The spectrum of Mira changes from M5e at maximum to M9e at minimum. The fact that the companion is also variable may account somewhat for the difference in depth of the minima of the system, which vary from about 8.5 to 9.5. See VARIABLE STAR.

[MARGARET W. MAYALL]

Mizar

A multiple star in the constellation Ursa Majoris. Mizar (ζ Ursa Majoris) is located at the bend in the handle of the Big Dipper. A very wide visual double, Mizar (the horse) and Alcor (the rider), is visible to the naked eye, separated by a distance of 12″. Mizar is also a telescopic double star, separation 14″; in addition, both components are also spectroscopic binary stars. Mizar was the first spectroscopic binary to be discovered (1889). It showed spectral lines from both stars which, because of the Doppler effect, shifted from violet to red and back, in a period of 20 days, as the two stars revolved about their common center of gravity. The amplitude and period together permit computation of the masses of the stars. Mizar is a member of the moving cluster of stars in Ursa Major. See CONSTELLATION; URSA MAJOR.

[JESSE L. GREENSTEIN]

Month

Any of several units of time based on the revolution of the Moon around Earth.

The calendar month is one of the 12 arbitrary periods into which the calendar year is divided. See CALENDAR.

The synodic month is the average period of revolution of the Moon with respect to the Sun, the same as the average interval between successive full moons. Its duration is 29.531 days.

The tropical month is the period required for the mean longitude of the Moon to increase 360°, or 27.322 days.

The sidereal month, 7 sec longer than the tropical month, is the average period of revolution of the Moon with respect to a fixed direction in space.

The anomalistic month, 27.555 days in duration, is the average interval between closest approaches of the Moon to Earth.

The nodical month, 27.212 days in duration, is the average interval between successive northward passages of the Moon across the ecliptic. See TIME.

[GERALD M. CLEMENCE]

Moon

The Earth's natural satellite. United States and Soviet spacecraft have obtained lunar data and samples, and Americans have orbited, landed, and roved upon the Moon (Fig. 1). Though the first wave of exploration has passed, it left a store of information whose meanings are still being deciphered. Many of the Moon's properties are now well understood, but its origin and relations to other planets remain obscure. Theories of its origin include: independent condensation and then capture by the Earth; formation in the same cloud of preplanetary matter with the Earth; and fission from the Earth. Because many of the Moon's geologic processes stopped long ago, its surface preserves a record of very ancient events. However,

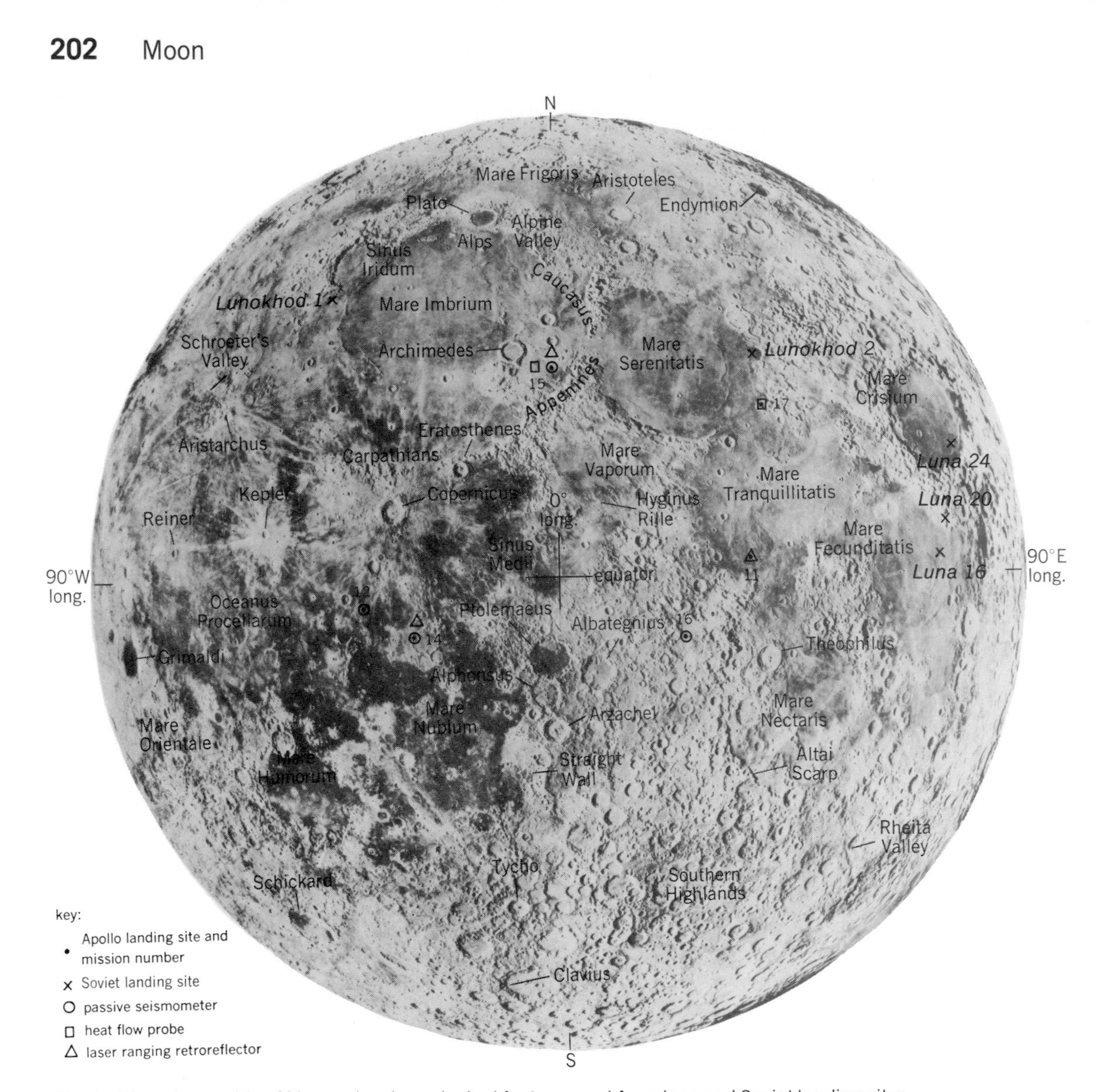

Fig. 1. Map of near side of Moon, showing principal features and American and Soviet landing sites.

because the Moon's rocks and soils were reworked by geochemical and impact processes, their origins are partly obscured, so that working out the Moon's early history remains a fascinating puzzle. Major characteristics of the Moon are listed in Table 1.

The apparent motions of the Moon, its waxing and waning, and the visible markings on its face (Fig. 1), are reflected in stories and legends from every early civilization. At the beginning of recorded history on the Earth, it was already known that time could be reckoned by observing the position and phases of the Moon. Attempts to reconcile the repetitive but incommensurate motions of the Moon and Sun led to the construction of calendars in ancient Chinese and Mesopotamian societies and also, a thousand years later, by the Maya. By about 300 B.C., the Babylonian astronomer-priests had accumulated long spans of observational data and so were able to predict lunar eclipses. Major events in the subsequent development of human knowledge of the Moon are summarized in Table 2.

Space flight experiments have now confirmed and vastly extended understanding of the Moon; however, they have also opened many new questions for future lunar explorers.

Motions. The Earth and Moon now make one revolution about their barycenter, or common center of mass (a point about 4670 km from the Earth's center), in $27^d\ 7^h\ 43^m\ 11.6^s$. This sidereal period is slowly lengthening, and the distance (now about 60.27 earth radii) between centers of mass is increasing, because of tidal friction in the oceans of the Earth. The tidal bulges raised by the Moon are dragged eastward by the Earth's daily rotation. The displaced water masses exert a gravitational force on the Moon, with a component along its direction of motion, causing the Moon to spiral slowly outward. The Moon, through this same tidal friction, acts to slow the Earth's rotation, lengthening the day. Tidal effects on the Moon itself have caused its rotation to become synchronous with its orbital period, so that it always turns the same face toward the Earth.

Tracing lunar motions backward in time is very difficult, because small errors in the recent data

Table 1. Characteristics of the Moon

Characteristics	Values and remarks
Diameter (approximate)	3476 km
Mass	1/81.301 Earth's mass, or 73.49×10^{24} g
Mean density	0.604 Earth's, or 3.34 g/cm^3
Mean surface gravity	0.165 Earth's, or 162 cm/sec^2
Surface escape velocity	0.213 Earth's, or 2.38 km/sec
Atmosphere	Surface pressure 10^{-12} torr (1.3×10^{-10} Pa); hints of some charged dust particles and occasional venting of volatiles
Magnetic field	Dipole field less than $\sim 0.5 \times 10^{-5}$ Earth's; remanent magnetism in rocks shows past field was much stronger
Dielectric properties	Surface material has apparent dielectric constant of 2.8 or less; bulk apparent conductivity is 10^{-5} mho/m or less
Natural radioactivity	Mainly due to solar- and cosmic-ray-induced background (about 1 mr/hr for quiet Sun)
Seismic activity	Much lower than Earth's; deep moonquakes occur more frequently when Moon near perigee; subsurface layer evident
Heat flow	3×10^{-2} W/m^2 (*Apollo 15* site)
Surface composition and properties	Basic silicates, three sites (Table 4); some magnetic material present. Soil grain size is 2–60 μm; 50% less than 10 μm. Soil-bearing strength 1 kg/cm^2 at depth of a few centimeters
Rocks	All sizes up to tens of meters present, concentrated in strewn fields. Rock samples from Mare Tranquillitatis include fine- and medium-grained igneous and breccia
Surface temperature range	At equator 400 K at noon; 80–100 K night minimum; 1 m below surface, 230 K; at poles ~150 K

propagate through the lengthy calculations, and because the Earth's own moment of inertia may not have been constant over geologic time. Nevertheless, the attempt is being made by using diverse data sources, such as the old Babylonian eclipse records and the growth rings of fossil shellfish. At its present rate of departure, the Moon would have been quite close to the Earth about 4.6×10^9 years ago, a time which other evidence suggests as the approximate epoch of formation of the Earth.

The Moon's present orbit (Fig. 2) is inclined about 5° to the plane of the ecliptic. Table 3 gives the dimensions of the orbit (in conventional coordinates with origin at the center of the Earth, rather than the Earth-Moon barycenter). As a result of differential attraction by the Sun on the Earth-Moon system, the Moon's orbital plane rotates slowly relative to the ecliptic (the line of nodes regresses in an average period of 18.60 years) and the Moon's apogee and perigee rotate slowly in the plane of the orbit (the line of apsides advances in a period of 8.850 years). Looking down on the system from the north, the Moon moves counterclockwise. It travels along its orbit at an average speed of nearly 1 km/sec or about 1 lunar diameter per hour; as seen from Earth, its mean motion eastward among the stars is 13°11′ per day.

As a result of the Earth's annual motion around the Sun, the direction of solar illumination changes about 1° per day, so that the lunar phases do not repeat in the sidereal period given above but in the synodic period, which averages $29^d\ 12^h\ 44^m$ and varies some 13 hr because of the eccentricity of the Moon's orbit. *See* EARTH ROTATION AND ORBITAL MOTION; ORBITAL MOTION.

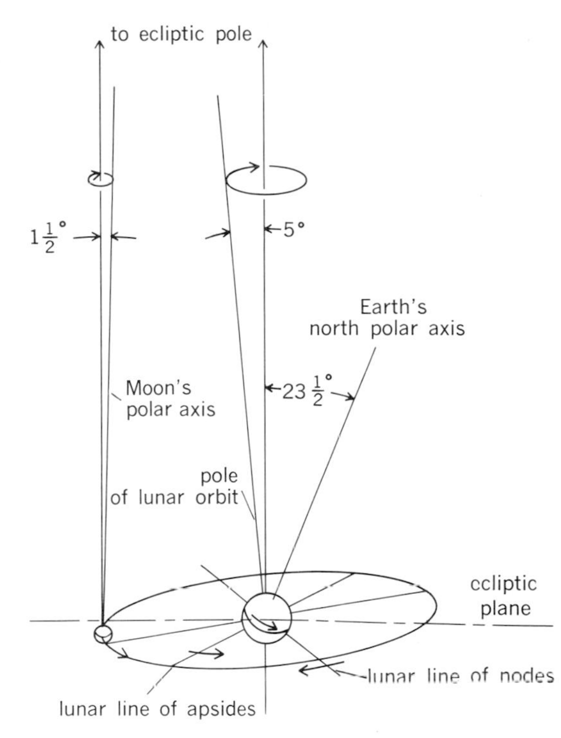

Fig. 2. Sketch of Moon's orbit.

When the lunar line of nodes (Fig. 2) coincides with the direction to the Sun and the Moon happens to be near a node, eclipses can occur. Because of the 18.6-year regression of the nodes, groups of eclipses recur with this period. When it passes through the Earth's shadow in a lunar eclipse, the Moon remains dimly visible because of the reddish light scattered through the atmosphere around the limbs of the Earth. When the Moon passes between the Earth and Sun, the solar eclipse may be total or annular. As seen from Earth, the angular diameter of the Moon (31′) is almost the same as that of the Sun, but both appar-

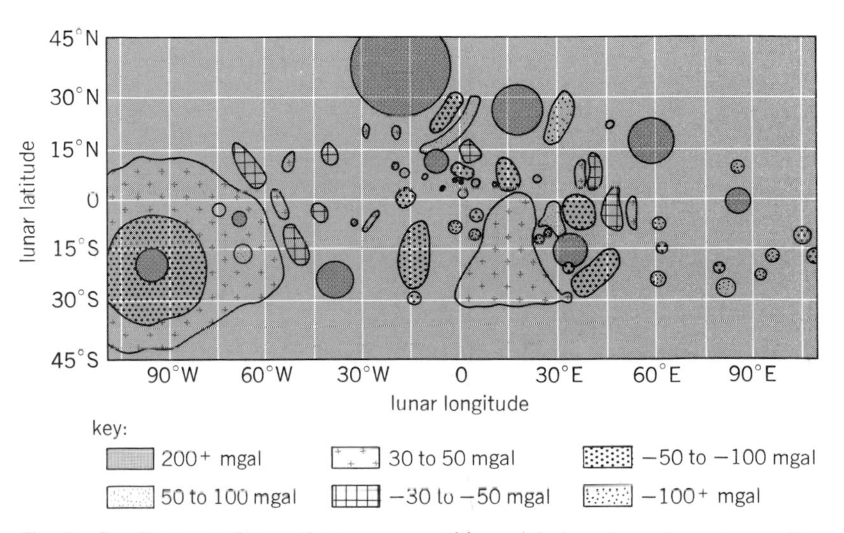

Fig. 3. Gravity anomalies on the lunar near side and limb regions. 1 mgal = gravitational acceleration of 10^{-5} m/s^2. Circular areas correspond to mass concentrations in circular maria. (*W. L. Sjogren, Jet Propulsion Laboratory, NASA*)

Table 2. Growth of human understanding of the Moon

Date	Event
Prehistory	Markings and phases observed, legends created connecting Moon with silver, dark markings with rabbit (shape of maria) or with mud.
~300 B.C.	Apparent lunar motions recorded and forecast by Babylonians and Chaldeans.
~150 B.C.	Phases and eclipses correctly explained, distance to Moon and Sun measured by Hipparchus.
~A.D. 150	Ancient observations compiled and extended by Ptolemy.
~700	Ephemeris refined by Arabs.
~1600	Empirical laws of planetary motion derived by Kepler.
1609	Lunar craters observed with telescopes by Harriot and Galileo.
1650	Moon mapped by Hevelius and Riccioli; features in system named by them still in use.
1667	Experiments by Hooke simulating cratering through impact and vulcanism.
1687	Moon's motion ascribed to gravity by Newton.
1692	Empirical laws of lunar motion stated by Cassini.
1700–1800	Lunar librations measured, lunar ephemeris computed using perturbation theory by Tobias Mayer. Secular changes computed by Lagrange and Laplace. Theory of planetary evolution propounded by Kant and Laplace. Many lunar surface features described by Schroeter and other observers.
1800–1920	Lunar motion theory and observations further refined, leading to understanding of tidal interaction and irregularities in Earth's rotation rate. Photography, photometry, and bolometry applied to description of lunar surface and environment. Lunar atmosphere proved absent. New disciplines of geology and evolution applied to Moon, providing impetus to theories of its origin.
1924	Polarization measured by B. F. Lyot, showing surface to be composed of small particles.
1927–1930	Lunar day, night, and eclipse temperatures measured by E. Pettit and S. B. Nicholson.
1946	First radar return from Moon.
1950–1957	Renewed interest in theory of lunar origin, new methods (for example, isotope dating) applied to meteorites, concepts extended to planetology of Moon. Low subsurface temperatures confirmed by Earth-based microwave radiometry.
1959	Absence of lunar magnetic field (on sunlit side) shown by *Luna 2*.
1960	Eastern far side photographed by *Luna 3*. Slower cooling of Tycho detected during eclipse.
1961	United States commitment to manned lunar flight.
1962	Earth-Moon mass ratio measured by *Mariner 2*.
1964	High-resolution pictures sent by *Ranger 7*. Surface temperatures during eclipse measured by Earth-based infrared scan.
1965	Western far side photographed by *Zond 3*.
1966	Surface pictures produced by *Luna 9* and *Surveyor 1*. Radiation dose at surface measured by *Luna 9*. Gamma radioactivity measured by *Luna 10*. High-resolution, broad-area photographs taken by *Lunar Orbiter 1*. Surface strength and density measurements made by *Luna 13*.
1967	Mare soil properties and chemistry measured by *Surveyor 3, 5*, and *6*. Whole front face mapped by *Lunar Orbiter 4*, sites of special scientific interest examined by *Lunar Orbiter 5*. Particle-and-field environment in lunar orbit measured by *Explorer 35*.
1968	Highland soil and rock properties and chemistry measured by *Surveyor 7*. Mass concentrations at circular maria discovered.
1968	Astronauts orbit Moon, return with photographs.
1969	Astronauts land and emplace instruments on Moon, return with lunar samples and photographs.
1969–1972	Lunar seismic and laser retroreflector networks established. Heat flow measured at two sites. Remanent magnetism discovered in lunar rocks. Geologic traverses accomplished. Orbital surveys of natural gamma radioactivity, x-ray fluorescence, gravity, magnetic field, surface elevation, and subsurface electromagnetic properties made at low latitudes. Metric mapping photos obtained. Samples returned by both crewed (United States) and automated (Soviet) missions; sample analyses confirmed early heating and chemical differentiation of Moon, with surface rocks enriched in refractory elements and depleted in volatiles. Age dating of lunar rocks and soils showed that most of the Moon's activity (meteoritic, tectonic, volcanic) occurred more than 3×10^9 years ago.
1975–to date	The multiplicity of theories of lunar origin, though much more closely limited by Apollo and Luna data, continues, with no model being accepted as unique.

ent diameters vary because of the eccentricities of the orbits of Moon and Earth. Eclipses are annular when the Moon is near apogee and the Earth is near perihelion at the time of eclipse. A partial solar eclipse is seen from places on Earth that are not directly along the track of the Moon's shadow.

The Moon's polar axis is inclined slightly to the pole of the lunar orbit (Fig. 2) and rotates with the same 18.6-year period about the ecliptic pole. The rotation of the Moon about its polar axis is nearly uniform, but its orbital motion is not, owing to the finite eccentricity and Kepler's law of equal areas, so that the face of the Moon appears to swing east and west about 8° from its central position every month. This is the apparent libration in longitude. The Moon does rock to and fro in a very small oscillation about its mean rotation rate; this is called the physical libration. There is also a libration in latitude because of the inclination of the Moon's polar axis. The librations make it possible to see

Table 3. Dimensions of Moon's orbit

Characteristics	Values
Sidereal period (true period of rotation and revolution)	(27.32166140 + 0.000000167T) ephemeris days, where T is in centuries from 1900
Synodic period (new Moon to new Moon)	(29.5305882 + 0.000000167T) ephemeris days
Apogee	406,700 km (largest); 405,508 km (mean)
Perigee	356,400 km (smallest); 363,300 km (mean)
Period of rotation of perigee	8.8503 years direct ("direct" meaning that the motion of perigee is in the direction of Moon's motion about the Earth)
Period of regression of nodes	18.5995 years
Eccentricity of orbit	0.054900489 (mean)
Inclination of orbit to ecliptic	5°8'43" (oscillating ±9' with period of 173 days)
Inclination of orbit to Earth's Equator	Maximum 28°35', minimum 18°21'
Inclination of lunar equator	
to ecliptic	1°32'40"
to orbit	6°41'

about 59% of the Moon's surface from the Earth.

The lunar ephemeris, derived from precise astronomical observations and refined through lengthy computations of the effects perturbing the movements of the Moon, has now reached a high degree of accuracy in forecasting lunar motions and events such as eclipses. Laser ranging to retroreflectors landed on the Moon, aided by radio ranging to spacecraft, provides measurements of Earth-Moon distances to a precision of the order of meters. *See* LUNAR LASER RANGING; PERTURBATION.

Selenodesy. The problem of determining the Moon's true size and shape and its gravitational and inertial properties has been under attack by various methods for centuries (Tables 1 and 2). However, results from space flights have invalidated some of the premises on which the earlier methods were based, and have revealed discrepancies in the older data. The relation between the Moon's shape and its mass distribution is very important to theories of lunar origin and the history of the Earth-Moon system. Radio-tracking data from Lunar Orbiters indicate that the Moon's gravitational field is ellipsoidal, with the short axis being the polar one (as expected for any rotating body), and with the equatorial section being an ellipse possibly slightly elongated in the Earth-Moon direction. But the Earth-based radar measurements and tracking data from Rangers and Surveyors showed that the Moon's actual surface at the points of landing is about 2 km farther from the Earth than expected. Further evidence of an anomalous relationship between mass and shape for the Moon is provided by the mass concentrations in circular maria, discovered through analysis of short-term variations in the Lunar Orbiter tracking data and then mapped in detail by Apollo tracking (see Fig. 3). By radio altimetry, Apollo confirmed that the Moon's surface on the far side is higher on the average than the near side; that is, the center of mass is offset from the center of figure. The offset is about 2 km toward the Earth. These observations suggest that the Moon's crust is thicker on the far side than on the near side, as shown (not to scale) in Fig. 4.

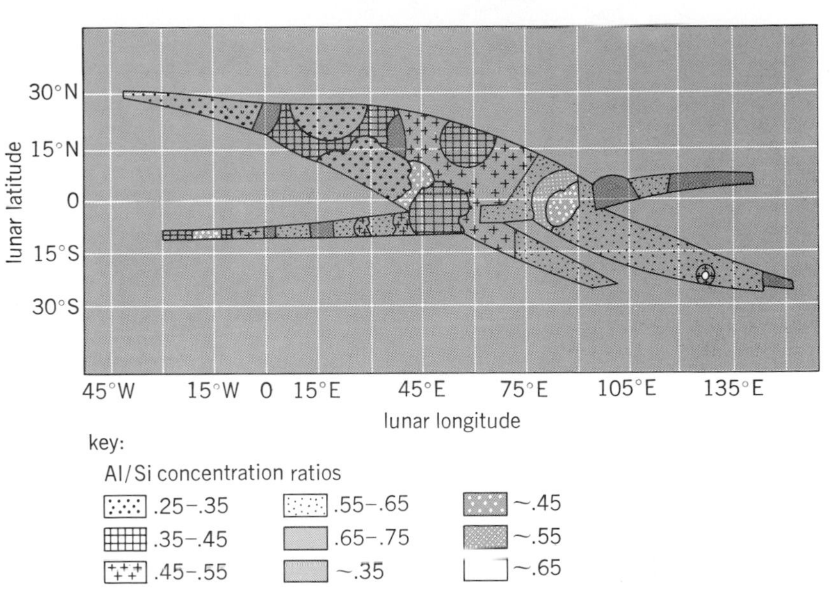

Fig. 5. Aluminum-silicon concentration ratios as detected by x-ray experiments on *Apollo 15* and *16*. (*I. Adler, University of Maryland*)

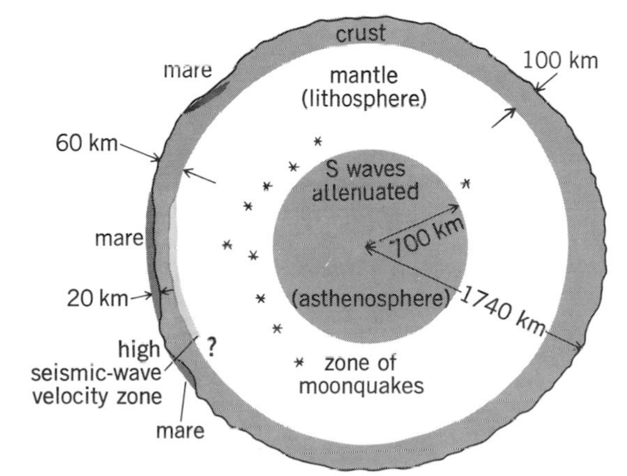

Fig. 4. Schematic diagram of lunar structure. The near side of the Moon is to the left of the figure. (*From S. R. Taylor, Lunar Science: A Post Apollo View, © 1975 by Pergamon Press Ltd.; reprinted with permission*)

Body properties. The Moon's small size and low mean density (Table 1) result in surface gravity too low to hold a permanent atmosphere, and therefore it was to be expected that lunar surface characteristics would be very different from those of Earth. However, the bulk properties of the Moon are also quite different—the density alone is evidence of that—and the unraveling of the Moon's internal history and constitution is a great challenge to planetologists. *See* EARTH.

The Earth, with its dense metallic fluid core, convective mantle, strong and variable magnetic field with trapped radiation belts, widespread seismic tremors, volcanoes and folded mountain ranges, moving lithospheric plates, and higher differentiated radioactive rocks, is plainly a planet seething with inner activity. Is the Moon also an active, evolving world or is it something very different? The answer lies in a group of related ex-

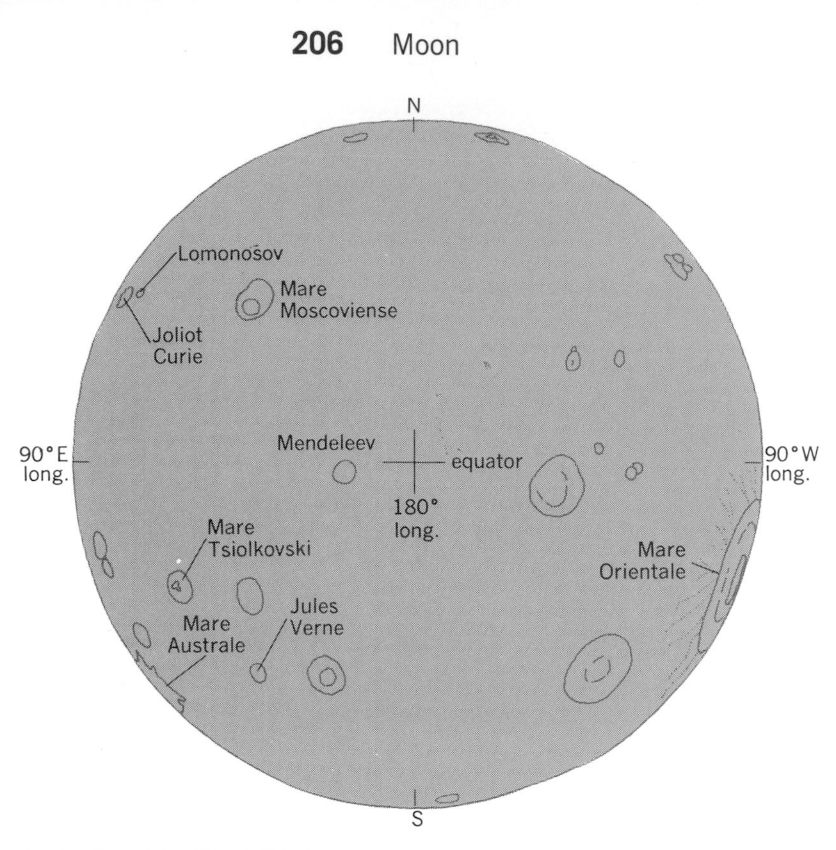

Fig. 6. Map of far side of Moon.

periments: seismic investigations, heat-flow measurements, surface magnetic and gravity profiles, determination of abundances and ages of the radioactive isotopes in lunar material, and comparison of the latter with those found in the Earth and meteorites. Present theories and experimental data yield the following clues to the problem.

1. The Moon is too small to have compressed its silicates into a metallic phase by gravity; therefore, if it has a dense core at all, the core should be of nickel-iron. But the low mass of the whole Moon does not permit a large core unless the outer layers are of very light material; available data suggest that the Moon's iron core may have a diameter of at most a few hundred kilometers.

2. The Moon has no radiation belts, and behaves as a nonconductor in the presence of the interplanetary field. Moon rocks are magnetized, but the source of the magnetism remains a mystery as there is now little or no general lunar magnetic field.

3. The Moon's natural radioactivity from long-lived isotopes of potassium, thorium, and uranium, expected to provide internal heat sufficient for partial melting, was roughly measured from orbit by *Luna 10*, and the component above the cosmic-ray-induced background radiation was found to be at most that of basic or ultrabasic earthly rock, rather than that of more highly radioactive, differentiated rocks such as granites. *Apollo 11* and *12* rock samples confirmed this result; *Apollo 15*, *16*, and *17* mapped lunar composition and radioactivity from orbit (Fig. 3). X-ray experiments showed higher aluminum-silicon concentration ratios over highland areas and lower values over maria (Fig. 5), while magnesium-silicon ratios showed a converse relationship—higher values over maria and lower values over highlands.

4. The Moon is seismically much quieter than the Earth. Moonquakes are small, many of them originate deep in the interior (Fig. 4), and activity is correlated with tidal stress: more quakes occur when the Moon is near perigee.

When all of the Apollo observations are taken together, it is evident that the Moon was melted to an unknown depth and chemically differentiated about 4.5×10^9 years ago, leaving the highlands relatively rich in aluminum and an underlying mantle relatively rich in iron and magnesium, with all known lunar materials depleted in volatiles. The subsequent history of impacts and lava flooding includes further episodes of partial melting until about 3.9×10^9 years ago, with on the final result being a thick, rigid crust with only minor evidence of recent basaltic extrusions. The temperature profile and physical properties of the Moon's deep interior are, despite the Apollo seismic and heat-flow data, under active debate. Figure 4 shows a rough sketch of the Moon as revealed by the data.

Fig. 8. The crater Copernicus, showing the central peaks, slump terraces, patterned crater walls, and (background) slopes of the Carpathian Mountains. (*Langley Research Center, NASA*)

Fig. 7. Mare Tsiolkovski on far side of Moon. Crater, partly flooded by dark mare material, is about 200 km across. (*Langley Research Center, NASA*)

Large-scale surface features. As can be seen from the Earth with the unaided eye, the Moon has two major types of surface: the dark, smooth maria

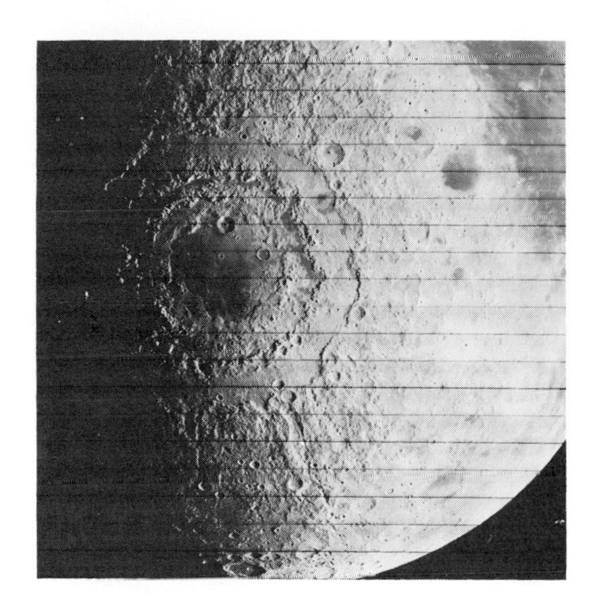

Fig. 9. Mare Orientale. (*Langley Research Center, NASA*)

and the lighter, rougher highlands (Figs. 1 and 6). Photography by spacecraft shows that, for some unknown reason, the Moon's far side consists mainly of highlands (Fig. 7). Both maria and highlands are covered with craters of all sizes. Craters are more numerous in the highlands than in the maria, except on the steeper slopes, where downhill movement of material apparently tends to obliterate them. Numerous different types of craters can be recognized. Some of them appear very similar to the craters made by explosions on the Earth; they have raised rims, sometimes have central peaks, and are surrounded by fields of hummocky, blocky ejecta. Others are rimless and tend to occur in lines along cracks in the lunar surface. Some of the rimless craters, particularly those with dark halos, may be gas vents; others may be just the result of surface material funneling down into subsurface voids. Most prominent at full moon are the

Fig. 10. Schroeter's Valley, a large sinuous rille, and craters Herodotus and Aristarchus. Aristarchus, brightest of the ray craters, is about 40 km across and 3600 m deep. (*Langley Research Center, NASA*)

bright ray craters (Fig. 1) whose grayish ejecta appear to have traveled for hundreds of kilometers across the lunar surface. Observers have long recognized that some erosive process has been and may still be active on the Moon. For example, when craters overlap so that their relative ages are evident, the younger ones are seen to have sharper outlines than the older ones. Bombardment of the airless Moon by meteoritic matter and solar particles, and extreme temperature cycling, are now considered the most likely erosive agents, but local internal activity is also a possibility. Rocks returned by the Apollo astronauts are covered with tiny glass-lined pits, confirming erosion by small high-speed particles.

The lunar mountains, though very high (8000 m or more), are not extremely steep, and lunar explorers see rolling rather than jagged scenery (Fig. 8). There are steep slopes (30–40°) on the inside walls and central peaks of recent craters, where the lunar material appears to be resting at its maximum angle of repose, and rocks can be seen to have rolled down to the crater bottoms.

Though a widespread network of fault traces is

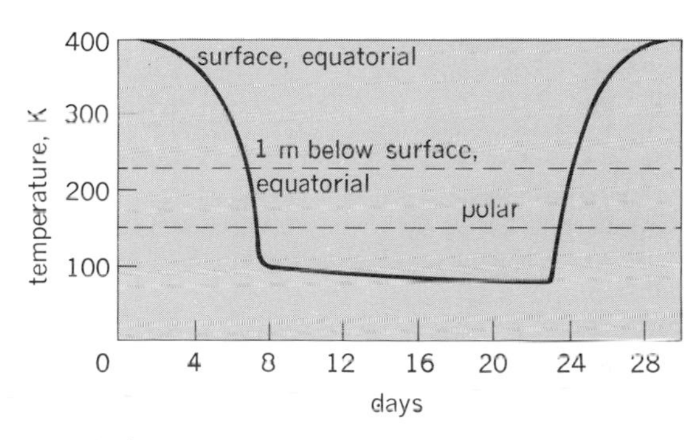

Fig. 11. Lunar surface and subsurface temperatures.

visible, there is no evidence on the Moon of the great mountain-building processes seen on the Earth. There are some low domes suggestive of volcanic activity, but the higher mountains are all part of the gently rolling highlands or the vast circular structures surrounding major basins. Figure 9 shows one of these, the Mare Orientale, as revealed by *Lunar Orbiter 4*. This large concentric structure is almost invisible from Earth because it lies just past the Moon's western limb; at favorable librations parts of its basin and mountain ramparts can be seen. The great region of radial sculpture surrounding the Orientale basin strongly suggests a catastrophic origin, with huge masses of matter thrown outward from the center. Note, however, the gentle appearance of the flooding by the dark mare material, which seems to lie only in the lowest parts of the concentric rings. Other basins, namely, Imbrium, Serenitatis, and Crisium, appear more fully flooded (Fig. 1). These were maria created by giant impacts, followed by subsidence of the ejecta and (probably much later) upwelling of lava from inside the Moon. Examination of small variations in Lunar Orbiter motions has revealed that each of the great circular maria is the site of a positive gravity anomaly (excess mass), shown in-

Fig. 12. Lunar patterned ground, a common feature on moderate slopes. (*Langley Research Center, NASA*)

Fig. 3. The old argument about impact versus vulcanism as the primary agent in forming the lunar relief, reflected in lunar literature over the past 100 years, appears to be entering a new, more complicated phase with the confirmation of extensive flooding of impact craters by lava on the Moon's near side, while on the far side, where the crust is thicker, the great basins remain mostly empty.

In some of the Moon's mountainous regions bordering on the maria are found sinuous rilles (Fig. 10). These winding valleys, some of them known since the 18th century, were shown in Lunar Orbiter pictures to have an exquisite fineness of detail. Some of them originate in small circular pits and then wriggle delicately across the Moon's gentle slopes for hundreds of kilometers, detouring around even slight obstacles, before vanishing on the plains. Though their resemblance to meander-

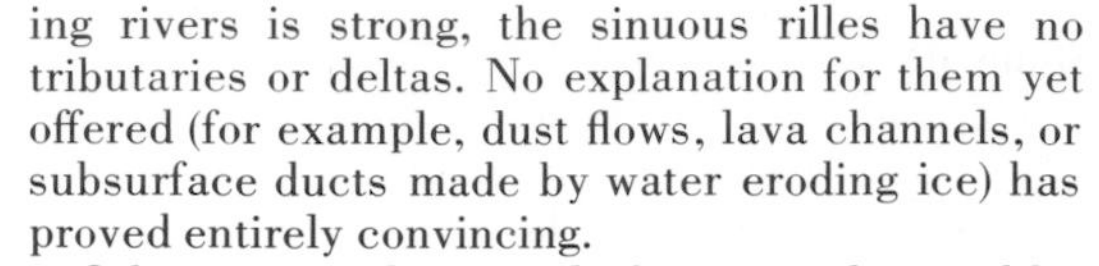

ing rivers is strong, the sinuous rilles have no tributaries or deltas. No explanation for them yet offered (for example, dust flows, lava channels, or subsurface ducts made by water eroding ice) has proved entirely convincing.

Other strange large-scale features, observed by telescope and then revealed in more detail by spacecraft cameras, are the ghost craters, circular structures protruding slightly from the maria, and the low, ropy wrinkle ridges that stretch for hundreds of kilometers around some mare borders.

Small-scale surface features. Careful observations, some of them made decades before the beginning of space flight, revealed much about the fine-scale nature of the lunar surface. Since the smallest lunar feature telescopically observable from the Earth is some hundreds of meters in extent, methods other than direct visual observation had to be used. Photometry, polarimetry, and later radiometry and radar probing gave the early fine-scale data. Some results of these investigations

Fig. 14. Crater showing appearance of upwelling on its floor. (*Langley Research Center, NASA*)

Fig. 13. Lunar soil and rocks, and the trenches made by *Surveyor 7*. (*Jet Propulsion Laboratory, NASA*)

suggested bizarre characteristics for the Moon. Nevertheless, many of their findings have now been confirmed by spacecraft. The Moon seems to be totally covered, to a depth of at least tens of meters, by a layer of rubble and soil with very peculiar optical and thermal properties. This layer is called the regolith. The observed optical and radio properties are as follows.

1. The Moon reflects only a small portion of the light incident on it (the average albedo of the maria is only 7%, darker than any familiar object except things like soot or black velvet).

2. The full moon is more than 10 times as bright as the half moon.

3. At full moon, the disk is almost equally bright all the way to the edge; that is, there is no "limb darkening" such as is observed for ordinary spheres, whether they be specular or diffuse reflectors.

4. Color variations are slight; the Moon is a uniform dark gray with a small yellowish cast.

Some of the maria are a little redder, some a little bluer, and these differences do correlate with large-scale surface morphology, but the visible color differences are so slight that they are detectable only with special filters. (Infrared spectral differences are more pronounced and have provided a method for mapping variations in the Moon's surface composition.)

5. The Moon's polarization properties are those of a surface covered completely by small, opaque grains in the size range of a few microns.

6. The material at the lunar surface is an extremely good thermal insulator, better than the most porous terrestrial rocks. The cooling rate as measured by infrared observations during a lunar eclipse is strongly variable; the bright ray craters cool more slowly than their surroundings.

7. The Moon emits thermal radiation in the radio wavelength range; interpretations of this and the infrared data yield estimates of surface and shallow subsurface temperatures (Fig. 11).

8. At wavelengths in the meter range, the Moon appears smooth to radar, with a dielectric constant lower than that of most dry terrestrial rocks. To centimeter waves, the Moon appears rather rough, and at visible light wavelengths (a conclusion from observations 1–5 above), it is extremely rough.

These observations all point to a highly porous or underdense structure for at least the top few millimeters of the lunar surface material. The so-called backscatter peak in the photometric function, which describes the sudden brightening near full Moon, is characteristic of surfaces with deep holes or other roughness elements which are shadowed when the lighting is oblique.

The Ranger, Luna, Surveyor, and Lunar Orbiter missions made it clear that these strange electromagnetic properties are generic characteristics of the dark-gray, fine soil that appears to mantle the entire Moon, softening most surface contours and covering everything except occasional fields of rocks (Figs. 12 and 13). This soil, with a slightly cohesive character like that of damp sand and a chemical composition similar to that of some basic silicates on the Earth, is a product of the radiation, meteoroid, and thermal environment at the lunar surface. Figure 12 shows a surface texture, called patterned ground, that is common on the moderate slopes of the Moon. This widespread phenomenon is unexplained, though there are some similar surfaces developed on the Earth when unconsolidated rock, lava, or glacial ice moves downhill beneath an overburden. At many places on the Moon, there is unmistakable evidence of downward sliding or slumping of material and rolling rocks. There are also a few instances of apparent upwelling (Fig. 14), as well as numerous "lakes" where material has collected in depressions. Figures 15, 16, and 17 show, at different scales, the landing site of *Surveyor 7* near the great ray crater Tycho (Fig. 1). *Surveyor 7* found the soil of the highlands near Tycho to be rockier than, and slightly different in chemical composition to, the mare materials sampled by earlier Surveyors. Figure 13 shows trenches made in the lunar soil during soil mechanics experiments on *Surveyor 7*. Magnets on the Surveyors collected magnetic particles from the soil, demonstrating the presence of either meteoritic or native iron minerals at the sites examined. Meteoroid experiments on the Lunar Orbiters showed about the same flux of small particles as is observed at the Earth, so that the lunar soil would be expected to contain a representative sample of meteoritic and possibly also cometary matter. Apollo results confirmed and extended the Surveyor data and also indicated that glassy particles are abundant in and on the soil. Evidence of micrometeoroid bombardment is seen in the many glass-lined microcraters on lunar rocks (Fig. 18).

Fig. 15. Region near crater Tycho, including *Surveyor 7* landing site. (*Langley Research Center, NASA*)

Chemical, mineral, and isotopic analyses of minerals from the *Apollo 11* site showed that mare rocks there are indeed of the basic igneous class and are very ancient ($3-4 \times 10^9$ years). The *Apollo 12* samples are significantly younger, suggesting that Mare Tranquillitatis and Oceanus Procellarum were formed during a long and complex lunar history. The *Apollo 12* astronauts visited *Surveyor 3* and brought back parts of that

Fig. 16. *Orbiter* photograph of *Surveyor 7* site. (*Langley Research Center, NASA*)

Fig. 17. Surface view at *Surveyor 7* site. (*Jet Propulsion Laboratory*)

spacecraft to permit analysis of the effects of its 2½-year exposure on the surface of the Moon. The lunar rock and soil samples returned by the Apollo and Luna missions have yielded much new information on the composition and history of the Moon. Among the dominant characteristics of these rocks are enrichment in refractories, depletion in volatiles, much evidence of repeated breaking up and rewelding into breccias, and ages since solidification extending back from the mare flows of $3-4 \times 10^9$ years ago into the period of highland formation more than 4×10^9 years ago, but not as yet including the time of the Moon's original accretion. Some characteristics of the lunar samples are summarized in Table 4.

As the Apollo missions progressed, each new landing site was selected with the aim of elucidating more of the Moon's history. A main objective was to sample each of the geologic units mapped by remote observation, either by landing on it or by collecting materials naturally transported from it to the landing site. Although this process did result in collection of both mare and highland materials with a wide range of ages and chemical compositions, it did not result in a complete unraveling of the history of the Moon. Apparently, the great impacts of $3-4 \times 10^9$ years ago erased much of the previous record, resetting radioactive clocks and scrambling minerals of diverse origins into the complicated soils and breccias found today. *See* ASTEROID; METEOR; METEORITE; SPACE PROBE.

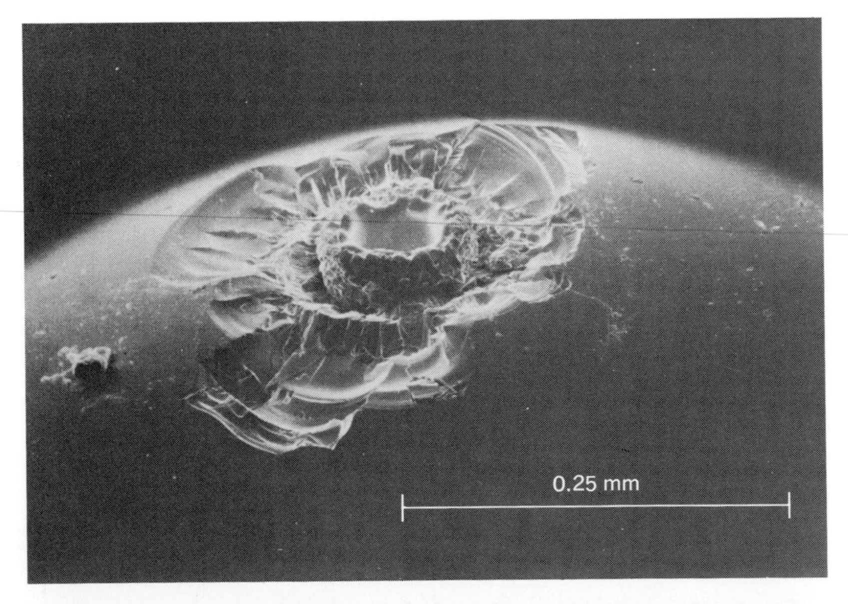

Fig. 18. Example of microcratering, caused by hypervelocity impact of tiny particles, on a dark-brown glass sphere. The diameter of the sphere is approximately 0.75 mm, and the diameter of the inner crater, inside the raised rim, is about 50 μm. This photograph was taken through a scanning electron microscope.

Atmosphere. Though the Moon may at one time have contained appreciable quantities of the volatile elements and compounds (for example, hydrogen, helium, argon, water, sulfur, and carbon compounds) found in the meteorites and on the Earth, its high daytime surface temperature and low gravity would cause rapid escape of the lighter elements. Solar ultraviolet and x-ray irradiation would tend to break down volatile compounds at the surface, and solar charged-particle bombardment would ionize and sweep away even the heavier gas species. Observations from the Earth, looking for a twilight glow of the lunar atmosphere just past the terminator on the Moon, and watching radio-star occultations have all been negative, setting an upper limit of 10^{-12} times the Earth's sea-level atmospheric density for any lunar gas envelope. Therefore, either the lunar volatile compounds have vanished into space or they are trapped beneath the surface. The samples returned by Apollo are enriched in refractory elements, depleted in volatiles, and impregnated with rare gases from the solar wind. No water appears to have ever been present at any Apollo site, and carbonaceous materials were present, if at all, only in very small amounts.

Occasional luminescent events reported by reliable observers suggest that some volcanic gases are vented from time to time on the Moon, particularly in the regions of the craters Aristarchus and Alphonsus. A slight, transient atmosphere does exist on the night side of the Moon as a result of the trapping and release of gas molecules at the very low temperatures prevailing there; also frozen liquids or gases could exist in permanently shadowed crater bottoms near the lunar poles. No experiment to detect such accumulations of volatiles has been made. *Lunokhod 2*, a Soviet roving spacecraft, measured a slight glow attributed to a very thin cloud of small particles near the surface, which could explain the Surveyor observations of a slight horizon glow after sunset. Also, the ALSEP (Apollo lunar surface experiments packages) experiments landed by Apollo have occasionally detected small gas emanations, including water, from unknown sources.

Lunar resources. Enough is known about the Moon to show that it is a huge storehouse of metals, oxygen (bound into silicates), and other materials potentially available for future human use in space. Because of the Moon's weak gravity, lunar

Table 4. Some selected data from Apollo and Luna missions

Mission	Main sample properties	Other data
Apollo 11	Mare basalts, differentiated from melt at depth 3.7×10^9 years ago. Some crystalline highland fragments in soils. Unexpected abundance of glass. Much evidence of impact shock and microcratering. No water or organic materials.	Study of seismic properties showed low background, much scattering, and low attenuation.
Apollo 12	Basalts 3.2×10^9 years old. One sample 4.0×10^9 years old includes granitic component. Some samples with high potassium, rare-earth elements, and phosphorus (KREEP) may be Copernicus crater ejecta.	Surveyor parts returned showed effects of solar and cosmic bombardment.
Apollo 13	Spacecraft failure—no samples.	Despite emergency, some lunar photos returned.
Luna 16	Basalt 3.4×10^9 years old, relatively high Al content.	
Apollo 14	Shocked highland basalts, probably Imbrium ejecta, 3.95×10^9 years old, higher Al and lower Fe than mare materials.	Deep moonquakes.
Apollo 15	Highland anorthosites including one sample 4.1×10^9 years old, mare basalts similar to *Apollo 11* samples.	Orbital remote sensing began mapping of surface compositions.
Luna 20	Possibly Crisium ejecta, 3.9×10^9 years old.	
Apollo 16	Highland anorthosite breccias $3.9-4 \times 10^9$ years old, also possibly Imbrium ejecta.	Seismic network began recording locations of impacts and deep moonquakes; orbital compositional mapping extended.
Apollo 17	Variety of basalts and anorthosites $3.7-4 \times 10^9$ years old, possibly volcanic glass, few dunite fragments 4.48×10^9 years old, possibly surviving from before the great highland bombardment.	Orbital mapping, and study of seismic, particle-and-field, and subsurface electrical properties yielded comprehensive (but still unexplained) picture of Moon.
Luna 24	Very low titanium basalt from Mare Crisium. Sample includes rock 3.3×10^9 years old.	

materials could be placed into orbit at less than one-twentieth of the energy cost for delivering them from Earth. At the sites so far explored, no water exists, and the only available hydrogen is the small amount implanted in soil by the solar wind. It will be a task for future explorations to find the polar ices if they exist, to discover concentrations of meteoritic or cometary materials, and to investigate atypical geologic phenomena such as the seemingly volcanic regions. Any of these sites might yield additional treasures, but it is already known that the Moon could be an important resource for humanity in space. [JAMES D. BURKE]

Bibliography: *Apollo 15 Preliminary Science Report*, NASA SP-289, 1972, *Apollo 16*, NASA SP-315, 1972, *Apollo 17*, NASA SP-330, 1973; Apollo Preliminary Examination Team, Preliminary examination of lunar samples from *Apollo 11*, *Science*, 165(3899):1211, 1969; J. R. Arnold, Ice at the lunar poles, *J. Geophys. Res.*, 84(B10):5659–5668, Sept. 10, 1979; J. R. Arnold (ed.), *Workshop on Near-Earth Resources*, La Jolla, NASA Conf. Pub. 2039, 1978; R. B. Baldwin, *The Face of the Moon*, 1949; V. de Callatay, in A. de Visscher (ed.), *Atlas of the Moon*, trans. by R. G. Lascelles, 1964; A. Dollfus (ed.), *Moon and Planets II: A Session of the 10th Plenary Meeting of Cospar, London*, July 26–27, 1967; F. El-Baz, The Moon after Apollo, *Icarus*, 25:495–537, 1975; W. Hess (ed.), *The Nature of the Lunar Surface: IAU-NASA Symposium*, 1965; Z. Kopal and Z. K. Mikhailov (eds.), *The Moon*, 1962; L. J. Kosofsky and F. El-Baz, *The Moon as Viewed by Lunar Orbiter*, NASA SP-200, 1970; R. A. Lyttleton, *Mysteries of the Solar System*, 1968; *The Moon: A New Appraisal*, Royal Society of London, 1979; A. Pannekoek, *A History of Astronomy*, 1961; Proceedings of the Lunar Science Conferences, *Geochim. Cosmochim. Acta*, suppl. 1–10, 1970–1980; *Surveyor 7: A Preliminary Report*, NASA SP-173, May 1968; S. R. Taylor, *Lunar Science: A Post-Apollo View*, 1975; S. R. Taylor, Structure and evolution of the Moon, *Nature*, 281:105, 1979; U.S.S.R. Academy of Science, *Atlas of the Far Side of the Moon*, pts. 1 and 2, 1960, 1967; U.S.S.R. Academy of Science, *First Panoramas of the Surface of the Moon*, 1966; F. L. Whipple, *Earth, Moon and Planets*, 3d ed., 1968.

Nebula

Originally, any fixed, extended, and usually fuzzy luminous object seen in a telescope. Nebulae are now distinguished from star clouds that can be resolved into individual stars, but earlier workers were unable to differentiate between white nebulae, which are stellar systems so remote as to show no individual stars, and gaseous or diffuse nebulae in the Milky Way Galaxy.

Extragalactic nebulae are stellar systems comparable with the Milky Way Galaxy or the Magellanic Clouds in size and number of stars, and are more properly termed external galaxies. They are grouped as spirals, ellipticals, or irregulars, and various classification systems have been devised. *See* GALAXY, EXTERNAL.

Types of nebulae. This article deals with gaseous nebulae. This class of objects includes diffuse nebulae which contain dust and gas of the interstellar medium, excited and caused to fluoresce by embedded stars—for example, the Great Orion Nebula, supernova remnants such as the

Network or Veil nebula in Cygnus, and the so-called planetary nebulae. Gaseous nebulae are members of the Milky Way galactic system, and small compared with its overall dimensions. Various types of gaseous nebulae have been identified.

Diffuse nebulae. One type of nebula ranges from huge masses of relatively high surface brightness, such as the Orion Nebula, down to faint, milky structures a hundred times less dense that are detectable only with long exposures and special filters. Diffuse nebulae may contain both dust and gas or may be purely gaseous, such as the California Nebula. The nebulosity found in the Pleiades and elsewhere consists of dust with no luminous gas, although it is probable that there is also a great quantity of neutral hydrogen. *See* ORION NEBULA.

Variable nebulae. Variable-brightness nebulae are associated with abnormal variable stars and are frequently fan-shaped in appearance. The best-known example is Hubble's Variable Nebula. Nebulae of this type are associated with T Tauri variables, which are believed to be stars in the process of formation.

Planetary nebulae. Planetary nebulae are so denoted because they often show small greenish disks in the telescope, not unlike the images of the planets Uranus and Neptune. The best-known object of this class is the Ring Nebula in Lyra, M 57 or NGC 6720 (Fig. 1). The energy emitted by planetary nebulae is derived from the rich ultraviolet radiation of stars embedded within them.

Supernova remnants. The detonation of a star in a supernova event causes the ejection of the outer layers into the surrounding interstellar medium. In early stages as in the Crab Nebula, the radiating material consists of ejecta from the star. In the later stages this rapidly moving material is slowed down as it mixes with the surrounding dust and gas of the interstellar medium. Such a phase is illustrated, for example, by the Network Nebula in Cygnus or NGC 433. Supernovae remnants characteristically emit nonthermal radio-frequency emission, whereby they are often detected in nearby galaxies as well as in the Milky Way system. *See* CRAB NEBULA; SUPERNOVA.

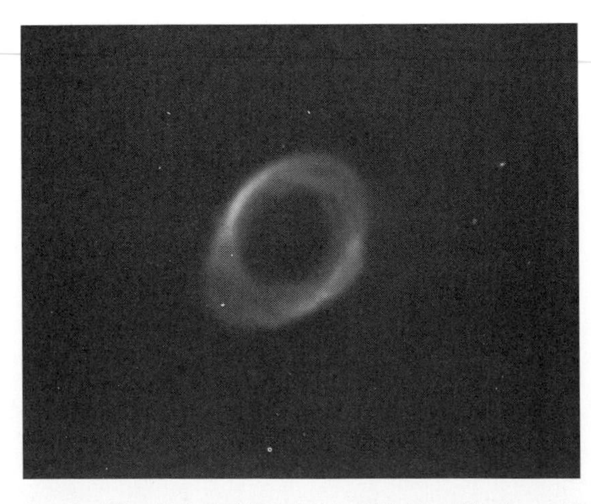

Fig. 1. Ring Nebula NGC 6720. (*U.S. Navy Electrograph made with 61-in. astrometric reflector at Flagstaff, Ariz.*)

Although the Carina Nebula appears to be a normal object, η-Carinae itself is a remarkable object which appears to resemble a nebula more closely than a star.

Catalogs. Nebulae are cataloged according to various systems. The first list was made by C. Messier. A much more complete list was given by J. L. E. Dreyer in the *New General Catalogue* (abbreviated NGC) and the two Index catalogs. These lists include also galaxies and star clusters. A revised catalog has been compiled by J. W. Sulentic and W. G. Tifft. Special lists have been published for planetary nebulae (L. Perek and L. Kohoutek), diffuse nebulae and HII regions (H. M. Johnson), and supernova remnants (K. R. Lang). *See* MESSIER CATALOG.

Methods and types of observations. Measurements of positions and sizes are straightforward, except for irregular structures that are hard to describe. Early observations were visual, but by the unaided eye only the brightest nebulae or their most conspicuous features could be detected. Photography has contributed greatly to nebular observation. Most emission nebulae radiate strongly in the red hydrogen line Hα. Hence, by using red-sensitive plates and narrow-bandpass filters, it is possible to suppress the sky background and register nebulosities of low surface brightness. Gaseous nebulae of both thermal and nonthermal types are observable with radio telescopes. Energy fluxes from nonthermal sources persist and even rise as one goes to the very lowest frequencies, while radiation from thermal sources eventually declines with decreasing frequency in accordance with Rayleigh-Jeans law. Infrared measurements have revealed a number of important emission lines of ions whose presence cannot be detected in ordinary spectral regions; they often show that great quantities of thermally emitting dust are found in diffuse nebulae as well as in planetaries. Ultraviolet and x-ray observations provide invaluable supplements to data obtained from ordinary spectral regions, sometimes revealing attenuated gases at temperatures of hundred of thousands of degrees.

Brightness. Because surface brightness is independent of the distance as long as the eye perceives the object as an extended area, no advantage is gained on objects such as Orion or the Trifid nebula by using large telescopes, unless one wishes to examine small details. For small diffuse nebulae and planetaries, a large telescope has considerable advantage. For monochromatic radiation, the surface brightness may be expressed in terms of ergs/(s)(cm^2)(unit solid angle), although other units such as SI units [W/(cm^2)(sr)] or even magnitudes per square minute of arc have also been used.

The brightness of a nebula can also be measured in the radio-frequency region, although it is necessary to take into consideration the limited resolving power of such telescopes. Surface brightness may be measured by photographic photometry, but the most accurate work is done by photoelectric methods, using a spectrum scanner or narrow-bandpass filters to select monochromatic radiations.

The measurement of the brightness of a nebula is more complicated than that of a star. The nebula is an extended surface of nonuniform brightness; hence, the complete description of a nebula in monochromatic radiation would consist of a set of isophotic contours calibrated in terms of intensity units. Gaseous and diffuse nebulae show a huge range in surface brightness from objects like Orion to faint wisps barely visible on long exposures with narrow-bandpass filters.

Distances. If the nebula is associated with a star or a star cluster, its distance may be found by measuring the stellar distances. For example, the distance of the Orion Nebula is found by establishing the absolute luminosities of the illuminating stars and the amount of space absorption. Then the distance is found from a comparision of the apparent and intrinsic brightnesses of the stars. Similar methods may be applied to the Lagoon and Trifid nebulae.

In some instances, such as the Network Nebula in Cygnus, it is possible to measure the angular rate of expansion and also the velocity in the line of sight, which gives the radial rate of expansion in kilometers per second. The method cannot be applied indiscriminately to the planetaries because the rates of expansion may represent not just a lateral motion of material but a change in size of the ionized volume. Often, statistical methods are used in which radial velocities and proper motions are compared, together with correlation between angular diameter and distance. Direct determinations are possible for a few objects—the nucleus of NGC 246 which has a dwarf comparison of absolute magnitude $M = 7.0$, the planetary nebula in the globular cluster M 15, the planetaries near the central bulge of the Galaxy, and those in external galaxies such as the Magellanic Clouds, in the Andromeda Spiral and its companions, and in other members of the local group. Astrophysical methods are often used for planetaries; often these procedures involve some assumption such as constancy of the mass within the emitting volume or constancy of luminosity.

Spectra. When small gaseous nebulae (which have diameters of a few seconds of arc) are observed with a slitless spectrograph, an image of the nebula is formed in each of its monochromatic radiations. It is found that the radiations of ions of higher excitation, such as neon, Ne^{4+}, are always concentrated closer to the central star than are the radiations of ions of lower excitation, such as oxygen, O^{+}. The reason is that the higher-energy quanta capable of producing highly ionized atoms are exhausted before they reach the outer layers. *See* ASTRONOMICAL SPECTROSCOPY.

For the spectroscopic studies of large nebulae, weak lines, or lines that fall close together, it is necessary to use a slit spectrograph, equipped with appropriate detectors.

The spectra of the gaseous nebulae show the recombination lines of hydrogen and helium which are observed in optical ultraviolet and radio-frequency ranges. The strongest lines in the optical region are often some that have never been produced in any terrestial laboratory. These are the so-called forbidden lines of ions of various abundant elements. They represent transitions between the metastable levels of the ground configuration. Figure 2 illustrates these transitions for the ions argon [Ar III] and chlorine [Cl III], where numeral III designates a doubly ionized atom, the ionized atoms having ground configurations $3p^3$ and $3p^4$, respectively. Transitions of the type ^2P-^2D in p^3 ions or ^1S-^1D in p^2 or p^4 ions are called auroral transitions, because this type of forbidden transition is the most important in the Earth's aurora. Transitions between the middle metastable term and the ground term give the so-called nebular lines, while jumps from the highest metastable terms to the ground term give transauroral lines.

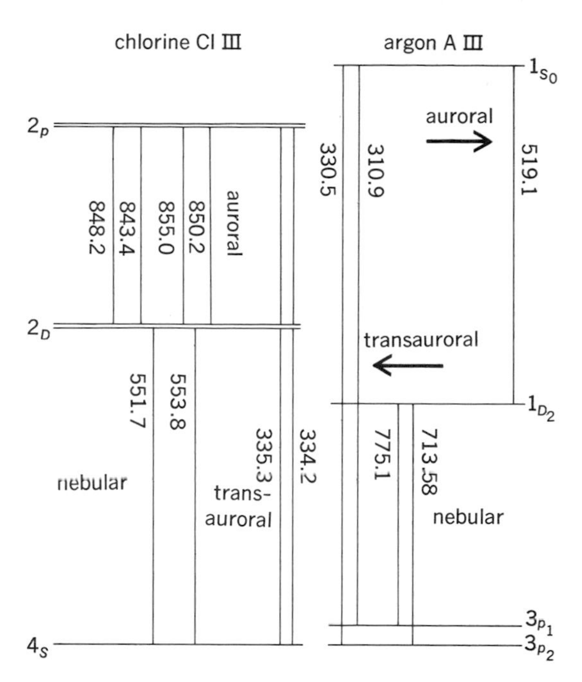

Fig. 2. Forbidden-line transitions in doubly ionized chlorine and argon with the energy of each transition given in wave number units. The nebular transitions of [Cl III] are observed in a large number of planetaries. Numbers indicate wavelengths in nanometers.

Weak recombination lines of oxygen, nitrogen, and carbon are observed in gaseous nebulae, but (aside from H and He) the strongest permitted lines are certain O III transitions observed in high-excitation planetaries. The O III lines are produced by a remarkable fluorescent mechanism discovered by I. S. Bowen. Ions of O^{2+} in the $2p^2\ ^3P_2$ level of the ground configuration absorb the 30.378-nm resonance line of ionized helium and are excited to the $3d^3\ P_2$ level, from which they cascade downward with the emission of observable lines.

The visible and radio-frequency lines of hydrogen and helium are produced by a process of photoionization from the ground level, followed by recombination in one of the highly excited levels with subsequent cascade and the emission of observable lines. For example, the red hydrogen line Hα may be produced by the recapture of an electron on the third level with a subsequent jump from the third to the second level. Finally, the atom goes from the second level to the ground level.

On the other hand, the forbidden lines are excited by electron impacts which cause the atoms to rise from the ground term to one of the nearby metastable terms. They return to the ground level with the emission of a forbidden line. The transition probabilities, such as the probability of an atom making the jump in a unit time, are very low. An atom may remain in a metastable level for seconds or even minutes, whereas it will leave an ordinary excited level in 0.0000001 s or even less time. The forbidden lines attain such great strength in gaseous nebulae because of their vast extent and because the processes operating to produce the ordinary permitted lines are enormously reduced in efficiency. In a typical planetary nebula the green forbidden lines of oxygen [O III] may be 10 times as strong as the hydrogen Hβ line, yet the concentration of hydrogen ions per unit volume may be 10,000 times as great as the concentration of O^{2+} ions.

The diffuse galactic nebulae always show relatively low excitation, the Balmer lines are strong, the [O II] lines are often more intense than the [O III] lines, and lines such as [Ar IV] and [Ne V] are absent. In nebulae of low surface brightness, only a few lines can be observed: Hα, and sometimes the 372.7-nm [O II], and occasionally [O III]. In external galaxies there sometimes exist high-excitation, extended gaseous nebulae. In the so-called Seyfert galaxies, broad high-excitation lines are observed.

Quasistellar sources show emission lines which resemble those of ordinary planetary nebulae, but strongly shifted toward the red. Lines which fall normally in the far ultraviolet are often observed in these objects, for example, Lyα of hydrogen. *See* QUASARS.

Masses, densities, and temperatures. The radius R of a nebula of angular radius R'' is obtained as soon as its distance r is known. Consider a uniform spherical nebula, where N_ϵ is the electronic density, $N(H^+)$ is the density of hydrogen ions, and T_ϵ is the electron temperature. The emission per unit volume in the Hβ line can be written as Eq. (1). Here A^0 is a factor that depends solely

$$E(H\beta) = N(H^+)N_\epsilon A^0\, 10^{-25} \quad \text{cgs units} \quad (1a)$$

$$E(H\beta) = N(H^+)N_\epsilon A^0\, 10^{-44} \quad \text{SI units} \quad (1b)$$

on temperature, with the value 2.22 at 5000 K, 1.24 at 10,000 K, and 0.66 at 20,000 K.

The entire energy emitted by the nebula in the Hβ line is $\frac{4}{3}\pi R^3 E(H\beta)$, so the flux passing through each unit area of surface is $RE/3$, and therefore the flux received by an observer at distance r will by $\frac{1}{3}RE(H\beta)(R/r)^2$.

To establish the electron density, one needs not only measurements of the surface brightness but also a knowledge of the temperature T_ϵ. One might estimate the electron temperature from the width of the spectral lines, but it would be necessary to separate the effects of the gas kinetic motion from those of the large-scale mass motion.

The best optical method involves the use of the relative intensities of the auroral and nebular transitions of a given ion, such as the ^1S-^1D (436.3-nm) and ^1D-^3P (500.7-nm, 495.9-nm) transitions of [O III]. The relative number of collisional excitations to the $^1D_2$ level and to the ^1S level depends on the velocity distribution of the electrons and, hence, on the temperature. If the target areas for collisional excitation and the transition probabilities are known, a relation involving temperature, density, and intensity ratio can be found. If the nebular and auroral lines of two ions, both occurring in the same region in the nebula, can be measured, both N_ϵ and T_ϵ can be found independently of the surface brightness measurements. If the nebula has a filamentary structure, the electron density found in this way will be greater than that found from the surface brightness, which represents an average over the space occupied by the nebula.

From radio-frequency observations carried out at different wavelengths, one can determine both the optical thickness and the gas kinetic temperature of the radiating gas. Additional information is provided by combining radio and optical observations. The radio-frequency observations are not affected by space absorption, but they give the temperatures in the cooler portions of the radiating gas and are affected by lack of spatial resolution.

The masses of the planetary nebulae turn out to be about a fifth that of the Sun, although this mass may be somewhat less in some objects. The diffuse nebulae often have masses several times that of the Sun, while the neutral hydrogen clouds in Orion (which are not ionized by the hot stars) have a total mass about 100,000 times that of the Sun.

Internal motions. The motions of the gases perpendicular to the line of sight have been found in only a few nebulae, notably the Network Nebula and the Crab Nebula, but motions along the line of sight can be detected by radial velocity measurements with a slit spectrograph and by a Fabry-Perot etalon. Use of a multislit consisting of a series of closely placed slits parallel to one another is the most efficient way to observe radial velocity shifts in small nebular regions. With this device O. C. Wilson has observed many planetaries. The planetaries appear to be expanding, the rate of expansion depending on the degree of ionization of the ion. Thus [Ne V] lines show the smallest expansion rate and [O II] lines show the largest, suggesting that because the degree of ionization depends on the distance from the central star, the material is accelerated on the outer side.

Studies of motions in diffuse nebulae, particularly Orion, show a variety of phenomena. Mass streaming motions and perhaps shock waves appear to occur. Also, the conventional theory of incompressible turbulence, which gives a definite relation between eddy size and velocity, cannot be applied. In some nebulae, internal motions are almost certainly complicated by the influence of magnetic fields.

Relation to illuminating stars. Except for the nonthermal radio-frequency sources, gaseous nebulae derive their energy from stars near or within them. If the star is relatively faint (like T Tauri), the nebula is small; large luminous nebulae are necessarily excited by high-temperature bright stars. The hot star ionizes the surrounding gas up to a boundary that is more or less sharp, depending

Fig. 3. Lagoon Nebula NGC 6523 and star cluster NGC 6530. The photograph was made with the Curtis Schmidt Telescope, University of Michigan; a red-sensitive filter and emulsion were used to isolate the spectral region about the red hydrogen line Hα. Note the irregular structure. The dark lanes and the globules are caused by solid grains in the neighborhood.

on the density inhomogeneities in the gas. Beyond that boundary the gas is neutral and nonluminous, although it may be detectable from its 21-cm radio-frequency emission. An example in point is the Lagoon Nebula NGC 6523, which is excited by the star cluster NGC 6530 (Fig. 3). The patchy luminous nebula appears to be surrounded by a much larger region of cold, neutral hydrogen. The bright O- and B-type stars that excite typical diffuse gaseous nebulae have effective temperatures of 25,000–40,000 K and luminosities 1000–10,000 times that of the Sun.

The central stars of planetary nebulae are by comparison dwarf stars, although their temperatures range from 25,000 K to perhaps 200,000 K. Their luminosities range from less than that of the Sun to about 200 times that of the Sun; precise figures cannot be given until the distances of the planetaries are accurately established.

Supernova remnants are excited by the dissipation of the kinetic energy of the ejected material. Spectra produced in shock phenomena differ from those caused by absorption of direct ultraviolet radiation from stars. Forbidden lines of ionized sulfur [S II] are usually prominent; sometimes in the wake of the shock as in the Cygnus loop, even lines of coronal excitation, [Fe X] can be produced.

Gaseous nebulae in external galaxies. Diffuse nebulae are observed in external galaxies, for example, in the Andromeda Spiral M31, NGC 6822, and Messier 101; the largest number has been cataloged in the Triangulum Spiral M33. The largest nebula in this galaxy, NGC 604, has an appearance similar to that of 30 Doradus in the Large Magellanic Cloud, a structure suggestive of the influence of magnetic fields.

As noted above, planetary nebulae are observed in nearby galaxies.

[LAWRENCE H. ALLER]

Bibliography: L. H. Aller, *Gaseous Nebulae*, 1958; S. A. Kaplan and S. B. Pikelner, *The Interstellar Medium*, 1970; K. R. Lang, *Astrophysical Formulae*, 1978; S. Maran (ed.), *The Gum Nebula*, NASA Spec. Publ. 332, 1973; B. Middlehurst and L. H. Aller (eds.), *Interstellar Medium and Gaseous Nebulae*, vol. 7 of *Stars and Stellar Systems*, 1968; D. E. Osterbrock, *Astrophysics of Gaseous Nebulae*, 1973; L. Perek and L. Kouhoutek, *Catalogue of Galactic Planetary Nebulae*, 1967; L. Spitzer, *Physical Processes in the Interstellar Medium*, 1978; J. W. Sulentic and W. G. Tifft, *Revised New General Catalogue of Non-Stellar Astronomical Objects*, 1978.

Neptune

The outermost of the four giant planets, a near twin of Uranus in size, mass, and composition. Its discovery in 1846 within a degree from the theoretically predicted position was one of the great achievements of celestial mechanics. Difficulties in accounting for the observed motion of Uranus by means of perturbations by the other known planets led early in the 19th century to the suspicion that a new planet, beyond the orbit of Uranus, might be causing the deviation from the predicted path. The difficult problem of deriving the mass and orbital elements of the unknown planet was solved independently in 1845–1846, first by J. C. Adams in Cambridge, England, and then by U. J. Leverrier in Paris. Adams's result did not receive immediate attention, so it was Leverrier's solution that led to the discovery of Neptune by J. G. Galle, in Berlin, who found the planet on Sept. 23, 1846, only 55′ from its calculated poisition. *See* CELESTIAL MECHANICS; PERTURBATION.

The planet and its orbit. The actual mass and orbit of Neptune differ considerably from the values predicted by Adams and by Leverrier, since both assumed that the mean distance of the planet to the Sun would be that predicted by the Titius-Bode relation, namely, 38.8 AU (astronomical units), whereas it is only 30.1 AU or 2.8×10^9 mi (4.5×10^9 km). The eccentricity of the orbit is only 0.009, the second smallest (after that of Venus) among the planets; the inclination is 1.8°; the period of revolution is 164.8 years; and the mean orbital velocity of Neptune is 3.4 mi/sec (5.45 km/sec). *See* PLANET.

Through a small telescope, Neptune appears as a tiny greenish disk, with a mean apparent diameter of about 2.1″ (the Moon has an apparent diameter of 31′). This corresponds to a linear diameter of 30,900 mi (49,400 km)—very similar to that of Uranus. The mass of Neptune is 17.21 times the mass of Earth, corresponding to a mean density of 1.66, somewhat above that of its sister planet. This suggests that the enrichment of heavy elements (when compared with Jupiter and Saturn, which exhibit a solar mixture) is somewhat greater in Neptune than in Uranus, but both must still be composed primarily of hydrogen and helium.

The apparent visual magnitude of Neptune at mean opposition, that is, when closest to Earth, is +7.8, too faint to be seen by the naked eye. The corresponding albedo is 0.4, a relatively high value characteristic of a planet with a dense atmosphere. Some observers have reported the presence of faint, dusky markings on the disk of Neptune, but these observations require confirmation. It is not

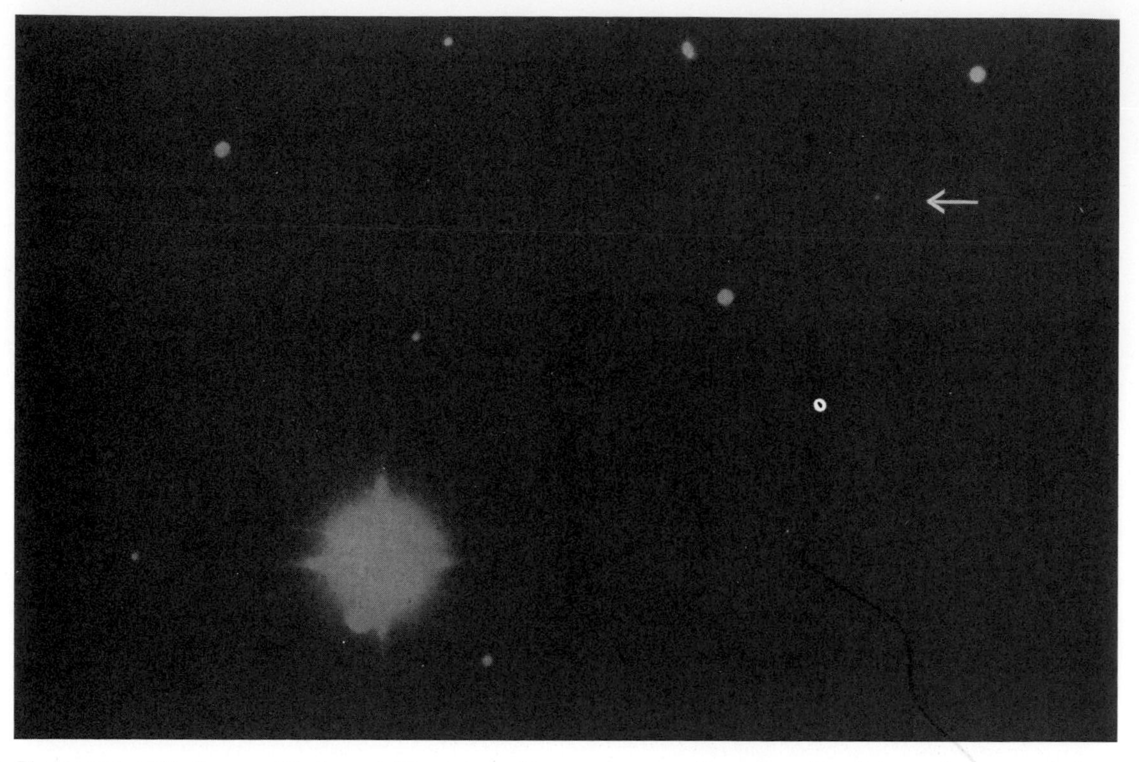

Photograph of Neptune and its two satellites taken with the 200-in. (5-m) telescope on April 17, 1950. The large satellite Triton appears as a white disk just to the left of the lower diffraction spike from the overexposed image of the planet. The small satellite Nereid is indicated by an arrow; the outer white dots are background stars. (*From G. P. Kuiper, Limits of completeness, in G. P. Kuiper and B. M. Middlehurst, eds., Planets and Satellites, ch. 18, University of Chicago Press, 1961*)

known what kinds of clouds or haze may be present in the planet's atmosphere. The period of rotation has been determined spectroscopically to be between 14 and 15 hr. A more precise value awaits the use of more modern techniques. The sense of rotation is direct, and the rotational axis is inclined at an angle of about 29°.

The temperature that this distant object would assume as a result of heating by solar radiation can be calculated as 45 K. But temperatures measured near a wavelength of 13 μm are very much higher, over 100 K. This suggests the presence of a region in the upper atmosphere where the temperature increases with height, a so-called thermal inversion. The temperature is probably associated with emission from ethane gas formed in this upper layer, as is known to be true for Jupiter and Saturn. There is also evidence for emissiom from methane at wavelengths near 8 μm. At radio frequencies, which permit observations of radiation from the lower atmosphere, one finds temperatures increasing with depth, the highest value thus far reported being 227±23 K at a wavelength of 6 cm. There is evidence from infrared and radio observations that Neptune has an internal energy source, as do Jupiter and Saturn. It is radiating more energy than it receives from the Sun, perhaps as a result of the conversion of gravitational potential energy to kinetic energy through a very slow contraction.

The optical spectrum of Neptune is dominated by strong absorption bands of methane. Hydrogen absorptions have also been detected, and the determination of the relative abundances of these two gases is a topic of active research. Normalized path lengths of 8 km for methane and 225 km for hydrogen have been reported; the comparable figure for the Earth's atmosphere is 8 km for oxygen and nitrogen combined. There must also be large amounts of helium and neon, but these gases are impossible to detect by ground-based observations. At lower, warmer levels of the atmosphere,

Orbital elements of Triton and Nereid

Satellite	Mean distance to Neptune, 10^3 mi	Sidereal period, days	Diameter, mi	Magnitude at mean opposition
Triton	220	5.877	2500	13.6
Nereid	3480	359.10	250	18.7

Solar eclipse; multiple-exposure photograph taken near St. Paul, Minnesota, on morning of June 30, 1954. (*M. Tinklenberg*)

(*a*) Large Magellanic Cloud; the large red patch is the 30 Doradus nebula, one of the largest H II regions known. (*b*) Small Magellanic Cloud. (*Photographs by R. J. Dufour*)

(a)

(b)

(c)

(*a*) M83, a prominent southern spiral galaxy. Blue patches are associations of very young stars; dark lanes and patches are produced by dust; red blobs are H II regions. (*b*) Galaxy NGC 5128. (*c*) NGC 6744, a barred spiral galaxy. (*Photographs by R. J. Dufour*)

Oblique view of Mars along Tharsis ridge from *Viking Orbiter1*, showing the three Tharsis volcanoes. (*NASA*)

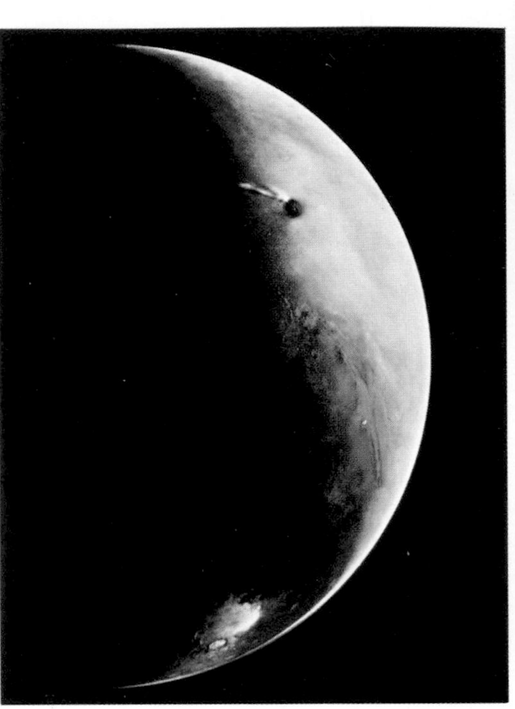

Dawn side of Mars from *Viking Orbiter 2*. Major features are visible, including the giant volcano Ascreus Mons (near top), the Valles Marineris (middle), and the large frost-covered crater basin Argyre (near bottom). (*NASA*)

Artist's conception of Olympus Mons wreathed in clouds. A well-defined cloud train extends several hundred kilometers beyond the mountain (upper left). (*NASA*)

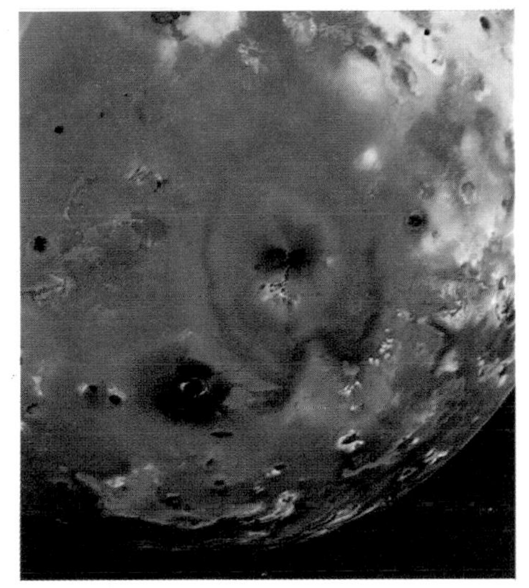

Io from *Voyager 1*. Reddish, white, and black areas are probably surface deposits. Many of the black spots are associated with volcanic craters. (*NASA*)

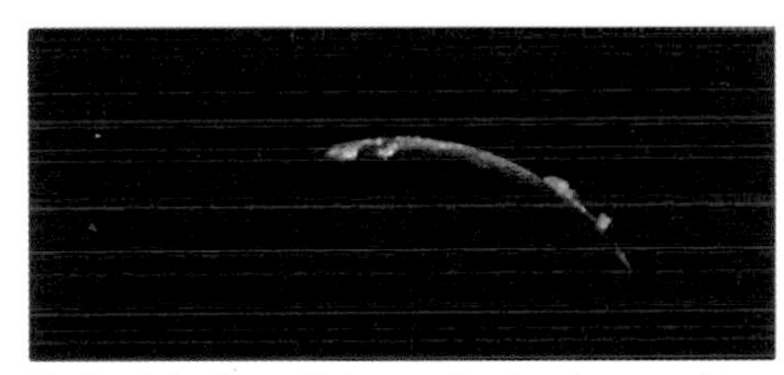

Limb of Io from *Voyager 2*, showing two blue volcanic eruption plumes about 100 km high. (*NASA*)

Jupiter from *Voyager 1*, with the Great Red Spot showing prominently below the center and Ganymede at the lower left. (*NASA*)

Jupiter from *Voyager 1*, showing the Great Red Spot (upper right) and a turbulent region immediately to the west. (*NASA*)

Crab Nebula in Taurus (*above*) is debris from a stellar explosion seen as a nova by Oriental astronomers in 1054. High-energy electrons cause the gas to glow. These colors are not visible because of low intensity; even sensitive color film requires exposures of several hours through 200-in. telescope. Great Nebula in Orion (*below*), visible as the middle "star" in the sword of Orion, is a gas cloud excited to incandescence by hot stars in its center. Stars may form in such a region. (*Photographs by William Miller, Carnegie Institution of Washington*)

Matter exists both as gas and as dust grains in the space between stars. This interstellar matter is observed in a variety of ways. For example, the North American Nebula in Cygnus (*to left*), which is a cloud of gas having a density possibly a thousand times greater than the average density of interstellar gas, fluoresces by ultraviolet light from adjacent stars; it is called an emission nebula. Dust in the space between the nebula and Earth scatters all but the red light. Density of this interstellar dust is about 30 dust grains per cubic kilometer. A much denser cloud of dust between the North American Nebula and Earth obscures portions of the emission nebula to create the appearances of the "Gulf of Mexico" and the "Atlantic Ocean." The Veil Nebula (*above*), also in Cygnus, consists of filaments of high-velocity gas ejected from an exploding star more than 50,000 years ago. As this gas collides with atoms in interstellar space, it ionizes and glows. High-velocity collisions along the leading edge create blue light; less-energetic collisions along the trailing edge create red light. Collisions also decelerate the gas filaments so that the nebula will cease to glow in another 25,000 years. Both photographs were made with the 48-in. Schmidt telescope, California Institute of Technology.

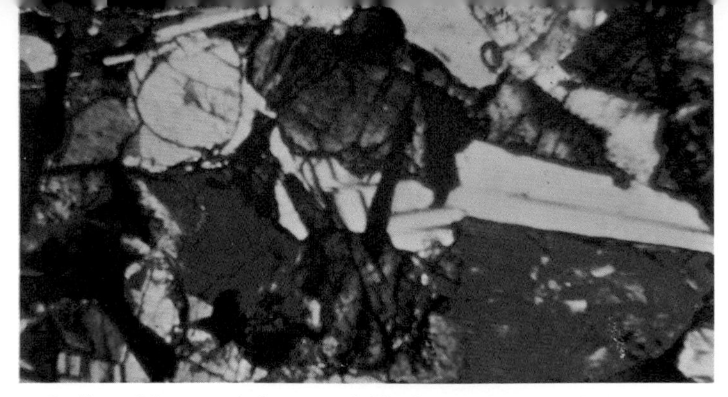

Section of lunar rock in crossed Nicols showing zoned pyroxene, plagioclase, and ilmenite. (*NASA*)

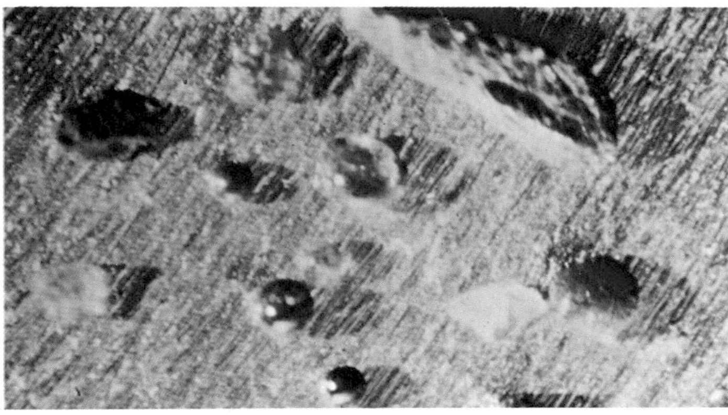

Microscopic view of spherules from lunar soil which was collected by the *Apollo 11* astronauts. (*NASA*)

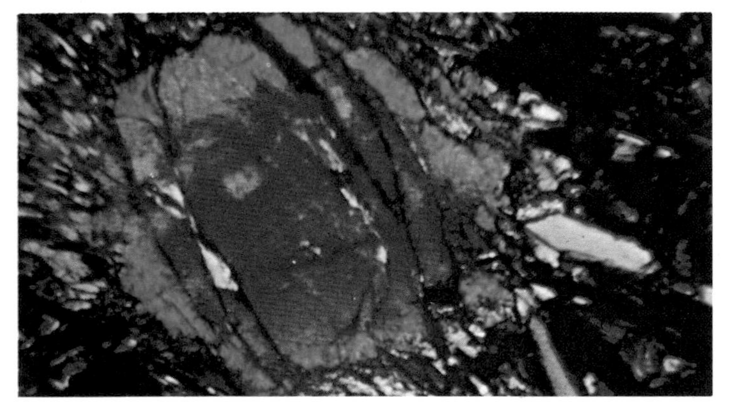

Thin section of lunar rock showing relatively large equant crystals. Porphyritic texture suggests differing rates of growth of the large crystals versus the groundmass minerals. Large crystals are olivine and the groundmass minerals are pyroxenes, feldspars, and metal compounds. Large crystals are approximately 1 mm across. The texture and mineralogy are both common in Earth volcanic rocks. (*NASA*)

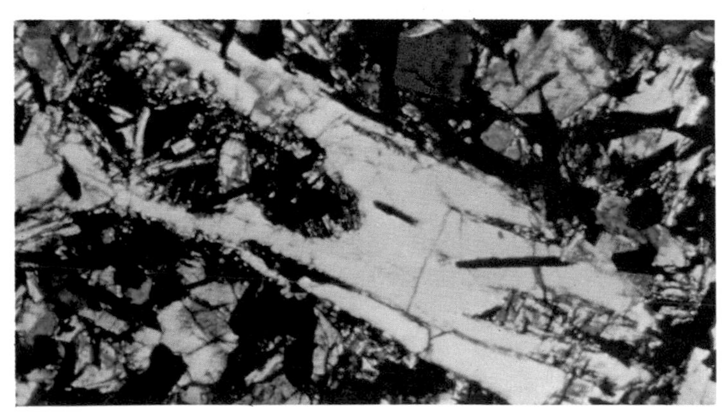

Photomicrograph of lunar sample. Light blue and white mineral is plagioclase; black mineral is ilmenite; blue and/or green and/or orange and/or yellow and/or red mineral is pyroxene. The large pyroxene is a phenocryst that had been partially resorbed. (*NASA*)

Rock from surface of the Moon. (*NASA*)

Photomicrographs of thin sections of lunar rock in polarized white light. Colors are caused by the interaction of the polarized light with the crystalline structure of the various minerals. Each color usually represents a different mineral. (*NASA*)

(*Below; top*) Chainpur, India, chondrite. Photomicrograph of a typical chondritic texture with a small intact chondrule composed of olivine and glass adjacent to two larger broken olivine-plus-pyroxene chondrules. The chondrules are set in a matrix of fine-grained silicate fragments.

(*Center*) Bununu, Nigeria, achondrite. Typical brecciated texture of the howardite achondrites. The dark angular glass fragments are mixed with mineral and rock fragments derived from basaltic crystalline rocks by impact brecciation.

(*Bottom*) Isna, Egypt, chondrite. The texture of a fine-grained carbonaceous chondrite displays chondrules, angular to rounded rock fragments, and black metal and sulfide particles set in a dark carbon-rich matrix.

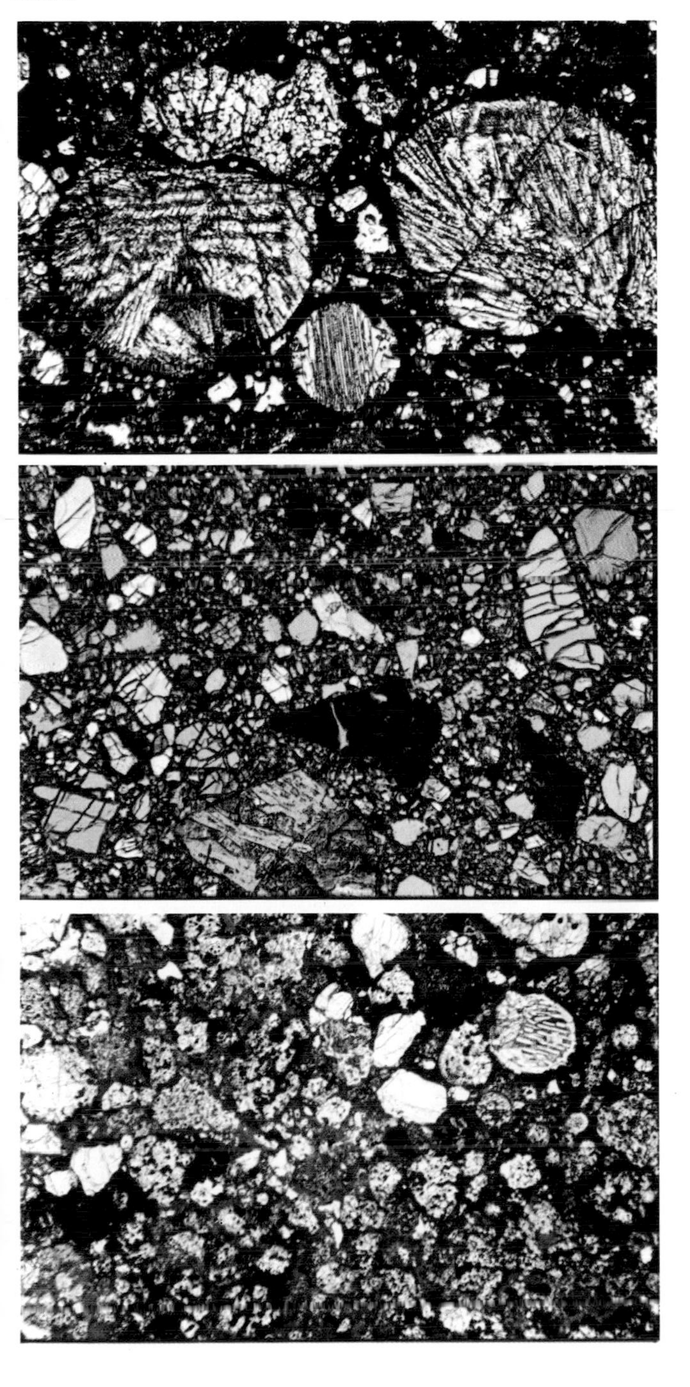

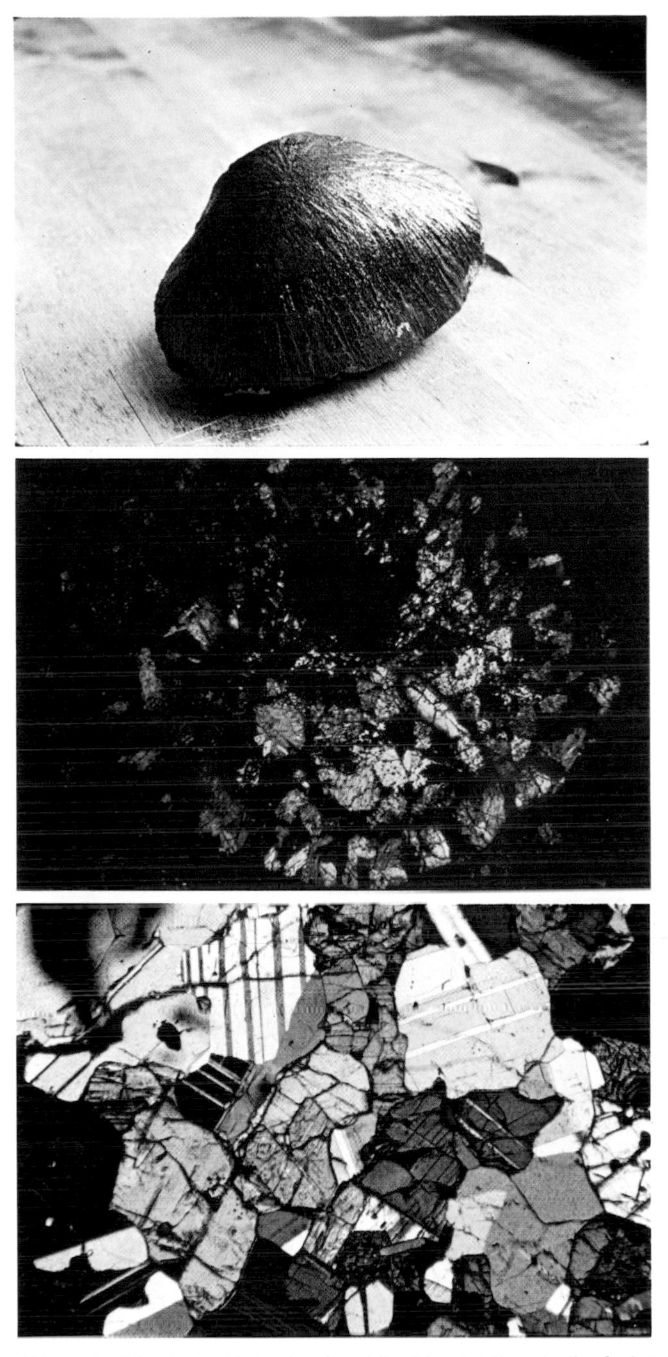

(*Above; top*) Luotolax, Finland, achondrite. The striations in the fusion crust covering this specimen show that it fell through the air in a fixed position.

(*Center*) Inclusion AL3S4 of the Allende carbonaceous chondrite. This inclusion is approximately 1.5 cm in diameter. It is composed of an outer rim of large melilite crystals and an interior of pyroxene, spinel, and an orthite. The pyroxene and spinel are highly enriched in nucleogenetic ^{16}O, but oxygen in the melilite is normal.

(*Bottom*) Moore County, North Carolina, achondrite. This coarse-grained eucrite, which contains interlocking grains of twinned plagioclase and pyroxene, is similar to many terrestrial basalts.

(*Photographs courtesy of Edward P. Henderson, except Allende chondrite courtesy of Dr. I. D. Hutcheon*)

(a)

(b)

(c)

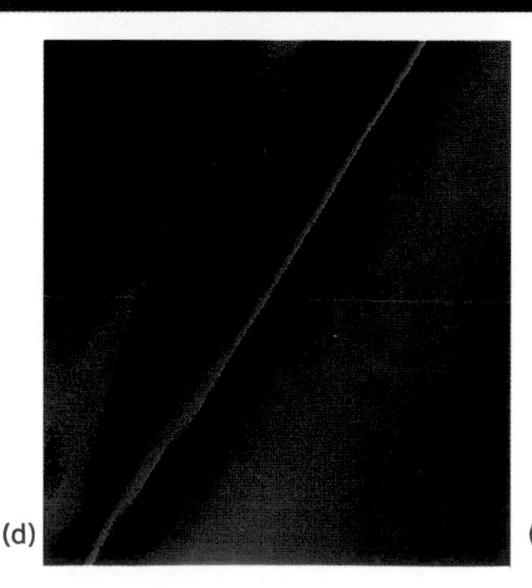

(d)

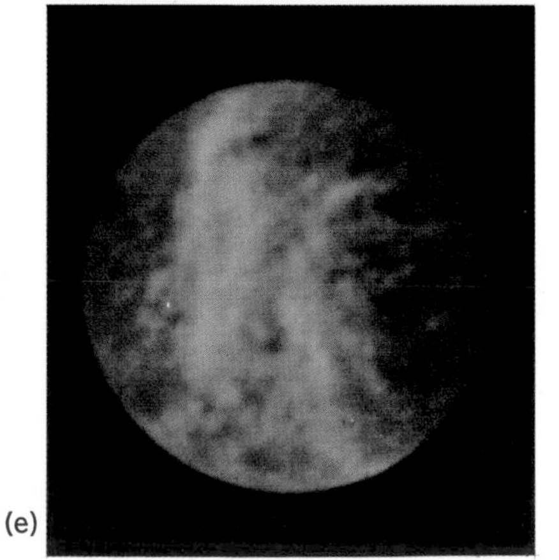

(e)

Images of Saturn system from *Voyager 1*. (*a*) The planet, color-enhanced. (*b*) Rings and their shadows. (*c*) The satellite Dione. (*d*) Layers of haze covering the satellite Titan; colors are false and are used to show details of the haze. (*e*) The satellite Rhea; colors have been exaggerated to bring out differences. (*NASA*)

Solar corona, during eclipse of the Sun, showing long equatorial streamers and short polar brushes typical of solar activity during the minimum portion of the sunspot cycle. (*Tersch Enterprises, Terry Schmidt*)

A few seconds before the Moon completely covers the Sun, the solar chromosphere flashes into view. This spectrogram taken at that time reveals the chemical elements in this region of the solar atmosphere. The phenomenon is called the flash spectrum. (*Tersch Enterprises, Terry Schmidt*)

Photograph of the Sun taken in visible light shows the majority of the solar activity on the northern hemisphere. Note the limb darkening and complexity of the spot groups. Eight spot groups are visible, containing a total of 210 individual spots. (*Tersch Enterprises, Terry Schmidt*)

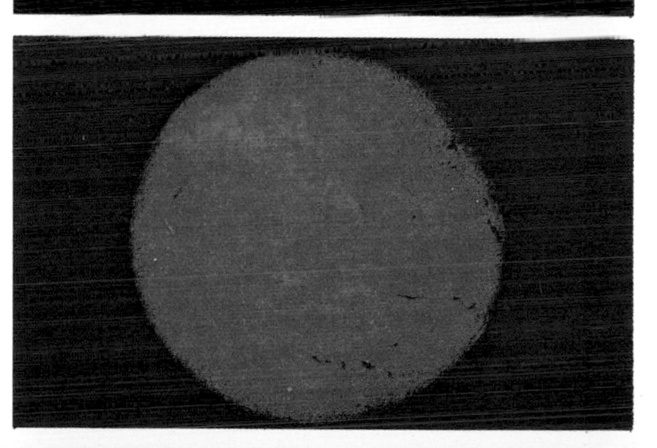

Spectroheliogram of the Sun in H-alpha light. Sunspots appear as dark centers, and hydrogen prominences projected against the solar disk as black filaments. At upper left, a large solar flare appears as a bright patch. The bright patches are hydrogen flocculi. (*Sacramento Peak Observatory, Air Force Cambridge Research Laboratories*)

Chromosphere of the Sun showing a very active surge prominence developing at the solar limb. (*Sacramento Peak Observatory, Air Force Cambridge Research Laboratories*)

These giant loop prominences on the surface of the Sun were photographed on June 28, 1957. (*Sacramento Peak Observatory, Air Force Cambridge Research Laboratories*)

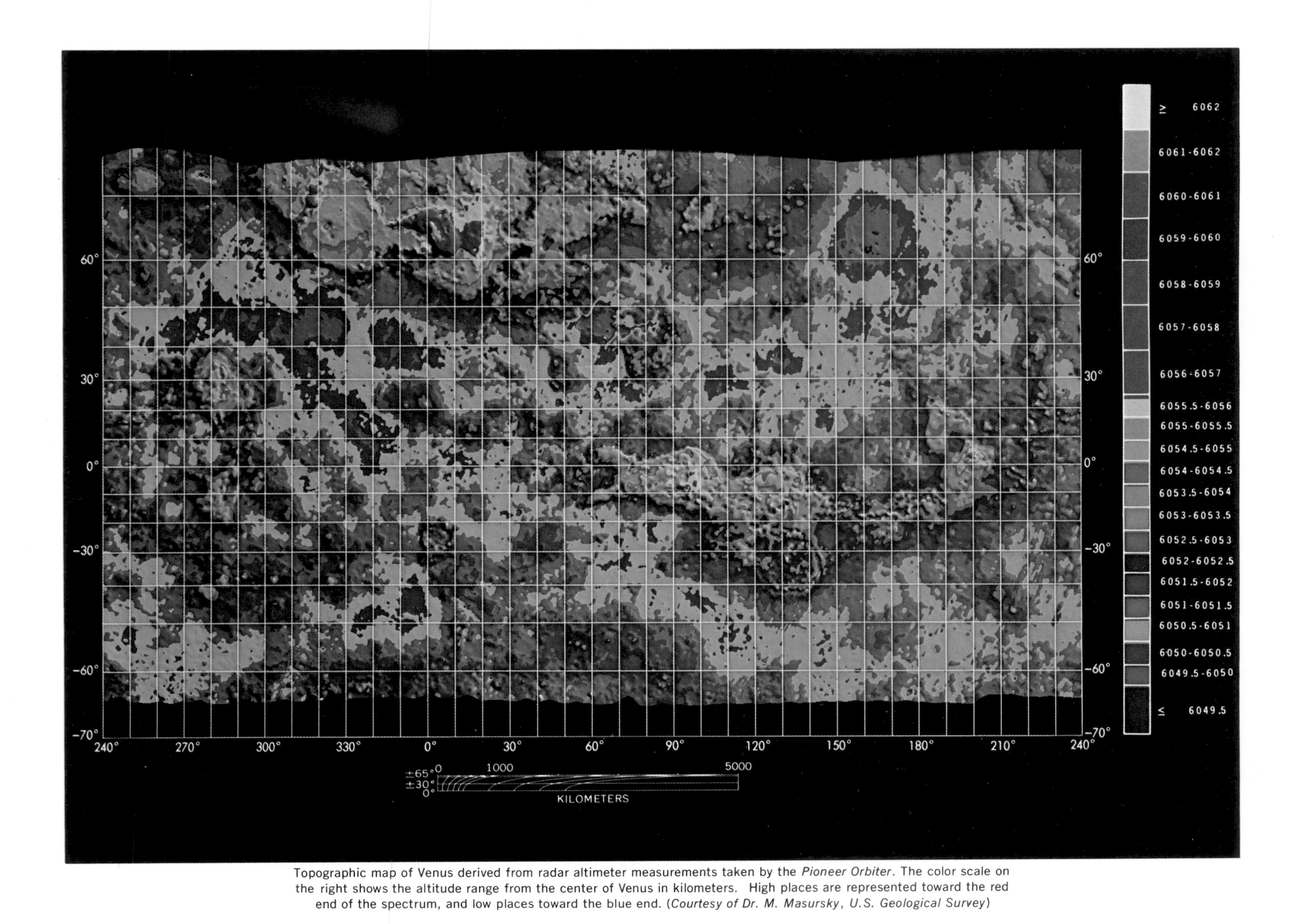

Topographic map of Venus derived from radar altimeter measurements taken by the *Pioneer Orbiter*. The color scale on the right shows the altitude range from the center of Venus in kilometers. High places are represented toward the red end of the spectrum, and low places toward the blue end. (*Courtesy of Dr. M. Masursky, U.S. Geological Survey*)

one expects ammonia, hydrogen sulfide, water vapor, and other reduced compounds to be present.

Satellites. Neptune has two known satellites (see illustration). The larger and brighter one, Triton, was found visually by W. Lassell in 1846; Nereid was discovered in a photographic search by G. P. Kuiper in 1949 (see table). The orbit of Triton is remarkably circular, but its inclination of 160° makes it an irregular satellite with retrograde motion about Neptune. It is too small to show a disk, so its diameter can only be estimated from its apparent brightness by assuming a value for the albedo (0.2 in this case). The situation for Nereid is even worse, since it is 100 times fainter. Nereid is also classed as an irregular satellite by virtue of the very high eccentricity (0.76—the highest of any known satellite) and relatively high inclination (27.5°) of its orbit. It may well be a captured object, in which case one might expect it to be very similar to the nucleus of a comet in structure and composition.

The relatively great brightness of Triton suggests a large size, but without a known mass or diameter, one can only speculate about its additional properties. The infrared spectrum of this object shows an absorption feature that may be caused by the presence of an atmosphere containing methane.

Cosmogony. The many similarities between Uranus and Neptune suggest a common mode of origin for these two bodies that was quite different from that for Jupiter and Saturn. The possibility that Pluto may once have been a satellite of Neptune is also a potential area of investigation. Neptune is now the only outer planet that does not have a known system of rings. This may simply be the result of the intrinsic difficulty in observing this distant object; several searches are planned. *See* JUPITER; PLUTO; SATURN, URANUS.

[TOBIAS C. OWEN]

Bibliography: W. K. Hartmann, *Moons and Planets*, 1972; G. P. Kuiper and B. Middlehurst (eds.), *Planets and Satellites*, 1961.

Neutron star

A star containing about 1½ solar masses of material compressed into a volume approximately 10 km in radius (1 solar mass $= 2 \times 10^{30}$ kg). Neutron stars are one of the end points of stellar evolution and are probably the final states of medium-mass and possibly of high-mass stars. The density of neutron star material is 10^{14} to 10^{15} times the density of water and exceeds the density of matter in the nuclei of atoms. Neutron stars are pulsars (pulsating radio sources) if they rotate sufficiently rapidly and have strong enough magnetic fields.

Neutron stars play a role in astrophysics which extends beyond their status as strange, unusual types of stellar bodies. The interior of a neutron star is a cosmic laboratory in which matter is compressed to densities which are found nowhere else in the universe. Precise measurements of the rotation of neutron stars can probe the behavior of matter at such densities. Neutron stars in double-star systems can emit x-rays when matter flows toward the neutron star, swirls around it, and heats up. Neutron stars are probably formed in supernova explosions, events in which a dying star becomes more luminous than an entire galaxy, up to 10^{12} times as powerful as the Sun. A few pulsars are found in double-star systems, and careful timing of the pulses they emit can test Einstein's general theory of relativity. *See* GRAVITATION; RELATIVITY.

Dimensions. The radius of a neutron star, while not precisely known, is about 10 km. A few neutron stars exist in double-star systems, wherein one can measure the strength of the gravitational pull of the neutron star on the other star in the system and hence determine the mass of each. The masses of these neutron stars are slightly more than the mass of the Sun, with the measured values ranging from 1.4 to 1.8 solar masses. *See* BINARY STAR.

The highest mass that a neutron star can have is about 2 solar masses. More massive neutron stars might exist if current ideas about the behavior of neutron star matter turn out to be wrong. If Einstein's theory of gravitation is the correct one, though, a neutron star with a mass larger than some limiting value will collapse catastrophically, because its internal pressure will be insufficient, and become a black hole. The exact value of this limiting mass is not known precisely, but lies between 3 and 5 solar masses. *See* BLACK HOLE.

Internal structure. The state of matter beneath the surface of a neutron star is completely different from matter on the surface of the Earth and cannot be reproduced in terrestrial laboratories. The outer crust of neutron stars is solid. The gaseous surface, only a few meters thick, is different from ordinary gases, for the structure of atoms is dominated by the strong magnetic field that certainly exists in all pulsars and probably is present in all neutron stars. The magnetic field distorts the shape of the electron orbitals which form the outer part of the atoms. A few meters below the surface, the material solidifies. The solid crust is about 1 km thick (see illustration) and is approximately 10^{17} times stiffer than steel. As one probes deeper, the densities increase, and a progressively greater fraction of the electrons are forced into atomic nuclei, where they combine with protons to form neutrons.

Most of the interior of a neutron star consists of matter which is almost entirely composed of neutrons. In the bulk of the star, this matter is in a superfluid state, where circulation currents can flow without resistance. In rotating superfluids maintained in physics laboratories, small vortices form; such vortices may also exist in neutron stars. This material is under pressure, since it must be able to support the tremendous weight of the overlying layers at each point in the neutron star. This pressure, called degeneracy pressure, is caused by the close packing of the neutrons rather than by the motion of the particles. As a result, neutron stars can be stable no matter what the internal temperature is, because the pressure that supports the star is independent of temperature.

The central regions of neutron stars are poorly understood. For example, the role played by strange, subnuclear particles such as the Δ-hype-

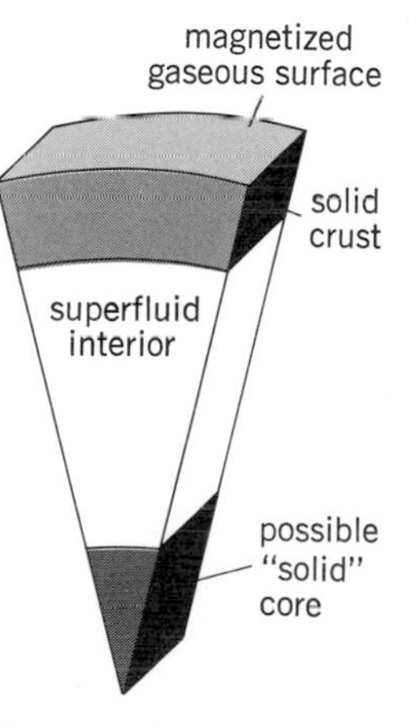

Cross section of a neutron star, with various layers.

ron is unknown. Some calculations indicate that pi mesons, particles which carry the strong nuclear force, are created by the strong forces between the neutrons and form some kind of condensation. It is remotely possible that quark matter, material composed of the postulated fundamental particles of matter, exists in the cores of neutron stars.

Observations. Theoretical calculations form the basis for the above description of a neutron star. A variety of astronomical observations may eventually confirm or disprove various aspects of this model. The pulses from a pulsar (a rotating neutron star) can be timed very precisely and can suggest the way that neutron stars change their shape as they spin more slowly. A single, isolated neutron star loses energy in the form of high-speed particles emitted from its surface, and its rotation rate slows down. A newly formed pulsar spins very fast—the pulsar in the Crab Nebula, formed a little less than 1000 years ago as observed from Earth, rotates 30 times every second. Eventually it will rotate more slowly, and its structure will readjust, occasionally producing abrupt changes in the rotation period which are called glitches. *See* CRAB NEBULA.

Some neutron stars exist in double-star systems, where matter flows from one star to the neutron star, forms a whirling disk around it, and causes it to rotate faster as it gains more mass. These neutron stars emit pulses of x-rays, which can be detected and timed by satellite observatories orbiting above the Earth's atmosphere. Timing observations of pulsars, in binary systems and isolated in space, have not provided any solid confirmation of detailed models of neutron stars. Sudden and irregular changes in pulse arrival times do demonstrate that some solid matter exists in neutron stars. *See* X-RAY ASTRONOMY.

Origin. Two neutron stars, the pulsar associated with the Crab Nebula and the one in the center of a supernova remnant in the constellation Vela, undoubtedly were created in supernova explosions, the violent death throes of massive stars. It is difficult to produce reliable models of such a violent event, so it is not clear how a neutron star is produced in the core of a star in its late evolutionary stages. One possibility is that the core collapses rapidly and violently, releasing energy that then explosively ejects the outer layers of the star into space. *See* PULSAR; STELLAR EVOLUTION; SUPERNOVA. [HARRY L. SHIPMAN]

Bibliography: I. Asimov, *The Collapsing Universe: The Story of Black Holes*, 1977; G. Baym and G. Pethick, Physics of neutron stars, *Ann. Rev. Astron. Astrophys.*, 17:415–445, 1979; K. Brecher and G. Caporaso, "Neutron" stars within the laws of physics, *Ann. N.Y. Acad. Sci.*, 302:471–481, 1977; H. L. Shipman, *Black Holes, Quasars, and the Universe*, 2d ed., 1980.

Nova

The brightening of a seemingly undistinguished star within a few days to thousands or even hundreds of thousands of times its previous luminosity. These events, which occur about 10 times a year in a large spiral galaxy like the Milky Way, are the classical novae (short for novae stellae, or new stars, so called because the pre-flare-up stars are too faint to have been seen in the days before telescopes). A much rarer and still more violent class of stellar flare-up is given the name supernova. *See* SUPERNOVA.

The classical novae are defined and distinguished from other kinds of explosive stellar events by the following observed properties: (1) brightening by more than a factor of 4000 in a few days; (2) noticeable fading in less than a few years (more often weeks to months); (3) a complex, changing spectrum, showing emission lines of common gases at high temperature; (4) ejection of gas at speeds of 100 to 5000 km/s; and (5) no historical record of a similar previous event involving that particular star.

The single most important step in understanding nova explosions occurred in 1964, when Robert Kraft showed that most previously observed novae had occurred in close (that is, stars not much farther apart than their own sizes) binary systems, containing one hot, compact star (like a white dwarf) and one cooler, more extended star (like a main sequence or subgiant star). The systems typically have large amounts of stray gas and masses near 1 solar mass (2×10^{30} kg). *See* BINARY STAR; STAR; WHITE DWARF STAR.

Characteristics of cataclysmic binaries. Several different kinds of systems sharing both a propensity for unpredicable behavior and the general structure that Kraft demonstrated for the classical novae are known. They include: (1) the recurrent novae, whose outbursts are similar to classical ones, but they have been seen to go off more than once, and the cool star is typically a giant; (2) the symbiotic stars, which have structures like the recurrent novae, but for which no outburst has been seen; (3) the dwarf novae, whose outbursts are much smaller (factors of 100 or so in brightness) and recurrent with periods of weeks or months; and (4) the novalike variables, with structures like the classical and dwarf nova systems, but no outbursts recorded. *See* VARIABLE STAR.

Such systems were collectively dubbed cataclysmic binaries by Sergei Gaposchkin and Cecilia Payne Gaposchkin. All have in common the evidence for two stars, hot and cool, and a stream of gas, typically passing from cool to hot component. Most also clearly display a disk of gas around the hot star, which contributes much of the total light of the system, and many have a hot spot on the disk, presumably where the gas stream hits it. Besides the major outbursts, most cataclysmic binaries vary in brightness more mildly on a variety of time scales, including minute-to-minute flickering, quasiperiodic oscillations of a few hours, and many others. There are also properties that differ systematically among the types of cataclysmic binaries. Novae and recurrent novae always blow off gas during outbursts. Dwarf novae never do, and their spectra show they do not get as hot as the others. The recurrent novae and the symbiotic stars have giant (rather than main-sequence) cool stars. And the strength of the gas stream decreases systematically from recurrent, to classical, to dwarf novae, averaging about 10^{-7}, 10^{-8}, and 10^{-10} solar mass per year in the three classes.

Most astronomers are now convinced that the origin of these systems and the causes of their assorted kinds of explosive behavior are understood reasonably well.

Origin of cataclysmic binaries. There exists a class of main-sequence (that is, relatively unevolved) binaries whose masses and separations are rather similar to the cataclysmic systems. These W Ursa Majoris variables were once thought to evolve fairly directly into novae and their relatives. This is no longer a tenable view, as further calculation has shown that transfer of gas between the two stars in a W Ursa Majoris system causes the two stars to merge (or, at least, one to become very small) long before either can evolve to a white dwarf of the type required for a cataclysmic system.

Instead, the evolution of a pair of stars whose initial separation is quite large compared to their sizes, and whose masses are rather larger than 1 solar mass, must be considered. In such a system the more massive star (which is always the more rapidly evolving), after 10^9 years or so of main sequence life, becomes a red giant, whose outer envelope engulfs the other star temporarily. The two stellar cores moving inside the single gaseous envelope transfer much of their orbital angular momentum to the envelope, driving it away from the system as the stars gradually spiral together. Thus, by the time the more massive star has become a degenerate dwarf, it is much closer to its companion than it was to begin with. *See* PLANETARY NEBULA.

Eventually (another 10^9 years or so) the less massive star begins its evolution away from the main sequence, gradually expanding its radius. In due course (just when depends on the separation of the two stars) it gets big enough to fill a critical volume, called the Roche lobe. The surface of this lobe is defined by the criterion that material inside it is pulled back toward the evolving star, while gas on or just outside it is free to flow through the system and stream down toward the degenerate star from the point on the lobe surface shown as L_1 in the illustration. Because of conservation of angular momentum, the gas stream does not fall directly onto the degenerate dwarf, but goes into an accretion disk around it, and slowly spirals inward, as indicated by the small arrows in the illustration. The impact of the stream hitting the disk provides additional energy and makes a hot spot. Thus there are formed precisely the components (cool star, hot compact star, accretion disk, gas stream, and hot spot) that observations of assorted kinds of cataclysmic binaries indicate must exist in these systems. *See* STELLAR EVOLUTION.

Outburst mechanisms. Clearly the complex arrangement shown in the illustration provides all sorts of opportunities for variable light output from the systems. If the cool star becomes a giant before filling its Roche lobe, the gaseous wind from its surface, hitting the white dwarf, may provide enough extra light and emission lines to make the system conspicuous. These are (at least some of) the symbiotic stars.

Once the lobe is filled, lumps, hot spots, and all sorts of instabilities in the stream, disk, and

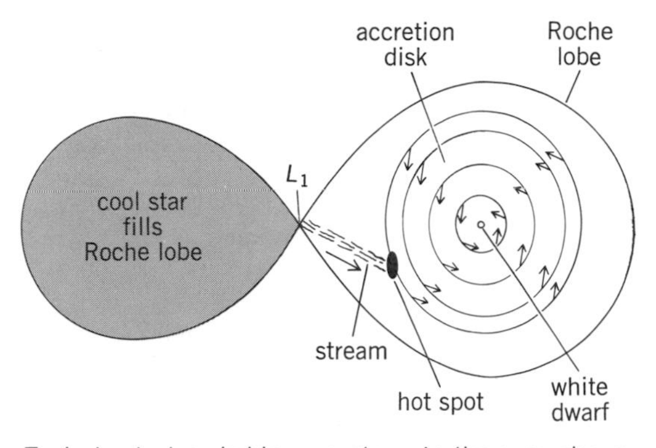

Typical cataclysmic binary system. In the case shown, the white dwarf is the more massive of the two stars, so that its Roche lobe is the larger; the opposite case also occurs. *(From V. Trimble, How to survive the cataclysmic binaries, Mercury, 9(1):8, Astronomical Society of the Pacific, 1980)*

interfaces between disk and star will give rise to both irregular and periodic (for example, the orbit period of a lump in the disk) variations in brightness, accounting for the many kinds of flickering and pulsations seen in the cataclysmic systems. Rare cases (called AM Her systems), where the recipient star has a strong magnetic field, are capable of additional interactions between gas and field and show additional complexity in spectra, light variation, and polarization.

A simple calculation reveals that accretion from the disk to the white dwarf releases more energy than is coming from the hot star itself (and, often, from either star). Thus changes in the accretion rate will change the brightness of the system. In particular, the flare-ups of dwarf novae can be convincingly modeled by a rapid increase in the accretion rate from perhaps 10^{-10} or 10^{-11} to 10^{-8} or 10^{-9} solar mass per year. It has not been determined whether the instability occurs in the disk itself or in the donor star, or why it occurs only at the low transfer rates found in dwarf novae and not among other cataclysmic systems.

The accreting gas, mostly hydrogen, since that is what the donor star is made of, gradually accumulates on the degenerate dwarf, perhaps mixing somewhat with its carbon-oxygen surface, until the bottom of the hydrogen layer is hot enough and dense enough for nuclear reactions to occur. The time required depends both on mass transfer rate and on the mass of the white dwarf. It can be as short as 30–100 years for a giant donor (whose transfer rate is large) and a massive receiver; more like 10^5 years for rates characteristic of old novae; and as long as 10^6 or 10^7 years for dwarf nova transfer rates. Nuclear reactions at densities as high as those in white dwarfs are nearly always explosive. Thus, according to detailed computer models, 10^{46} ergs (10^{39} J) or more can be released in a very short time, matching the energy release of typical classical novae. The energy shows up in two ways. Some is radiated as light, making the system suddenly brighten; and some blows off the outer, hydrogen-polluted layers of the star into what eventually becomes an observable expanding gas cloud. As the outer layers expand, they cool, turning off the nuclear reactions and bringing the explosive phase to an end. As the surface of the white dwarf becomes visible and as dust forms in the outgoing envelope, much of the light output is shifted to the ultraviolet and infrared parts of the spectrum, making the system seem to fade faster than it really does. When the fireworks have died away, accretion onto the dwarf resumes, and continues until another explosive layer (10^{-4} solar mass or so is required) has built up. If the time required is shorter than the span of modern astronomical observations (200 years or so), a recurrent nova results; otherwise a classical "one-shot" nova is observed.

In principle, accretion of ordinary interstellar material onto an isolated white dwarf could also produce such an explosive layer, and a nova event not in a binary system. However, the time required is billions of years, and such events must be very rare. It is possible that the conspicuous naked-eye nova in Cygnus in 1975 was of this type.

Ultimate fate of cataclysmic binaries. The continuous transfer of material to the degenerate dwarf in these systems raises the possibility that the dwarf's mass might eventually be raised above the Chandrasekhar limit. The star would then collapse to a neutron star configuration, releasing enough energy to power a supernova. This has been suggested as a possible way to produce the subset of supernovae (those of type I) that apparently occur among relatively old, low-mass stars. However, it now seems more probable that a nova explosion blows off at least as much gas as was accreted to produce it, so that the white dwarf mass is not increased. Thus mass transfer and periodic explosions will occur until the donor star is stripped down to a dense core and can no longer fill its Roche lobe, leaving behind a pair of degenerate dwarfs in a short-period orbit. Systems like this will not be very conspicuous or easy to observe, but one or two are known. Buildup of accreted material on the degenerate dwarf may still be possible when the accretion rate is much larger or smaller than average, giving rise to an occasional supernova from within the cataclysmic binary population. *See* NEUTRON STAR.

[VIRGINIA TRIMBLE]

Bibliography: J. S. Gallagher and S. Starrfield, Theory and observations of classical novae, *Annu. Rev. Astron. Astrophys.*, 16:171–214, 1978; H. van Horn and V. Weidemann (eds.), *White Dwarfs and Variable Degenerate Stars*, 1979; A. D. Mallama and V. L. Trimble, Novae vs. dwarf novae: Energy sources and systematics, *Quart. J. Roy. Astron. Soc.*, 19:430–441, 1978; C. H. Payne-Gaposchkin, *The Galactic Novae*, 1957.

Nutation

In mechanics, the term nutation refers to a bobbing or nodding up-and-down motion of a spinning rigid body, such as a top, as it precesses about its vertical axis. Astronomical nutation refers to irregularities in the precessional motion of the equinoxes caused by the varying torque applied to the Earth by the Sun and Moon. Astronomical nutation, which is sometimes called nutational wandering of the terrestrial poles, should not be confused with nutation as defined in mechanics; the latter is present even if the source of the torques is unvarying.

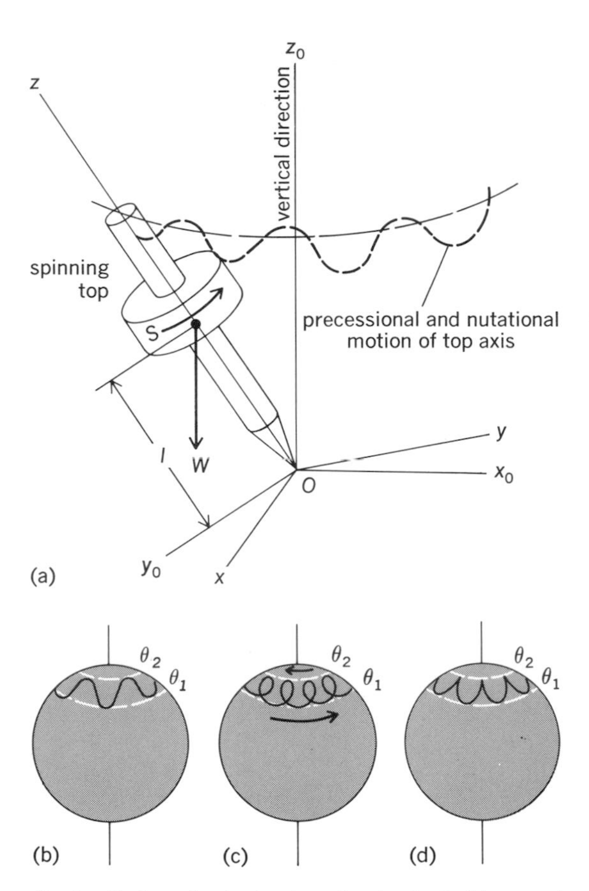

Fig. 1. Motion of spinning top showing typical traces of the top spin axis on unit spheres for different initial conditions. (*a*) General motion. (*b*) Top released with initial angular velocity in direction of precession. (*c*) Top released with initial angular velocity opposite to direction of precession. (*d*) Axis of spin released from rest.

Nutation of tops. The general motion of a spinning top, easily observed at low spin rates, consists of both precession and nutation. Figure 1*a* shows a symmetrical top spinning about a fixed point with its axis tracing out this general motion. Figure 1*b* shows the motion for the case when the axis of the spinning top is released with an initial angular velocity in the direction of precession; Fig. 1*c*, that with an initial velocity opposite to the precession: Fig. 1*d*, that with zero initial angular velocity (axis of spin released from rest).

The angular frequency of the nutation of a top axis at a high spin rate is given by Eq. (1), where I_z

$$\omega_n = \frac{I_z}{I_x} S \tag{1}$$

and I_x are moments of inertia about the z and x axes, respectively, and S is the angular velocity of spin. Furthermore, the rate of precession ω_p for the general motion is not uniform but varies harmonically with time with the same frequency as does the nutation, as shown in Eq. (2). The aver-

$$\omega_p = \frac{Wl}{I_z S}(1 - \cos \omega_n t) \tag{2}$$

age precessional frequency is given by Eq. (3).

$$(\omega_p)_{av} = \frac{Wl}{I_z S} \qquad (3)$$

As the spin rate S is increased the frequency of nutation increases, as shown by Eq. (1), and the nutational displacement $(\theta_2 - \theta_1)$ decreases very rapidly. Furthermore, as S is increased in Eq. (2), the frequency of the precessional variation increases, but from Eq. (3) the average rate of precession decreases. Therefore, in practice, for a sufficiently fast top, the nutation is so small and fast that it is damped out by the friction at the pivot and is unobservable. The top appears to precess uniformly about the vertical axis for this common case.

Nutation of projectiles. The motion relative to the centers of mass of bullets and shells stabilized by high rates of spin is identical to that of a spinning top relative to the fixed point of contact. Torques about the centers of mass due to aerodynamic forces acting on such bodies during flight cause precession and nutation to occur (Fig. 2). Furthermore, finned missiles, which usually rotate or spin rather slowly during flight unless prohibited from doing so by a suitable control system, also develop precessional and nutational angular velocities, similar to those of a spinning pendulum.

Astronomical nutation. The rotating Earth can be regarded as a spinning symmetrical top with small angular speed but large angular momentum, the latter due to its large mass. The gravitational attractions of the Sun and Moon cause the Earth's axis to describe a cone about the normal to the plane of its orbit. However, the magnitude of these gravitational attractions is continually varying, due to the changing positions in space of Sun, Moon, and Earth. The Moon's orbit is continually changing its position in such a way that the celestial pole undergoes a nodding (nutation) as well as a periodic variation in the rate of advance. The largest nutation is about 9.″2, and occurs in a period of a little less than 19 years; that is, the celestial pole completes a small ellipse of semimajor axis 9.″2 in about 19 years. *See* PRECESSION OF EQUINOXES.

There are lesser nutation effects which are due to the motion of the Moon's nodes, the changing declination of the Sun, and the changing declination of the Moon.

Nutation of gyroscopes. Still another example of the nutation of a spinning symmetrical body is given by the general motion of a gyroscope.

[RAY E. BOLZ]

Bibliography: E. P. Federov and M. L. Smith (eds.), *Nutation and the Earth's Rotation: Proceedings*, I.A.U. Symposium no. 78, Kiev, May 23–28, 1980; H. Goldstein, *Classical Mechanics*, 2d ed., 1980.

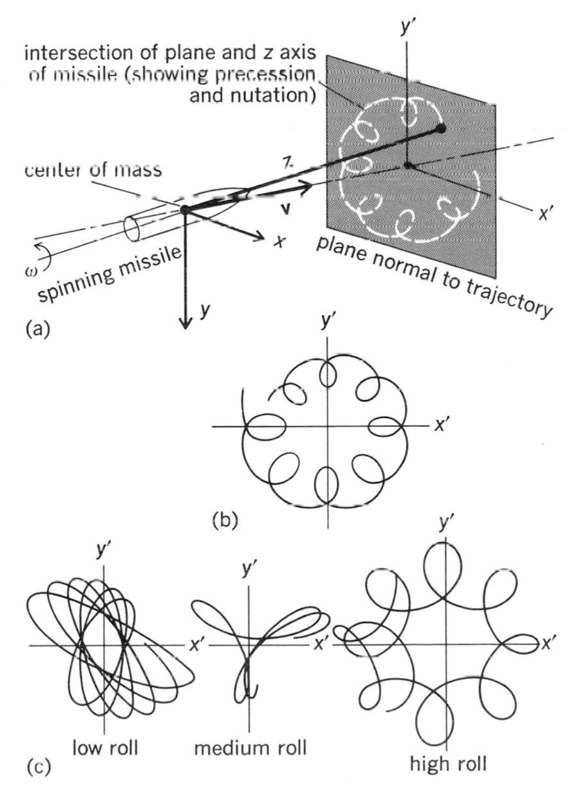

Fig. 2. Precessional and nutational displacements of missiles in flight. (*a*) Intersection of missile axis on a plane. The velocity of the center of mass of the missiles is indicated by **v**. (*b*) Typical motion of spin-stabilized missiles (bullets, shells, and the like). (*c*) Typical motion of rolling fin-stabilized missiles for low, medium, and large rates of roll, respectively.

Occultation

The apparent disappearance of a star or planet behind the surface of the Moon, or of a satellite behind the disk of the parent planet.

Occultations by the Moon are observed to determine the position of the Moon at the time of the phenomenon. They are also used for precise determinations of longitude. Because of the eastward motion of the Moon against the background of stars, immersion (or disappearance) of the star or planet occurs at the eastern limb, and emersion (or reappearance) at the western limb. A single occultation is visible only from a certain region of Earth. The parallels of latitude which enclose the region of visibility are called limiting parallels. Outside that region, the effect of parallax causes the star or planet to remain clear of the lunar disk. Location of the observing site within the limiting parallels is a necessary, but not sufficient, condition of visibility. It is also necessary that the Moon be above the horizon at the time of the occultation for the location considered.

Occultations of the four Galilean satellites of Jupiter occur at each of their superior conjunctions, with the exception of satellite IV, which occasionally passes clear of the planet's disk. In general, the immersions cannot be observed when Jupiter's shadow extends westward from the planet. Similarly, the emersions cannot be observed when the shadow extends eastward. In the first instance, the satellite is still eclipsed when the occultation begins, whereas in the latter, the satellite is in eclipse as the occultation ends. *See* JUPITER.

[SIMONE DARO GOSSNER]

Optical telescope

An instrument that collects light energy from a distant source and focuses it into an image that can then be studied by a number of different techniques. This definition of an optical telescope must be narrowed somewhat. Satellite-borne telescopes operated to study celestial x-rays or ultraviolet radiation and ground-based telescopes used to study radio radiation can be called optical telescopes in the sense that they operate according to the principles of geometrical optics. The following discussion stresses ground-based telescopes which are used to study radiation from celestial objects in the wavelength range from the Earth's atmospheric cutoff in the near ultraviolet to the infrared, that is, from 300 to 1000 nanometers. Such instruments may be classified as (1) large astronomical telescopes, used to study the nature of astronomical objects themselves, and (2) astronomical transit instruments, used to study positions and motions of astronomical objects and in the accurate determination of time. *See* RADIO TELESCOPE; TELESCOPE.

LARGE TELESCOPES

A large astronomical telescope is used by astronomers to study the fundamental problems in the field. The size required for a telescope to be considered large depends on its type, as discussed below.

Types of telescopes. There are basically three types of optical systems in use in astronomical telescopes: refracting systems whose main optical elements are lenses which focus light by refraction; reflecting systems, whose main imaging elements are mirrors which focus light by reflection; and catadioptric systems, whose main elements are a combination of a lens and a mirror. The most notable example of the last type is the Schmidt camera.

In each case, the main optical element, or objective, collects the light from a distant object and focuses it into an image that can then be examined by some means. Specific types of tools that are frequently employed to study astronomical objects are discussed below.

Refracting telescopes. The main optical element, or objective, of a refracting telescope is usually a long-focal-length lens. The objective lens is typically compound; that is, it is made up of two or more pieces of glass, of different types, designed to correct for aberrations such as chromatic aberration. Figure 1 shows a refractor lens imaging the light of two stars onto a photographic plate. To construct a visual refractor, a lens is placed beyond the images and viewed with the eye. To construct a photographic refractor or simply a camera, a photographic plate is placed at the position of the image. The characteristics of the components of a photographic lens may differ from those of a visual lens.

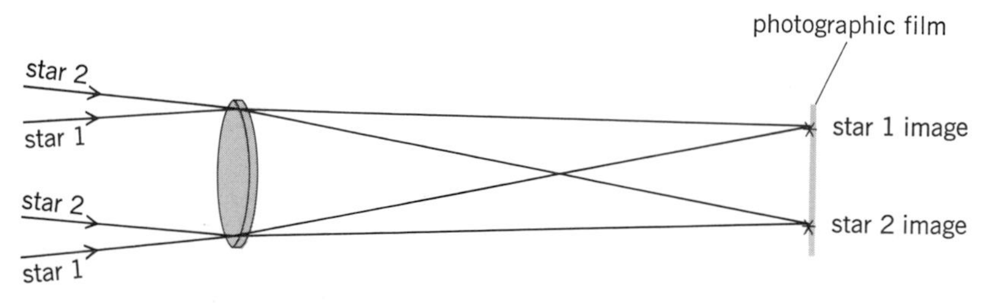

Fig. 1. Refracting optical system used to photograph a star field.

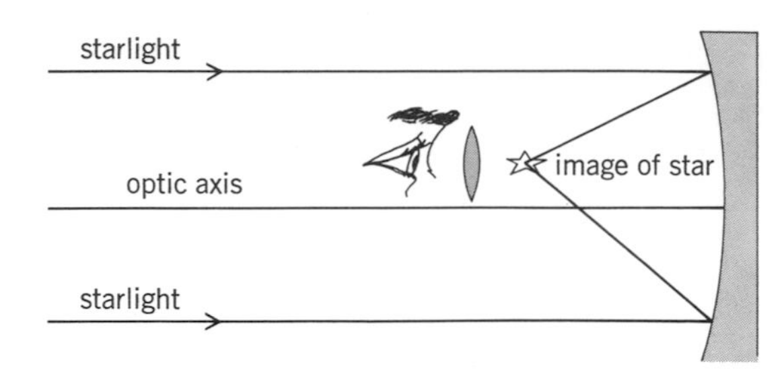

Fig. 2. Viewing a star with a reflecting telescope. In this configuration the observer may block the mirror, unless it is a very large telescope.

Generally, refracting telescopes are used in applications where great magnification is required, namely, in planetary studies and in astrometry, the measurement of star positions and motions. For example, most stellar parallaxes have been measured with refractors such as the Sproul Observatory 24-in. (61-cm) *f*/18 refractor. However, this practice is changing, and the traditional roles of refractors are being carried out effectively by a few new reflecting telescopes. This changing role has come about in part because of effective limitations on the size of refracting telescopes. The largest refractor, located at Yerkes Observatory at Williams Bay, WI, has a 40-in. (1.02-m) objective lens.

A refractor lens must be relatively thin to avoid excessive absorption of light in the glass. On the other hand, the lens can be supported only around its edge and thus is subject to sagging distortions which change as the telescope is pointed from the horizon to the zenith; thus its thickness must be great enough to give it mechanical rigidity. An effective compromise between these two demands is extremely difficult, if not impossible, for a lens over 1 m in diameter, making larger refractors unfeasible.

Reflecting telescopes. The principal optical element, or objective, of a reflecting telescope is a mirror. The mirror forms an image of a celestial object (Fig. 2) which is then examined with an eyepiece, photographed, or studied in some other manner.

Reflecting telescopes generally do not suffer from the size limitations of refracting telescopes. The mirrors in these telescopes can be as thick as necessary and can be supported by mechanisms which prevent sagging and thus inhibit excessive distortion. In addition, mirror materials having vanishingly small expansion coefficients (Cer-Vit, ultra-low-expansion-fused silica, and others), together with ribbing techniques which allow rapid equalization of thermal gradients in a mirror, have eliminated the major thermal problems plaguing telescope mirrors. Telescopes with mirrors up to 6 m in diameter have been built.

The reflecting telescope has other advantages which make it an attractive system. By using a second mirror (and even a third one, in some telescopes), the optical path in a reflector can be folded back on itself (Fig. 3*a*), permitting a long focal length to be attained with an instrument housed in a short tube. A short tube can be held by a smaller mounting system and can be housed in a smaller dome than a long-tube refractor, thus decreasing costs.

Finally, a large mirror has only one surface to be figured—that is, to be ground and polished to the desired optical shape—whereas a refractor lens usually consists of two pieces of glass that must be laboriously figured. The secondary mirror of a reflector is a second surface that needs to be figured, but it is much smaller than the primary and thus is much easier and cheaper to fabricate.

A variety of optical arrangements are possible in large reflecting telescopes, including the prime focus, the Newtonian focus, the Cassegrain focus, and the coudé focus.

The Newtonian focus is probably most widely used by amateur astronomers in reflectors having apertures on the order of 6 in. (15 cm; Fig. 3*b*). A flat mirror placed at 45° to the optical axis of the primary mirror diverts the focused beam to the side of the telescope, the image being formed by the paraboloidal primary mirror alone. An eyepiece, camera, or other accessories can be attached to the side of the telescope tube to study the image. In the largest telescopes, with apertures over 100 in. (2.5 m), provision is not usually made for a Newtonian focus. Instead, an observing cage is placed inside the tube, where the observer can take accessories to observe the image formed by the primary mirror. This prime focus is identical to the Newtonian focus optically, since the Newtonian flat does nothing more than divert the light beam. The modern reflectors have fast primary mirrors, so that the focal ratio at the prime focus is *f*/2.5 to *f*/6. Lower focal ratios permit shorter exposures on extended objects such as comets and nebulae.

A Cassegrain system consists of a primary mirror with a hole bored through its center, and a convex secondary mirror which reflects the light beam back through the central hole to be observed behind the primary mirror (Fig. 3*a*). Since the secondary mirror is convex, it decreases the convergence of the light beam and increases the focal length of the system as a whole. The higher focal ratios (*f*/8 to *f*/13) of Cassegrain systems permit the astronomer to observe extended objects, like planets, at higher spatial resolution and to isolate individual stars from their neighbors for detailed studies.

The classical Cassegrain system consists of a paraboloidal primary mirror and a hyperboloidal secondary mirror. The Newtonian focus, prime focus, and Cassegrain focus are not affected by spherical aberration. However, all of the systems are plagued by coma, an optical aberration of the paraboloidal primary. Coma causes a point source off the center of the field of view to be spread out into a comet-shaped image. To correct for the effect of coma, a corrector lens is often used in front of the photographic plate. The design of prime-focus corrector lenses is a major consideration in large telescope design, since the corrector

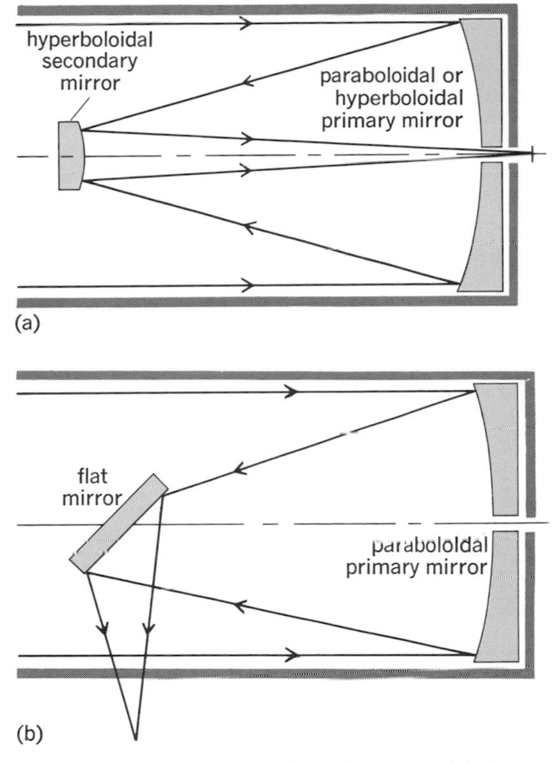

Fig. 3. Diagrams of reflecting telescopes. (*a*) Cassegrain telescope, with either classical or Ritchey-Chrétien optics. (*b*) Newtonian telescope.

lens itself can introduce additional aberrations. *See* ABERRATION; ASTRONOMICAL PHOTOGRAPHY.

To avoid complicated corrector lenses at the Cassegrain focus, the Ritchey-Chrétien system, an alternate design with both a hyperboloidal primary and a hyperboloidal secondary, is used in modern telescopes. This arrangement is not affected by coma or spherical aberration, so it has a wider field of view than the classical Cassegrain.

Schmidt camera. This is an optical system used almost exclusively for photographic applications such as sky surveys, monitoring of galaxies for supernova explosions, and studies of comet tails. The primary mirror of a Schmidt camera has a spherical shape, and therefore suffers from spherical aberration. To correct for this problem, the light passes through a thin corrector plate as it enters the tube, as illustrated in Fig. 4. Schmidt

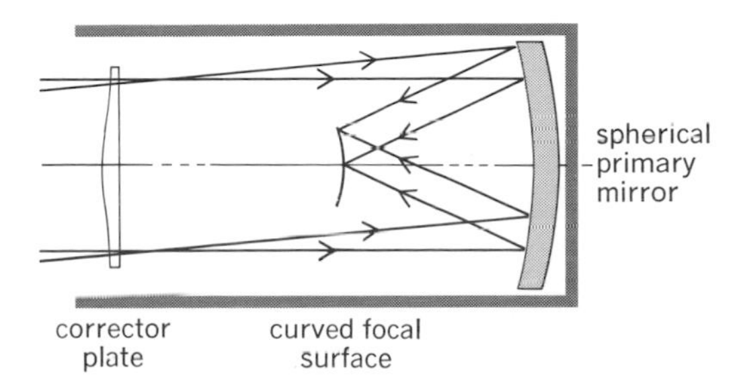

Fig. 4. Optics of the Schmidt system showing the axial and the extra-axial light paths.

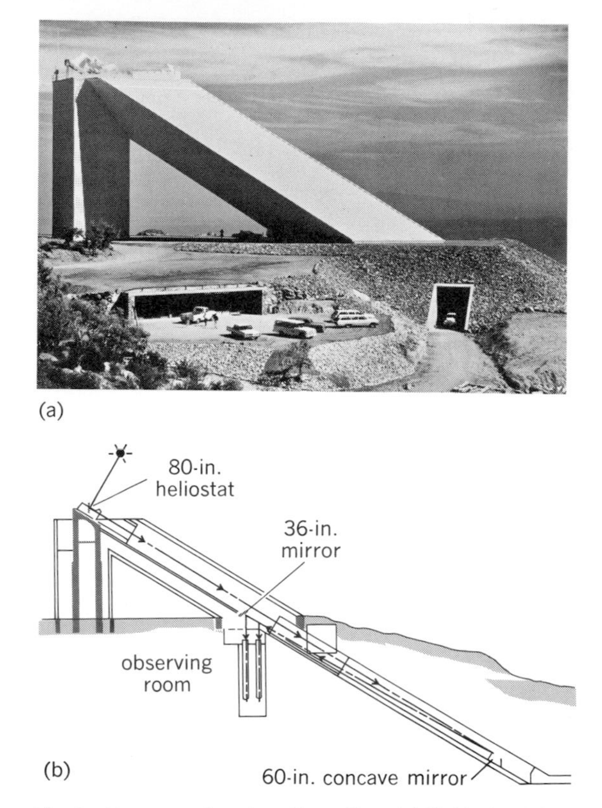

Fig. 5. Unconventional configuration. (*a*) McMath 60-in. solar telescope, Kitt Peak, AZ. (*b*) Diagram of optical elements in the telescope. 1 in. = 2.5 cm.

camera correcting plates are among the largest lenses made for astronomical applications. The special features of this system combine to produce good images over a far larger angular field than can be obtained in a Cassegrain. The Schmidt camera at the Hale Observatories, which carried out the National Geographic Society–Palomar Observatory Sky Survey, can photograph a 6° by 6° field. *See* SCHMIDT CAMERA.

Solar telescopes. Solar instrumentation differs from that designed to study other celestial objects, since the Sun emits great amounts of light energy. One solar instrument, the Robert R. McMath Solar Telescope located at Kitt Peak National Observatory near Tucson, AZ (Fig. 5), consists of an 80-in. (2.03 m) heliostat that reflects sunlight down the fixed telescope tube to a spectrograph. This spectrograph is evacuated to avoid problems that would be created by hot air currents. An alternate design has been used at the Sacramento Peak Observatory, in which the heliostat feeds light into a vertical telescope that is evacuated.

Site selection. Efficient use of a telescope requires a site with a clear, steady atmosphere. Site selection for a large telescope is given considerable attention, every effort being made to locate the instrument in an area that is climatologically and geologically favorable—preferably an area at high elevation, on solid footing, and having a proved record for number of clear nights per year and maximum mean atmospheric stability. The site should be far from the lights and polluted air of large population centers. Today, it is difficult to find excellent sites in the continental United States. The great observatories in California, for instance, all feel the impact of civilization. The most sought-after sites are in the high mountains of Chile and Hawaii (see the table).

Tools. Astronomers seldom use large telescopes for visual observations. Instead, they record their data for future study. Modern developments in

Large telescopes

Mirror diameter			
Meters	Inches	Observatory	Year completed
		World's largest reflecting telescopes	
6.0	236	Special Astrophysical Observatory, Zelenchukskaya, Crimea, Soviet Union	1976
5.1	200	Hale Observatory, Palomar Mountain, CA	1950
4.0	158	Kitt Peak National Observatory, AZ	1973
4.0	158	Cerro Tololo Inter-American Observatory, Chile	1976
3.9	153	Anglo-Australian Telescope, Siding Spring Observatory, Australia	1975
3.8	150	United Kingdom Infrared Telescope, Mauna Kea Observatory, HI	1978
3.6	144	Canada-France-Hawaii Telescope, Mauna Kea Observatory, HI	1979
3.6	142	Cerro La Silla European Southern Observatory, Chile	1976
3.2	126	NASA Infrared Telescope, Mauna Kea Observatory, HI	1979
3.0	120	Lick Observatory, Mount Hamilton, CA	1959
2.7	107	McDonald Observatory, Fort Davis, TX	1968
2.6	102	Crimean Astrophysical Observatory, Soviet Union	1960
2.6	102	Byurakan Observatory, Yerevan, Soviet Union	1976
2.5	100	Hale Observatory, Mount Wilson, CA	1917
2.5	100	Cerro Las Campanas, Carnegie Southern Observatory, Chile	1976
2.5	98	Royal Greenwich Observatory, United Kingdom	1967
6 × 1.8 meters		Multi-Mirror Telescope, Mount Hopkins, AZ	1979
		World's largest refracting telescopes	
1.02	40	Yerkes Observatory, Williams Bay, WI	1897
0.91	36	Lick Observatory, Mount Hamilton, CA	1888
0.83	32.7	Observatoire de Paris, Meudon, France	1893
0.80	32	Astrophysikalisches Observatory, Potsdam, Germany	1899
0.76	30	Allegheny Observatory, Pittsburgh, PA	1914

photoelectric imaging devices are supplanting photographic techniques for many applications. The great advantages of detectors such as charge-coupled devices is the fact that they are very sensitive, and the images can be read out onto a computer-compatible magnetic tape or disk for immediate analysis.

Light received from most astronomical objects is made up of radiation of all wavelengths. The spectral characteristics of the radiation emitted from or reflected by a body may be extracted by special instruments called spectrographs. Wide field coverage is not critical in spectroscopy of stellar objects, so spectrographs are mounted at the Cassegrain and coudé foci. The coudé position has the advantage that its focal point is fixed in position, regardless of where the telescope points in the sky (Fig. 6). Thus a spectrograph or other instrument that is too heavy or too delicate to be mounted on the moving telescope tube can be placed at the coudé focus. *See* ASTRONOMICAL SPECTROSCOPY.

Photoelectric imaging devices may be used in conjunction with spectrographs to record spectral information. Photoelectric detectors, usually photomultiplier tubes, are useful tools for classifying stars, monitoring variable stars, and quantitatively measuring the light flux from any source at which the telescope is pointing. The photocathode of the detector is placed just behind the focal plane of the telescope, preceded by a small diaphragm which permits a view of only a very small area of the sky. The phototube converts the incident light energy into an electrical signal, which is subsequently amplified and recorded. Photometry carried out with different filters yields basic information about the source with shorter observing time than that required for a complete spectroscopic analysis.

Limitations. The largest telescope in operation is the 236-in. (6 m) reflector in the Caucasus Mountains in the Soviet Union. For many applications the Earth's atmosphere limits the effectiveness of larger telescopes.

The most obvious deleterious effect of the Earth's atmosphere is image scintillation and motion, collectively known as "poor seeing." Atmospheric turbulence produces an extremely rapid motion of the image resulting in a smearing of the image. On the very best nights at ideal observing sites, the image of a star will be spread out over a 0.25-arc-second seeing disk; on an average night, the seeing disk may be between 0.5 and 2.0 second of arc. The theoretical resolving power of a 20-in. (51 cm) telescope is 0.25 arc second, so that any telescope with much over 20-in. aperture is limited by the Earth's atmosphere as far as the finest detail it can resolve.

Telescopes larger than about 50 cm are built, not for resolution, but for light-gathering power, which depends on the area of the primary mirror or objective lens. One of the chief uses of large telescopes is to study phenomena in quasars and galaxies in distant regions of the observable universe. Once again, however, the atmosphere limits what can be observed. The upper atmosphere glows faintly because of the constant influx of charged particles from the Sun. This airglow is a similar

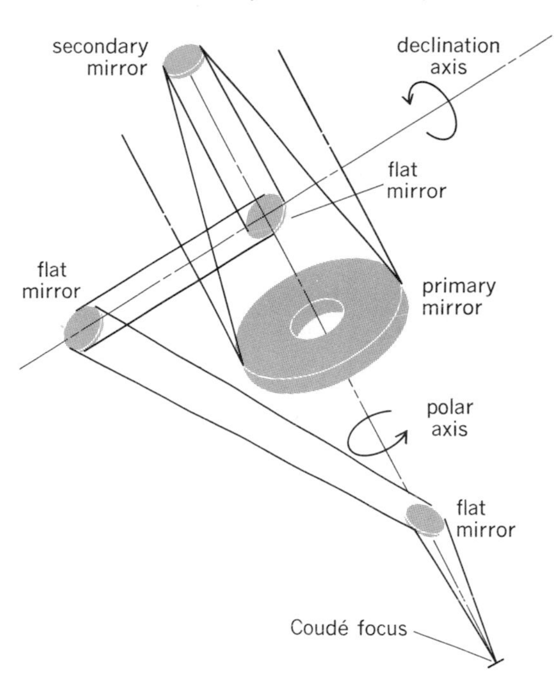

Fig. 6. Optical diagram for a common coudé configuration, showing axes of rotation of the telescope, viewed in isometric projection for clarity.

phenomenon as the aurora borealis, although it is typically fainter than a visible aurora display. Airglow adds a background exposure or fog to photographic plates that depends on the length of the exposure and the speed (*f*-ratio) of the telescope. The combination of the finite size of the seeing disk of stars and the presence of airglow means that a 10-m telescope could not see an object $(10/6)^2 = 2.8$ times fainter than the faintest object seen by a 6-m telescope. In fact, the gain is much less than that figure. On the other hand, a 10-m telescope might cost as much as 10 times more than a 6-m telescope. One solution is placing a large telescope in orbit above the atmosphere.

In practice, the effects of air pollution and light pollution from large cities outweigh the effect of airglow at most observatories in the United States. There are few unspoiled observatory sites left in the continental United States.

Notable telescopes. The definition of a large telescope depends on its type. For a refracting telescope to be considered a large telescope, its objective lens must be larger than about 24 in. (0.6 m), whereas a reflecting telescope will have to exceed 79 in. (2 m) to be considered large. The table lists a few of the largest telescopes. It is of interest to compare the dates of the reflecting telescopes and refracting telescopes. The limitations of refractors were recognized very early. A few telescopes that are notable for their historical importance, large size, or innovative design will be discussed. *See* ASTRONOMICAL OBSERVATORY.

Although Galileo's telescopes were small in size, they were notable because of the tremendously important discoveries made with them.

In 1790, Sir William Herschel built a telescope whose mirror was 48 in. (1.22 m) in diameter. Also, William Parsons, the third Earl of Rosse, built a telescope with a 72-in. (1.83 m) mirror about 1840. However, these telescopes were difficult to operate and led to relatively few discoveries. Both were built with alt-azimuth mountings which made it very difficult to follow celestial objects as the Earth rotated. In an equatorial mounting, found in almost all modern telescopes, rotation about one axis, the polar axis, compensates for the Earth's rotation. However, in an alt-azimuth mounting, motion in both altitude and azimuth is required to follow an object.

The 40-in. (1.02 m) refractor at the Yerkes Observatory was completed in 1897. The 40-in. objective lens has a focal length of 62 ft (18.9 m) and is housed in a 90-ft (23 m) dome. With its delicately balanced equatorial mounting and clock drives, the Yerkes refractor has been a major contributor to astronomy.

The 200-in. (5.08 m) Hale telescope at Palomar Mountain, CA, was completed in 1950. The primary mirror of this reflector is 200 in. in diameter with a 40-in. (1.02 m) hole in the center. Its focal length is 660 in. (13.97 m) and it has a paraboloidal figure. The focal ratio of the prime focus is *f*/3.3, and of the Cassegrain focus *f*/16. The Hale telescope was the first that was large enough to have a prime-focus cage where the observer could sit inside the main tube.

Fig. 7. The 158-in. (4.01 m) Mayall reflector of the Kitt Peak National Observatory. The dark tube at the upper end of the telescope is the prime focus cage. The large horseshoe bearing floats on oil pads, seen near each end of the mounting walkway. (*Kitt Peak National Observatory Photograph*)

Since the completion in 1950 of the Hale reflector, the number of telescopes over 100 in. (2.5 m) in aperture has steadily grown, and a dozen such instruments have been built since 1960. It is not feasible to describe each of these instruments in detail, although it is worth mentioning the modern instruments which represent the most advanced systems in operation.

The 158-in. (4.01 m) Mayall reflector at the Kitt Peak National Observatory was dedicated in 1973 (Fig. 7). The 158-in. mirror is made from a 24-in.-thick (61 cm) fused quartz disk which is supported in an advanced design mirror cell. It took three years to grind and polish the mirror to its *f*/2.7 hyperboloidal shape. The prime focus has a field of view six times greater than that of the Hale reflector. The first photographs with the 158-in. mirror showed its outstanding optical characteristics. An identical telescope was subsequently installed at Cerro Tololo Inter-American Observatory, in Chile.

Another notable instrument is the 236-in. (6 m) Soviet reflector. It is supported in an alt-azimuth mounting and like the great telescopes of Herschel and the Earl of Rosse, must be moved in both altitude and azimuth to compensate for the Earth's rotation. This delicately balanced instrument is driven by a computer which calculates the relative rates of motion in azimuth and in altitude required to follow an object, and controls the motors which provide the motion. However, there is yet a third motion required. In an alt-azimuth system, the sky rotates relative to the telescope tube as the telescope tracks a celestial object. This rotation would smear a photograph of the object. To avoid this smearing, the photographic plate (or other observing device) must be rotated by the computer at the proper rate. The Soviet astronomers estimate the alt-azimuth mounting cost half that of an equatorial mounting.

The Smithsonian Astrophysical Observatory/University of Arizona Multi-Mirror Telescope or MMT (Fig. 8) is one of the most innovative of the new telescopes. It uses six 1.8-m mirrors working together to provide light-gathering power equivalent to a 4.5-m mirror. The six images from the telescopes are brought to a common focus by a mirror system. The coalignment of the six optical systems is maintained by an active laser and computer system which continually moves the secondary mirrors to maintain alignment.

Observations with the MMT were begun in early 1979, and it quickly demonstrated its capabilities by producing a number of interesting discoveries about stars and quasars.

Two space-based telescopes have been highly successful. The *International Ultraviolet Explorer* (IUE) was launched into geosynchronous orbit in early 1978. It is not a giant telescope; its mirror is only 45 cm in diameter. The satellite hangs over the Atlantic Ocean, where it can be operated in real time from both the United States and Europe to obtain high-dispersion ultraviolet spectra of solar system objects, stars, nebulae, and extragalactic objects. The *Einstein Observatory (High Energy Astronomy Observatory 2)*, launched in November 1978, carries a telescope which images soft x-rays in the 0.3–5-nm wavelength range. The data that scientists have obtained for stars from the IUE and the Einstein are revolutionizing understanding of stellar physics. *See* SATELLITE ASTRONOMY; ULTRAVIOLET ASTRONOMY; X-RAY TELESCOPE.

Future developments. In 1985 the Space Telescope (ST) is to be launched into Earth orbit. The telescope for the ST will be a 2.4-m (94.5-in) Cassegrain reflector, which will feed light into a number of scientific instruments. The instruments on the ST will be able to study the celestial sphere with a resolution fully 10 times better than can be achieved on the ground. Concentration of the light into smaller images, and a lack of airglow, will allow the ST to see objects many times fainter than achieved so far. If the results from the small telescope on the IUE combined with the Einstein are any clue, the ST should begin a major revolution in understanding the universe. The scientific instruments to be flown on the ST include a wide-field and planetary camera, a faint-object camera, a faint-object spectrograph, a high-resolution spectrograph, and a high-speed photometer. After several years of operation the instruments will be changed out in orbit and replaced by innovative instruments. The second round of instruments will probably stress some infrared science. NASA plans to operate the ST through the remainder of the 20th century.

Astronomers are also considering the future of ground-based telescopes. An optical and infrared telescope is planned that will have considerably greater light-gathering power than the telescopes built in the 1970s. The concepts for the "next-generation telescope" look substantially different than the traditional designs of large reflectors. The Multi-Mirror Telescope on Mount Hopkins is clearly the first of a generation of multiple telescope systems that may approach the light-gathering power of a 1000-in. (25-m) telescope. These telescopes will probably be able to sense the distortion in the light waves from celestial objects caused by the Earth's atmosphere, and move the various mirrors relative to one another to correct for the distortions. In such a system, very-high-resolution imaging will also be possible. [ROBERT D. CHAPMAN]

Fig. 8. Multi-Mirror Telescope. (*Multi-Mirror Telescope Observatory, Smithsonian Institution/University of Arizona*)

ASTRONOMICAL TRANSIT INSTRUMENTS

Astronomical transit instruments are telescopic instruments adapted to the observation of the passage, or transit, of an astronomical object across the meridian of the observer. The astronomical transit instrument is the classic instrument of positional astronomy, the study of the positions and motions of astronomical objects and the related determination of positions by observation of these astronomical bodies from the Earth (the specific categories of astronomy concerned with these investigations are astrometry and celestial mechanics). The chief variants of the classic design include the meridian circle, the vertical circle, the horizontal transit circle, the broken or prism transit, and the photographic zenith tube.

The astronomical transit instrument was first

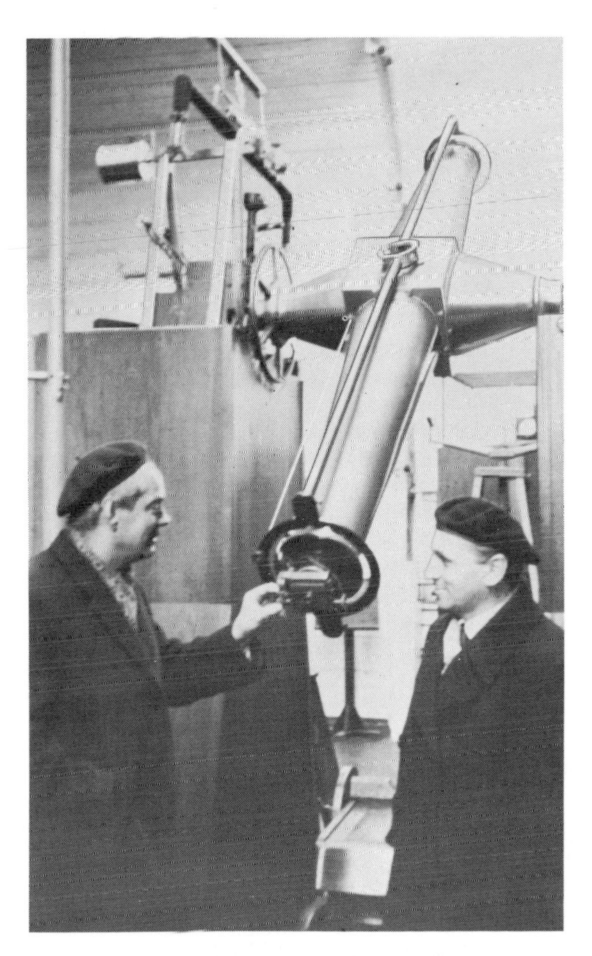

Fig. 9. Large transit instrument, Pulkovo Observatory, Soviet Union. (*Courtesy of B. L. Klock*)

developed by the Danish astronomer Ole Roemer. The modern transit instrument has a telescopic objective with a diameter of 6–10 in. (15–25 cm) and a focal length of 72–90 in. (180–230 cm). The instrument consists of a telescope mounted on a single fixed horizontal axis of rotation. The horizontal axis has a central hollow cube (sometimes a sphere) and two conical semiaxes ending in cylindrical pivots. The objective and eyepiece halves of the telescope are also fastened to the cube of the instrument, perpendicular to the horizontal axis. Rotation of the instrument in its bearings, or wyes, permits the optical axis to sweep only in the plane of the meridian. An accurate clock is the essential ancillary scale by which the transits of the astronomical objects are observed. *See* ASTRONOMICAL COORDINATE SYSTEMS.

The large transit instrument of Pulkovo Observatory, Soviet Union, in use since 1838, is seen in Fig. 9.

Applications. The astronomical transit instrument has three interrelated uses. (1) From a known position on the Earth, observations of transits of stars lead to the determination of their right ascension with respect to the astronomical coordinate system. (2) The determination of corrections to the clock may be made by the observation of stars of known position with an instrument situated at a known longitude. (3) Finally, with a knowledge of the positions of the stars observed and of Greenwich time, the longitude of the observer can be computed from observations of the time of transit of a star.

The astronomical transit instrument takes advantage of a special case of the astronomical triangle which is composed of arcs of great circles on the celestial sphere. Its vertices are, respectively, the north celestial pole, the zenith point of the observer, and the celestial object under observation. The angle at the north celestial pole represents the hour angle of the object; hence when the object is on the meridian, the hour angle is zero, and the triangle degenerates to a single arc, a segment of the meridian. At that instant the local sidereal time equals the right ascension of the celestial object.

A divided circle, graduated into fractions of a degree, is seated on the horizontal axis of the transit instrument and is used to set the instrument at the required zenith distance. The instrument has a fastening clamp also situated on the horizontal axis. The clamp may have a fine-motion adjustment mechanism to improve alignment for an observation.

The micrometer, or eyepiece part of the instrument, contains a movable wire, or pair of wires, and a stationary grid of vertical wires for use in registering the transit. At the Bordeaux, Perth, and Pulkovo observatories, photoelectric observations have been made, thereby replacing the observer and yielding higher-quality observations.

An accurate quartz crystal or atomic clock, with a rate of less than $0^{s}.001$ per day, is used as the scale for recording the transit data in conjunction with some form of data storage, such as punched tape, a printing chronograph, or a computer.

Corrections. It is extremely difficult to adjust the instrument to the point of perfection, where the mean wire will trace the true meridian as the instrument is rotated on its pivots; therefore corrections must be determined and applied to the observational data. The three principal instrument errors that require correction are azimuth, collimation, and level. These errors do not remain constant even though the instrument may be well constructed and mounted. The principal cause for their change is attributed to variations of temperature in the environment of the pavilion housing the instrument. This temperature is ambient since the roof of the pavilion must be open to make observations. Modern design of transit instruments attempts to utilize new advances in metallurgy to minimize the thermal influence on the instrument.

The azimuth correction is the horizontal angle between the axis of rotation and the true east-west direction. This correction is usually less than a few seconds of arc. It may be determined via observational data made on circumpolar stars in conjunction with artificial stars (marks) located several hundred feet to the north or to the south of the instrument pavilion.

The collimation correction is the angle between

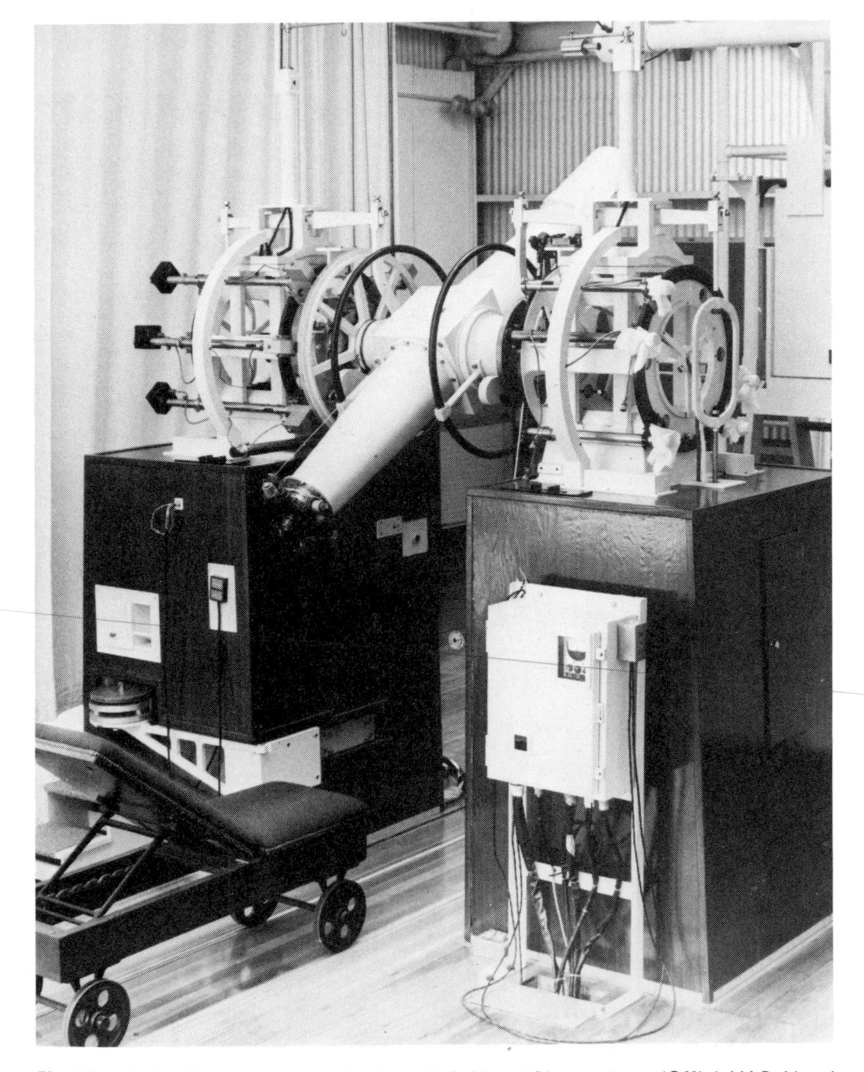

Fig. 10. Six-in. (152.4-mm) transit circle, U.S. Naval Observatory. (*Official U.S. Naval Observatory photograph*)

Fig. 11. Exterior view of the world's largest photographic zenith tube (65 cm), at the U.S. Naval Observatory. (*Courtesy of T. J. Rafferty*)

the line from the optical center of the telescope objective to the mean wire in the micrometer and the plane perpendicular to the horizontal axis of rotation. If this line intersects the horizon to the west of the south point, the stars appear to transit late and a negative correction is required. The reverse is true for stars which appear to transit early because of this error. The magnitude of this correction is generally no larger than 0ˢ010 – 0ˢ020. There are several methods of determining the collimation error. With large, permanently mounted instruments two horizontal collimating telescopes are placed 15 – 20 ft (4.6 – 6.1 m) to the north and to the south of the main instrument. These two telescopes have cross hairs in their focal planes, with a diffuse illumination source behind them. The transit instrument generally has a large hole in its central cube section that can be opened to permit the collimating telescopes to sight through it. The cross hairs of the collimating telescopes are then aligned on each other so that they lie in a plane passing approximately north and south of the transit instrument. The transit instrument is then pointed first toward one telescope and then the other. The position of the cross-hair image is observed in each case, and half the sum of their displacements is the collimation error. The collimation may also be determined by making observations on a distant mark and then reversing the telescope on its pivots and repeating the readings.

The level correction is the angle that the axis of rotation makes with the plane of the horizon. The value of this constant may be determined by observing the reflected images of the cross wires in a mercury horizon, when the telescope is pointed toward the nadir. The mercury basin is placed under the instrument only while the level correction is being determined. The level can also be determined through a striding level supported on a special frame designed to ride on, or hang from, each end of the horizontal axis. The magnitude of this correction is normally no greater than a few seconds of arc.

The clock correction of the transit instrument represents the error between the true sidereal time that the star should transit the local meridian and the time of transit recorded by the local sidereal clock. At one time the clocks had rates which were significant enough to warrant a correction; however, today they have sufficient accuracy so that this correction may be generally regarded as a constant for each night's work. This correction is evaluated through observations on bright stars whose positions are already well known.

Meridian circle. The major astrometrical observatories of the world have astronomical transit instruments called meridian or transit circles. These instruments are similar to the transit instrument previously described, except they have a micrometer eyepiece which has an extra pair of moving wires perpendicular to the vertical set. These wires are used to measure the zenith distance or declination of the celestial object in con-

Fig. 12. Photographic zenith tube, U.S. Naval Observatory. (*Official U.S. Naval Observatory photograph*)

junction with readings taken from a large, accurately calibrated circle attached to the horizontal axis. The circle may be read photographically, photoelectrically, or electronically.

The 6-in. (152.4-mm) transit circle of the U.S. Naval Observatory (Fig. 10) was designed and constructed at the end of the 19th century, but it has been improved continuously with the latest technological developments. It is the first transit circle in the world to have an electronic circle and the first to have the micrometer data read directly into an electronic computer. In addition to the electronic circle data, the divided glass circle is scanned photoelectrically with six scanning micrometers. The data from the scanners are entered into the computer for real-time processing. The probable error of a single set of circle readings is 0.″07. However, the declination of a star may contain uncertainties of 0.″25 – 0.″50 due to errors from other sources, such as atmospheric refraction, mechanical flexure of the instrument, and residual errors in the divided circle.

The 6-in. transit circle is used not only to observe the brighter stars (as faint as ninth magnitude) but also the Sun, Moon, planets, and several of the brighter asteroids. These observations are made visually, with the observer seated on a couch with an adjustable back to lend head support during an observation. The probable error of a single observation of the right ascension data of an equatorial star is about 0.ˢ012 with the use of the motor-driven micrometer. The observer presets the basic speed of the right-ascension wires and then adds or subtracts to this speed through pushbuttons on a hand keyboard.

The U.S. Naval Observatory has undertaken the refurbishing of its 7-in. (177.8-mm) transit circle with new instrumentation featuring an automatic micrometer, automatic setting, and a circle scanning system similar to the 6-in. (152.4-mm) transit circle system. When completed, the 7-in. transit circle will be used for the southern hemisphere in order to improve knowledge of the positions and motions of celestial objects in that part of the sky.

The Tokyo Astronomical Observatory at Mitaka completed construction of a new transit circle with automatic tracking and automatic reading of the circle in 1981.

Another configuration for the meridian circle is that proposed by E. Hög, of Brofelde Observatory in Denmark, in which the light from a star on the meridian is reflected by a 45° flat mirror to a horizontal reflecting telescope oriented east-west. A photoelectric micrometer and circle scanning system will complement the configuration.

The meridian (transit) circle is the instrument used to determine fundamental star positions in most of the world except the Soviet Union. There the philosophy is to separately measure each coordinate, right ascension, and declination, with its own particular kind of telescope. Right ascensions are measured with the transit instrument, and declinations are measured with a vertical circle. The vertical circle is similar to the meridian circle, except that its micrometer contains wires which permit measurement only in the vertical

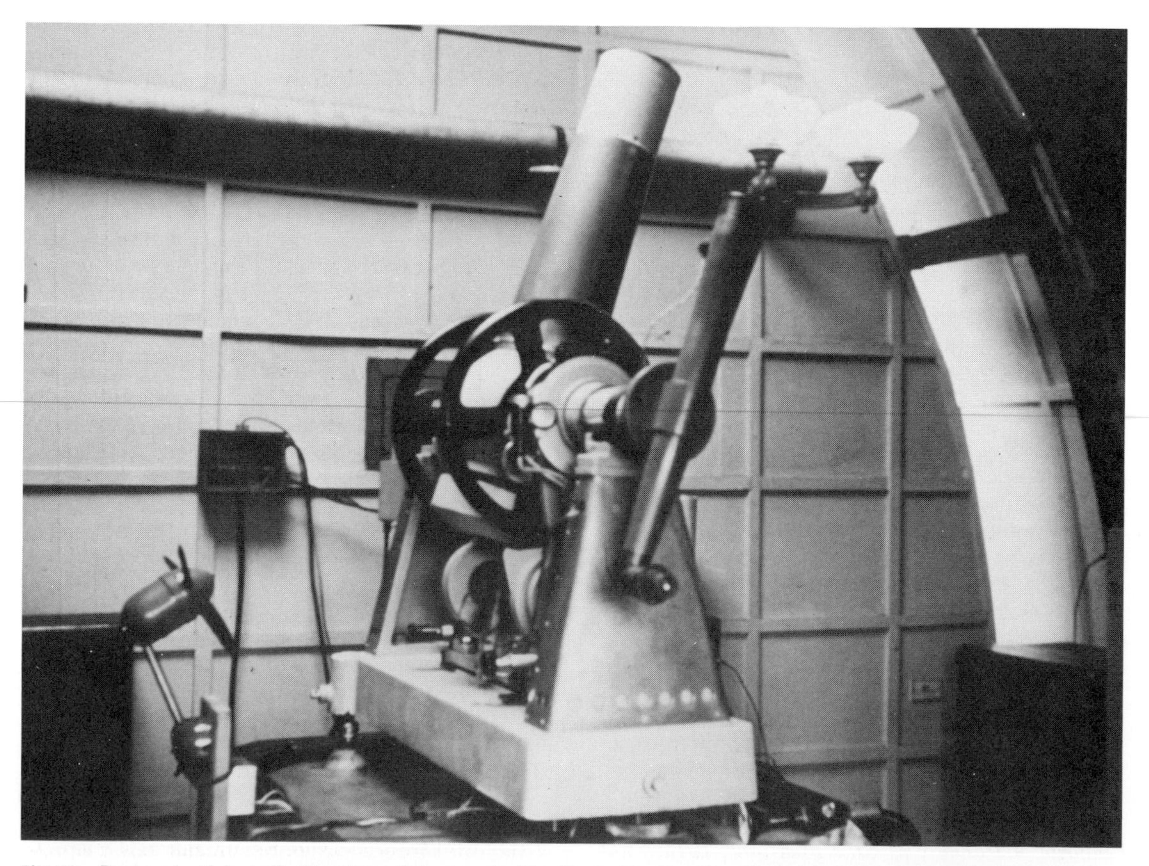

Fig. 13. Broken-back transit, Sternberg Astronomical Institute, Moscow, Soviet Union. (*Courtesy of B. L. Klock*)

plane. A photographic vertical circle has been constructed at Pulkovo Observatory in Leningrad. With this instrument the observations are made photographically rather than visually. The observatory Pulkovo also has a horizontal mirror transit circle.

Photographic zenith tube. This specially designed telescope is used for the accurate determination of time. The world's largest instrument of this kind (65 cm) is now operating at the U.S. Naval Observatory (Fig. 11). The telescope tube is mounted in a permanent vertical position (Fig. 12). Stars passing close to the zenith of the observatory form the selected list of observed objects.

The optical axis of this instrument is folded back upon itself by a pool of mercury, so that the image is formed just below the center of the objective. The objective is designed so that its second nodal point is slightly behind the last glass surface so the focal plane can be brought to this point by raising or lowering the pool of mercury. When the adjustment is accomplished, the time of passage of a star image through the nodal point becomes independent of the tilt of the instrument as the plate carriage is reversed through 180° midway through the observation. This reversal cancels the instrumental errors introduced into the images recorded before and after the reversal. The photographic plate is mounted in the carriage, which is motor-driven across the field at the exact speed of the star image. As the carriage proceeds, the plate is marked at several positions at known clock times. Thus, the position of the star image on the plate, with respect to the time marks, gives directly the correction to the clock.

Most of the major time services in the world, with the exception of the Soviet Union, use the photographic zenith tube (PZT). The time services at the Sternberg Astronomical Institute in Moscow and the Pulkovo Observatory make use of the broken-back, or prism, transit instrument. The U.S. Naval Observatory 65-cm PZT has a wide field to complement its natural ability to reach to fainter magnitudes.

Broken-back transit. In the broken-back, or prism, transit (Fig. 13) the telescope is bent at the axis of rotation by the insertion of a prism at the intersection of the optical and rotational axes. This places the eyepiece at one end of the rotation axis, where it remains in a stationary position except when the instrument is reversed on its bearings. The reversal is usually performed midway through an observation. This aids in the partial elimination of some of the instrumental corrections. The time services of the Soviet Union have adopted a special photoelectric micrometer to their broken-back transits which enables them to obtain clock corrections of the order of $0^{s}.005$ (here "s" denotes seconds of time). This is the same order of precision achieved with the photographic zenith tube.

[BENNY L. KLOCK]

Bibliography: A. N. Adams and D. K. Scott, *Publ. U.S. Nav. Observ.*, 2d ser., vol. 19, pt. 2, 1968; J. Cornell, Six new eyes peer from Mount Hopkins, *Sky Telesc.*, 38:23–34, 1979; D. L. Crawford (ed.), *The Construction of Large Telescopes*, 1966; H. Eichorn, *Astronomy of Star Positions*, 1974; First 4-meter photographs from Kitt Peak, *Sky Telesc.*, 46(1):10–13; P. D. Hemenway, Washington 6-inch transit circle, *Sky Telesc.*, February 1966; H. C. King, *The History of the Telescope*, 1955; B. L. Klock and F. S. Gauss, Instrumental parameters of the U.S. Naval Observatory automatic transit circle, *New Problems in Astronomy*, 1974; G. P. Kuiper and B. M. Middlehurst (eds.), *Telescopes*, 1960; J. B. Oke, Palomar's Hale telescope: The first 50 years, *Sky Telesc.*, 58:505–509, 1979; T. Page and L. W. Page, *Telescopes*, 1966; Photographic report from Kitt Peak, *Sky Telesc.*, 45(1): 10–17, 1973; A. G. Davis Philips, A visit to the Soviet Union's 6-meter reflector, *Sky Telesc.*, 47(5):290–295, 1974; V. V. Podobed, *Fundamental Astronomy*, 1965; W. T. Powers and R. S. Aikens, Image orthicon astronomy at the Dearborn Observatory, *Appl. Opt.*, 2(2):157–163, 1963; G. M. Sanger and R. R. Shannon, Optical fabrication techniques for the MMT, *Sky Telesc.*, 46(5):280–284, 1973; D. H. Schulte, Auxiliary optical systems for the Kitt Peak telescopes, *Appl. Opt.*, 2(2):141–151, 1963; C. L. Tichenor, Notes on modern telescope mountings, *Sky Telesc.*, 35(5): 290–295, 1968; A. B. Underhill, The international ultraviolet explorer satellite, *Sky Telesc.*, 46(6): 377–379, 1973; P. van de Kamp, *Principles of Astrometry*, 1967; E. W. Woolard and G. M. Clemence, *Spherical Astronomy*, 1966; The x-ray eyes of Einstein, *Sky Telesc.*, 57:527–534, 1979.

Orbital motion

In astronomy the motion of a material body through space under the influence of its own inertia, a central force, and other forces. Johann Kepler found empirically that the orbital motions of the planets about the Sun are ellipses. Sir Isaac Newton, starting from his laws of motion, proved that an inverse-square gravitational field of force requires a body to move in an orbit that is a circle, ellipse, parabola, or hyperbola.

Elliptical orbit. Two bodies revolving under their mutual gravitational attraction, but otherwise undisturbed, describe orbits of the same shape about a common center of mass. The less massive body has the larger orbit. In the solar system, the Sun and Jupiter have a center of mass just outside the visible disk of the Sun. For each of the other planets, the center of mass of Sun and planet lies within the Sun.

For this reason, it is convenient to consider only the relative motion of a planet of mass m about the Sun of mass M as though the planet had no mass and moved about a center of mass $M + m$. The orbit so determined is exactly the same shape as the true orbits of planet and Sun about their common center of mass, but it is enlarged in the ratio $(M + m)/M$. See PLANET.

Parameters of elliptical orbit. The diagram shows the elements or parameters of an elliptic orbit. Major axis AP intersects the ellipse AOP at the apsides; the extension of the major axis is the line of apsides. The body is nearest the center of mass at one apside, called perihelion P, and is farthest away at the other, called aphelion A.

Shape and size of an orbit are defined by two elements: length of semimajor axis and departure of the orbit from a circle. Semimajor axis a equals CP; this length is expressed in units of the mean

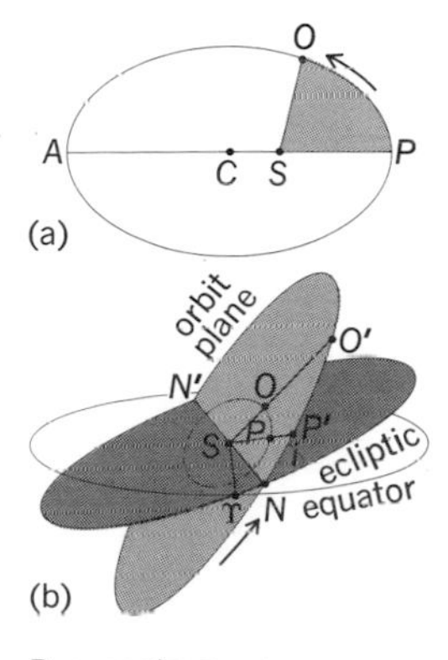

Parameters, or elements, of an elliptical orbit. (*a*) Relative orbit. (*b*) Orbit in space.

distance from the Earth to the Sun. Eccentricity e equals CS/CP where C is the center of the ellipse and S is a focus. For elliptical orbits e is always less than unity.

Position of a body in its orbit at time t can be computed if a, e, and time of perihelion passage p and period of revolution T are known. Let O be the position of a planet at time t and OSP be the area swept out in time $t-p$. From Kepler's area law, area OSP equals $(t-p)/T$ multiplied by the area of the full ellipse. *See* AREAL VELOCITY.

To describe the orientation of an orbit in space, several other parameters are required. All orbits in the solar system are referred to the plane of the ecliptic, this being the plane of the orbit of Earth about the Sun. The reference point for measurement of celestial longitude in the plane of the ecliptic is the vernal equinox ♈, the first point of Aries. This is the point where the apparent path of the Sun crosses the Earth's Equator from south to north. The two points of intersection of the orbit plane with the plane of the ecliptic (N and N') are called the nodes, and the line joining them is the line of nodes. Ascending node N is the one where the planet crosses the plane of the ecliptic in a northward direction; N' is the descending node. The angle as seen from the Sun S measured in the plane of the ecliptic from the vernal equinox to the ascending node is ♈SN; it is termed the longitude of the ascending node ☊ and fixes the orbit plane with respect to the zero point of longitude. The angle at the ascending node between the plane of the ecliptic and the orbit plane is called the inclination i and defines the orientation of the orbit plane with respect to the fundamental plane. The angle as seen from the Sun, measured in the orbit plane from the ascending node to perihelion, is NSP' and is referred to as the argument of perihelion; it defines the orientation of the ellipse within the orbit plane. The angle $NSP' +$ ☊, measured in two different planes, is called the longitude of perihelion $\tilde{\omega}$. Because dynamically the semimajor axis a and period T of a planet of mass m revolving under influence of gravitation G about the Sun of mass M are related by Eq. (1), only six elements,

$$\frac{4\pi^2}{T^2}=\frac{G(M+m)}{a^3} \qquad (1)$$

a, e, i, ☊, $\tilde{\omega}$, and p, are required to fix the position of a planet in space. Instead of these elements, however, a position vector x,y,z and the associated velocity vector $\dot{x},\dot{y},\dot{z}$ at a given instant of time would serve equally well to define the path of a planet in a rectangular coordinate system with origin at the Sun.

Orbital velocity. Orbital velocity v of a planet moving in a relative orbit about the Sun may be expressed by Eq. (2) where a is the semimajor axis,

$$v^2=G(M+m)\left(\frac{2}{r}-\frac{1}{a}\right) \qquad (2)$$

and r is the distance from the planet to the Sun. In the special case of a circular orbit, $r=a$, and the expression becomes Eq. (3). When the eccentricity

$$v^2=\frac{G(M+m)}{a} \qquad (3)$$

of an orbit is exactly unity, the length of the major axis becomes infinite and the ellipse degenerates into a parabola. The expression for the velocity then becomes Eq. (4). This parabolic velocity is

$$v^2=G(M+m)\left(\frac{2}{r}\right) \qquad (4)$$

referred to as the velocity of escape, since it is the minimum velocity required for a particle to escape from the gravitational attraction of its parent body.

Eccentricities greater than unity occur with hyperbolic orbits. Because in a hyperbola the semimajor axis a is negative, hyperbolic velocities are greater than the escape velocity.

Parabolic and hyperbolic velocities seem to be observed in the motions of some comets and meteors. Aside from the periodic ones, most comets appear to be visitors from cosmic distances, as do about two-thirds of the fainter meteors. For ease of computation, the short arcs of these orbits that are observed near perihelion are represented by parabolas rather than ellipses. Although the observed deviation from parabolic motion is not sufficient to vitiate this computational procedure, it is possible that many of these "parabolic" comets are actually moving in elliptical orbits of extremely long period. The close approach of one of these visitors to a massive planet, such as Jupiter, could change the velocity from parabolic to elliptical if retarded, or from parabolic to hyperbolic if accelerated. It is possible that many of the periodic comets, especially those with periods under 9 years, have been captured in this way. *See* CELESTIAL MECHANICS; COMET; GRAVITATION; PERTURBATION; STELLAR ROTATION. [RAYNOR L. DUNCOMBE]

Bibliography: G. O. Abell, *Exploration Of The Universe*, 1975; P. Herget, *The Computation Of Orbits*, 1948; F. R. Moulton, *Introduction To Celestial Mechanics*, 2d rev. ed., 1970.

Orion

The Warrior, in astronomy, and undoubtedly the finest of all constellations in the sky. Orion is a winter group near the celestial equator. Four of the

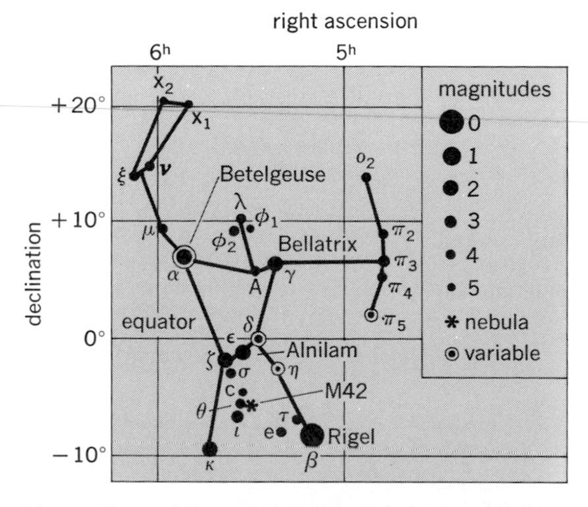

Line pattern of the constellation Orion. The grid lines represent the coordinates of the sky. The apparent brightness, or magnitude, of the stars is shown by the size of the dots, graded by appropriate numbers.

most prominent stars, α, γ, β, and κ, form a huge crude rectangle (see illustration). The group is pictured as the figure of a warrior, holding a shield with his left hand and swinging a club with his raised right arm ready to strike the charging Bull. Betelgeuse (meaning armpit), one of the largest stars known, is the red star at the right shoulder, Bellatrix is at the left shoulder, and Rigel, the blue-white star, is at the left leg. Three bright stars, δ, ϵ, and κ, in a straight line in the middle of the rectangle represent the warrior's belt. The center star is Alnilam. These four stars just cited are all navigational stars. Three faint stars below the belt form the Sword of Orion, the middle one of which is actually four very hot stars, the Trapezium, embedded in the Great Nebula of Orion (M42). *See* CONSTELLATION; TAURUS.

[CHING-SUNG YU]

Orion Nebula

A portion of a vast complex of ionized hydrogen, cool molecular clouds, and solid grains which is found in the sword of the constellation of Orion, and which shines because of radiation received from hot stars (surface temperatures of about 30,000 K) embedded within it (see illustration). Designated Messier 42 or NCG 1976, it is the brightest and perhaps one of the densest ionized hydrogen (HII) regions known. The distance of the Orion Nebula is about 500 parsecs or 1600 light-years (1.5×10^{19} m); radius of the nebula is roughly 10 light-years (10^{17} m). The Orion Nebula is a pygmy compared with the 30 Doradus Nebula, in the Large Magellanic Cloud or the W49 complex in the Milky Way Galaxy, which is observable only in the radio-frequency range. *See* MAGELLANIC CLOUDS.

Radio-frequency and infrared observations have shown the Orion Nebula to be extremely intricate, with dense cool clouds in which numerous, often complex molecules abound. The hot Trapezium stars, which excite the nebula, are very young and only recently formed from the interstellar medium. This and various other lines of evidence show the luminous Orion Nebula to be very young, less than a million years old, and possibly only a few tens of thousands of years of age. Radio observations indicate that the Orion Nebula and the entire complex of hot stars, gas, and grains are surrounded by an expanding shell of neutral hydrogen with a radius of 8.5°, or about 68 parsecs (2.10×10^{18} m), and a mass equivalent of 100,000 suns. *See* INTERSTELLAR MATTER; NEBULA. [LAWRENCE H. ALLER]

Parallax

The difference in direction to a celestial object from two separated points of observation. The distance between observational points is the base line. The base line may be provided by the diurnal motion of Earth around its axis, in which case the parallax is diurnal or geocentric; it may be provided by the motion of the Earth in its orbit about the Sun, in which case the parallax is annual or heliocentric; or it may be provided by the motion of the solar system within the local galaxy, in which case the parallax is secular, as illustrated.

Diurnal parallax. The size of Earth produces a geocentric parallax noticeable in observations of objects within the solar system. Horizontal parallax p_h is defined as the angle between the directions to the object S at the horizon as seen from the center C of Earth and from the observer's location O on the surface of Earth. The mean equatorial horizontal parallax refers to Earth's equatorial radius (6378 km) as seen from the mean distance of the celestial object. Thus defined, the diurnal parallax of the Moon is 57′2″.7. Of great historical importance is the measurement of solar parallax; the most accurate measurements give 8″.7941. This value provides the so-called astronomical unit of distance, which is the mean distance from Earth to Sun; it is 149,597,893 km from radar observations of Venus. The diurnal parallax of stars is negligible because of their great distances.

Stellar parallax. The size of the Earth's orbit produces a heliocentric parallax noticeable in observations of the nearer stars. This parallax plays a fundamental role in the determination of the distances to stars and is of basic importance in the study of the physical properties of stars. Stellar parallax or annual parallax is the maximal angle subtended by one astronomical unit at the star's location, hence the simple relation $p = 1/r$ where p is the stellar parallax expressed in seconds of arc, and r the distance expressed in parsecs. One parsec equals 3.26 light-years, or 206,265 astronomical units, or 30.86×10^{12} km.

Because of the great distances of the stars, annual parallax of a star was not successfully measured until 1838; accurate determinations became possible in the 20th century through photography with long-focus telescopes. *See* ASTROMETRY.

Photographic measurement. The observed annual parallactic path of a star on the sky is measured

ORION NEBULA

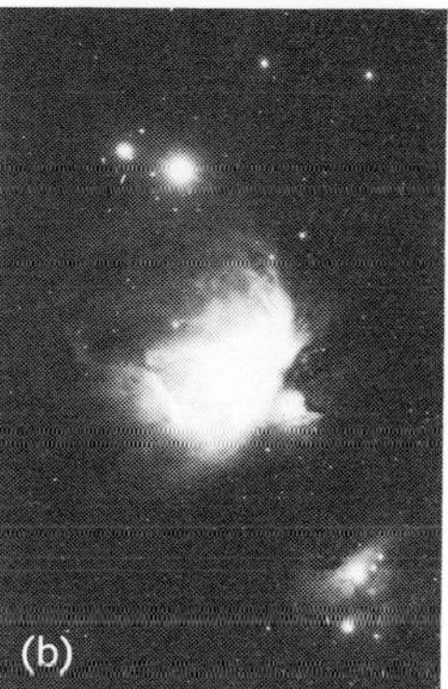

Orion Nebula, NGC 1976, photographed through Curtis Schmidt telescope at the University of Michigan Observatory. Image produced by *(a)* green nebular lines of oxygen [O III], *(b)* ultraviolet region, chiefly oxygen [O II].

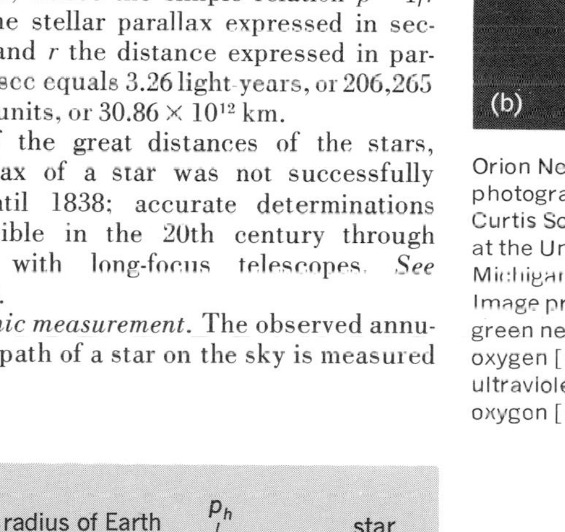

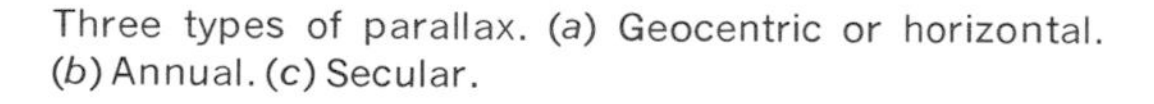

Three types of parallax. *(a)* Geocentric or horizontal. *(b)* Annual. (c) Secular.

against a background of three or more faint reference stars, at much greater distances, whose minute parallactic displacements may be allowed for. The photographic plates are commonly measured on a long-screw precision measuring engine and are reduced by conventional algebraic methods to allow for differences in scale, orientation, and origin of the different plates taken of any one star. Other parameters such as magnitude and color ("coma") effects may also be taken into account. The first parallax determinations made in this way were commonly based on some 20 plates, each with two or three exposures, spread over an interval of 2 or 3 years, the observations being centered near maximum parallactic displacement; an accuracy of ±0″.01 is thus reached. For a time there was a tendency to increase the number of plates to reach higher accuracy; however, there is a definite limitation in accuracy, that is due not so much to the photographic plate and the measuring machine as to systematic errors caused by the telescope and by the atmosphere. The *General Catalogue of Trigonometric Stellar Parallaxes* lists the parallaxes of over 6000 stars; for the nearer stars the percentage error is often well below 5%. For example, the parallax of Sirius is 0″.375 ± 0″.004, its distance 2.67 ± 0.03 parsecs or 8.7 ± 0.1 light-years. On the other hand, the trigonometric parallax method begins to fail for stars beyond a distance of 20 parsecs.

Use of data. Parallax measurements for the hundreds of nearer stars yield basic astronomical information. Knowledge of the spatial properties of the universe is based on accurate geometric measurements made in the Earth's immediate cosmic neighborhood. Annual parallax is a prerequisite for determining the space velocity of a star and a star's luminosity. The latter, expressed in absolute magnitude M, is related to the apparent magnitude m and the parallax p by $M = m + 5 + 5 \log p$.

In the case of binary stars, parallax is a prerequisite for measuring the total mass $(M_1 + M_2)$ of the binary system. The sum of the masses $(M_1 + M_2)$ is expressed in terms of the Sun's mass, which is related to the space-time dimensions of the double-star orbit through the relation $M_1 + M_2 = (a^3/p^3)\ (1/P^2)$. Here a is the semimajor axis of the double-star orbit, p the parallax; both a and p are expressed in seconds of arc; P is the binary period expressed in years. *See* BINARY STAR.

Secular parallax. The motion of the solar system seems attractive for the purpose of measuring stellar distances, because time alone provides an indefinite extension of the base line. However, there are severe limitations because of the motions of the stars themselves; hence only values of the average parallax for groups of stars, called mean secular parallaxes, may be measured. The method has proved important to extend our geometric knowledge of the Milky Way system, in a statistical fashion, by measuring average distances up to several thousand light-years for groups of stars. The method also has been important in obtaining information about the average luminosity for stars whose parallaxes are too small to be measured by annual parallax. Closely related to mean secular parallax is the mean parallax of a group of stars obtained from a comparison of their average radial velocity with average values of their proper motions or components thereof.

Distance of remote stars. Geometric measurements of parallax yield the distances of stars. From these distances and the apparent magnitudes of the stars, their absolute magnitudes or luminosities are determined. Spectral observations show that stars can be grouped into classes of like character. On the assumption that all stars of like spectral emission have equal intrinsic luminosity or absolute magnitude M, the apparent magnitude m of a remote star can be measured photometrically and the parallax deduced from $5 \log p + 5 = M - m$. This photometric method is limited by loss of light in space due to scattering or absorption. Another means for obtaining absolute magnitudes for use in photometric determination of parallax is from observed periods of cepheid variables. *See* STAR; VARIABLE STAR.

The spectroscopic method is calibrated from nearby stars whose geometric parallaxes are known; the method permits, in principle at least, the determination of stellar parallaxes up to any distance, as long as the star's spectrum can be observed. The method works better for stars of certain spectral type than for others.

The period-luminosity relation of variable stars has yielded knowledge of the tremendous distances of globular clusters and galaxies because of the appearance of cepheids in those distant systems. The use of this approach, and that of other highly luminous distance indicators, is qualified by their apparently nonunique behavior, as is illustrated by the existence of two types or populations of stars which results in two parallel period-luminosity relations. *See* CEPHEIDS.

Still another method is that of dynamical parallaxes, based on the fact that there is a comparatively small dispersion in stellar masses. From the known apparent size of a double-star orbit and its period of revolution, it is possible—with a reasonable assumption of the sum of the masses—to arrive at a good estimate of the parallax, whose percentage error is only one-third of any percentage error in the assumed combined mass.

[PETER VAN DE KAMP]

Bibliography: G. Abell, *Exploration of the Universe*, 3d ed., 1975; K. Aa. Strand, *Basic Astronomical Data*, 1963, reprint 1980; K. Aa. Strand, Determination of stellar distances, *Science*, 144:1299–1309, 1964; W. van Altena, *General Catalogue of Trigonometric Stellar Parallaxes*, 3d ed., 1981; P. van de Kamp, Elements of long-focus photographic astrometry, *Photogramm. Eng.*, 22(2):32–43, 1956.

Parsec

A unit of measure of astronomical distances. One parsec is equivalent to 3.084×10^{13} kilometers, or 1.916×10^{13} miles. There are 3.26 light-years in 1 parsec. The parsec is defined as the distance at which the semimajor axis of Earth's orbit around the Sun (1 astronomical unit) subtends 1 second of arc. Thus, because the angle is small, the equation below holds. A parsec is then 206,265 astronomical

$$\frac{1 \text{ astronomical unit}}{1 \text{ parsec}} = 1 \text{ second} = \frac{1}{206{,}265}$$

units; its accuracy depends on the precision with which the distance from Earth to Sun is measured. At a distance of 1 parsec, the parallax is 1 second of arc. The nearest star is about 1.3 parsecs distant; the farthest known galaxy is several billion parsecs. *See* PARALLAX.

[JESSE L. GREENSTEIN]

Pegasus

The Winged Horse, in astronomy, an autumnal constellation. Pegasus is usually identified by the four bright stars α, β, γ, and α situated on the corners of a large square known as the Great Square in Pegasus (see illustration). The constellation is represented by a winged horse. Markab (the Saddle), a navigational star, occupies the southwestern corner of the square. The star Alpheratz at the opposite corner is really in the constellation Andromeda. The star at the northwestern corner is a red star, known as Scheat, a giant irregular variable. Diagonally opposite on the southeastern corner of the square is Algenib. Enif, another navigational star, lies in the nose of the horse. *See* CONSTELLATION.

[CHING-SUNG YU]

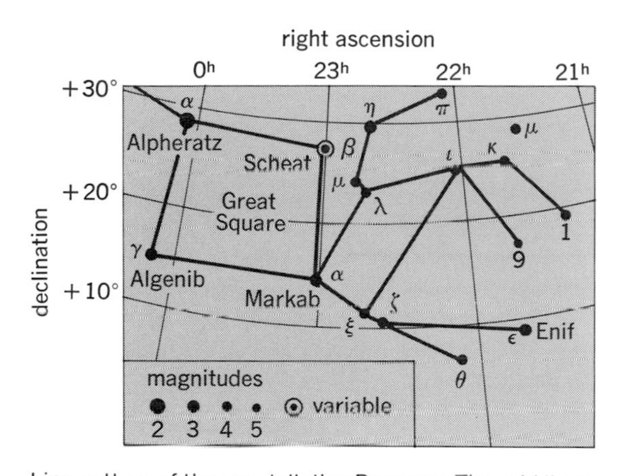

Line pattern of the constellation Pegasus. The grid lines represent the coordinates of the sky. The apparent brightness, or magnitude, of the stars is shown by the sizes of the dots, graded by appropriate numbers.

Perigee

The point nearest the Earth in the orbit of the Moon or of an artificial satellite. At perigee the Moon is 5% closer to Earth than at its mean distance, the orbital eccentricity being 0.055. Because, on the average, the Moon and Sun subtend nearly equal angles, a solar eclipse near perigee lasts about 5 min; an eclipse near apogee is annular. The line perigee-Earth-apogee is the major axis of the orbital ellipse on the line of apsides. The differential attraction of the Sun on Earth and Moon causes the line of apsides of the Moon to move forward in the orbital plane with a period of 8.85 years. *See* MOON; PERIHELION.

[GERALD P. KUIPER]

Perihelion

In astronomy, that point at one extremity of the major axis of the elliptical, parabolic, or hyperbolic orbit of a planet or comet about the Sun where the planet or comet is closest to the Sun. The instant when a planet or comet is at perihelion is referred to as the time of perihelion passage. For Earth this occurs about Jan. 3, at which time Earth is some 2.5×10^6 km closer to the Sun than its mean distance of 149.6×10^6 km. *See* CELESTIAL MECHANICS; ORBITAL MOTION.

[RAYNOR L. DUNCOMBE]

Perseus

A compact circumpolar constellation of the northern sky, like its neighbor, Cassiopeia, on the east. Both constellations lie in a brilliant part of the Milky Way. The prominent stars in Perseus form the capital script letter *A* (see illustration). This group is represented by the figure of the hero Perseus. The conspicuous curved arc of stars, bright and easy to identify, is commonly known as the Segment of Perseus. Mirfak, a navigational star, lies in the right shoulder. The constellation is noted for its clusters of stars. Just above the head are the famous double clusters *h* and χ in Perseus. Algol, the Demon Star, which is an eclipsing variable, is located in this constellation. *See* CASSIOPEIA; CONSTELLATION.

[CHING-SUNG YU]

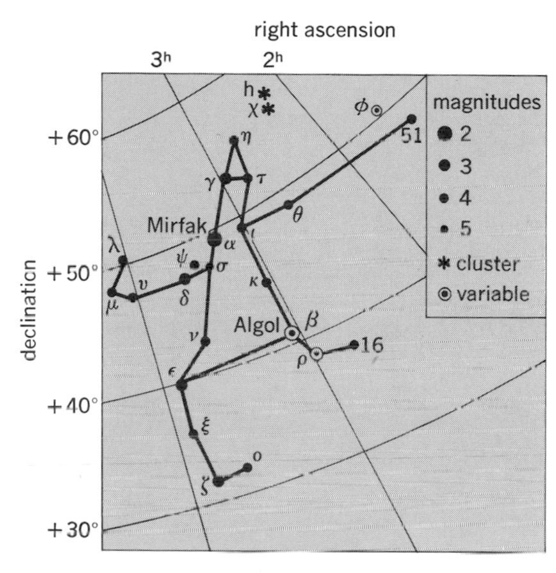

Line pattern of the constellation Perseus. The grid lines in the chart represent the coordinates of the sky. The apparent brightness, or magnitude, of the stars is shown by the size of the dots, which are graded by appropriate numbers as indicated.

Perturbation

Departure of a celestial body from the trajectory it would follow if moving only under the action of a single central force. Perturbations may be caused by either gravitational or nongravitational forces.

Corrections to elliptic orbits. In the solar system, orbits of planets may be adequately represented by mean elliptical elements to which are added small corrections due to the mutual planetary attractions. Although such motion is referred to as disturbed, it is as much a consequence of the law of gravitation as is undisturbed elliptic motion.

Another method of representing perturbed motion is to augment the position derived from the mean ellipse by the actual displacements in the coordinates due to the disturbing forces. These perturbations of the elements, and perturbations of the coordinates, are represented by infinite series; usually many terms are required to represent the disturbed motion accurately. These analytical expressions are referred to as general perturbations and, with their associated mean elements, form a general theory of the motion. In some instances, such as the orbit of an outer satellite of Jupiter or the motion of a comet moving with nearly parabolic velocity, the analytical expressions representing the perturbing function become so involved (mainly because of lack of convergence of the Fourier series) that general perturbations are not attempted. Instead, the perturbed positions are computed from a step-by-step numerical integration of the equations of motion; this is known as the method of special perturbations.

Long- and short-term disturbances. Planetary orbits are subject to two classes of disturbances: secular, or long-term, perturbations; and periodic, or relatively short-term, perturbations. Secular perturbations, so called because they are either progressive or have excessively long periods, arise because of the relative orientation of the orbits in space. They cause slow oscillatory changes of eccentricities and inclinations about their mean values with accompanying changes in the motions of the nodes and perihelia. The periods of time involved in these oscillations may extend from 50,000 to 2,000,000 years. Periods and major axes of orbits are not affected by secular changes. For the orbit of the Earth, the present inclination to the invariable plane is 1°35′. This will diminish to a minimum of 47′ in approximately 20,000 years. The eccentricity, presently 0.017, is diminishing also and will reach a minimum of 0.003 in about 24,000 years.

Periodic perturbations arise from the relative positions of the planets in their orbits. When the disturbed and disturbing planets are aligned on the same side of the Sun, the perturbation reaches a maximum, and reduces to minimum when alignment is reached on opposite sides of the Sun. The size of a periodic perturbation is a function of the mass of the disturbing body and of the length of time the two planets remain near the point of closest approach. Periodic perturbations continually shift a planet away from the position it would occupy in undisturbed motion, moving it above or below the orbital plane, nearer to or farther from the Sun, and forward or backward in the orbit.

Commensurable motions. If the mean motion of the disturbed planet were exactly a submultiple, say 1/2, of the mean motion of the disturbing planet, the maximum perturbation produced by their close approach would always occur in the same part of the disturbed orbit. The displacement in position of the disturbed planet would increase with each coincidence until the character of the orbit became modified to the point where exact commensurability of the mean motions would cease to exist.

Because the solar system is middle-aged, cosmically speaking, few examples of commensurability of mean motions exist today. None is found in the motions of the major planets. Cases of near commensurability exist which give rise to long-period periodic terms of large amplitude. As an example, the periods of Jupiter and Saturn are nearly in the ratio of 2:5. Thus, after nearly five revolutions of Jupiter, the two planets return to approximately the same juxtaposition. Their line of coincidence, however, sweeps slowly around Jupiter's orbit, completing a circuit in about 850 years and thus producing a perturbation of this period.

Among the four inner planets, the periodic perturbations are small, amounting in orbital longitude at most to 0′.25 for Mercury, 0′.5 for Venus, 1′ for Earth, and 2′ for Mars. Periodic perturbations of the outer planets are larger, reaching in the case of the long-period terms to 30′ for Jupiter, 70′ for Saturn, 60′ for Uranus, and 35′ for Neptune.

Because the amplitude of a periodic perturbation depends on the mass of the disturbing planet, observational measurement of this amplitude affords a method of determining the disturbing mass. For the planets Mercury, Venus, and Pluto, which do not have satellites, this is the only method of determining the mass. As a consequence of the mutual perturbations of the planets, the distance of a planet from the Sun is, on the average, decreased by the action of planets closer to the Sun, and increased by planets farther from the Sun; this mean effect represents a perturbation of the radius vector with a constant value.

The orbits of the minor planets are affected in varying degree by the attractions of the major planets. Those orbits passing close to Jupiter suffer large perturbations which, if the mean motions were commensurable with that of Jupiter, would be augmented at each close approach until the trajectories were sufficiently altered to reduce the commensurability. In the overall distribution of mean motions of the minor planets there are noticeable gaps near the points where the period would be an exact submultiple (1/2, 1/3, 2/5, . . .) of the period of Jupiter. In cases of near commensurability, observational determination of the amplitude of the long-period perturbation affords a method for measuring the mass of Jupiter. A small group of minor planets, called the Trojan asteroids, has been so completely captured by Jupiter that they oscillate about the 60° points which form equilateral triangles with Jupiter and the Sun. *See* TROJAN ASTEROIDS.

Effect on comets. Planetary perturbations also affect the orbits of comets. Studies of the motion of Halley's Comet indicate that the time from one perihelion passage to the next has varied by almost 5 years because of perturbations. Most comets approach the Sun at nearly parabolic speeds in randomly oriented orbits, but if a comet approaches close to one of the more massive major planets, the planet may so alter the trajectory that the comet pursues an elliptical orbit thereafter. A number of short-period comets whose orbits agree only in that they all pass close to Jupiter illustrate the perturbing effect of this planet on cometary orbits.

Nongravitational causes. Material forming the tails of comets is subject to a nongravitational type of perturbation. This rarefied matter given off by the head of the comet is forced into a trajectory away from the Sun by the pressure of solar radiation.

Associated with many of the periodic comets are swarms of smaller particles which appear as meteors upon collision with the upper atmosphere of the Earth. The density of these swarms is so tenuous that they cannot hold themselves together by their own gravitation, and planetary perturbations of speed and direction soon spread the components completely around the orbit. The annual meteor showers, such as the Perseids, reflect this dispersal of particles along the orbit. The effect of the Earth's attraction on a meteor trajectory depends on the relative velocity, that is, whether the Earth is overtaking the meteor or meeting it head on. Once the meteor enters the upper reaches of the Earth's atmosphere its motion is subject to a nongravitational perturbation caused by atmospheric drag. This resistance to the passage of the particle is evidenced by the trail of incandescent gas and vapor which forms until the particle is consumed or continues in its trajectory greatly decelerated. *See* METEOR.

Perturbations of satellite orbits. The motions of planetary satellites, natural and artificial, reflect both gravitational and nongravitational perturbations. The centrifugal force arising from the rotation of a planet causes a deformation or oblateness of figure. In such a case the central mass does not attract as if it were concentrated at its center. For a close satellite the principal perturbation arises from the attraction of this equatorial bulge. The effect of this attraction on an otherwise undisturbed satellite orbit is a gradual regression of the line of nodes on the equatorial plane and a rotation of the line of apsides. Both rotations vary with the inclination of the satellite orbit. Nearer to the primary, the tidal forces may become so great that a satellite would be literally torn to pieces. For a fluid satellite of the same density as the planet, the limit within which this disruptive perturbation occurs is about $2\frac{1}{2}$ times the radius of the planet. *See* SATURN.

Satellite motions are also disturbed by the direct attraction of other satellites, the Sun, and, to a lesser amount, by other planets. Observation of the orbital displacements caused by the mutual perturbations of satellites in the systems of Jupiter and Saturn makes possible the determination of the masses of these satellites. The solar attraction is significant in the orbits of the outer satellites of Jupiter and Saturn, reaching to one-ninth the planet's attraction for the eighth satellite of Jupiter. So greatly disturbed is this satellite that it is not possible to derive a general theory for its motion.

The orbit of the Moon is disturbed mainly by the Sun, with some changes in motion due to the oblateness of the Earth, the figure of the Moon, and smaller perturbations caused by the planets. The attraction of the Sun on the Moon is more than twice the Earth's attraction, but because both the Earth and Moon are free to move it is only their relative acceleration with respect to the Sun which determines the motion. This relative acceleration toward the Sun is always less than 1/80 of the acceleration of the Moon toward the Earth. The eccentricity and inclination of the Moon's orbit oscillate slowly about their mean values, while the line of apsides advances with an average period of almost 9 years and the nodes regress through one revolution in 18.6 years.

The observed motion of the lunar node and perigee affords one means of measuring the oblateness of the Earth. The present lunar theory incorporates the value 1/294. The *International Astronomical Union System of Astronomical Constants* (1976) contains a reference ellipsoid of revolution for the Earth having a flattening of 1/298.257. This value has been derived mainly from measures of the motions of the nodes and apsides of artificial Earth satellites. Lunar and solar perturbations of artificial Earth satellite orbits are minor for orbits 500 mi above the surface but grow with increasing distance from the Earth. Atmospheric drag perturbations are significant at this altitude, but decrease with increasing altitude. *See* CELESTIAL MECHANICS.

[RAYNOR L. DUNCOMBE]

Bibliography: G. Brouwer and G. M. Clemence, *Methods of Celestial Mechanics*, 1961; International Astronomical Union, *Inform. Bull.*, no. 37, January 1977; A. E. Roy, *Orbital Motion*, 1978.

Phase

In astronomy, the changing appearance of the Moon, inner planets, and Mars due to the angular difference between the incident light from the Sun and the viewing direction of the observer. Phases of the Moon are a familiar sight. During a lunar month (29.53 Earth days), the Moon completes a cycle of appearances or phases: dark of the moon, or new moon, during which the Moon is nearer the Sun than is the Earth; crescent until first quarter (about a week after new moon); half an illuminated disk, the Moon continues to wax, being gibbous, until it is full; it then wanes through third quarter and completes the cycle as illustrated. A solar eclipse by the Moon can occur only at dark of the moon; a lunar eclipse when the Earth's shadow falls on the Moon can occur only at full moon.

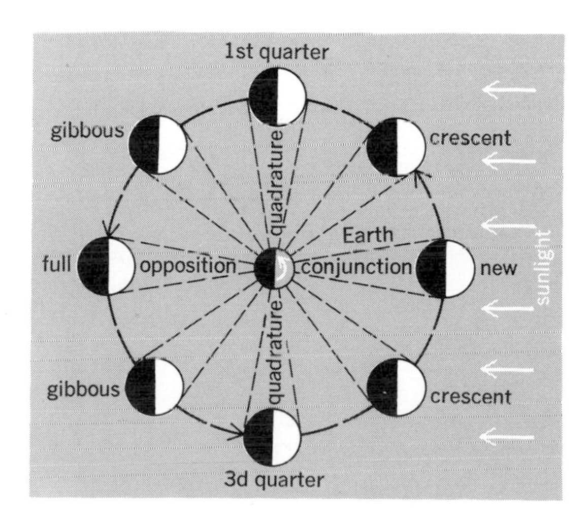

Mechanism of Moon's phases. Dashed lines indicate hemisphere of Moon that is visible from Earth at various positions in Moon's orbit.

Mercury and Venus show phases like those of the Moon. Mars varies in appearance from full to gibbous. The more remote planets are so far from Earth and Sun that they are viewed from substantially the same angle as that from which they are illuminated and thus show no phase change. *See* MARS; MOON.

[GERARD P. KUIPER]

Photosphere

The photosphere of the Sun (the visible surface of the Sun or other stars) is a gaseous layer a few hundred kilometers thick with an average effective temperature of 5780 K, determined from the total radiation per square centimeter. The temperature is maintained by convection, which brings hot material from the opaque solar interior to the surface in the form of rising columns of gas. The convection produces a small-scale granular texture which is visible through a projection telescope. In areas where strong magnetic fields inhibit the convection, the photosphere cools and dark sunspots appear. *See* STELLAR EVOLUTION; SUN.

[JOHN W. EVANS]

Pisces

The Fishes, in astronomy, a zodiacal constellation appearing in the autumn evening sky. Pisces is the twelfth and last sign of the zodiac. It is inconspicuous, having no star brighter than the fourth magnitude. But it is an important constellation because the vernal equinox, which marks the beginning of the astronomical year, is now located in it. Its most distinctive feature is a V-shaped figure, with the fishes' tails toward the point of the V tied together by a ribbon (see illustration). The northern fish is poorly defined, but the western one is marked with a group of stars forming an irregular pentagon, known as the Circlet in Pisces. *See* CONSTELLATION.

[CHING-SUNG YU]

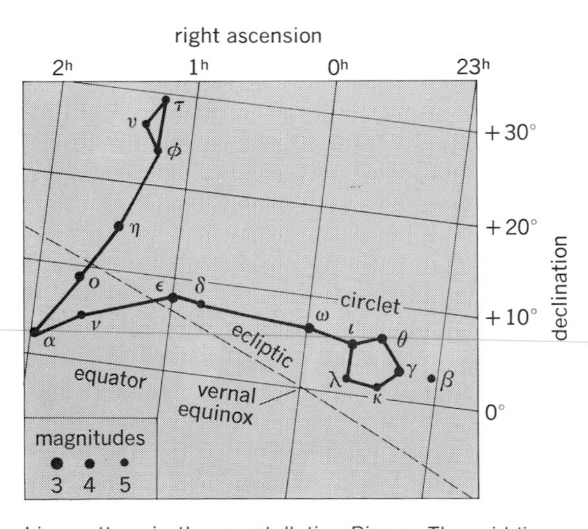

Line pattern in the constellation Pisces. The grid lines represent the coordinates of the sky. The apparent brightness, or magnitude, of the various stars is shown by the sizes of the dots, which are graded by appropriate numbers as indicated.

Planet

A relatively small, solid celestial body moving in orbit around a star, in particular the Sun. Besides Earth, the eight known planets of the solar system are Mercury, Venus, Mars, Jupiter, Saturn, Uranus, Neptune, and Pluto; in addition, over 2000 minor planets, or asteroids, mostly located between the orbits of Mars and Jupiter, are known (Fig. 1).

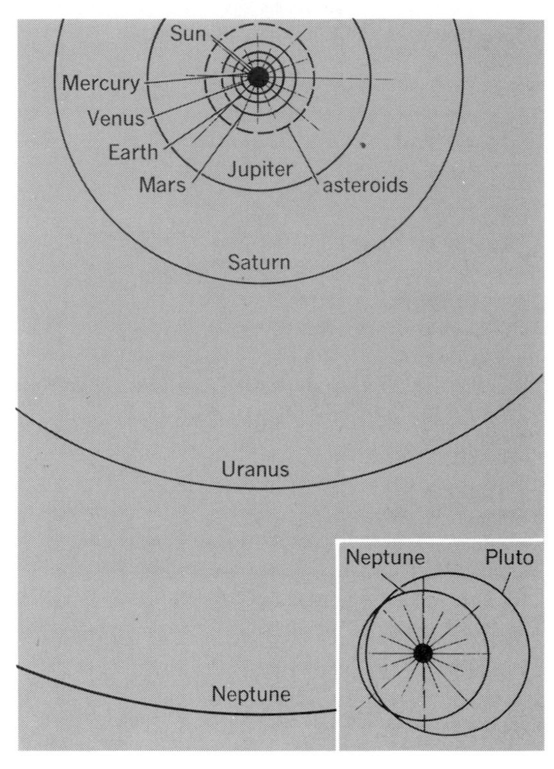

Fig. 1. Plan of the solar system. (*From L. Rudaux and G. de Vaucouleurs, Astronomie, Larousse, 1948*)

Classification. There are two basic groups of planets: the small, dense, terrestrial planets—Mercury, Venus, Earth, Mars, and Pluto—and the giant or Jovian planets—Jupiter, Saturn, Uranus, and Neptune. With the exception of Pluto, the terrestrial planets are all located within the inner solar system. The low-density Jovian planets extend outward from Jupiter to the remote outer reaches of the solar system. This distribution is not accidental, but is related to the fractionation of rocky, icy, and gaseous materials during the early stages of formation of the solar system.

Each of the main planets from the Earth to Neptune is accompanied by one or more secondary bodies called satellites. Although now traveling in its own orbit around the Sun, it has been suggested that Pluto may once have been a satellite of Neptune. *See* SATELLITE.

The planets may also be divided into inferior planets, Mercury and Venus, located inside the Earth's orbit, and superior planets, from Mars to Pluto, circulating outside the Earth's orbit.

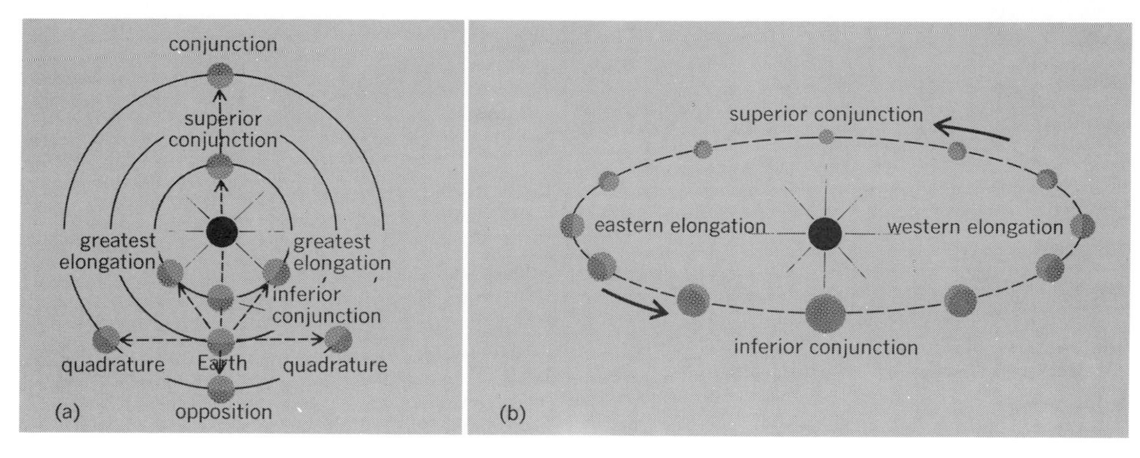

Fig. 2. Planetary configurations and phases. (*a*) Positions of Earth and other planets relative to Sun. (*b*) Phases of inferior planets (Mercury and Venus). (*From L. Rudaux and G. de Vaucouleurs, Larousse Encyclopedia of Astronomy, Prometheus Press, 1959*)

Kepler's laws. The motions of the planets in their orbits around the Sun are governed by three laws discovered by J. Kepler at the beginning of the 17th century.

First law: The orbits of the planets are ellipses of which the Sun occupies a focus.

Second law (law of areas): Equal areas of the ellipse are described by the radius vector from the Sun to the planet in equal intervals of time.

Third law (harmonic law): The squares of the periods of revolution P are proportional to the cubes of the semimajor axes of the orbits a; that is, for all planets the ratio P^2/a^3 is equal to a constant. The constant is equal to unity if a is given in astronomical units and P in sidereal years. One astronomical unit (AU) is the mean distance from Earth to the Sun and is approximately equal to 149.6×10^6 km (93.0×10^6 mi). *See* ASTRONOMICAL UNIT.

Otherwise, the constant of the harmonic law is given by Newton's law of gravitation as $G(M+m)/4\pi^2$, where M and m are the masses of the Sun and the planet, and G is the constant of gravitation. *See* GRAVITATION.

Kepler's laws are true only when the mutual perturbations of the motions of the planets by the others are neglected.

Planetary configurations. In the course of their motions around the Sun, Earth and other planets occupy a variety of relative positions or configurations (Fig. 2*a*), the principal of which are designated as follows: The inferior planets are in conjunction with the Sun when closest to the Earth-Sun direction, either between the Earth and the Sun (inferior conjunction) or beyond the Sun (superior conjunction). On rare occasions when the planet is very close to the plane of the Earth's orbit at the time of an inferior conjunction, a transit in front of the Sun is observed. *See* TRANSIT.

Between conjunctions, the geocentric angular distance from the planet to the Sun, or the elongation, varies up to a maximum value; the greatest or maximum elongations of Mercury and Venus are 28° and 47°, respectively. The superior planets are not so limited, and their elongations can reach up to 180° when they are in opposition with the Sun; when the elongation is ±90°, they are in quadrature (eastern or western) with the Sun.

The telescopic aspect of the disks of the planets varies according to their configurations, which determine the angle between the directions of illumination and observation, or the phase angle. Between inferior conjunction and greatest elongations, the interior planets show crescent phases, like the Moon between new moon and first or last quarters (Fig. 2*b*); between greatest elongations and superior conjunction they show a gibbous phase, like the Moon between quarters and full moon. At superior conjunction, they show a circular disk, fully illuminated and seen face on, while during transits, the dark side is profiled against the Sun. The superior planets show their full phase at both conjunction and opposition and a gibbous phase near quadrature, at which time the unilluminated portion of the disk is at a maximum.

Apparent motions. The combinations of the orbital motions of Earth and of any other planet give rise to complicated apparent motions of the planets as seen from the Earth. Because the orbits of the main planets are, except for Pluto, only slightly inclined to the plane of the orbit of Earth, the apparent paths of the planets (except Pluto) are restricted to the zodiac, a belt 16° wide centered on the ecliptic. The ecliptic is the path in the sky traced out by the Sun in its apparent annual journey as the Earth revolves around it. Along this path, the apparent motions of the inferior planets with respect to the Sun are alternatively westward, from greatest elongation through inferior conjunction to greatest elongation, then eastward, from greatest elongation through superior conjunction to greatest elongation (Fig. 2). The motion of the superior planets is always westward. *See* ASTRONOMICAL COORDINATE SYSTEMS.

The apparent motions with respect to the celestial sphere, that is, to the fixed stars, appear for the inferior planets as oscillations back and forth about the position of the Sun steadily moving eastward among the stars. For the superior planets, the apparent motion is generally eastward or direct, but for short periods near the time of opposition it

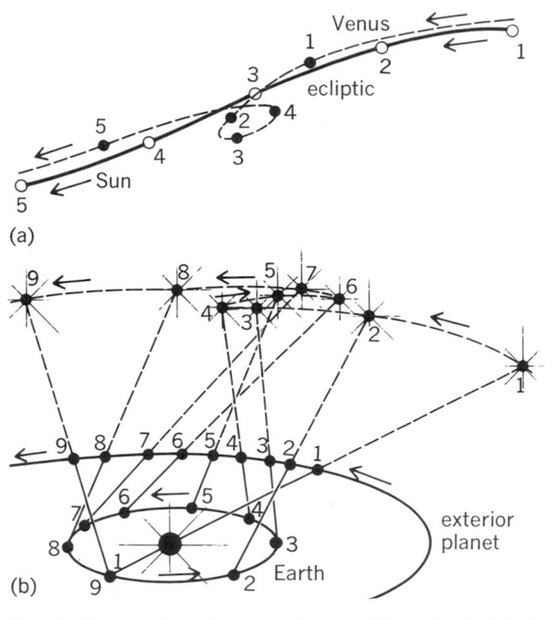

Fig. 3. Apparent motions as observed from Earth (*a*) of an interior planet with respect to the Sun and (*b*) of an exterior planet with respect to the fixed stars. (*From L. Rudaux and G. de Vaucouleurs, Larousse Encyclopedia of Astronomy, Prometheus Press, 1959*)

is westward or retrograde (Fig. 3). At times when the direction of the apparent motion on the sphere reverses, the planet appears to be stationary.

The mean interval of time between successive returns to the same place with respect to the stars is the sidereal period, which is established by the true motion of revolution of the planet in its orbit around the Sun. The mean interval of time between successive returns of the same configuration with respect to the Sun (for example, conjunctions or oppositions) is the synodic period, which governs the apparent motion of the planet as seen from Earth (Table 1).

Elliptic motion. The motion of a planet having an elliptical orbit of semimajor axis a, with the Sun at the focus S, brings it each revolution to the perihelion A and to the aphelion A', the points of the orbit respectively nearest to and farthest from S. If C is the center of the ellipse, the semimajor axis is $a = CA = CA'$; the eccentricity of the ellipse is $e = CS/CA = CS/a$, whence $SA = a(1-e)$, $SA' = a(1+e)$. The distance $SP = r$ of the planet to the Sun at any other point is $r = a(1-e^2)/(1+e\cos\theta)$, where the angle $\theta = \angle ASP$ is the true anomaly. If P' is the point on the principal circle of radius $CA = a$ whose projection in the ellipse is P (Fig. 4), the eccentric anomaly is the angle $\theta' = \angle ACP'$, so that $r = a(1 - e\cos\theta')$. If the planet is at perihelion at time T and returns to it at time $T + P$, the mean angular velocity (or mean motion) is $n = 2\pi/P$, and the mean anomaly at any time t is $M = n(t - T)$.

The relation between the mean and eccentric anomalies, $\theta' - e\sin\theta' = M$, is known as Kepler's equation; its solution gives θ' and, consequently, r at any time t when the orbital elements a, e, n, T are known. *See* KEPLER'S EQUATION.

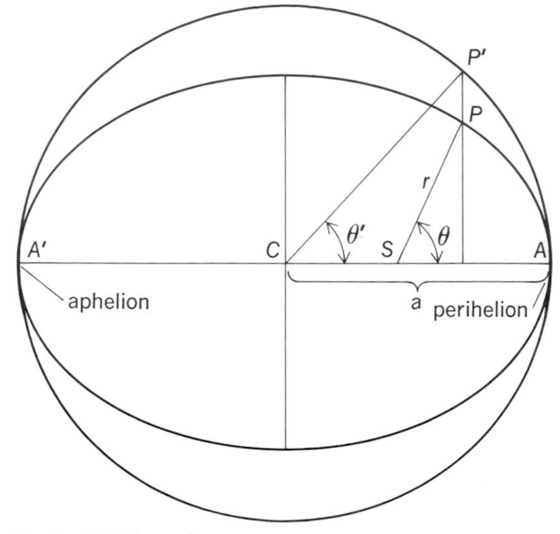

Fig. 4. Elliptic motion.

Orbital elements. The position of a planet in its orbit and the orientation of the orbit in space are completely defined by seven orbital elements (Fig. 5): (1) the semimajor axis a, (2) the eccentricity e, (3) the inclination i of the plane of the orbit to the plane of the ecliptic, (4) the longitude ☊ of the ascending node N, (5) the angle ω from the ascending node N to the perihelion A, (6) the sidereal period of revolution P, or the mean (daily) motion $n = 2\pi/P$, and (7) the date of perihelion passage T, or epoch E.

Table 1. Elements of planetary orbits

Planet	Symbol	Mean distance (semimajor axis)		Sidereal period of revolution		Synodic period, days	Mean velocity, km/sec	Eccentricity	Inclination
		AU	10^6 km	Years	Days				
Mercury	☿	0.387	57.9	0.241	87.97	115.88	47.89	0.206	7°00′
Venus	♀	0.723	108.2	0.615	224.70	583.92	35.03	0.007	3°24′
Earth	⊕	1.000	149.6	1.000	365.26		29.79	0.017	0°00′
Mars	♂	1.524	227.9	1.881	686.96	779.94	24.13	0.093	1°51′
Jupiter	♃	5.203	778.3	11.862	4332.59	398.88	13.06	0.048	1°18′
Saturn	♄	9.539	1427.	29.458	10759.	378.09	9.64	0.056	2°30′
Uranus	♅	19.18	2870.	84.014	30685.	369.66	6.81	0.047	0°46′
Neptune	♆	30.06	4497.	164.79	60189.	367.49	5.43	0.009	1°47′
Pluto	♇	39.44	5900.	247.7	90465.	366.73	4.74	0.250	17°09′

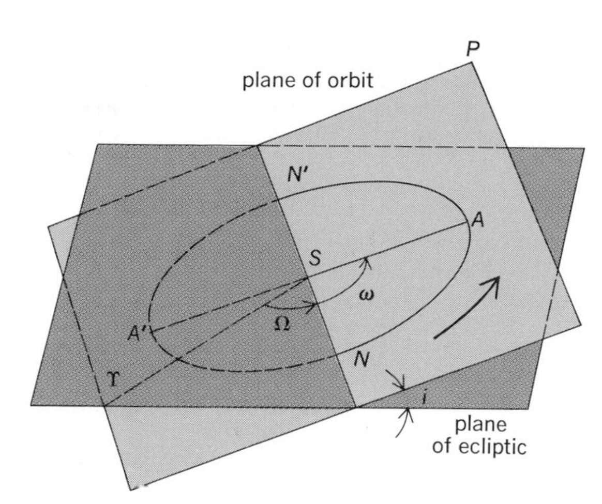

Fig. 5. Orbital elements.

If the plane of a planet's orbit is inclined to the plane of the ecliptic, their intersection NN' is the line of nodes; in its motion, the planet crosses the plane of the ecliptic from south to north at the ascending node N and from north to south at the descending node N'. The longitude of the ascending node is the angle ☊ $= \angle$♈ SN, measured in the plane of the ecliptic from the vernal equinox ♈. The longitude of perihelion is $\tilde{\omega}=$ ☊ $+\omega=\angle$♈ $SN+\angle NSA$, the second angle being measured in the plane of the planet's orbit (Fig. 5). The location of the plane of the orbit in space is defined by i and ☊, the orientation of the ellipse in this plane by ω, its form by e, and its size by a, and the position of the planet on the ellipse by P and T (and by the time t). See ORBITAL MOTION.

Determination of orbital elements. Accurate observations of the positions of the planets with respect to the stars (for example, as measured on photographs) or with respect to the celestial coordinates (for example, by means of the meridian circle instrument) are used to determine the elements of their orbits. In principle, three observations of two coordinates (right ascension and declination) and the laws of elliptic motion are sufficient to determine the six independent elements of a planetary orbit, since by Kepler's third law $a^3 \propto P^2$. In practice, as many observations as possible are combined, and the equations solved by the method of least squares; the elements for a given epoch so obtained are subject to variations and corrections allowing for planetary perturbations. Tables of the motions of the planets for several centuries past and future have been established, from which the yearly ephemerides are extracted in a form convenient for immediate use. See EPHEMERIS.

The elements of the planetary orbits are given in Table 1.

Planetary sizes. The apparent diameter of a planet may be determined visually by means of a filar micrometer or, preferably, a birefringent or double-image micrometer attached to a telescope, or it may be measured on large-scale photographs taken through telescopes or by planetary spacecraft. If the apparent diameter of a planet is d'' when its distance to the Earth is Δ, the linear diameter is $D=\Delta \sin d''=\Delta d''/206{,}265$, where d'' is measured in seconds of arc. The linear diameter is expressed in the same units as Δ, which is given by the ephemerides in astronomical units; conversion to kilometers or miles is given by the adopted value of the astronomical unit: 1 AU $=149.6\times10^6$ km $=93.5\times10^6$ mi.

When polar flattening is perceptible, both the polar and equatorial radii r_p, r_e can be determined or, as in Table 2, the mean radius $r=(r_p+r_e)/2$ and the ellipticity $\epsilon=(1-r_p/r_e)$. The mean radius may also be expressed in terms of the mean radius of the Earth (6371 km) as a unit. The relative area is then very nearly equal to r^2 and the relative volume to r^3. For the nearer planets with a solid surface (Mercury, Venus, and Mars), the linear diameter can be determined more precisely by radar, using the time-delay technique. See RADAR.

Masses, gravity, and density. The mass of a planet is found easily if it has one or more satellites. If a is the mean distance (semimajor axis) of the satellite's orbit and P its period of revolution expressed respectively in astronomical units and sidereal years, the mass m of a planet is given through Newton's law of gravitation by $m=A^3/P^2$, in terms of the mass of the Sun as a unit. This assumes that the mass of the satellite relative to that of the planet, and of the Earth relative to that of

Table 2. Physical elements of planets

Planet	Equatorial radius, r_e (Earth = 1)	Equatorial radius, r_e km	Ellipticity	Volume (Earth = 1)	Mass (Earth = 1)	Density, g/cm³	Escape velocity, km/sec	Rotation period	Inclination of axis*
Mercury	0.38	2425	0.000	0.055	0.053	5.6	4.2	58.65d	7
Venus	0.95	6051	0.000	0.87	0.815	5.2	10.3	243d	176.6
Earth	1.00	6378	0.0034	1.00	1.000	5.52	11.2	23h56m22.7s	23.45
Mars	0.53	3395	0.009	0.150	0.107	3.95	5.0	24h37m22.6s	24
Jupiter	11.20	71600	0.064	1317.	317.89	1.33	59.5	9h50m30s†	3.1
Saturn	9.47	60000	0.102	762.	95.15	0.69	35.6	10h14m‡	26.75
Uranus	4.06	25900	0.01	67.	14.54	1.20	21.2	10h49m	97.9
Neptune	3.87	24700	0.03	58.	17.23	1.64	23.6	18h24m	24
Pluto	0.5?	3200?	?	0.12?	0.002	0.8?	8?	6.3d	115

*To perpendicular to orbit, in arc degrees.
†Latitude < 12° (system I); 9h55m40.6s, latitude > 12° (system II).
‡Near Equator, 10h38m at intermediate latitudes.

the Sun, may be neglected, which is nearly always the case within the accuracy of the data. Since the ratio m_e/M_s of the masses of the Earth and the Sun can be determined similarly, planetary masses are also known in terms of m_e.

If the planet has no satellite (Mercury and Venus), its mass can be derived only from the perturbations it causes in the motions of the other planets and occasionally of comets. Since the perturbations are small, the masses so obtained are generally of low accuracy. More precise mass values for Mercury, Venus, Mars, Jupiter, and Saturn, however, have been derived from the perturbations of the orbits of space probes. *See* PERTURBATION.

Once the mass m and the radius r of a planet are known in terms of the Earth's mass and radius, its surface gravity and mean density relative to the Earth are given by $g = m/r^2$ and $\rho = m/r^3$, respectively. Multiplication by 981 and 5.552 gives the corresponding values in cgs units.

From r and m follows also the escape velocity V_1 that permits a projectile (or a molecule) to leave the planet on a parabolic orbit: $V_1 = (2Gm/r)^{1/2}$; this is $\sqrt{2}$ times the velocity of an hypothetical satellite moving in a circular orbit close to the surface of the planet. These elements are listed in Table 2.

Rotation periods. The period of rotation of a planet can be determined by several methods: (1) Direct telescopic observation of the permanent markings on its surface (Mars, Mercury) or of the semipermanent cloud formations in its atmosphere (Jupiter, Saturn) is the best method; (2) the spectroscopic determination of the velocity difference between the opposite equatorial limbs can give, in combination with the linear diameter, an approximate value of the rotation period (Uranus, Neptune); (3) radar observations of the broadening of the signal reflected by the planet due to the different line-of-sight velocities of the various parts of its disk (Venus) are in effect a much more precise radio analog of method (2); and (4) when the apparent diameter of the disk is too small for any of these methods (Pluto, asteroids) a determination of the periodicity of the light variations, if any, due to the changing presentation of bright and dark regions of the surface may give a fairly accurate value of the rotation period. The rotation periods of all the main planets are now well determined (although less so for Uranus and Neptune), as indicated in Table 2.

Planetary radiations. The electromagnetic radiation received from a planet is made up of three main components: the visible reflected sunlight, including some ultraviolet and near-infrared radiation; the thermal radiation due to the planet's heat, including both infrared radiation and ultrashort radio waves; and the nonthermal radio emission due to electrical phenomena, if any, in the planet's atmosphere or in its radiation belts.

Planetary brightness. The apparent brightness of a planet can be measured by visual, photographic, and photoelectric photometry and is usually expressed in the stellar magnitude scale; it varies in inverse proportion to the squares of the distances r from the Sun and Δ from the Earth. The fraction of the incident light reflected at full phase compared with the fraction that would be reflected under the same conditions by an equivalent perfect diffuse reflecting disk is called the geometrical albedo. It is a measure of the backscattering reflectivity of the planet's visible surface. The visual albedos of the planets vary between 5 and 70%. *See* ALBEDO.

Thermal radiation. The thermal radiation from a planet can be measured either with a radiometer at wavelengths of 8–14 μm, 17–25 μm and 30–40 μm (which are partially transmitted by the Earth's atmosphere) or with a radio telescope at wavelengths between 1 mm and 30 cm. In either case, the amount of energy corresponds to that which would be received under the same conditions from a perfect radiator of the same size at a certain temperature T, called the blackbody temperature of the planet. Its relation to the actual temperature depends on the thermal and radiative properties of the atmosphere and surface of the planet. Jupiter has been found to radiate twice as much energy as it receives from the Sun. From this it is inferred that the planet has an internal source, perhaps primordial heat continuing to escape long after Jupiter's formation 4.6×10^9 years ago. Saturn has also been observed to emit more energy than it receives from the Sun, as verified both from Earth and by the *Pioneer 11* spacecraft in 1979. The output is some two to four times higher than incoming sunlight. Unlike Jupiter, however, Saturn may be too old to still possess a primordial heat source. While gravitational separation of the planet's hydrogen and helium has been suggested as one possible mechanism for this excess energy, no single solution appears to account for the entire energy outflow. *See* JUPITER; SATURN.

Nonthermal radiation. Nonthermal radio emission at decimeter and dekameter wavelengths has been received from Jupiter by means of large radio telescopes. The dekameter emission takes the form of irregular bursts of noise originating in the planet's atmosphere from subjacent sources. Voyager spacecraft revealed that powerful electric currents exist inside the Jovian magnetosphere, particularly one called a flux tube linking higher latitudes on the planet with satellite Io. Since the observed dekametric radiation is modulated by the orbital position of Io, this current loop may be responsible for the outbursts, as well as for auroral discharges observed by the spacecraft in the planet's polar regions.

Planetary atmospheres. The principal constituents of the atmospheres of the terrestrial planets (Mercury has no atmosphere; Pluto, a special case, is discussed below) are carbon dioxide, argon, nitrogen, water, and (on Earth only) oxygen. The atmospheres of the giant planets are composed primarily of hydrogen and helium, with lesser amounts of methane, ammonia, and water. Atmospheric motions are driven by temperature gradients—in general, those existing between the warm equatorial regions and the cooler polar areas. An atmosphere thus tends to redistribute heat over the planetary surface, lessening the temperature extremes found on airless bodies.

On planets having relatively dense atmospheres, heat from the Sun is trapped by the greenhouse effect, wherein visible radiation from the Sun passes readily through the atmosphere to heat the planetary surface, but infrared radiation reemitted from the surface is constrained from escaping back to space by the lower transparency of the atmosphere to longer wavelengths. Although Venus absorbs approximately the same amount of energy from the Sun as does Earth, the greenhouse effect is responsible for heating the surface of Venus to a much higher temperature, approximately 750 K (900° F). *See* VENUS.

Directly through wind erosion, or indirectly by carrying water vapor which can precipitate as rain, the atmospheres of the terrestrial planets are a major factor in modifying surface structure and rearranging the distribution of surface materials. Mercury, being an airless planet, exhibits a relatively unmodified surface, very similar in appearance to that of the Moon. Pluto, however, is covered with a methane ice frost, and thus maintains a tenuous atmosphere of methane vapor when in the part of its orbit nearest the Sun. *See* MERCURY; PLUTO.

Possible unknown planets. During the 19th century, an unexplained irregularity in the motion of Mercury was thought by some investigators to be caused by an unknown planet circulating between the Sun and Mercury, called Vulcan, which was looked for in vain. This irregularity was satisfactorily explained in 1915 by Einstein's general theory of relativity. It is now certain that no intra-Mercurial planet of size comparable to the terrestrial planets exists. *See* RELATIVITY.

The possibility of one or more planets circulating beyond the orbits of Neptune and Pluto has also been discussed, but there is no compelling evidence for the existence of such planets.

Planets outside solar system. Minute perturbations in the motion of some nearby stars have indicated the existence of minor components of small mass. The masses of these satellite bodies, although larger than planetary masses in the solar system, are considered to be too small to be self-luminous; they are consequently more like planets than dwarf stars. From this evidence it is inferred that planetary systems are not as uncommon in the universe as was previously believed, and studies of multiple-star systems suggest that one star in every three may be attended by planetary companions. Whether some of these planets are inhabited by advanced beings or sustain lower life forms is unknown, but it would perhaps be more surprising to learn that life on Earth is unique. Presumably, whenever temperature, composition, and other conditions are favorable, the chemical evolution of complex organic molecules will begin, and the crucial transition from chemical to biochemical evolution becomes just a matter of time. *See* ASTEROID; CELESTIAL MECHANICS; CERES; EARTH ROTATION AND ORBITAL MOTION; EROS; MARS; NEPTUNE; OPTICAL TELESCOPE; RADIO ASTRONOMY; SOLAR SYSTEM; TROJAN ASTEROIDS; URANUS.

[J. KELLY BEATTY]

Bibliography: J. K. Beatty, B. O'Leary, and A. Chaikin, *The New Solar System*, 1981; C. R. Chapman, *The Inner Planets*, 1977; R. M. Goody and J. C. G. Walker, *Atmospheres*, 1972; W. K. Hartmann, *Moons and Planets: An Introduction to Planetary Science*, 1972; J. A. Wood, *The Solar System*, 1979.

Planetarium

A projection device which accurately portrays the stars and planets at any time in the past, present, or future from any point on the Earth or the near region of space. The modern planetarium instrument is a mechanical-electrical analog of space.

Design features. Planetarium projectors are basically designed around three major components: a star lamp that has the intrinsic brilliance of stars in nature; a planet-projection system that allows for freedom of motion on or off the Earth; and a computer that solves all of the mechanical and mathematical problems incurred with modern planetarium programming.

Star lamp. The star lamp is a high pressure xenon arc which produces a tiny, intense-point source of light. This light is focused through thousands of individual lenses and pinholes and is projected to the dome. When the star projector is put in motion, the sky appears to move. There are two star balls, one for the northern hemisphere and one for the southern. Since the two hemispheres are off-center in the projection hemisphere, parallax correction and star-field overlaps are resolved by the engineer during the star-ball construction. The lens positions and the pinholes are drilled by a computer-controlled machine that is programmed by data from the Yale Star Catalog. The parallax correction is engineered into the control tape. A self-leveling (liquid level) cutoff maintains the horizon line so that the stars do not appear on the walls and floor of the chamber.

Projection system. The planets are projected through a system of individual, mechanical analogs. With modern instrumentation the five planets visible to the naked eye are shown in addition to the Moon and the Earth as seen from the Moon. The traditional method of projecting planets with the stars has been through the use of gear trains. Today's designs have resulted in electromechanical analogs for greater flexibility, related to position in time and in space. The mechanical analog is a miniature model of a planet's orbital characteristics (one analog for each planet projector). Earth, Sun, and planet positions are mechanically represented. The output shaft of the analog controls a two-axis mirror (on a gear-driven angle head), and the mirror position determines the placement of the planet in the star field. High-speed motors and individual controls allow for rapid setting of the planet position. The operator can select either an Earth view or a Sun (heliocentric) view of planetary motions.

Computer use. The computer is used to interconnect the three axes of motion in a manner that permits the operator to rotate about any point in the celestial sphere, thus allowing observation of the sky from any planet in the solar system or any contrived point in space. This system demonstrates the normal Earth view of the sky through the concepts of Copernicus or Galileo and generates entirely new motions for the observer to analyze. The opportunity to freely change perspective gives the casual observer the perspective abstracted by the involved astronomer.

Size. Planetarium chambers range from 24 ft (7.3 m) (for schools) to 80 ft (24 m) in diameter for

Fig. 1. The Spitz STP public-oriented planetarium, which is also used in a 40-ft (12-m) diameter dome for school planetariums. This type of planetarium is computer-controlled. (*Denver Museum of Natural History*)

large-city installations (Fig. 1). The planetarium dome is constructed of perforated aluminum sheets, and for a dome that is 50 ft (15 m) in diameter, there are roughly 56,000,000 holes. With the large number of perforations, the planetarium dome can function as a theatrical scrim. This allows the instrument to project the stars on the surface of the dome and also allows for projection through the dome (Fig. 2).

With computer-controlled instruments, seating arrangements in planetariums have changed: The entire audience can now face in one direction. This not only stops the head twisting, but also gives everyone in the chamber the same reference meridian. When demonstrating motions such as roll, pitch, and yaw, a common reference meridian is essential.

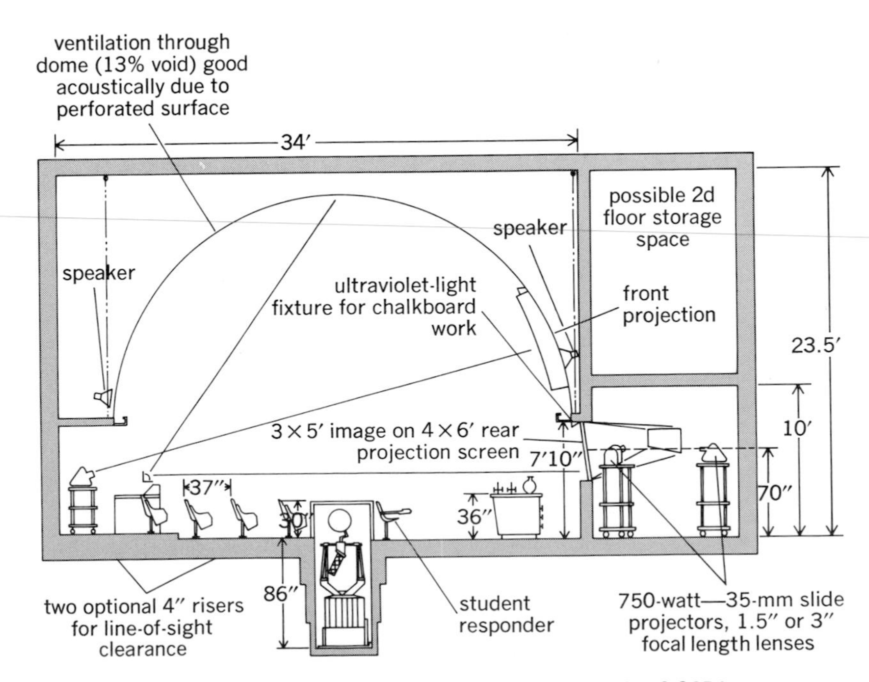

Fig. 2. Section view of classroom planetarium. 1′ = 0.3048 m; 1″ = 0.0254 m.

Educational value. Planetarium programming has changed markedly since the installations of the 1930s. Samples of programming can vary from basic astronomy to ecological systems. Since the universe is an umbrella that covers most of nature, the stars can provide a mood that allows for discussions of biology, geology, chemistry, physics, mathematics, history, art, and general environmental experiences. It is not uncommon to have lightning storms, sounds of a forest, and the crashing surf in addition to the planetarium experience of rocket flight or travel through time and space (Fig. 3). In order to create these additional effects, there are hundreds of projectors spaced around the dome, in the center of the room, and in special projection ports. Projectors showing binary stars in motion, comets, meteors, auroras, satellites, and horizon-line panoramas represent some of the units that supplement the star and planet projection system.

There are approximately 1000 planetariums in the United States as of 1981, the greatest majority having domes 40 ft (12 m) or smaller and located in

Fig. 3. A view from an operator console showing a panorama of the surface of the Moon, a view of the Earth over the lunar horizon, and lunar and command modules in flight (*Denver Museum of Natural History*)

schools. There are approximately 70 major planetariums 50 ft (15 m) in diameter and above.

[DONALD M. LUNETTA]

Planetary nebula

A gaseous shell thrown off by a star during the later stages of its evolution. Planetary nebulae are so called because many display small greenish disks, akin to those of the planets Uranus and Neptune in small telescopes. They have nothing to do with planets; rather they are associated with late stages of stellar evolution, a planetary nebula being the shell thrown off by a dying star just before it settles down to become a degenerate white dwarf.

These nebulae do not congregate in spiral arms, but show a spatial distribution similar to that of most stars, that is, a concentration toward the galactic center and some preference for the galactic disk. Well over a thousand have been cataloged, and the total number in the Galaxy must be about 20,000. They are also found in neighboring galaxies, such as M31 and M32. *See* GALAXY; GALAXY, EXTERNAL.

Determinations of distances of individual galactic planetaries is extremely difficult. All are too remote for trigonometric parallaxes; proper motions give only statistical results. A few nebular central stars have companions of known luminosity. Astrophysical methods based, for example, on the assumption that ejected shells have constant masses are likely to fail for individuals.

Although a planetary nebula is often symmetrical about the central star, which is usually seen, a great variety of forms appear. These include double rings, bilaterally symmetrical envelopes, amorphous structures, and even apparent central holes. Delicate filamentary wisps and knotlike concentrations are the rule, indicating that the ejected shells must be usually very inhomogeneous.

Spectra of planetary nebulae are dominated by

Chemical composition of a typical planetary nebula compared with the Sun*

	Nebula	Sun		Nebula	Sun		Nebula	Sun
He	102,000	68,000	Ne	115	105	Cl	0.18	0.3
C	1200	420	Na	1.7	1.9	Ar	2.8	1
N	93	87	P	0.2	0.3	K	0.08	0.15
O	460	690	S	12	16	Ca	0.11	2.25

*Numbers of atoms on scale H = 1,000,000.

emission lines of abundant elements, in various ionization stages, all excited because of the rich ultraviolet radiation emitted by the central star. In the usual optical range, the principal lines are those of H and He (produced by recombination and subsequent cascade) and forbidden lines of ions of N, O, Ne, S, and Ar. Prominent lines of some elements such as carbon appear only in the ultraviolet. There is also a continuous spectrum due mainly to recombination of protons and electrons, double photon emission, and bremsstrahlung. Many objects show a strong infrared continuum caused by thermal radiation of solid grains, most likely soot.

The chemical composition of the nebular gas (see table) is predestined by the material from which the parent stars were formed, the effects of element-building processes in those stars, and the removal of refractory elements by grains. The nebular carbon/oxygen ratio exceeds the solar value because carbon is generated in cores of parent stars, but calcium is less abundant in the nebular gas because it is trapped in grains. *See* ELEMENTS AND NUCLIDES, ORIGIN OF.

Possibly most stars with masses between one and five times that of the Sun produce planetary nebulae. When hydrogen becomes exhausted in the core of a solar-type main-sequence star, it evolves into a red giant whose outer envelope is eventually ejected. As the shell density declines, the shell is exposed to energetic photons of the hot residual core. Hence, the characteristic bright-line nebular spectrum is produced. As the nebula expands further, this spectrum gradually fades away. The hot, blue-white core, deprived of all nuclear fuel, slowly evolves to a white dwarf. *See* STELLAR EVOLUTION; WHITE DWARF STAR.

The chemical composition of shells ejected by low-mass stars reflects that of the interstellar material from which they were formed. In more massive stars, however, products of nuclear reactions in the core get blended in with surface material, and the resultant planetary nebula is enriched in He, C, and N. Some of the ejecta remains as a gas, and some must condense into solid grains in the inhomogeneous escaping shell, which eventually returns to the interstellar medium. The cool blobs of material contain molecules as well as solid particles and neutral gases. Planetary nebulae appear to be important suppliers of dust and gas to the interstellar medium. *See* INTERSTELLAR MATTER; NEBULA. [LAWRENCE H. ALLER]

Bibliography: L. H. Aller, *The Planetary Nebulae*, 1971; Y. Terzian (ed.), *Planetary Nebulae*, 1978.

Planetary physics

The study of the structure, composition, and physical and chemical properties of the planets of the solar system, including their atmospheres and their immediate cosmic environment.

Extent of knowledge. While the atmosphere and the oceans have been explored quite thoroughly, probes in the form of deep holes have barely sampled the outermost layer of the Earth's crust. A great deal has been deduced about the interior of the Earth as the result of the application of principles of physics and chemistry and knowledge of the properties of materials. Knowledge of the structure of the deep interior of the Earth principally depends upon studies of the properties of the seismic (earthquake) waves which penetrate through the innermost regions and provide knowledge of conditions there. The Earth's magnetic field also has very deep roots in the interior, and the knowledge obtained from studies of the variations in the magnetic field are supplementary to that obtained from seismic waves.

Knowledge of the other planets in the solar system is much less than that of the Earth. Quite a bit is known about the Earth's Moon as the result of the lunar samples which have been brought back for study in terrestrial laboratories, and of the instruments which astronauts left operating upon the surface of the Moon, which continued to gather a variety of physical information for a period of years following the lunar landings in the Apollo program. A great deal is known about the surface features of Mars, owing to the long series of images taken of the surface by Mars-orbiting spacecraft such as *Mariner 9* and the two Viking orbiters. Both the Soviet Union and the United States have landed spacecraft upon the surface of Mars, but so far these have given relatively little information about the interior of the planet. In the case of Venus, the very thick hazy atmosphere obscures the surface of the planet, and a detailed map of the surface features awaits the arrival of a spacecraft with a synthetic-aperture radar capability, such as is planned by the United States. Very-low-resolution maps of the surface have been provided by using radar antennas on the Earth, and improved but still low-resolution topographic information is being provided by a radar altimeter on the Pioneer Venus orbiting spacecraft and by the Arecibo radio telescope. Both the United States and the Soviet Union have landed spacecraft on Venus, but these have primarily been intended to investigate the atmosphere, and have given practically no information about the interior. For Mercury, reasonably good images of one of the hemispheres were taken by a fly-by spacecraft. For the outer solar system, images of Jupiter and its four large Galilean satellites, and of Saturn and its larger satellites and ring system, have been returned by the Voyager spacecraft. *See* SPACE PROBE.

Construction of models. Planetary scientists attempt to synthesize their information about the structure and properties of each of the planets by constructing models of them. These models make use of the laws of physics and chemistry and are considered to be successful when they reproduce all of the known measured information about a planet. Sometimes it is possible to fit all of this information using more than one model, indicating a considerable uncertainty about the interior properties of the planet, but nevertheless such model building is a useful exercise, because it tends to limit many of the properties of the planet to certain ranges of values.

The most obvious gross properties of the planet are its mass and its radius. In constructing a model, it is required that the model be in hydrostatic equilibrium. This means that at any interior point in the model, the pressure must be great enough to sustain the weight of the overlying mass of material. Thus, given the mass and the radius, estimating the interior pressures through the principles of hydrostatic equilibrium, and knowing something about the compressibilities of materials, it is generally possible to place constraints upon the interior composition of the planets.

All of the planets rotate to some degree, and some of them spin quite rapidly. The effect of the spin is equivalent to a force pressing outward in the equatorial plane of the planet, which helps to sustain the weight of the material in that plane, producing an equatorial bulge. Knowledge of the equatorial bulge and the rate of spin provides additional information about the distribution of mass in the interior of the planet.

Another property which is difficult to measure, but which is known in some detail for the Earth and very crudely for the Moon, is the rate of heat flow from the interior of the planet. For the Earth and the Moon this heat flow appears to result from radioactive heating of the interior of the body, and the knowledge of this heat flow gives information about the distribution of temperature in the interior. In the case of the Earth and the Moon, this interior heat flow is very small compared to the heat which is received from the Sun. In the case of Jupiter and Saturn, the heat flow from the interior is comparable to the heat received from the Sun, so that it has been possible fairly easily to measure this heat flow for these planets because their temperatures are significantly greater than would be expected on the basis of the heat received from the Sun alone. *See* EARTH.

Classes of chemical composition. The planets in the solar system have an extremely wide range of properties. This distribution of characteristics can be understood in part from a knowledge of the more abundant elements in nature and their volatility properties.

Approximately 98% of matter in the Sun, and therefore also presumably in the matter from which the Sun and the solar system were formed, consists of the gases hydrogen and helium. Most of the remaining material consists of carbon, nitrogen, and oxygen, which in the presence of very large amounts of hydrogen tends to form methane, ammonia, and water. These substances are collectively called ices, and they evaporate at relatively low temperatures. Both the light gases hydrogen and helium and the ices are of quite low abundance on the Earth and the other inner planets in the solar system. What comprises the bulk of the material in these planets is the rocky material, constituting only about 3 parts in 1000 of the solar mix of elements, and among the rocks the most abundant elements are magnesium, silicon, iron, aluminum, calcium, and chromium, all of which are present in rocks predominantly in the form of oxides. In the relatively undifferentiated rocks which fall to Earth from outer space, called meteorites, iron appears both in the form of oxide and as the metal. In the surface rocks of the Earth, iron is almost entirely in the form of the oxide, since the metallic iron has predominantly collected near the center of the Earth to form the core. *See* ELEMENTS, COSMIC ABUNDANCE OF; METEORITE.

The differences in the volatilities of these materials, which correlate with the properties of the planetary bodies in the solar system, give information about the properties of the environment in which the planets formed in the solar system. The inner planets, composed predominantly of rocks, evidently formed in a rather hot environment, so that the volatile gases and ices were not condensed and did not collect along with the rocky material, which presumably was condensed. The comets, residing at very large distances from the Sun in the solar system, appear to be mixtures of rocky materials and of the ices. The outer giant planets, Uranus and Neptune, appear to be primarily composed of materials heavier than hydrogen and helium, probably mixtures of rocky and icy materials. The two largest planets in the solar system, Jupiter and Saturn, are much closer in composition to that of the Sun itself, although studies of their interior structures tend to indicate that there is some degree of enrichment in the heavier elements. These differences in composition thus indicate that the tendency to collect hydrogen and helium depends not upon the ability to condense hydrogen and helium into solid form, which would require extremely low temperatures, but rather upon the size of the body which has formed, the larger bodies being more successful in gravitationally capturing the elusive hydrogen and helium. Comets are quite small and are devoid of these materials; Uranus and Neptune are intermediate in mass between the Earth on the one hand and Jupiter and Saturn on the other, and they have been only moderately successful in obtaining hydrogen and helium. However, Jupiter and Saturn were very successful in obtaining these gases. *See* COMET.

Since these compositional classes provide a natural means for dividing the planetary objects within the solar system into separate groups, the structure of the various planetary groups will be discussed in turn.

Giant planets. The giant planets are Jupiter, Saturn, Uranus, and Neptune.

Jupiter. Jupiter, the most massive planet in the solar system, is only about 1/1000 of the mass of the Sun. It comes closest in composition to that of the Sun itself. If the composition of Jupiter truly matches that of the Sun, then it would contain in its total mass the equivalent of about one earth mass of rocky material. However, the best attempts to construct models of the interior of Jupiter indicate the amount of material heavier than hydrogen and helium is significantly in excess of that which would be expected for the solar composition. There are probably something like 10 to 20 earth masses of rock and ice in the interior of Jupiter, which is an enrichment of a factor of three to six over the solar composition if the ice-to-rock ratio in the interior of Jupiter is the solar ratio, which is not known. Even this enhanced amount of material amounts to only a few percent of the total mass of Jupiter. The considerable uncertainty in the amount of heavy-element enrichment in the Jovian interior results from the uncertainties in the extrapolation of the properties of hydrogen and helium to very high pressures and temperatures such as those found in the interior of Jupiter. It is not even clear whether these heavier materials have settled to the center of Jupiter, or whether they are suspended in the atmosphere which is being continually mixed throughout the different interior levels of Jupiter due to convective motions.

One of the interesting properties of hydrogen at higher pressures is its tendency to form a conducting metal, metallic hydrogen. Because hydrogen is a simple substance, the physical calculations that lead to the expected transformation from molecular to metallic hydrogen are reasonably certain, but the precise pressure at which this transformation takes place is still quite uncertain. It appears to be somewhat in excess of 10^6 atm (10^{11} Pa). Most of the mass of Jupiter exists at a pressure considerably in excess of this amount, so that metallic hydrogen is anticipated to form a substantial portion of the interior mass of the planet. *See* JUPITER.

Saturn. Saturn has about only one-third of the mass of Jupiter, but nevertheless it also is predominantly composed of hydrogen and helium, and in this case it is definitely clearer that there are heavier elements in excess of solar composition within the interior of Saturn. Again, it is not known whether these heavy elements maintain the solar composition ratio between the ices and rocky materials, and the precise amount of enrichment is therefore uncertain, depending upon this ratio. However, the total amount of heavy materials in the interior of Saturn is comparable to the excess amount in Jupiter. *See* SATURN.

Heat flow and helium segregation. Attempts have been made to construct evolutionary sequences of models of Jupiter and Saturn which would follow the changes in structure that take place as the planets cool off after their formation. The research has suggested that Jupiter should still be radiating away its interior heat of formation at about the rate which is actually observed as an excess heat flow from the interior, whereas the amount of primordial heat still emerging from Saturn is expected to be much less than is observed. The explanation of this discrepancy may lie in another interesting property expected for a mixture of helium and hydrogen at higher pressures. Below some temperature which is still quite uncertain, it is expected that helium will collect to form small bubbles within the hydrogen; these bubbles, being heavier, will then sink through the hydrogen toward the center of the planet. Not only does this lead to a greater mass concentration toward the center of the planet, but it also releases additional gravitational potential energy, thereby enhancing the heat flow from the interior. It has been suggested that the interior of Jupiter is still sufficiently hot to have prevented this segregation of helium from hydrogen, whereas the interior of Saturn is sufficiently cooler so that a significant amount of such segregation has and is continuing to occur, thus leading to the observed heat outflow from Saturn.

Uranus and Neptune. Uranus and Neptune are quite similar planets, being 14.5 and 17.2 times the mass of the Earth, respectively. Approximately three-quarters of this mass is expected, on the basis of model building, to consist of materials heavier than hydrogen and helium. The precise numbers will depend upon whether these materials are in the solar ratio of ices to rock, which is not known. If one assumes this ratio to be valid, then each of the planets contains approximately four earth masses of rock and approximately twice that much in the form of ices. The remaining hydrogen and helium form a very deep atmosphere. *See* NEPTUNE; URANUS.

Physical composition. Nowhere in the interiors of the giant planets can anything resembling a solid surface be expected. The temperatures in the interiors are very uncertain and can be estimated only as the result of model construction, but they tend to be thousands to tens of thousands of degrees Celsius. The pressures range up to tens of millions of bars (10^7 bars = 10^{12} Pa) and higher. Under these circumstances all materials behave like fluids. There may be a certain amount of compositional stratification, with denser fluids underlying lighter ones.

This issue of stratification is significant in connection with one of the interesting properties of the interior, the transport of gravitational potential energy released in the deep interior to the surface. The thermal conductivity within the interiors of these planets appears to be much too small to do this job efficiently, even in regions of metallic hydrogen. Conduction may be required to transport heat from a layer of one composition to a neighboring layer of different composition. But within a layer of any given composition, the transport of heat appears to require convection. Convection consists of an irregular pattern of overturning motions within a fluid, similar to that which occurs when one boils water within a pot. It has been argued on this basis that the interiors of the giant planets are primarily engaged in convective motions which transport heat outward.

Terrestrial planets. The terrestrial planets include Mercury, Venus, Earth, and Mars. The Earth's Moon may also be considered a terrestrial planet.

Earth. The prototype for the terrestrial planets, and the one about which the most is known, is the Earth. The Earth consists of a thin upper crust composed of rocks of relatively low density and low melting points, overlying a much thicker mantle composed predominantly of metallic silicates and oxides, which in turn overlies a substantial core, which is composed of much denser materials, believed predominantly to be iron with other elements, either alloyed or in solution. Most of the core is liquid, but there is a smaller inner core which appears once again to be solid, and which probably has some compositional differences relative to the outer core.

On the scale of volatility, the Earth is a very refractory place. Most of the materials in its composition condense at quite high temperatures in a gas of solar composition, usually considerably in excess of 1200°C. Under such circumstances, most of the iron is expected to be metallic, and since metallic iron is so much heavier than other typical rocky material, such as magnesium silicates, it is natural for the metallic iron to collect at the center of the planet. The detailed seismic evidence indicates that the core of the Earth is not pure iron, but also has some admixture of lighter elements, probably some combination of oxygen, silicon, and sulfur. Several percent of the core must also be nickel, which has properties very similar to that of iron.

The overlying mantle is composed of the oxides and silicates of the metals which are more abundant in nature. Many phase changes take place as such material is subjected to increasing pressure, and some of the increasing density with depth in the Earth's mantle is due to such phase changes.

Among the many different mineral phases which are present within the Earth, there is a natural sorting process for those minerals which combine a relatively low melting point with low density. Such minerals melt easily and tend to find their way to the surface of the Earth through such cracks or pores as become available. In this way the crust of the Earth is formed predominantly of such materials through tectonic activity.

One of the major revolutions in thinking in the earth sciences has come with the realization that the Earth is a very dynamic place. The position of the Earth's pole has changed dramatically in location with respect to the surface throughout the history of the Earth, and the land masses themselves have drifted about from one part of the surface to another. This continental drift is rendered somewhat easier by the relatively large mass of the Earth and hence the fairly rapid rate with which the temperature increases into the Earth's interior, thereby weakening the materials and allowing them to deform and flow more easily.

Venus. The next most massive planet within the inner solar system is Venus, which has slightly more than four-fifths of the mass of the Earth. Venus has a very thick atmosphere, and the temperature at its surface is very much higher than is typical of the Earth's surface. The conditions make it very difficult to land spacecraft which can operate for appreciable lengths of time such as would be required to obtain seismic signals from the interior of the planet. On these grounds it can only be conjectured that the interior of the planet is probably much like that of the Earth, with a core, a mantle, and a crust. The Pioneer Venus orbiter radar altimeter has found some major structural features on the surface of the planet, suggestive of extensive tectonic activity, but also, to the extent that some of the features are correctly determined to be large craters, indicating that surface weathering processes take place very slowly. The extent to which the crust of Venus is subject to extensive continental drift motions is quite unknown. *See* VENUS.

Mars. The mass of Mars is approximately one-tenth that of the Earth, and hence significant differences in the internal structure are to be expected. There appears to be less of a density contrast between the core of Mars and that of its mantle, suggesting that the amount of lighter material allied with the core, such as possibly sulfur, is increased relative to the Earth. Because the planet is smaller, the temperature increases less rapidly with depth than in the case of the Earth, and hence Mars should have a somewhat more rigid outer mantle and crust than the Earth. There is no indication that large amounts of continental drift have taken place on Mars. On the other hand, tectonic activity has clearly played a large role in the history of Mars, since the planet can be roughly divided into a hemisphere which is of predominantly ancient and heavily cratered terrain, and another hemisphere which is of much younger and less heavily cratered material. The density of craters on the surface of a planet such as Mars, with so little atmosphere that incoming massive bodies are not significantly impeded in striking the planet, is a measure of the relative age of the surface which has been exposed to space. Since the cratering rate apparently fell off rapidly throughout the first few hundred million years of the history of the inner solar system, differences in crater density frequently represent age differences of some few hundred million years back in the heavy cratering epoch. *See* MARS.

Mercury. Mercury has only about half the mass of Mars, but has several distinct planetary characteristics. The mean density of Mercury is very high, indicating that Mercury probably has an abnormally large core predominantly composed of metallic iron. There is much evidence of extensive tectonic activity, although, like Mars, the increase of temperature below the surface of Mercury probably occurs sufficiently slowly so that the crust and upper mantle are relatively rigid, and nothing resembling continental drift has probably taken place. Mercury is a very heavily cratered planet, with the craters of a given size apparently having been produced by smaller projectiles than in the case of Mars. The reason is that at the distance of Mercury from the Sun such infalling projectiles tend to have higher velocities than they do near the orbit of Mars, so that the resulting impacts are more energetic. *See* MERCURY.

Moon. Although the Earth's Moon is technically a satellite, it makes sense to describe it as a planetary body, and planetary scientists consider the twin bodies of the Earth and the Moon as interesting examples of the extremes of planetary physics ranging from relatively large bodies to relatively small but still chemically differentiated objects. The Moon has a history which includes extensive episodes of melting and differentiation, much of which can be reconstructed on the basis of the returned lunar samples. The upper layers of the Moon, which is only just over 1% of the mass of the Earth, are quite rigid, and there is no evidence for extensive horizontal motions of the structural units.

The Moon is unique in the solar system in having a relatively low density among the inner planets, and at best a very small core, indicating that the planet is practically devoid of metallic iron. Relative to the Earth, it is also highly depleted in the more volatile elements. This unusual compositional pattern presumably requires an explanation in the mechanisms which resulted in the formation of the Moon, about which there has been much controversy. *See* MOON.

Galilean satellites. The four Galilean satellites of Jupiter—Io, Europa, Ganymede, and Callisto—have masses which are all roughly comparable to the mass of the Earth's Moon. It is therefore quite clear that they should be considered as planetary bodies in their own right by planetary scientists. The detailed images of these satellites returned by the Voyager spacecraft which passed through the

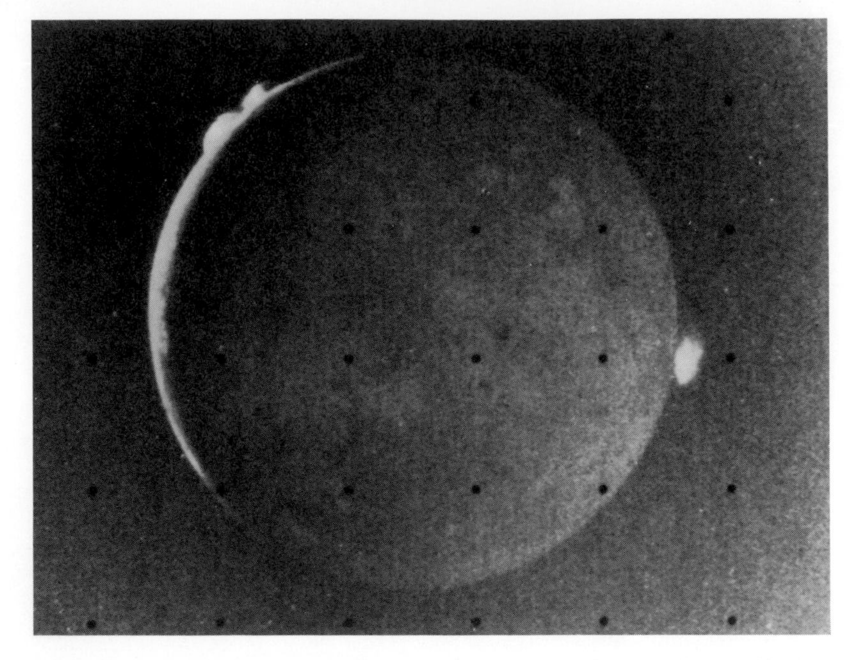

Photograph of Io taken by *Voyager 2* on July 10, 1979, from distance of 1,200,000 km. Three volcanic eruption plumes are visible on the limb, all previously seen by *Voyager 1*, 4 months earlier. (*NASA*)

Jupiter system revealed them to be very interesting places with many rich, complex, and exotic properties.

The most spectacular of these planetary bodies is undoubtedly Io. This satellite has a surface characterized by large deposits of sulfur and sulfur dioxide, which is in a state of continual change. It appears to have at any time several active volcanos, each of which is likely to be spewing a stream of gas and entrained rocky particles about 100 km or so above the surface (see illustration). Such volcanic plumes spread the gases and rocky material from the volcano over a considerable portion of the surrounding terrain. This vigorous tectonic activity is understood to arise from a combination of orbital perturbations of Io by the other Galilean satellites and tidal damping by Jupiter, which results in the dissipation in the interior of Io of very large amounts of heat.

The Galilean satellites appear to represent a composition class which is slightly more volatile-rich than the pure rocky materials characteristic of the inner solar system. In particular, the sulfur content is likely to be considerably higher. The water and carbonaceous contents may also be much higher. If Io ever had much water, it appears to have been lost from the body quite early in its history. It has been suggested that Io has a sufficiently large reservoir of sulfur that this may form an effective fluid layer, or ocean, underlying the solid surface crust.

Europa, Ganymede, and Callisto all appear to have outer crusts composed of water ice. There are a variety of surface markings which indicate a history of cracking, cratering, and in some instances renewal of the icy surfaces. The mean density of Europa is sufficiently high that the planetary body is probably primarily composed of rock. On the other hand, both Ganymede and Callisto have a significantly reduced density, suggesting that quite thick layers of an icy mantle are likely to be part of those planetary bodies. It would not be surprising if the amount of tidal and radioactive heating in the interiors of these bodies was sufficient to maintain a substantial portion of this icy mantle in the form of a liquid brine. Nothing is known about the character of the underlying rocky core.

Oceans. With the possible exception of the Galilean satellites, the Earth is the only planet in the solar system having oceans. Mars is too cold for substantial bodies of liquid water to exist upon its surface, although there is evidence in its surface features that water once ran through a number of channels for at least short periods of time. Mars also has a substantial amount of ice in its polar caps. The atmosphere of Venus is sufficiently hot that if any liquid water were to be placed upon its surface, it would quickly be evaporated into steam. Venus contains very little water in its atmosphere, giving rise to the question of whether Venus has never had substantial amounts of water at any time in its past, or whether it has found some mechanisms for getting rid of the bulk of it. Both the Moon and Mercury are very dry.

Within the oceans of the Earth, a complex set of currents and motions takes place. Many of these currents are driven by slight differences of density within the oceans, which in turn arise because of variations of the amount of dissolved salt, or salinity. Some differences in salinity arise from the evaporation of water into the atmosphere from the oceans, followed by rain upon the land and the runoff of salt-depleted water from the rivers. Other differences in salinity arise from the formation of ice from the oceans, which results in a concentration of salt within a liquid phase.

Large-scale currents are also set up in the oceans as the result of temperature differences within the ocean waters, resulting from the preferential heating of the water in the tropical regions of the Earth. The oceans play an important role in the transport of heat from the Equator toward the poles of the Earth.

Atmospheres. There are certain general principles which govern the structure and dynamics of planetary atmospheres. In most cases these atmospheres receive their primary heat input from above, resulting from heating due to the Sun. Most atmospheres contain some form of haze or condensed layers in the form of clouds, which results in a reflection of a portion of the incident sunlight back into space where it has not contributed to the deposition of heat within the atmosphere. The remainder of sunlight is either absorbed within the atmosphere or transmitted or scattered downward to the ground where absorption takes place. The heat thus received by the ground must be reradiated into the atmosphere, which will transmit some of it and absorb some of it. The absorbed radiation from the ground will in part be reradiated by the atmosphere toward the ground, adding to the heating effect that has taken place as the result of the original receipt of the corresponding energy. This enhancement of the heating effect is commonly called the greenhouse effect, despite the fact that the mechanism by which a greenhouse keeps warm is somewhat different.

The temperature at the surface of the planetary body therefore depends in a complex manner on the properties of the overlying atmosphere, as well as upon the distance of the planet from the Sun. The atmosphere of Venus is very much hotter relative to the Earth than would be expected purely on the basis of the relative distances from the Sun. The difference appears to arise from the extensive operation of the greenhouse effect within the very thick atmosphere of Venus. Soviet spacecraft landed upon the surface of Venus have found quite large amounts of illumination by sunlight there, indicating that significant amounts of solar energy do manage to penetrate to the ground of that planet.

The only terrestrial planets with atmospheres are Venus, Earth, and Mars. Both Mars and Venus have atmospheres composed predominantly of carbon dioxide. If all of the carbonate rocks of the Earth had the carbon dioxide extracted from them, the Earth also would have a thick predominantly carbon dioxide atmosphere very similar to that of Venus. Thus the difference between these two planets arises to a large extent from the ability to form carbonate rocks, which is a function of temperature. At the high temperature of the ground surface of Venus, carbonate rocks are broken down, and the carbon dioxide is released to the atmosphere; the thick atmosphere of Venus is therefore the stable state under the circumstances. In the case of the Earth, it appears that water has played an important role in the formation of carbonate rocks from carbon dioxide, and such water has not been present significantly in liquid form on Mars. There is no evidence that Mars has a substantial reservoir of carbon dioxide in the form of carbonate rocks.

The element of next greatest abundance in the atmospheres of Mars and Venus is nitrogen. This happens to be the predominant element in the atmosphere of the Earth. The next most abundant element in the terrestrial atmosphere is oxygen, which is maintained there predominantly as the result of the operation of life upon the surface of the Earth. Any planet with a large content of oxygen in the atmosphere is very likely to be extensively populated by living organisms.

The atmospheres of the terrestrial planets appear to be substantially mixed as a result, in part, of convective processes which transport heat, and in part from winds which are produced by pressure differences and which cause mixing by stirring up the atmosphere. At a sufficiently great height in the atmosphere, mixing is no longer effective, and a gravitational stratification of the components of the atmosphere takes place, with the lighter components of the atmosphere extending to greater heights. At these great altitudes, solar ultraviolet radiation produces an extensive amount of ionization of these atmospheric constituents, producing a plasma layer at the top of the atmosphere which is called the ionosphere. At a sufficiently great height the molecules of the atmosphere are in free ballistic trajectories; this region is called the exosphere.

The same principles of physics and photochemistry also apply to the atmospheres of giant planets, but the details are considerably different, because the predominant constituents are hydrogen, helium, and methane. At significantly lower levels in the atmosphere, a substantial amount of ammonia appears, and this forms a layer of ammonia clouds, probably contaminated by some amount of hydrogen sulfide which forms a compound with ammonia. At a still lower level in the atmosphere it is expected that water clouds will be present.

In the case of the giant planets, the rapid rotation gives rise to a distinctive banded structure along parallels of latitude within the atmosphere, because Coriolis forces make it difficult for the convective motions within the atmosphere to transport material significantly across parallels of latitude.

Magnetospheres. Some of the planets contain substantial magnetic fields; others do not. Within the inner solar system, the Earth possesses a relatively strong field, Mercury a relatively weak one, and if Venus and Mars contain significant intrinsic fields, they are sufficiently weak that they have not been confirmed. On the other hand, it is known by direct measurement that Jupiter and Saturn have very strong magnetic fields.

The generation of planetary magnetic fields appears to depend upon a combination of planetary rotation with an inner convecting layer having significant electrical conductivity. These conditions appear to be met in the core of the Earth and in the metallic hydrogen mantles of Jupiter and Saturn.

One of the most striking features of planetary magnetospheres is the trapping of energetic particles within them. This gives rise to a great variety of phenomena within the Earth's magnetic field, but in the case of Jupiter the effects are so strong that radiation damage significantly affects the operational lifetime of any space probe inserted into the magnetosphere. *See* MAGNETOSPHERE; PLANET.

[A. G. W. CAMERON]

Bibliography: T. Gehrels (ed.), *Jupiter*, 1976; H. Jeffreys, *The Earth*, 5th ed., 1970; *The Solar System* (articles from *Scientific American*, September 1975), 1975.

Pleiades

A beautiful group of stars resembling a little dipper, in the constellation of Taurus, known since earliest records. The Pleiades is a typical open cluster (see illustration); it contains several hundred stars within a radius of 1° from Alcyone. Its distance is 420 light-years (1 ly = 9.46×10^{12} km), its linear diameter about 15 light-years, its age 5×10^7 years. The brightest stars are blue, of B type. The cluster is permeated with diffuse nebulosity. Though early accounts refer to the

The Pleiades. (*Lick Observatory photograph*)

Pleiades in terms of seven stars, only six are now conspicuous to the unaided eye, which raises a theory that one, the lost Pleiad, has faded. *See* CONSTELLATION. [HELEN S. HOGG]

Pluto

The most distant known planet in the solar system. Pluto was discovered on Feb. 18, 1930, by C. W. Tombaugh at the Lowell Observatory, Flagstaff, AZ, on photographic plates taken as part of a systematic search. The presence of a planet beyond Neptune's orbit had been predicted independently by P. Lowell and W. H. Pickering (among others) on the basis of analyses similar to those which had led U. J. Leverrier and J. C. Adams to the prediction of Neptune. In this case, apparent perturbations of Neptune's motion seemed to require the presence of another planet. Pluto was found surprisingly near its predicted position, since modern investigations have shown that the mass of Pluto is far smaller than was assumed in the calculations.

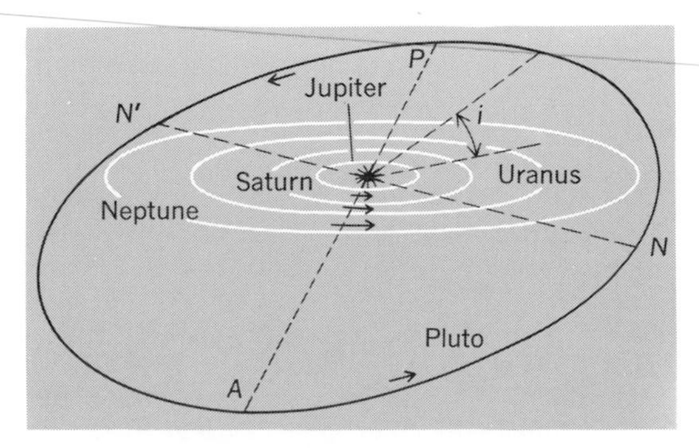

Orbit of Pluto. A perspective view to show the inclination *i* and eccentricity of the orbit. *A*, aphelion; *P*, perihelion; NN', line of nodes. (*From L. Rudaux and G. de Vaucouleurs, Astronomie, Larousse, 1948*)

Pluto's orbit (see illustration) has a semimajor axis (mean distance to the Sun) of 3.7×10^9 mi (5.9×10^9 km), an eccentricity of 0.25, and an inclination of 17.2°. The inclination and eccentricity are the largest of any of the planets. At a mean orbital velocity of 2.96 mi/s (4.7 km/s), it takes Pluto 248.4 years to make one revolution around the Sun. The large orbital eccentricity means that at perihelion Pluto is closer to the Sun than Neptune, but the orbits of the two planets do not intersect. *See* NEPTUNE; PLANET.

Pluto is visible only through fairly large telescopes, since its visual magnitude at mean opposition is 14.7. Periodic variations in its brightness demonstrate that the surface of Pluto is covered with bright and dark markings and indicate a period of rotation of 6.3 days. This is the longest rotational period of any outer planet and is similar to the *orbital* period of Triton about Neptune. This fact and the high inclination and eccentricity of Pluto's orbit around the Sun have led several scientists to propose that Pluto was originally a satellite of Neptune which escaped from the control of the planet—perhaps by a near collision with Triton, which then assumed its present retrograde orbit.

Pluto has a small satellite, discovered in 1978, in a synchronous orbit; that is, the satellite's period is identical to the rotational period of the planet. This discovery made possible a determination of Pluto's mass, which was found to be about one-seventh that of the Earth's Moon. With a diameter in the range of 3000–4000 km (it is still poorly determined), Pluto must have a mean density close to 1 gm/cm³. It is therefore a very different object from the rocky terrestrial planets. The Earth's Moon has a mean density of 3.3, consistent with a composition of silicates. In contrast, Pluto must be composed largely of ices, a conclusion that gains support from the detection of an infrared absorption band of solid methane in its spectrum. Methane gas may also be present.

The size and mass of Charon—the provisional name for Pluto's satellite—are even less well determined. However, current observations suggest that this object may be almost half as large as Pluto itself, which would make this system virtually a double planet. Observations with the Space Telescope in 1986 should greatly improve knowledge about these bodies.

The average temperature of Pluto must be close to 50 K, which is consistent with the presence of solid methane on the planet's surface, in equilibrium with a small amount of methane gas. But the temperature varies with the position of the planet in its eccentric orbit. Since Pluto reaches perihelion in 1989, it is now actually closer to the Sun than Neptune, and the surface temperature is several degrees above the average value. This will maximize the amount of methane in the atmosphere. A heavier gas, such as nitrogen or argon, must also be present to prevent the methane from escaping. Like the diameter, the temperature has not yet been measured directly. Thus there is a possibility that it could be slightly higher than the predicted value if additional gases that can absorb infrared radiation are present to produce a small greenhouse effect. Once again, new observations are needed before a fundamental question can be answered about this distant object. *See* PLANETARY PHYSICS; VENUS.

Tombaugh continued his photographic survey at Lowell after the discovery of Pluto, with negative results. Planets at still larger distances must be fainter than sixteenth magnitude, and a massive effort would probably be required to detect them. [TOBIAS C. OWEN]

Bibliography: W. K. Hartmann, *Moons and Planets*, 1972; G. P. Kuiper and B. Middlehurst (eds.), *Planets and Satellites*, 1961.

Polaris

Alpha Ursae Minoris, the North Star, or polestar. Polaris is located approximately 1° of arc from the north celestial pole, the point where the Earth's axis of rotation now pierces the celestial sphere. Two bright stars at the end of the bowl of the Big Dipper point toward Polaris.

Polaris is a high-luminosity star of spectral type late F. It is one of the brightest cepheid variables, but is atypical in having the small light range of only 0.1 magnitude. The period is 4 days, the luminosity 3000 times that of the Sun, and the distance 200 parsecs. Polaris has a faint, main-sequence companion. *See* STAR.

[JESSE L. GREENSTEIN]

Precession of equinoxes

A slow change in the direction of orientation of Earth's axis of rotation which results in a gradual westward motion of the equinoxes. There are two types, known as lunisolar precession and planetary precession. The total precession is the sum of these two. The phenomenon of lunisolar precession, which is by far the more important, was discovered by Hipparchus about 125 B.C., and was first explained by Isaac Newton.

The term equinox has a dual meaning. It refers to (1) either of the two imaginary points at which the ecliptic (the apparent annual path of the Sun among the stars) crosses the celestial equator (the great circle in which the plane of Earth's equator intersects the celestial sphere), or (2) the date when such a crossing occurs. During its annual journey around the ecliptic the Sun crosses the celestial equator twice. There are therefore two equinoxes each year; the vernal equinox occurs

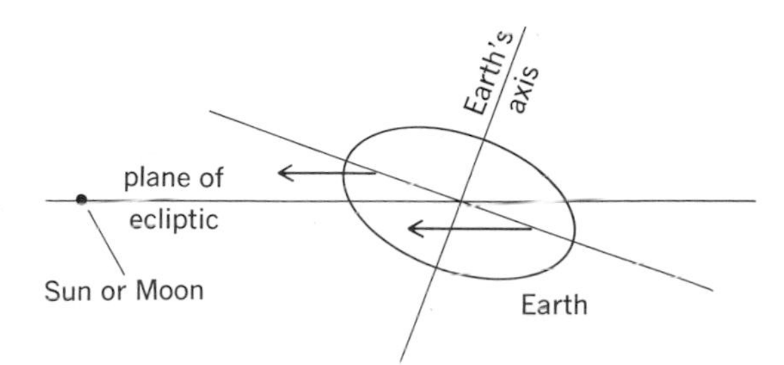

Fig. 1. Gravitational attraction of Sun or Moon on Earth. A torque results which tends to tip Earth's axis.

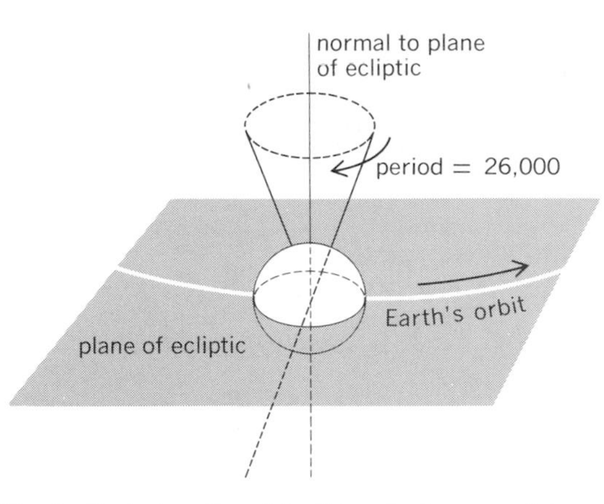

Fig. 2. Conical motion of Earth's axis. One complete cycle of precession requires 26,000 years.

about March 21, and the autumnal equinox occurs about September 23. The annual westward motion of the equinoxes caused by precession amounts to about 50."27. The general motion of Earth's axis of rotation consists of precession plus a second angular motion of the axis which is called nutation. For general discussions of these phenomena *see* NUTATION. *See also* ASTRONOMICAL COORDINATE SYSTEMS; EARTH ROTATION AND ORBITAL MOTION.

Physical causes. The lunisolar precession of the equinoxes is caused by the gravitational attraction of the Sun and the Moon which, as a result of the polar flattening of Earth and the inclination of Earth's axis, gives rise to a small turning moment, or torque, on the Earth in its orbit (Fig. 1). This torque is not constant because of the varying positions in space of the three bodies involved. Because of the closer distance of the Moon, the magnitude of the torque due to the Moon is greater than that due to the Sun. As a result of this torque, Earth's axis describes a cone about the normal to the plane of its orbit (Fig. 2); the period of this precession is approximately 26,000 years in a direction opposite to that of Earth's rotation; that is, the equinoxes precess from east to west. Because the angular momentum of the spinning Earth is so large, the rate of precession is extremely slow.

Planetary precession. This is a comparatively small eastward motion of the equinoxes caused by the action of the other planets in altering the plane of Earth's orbit. Planetary precession also causes a small change in the rate of lunisolar precession.

Effects. As a result of precession, Polaris will not always be the pole star. Vega will be nearer to the north celestial pole in about 12,000 years; α Draconis was the pole star about 4600 years ago. Another effect of precession is that the signs of the zodiac no longer correspond to their respective constellations. The first degree of Aries is now in Pisces and is approaching Aquarius; hence the term Age of Aquarius.

[J. ALLEN HYNEK]

Bibliography: W. H. Munk and G. J. F. MacDonald, *Rotation of the Earth: A Geophysical Discussion*, 2d ed., 1975; F. D. Stacey, *Physics of the Earth*, 2d ed., 1977; S. P. Wyatt, *Principles of Astronomy*, 3d ed., 1977.

Procyon

Alpha Canis Minoris, a nearby bright star, 0.3 magnitude, of spectral type F5. Located at a distance of 3.5 parsecs, Procyon has an absolute magnitude of +2.7, which is slightly brighter than a normal main-sequence F5 star. Thus Procyon may be an old star that has begun to evolve off the main sequence. Its great age is indicated by a close faint companion of absolute magnitude +13, yellowish in color, that is, a white dwarf star, which is nearing the end of its evolution. The visual binary orbit, of 40-year period, gives Procyon a mass of 1.76, and its companion 0.65, that of the Sun. *See* STAR.

[JESSE L. GREENSTEIN]

Proton-proton chain

A group of nuclear reactions involving fusion of light nuclei that converts hydrogen into helium. It is believed to be the principal source of energy in main sequence stars of a little more than a solar mass and of less-massive stars. Completion of a chain results in the consumption of four protons (hydrogen-1 nuclei), the synthesis of a helium (^{4}He) nucleus and two neutrinos (ν), and the release of 26.73 MeV of energy. The energy E reflects the mass difference between the four protons and the helium nucleus, and is calculated on the basis of the Einstein mass-energy equation $E = mc^2$, where m is the mass difference and c^2 is the square of the velocity of light. Because hydrogen is the fuel consumed in the process, it is referred to as hydrogen burning by means of the proton-proton chain.

Reactions. The first reaction involves the fusion of two protons to form a nucleus of heavy hydrogen (the deuteron, ^{2}H) with the release of a positron (e^+) and a neutrino (ν). This is followed by the fusion of a proton and a deuteron to form a helium nucleus of mass 3 (^{3}He) with the release of a 5.494-MeV photon (γ) of electromagnetic energy. Most often, the chain is completed by fusion of two ^{3}He nuclei to form a ^{4}He nucleus with the release of two protons. These reactions are represented by (1), where the energy release of each reaction is indicated following the reaction.

$$^1H + ^1H \rightarrow ^2H + e^+ + \nu \quad \text{(1.442 MeV, of which 1.192 is exclusive of neutrino energy)} \qquad (1a)$$
$$^1H + ^2H \rightarrow ^3He + \gamma \quad \text{(5.494 MeV)} \qquad (1b)$$
$$^3He + ^3He \rightarrow ^4He + 2\,^1H \quad \text{(12.860 MeV)} \qquad (1c)$$

If there is sufficient ^{4}He already present at the site of hydrogen burning, it is possible to complete the proton-proton chain by reactions (2), or at higher hydrogen-burning temperatures by reactions (3).

$$^3He + ^4He \rightarrow ^7Be + \gamma \quad \text{(1.586 MeV)} \qquad (2a)$$
$$^7Be + e^- \rightarrow ^7Li + \nu \quad \text{(0.862 MeV, or 0.049 exclusive of neutrino energy)} \qquad (2b)$$
$$^7Li + ^1H \rightarrow 2\,^4He \quad \text{(17.348 MeV)} \qquad (2c)$$

$$^7Be + ^1H \rightarrow ^8B + \gamma \quad \text{(0.137 MeV)} \qquad (3a)$$
$$^8B \rightarrow ^8Be^* + e^+ + \nu \quad \text{(15.04 MeV, or 7.41 MeV exclusive of neutrino energy)} \qquad (3b)$$
$$^8Be^* \rightarrow 2\,^4He \quad \text{(3.03 MeV)} \qquad (3c)$$

Reaction (2*b*) represents the capture of an electron by the beryllium-7 nucleus to produce a lithium-7 nucleus and a neutrino. Boron-8 is unstable to β^+ decay and is shown releasing a positron (e^+) and a neutrino in its decay to an excited state in ^{8}Be (the asterisk on ^{8}Be* shows that it is an excited state rather than the ground state of that nucleus). The ^{8}Be* then decays [reaction (3*c*)] to two alpha particles (^{4}He nuclei) releasing 3.03 MeV, which represents the 2.94-MeV excitation energy plus the 0.092 MeV by which the mass of two alpha particles differs from that of the ground state of ^{8}Be.

Using the rates of the nuclear reactions in the chain under conditions believed to be prevalent in the hottest part of the core of the Sun, R. W. Kavanagh calculated the probable percentages of the different modes of completion of the *p-p* chain. He determined that about 69% of the reaction chains in the Sun are completed through the ^{3}He + ^{3}He reaction, and about 99.7% of the remainder go by the electron capture by ^{7}Be.

Neutrino emission. The photons that are produced in these nuclear reactions in the central regions of a star interact with the stellar medium, and radiation of resulting energy from the surface of the star is delayed. In the Sun, for example, millions of years are required for the energy to reach the surface. In contrast, the neutrinos that are produced interact only weakly with matter and thus travel directly out of the Sun at the velocity of light. Although the neutrinos carry only a small percentage of the energy produced in the nuclear reactions, their penetrability provides a means of determining what is happening at the center of the Sun now, rather than millions of years ago. R. Davis and his collaborators devised an ingenious experiment to measure the influx of these neutrinos reaching the Earth from the Sun. Their detector is particularly sensitive to the relatively high-energy neutrinos of the ^{8}B decay, so the analysis of the importance of that infrequent branch of the chain is of particular interest. Scientists are puzzled by the fact that the flux of neutrinos predicted by theories of the present structure and evolutionary state of the Sun is several times larger than that found in Davis's very careful measurements. Astrophysicists are reexamining all aspects of their theories of stellar structure and conditions of the core of the Sun in their attempts to solve the solar neutrino problem. *See* CARBON-NITROGEN-OXYGEN CYCLES; STELLAR EVOLUTION; SUN.

[GEORGEANNE R. CAUGHLAN]

Bibliography: C. A. Barnes, Nucleosynthesis by charged particle reactions, *Advan. Nucl. Phys.*, 4: 133–203, 1971; W. A. Fowler, *Nuclear Astrophysics*, 1967; R. W. Kavanagh, Reaction rates in the proton-proton chain, in F. Reines (ed.), *Cosmology, Fusion, and Other Matters*, George Gamow Memorial Volume, pp. 169–185, 1972; P. D. Parker, J. N. Bahcall, and W. A. Fowler, Termination of the proton-proton chain in stellar interiors, *Astrophys. J.*, 139:602–621, 1964.

Pulsar

A celestial radio source producing intense short bursts of radio emission. The unforeseen discovery of pulsars was made by A. Hewish, J. C. Bell, and colleagues at the Mullard Radio Astronomy Observatory of Cambridge University in early 1968. Since that time, a total of about 330 pulsars have been found, and it has become clear that hundreds of thousands of pulsars must exist in the Milky Way Galaxy—most of them too distant to be detected with existing radio telescopes. *See* RADIO ASTRONOMY; RADIO SOURCES.

Pulsars are distinguished from all other types of celestial radio sources in that their emission, instead of being constant over time scales of years or longer, consists of periodic sequences of brief pulses. The interval between pulses, or pulse period, is nearly constant for a given pulsar, but for different sources ranges from approximately 0.03 to 4 s. The bursts of emission are generally confined to a window whose width is a few percent of the interpulse period. A tracing of the signal received from a pulsar by a large radio telescope is shown in Fig. 1. Individual pulses can vary widely in intensity; however, their periodic spacing is accurately maintained.

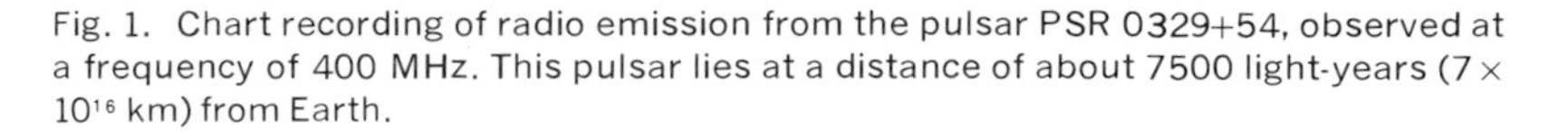

Fig. 1. Chart recording of radio emission from the pulsar PSR 0329+54, observed at a frequency of 400 MHz. This pulsar lies at a distance of about 7500 light-years (7×10^{16} km) from Earth.

Neutron stars. Compelling evidence exists that pulsars are neutron stars, the collapsed cores left behind when certain types of moderate- to high-mass stars become unstable and undergo supernova explosions. Basic models of neutron stars were first produced in the 1930s, when the nuclear physics relevant to stellar interiors was first being studied. It was generally thought then that neutron stars, because of their extremely small size (radius less than about 20 km), would be utterly invisible at astronomical distances. At about the time of the discovery of pulsars, it was realized that conservation of angular momentum and magnetic flux required that a newly formed neutron star must spin rapidly (perhaps many times per second) and have an extremely strong magnetic field strength, of order 10^{12} gauss (10^8 teslas) at the surface. This combination of properties evidently gives rise to highly directive radio emission in the centimeter and meter wavelength range. The neutron star's spin causes the beamed emission to sweep around the sky like a rotating searchlight, thereby causing the periodic pulses observed from a particular location such as the Earth.

Evidence in favor of the rotating neutron star model for pulsars is wide-ranging and varied. Perhaps the single most important observed fact is the remarkable stability of the basic pulsation periods, which typically remain constant to within a few tens of nanoseconds per year. Such stability seems natural when ascribed to the rotation in free space of a compact, rigid object like a neutron star, but is extremely difficult to achieve by any other known astrophysical mechanism—especially over the necessary range of more than a hundredfold in period. The small size of a neutron star is also required by the observed rapid intensity variations within pulses. One expects a rapidly spinning, highly magnetic star to gradually slow down as it radiates part of its energy away, and such behavior has been observed for all pulsars where observations of sufficient precision have been made. A final, telling piece of evidence was the discovery of a pulsar at the center of the Crab Nebula, the gaseous, filamentary remnant of a supernova explosion recorded by Far Eastern astronomers in the year 1054. This pulsar has the shortest period of all pulsars known, and is evidently one of the youngest. Since supernovas are predicted to be the birthplaces of neutron stars, it is reassuring to find that they are also the birthplaces of pulsars. *See* CRAB NEBULA; NEUTRON STAR; SUPERNOVA.

Pulsed emission. The waveform of a pulsar's periodic emission, averaged over several hundred pulses or more, has a distinctive shape characteristic of each individual source. Several examples of average pulse profiles are shown in Fig. 2. Variations in pulse shape from one pulsar to another are probably the result of differences in detailed structure of the neutron star's magnetic field, or of a different orientation of the rotating beam relative to the Earth, or both. Pulse shapes generally exhibit approximate mirror symmetry about a central point. The emission is usually highly polarized, and observations with high time resolution show that the angle of the polarization vector (projected onto the plane of the sky) rotates smoothly across the pulse profile. The total change in angle is always less than or about 180°, and this behavior is taken to be the straightforward result of the changing projection of the pulsar's magnetic field lines as the star spins relative to the line of sight.

Individual pulses from many of the strong, relatively nearby pulsars have been studied in great detail, and they are found to vary rapidly in intensity, shape, and polarization. Intensity fluctuations on time scales down to 300 μs are shown for one pulsar, PSR 1133+16, in Fig. 3. Individual pulses normally consist of one or more subpulses, and the subpulses may in turn contain "microstructure" of even shorter duration. In some pulsars, subpulses vary in intensity in an obviously nonrandom way, and may even exhibit secondary periodicities incommensurate with the basic pulsar period. Individual pulses are more highly polarized than the average pulse profiles, and sometimes undergo abrupt changes of polarization angle of almost exactly 90°. Many attempts have been made to incorporate such complications as these into models of the pulse emission mechanism, but no

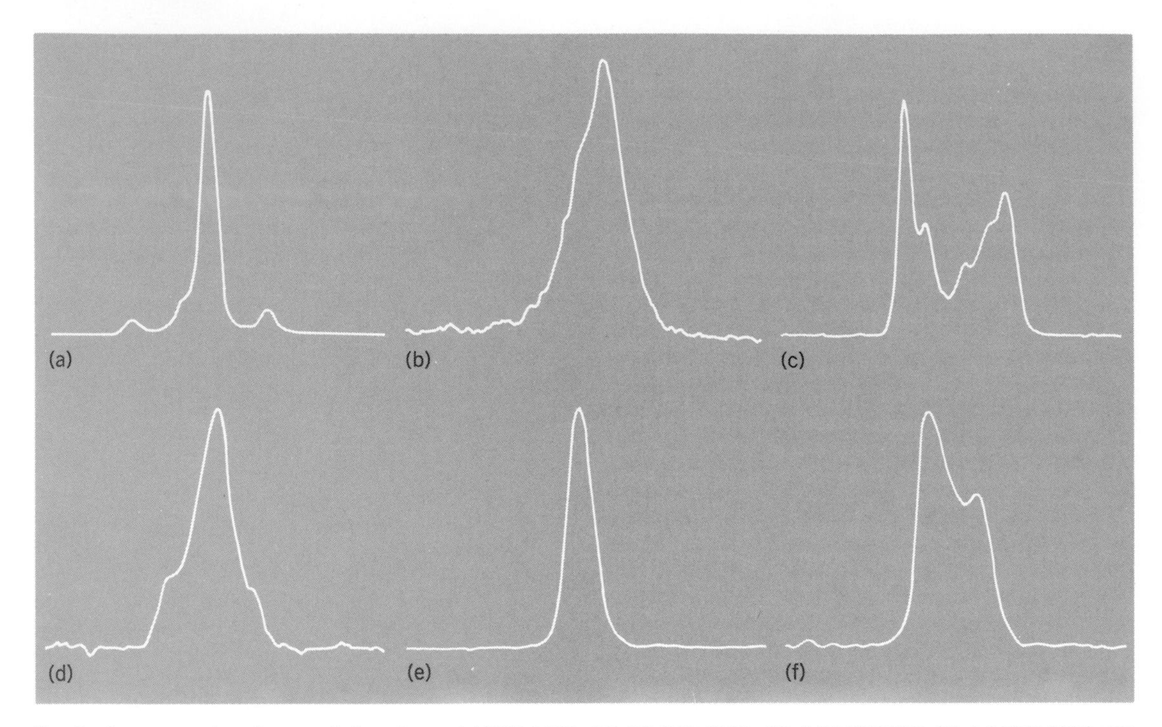

Fig. 2. Average pulse shapes of six pulsars. (*a*) PSR 0329+54. (*b*) PSR 0950+08. (*c*) PSR 1237+25. (*d*) PSR 1508+55. (*e*) PSR 1642–03. (*f*) PSR 1919+21.

one model has won general acceptance, and many uncertainties remain.

Spectra. With a few notable exceptions, pulsars are observable only at radio frequencies, typically between 100 and 10,000 MHz. Over this range the average flux density decreases rapidly, usually in approximate proportion to f^{-1} or f^{-2}, where f is the observing frequency. At frequencies below a few hundred megahertz, a combination of galactic background radiation and terrestrial interference makes observations very difficult; at frequencies above a few thousand megahertz, most pulsars become too weak to observe with existing instruments.

When extrapolated to optical or higher frequencies, the steep radio-frequency spectra of pulsars predict flux densities far below detectable limits. However, the pulsar in the Crab Nebula and another young pulsar, PSR 0833 – 45, have been detected at optical, x-ray, and γ-ray frequencies. Evidently another emission mechanism (or mechanisms) is responsible for the high-frequency radiation from these pulsars. Several dozen pulsarlike objects have also been observed in x-rays, but not at radio frequencies. Invariably these sources are found to be in gravitationally bound orbits around large, evolving stars which are transferring some of their atmospheric gas onto their companions, presumably neutron stars. The transferred gas becomes so hot that it emits thermal x-rays, but at the same time interferes with the radio emission process and prevents the neutron star from becoming an observable radio-frequency pulsar. *See* X-RAY ASTRONOMY.

Interstellar medium. Pulsars have provided astronomers with a unique set of probes for the investigation of the diffuse gas and magnetic fields in interstellar space. Measurement of absorption at 1420 MHz, the frequency of the hyperfine transition in ground-state neutral hydrogen atoms, gives information on the structure of gas clouds, and in many cases provides an estimate of the pulsar distance. Because the gas density in interstellar space is so low, atomic recombination times are large and a significant fraction of the gas is ionized. Consequently the index of refraction for radio waves is strongly frequency-dependent, and low-frequency signals propagate more slowly than high. The broadband, pulsed nature of pulsar signals makes them ideal for measurements of this dispersion. When independent estimates of individual pulsar distances are available, the dispersion may be then used to determine the mean interstellar electron density. Conversely, when a value for the mean density is assumed, individual pulsar distances may be derived from observed dispersion measures. It has been found in this way that the average interstellar electron density is approximately 0.03 cm^{-3} in the galactic disk, and that most of the known pulsars lie within 10,000 light-years (1 light-year = 9.46×10^{12} km) of the Earth.

Because of the large-scale galactic magnetic field, the index of refraction in the interstellar medium is different for the two circular components of a linearly polarized wave. For this reason, the plane of polarization rotates along the propagation path, an effect known as Faraday rotation. Measurements of this effect for some 60 pulsars have shown that the magnetic field strength in the solar neighborhood of the Galaxy is approximately 2×10^{-6} gauss (2×10^{-10} teslas) in a direction along the local spiral arm. *See* GALAXY; INTERSTELLAR MATTER.

Galactic distribution. Over 90% of the sky has been searched for pulsars down to a uniform minimum flux density of approximately 10^{-28} W m^{-2} Hz^{-1}. Some 290 pulsars have been detected above this limit, and their distribution in the sky is shown in the galactic coordinates map in Fig. 4. The figure also shows 40 pulsars with flux densities as much as 10 times weaker, detected in a limited deep survey of the galactic plane at longitudes 40–60°. The obvious concentration of pulsars within ±10° of the galactic equator and in the hemisphere toward the galactic center (longitudes 0–90° and 270–360°) demonstrates that pulsars are members of the Milky Way Galaxy and that their distances are large compared to the thickness of the galactic disk. When distance estimates are combined with the two-dimensional information in Fig. 4, a more elaborate analysis shows that pulsars occupy a disk-shaped region of space approximately 2000 light-years thick and 40,000 light-years in radius. (The Sun, at a distance of about 30,000 light-years from the galactic center, lies near the outer edge of the pulsar distribution.) The number of pulsars per unit volume is several times larger in the inner portion of the disk than near the outer edge, and the total number of galactic pulsars is inferred to be roughly 5×10^5.

The massive, luminous stars which are believed to be the progenitors of pulsars have a galactic distribution similar to that of the pulsars, except that the thickness of their disk is only about 500 light-years. It appears that pulsars are found at much larger distances from the galactic plane because most pulsars acquire a substantial velocity (100 km s^{-1} or more) at birth, as a result of the supernova explosion. During a typical pulsar's active lifetime, roughly 5×10^6 years, it has time to move as much as 1500 light-years away from its birthplace. The equivalent thickness of the pulsar disk will thus be several times greater than that of the parent stars.

If pulsars typically remain active for 5×10^6 years, and if the present inferred galactic population of 5×10^5 pulsars is to be maintained, then a new pulsar must be formed somewhere in the Galaxy approximately every 10 years. This birth rate is reasonably close to the estimated rate of occurrence of galactic supernovas, if allowance is made for those that are optically obscured from view by interstellar dust.

Binary pulsars. Since approximately half of the stars in the Galaxy are members of binary or more complicated orbiting systems, it is noteworthy that radio pulsars are, in general, solitary objects. For reasons mentioned earlier, it appears that neutron stars with ordinary stars as nearby orbiting companions are prevented from being observable pulsars by the unfavorable interaction of atmospheric gases with the pulsar magnetosphere. Furthermore, the probability of such a system surviving a second supernova explosion, thereby creating a pair of pulsars in close orbit, is probably not more than a few percent. Pairs of orbiting stars have always been very important in astronomy, because measurements of their mutual interactions under the force of gravity provide the only method of obtaining stellar masses. Thus the three pulsars which have been shown to be members of orbiting systems have been the subject of much interest.

One of these objects, PSR 0820+02, is in a large orbit with a period of more than 3 years. The nature of its companion star is not yet known. A second binary pulsar, PSR 0655+64, is in a much closer orbit with a period of 1.029 days. Again the identity of the companion has not been established, but in this case it is likely to be a compact object, perhaps a white dwarf, because a large main-sequence star would scarcely fit inside the orbit. Detailed observations of both of these systems will probably yield important information on the evolution of binary star systems and on the generic relationship of pulsars to other classes of stars.

The remaining binary pulsar system, PSR 1913+16, was the first of the three to be discovered. Observations made in 1974 revealed that this pulsar moves in a very close orbit with a period of only $7^h\ 45^m$, and an orbital velocity approximately one-thousandth the speed of light. In this case the companion star is certainly a compact object, and the large velocity and intense gravitational fields in

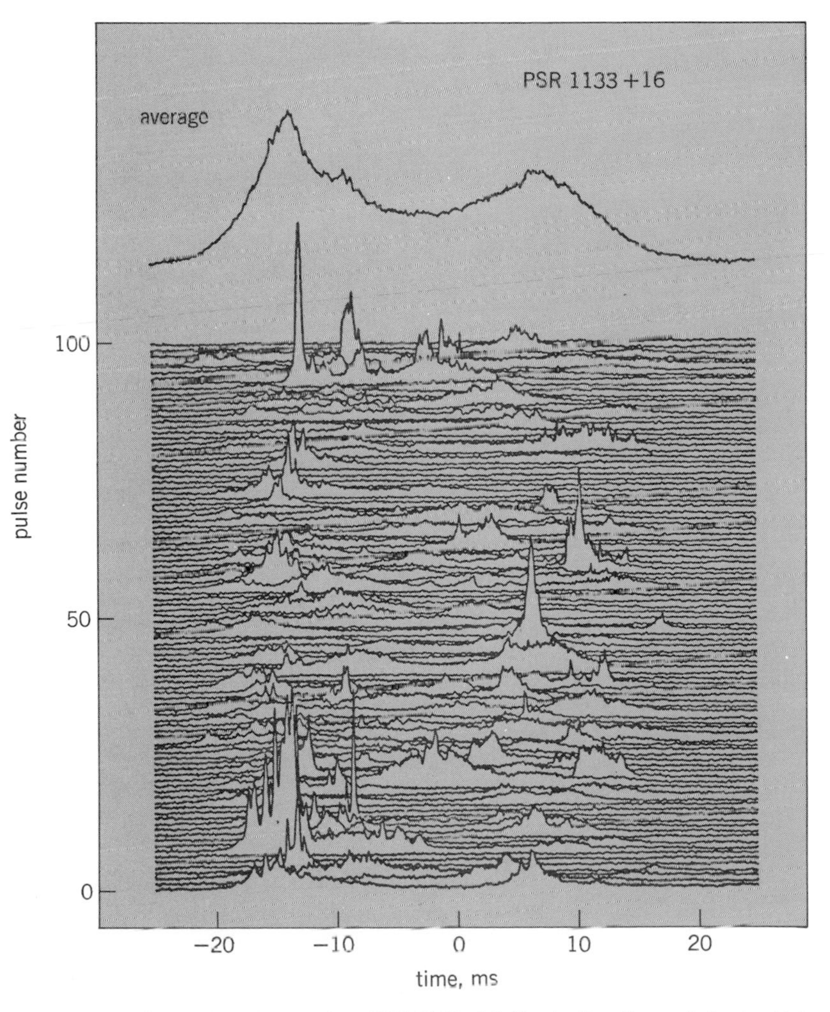

Fig. 3. Individual pulses from pulsar PSR 1133+16, illustrating the variations which take place within pulses and from one pulse to the next.

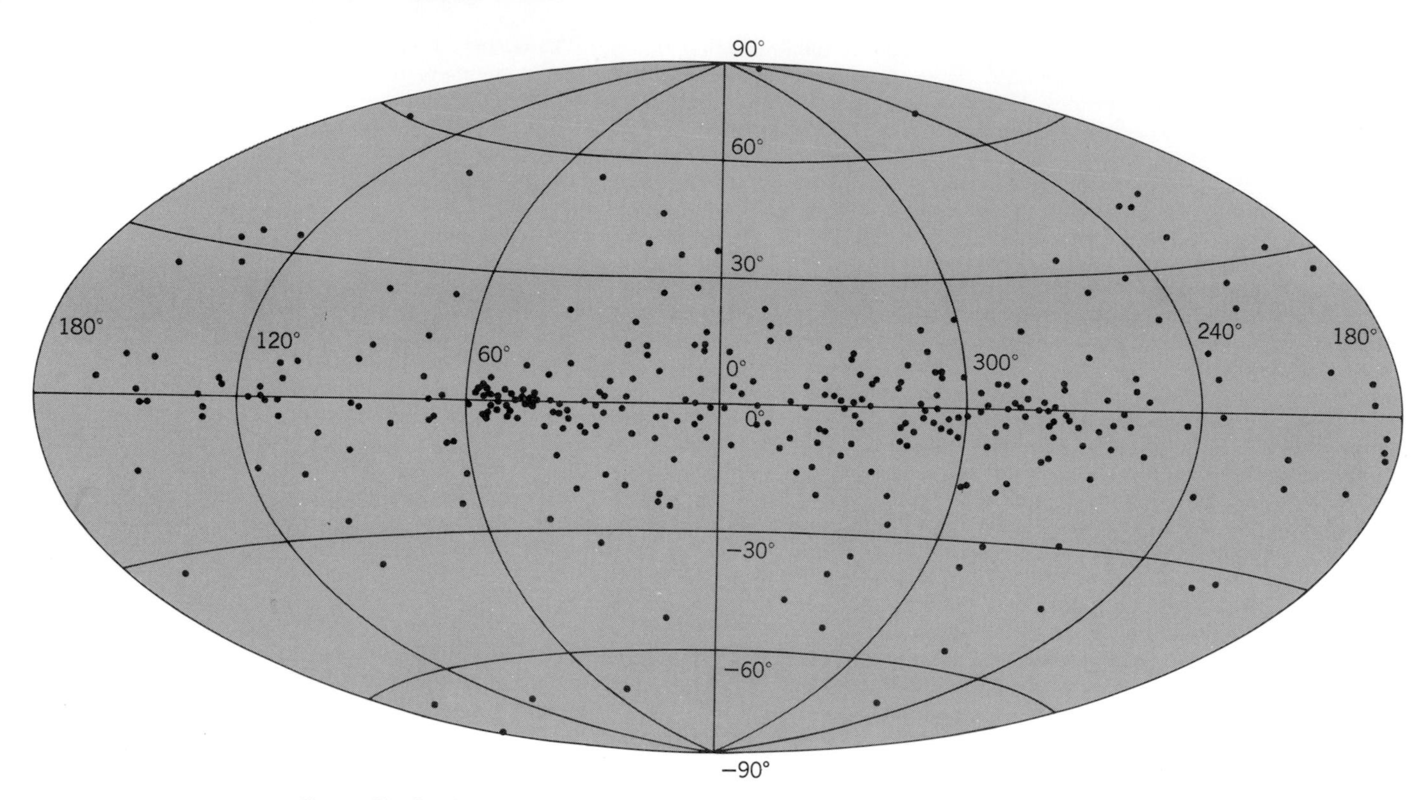

Fig. 4. Distribution of known pulsars in galactic coordinates.

the system, together with the accurate clock mechanism of the pulsar, make it a nearly ideal testing ground for relativistic gravitation theories. One especially important prediction of the general theory of relativity is that a close binary star system should gradually lose energy by the radiation of gravitational waves, and consequently the two stars should slowly spiral closer together. Observations of Doppler shifts of the periodic pulsar signals over more than 5 years have established that the two masses in the PSR 1913+16 system are each approximately 1.4 times the mass of the Sun. The quantitative prediction of general relativity is, then, that the orbital period should diminish by about 10^{-7} s per orbit, amounting to a cumulative orbital phase shift of 1.2 s after 5 years. As shown in Fig. 5, just such an effect has been found, and the observations provide the first (and only) experimental evidence in support of the existence of gravitational waves. *See* BINARY STAR; GRAVITATION; RELATIVITY.

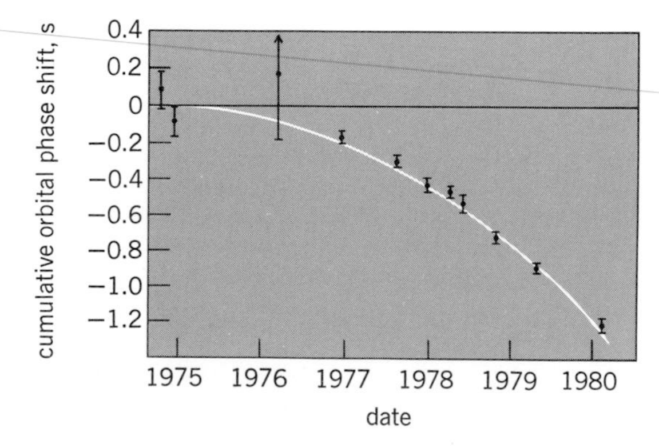

Fig. 5. Accumulating phase shift in the orbit of binary pulsar PSR 1913+16. The parabolic curve gives the prediction of general relativity for energy loss by the radiation of gravitational waves.

[JOSEPH H. TAYLOR, JR.]

Bibliography: R. N. Manchester and J. H. Taylor, *Pulsars*, 1977.

Quartz clock

A clock that uses the piezoelectric property of a quartz crystal. When a quartz crystal vibrates, a difference of electric potential is produced between two of its faces. The crystal has a natural frequency of vibration, depending on its size and shape, and if it is introduced into an oscillating electric circuit having nearly the same frequency as the crystal, two effects take place simultaneously: The crystal is caused to vibrate at its natural frequency, and the frequency of the entire circuit becomes the same as the natural frequency of the crystal.

In the clock the alternating current from the oscillating circuit is amplified, and the frequency is subdivided in steps such as from 100 kHz to 1 kHz. This setup finally drives a synchronous motor and gear train to display time by hands on a clock face.

The natural frequency of a quartz crystal is nearly constant if precautions are taken in cutting and polishing it, and if it is maintained at nearly constant temperature and pressure. After a crystal has been placed in operation, the frequency changes slowly as a result of physical changes. When allowance is made for such changes, the best crystals run for a year with accumulated errors of less than 0.1 sec.

The practice at astronomical observatories that determine time is to rely on the average indication of several quartz clocks to carry the time from one astronomical determination to the next, and to smooth out the accidental errors of the astronomical observations. In this way the mean solar time is known at any instant with a precision of a few thousandths of a second.

[GERALD M. CLEMENCE]

Quasars

A class of astronomical objects that appear starlike on a photographic plate but possess many other characteristics, such as a large red shift, that prove they are not stars. The name "quasar" is a contraction of the term "quasistellar object" (QSO) which was originally given to these objects for their photographic appearance. The objects appear starlike because their angular diameters are less than about 1 second of arc, which is the resolution limit of ground-based optical telescopes imposed by atmospheric effects. Stars also have angular diameters much less than this, and so they too appear unresolved or pointlike on a photograph. *See* RADIO SOURCES.

Location techniques. Quasars were discovered in 1961 when it was noticed that very strong radio emission was coming from a localized direction in the sky that coincided with the position of a starlike object. Prior to this time, it was believed that strong radio emission originating beyond the solar system came only from the direction of certain exploding galaxies (radio galaxies) or from previously exploded stars (supernova remnants). Two techniques were used by radio astronomers to measure within a few arc-seconds the direction of the radio emission. The more common technique, called interferometry, makes use of two (or more) antennas pointing in the direction of the source. After the signals are amplified at each antenna, they are combined via cables or a microwave radio link. Another method, very-long-baseline interferometry, VLBI, records the signals on magnetic tapes which are then synchronously played back so that the signals can be combined. The combined signals yield sinusoidally varying output voltages called fringes. Fringes represent the beating in and out of phase of the incoming radio waves arriving at different times at the two antennas. The phase of the fringes yields an accurate determination of the direction of arrival of the radio waves. *See* SUPERNOVA.

The second technique makes use of the motion of the Moon through the sky in its orbit around the Earth. Occasionally the Moon will pass in front of a radio source and block its emission from reaching the Earth. The exact time of such an occultation is observed using a radio telescope and compared with the known position of the edge of the Moon at that instant. When the source reappears from behind the Moon several minutes later, the time of reappearance is also measured and compared with the location of the Moon's edge. These two fixes on the radio source are usually sufficient to permit its position to be measured accurately.

When these techniques were used to make accurate position measurements of small-angular-diameter radio sources, the coincidence with starlike objects on optical photographs led to the discovery of a new, hitherto unsuspected class of objects in the universe, the quasars. The full significance of the discovery was not appreciated until 1963 when it was noted that the hydrogen emission lines seen in the optical spectrum of the quasar 3C 273 were shifted by about 16% to the red from their normal laboratory wavelength. This red shift of the spectral lines is characteristic of galaxies whose spectra are red-shifted because of the expansion of the universe and is not characteristic of stars in the Milky Way Galaxy, whose spectra are only slightly shifted to the red or blue from the Doppler effect of their small velocities relative to the Earth and Sun.

Optical characteristics. The color of quasars is generally much bluer than that of most stars with the exception of white dwarf stars. The optical continuum spectrum of stars has a blackbody distribution, while the energy distribution of quasars is generally a power law of the form f^{α}, where f is the frequency and α is the spectral index and is usually negative. The blueness of quasars as an identifying characteristic led to the discovery that many blue starlike objects have a large red shift and are therefore quasars. The quasistellar objects (QSOs) discovered this way turned out to emit little or no radio radiation and to be about 100 times more numerous than the radio-emitting quasistellar radio sources (QSSs). Why some should be strong radio emitters and most others not is unknown at the present time. The orbiting Einstein x-ray observatory has found that most quasars also emit strongly at x-ray frequencies. The continuous emission of quasars also increases abruptly at infrared wavelengths longer than a few micrometers. When the sensitivity of long-wavelength infrared detectors is improved, it is expected that quasars will be found to emit large amounts of energy through the infrared region of the spectrum as has been observed in the center of the Earth's and other nearby galaxies. The optical continuum emission is also highly variable, with time scales ranging from fractions of a day to several years. The rapid fluctuations indicate that there are some components in quasars that have diameters less than a light-day or about 10^{15} cm, the size of the solar system. The radiation mechanism responsible for the optical and x-ray continuum emission is not known, but the most likely possibilities are either synchrotron emission or inverse Compton radiation. *See* GALAXY; GALAXY, EXTERNAL.

Emission lines characteristic of a hot ionized gas are superimposed on the continuum. The emission lines are highly Doppler-broadened, indicating internal velocities of hundreds to thousands of kilometers per second. The most common permitted lines seen are those of H, He, C IV, Si IV, and Mg II, while forbidden lines originating from less dense gaseous regions in the quasar are emitted by the ions [C III], [O II], [O III], [Ne V], and others.

The number of emission lines in a quasar spectrum ranges from none to about a dozen. Those with no lines are called lacertids after a well-known variable "star," BL Lacertae, which turned out to be a quasar. In general the emission from the lacertid quasars tends to be more highly variable than the others. In addition to the emission lines, many quasars, especially those with large red shifts, show very narrow absorption lines. In contrast to the emission line regions, the absorbing regions have internal velocities of about 10 km/s. Frequently, one finds many absorption line systems with differing red shifts. The absorption-line red shifts are typically less than the emission-line red shifts by amounts corresponding to relative velocities of several thousand kilometers per second. In some quasars one does find absorption-line red shifts substantially less than the emission-line red shift. The most likely interpretation of the latter is that the light from the quasar is passing through and being absorbed by a distant galaxy that just happens to lie along the line of sight. The more numerous other lines could be absorptions by clouds ejected from the quasar, but it is a mystery as to why these lines would show so little turbulent motion within the ejected clouds.

For the approximately 500 quasars measured, the emission-line red shift z ranges from near 0 to 3.53. For quasars having a z greater than about 0.5, one sees many normally ultraviolet spectral lines that are red-shifted into the visible region of the spectrum. These lines, such as the Lyman α line at 1216 A (121.6 nm)—seen when $z > 1.6$—cannot otherwise be seen in astronomical objects observed from the surface of the Earth because of the atmospheric absorption of all radiation less than about 3100 A (310 nm).

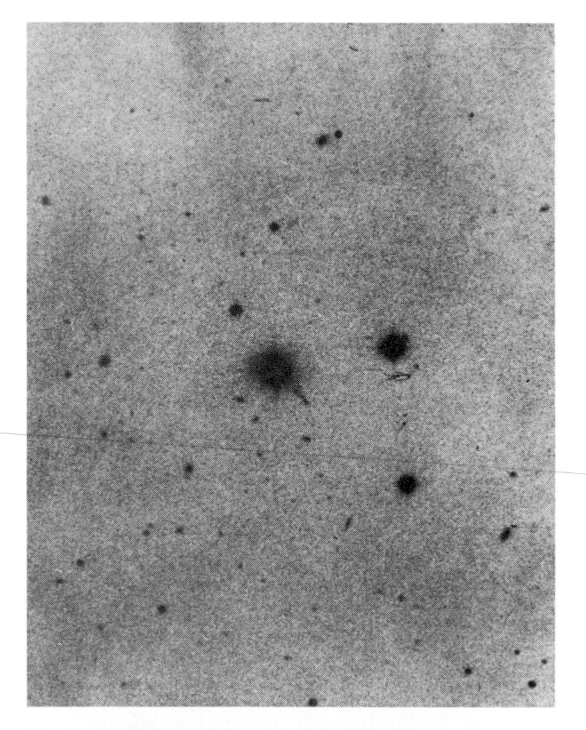

Fig. 1. A photographic negative of the brightest quasar, 3C 273. The cross-shape pattern is due to diffraction within the telescope. The jet extends toward the lower right. (*Photograph taken by A. R. Sandage with the 200-in. Mount Palomar telescope; R. A. Schild and E. L. Schucking, eds., Quasi-Stellar Sources and Gravitational Collapse, University of Chicago Press, 1965*)

Not all quasar images are entirely point sources of light. Some show faint nebulosity surrounding the starlike image. A measurement of this nebulosity around 3C 48 shows that it has an emission-line red shift that is the same as that of the central quasar. Quasar 3C 273 (Fig. 1) has an elongated jet of optical emission 20 arc-seconds from the quasar but pointing radially away from it as if the gas in the jet had been ejected from the quasar. The image of the quasar has been intentionally overexposed to reveal the ejected jet of material. Furthermore, objects having quasar characteristics but with small red shifts are sometimes found surrounded by a faint distribution of stars. These are called N galaxies and may be similar to the Seyfert galaxies, which are spiral galaxies having a very bright quasarlike nucleus. Both of these types of galaxies suggest that the quasars may have an associated galaxylike component but that they are too far away for this fainter component to be seen.

Distances of quasars. When the quasar red shift is plotted against the quasar's brightness or magnitude, as is done for galaxies, one does not find a strong correlation between large z and faint magnitudes. Such a correlation is observed for galaxies and indicates that the fainter, more distant galaxies have progressively larger red shifts. This correlation, called Hubble's law, is considered to be proof that the universe is expanding. In the case of the quasars the scatter in the plot is so great that it does not prove that the red shift can be used as a distance indicator. This does not prove that the large-z quasars are not the most distant objects known, since the scatter in the diagram could easily be due to large differences in the intrinsic luminosities of the quasars. *See* HUBBLE CONSTANT; RED SHIFT.

If the quasars exhibited the Hubble law relationship between their magnitude and red shift, then their red shifts would unambiguously show them to be the most distant known objects in the universe, ranging up to about 10×10^9 light-years away. The statistics of the red-shift number distribution could then be interpreted as showing a greater abundance of quasars at earlier epochs in the universe. However, the paucity of known quasars with z greater than 3 implies that they may have been absent at even earlier epochs when the density of the universe was 27 times greater than its present density.

Some astronomers have argued that the red shifts are not cosmological in origin and hence do not indicate the distances of quasars. One of these arguments rests on the observations that there are examples where two quasars of very different red shift are seen closer together than one would expect by chance if the two quasars were not physically related. On the other hand, the red shifts of a few QSOs seen in nearby clusters of galaxies have been found to agree with the measured red shifts of the associated galaxies, implying that the red shifts are cosmological. Models postulating non-cosmological red shifts, such as gravitational or Doppler red shifts, have met with difficulties. If the red shifts are not cosmological, then they are probably a result of some new unknown physical effect. *See* COSMOLOGY.

Radio characteristics. Some quasars emit a significant fraction of their radiated energy at radio frequencies ranging from about 30 MHz (wavelength, 10 m) to 100 GHz (wavelength, 3 mm). There appear to be two general categories of radio emitters. Those that emit predominantly at the lower radio frequencies (less than 1 GHz) have spectral indices α of about $-.8$, showing that the flux of radio emission is rapidly decreasing with increasing frequency. Interferometers have shown that the radio emission from these types of QSSs does not originate from the optical quasar but comes from two considerably distant regions symmetrically displaced on either side of the optical object. Such radio-frequency characteristics are nearly identical to those of radio galaxies, which suggests that the two may be generically related.

The other, more interesting type of QSS emits primarily at centimeter and millimeter wavelengths (1 GHz to 100 GHz) and originates in the same region as the optical quasar. The flux of emission of these sources is nearly independent of frequency except that there is an abrupt drop-off in the emission spectrum at the low-frequency end. There are a few examples of QSSs that exhibit both sets of characteristics, suggesting that one type may evolve into another. In 1965 large variations in the flux of radio emission from the second type of QSS were discovered, indicating that, since these types are active, they are young objects while those of the first type are probably decaying remnants. Figure 2 shows an example of measurements of the active galaxy source 3C 120 made at a frequency of 15.5 GHz (1.9 cm) with the 120-ft radio telescope of the Haystack Observatory in Massachusetts. Sudden outbursts of emission are evident, particularly the pair in late 1972 and early 1973. Outbursts such as these have enormous luminosities sometimes approaching 10^{45} ergs/s, nearly 10^{12} times the luminosity of the Sun.

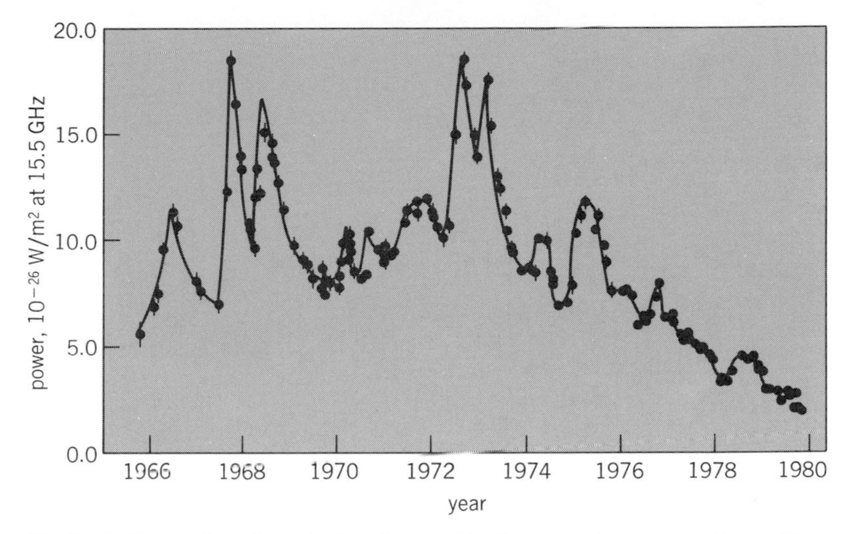

Fig. 2. Outbursts in radio emission observed in the quasarlike source 3C 120. (*Data taken by W. A. Dent using the 120 ft Haystack antenna*)

The nature of the origin of this energy is not known and represents one of the most important unanswered questions concerning quasars. The radio emission is most probably synchrotron radiation produced by electrons with velocities very near the speed of light that are spiraling in magnetic fields. The electrons can absorb their own photons at lower frequencies, producing the drop-off in their emission spectrum. From the observed frequency of this self-absorption, the magnetic field strengths can be estimated as 0.01 to 1 gauss.

In the past few years there have been VLBI observations of some of the active quasars. These observations show that very compact radio components with angular diameters of about 0.001 arcsecond exist in quasars of the second type. Furthermore, successive observations seem to show that these compact components are expanding rapidly and separating at speeds greater than the speed of light, if cosmological distances are assumed. A number of theories have been advanced to explain the phenomena without requiring that the quasars be relatively nearby. The most likely explanation is that one of the separating radio components is moving almost directly toward the Earth with a velocity close to the speed of light. At high velocities of approach, effects predicted by the special theory of relativity become important. Intervals of time in a rapidly approaching object would appear to be much shorter, giving the appearance that the object is moving much faster than it really is. *See* RELATIVITY.

Theories of quasars. The many similarities of the observed characteristic of quasars with radio galaxies, Seyfert galaxies, and BL Lacertae objects strongly suggest that quasars are also active nucleii of galaxies. There is good statistical evidence which shows that quasars with large red shifts are spatially much more numerous than those with small red shifts. Because high-red-shift objects are very distant and emitted their radiation at an earlier epoch, quasars must have been much more common in the universe about 10^{10} years ago. Since this is about the same epoch when galaxies are thought to form, it is possible that quasars may be associated with the birth of some galaxies.

More than 10^{60} ergs (1053 J) of energy are released in quasars over their approximately 10^6-year lifetime. Such a prodigious amount of energy must involve a quantity of matter more than 10^9 times the mass of the Sun, and contained within a volume similar in size to the solar system. The source of this energy and the mechanism by which particles are accelerated to velocities very near the speed of light are not known. One theory proposes supernovalike explosions of supermassive objects following their gravitational collapse; another theory postulates that a massive black hole is somehow driving the quasar; still others assume the particles are accelerated by magnetic fields associated either with a giant rotating object or by many smaller rotating neutron stars. Yet another theory postulates that the energy is released in collisions between rapidly moving stars in a very dense cluster at a galactic nucleus. *See* ASTRONOMICAL SPECTROSCOPY; BLACK HOLE; INFRARED ASTRONOMY; NEUTRON STAR; RADIO ASTRONOMY; SUPERMASSIVE STARS.

[WILLIAM DENT]

Bibliography: G. Burbidge and M. Burbidge, *Quasi-Stellar Objects*, 1967; F. D. Kahn and H. P. Palmer, *Quasars*, 1967; H. L. Shipman, *Black Holes, Quasars, and the Universe*, 1980.

Radio astronomy

The study of celestial objects by measurement and analysis of the electromagnetic radiation they emit in the wavelength range from 1 mm to 30 m. Karl Jansky was the first to determine that radio emission from the heavens could be detected. At the Bell Telephone Laboratories in 1932 Jansky built a rotating antenna array operable at 14.6-m wavelength and attempted to investigate the source of short-wave interference. He found that, in addition to the intermittent interference resulting from thunderstorms, there was a steady hiss-type static that was most intense when his antenna was directed toward the center of the Galaxy. With these and subsequent observations Jansky concluded that the Galaxy itself was an intense source of cosmic radio radiation at frequencies of a few tens of megahertz.

Jansky's work was not continued seriously until 1940 when an electronics engineer, Grote Reber, built a 9.4-m parabolic reflector in his backyard in Wheaton, IL, and studied the galactic radio emission at 162 MHz. He confirmed Jansky's impression that the radio emission was strongest toward the galactic center; he further succeeded in detecting radio emission from the Sun. Solar radio emission was also discovered during World War II by radar researchers in Britain, who found that the severe interference they had noted each day at dawn was in fact emanating from the rising Sun. After World War II the radar scientists made use of their expertise in astronomical pursuits, and a concerted study of astronomical objects at radio wavelengths began. Today radio astronomers study an entire range of celestial objects, including the normal stars, planets, galaxies, and the exotic quasars, pulsars, and x-ray sources.

Radio universe. Since the late 1940s, radio telescopes have been used to map the skies and determine the positions and intensities, or fluxes, of individual sources of radio emission. Such maps have been made with increasing sensitivity and angular resolution; the latter property enables astronomers to determine the position of the radio sources accurately. Knowing the position, astronomers can refer to optical photographs of the sky and establish precisely which object is emitting radio waves. This procedure has led to the identification of radio sources with many bright galaxies and even with the most distant objects in the universe, the quasistellar objects (or quasars). However, nearly one-fifth of all radio sources are unidentified, that is, excellent photographs taken at the radio source positions show no object at all from which the radiation could arise. One concludes from this that these unidentified sources are "normal" galaxies and quasars at such great distances that they cannot be seen optically. Hence, radio astronomy seems to provide a more complete sampling of objects in the universe than does optical astronomy; as such it may be a tool for investigating cosmological questions, in particular questions about the origin of the universe. *See* RADIO SOURCES; RADIO TELESCOPE.

One can address cosmological questions in radio astronomy by a simple procedure that makes use only of the fact that the further away a radio source is, the weaker it appears. Specifically, the measured radio flux of a source decreases inversely with distance R squared, $1/R^2$. If the universe is filled uniformly with radio sources, then as one looks to greater and greater distances one can sample more and more sources, that is, the volume of space surveyed by the observation increases as the distance cubed, R^3. Therefore, if one simply counts the number of sources at a given flux level S, the number should increase as the inverse 3/2 power of the received flux, that is, the number of sources is proportional to $S^{-3/2}$. This proportionality is not found. Rather it appears that there are more sources at great distances per unit volume of space than there are nearby, a result that means the universe is expanding, and that those distant sources are, in general, stronger than ones nearby. These two conclusions are fundamentally linked, and together they mean that further work on cosmological problems must await a clearer understanding of the nature and evolution of individual radio sources. *See* COSMOLOGY.

Galaxies and radio galaxies. The morphological structure of most galaxies resembles either a spiral (or whirlpool) or an ellipse. Spiral galaxies, such as the Milky Way Galaxy, are seen to be laced with gas, dust, and newly formed stars; in addition they are often radio sources, although most are quite weak. Elliptical galaxies, on the other hand, consist of older stars and show no evidence of gas and dust; they are usually not radio sources. However, the few elliptical galaxies that are radio sources are very spectacular ones, being among the most energetic radio objects in the sky. These are known as the radio galaxies.

The radio emission from spiral galaxies is typically confined to a small nuclear region supplemented by much weaker extended emission from the disk of the galaxy. In elliptical radio galaxies, on the other hand, the radio emission emanates not from small radio sources within the galaxy but from the whole volume of the galaxy and, in most cases, a huge volume outside the galaxy. Frequently there are two large regions 10^6 light-years in diameter (1 light-year $= 9.46 \times 10^{12}$ km) on opposite sides of the galaxy from which most of the radio energy is emitted. An example of a bright radio galaxy is Fornax A, shown in Fig. 1. The two radio-emitting lobes of this galaxy are separated by more than 500,000 light-years.

The radio emission from galaxies and radio galaxies is almost certainly generated by the electron synchrotron process, in which relativistic electrons spiral around magnetic field lines and emit a continuous radiation spectrum throughout the band accessible to radio astronomers. The total energy in the form of magnetic fields and relativistic particles that is contained within the immense radio-emitting regions which surround radio galaxies must be enormous. The events that produce radio emission from galaxies, the sources of their energy, and their ultimate fate are poorly understood. *See* GALAXY, EXTERNAL.

Quasars. Quasars are perhaps the most enigmatic objects in the universe. To the radio astronomer they are essentially indistinguishable from the radio galaxies, with regard to both their spectrum and apparent radio flux. However, optical measurement of the red shift of the quasars reveals that they are receding from Earth at velocities approaching that of light. This means, according to the work of Edwin Hubble, who showed that the distance to extragalactic objects is proportional to their velocity, that the quasars must be the most distant objects in the universe. If the quasars are so much more distant than the radio galaxies, yet produce radio emission that is observed at a flux level comparable to nearby galaxies, their intrinsic radio power must be considerably greater than that of the already very energetic radio galaxies. *See* RED SHIFT.

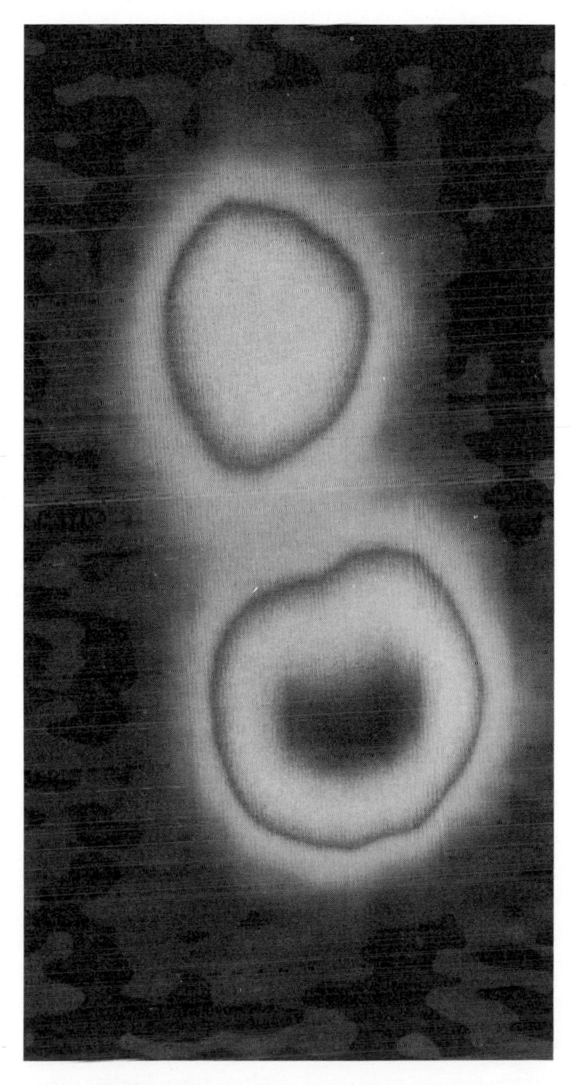

Fig. 1. Radio picture of the giant radio galaxy Fornax A: This picture was made at 5000 MHz with the 43-m radio telescope at Green Bank, WV. (*National Radio Astronomy Observatory*)

Further puzzles regarding the quasars come from radio observations that use the technique of very-long-baseline interferometry (VLBI). These observations have revealed that the size of the radio-emitting region in quasars is very small compared with the size of the radiating region in radio galaxies: the quasar emission arises in a region approximately 1/1000 of the extent of that of the radio galaxies. How quasars can emit such vast amounts of energy in such a small volume, and by what mechanism, is one of the outstanding unanswered questions in radio astronomy. *See* QUASARS.

Big bang. The universe itself, as distinguished from all of the discrete objects individually or collectively within it, is also a source of radio radiation. In 1965 Arno Penzias and Robert Wilson of the Bell Telephone Laboratories measured the radio temperature of the sky to be 3 K and showed that the universe is uniformly permeated by an electromagnetic flux of radiation at this temperature. This radiation is a relic of the initial evolutionary phase of the universe. When the universe was a small fraction of its present size, the "heat" of its formation was contained in subnuclear particles and electromagnetic waves. As time passed, the particles coalesced into atoms and ultimately into galaxies and stars, whereas the radiation simply cooled by the expansion of the universe to its present temperature near 3 K. The detection of this radiation along with its interpretation as direct evidence for the "big bang" origin of the universe is the most significant achievement of radio astronomy. *See* COSMIC MICROWAVE RADIATION.

Solar system astronomy. The Sun is an intense radio source, but only because it is so close to Earth. If it were at the distance of the nearest stars, its radio emission could not be detected. Solar radio emission tends to be intermittent: solar flares that produce cosmic rays and plasma streams that interact with Earth are visible as radio bursts; these are most frequent during the peak of the 11-year solar cycle. *See* COSMIC RAYS; SUN.

Radio observations of the planets have revealed a great deal. Radio astronomers provided the first indication that Mercury does not keep one face constantly toward the Sun, as had been believed for nearly a century by optical astronomers, by showing that the dark side of the planet had a high temperature. Radar astronomers subsequently proved this point conclusively. Venus was first discovered to have an extremely hot (roughly 1000°C) surface by radio measurements of its temperature; measurements at different wavelengths have yielded information on the atmospheric composition of Venus. High-resolution measurements of Venus and the Moon by large radio telescopes and interferometers have provided information on the surface composition and roughness, as well as on the distribution of temperature on the surface. *See* MERCURY; MOON; VENUS.

Jupiter is a most interesting planet, being a much stronger radio source than had been expected from estimates of its surface temperature by optical astronomers. Most of its radio emission is caused by electron synchrotron emission in its very strong magnetic field. The very-long-wavelength emission of Jupiter is impulsive, and its strength depends upon the position of Io, one of Jupiter's moons. These processes are not completely understood. Similar impulsive long-wavelength bursts have also been discovered from Saturn. *See* JUPITER; SATURN.

Radio stars. Several nearby, apparently normal stars are detectable at radio wavelengths. Such stars as Algol, β Persei, and AR Lacerta are multiple star systems, long studied by optical astronomers, that have been found to be radio sources. The radio emission from these stars is dominated by radio bursts in which the radio fluxes may increase by a factor of 100 or more. It is clear that the radio bursts are initiated by or are a product of mass exchange processes going on between (at least) two closely bound stars.

An extreme example of radio emission from stars comes from stars that are also x-ray sources. Again, these objects are usually binary systems in which mass exchange plays a deciding role in their continuing evolution, but with x-ray sources one of the component stars appears to be a star that has exhausted its reservoir of nuclear fuel and is collapsing to its final state. *See* BINARY STAR; X-RAY ASTRONOMY.

Pulsars. In the spring of 1968 a completely new class of continuum radio sources was discovered. Instead of emitting energy continuously, these sources emit bursts of energy in extremely regular intervals. Precise measurements have shown that the pulse period is highly constant: better than 1 part in 10^{12}, but the pulse intensity varies unpredictably with time. In all pulsars the pulses are stronger at the longer wavelengths; often the pulse strength varies rapidly and unpredictably with wavelength. The length of individual pulses ranges from 1 to about 20 msec, depending on the particular pulsar; time structure of less than 1 msec also appears and exhibits both linear and circular polarization. The arrival time of the pulses at long wavelengths is later than at short wavelengths, because of dispersion effects caused by ionized gas between the pulsar and Earth; analysis of such data yields otherwise unattainable information concerning the electron density, magnetic field, and cosmic-ray intensity in interstellar space.

Pulsars appear to be very highly evolved stars that have exhausted their inner energy sources and are slowly collapsing. The size inferred for these stars (10 km) and their mass (10^{33} g) means that they are so dense that individual protons and electrons cannot exist as separate entities. Rather, the extreme pressure inside the pulsars will cause the proton and electrons to "coalesce" into neutrons. Pulsars are therefore true neutron stars with enormous magnetic fields, $B = 10^{12}$ gauss (or 10^8 tesla), which emit radio radiation most likely by cyclotron radiation from electrons moving in these intense magnetic fields. *See* PULSAR.

Supernova remnants. The final neutron star phase of stellar evolution is thought to be preceded by a phase of catastrophic stellar instability in which the core of the star suddenly collapses, causing the outer envelope of the star to be explosively ejected. During this process the luminosity of the star increases briefly by a factor of 10^{12} or more, and the star is said to be a supernova. The atmospheric envelope of the star that is ejected in this process becomes a rapidly expanding cloud of relativistic particles, magnetic field, and filaments of ionized gas. These conditions are precisely those necessary for the generation of radio emission through electron synchrotron radiation, and very intense radio sources indeed exist at the positions of old supernovae chronicled by ancient astronomers.

One of the most interesting of the radio sources associated with a supernova remnant is the Crab Nebula. It first appeared as a visible supernova in A.D. 1054 when its position and brightness were recorded by Chinese and Japanese astronomers; today the Crab Nebula is one of the strongest radio sources in the sky, although it is slowly dimming as a result of its expansion. The polarization of the radio and optical radiation indicates that both are produced by synchrotron radiation. The polarization and intensity of the optical radiation vary with time; such variations are most pronounced at the position of the pulsar (presumably the remains of the star that exploded producing the supernova remnant) in this nebula. Evidently the pulsar supplies much of the energy for the nebular radio emission in the form of relativistic particles.

At the location of the pulsar in the Crab Nebula, one finds both an x-ray source and an intense discrete radio source discernible only at wavelengths longer than 10 m. Both the low-frequency radio source and the x-ray source have pulsed components that must be direct manifestations of the pulsar, but the precise nature of the correspondence between the pulsar and the radiation from the nebula as a whole is not fully understood. *See* CRAB NEBULA; SUPERNOVA.

HII regions. An HII region (a region of ionized hydrogen) is a large cloud of interstellar gas that has been ionized and heated by one or more bright, hot stars located within. These nebulae are sources of both continuum and line energy at radio and optical wavelengths. Since cosmic matter consists mostly of hydrogen, the ionized gas consists mainly of protons and electrons that emit continuum energy by the bremsstrahlung process. *See* NEBULA.

The gas also emits recombination line radiation. In the radio region these lines arise from transitions between two high quantum levels in the hydrogen atom. Since many such levels exist, many lines at different wavelengths appear. Similar transitions exist for elements other than hydrogen; comparison of the intensities of lines from two elements yields a direct measure of their relative abundance.

Measurement of the line wavelength usually reveals that it is displaced slightly from the value that would be arrived at in a terrestrial laboratory. This difference arises from the Doppler effect, caused by relative motion of the HII region and Earth. In a similar way the wavelength width of the line yields information on the relative motion among the hydrogen atoms within the HII region; if there were no relative motion, the width of the line would be very small. This motion arises from a combination of thermal motions of individual atoms, turbulence, and large-scale velocity fields within the nebula. *See* DOPPLER EFFECT.

Comparison of the strengths of lines arising from different levels, along with comparison of these strengths with the continuum intensity, yields information on the number of atoms per unit volume, the temperature, and the degree of fluctuation of both of these quantities within the nebula. It is of course of great value to be able to measure the variation of these quantities with position within an HII region; but this is a difficult task requiring many observations with the largest telescopes to extract a reasonably complete physical picture. *See* INTERSTELLAR MATTER.

Hydrogen line. In 1944 Henk van de Hulst suggested that radiation from individual hydrogen atoms could be detected at 21.1-cm wavelength. The radiation van de Hulst had in mind was radiation in a hyperfine spectral line, which arises when a hydrogen atom with the proton and electron spinning in opposite directions suddenly rearranges itself so that the electron and proton spin in the same direction; a discrete radio-frequency quantum or photon is emitted at 21.1 cm in this rearrangement process. In 1951 H. I. Ewen and E. M. Purcell at Harvard succeeded in detecting this spectral line and introduced a new field of research into the nature of the gaseous content of the Milky Way Galaxy and other galaxies.

Study of the 21-cm line of neutral atomic hydrogen has been exceptionally rewarding in its contribution to the knowledge of galactic structure and of the physical characteristics of interstellar gas. Line intensity normally reflects the amount of gas in the line of sight; line wavelength and width indicate the line-of-sight velocity of the gas and the state of internal motion, just as with recombination lines in HII regions. If the gas overlies a strong radio source, the gas temperature can be inferred by observing the 21-cm line in absorption.

The structure of the Galaxy has been elucidated by the study of the amount and velocity of the hydrogen within it. A prime advantage of this method is that the very distant gas is just as visible as nearby gas, whereas optical studies of the whole Galaxy are impossible because the very distant stars are made invisible by intervening clouds of dust. The results of these radio studies indicate that the Milky Way Galaxy is a spiral, as are many of the galaxies in the sky. About 5% of the total mass of the Galaxy exists in the form of interstellar gas; it is spread throughout most of the Galaxy, resulting in an average density of about one hydrogen atom per cubic centimeter. The hydrogen in other galaxies can also be studied by its 21-cm line radiation. The structure of hydrogen in this Galaxy is similar to that in other spiral galaxies; an important result is that, since spiral galaxies have a deficiency of gas in their central portions, the gas distribution is doughnutlike. Elliptical galaxies contain very little hydrogen gas. *See* GALAXY.

Study of the small-scale structure of interstellar gas in the Galaxy has revealed that the gas is clumped into structures of various sizes and shapes. These include clouds that are tens of light-years in diameter, smaller cloudlets, and larger flat sheets of gas. The intensity of the magnetic field has been measured in some of these clouds by observations of the Zeeman effect in the 21-cm line.

Molecular lines. It has long been believed that simple molecules could not exist in the tenuous gas between stars, because the starlight radiation field would be sufficiently intense to break apart even the simplest molecular species. In spite of these arguments, by 1968 radio astronomers had found rotational transitions of three simple molecules, OH, H_2O, and NH_3. These molecules were found in dark clouds of gas and dust usually associated with HII regions; such dark clouds are believed to be the sites of recent and continuing star formation.

The OH and H_2O emission lines were also found to be peculiar in two respects: first, they were unexpectedly strong; and second, they were composed of many components separated slightly in frequency, presumably as a result of the Doppler effect (see Fig. 2). VLBI observations of these individual line components show that each component arises in a discrete region spatially separated from the others, and moreover that the diameter of each of these discrete regions is quite small, usually only a few astronomical units. (An astronomical unit is the distance between the Sun and Earth, 1.5×10^8 km.) The great intensity of these sources combined with their small sizes leads one to infer that the equivalent temperatures of the emitting gas is unreasonably high; it is on the order of 10^{13} K. Since this cannot represent a real temperature, the line emission must be amplified in the source; that is, microwave amplification by stimulated emission of the OH and H_2O lines is being directly observed. These lines are interstellar masers. The ultimate energy source which energizes or "pumps" the masers may be collisional processes with H_2 or free electrons; or it may be intense infrared, optical, or ultraviolet radiation.

Since 1968 dozens of molecular species have

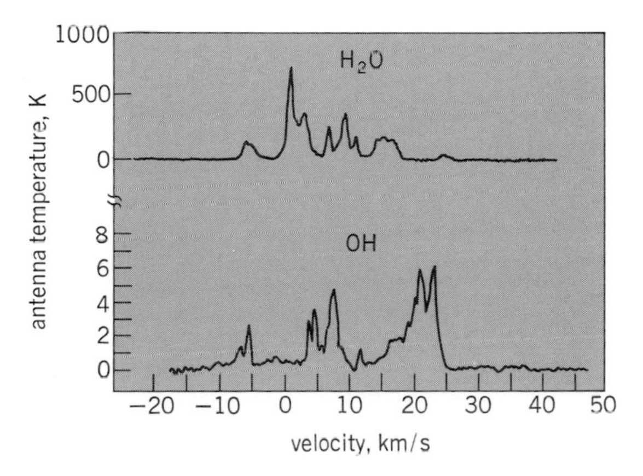

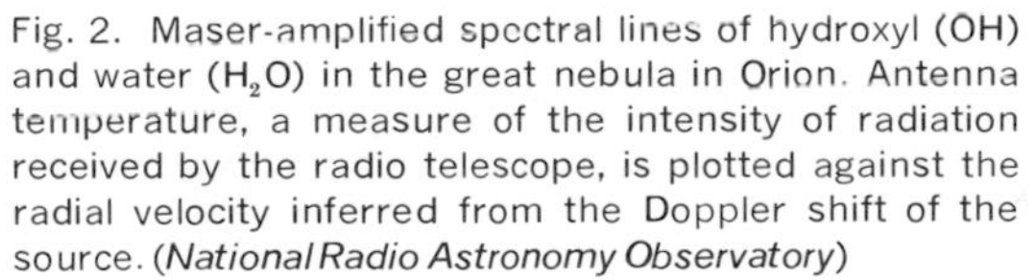
Fig. 2. Maser-amplified spectral lines of hydroxyl (OH) and water (H_2O) in the great nebula in Orion. Antenna temperature, a measure of the intensity of radiation received by the radio telescope, is plotted against the radial velocity inferred from the Doppler shift of the source. (*National Radio Astronomy Observatory*)

Molecules discovered in space by radio astronomers

Year	Molecule	Symbol	Wavelength
1963	Hydroxyl	OH	18 cm
1968	Ammonia	NH_3	1.3 cm
1968	Water	H_2O	1.3 cm
1969	Formaldehyde	H_2CO	6.2 cm
1970	Carbon monoxide	CO	2.6 mm
1970	Cyanogen radical	CN	2.6 mm
1970	Hydrogen cyanide	HCN	3.4 mm
1970	Formylium	HCO^+	3.4 mm
1970	Cyanoacetylene	HC_3N	3.3 mm
1970	Methyl alcohol	CH_3OH	36 cm
1970	Formic acid	$CHOOH$	18 cm
1971	Carbon monosulfide	CS	2.0 mm
1971	Formamide	NH_2COH	6.5 cm
1971	Carbon sulfide	OCS	2.7 mm
1971	Silicon monoxide	SiO	2.7 mm
1971	Methyl cyanide	CH_3CN	2.7 mm
1971	Isocyanic acid	$HNCO$	3.4 mm
1971	Hydrogen isocyanide	HNC	3.3 mm
1971	Methyl acetylene	CH_3CCH	3.5 mm
1971	Acetaldehyde	CH_3CHO	20.0 cm
1972	Thioformaldehyde	H_2CS	9.5 cm
1972	Hydrogen sulfide	H_2S	1.8 mm
1972	Methanimine	CH_2NH	5.7 cm
1972	Sulfur monoxide	SO	3.0 mm
1973	Methyldadyne	CH	9.0 cm
1973	Methylamine	CH_3NH_2	3.5 mm
1973	Ethynyl	C_2H	3.5 mm
1974	Dimethyl ether	H_3COCH_3	3.3 mm
1974	Silicon monosulfide	SiS	2.8 mm
1974	Diazenylium	N_2H^+	3.2 mm
1974	Ethyl alcohol	CH_3CH_2OH	3.2 mm
1975	Vinyl cyamide	CH_2CHCN	21.8 cm
1975	Methyl formate	$HCOOCH_3$	18.6 cm
1975	Sulfur dioxide	SO_2	3.6 mm
1975	Nitrogen monosulfide	NS	2.6 mm
1975	Cyanamide	NH_2CH	3.0 mm
1976	Formyl radical	HCO	3.5 mm
1976	Cyanodiacetylene	HC_5N	2.8 cm
1977	Nitroxyl hydride	HNO	3.7 mm
1977	Cyanoethynyl radical	C_3N	3.4 mm
1977	Cyanohexatriyne	HC_7N	1.3 cm
1977	Ketene	H_2CCO	3.0 mm
1977	Ethyl cyanide	CH_3CH_2CN	3.4 mm
1978	Nitric oxide	NO	2.0 mm
1978	Cyanooctatetrayne	HC_9N	2.1 cm
1978	Butadiynyl radical	HC_4	3.1 mm
1979	Isothiocyanic acid	$HNCS$	3.7 mm
1979	Formaldehyde-D	$HDCO$	3.4 mm
1979	Methyl mercaptan	CH_3SH	2.0 mm
1980	Ethynyl	C_2H	2.2 mm
1980	Cyanoethynyl	C_3N	2.2 mm
1980	Cyanoacetylene-D	DC_3N	2.2 mm
1981	Thioformyl ion	HCS^+	3.0 mm

been detected in interstellar space by observations of millimeter rotational lines or lines arising from the interaction of the rotation of the molecule's nuclei with the spin of its electrons (lambda-doubling transitions; see the table). Nearly all of these discoveries were made on the highly precise 36-ft (11 meter) millimeter-wave telescope on Kitt Peak, AZ, operated by the United States National Radio Astronomy Observatory. Observations of the molecular species may make it possible to establish the chemistry and thermodynamics in the interstellar clouds from which, ultimately, stars, planets, and life itself must form.

[ROBERT L. BROWN]

Bibliography: J. S. Hey, *The Radio Universe*, 1971; R. N. Manchester and J. H. Taylor, *Pulsars*, 1977; G. L. Verschuur, *The Invisible Universe*, 1974; G. L. Verschuur and K. I. Kellerman, *Galactic and Extragalactic Radio Astronomy*, 1974.

Radio sources

Virtually every object in the sky is a source of at least weak radio emission, but this discussion will consider only those sources that produce significant detectable radiation at the Earth. These are categorized into three main locations in space: first, within the solar system; second, outside the solar system but within the Milky Way Galaxy; and third, extragalactic. *See* RADIO ASTRONOMY.

Solar system. The most prominent source of radio emission in the sky is the Sun itself. Although it is not an intrinsically intense star, its close proximity to the Earth makes it appear very bright to a radio telescope just as it does visually. The far outer atmosphere or corona of the Sun is extremely hot—about 1,000,000°C—and this gas produces a continuous glow at all wavelengths, which depends on the temperature of the region. Such emission is called thermal radiation. *See* SUN.

In addition, there are often violent storms on the Sun, called solar flares, which spew large numbers of high-energy electrons and protons out into the atmosphere. The radio emission associated with this solar activity, first discovered as serious interference on World War II radar sets, can show significant variations in both intensity and frequency in a matter of a few seconds. As the particles ejected from the active region rush out through the solar atmosphere, they excite the material there to oscillate and thus radiate at a characteristic frequency, called the plasma frequency. The plasma frequency depends on the density of the medium so that, as the particles pass outward through the continually decreasing density of the solar atmosphere, a drift in the frequency of the radiation is observed. Some of the observed emission is also produced by synchrotron radiation, so called because it was first observed in giant particle accelerators, called synchrotrons. Relativistic electrons, traveling at nearly the speed of light, are trapped in the strong magnetic fields of the Sun and radiate large amounts of radio energy. As shall be seen below, this synchrotron process is responsible for the emission observed from almost all the objects outside the Galaxy and many sources within the Galaxy as well.

The planets are heated by the Sun and then, because they are hot, radiate thermal blackbody emission. For cloudy Venus, it is possible to "see" through the opaque cloud layers at radio wavelengths and actually measure the temperature of the surface. The incredibly high value found of 475°C is attributed to the insulating effect of the thick cloud blanket. Jupiter also appears too hot, but in this case it is because the planet is so large that it produces internal heat by contraction under its own self gravitation. In addition, Jupiter has vast Van Allen radiation belts of relativistic particles trapped in the equatorial plane of its magnetic field. These particles produce synchrotron radiation and also appear responsible for short bursts of emission from the planet at low radio frequencies. The repetition of the bursts is correlated not only with the rotation of Jupiter's magnetosphere but with the position of one of Jupiter's moons, Io, in its orbit around the planet. Apparently Io sweeps particles out of the radiation belts to produce the burst radiation as they are dumped into the atmosphere of Jupiter. There is some evidence that Saturn has similar but much weaker radiation belts. *See* JUPITER; SATURN; VENUS.

Radio emission has been detected from several molecules in comets. As the "dirty snowball" composing the nucleus of a comet approaches the hot Sun, the material evaporates and trails off away from the Sun. The molecules in this released gas will be highly excited and, as they relax, vibrations and rotations of their constituent atoms will cause radiation to be emitted only at certain discrete frequencies characteristic of each molecule. These are called spectral lines and can be observed with high-resolution radio spectrometers, which divide the emission over a broad frequency range into a number of adjacent smaller channels. *See* COMET.

The Galaxy. Most of the physical processes responsible for emission from radio astronomical sources were discussed above. Further discussion will be limited to radio sources themselves.

One widely distributed class of radio source in the Milky Way Galaxy consists of large clouds of hydrogen and other elements. These clouds often contain compact regions of strong spectral-line emissions from various organic molecules, which, to date, range up to eleven atoms in complexity. Mixed in with the molecules there is usually dust that can block the light from the stars and cause condensation of the cloud material into new protostars, which will eventually contract sufficiently to allow nucleosynthesis in their cores, thus becoming normal stars. Often at the centers of these protostars are found bright sources of infrared radiation and also maser emission. A maser (the short term for microwave amplification by stimulated emission of radiation) is apparently created when the infrared radiation is absorbed by molecules, putting them into higher energy states than they normally have. The overenergetic molecules are then stimulated by small amounts of incoming radiation to create an avalanche of spectral-line emission as the molecules fall back down to their normal levels. This pumping and stimulation is possible only when the density of the material is nearing that found in the stars. *See* INTERSTELLAR MATTER.

As the newly formed stars turn on, they can heat and ionize the surrounding cloud. The heated gas produces thermal radiation similar to that from the Sun's corona, but in this case at a temperature of about 10,000°C, and also spectral lines associated with the recombination of free electrons and protons to form hydrogen atoms.

The normal stars are not detectable sources of radio emission because of their great distances, although a few spectacular flares have been detected from selected stars. Some stars in pairs, however, are known to have very extended atmospheres and large amounts of gas between them. With highly sensitive modern instruments, one can sometimes detect the thermal emission from this material.

After they use up most of their nuclear fuel, completing the main phase of the evolution, most stars will swell to become giants, and conditions for maser activity prevail in the cool atmospheric envelopes. Maser emission from such stars has been observed in the spectral lines of water vapor, the hydroxyl radical, and silicon monoxide. After the giant stage, very massive stars literally explode and spew a large fraction of their material out into the surrounding space at speeds of up to about 30,000,000 km/h. At the time of the explosion these objects appear as bright new stars called supernovae, but they quickly fade (with time scales of generally less than a year) as the outflowing material is diluted by filling an ever-increasing volume. The magnetic fields and relativistic particles in the surroundings can be compressed and amplified by interaction with this expanding gas to produce large amounts of synchrotron emission which appears in a generally irregular shell around the position of the original star. It has been calculated that this emission is often visible to radio telescopes for up to 100,000 years. As the material gradually merges with its surroundings, it may be a source of the galactic cosmic rays and also the general synchrotron radiation which appears to emanate weakly from everywhere within the Milky Way but with a concentration toward the center of the Galaxy. *See* SUPERNOVA.

A second product of a supernova is an extremely condensed stellar core called a neutron star. As this star, with a mass comparable to that of the Sun contracted to a diameter of perhaps 10 km, it carried with it all the angular momentum and magnetic field of the original star so that it ended up spinning some tens of times each second and carrying along a tremendous magnetic field. Once each rotation, when the magnetic field is properly aligned toward the Earth, such an object produces a short pulse of radiation; hence the name "pulsar," *See* NEUTRON STAR; PULSAR; STAR; STELLAR EVOLUTION.

One final form of radio emission from the Galaxy is the 21-cm spectral line of hydrogen which is caused by a flip in the spin of the electron as it orbits around the proton within an atom. The tremendous abundance of hydrogen makes this line visible everywhere in the Galaxy, and by use of the Doppler principle (the observed frequency of a spectral line is shifted in proportion to its velocity with respect to an observer) it has become the most powerful tool available in analyzing the general rotation and other motions within the Milky Way galactic system. *See* GALAXY.

Extragalactic sources. The simplest radio sources outside the Milky Way are the normal galaxies, so called because of their resemblance to it. A number of these galaxies are close enough to be studied in considerable detail with modern aperture-synthesis telescopes. The results show similar forms of stellar birth and death activity as well as the general synchrotron radiation and hydrogen-line emission found in the Milky Way Galaxy. *See* GALAXY, EXTERNAL.

Some galaxies, however, have radio luminosities that can be thousands of times brighter than those of the normal galaxies. The synchrotron emission from these radio galaxies often extends far beyond the optical outline of the galaxy (up to 18,000,000 light-years for the largest one discovered to date) and generally occurs in two components on either side of the central galaxy. The illustration is a "radiophoto" (that is, a visual representation of the brightness recorded by a radio telescope) of such a galaxy. The source of the emission has been very strongly directed as it has moved out from the galaxy.

In some sources, fine-scale components with sizes smaller than 1/1000 arc-second have been found. These features generally have an excess brightness at centimeter wavelengths and also vary in radio intensity on time scales of weeks to months. Many of these sources have been identified with quasars, bright "point" sources that appear to be receding from the Earth at large fractions of the speed of light, as based upon measurements of optical spectral lines. The predominating opinion is that these quasars are very distant objects that are participating in the general expansion of the universe, but other explanations of the large velocities, such as self-gravitational collapse, cannot be ruled out. In either event, the quasars are very small, perhaps only the size of the solar system, yet produce hundreds of times more synchrotron emission than whole normal galaxies. The source of the energy may well be the release of gravitational potential energy as material falls onto a compact object at the center of the system, but other unknown possibilities may also contribute. See QUASARS.

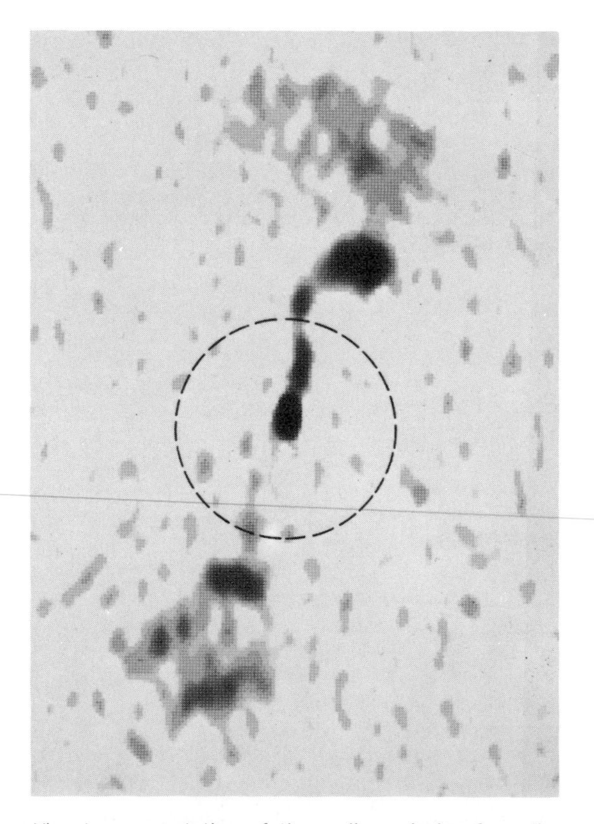

Visual representation of the radio emission from the radio galaxy B 0844+31. This is a negative presentation so that the bright features are shown dark. The outline of the visual galaxy is shown by the dashed line. The data were obtained with the Westerbork Synthesis Radio Telescope in the Netherlands. (*Courtesy of W. van Breugel*)

A final source of extragalactic radio emission is the all-pervasive "2.7 K background". This is the remnant of the fireball of the primeval universe. This early universe cooled rapidly as it expanded, quickly reaching the equilibrium temperature now found in any observed direction. *See* COSMIC MICROWAVE RADIATION.

Intelligent extraterrestrial signals. No discussion on radio astronomy would be complete without mention of the search for intelligent signals from beyond the Earth. As of 1980, no such signals have been detected, although a number of prominent radio astronomers are spending some of their time in such activities. The negative results certainly do not disprove the existence of other intelligent beings, however, but could arise merely from accidentally observing in the wrong place at the wrong time or the lack of understanding of differences in other possible technologies. *See* RADIO TELESCOPE. [JOHN R. DICKEL]

Bibliography: E. Chaisson, Gaseous nebulas, *Sci. Amer.*, 239(6):164–180, December 1978; D. F. Dickinson, Cosmic masers, *Sci. Amer.*, 238(6): 90–105, June 1978; L. C. Green, Radio galaxies and quasars, *Sky and Telescope*, 54:384–389, November 1977; J. S. Hey, *The Radio Universe*, rev. 2d ed., 1977; P. Morrison, J. Billingham, and J. Wolfe (eds.), *The Search for Extraterrestrial Intelligence*, NASA SP419, 1977.

Radio telescope

An instrument used in astronomical research to detect and measure the radio-frequency power coming from various directions in the sky. It consists of three complementary parts: the large reflecting surface that collects and focuses the incident radiation; the electronic receiver that amplifies and detects cosmic radio signals; and a data display device. From the ground, observations with radio telescopes must be made at wavelengths shorter than 30 m, because of ionospheric attenuation, and longer than 1 mm, because the very-short-wavelength radio radiation is absorbed by atmospheric H_2O, CO_2, and O_3. Large radio telescopes and arrays are listed in the table.

Principle of operation. The fundamental principle of a radio telescope is identical to that of a reflecting telescope used at visual wavelengths. The incoming waves (radio or optical) are intercepted by a precise mirror and reflected to a common focal point. The shape of the reflecting surface or "dish" is important: the radio waves must arrive "in phase" at the focal point following their reflection from the dish; that is, the path length from the point of reflection to the focus must be exactly the same for all points on the dish. This restriction can be most simply satisfied if the shape of the reflecting surface is made paraboloidal; consequently, most modern radio telescopes have this shape (Fig. 1). *See* OPTICAL TELESCOPE; TELESCOPE.

The requirement that the radio waves arrive at the focus in phase means that the entire surface of the dish must be very accurate. As a general rule, observations at a particular radio wavelength λ can

Large radio telescopes and synthesis arrays

Institution	Location	Size of reflector, m
Radio telescopes for meter and centimeter wavelengths		
Fully steerable paraboloids		
Max Planck Institut für Radioastronomie	Effelsberg, West Germany	100
Nuffield Radio Astronomy Laboratory	Jodrell Bank, England	76
CSIRO	Parkes, N.S.W., Australia	64
Jet Propulsion Laboratory	Goldstone, CA	64
Algonquin Radio Observatory	Lake Traverse, Ontario	46
National Radio Astronomy Observatory	Green Bank, WV	43
California Institute of Technology	Big Pine, CA	40
Haystack Observatory	Westford, MA	37
Crimean Astrophysical Observatory	Crimea, Soviet Union	22
Limited-tracking transit telescopes		
Special Astrophysical Observatory	Zelenchukskaya, Soviet Union	10 × 1885
Tata Institute	Ootacamund, India	30 × 529
National Astronomy and Ionosphere Center	Arecibo, Puerto Rico	305
Observatory of Paris	Nancay, France	40 × 200
National Radio Astronomy Observatory	Green Bank, WV	91
Radio telescopes for millimeter wavelengths		
Onsala Observatory	Gothenburg, Sweden	20
University of Massachusetts	Amherst, MA	14
National Radio Astronomy Observatory	Kitt Peak, AZ	11
California Institute of Technology*	Big Pine, CA	10
University of Texas	Fort Davis, TX	5
Synthesis arrays		
National Radio Astronomy Observatory (very large array, VLA)	Socorro, NM	Resolution 0.″1
Mullard Radio Astronomy Observatory (5-km array)	Cambridge, England	Resolution 0.″5
Westerbork Radio Observatory (WSRT)	Westerbork, Netherlands	Resolution 1″

*Also three-element millimeter-wave interferometer.

be made efficiently with a particular radio telescope if the mean deviation of the surface of that telescope from a perfect parabola is no greater than λ/10 at each point on the surface. Hence, for example, the 100-m-diameter telescope shown in Fig. 1, which is designed for observations at 1-cm wavelength, must have a surface that is paraboloidal to better than 1 mm across the whole dish. These restrictions can place severe demands on the design, construction, and financing of large telescopes. Of course, the smaller the telescope, the more feasible it is to make the surface highly precise; thus, one radio astronomy observatory may have several telescopes of different sizes for use at different wavelengths. *See* ASTRONOMICAL OBSERVATORY.

Once the radio waves are collected and brought together at the focal point of the telescope, they are in general still extremely weak; power levels near 10^{-20} W are common for astronomical objects. The incoming radio-frequency (rf) signals are first amplified at the focus 10 to 1000 times and then converted to a lower frequency, the intermediate frequency (i-f), that can be easily transmitted by cables from the focal point to the telescope-control building. There the i-f is further amplified, and the signal is detected and displayed in the manner most suited to the investigation.

Astronomical considerations. The types of astronomical objects that emit radio-frequency radiation and hence can be studied by radio astrono-

Fig. 1. The 100-m-diameter radio telescope operated by the Max Planck Institut für Radioastronomie at Effelsberg, West Germany. (*Max Planck Institut für Radioastronomie*)

mers are of such a diverse nature that a variety of radio telescopes and receiving equipment are necessary for a modern radio observatory. Two general astronomical considerations dictate what instruments are needed: first, radio telescopes should have the highest possible angular resolution so that the small-scale details of radio sources can be studied; second, the radio receivers should be extremely sensitive to the very weak signals emitted by cosmic radio sources.

The angular resolution obtainable by a particular radio telescope is limited by diffraction, a blurring of the image that will, for example, cause two radio sources that are close together to appear as if they were a single source. In the presence of this diffraction limit, the minimum angular separation between two sources that a radio telescope would resolve as two sources and not as a single blurred image depends only on the wavelength and the diameter D of the telescope as λ/D (radians). To obtain better angular resolution, one may observe at a shorter wavelength (reduce λ) or employ a larger telescope (increase D). There are, of course, practical limitations to both of these proposals: there is a finite limit to how large a single telescope could be built (as well as concomitant financial limitations); whereas at shorter wavelengths one requires more precise telescope construction, and such instruments must necessarily be smaller. The highest angular resolution obtainable by single-dish radio telescopes is about 1 minute of arc.

Since nearly all sources of radio emission have structure on a scale much smaller than this, it is necessary to realize better angular resolution: this can be done by employing the principles of interferometry. One can imagine replacing a single large radio telescope with many small dishes connected electrically by cables and separated by distances ranging up to the diameter of the original large dish. This concatenation of small dishes set up to simulate the single large dish is called an array. An array provides an angular resolution defined by λ/d, where d is the maximum separation between two of the array members. Since two such small dishes can be separated by reasonable distances (1 to 10 mi, or 1.6 to 16 km, is common) and connected electrically by cables or waveguides, very high angular resolution on the order of 1 second of arc is obtainable.

Fig. 2. Very large array (VLA) near Soccoro, NM, with 27 separate radio telescopes along the arms of a Y. Each arm is 21 km long. (*National Radio Astronomy Observatory*)

To adequately reproduce a single large dish with an array would require a very large number of small dishes separated by roughly the diameter of each small dish, which is not feasible. However, the simulation of such an array using a small number of elements is feasible with the techniques of Earth-rotation aperture synthesis, a concept pioneered by Sir Martin Ryle at Cambridge, who was awarded the Nobel Prize in Physics in 1974 for this work. Imagine an array with a small number of radio telescopes seen from a fixed radio source. As the Earth rotates, the relative positions and displacements of the individual dishes seem (to the radio source) to be changing. If the original array had a reasonable number of elements with various initial separations, the net effect after 12 hr of observing would be to nearly simulate an observation by a single very large telescope. These aperture synthesis techniques provide extremely detailed pictures of radio sources. The latest generation of arrays, such as the very large array (VLA) in New Mexico, Fig. 2, are designed to provide radio pictures which have an angular resolution comparable with that of the largest optical telescopes.

The ultimate in angular resolution is achieved by the technique of very-long-baseline interferometry (VLBI), in which one simultaneously utilizes radio telescopes separated by thousands of miles. It is clearly impossible to connect such telescopes by cables to operate them as a single interferometer, and so data are acquired independently at each telescope and recorded on video tape. Precise time markings are also made on the tape using hydrogen maser clocks that are so accurate that they neither gain nor lose more than 1 sec in 1,000,000 years. After the data are recorded, the video tapes from the separate telescopes are brought together; the time markings on the individual tapes are aligned, and the data taken at precisely the same times can be compared and analyzed. Such VLBI techniques have achieved angular resolutions of about 0.0003 seconds of arc.

Electronics. The radio receiver that is mounted on the telescope and used to detect the incoming radio waves is generally composed of two parts, both of which are optimized so as to enable one to detect extremely weak signals. The "front end" of the receiver is a sensitive preamplifier mounted at the telescope focus that amplifies the incoming rf signal 10 to 1000 times. The rf is then mixed with the lower i-f, and the i-f is further amplified in the "back-end" of the receiver.

The sensitivity of a radio receiver depends critically on the quality of the front-end preamplifier. The front end generates its own radio noise, which is added to the radio power from the sky; since the receiver noise far exceeds the power coming from the sky, the sensitivity of the receiver is increased if the front-end noise is reduced as much as possible. A variety of low-noise amplifiers have been developed that are suitable as radio astronomy front ends. These amplifiers utilize negative-resistance circuits: the two most common ones are parametric amplifiers and maser amplifiers. Parametric amplifiers rely on variable-capacitance diodes to achieve amplification, whereas maser amplifiers make use of the quantum nature of the energy states of atomic particles to provide amplification. In either case, the amount of radio noise generated by the amplifiers is further reduced by cooling them cryogenically with liquid helium.

The back-end configuration of the receiver may be varied to suit the purposes of the type of measurement undertaken. For continuum observations in which one merely wants to identify the total radio power emitted by a source in the sky, the back end consists of a calibrated laboratory source of radio power and a switch enabling one to compare the difference between the source in the sky and the laboratory source. However, when a spectral line is being observed, it is necessary to measure the radio power at many frequency points at and around the frequency of the line. Thus a spectral-line receiver must perform all the functions of a continuum receiver, but it must do so simultaneously at many closely spaced frequency points (typically 100 to 1000). The output of the radio receiver is recorded on magnetic tape suitable for later processing with an electronic computer. *See* RADIO ASTRONOMY.

[ROBERT L. BROWN]

Bibliography: N. G. Basov (ed.), *Radio, Submillimeter, and X-ray Telescopes*, 1976; J. D. Kraus, *Radio Astronomy*, 1966; G. L. Verschuur, *The Invisible Universe*, 1974.

Red dwarf star

A red star of low luminosity, so designated by E. Hertzsprung. Dwarf stars are commonly those main-sequence stars fainter than an absolute magnitude of about +1. Red dwarfs are the faintest and coldest of the dwarfs. They are present among both old and young stars. Red dwarfs are the most numerous class of stars in space, although they are so faint that their presence in remote parts of the Galaxy must be inferred from their frequency near the Sun and their contribution to the total mass of the Galaxy. They have a low-energy output per unit mass and a long nuclear lifetime.

Theoretical work has shown it is improbable that red dwarfs will have appreciable energy generation or be on the main sequence if their masses are less than 0.06 times that of the Sun. Furthermore, infrared observations have defined the total energy emitted, which in cool red dwarfs near 2000 K is mostly in the infrared. Thus, the total luminosity of the faintest and lowest-mass stars is about 3×10^{-4} times that of the Sun. *See* DWARF STAR; STAR. [JESSE L. GREENSTEIN]

Red shift

A systematic displacement toward longer wavelengths of lines in the spectra of distant galaxies, and also of the continuous part of the spectrum. First studied systematically by E. Hubble, red shift is central to observational cosmology, in which it provides the basis for the modern picture of an expanding universe. There are two fundamental properties of red shifts. *See* COSMOLOGY.

First, the fractional red shift $\Delta\lambda/\lambda$ is independent of wavelength. ($\Delta\lambda$ is the shift in wavelength of radiation of wavelength λ). This rule has been verified from 21 cm (radio radiation from neutral hydrogen atoms) to about 6×10^{-5} cm (the visible region of the electromagnetic spectrum) and leads to the interpretation of red shift as resulting from a recession of distant galaxies. Though this interpretation has been questioned, no other mechanism is known that would explain the observed effect.

Second, red shift is correlated with apparent magnitude in such a way that when red shift is translated into recession speed and apparent magnitude into distance, the recession speed is found to be nearly proportional to the distance. This rule was formulated by Hubble in 1929, and the constant of proportionality bears his name. Hubble's constant is currently estimated to lie between 20 and 35 km/(s) (10^6 light-years) or 1.9 and 3.3 $\times$ $10^{-18}s^{-1}$. *See* HUBBLE CONSTANT.

Until 1975 the largest red shifts that could be routinely measured for optical galaxies were around $\Delta\lambda/\lambda = 0.2$. Since that time the use of image tubes and techniques for automatically subtracting the sky background have made it possible to measure red shifts of order $\Delta\lambda/\lambda = 1$ for bright spiral and elliptical galaxies. Quasars, being brighter than ordinary galaxies, can be observed at considerably greater distances and correspondingly greater red shifts. The largest measured red shifts of quasars are around 3.5. In quasars the far-ultraviolet spectrum is shifted into the infrared. Although most astronomers interpret the displacement of lines in the spectra of quasars as cosmological red shifts, the view has also been advanced that they are Einstein shifts resulting from enormously strong gravitational fields at the surfaces of these objects. *See* IMAGE TUBE; QUASARS.

The recession speed indicated by the red shift in the spectrum of a given galaxy is not the current value for that galaxy but the value appropriate to the epoch when the light now reaching the Earth was emitted. Consequently, the observed relation between red shift and apparent magnitude contains information about past values of Hubble's constant, as well as about the present value. If this information could be extracted from the record it would enable astronomers to choose among various model universes that have been proposed by cosmologists. At present, however, this cannot be done.

[DAVID LAYZER]

Relativity

A general theory of physics, primarily conceived by Albert Einstein, which involves a profound analysis of time and space, leading to a generalization of physical laws, with far-reaching implications in important branches of physics and in cosmology. Historically, the theory developed in two stages. Einstein's initial formulation in 1905 (now known as the special, or restricted, theory of relativity) does not treat gravitation, and one of the two principles on which it is based, the principle of relativity (the other being the principle of the constancy of the speed of light), stipulates the form invariance of physical laws only for inertial reference systems. Both restrictions were removed by Einstein in his general theory of relativity developed in 1915, which exploits a deep-seated equivalence between inertial and gravitational effects, and leads to a successful "relativistic" generalization of Isaac Newton's theory of gravitation. Because of the extreme weakness of gravitation compared with all other known forces, the non-newtonian consequences of the general theory have been verified only in a few instances, and their applications are largely confined to cosmology, the structure of neutron stars and black holes, and the motion of bodies in the solar system. For the same reason, where it can be applied, the special theory suffices in the treatment of relativistic effects in atomic, nuclear, and high-energy physics, and its confirmation in numerous instances of this kind renders it among the most securely established of modern scientific theories.

SPECIAL (RESTRICTED) THEORY

The scientific developments that led to the birth of special relativity arose from a dilemma confronting physicists in the latter part of the 19th century in their attempts to reconcile Maxwell's electromagnetic equations with the principles of newtonian mechanics. The latter were then widely held to have universal validity, but Maxwell's equations do not preserve their form under galilean transformations, such as that given by Eqs. (1) [see Fig. 1], that is, they do not satisfy the gali-

$$x' = x - vt \qquad y' = y \qquad z' = z \qquad t' = t \qquad (1)$$

lean principle of relativity. Hence, it appeared necessary to assume that these equations apply only in one special inertial frame, one in which the electromagnetic, or luminiferous, ether is at rest, and which, it was also natural to suppose, served to define Newton's absolute space and absolute velocity. In particular, the absolute velocity of the Earth in its orbital motion around the Sun could be expected to reveal itself in appropriate electromagnetic or optical experiments. However, all such experiments yielded negative results.

The most famous of these, the interferometer experiment of A. A. Michelson and E. W. Morley in 1881, presented the greatest challenge. It was the first and, until the beginning of the 20th century, the only experiment whose negative result involved an effect of order v^2/c^2 (second-order effect), v being the average orbital speed of the Earth and c the speed of light (so that $v^2/c^2 \sim 10^{-8}$). The simplest explanation—that this negative result arises from the dragging of the ether by the Earth, a hypothesis proposed in another connection earlier (1845) by G. G. Stokes—was shown by H. A. Lorentz to be untenable because of difficulties in explaining stellar aberration. Lorentz was therefore led in 1892 to postulate that a body whose absolute speed of translational motion is v is contracted in the direction of motion by the factor $\sqrt{1-v^2/c^2}$. This hypothesis was independently suggested by G. F. FitzGerald, and is thus known as the Lorentz-FitzGerald contraction hypothesis. With its aid, Lorentz succeeded in accounting for the Michelson-Morley experiment, and he also explained it on the basis of ideas from his theory of electrons. However, other second-order experiments performed at the turn of the 20th century, as well as critical methodological considerations, led Lorentz to arrive gradually, in 1899 and 1904, at a general solution of the problem posed by all these unexpected negative ether-drift results. This solution involved replacing galilean transformations (1) by space-time transformation equations (2), and introducing associated transformation

$$\begin{aligned} x' &= \frac{x - vt}{\sqrt{1-v^2/c^2}} \\ y' &= y \\ z' &= z \\ t' &= \frac{t - vx/c^2}{\sqrt{1-v^2/c^2}} \end{aligned} \qquad (2)$$

equations for the electromagnetic field quantities. However, the theory was partly logically flawed, and it was Henri Poincaré who corrected and completed it in 1905 and 1906. Poincaré named Eqs. (2) Lorentz transformations, and he was the first to discuss fully the complete group of Lorentz transformations, including rotations of the spatial coordinate axes.

Although the mathematical structure of the Lorentz-Poincaré theory is the same as that of the theory advanced by Einstein in 1905, independently of Lorentz's work of 1904 and of Poincaré's work, there is nevertheless a vast difference in the underlying conceptual framework. Lorentz retained the prevailing notions of absolute time and space, and he clung to the belief that an absolutely stationary ether is necessary for a consistent explanation of electromagnetic phenomena. Einstein, on the contrary, started with the idea that the absoluteness of time and space is imposed not by nature but by custom, and that therefore, in particular, the notion of simultaneity must be reexamined critically. In addition, he did not consider the existence of an electromagnetic ether essential to electrodynamics. This intellectual independence enabled him to develop an exceptionally viable formulation of what is in consequence commonly referred to as Einstein's theory of relativity.

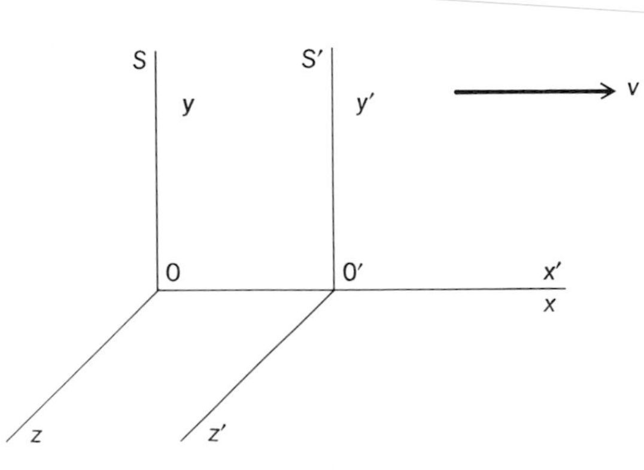

Fig. 1. Inertial reference frames S, S' with parallel coordinate axes; **v** is the velocity of S' relative to S.

Einstein's formulation. The theory is erected on two fundamental postulates and two definitions. The postulates, which Einstein called the principle of relativity and the principle of the constancy of the velocity of light, can be stated respectively as follows: (I) The analytical form of physical laws is the same in all inertial reference systems. (II) The speed of light in vacuum is a universal constant.

Postulate I is a generalization of the Galilean principle of relativity to embrace all phenomena rather than only those of mechanics (excepting, however, gravitational phenomena). It is II that, taken together with I, clearly represents a radical breaking away from traditional thinking. The new formulation of simultaneity is also intimately connected with II. Two spatially separated localized occurrences (or events) are simultaneous when the readings of two identical clocks adjacent to the events are the same, and it is known that the clocks are synchronized. However, when the clocks are not near each other, their synchronism must be defined.

The definition given by Einstein is tied to II, and can be stated as follows: (A) Two identical clocks C, C', situated at two distant points P, P' fixed in a given inertial frame S, synchronize in S, when the respective C and C' times t_1, t_1' of the sending of a light signal at P and its arrival at P' are connected by the formula $t_1' - t_1 = t_2 - t_1'$ with the C time t_2 of its return to P after reflection at P' back to P.

As will be shown below, two events that are simultaneous in terms of clocks synchronized in S according to definition A are not simultaneous when referred to identical clocks synchronized according to A in a reference frame that is moving relative to S. In other words, simultaneity is a relative concept: it depends on the reference system under consideration. This relativity is at the heart of many relativistic phenomena.

Einstein's second definition clarifies the operational meaning of the length of a moving rod, and can be stated as follows: (B) The length, as measured in a given inertial frame S, of a rigid rod that is in uniform motion relative to S is the length between the instantaneous positions of the ends of the rod, where the instantaneousness and the length are determined respectively by clocks synchronized in S and by meter sticks at rest in S.

With the aid of postulates I and II and definition A, Einstein succeeded for the first time in deriving the Lorentz transformations. Then, using B and I he deduced the Lorentz-FitzGerald contraction: Suppose a rigid rod is at rest in the system S' (of Fig. 1) along the x' axis, its ends having the coordinates x_1', x_2' ($x_1' < x_2'$), then by Eq. (2) for a constant value t_0 of t, $x_k' = \gamma(x_k - vt_0)$, $k = 1, 2$, where $\gamma \equiv 1/\sqrt{1 - v^2/c^2}$, so that $x_2' - x_1' \equiv \Delta x' = \gamma \Delta x$, $\Delta x = \Delta x'/\gamma \equiv \sqrt{1 - v^2/c^2}\ \Delta x'$, which is the required result, since $\Delta x'$ is by I the same as the length of the rod when at rest in S and measured in S. By transformation equations (3), inverse to Eqs. (2), a similar result is obtained when the roles

$$\begin{aligned} x &= \gamma(x' + vt') \\ y &= y' \\ z &= z' \\ t &= \gamma(t' + vx'/c^2) \end{aligned} \qquad (3)$$

of S and S' are interchanged, as is in any case required by I. It can also be concluded from Eqs. (2) that the relativistic contraction of a body takes place only in the direction parallel to that of the relative motion of S and S'.

Time dilation. Another remarkable result concerns the relativity of time. Setting x' = constant in Eqs. (3), one finds that $\Delta t = \gamma \Delta t'$, and hence that Eq. (4) is true. Since Eqs. (2) involve of course the

$$\Delta t' = \sqrt{1 - v^2/c^2}\ \Delta t \qquad (4)$$

assumption that the clocks in S and S' (and the meter sticks) are identical, Eq. (4) states that as measured in S', the time interval between the given pair of events is smaller by the factor $\sqrt{1 - v^2/c^2}$ than it is as measured in S. In this respect, too, there is complete reciprocity, as there must be, between S and S': the time interval between two events is a relative concept. Einstein pointed out also the following corollary of Eq. (4): If one of two identical clocks, initially synchronized and adjacently at rest in an inertial frame, makes a round trip, it will lag behind the other upon its return. This is proved by integrating Eq. (4) over the closed path of the moving clock, on the assumption that in case of accelerated motion relative to an inertial frame, the kinematic relativistic results hold during each infinitesimal interval—an assumption, usually tacit, which is involved in many relativistic topics. If the clocks are replaced by the beating hearts of identical twins, it follows that the traveling one will upon return be younger than the stay at-home. Convincing verification of the effect indicated by Eq. (4) is provided by the measured lifetimes of fast mesons, whose increase with the speed of the particles conforms to Eq. (4), and by more precise experiments employing the Mössbauer effect. The Michelson-Morley-type experiment, as well, has been repeated with greater precision using laser light.

Composition of velocities. If $\mathbf{u}'$ is the velocity of a projectile relative to S' as measured in S', and $\mathbf{u}$ is its velocity relative to S as measured in S, one finds by Eqs. (3) that the components of the two velocities are related by Eqs. (5). If $\mathbf{u}'$ is taken to

$$\begin{aligned} u_x &= (u_x' + v)/(1 + vu_x'/c^2) \\ u_y &= u_y'/\gamma(1 + vu_x'/c^2) \\ u_z &= u_z'/\gamma(1 + vu_x'/c^2) \end{aligned} \qquad (5)$$

represent, instead, the velocity of the projectile relative to S' as measured in S, then $u_x = u_x' + v, u_y = u_y'$, $u_z = u_z'$, as in newtonian physics. In any case, Eqs. (5) reduce to the latter equations when vu_x'/c^2 and v^2/c^2 are negligible. On the other hand, if $u_x' = c$ (and hence, $u_y' = u_z' = 0$), then $u_x = c$, $u_y = u_z = 0$, in agreement with postulate II.

The preceding discussion involves the tacit assumption that particle speeds cannot exceed c. That c does indeed represent the maximum speed of energy propagation is indicated by an argument of Einstein (1907) employing Eqs. (5), which shows that in the contrary case it would be possible to transmit information into the past. Such a situation, while perhaps not entailing a formal logical contradiction, is unacceptable to the great majority of physicists, although the possibility of the existence of particles with speed greater than c, dubbed tachyons, has been suggested. However, such particles would represent a category quite distinct from known particles, since the latter cannot be accelerated to speeds greater than c.

Minkowski's formulation. Hermann Minkowski's geometric approach, formulated in 1908, which was an important link in the development of general relativity, centers on absolute quantities in the special theory, and provides it with the efficient formalism of tensors in space-time.

Minkowski's space-time, his "world," is the continuum of all events (Minkowski's "world points"), the primitive absolute quantities of relativistic kinematics. It is a four-dimensional, quasi-euclidean space, with the line element ds^2 (the distance squared between two neighboring events) given by Eq. (6), defining the metric of the space the "Minkowski metric."

$$ds^2 = \sum_{\alpha,\beta=0}^{3} \eta_{\alpha\beta}\, dx^\alpha\, dx^\beta \equiv (dx^0)^2 - \sum_{i=1}^{3} (dx^i)^2 \qquad (6)$$

$$x^0 = ct \qquad x^1 = x \qquad x^2 = y \qquad x^3 = z$$

$$\eta_{\alpha\beta} = 0 \text{ (if } \alpha \neq \beta)$$

$$\eta_{00} = -\eta_{11} = -\eta_{22} = -\eta_{33} = 1$$

The tensors of special relativity represent absolute quantities, which are determined by their components, relative quantities, obeying the appropriate transformation rules associated with the Lorentz group. Because the metric of Eq. (6) is not definite ($\eta_{\alpha\beta} \neq \delta_{\alpha\beta}$, since $\eta_{ii} = -1$ for $i = 1, 2, 3$) one must distinguish between contravariant, covariant, and mixed components of a tensor, which differ, however, at most in sign; for example, $\eta_{\alpha\beta}$ and dx^α are covariant and contravariant components of the respective tensors, and $\eta_{\alpha\beta} = \eta^{\alpha\beta}$ while $dx_i = -dx^i$ ($i = 1, 2, 3$). One usually attaches the prefix "four" to relativistic tensors. Thus, the scalar of Eq. (6) is a four-scalar, and the dx^α represent a four-vector.

One may replace dx^α in Eqs. (6) by finite Δx^α, and according as Δs^2 is greater than 0, less than 0, or equal to 0, one has a timelike, spacelike, or null interval (displacement four-vector). The first and last cases correspond to possible displacements of a particle with $m_0 \neq 0$ or $m_0 = 0$. Considering the motion of a particle ($m_0 \neq 0$), one finds from Eq. (6) that $ds^2/c^2 \equiv d\tau^2 = dt^2(1 - v^2/c^2)$, $\mathbf{v} = d\mathbf{x}/dt$. The integral $\int d\tau$ along the particle's space-time path (or "world line") gives the time as measured by a clock accompanying the particle: the "proper time" on the world line.

Applications. The branches of classical physics can usually be relativistically generalized by applying postulate I in Minkowski's formulation of expressing the physical laws in tensor form.

Mechanics. Minkowski's method is successful in the case of a particle moving under the action of an external field of force, and important but limited results exist for systems of interacting particles. A dramatic consequence is Einstein's mass-energy equivalence, which has implications ranging from nuclear reactor energy production to atomic bombs.

Continuum mechanics and fields. The space-time formulation is here especially effective. Maxwell's equations of the free electromagnetic field are already Lorentz-invariant, and their four-tensor form is immediate. Using relativistic-mechanics results, the four-tensor reformulation of the Maxwell-Lorentz equations for the electromagnetic field generated by a system of elementary charges and for the motion of the charges in the electromagnetic field, and of the phenomenological equations of polarizable and magnetizable media, first developed by Minkowski in 1909, present few difficulties. Of central importance in this area is the energy-momentum tensor, whose components T^{00}, T^{ij}, T^{0i}, and T^{i0} ($i = 1, 2, 3$) represent the respective densities of energy, momentum flux (stress), (energy flux)/c, and $c \cdot$ (momentum). The symmetry of $T(T^{\alpha\beta} = T^{\beta\alpha})$ implies, among other things, the inertia of energy. For closed systems, the equations of motion assume the compact form

$$\sum_{\beta=0}^{3} \partial T^{\alpha\beta}/\partial x^\beta = 0 \qquad (\alpha = 0, 1, 2, 3)$$

and the integrals $\int T^{\alpha 0} dV$ over space for any fixed t give the conserved total energy and momentum of the system in energy units.

Optics. From the four-scalar property of the phase function $2\pi\nu\ (t - \mathbf{n} \cdot \mathbf{x}/c)$ of a plane light wave, it follows that if a light ray is emitted in S' in a direction making in S' the angle θ' with the x' axis, then in S its direction with the x axis is given by the angle θ such that $\cos\theta = (\cos\theta' + v/c)/1 + v\cos\theta'/c)$; and if the frequency of the emitted light is ν' in S', then in S it is $\nu = \nu'/\gamma(1 - v\cos\theta/c)$. The two formulas are respectively the relativistic generalizations of the laws governing the aberration of light and the Doppler effect. *See* ABERRATION; DOPPLER EFFECT.

Statistical physics. The relativistic extension of classical thermodynamics presents some conceptual difficulties. Unlike Einstein's definition of the length of a moving body, the natural operational definition of, say, the temperature of a moving body is neither obvious nor always independent of the thermodynamic state of the body. This circumstance explains in part the paradoxical situation which arose regarding the conceptual framework of relativistic thermodynamics, when three different sets of Lorentz transformation formulas were being advocated. No such difficulties attach to statistical mechanics and kinetic theory, and the simplest topics of these subjects were treated relativistically not long after the rise of relativity. Considerable progress has been achieved in studies of relativistic kinetic theory and statistical mechanics of both reversible and irreversible processes, involving proper choices of the energy-momentum tensor of the thermodynamic systems. These studies have been mostly motivated by astrophysical problems presented by superdense stars such as pulsars and certain white dwarfs and by some speculative cosmological theories. In most of these applications intense gravitational fields are involved that require general-relativistic treatment. However, the special-relativistic formulations represent an essential step. *See* PULSAR; WHITE DWARF STAR.

[H. M. SCHWARTZ]

GENERAL THEORY

General relativity is the geometric theory of gravitation developed by Einstein in 1915. It is a generalization of special relativity, and includes the classical gravitational theory of Newton as the limiting case when the gravitational fields involved are weak and the velocities of all the bodies involved are small compared to the speed of light *c*. The most important applications of the theory are to the structure of neutron stars and black holes, the large-scale cosmological description of the universe, and the motion of bodies in the solar system. *See* GRAVITATION.

Need for relativistic theory. The special theory of relativity proposed in 1905 gained acceptance rapidly among physicists by virtue of its theoretical elegance and experimental success, and was rather well established by 1915. One of the basic tenets of special relativity is that no physical effect can propagate with a velocity greater than the speed of light, and so *c* represents a universal speed limit.

On the other hand, classical gravitational theory describes the gravitational field of a body throughout space as a function of its instantaneous position, which is equivalent to the assumption that gravitational effects propagate with an infinite velocity; that is, classical gravitational theory is an action-at-a-distance theory. Thus, special relativity and classical gravitational theory are inconsistent, and a modified theory of gravity is necessary. This is the theory searched for and found by Einstein from 1905 to 1915, following his discovery of special relativity.

Principle of equivalence. It had long been considered a fundamental and puzzling question why bodies of different mass fall with the same acceleration in a gravitational field, or equivalently why the trajectory of a test body is independent of its mass. This situation was explained by Newton with the statement that both the gravitational force on a body and its inertial resistance to acceleration are proportional to its mass. Thus the mass cancels out of the mathematical description of the motion. In laboratory experiments early in the 20th century by L. von Eötvös, it was found that this cancellation is true to a few parts in 10^8. Later work by R. Dicke improved the accuracy to a few parts in 10^{11}, and V. Braginsky obtained an accuracy of a few parts in 10^{12}. Thus the independence of the motion of a test body on its mass is one of the most accurately tested experimental facts in physics.

The explanation by Newton is not very profound and is more in the nature of an ad hoc description. A deeper and more natural explanation occurred to Einstein. In physics there are numerous examples of forces other than gravitation which are mass-proportional; these generally arise due to the use of accelerated coordinate systems to describe the motion. One well-known example is the centrifugal force encountered in a rotating coordinate system. Consider one observer in the gravitational field of the Earth and another in an accelerating elevator or rocket in free space (Fig. 2). If both drop a test body, they will observe it to accelerate relative to the floor. According to classical theory, the Earth-based observer would attribute this to a gravitational force and the elevator-based observer would attribute it to the accelerated floor overtaking the uniformly moving body. However, Einstein reasoned that the effects are identical and the theory of gravity should provide an equivalent description of the two systems. This is the famous principle of equivalence that Einstein made the physical cornerstone of general relativity; it states that on a local scale the physical effects of a gravitational field are indistinguishable from the physical effects of an accelerated coordinate system. From the point of view of the principle of equivalence, it is evident why the motion of a test body in a gravitational field is independent of its mass.

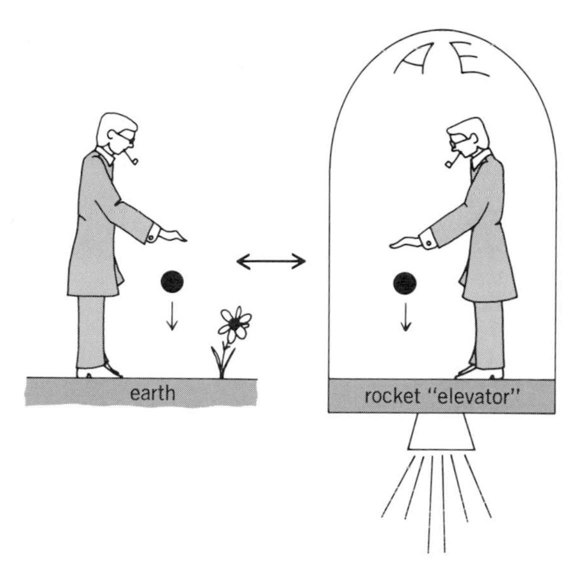

Fig. 2. Einstein's elevator, illustrating the principle of equivalence. The elevator acceleration is equal to *g*, the acceleration of gravity.

The principle of equivalence is strictly local and applicable only to a region of space and time sufficiently small that inhomogeneities in the gravitational field can be ignored. There is an intrinsic difference between gravitational and accelerative effects on a finite scale. This is well illustrated by considering in a nonuniform gravitational field two nearby test bodies, which, being in slightly different parts of the field, follow slightly different trajectories. The relative deviations of the trajectories characterize the inhomogeneities of the field. They intrinsically distinguish the effects of gravity and acceleration, which is impossible on a strictly local scale according to the principle of equivalence.

The principle of equivalence is heuristic and somewhat imprecise; despite its logical imprecision, it has played a very important historical role.

Tensor field equations. The close connection between gravity and accelerating coordinate systems suggests that the equations describing the gravitational field be cast in a form that is manifestly independent of the coordinate system in order to achieve the maximum simplicity and elegance. The study of systems of equations that are independent of coordinate systems was begun by K. Gauss in connection with his study of geometry on curved two-dimensional surfaces, and was carried to a high state of development by B. Riemann and T. Levi-Civita in the tensor calculus. Tensors are a basically simple generalization of vectors.

Only tensors in the four-dimensional space-time of relativity need be considered. Suppose that the points in four-dimensional space-time are labeled by two essentially arbitrary systems of coordinates x^μ and x'^ν (the indices μ and ν range from 0 to 3) and that the two coordinate systems are related to each other by an infinitely differentiable set of four functions, $x'^\mu = x'^\mu(x^\alpha)$ and $x^\beta = x^\beta(x'^\nu)$. Then contravariant tensors of rank 0, rank 1, rank 2, etc. are defined as sets of 1, 4, 16, etc. numbers or functions whose values at the same point in the two systems of coordinates are related by transformation equations (7). Covariant tensors are similarly defined by Eqs. (8). In these equations the indices

$$\phi' = \phi \quad \text{rank 0}$$

$$\eta'^\mu = \frac{\partial x'^\mu}{\partial x^\alpha}\eta^\alpha \quad \text{rank 1} \tag{7}$$

$$T'^{\mu\nu} = \frac{\partial x'^\mu}{\partial x^\alpha}\frac{\partial x'^\nu}{\partial x^\beta}T^{\alpha\beta}, \text{ etc.} \quad \text{rank 2}$$

$$\sigma'_\mu = \frac{\partial x^\alpha}{\partial x'^\mu}\sigma_\alpha \quad \text{rank 1} \tag{8}$$

$$\omega'_{\mu\nu} = \frac{\partial x^\alpha}{\partial x'^\mu}\frac{\partial x^\beta}{\partial x'^\nu}\omega_{\alpha\beta} \quad \text{rank 2}$$

that appear twice in one expression are to be summed over; this is the Einstein summation convention. Its use allows a great saving of effort and space by avoiding numerous explicit summation signs Σ. Tensors of rank 0 are also called scalars or invariants since they have the same value in all coordinate systems and are thus of great importance in the description of physical quantities. Tensors of rank 1 are also called four-vectors. One example of a scalar is the inner product of a contravariant four-vector ξ^α and covariant four-vector η_α defined as $\xi^\alpha\eta_\alpha$.

The space-time of relativity contains one covariant second-rank tensor of particularly great importance, called the metric tensor $g_{\mu\nu}$, which is a generalization of the Lorentz metric of special relativity $\eta_{\alpha\beta}$, introduced in Eqs. (6). Nearby points in space-time, known as events, which are separated by coordinate distances dx^μ have an invariant physical separation whose square, the line element, is given by Eq. (9). This quantity is a general-

$$ds^2 = g_{\mu\nu}\, dx^\mu\, dx^\nu \tag{9}$$

ization of the line element in special relativity and has the same relation to the concept of proper time; that is, when ds^2 is positive, it represents the square of the time interval between nearby events (multiplied by c^2) measured by an observer who moves at approximately constant velocity in such a way as to be present at both nearby events. (A space with a quadratic metric form as in Eq. (9) is called a Riemann space by mathematicians.)

Tensor equations are equations in which one tensor of a given rank and type (contravariant or covariant) is set equal to another of the same type, for example, $T^\mu = S^\mu$. In a different coordinate system, denoted by a prime, the definition of a tensor implies that both sides transform in the same way so that the partial derivatives of the transformation cancel from the equation and it has the same form as in the original system, for example, $T'^\mu = S'^\mu$. Tensor equations are thus called form-invariant, or covariant.

The field equations of general relativity are tensor equations for the metric tensor, which completely describes the geometry of the space. To present the field equations, a Cristoffel symbol (not a tensor) is first defined by Eq. (10). Here $g^{\mu\tau}$ is the

$$\begin{Bmatrix}\mu\\ \alpha\beta\end{Bmatrix} = \frac{1}{2}g^{\mu\sigma}\left(\frac{\partial g_{\sigma\alpha}}{\partial x^\beta} + \frac{\partial g_{\sigma\beta}}{\partial x^\alpha} - \frac{\partial g_{\alpha\beta}}{\partial x^\sigma}\right) \tag{10}$$

inverse of $g_{\mu\tau}$, considered as a matrix. The Riemann tensor (or curvature tensor) is defined by Eq. (11). This tensor plays a central role in the geo-

$$R^\alpha_{\mu\beta\nu} = \frac{\partial}{\partial x^\nu}\begin{Bmatrix}\alpha\\ \beta\mu\end{Bmatrix} - \frac{\partial}{\partial x^\mu}\begin{Bmatrix}\alpha\\ \beta\nu\end{Bmatrix} + \begin{Bmatrix}\alpha\\ \tau\nu\end{Bmatrix}\begin{Bmatrix}\tau\\ \beta\mu\end{Bmatrix} - \begin{Bmatrix}\alpha\\ \tau\beta\end{Bmatrix}\begin{Bmatrix}\tau\\ \mu\nu\end{Bmatrix} \tag{11}$$

metric structure of a space; if it is zero, the space is termed flat and has no gravitational field; if nonzero, the space is termed curved. In terms of the contracted Riemann tensor, a Riemann tensor summed over $\alpha=\beta$, the Einstein field equations for free space are given by Eq. (12). In a region of

$$R^\alpha_{\mu\alpha\nu} \equiv R_{\mu\nu} = 0 \tag{12}$$

space containing matter or energy, the zero on the right side of this equation is replaced by a tensor representing the energy content of space, usually written $-(8\pi G/c^2)(T_{\mu\nu} - 1/2 g_{\mu\nu}\, T^\alpha_\alpha)$, where G is the gravitational constant (equal to 6.670×10^{-11} N · m² · kg⁻²), and $T_{\mu\nu}$ is called the energy-momentum tensor.

The field equations are a set of 10 second-order partial differential equations since the four-by-four symmetric tensor $R_{\mu\nu}$ has 10 independent components; they are to be solved for the metric tensor. A solution in a given coordinate system defines an Einstein space-time. The curvature of this space corresponds to the intrinsic presence of a gravitational field. That is, the concept of a field of mechanical force in classical gravitational theory is replaced by the geometric concept of curved space in relativity theory.

Motion of test bodies. Within the context of tensor notation and the description of gravity by means of curved space, the equations of motion are almost obvious. The path of a test body is a generalization of a straight line in euclidean space; it is the shortest "distance" (in terms of intervals ds) between points in space-time, known as a geodesic. The differential equation of a geodesic, Eq. (13), involves derivatives of the coordinates along the path and the Christoffel symbols. In

$$\frac{d^2x^\mu}{ds^2}+\left\{\begin{matrix}\mu\\ \alpha\beta\end{matrix}\right\}\frac{dx^\alpha}{ds}\frac{dx^\beta}{ds}=0 \qquad (13)$$

the special case of no gravitational field, with the metric equal to the constant Lorentz metric, the Christoffel symbols vanish and this is indeed the equation of a straight line. Basically the scenario for gravitational influence in the general theory of relativity is that matter curves space in its vicinity, in accord with the Einstein field equations, and in this curved space test bodies move on geodesics, or generalized straight lines. The motion is thereby clearly independent of the mass of the test body, consistent with the principle of equivalence.

The basic equations of relativity were thought at first to be the Einstein field equations and the separately postulated geodesic equations of motion. However, general relativity theory possesses an extraordinary property; because the field equations are nonlinear, unlike newtonian theory, the motion of a test body in a gravitational field is not arbitrary since the body itself has mass and contributes to the field. Indeed, it turns out that the field equations are so restrictive that the geodesic equation of motion is a necessary consequence and need not be treated as a separate postulate. This has been shown by Einstein, L. Infeld, and B. Hoffman, by Levi-Civita, and by numerous others. This remarkable property of the field equations is unique to general relativity among the accepted theories in physics.

Test bodies are defined as bodies without significant extent or structure, and do not materially affect the field. Bodies with nonnegligible size should not be considered as test bodies, and indeed have more complicated equations of motion.

Schwarzschild solution. A very important solution of the field equations was obtained by K. Schwarzschild in 1916, surprisingly soon after the inception of general relativity. This solution represents the field in free space around a spherically symmetric body such as the Sun. It is the basis for a relativistic description of the solar system and all of the experimental tests of general relativity which have been carried out. In spherical coordinates r, θ, ϕ and time coordinate $x^0=ct$ the solution is represented by a line element (valid only outside the body) given by Eq. (14). Here M is the

$$ds^2=\left(1-\frac{2GM}{c^2r}\right)c^2\,dt^2-\left(1-\frac{2GM}{c^2r}\right)^{-1}dr^2 - r^2(d\theta^2+\sin^2\theta\,d\phi^2) \qquad (14)$$

mass of the body. In the limit of $M=0$ this represents the line element of special relativity written in spherical coordinates. This solution is the relativistic analog of the classical gravitational potential field $\Phi=-GM/r$. Indeed, there is an important approximate relation between Φ and g_{00}, $g_{00}=1+2\Phi/c^2$, which is necessary and sufficient to show that general relativity has classical gravitational theory as its limit for very weak fields and low velocities of test bodies.

Gravitational red shift. There are a number of ways to show that electromagnetic radiation of a given frequency emitted in a gravitational field will appear to an outside observer to have a lower frequency; that is, it will be red-shifted. For the Schwarzschild solution the fractional decrease in frequency is equal to the difference in GM/c^2r between emission point and observation point. The red shift was first verified for radiation of optical frequency emitted by atoms on the Sun. However, nongravitational effects on such radiation make the measurement of the 1-part-in-10^6 effect difficult and somewhat uncertain. More accurate and dependable measurements involve the use of the Mössbauer effect in terrestrial experiments. Certain radioactive nuclei, such as ^{57}Fe, in crystals emit and absorb gamma radiation in extremely narrow frequency bands of fractional width about 10^{-12}. A gamma ray emitted upward by such a crystal will be red-shifted about 1 part in 10^{15} in 100 ft (30 m) and will not be resonantly absorbed by a receiving crystal, unless the receiving crystal is given a small downward velocity to compensate for the gravitational red shift with a Doppler shift; the velocity necessary to reestablish absorption provides a measurement of the red shift. With this technique the red shift has been measured by R. Pound and G. Rebka to be within about 1% of the value predicted by general relativity theory.

The most accurate test of the red shift to date was performed using a hydrogen maser on a rocket that reached an altitude of about 10^4 km. Comparison of the maser frequency with Earth-based masers by R. Vessot gave a measured red shift in agreement with theory to about 1 part in 10^4.

The red shift can be derived from the principle of equivalence without the use of the Schwarzschild solution, so these experiments do not test the Schwarzschild solution or the Einstein field equations. *See* EINSTEIN SHIFT.

Perihelion shift. The equations of motion can be solved for a planet considered as a test body in the Schwarzschild field of the Sun. As should be expected, the orbits obtained are very similar to the ellipses of classical theory. Small differences occur, however, the most interesting of which is that the ellipse rotates very slowly in the plane of the orbit so that the perihelion, the point of closest approach of the planet to the Sun, is at a slightly different angular position on each orbit. This shift is extremely small. It is greatest in the case of the planet Mercury, whose perihelion advance is predicted to be only 43 seconds of arc in a century. This is in excellent agreement with the value for the discrepancy between classical theory and observation, which was well known and unaccountable for many years before the discovery of general relativity.

Modern tests of the perihelic motion involve the planet Mars in addition to Mercury and also the asteroid Icarus in a comprehensive model of the solar system. Very precise planetary and asteroid position measurements using radar have been made by I. Shapiro, and the results are consistent with general relativity theory. In particular the perihelion shift of Mercury has been determined to an accuracy of about half a percent and agrees with theory.

Some question still exists about the interpretation of these experimental results. The reason is that the Sun's quadrupole moment is not known precisely, and such a quadrupole moment would produce a perihelion advance analogous to the relativistic effect. *See* CELESTIAL MECHANICS.

Deflection of star light. The principle of equivalence suggests an extraordinary phenomenon of gravity. Light or other electromagnetic radiation crossing the Einstein elevator horizontally will appear to be deflected downward in a parabolic arc because of the upward acceleration of the elevator. The same phenomenon must occur for light in the gravitational field of the Sun; it must be deflected toward the Sun. A calculation of this deflection gives 1.75 seconds for the net deflection of starlight grazing the edge of the Sun. A star near the edge of the Sun viewed from the Earth will appear to be artificially displaced away from the Sun. Early measurements of this effect, notably by A. S. Eddington, were made by photographing stars near the Sun during total solar eclipses and comparing their positions with those when the Sun is in a different part of the sky. They verified the qualitative correctness of the deflection phenomenon. Modern measurements are made by tracking quasars as they pass near or behind the Sun. With these techniques, which are free of the limitations imposed by infrequent eclipses and photographic problems, the deflection has been measured to be within 1% of the value predicted by general relativity. *See* QUASARS; RADIO ASTRONOMY.

Radio time delay. In the curved space around the Sun the distance between points in space, for example between two planets, is not the same as it would be in flat space. In particular, the round-trip travel time of a radar signal sent between the Earth and Venus will be measurably increased by the curvature effect when the Earth, the Sun, and Venus are approximately lined up. Although the maximum time delay is only a few hundred microseconds, it has been accurately measured and found to agree with the predictions of general relativity to within 4%. Subsequent experiments used the Viking spacecraft, which contained a transponder, instead of the planet Venus, and were able to achieve an accuracy of about one-half of 1%. These measurements provide the first qualitatively new test of general-relativity theory in 50 years.

Precession of a gyroscope. A particularly interesting test of general-relativity theory involves the motion of a highly accurate and stable gyroscope in orbit around the Earth. Relativity theory predicts that such a gyroscope will precess at a rate of about 7 seconds of arc per year.

Most of this relativistic precession can be understood as due to the curvature of space described by the Schwarzschild metric in Eq. (14); this is called the geodetic precession. However, since the Earth spins, the Schwarzschild metric does not describe its field completely, and small additional terms occur in the metric. Loosely speaking, these terms arise because in relativity theory a spinning body partially drags space around with it. This is known as the dragging of inertial frames. It is sometimes also referred to as the Lense-Thirring effect after its discoverers J. Lense and H. Thirring. There is no classical analog; the classical gravitational field of a spinning spherical body is identical to that of a nonspinning spherical body.

A small part, about 1%, of the precession of the gyroscope is due to the dragging of the inertial frame. Moreover, for a satellite in a polar orbit the precession due to the Lense-Thirring effect is at right angles to the dominant geodetic precession. This fact makes it possible to separate and measure the geodetic effect and the Lense-Thirring effect in the same experiment. Thus, if sufficiently accurate, the gyroscope experiment can provide a test of this novel effect of general-relativity theory. A gyroscope and associated experimental apparatus have been developed that should allow the precession due to the Lense-Thirring effect to be measured to an accuracy of about 2%, and the geodetic precession to about 1 part in 10^4.

Neutron stars and black holes. The astronomical sources of regularly spaced pulses of electromagnetic radiation, known as pulsars, are believed to be small, rapidly spinning stars of extremely high density known as neutron stars.

For ordinary stars, such as the Sun, the gravitational field is sufficiently weak that classical gravitational theory is an adequate approximation for studying the internal structure. This is not true for neutron stars, which contain a core composed largely of neutrons and other elementary particles at nearly nuclear density. The gravitational field in these stars can be quite large, and it is necessary to describe them using hydrodynamic equations derived from the Einstein field equations in the presence of matter. *See* NEUTRON STAR.

One of the most remarkable theoretical properties of neutron stars is that they have an upper mass limit of about 2 solar masses. A nonrotating, spherically symmetric neutron star with a larger mass is not stable and undergoes a process known as gravitational collapse, one of the most exotic concepts of relativity theory. In such an unstable neutron star the pressure produced by compression of the star is not sufficient to balance the inward force of gravity, and the star shrinks in size; the remarkable fact is that it shrinks indefinitely, the outward pressure never being able to balance the force of gravity, no matter how large the pressure becomes. Viewed by an observer far outside the very strong gravitational field region, such a star appears to asymptotically approach a sphere of constant radius $r = 2GM/c^2$, which is known as the Schwarzschild radius. This is the radius at which the Schwarzschild solution becomes singular, and is equal to about 1.5 km for a star with the mass of the Sun. Light emitted by a source at this radius suffers a red shift to zero frequency and thus cannot be observed. As a consequence, the collapsing star will asymptotically approach a radius where it is invisible, and becomes what is known as a black hole. In fact, not only is light unable to escape from the black hole, but no physical effect is able to reach the external world from the surface or interior of the black hole. Matter

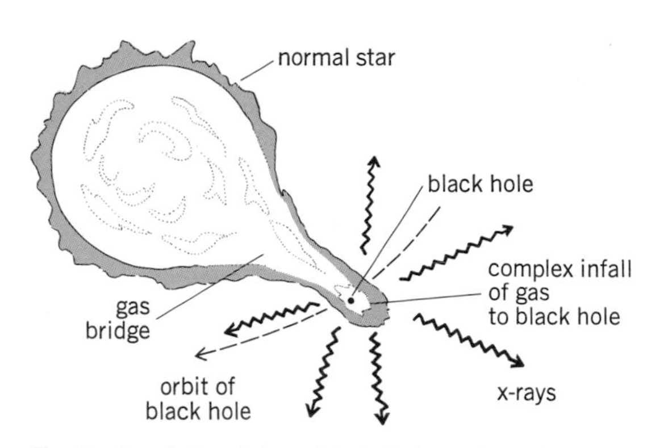

Fig. 3. Gas falling into a black hole and producing x-rays. (*From R. J. Adler, M. Bazin, and M. M. Schiffer, Introduction to General Relativity, McGraw-Hill, 1975*)

and radiation can fall into the black hole, but can never emerge. The surface acts as a one-way membrane that separates space into two disjoint parts; it is impossible to communicate with the outside world from the surface or interior of a black hole.

Among the end products of the evolution of stars it is expected that occasionally a configuration should occur that leads to gravitational collapse and a black hole. Such stellar black holes may have already been observed. It is believed by many astronomers that the x-ray source known as Cygnus X-1 is a binary system of an ordinary star and a black hole in orbit, and that gas approaching the black hole from the ordinary star is strongly heated by gravitational compression, emits x-rays in very short random bursts, and ultimately falls into the black hole (Fig. 3).

If a collapsing star is rotating, as is most likely to be the case, the asymptotic state is believed to be a generalization of the Schwarzschild solution to rotating systems discovered by R. Kerr. Like the Schwarzschild solution, the Kerr solution has a spherical singular surface and describes a black hole, but the black hole is surrounded by an annular region known as the ergosphere. The ergosphere can contain an enormous concentration of energy, equal to a significant fraction of the total rest energy of the star. It is possible that this phenomenon is of great astrophysical importance if the energy can be liberated by naturally occurring astronomical processes.

Studies by S. Hawking indicate that black holes are not totally stable but essentially evaporate away by emitting radiant energy. For stellar-mass black holes the rate of evaporation is completely negligible, but if very-low-mass black holes exist, possibly formed at the birth of the universe, they would evaporate explosively with large energy release. Such bursts of radiation have not yet been detected. *See* BLACK HOLE.

Cosmology. The geometric viewpoint of relativity provides an elegant description of the shape, origin, and evolution of the universe on the cosmological scale; this scale is so large that entire galaxies and even clusters of galaxies are treated in theory like atoms in a gas. On this scale the universe appears observationally to be spatially isotropic and homogeneous, that is, it is substantially the same in all directions and at all points in space. It is thus assumed that the large-scale three-dimensional geometry of the universe is isotropic and homogeneous. This assumption leads to a type of three-dimensional metric known as the Robertson-Walker metric, after its discoverers. There are three subtypes of Robertson-Walker metric: (1) hyperspherical, (2) hyperplane, (3) hyperpseudospherical. These are precisely the three-dimensional analogs of (1) the surface of a sphere, (2) a plane surface, (3) a negative-curvature two-dimensional surface studied in the last century by N. Lobachevski; it unfortunately cannot be simply visualized.

The "galactic atoms" in this picture are at fixed three-dimensional coordinate positions, but the three-space has a single arbitrary metrical scale factor, generally time-dependent, which relates the constant coordinate distance between galaxies to the physical distance. This description is based on observationally motivated geometrical assumptions, and is independent of the Einstein field equations. The hyperspherical universe is of particular philosophical interest since it represents a three-dimensional model of the universe which has no boundaries but is of finite volume, analogous to the surface of a sphere, which has no boundaries but is of finite area. It is thus a closed universe.

The Einstein field equations enter the cosmological problem by providing a differential equation for the metrical scale factor. They cannot determine the geometrical type of the metric, which is an observational problem. A. Friedman in 1922 obtained the solutions of the field equations representing the special case in which the pressure of the "galactic gas" is negligible compared to its density. This is realistic at the present epoch, but not for very early epochs. For the three subtypes of Robertson-Walker metric the solutions have the following properties: (1) For the hyperspherical case, the scale factor increases from zero, reaches a maximum, and then contracts again to zero. That is, the average intergalactic distance expands from zero to a maximum and then contracts to zero. (2) For the hyperplane case, the scale factor increases from zero and becomes indefinitely large. (3) For the hyperpseudospherical case, the scale factor increases from zero, becoming indefinitely large, more rapidly than in the hyperplane case. In all cases the universe begins with an explosive "big bang" (here ignoring details) and then expands. For sufficiently early epochs the behavior of the scale factor is in fact nearly the same for the three cases, and at the present epoch it is difficult to distinguish between these three models.

It is remarkable that Einstein's first attempts at cosmological solutions of the field equations yielded time-evolutionary solutions, some years before the observational discovery of the expansion of the universe as evidenced by the Doppler shift of distant galaxies. He was forced to introduce a new cosmological term in the field equations to obtain static solutions which he thought were necessitated by observation; the subsequent discovery of the universal expansion removed the need for static solutions and made the cosmological term unnecessary.

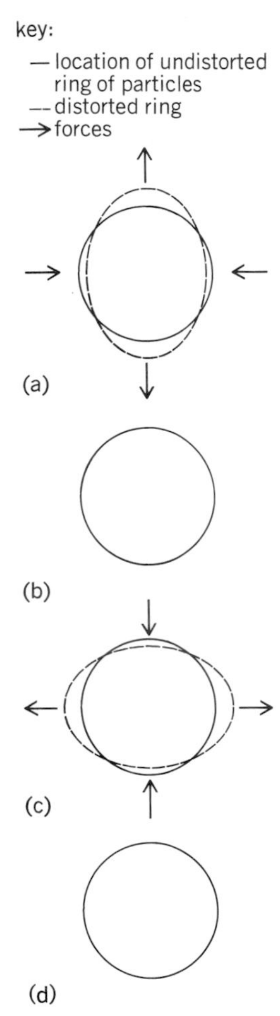

Fig. 4. The forces and distortions produced by a gravitational wave. (*a*) Begin cycle. (*b*) 1/4 cycle (forces equal zero). (*c*) 1/2 cycle (forces reverse). (*d*) 3/4 cycle (forces equal zero).

Observational cosmology mainly involves the measurement of two important numbers. The first is Hubble's constant H, with the units of inverse time; the fractional change in wavelength of light from a galaxy at distance L is given by HL/c. This number H can be related to the scale factor of the Friedmann model universes. The observational value for $1/H$ of about 2×10^{10} years is a rough measure of the age of the universe. The second number is the deceleration parameter q_0, which determines whether the expansion of the universe is decelerated $q_0 > 1/2$, uniform $q = 1/2$, or accelerated $q < 1/2$. These three cases correspond to the three subtypes of Robertson-Walker metric, hyperspherical, hyperplane, or hyperpseudospherical. Measurements of q_0 involve very distant galaxies and are quite difficult and subject to large uncertainties. The difficulty stems from lack of understanding of the evolution of galaxies and their intrinsic luminosity. As a result, there is considerable controversy over the observational value of the deceleration parameter and thus over the question of whether the universe is closed (hyperspherical) or open (hyperplane or hyperpseudospherical). *See* GALAXY, EXTERNAL; HUBBLE CONSTANT.

There is another method of determining whether the universe is open or closed. It is a remarkable consequence of the Einstein equations that, given the assumption of a Robertson-Walker metric, if the average density of mass and energy in the universe exceeds a critical value of about 2×10^{-29} g/cm^3, then the universe must be closed. The density of visible material, such as that contained in stars and luminous gas, is observed to be only about 2×10^{-31} g/cm^3. However, it does not follow from this that the universe is open, since there could well be a great deal of mass and energy that is not visible. The density of material in the form of very dim stars, dark gas and dust in galactic halos, black holes, neutrinos, gravitational radiation, and so forth cannot easily be determined. Studies of the gravitational dynamics of galaxies do in fact indicate the presence of large amounts of invisible material, often called the missing mass. This is an active area of research, and it is not yet clear if sufficient density is present to indicate a closed universe. *See* COSMOLOGY; GALAXY.

Gravitational radiation. Gravitational radiation is closely analogous to electromagnetic radiation; whereas electromagnetic radiation is emitted by charges in accelerated motion, gravitational radiation is emitted by masses in accelerated motion. Such motion produces small ripples or waves in the gravitational field that propagate at the velocity of light. The usual method of studying these ripples is with linearized general-relativity theory; in this theory the metric tensor is expressed as the constant Lorentz metric of special relativity plus a small perturbation term representing small gravitational ripples in an almost flat space. Quantities of higher than first-order in this perturbation are discarded in the field equations. In this manner the nonlinear field equations become linear approximate equations. This simplifies the mathematics enormously and, in view of the small amplitude of gravitational radiation, it is an excellent approximation. The forces exerted by plane gravitational waves are at right angles to the direction of propagation of the wave; that is, they are transverse. The directions of these forces are such as to produce the distortions shown in Fig. 4 in a circular ring of particles for a periodically varying wave moving perpendicular to the paper. As with electromagnetic waves, there are two polarizations possible for gravitational waves. Both polarizations exert forces as illustrated in Fig. 4, but the force field of one polarization is rotated 45° to the other.

The only known sources of gravitational radiation strong enough to be directly detectable are violent astrophysical events such as supernova explosions or the gravitational collapse of stars. The number of such catastrophic events that occur in the Milky Way Galaxy can only be roughly estimated. Supernovae are seen only about once every 100 years in the Milky Way, but others may be obscured by dust or gas, or may involve too little visual display to be seen on Earth. Thus the frequency of about 1 per 100 years should be taken as a rough lower limit. This number is consistent with the observed frequency of supernovae seen in other galaxies. For example, in the Virgo cluster of about 2000 galaxies there is about one supernova seen per month. *See* SUPERNOVA.

The number of collapsing stars in the Galaxy can be estimated from the number of pulsars observed, which is about 450, and the average lifetime of a pulsar. Unfortunately, characteristic lifetimes of pulsars vary greatly from about 10^3 to 10^8 years. If a lifetime of 5000 years is adopted, somewhat arbitrarily, a rate of neutron star formation of about 1 per 10 years is obtained. The frequency of catastrophic events in the Galaxy is thus very roughly 1 every 10 or 100 years, a disappointingly small number.

On the other hand, it is possible that intense gravitational radiation is produced in extragalactic sources such as the collapse of entire galactic nuclei, the quasars, or the collisions of black holes. Thus, the amount of gravitational radiation that impinges on the Earth cannot be accurately predicted, and the only reliable procedure is to make observational searches.

When gravitational radiation passes over a solid body, it exerts forces that slightly distort the body. These distortions can be utilized to make massive bodies act as antennas. For example, a large metallic cylinder will ring like a bell when struck with a pulse of gravitational radiation, and the small displacements can in principle be measured. A number of experiments have already been performed with such cylinders, but no reproducible and definitive evidence for gravitational radiation has been obtained. To improve the sensitivity of such antennas, it is necessary to use low-temperature techniques, with cylinders of very high Q cooled to only a few kelvins. The sensitivity should soon approach the limits set by the uncertainty principle of quantum theory, and it is expected that by 1990 many groups of experimentalists will possess antennas capable of detecting most gravitational-radiation-producing events occurring in the Milky Way. If the frequency of such events is in the optimistic end of the range discussed above (about one every 10 years), there could well be an unambiguous detection of gravitational radiation.

The prospects for detecting extragalactic sources, such as supernovae in the Virgo cluster, are much less promising since the radiation from such distant sources would be about 10^6 times weaker than from the Milky Way. Very sophisticated means for obtaining great sensitivity would be necessary, but the high frequency of events makes it an intriguing prospect.

Whereas the direct detection of gravitational radiation is probably many years in the future, there exists very convincing indirect evidence of such radiation. The pulsar PSR 1913+16, discovered by J. Taylor and R. Hulse in 1975, appears to be one component of a close binary system with an orbital period of only 8 h. By using the short pulses from the pulsar as calibrations, observers have been able to obtain very precise information about its orbit, in particular the decrease in orbital period. This decrease is interpreted as due to orbit decay via the emission of energy in the form of gravitational radiation. The numerical value agrees with the predictions of the theory within an accuracy of about 10%.

Relation to electromagnetism. General relativity is the only fundamentally geometric theory in physics and has stood apart from other theories since its inception. Its geometric nature is intimately connected with the principle of equivalence and the independence of the trajectory of a test body on the properties of the body such as its mass. There is no analog of the principle of equivalence in electromagnetism: the motion of a charged test body in an electromagnetic field depends directly on its charge and mass. The lack of such a principle has hindered the development of a true geometric theory of electromagnetism. Many attempts at a classical unified field theory of gravitation and electromagnetism were made by Einstein, H. Weyl, and others, but the results were not very convincing. These attempts seem less interesting since the discovery of other forces in nature besides gravitation and electromagnetism. It would be desirable ultimately to unify the strong and weak nuclear forces, electromagnetism, and gravity, and such a grand unification would necessarily involve quantum theory.

Although there is no convincing classical unified field theory, the Maxwell equations of electromagnetism, when expressed in covariant form, are completely consistent with the ideas and equations of general relativity, and no geometric interpretation of the electromagnetic field is logically necessary. In this view the electromagnetic field operates conventionally in the curved space described by the gravitational field equations. In turn, the electromagnetic field contains energy and is thus the source of some of the curvature of the space. Much work has gone into the elucidation of the properties of the coupled Einstein-Maxwell equations in vacuum, sometimes referred to as already-unified field theory, and interesting formal results and interpretations have emerged.

Constancy of gravity. The ratio of the electrostatic to the gravitational force between the electron and the proton in a hydrogen atom is about 10^{39}. Such an enormous number presents a manifest challenge for theoreticians to explain on some fundamental level; moreover, it is hard to see how a theoretical unification of gravity with the other forces in nature could be accomplished without some understanding of the extreme weakness of gravity.

P. Dirac suggested an interesting approach to this problem. A natural unit for the measurement of time is the time interval required for light to cross an elementary particle diameter, about 10^{-23} s. In terms of this unit the lifetime of the universe is about 10^{40}, roughly the same as the ratio of electrostatic to gravitational forces in the hydrogen atom. Dirac argued that the rough equality of such enormous dimensionless numbers must be more than a coincidence, and suggested that the numbers may be actually equal at all times. This is referred to as the large-number hypothesis. It implies that the inverse of the gravitational constant G may be proportional to the lifetime of the universe, and not be a constant at all. G should decrease by roughly 1 part in 10^{10} per year. That is, gravity weakens with time.

Many approaches to searching for a variation in G have been made using geology, paleontology, celestial mechanics, and so forth, but the only positive result is somewhat uncertain and controversial. The upper limit on the variation of G is now about 1 part in 10^{10} per year, and work is continuing in order to improve the accuracy of the observations.

Relation to quantum theory. Fusing the ideas of general relativity and those of the quantum theory is a difficult problem that is the subject of extensive research. The most naive approach to this problem is to assume that some invariant generalizations of the conventional quantum equations are valid in the curved space specified by the Einstein equations. This approach has led to many important results and insights such as the evaporation of black holes, discussed above. However, it does not provide a natural way to quantize the gravitational field itself and thus is not a consistent approach. For example, the uncertainty principle of quantum theory could be subverted, in principle, by gravitational measurements, which presents a manifest inconsistency.

A more fruitful approach is to deemphasize the geometric interpretation of gravity and develop the theory in analogy with successful quantum field theories, in particular quantum electrodynamics. Quantum electrodynamics accurately describes the interactions of charged particles, such as electrons, with photons, which are the quanta of the electromagnetic field. This theory has already served as a model for the development of quantum field theories of both the weak and strong nuclear interactions, and the two theories of quantum electrodynamics and the weak interactions have been combined by S. Weinberg and A. Salam in a very successful unified theory. The relatively successful theory of the strong interactions, known as quantum chromodynamics, describes the interactions of quarks with the quanta of the strong nuclear force, called gluons. Much attention has been focused on grand unified theories, which attempt to describe in a coherent way the strong, electromagnetic, and weak interactions—that is, all the fundamental forces of nature except gravity.

The basis of much of the success of these quantum field theories has been their renormalizability.

This means that certain divergences or infinite results that occur during a calculation do not appear in the final answer. Thus the predictions for physically measurable effects are finite. The divergences disappear by rather subtle cancellations.

When a quantum theory of gravity is constructed by analogy with quantum electrodynamics, it is convenient to abandon or at least deemphasize the geometric interpretation of the gravitational field as necessary to the basic physics. Then the formal construction of the theory is in fact not difficult, and the quanta of the field, called gravitons, can be studied. However, quantum general relativity is not renormalizable (as yet) and can predict infinite results for real experiments, which is manifestly unacceptable.

Some progress has been made in this problem with a generalization of the theory known as supergravity. In supergravity the quanta of the gravitational field are grouped into a family with the quanta of another field, and some of the divergences of the theory do indeed vanish. However, supergravity and variations are still in a preliminary stage of development and as yet offer only hopes and hints as to how theoretical study might proceed in the future.

Quantum field theories at present are based on the assumption that a space-time continuum exists down to indefinitely small distances and short times. Many physicists have noted that this assumption cannot in fact be justified by experience and that difficulties with the concept probably occur for distance scales of about 10^{-31} cm and times of 10^{-42} s. The geometric view of gravity suggests that the energy density of quantum fluctuations at such a scale should distort space-time out of all recognition and usefulness. Field theories other than gravity may avoid such difficulties by the process of renormalization, but it may be that a successful quantum theory of gravity will require a much deeper understanding of space and time at very small scales. [RONALD J. ADLER]

Bibliography: R. J. Adler, M. Bazin, and M. M. Schiffer, *Introduction to General Relativity*, 1975; A. Einstein, *The Meaning of Relativity*, 1956; N. D. Mermin, *Space and Time in Special Relativity*, 1968; C. Misner, K. Thorne, and J. Wheeler, *Gravitation*, 1973; W. Pauli, *Theory of Relativity*, 1958; W. Rindler, *Essential Relativity*, 1969; H. M. Schwartz, *Introduction to Special Relativity*, 1968, corrected reprint, 1977; S. Weinberg, *Gravitation and Cosmology*, 1972.

Retrograde motion

In astronomy, either an apparent east-to-west motion of a planet or comet with respect to the background stars or a real east-to-west orbital motion of a comet about the Sun or of a satellite about its primary. The majority of the objects in the solar system revolve from west to east about their primaries. However, near the time of closest approach of Earth and a superior planet, such as Jupiter, because of their relative motion, the superior planet appears to move from east to west with respect to the background stars. The same apparent motion occurs for an inferior planet, such as Venus, near the time of closest approach to Earth.

Actual, rather than apparent, retrograde motion occurs among the satellites and comets; the eighth and ninth satellites of Jupiter and the ninth satellite of Saturn are examples. See ORBITAL MOTION.

[RAYNOR L. DUNCOMBE]

Rigel

Beta Orionis, a blue supergiant of spectral type B8. Although Rigel is one of the apparently brightest stars in the sky (0.2 visual magnitude), it is too distant to have a measurable parallax or proper motion. Its luminosity can be estimated as -7^m or -8^m because it has a faint companion with which it is probably physically connected. So high a luminosity, about 60,000 times that of the Sun, means that Rigel is an exceptionally young star in rapid evolution, with a life span of only a few million years. See STAR. [JESSE L. GREENSTEIN]

Roche limit

The closest distance which a satellite, revolving around a parent body, can approach the parent without being pulled apart tidally. The simplest formal definition is that the Roche limit is the minimum distance at which a satellite can be in equilibrium under the influence of its own gravitation and that of the central mass about which it is describing a circular orbit. If the satellite is in a circular orbit and has negligible mass, the same density as the primary, and zero tensile strength, the Roche limit is 2.46 times the radius of the primary.

The concept of an equilibrium state was first given by Édouard Roche in 1849. The radius r of the Roche limit appropriate to satellite formation is given by the equation below, where B is the

$$r = 2.46\left(\frac{\rho}{\rho_s}\right)^{1/3} B$$

radius and ρ the mean density of the parent planet, and ρ_s is the mean density of the satellite. This is the radius at which the surface of a homogeneous, fluid satellite can remain in stable hydrostatic equilibrium.

The popular definition of Roche's limit, that is, the distance from the planet at which a satellite would suffer tidal loss of particles, depends on properties of the material and the shape and density of the satellite. If, for instance, the material can flow to adopt the hydrostatic equilibrium shape, then the body will be elongated in the direction of the radius vector. For a binary system, the Roche limit is defined by the value of the Jacobi constant such that the zero velocity surfaces around the two bodies intersect in the lagrangian libration point L_2 between the finite masses. This implies an upper limit to the size of the components. While magnetohydrodynamic forces act on the outer layers of stars and the theory holds only for the circular restricted problem with the masses acting as point masses, the Roche limit seems to correspond to the observational data. When a star has exhausted the supply of hydrogen in its core, its radius will increase by a factor of 10 to 100. A star in a binary system may then exceed its Roche limit, material will thus escape from that star, and its companion will receive the excess material. See BINARY STAR; CELESTIAL MECHANICS.

[P. K. SEIDELMANN]

Rocket astronomy

The discipline comprising measurements of the electromagnetic radiation from the Sun and other celestial bodies of wavelengths that are almost completely absorbed between the 250-km level reached by sounding rockets and 40-km level attained by balloons. Beginning in 1946 with captured German V-2 rockets, rocket astronomy has developed from the simplest instrumentation to the extremely complex. Countries involved are the United States, the United Kingdom, France, West Germany, Japan, the Soviet Union, and others. The vehicle in longest and most extensive use is the Aerobee, a liquid-propelled, solid-boosted rocket. Also in general use is the Black Brant.

The Aerobee-170 and the Black Brant V-C lift several hundred pounds to 250–300 km. Recovery of scientific payloads by parachute makes possible the retrieval of photographic and other nontelemetered data as well as reflight of the same equipment. Although the first instruments were unpointed, the early results opened the entire field through the extreme ultraviolet (xuv) into x-rays. Since 1952, the biaxial pointing control (BPC, 1-arc-minute stabilization) has made possible much of the most important solar research from rockets. After about 1970 these systems were replaced by the triaxial solar pointing and roll control system (SPARCS). By means of small gas jets, SPARCS points and stabilizes the entire rocket. The error signals that control yaw and pitch are produced by a solar sensor; for roll they are from a gyro. The solar image can be held stable to better than 1 arc second.

A further development enables the scientist on the ground both to observe and to change the exact pointing of the rocket during unpowered flight. This is accomplished by viewing a solar image televised from the rocket and by radio command of the SPARCS. The first system for pointing at stars and other celestial sources was the attitude control system (ACS) that enabled pointing during flight at five positions in the sky, with the use of gas jets controlled by gyros and a computer; several advanced systems of this type are now available.

Solar rocket astronomy. Rocket astronomy's principal subdiscipline is the Sun. The known solar spectrum was extended in 1946 from 290 nm, the limit reached from the Earth, to 210 nm. Now the spectrum is known to about 1.5 nm from photographic as well as telemetered photoelectric rocket spectroscopy using all-reflective diffraction grating spectrographs. The x-ray spectrum, discovered in 1948 with broadband detectors, now is studied with Bragg crystal spectrometers to 0.15 nm and to shorter wavelengths with pulse-height analyzer detectors. It has been found that the continuum with Fraunhofer (absorption) lines, typical of the visible, extends to 208.5 nm; at shorter wavelengths it changes to a continuum with fewer absorption lines and becomes ever more densely overlaid with emission lines; below 153 nm there are few, if any, absorption lines. Between those wavelengths the continuum originates from the coolest part of the Sun's atmosphere, where the temperature falls to a minimum of about 4500 K. The emission lines, however, arise from the higher atmosphere, where the temperature increases rapidly with height. Short of 70 nm the spectrum consists mainly of emission lines from the chromosphere, active regions, the transition layer, and the 1,000,000 K corona.

Solar flares, because of their extremely high temperatures, emit many additional lines from atoms that are highly ionized. Although flare lines occur throughout the solar spectrum, they become much more abundant with decreasing wavelength through the xuv to 0.15 nm in x-rays. From analyses of flare spectra, temperatures within a flare have been shown to rise at least to 25,000,000 K, and the temperature history of a flare can be derived from the change in its spectrum with time.

Monochromatic images of the entire Sun are obtained with spectroheliographs. Spectroheliograms in the xuv and in x-rays show that the Sun's appearance depends on the origin of the particular emission line in which it is imaged. In chromospheric lines there is great detail of high contrast; in lines from regions of still higher temperature the detail and contrast become less; and in coronal lines the emission extends well above the limb and is diffuse. The principal xuv rocket spectroheliograph is the single, objective grating combined with a filter of aluminum foil 100 nm thick, first flown in 1963; this covers the range 63 to 17.1 nm. In x-rays the first images were photographed with a simple pinhole camera and a proper filter. Spectacular, high-resolution solar images in x-rays have been obtained, and coronal holes have been discovered with the Wolter double-cylindrical mirror system operating at grazing incidence combined with suitable filters. *See* ASTRONOMICAL PHOTOGRAPHY; ASTRONOMICAL SPECTROSCOPY; X-RAY TELESCOPE.

An important advance in solar xuv spectrographs is the double-dispersion, high-resolution stigmatic spectrograph (HRTS). Flown first in 1975 on a Black Brant V-C equipped with SPARCS and the command control system, this instrument has photographed spectra of areas of the Sun's surface measuring 1 arc-second square. Some of these extremely small solar regions exhibit spectra that are most unusual and not yet well understood.

The Sun's outer, white-light corona that can be seen from the ground only during a total eclipse can be recorded by a coronagraph on a rocket at an altitude above the veiling light of the daytime sky. Coronal white light images to 10 solar radii were obtained in this way from 1963 to 1972. These images show that the outer corona changes dramatically, even from one day to the next. With a similar coronagraph placed in orbit, the first observation of a coronal transient was made on December 14, 1971; images made at intervals of 11 min showed a large mass of plasma leaving the Sun at 1000 km per second; it reached the Earth 2 days later and produced a magnetic storm.

The results from solar rocket astronomy have changed and refined the model of the solar atmosphere, have thrown new light on the role of magnetic fields in the Sun's atmosphere, and have led to a new understanding of solar flares. Coronal holes have been discovered and shown to be related to high-speed particle streams in the solar wind and to ionospheric and geomagnetic effects on Earth. Solar rocket spectroscopy has led to the stimulation of spectroscopic research in the laboratory, much in close relation to spectroscopic work involved in research on fusion. Because the Sun is the only star that can be studied in detail, solar rocket-spectroscopic results are essential in guiding the development of theories of stellar atmospheres. *See* SUN.

Other celestial bodies. Rocket astronomy is more difficult for the study of celestial bodies other than the Sun because of the low intensity of their emissions. Early research with broadband or very-low-dispersion systems served to obtain spectra of certain stars, especially those that are very hot. This has been of great value in improving the models of stellar atmospheres. The first stellar spectra of sufficient resolution to distinguish lines were obtained in 1965, using a photographic grating spectrograph. The Doppler shift of xuv emission lines, observed for the hot supergiants in Orion, was evidence of a rapid loss of mass, required for their eventual evolution into dwarfs. Below 91.2 nm, the long-wavelength limit of the Lyman continuum of hydrogen, because of strong absorption by interstellar hydrogen, it is unlikely that xuv stellar radiation will be observed until much shorter wavelengths are reached. Molecular hydrogen, on the other hand, is relatively transparent; its presence in interstellar space was first detected in 1970 through faint xuv absorption bands, observed in spectra of ξ-Persei photographed from a rocket with an electronographic spectrograph. X-ray radiation from celestial bodies, however, is able to reach the solar system, and was discovered with instruments carried on Aerobee rockets. The first detection was of a diffuse cosmic background in 1956. The first discrete source was discovered near the center of the Galaxy in 1962, and the most intense source, Scorpius X-1, in 1963. Thus was opened the now extremely active subdiscipline x-ray astronomy. *See* X-RAY ASTRONOMY.

Solar-terrestrial relationships. Closely connected with and forming a part of rocket astronomy is the study of the ionosphere, the airglow, and the composition of the Earth's outer atmosphere by measuring the attenuation of solar xuv radiation with rocket-borne instrumentation. This work is often placed in the interdisciplinary fields of solar-terrestrial relationships, aeronomy, and ionospheric physics. As a result, the composition and chemistry of the upper atmosphere, including the dissociation of molecular oxygen and nitrogen and the formation of ozone, nitric oxide, and various ionic species, are becoming fairly well understood. The origin of the ionosphere and changes produced in it by changes in the x-ray output of the Sun are also understood. Still controversial, however, and considered of great importance are effects produced in the ozone layer by introduction of nitrogen oxides from jet aircraft exhausts, and of fluorocarbons. This places new emphasis on the need for research on the Earth's atmosphere with instruments borne on rockets.

Orbiting observatories. Astronomical research from space vehicles is conducted now to a greater degree from orbiting observatories than from rockets. The change has been gradual. Observations of the Sun began in 1960, and of other celestial bodies in 1968. Future plans are based principally on the use of the space shuttle, which is designed to carry astronomical equipment to be used by the crew, and to launch, service, and retrieve uncrewed observatories. Nevertheless, rocket astronomy will continue. Historically, it has been from experiments carried on rockets that most of the breakthrough discoveries in space astronomy have come. These results prepared the way for the far more complicated experiments that were later flown in satellites. Rockets will continue to provide means for making new discoveries and for proving experiments for satellites at costs that are modest. *See* SATELLITE ASTRONOMY.

[RICHARD TOUSEY]

Sagittarius

The Archer, in astronomy, a zodiacal and summer constellation, the major portion of which lies directly in the Milky Way. Sagittarius is the ninth

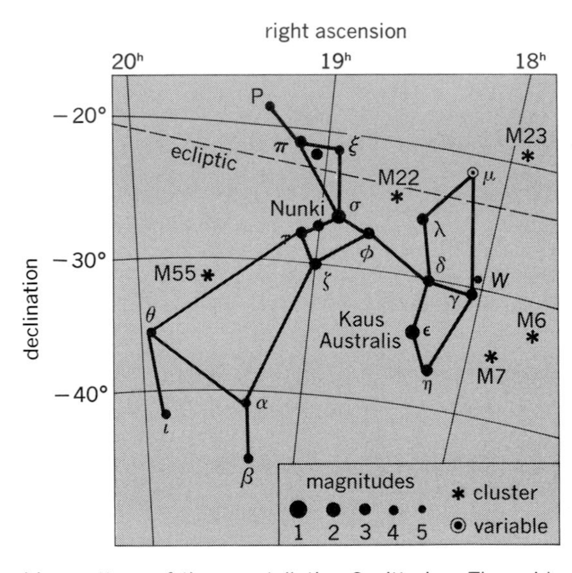

Line pattern of the constellation Sagittarius. The grid lines in the chart represent the coordinates of the sky. The apparent brightness, or magnitude, of the stars is shown by the sizes of the dots, which are graded by appropriate numbers as indicated.

sign and the southernmost constellation of the zodiac. In mythology, it is represented by a centaur, Chiron, drawing his bow to release an arrow. Its most prominent feature is a star group commonly called the little Milk Dipper. It is an inverted dipper with four stars to form the bowl and one to form the handle (see illustration). The Milky Way in Sagittarius is very bright, containing rich star fields and clusters, because its direction lies in the center of the Milky Way stellar system. *See* CONSTELLATION.

[CHING-SUNG YU]

Satellite

A relatively small celestial body moving in orbit about a planet. The Moon is the only natural satellite of Earth. In all, the solar system contains 43 known satellites distributed among the nine planets as follows: Jupiter, 15; Saturn, 17; Uranus, 5; Mars, 2; Neptune, 2; Earth, 1; and (probably) Pluto, 1. Venus and Mercury have no satellites.

The largest satellites reach planetary dimensions, with four (Ganymede and Callisto of Jupiter, Titan of Saturn, and possibly Triton of Neptune) having diameters greater than 5000 km (3000 mi), that is, exceeding that of the planet Mercury. The smallest known satellites, found around Jupiter, Saturn, and Mars, are less than 50 km (30 mi) in diameter, and some are as small as 10 km (6 mi).

The surface gravity of most satellites is too feeble to retain an atmosphere having a density comparable to Earth's. An exception is Titan, with an atmosphere composed primarily of hydrogen and methane and having a density about equal to Earth's. The reddish color of Titan as seen in the telescope has been ascribed by some astronomers as being due to clouds of complex organic molecules floating in the satellite's atmosphere. Among Jupiter's satellites, Io is known to have an extremely tenuous atmosphere, less than one-hundred-millionth that of Earth's. In 1979 the Voyager spacecraft discovered that Io had numerous powerful fissures spewing material for hundreds of kilometers out into space. It is the only body in the solar system, in addition to Earth, known to have volcanolike eruptions.

Some of the larger satellites, such as the four Galilean satellites of Jupiter (Io, Europa, Ganymede, and Callisto), were formed along with, and at about the same time as, the planet about which they now revolve. Except for Io, these contain large amounts of water in the form of ice. Others, such as the small outer satellites of Jupiter, may well be rocky asteroidal fragments captured long after the formation of the inner Jovian system. In many other cases, the early histories of satellites are uncertain. The origin of the Moon, for example, remains in doubt, even though the Moon has been studied more extensively than any other extraterrestrial body. *See* JUPITER; MARS; MOON; NEPTUNE; PLANET; PLUTO; SATURN; URANUS.

[J. KELLY BEATTY]

Bibliography: C. R. Chapman, *The Inner Planets*, 1977; R. M. Goody and J. C. G. Walker, *Atmospheres*, 1972; W. K. Hartmann, *Moons and Planets: An Introduction to Planetary Science*, 1972; B. O'Leary and J. K. Beatty, *The New Solar System*, 1981; J. A. Wood, *The Solar System*, 1979.

Satellite astronomy

The study of astronomical objects using detectors mounted on Earth-orbiting satellites or deep-space probes so that observations unobstructed by the Earth's atmosphere can be made. Many astronomical satellites carrying detectors to record electromagnetic radiation at wavelengths shorter than visible light (ultraviolet, x-, and gamma rays) were launched in the 1960s and 1970s. The most noteworthy were the OSO, *Uhuru* (*SAS 1*), OAO, *SAS 2*, ANS, *SAS 3*, IUE, and HEAO satellites. In general, one or two wavelength regions were observed from each satellite. Successive satellites incorporated significant developments in operating modes and improvements in detector sensitivity and resolution such that new discoveries were possible.

Ultraviolet astronomy. Since the Earth's atmosphere is opaque to ultraviolet (uv) light with wavelengths shorter than about 310 nm, ultraviolet astronomy had to await the space age. In the 1940s Lyman Spitzer of Princeton University pointed out some of the astrophysical problems that could be addressed with a large telescope above the Earth's atmosphere. Many of these problems have now been explored with the several ultraviolet detector systems flown on the OSO (Orbiting Solar Observatory) and OAO (Orbiting Astronomical Observatory) satellite series in the 1960s and early 1970s, as well as with the subsequent IUE (International Ultraviolet Explorer) satellite. A much more powerful tool will become available with the launch of the Space Telescope.

Solar ultraviolet astronomy. The OSO satellites were the first to be devoted primarily to astronomical ultraviolet and x-ray observations. Eight satellites (*OSO 1–8*) were launched over a 12-year period beginning in 1963. Each was of a similar design and contained a pointed section mounted above and across a wheel section which spun and maintained stable pointing of the satellite toward the Sun. The experiments mounted in the wheel section scanned a great circle in the sky including the Sun.

The solar ultraviolet experiments included both low- and high-spectral-resolution spectrometers. Small entrance slits were used so that high-spatial-resolution (several arc-sec) raster scans could be made to map the entire Sun in the light of a given ultraviolet emission line. Different spectral lines, or narrow wavelength intervals, could then be chosen by changing the tilt of the spectrograph grating. The resulting detailed maps of the Sun greatly increased understanding of the temperature versus height profile of the solar atmosphere. This, in turn, enabled much more detailed theoretical models of the transfer of radiation out through the solar atmosphere. *See* ASTRONOMICAL SPECTROSCOPY.

Time-resolved spectra and images were also recorded so that the first ultraviolet observations of solar flares were conducted. Although superseded in quality by the higher-resolution observations made from Skylab in 1972 and 1973, the first solid evidence for the existence of coronal holes came from experiments on *OSO 4*. These are apparent holes in the solar corona, or hot (10^6 K) outer low-density region of the solar atmosphere, through which the solar wind is escaping. Another fundamental discovery made from *OSO 8* was that the solar corona could not be heated by acoustic waves driving up from below (in the chromosphere) as had been commonly assumed. *See* SUN.

Galactic ultraviolet astronomy. Extensive ultraviolet observations of stars and nebulae in the Galaxy but outside the solar system were first conducted with ultraviolet telescopes on the Orbiting Astronomical Observatory satellites. Two of these spacecraft were successfully launched. *OAO 2*, launched in December 1968 (*OAO 1* did not achieve orbit), carried two principal experiments—a spectrophotometer with 2-nm resolution to study ultraviolet emission lines from bright stars (as well as the first ultraviolet absorption lines from interstellar gas) and a broadband photo-

Fig. 1. International Ultraviolet Explorer spacecraft in geosynchronous orbit. (*IUE Observatory, NASA Goddard Space Flight Center*)

meter with a 31-cm telescope to measure the ultraviolet fluxes in broader bands of stars down to fainter magnitudes. More than 5000 stars were observed at 140 nm and 200 nm by the latter instrument, which provided the first large sample of ultraviolet observations of stars for comparison with predictions from model stellar atmospheres derived from visible-light observations.

Whereas *OAO 2* operated for about 16 months (and the individual OSO satellites typically for 1–2 years), the *OAO 3* satellite, called *Copernicus*, was still functioning as an ultraviolet and x-ray observatory in 1981 after a 1972 launch. This satellite contains an 80-cm-aperture ultraviolet telescope and high-resolution spectrograph with sensitivity and resolution such that fractional angstrom (1 A = 0.1 nm) resolution can be obtained on stars brighter than about seventh magnitude. The spectrograph is actually a scanning monochromator (as on the OSOs) so that detailed line profiles produced by absorption from the fractional percent composition of heavy elements (carbon, nitrogen, oxygen, silicon, and so forth) in the interstellar medium can be studied. Surprisingly, these heavy elements and "metals" were found to be much lower in their total apparent abundance in the interstellar medium than the average cosmic composition which they might otherwise be expected to show. The explanation seems to be that the heavier elements are "locked up" in the interstellar grains. These grains, which are typically 10-μm particles, could not themselves produce the ultraviolet absorption spectra seen (as would single atoms); instead they can totally absorb the ultraviolet and reradiate it as heat or infrared radiation. *See* INTERSTELLAR MATTER.

Extragalactic ultraviolet astronomy. The International Ultraviolet Explorer (IUE) was launched in January 1978 into a geosynchronous orbit such that it can be continuously viewed from both the NASA Goddard Space Flight Center and the European Space Agency (ESA) Operations Control Center near Madrid, Spain. United States astronomers use the facility 16 hours per day whereas ESA controls the satellite for 8 h. The IUE spacecraft (Fig. 1), like *OAO 3*, is three-axis-stabilized and can be pointed at any object outside a 45° cone centered on the Sun.

The IUE telescope (Fig. 2) is a 45-cm-aperture

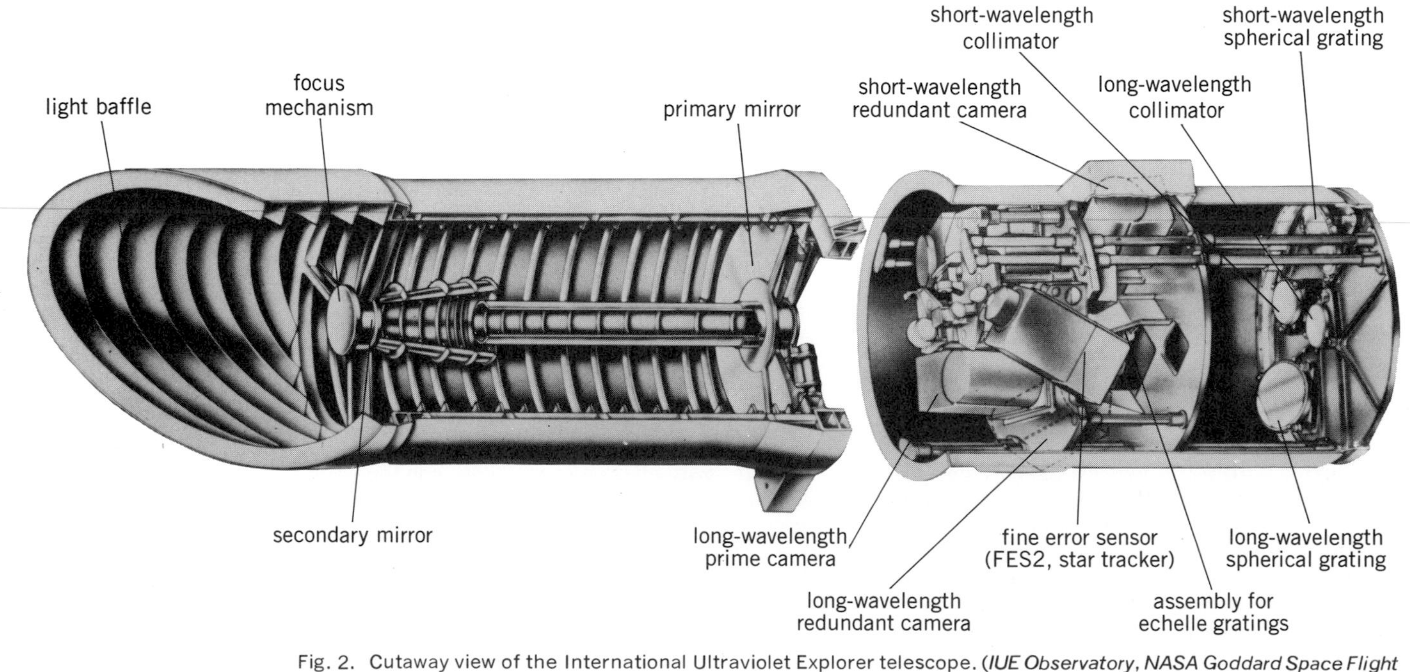

Fig. 2. Cutaway view of the International Ultraviolet Explorer telescope. (*IUE Observatory, NASA Goddard Space Flight Center*)

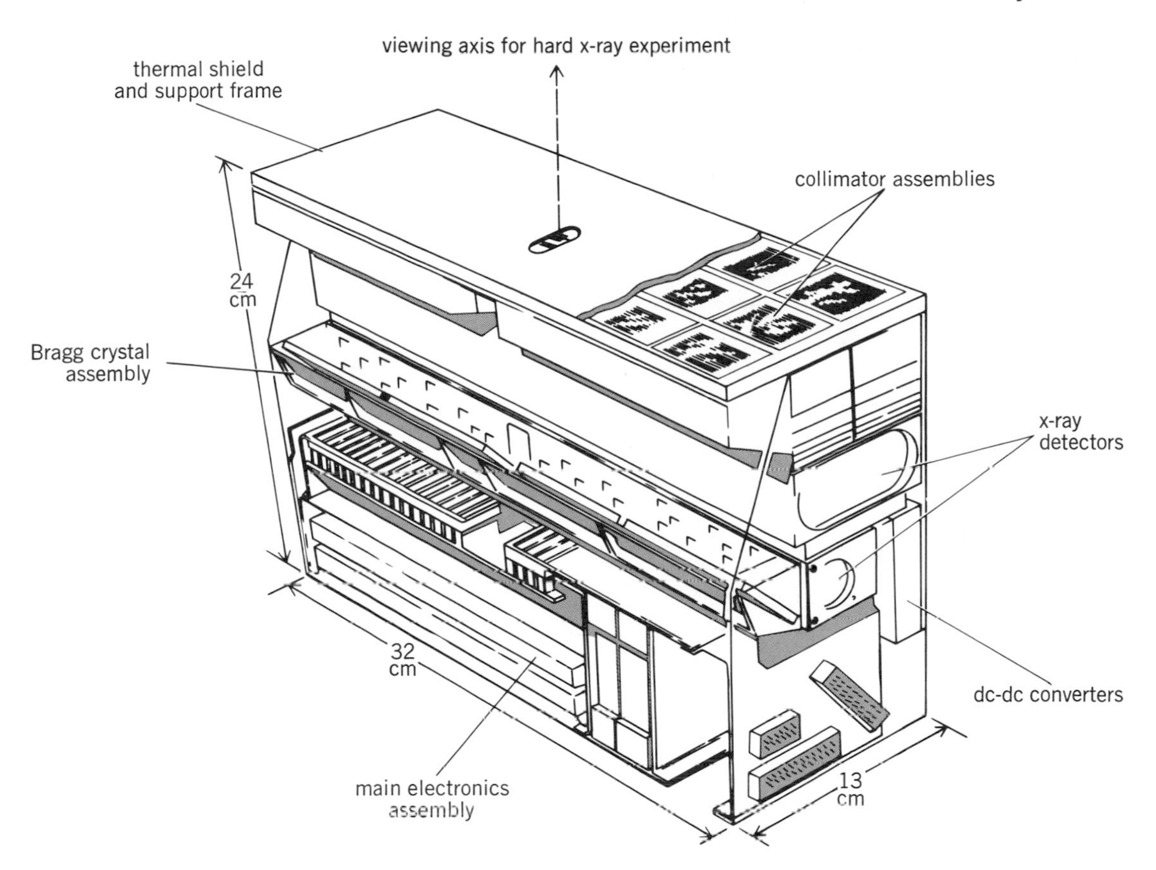

Fig. 3. United States hard x-ray (1 – 20 keV) experiment on the Astronomical Netherlands Satellite.

Ritchey-Chrétien with an effective focal ratio of *f*/15 and image quality of about 3 arc-sec within a 16-arc-min field. At the telescope focus is an echelle spectrograph, which allows very high spectral resolution ($\lambda/\Delta\lambda = 10^4$) over a large spectral range (120 nm) by stacking multiple orders of the dispersed spectrum on a two-dimensional detector.

IUE has, like *OAO 3*, been used for a great many observations of objects within the Galaxy, such as galactic x-ray binaries, hot stars, and globular clusters, to name but a few. However, its greater sensitivity has also permitted for the first time ultraviolet observations of extragalactic objects. Active galaxies and nearby bright quasars such as 3C 273 have been some of the primary targets. These observations have enabled the first studies of the ultraviolet emission line spectrum in nearby quasars for comparison with the many ground-based optical observations of these same lines from higher red-shift (more distant) quasars. This in turn allows the evolution of quasars, which are the most distant objects known, to be studied. *See* GALAXY, EXTERNAL; QUASARS; ULTRAVIOLET ASTRONOMY.

X-ray astronomy. X-ray astronomy can only be done from above the Earth's atmosphere. The first astronomical x-ray detectors, launched with captured V2 rockets in the late 1940s, discovered x-ray emission from the Sun. Cosmic (that is, nonsolar) x-ray sources were first discovered in 1962 by detectors carried above the Earth's atmosphere on sounding rocket flights of, typically, 5 minutes duration.

Galactic x-ray astronomy. X-ray astronomy really came of age with the launch of the first Small Astronomy Satellite (*SAS 1*) on December 12, 1970, designated *Uhuru*. It carried two proportional-counter x-ray detectors in which a pulse of electric charge proportional to the energy of the incident x-ray photon is detected. The satellite was spinning so that the two detectors, with differing fields of view, alternately scanned over a given source, and the first survey of the entire sky could be conducted. *Uhuru* enabled the key discovery of x-ray binary systems in which a collapsed object (neutron star or black hole) accretes gas from the atmosphere of a "normal" companion star. These systems are often 10,000 times as luminous in x-rays (alone) as the entire output of the Sun, and they are probably involved with most (if not all) of the bright galactic x-ray sources. *Uhuru* also made the initial discoveries that active galaxies (Seyferts) and galaxy clusters are also prodigious sources of x-rays. *See* BINARY STAR; BLACK HOLE; NEUTRON STAR.

The OSO satellites (described above) also carried cosmic x-ray detectors (proportional counters similar to *Uhuru*) and contributed much to the detailed understanding of individual sources. Qualitatively different cosmic x-ray satellite experiments, however, were launched in 1974 and 1975 with the Astronomical Netherlands Satellite (ANS) and *SAS 3* satellite, respectively.

ANS was the first x-ray observatory: it was a pointed instrument carrying a variety of x-ray detectors that could be operated by an on-board computer in a variety of modes. A schematic view of the United States experiments on ANS is shown in Fig. 3. (ANS, a joint United States–Netherlands venture, also carried two Dutch experiments.) This experiment consisted of two proportional counters with fields of view slightly inclined to each other (so that accurate source locations in one dimension could be derived from the ratio of fluxes recorded in the two detectors) and two Bragg crystal spectrometers. The Bragg spectrometer was tuned to reflect (from the planes of atoms in a crystal) onto proportional counter detectors only x-rays at the wavelengths (about 0.67 nm) produced by *K*-shell transitions in the element silicon stripped of all but one or two electrons—that is, Si(XIV) or Si(XIII), respectively). The upper limits found for silicon-line emission from a number of bright x-ray sources were able to restrict the physical conditions in the source. The most significant discovery made with ANS, however, was that of the x-ray burst sources. Intense bursts of x-rays were found to be occasionally emitted by certain strong x-ray sources, which are often located in globular clusters. Although it was originally thought that these might be produced by gas falling into a massive black hole, subsequent studies indicated that the sources are probably (again) binary systems containing a low-mass black hole or a neutron star. The discovery of the comparatively rare x-ray bursts was due in large part to the ability of ANS to point continuously and thus observe an object for much longer times than the earlier scanning experiments.

The *SAS 3* satellite, launched in May 1975, contained a variety of x-ray detectors, including a low-energy x-ray flux concentrator (a similar device was included on ANS as one of the two Dutch experiments), which detected x-rays from "normal" stars, and a rotating modulation collimator (RMC). The RMC was the first to be flown on a satellite. It was able to determine relatively precise x-ray source positions (to about 30 arc-sec) by measuring, on a proportional-counter detector, the modulation of the detected x-ray flux due to the shadow cast by a rotating grid of closely spaced wires. These precise source positions in turn allowed many x-ray sources to be identified optically for the first time. *SAS 3* was able to both scan and point, and detailed timing studies of x-ray pulsars and bursters were emphasized.

Extragalactic x-ray astronomy. A major increase in sensitivity for x-ray astronomy was achieved with the High Energy Astronomical Observatory (HEAO) satellite. *HEAO 1* carried four major experiments: a large-area (nearly 1 m^2) proportional counter; a broad-energy-range and wide-field-of-view set of proportional counters; a modulation collimator; and a crystal scintillator-detector system to extend the energy range to much higher values (that is, to low-energy gamma rays). The total number of x-ray sources known was quadrupled to some 1500 as a result of *HEAO 1*, and high-quality continuum spectra were obtained for many sources. The increased sensitivity meant that fainter objects (primarily extragalactic) such as active galaxies and quasars could be well observed for the first time. *HEAO 1* was a scanning instrument (though some pointed observations were also done) designed to survey the entire sky.

Finally, the *HEAO 2* satellite, or Einstein Observatory, marked the start of a new era in x-ray astronomy and in satellite astronomy in general. Launched in November 1978, Einstein has provided the first x-ray images of celestial objects and detected thousands of new sources. The satellite carries the largest x-ray telescope yet constructed (with 0.6-m aperture), and the first to be used exclusively for studies of cosmic x-ray sources (a smaller x-ray telescope was included on the Skylab satellite and used to take x-ray pictures of the Sun). The overall length of the satellite is about 7 m. Einstein, like IUE, is used extensively by guest observers as if it were a ground-based astronomical observatory.

In the grazing-incidence x-ray telescope on *HEAO 2*, x-rays are reflected from highly polished, nickel-coated surfaces of first a hyperboloid and then a paraboloid. The incidence angles must be small (less than 1°) for x-rays to reflect from, and not be instead absorbed by, the mirror surface. Thus, although the *HEAO 2* telescope has a large surface area, its projected area for imaging is small, only several hundred square centimeters. However, a true image of the object (extended or point source or sources) being observed is formed, and thus the detected background is very low in a small-image pixel element. Thus, the sensitivity of *HEAO 2* or any imaging detector is very much greater than a nonimaging detector of the same size. In fact, the sensitivity increase achieved in the 16 years of x-ray astronomy from the first rocket-launched detectors to the *HEAO 2* instrument is comparable to that achieved in the 300 years of optical astronomy from the first telescope of Galileo to the 5-m telescope on Palomar Mountain. *See* X-RAY TELESCOPE.

The Einstein Observatory includes four different detectors which can be individually positioned at the focus of the telescope (Fig. 4). These are: a low-resolution (arc-min) and high-resolution (arc-sec) imaging detector (that is, "camera") for recording x-ray images with and without spectral information and a moderate-resolution and high-resolution spectrometer for measuring x-ray spectral lines in a broad versus narrow spectral range. Major discoveries have been made by each of these instruments. Perhaps foremost of these discoveries is the finding that even the most distant quasars known (and many previously unknown) are detected as strong x-ray sources and that collectively these may contribute most of the mysterious cosmic x-ray background. Another important result is that "normal" stars can be much more powerful x-ray emitters than was expected from comparisons with the Sun. A permanent x-ray observatory (the Advanced X-ray Astrophysics Facility, or AXAF), employing a much larger (1.2-m aperture) imaging telescope, may be launched in 1988–1990 to carry on these highly productive studies. *See* X-RAY ASTRONOMY.

Gamma-ray astronomy. Gamma rays, which are more energetic than x-rays, still do not penetrate the Earth's atmosphere and, except at the very highest energies, can only be detected from high-

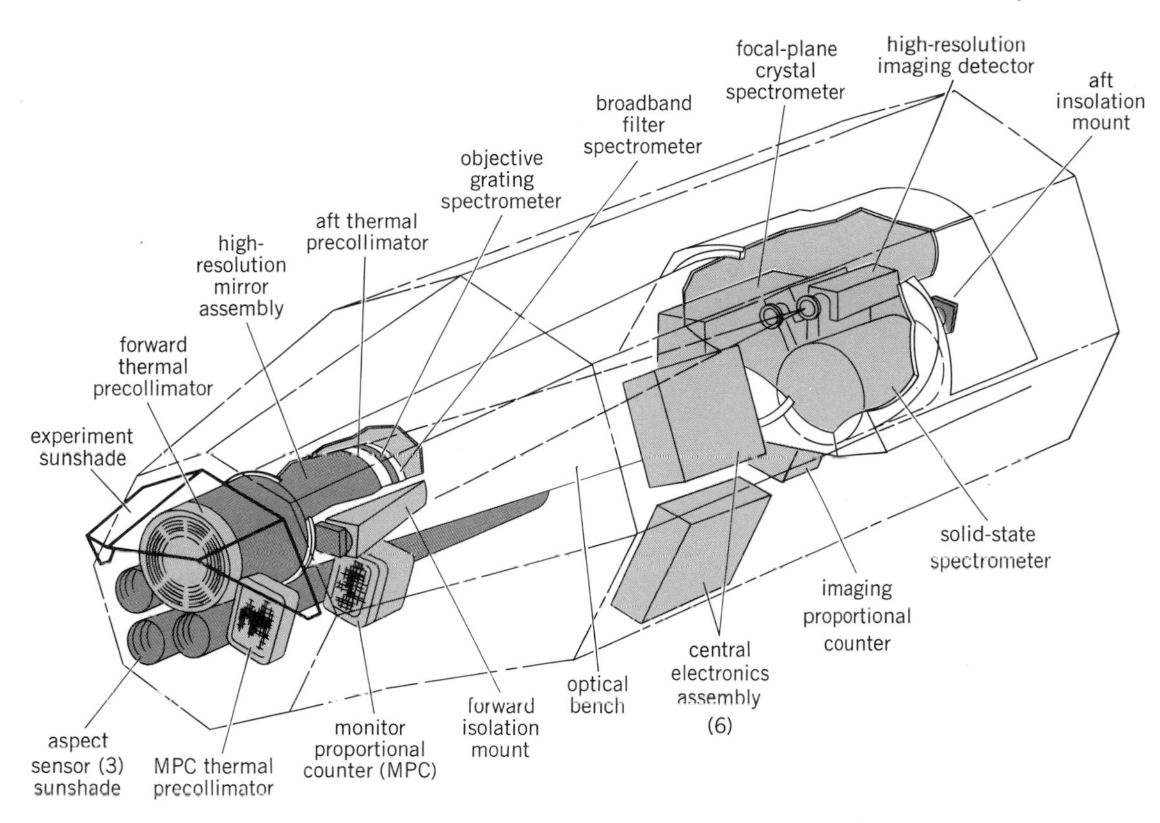

Fig. 4. Experiment configuration and overall layout of the *HEAO 2* (Einstein Observatory).

altitude balloons, rockets, or satellites. The first cosmic gamma-ray detectors were flown in the early 1960s on the *Explorer 11* satellite and a Ranger spacecraft; the first to detect the clear signature of a cosmic gamma-ray source (in this case, gamma rays from the disk of the Galaxy), was an experiment carried on the *OSO 3*. The *SAS 2*, launched in 1972, first established the existence of gamma-ray point sources (such as the Crab and Vela pulsars) at energies of about 100 MeV. The detector used was a digitized spark chamber in which gamma rays are detected by the secondary electron pair they produce upon interacting in the detector. *SAS 2* conducted pointed observations but, unfortunately, operated for only about 6 months.

A similar but much longer-lived (and slightly more sensitive) spark chamber gamma-ray telescope was launched in 1975 by the Europeans as the COS-B mission. It has discovered more than 40 point sources of gamma rays in the plane of the Galaxy. Most of these have not yet been identified with objects detected at x-ray or optical wavelengths. This pioneering phase of gamma-ray astronomy will be greatly expanded with the launch (in about 1986) of the Gamma Ray Observatory (GRO), containing detectors of much greater sensitivity. *See* GAMMA-RAY ASTRONOMY.

[JONATHAN E. GRINDLAY]

Space Telescope. The Space Telescope is a large telescope to be operated in orbit above the disturbing effects of the Earth's atmosphere. It is being developed by the National Aeronautics and Space Agency with European Space Agency participation, and builds on experience gained in previous astronomical satellite programs such as the SAS and OAO series. The Space Telescope program involves several major departures and advances relative to previous satellites. The telescope provides broad wavelength coverage and can be used to collect light with wavelengths from 115 nm in the ultraviolet, through the optical (or visible) portion of the spectrum, and on to wavelengths of 1 mm in the far infrared. With its 2.4-m primary mirror diameter, it will be the largest astronomical telescope ever placed above the atmosphere. The Space Telescope is a long-lived mission. After a shuttle launch in 1983, periodic visits by the shuttle will allow in-orbit replacement of failed or obsolete components, and if necessary, the telescope may be returned to the Earth for major refurbishment followed by relaunch. The telescope will be managed as a guest observer facility, and the observations actually performed will be selected on the basis of scientific merit from observing proposals submitted by astronomers.

Optics. The configuration of the Space Telescope is shown in Fig. 5. Its overall length is about 12 m. The telescope optics are of the Ritchey-Chrétien type with an $f/24$ Cassegrain focus. Light enters the telescope through the aperture door, strikes the 2.4-m-diameter primary mirror, and is reflected to the secondary mirror where it is reflected again, back through a hole, to a focus behind the primary mirror.

Instruments. At the focal plane, the light is shared among at most eight instruments, and only four may be operating simultaneously. There are

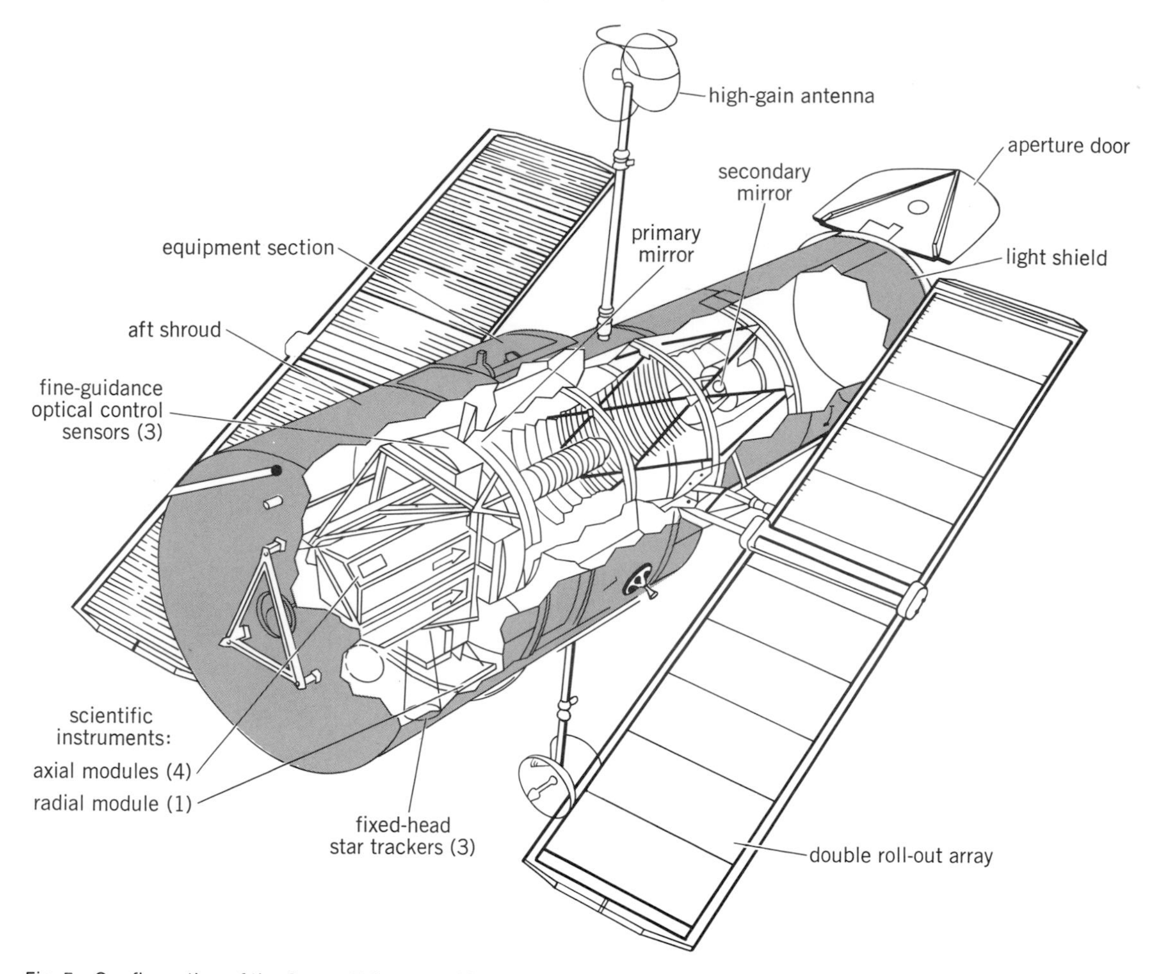

Fig. 5. Configuration of the Space Telescope. (*Courtesy of J. N. Bahcall and C. R. O'Dell*)

four radial instrument modules and four axial modules. All of the modules are replaceable in orbit so that instruments may be changed in the event of component failure or as instrument technology improves. Three of the radial modules are occupied by fine-guidance sensors. During an observation, with any instrument, at least two fine-guidance sensors will lock onto guide stars and use these stars to generate error signals. The error signals are processed by the pointing control system, which generates small corrections to the telescope pointing in order to keep a target object accurately centered in the field of view of one of the other instruments. The point stability achieved by this system will be 0.007 arc-sec, much smaller than has been achieved with any previous spacecraft or telescope.

The remaining radial module and the four axial modules are available for scientific instruments. The initial set of scientific instruments includes two cameras, two spectrographs, and a photometer, thus forming a powerful complement of instruments capable of carrying out most types of observations performed by ground-based observatories. In addition, the third fine-guidance sensor (the one not being used to supply pointing information) can be used to measure accurate relative positions of stars, thus providing the Space Telescope with an astrometric capability. All of the instruments take advantage of the ultraviolet capability of the telescope, but due to detector limitations, the maximum wavelength that can be reached by any instrument is only 1 μm. NASA intends to develop an infrared instrument, capable of exploiting the Space Telescope infrared capability, at a later date when infrared detector technology has improved. *See* INFRARED ASTRONOMY.

Other equipment. The rest of the spacecraft includes computers, communications equipment, star trackers, batteries, solar arrays, and all the other equipment needed to operate a spacecraft by remote control from the ground.

Capabilities. A large telescope above the atmosphere is a very powerful tool for astronomy. Although the Space Telescope is not as large as the largest ground-based telescopes—for example, there are 10 United States telescopes as large or larger than the Space Telescope—the absence of the atmosphere means that for many problems the Space Telescope can substantially outperform the largest United States ground-based telescope, the 5-m Hale Telescope on Palomar Mountain, even though the Hale Telescope has about four times the light-gathering power of the Space Telescope.

Above the atmosphere, the Space Telescope can observe at wavelengths where the atmosphere is opaque or only partially transparent, as in the

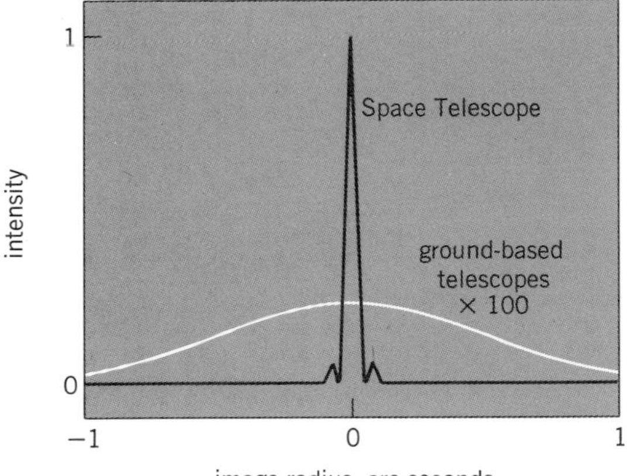

Fig. 6. Comparison of stellar image profiles produced by the Space Telescope (narrow profile) and a ground-based telescope. The ground-based profile is so spread out that it had to be multiplied by 100 times to be seen on this diagram.

ultraviolet and the infrared. Very hot objects and very cool objects radiate most of their energy in the ultraviolet and infrared, respectively. Ground-based studies of these objects, made with visible light, can only see the "tip of the iceberg." The atmosphere itself is a source of background light, especially in the infrared. Absence of this background will make it easier to study very faint objects. Furthermore, the atmospheric absorption and background light fluctuate with time, making it difficult to determine whether observed changes in the brightness of an object are due to the atmosphere or to intrinsic changes in the object.

All of the foregoing provide a strong case for operating a telescope in space. But by far the most compelling reason to place a telescope above the atmosphere is the greatly improved angular resolution that can be achieved. As a telescope is made larger, it is capable of forming smaller images of stars (in inverse proportion to the diameter of the primary mirror). Yet this capability is never used on ground-based telescopes because irregularities in the atmosphere distort the light and smear the images—a phenomenon called astronomical seeing. Under good conditions, images with a diameter of about 1 arc-sec may be formed with ground-based telescopes. At visible wavelengths, the Space Telescope will form images with a diameter of 0.07 arc-sec (Fig. 6). When the Space Telescope observes a faint star, the area of the image occupied by the star will be 200 times smaller and include 400 times less background light (because the sky is twice as dark in space) than if the star is observed by the same-sized ground-based telescope. This means that the Space Telescope can detect stars 20 times fainter than the faintest stars that can be detected by a similar ground-based telescope and 10 times fainter than can be detected with the Hale Telescope.

When the Space Telescope is used to observe objects which are not points, but which are extended and have observable structure such as planets or galaxies, the improved angular resolution will provide a much clearer and more detailed image (Fig. 7).

Several conferences have been held to discuss and plan the observations which may be carried out by the Space Telescope. It is straightforward to assess its capabilities and determine how it will provide answers to current problems in astronomy; it is much more difficult and probably impossible to predict what new discoveries it will lead to. In the history of astronomy, there have been only two advances as dramatic as that which will occur when the Space Telescope begins observing in 1983; when Galileo constructed the first tele-

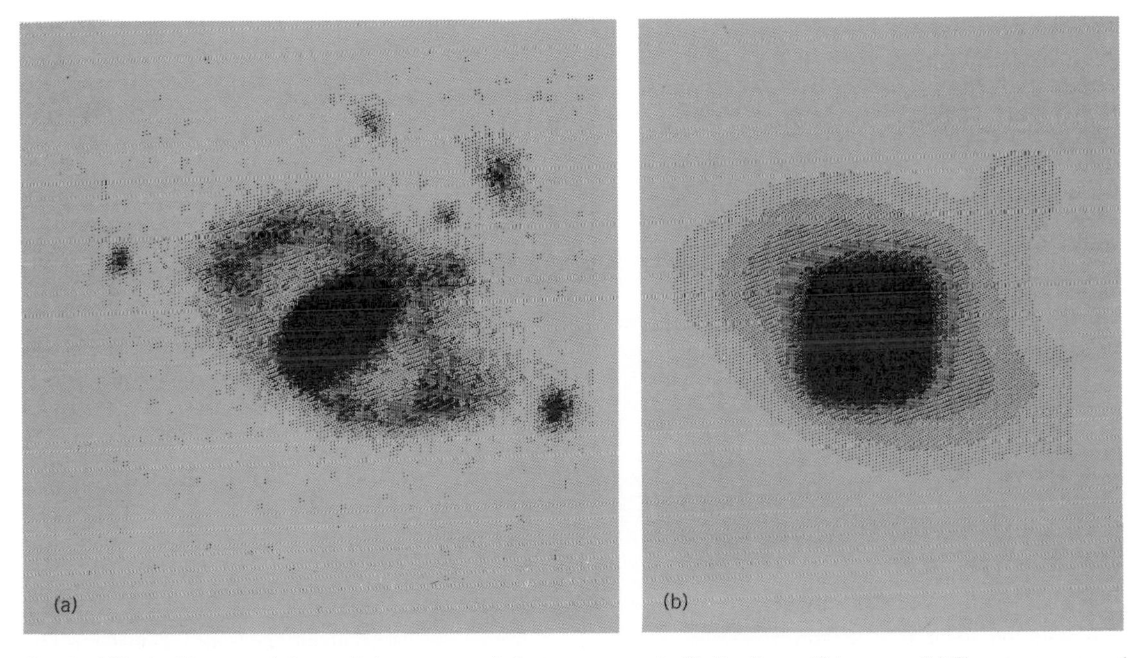

Fig. 7. Effect of improved Space Telescope resolution. (*a*) A galaxy (and some other objects) as it might be observed with the Space Telescope. (*b*) The same group of objects degraded to ground-based resolution.

scope and discovered the satellites of Jupiter in 1610, and when the 2.5-m Hooker Telescope on Mount Wilson began operation in 1917. Thus, one can predict with confidence that many important discoveries will result from the Space Telescope Program. *See* OPTICAL TELESCOPE; TELESCOPE.

[EDWARD J. GROTH]

Bibliography: J. N. Bahcall and C. R. O'Dell, The Space Telescope observatory, *J. Astronaut. Sci.*, 28(2):107–121, 1980; R. Giacconi, The Einstein x-ray observatory, *Sci. Amer.*, 242(2):80–102, February 1980; J. Grindlay, New bursts in astronomy, *Mercury*, 6(5):6–11, 1977; D. S. Leckrone, The Space Telescope scientific instruments, *Publ. Astron. Soc. Pacific*, 92:5–21, 1980; M. S. Longair and J. W. Warner (eds.), *Scientific Research with the Space Telescope*, 1979; F. Macchetto, F. Pacini, and M. Tarenghi (eds.), *Astronomical Uses of the Space Telescope*, 1979; S. Maran and A. Boggess III, Ultraviolet astronomy enters the eighties, *Phys. Today*, 33(9):40–46, September 1980.

Saturn

The second-largest planet in the solar system and the sixth in order of distance to the Sun. The outermost planet known prior to 1781, Saturn is surrounded by a beautiful system of rings. Despite the planet's huge size, its mean density is so low it would float in water. Saturn is also the only planet that has a satellite (Titan) with a dense atmosphere. This distant planetary system has been visited by three NASA spacecraft: a preliminary survey by *Pioneer 11* in September 1979, and a more sophisticated reconnaissance by *Voyager 1* in November 1980 and *Voyager 2* in August 1981.

Orbit and physical elements. The orbit of Saturn has a semimajor axis or mean distance to the Sun of 8.95×10^8 mi (1.43×10^9 km), an eccentricity of 0.056, and its plane is inclined to the plane of the ecliptic at an angle of 2.5°. With a mean orbital velocity of 6 mi/s (9.65 km/s), Saturn makes one revolution about the Sun in 29.46 years. *See* PLANET.

Fig. 1. Saturn viewed from *Voyager 1*. The soft, velvety appearance of the low-contrast banded structure is due to scattering by a haze layer above the planet's cloud deck. (*NASA*)

The equatorial diameter of Saturn is about 75,000 mi (120,800 km), and the polar diameter about 67,700 mi (109,000 km). The volume is 769 (Earth = 1) with a few percent uncertainty. The polar flattening caused by the rapid rotation is the largest of all the planets, and the ellipticity, $(r_e - r_p)/r_e = 0.096$, is almost 30 times the value for Earth (r_e is the equatorial radius and r_p is the polar radius).

The mass, about 95.2 (Earth = 1) or 1/3500 (Sun = 1), is accurately determined from the motions of the planet's brighter satellites. The mean density is 0.68 g/cm^3, the lowest mean density of all the planets. The corresponding value of the mean gravity at the visible surface (superimposed cloud layers) is 1.14 (Earth = 1) or 11.2 m/s^2; however, because of the planet's rapid rotation, the centrifugal acceleration at the equator is 1.76 m/s^2, and so the effective gravity is reduced to about the same value as on Earth.

Photometric properties. The apparent visual magnitude of Saturn at mean opposition is +0.7 when the ring is seen edgewise, and the corresponding value of the reflectivity (geometric albedo) is about 0.44. *See* ALBEDO.

Appearance. Observed through a telescope, Saturn appears as an elliptical disk, darkened near the limb and crossed by a series of barely discernible bands parallel to the equator. Frequently only the bright equatorial zone and the two darker tropical bands on either side of it are visible (Fig. 1). The rings may be seen even with a relatively small telescope, but their visibility changes with the position of the planet in its orbit, because the rotation axis (of both the planet and the rings) is inclined 26°45′ to the perpendicular to the orbital plane. Thus Saturn has seasons similar to those on Earth, but much longer in duration (Fig. 2).

The rotation period of Saturn's interior is 10^h40^m, as determined by radio emissions controlled by the planet's magnetic field. The disk of Saturn is much more homogeneous than that of Jupiter, even when seen at close range with the cameras of the Pioneer and Voyager spacecraft. There is no feature comparable to the Great Red Spot, and the contrast of the features that are visible is very low (Fig. 1). By studying the movement of these features, it has been possible to determine that the circulation patterns on Saturn are also very different from those on Jupiter. At a latitude of ±40°, the atmosphere rotates with the same velocity as the interior, but wind velocities increase smoothly toward the equator, where they reach a value of 500 m/s, about four times faster than Jupiter's equatorial jet. There is no alternation of easterly and westerly currents corresponding to transitions between belts and zones, as there is on Jupiter, except at latitudes above ±40°. This difference between Saturn and Jupiter represents a fundamental difference in global circulation, perhaps related to the relative sizes of the cores of the two planets. *See* JUPITER.

Atmosphere. The optical spectrum of Saturn is characterized by strong absorption bands of methane (CH_4) and by much weaker bands of ammonia (NH_3). Absorption lines of molecular hydrogen (H_2) have also been detected. The estimated quantities of these gases that are present above the clouds are equivalent to STP path-lengths of about 70 km (H_2), 60 m (CH_4), and 2 m (NH_3). Here STP refers to standard temperature (273 K) and pressure (1 atm or 1.01325×10^5 Pa). In these units, the entire atmosphere of the Earth would be equivalent to an 8-km path length. The presence of about 4 km of helium (He) has been deduced indirectly from infrared observations of pressure-broadened hydrogen emission lines.

The temperature the planet should assume in response to solar heating is calculated to be about 76 K, somewhat lower than the measured value of 92 K. This suggests that Saturn has an internal heat source of roughly the same magnitude as that on Jupiter. Relatively intense thermal emission near a wavelength of 12 μm has been identified as resulting from ethane (C_2H_6) formed in the upper atmosphere from the dissociation of methane. As in the case of Jupiter, a thermal inversion exists in this region of Saturn's atmosphere. Emission bands of methane and phosphine have also been detected here. This region is well above the main cloud layer, which is thought to consist primarily of frozen ammonia crystals, with an admixture of some other substances to provide the yellowish color sometimes observed in the equatorial zone. The ammonia cirrus on Saturn is apparently denser and more ubiquitous than on Jupiter, since one does not see through it to lower cloud layers (Fig. 1). This difference probably results from the lower atmospheric temperature and smaller gravity of Saturn, which will act together to spread out and increase the density of the cloud layer.

Internal structure and radiation belts. Observations of Saturn at radio frequencies indicate that the temperature steadily increases with depth into the atmosphere. Theoretical models for the internal structure of Saturn are similar to those for Jupiter, that is, a dense core surrounded by hydrogen compressed to a metallic state which gradually merges into an extremely deep atmosphere. The fact that the two planets radiate comparable amounts of energy despite their difference in size means that smaller Saturn must have some additional energy source besides gravitational contraction. The gradual solution of helium in the liquid hydrogen surrounding the core would suffice, and would also explain the smaller abundance of helium (relative to hydrogen) found in Saturn's atmosphere. The existence of a magnetic field and belts of trapped electrons was initially deduced from observations of nonthermal radiation at dekameter wavelengths and was mapped out in detail by the Pioneer and Voyager spacecraft. Saturn's magnetic field has the same polarity as Jupiter's; a terrestrial compass taken to either planet would point south instead of north. The magnetic moment of Saturn is 4.3×10^{28} gauss cm^3 (4.3×10^{18} teslas $\cdot$ m^3), 500 times Earth's and 34 times smaller than Jupiter's. But the magnetic field at the equator is only 0.2 gauss (2×10^{-5} tesla), two-thirds the value of Earth's field, because

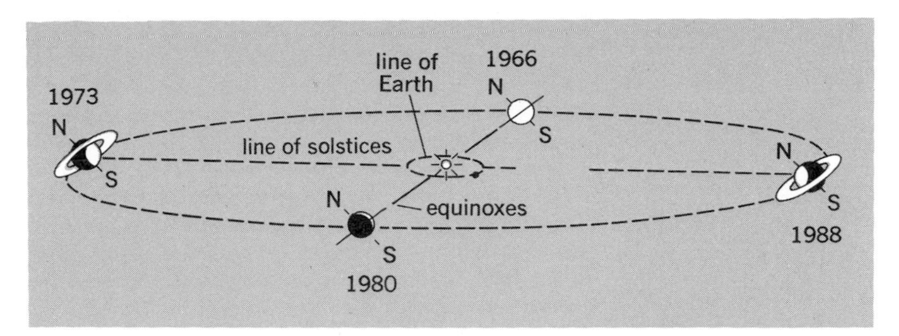

Fig. 2. Presentations of Saturn's ring system. *(Adapted from L. Rudaux and G. de Vaucouleurs, Larousse Encyclopedia of Astronomy, Prometheus Press, 1959)*

of Saturn's greater size. The magnetosphere contains a plasma of charged particles, but no distinct torus like the one associated with Io in the Jupiter system. The charged particle belts are absent in the region of the rings. However, radio signals resembling radiation from lightning discharges in the rings were detected by the Voyager spacecraft.

Jupiter and Saturn are relatively similar bodies. Both seem to have bulk compositions close to that of the Sun and the other stars, and are rich in hydrogen and helium. In that sense, they may represent the primitive material from which the entire solar system was formed, whereas the other planets have undergone fractionation processes resulting in the loss of most of the light gases. *See* PLANETARY PHYSICS.

Ring system. The most remarkable feature associated with Saturn is the complex ring system that surrounds the planet (Fig. 3). The system is divided into six main regions, designated A

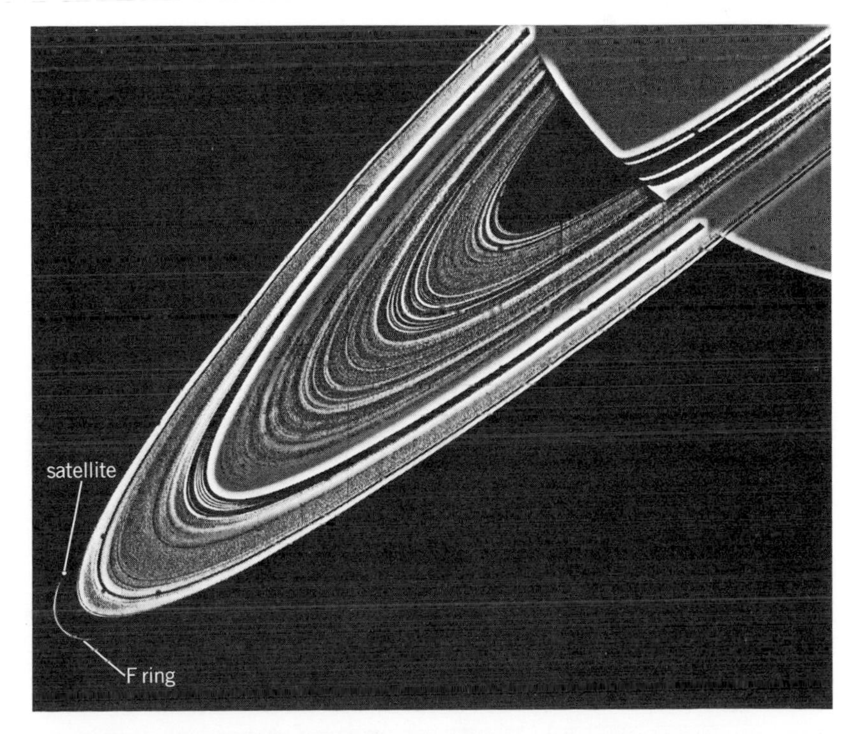

Fig. 3. Saturn's rings, viewed from *Voyager 1*. Approximately 95 individual concentric features are visible. One of the satellites (arrow) discovered by *Voyager 1* is visible just inside the narrow F ring. *(NASA)*

through F. Observations by several instruments on the Voyager spacecraft demonstrated that each of these six regions is subdivided into many individual "ringlets," so that Saturn is actually surrounded by thousands of rings.

Both theory and observations prove that the ring system is made up of myriad separate particles that move independently in flat, mostly circular orbits in Saturn's equatorial plane. The discontinuous, meteoric nature of the rings is demonstrated directly by spectroscopic observations, which show that the inner edge of the ring rotates faster than the outer edge and precisely with the velocity that independent satellites would have at the same distance from the planet.

Appearance from Earth. As viewed with telescopes from Earth, the nearly circular ring system looks like an ellipse whose appearance changes with the relative positions of Earth and Saturn (Fig. 2). The maximum opening of the rings occurs when the system is tilted 27° to the line of sight. The rings are seen as a thin line when Earth crosses Saturn's equatorial plane. The slightly variable point of view, depending on the position of Earth in its orbit, causes a slight annual oscillation of the tilt angle. Thus when the tilt is near 0°, Earth may cross the plane of the ring system either once or three times. During these edge-on configurations, it is possible to search for faint inner satellites and distant rings that ordinarily are hidden by the bright light reflected by the main ring system.

Nomenclature and structure. The bright outer ring, A, has an outside diameter of 170,000 mi (274,000 km) and an inner diameter of 149,000 mi (240,000 km). In the high-resolution pictures obtained by the Voyager cameras, this ring is seen to contain many wavelike patterns, perhaps indicating gravitational disturbances by the satellites. A dark gap in the A ring, often called the Encke division, was found to contain a narrow, elliptical, discontinuous and kinky ring. The broad (2500 mi or 4030 km wide) division separating ring A from ring B is called the Cassini division after its discoverer. It too was found by the Voyager instruments to contain material comprising at least five discrete rings, each of which shows some internal structure.

Ring B, which is the brightest segment of the system, has an outer diameter of 144,000 mi (232,000 km) and an inner diameter of 112,000 mi (180,000 km). It is slightly less bright in its inner regions, where it is also more transparent.

Ring C, sometimes called the crepe ring, is much fainter and even more transparent. It appears as a dusky band in projection against the disk of the planet and only faintly against the sky. Its outer diameter is 111,000 mi (178,000 km), and its inner diameter 89,950 mi (144,000 km). This means the inner edge of ring C is just 7500 mi (11,500 km) above the visible cloud deck on the planet. This may be compared with the mean distance of 240,000 mi (384,000 km) between Earth and its Moon.

A fourth zone, D, of the ring of Saturn between C and the globe was discovered by P. Guerin at Pic-du-Midi Observatory in October 1969. It is fainter than the crepe ring, but its reality has been confirmed by Voyager pictures.

The E ring has been observed well only when the Earth passed through the ring plane in 1966 and especially again in 1980 (Fig. 2). It begins inside the orbit of Enceladus at a distance of about 180,000 km from Saturn, and extends outward past the orbit of Dione, fading from view at about 480,000 km from the planet. The ring is very much brighter just at the orbit of Enceladus, suggesting that this satellite is somehow responsible for the production of the material seen in the ring.

The F ring was discovered by the Pioneer spacecraft and seen in detail by Voyager (Fig. 4). It lies just outside the A ring and actually consists of more than five separate strands, which may not all be in the same plane. The two brightest strands deviate markedly from ellipses, showing the effects of nongravitational forces and perhaps the influence of embedded moonlets. The F ring is held in place by two small satellites, one inside and one outside. A third guards the outer edge of the A ring.

Pictures obtained by the Voyager spacecraft showed that in addition to these nearly circular rings, transient radial features exist in the B ring. These features, called spokes, appear to be clouds of micrometer-size charged particles that are initially controlled by the planet's magnetic field but soon begin moving in keplerian orbits like the larger particles.

The G ring was established by the Voyager spacecraft to lie at a distance of about 19,000 mi (30,000 km) outside the A ring, placing it well inside the orbit of Mimas. It is a diffuse, faint ring like the E ring, but no satellites have been discovered in association with it.

Origin and nature. E. Roche proved in 1849 that a liquid satellite of a planet with the same density

Fig. 4. Saturn's F ring viewed from *Voyager 1*. Two narrow, braided bright rings that trace distinct orbits, and a broadened, very diffuse component about 35 km in width, are visible. Also visible are knots, which probably are local clumps of ring material, but which may be minisatellites. (*NASA*)

cannot form if it is closer to its planet than 2.44 times the planetary radius. Within this distance, disruptive tidal forces (the gradient of the planet's gravitational field) will be greater than the self-gravity of the satellite. But a rocky or icy satellite is held together by stronger forces, and will, therefore, not be disrupted by the planet's field outside a distance of 1.35 radii. In fact, the F ring and its guardian satellites are at 2.33 radii, just inside Roche's limit.

In 1859 J. C. Maxwell was able to show that a ring system of small mass, formed of a large number of particles, is quite stable against external perturbations such as those caused by the larger satellites. The aggregate mass of the ring system is quite small, probably less than one-half the mass of the Earth's Moon. Periodic perturbations by the major satellites are responsible, in part, for the main divisions of Saturn's rings in the same way that perturbations by Jupiter cause the Kirkwood gaps in the asteroid belt. The cause of the many small divisions discovered by Voyager (Fig. 3) is still obscure. Following a theory of P. Goldreich and S. Tremaine, it has been proposed that tiny moonlets may be present in these gaps, stabilizing individual rings even as the F ring is stabilized by its two satellites. But a careful search has failed to reveal any small satellites (with diameters of 5–10 km) in any of the gaps. *See* ASTEROID; PERTURBATION.

G. P. Kuiper demonstrated in 1952 that the reflection spectrum of the ring was identical with that of water ice, a result that has been confirmed by subsequent measurements. In 1974 the ring system was found by R. M. Goldstein to be a surprisingly good reflector of radar waves, suggesting that some relatively large (diameter approximately 1 km) particles may be present. The scattering of radio waves from the Voyager spacecraft by the C ring indicated an effective particle size of 1 m. The mean thickness of the ring system is not well determined, but Voyager observations of the occultation of a star by the rings demonstrated that the outer edge of the A ring is only 500 ft (150 m) thick.

Satellites. Saturn has 21 known satellites, more than any other planet (see table). The largest and brightest, Titan, was discovered by C. Huygens in 1655 and is visible with small telescopes; the other satellites are much fainter. The outermost satellite was discovered photographically by W. H. Pickering in 1898; two inner satellites (S-10 and S-11) were discovered in 1966, when Saturn's rings were seen nearly edgewise, but it was not until the next such configuration, in 1980, that the true nature of these objects was determined: they are very nearly in the same orbit. Three additional satellites were discovered from the ground in 1980 and 1981; one of them is in the leading lagrangian point of Dione's orbit, and two are in the lagrangian points of Tethys's orbit. The three satellites associated with the rings that were discovered by Voyager have already been mentioned. The four newest satellites discovered on Voyager pictures consist of two in the langrangian points of Mimas's orbit, an additional member of the Dione family, and a satellite in orbit between Tethys and Dione. All of these objects are small (less than 60 mi or 100 km long) and irregular in shape. *See* CELESTIAL MECHANICS.

The small outermost satellite, Phoebe, moves in a retrograde direction (opposite to that of the inner 20 satellites and to the direction of the planets about the Sun) in an orbit of relatively high eccentricity (0.16). These two characteristics place Phoebe in the class of irregular satellites that have probably been captured by the planet they orbit, instead of being formed with it from the original solar nebula. Phoebe's rotation period is only 9 h, completely out of synchrony with its 550-day period of revolution. *See* RETROGRADE MOTION.

Titan shows a measurable disk in large telescopes; the mean apparent diameter corresponds to a linear diameter of approximately 3600 mi (5800 km). But this diameter refers to the satellite's atmosphere, which is filled with a dense aerosol produced photochemically by incident sunlight. The solid surface of Titan has a diameter of 3128 mi (5120 km), making this satellite larger than Mercury but smaller than Jupiter's giant Ganymede. This large satellite has a mass about two times the mass of the Moon, with a corresponding mean density of 1.9 g/cm³. The low density (Moon = 3.3) means that this object contains a large fraction of icy material and is thus quite different from the Moon or the inner planets in composi-

Satellites of Saturn*

Satellite	Mean distance from Saturn, 10^3 km	Period of revolution, days	Mean density, g/cm³	Diameter, km	Visual magnitude at mean opposition
I Mimas	186	0.942	1.2 ± 0.1	390 ± 10	12.1
II Enceladus	238	1.370	1.1 ± 0.6	500 ± 20	11.8
III Tethys	295	1.888	1.0 ± 0.1	1050 ± 20	10.3
IV Dione	377	2.737	1.4 ± 0.1	1120 ± 20	10.4
V Rhea	527	4.518	1.3 ± 0.1	1530 ± 20	9.7
VI Titan	1222	15.945	1.9 ± 0.1	5120 ± 60	8.4
VII Hyperion	1481	21.277	?	290 ± 40	14.2
VIII Iapetus	3560	79.331	1.2 ± 0.5	1440 ± 40	10.2–11.9
IX Phoebe	12,930	550.45	?	(100)	16.5
S-10	151.4	0.750	?	(200)	(14)
S-11	151.4	0.750	?	(150)	(15)

*Values in parentheses are still uncertain. Names and roman numerals have not yet been assigned to S-10 and S-11. No data are given here for S-12–17, as even the designations are not yet secure.

tion. Furthermore, it is large and cold enough to retain an atmosphere of gases with relatively high molecular weights. In fact, the existence of a methane atmosphere about Titan was established through spectroscopic observations by Kuiper in 1944. Subsequent studies have indicated that a thermal inversion exists in the satellite's upper atmosphere, where temperatures as high as 175 K have been measured, produced by absorption of solar ultraviolet radiation in the aerosol. The main constituent of this atmosphere is molecular nitrogen (N_2), which produces a surface pressure of 1.5 bars (1.5×10^5 pascals), or 1.5 times the sea-level pressure on Earth. The surface of Titan is so cold (95 K) that methane can liquefy. Hence, this substance may play the same role on Titan as water does on Earth. The presence in the atmosphere of a variety of organic compounds such as cyanoacetylene (HC_3N), hydrogen cyanide (HCN), and propane (C_3H_8) in addition to the aerosol should allow future investigators to use this satellite as a natural laboratory for testing ideas about chemical evolution on the primitive Earth.

The other satellites encompass a variety of characteristics. All have densities near unity, indicating a predominantly icy composition. Yet the small variations in density that do occur appear random, rather than showing a radial trend as is the case for the Jupiter system. It has frequently been suggested that these small, icy satellites are similar in composition to the nuclei of comets. The slight differences in density would reflect differences in the amount of rocky material embedded in the ices. The surfaces of these objects are covered with impact craters, with the exception of Enceladus and Iapetus. Large regions of the surface of Euceladus are free of craters, indicating reworking of the surface in recent times. Coupled with the unusually high reflectivity of this satellite (close to 100%), the modified surface suggests internal activity leading to partial melting and the production and expulsion of the tiny ice grains that populate the E ring. Iapetus is unique in the solar system in that its trailing hemisphere is six times brighter than the leading one. The cause for this anomaly is not known. All of these satellites except Phoebe appear to keep the same hemisphere facing Saturn as they revolve around it, meaning that their rotational periods are equal to the periods of revolution as a result of tidal frictions, just as is the case for the Moon. *See* COMET; MOON.

[TOBIAS C. OWEN]

Bibliography: D. Goldsmith, *The Evolving Universe*, 1981; W. K. Hartmann, *Moons and Planets*, 1982; Special issue on *Voyager 1* encounter with the Saturnian system, *Science*, 212(4491):159–243, 1981; special issue on *Voyage 2* encounter, *Science*, 215(4532):499–594, 1981.

Schmidt camera

An optical system consisting of a spherical mirror with a corrector plate near its focus. Cameras of the Schmidt type are of great value in direct astronomical photography and also as parts of spectroscopes.

The corrector plate is a figured plate with one plane surface, its second surface being aspherical, slightly deviating from a plane, such that the aperture aberration of the mirror is balanced, as shown in the illustration. The curvature of the plate at the axis is zero.

The Schmidt camera is a catadioptric system which is free from aperture aberration, asymmetry, and chromatic aberrations. It can therefore be used at the very large aperture of *f*/2.0 or more. The image, although sharp, has curvature of field, but the field can be flattened by adding a second mirror to produce what could be called a Schmidt-Cassegrain system.

A Newtonian type of mirror with an additional concentric lens in place of the irregular corrector plate was suggested at about the same time (1941) in four different countries. In 1944 a proposal was made to replace the concentric lens by a meniscus lens free from color aberrations.

The Cassegrain type of mirror has also been used in combination with an additional, more or less afocal system, to increase the aperture and correct the monochromatic errors of the mirror, without introducing large color aberrations. Additional systems, which are quite complex, have been used.

The Cassegrain type of Schmidt camera has the disadvantage of considerable vignetting, and thus of a very small field. It requires careful baffling besides to prevent undesirable light from going through the instrument. Frequently, the Cassegrain mirrors are oriented at 90° with respect to each other by means of a prism or a mirror.

The success of the Schmidt camera, which at first was used mostly for astronomical purposes, has led to numerous designs of catadioptric systems for other purposes.

The application of such systems is very successful in microscope optics, where large aperture and freedom from color aberrations combined with a small field are needed. Most microscopic work in the ultraviolet is done with catadioptric systems. Catadioptric systems have also been designed by A. Bouwers and others for use as camera lenses, especially when a great focal length is required.

The disadvantage of catadioptric systems is that the mirrors and lenses are in one another's way, thus vignetting the center of the aperture. The diffraction effect connected with this cutting out of the central part of the beam may jeopardize the recognition of very fine detail. *See* ASTRONOMICAL PHOTOGRAPHY.

[MAX HERZBERGER]

Bibliography: M. Born and E. Wolf, *Principles of Optics*, 6th ed., 1980; D. F. Horne, *Optical Instruments and Their Applications*, 1980; F. A. Jenkins and H. E. White, *Fundamentals of Optics*, 4th ed., 1976.

Scorpius

The Scorpion, in astronomy, one of the most beautiful and vivid constellations in the sky. Scorpius is the eighth sign of the zodiac. The constellation resembles a scorpion even to the sting (see illustration). The bright red star Antares is situated at the heart. Its name (Ant-Ares) means the Rival of Mars, since both the plant and the star are bright and red in color, and the two are often found near each other. Antares is one of the largest stars

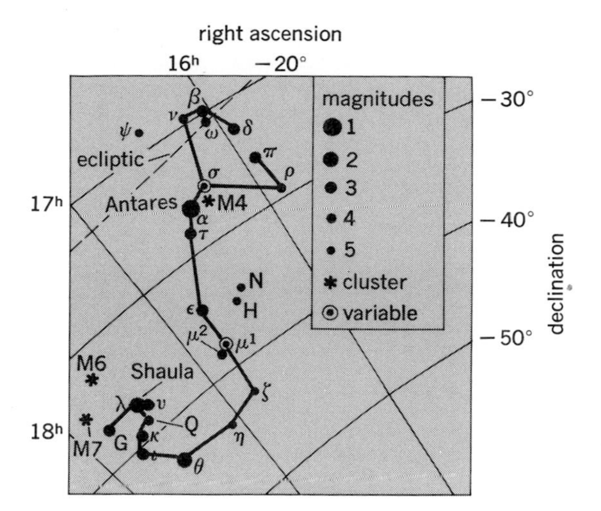

Line pattern of the constellation Scorpius. The grid lines represent the coordinates of the sky. The apparent brightness, or magnitudes, of the stars is shown by the sizes of the dots, which are graded by appropriate numbers as indicated.

known, having a diameter over 450 times that of the Sun. As in Sagittarius, the Milky Way in Scorpius is bright and rich in star clouds and clusters. *See* CONSTELLATION.

[CHING-SUNG YU]

Sirius

Alpha Canis Majoris, the brightest star visible from Earth. From the Northern Hemisphere, Sirius appears in the south in midwinter. Its apparent magnitude, −1.5, makes it a rival of the planets in brightness. Its distance, 2.7 parsecs (8.3×10^{6} m), makes it the sixth nearest star to the Sun. Absolute magnitude +1.4 and spectral type Al are representative of early type A stars, with an effective temperature near 10,000 K, strong hydrogen lines, and large Balmer continuous absorption. Special interest attaches to the system of Sirius because of the close companion, of the ninth magnitude (more than 10^4 times fainter)—the white dwarf α CMaB. The period of the binary system is 50 years; the mass of α CMaA is 1.05 M⊙, that of α CMaB, 2.14 M⊙. Stars of solar mass usually have spectral type G and yellow color, but α CMaB is bluish and shows hydrogen lines; that is, it is hotter, and much smaller, than the Sun.

Although the faint companion is buried under the brilliant light of its primary, indirect measurements of its temperature and radius have been obtained; it is about 27,000 K at the surface and has a radius of almost 5000 km, as contrasted with that of the primary, which is 10^6 km. The spectrum of the companion shows pronounced pressure-broadened lines of hydrogen.

A special peculiarity has been found in the composition of the bright star α CMaA. Certain chemical elements are present in excessive amounts, compared to their presence in the Sun or other stars like Vega. In addition, still other elements are deficient. This makes Sirius a member of the so-called metallic-line star class. It is of some interest that it is a binary, since this peculiarity is very common among close binaries. The white dwarf was once the more massive of the two. *See* BINARY STAR; STAR; WHITE DWARF STAR.

[JESSE L. GREENSTEIN]

Solar constant

The rate at which energy is received from the Sun just outside Earth's atmosphere. At Earth's mean distance from the Sun, the solar constant is 1.36×10^{6} ergs/(cm^2)(sec). Depending on the Sun's distance from the zenith, up to a third of this energy may be scattered in Earth's atmosphere. From the measured solar constant, the total radiation of the Sun is 3.86×10^{33} ergs/sec.

The accuracy with which the solar constant is known at present is no better than about 2% because of observational difficulties due to variable atmospheric absorption. Solar variations within this limit have been suspected, but are questionable. Observations from above the atmosphere by satellite-borne instruments will eventually both improve the accuracy and settle the question of variation within the improved accuracy. *See* SUN.

[JOHN W. EVANS]

Solar corona

The outermost extension of the atmosphere of the Sun which is normally visible only at a total solar eclipse. The corona consists of gas at a density of about 10^{-15}g/cm^3 (or 10^{-12} atmospheric density) and a normal kinetic temperature between 1 and 2×10^{6} K. It is easily observable at an eclipse to a height of 2–3 solar diameters. The outer envelope often consists of beautiful radial streamers composed of electrons and protons ejected by the Sun. *See* SUN.

[JOHN W. EVANS]

Solar magnetic field

The magnetic field that pervades the ionized and highly conducting gas composing the Sun. The field is observable by optical means at the Sun's surface; subsurface fields are inferred from the laws of magnetohydrodynamics and from the behavior of sunspots; the effects of magnetic fields external to the Sun are seen in solar prominences, flares, and coronal streamers, and in relation to the solar wind, an outflow of tenuous plasma into interplanetary space. *See* SOLAR WIND; SUN.

Direct measurement of the strength and polarity of the complex patterns of magnetic fields at the Sun's surface is accomplished by means of an instrument known as the solar magnetograph that utilizes the Zeeman effect. It comprises a telescope to form an image of the Sun plus a differential analyzer for circular polarization, a spectrograph in which magnetically sensitive spectrum lines are selected for measurement of Zeeman splitting, and photoelectric or photographic means of differencing and recording. The photoelectric instrument is capable of measuring local fields as weak as a fraction of 1 gauss.

The small-scale features of the Sun's magnetic pattern are continuously changing with a degree of randomness resulting from turbulence and convection in the outer layers of the Sun, but in their regularity of position and other characteristics they reflect systematic dependence on a large-scale hydromagnetic cycle operating within the differentially rotating Sun. This magnetic cycle, which is most evident in the varying frequency of occurrence of sunspots and in their progressively decreasing heliographic latitude (Spörer's law), as well as in their characteristic magnetic polarity (Hale's law), has a duration of about 22 years.

In sunspots, the magnetic-field lines are concentrated so that the field intensity reaches 1000–3000 gauss. Such fields are strong enough to inhibit turbulence, diminish the outflow of energy to the surface, and lower the temperature and darken the appearance of the spot. *See* SUNSPOT.

Sunspot groups are thought to arise through internal amplification of submerged toroidal bands of magnetic flux, of opposite polarity on the two sides of the Sun's equator. This amplication results from the differential rotation of the Sun. Loops of these flux bands, rising to the surface, produce bipolar sunspot groups. Such groups invariably disappear by expansion and consequent weakening of the magnetic field, with ultimate expulsion of flux loops into the solar atmosphere. Poleward migration of the trailing parts of bipolar regions produces accumulations of flux, of opposite polarity in the two heliographic polar caps; these persist until they are neutralized and reversed in the subsequent half of the 22-year cycle.

[HORACE W. BABCOCK]

Bibliography: H. W. Babcock, The Sun's magnetic field, *Annu. Rev. Astron. Astrophys.*, 1:41, 1963; H. W. Babcock, The topology of the Sun's magnetic field and the 22-year cycle, *Astrophys. J.*, 133:572–587, 1961; H. W. Babcock, The Zeeman effect in astrophysics, *Physica*, 33:102–121, 1967; R. Howard, Magnetic field of the Sun (observational), *Annu. Rev. Astron. Astrophys.*, 5:1, 1967; E. N. Parker, The Sun, *Sci. Amer.*, 233(3):42–50, 1975.

Solar neutrinos

Neutrinos produced in nuclear reactions inside the Sun. The first direct test of how the Sun produces its luminosity (observed most conspicuously on Earth as sunlight) has been carried out by observing these particles. The results of this experiment are in disagreement with theoretical predictions based upon the supposedly well-established theory of stellar evolution. This discrepancy between theory and observation suggests that either the process by which the Sun shines is not understood as well as previously believed, or some modification in the classical physical theory of neutrinos is required.

Nuclear fusion in the Sun. The Sun shines because of fusion reactions similar to those envisioned for terrestrial fusion reactors. The basic solar process is the fusion of four protons to form an alpha particle, two positrons (e^+), and two neutrinos (ν); that is, $4p \rightarrow \alpha + 2e^+ + 2\nu_e$. The principal reactions are shown in the table, with a column indicating in what percentage of the solar terminations of the proton-proton chain each reaction occurs. The rate for the initiating proton-proton (PP) reaction, number 1 in the table, is largely determined by the total luminosity of the Sun. Unfortunately, these neutrinos are below the threshold, which is 0.81 MeV, for the first experiment to detect solar neutrinos (with ^{37}Cl). Several of the proposed experiments, especially a ^{71}Ga experiment, are designed to be primarily sensitive to neutrinos from the PP reaction.

The proton-electron-proton PEP reaction (number 2), which is the same as the familiar PP reaction except for having the electron in the initial state, is detectable in the ^{37}Cl experiment. The ratio of PEP to PP neutrinos is approximately independent of which solar model is used for the solar predictions. Two other reactions in the table are of special interest. The capture of electrons by ^{7}Be (number 6) produces detectable neutrinos in the ^{37}Cl experiment. The ^{8}B decay (number 9) was expected to be the main source of neutrinos for the ^{37}Cl experiment because of their relatively high energy (14 MeV), although it is a rare reaction in the Sun. There are also some less important neutrino-producing reactions from the carbon-nitrogen-oxygen (CNO) cycle, which will not be discussed here in detail since this cycle is believed to play a rather small role in the energy-production budget of the Sun. *See* CARBON-NITROGEN-OXYGEN CYCLES; PROTON-PROTON CHAIN.

Experimental test. The first solar neutrino detector was based on the reaction $\nu_{solar} + {}^{37}\text{Cl} \rightarrow$

Proton-proton chain in the Sun

Number	Reaction	Solar terminations	Maximum neutrino energy, MeV
1	$p + p \rightarrow {}^2\text{H} + e^+ + \nu$	99.75	0.420
	or		
2	$p + e^- + p \rightarrow {}^2\text{H} + \nu$	0.25	1.44 (monoenergetic)
3	${}^2\text{H} + p \rightarrow {}^3\text{He} + \nu$	100	
4	${}^3\text{He} + {}^3\text{He} \rightarrow {}^4\text{He} + 2p$	86	
	or		
5	${}^3\text{He} + {}^4\text{He} \rightarrow {}^7\text{Be} + \nu$	14	
6	${}^7\text{Be} + e^- \rightarrow {}^7\text{Li} + \nu$		0.861 (90%), 0.383 (10%) (both monoenergetic)
7	${}^7\text{Li} + p \rightarrow 2\,{}^4\text{He}$		
	or		
8	${}^7\text{Be} + p \rightarrow {}^8\text{B} + \nu$	0.02	
9	${}^8\text{B} \rightarrow {}^8\text{Be}^* + e^+ + \nu$		14.06
10	${}^8\text{Be}^* \rightarrow 2\,{}^4\text{He}$		

$^{37}Ar + e^-$, which is the inverse of the electron-capture decay of ^{37}Ar. This reaction was chosen for the first experiment because of its unique combination of physical and chemical characteristics, which were favorable for building a large-scale solar neutrino detector. Neutrino capture to form ^{37}Ar in the ground state also has a relatively low energy threshold (0.81 MeV) and a favorable cross section, nuclear properties that are important for observing neutrinos from ^{7}Be, ^{13}N, and ^{15}O decay and the PEP reaction.

The ^{37}Cl detector was built deep underground to avoid the production of ^{37}Ar in the detector by cosmic rays. The final detector system consists of an approximately 400,000-liter tank of perchloroethylene, a pair of pumps to circulate helium through the liquid, and a small building to house the extraction equipment, all in a deep gold mine (Homestake) about 1500 m below the surface.

A set of about 40 experimental runs carried out in the ^{37}Cl experiment during the 1970s showed that the ^{37}Ar production rate in the tank is 0.50 ± 0.06 ^{37}Ar atoms per day. Even though the tank is nearly a mile (about 1.6 km) underground, a small amount of ^{37}Ar is produced by cosmic rays. This background rate has been estimated to be of order 0.1 ^{37}Ar atoms per day, but could be larger.

If the above-mentioned background rate is assumed, then a positive signal of (2.2 ± 0.4) solar neutrino units (SNU) is inferred; 1 SNU $= 10^{-36}$ captures per target particle per second.

The predicted capture rates for the ^{37}Cl experiment (in SNU) for one solar model are: PP reaction: 0; ^{8}B beta decay: 6; PEP reaction: 0.2; ^{7}Be electron capture: 1; ^{13}N decay: 0.06; and ^{15}O decay: 0.2. The total theoretical prediction is 7.5 SNU. Further investigations have been undertaken of the best value to use for various nuclear parameters, and the total predicted rate may well differ from 7.5 SNU by 1 or 2 SNU.

Observational implications. The ^{37}Cl experiment tests theoretical ideas at different levels of meaning, depending on the counting rate being discussed. The value of 28 SNU that would be expected if the Sun shone because of the fusion reaction based on the carbon-nitrogen-oxygen cycle is ruled out. More surprisingly, the best models based on standard theory, which imply approximately 6 to 8 SNU, are also inconsistent with the observations. This disagreement between standard theory and observation has led to many speculative suggestions of what might be wrong. One such suggestion, that in the solar interior the heavy-element abundance is at least a factor of 10 less than the observed abundance, leads to an expected counting rate of 1.5 SNU, which is about as low a prediction as can be obtained from solar models without seriously changing current ideas about the physics of the solar interior.

Present and future versions of the ^{37}Cl experiment are not likely to reach a sensitivity as low as 0.3 SNU, the minimum counting rate (from reaction number 2 of the table) that can be expected if the basic idea of nuclear fusion as the energy source for main-sequence stars is correct. Certain developments of grand unified theories (unifying weak, electromagnetic, and strong interactions) suggest that it is not implausible that the neutrinos originating in the Sun may arrive at the Earth in a form that is undetectable in the ^{37}Cl experiment.

Further experiments. Another experiment is required to settle the issue of whether present concepts in astronomy or physics are at fault. Fortunately, a testable distinction can be made. The flux of low-energy neutrinos from the PP and PEP reactions (numbers 1 and 2 in the table) is almost entirely independent of astronomical uncertainties and can be calculated from the observed solar luminosity, provided only that the basic physical ideas of nuclear fusion as the energy source for the Sun and of stable neutrinos are correct. If these low-energy solar neutrinos are detected in a future experiment, then it will be known that the present crisis is caused by a lack of astronomical understanding. If the low-energy neutrinos are absent, then it will be established that the present discrepancy between theory and observation is due, at least in part, to faulty physics, and not just to poorly understood astrophysics.

The preferred experiments, based on feasibility and new information provided, would use detectors of ^{2}H, ^{7}Li, ^{71}Ga, ^{115}In, or electron-neutrino scattering. The two leading detectors considered for the first additional experiment are ^{71}Ga and ^{115}In, since both of these detectors are primarily sensitive to the neutrinos from the basic PP reaction (number 1 in the table). A modular demonstration experiment using 1.4 tons (1.3 metric tons) of gallium has been undertaken. *See* STELLAR EVOLUTION.

[JOHN N. BAHCALL]

Bibliography: J. N. Bahcall, Solar neutrinos: Theory versus observation, *Space Sci. Rev.* 24: 227–251, 1979; J. N. Bahcall and R. Davis, Jr., Solar neutrinos: A scientific puzzle, *Science*, 191: 264–267, 1976; J. N. Bahcall et al., Proposed solar-neutrino experiment using ^{71}Ga, *Phys. Rev. Lett.*, 40:1351–1354, 1978; R. Davis, Jr., Results of the ^{37}Cl experiment, *Proceedings of the Brookhaven Solar Neutrino Conference*, BNL 50879, 1:1–54, 1978.

Solar radiation

The electromagnetic radiation and particles (electrons, protons, and rarer heavy atomic nuclei) emitted by the Sun. Electromagnetic energy has been observed over the whole spectrum with wavelengths varying from 0.01 nm to 30 km. The bulk of the energy is in the spectrum of visible light (400–800 nm). The solar spectrum doubtless extends far beyond the observed limits in both directions. The total power is 3.86×10^{33} ergs/sec.

The Sun also emits a continuous stream of electrons with shorter bursts of electron and proton showers sufficiently intense to affect the ionization of the upper terrestrial atmosphere. These sporadic particles have energies from a few thousand to a few billion electron volts. The lower-energy particles are much more abundant, but those of high energy are sufficient to occasionally damage the solid-state circuitry of spacecraft. The physical mechanism of high-energy particle emission is not understood, but is closely associated with the more energetic forms of solar activity. *See* SOLAR WIND; SUN.

[JOHN W. EVANS]

Solar system

The Sun and the bodies moving about it. There are more than 10^8 stars in the Galaxy, but only in the case of the solar system is a star with an accompanying planetary system known. The other stars are all so far away that there has been no means of detecting any possible accompanying planets. The Sun is about 3×10^7 light-years (3×10^{20} km) from the center of the Galaxy, part of the flattened distribution of stars forming the disk of the Galaxy. The Galaxy also has a substantial spheroidal content of stars which are believed to have been the first to form when the Galaxy was formed. The disk stars, such as the Sun, are believed to have formed later, and many examples of the process of star formation are observed continuing within the galactic disk today. By observing how such star formation occurs, it may be possible to gain clues as to the way in which the Sun, and possibly its accompanying planetary system, was formed. *See* GALAXY; SUN.

COMPOSITION OF THE SOLAR SYSTEM

The planetary system contains an inner group of four planets, Mercury, Venus, Earth, and Mars, which are primarily composed of rocks and iron. A large number of very small bodies, typically only a few kilometers in radius or less, called asteroids, occur in orbits beyond that of Mars. These too are primarily rocky in composition. *See* ASTEROID; EARTH; MARS; MERCURY; VENUS.

Beyond the asteroids lies the largest planet in the solar system, Jupiter. Jupiter is primarily composed of the light gases hydrogen and helium. It is believed to be somewhat more enriched in the heavier elements than is the solar mixture relative to hydrogen and helium, but the total content of these heavier elements probably amounts to only a few percent of the mass of the planet. Saturn, the second most massive planet in the solar system, lying beyond Jupiter, is basically similar in composition and structure. *See* JUPITER; SATURN.

Beyond Saturn lie two intermediate-mass planets, Uranus and Neptune. While these planets contain substantial amounts of hydrogen and helium, the majority of their mass appears to be composed of heavier elements, probably some combination of rocky materials plus "ices," water, ammonia, and methane. *See* NEPTUNE; URANUS.

Still farther out is a very small planet, Pluto, considerably less massive than the Earth and apparently composed primarily of rocks and frozen ices. *See* PLANET; PLUTO.

Most of the planets in the solar system have accompanying satellites. The most massive satellite in relation to its primary planet is the Earth's Moon. Mars has two very tiny satellites, Phobos and Deimos. Jupiter has four major satellites, each comparable in mass to the Moon, Io, Europa, Ganymede, and Callisto. It has a substantial number of additional small bodies rotating about it. Saturn has a similar large swarm of satellites, although only Titan has a mass comparable to that of the Moon. Titan is unique in that it is the only satellite within the solar system having a substantial atmosphere. Uranus has many small satellites, Neptune has two, and Pluto has a satellite which is of substantial mass in relation to its primary planet. *See* MOON; SATELLITE.

Well beyond the region of the planets lies a huge region of space, out to about 100,000 astronomical units (1.5×10^3 km) from the Sun, which appears to contain a very large number of small, frozen bodies called comets. This reservoir of cometary bodies is usually called the Oort cloud. From time to time passing stars cause gravitational deflections of the orbits of the comets in the Oort cloud, so that some of these comets penetrate far into the solar system, where they become visible for the first time. Gravitational perturbations by the planets greatly decrease the periods and orbits of a few of these comets, causing them to stay within the inner solar system. These are the short-period comets. The space within the solar system is also filled with much smaller bodies. Some of these weigh many kilograms. These are the meteoroids, which become bright meteors or fireballs after entering the Earth's atmosphere, and are called meteorites when they are found on the ground. Apart from the samples returned from the Moon, the meteorites are the only samples of extraterrestrial material which are available on the Earth. They have been subjected to extremely delicate and sensitive tests and measurements in laboratories, giving a great deal of information about conditions which were present early in the history of the solar system. Unfortunately, it is not known from which bodies within the solar system the meteorites have come; this has hampered evaluation of the relevance of the meteorite clues to the early history of the solar system. *See* COMET; METEOR; METEORITE.

Many of the particles are very tiny, much less than 1 mm in size. Such particles are particularly prevalent along the orbits of comets, and it is generally believed that such dust particles are cometary debris. *See* INTERPLANETARY MATTER.

Also pervading the solar system is a very tenuous ionized gas, or plasma. This gas is expanding away from the Sun at a velocity of several hundred kilometers per second. It represents the expansion of the outermost fringes of the solar atmosphere, and is called the solar wind. *See* SOLAR WIND.

ORIGIN OF THE SOLAR SYSTEM

There is a long history of speculation about the origin of the solar system, extending over more than 300 years. These theories can be classed as monistic, in which usually the Sun and planets are thought to have formed as an isolated system, and dualistic, in which the planets are generally thought to have been formed as a result of an interaction between two stars. It is now generally believed that some form of monistic theory is required to understand the origin of the solar system. Some of the results of research activity bearing on these problems and some of the conceptual issues that stimulate much debate will be discussed.

Solar system chronologies. It is generally believed that the heavier elements are formed in stars, particularly in supernova explosions, in which a star blows up and distributes into space the products of nuclear reactions which have been going on in its interior and creating heavier elements. Many of these products are unstable, or radioactive, and decay with half-lives ranging from seconds to billions of years. The longer-lived radioactivities may still be present as such in material which formed the solar system, and measurements of the time during which the radioactivities were still present provide a tool to determine the times for various things to happen during the history of the solar system. *See* ELEMENTS AND NUCLIDES, ORIGIN OF; SUPERNOVA.

The decay of the longer-lived activities, uranium, thorium, and potassium, provides a fairly accurate date of 4.5×10^9 years for the age of the Earth and the meteorites, and hence presumably for the solar system itself. These elements are also responsible for almost all of the internal heat generation within the Earth. *See* EARTH.

A considerable amount of information more relevant to the early history of the solar system is provided by two extinct radioactivities: a plutonium isotope with a mass number of 244 and a half-life of 8.3×10^7 years, and an iodine isotope with a mass number of 129 and a half-life of 1.7×10^8 years. Evidence indicates that both of these isotopes were present when the solar system was formed, and from the amount of their decay products which are present in various materials, it is possible to date the formation of those materials after the formation of the solar system to an accuracy of a few millions or tens of millions of years.

Still more startling has been the discovery that two other isotopes of still shorter half-life were present in the early solar system. These are a palladium isotope with a mass number of 107 and a half-life of 6.5×10^6 years, and an aluminum isotope with a mass number of 26 and a half-life of 7.3×10^5 years. The latter isotope in particular indicates that not more than a few million years can have passed between the time that the radioactive aluminum was produced in a supernova explosion and the time that certain meteorite materials were formed.

Star formation. A substantial portion of the mass in the disk of the galaxy is contained not in stars, but in the form of gas and dust spread out among the stars. This interstellar medium has an intricate structure of its own. Some of it is hot, tenuous, and substantially ionized. Some of it is clumped together in cooler, denser clouds. Many of these denser clouds are clumped together into cloud complexes of substantial mass. *See* INTERSTELLAR MATTER.

It is precisely in these cool, dense cloudy regions of the Galaxy that star formation appears to take place. Many of the cloud complexes lie along the spiral arms of the Galaxy. The galactic matter in the spiral arms exerts an additional gravitational pull upon matter entering the arms, causing it to be compressed with a significant increase in pressure.

These dense interstellar clouds typically have masses which are hundreds or thousands of times as great as the Sun. Ordinarily their internal pressures are sufficient to prevent them from contracting or collapsing to form stars. However, from time to time events cause local increases in the pressure in the interstellar medium, producing additional compression of interstellar clouds. This compression occasionally brings these clouds to the threshold of gravitational collapse, during which time the internal gravitational attraction of the matter is greater than any tendency to expand due to the internal pressure. Among the events which can produce these local increases in pressure are several stages in the lifetimes of massive stars. When such stars are in the main sequence, burning hydrogen into helium in their central regions, they emit a great deal of ultraviolet radiation which ionizes the surrounding gas in the interstellar medium and causes a substantial increase in the local pressure. It is believed that later these massive stars undergo supernova explosions, in which large amounts of hot gas are expelled in a stellar catastrophe, increasing the local gas pressures for a few tens of light-years away from the site of the explosion. The presence of radioactive aluminum-26 in the early solar system has led to the suggestion that it was a supernova explosion of this kind which triggered the collapse of an interstellar cloud and led to the formation of the solar system.

The detailed behavior of the matter in a collapsing interstellar gas cloud is not well known. The collapsing gas is expected to be very cold and quite turbulent. It is conjectured that the gas undergoes a great deal of fragmentation into separately gravitating fragments, since the masses of stars are so much less than the masses of the interstellar clouds. This conjecture has observational support, since many examples are known in the Galaxy of freshly formed clusters of massive stars which have obviously just recently been formed and which are expanding away from a common central region. These young stellar associations will become unstable against expansion in this way if the formation of massive stars, as a part of the collapsing fragmentation process, results in heating of the residual gas, so that it escapes from the region and no longer contributes to the gravitational forces binding the forming stars together. *See* STELLAR EVOLUTION.

Primitive solar nebula. Most modern theories of formation of the solar system regard the Sun as a product of an interstellar cloud fragment. The Sun today has very little angular momentum, and even if the present-day Sun were to rotate very rapidly upon its spin axis, as presumably it did early in its history, the angular momentum which it would possess would be very much less than an interstellar cloud fragment is expected to have. From this general consideration it is expected that the gas of the collapsing fragment cannot have fallen together directly to form a star, which became the Sun, but that the gas formed first a rotating flattened disk of substantial dimensions, probably at least comparable to the dimensions of the planetary system. This flattened disk of gas and dust is frequently called the primitive solar nebula.

Much has been conjectured about the properties of the primitive solar nebula, and many of the measurements made of the properties of minerals within meteorites have been interpreted in terms of the temperatures and pressures in the meteorite regions of formation within the primitive solar nebula. The pressures range from about 10^{-6} atm (10^{-1} Pa) to as high as 1 atm (10^5 Pa), and the temperatures range from room temperature to about 1500°C or slightly higher. However, lack of knowledge of the types of bodies from which the meteorites come makes it impossible to know where in the solar nebula the inferred pressures and temperatures occurred. A further complication is that practically nothing is known about the history of the primitive solar nebula, and the extent to which the temperature and pressure varied with time at one part or another of the nebula.

Nor is it known how the primitive solar nebula dissipated. Substantial amounts of its mass were presumably accumulated into the Sun. A smaller amount of the mass went into the planets and smaller objects within the solar system. The rest of the mass was probably dissipated into space. It is known that young stars which have not yet evolved to the point of burning hydrogen into helium in their central regions are frequently accompanied by substantial outflows of matter. These young stars are often called T Tauri stars. It is possible that the Sun had a similar strong gas outflow, more vigorous than the present-day solar wind by a large factor, and that the primitive solar nebula was blown away into space through interaction with this intense early solar wind. It is also possible that the primitive solar nebula formed its own "solar wind," causing some of its mass to flow off into space, while the remainder flowed inward to accrete into the Sun, releasing gravitational energy, some of which went into the gas outflow.

If there were substantial amounts of mass in the primitive solar nebula, particularly in comparison to the amount of mass which had been accumulated into the central Sun at any stage in the evolution of the system, then it is very likely that the gas of the primitive solar nebula would have undergone gravitational instabilities of its own. These instabilities are likely to take the form of doughnut-shaped rings extending around the spin axis of the system, into which the gas in the primitive solar nebula would become clumped. Such rings are themselves unstable, and break up into a small number of major gaseous blobs, which are called giant gaseous protoplanets.

Formation of planets. There exists a substantial body of calculations investigating the following sequence of events. It is assumed that the primitive solar nebula became quiescent enough for small, solid dust particles within it to settle downward toward the central plane of the nebula. When sufficiently thin layers of these dust particles had collected at the central plane of the nebula, gravitational instabilities in the dust layer itself would cause it to break up into a large number of solid bodies, typically having dimensions of the order of a kilometer. In the inner solar system these small bodies would be primarily rocky in composition, whereas in the outer solar system it is presumed that the temperature would be lower and these bodies would have a substantial component of primitive ices. This is a natural way for rocky asteroidal bodies to form in the inner solar system, and cometary bodies in the outer solar system.

The problem is how these bodies grew from the size range of a kilometer to the planetary size range, typically many thousands of kilometers. This requires that the huge swarm of smaller bodies is collected into a few larger ones. This can occur only as the result of a great many collisions among the smaller bodies, collisions which take place gently enough so that the bodies grow in size by amalgamation rather than being smashed into pieces by the violence of the collision. A problem with the process is that it appears that collisons in the asteroidal belt today are primarily destructive rather than amalgamative.

The time required for planets to form by schemes of this kind is quite substantial: several hundred million years. This time would be unconvincingly long for the outer planets, because it must be presumed that if the outer planets depended upon this mechanism, they must grow by collisional amalgamation of solid bodies to a sufficient size to be able to capture from the surrounding primitive solar nebula large amounts of hydrogen and helium. In turn, this would require that the gas of the primitive solar nebula should stay around for several hundred millions of years, which appears to be much longer than the ages of young stars in their T Tauri phases.

If instabilities within the primitive solar nebula have led to the formation of giant gaseous protoplanets, an additional set of mechanisms of planet formation are provided in addition to those described above. Giant gaseous protoplanets are expected to have masses within a factor of a few of that of Jupiter. They are of very large dimensions, initially more than an astronomical unit in radius. They can fit into the region of the inner solar system only by being formed very early in the development of the primitive solar nebula, when there is relatively little mass near the spin axis, and hence the gravitational bonds between objects are considerably smaller than those which will pertain at a later time when the objects such as the central Sun have grown in mass. When that happens, the giant gaseous protoplanets become too large to fit properly into the inner solar system, and they undergo tidal disruption.

After a giant gaseous protoplanet has been formed, it evolves in a manner similar to a newly formed star, contracting with an increase of pressure and temperature in its central regions. In the case of the protoplanet, these pressures and temperatures were initially very low. In at least some of the protoplanets, it is likely that the internal temperatures and pressures become high enough so that small, solid dust particles immersed within the gas of the protoplanet melt, forming liquid droplets which collect together and rain out toward the center of the protoplanet, much as water droplets do in the clouds in the Earth's atmosphere. This can result in the formation of a substantial molten rocky core of the protoplanet before the tidal stripping of the envelope takes place. These cores, if formed in this way, become powerful gravitating centers which attract and collect the large swarm of smaller bodies likely to be formed by the gravitational instability mechanism described previously. The time required for planet formation in this way is greatly reduced compared to that required to build up a planet entirely from a swarm of small bodies.

In the outer solar system, the gaseous envelope of a giant gaseous protoplanet would not be tidally stripped away. Much of it may be thermally evaporated away, leaving behind the less volatile constituents. These may form cloud layers or may settle as solids toward the interior. If the giant planets of the solar system are formed from protoplanets in this way, not only is the formation time for these planets greatly reduced as compared to that required for growth entirely from a swarm of small bodies, but there is no need to collect additional gas from the primitive solar nebula, which therefore need not be present for a long period.

Prospects. Investigations of the other planets in the solar system by means of interplanetary spacecraft have provided a wealth of data pertaining to the origin and history of the solar system, and have stimulated a great deal of thinking upon the subject. It is very likely that new knowledge and calculations will greatly clarify the subject matter discussed in this article. [A. G. W. CAMERON]

Bibliography: T. Gehrels (ed.), *Protostars and Planets*, 1978; D. Ter Haar and A. G. W. Cameron, Historical review of theories of the origin of the solar system, in R. Jastrow and A. G. W. Cameron (eds.), *Origin of the Solar System*, 1963.

Solar wind

The continuous outward flow of ionized solar gas and a "frozen-in" remnant of the solar magnetic field through the solar system. This flow arises from strong outward pressure in the solar corona, becomes supersonic at a few solar radii (1 solar radius = 6.96×10^{10} cm) above the visible surface of the Sun (the photosphere), and attains speeds in the range 250–750 km s^{-1} in interplanetary space. The solar wind is believed to remain supersonic out to a distance from the Sun of 50–100 astronomical units (AU), where it is slowed by interaction with the interstellar gas and magnetic field.

Studies of geomagnetic activity led to the first suggestions (as early as the 1850s) that the Sun emits particles in connection with solar activity. In the late 1950s E. N. Parker demonstrated that the extended, high temperature of the solar corona (above a million degrees) precluded the nearly static equilibrium state that prevails in the atmospheres of planets like the Earth. Rather, the corona attains equilibrium by expanding continuously into interplanetary space. Simplified theoretical models of this expansion predicted a transition from very slow motions in the lower corona to a supersonic outward flow at 200–1000 km s^{-1} in the interplanetary region.

Continuous emission. This concept of a continuous emission of particles (as an ionized but electrically neutral plasma), or a "solar wind," unified many of the ideas about solar particle emission that had been deduced from indirect sources. By 1960 direct observations had led to a fairly complete and detailed description of the basic characteristics of the solar wind near the orbit of Earth. The range of observed values and the long-term averages of most important solar wind properties are given in the table.

During the 1970s there were interesting developments in the focus of solar wind research. In the early years of the decade, much attention was given to the study of inhomogeneities in the flow, in particular of streams of abnormally fast plasma. In the mid-1970s, studies of coronal holes, regions of exceptionally low density in the corona, led to the identification of the solar sources of these streams. Further understanding of the relationship between holes and streams and of both to the solar magnetic field led to a remarkably unified view of the three-dimensional structure of the corona and interplanetary medium. It also led to a reevaluation of the earlier comparisons of theoretical models of the coronal expansion with observed average solar wind properties and to a new set of difficulties to be confronted by these models. *See* INTERSTELLAR MATTER; SOLAR MAGNETIC FIELD.

Solar wind streams and magnetic sectors. In 1962, sampling instruments on the *Mariner 2* spacecraft revealed an important pattern in the variations of solar wind speed with time. The observed speed rose systematically from low values, 300–400 km s^{-1}, to high values, 600–700 km s^{-1}, in 1 or 2 days and then returned to low values during the next 3 to 5 days (Fig. 1*a*.) Each of these high-speed streams tended to be seen at approximately 27-day intervals or to recur with the rotation period of the Sun as viewed from the spacecraft. A similar recurrence tendency had been noted in geomagnetic activity, and was widely interpreted as the effect of localized, long-lived streams of particles emitted from the Sun and swept past the Earth once during each solar rotation. The high-speed solar wind streams have been linked to recurrent geomagnetic activity and found to be prominent features of the solar wind much of the time since 1962. *See* MAGNETOSPHERE.

Interplanetary magnetic sectors. Variations in numerous other solar wind properties have been found to be organized within these streams. For

Ranges and average values of observed solar wind properties

Property	Range*	Average value
Flow speed	300–700 km s^{-1}	450 km s^{-1}
Flow direction	±5° of direction from Sun	Within 2° of direction from Sun
Proton density	3–20 cm^{-3}	8 cm^{-3}
Proton temperature	$0.1-3 \times 10^5$ K	1×10^5 K
Electron temperature	$0.9-2 \times 10^5$ K	1.5×10^5 K
Magnetic field intensity	$2-10 \times 10^{-5}$ gauss ($2-10 \times 10^{-9}$ tesla)	6×10^{-5} gauss (6×10^{-9} tesla)

*Determined by excluding the lowest and highest 5% of observed values.

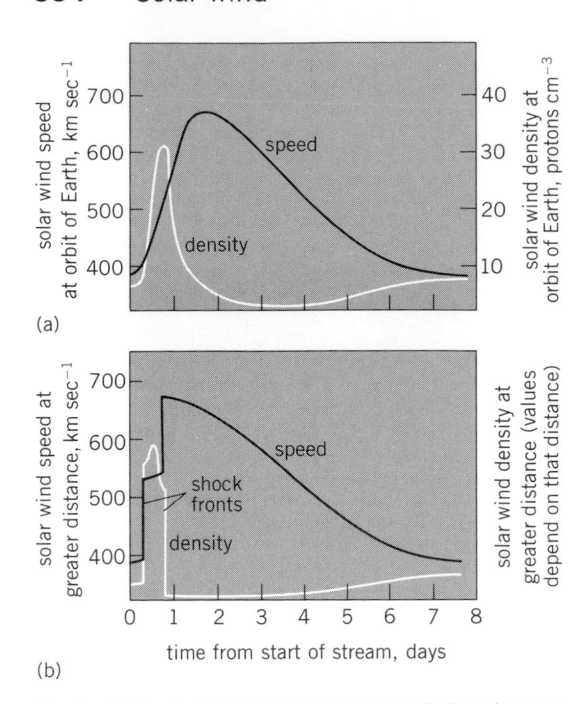

Fig. 1. Pattern of speed and density variations in solar wind streams (*a*) observed near orbit of Earth and (*b*) predicted and later observed farther from the Sun.

example, the interplanetary magnetic field points predominantly toward or away from the Sun within a stream, normally changing this polarity in the low-speed wind between streams. Indirect, geomagnetic evidence strongly suggests the presence of these interplanetary magnetic sectors for the past five 11-year sunspot cycles. Another important, stream-organized variation involves the density. The solar wind particle density tends to increase near the front of a stream, attain a high, maximum value where the speed is increasing rapidly, and then decrease to abnormally low values as the speed peaks and falls, as shown in Fig. 1.

Stream steepening. Straightforward physical arguments suggest that high-speed streams will evolve as they move outward from the Sun through a process known as stream steepening. The fast-moving material at the crest of the stream should overtake the slower-moving material in front of it and outrun the slower-moving material behind it. This will result in a fast-rise, slow-decay profile for the speed variation, and produce a compression of the plasma ahead of the stream crest and a rarefaction in the plasma behind the crest. The typical variations in solar wind speed and density observed near the orbit of Earth (as in Fig. 1) follow these patterns and are widely interpreted as evidence that the streams have evolved in this fashion between the Sun and 1 AU.

Smoothing process. The rearrangement of material (compression and rarefaction) produced by stream steepening implies an additional effect that resists further steepening and that must ultimately smooth out the speed variations at the root of the process. The compressed plasma between the stream front and crest will be at an elevated pressure, and thus pressure forces will act away from this region on both sides. These forces will accelerate the slow-moving material near the front of the stream and decelerate the fast-moving material near the crest, tending to diminish the speed difference that produced the compression. An important complication arises from the large amplitudes, 100–400 km s^{-1}, of the speed variations in solar wind streams. As these differences are much larger than the speed of sound in the interplanetary plasma (typically about 40 km s^{-1} near 1 AU), the steepening will continue until very large spatial gradients are produced; dissipative processes at these gradients lead to the formation of shock fronts on both sides of the compressed region, as shown in Fig. 1*b*. Strong heating and the acceleration and deceleration of the plasma will then occur mainly at these fronts.

Evolution of high-speed streams. These rather complex steepening and smoothing processes combine to determine the evolution of high-speed streams as they move outward through the solar system. Detailed theoretical models of this evolution, applied to the streams actually observed in the solar wind during the mid-1970s, predicted that the steepening would dominate until shock fronts formed between 1 and 2 AU and that the smoothing out of the streams would then occur very slowly, over a distance scale of about 10 AU. Thus shells of highly compressed plasma, bounded by shock fronts as illustrated in Fig. 1*b*, should arise in each recurrent solar wind stream, and these shells should be the dominant structure in the interplanetary plasma between about 2 and 10 AU from the Sun.

The flights of Pioneer and Voyager spacecraft to the outer solar system during the mid and late 1970s provided opportunities to test the general concept and detailed models of stream evolution. The solar wind structure observed as these spacecraft moved slowly outward from the orbit of Earth showed an increasing concentration of plasma and frozen-in magnetic field in the stream-associated compressions. By 2–3 AU, most streams showed the predicted pair of shock fronts, and these shocks persisted as the spacecraft moved to the orbit of Jupiter (at 5 AU) and beyond. The solar wind observed near Jupiter was far more inhomogeneous than that near 1 AU, as a result of this expected evolution of high-speed streams. In addition to confirming these predictions and determining the details of the resulting solar wind structure that fall beyond the capabilities of theoretical prediction, these spacecraft have also found that the stream-associated compressions and shock waves are unexpectedly important sources of energetic particles.

Streams, coronal holes, and solar magnetism. A number of solar features were proposed as sources of the particle streams thought to be responsible for recurrent geomagnetic activity. The early in-place observations established the basic characteristic of the streams, that is, abnormally fast flow speeds. By the early 1970s, it could be reasonably argued that high-speed streams emanated from the central portions of open magnetic structures in the corona, with the associated magnetic sector pattern resulting from the basically

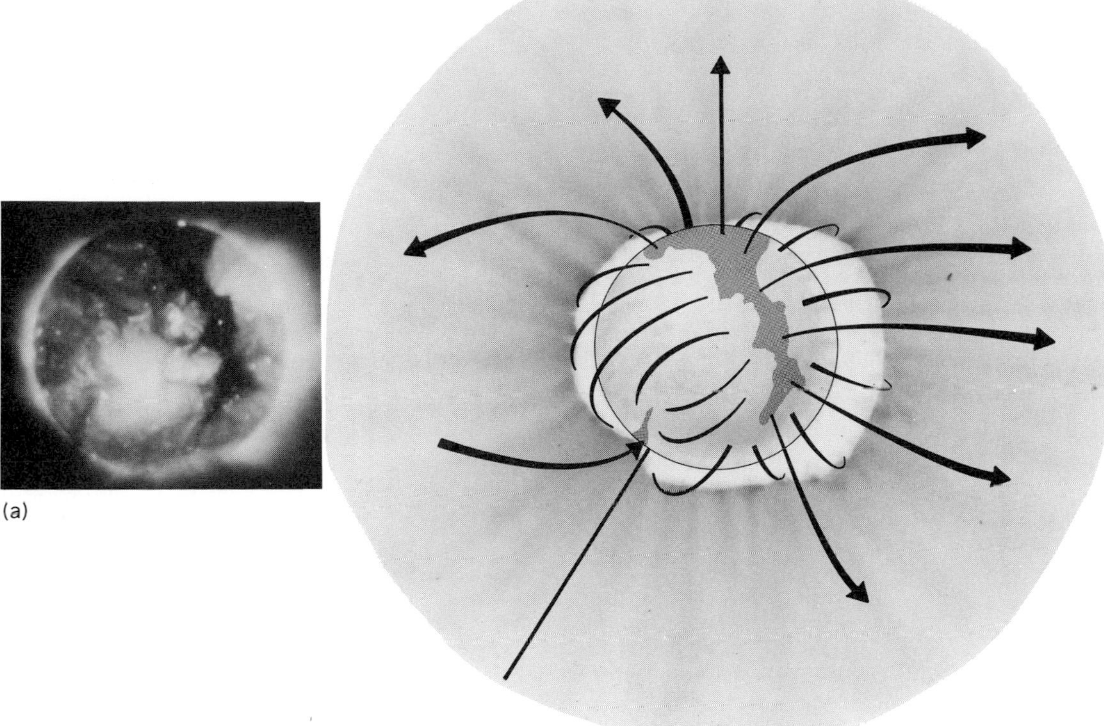

Fig. 2. Coronal holes and solar magnetic geometry. (*a*) X-ray image of the Sun obtained by an instrument on the Skylab mission. The dark regions near the north pole of the Sun and extending in a long north-south lane to the right of center are coronal holes (*American Science and Engineering Co.*). (*b*) Magnetic geometry associated with coronal holes. Open magnetic field lines extend from the holes (with a polarity as indicated by the arrows), while the neighboring regions are threaded by closed magnetic field lines.

unipolar nature of the magnetic field underlying and permeating such a region.

Coronal holes. In the early 1970s x-ray and ultraviolet images of the Sun brought attention to the features known as coronal holes. These are regions of abnormally low coronal density that appear dim in any radiation emitted (x-rays and the ultraviolet) or scattered (the white-light corona observed at eclipse) from the corona. For example, in the x-ray image (Fig. 2*a*) made at wavelengths of nanometers (the characteristic radiation of the million-degree corona), holes appear as dark patches outlined by emission from neighboring, dense, hot regions. Detailed examination of these images suggests that the holes are regions where the solar magnetic field is open (Fig. 2*b*), with field lines reaching from their visible roots in the photosphere out into interplanetary space. In contrast, the structure seen in the bright or dense corona suggests closed magnetic fields, with field lines connecting separated locations in the photosphere. Numerous studies, using coronal holes observed by a variety of techniques, have established that the appearance of a hole near the equator of the Sun leads to the appearance in the solar wind observed in the ecliptic plane (near the solar equator) of a high-speed stream, embedded in a magnetic sector with the polarity matching that observed in the photospheric region underlying the hole. The low density of the holes can be attributed directly to their open magnetic character and the resulting escape of coronal material to interplanetary space. In contrast, the closed magnetic fields in other parts of the corona inhibit the escape of plasma and lead to a confined, dense corona.

Polar holes. The largest and longest-lived coronal holes have been observed to occur in the polar regions of the Sun (as defined by its rotation axis); conspicuous polar holes existed for at least 8 years during the past 11-year sunspot cycle. These polar holes are related to the weak, dipolelike, general magnetic field of the Sun and are thus of opposite polarity in the two hemispheres. Near-equatorial holes of a given magnetic polarity tend to occur in the same hemisphere as the polar hole of the same polarity; at some time during their lifetime most near-equatorial holes "connect" to the polar hole of the same polarity and thus appear as an equatorward extension of a polar hole. This suggests a rather simple large-scale geometry for magnetically open regions of the corona and their extension into interplanetary space—two regions of opposite magnetic polarity, each arising from a polar hole and its equatorward extensions, separated by a neutral sheet or surface at which the outwardly extended magnetic field changes its sign or polarity. This neutral sheet must separate all holes of opposite polarity.

Sector boundaries. Were the solar magnetic field a simple dipole aligned with the axis of rotation, the northern and southern hemispheres would be expected to contain the field lines of opposite polarity, each arising from a polar hole, and these hemispheres to be separated by a neutral sheet in the equatorial plane. The complexities of the actual solar magnetic field distort this idealized geometry, producing a neutral sheet that is warped out of the equatorial plane as in Fig. 3. The major warps in the sheet would correspond to the positions of equatorward extensions of the polar holes (and perhaps other near-equatorial holes). The rotation of this spatial structure with the Sun then produces the changing pattern of magnetic polarity seen by a stationary interplanetary observer. "Sector boundaries" occur when the neutral sheet sweeps past the observer, with the unipolar magnetic sectors between such boundaries corresponding to residence on one side of the neutral sheet.

Three-dimensional model. The observed pattern of recurrent variations in solar wind speed (that is, high-speed streams) can be simply related to the same magnetic spatial structure if the empirical findings that fast wind comes from near-equatorial holes while slow wind occurs near sector boundaries is assumed to hold in three dimensions. That is, low speeds occur everywhere near the neutral sheet, high speeds everywhere sufficiently far from this three-dimensional surface. Under this assumption, the pattern of alternating fast and slow solar wind flow familiar from in-place observations performed in the ecliptic plane (and hence within approximately 7° of the solar equatorial plane) would then be characteristic only of near-equatorial latitudes. The polar latitudes, sites of the large, long-lived polar holes, would be regions of uniform high-speed solar wind flow.

This view of a large-scale, three-dimensional solar wind structure organized about a simple magnetic geometry implies a significant revision of many earlier ideas about the solar wind. On the purely phenomenological side, the dominance of the polar influence on the three-dimensional structure and hence the pattern of solar wind variations even in the near-equatorial wind is a radical departure from the previous orthodox model. This revision extends to the view of the basic physical characteristics and origins of the wind. The average solar wind properties determined in the late 1960s and used as the touchstone of theoretical models may deviate markedly from true global averages wherein the high-speed flows from the polar regions would be more heavily weighted. The much higher speeds and energies in the high-speed wind have become the most important, and difficult, characteristics of the solar wind, which will require explanations by future models. *See* SOLAR MAGNETIC FIELD; SUN.

[A. J. HUNDHAUSEN]

Bibliography: A. J. Hundhausen, Solar activity and the solar wind, *Rev. Geophys. Space Phys.*, 17: 2034–2048, 1979; E. J. Smith and J. H. Wolfe, Fields and plasmas in the outer solar system, *Space Sci. Rev.*, 23:217–252, 1979; J. B. Zirker (ed.), *Coronal Holes and High-Speed Wind Streams*, 1977.

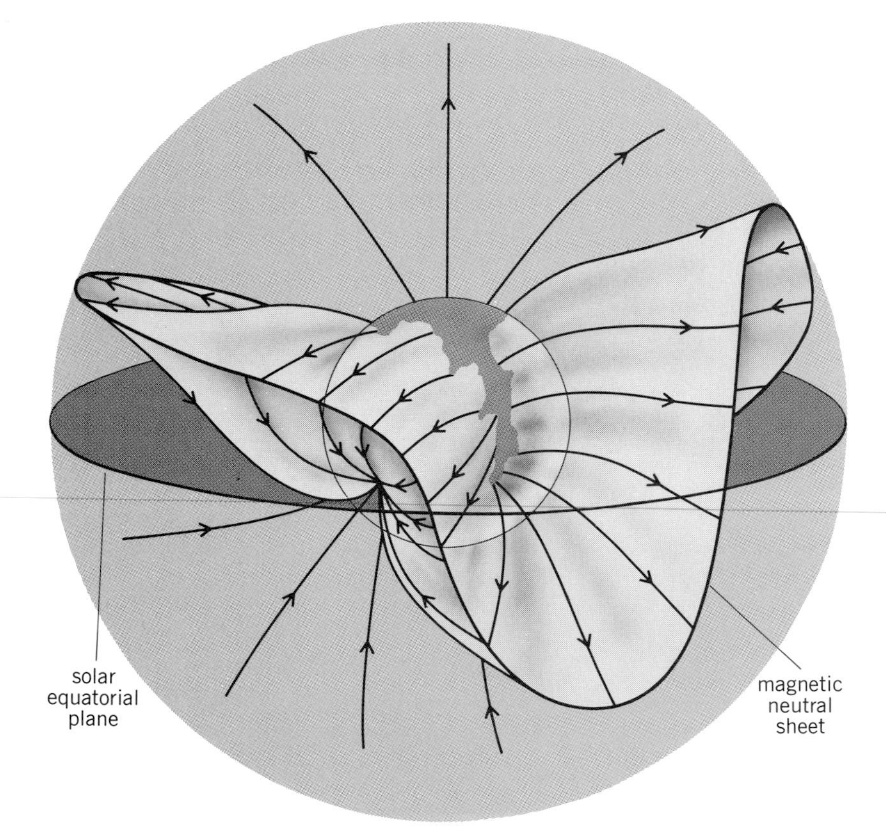

Fig. 3. Magnetic neutral sheet separating outward-pointing magnetic field lines (positive magnetic polarity) from inward-pointing magnetic field lines (negative magnetic polarity). In this case the positive polarities originate in the north polar hole and its equatorward extensions. The latter warp the neutral sheet from its idealized equatorial location to produce the magnetic polarity or sector structure observed in the near-equatorial solar wind.

Space probe

An instrumented vehicle, the payload of a rocket-launching system, designed specifically for flight missions to other planets, the Moon, and into deep space, as distinguished from Earth-orbiting satellites. Some of the important space probes are given in Table 1.

The space probe is used primarily for scientific purposes, and the payload is designed to explore planetary geology, geography, geodesy, atmospheric physics, exobiology, other specialized aspects of planets, their satellites, comets, and, lastly, the charged particle and electromagnetic field environment about planets and in deep space. The task is usually designated as the mission, which, for exploration to celestial bodies, is customarily divided into the cruise and encounter phases. The instrumentation is usually distinct for the two parts of the mission. Instruments carried are cosmic-ray telescopes, plasma detectors, magnetometers, neutral gas detectors, television cameras, gamma-ray telescopes, various specialized surface scanners, and micrometeoroid detectors. Cruise-mode operation entails interplanetary physics such as the measurement of solar-wind and cosmic-ray propagation and diffusion, whereas the encounter mode is occupied with exploration and measurements of the properties of the planet.

Table 1. Important space probes

Name	Launch date	Comments
Pioneer 1	Oct. 11, 1958	Magnetic fields, cosmic rays; launched to Moon; returned data to 14.3 earth radii
Pioneer 4	Mar. 3, 1959	Cosmic rays; passed 37,300 mi (60,000 km) from Moon
Luna 2	Sept. 2, 1959	Impacted Moon
Luna 3	Oct. 4, 1959	Photographed far side of Moon
Pioneer 5	Mar. 11, 1959	First deep-space probe; magnetic fields and cosmic rays
Mariner 2	Aug. 26, 1962	Venus probe
Ranger 7	July 28, 1964	Lunar impact and approach photography
Mariner 4	Nov. 28, 1964	Mars encounter; photography, magnetic fields, cosmic rays
Ranger 8	Feb. 17, 1965	Lunar impact and approach photographs
Ranger 9	Mar. 21, 1965	Lunar impact in Alphonsus approach photography
Zond 3	July 18, 1965	Photographs from lunar encounter
Pioneer 6	Dec. 16, 1965	Interplanetary probe
Luna 9	Jan. 31, 1966	First photographs of lunar surface
Luna 10	Mar. 31, 1966	Lunar and interplanetary data
Lunar Orbiter 1	Aug. 10, 1966	Lunar photographs
Pioneer 7	Aug. 17, 1966	Interplanetary probe
Luna 11	Aug. 24, 1966	Lunar data
Surveyor 1	May 30, 1966	Soft landing on Moon; environmental data and photography
Luna 12	Oct. 22, 1966	Lunar orbital photography and other data
Lunar Orbiter 2	Nov. 6, 1966	Lunar orbital photography
Luna 13	Dec. 21, 1966	Lunar surface photography and soil information
Mariner 5	June 14, 1967	Venus probe; atmospheric and magnetospheric data
Venera 4	June 12, 1967	Analysis of Venus atmosphere; first instrumented landing on another planet
Lunar Orbiter 3	Feb. 5, 1967	Lunar orbital photography
Lunar Orbiter 4	May 4, 1967	Lunar orbital photography
Surveyor 3	Apr. 1967	Lunar surface photography and surface properties
Lunar Orbiter 5	Aug. 1, 1967	Lunar orbital photography
Surveyor 5	Sept. 8, 1967	Lunar surface photography and surface properties, including elemental analysis of surface
Surveyor 6	Nov. 7, 1967	Same as *Surveyor 5*; landing in Sinus Medii
Surveyor 7	Jan. 7, 1968	Same as *Surveyor 5*
Pioneer 9	Nov. 8, 1968	Interplanetary probe
Zond 6	Nov. 10, 1968	Interplanetary probe
Venera 5	Jan. 5, 1969	Same as *Venera 4*
Venera 6	Jan. 10, 1969	Same as *Venera 4*
Mariner 6	Feb. 25, 1969	Photography and analysis of surface and atmosphere of Mars
Mariner 7	Mar. 27, 1969	Same as *Mariner 6*
Luna 15	July 14, 1969	Lunar reconnaissance (crashed during attempted lunar landing)
Zond 7	Aug. 8, 1969	Reentered Aug. 14, 1969; third uncrewed circumlunar flight; recovered in the Soviet Union
Venera 7	Aug. 17, 1970	Lander capsule transmitted 23 min from surface of Venus Dec. 15, 1970
Luna 16	Sept. 12, 1970	Reentered Sept. 24, 1970; unmanned Moon lander touched down on Sea of Fertility Sept. 20, 1970; returned lunar soil samples
Zond 8	Oct. 20, 1970	Circled Moon; recovered Oct. 27, 1970
Luna 17	Nov. 10, 1970	Landed on Moon Nov. 17, 1970; unmanned Moon rover
Mars 2	May 19, 1971	First Soviet Mars landing
Mars 3	May 28, 1971	Mars probe
Mariner 9	May 30, 1971	Mars probe
Luna 18	Sept. 2, 1971	Impacted Moon Sept. 11, 1971
Luna 19	Sept. 28, 1971	Lunar photography mission
Luna 20	Feb. 14, 1972	Recovered Feb. 25, 1972; returned lunar sample
Pioneer 10	Mar. 2, 1972	Jupiter encounter; transjovian interplanetary probe
Venera 8	Mar. 27, 1972	Venus landing July 22, 1972
Luna 21	Jan. 8, 1972	Moon landing Jan. 16, 1972, with Lunikhod rover
Pioneer 11	Apr. 5, 1973	Jupiter encounter and transjovian interplanetary probe; also Saturn encounter
Mars 4	July 21, 1973	Mars orbiter
Mars 5	July 25, 1973	Mars orbiter
Mars 6	Aug. 5, 1973	Mars lander
Mars 7	Aug. 9, 1973	Mars lander
Mariner 10	Nov. 3, 1973	Venus and Mercury encounter
Luna 22	May 29, 1974	Lunar probe
Helios	Dec. 10, 1974	Close solar probe
Venera 9	June 8, 1975	Venus probe
Venera 10	June 14, 1975	Venus probe
Viking 1	Aug. 20, 1975	Mars lander and orbiter
Viking 2	Sept. 9, 1975	Mars lander and orbiter
Voyager 2	Aug. 20, 1977	Same as *Voyager 1* (below); continuing to Uranus and Neptune (?)
Voyager 1	Sept. 5, 1977	Returned encounter information concerning Jupiter, Saturn, and their satellites and rings
Pioneer Venus Orbiter	May 20, 1978	Returning atmospheric, surface, and particle and field information

Table 1. Important space probes (cont.)

Name	Launch date	Comments
Pioneer Venus Multi-Probe Bus	Aug. 8, 1978	Penetration of Venus atmosphere by four probes; returned atmospheric data
Venera 11	Sept. 8, 1978	Lander; detected lightning and sounds believed to be thunder
Venera 12	Sept. 14, 1978	Same as *Venera 11*.

Research on planetary surfaces usually includes instruments devoted to chemical examination of the surface and atmosphere, as well as the surface geology and meteorology. Exobiological instrumentation may be included.

DESIGN CONSIDERATIONS

The spacecraft is a self-contained automatic device with a self-sustaining power generation system deriving energy from the Sun or a nuclear thermalelectric generator converting this energy to usable form, and distributing electricity to the various electronic subsystems. The other subsystems of the spacecraft are: communication, which includes command receivers, data encoder and transmitter, and antennas; attitude control, which may be passive, using spin angular momentum or solar pressure vanes; a frame or structure which must maintain its integrity throughout the mission; and a thermal control system to provide the proper operating temperatures in the spacecraft interior. Spacecraft which are designed to enter a planetary atmosphere or land on a planetary surface must also include an aeroshield and a landing system. Viking also included a bioshield to maintain a sterile environment during Earth launch.

Ideally, a spacecraft will have an extensive system of monitoring sensors for determining the engineering status of the various subsystems during flight and relaying this information back to Earth via the telemetry link. Propulsion is provided in those spacecraft where midcourse trim of the velocity is needed; such instances are required when a target planet is to be encountered. Lastly, the payload comprises the package to be delivered to a specified point in space in operational condition so measurements can be taken and signals returned to Earth.

Mission. Although some missions require only a trajectory into interplanetary space, others are directed to specified bodies in the solar system. Such trajectories invariably require that during the cruise mode (midcourse) the velocity be corrected, since the launch booster rocket alone seldom can attain sufficient accuracy of guidance to intercept the target planet with acceptable precision. The correction is applied by firing a small vernier rocket which is a part of the spacecraft. The thrust is in a direction computed so that the correct velocity to be gained is attained by vectorially adding the increment to the precorrection cruise velocity. To determine the correction requires tracking from one to several days to ensure accuracy.

Some planetary probes are designed for encounter only (flyby); others are designed for capture into orbit as an artificial satellite; and, finally, there are those which are decelerated to land. All these require special on board propulsion. In cases where an atmosphere is present, aerodynamic devices are useful to reduce the entry speed. These can take the form of ablating nose cones, balutes, retrorockets, pure parachutes, or some combination of these. The Venera spacecraft were aided by a parachute in their descent into the Cytherian (Venusian) atmosphere. Additional devices are available for descent onto atmosphere-free bodies such as the Moon, where retrorocket thrust is the primary deceleration method. The final part of the landing is then aided by a landing gear system (Surveyor) or, for semihard landings such as *Ranger Block 1*, by a balsa wood spherical impact absorber surrounding the payload during the final drop onto the surface (*Luna 9*).

The Moon has been targeted many times by the United States and the Soviet Union. The technique used is representative of missions which might be planned to planets. The Moon literally flies into the spacecraft which is proceeding along a predetermined course. The geometry of the collision discloses that the final descent onto the surface is not vertical and that a residue of lateral velocity is present. This must be removed by trim rockets which are controlled by Doppler velocity radar using reflections from the surface. It is also possible to drop onto the surface of a body from orbit rather than directly as in the case of Ranger and Surveyor. *See* MOON.

Interplanetary spacecraft are somewhat simpler in a relative sense; complexity is nevertheless an important characteristic of their design. Such vehicles need not be fitted with midcourse propulsion, and the system operates in only an interplanetary mode.

The space probe is a self-contained entity capable of operating for long periods and returning a continuous scientific log and status report of its operation. On-board data storage, buffering, and processing equipment are sometimes included to prepare data for optimum transmission.

General background. Space probes are among the most sophisticated engineering devices in existence. They encompass all the technology connected with computers, electrical power systems, and scientific instruments of a wide variety, as well as rocket technology and a great deal of mechanical engineering. Their mode of operation in space ties them closely to the Earth; control is always needed, and the exact operational decisions to be made at a given time generally require that a complicated series of mathematical calculations be carried out. These are often associated with

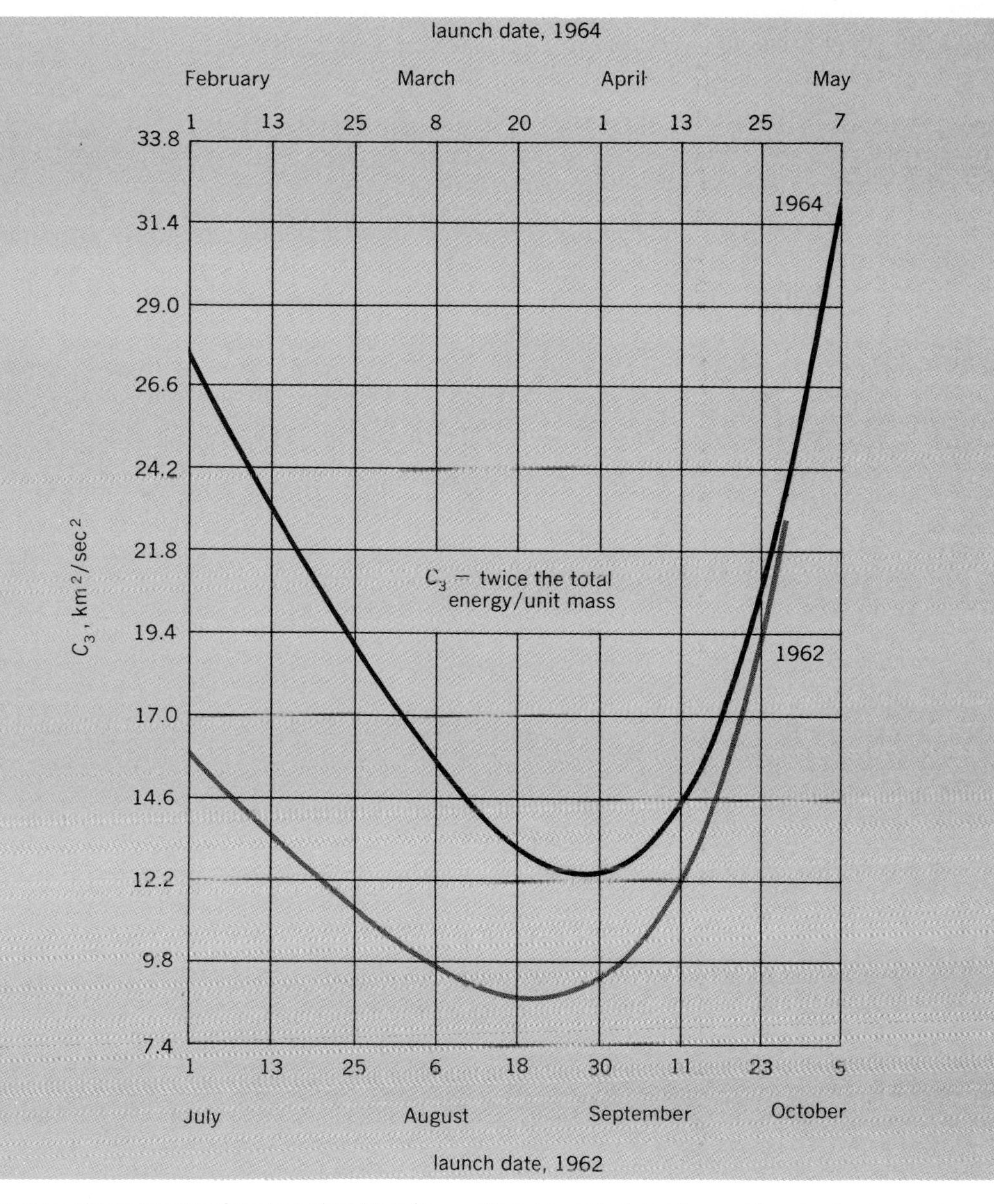

Fig. 1. Energy requirement curve for a launch to Mars in the spring of 1962 and 1964. Differences are due to changes in the inclination of the orbit planes of Earth and Mars between the two launch opportunities. The minimum energy point is the optimum time for firing, but the exact day cannot usually correspond to this since any schedule slip would make the launch impossible from an energetic standpoint. These graphs satisfy the requirement that radius vectors to the Sun between the launch and encounter make an angle measuring less than 180°. (*JPL/NASA*)

radio tracking and desired changes in the trajectory. For example, the initial trajectory is seldom sufficiently accurate for a planetary target to be reached without augmentation of the velocity in midcourse. Such changes require refined and fast calculations followed by an involved coded command stream to the spacecraft. The transit time delay for light becomes important at great distance, for example, at Jupiter (approximately ½ h one-way), where automation involved in modifying spacecraft operation is essential for a total of 1½ h, taking into consideration the total time from initial transmission from the spacecraft to receipt of a command change and verification at Earth. Also, the tracking by antenna requires that antenna pointing be known and predictable, again requiring the use of Earth computers. The spacecraft usually carries electronic equipment of great complexity capable of coding and decoding commands from Earth as well as internally generated commands. All these are processed by using computer technology; the spacecraft includes much of the methodology of computers.

In addition to the spacecraft a large array of ground equipment, without which the spacecraft could not continue to function, is necessary. Foremost is the radio link for passing information to Earth, but almost equally important is the up-link that sends instructions to the spacecraft. The ground complement includes large antennas, radar transmitters and receivers of the greatest complexity, on-line computers of the largest types, and a worldwide communication system which connects the command center on Earth to the antenna range complex that encompasses the whole Earth.

In addition to these operational parts of the whole system, there is the vital and extremely complicated launching operation. Its store of equipment includes the launch complex itself and a distinct computer-radio system for controlling the launch phase and transferring control to the space network. Lastly, there is the test equipment for proving that the spacecraft will operate properly in space. The test equipment or GSE (ground support equipment) together with the simulation gear that provides vacuum, thermal, vibration, and other tests is, in itself, the subject of considerable thought and design effort. It involves a large amount of planning and the expenditure of a large fraction of the total costs of a spacecraft program.

Orbits. For simplicity of discussion, orbits can be categorized into four general cases; in reality, however, where the perturbations of the planets are taken into account, they are all treated in numerical computation as a many-bodied problem. Most simply, the lunar transfer orbit can be taken as a restricted three-body problem where the spacecraft mass is trivial. Interplanetary orbits are elliptic with minor perturbations resulting from certain planets. Planetary trajectories are the most complicated, since they usually involve midcourse propulsive maneuver corrections and must be designed specifically with regard to the type of encounter. The reason is that the terminal velocity vector defines atmospheric entry conditions or, in a near miss, the impact parameter, the illumination available on the planetary surface, and so on.

The final velocity and position attained by a specific booster determine the interplanetary orbit of a spacecraft. Since the Earth's orbit plane is almost invariably tilted with respect to that of the target planet, it is usually necessary to launch the spacecraft with a component of velocity out of the ecliptic. Very special exceptions take place when the target planet-orbit plane-ecliptic line of nodes is coincident with the target arrival date of the spacecraft. Such events occur at infrequent intervals but, when available, decrease the energy and guidance accuracy requirement since the launch is into the ecliptic plane. Such a condition existed for the Venus launch opportunity in 1959. Generally, the energy requirement involved is so great that only very restricted times are available for planetary launches. Such times are called windows, and even then the required velocity is a function of launch time. Figure 1 shows the velocity in units of energy/mass for injection into a transfer orbit to Mars for 1962 and 1964. The latter case is considerably less advantageous because of the longer required component of velocity out of the ecliptic. Thus, a minimum-energy orbit is a unique minimum for each window.

An important constraint upon the final available energy is the way in which a spacecraft is boosted into orbit. The best available launch site for planetary-interplanetary flight in the United States is Cape Kennedy (Cape Canaveral), Fla. This location is favored because it makes good use of the rotation of the Earth to add velocity. Even for this site, it is seen upon examination that the component of Earth velocity available is $V_e \sin \psi$, where V_e is the surface velocity of the Earth at Cape Kennedy and ψ the compass heading. Thus, a firing due east would maximize the added velocity. However, only when $V_e \sin \psi$ is also parallel to the desired interplanetary orbit can this maximum value be realized. This happens twice per day but, unless the Earth's axis is also tilted correctly, the maximum increment of Earth velocity to be gained still cannot be utilized. Since the time of day and seasonal tilt of the Earth's axis are both critical to this problem, these factors become important constraints.

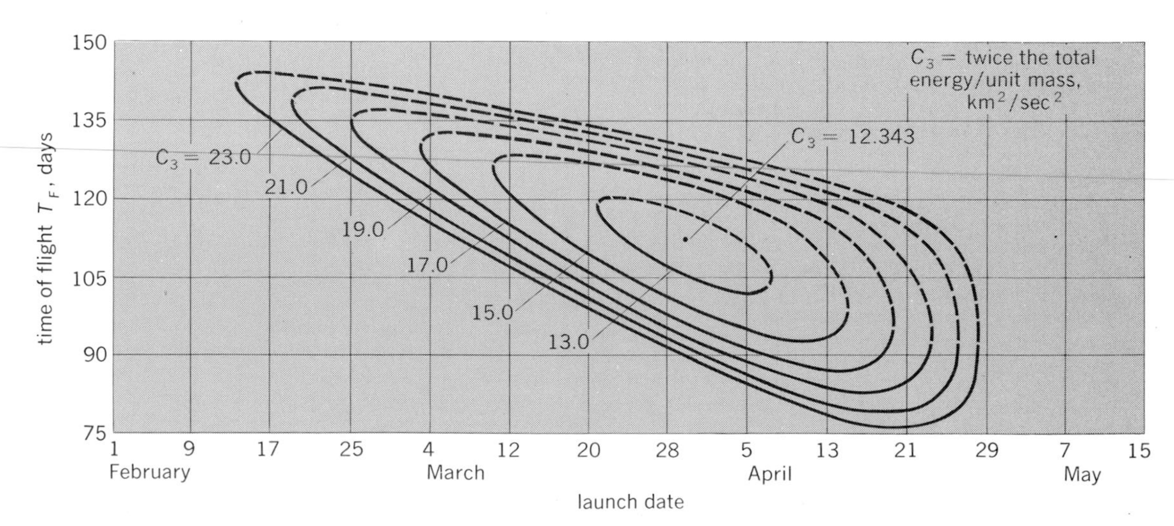

Fig. 2. Graph of trip time versus launch dates for spring, 1964. The multiple values correspond to different starting energies. The dotted loci correspond to trajectories where the apsidal angle of the interplanetary transfer ellipse is greater than 180°. (*JPL/NASA*)

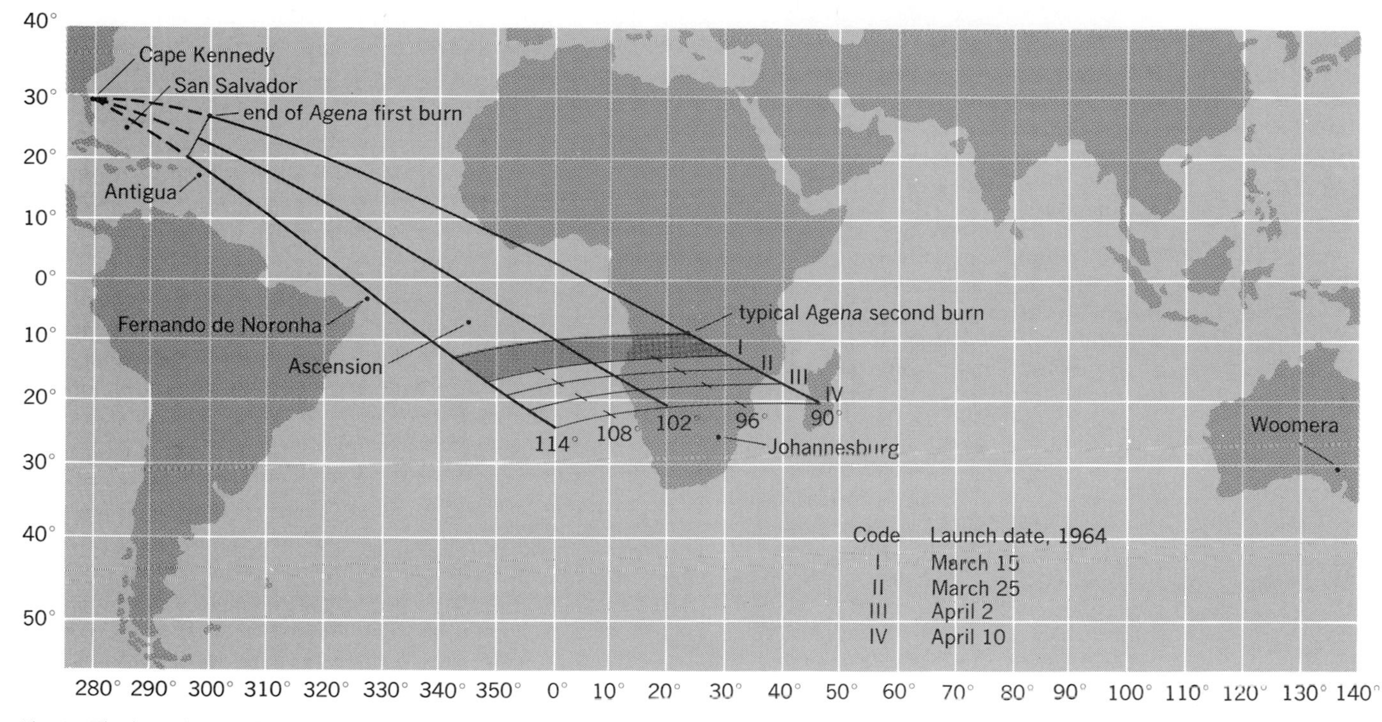

Fig. 3. The injection loci for a hypothetical 1964 Mariner Venus launch from Cape Kennedy, Fla. The variation of launch azimuth with reference to the time of launching is shown. (*JPL/NASA*)

There is a means for increasing the Earth-added velocity when the launching system is capable of being restarted in orbit. Typically, in configurations utilizing an Agena or other advanced upper stage, the burning period is followed by a time of free coasting during which the rocket is reoriented. At a precisely determined time the rocket is restarted so as to give a velocity increment in the proper direction. Since the rocket is essentially free of the Earth, this vector may be in an arbitrary direction. Some of the loss incurred by not having all of V_e available is recouped by this means. Figure 2 shows a set of interplanetary trip times for a typical launch, using an Agena upper stage, from Cape Kennedy for Mars for both 1962 and 1964. Additional constraints such as range safety and downrange communication are not discussed here; the extent of azimuthal firing lane shown in Fig. 3 is a result of these factors.

Additional constraints upon planetary trajectory shaping come from reliability and communication requirements. The time of flight is altered drastically by small changes in final injection velocity because the greatest proportion of booster energy is utilized in escaping the gravitational potential of Earth. Consequently, only very small fractional changes in final velocity result in enormous changes in transit time as shown in Fig. 2, the companion graph to Fig. 1. These times enter into the design reliability philosophy, which often has a critical bearing on the configuration of the scientific experiments. In this guise the problem of communication is of great importance, since the distance over which the return link from the spacecraft to Earth must operate (telemeter) determines the information rate capacity; or, conversely, the communication system, for extended distances, must utilize a greater share of the power, weight, and space available within the spacecraft.

Generally, the energy required to launch toward the Sun has a simple relation to that required to launch in the antisolar direction. The Sun's gravitation potential energy at Earth's orbit corresponds to an orbital velocity (30 km/sec), that is, nearly half of the solar system escape velocity (70 km/sec). The launch energy for inward (< 1 astronomical unit, or AU; 1 AU = 1.496×10^8 km) and outward (> 1 AU) missions can be shown in a simplified manner (Fig. 4). It can be seen that the very close inward mission (~0.2 AU) requires more en-

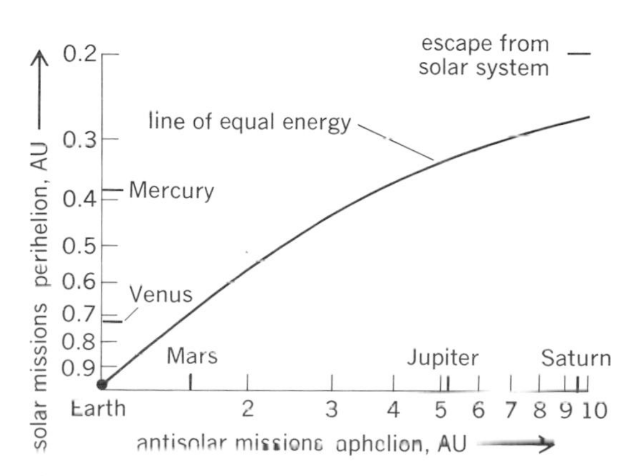

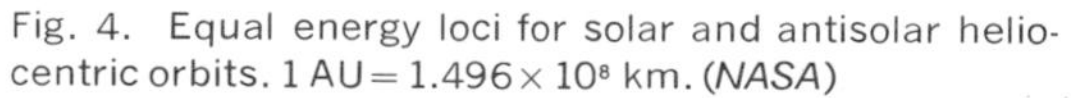

Fig. 4. Equal energy loci for solar and antisolar heliocentric orbits. 1 AU = 1.496×10^8 km. (*NASA*)

ergy, for example, than a mission to Saturn.

As seen from Earth, the interplanetary probe displays a complicated trajectory. It is useful to utilize an Earth-fixed coordinate system, that is, one rotating about the Sun, since many important effects are emphasized. Figure 5 shows such representative solar and antisolar trajectories in the plane of the ecliptic. The solar trajectory shows the characteristic lag in orbital velocity behind Earth at launch, with an overtaking and crossing of the Earth-Sun line (inferior conjunction) as a consequence of the conservation of angular momentum of the spacecraft about the Sun. Outbound or antisolar trajectories are considerably simpler. In this case the initial excess velocity in the direction ahead of Earth causes the spacecraft to assume an orbit greater than 1 AU, resulting, in turn, in a reduced angular velocity as the radius vector to the Sun increases.

Booster or launch vehicle performance is a critical factor in the mission design. The final spacecraft weight and the maximum (aphelion) or minimum (perihelion) solar distance, depending on whether the mission is antisolar or solar, are geared to the booster.

Descent maneuver. The very large velocity of a typical spacecraft just after burnout of the propulsion rockets decreases as the spacecraft ascends from the Earth's gravitational field. During cruise to another planet the speed is that left after subtracting the required velocity of escape from the Earth. For lunar trajectories the escape is not complete, but the Moon is high on the gravitational potential well of the Earth and most of the velocity of escape is required. Although the speed in space is substantially decreased as measured in an Earth-centered coordinate system, compared to just after burnout, the velocity with respect to the Sun is still large. That is, the speed is usually not greatly different from the speed of the Earth in its orbit, about 30 km/sec. The interplanetary speed determines the trip time to another planet. Since most of the original speed measured in the Earth frame is subtracted in escape from the Earth, it is clear that the residue is very important in determining the trip time; small errors in the burnout speed are magnified, as the errors apply in full measure to the interplanetary speed. A small error can cause the planetary probe to arrive at the calculated planetary position early or late by a substantial margin. Thus, the targeting requires especially good knowledge of the speed. That is why small midcourse corrections must be applied to the spacecraft velocity during interplanetary flight, after sufficient time has elapsed for an accurate determination of the speed using two-way Doppler radar.

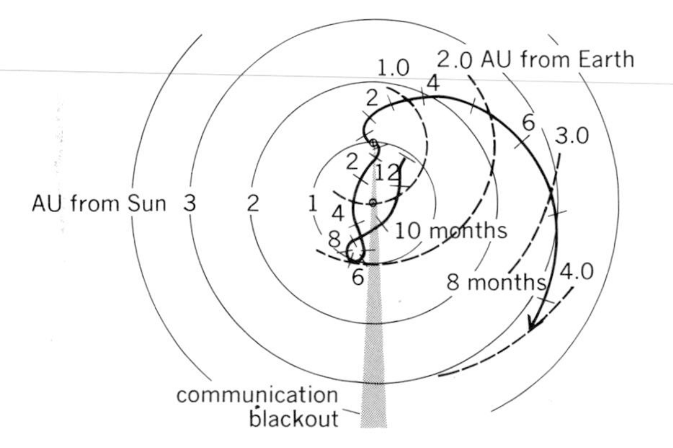

Fig. 5. Representative solar and antisolar trajectories from Earth shown in coordinates which rotate with Earth's yearly motion about the Sun. Marks on the trajectories denote months after launch. 1 AU = 1.496×10^8 km. (*NASA*)

The descent phase of a trajectory adds new problems to the total mission. Although targeting is important for correct arrival, an even more serious requirement is that the correct aiming point in the neighborhood of the target planet be reached at a specified time if accurate descent is to be achieved. The descent requires sufficient velocity to be subtracted from the spacecraft so that impact with the planet takes place at the correct latitude and longitude. Otherwise, such mission requirements as photography are seriously compromised because of improper solar illumination. To decrease the spacecraft speed so that it is correct in the planet frame of reference, deboost rockets are used. This is best done by capture of the spacecraft into a fixed orbit; then an additional deboost is applied at some appropriate point in orbit so that the aiming point on the surface is reached.

The final or terminal phase of the descent in the case of atmospheric entry employs other deboost aids. It is far more conservative of weight to use aerodynamic entry devices to reduce the speed, and these are customarily employed in design. The Venera spacecraft used such a device to enter into a descent trajectory to Venus; their descent into the planet's atmosphere was aided by a parachute.

In the case of bodies not endowed with a sufficiently dense atmosphere, such as the Moon, it is necessary to use retrorocket propulsion to the surface with the final shock taken up by a landing-gear system as in Surveyor. (A similar system is used in the manned Apollo program.) This terminal phase is especially delicate, as it is basically a launch in reverse without the addition of all the aids present at the usual launch site.

Communication. Two-way communication has been standard on all space probes since the flight of *Pioneer 1*. It is customary to include aboard the spacecraft a matrix of Earth-activated commands which are the end link of a command loop. This configuration consists of an Earth-based transmitter, antenna, encoding equipment, and one or more command receivers aboard the spacecraft. Usually the command carrier is phase-locked to the spacecraft transmission system, and a coherent Doppler ranging system is thereby established. By appropriate carrier modulation, both range and range-rate information can be obtained. Such information is crucial to successful midcourse velocity trim, especially for planetary missions, and is of great importance in the study of parameters which are associated with the properties of the spacecraft orbit.

The primary design problem of spacecraft communication rests in the area of telemetry rather

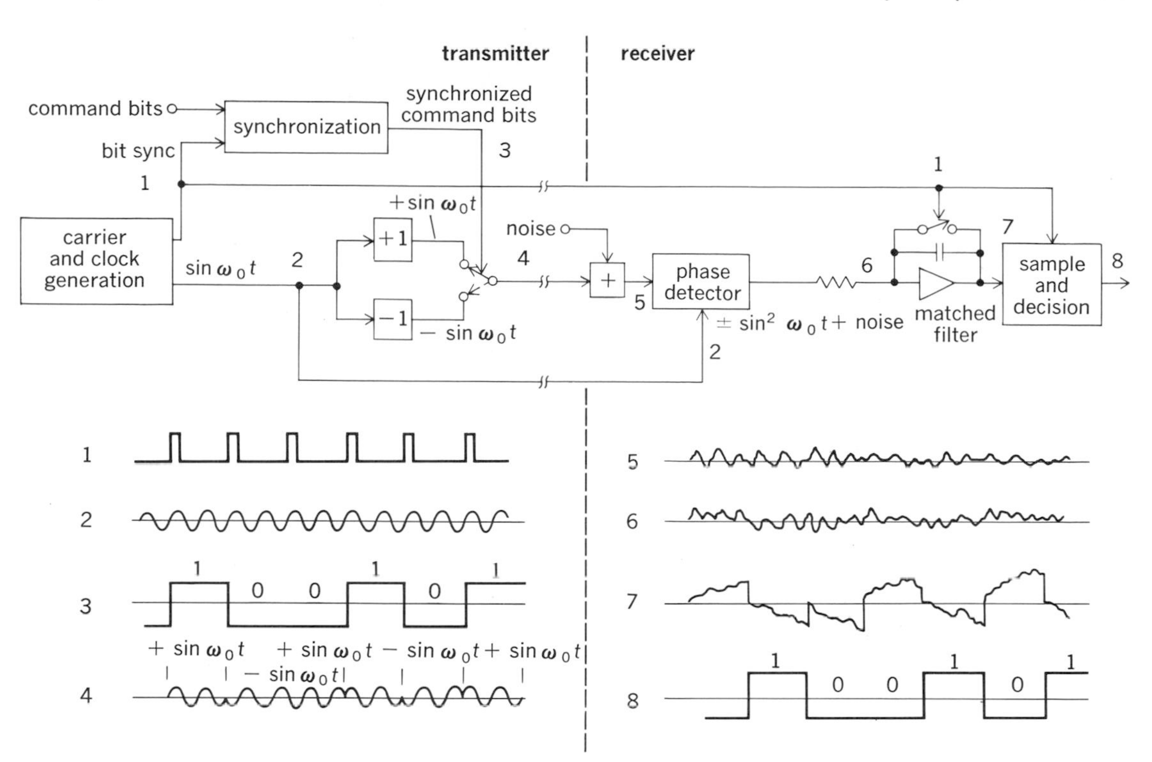

Fig. 6. Basic waveforms of phase-shift-key (PSK) transmitter and receiver. Both a bit-sync and sin $\omega_0 t$ (data carrier) waveform are generated at transmitter. Sin $\omega_0 t$ is also made available as $-\sin \omega_0 t$ according to whether a 1 or 0 is to be carried. The PSK output is phase detected by comparison with unmodulated sin $\omega_0 t$ at the receiver. Transmission of bit-sync signal provides timing for filtering and sampling to reconstruct data to form of original synchronized command. (*Space/Aeronautics*)

than command, since signal strength versus system noise is the fundamental limitation; that is, the spacecraft's power capacity is limited. In command systems power can generally be raised arbitrarily, since the transmitter is on Earth. The design of the return link is a trade-off between spacecraft transmitter power, transmitter antenna gain, receiver noise level, and antenna size. For the spacecraft portion of the communication link the quoted parameters are extremely sensitive to the basic spacecraft design. For example, the method of spacecraft stabilization determines the antenna configuration and therefore the maximum gain which can be employed. The procedure for increasing gain emp oyed in the Pioneer program involved the use of a stack of dipoles (Franklin array) to generate a beam having cylindrical symmetry. With the symmetry axis coincident with the spacecraft spin axis and these axes normal to the ecliptic, the Earth always "sees" the main lobe of the antenna signal independent of spacecraft spin. A further variant is that used for some communication satellites, which consists of a group of Franklin arrays arranged about a common center. The patterns are phased so that the cylindrical symmetry is destroyed and a pencil beam is produced. The phasing is made to rotate counter to the spacecraft spin and at the same rate so that the beam is fixed inertially. In this way the 8–10-dB gain of the cylindrical array can be raised to perhaps 18 dB. The most advanced space probes, such as Voyager, Galileo, and International Solar-Polar Mission (ISPM), are continually directed toward Earth by using conical scanning for attitude correction. This permits use of a very large antenna.

Returning of data to Earth—the primary objective—entails the encoding of information in addition to selecting the most efficient transmission system. The information forwarded to the telemetry system for transmission must be serial and is usually digitally encoded. It is common to utilize a binary logic. Consequently, the information must be both digitized and converted to binary form, as well as properly tagged and synchronized with the spacecraft master timing system. Digitization of data from a particular experiment is carried out either as part of the experiment or else is incorporated into the encoder which samples, orders, stores, and tags the data. Thus, the encoder contains some of the intrinsic properties of a computer.

Phase-shift keying (PSK), a commonly used digital modulation technique in which the information is represented by discrete shifts in phase of the rf carrier, is shown schematically in Fig. 6. The system is symmetric; that is, it is applicable to either the command (up) link or telemetry (down) link of the spacecraft. The sin $\omega_0 t$ signal carries the information within its phase (±) which is chosen by the subcarrier modulator. The sin $\omega_0 t$ is convoluted with the command signal and transmitted to the receiver. Three information lines are required, that is, carrier, subcarrier (sin $\omega_0 t$), and synchronizing code.

As discussed previously, the communication

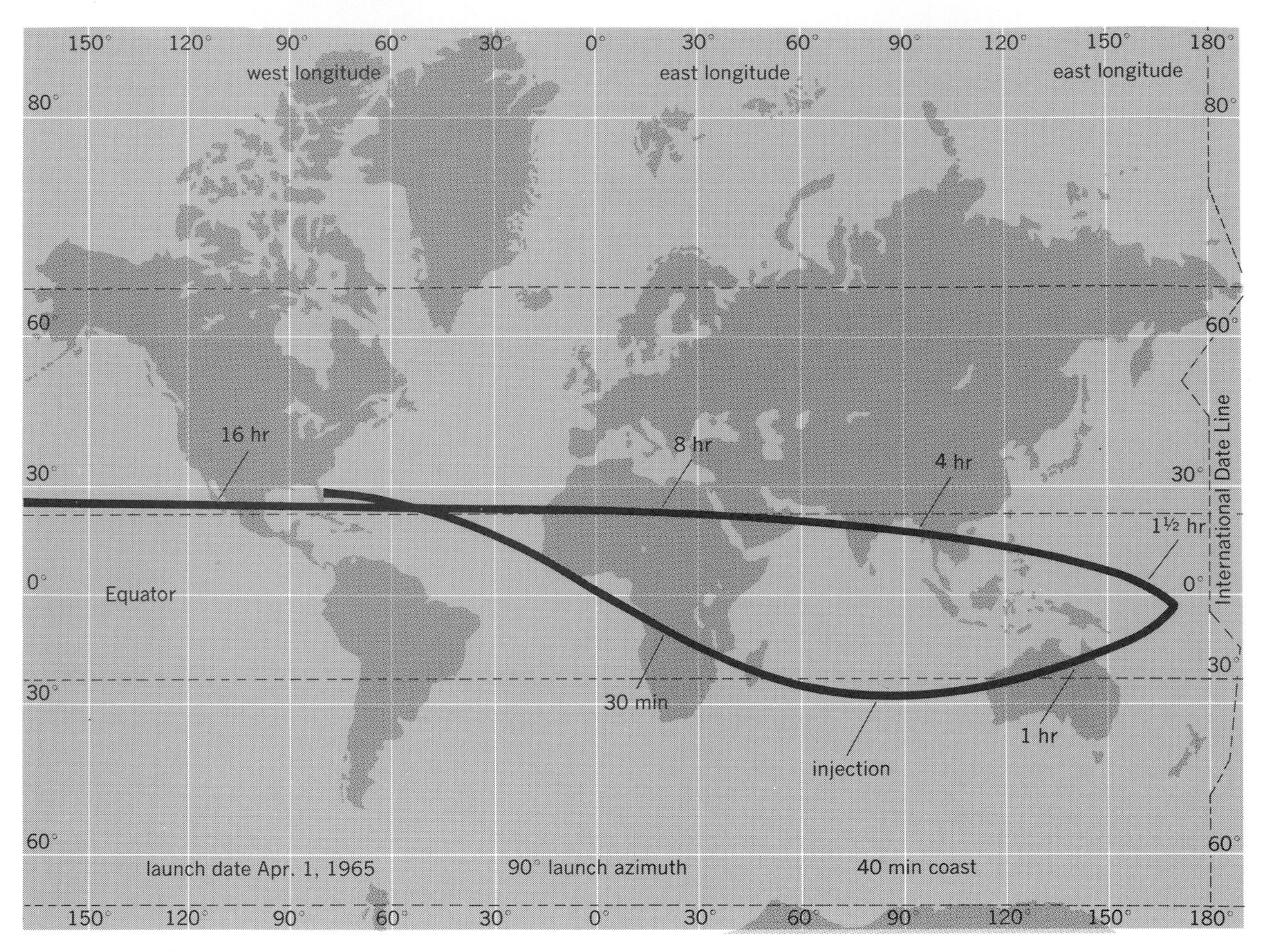

Fig. 7. Ascent trajectory projected on Earth for typical solar orbit to 0.8 AU (from Pioneer). Retrograde motion results from slowdown in angular velocity as spacecraft ascends (conservation of angular momentum). 1 AU = 1.496×10^8 km. (*NASA*)

system is inseparably linked to the design of the orbit. First, the problem of tracking the spacecraft must be a judicious choice between ground-station locations, Earth rotation, and launch or ascent phase. An illuminating way of displaying the orbit for tracking purposes is to plot the locus of the orbit as seen from the rotating Earth. As the spacecraft climbs, its angular velocity lags that of Earth so that finally its motion becomes retrograde, as shown in Fig. 7. This condition is for an interplanetary, Pioneer-type orbit with launching strictly eastward and with the ascent divided by a coast phase. It is also convenient to consider the decrease in communication-channel capacity (defined as bits/sec for some mean error rate) as a function of time in orbit. All interplanetary spacecraft orbits show such a decrease as distance increases; the loss is nonlinear because of the complicated manner in which radial distance varies. The telemetry profile or channel capacity for standard Pioneer solar (perihelion 0.8 AU) and antisolar (aphelion 1.2 AU) orbits is shown in Fig. 8, which displays channel rate versus time.

Attitude. Spacecraft orientation or attitude control is a special area of design which affects many other systems. The two general methods of atti tude stabilization are spin and active control, which uses gas jets, inertia systems, or combinations of these. Only certain early Soviet Lunik probes were allowed to tumble at a rate determined by final booster separation anomalies; even then the roll rate was sometimes used.

Since many experiments are designed to scan either the celestial sphere or a planetary disk, the means of attitude stabilization must be compatible with the experiment package. For simple scanning, it is often most appropriate to use a spinning spacecraft. Systems stabilized in inertial space do not lend themselves to the kind of scanning required for many interplanetary experiments. The situation for planetary scanning, however, is distinctly less clear. Though the equivalent of a simple flying spot scanner has been employed in spinning spacecraft, present practice is to use the more conventional television procedure. In this method the image field of a telescope is scanned electronically, and the system is mounted upon an active attitude-controlled spacecraft. The most sophisticated spacecraft, such as Voyager and Galileo, use dual-spin, with the image system mounted on an inertially stabilized section of the spacecraft with particle experiments on the spinning half.

The type of attitude control employed is also

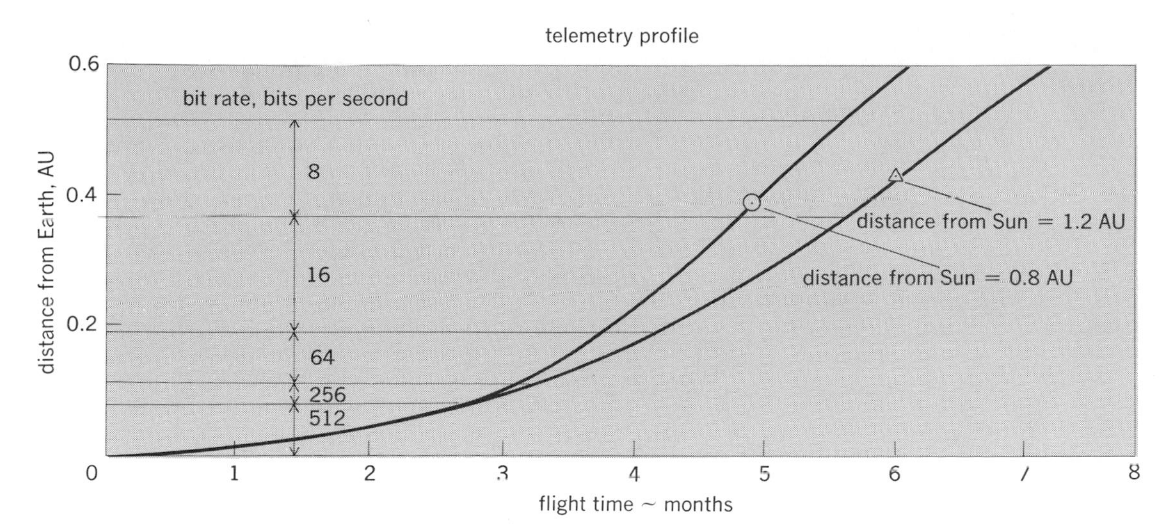

Fig. 8. Communication channel capacity, or telemetry profile, in bits/sec at various flight times for solar 0.8-AU perihelion and antisolar 1.2-AU aphelion (from the Pioneer program). 1 AU = 1.496 × 10⁸ km. (*NASA*)

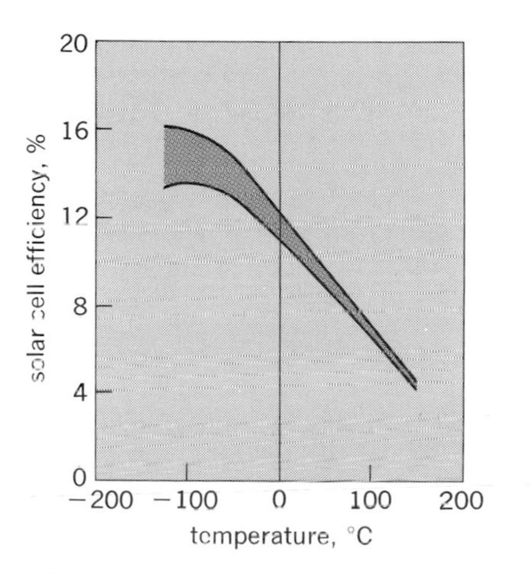

Fig. 9. Silicon *n/p* solar-cell efficiency as a function of temperature. (*NASA*)

influenced by guidance and communication constraints. A factor in communications is that more flexible antenna design is permitted by a nonspinning orientation; a consideration in guidance is connected with the greater ease of performing midcourse velocity trimming on a spacecraft which is nonspinning. Although this last comment is generally held to be true, it is recognized that maneuvering can also be done on a spinning spacecraft. Indeed, the Pioneer spacecraft oriented themselves by torque-induced precession. In addition to the constraint of communication antennas, the attitude-control system is strongly influenced by the power system in cases where the energy source is solar. Solar-cell array yield varies as the cosine of the incidence angle; attitude-stable spacecraft are favored for this reason, though they are not essential, as seen by the extensive use of solar-cell configurations on spinning spacecraft. More exotic solar power systems require nonspinning platforms. For these later-generation power systems, spinning spacecraft cannot be utilized.

A basic means of storing angular momentum is a flywheel. The spacecraft is linked electromagnetically to the flywheel, and control of coupling takes place electronically. Friction and momentum saturation limit this type of system, and it must be augmented by a gas-jet arrangement. Gas-jet nozzles supply the torques for Mariner-type spacecraft, with the *Mariner 4* being supplemented by solar pressure vanes. Since the gas-jet system alone has no friction, it is conditionally stable and can never be brought to rest. The spacecraft continually rolls between limit stops.

For interplanetary orbits, nonspinning systems are first oriented to the Sun and then roll-fixed about the axis pointing to the Sun. The latter requires reference to a celestial body other than the Sun. Earth is customarily utilized for this purpose. The Earth scanner has a cylindrical acceptance geometry and must be designed with a threshold so as not to lock inadvertently on the Moon or a bright star. For orbits exterior to 1 AU such a system will not suffice because in some cases the Earth and Sun are unfavorably located (inferior conjunction). In these instances a star (Canopus) is

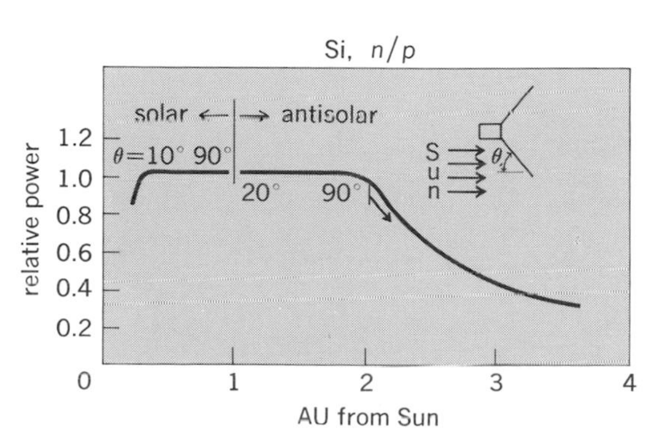

Fig. 10. Programmed solar-cell panel system efficiency using sweepback (canted angle of incidence of solar radiation) and blue-red filter glass for ultraviolet and infrared protection. 1 AU = 1.496 × 10⁸ km. (*NASA*)

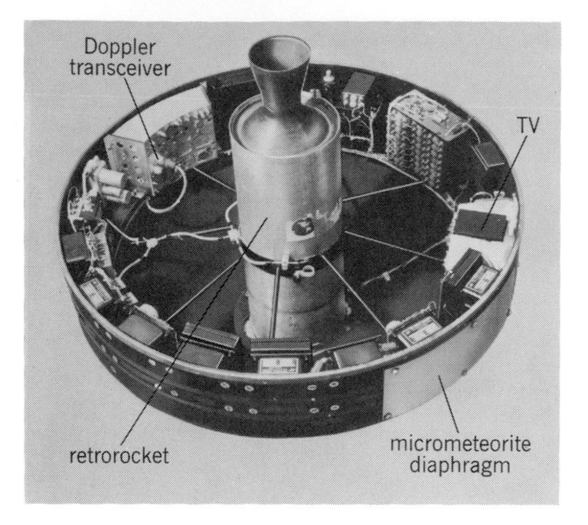

Fig. 11. *Pioneer 1*. Originally intended as a lunar orbiter, it passed through the Van Allen belts, the magnetosphere, and into the solar wind. (*TRW Systems*)

used as the roll reference. Clearly, as the spacecraft moves in orbit, a slow roll takes place for cases interior to 1 AU, since the Earth moves. For a Canopus seeker this is not so, as the star is celestially fixed. Since attitude-locked spacecraft such as Mariner have hinged antennas, the orientation must continually be modified. This is doubly complex for the orbit interior to 1 AU, since the spacecraft is in slow roll because of Earth motion.

Power. Electrical power is required for all spacecraft functions. Most probes employ solar-cell arrays that generate power from direct illumination by the Sun. Arrays of solar cells are generally mounted in plane configurations which are deployed sometime after the aerodynamic shroud used during launch has been jettisoned. On nonspinning spacecraft an important advantage is gained because the arrays can always be directed normal to the incident sunlight. Spinning spacecraft suffer a distinct disadvantage in this respect. Solar cells have an efficiency which is strongly temperature-dependent; they are useful only to distances of 2–3 AU from the Sun before their efficiency falls to a very low level. Figure 9 shows silicon *n*/*p* cell efficiency versus temperature. To maintain temperature to an adequate level near the Sun, one technique is to cant the cells. The high solar constant still provides an adequate level of power as close as a few tenths AU, although heating the cells degrades operation. In the case of antisolar missions, the low solar constant eventually degrades operation. The curves in Fig. 10 show both cases, where the progressive change in angle of incidence is used to extend the useful range of the cell array.

There have been advances in nuclear power supplies. In particular, SNAP-19 has been used on a

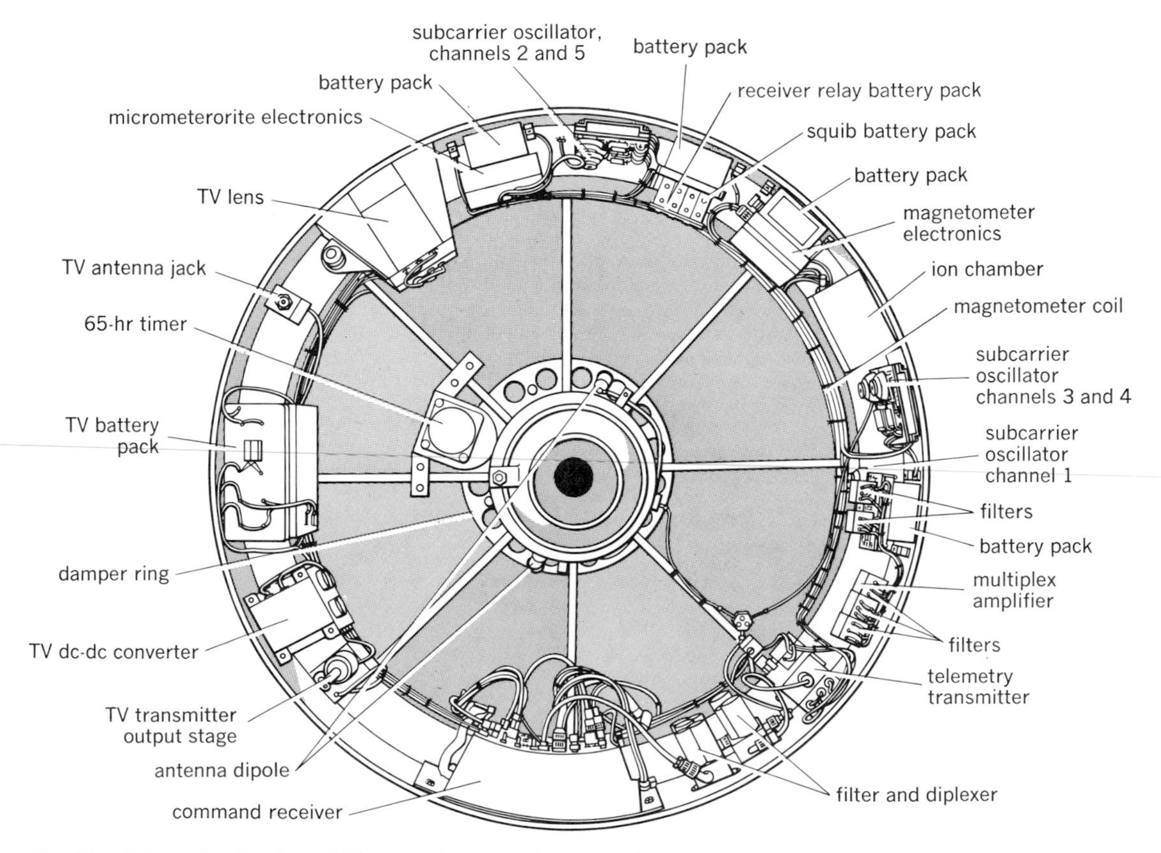

Fig. 12. Schematic drawing of *Pioneers 1* and *2* with primary subsystems. Total weight was 95 lb with about half assigned to the rocket. These spacecraft carried a TV system and experiment packages. (*NASA*)

number of spacecraft missions. This device is a thermalelectric generator. Electricity is generated by the thermal gradient across dissimilar metals; the heat is supplied from a radioactive isotope, for example, ^{238}Pu. Such systems were used extensively in the Apollo program to power the Apollo Lunar Surface Experiments Package (ALSEP) central station. There the fuel was loaded into the generator by the astronauts on the lunar surface. Thus the safety requirement during launch could be more easily met. The fuel was contained in a failure-proof special entry vehicle in the event of mission abort resulting in return to Earth.

For spacecraft use, the fuel must be inserted prior to launch. In this case, the design considerations are considerably more complex, as there are problems of decay and decay-gas products and production buildup, as well as material degradation from nuclear damage during the final checkout, prelaunch procedures, and launch. The SNAP-19 generator has been used successfully on *Pioneers 10* and *11* and will probably be used on all probes which are outbound, that is, have outer-planet trajectories.

SPACECRAFT CLASSIFICATION

A number of families of spacecraft, with similar missions and often similar designs, have arisen. Some of the important types of space probes are listed below.

Pioneer. This family of spacecraft, used mainly for interplanetary physics, constitutes a series which has radically changed in appearance over the course of a decade. The first group, *Pioneers 1* and *2* (Figs. 11 and 12) [also *Ables 3, 4*, and *5* where launch failures occurred], were intended to explore either the Moon or Venus or serve as a test-bed for instruments for later probes. *Pioneers 1* and *2* were both sterilized to satisfy the life-science requirements for a possible, but unplanned, lunar impact. They carried spot-scanning TV systems and a complement of particle and field experiments as well as a micrometeoroid detector. There were no solar paddles. *Pioneer 1* ceased transmission upon battery failure after lunar injection was missed because of error in the Thor guidance system. Premature battery failure occurred because of low temperature, which resulted from solar orientation misdirection due to booster guidance angle error. Data were received out to some 15 earth radii, providing the first pass through the Van Allen radiation belt in nonpolar latitudes and through the magnetosphere into the solar wind.

All Pioneer spacecraft are spin-stabilized using the angular momentum of the spacecraft following spin motor firing some time in the launch sequence. Perhaps the archetype of interplanetary spacecraft of the spinning type is the *Pioneer 5* (Fig. 13), which set a record early in 1960 for deep-space communication and the first long-term measurement of various interplanetary particle and field effects. The interior contains a complement of instruments for investigation of the solar wind, two transmitters, one of 5-watt and the other 150-watt power. The latter was used only in communication tests; it was planned for a Venus probe for a later flight in the series. The spacecraft con-

Fig. 13. *Pioneer 5* in flight configuration. Complicated paint pattern is for thermal control; solar panels are erected as they would be in space; small command reception antenna is shown at the bottom. (*TRW Systems*)

tained a solar-driven power supply, data processing and encoding, and power distribution. The attitude was fixed by the intrinsic spin imparted during launch. The antennas were of two types, one for up-link command reception and the other for data transmission to Earth. The antenna patterns were approximately omnidirectional so that the changing aspect of spacecraft to Earth would not severely modulate the signal strength. Solar power varied during the year because of changing aspect with respect to the Sun, requiring extra cell area when aspect was nonoptimum. *Pioneer 5* was launched in the general direction of Venus so that the orbit resembles later minimum-energy flights to Venus.

A considerably more complex Pioneer-type spacecraft with the total weight increased to some 350 lb (159 kg) from the early Pioneer class of 100 lb (45 kg) is shown in Fig. 14. This class is representative of the Pioneer design philosophy circa 1960. A motor was included for midcourse trim and deboost into lunar injection. The complement of experiments was considerably enhanced over that of the earlier spacecraft; the general instrument configuration still included a preponderance of particle and field sensors. Three spacecraft of this series were launched. In all firings the launching system (Atlas/Able) failed (in one case, the spacecraft shroud failed in ascent).

Attempts using the Pioneer were revived in 1962 (Fig. 15*a* and *b*) when a new Pioneer configuration was planned using orientation of the spinning spacecraft after insertion into interplanetary orbit. There are presently four such spacecraft in interplanetary orbit about the Sun. Solar-cell paddles are no longer used, the cells being distributed over a cylindrical outer spacecraft shell. Perhaps the most basic system change affecting many of the design features of this spacecraft is that the spin axis is pointed in the direction perpendicular to

Fig. 14. *Atlas 4-Able* spacecraft, intended as a lunar orbiter, mounted and in the gantry prior to attachment of solar panels and shroud. (*TRW Systems*)

the plane of the ecliptic some hours or days subsequent to the launch; this is carried out by use of cold gas jets that apply a torque to the spacecraft. (The spacecraft is similar dynamically to a spinning top.) The torque maneuver takes some hours to complete, after which the spin axis is essentially invariant. The solar cell configuration provides constant power throughout the orbit except for variations in the radial distance from the Sun. The thermal system consists of heat rejection through the bottom of the spacecraft, using louvers driven by bimetal spring motors. The direction in which heat is rejected is fixed and perpendicular to the Sun at all times, which makes this design particularly appealing. Antenna directionality is constant, with a beam geometry of pancake form achieved by using a stack of dipolar radiators (Franklin array); the pancake lies in the plane of the ecliptic and therefore always intersects the Earth. The beam is used to aid in the orientation maneuver. Scanning in the plane of the ecliptic is especially convenient for plasma instruments which must sweep through the solar wind. Since its beginnings early in the space program, Pioneer has undergone significant system changes, as has the Mariner spacecraft. The key differences are still the use of spin angular momentum to stabilize the Pioneer and an attitude-control system to maintain orientation of the Mariner. The Pioneer-Jupiter spacecraft (*Pioneers 10* and *11*) encountered (periapsis) the planet Jupiter on Dec. 3, 1973, and Dec. 3, 1974, respectively. *Pioneer 10* was sent on a solar-

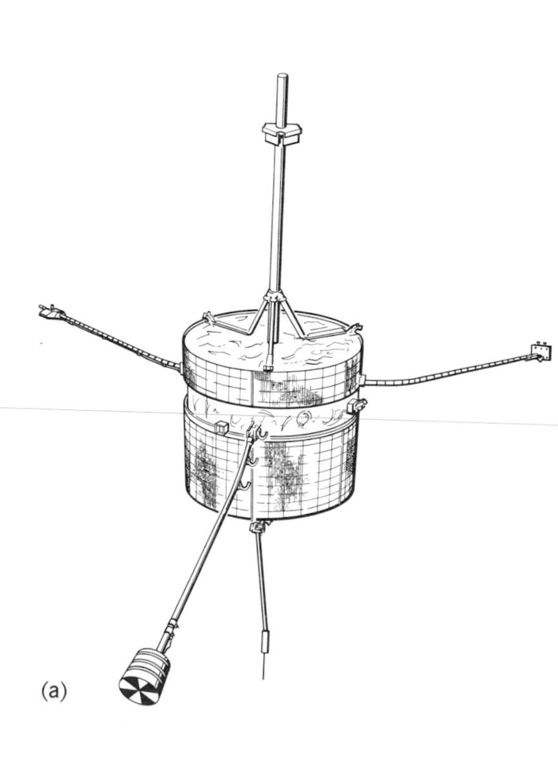

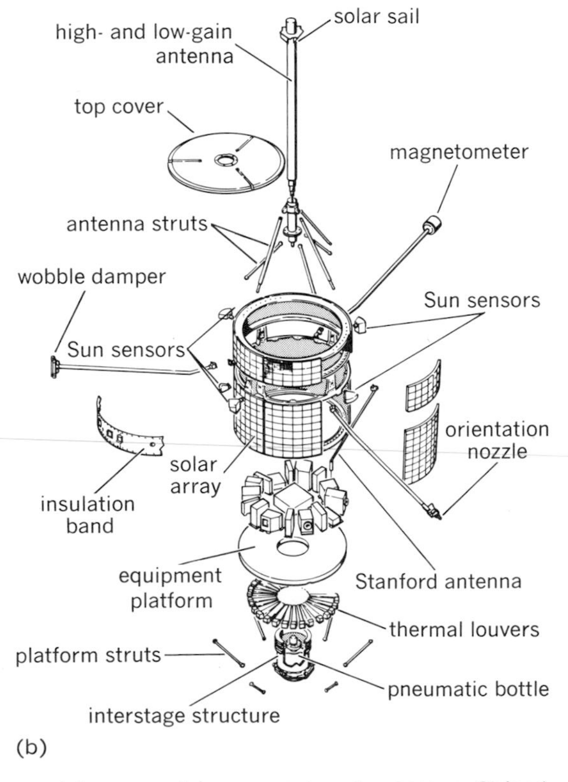

Fig. 15. Pioneer spacecraft. (*a*) View of the assembled *Pioneer 6* spacecraft with booms deployed. Vertical structure is stacked dipole (Franklin array) for telemetry transmission to Earth. Small vanes on antenna are used to maintain orientation by solar radiation pressure. Forward boom contains magnetometer sensor. Striped coating is for thermal control. Folded boom at bottom is for Stanford two-frequency radar experiment, and boom at right contains jets used to apply torque. (*b*) Exploded view of Pioneer showing main subsystems. (*NASA*)

system escape trajectory. *Pioneer 11* encountered Saturn in September 1979. *See* JUPITER; SATURN.

The most obvious change in the Pioneer from earlier versions is the antenna configuration, which was changed from a Franklin array transmitting a cylindrically symmetric beam to a 9-ft (2.7-m) paraboloid (Fig. 16). The orientation of earlier Pioneers perpendicular to the plane of the ecliptic or the spacecraft orbit would not work for a paraboloid, and the spacecraft orientation had to have the spin vector lying in the orbital plane. In turn, this implied a periodic updating of the direction of the spin axis and antenna-beam axis so that the peak beam strength would be directed toward Earth. The updating was carried out using a gas system for applying small precessional torques to the spacecraft. Spin orientation was also required for this and was obtained using a star seeker pointed at Canopus, as used for Mariner.

A second profound change in the spacecraft is in the power system. Since the distances from the Sun are great, solar energy would have been insufficient. Thus for the first time, other than for Apollo (where fuel loading was carried out manually on the Moon), radioisotope thermoelectric generators (RTGs) were employed as the source of electrical energy. These provide electrical power by the heat generated from the decay of ^{238}Pu, which drives an array of thermoelectric junctions consisting of positive elements of antimony and negative elements of an alloy of tellurium, silver, and germanium. A key consideration in the power subsystem is the half-life of ^{238}Pu, which is 50 years, and reflects upon the length of the Pioneer mission. Additional considerations include the production of decay products, especially He, and containment of the source between manufacturing and flight.

The design of the spacecraft is seen in Fig. 17, which shows the RTG and magnetometer booms. RTG deployment took place after launch and subsequent to boom deployment. These devices were run out on cables with drag braking. The scientific payload consisted of 11 experiments on *Pioneer 10* and an additional magnetometer on *Pioneer 11*. Space used for experiments included the backside of the antenna where meteorite detectors of the pressure-cell type were attached. A spin-scanning polarimeter was also carried which used the combination of spin motion and the nodding motion of the telescope attached to the side of the spacecraft bus. Pictures of Jupiter were reconstructed from the spin-scan images, some of which were four times better than the best obtained from Earth. Although this factor appears modest, the change gave a vivid new view of the cloud structure of the planet (see Fig. 18). From the standpoint of the scientific experiments, key considerations for a mission of this type were the extreme lifetime required and operation in the presence of radiation from both the radiation belts of Jupiter and the RTGs.

From the standpoint of navigation, Pioneer-Jupiter represented the most complex uncrewed mission to that time with the exception of Viking. Figure 19 shows the Jovian close-approach trajectory of the two spacecraft complicated by the presence of the numerically large satellite system, about which scientific information was also recovered.

Fig. 16. Pioneer-Jupiter spacecraft in test base. To the left is the Centaur shroud. The large paraboloid is shown at the top of the spacecraft with the feedhorn and lower-gain antennas mounted above. Instrument bays are below the antenna and are covered with thermal protection. (*Ames/NASA*)

Among the more important results are density determinations obtained from analysis of orbital perturbations and radio occultations.

The communication, tracking, and data formatting are done compatibly with the Jet Propulsion Laboratory (JPL) Deep-Space Network, and this system is used for all tracking functions. The early operational control originates at Kennedy Space Flight Center (KSC) with a handover to JPL, and finally to Ames Research Center, which operates the spacecraft through the Deep-Space Network.

Ranger. This series of spacecraft began with an attempt to land a single-axis seismometer upon the lunar surface. The technique included deboost into lunar trajectory and a semisoft landing, where the residual impact was removed using a balsa wood absorbing sphere. This design was abandoned in favor of a hard landing using television coverage during the approach to obtain photogeological information of the surface. An early Ranger of this type (Block 1) is shown in Fig. 20. The use of an impact absorber (the ball in Fig. 20) is possibly of more than historical interest, since it appears that

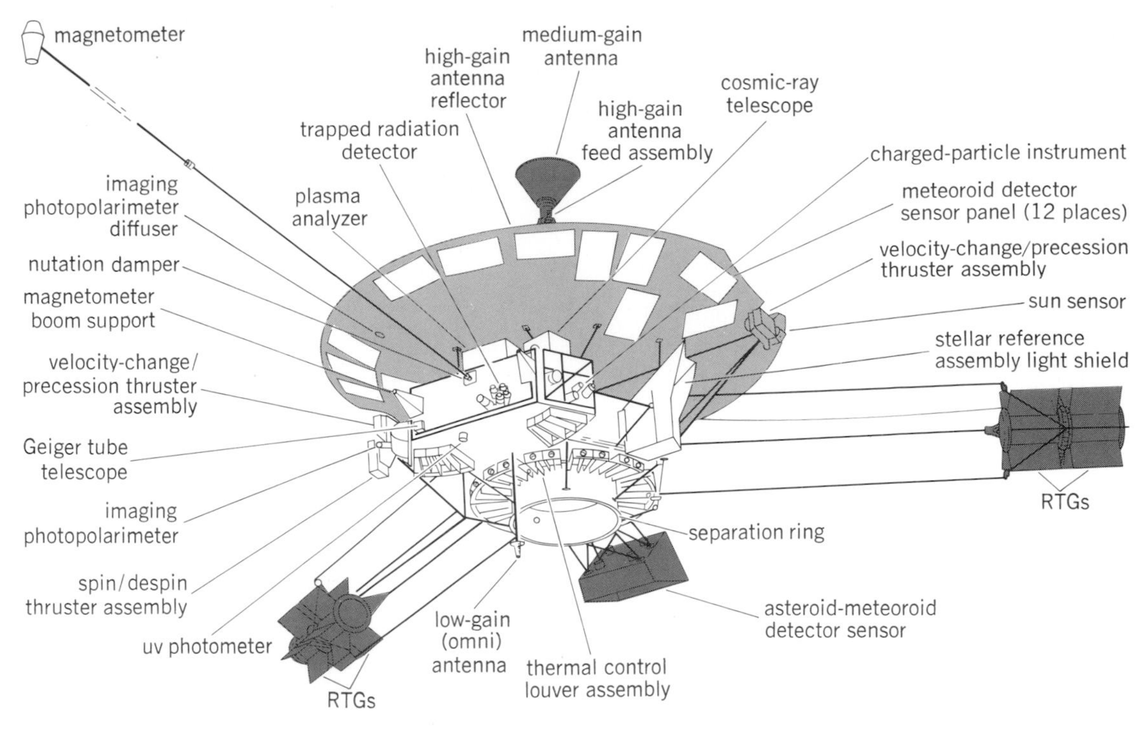

Fig. 17. Oblique view of *Pioneers 10* and *11* (Pioneer-Jupiter) spacecraft giving the various subsystems. The magnetometer and RTGs are shown in deployed configuration. (*From C. F. Hall, Pioneers 10 and 11, Science, 188:445–446, 1975*)

some of the Soviet Moon probes were landed by a technique of this type. However, United States experience ended because of a series of failures leading to redirection of the program to the *Ranger Block 3* with a television system.

The consequences of Ranger program redirection was a primarily photogeological spacecraft boosted by an Atlas-Agena rocket. The whole Ranger program included nine launches (Table 2). The basic spacecraft system followed the guidance and control philosophy developed at JPL, similar to that of Mariner and Surveyor. It was based upon solar and Earth or Canopus attitude control together with short-time inertial stabilization for maneuvers. Ranger can be considered as a test-bed for these later programs. The attitude system using the Sun and Canopus or Earth is a basic means of in-flight, nearly inertial, stabilization of interplanetary spacecraft.

The Doppler tracking system on Ranger was similar to that used in the other spacecraft of the JPL series. There were differences in propulsion, since the Ranger was not required to carry out the extensive maneuvers used later for Surveyor. The Block 3 payload consisted of a television system,

Fig. 18. Spin-scan image of Jupiter taken by *Pioneer 11* on Dec. 4, 1974, showing the Giant Red Spot. The image is in blue light. (*NASA*)

Table 2. Ranger firings

Designation	Launched	Remarks
1 and 2 (Block 1)	Fall, 1961	Launch vehicle failure; low-altitude Earth orbit
3 (Block 2)	1962	High launch vehicle dispersion; spacecraft failure
4 and 5 (Block 2)	1962	Spacecraft failure; tracked capsule transmitter to impact (*Ranger 4*) and interplanetary for 11 days (*Ranger 5*)
6 (Block 3)	Jan. 30, 1964	Lunar impact; 9.3° N, 21.5° E; television failure
7	July 28, 1964	Lunar impact; 20.6° W, 10.6° S; extensive photo coverage
8	Feb. 17, 1965	Lunar impact; 2.7° N, 24.8° E; extensive photo coverage
9	Mar. 21, 1965	Lunar impact; 13.1° S, 2.4° W; continuation of photo coverage

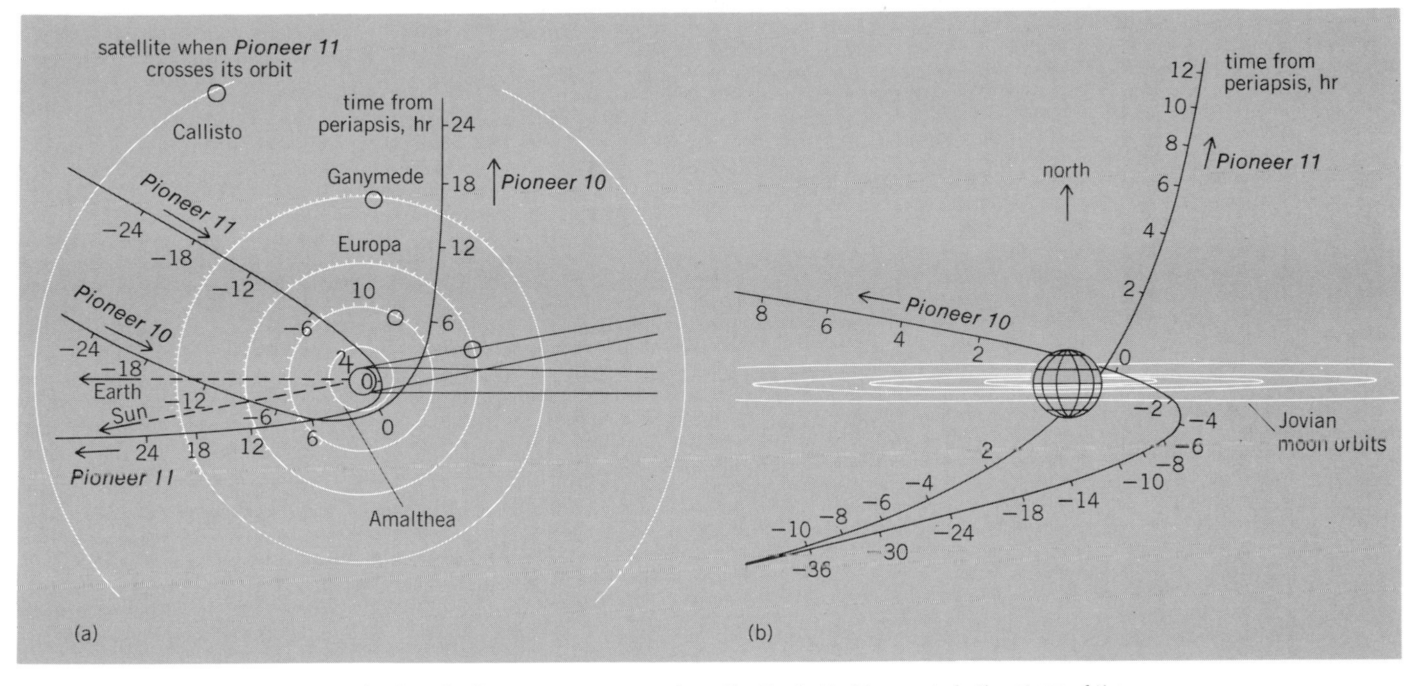

Fig. 19. *Pioneer 10* and *11* trajectories in a Jupiter-centered coordinate system. (*a*) The trajectories resolved into the plane parallel to the ecliptic. (*b*) The trajectories as seen from the Earth, that is, nearly in the plane of the ecliptic. (*From C. F. Hall, Pioneers 10 and 11, Science, 188: 445–446, 1975*)

including a complex of cameras designed for variable resolution photography of the lunar surface. The motion of the spacecraft had to be taken into account in the photographic sequence, and this formed an important aspect of the operation. As the spacecraft approached the Moon, the area covered by the camera decreased; with the approach of the spacecraft, lateral motion became significant since the spacecraft did not head radially into the planet. Thus the picture sequence was as shown schematically in Fig. 21. The final result, showing one close photograph from Ranger covering about 18 mi (29 km) across the photograph, is shown in Fig. 22.

The Ranger system required midcourse correction and steering similar to that for Surveyor. The spacecraft was steered by main booster propulsion into a final burnout trajectory, after which the free-flight trajectory permitted a determination of the lunar impact or miss using two-way Doppler radar. Then fast computation was carried out on Earth to determine the velocity to be added for an impact at the nominal point on the Moon which would provide appropriate solar illumination. This correction was inserted into the spacecraft control system in the following manner. The spacecraft was reoriented away from the Sun along the direction of the vectorial velocity to be added, and the vernier or midcourse rocket was fired for a predetermined time. Then reorientation toward the Sun was carried out. This placed the spacecraft into a lunar impact trajectory. An example of the effects of the velocity gained upon the lunar approach trajectory for the *Ranger 8* flight is shown in Fig. 23. As shown, this vehicle would have missed the Moon altogether without the midcourse trim. For *Ranger 9*, near terminator coverage was desired to increase the picture contrast. This, in turn, reflected upon the targeting accuracy. Terminal trajectory for *Ranger 9*, typifying the Ranger lunar approach, is shown in Fig. 24. There was a total of six cameras; the F channel contained two fully

Fig. 20. *Ranger Block 1*. At top is the impact absorber system with command antenna. Directly under is the solid-propellant retromotor for primary lunar deboost. (*NASA*)

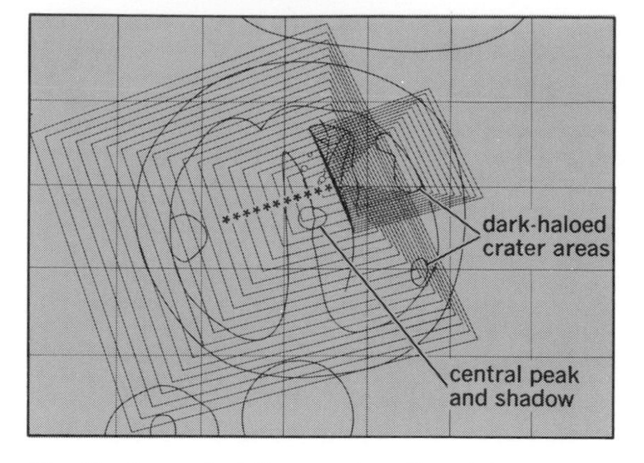

Fig. 21. Conceptual diagram of nesting of photographs as taken by *Ranger Block 3* spacecraft approaching Moon. Nesting is due to a combination of contraction of the field of view as the Moon is approached, together with the lateral motion of the spacecraft as it bears down slantwise on the Moon. (*JPL/NASA*)

Table 3. Summary of Ranger camera characteristics

Characteristic	Camera F_A	Camera F_B	Cameras P_1 and P_2	Cameras P_3 and P_4
Aspect ratio	1:1	1:1	1:1	1:1
Active scan area, in. (mm)	0.44 x 0.44 (11.2 x 11.2)	0.44 x 0.44 (11.2 x 11.2)	0.11 x 0.11 (2.8 x 2.8)	0.11 x 0.11 (2.8 x 2.8)
Line rate, Hz	450	450	1500	1500
Horizontal line time, msec	2.22	2.22	0.6666	0.6666
Horizontal blanking time, msec	0.22	0.22	0.1111	0.1111
Frame rate, Hz	0.39	0.39	5	5
Frame time, sec	2.56	2.56	0.2	0.2
Vertical blanking time, msec	46.6	46.6	6.6	6.6
Group vertical blanking time, msec	–	–	40	40
Video bandwidth, kHz	177	177	144	144
Lens optics	25 mm, $f/1$	75 mm, $f/2$	25 mm, $f/1$	75 mm, $f/2$
Shutter speed, msec	4	4	2	2
Horizontal erase rate, Hz	25	25	3000	3000
Number of TV lines scanning	1130	1130	280	280

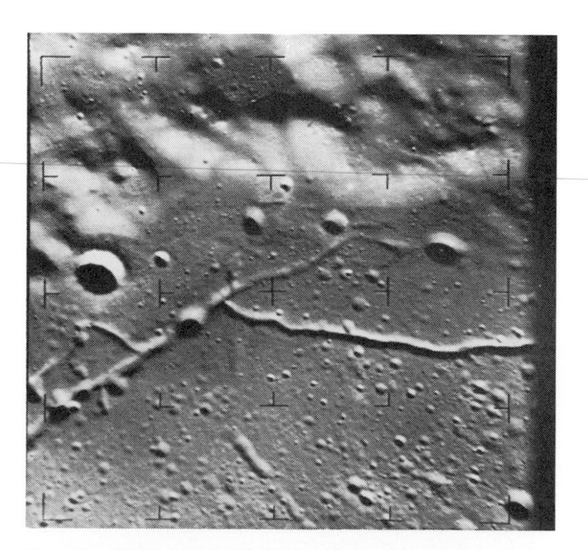

Fig. 22. *Ranger 9* photograph of surface of Moon taken near crater Alphonsus. Most prominent and striking features are rills or valleys across photograph. (*NASA*)

scanned vidicons and the P channel the remaining four, partially scanned. Technical details of the cameras are summarized in Table 3.

Surveyor. This spacecraft (Fig. 25) was a complicated system for delivering a payload of exploration and scientific instruments to the surface of the Moon and providing telemetry, power, and control after landing. The original design was planned to permit operation through the first lunar night following landing. Because of Centaur rocket payload limitations, the spacecraft design restricted operation to the first lunar day. However, the electrical system on some spacecraft survived the lunar night, and partial operation was possible into the second day.

Surveyor included all equipment for midcourse or cruise corrections to the trajectory, as well as propulsion, guidance, and control for the final descent to the surface. The basic portions of the spacecraft were the frame or structure, the power system, communication, attitude control, thermal system, propulsion, and the payload which, on early missions, was restricted to a television camera and numerous engineering sensors. These were sufficiently broad in scope of measurement that study of the environmental conditions on the lunar surface aided the lunar surface science program considerably. This often occurs in the early phase of a planetary exploration mission and forms a major link between the strictly scientific and engineering data needs.

The operation of the spacecraft is best seen by examining a model flight profile. The launching rocket system was an Atlas-Centaur. Ascent trajectory was similar to the general class discussed previously. Separation of the spacecraft from booster took place after final thrust termination of the Centaur. The profile of the cruise flight is shown in Fig. 26. The Sun sighting was acquired prior to separation, and sighting on the star Canopus, acquired afterward, provided roll control. A midcourse correction to the desired trajectory was generally required to fix the spacecraft properly upon the target area on the Moon. At a predetermined aiming point in space, Canopus pointing was unlocked, the spacecraft put on gyro control, an Earth-programmed two-axis roll instituted, and the retrorocket system fired. This committed the vehicle to a landing area. The spacecraft was returned to proper orientation, and the remaining operations of landing and operation on the surface were carried out.

To carry out the landing operations, the spacecraft contains numerous systems not found on vehicles used for interplanetary missions, including a sophisticated propulsion system programmed with a complex series of internally generated instructions. For the main deboost maneuver, Surveyor carried a large spherical solid-propellant motor of 8000–10,000-lb (35–45-kN) thrust which burned for some 41 sec. This reduced the speed of the vehicle measured in lunar coordinates from 6100 to 250 mph (2700 to 110 m/sec), removing the major fraction of the descent speed and leaving only a minor residue. The start of the main engine was by a radar altimeter placed in the nozzle of the motor; the altimeter was blown away by the start,

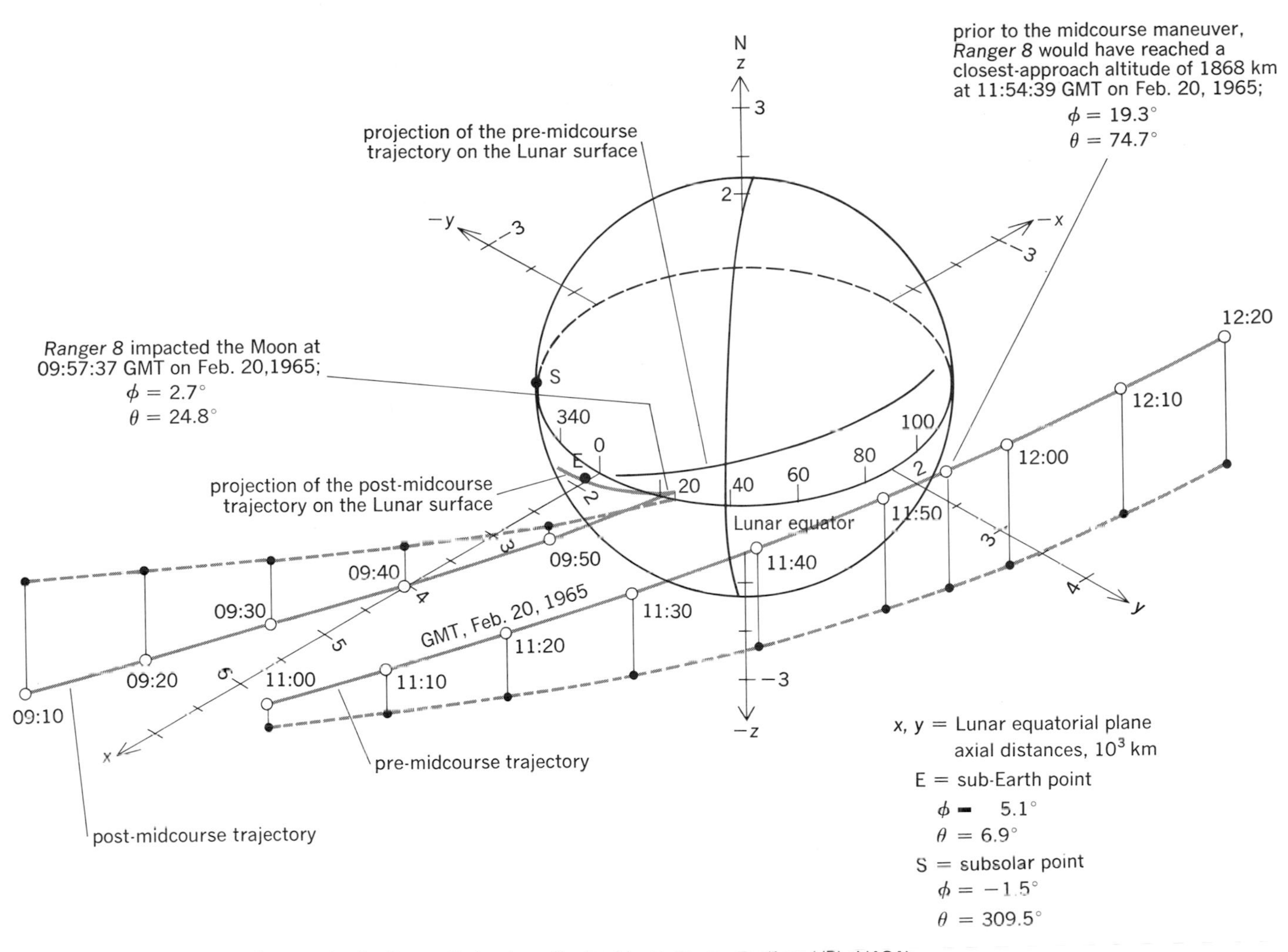

Fig. 23. Pre-midcourse and encounter for *Ranger 8* showing effects of trajectory correction. (*JPL /NASA*)

which took place at about 60 mi (100 km) slant range from the surface. It is important to bear in mind that the primary retrothrust was along the trajectory and in the direction opposed to the velocity vector. Thus, there remained a velocity laterally along the surface as well as the residual 250 mph (110 m/sec) of vertical velocity which was removed from the system prior to touchdown.

The final reduction of the lateral and normal components of approach velocity was attained by use of three small liquid-fueled vernier rocket motors. Two were rigidly attached to the frame of the spacecraft whereas the third was gimbaled about two axes. Doppler radar together with the vernier rocket system provided attitude stabilization and lateral and vertical thrust, and was used during the main motor firing for attitude control. The vernier system shut down 14 ft (4.3 m) off the surface, and the final landing speed was some 10 mph (4.5 m/sec), which was absorbed by the landing gear. The latter formed one of the distinctive features of Surveyor and is shown in Fig. 27 with the vernier propulsion system delineated against a background of the spacecraft. This rocket system used two liquids which ignite pyrogallically (without aid). Fuel tanks were pressurized using helium and pressure bladders so that the firing could take place in zero gravity for midcourse corrections.

The TV camera operated with a programmable mirror ahead of the optical train so that panoramic coverage over 360° in azimuth and from 40 to −60° could be obtained. The camera used a vidicon tube and the scan was 600 lines, with a backup mode of 200 lines in the event of failure of the high-gain

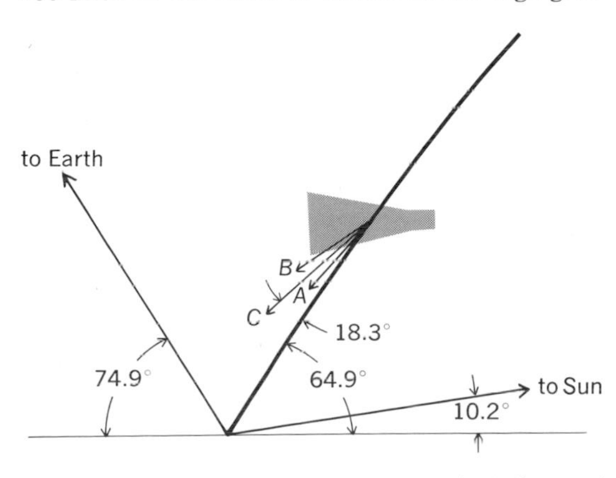

Fig. 24. *Ranger 9* terminal trajectory orientation and glide path. (*NASA*)

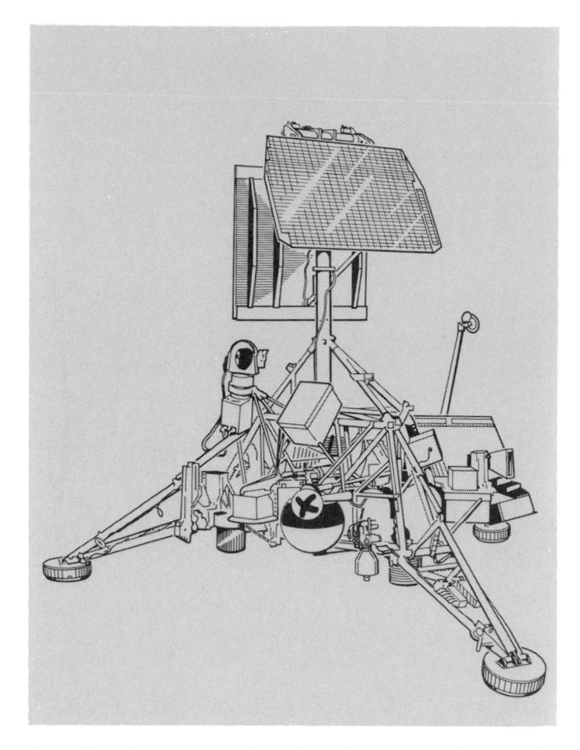

Fig. 25. Surveyor full-scale science training model which is used in operational simulation of the instrument complement. (*NASA*)

antenna. The frame rate of the primary system was once every 3.6 sec. A cutaway view of the camera is shown in Fig. 28. The color system, composed of a three-filter wheel, was available for special spectrophotometry of the surface.

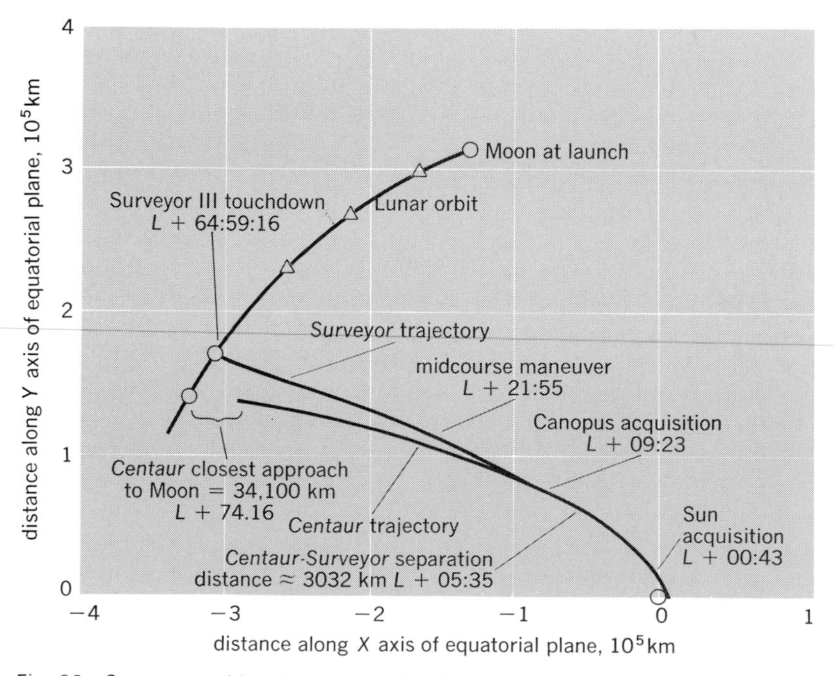

Fig. 26. Surveyor and booster trajectories. Solar acquisition points spacecraft and booster in direction of Sun; then booster is separated and Canopus acquisition begins for Surveyor. This provides complete attitude control during flight to Moon. The Centaur booster is no longer of interest after separation except that a slight boost is given it to ensure that impact with Moon does not take place. (*JPL/NASA*)

The alpha backscatter experiment for elemental analysis of the surface soil was a novel device which formed the basic chemical probe used on Surveyor. The experiment combined elastic backscatter of α-particles from target nuclei and protons from (α, p) reactions on surface elements. The α-scatter measurements were made at fixed angle. An energy-sensitive detector measured the energy spectrum of the scattered α-particles which for a fixed angle depended on the nuclear charge Z of the target material. The (α, p) reactions on certain light elements, where the α-scattering efficiency dropped off, were intended to provide low Z composition.

Surveyor 7 was the only spacecraft targeted to the highlands region of the Moon; it landed some 18 mi (29 km) north of the crater Tycho. From composition measurements there appear to be certain differences from the maria composition measured on *Surveyors 5* and *6* as well as different surface morphology. The area in which *Surveyor 7* landed is composed of a much rougher surface, characteristic of a rubble-strewn field from secondary ejecta from a large crater, possibly Tycho. Despite the more severe landing conditions the operation was successful, and the photogeology, forming one of the primary scientific functions of Surveyor, was extended from the case of the maria. A panoramic photograph taken by the television system of *Surveyor 7* is shown in Fig. 29.

Lunar orbiter. This was a spacecraft specifically designed to place an instrument package in a stable orbit about the Moon. It was a solar-powered vehicle with the customary two-way Doppler tracking system, a frame, thermal system, and data transmission link. The spacecraft was modular. The photographic instruments including cameras and automatic development and video scanning system were carried in a sealed container which was removable and could be replaced by other instruments, though only photographic missions were flown. There were five flights in the series, all of which produced high-quality photographs. The spacecraft weight was approximately 850 lb (385 kg), of which 150 lb (68 kg) was assigned to the photographic system and 262 lb (119 kg) was propellant for the midcourse engine. The camera was equipped with two lenses, one with a focal length of 80 mm and the other 610 mm.

The spacecraft was placed in an elliptic orbit with perilune (closest approach to the Moon's center, also called periselene) at 120 mi (193 km) and apolune (furthest orbital point from the Moon, also called aposelene) at 1150 mi (1850 km). This ellipticity was decided on to provide sufficient time in sunlight for complete battery recharging. Actually, some of these spacecraft were steered as close as 28.5 mi (45.9 km) from the surface. The high surface speed (4300 mph or 1900 m/sec) at perilune meant that the long photographic exposure times, 1/25 to 1/100 sec, would have caused exposure blurring if the camera had not been equipped with a motion compensation mechanism.

Orbiter returned 1950 photographs of the Moon which permitted selection of five prime Apollo landing sites; the photographs have proved extremely useful in interpretive studies of the geology of the lunar surface. A good fraction of the back side of the Moon was photographed. *Orbiters 4* and *5* were placed in polar orbits, providing coverage not attainable from the Earth even with rectification to correct for the extreme obliquity of limb photographs.

Mariner. Mariner is the archetype of planetary explorers in the United States space program. There were seven missions flown in the 1960s. Five were successful, two to Venus and three to Mars. The Mariner spacecraft is solar-powered, has an attitude-controlled stabilization system with Sun and Canopus or Earth pointing, and uses two-way Doppler compatible with the Deep Space Network for tracking and trajectory determination. A gradual evolution of design has taken place since the first firing to Venus in 1962. Generally, the tendency has been in the direction of a compact structure with a thermal system common to the whole frame, a departure from the Surveyor and Ranger designs. The antennas and solar-cell structures have also evolved in time. However, the configuration does not grossly differ from that of *Mariner 4* (Fig. 30). *Mariner 2* weighed 449 lb (204 kg), *Mariner 4* weighed 575 lb (261 kg), and *Mariners 6* and *7* each weighed 850 lb (385 kg). Miniaturization of equipment allowed for much greater efficiency in the later spacecraft. Except for the addition of a rocket engine, the 1971 Mariner Mars orbiter spacecraft, *Mariner 9*, was in many respects similar to *Mariners 6* and *7*.

These spacecraft, which are combined interplanetary probes and planetary explorers, carry a complement of equipment intended to carry out primary research in cosmic rays and the solar wind, the latter requiring both a magnetometer and plasma detector. Planetary measurements have included investigation of the interaction of the solar wind with the planets as well as atmospheric and surface measurements. Atmospheric studies include radio probing of the upper atmosphere, using variation from nominal in the transmission of the tracking signals sent to and received from the spacecraft. For surface observation a number of sensors have been used. Much information has been found using television coverage in the case of Mariner Mars series, where the photographic scanning of the surface returned the first details of the planet's surface. The photographic coverage proved, following the Ranger and Surveyor lunar missions, that television is likely to become a permanent feature of the design of planetary probes in the future. From *Mariner 4* the results were especially novel since, for the first time, intense cratering was seen on the surface of a body other than the Moon. Some 300 craters were identified in these first closeup pictures of Mars, with another 300 possible, ranging from 3 to 180 km in diameter. The central picture in the scan sequence of *Mariner 4* shows perhaps the most visible crater field. This photograph (Fig. 31), covering a field of view of some 175 km in the east-west and north-south directions, is thought to include primarily the Mare Sirenum.

Better-quality pictures from *Mariners 6* and *7* mostly showed similar terrain, although erosion was more pronounced, especially in the southern

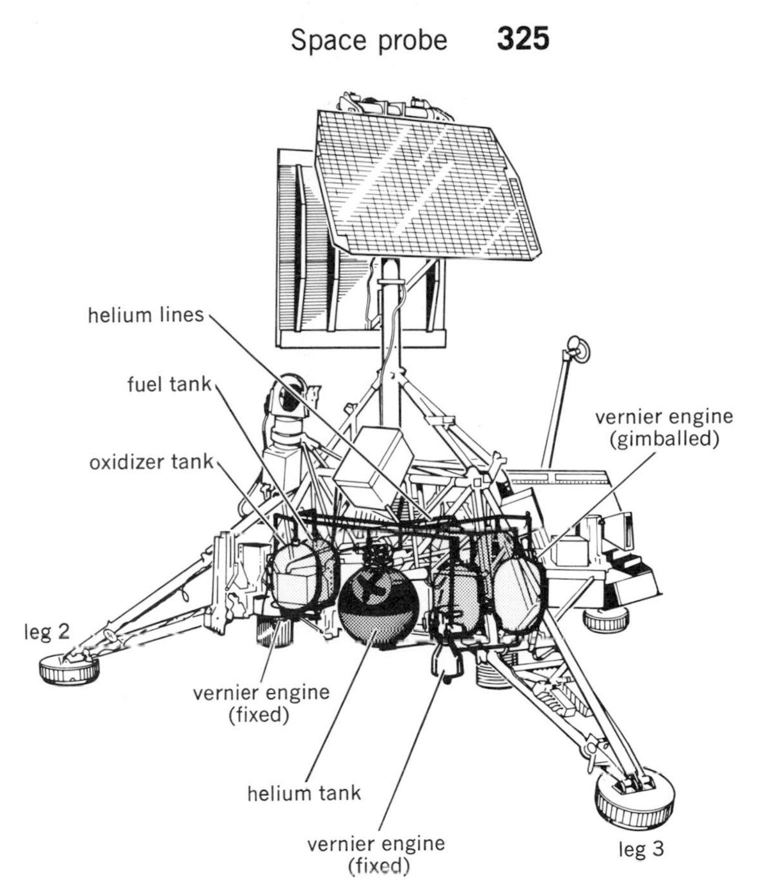

Fig. 27. Partial schematic of Surveyor to show part of propulsion system associated with vernier engines which provide midcourse trim, attitude control during firing of large retrorocket, and final velocity removal near lunar surface. (*JPL/NASA*)

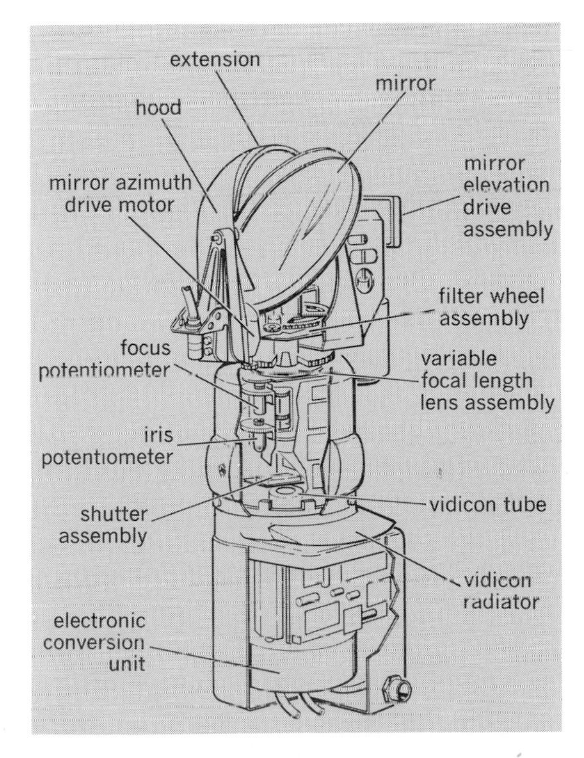

Fig. 28. Cutaway view of a Surveyor television camera, including filter wheels for color photography, focus control, and directional control using top mirror. (*NASA*)

Fig. 29. Photo mosaic by *Surveyor 7* of rubble field associated with secondary crater some 1½ mi (2.4 km) wide. (*NASA*)

hemisphere. A "jumbled" terrain, unlike any found on the Earth or Moon, was evident in some pictures. *Mariner 7* pictures of the southern polar region showed some drifting of "snow." Each picture transmitted by *Mariner 4* was composed of 200 200-element lines, whereas each picture transmitted by *Mariners 6* and *7* was composed of 704 945-element lines. *See* MARS.

The scan sequence of the Mariner spacecraft required the utmost planning and was, in some ways, indicative of the complexity of the engineering task in preparing a planetary spacecraft. It is informative to examine the evolution of the operation. The combination of constraints in the targeting included a requirement that the probability of accidental impact upon the planet be less than 1: 10^4 to satisfy a low probability of biological contamination of the planet. Lighting upon the surface was critical for attainment of sufficient relief and for the scan needed to cover a reasonably interesting region representative of a large part of the planet. Other experiments also needed special orbit characteristics, for example, to search for any trapped radiation. From the engineering standpoint it was necessary to ensure that solar pointing would be maintained so that power would always be available; Earth occultation became a requirement late in the program to carry out limb radar propagation studies through the atmosphere; lastly, the Goldstone tracking station was to be used during the encounter because of the superior communication link with the operation facility in Pasadena, Calif. These factors together determined a matrix of target points in the vicinity of the planet. A representation of the aiming problem can be graphically shown using the plane normal to the approach vector of the spacecraft prior to the beginning of influence of the Martian gravitational field. This plane, essentially normal to the ecliptic, defines the point in space near the planet which is aimed at, the locus of the boundary of probability 10^{-4} that an accidental impact would be made on the planet destroying its pristine biological environment, and three angles, Canopus-probe-Mars (CPM), Earth-probe-Mars (EPM), and Sun-probe-Mars (SPM). These quantities are considered jointly in arriving at the best compromise of aiming or target point in space. In Fig. 32, CPM angles less than 36° cause loss of Canopus lock and result in random roll of the spacecraft. The boundaries SPM = 0 define the bounds within which solar occultation takes place, indicating dark side coverage, that is, solar eclipse by Mars; EPM = 0 defines Earth occultation by Mars, permitting Doppler radio transmission through the atmospheric limb to be carried out. A complex choice of possibilities including camera illumination, pristine ecology, radio occultation, and arrival at Mars when the probe would be seen from the prime DSN station at Goldstone resulted in the sample arrival target point shown in Fig. 32 for the arrival date of July 15, 1964.

Mariner 9 was designed to map 70% of the Martian surface within 90 days with high-resolution cameras, and at the same time study the atmosphere and surface via ultraviolet and infrared spectroscopy and infrared radiometry. S-band occultation was also studied by using the spacecraft radio experiment, and celestial mechanics by using orbital information. *Mariner 9* used the basic octagonal structure of the Mariner class with four solar panels, a louvered thermal control system, and a 300-lb (135-kg) rocket engine for maneuvering, that is, deboost into orbit. This spacecraft was launched using the high-energy Centaur upper stage. The spacecraft is fully attitude-controlled via gas jets and carries a sophisticated complement of transmitters and receivers. A typical science sequence of recording at the Goldstone station is shown in Fig. 33. Initial results of the imaging experiment from this spacecraft were expected to be poor because of the onset of long-term global dust storms on Mars, which caused a profound decrease in object contrast. Since the image systems were designed to record an extremely wide range of light intensities, digital processing permitted scale expansion which resulted in a remarkable series of geophotographic maps of nearly the whole planetary surface. This was aided by the extension of the lifetime to about a year. Figures 34 and 35 show, respectively, two of the most remarkable discoveries made, the graded surface near the northern polar cap and the volcano Olympus Mons in the Tharsis region, the largest known volcano in the solar system.

Mariner 10 was the first in the long series of Mariner spacecraft to encounter the planet Mercury. The appearance of the spacecraft is similar to earlier Mariners, but significant changes were made in the solar panel configuration and thermal control. Since *Mariner 10* approached close to the Sun, the energy input rose steeply, but as shown in Fig. 10, this is mitigated by the increase in temperature. An adjustment for this was made by designing the solar panels to rotate away from the direction of the Sun. Since the solar input was so high, only two solar panels were included in this spacecraft.

Mariner 10 used the swing-by technique of celestial mechanics and navigation both at Venus and Mercury. The swing-by technique at Venus aided fulfillment of the energy requirement and permitted experiments to be carried out in the neighborhood of that planet, while that at Mercury put the spacecraft into a solar orbit, resonant with that planet. Three encounters were made, terminated eventually by depletion of attitude-control gas.

Mariner 10 carried a complement of combined planetary and interplanetary experiments, among them various field and particle detectors, an infrared radiometer, and an ultraviolet spectrometer, as well as a sophisticated set of two-imaging cameras. The unique role of spacecraft in planetary science was again demonstrated by *Mariner 10* with its close view of the dynamical structure of the upper cloud layer of Venus (Fig. 36), the surface of Mercury (Fig. 37), and finally the quite unexpected discovery of a global magnetic field at Mercury. The cloud pattern of Venus shown in Fig. 36 is indicative of planetary winds tending to blow away from the sunlit hemisphere. This is the first view of such a cloud pattern on a planet. The cratered plains of Mercury (Fig. 37) have at least a superficial similarity to the Moon and show a heavily bombarded early surface. *See* MERCURY; VENUS.

Viking. This spacecraft (Fig. 38) was the most ambitious uncrewed probe attempted. Two launches of a combined orbiter-lander took place in Au-

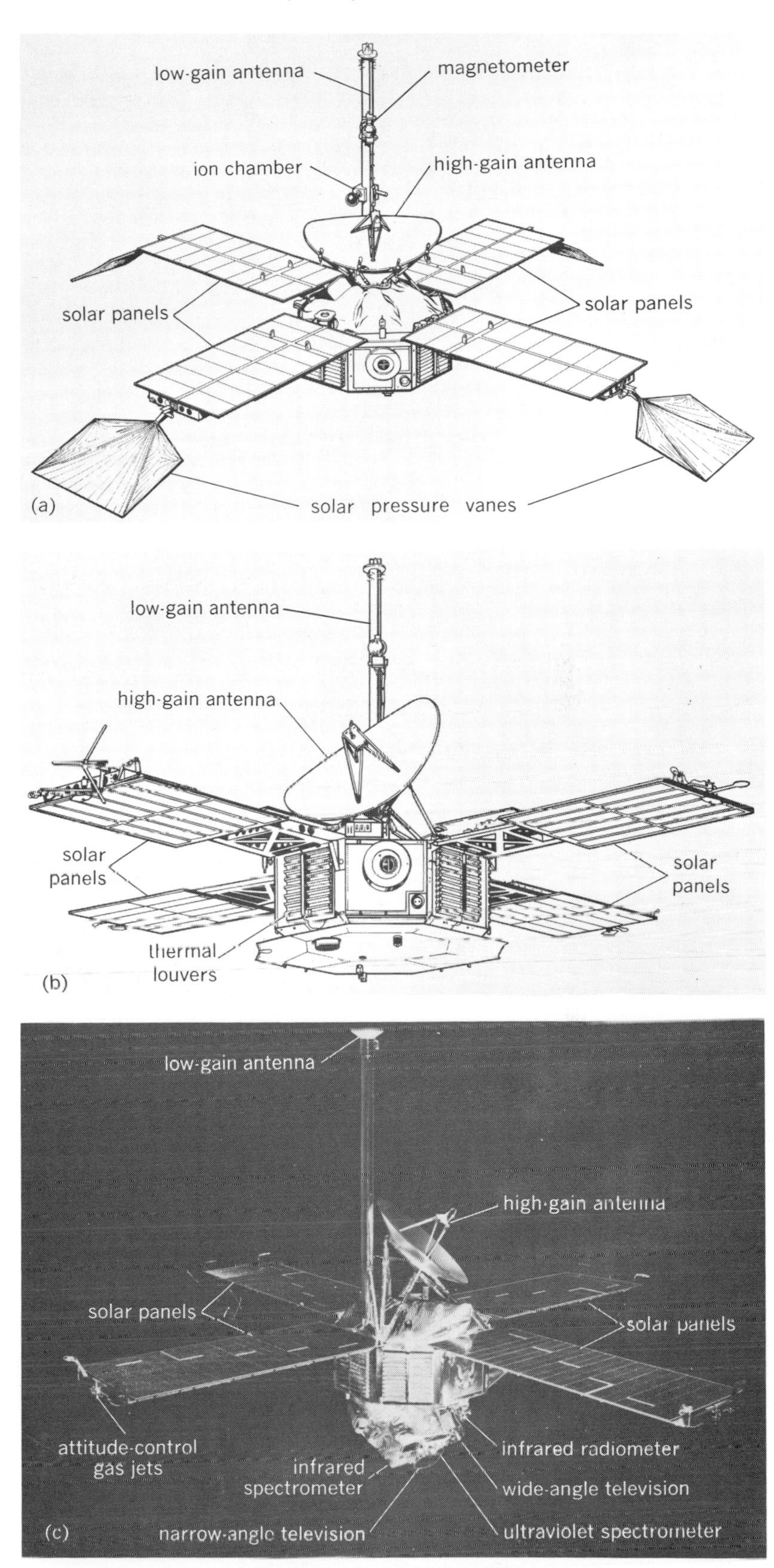

Fig. 30. Evolution of the Mariner spacecraft. (*a*) *Mariner 4*. (*b*) *Mariner 5*. Active thermal louvers, shown on sides of main body, are driven by temperature-sensitive motors which control degree of opening, and thus heat-emission rate from spacecraft. Many changes from *Mariner 4* are not seen in this gross view from the outside. (*c*) *Mariners 6* and *7*. Although only about 300 lb (135 kg) heavier than *Mariners 4* and *5*, this type of spacecraft is much more complex because of advances in electronics. (*NASA*)

Fig. 31. Photograph from the central part of the track of the television camera from *Mariner 4*, nearly 6000 mi (9700 km) above the surface of the planet Mars. Area covered is approximately 150 by 150 mi (240 by 240 km). (*NASA*)

gust and September of 1975, with arrival at Mars in June and August 1976. The principal motivation for these missions was the search for biological activity on the Martian surface. Besides these experiments, some additional instruments were carried for geophysical and geochemical exploration, and comprehensive imaging systems were borne on both the orbiter and lander. The system on the orbiter was designed to reconnoiter landing sites prior to deboost of the lander, as well as to carry out a global mapping program. The launch vehicle for these missions was the Titan 3-E/Centaur, but additional propulsion was carried for orbit insertion, deboost operation, and reduction in lander velocity after parachute separation.

Conceptually, the Viking lander (Fig. 39) borrowed heavily from lunar programs. Radar altimeter and Doppler velocity devices were used for final correction prior to touchdown, but the atmosphere of Mars made the overall descent different. Considerable saving in the deboost energy requirement resulted from the use of aerodynamic braking during the first phase of descent, while a parachute aided the drag after the lander slowed to below Mach 1 (Fig. 40). Besides this major difference from lunar landings, there was the severe requirement for biological cleanliness, which made the design and operation of Viking quite different from earlier space missions. The entire lander was subject to this requirement while on Earth; a bioshield covered it during the launch and was jettisoned only after the Centaur second burn.

Solar panels were used to supply power for the cruise to Mars and for the orbiter after separation of the lander stage, while the lander power supply used radioisotope thermoelectric generators of the SNAP-19 type similar to Pioneer-Jupiter. In the case of Viking, only two were required: batteries were used for peak load demand. Excess heat from the RTGs was used to warm the instruments or dumped as required. Requirements for thermal design were severe as in the case of lunar missions, but the lunar temperature extremes are greater. The shorter Martian night also makes some difference in the thermal requirements.

Sophisticated telemetry and control systems were included in the lander along with an on-board computer. Lander sequencing was complex, as soil sampling took place both for biological and geochemical purposes.

Helios. This is a German-built interplanetary probe which because of its close approach to the Sun (0.31 AU) is called a solar probe. The general appearance of this spacecraft (Fig. 41) is that of a giant bobbin, that is, a 16-sided cylinder with ends which are flared out. A cutaway expanded view is shown in Fig. 42. The spacecraft has a mass of 370 kg. The solar arrays are mounted on the flared end-skirts, with the antenna system extending along the axis of symmetry. The particular configuration is dictated partially by the high solar flux encountered. Heat is rejected from the spacecraft by radiation along the axes, which are normal to the plane of the solar ecliptic or solar equator. The upper and lower surfaces of the cylinder never see the Sun. Additional heat rejection is attained by mirrors that are interspersed between the solar cells which both reflect the external heat and radiate heat from the cell regions.

The spacecraft uses an antenna which is mechanically despun, that is, rotates counter to the spacecraft motion, for the primary communication system, with backup by two antennas of lesser gain which radiate much in the manner of the Pioneer spacecraft, that is, into the plane of the spacecraft orbit. Instead of gas, a dc motor is used to rotate the high-gain antenna with respect to the spacecraft, so that preferential pointing toward Earth is achieved. Since this spacecraft is an interplanetary probe, the experiments are directed toward phenomena taking place in the solar wind and consist mostly of magnetometers and particle detectors. However, the occultation of *Helios* by the Sun permits the observation of radio waves passing near to and through the solar atmosphere.

The orbit of *Helios* is solar-centered (Fig. 43). It is of the high eccentricity required to be able to approach the Sun closely while holding the energy requirement to a minimum. The unique factor of the *Helios* orbit is its approach to within 0.31 AU of the Sun.

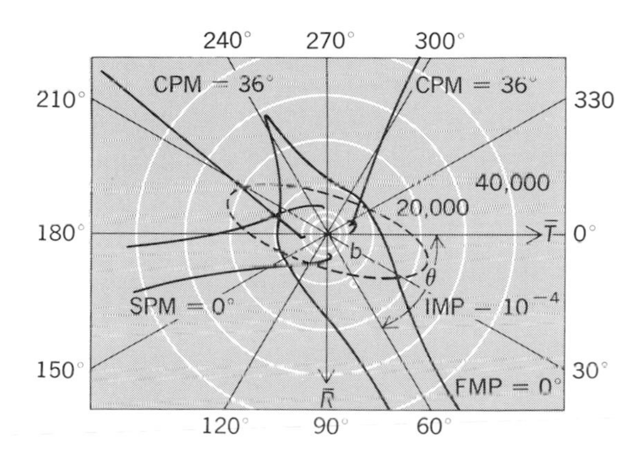

Fig. 32. Target aiming plane perpendicular to the velocity vector of *Mariner 4* just prior to the beginning of Mars encounter. The various notations and their meaning are discussed in the text. This target plane provides the constraints on spacecraft operation used in planning the detailed encounter trajectory. (*NASA*)

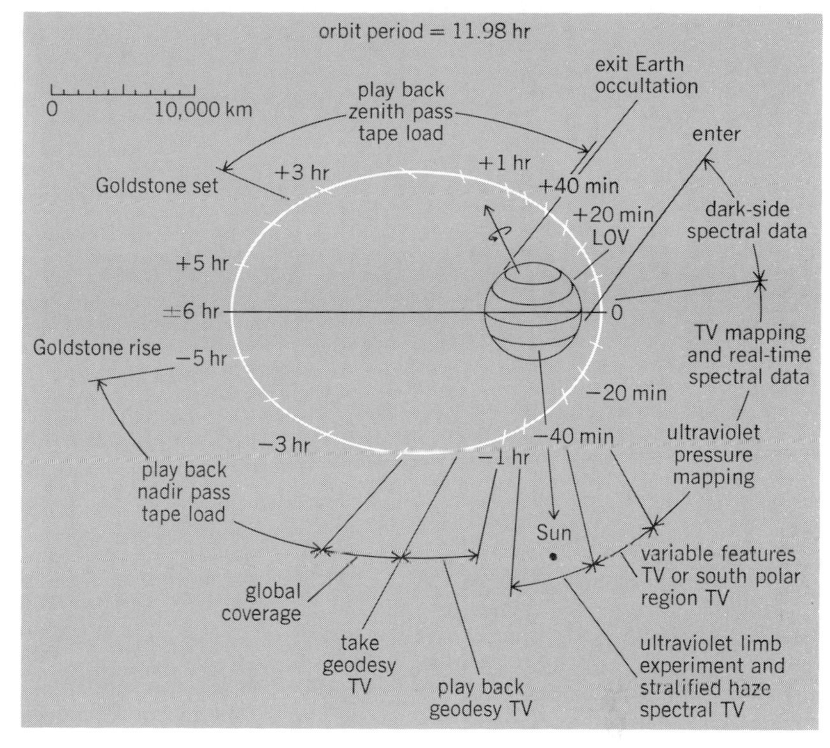

Fig. 33. Time division of scientific activities during typical orbit of *Mariner 9* about Mars. (*Mission to Mars, JPL, 1971*)

Fig. 34. Martian northern polar frost cap taken by *Mariner 9* about half a Martian month after summer solstice. The cap is about 1000 km across. (*JPL/USGS/NASA*)

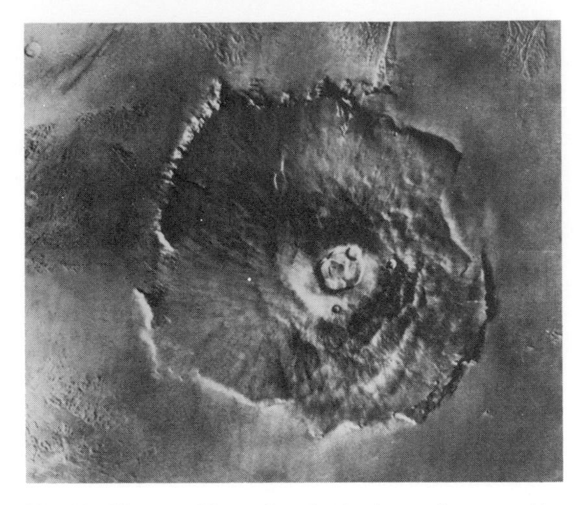

Fig. 35. Olympus Mons, the giant volcano discovered by *Mariner 9*. (*JPL/USGS/NASA*)

Pioneer Venus. This mission consisted of two spacecraft, an orbiter and a bus carrying a large (315 kg) and three small probes (90 kg each). The orbiter mass was 553 kg, and the bus with probes on board 875 kg. Launch of the bus probe carrier used the Atlas SLV-3D/Centaur D-1 AR. Telemetry rate was 2048 bits per second to the Deep Space Network. Separation of the large probe (Figs. 44 and 45) was done about 24 days prior to entry into the Venus atmosphere, and small probe separation at −20 days. Communication between bus and Deep Space Network was via S-band (2.115 and 2.295 GHz). Power was derived from solar cells placed about the main body as used earlier in *Pioneers 6–9*. Peak deceleration of the large probe was 280 *g*, indicating a fast entry. Phenolic forebody construction was used for heat control via ablation.

Fig. 36. Closeup of Venus taken by *Mariner 10* on its swing-by trajectory. (*NASA*)

A parachute deployed subsequent to entry slowed the large probe, with separation of the parachute at 47 km above the surface. Communication was direct to Deep Space Network via a 40-W transmitter using a crossed dipole on the afterbody. Because the scientific lifetime was short, power was derived from a 40 ampere-hour silver-zinc battery. During entry the data rate was 256 bits/s, except for short pre- and postentry times when it was reduced to 128 bits/s. A 3072-bit memory was included for information during entry blackout (due to shock-wave ionization). These data were then transmitted after blackout.

The small probes (Fig. 45) encountered deceleration peaks of 223 to 458 *g*, depending upon the individual entry profiles. No parachutes were used, and the heat shields were retained until surface contact, in contrast to the large probe where shield jettison took place during descent. Communication power was 10 W, and power was from 11-A-h silver-zinc batteries. Data rate was 64 bits/s to 29 km altitude, then 16 bits/s. Of the small probes, the "day" probe survived on the surface for 67 min. Since no scientific instrumentation was carried on the bus, it was not redirected after separation of the probes. It entered the Venus atmosphere and, without thermal protection, burned on entry.

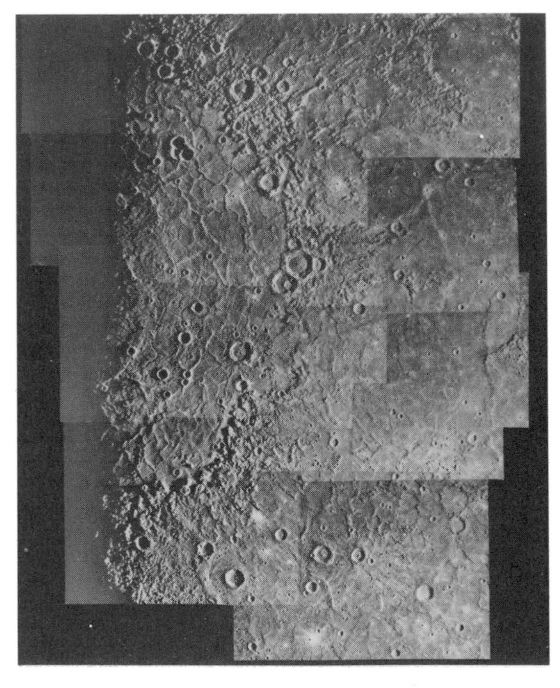

Fig. 37. View of Mercury taken by *Mariner 10*, with the Caloris basin to the left. (*NASA*)

Since the Venus atmosphere is at a high temperature and pressure, special protection was required in the probe design.

In contrast to the bus and probes, the Pioneer Venus orbiter was designed for an extended mission time. Similar mechanical and power construction was used on the bus and orbiter, though the communication system and attitude control varied greatly with the differing requirements. S-band is employed to the Deep Space Network by using a despun antenna on the spacecraft. Sixteen scientific experiments are carried. The 24-h orbit of the spacecraft, after insertion into Venus orbit, was maintained, and the spacecraft was placed in an extended mission mode. Attitude control is by spin; propulsion includes hydrazine monopropellant used for orbit trim and attitude changes. Data can be transmitted at rates from 8 to 4096 bits/s to the Deep Space Network. Data storage on the spacecraft is available from two memories, each of 524,288 bits. Uplink commands from the Deep Space Network take place at 4 bits/s. Although the main channel is via S-band, a transmitter (in addition to the four S-band transmitters) is carried at X-band (11/3 of the S-band frequency). Primary power is supplied from a solar cell system about the center section of the spacecraft, as for the bus, backed up by 7.5-A-h nickel/cadmium batteries.

Many design concepts can be traced to the earlier Pioneer series as is seen from the familiar structural configuration of the spacecraft. Significant changes are, however, apparent from the requirements of the bus and the introduction of the

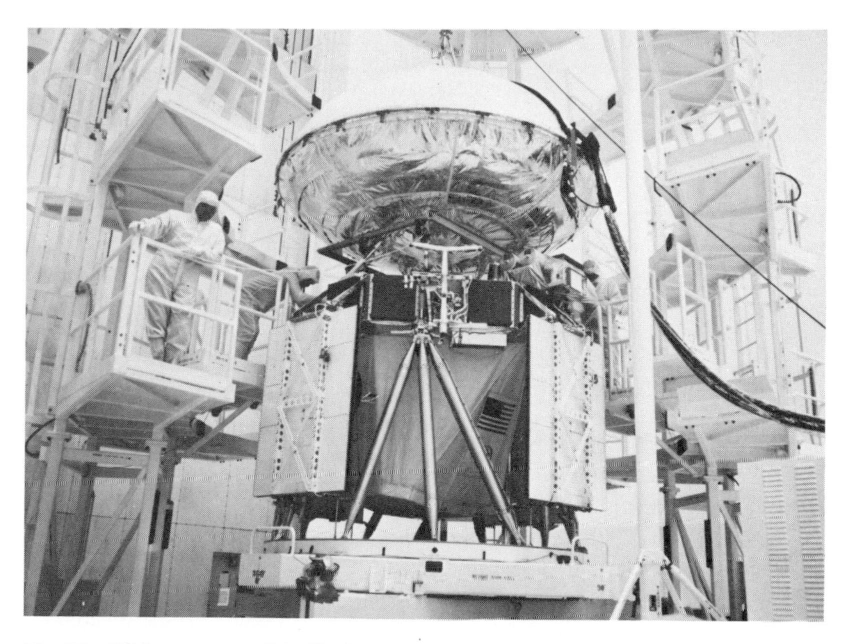

Fig. 38. Viking spacecraft in final test configuration. The upper ovoid is the sterile shield surrounding the lander. (*Langley/NASA*)

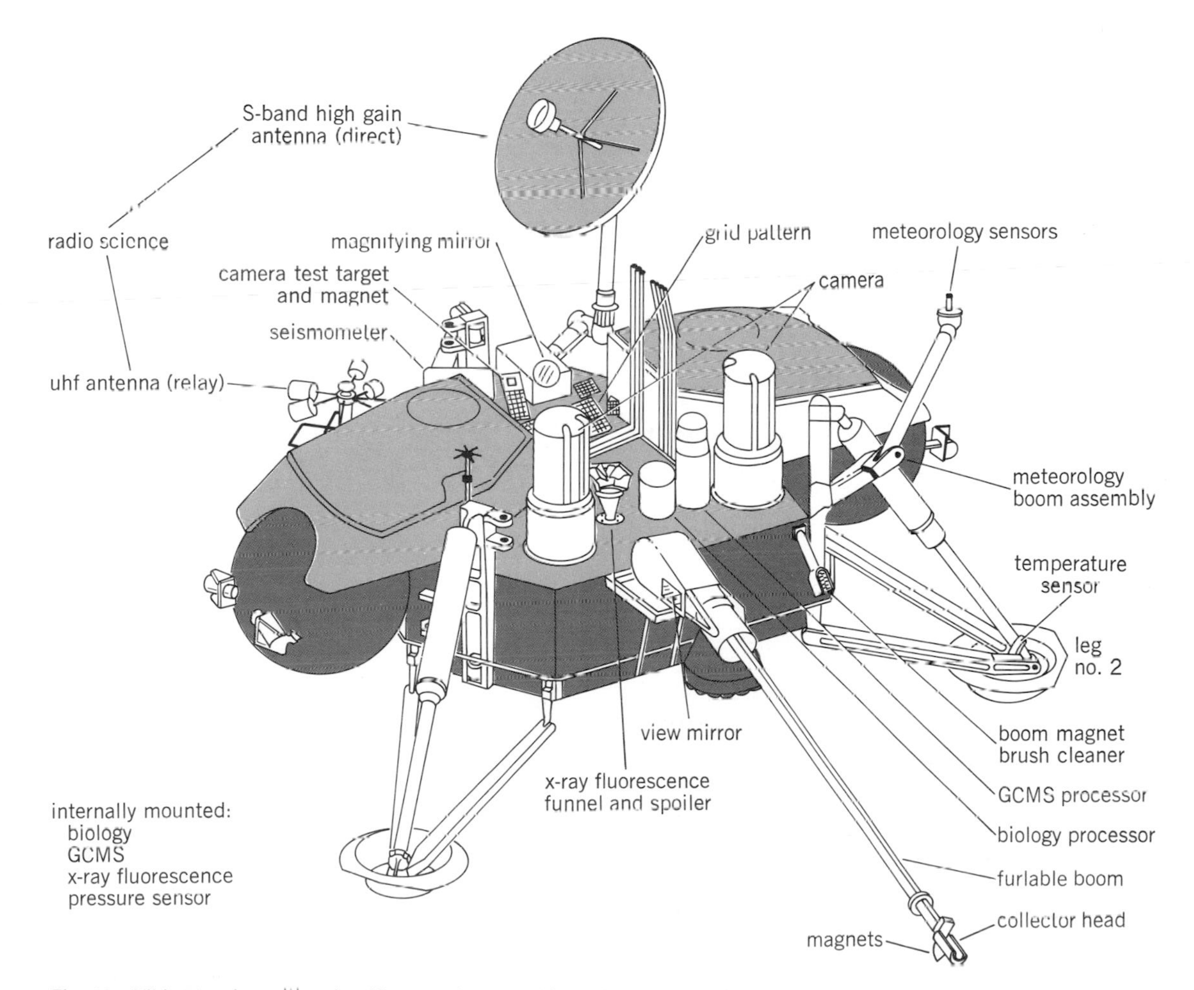

Fig. 39. Viking lander, with scientific experiments. (*Viking Press Kit, NASA News Release no. 75-183, 1975*)

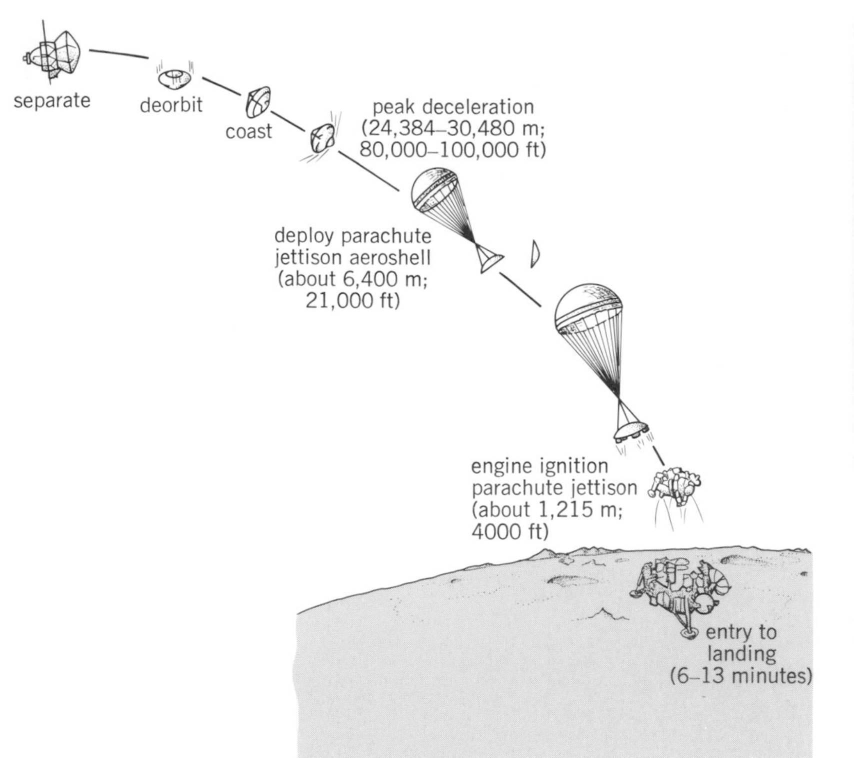

Fig. 40. Descent profile for the Viking spacecraft from separation to landing. (*Viking Press Kit, NASA News Release no. 75-183, 1975*)

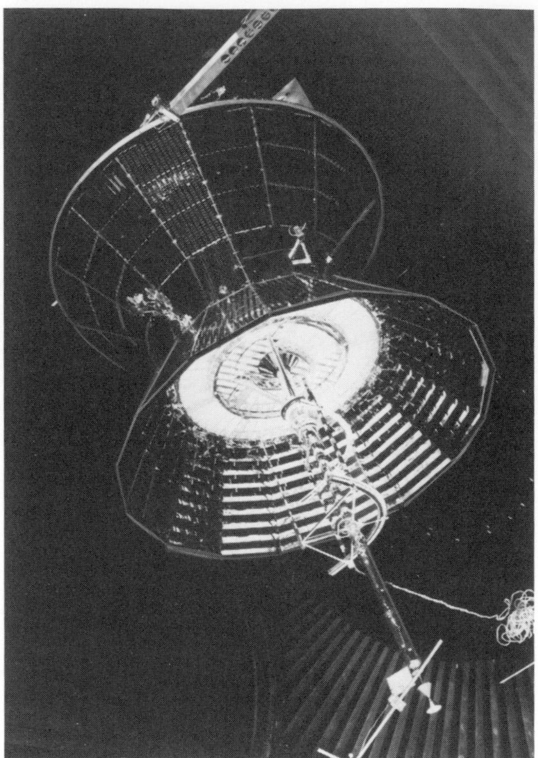

Fig. 41. The *Helios* spacecraft. (*NASA*)

despun antenna. Although the main communication system is described above, the radar experiment on Pioneer Orbiter adds a component of rf radiation. This experiment for study of the Venus topography through the cloud layer radiates 10 W at a frequency of 1757 MHz. Tracking of the probes for positional determination during entry was done directly from the Deep Space Network. In addition, gravity sounding of the planet is carried out from orbit determinations, also using tracking, in this case, of the orbital characteristics of the Orbiter.

Voyager. Voyager (Fig. 46) is a dual-spacecraft mission to the outer planets. It encountered Jupiter and several of the Jovian satellites, providing the first close-up information on these objects, includ-

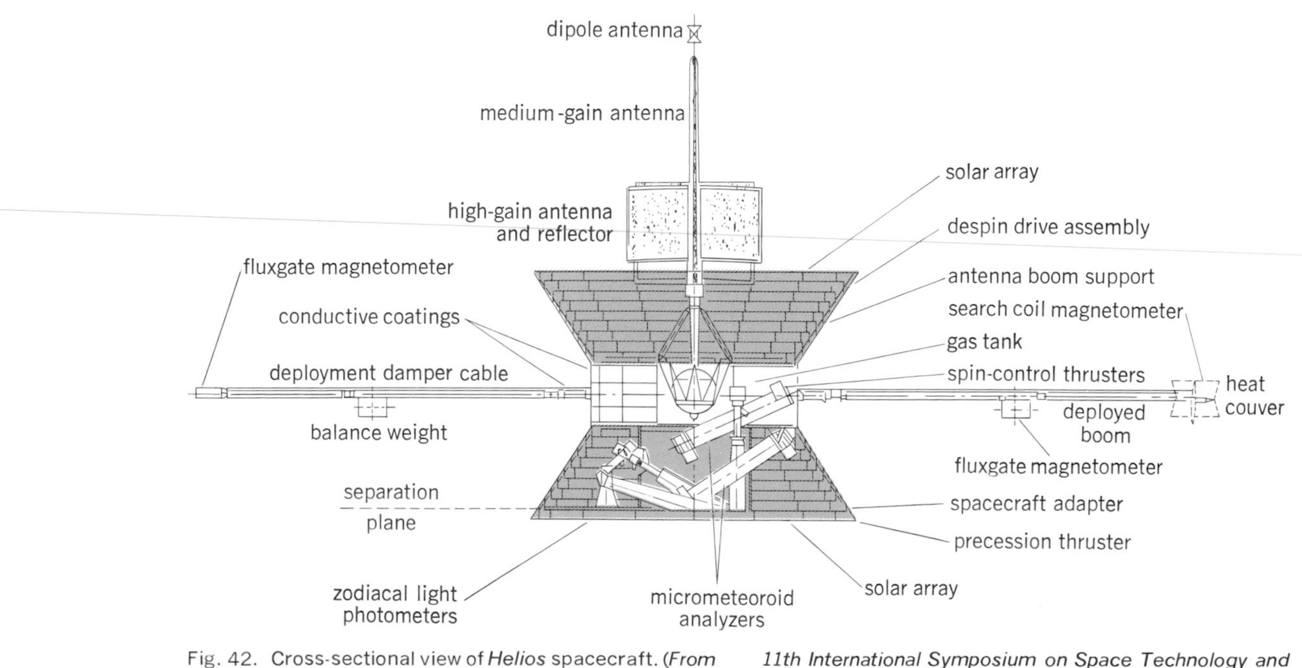

Fig. 42. Cross-sectional view of *Helios* spacecraft. (*From G. Ousley and A. Kutzer, Helios* (*Cooperative Solar Probe*), *11th International Symposium on Space Technology and Science, Tokyo, 1975*)

ing images of the Jovian satellites. Subsequent to the Saturn encounters, which took place respectively in November 1980 (*Voyager 1*) and August 1981 (*Voyager 2*), these spacecraft were turned to the outer parts of the solar system. *Voyager 1* will eventually move toward the apex of the heliosphere (the direction of the motion of the solar system with respect to the local star system), while *Voyager 2* is to be aimed first to Uranus, encountering this planet in January 1986, and finally reaching Neptune in August 1989 (Fig. 47).

Voyager belongs to the same general class of spacecraft as Galileo, discussed below, with respect to telecommunication, attitude control, imaging, power, propulsion, thermal control, radiation hardening, and so forth, but there are significant differences because Voyager spacecraft were designed earlier, and their functions were specialized for encounter rather than orbiting. Coupling to the main ground communication system of NASA (the Deep Space Network) results in key similarities in telecommunication, while the use of RTG power supplies is nearly obligatory for outer-planet spacecraft design, and similar star and Sun sensor design is used. A significant distinction is the fully inertial design of Voyager, whereas Galileo uses dual spin to enhance particle data. All spacecraft which encounter Jupiter must also take into consideration radiation-hardening requirements, since all approach Jupiter as close as Io's radius at least once (Fig. 48). Thus the radiation environment is especially severe. The role of spacecraft electrical charging has become recognized even for the Earth's magnetosphere. In Jupiter's environment, the situation is compounded by the energetics of that magnetosphere, and care must be taken to ensure that a common electrical potential is provided for over the whole spacecraft. Otherwise, large electrical fields can be acquired, discharging through the spacecraft with damage to solid-state devices.

Galileo. Galileo is a mission designed to launch two spacecraft toward Jupiter, one to orbit the planet over a period of up to 20 months, together with an entry probe transported to Jupiter on an independent spacecraft (bus), with launches scheduled in February 1984. The configuration of the Galileo mission has been changed because of fiscal constraints; the current plans are for a dual launch, whereas originally the Jovian entry probe was to be separated from a single spacecraft. Arrival at Jupiter is between May and August 1986, with an approach to Mars (possibly as close as 275 km) en route to Jupiter. Galileo is a massive spacecraft system, totaling 3770 kg when fully loaded with propellant. The orbiter mass is 3660 kg, with the remaining 1050 kg allocated to the probe and probe carrier. The orbiter spacecraft carries a despun section as in the case of the International Solar Polar Mission (ISPM). Propellant is used for midcourse maneuvers at Mars and other positions in the transfer trajectory, as well as for injection into Jovian orbit attitude control. Two engine burns at Mars (736 m/s and 221 m/s velocity change) are planned, with further use of the propellant for Jovian orbit insertion (674 m/s) together with subsequent orbit changes optimizing the geometry for

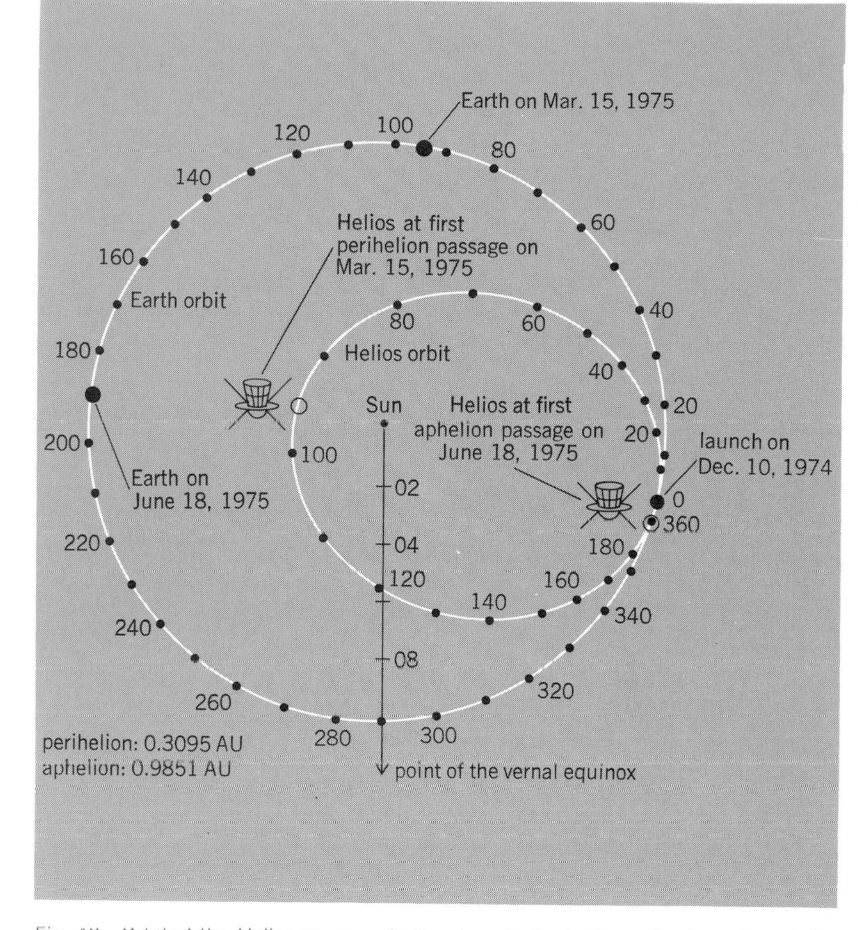

Fig. 43. Orbit of the *Helios* spacecraft. Numbers indicate days after launch and distance from Sun in astronomical units (AU). (*From G. Ousley and A. Kutzer, Helios (Cooperative Solar Probe), 11th International Symposium on Space Technology and Science, Tokyo, 1975*)

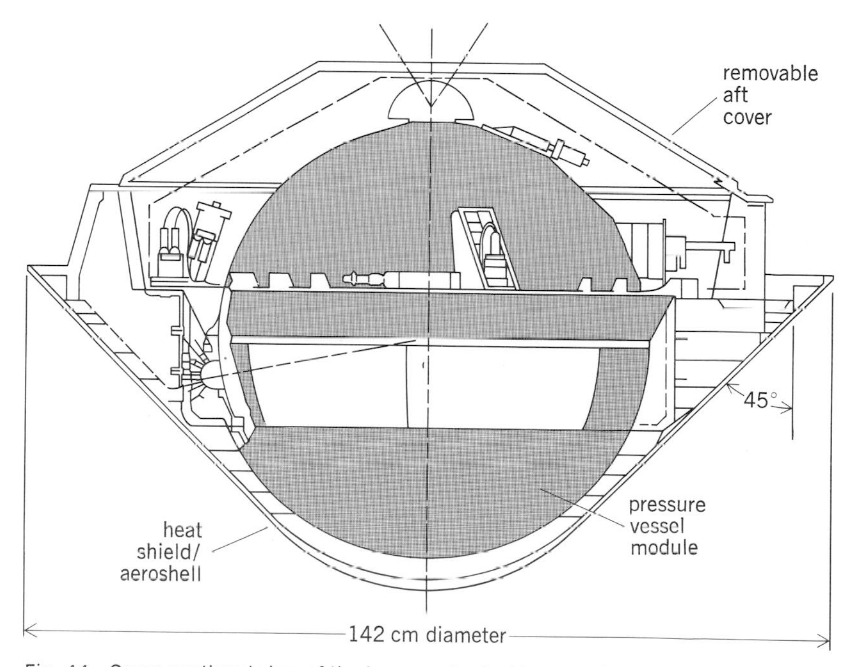

Fig. 44. Cross-sectional view of the large probe for Venus entry, part of the Pioneer-Venus spacecraft. (*From L. J. Nolte and S. C. Sommer, Probing a planetary atmosphere: Pioneer-Venus spacecraft description, 1975*)

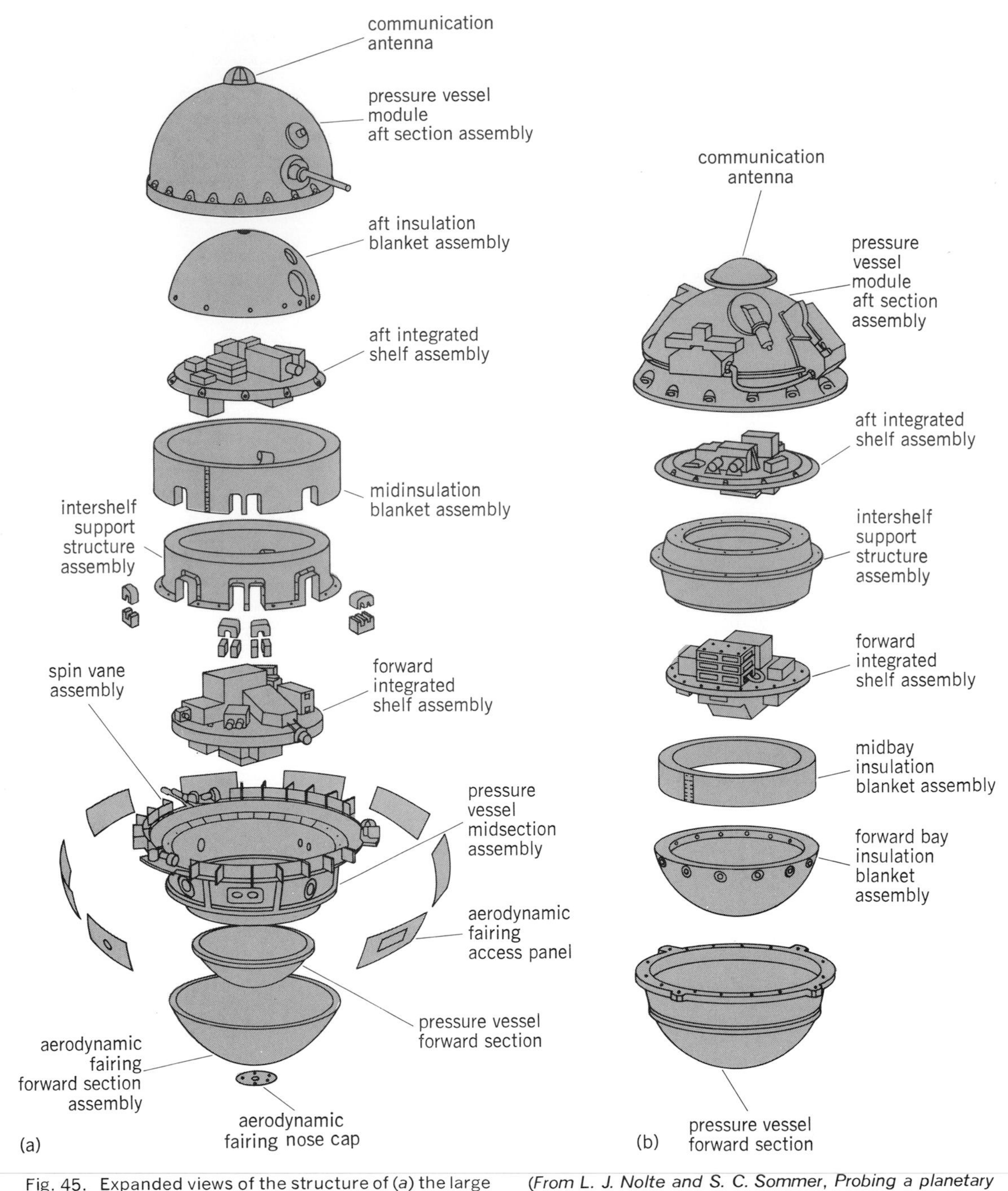

Fig. 45. Expanded views of the structure of (*a*) the large probe and (*b*) one of the smaller probes to be launched into the Venus atmosphere from the Pioneer-Venus bus. (*From L. J. Nolte and S. C. Sommer, Probing a planetary atmosphere: Pioneer-Venus spacecraft description, 1975*)

scientific contact with satellites for imaging and environmental measurements. The orbiter carries a sophisticated telecommunication system making use of radiation-hardened microprocessor technology. It has a downlink (to Earth) capacity of 115.2 kilobits/s. There is also 9×10^8-bit on-board magnetic tape storage for use during special periods such as Earth occultations, encounters, orbital maneuvers, and nontracked cruise times. Telecommunication with Earth is over S-band (uplink) and both S- and X-band (downlink). Thus in addition to normal communication requirements, differential Doppler (as well as ranging) can be carried out from the orbiter.

Internal control of the spacecraft uses six microprocessors in a configuration similar in some respects to interconnected and distributed computation and data transfer used on Earth. Special provisions are made for data verification.

The orbiter spacecraft is a dual-spin device with the spinning section or rotor carrying booms and fuel tanks (Fig. 49). This section is designed from the scientific standpoint to provide for full spherical coverage for particle experiments. The stationary section of the spacecraft carries an articulated science platform for the various imaging experiments. The two parts of the spacecraft are connected via the spin bearing assembly (SBA;

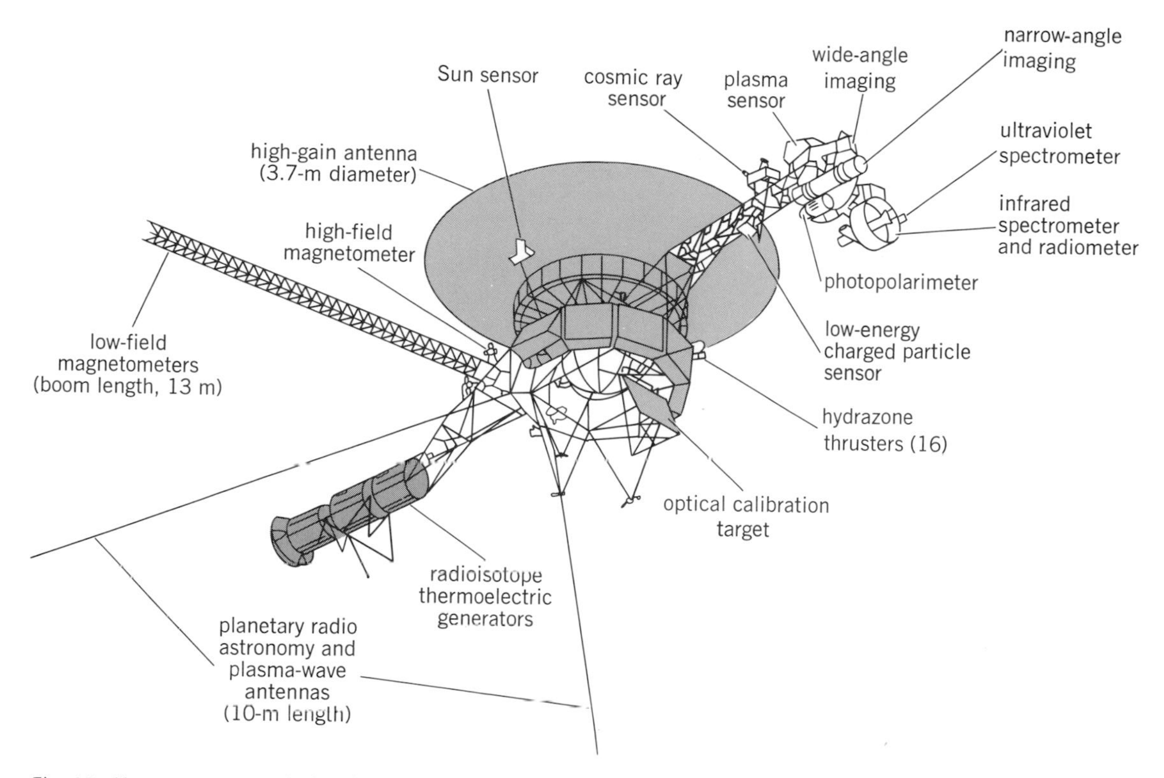

Fig. 46. Voyager spacecraft showing primary subsystems. (*NASA*)

Fig. 50), which must also provide the main channel for communication and power transfer between the two spacecraft sections. Since the two sections have variable relative spin rates, a brushless dc torque motor is incorporated along with 100 roll rings and a 16-bit optical encoder. Provision is also included for mechanical support of the main engine. Attitude-sensing information is obtained from a combination of Sun scanner and star tracker.

The probe is to be carried on a separate spacecraft but with tandem launch from the shuttle. The Jovian probe design is a major consideration in the Galileo project, as the entry energy loss exceeds that of any design so far dealt with. The entry velocity is 48 km/s. With a probe mass of 325 km, the total energy loss is nearly 375,000 megajoules. The rate of loss of energy during entry depends upon the shape of the trajectory. Peak loss would take place for vertical entry, but would take place most rapidly. Assuming a 100-s period of significant heating due to entry, the average energy loss becomes 3750 megajoules/s, over a surface which is approximately 12,000 cm². Thus, the surface heating rate becomes 300 kW/cm², corresponding to a blackbody at a temperature of about 16,000 K. If the peak heating takes place in 10 s, the rates are increased correspondingly and the equivalent temperature rises to 19,000 K. Design of the aero shield (heat shield) is thus complicated by the need for extreme accuracy in entry positioning; this becomes a key factor in the design of the probe carrier attitude system consistent with a permissible total energy and heating rate. Transmission of data from the probe is via relay to the Galileo orbiter. Storage of data in the probe is done using about 41 kilobytes of memory, necessitated by the ionization-induced blackout during entry (Fig. 51). Parachute deployment follows the entry cycle; approximately 1 h of descent is envisioned for the data taking and radio sequence.

International Solar-Polar Mission. This is a mission, jointly sponsored by NASA and the European Space Agency (ESA), to place two spacecraft into antipodal heliographic polar orbits about the Sun, with launch scheduled for 1985. The trajectory makes use of a gravity assist from the planet Jupiter. Thus, the novel feature is launch outward away from the Sun to Jupiter with orbits swinging back to the Sun at high heliographic latitude, passing over the solar poles at about 2 AU distance

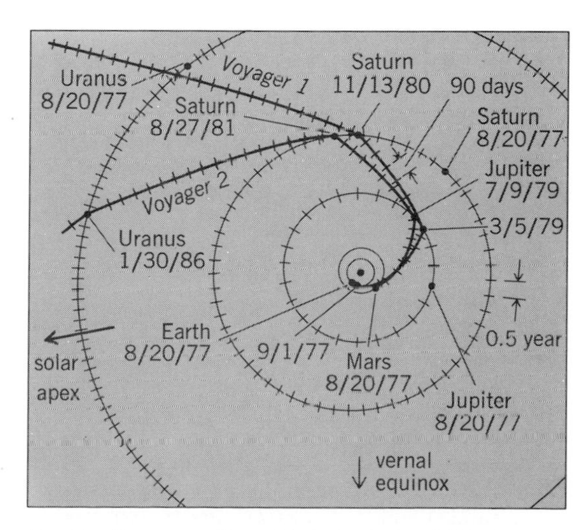

Fig. 47. Voyager heliocentric trajectories. (*NASA*)

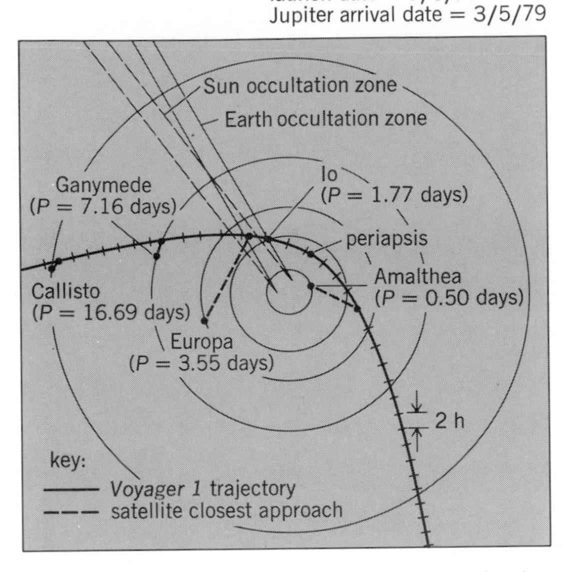

Fig. 48. *Voyager 1* Jupiter encounter trajectory showing secondary encounters to selected satellites. View is from direction of Jupiter's north pole. *P* is the satellite orbital period. (*NASA*)

some 3–4 years later. Perihelion is at 1 AU and does not correspond to solar-polar passage. Although originally the two spacecraft were to be launched in tandem from the space shuttle, present plans call for separate launches. As with all current programs, shuttle launch is planned. The additional feature of the Jovian encounter, not used hitherto, is the modification of the out-of-the-ecliptic component of velocity, enabling high heliographic latitudes to be attained. The close approach to Jupiter (within the orbit of Io) suggests the possibility of obtaining Jovian scientific data. The orbits are nonecliptic, both before and after Jovian encounter. Post-Jovian orbit plane angles to the ecliptic are 76° for the north trajectory and 81° for the south trajectory.

Designs of the two spacecraft are similar in gross aspects; thermal considerations, trajectory, and communication are necessarily similar, being dictated by external environmental conditions. However, the different scientific instrument payloads require distinctions in design which lead to significant departures from commonality. Insofar as possible, design and components are borrowed from previous programs.

The communication system for the NASA spacecraft includes the Viking 1.477-m parabolic antenna used with Voyager X-band traveling-wave tubes. Other subsystem design spinoffs from earlier programs are Mariner low-gain antennas, the Voyager power amplifiers and propellant tanks, and the OSO (Orbiting Solar Observatory) spin bearing assembly. The NASA spacecraft is called a dual spinner, since the platform consists of two sections: one spins at a rate dictated by scientific instrument scanning needs and spacecraft total angular momentum stability requirements, while the second section is "despun" to provide a stable platform for instruments intended to scan the Sun, such as the coronagraph. RTG power is used because solar cells could not supply electrical needs over the wide-ranging trajectory. Sun and star sensors are used to provide attitude control information, both for orientation of the spin axis and for roll rate.

Propulsion is provided by monopropellant (hydrazine) for both midcourse velocity corrections and attitude and precession control of the spacecraft.

Attitude control for this spacecraft is especially complicated by the numerous booms (for the RTG and magnetometers), the scan platform, and the despun section. Attitude information is obtained from a combination of solar and celestial sensors. Actual control, determined by on-board computation, is carried out by using eight monopropellant thrusters for spin and attitude control. A common propellant (hydrazine) source consisting of two tanks is used for both attitude control and midcourse vector velocity correction. Total available velocity correction is 160 m/s, which determines the fraction of fuel partitioned for this purpose. The remainder is allocated to attitude control. Attitude control of the spacecraft is dictated by a combination of requirements generated from accuracy needs for velocity correction and scientific pointing accuracy, in turn mostly forced by the coronagraph (30 arc-seconds).

Telecommunication has the usual multiple function of spacecraft command and data transmission, and in addition possible propagation experiments aimed at study of the medium, astrometry of the solar system, and even relativity, though the rela-

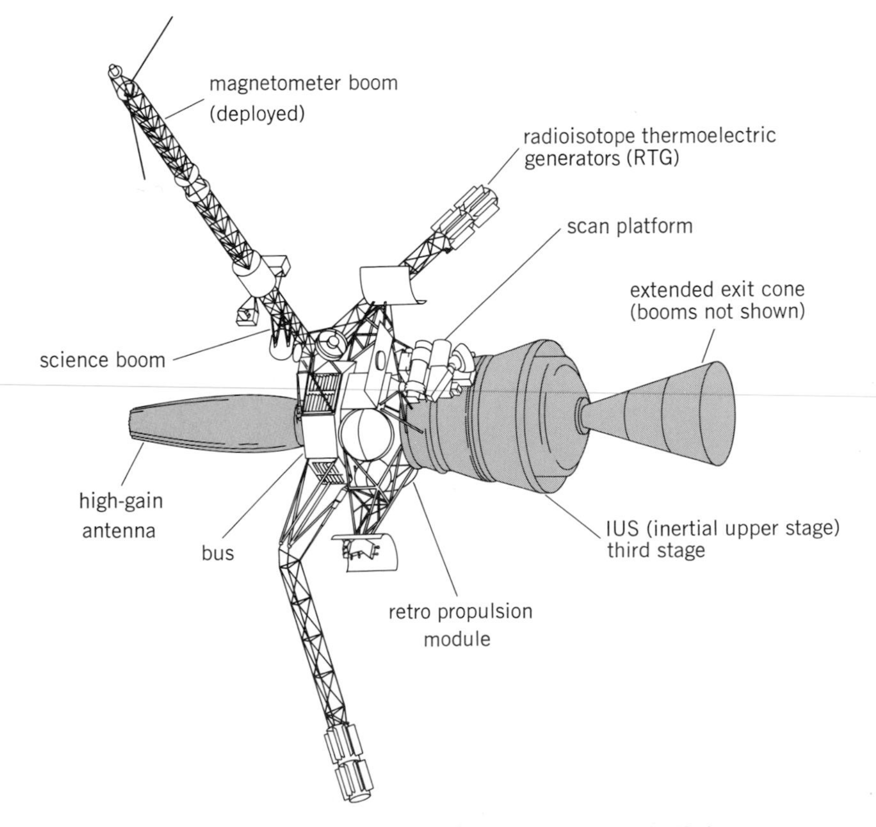

Fig. 49. Galileo dual-spinner spacecraft with main subsystems. (*NASA*)

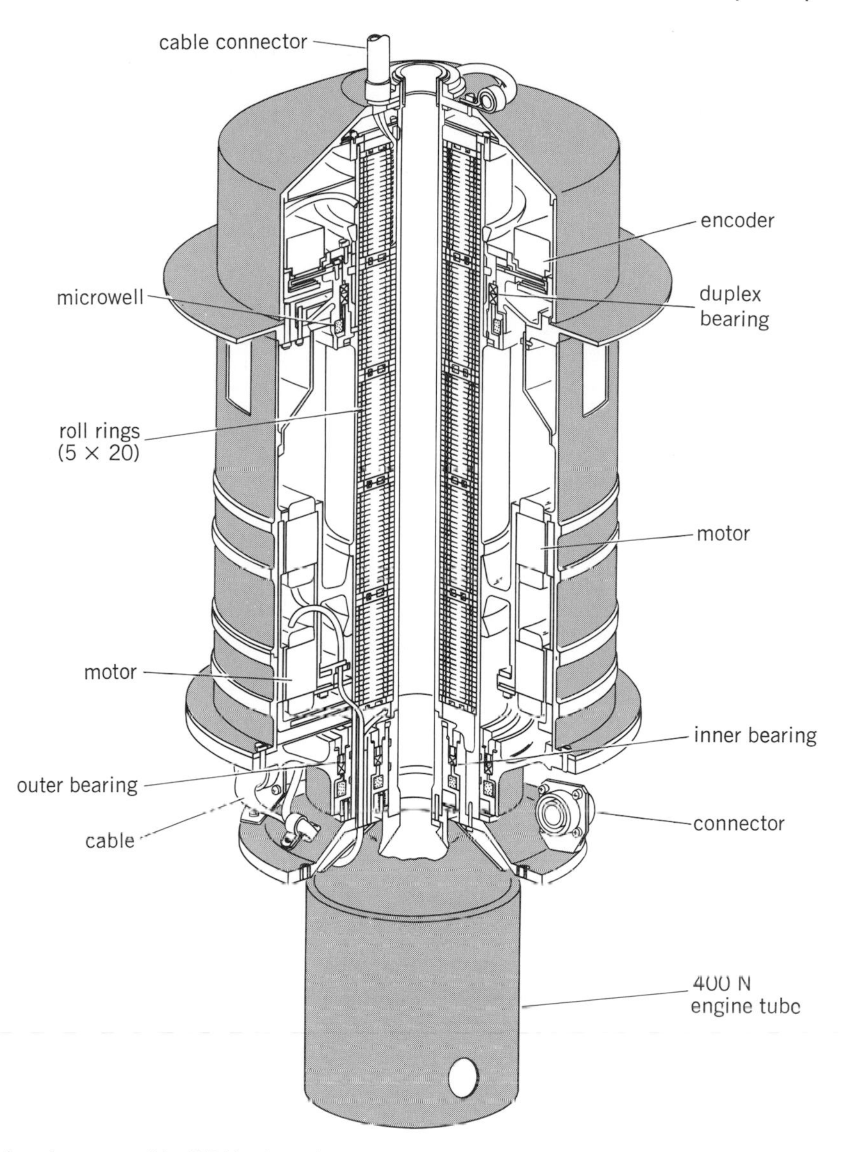

Fig. 50. Spin bearing assembly (SBA), the primary mechanical and electrical interface between rotor and stator sections of the Galileo orbiter spacecraft. (*NASA*)

tivity experiments are currently thought to be perhaps marginally possible at best. Uplink (command) is S-band (2115 MHz), while downlink is dual coherent S (2295 MHz) and X (8415 MHz) band. Design limits are 6 kilobits/s at 3 AU using the 34-m Deep Space Network antenna and coherent transmission, decreasing to 3 kilobits/s for incoherent transmission at 6 AU.

Total mass of the NASA spacecraft is about 360 kg. Even in the shuttle bay, the envelope is some 4 m across, while with erection the magnetometer particle instrument boom extends for 15 m from the center, and the radio astronomy antenna projects to 50 m.

The ESA spacecraft, forming the dual to the NASA vehicle, is fully spin-stabilized with antenna pointing to Earth as in the case of the advanced *Pioneer 10* and *11*. Communication is similar to that on the NASA spacecraft, consisting of uplink S-band and dual S- and X-band downlink. Total mass is about 326 kg. Figure 52 shows a cutaway of the NASA spacecraft with a portion of the magnetometer/particle boom.

Soviet spacecraft. Much less can be said technically about the Soviet Union space probes than about United States vehicles. It is known with reasonable assurance that the Soviet efforts have resulted in considerably more planetary missions. Technical requirements are similar in the two competing countries, but differences would have to result from the wholly different tracking station networks and from the very different latitudes of firing; in the Soviet case this would mean that to obtain the maximum possible thrust a more "dog-leg" trajectory into the plane of the ecliptic would be required, and the velocity to be gained from Earth rotation cannot assist the Soviet launches as greatly as those of the United States.

The Soviet Union spacecraft often display greater weight. It appears that attitude stabilization is used, though in some early cases it is possible that this was done only over part of the trajec-

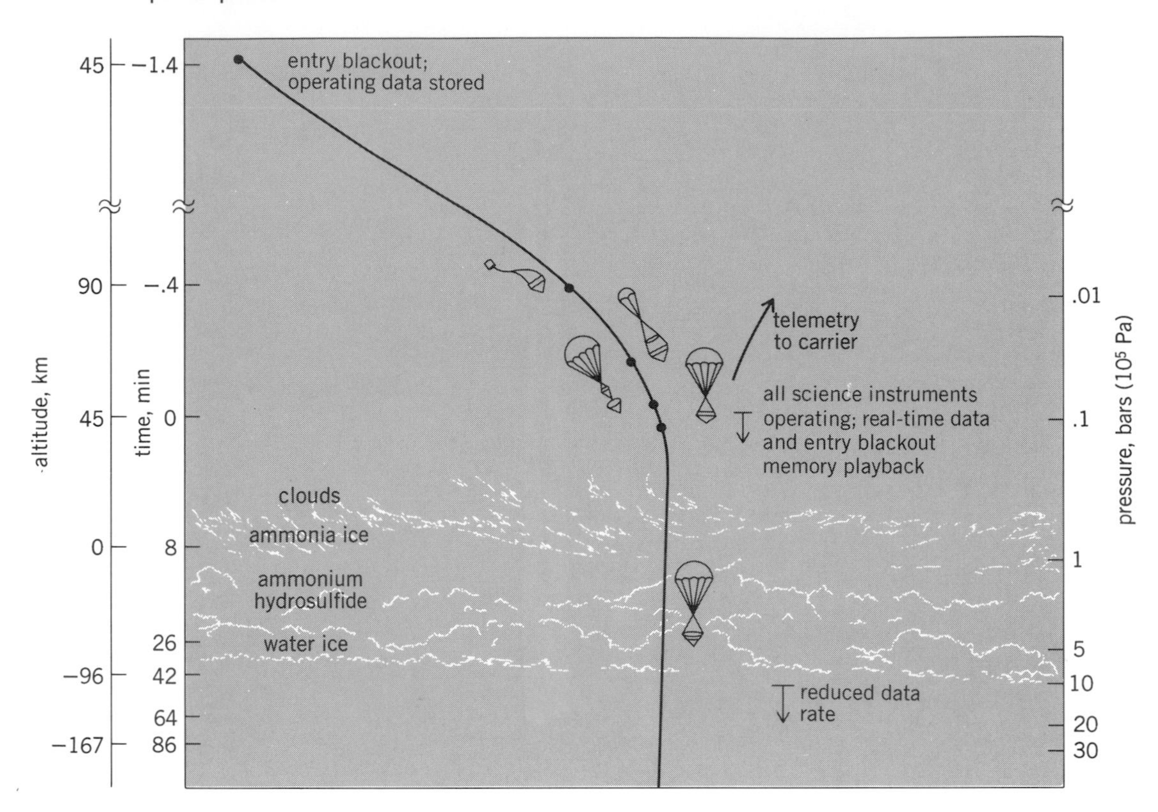

Fig. 51. Galileo probe entry at Jupiter. (*NASA*)

tory, perhaps to conserve control gas. The earlier spacecraft appear to have used radio frequencies much lower than those of United States vehicles. This has the double disadvantage that a larger transmitting antenna is required for the same beamwidth or gain and the cosmic noise background from the Galaxy is higher at lower frequencies. The frequency for *Sputnik 8* was raised to 922.8 MHz, and later probes presumably used the same or still higher values. Some of the early Sovi-

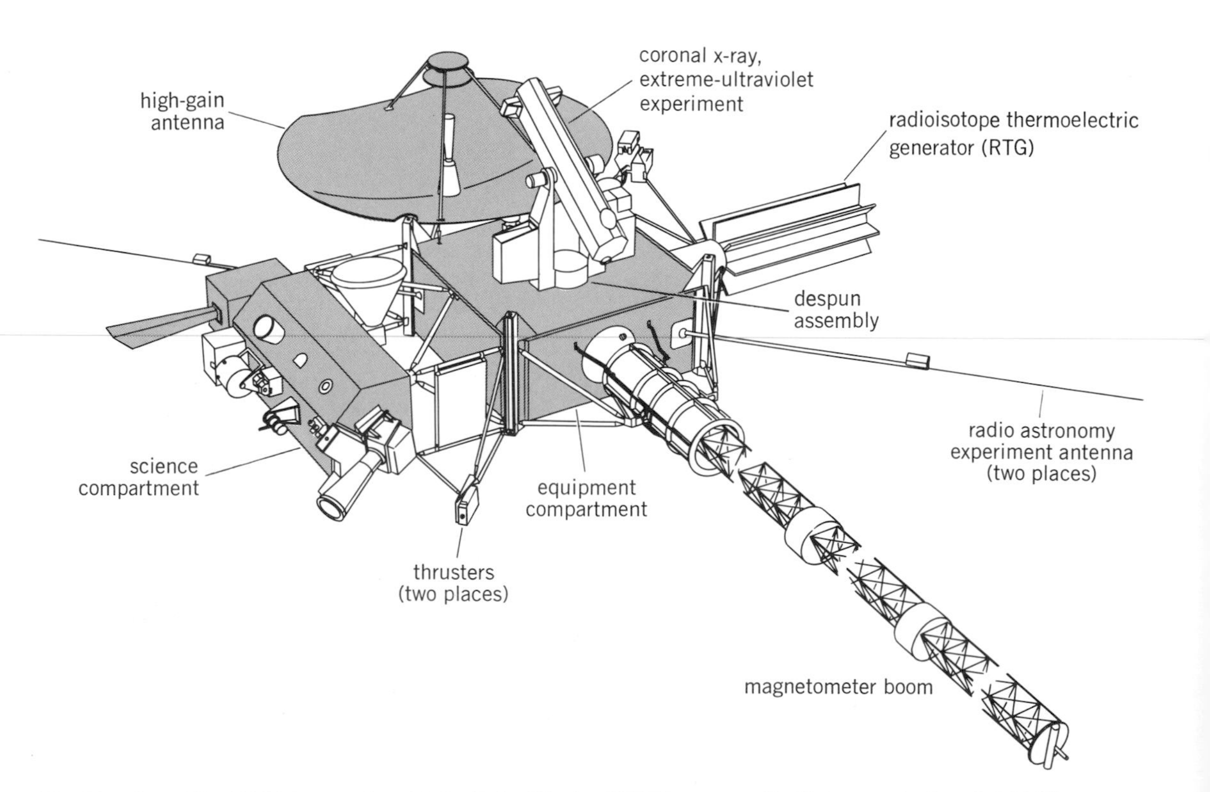

Fig. 52. View of the NASA International Solar-Polar Mission (ISPM) spacecraft with booms deployed. (*NASA*)

et Union spacecraft used pressurized containers. Very few United States spacecraft have done so. In *Pioneer 5*, the 150-watt transmitter required a sealed compartment to facilitate the cooling of the transmitting final-stage tube. Today more sophisticated schemes are used on United States spacecraft and components are individually sealed if required. Since *Sputnik 1*, the first Earth satellite, operated for 23 days and *Sputnik 2* for 7 days, it seems likely that these spacecraft were battery-powered exclusively; however, the next Soviet satellite, *Sputnik 3*, lasted for nearly 2 years, indicating the use of solar cells. The first space probes, the Lunik series, are thought to have used batteries for the first two launches and solar power for the third, the vehicle which first photographed the rear side of the Moon. The length of mission implies, and photographic evidence shows, that solar cell technology is utilized in Soviet Union probe design in the same manner as in United States attitude-controlled spacecraft. The thermal system in Soviet spacecraft deviates from United States design in at least some cases. Circulating fluid–radiator cooling has been used, and the *Luna 13* lunar lander used a boiling-water system dependent upon the heat of vaporization to maintain sufficiently low temperature for operation.

The lesser descriptive material regarding the Soviet space program is a result of more closely guarded technical reporting; in most cases where direct competition exists between the United States and the Soviet Union, the latter has been the first to attain a given goal. The relative failure rates are comparable in the lunar program if the early Rangers are included; the planetary programs appear to favor the United States from the standpoint of reliability, though if all the firings are included from the early days of the space program this advantage is somewhat diminished. The landings of the *Venera 4*, *5*, and *6* capsules on Venus were a definite leap forward technologically.

Soviet spacecraft have designations that generally are associated with their eventual mission. Sputniks are Earth satellites. The Luniks are lunar probes, the first also being called *Mechta* (dream). *Lunik 1* (*Mechta*) appears to have been instrumented primarily for particle and field measurements (Fig. 53). *Lunik 2* impacted the Moon in about 35 hr, indicating a much higher than minimum energy orbit. Here the apparent great excess in booster thrust permitted a much lessened degree of accuracy in launch than in the early United States probes where the greatest care was needed in guidance. *Lunik 3* was the first orbital lunar flight, technically a circumlunar orbit with return to Earth. Its prime function was back-side photography of the Moon, about 70% of the side away from the Earth being photographed. This spacecraft was also referred to as the AES (automatic interplanetary station).

In February, 1961, the Soviet Union launched the first of a series of planetary probes using a tandem launch which included a Sputnik. The Sputnik is thought to have weighed over 7 tons (6350 kg) and the Venus probe was 1419 lb (644 kg). Radio contact was lost some 15 days after launch. The next interplanetary attempt was on Aug. 25, 1962, when *Sputnik 27* was launched together with a Venus probe. It appears possible that the Venus probe was about the same design as that of the earlier attempt since the weight is quoted as the same. Insertion into interplanetary orbit from parking orbit about the Earth failed. Two other attempts to Venus were made on Sept. 1 and 12, 1962, followed by launches to Mars on Oct. 24, Nov. 1, and Nov. 4, 1962. The Mars probe design was different from the earlier Venus probes in configuration, but little interior detail is available.

Lunas, *5*, *6*, *7*, and *8* were attempts to soft-land and or semisoft-land payloads on the Moon: *Luna 9* was the first successful attempt in this series and returned the first pictures ever taken from the surface of the Moon. Most of the following Lunas returned data from lunar orbit. *Luna 15* crashed in what appears to have been an attempt to soft-land on the Moon.

Flight attempts to Venus continued, and *Venera 4* landed a capsule on the surface of the planet on Oct. 18, 1967. The final part of the descent from the spacecraft was aided by parachute; the landing was on the night side of the planet some 1500 km from the terminator and close to the equator. The instrumentation of the landing probe was oriented primarily to sampling atmospheric constituents and forms the first basis for models of the Venus atmosphere based upon direct measurements. On May 16 and 17, 1969, *Veneras 5* and *6* made similar descents through the atmosphere of Venus. Each of these spacecraft weighed 2492 lb (1130 kg). Because of the extreme pressure and temperature of the Venusian atmosphere, all three spacecraft ceased transmission of data before reaching the surface of Venus.

Little information of a technical nature has been released regarding the Soviet program. *Veneras 7*, *8*, *9*, and *10* continued the exploration of Venus,

Fig. 53. Model of *Lunik 1* (*Mechta*). Long metal strips are antennas, tapered post is the magnetometer boom, and raised windowlike apertures are ion detectors.

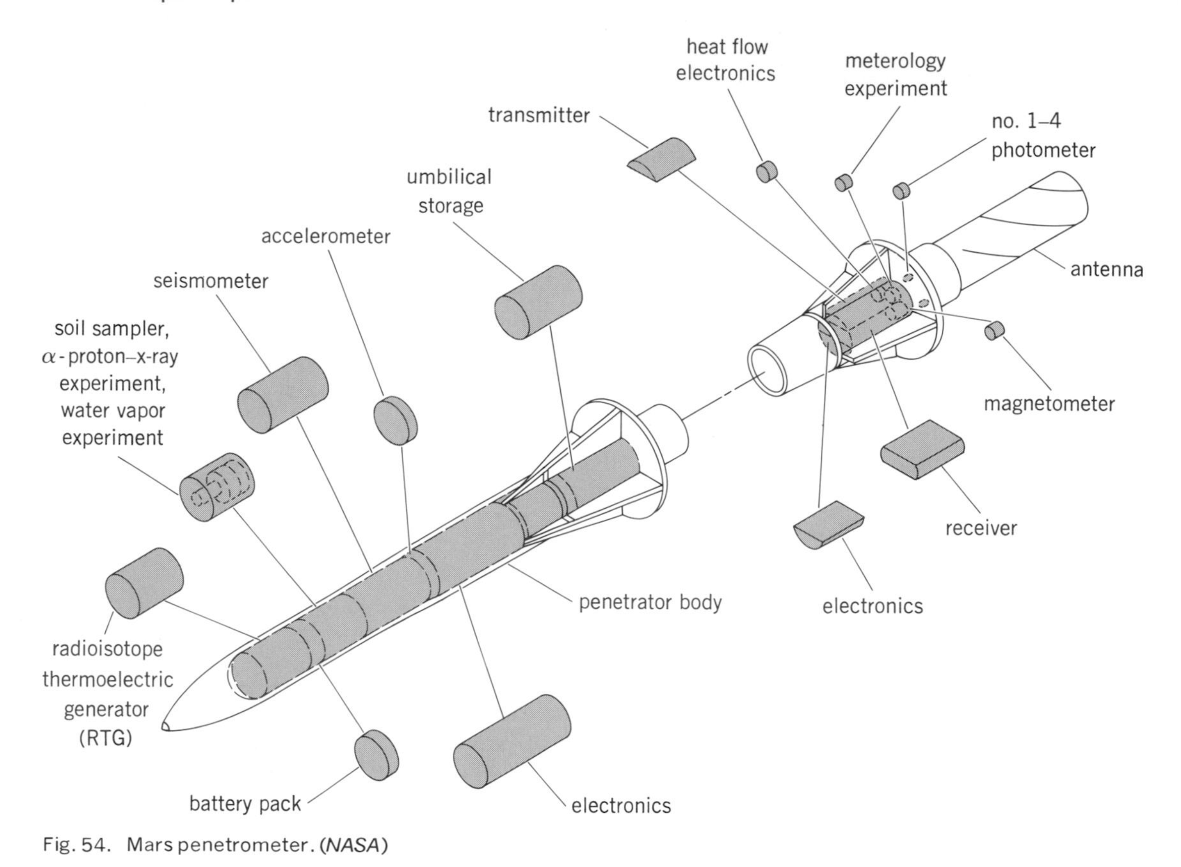

Fig. 54. Mars penetrometer. (*NASA*)

including nuclear counters for the detection of the surface radioactivity. Nuclear pulse counting suggests a high level of long-lived radionuclides based upon expected solar system abundances and experience from the Earth and Moon. These levels suggest a highly differentiated planet with a classical outward transport of the large-ion, high-Z uranium and thorium nuclides. A most important result comes from the imaging device used in this program to show the surface of Venus. The fact that television imaging could be carried out is indicative of a much higher light level than expected below the cloud deck of Venus. Also, one photograph shows a broken surface similar to that of the lunar regolith, while another shows a more fine-grained physical appearance suggesting aeolian erosion. To all appearances, Venus entry technology, involving extreme pressure and temperature, is well in hand. It seems reasonable to infer that this technology is similar in major respects to that designed for the United States Pioneer-Venus (PV) probe.

Advanced probes. Design has been undertaken on several missions, including the Mars lander and comet and asteroid encounters. A Venus Orbiting Imaging Radar (VOIR) probe, which may also carry some other Venus-related experiments, has been considered for flight missions. VOIR is a large spacecraft, with a mass of 4000–5000 kg. Study shows that deboost is expensive; one alternative mode of entry is aerobraking by using repeated shallow grazing of the atmosphere to decrease the angular momentum and energy of the spacecraft. Launch from Earth would use the shuttle, as is the case for all other projected missions.

Conceptual studies have also been made of Mars landers, including a surface roving vehicle and sample return. Studies of more economical alternate vehicles have also been undertaken including surface penetrometers which are launched under power from a bus spacecraft. A global geophysical network is envisioned by using a set of penetrometers. The penetrometer (Fig. 54) must enter the atmosphere of Mars, decelerating hypersonically by using an umbrella for the first velocity decrement. A final mass of about 10 kg must be brought to rest in the Martian soil. The penetrometer subsystem includes structure, data processing (via microprocessor), communication to the orbiter, power, and thermal control.

Further work has also been done on atmospheric entry, for example, into the atmospheres of the outer planets, and even the satellite, Titan. These all borrow from existing technology, with improvements.

For very deep penetrations of space, for example, Jupiter, electric propulsion becomes competitive with chemical propulsion. This scheme uses an ion engine which thrusts continuously. Although the thrust level is low, the continual application of thrust means that the eventual velocity increment obtained can exceed that from a chemical engine of equivalent weight, decreasing trip times substantially. The electric engine performs well because the specific impulse is higher than for any chemical system. Specific impulse is the force per unit mass expelled from the system. It is well known in rocket engineering that the mass to be expelled from the spacecraft carries a profound significance for the efficiency of the overall vehicle. The reason is that mass must be expelled in order to provide the forward momentum; the sum of the expelled mass and spacecraft momentum is maintained at zero; loss of momentum through the emission of hot gas means that the spacecraft proceeds in the opposite direction. This is done most efficiently by increasing the exhaust velocity, thereby decreasing the required mass. This, in turn, means that less fuel mass must be carried on board. Since the fuel is part of the mass to be accelerated until that fraction of fuel is burned, it is clear that the less the weight of fuel needed, the lighter the mean mass of the spacecraft and therefore the higher the acceleration given at any one instant of time. Thus the efficiency of a rocket engine is given in terms of unit impulse per unit mass of fuel. Since for ion propulsion the impulse to fuel mass ratio can exceed that of chemical fuels by several orders of magnitude, a potentially significant gain can be achieved in the performance. This gain can be measured either by a decrease in the overall weight of the spacecraft or an increase in the payload. Generally, such electric systems demand a large solar-cell array so that the large power requirement can be met.

[CHARLES P. SONETT]

Bibliography: L. Colin, Pioneer Venus program, *J. Geophys. Res.*, 85:7575–7598, 1980; W. R. Corliss, *Scientific Satellites*, NASA SP-133, 1967; W. R. Corliss, *Space Probes and Planetary Exploration*, 1965; J. W. Dyer, Pioneer Saturn, *Science*, 207:400–404, 1980; J. James, The voyage of *Mariner 4*, *Sci. Amer.*, 214:42–52, March 1966; Jet Propulsion Laboratories (JPL), *Mariner Mars 1964 Project Report: Television Experiment*, Tech. Rep. no. 32-884; JPL, *Ranger VI Mission and Analysis*, Tech. Rep. no. 32 699; JPL, *Ranger VII Mission and Analysis*, Tech. Rep. no. 32-700; JPL, *Surveyor III Mission Report*, Tech. Rep. no. 32-1177; JPL, *Surveyor V Mission Report*, Tech. Rep. no. 32-1246; C. E. Kohlase et al., The 1969 Mariner mission to Mars, *Astronaut. Aeronaut.*, 7(7):80–96, 1969; D. P. LeGalley (ed.), *Space Science*, 1963; D. P. LeGalley and A. Rosen (eds.), *Space Physics*, 1964; R. B. Leighton, The surface of Mars, *Sci. Amer.*, 222(5):26–41, 1970; E. Leven, D. D. Viele, and L. B. Eldrenkamp, The lunar orbiter missions to the moon, *Sci. Amer.*, 218:59–78, May 1968; *McGraw-Hill Encyclopedia of Space*, rev. 1970; NASA, *Aeronautics and Space Report to the President*, published annually; NASA, *Surveyor V: A Preliminary Report*, NASA SP-163, 1967; NASA, *Surveyor VI: A Preliminary Report*, NASA SP-166, 1968; The solar system, *Sci. Amer.*, Special Issue, 233(3):22–173, September 1975; E. C. Stone and A. L. Lane, *Voyager 2*: Encounter with the Jovian system, *Science*, 206:925–927, 1979; E. C. Stone and E. D. Miner, Encounter with the Saturnian system, *Science*, 212:159–162, 1981; *TRW Systems and Space Log*, published periodically; Viking voyage to Mars, *Astronaut. Aeronaut.*, 7(11):30–59, 1969.

Spectrohelioscope

An instrument for the monochromatic visual observation of the Sun. A telescope projects an image of the Sun on the first slit of a powerful spectroscope (Fig. 1). The resulting spectrum is imaged in the plane of a second slit which permits only a

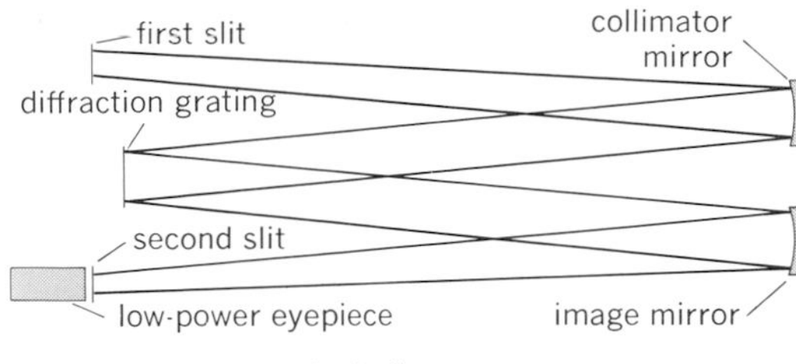

Fig. 1. The Hale spectrohelioscope.

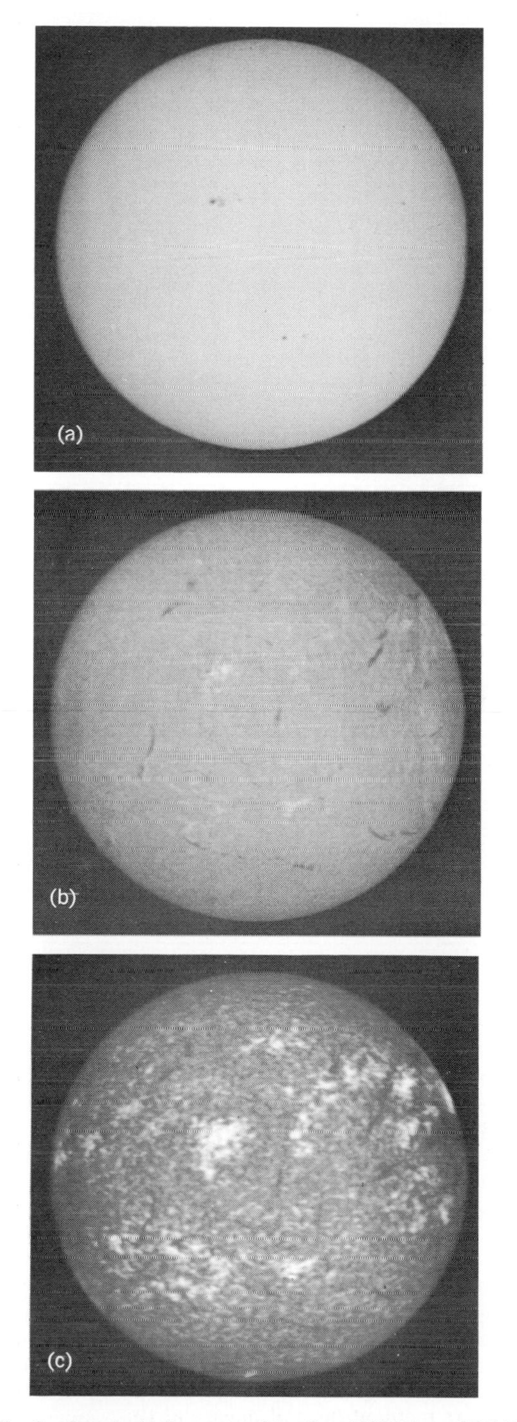

Fig. 2. Spectroheliograms of the Sun (1958 June 20ᵈ12ʰ 10ᵐ UT) formed of line elements 0.3 A wide. (*a*) At 6600 A. (*b*) At 6563 A (Hα). (*c*) At 3933 A (K). The slits were moved at a slow rate. 1 A = 0.1 nm.

single line element of the spectrum to emerge from the instrument. The emergent line element is a monochromatic image of that part of the Sun that falls on the first slit. The widths of the slits are generally chosen to isolate a spectral interval 0.5 angstrom (0.05 nanometer) or less in width. When the two slits are vibrated synchronously at high frequency, persistence of vision permits monochromatic observation of an area of the solar surface. The slits may also be moved at a slow rate and the image recorded photographically (Fig. 2). This modification of the spectrohelioscope is a simple form of the spectroheliograph.

[ROBERT R. MC MATH/JOHN W. EVANS]

Spica

Alpha Virginis, one of the brightest and nearest of the hot, main-sequence stars, spectral type B1. Located at a distance of 75 parsecs, Spica is 1500 times brighter than the Sun and has a temperature of nearly 20,000 K. Spica is a spectroscopic binary of 4-day period consisting of two nearly identical stars. Rapidly rotating, and relatively close, they are nearly unstable and are connected by streams of matter. The relative intensities of the spectral lines of the two components vary around the orbit because of these circumstellar streams. *See* STAR.

[JESSE L. GREENSTEIN]

SS 433

A remarkable stellar object with unique properties: it shows evidence of ejection of two narrow streams of cool gas traveling in oppositely directed beams from a central object at a velocity of almost one-quarter the speed of light—the beams executing a repeating, rotating pattern about the central object once every 164 days.

Discovery. SS 433 is an example of an astronomical object discovered, forgotten, and rediscovered several times over a period of two decades. The initial observation of the object was made in 1959 by C. B. Stephenson and N. Sanduleak during a survey of the Milky Way for peculiar stars. The Stephenson-Sanduleak (SS) published survey lists this star as the 433d entry, thus the nomenclature SS 433. The SS survey searched for objects with a specific anomaly in their spectrum—namely, bright emission lines. Ordinarily, when starlight is dispersed via a prism or grating into its component colors, the resulting spectrum is relatively smooth, with comparable amounts of light reaching Earth at every wavelength (color). However, occasionally stars are found which have intense peaks in their spectra at very specific wavelengths. When seen on a spectrum recorded on a photographic plate, this phenomenon is called an emission line, because the convergence of light at this particular wavelength causes a linear thickening of the developed emulsion at a specific location. Emission lines in stars are caused by light emitted from atoms in the star which have become excited due to collisions with other atoms or with electromagnetic radiation (light). The relaxation from this excitation is normally accompanied by emission of light at certain specific, unchanging wavelengths characteristic of the chemical element involved. Thus the detection of emission lines is a valuable insight into the nature of the star, since they provide a "fingerprint" enabling identification of chemical elements present in the object. Perhaps 5–10% of all stars show emission lines in their spectra, so this characteristic of SS 433 was not in itself sufficiently unusual to provoke more detailed observations of the object. *See* ASTRONOMICAL SPECTROSCOPY.

During the 1960s and 1970s a variety of astronomers mapping the skies at x-ray and radio wavelengths unknowingly rediscovered SS 433, as they found that intense x-ray and radio emissions were emanating from this region of sky. However, the precision of these observations was insufficient to permit these workers to associate the source of the emissions specifically with SS 433, as opposed to numerous other nearby stars (Fig. 1). Finally in 1978 three independent groups of English and Canadian astronomers recognized that the visible object SS 433 and the previously cataloged sources of radio and x-ray emission were in fact all the same object. This is extraordinary, as only the tiniest fraction of all stars emit detectable amounts of either radio or x radiation. The first modern spectroscopy of SS 433 confirmed the results of the SS catalog by showing the object to have extraordinarily intense emission lines of hydrogen and helium. These workers also pointed out that SS 433 is surrounded in the sky by a large, diffuse glow of radio emission, itself a previously cataloged object, termed W50. The structure of W50 has led most astronomers to conclude that it is the remnant of an ancient exploded star, or supernova, in this case probably occurring more than 100,000 years ago. The central location of SS 433 within W50 leads to the speculation, appealing but unproved, that the two objects are in fact associated. *See* RADIO ASTRONOMY; X-RAY ASTRONOMY.

Jets of matter. The most peculiar characteristics of SS 433 have been revealed by an intensive series of spectroscopic observations. These observations show that the spectrum possesses not only a set of emission lines due to hydrogen and helium, but two further sets of lines, one displaced to longer (redder) wavelengths from the familiar lines, and the second displaced to shorter (bluer) wavelengths. These displacements can be understood in terms of the Doppler effect, a familiar mechanism which lengthens the apparent wavelength of any wave phenomenon (including light or sound) when there is a recessional motion between the source and the observer, and shortens the wavelength if there is approach. Thus the observations imply that in addition to a stationary object, SS 433 possesses some gas (a mixture of hydrogen and helium) approaching the Earth, while some presumably different patch of gas recedes. The remarkable property is the velocity of approach and recession, calculated simply from the magnitude of the observed spectral Doppler shifts. The velocity of this gas in the initial observations was found to be up to 50,000 km/s, that is, about 16% of the speed of light. Because the escape velocity from the Milky Way Galaxy is only a few hundred kilometers per second, one never observes stellar objects with velocities in excess of this, since they would rapidly leave the Galaxy.

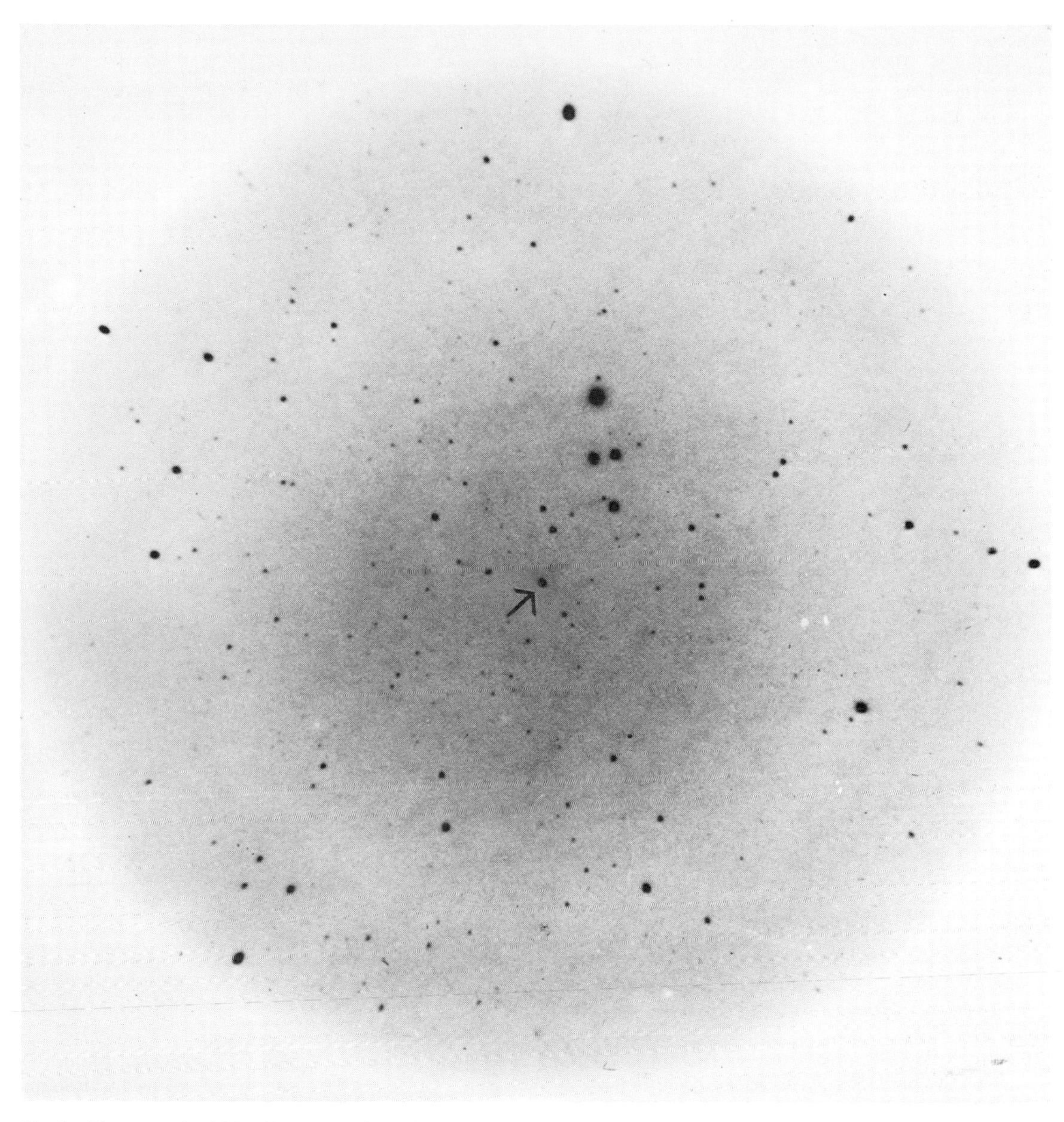

Fig. 1. Photograph of SS 433 made at Lick Observatory, using the 36-in (0.9-m) refractor telescope. The obvious lack of any features which distinguish the object from its numerous neighbors in this dense portion of the Milky Way helps explain why the star was overlooked for so long.

Further spectral monitoring of SS 433 has revealed spectacular changes in these two sets of Doppler-shifted lines. The wavelengths of the lines change every night in a smoothly progressing pattern, indicating that the velocity of the emitting regions is also changing. Each set of lines proves to cycle in a regular pattern between a recessional velocity of 50,000 km/s and an approach velocity of 30,000 km/s, with the cycle lasting approximately 164 days (Fig. 2). The pattern then begins again. The currently accepted interpretation of this periodic behavior is that the "moving" emission lines are due to light from two narrow streams or jets of matter ejected from a central object in opposite directions. A slow rotation of the axis of these jets, once every 164 days, is then responsible for the changing velocities observed at the Earth. This rotation is probably caused by precession, a wobbling motion of the star. Different velocities are seen on different days because the moving axis of the jets may be more or less directly pointed toward Earth at a given time. Interpretation of the observations shown in Fig. 2, using this concept, shows that the true velocity of the ejected beam remains constant throughout the 164-day cycle at a value of about 80,000 km/s, about one-quarter of the velocity of light. A velocity this high is never directly observed, since the Earth would have to be fortuitously located exactly in the conical surface of the rotation pattern to have the jets point exactly toward and away from it. The tremendous beam velocity inferred implies a huge energy source to accelerate a substantial amount of gas to this speed: the kinetic energy in the beams is approximately a million times as large as the total amount of light energy radiated by the Sun.

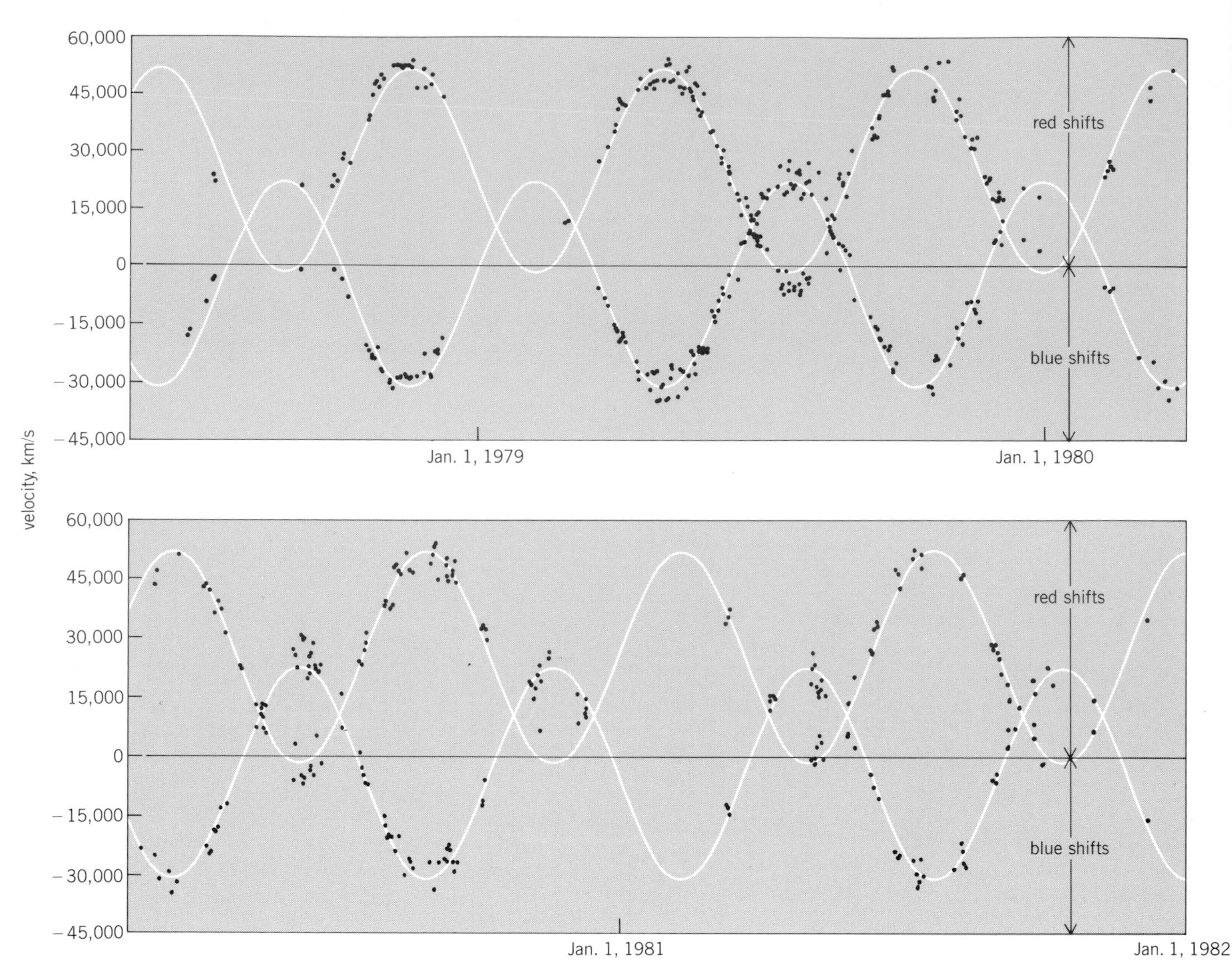

Fig. 2. Graph of the value of red shifts and blue shifts of SS 433 over a period of about 3 years. Large gaps in the data are caused when the object is too close to the Sun for nighttime observation; smaller gaps are due to the proximity of the Moon, making observations difficult. Curves show the predicted behavior of the Doppler-shifted gas is located in two narrow, oppositely directed beams which rotate every 164 days.

Time dilation. Especially intriguing is the observation (Fig. 2) that on a given night the average velocity of the approaching and receding beams is not zero, but rather a large positive value, about 12,000 km/s, despite the fact that SS 433 is approximately stationary with respect to the Earth. This proves to be a direct consequence of Einstein's special theory of relativity. An outside observer perceives a change in measured times and lengths of a system moving at very large velocity. The beam velocity in SS 433 is large enough that special relativity is important, and this "time dilation" effect causes a permanent red shift of the spectral lines of 12,000 km/s; the 164-day rotational component is then superposed on top of this underlying effect. *See* RELATIVITY.

Radio maps. Elegant confirmation of these indirect inferences concerning the nature of SS 433 has been provided by detailed maps of the radio structure of the central object. These observations, which utilized the Very Large Array, an ensemble of 27 radio astronomy antennas in New Mexico, indicate that this radio structure also changes cyclically every 164 days, as radio-emitting blobs of gas coast outward after ejection from the jets. The radio maps directly reveal the spirallike trail created by the jet rotation, thus confirming the concept of SS 433 developed from the optical spectroscopy. The radio observations also provide clues enabling an estimate of the distance from the Earth to SS 433 as 15,000 light-years (1.4×10^{17} km), about one-sixth the diameter of the Milky Way Galaxy.

Star type and matter source. There has been much speculation as to the type of star present in SS 433, with many astronomers now agreeing that the enormous velocities in the beams require a highly collapsed, compact star with a strong gravitational field. A neutron star, the same end point of stellar evolution responsible for pulsars, or possibly a black hole could satisfy this requirement. The source of matter ejected through the beams is also a problem; it seems probable that this may be supplied by a relatively normal, nearby companion star trapped in an orbit about SS 433. The emission lines in SS 433 have also been shown to cycle through a very small-amplitude period every 13 days, implying that the mutual orbital motion of SS 433 and this unseen companion may have been detected. *See* BINARY STAR; BLACK HOLE; NEUTRON STAR; PULSAR.

Perhaps the most vexing question of all is why there is only one object like SS 433 known in a galaxy of 10^{11} stars. Only the accidental discovery of a second such star can determine whether this is a possible end point of stellar evolution for a certain class of binary star, or whether SS 433 is truly a unique and obscure stellar accident.

[BRUCE MARGON]

Bibliography: B. Margon, The bizarre spectrum of SS 433, *Sci. Am.*, 243(4):54–65, October 1980; B. Margon, Relativistic jets in SS 433, *Science*, 215:247–252, 1982; D. Overbye, Does anyone understand SS 433?, *Sky Telesc.*, 58:510–516, 1979.

Star

A celestial body, consisting of a large, self-luminous mass of hot gas held together by its own gravity. The Sun is a typical star; its physical parameters are

$$\text{Radius } R_\odot = 6.9 \times 10^{10} \text{ cm} = 6.9 \times 10^{8} \text{ m}$$
$$\text{Mass } M_\odot = 2 \times 10^{33} \text{ g} = 2 \times 10^{30} \text{ kg}$$
$$\text{Luminosity } L_\odot = 4 \times 10^{33} \text{ ergs/sec} = 4 \times 10^{26} \text{ W}$$

The Sun's mean density is 1.45 g/cm³, and its central density is about 120 g/cm³. The surface temperature is 5750 K, the mean temperature about 5×10^6 K, and the central temperature 13×10^6 K. In spite of the high density, the gas in the Sun is almost completely ionized from surface to center.

Composition and distribution. The composition by weight of the average star is about 70% hydrogen, 28% helium, 1.5% carbon, nitrogen, oxygen, and neon, and 0.5% iron group and heavier elements.

The stars contain by far the largest fraction of the mass of the universe. The parameters given above describe their average condition. Stars are born, produce nuclear energy, evolve, and eventually die. Their life-spans range from 10^6 years, for a star of high luminosity, to 10^{10} for the Sun, and up to 10^{13} years for the faintest main-sequence stars. The oldest known in the Milky Way Galaxy are over 10^{10} years old.

The nomenclature for the identification of stars gives their general location and brightness. The sky was subdivided into constellations and the brighter stars were named; Greek alphabet letters generally describe them, with α representing the brightest star, visually, in a given constellation.

Table 1. The 26 nearest stars (from P. van de Kamp)[a]

Name	Parallax, seconds of arc	Distance, light-years	Annual proper motion, seconds of arc	Radial velocity, km/sec	Transverse velocity, km/sec	Apparent magnitude and spectrum	Absolute magnitude
Sun						−26.7 G2	+4.8
α Centauri[b]	0.760	4.3	3.68	−25	23	+0.3 G2(1.7 K5)	4.7(6.1)[e]
Barnard's star	.545	6.0	10.30	−108	90	9.5 M5	13.2
Wolf 359	.421	7.7	4.84	+13	54	13.5 M6e, *v*[f]	16.6
Luyten 726–8	.410	7.9	3.35	+29	38	12.5 M6e (13.0 M6e)*v*	15.6(16.1)
Lalande 21185	.398	8.2	4.78	−86	57	7.5 M2	10.5
Sirius[c]	.375	8.7	1.32	−8	16	−1.5 A0(8.7 DA)	1.4(11.6)
Ross 154	.351	9.3	0.67	−4	9	10.6 M5e	13.3
Ross 248	.316	10.3	1.58	−81	23	12.2 M6e	14.7
ε Eri	.303	10.8	0.97	+15	15	3.8 K2	6.2
Ross 128	.298	10.9	1.40	−13	22	11.1 M5	13.5
61 Cyg	.293	11.1	5.22	−64	84	5.6 K6 (6.3 M0)	7.9(8.6)
Luyten 789–6	.292	11.2	3.27	−60	53	12.2 M6	14.5
Procyon	.288	11.3	1.25	−3	20	0.5 F5 (10.8 DA?)	2.8(13.1)
ε Indi	.285	11.4	4.67	−40	77	4.7 K5	7.0
Σ 2398	.280	11.6	2.29	+1	38	8.9 M4 (9.7 M4)	11.1(11.9)
Groombr. 34	.278	11.7	2.91	+14	49	8.1 M2e (10.9 M4e), *v*	10.3(13.1)
τ Cet	.275	11.8	1.92	−16	33	3.6 G4	5.8
Lacaille 9352	.273	11.9	6.87	+10	118	7.2 M2	9.4
+5° 1668	.263	12.4	3.73	+26	67	9.8 M4	12.0
Lacaille 8760	.255	12.8	3.46	+23	64	6.6 M1	8.6
Kapetyn's star	.251	13.0	8.79	+242	166	9.2 M0	11.2
Ross 614	.251	13.1	0.97	+24	18	11.1 M5e(14.8)	13.1(16.8)
Kruger 60	.249	13.1	0.87	−24	16	9.9 M4 (11.4 M5e)	11.9(13.4)
−12° 4523	.244	13.4	1.24	−13	24	10.0 M5	11.9
vMa 2[d]	.236	13.8	2.98	+70	59	12.3 DG	14.2

[a]In 26 listings there are 36 individual stars, and some have still undiscovered faint companions.
[b]α Centauri has a second companion (dM5e) of absolute magnitude + 15.4.
[c]Sirius and Procyon each have a white dwarf companion.
[d]vMa 2 is a white dwarf.
[e]Parentheses indicate uncertain value.
[f]Here *v* = variable flare star which may emit bursts of light and even radio noise.

Thus Betelgeuse, or α Orionis, is the visually brightest star in Orion. Fundamentally a star is defined by its coordinates on the celestial sphere, right ascension and declination, and its brightness, or apparent magnitude. Because of the precession of the equinoxes, celestial coordinates must be specified for a given epoch. About 6000 stars are visible to the naked eye, but over 10^{12} exist in the Milky Way Galaxy. Catalogs give positions, brightnesses, motions, parallaxes, spectral types, velocities, and other properties of the stars. Many of these catalogs are now available in magnetic tape format. Several hundred thousand stellar positions are also available, with much of the data on tape suitable for computer search. Some fainter stars are tabulated by the name of the discoverer of any interesting property and are listed by many unrelated and overlapping names. *See* CELESTIAL SPHERE.

Near stars. The largest stars have been so identified by their large angular proper motions in the plane of the sky and by subsequent measurements of parallax. The 26 nearest stars are listed in Table 1 together with relevant data. A few more, intrinsically faint stars may exist at or within this limit of distance. Table 1 also gives the transverse motions, in kilometers per second, derived from the proper motion and distance of the star, and the radial velocity, which is from the measured Doppler shift. Many stars have a total space motion of from 100 to 500 km/sec with respect to the Sun. Such objects are called high-velocity stars and belong, according to the nomenclature of Walter Baade, to population II, that is, stars found in the spheroidal halo of a galaxy. The more slowly moving stars are younger, and may be members of population I like the Sun; such stars are found in the spiral regions or flattened disk of a galaxy. The absolute visual magnitudes show that 36 of the 40 nearby stars are intrinsically fainter than the Sun. In addition, about half are in multiple systems, doubles or triples. Most of the stars in Table 1 have spectral types which put them on the main se-

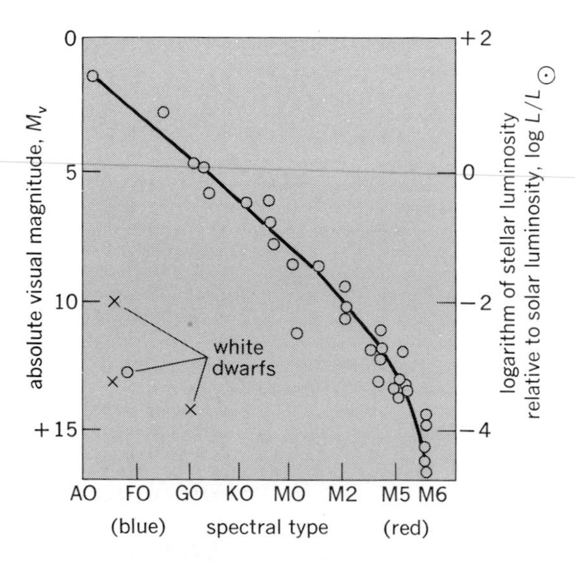

Fig. 1. Hertzsprung-Russell diagram for nearby stars showing main sequence and white dwarfs.

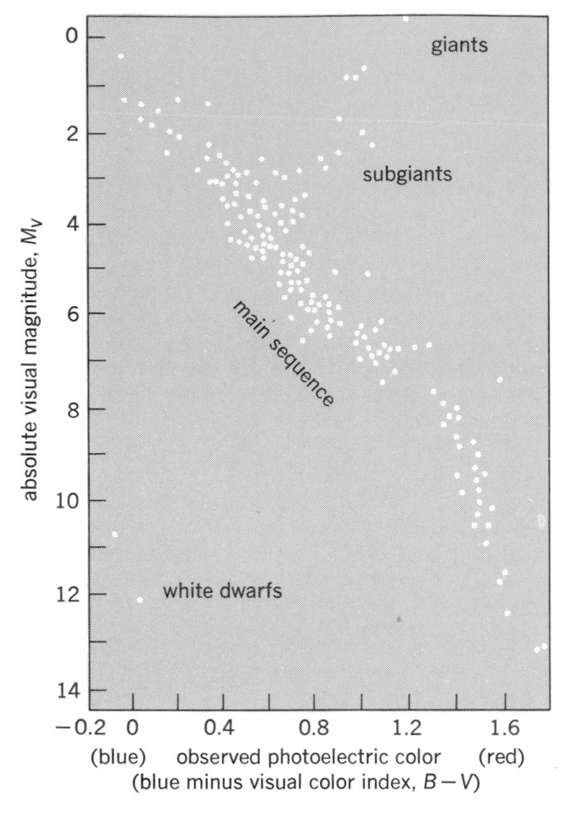

Fig. 2. Main-sequence stars, subgiants, and giants.

quence or dwarf branch. No red giant or supergiant is included, although there are three white dwarfs in the immediate neighborhood of the Milky Way Galaxy. Figure 1, a Hertzsprung-Russell or H-R diagram, is a means for plotting the relation between the luminosity and surface temperature or other temperature-dependent parameter, such as color or spectral type of the stars. Along the main sequence, through which the line is drawn in the diagram, the stars are distinguished chiefly by mass; luminosity is a steep function of the mass. Nearby stars, for which data are most accurate, do not include all interesting types. To include a greater variety of stars, the less accurate data for stars out to 20 parsecs (6.2×10^{17} m or 65 light-years) are plotted. The relation between absolute visual magnitude and observed photoelectric color for stars closer than 20 parsecs shows the main sequence in more detail for the somewhat brighter stars and also the red giants and subgiants (Fig. 2). The use of photoelectric color, like that of spectral type, is essentially an arrangement by surface temperature and is convenient and accurate. *See* HERTZSPRUNG-RUSSELL DIAGRAM; PARALLAX; PARSEC.

Bright stars. The apparently brighter stars are listed in Table 2. Because stars of high intrinsic luminosity can be seen at great distances, Table 2 includes many such stars, including giants and supergiants. For many of these the parallax is too small to be measured directly and the luminosities are only approximate. Similarly, physically close visual doubles would be missed. The H-R diagram for the brighter stars illustrates the existence of

Table 2. The 25 brightest stars (from H. L. Johnson)*

Star	Name	Spectrum	Absolute visual magnitude, M_v	Visual brightness, V	Color index, B−V	Remarks
α CMa	Sirius	A1 V	+1.4	−1.43	0.00	
α Car	Canopus	F0 Ia	−4.5	−0.73	+0.15	
α Cen		G2 V	+4.7	−0.27	+0.66	Double
α Boo	Arcturus	K2 IIIp	−0.1	−0.06	+1.23	
α Lyr	Vega	A0 V	+0.5	+0.04	0.00	
α Aur	Capella	G0 IIIp	−0.6	+0.09	+0.80	Spectroscopic binary, double
β Ori	Rigel	B8 Ia	−7	+0.15	−0.04	Double
α CMi	Procyon	F5 IV-V	+2.7	+0.37	+0.41	
α Eri	Achernar	B3 V	−2	+0.53	−0.16	
β Cen		B0.5 V	−4	+0.66	−0.21	Double
α Ori	Betelgeuse	M2 Iab	−5	+0.7	+1.87	Variable
α Aql	Altair	A7 IV-V	+2.2	+0.80	+0.22	
α Tau	Aldebaran	K5 III	−0.7	+0.85	+1.52	Variable, double
α Cru		B0.5 V	−4	+0.87	−0.24	Double
α Sco	Antares	M1 Ib	−4	+0.98	+1.80	Double, variable
α Vir	Spica	B1 V	−3	+1.00	−0.23	Spectroscopic binary
α PsA	Fomalhaut	A3 V	+1.9	+1.16	+0.09	
β Gem	Pollux	K0 III	+1.0	+1.16	+1.01	
α Cyg	Deneb	A2 Ia	−7	+1.26	+0.09	
β Cru		B0.5 IV	−4	+1.31	−0.23	
α Leo	Regulus	B7 V	−0.7	+1.36	−0.11	Double
ε CMa	Adhara	B2 II	−5	+1.49	−0.17	
α Gem	Castor	A0	+0.9	+1.59	+0.05	Double, spectroscopic binary
λ Sco	Shaula	B2 IV	−3	+1.62	−0.23	
γ Ori	Bellatrix	B2 III	−4	+1.64	−0.23	

*The spectra are from H. L. Johnson and Morgan; colors and magnitudes are photoelectric, V being the equivalent of a visual brightness and B−V a blue minus visual color index. The absolute visual magnitudes M_v are based on measured parallaxes; when only one significant figure is given, however, they are only estimates.

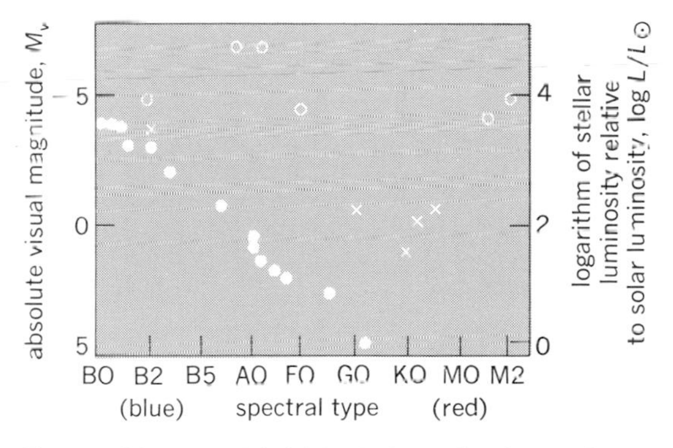

Fig. 3. Diagram of brightest stars showing main sequence as in Fig. 1 and branches of other types of stars.

the other branches in addition to the main-sequence and red-giant branch; white dwarfs are missing (Fig. 3).

Stellar spectra. The spectrum of a star in a large majority of cases shows absorption lines superposed on a continuous background. The interior of the star is at high temperature and pressure; its spectrum is nearly that of a blackbody. The star shades off into space through a reversing layer or stellar atmosphere in which the continuous spectrum and absorption lines are formed. In the Sun this reversing layer is about 250 km thick; its base is about 5750 K, and its outer layer is almost 4200 K. The continuum is formed at a depth equal to the mean free path for an average quantum. The emergent continuum resembles, but is not identical with, a blackbody. The atoms in the reversing layer produce absorption lines because of the existence of this temperature gradient. Outside the normal reversing layer in many stars there may be a temperature inversion in the low-density chromosphere and corona. The temperature of the corona eventually reaches 1,500,000 K. The emission lines of these outer layers affect only slightly the spectrum of the integrated light of a star and carry about one-millionth of the total energy. The x-ray radiation of the Sun is highly variable. Some stars have been found by their strong (and variable) x-ray emission. *See* SUN; X-RAY ASTRONOMY; X-RAY STAR.

Stellar spectra are normally obtained with a slit spectrograph by black and white photography.

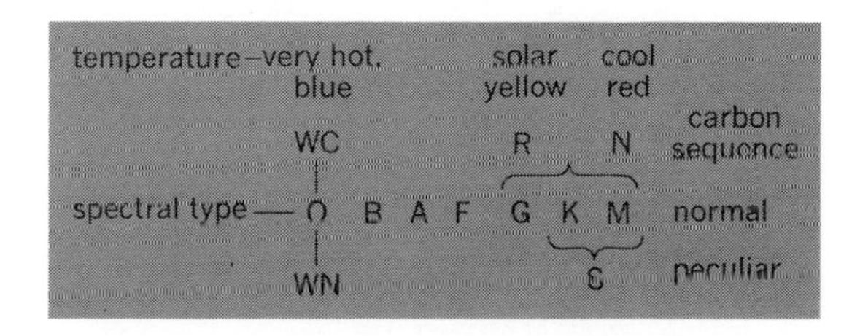

Fig. 4. Spectral classification from Draper catalog.

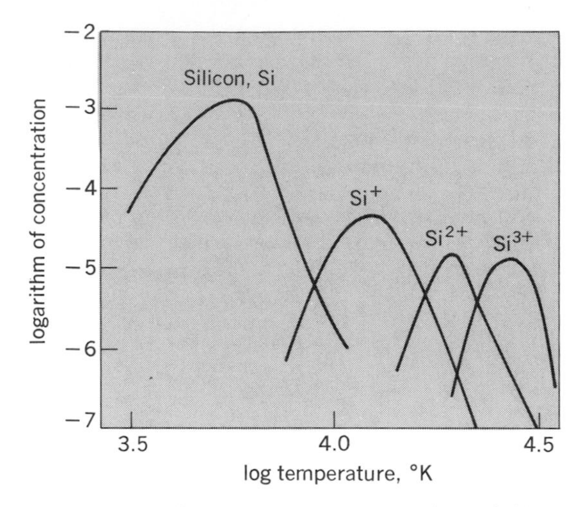

Fig. 5. Effect of temperature on spectral lines of silicon in various states of ionization.

Astronomical spectrographs largely employ plane gratings, operate in the wavelength region from 300 to 700 mm, and provide dispersions ranging from 1 to about 400 A/mm (0.1 to about 40 nm/mm). Wavelength standards are provided by a comparison spectrum impressed during the exposure of a plate, usually by a laboratory source imaged at both ends of the spectrograph slit. Standards for photometry of the lines are provided by plate calibration devices. Because a point source, like a star, gives only a narrow streak of spectrum with the stigmatic spectrographs used, the spectra are suitably widened by allowing the star image to trail up and down the slit. Low-resolution spectra and some high-resolution spectra of bright stars and of the Sun can be obtained by photoelectric scanning at the telescope. The infrared is best observed with cryogenic solid-state bolometers. Great advances in sensor technology have permitted moderate-resolution spectrometry of many cooler stars, down to 1500K temperatures. Even cooler stars have been detected at low spectral resolution. Image-intensifier tubes have been used for low-resolution spectra, providing an information-rate gain of over 20. *See* INFRARED ASTRONOMY; IMAGE TUBE.

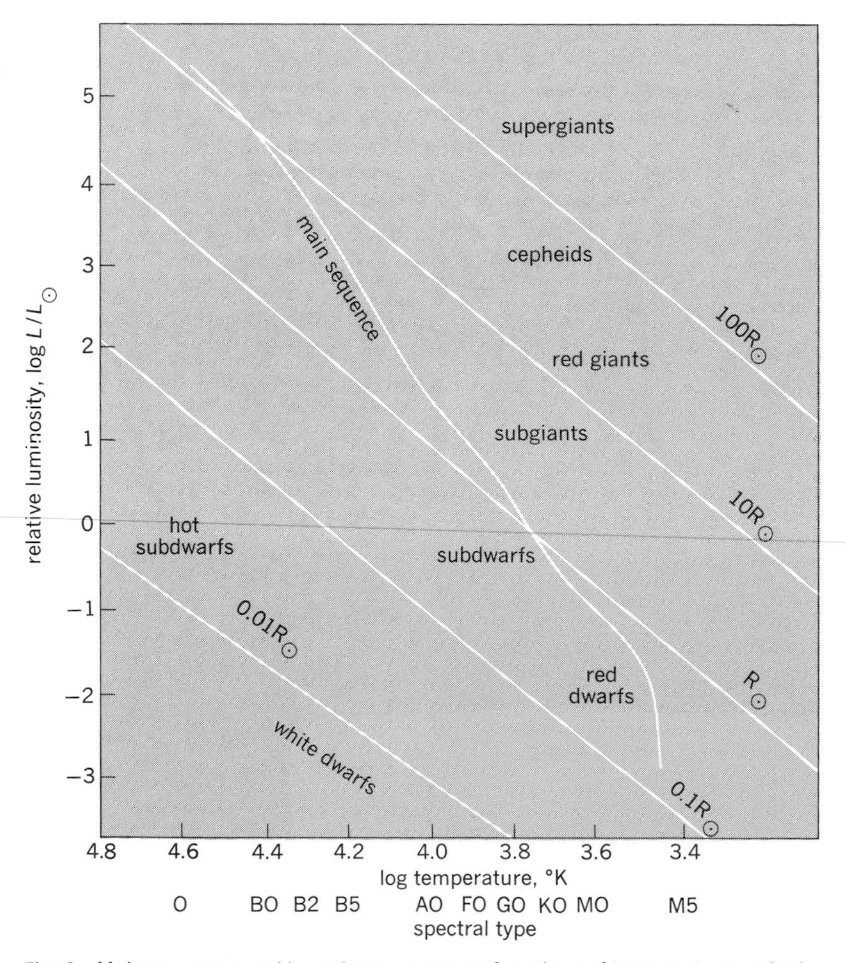

Fig. 6. Main sequence and branch sequences as functions of temperature and relative luminosity; approximate spectral types listed under temperature scale.

Spectral classification. The spectral classification of a star gives in a simple symbolic form the essential features of its complex spectrum. By inspection, at dispersions ranging from 100 to 400 A/mm (10 to 40 nm/mm), many features are found to vary in a smooth way from one star to another. This variation is correlated with the colors of the stars. As a result, spectral classification includes a vast majority of the stars and represents a sequence of decreasing reversing-layer temperature, from the left to the right (Fig. 4). There are several side sequences whose temperatures approximately correspond with those of normal stars, in which apparent abundance differences exist. Both dwarf and giant stars exist over a wide range of spectral type. Decimal subdivisions are used; in more refined analyses prefixes and suffixes are added, such as e for emission lines, n for broad lines because of rotation, q for novalike, p for peculiar, d for dwarf, D for white dwarf, sg for subgiant, g for giant, and c for supergiant. Almost 300,000 stars have been classified on this system.

Peculiar spectra exist, such as the carbon stars (old name R and N, new type C) and S stars, which have about the same temperature as K or M stars. Another very important type of peculiarity has been found associated with stars with magnetic fields. *See* STELLAR MAGNETIC FIELD.

The percentage of stars of different spectral types in the Henry Draper catalog is roughly as follows: B, 3%; A, 27%; F, 10%; G, 16%; K, 37%; and M, 7%; other types are rare. This is a selection of stars by apparent brightness and does not represent their true distribution in space, which from the data in Table 1 heavily favors the late-type M dwarfs.

A more refined system of spectral classification developed by W. W. Morgan, with P. C. Keenan and other collaborators, has been widely used. In this system, employing spectra at 120 A/mm (12 nm/mm), a two-parameter set of criteria provides by inspection both spectral type and estimated luminosity. The luminosity is indicated by a suffix ranging from Ia, extremely bright supergiant, to III, normal giant, to V, main-sequence or dwarf star. Thus the designation G2 V represents a star like the Sun, G2 III a giant of nearly the same temperature, with certain luminosity-sensitive features enhanced.

Temperature and luminosity. Most differences in appearance of stellar spectra are caused by changes in the surface temperature. The degree of ionization and excitation of the atoms at a given temperature dictates whether an element will have an appreciable concentration of atoms in the lower atomic level that produces the absorption line. For example, at very high temperature, helium (He) is all He^{++} and thus has no lines; at about 35,000 K, He^{+} dominates, and below 20,000 K, He exists. Below about 12,000 K insufficient atoms of He are excited to the states at 19 electron volts which produce lines in the normal spectral region; no helium lines will be seen (except in the inaccessible ultraviolet) in cooler stars. Stars with He^{+} lines exist and are of type O; He lines occur in type B stars. Stellar temperatures can be accurately determined by quantitative application of these methods, involving the Saha ionization and the Boltzmann excitation equations (Fig. 5).

Absolute magnitude effects occur because the ionization equation depends on pressure, the electron concentration setting the recombination rate. Consequently, stars of low surface gravity, which have lower pressures, will show a given percentage of ionization at a temperature about 500 K lower than those of the main sequence. A red giant of the same temperature as the Sun, 100 times brighter, has 10 times the radius and about 0.04 times the surface gravity (allowing for the larger mass of the giant). The lower gravity and pressure result in an increased level of ionization of sensitive elements. Thus the luminosity classification of a star, a second parameter, is possible after a temperature classification has been made. These effects are calibrated by stars of otherwise known luminosity.

The luminosity of a star is its energy output, either in ergs per second, or in watts, or in units of solar luminosity, or in absolute magnitude. First, apparent magnitude and distance in parsecs or parallax must be known. Let a star have luminosity L, radius R, and effective temperature T. Then from Stefan's law and the areas of the star Eq. (1)

$$\frac{L_{\text{star}}}{L_{\odot}} = \left(\frac{R_{\text{star}}}{R_{\odot}}\right)^2 \left(\frac{T_{\text{star}}}{T_{\odot}}\right)^4 \quad (1)$$

is obtained. If the temperature corresponding to a given spectral type or color is known, one of a variety of types of H-R diagrams can be plotted, connecting luminosity, or absolute magnitude (or apparent magnitudes if all stars of a group are located at the same distance) as ordinates and temperature, or spectral type or color as abscissas. In a diagram in which log L and log T are used, loci of constant radius are straight lines (Fig. 6). Such a diagram, with main-sequence and giant branch, makes clear the location of many of the various sequences of stars.

Age, evolution, and mass. A group of color-magnitude curves for clusters, open (population I) and globular (very old population II), shows that, although the fainter end of the main sequence is essentially the same in all groups of stars, the brighter ends vary from one group to another in accordance with differences in age and composition (Fig. 7). Stellar evolution causes such variations.

These H-R diagrams are most significant for the study of the ages, nuclear energy sources, and evolution of the stars. The location of a star in an H-R diagram is completely determined by its mass and chemical composition, if the latter is specified in detail throughout the star. The effects of rotation and moderate magnetic field are observable but small. Large rotation, near instability, causes hot stars to change luminosity, color, and spectrum.

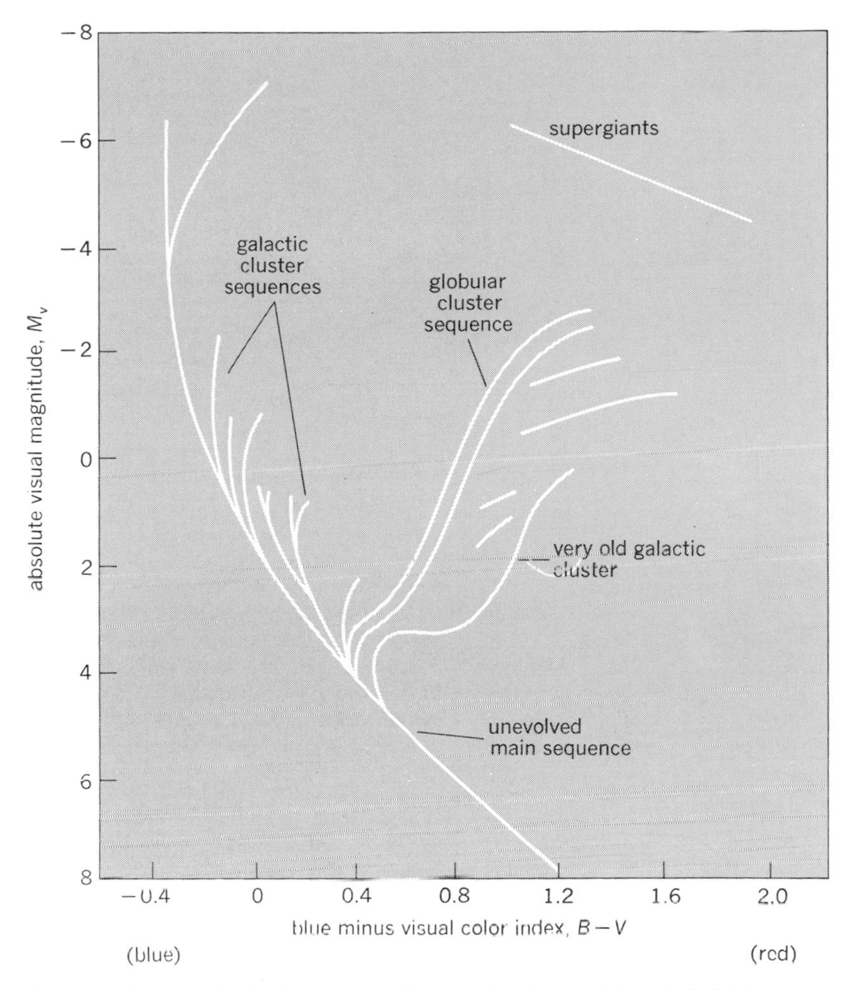

Fig. 7. Color-magnitude diagram for clusters showing variations in brightness for groups of different ages and compositions. Clusters are arranged chronologically, the oldest being at the bottom of diagram.

A point in an H-R diagram, specified by L and R, does not uniquely determine the mass, however, because of the possible variation of composition with depth. (Such variation is caused by the consumption of hydrogen and its conversion to helium by thermonuclear processes.)

The masses of the stars are determined for stars which are members of a double or multiple system, either visual or spectroscopic binaries. The application of Newton's laws provides the mass, although often with considerable uncertainty. If a pair of stars rotates in an orbit whose plane includes the line of sight, so that the star is also an eclipsing binary, and if the lines of both stars are visible in the composite spectrum and show measurable Doppler shifts, it is a two-line spectroscopic binary. Then the size of the orbit, its inclination to the line of sight, the relative masses, and the actual masses are determinable. Such completely

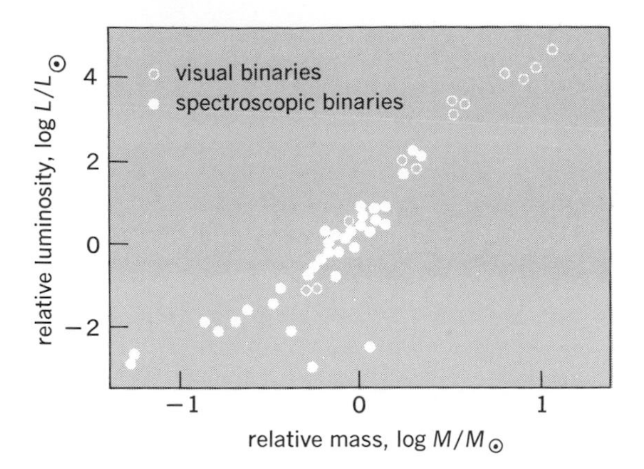

Fig. 8. Mass-luminosity relation for main-sequence stars. (*From observations by K. A. Strand*)

observed cases are rare but provide what quantitative information there is about masses, radii, and surface temperatures. Visual binaries with well-observed absolute orbits also give masses. The number of favorable circumstances required makes it difficult to measure directly the luminosities, masses, and radii of all the interesting types of stars. Thus many of the desirable relationships between parameters of a star are statistical in nature. The mass-luminosity relation is of this nature. However, a few classic, well-observed cases provide accurate and complete data for theoretical analysis. *See* BINARY STAR; MASS-LUMINOSITY RELATION.

The stars of the main sequence, for which stellar evolution has not yet been a substantial factor in displacement from their normal positions in an H-R diagram, obey a fairly well-established mass-luminosity relationship (Fig. 8). However, because of stellar evolution, stars move off the main sequence. If they belong to population I, they essentially move horizontally in an H-R diagram, so that they still obey the mass-luminosity relationship. Population II stars with low metal abundances brighten by factors up to a hundred in the subgiant and red-giant stage. A few such cases of clear violation of the mass-luminosity relation are known: for example, the population II visual binary, ζ Herculis A, for which the brighter component is 4 times brighter than the Sun, although it weighs only 1.07 times as much as the Sun. Other types of stars which deviate from the mass-luminosity relationship are the white dwarfs, the fainter components of Algol-type eclipsing variables, and some close O-type binaries.

Table 3 summarizes surface temperature *T*, lu-

Table 3. Approximate physical parameters of the stars

Properties of the Sun*:
G2, $L_\odot = 3.9 \times 10^{33}$ ergs/sec $= 3.9 \times 10^{26}$ W $M_\odot = 2 \times 10^{33}$ g $= 2 \times 10^{30}$ kg, $T = 5750$ K, $M_v = +4.64$, $R_\odot = 6.96 \times 10^{10}$ cm $= 6.96 \times 10^{8}$ m, $\log g = +4.44$, $\log \rho = +0.16$

Type	Color B−V	Absolute visual magnitude, M_v	$\log L/L_\odot$	T/1000 K	$\log R/R_\odot$	$\log M/M_\odot$	Remarks
Main sequence							
O8	−0.3 :†	− 5	+5.05	35	+0.96	+1.25	Uncertain, wide range
B0	−0.32:	− 4.3	+4.66	25	+1.05	+1.15	Uncertain
B1	−0.28	− 3.5	+4.06	22	+0.85	+1.05	
B2	− .24	− 2.8	+3.72	20	+0.76	+0.95	
B5	− .16	− 1.3	+2.96	15	+ .62	+ .78	
A0	0.00	+ 0.8	+1.73	11	+ .31	+ .45	
A5	+ .19	+ 1.9	+1.21	8.7	+ .24	+ .25	
F0	+ .37	+ 2.5	+0.85	7.6	+ .18	+ .14	
F5	+ .47	+ 3.5	+ .49	6.6	+ .12	+ .08	
G0	+ .60	+ 4.2	+ .21	6.0	+ .06	+ .04	
G5	+ .70	+ 5.2	− .19	5.5	− .06	− .17	
K0	+ .86	+ 6.1	− .55	5.1	− .18	− .22	
K5	+1.24	+ 7.5	− .84	4.4	− .18	− .25	
M0	+1.45:	+ 9.0	−1.23	3.6	− .22	− .30	
M2	+1.5 :	+10.0	−1.48	3.2	− .24	− .40	
M4	+1.6 :	+12.0	−1.91	3.1	− .42	− .55	Uncertain, because of
M6	+1.8 :	+15.0	−2.71	2.9	− .76	−0.90	steepness of main sequence
Giant stars							
G0		+ 0.7	+1.65	5.3	+0.90		Masses, to 4 M☉
K0		+ 0.2	+2.04	4.2	+1.30		Masses, to 4 M☉
M0		− 0.4	+2.72	3.3	+1.84		Masses, to 4 M☉
Supergiant stars							
B0		− 7	+5.66	25	+1.55		All data have wide range
A0		− 7	+4.77	10	+1.90		All data have wide range
K0		− 7	+5.26	3.6	+3.03		All data have wide range
M0		− 7	+5.66	3.0	+3.39		All data have wide range

*g = surface gravity in g/cm^2; ρ = mean density in g/cm^3. †Colons indicate discordant determinations.

minosity L, and radius R, and mass M of typical stars on the main sequence as a function of spectral type and gives, with a good deal less certainty, similar data for giants and supergiants. From the data surface gravity $g = GM/R^2$ (where G is the gravitational constant) and mean density $\rho = 3M/4\pi R^3$ can be obtained. A few direct measures of the angular diameters of red supergiants were made by stellar interferometers at the Mount Wilson Observatory and agree approximately with the estimates in Table 3. An interferometer has been devised by R. H. Brown and used at Narrabri, Australia, for a very accurate measurement of angular diameters of about 20 nearby stars of high temperature. This correlation-type interferometer, while of limited applicability, yields radii if the stars have accurate parallaxes. Direct effective temperatures are obtained from their luminosities. The radius of Sirius is 1.76 times that of the Sun, slightly smaller than that given in Table 3. The radius of Vega, however, which also is of spectral type AO, is 3.03 $R_{\odot}$. The calibration of the mean effective temperature of AO stars by Brown gives 10,250 K. Other methods have given a slightly lower temperature. The bright F5 star α CMi has 2.17 $R_{\odot}$, appreciably larger than given in Table 3; on the other hand, α CMi has, in fact, evolved above the main sequence; its temperature is about 6450 K.

Increased resolving power has been made possible by technical developments. Fast recording of the signal as a star is occulted by the limb of the Moon shows the diffraction pattern; the nulls in the diffraction are partly filled in, if the star has appreciable angular diameter. Still another technique, speckle interferometry, records electronically the image blurred by the Earth's atmosphere, in a very short time interval. Computer analysis restores the image to its diffraction-limited state. Major questions concerning radius and temperature still exist for the hotter stars. *See* ASTRONOMICAL PHOTOGRAPHY.

Surface temperature. The determination of surface temperature of stars is carried out by various techniques. Measurements of color, because the star's radiation is approximately that of a blackbody, provide a color temperature. But when combined with the presently highly developed theory of stellar atmospheres, the knowledge of the opacity of stellar material and detailed measures of the energy distribution at many wavelengths yield good measurements of the effective temperatures of stars. Late-type stars with strong molecular bands, however, have highly distorted energy curves, and temperatures can be deduced best from infrared magnitudes, measured with solid-state bolometers out to wavelengths of 20 μm.

Because all blackbody curves have essentially the same shape on the long-wavelength side of their energy maxima, all hot stars have essentially the same blue color, because their energy maxima are in the vacuum ultraviolet. In addition, hot stars are usually of high luminosity so that interstellar absorption and reddening impedes the analysis of their colors and luminosities.

Rockets and the first successful orbiting observatory provide the energy distribution in the far-ultraviolet and some information on the line spectra of the hottest and apparently brightest stars. Interstellar reddening does not increase as rapidly toward the violet as from the red to blue. Nevertheless, the flux is less than that predicted from model atmospheres and the standard temperature scale. Absorption lines are of highly ionized elements. In addition, mass loss from hot B stars has been established by displaced absorption lines, formed in an expanding cloud near the star.

Satellite observations of the ultraviolet flux have greatly increased the reliability of the temperature scale for hot stars. These include the Netherlands ANS satellite, the United Kingdom TD1, and the International Ultraviolet Explorer. As a result, the ground-based scale has been largely substantiated. Theoretical model atmospheres fit down to 110 nm. In binary stars, however, interaction may drive mass exchange, producing an optically thick accretion disk and excess ultraviolet radiation.

Strengths of absorption lines of different stages of ionization and excitation may be used to determine the temperature in hot stars. Certain of the hottest O stars, the Wolf-Rayet stars, and the nuclei of planetary nebulae show emission lines excited by a fluorescent conversion of their far-ultraviolet into visible radiations. The process is one of photoelectric ionization followed by recombination and emission of subordinate lines. Temperatures for such objects can be determined by the Zanstra method. The range of stellar temperatures is large. The hottest stars have effective temperatures near 50,000–100,000 K; the cooler stars are near 1500–2000 K. Although these extremes are somewhat unreliable, the temperatures given in Table 3 serve as a guide to the physical conditions in average stars. The x-ray stars observed optically give evidence for envelopes as hot as 50,000,000 K.

Stellar rotation. Another important property of stars is their rotation on their axes. The prevalence of wide double stars and of close spectroscopic binaries indicates that a large amount of angular momentum is often contained in the material that condensed to form the stars. In close double systems, revolution and rotation are often synchronous. In single stars, especially those of early spectral type, rotation is rapid.

The Sun has an equatorial velocity v of only 2 km/sec. Measurements of the rotational broadening of spectral lines give results contained in Table 4, based largely on the work of A. Slettebak. The large rotation of the early type B stars often results in instability, in the form of the ejection of matter from the rapidly rotating equatorial regions. As a result, such stars are often surrounded by disklike rings of low-density material about 10 times the

Table 4. Mean rotational surface velocities, in km/sec

Type	Dwarfs	Giants
B1–B3	200	127
B5–B7	257	163
B8–A2	177	93
A3–A7	173	202
A9–F2	87	125
F3–F6	31	67
F7–G0	<25	34

radius of the star. These are detected by the presence of emission lines and by absorption lines if the disk contains the line of sight. The rotation drops rapidly down the main sequence and usually is too small to be detected in single stars of types later than G0. *See* STELLAR ROTATION.

Composition. The chemical composition of a star can be deduced from the spectrum of its atmosphere or from the theory of its internal structure. There is evidence that stars need not be chemically homogeneous. Products of nuclear reactions may concentrate in the center so that the ratio of helium to hydrogen increases inward. Very slow mixing, however, counteracts the tendency toward complete diffusive separation by gravity of heavy from light elements.

Detailed studies of the composition of stars can be made only in their atmospheres. The ratio of hydrogen to heavy elements is about 8000 – 12,000 to 1, by number of atoms, for the Sun and young population I stars. For extreme high-velocity stars, old objects of population II, the abundance of the metals may be up to 400 times lower. Certain elements are unobservable in the spectra of stars, and the abundances of heavy and generally rare elements, as well as the isotopic abundances, are best obtained from the crust of Earth or meteorites. Table 5 is a composite résumé of current determinations of the abundances. *See* ELEMENTS, COSMIC ABUNDANCE OF.

Table 5. Abundances of elements in normal stars and Sun*

Logarithm of number of atoms = log N		
Atomic number	Symbol	log N
1	H	12.0
2	He	11.0
3	Li	1.0
4	Be	1.1
5	B	2.1
6	C	8.7
7	N	8.0
8	O	8.9
9	F	5.4
10	Ne	8.5
11	Na	6.3
12	Mg	7.5
13	Al	6.2
14	Si	7.5
15	P	(5.5)
16	S	(7.3)
17	Cl	(5.4)
18	Ar	(6.8)
19	K	4.7
20	Ca	6.2
21	Sc	3.2
22	Ti	4.3
23	V	3.7
24	Cr	5.4
25	Mn	4.9
26	Fe	7.5
27	Co	4.3
28	Ni	5.1
29	Cu	3.9
30	Zn	4.0
31	Ga	2.4
32	Ge	3.3
37	Rb	2.5
38	Sr	3.1
39	Y	2.8
40	Zr	2.3
41	Nb	2.0
42	Mo	2.1
44	Ru	1.8
45	Rh	1.7
46	Pd	1.8
47	Ag	0.8
48	Cd	2.0
49	In	1.7
56	Ba	2.9

*Parentheses indicate poor values. The rare earths have been omitted; they average about 1.0 in log N.

Stellar or solar abundances are determined by a set of measurements and interpreted by theory. Wavelengths of lines yield identification of the elements, and line intensities can be interpreted in terms of the numbers of atoms in the atomic levels producing the lines. However, to do this, the transition probability in the line must be known, either from quantum mechanics or laboratory measurements. Laboratory data on atomic transition probabilities are still in flux. In addition, corrections for the state of ionization and excitation must be applied to permit computation of the total concentration of the element, and these corrections depend strongly on the temperature. This is especially true for atoms whose visible lines arise from levels of high excitation potential; for example, in the Sun only one atom of helium in 10^{20} is likely to be excited into a state capable of producing a visible absorption line. Astrophysical abundance determinations are severely limited by Earth's atmosphere, which prevents observation of the far-ultraviolet spectra, where the resonance lines of many important elements are located. Satellite observations now permit such determinations. The subject of stellar atmospheres is an important branch of modern astrophysics. It relies heavily on the mathematical solution of problems of the diffusion of light outward through an absorbing and emitting medium and on physical information about the shape of atomic absorption lines, the amount of continuous absorption of light (which limits the depth to which we can see), and the theories of line broadening.

Motions of stars. Galactic dynamics and kinematics are an important part of the subject of stellar statistics and are intimately connected with the distribution of the stars in space. The Milky Way galaxy has a mass of about 6×10^{11} Suns, of which only a small fraction is visible from Earth. These stars move in orbits around their common center of gravitation, located about 8500 parsecs (2.6×10^{20} m) from Earth in the constellation Sagittarius. The galaxy is highly flattened by its systematic rotation, the linear rotational velocity being approximately 220 km/sec at the Sun's distance from the galactic center; this corresponds to a rotation period of 200,000,000 years. *See* GALAXY; GALAXY, EXTERNAL.

Differential rotation. Within the Milky Way galaxy, the rotation is described with respect to an external stationary frame of reference, from which it would also appear as a differential rotation, with angular velocity varying with distance from the galactic center (Fig. 9). This rotation curve is established by the study of the relative velocities of distant stars, and of clouds of interstellar hydrogen seen by their 21-cm radio-frequency radiation, both measured with respect to the Sun. Were the galaxy to rotate as a rigid body, there would be no relative velocities of approach or recession, that is, no radial velocity of a systematic character. A differential rotation, however, is detectable in the transverse motions of the stars, with respect to an outside frame of reference, even in the rigid body case. In nonrigid body rotation, the distant stars show systematic velocities in certain preferred directions. The relative radial velocity $V(r, l)$ is given by Eq. (2), where l is the azimuthal coordi-

$$V(r, l) = R_0[\omega(R) - \omega(R_0)] \sin l \qquad (2)$$

nate, in the galactic plane, of the star, measured from the galactic center, R_0 is the Sun's distance to the galactic center, and R that of the star; the motion is assumed to be in the galactic plane. To a first approximation, for small distances r measured from the Sun to the star, the above formula can be written as Eq. (3), where A is the first-order galac-

$$V(r,l) = 2rA \sin 2l \qquad (3)$$

tic rotation constant, which lies between about 13 and 17 km/sec per 1000 parsecs (1 parsec $= 3.086 \times 10^{16}$ m). This double wave is visible in the mean radial velocities of the stars in the range of distance 300 to 3000 parsecs; at larger distances higher-order terms must be included. The cosmological expansion, seen on a large scale in the red shifts of distant galaxies, does not operate within the Milky Way or within clusters of galaxies.

Peculiar velocity. Superposed on the systematic rotation of the galaxy are individual motions of the stars. Each star moves in a somewhat elliptical orbit and therefore shows a peculiar velocity with respect to the local standard of rest, the standard moving in a circular orbit around the galactic center. The Sun has an orbit of small ellipticity and inclination, so that solar motion with respect to the mean of neighboring stars can be detected by analysis of either radial velocities or proper motions of nearby stars.

Stars can be considered to be particles in a gas which has no collisions, with gravitational distant encounters between pairs of stars only slightly altering their orbits. The velocity dispersion of the stars with respect to their local standard of rest is essentially Maxwellian, except that they display greater mobility in the directions toward and away from the galactic center than in other directions.

The velocity distribution of stars in Earth's neighborhood is not a single Maxwellian one, but is characterized by at least two different dispersions. One young group (population I) is characterized by a small dispersion of space motion with respect to the local standard of rest, for example, about ±8 km/sec for B stars, increasing to +20 km/sec for M stars. Most stars have a higher dispersion, about ±45 km/sec, and belong to the so-called old-disk population. Such stars are typically 10^{10} years old, but have a chemical composition resembling that of young population I. Very low metal abundances characterize another group, the oldest population II stars, which have velocity dispersions ranging from ±30 to ±150 km/sec. The high-velocity popu-

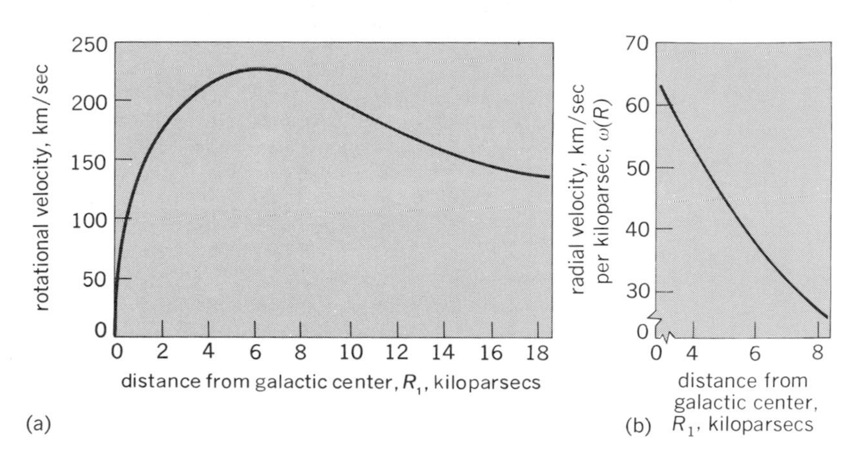

Fig. 9. Velocity curves. (*a*) Rotational velocity varies with distance from galactic center. (*b*) If the galaxy were a solid, the angular velocity would be constant instead of decreasing with radius as it does.

lation II stars are moving in galactocentric orbits of high eccentricity and inclination to the galactic plane. If they were moving with a velocity vector measured with respect to the local standard of rest in the same direction as the galactic rotation, their kinetic energy would be so great compared to the gravitational force of the galaxy as to exceed escape velocity. Thus stars with large velocity vectors, greater than 65 km/sec with respect to the Sun, are absent in about one quadrant of the sky, in the forward direction of galactic rotation.

Solar motion. The solar motion is defined by a vector giving the direction of A, the apex of solar motion, and U, the velocity with respect to the local standard of rest. The radial velocity of the Sun with respect to a direction located at angular distance λ from the apex is a differential motion, $\delta V = -U \sin \lambda$. By averaging over the observed radial velocities of stars of a given type, at moderate distances from the Sun (or correcting separately for galactic rotation if necessary), both U and λ can be determined.

The apex of solar motion with respect to the nearby population I stars is at right ascension 18^h, declination +29°, with U about 20 km/sec. With respect to the population II stars, the direction of the apex is strongly dependent on the velocity dispersion.

Proper motion. The transverse motions of the stars with respect to a fixed coordinate system on the plane of the sky are seen as proper motions, which are usually given in seconds of arc per year. The total proper motion μ is derived from the component μ_α, in right ascension, in seconds of time, and μ_δ, in declination, in seconds of arc, by Eq. (4). Proper motion of a star is measured by

$$\mu^2 = (15\, \mu_\alpha \cos \delta)^2 + \mu_\delta^2 \qquad (4)$$

displacements on photographic plates taken at a sufficiently long interval of time. The standard of rest is usually established by stars of accurately known absolute positions and motions. In principle, an ideal standard of rest would be provided by the faint extragalactic nebulae, and work is in progress with this technique. Normally, however, absolute positions of reference stars are deter-

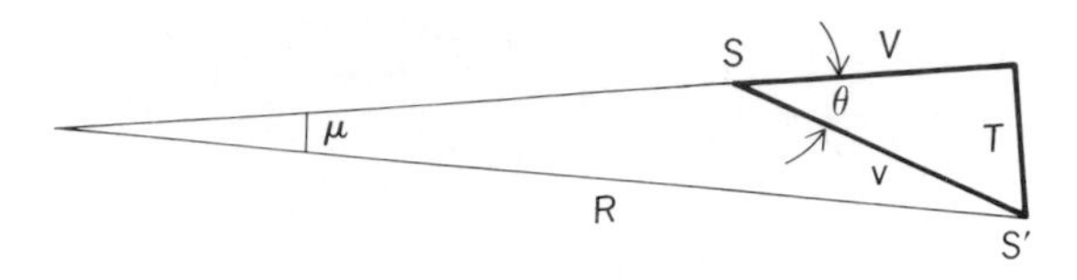

Fig. 10. Vector diagram of components of proper and radial motions of a star as observed from the Sun.

mined by visual observations, accurately timed, of the transits of stars, using meridian circles. About 200 stars have motions exceeding 1″ per year; motions down to 0.″005 per year are moderately dependable.

If the Sun is considered at rest, components of the peculiar motion of a star as observed from the Sun form a vector diagram (Fig. 10). If, in a year the star moves from S to S' at velocity v, the velocity in the line of sight is radial velocity V, and the velocity across the line of sight is proper motion μ, in seconds of arc per year. Parallax p is also expressed in seconds of arc; then tangential velocity T is $T=4.74\ \mu/p$, the space velocity v is $v_2=V^2+T^2$, and the radial and tangential velocities are $V=v\cos\theta$ and $T=v\sin\theta$. All velocities are expressed in kilometers per second.

Conversely, if the stars were all standing still and the Sun alone were moving, or if the stars' motions are averaged out, the formula above for the tangential velocity indicates that a star would show a proper motion due to the projection of the solar motion on the plane of the sky. This permits the determination of the solar motion from the proper motions of nearby stars for which p is not too small. Similarly, if the observed motions of a group of stars, after correction for solar motion, are used, an estimate is obtained of the average parallax. Special catalogs give accurate positions and proper motions.

Radial velocity. Spectroscopic measures of Doppler shifts give radial velocity. A general catalog of stellar radial velocities of 15,000 stars is available. Stellar radial velocities are determined with accuracies up to ±0.1 km/sec; the largest velocities are about 400 km/sec, but only 4% are greater than 60 km/sec.

Galactic dynamics has as its main result the interpretation of the space motions of stars in terms of galactic rotation, orbits, and the distribution of mass in the Milky Way galaxy. From the rotational-velocity curve for circular orbits, the mass is determined; this method has been applied to a few extragalactic nebulae, with the general result that masses of galaxies range from 10^9 to 2×10^{11} Suns. The dynamics of such features as the spiral arms are not understood without reference to the motion of the interstellar gas, galactic magnetic fields, and the effects of local perturbations of the gravitational potential field. Some features of the spiral arms have been simulated by computer experiments; however, the arms do not survive more than a few rotations.

Spatial distribution. The distribution of stars in space is the subject of studies of stellar statistics and galactic structure. The Milky Way is the dominant feature of the galaxy, even to the eye, and represents the mean plane in which the stars are concentrated. It is nearly a great circle, indicating that the Sun lies near the galactic plane.

The number of stars at a given apparent magnitude is a function of galactic latitude and longitude; in the galactic plane the complex structure of interstellar clouds of dust that absorb light and the irregularities of spiral structure produce a somewhat irregular, patchy appearance, with a maximum number of faint stars in the direction of the galactic center in Sagittarius, with subsidiary maxima in Cygnus and Carina. These maxima are probably caused by spiral arms seen lengthwise.

For an average over galactic longitude, the latitude dependence is shown in Table 6, which gives $\log N_{m,b}$, where $N_{m,b}$ is the number of stars brighter than apparent magnitude m, per square degree, at latitude b. The light of the entire sky is equivalent to that of one star of magnitude −6.6. The number of stars increases very rapidly with m, about threefold per magnitude, in the galactic plane; the rate of increase is slower for fainter stars and at higher galactic latitudes. The galactic concentration of stars increases for fainter stars, a natural result because the faint stars are on the average more distant.

Table 6. Apparent distribution $\log N_{m,b}$ of stars as a function of galactic latitude

m \ b	0°	10°	25°	50°	90°
6	−0.89	−0.97	−1.16	−1.35	−1.43
8	0.00	−0.08	−0.26	−0.45	−0.56
10	+0.89	+0.79	+0.59	+0.40	+0.26
12	+1.74	+1.63	+1.41	+1.18	+1.00
14	+2.57	+2.43	+2.17	+1.88	+1.65
16	+3.33	+3.19	+2.84	+2.48	+2.21
18	+4.01	+3.87	+3.42	+2.98	+2.68
20	+4.60	+4.46	+3.90	+3.38	+3.07

Certain types of objects are highly concentrated toward the galactic plane, particularly cepheid variables, luminous O and B stars, and galactic clusters. Others, like M dwarfs and long-period variables, are less concentrated, while globular clusters and RR Lyrae variable stars show almost no concentration.

Part of this effect is caused by luminosity differences; for example, M dwarfs are so intrinsically faint that they cannot be seen to great distances. On the other hand, O and B stars are intrinsically highly concentrated to the galactic plane, while the globular clusters and other extreme population II high-velocity stars are found at great heights, up to 15,000 parsecs from the plane.

In the galactic plane the frequency distribution of the types of stars varies from point to point. Interstellar gas and dust are highly concentrated in the spiral arms, together with O and B stars and other population I objects of high luminosity. In the neighborhood of the Sun, the luminosity function given in Table 7 is a useful average value. It gives the density of stars per cubic parsec as a function of absolute visual magnitude $\phi(M_v)$ within range $M_v+1/2$ to $M_v-1/2$. However, outside the spiral arms, and at heights greater than 100 par-

Table 7. Frequency log $\phi(M_v)$ of stars in the neighborhood of the Sun, as a function of absolute visual magnitude

M_v	log $\phi(M_v)$	M_v	log $\phi(M_v)$	M_v	log $\phi(M_v)$
−6	−8.4	0	−4.0	+6	−2.4
−5	−7.4	+1	−3.5	+8	−2.3
−4	−6.6	+2	−3.2	+10	−2.1
−3	−6.0	+3	−3.0	+12	−1.9
−2	−5.4	+4	−2.7	+14	−1.8
−1	−4.7	+5	−2.5	+16	−2.0

secs from the galactic plane, the high-luminosity end of $\phi(M)$ is cut off.

At great heights, stars with M_v less than +3 are rare. The space density of all types of stars together is about 0.1 solar masses per cubic parsec in the galactic plane and decreases rapidly with height above the plane, following an approximately exponential law, with a scale height h_0, which varies with the type of star. The B stars and the interstellar gas clouds have h_0 about 100 parsecs; RR Lyrae variables have h_0 in the order of 2500 parsecs. This difference reflects the different kinetic energies of slow- and fast-moving stars. The space density of stars increases greatly toward the center of the Milky Way galaxy, possibly by a factor of 100. The thickness of the galaxy also increases; the bulge surrounding the galactic center is composed mainly of stars of population II, which show a very large h_0. The center, obscured in visual light, has a point source of infrared and radio waves, presumably coincidental with a density singularity. Other galaxies show starlike objects at their centers. Such a center, or nucleus, may be a cluster of 100,000,000 stars within a few light-years. These nuclei are often the seat of violent explosions resembling the quasars. Considerable evidence suggests that a massive black hole is present in most galaxies, including the Milky Way Galaxy [JESSE L. GREENSTEIN]

Bibliography: L. H. Aller, *The Abundance of the Elements*, 1961; L. H. Aller and D. B. McLaughlin (eds.) *Stellar Structure*, 1965; A. Blaauw and M. Schmidt (eds.), *Galactic Structure*, 1965; E. M. Burbidge et al., Synthesis of the elements in stars, *Rev. Mod. Phys.*, 29:547–650, 1957; R. Burnham, Jr., *Burnham's Celestial Handbook*, vols. 1–3, 1978; A. J. Cannon and E. C. Pickering, *The Henry Draper Catalogue of Stellar Spectra*, Harvard Observatory Annals, vols. 91–100, 1918–1936; W. Gliese, *Catalogue of Nearby Stars, Edition 1969*, 1969; J. L. Greenstein (ed.), *Stellar Atmospheres*, 1960; D. Hoffleit, *Catalogue of Bright Stars*, 3d ed., 1964; L. F. Jenkens, *General Catalogue of Stellar Parallaxes*, 1963; J. Lequeux, *Structure and Evolution of Galaxies*, 1969; W. W. Morgan, P. C. Keenan, and E. Kellman, *An Atlas of Stellar Spectra*, 1943; *Star Catalog*, Smithsonian Astrophysical Observatory, 1971; K. A. Strand (ed.), *Basic Astronomical Data*, 1965; R. E. Wilson, *General Catalogue of Stellar Radial Velocities*, Carnegie Inst. Wash. Publ. no. 601, 1953.

Star clouds

Aggregations of thousands or millions of stars spread over hundreds or thousands of light-years in space. The Milky Way is composed of such star clouds, the heaviest clouds being in the richest parts, such as Cygnus, Sagittarius, Carina, and Scutum (see illustration). The stars in such clouds may appear unevenly distributed because of the presence of obscuring interstellar dust and gas. *See* GALAXY.

The great star cloud in Scutum. *(Yerkes Observatory)*

[HELEN S. HOGG]

Star clusters

Groups of stars held together by gravitational attraction. The two chief types are open clusters, containing from a dozen up to many hundreds of stars, and globular clusters, composed of thousands to hundreds of thousands of stars. A relative of the star cluster is the stellar association, a group of dozens or hundreds of relatively young stars spread loosely over a large volume of space. Star clusters are important in outlining the shape and extent of the Milky Way Galaxy and in deriving theories of stellar evolution on the assumption that stars of a given cluster were formed at the same time.

Open clusters. Open clusters, formerly called galactic clusters, lie along the backbone of the Milky Way Galaxy, strongly concentrated to the central plane of the Milky Way. A dozen are visible to the unaided eye, over 1000 are cataloged, and many more must exist. Most open clusters have an asymmetrical appearance (Fig. 1).

Distances and dimensions. The distances to open clusters range from 25 parsecs (1 parsec = 3.09×10^{13} m) for the sparse Ursa Major cluster up to 5000 parsecs for the faintest detectable against a rich stellar background. Distances may be determined by geometric methods, including trigonometric parallaxes or stellar motions, and photometric methods. *See* PARALLAX.

Angular diameters of open clusters range from several degrees down to several minutes of arc, and linear diameters range from 15 parsecs to 2 parsecs. From a study of the way in which linear diameters appeared to increase with increasing

Fig. 1. Double cluster in Perseus. *(Yerkes Observatory)*

distance, the absorption of light in space was deduced by R. J. Trumpler in 1930.

Spectral characteristics. The brightest stars in some clusters, like the Pleiades, are blue, of spectral type B; in others, like the Hyades or Praesepe, they are yellow or red. Some stars with luminosities brighter than absolute magnitude −3 are found, and sometimes supergiants up to −7. *See* HYADES; PLEIADES.

Fig. 2. The great globular cluster in Hercules, Messier 13. Photographed with 200-in. telescope. *(Hale Observatories)*

Most stars fall along the highly populated branch of the spectrum-luminosity diagram known as the main sequence. The point where this sequence starts furnishes a criterion of the age of the cluster. Open clusters show great diversity in their spectrum-luminosity diagrams. They may be classified on this basis, as well as by richness and central concentration. They may contain such types of stars as bright O stars, visual and spectroscopic binaries, certain kinds of variables, and white dwarfs. Some clusters, like the Pleiades, contain amounts of nebulosity equivalent to many solar masses.

Motions. Measures of proper motion and radial velocity show the cluster stars to be sharing a common motion in space, with velocities up to tens of kilometers per second. Open clusters whose most prominent characteristic is a large common proper motion of the stars are called moving clusters. The diagram of proper motions in a cluster shows a conspicuous convergent point, where their parallel motions appear to meet in space. This category includes the Taurus and Ursa Major moving clusters. *See* CONSTELLATION.

Age and dissolution. By comparing the spectrum-luminosity diagram with a standard main sequence, the age of a cluster may be determined. At one extreme are young clusters, formed in recent geologic times, like the Orion Nebula cluster with an age of about 10^6 years. At the other extreme is NGC 188, with an age of 5×10^9 years, comparable with that of old systems like globular clusters. The lifetime of a cluster depends on a balance between its mass and its radius. High-velocity stars may "evaporate" from dense clusters; encounters with interstellar clouds disrupt loose clusters.

Stellar associations. Systems in which early-type (O to B2) stars are more numerous than in the surrounding field are known as stellar associations. About 70 are cataloged. The radii range up to 200 parsecs. Stellar associations are perishable, lasting perhaps 10^7 years.

Globular clusters. Groups of thousands to hundreds of thousands of stars in globular symmetry constitute globular clusters. Though they are scattered widely in galactic latitude, their strong concentration toward the region of Sagittarius-Scorpio led Harlow Shapley in 1917 to postulate this as the center of the Milky Way Galaxy. Several are visible to the unaided eye, like Messier 13, the great cluster in Hercules (Fig. 2). A total of 131 have been cataloged in the Milky Way Galaxy, including several so far distant that they are really intergalactic. A few dozen more may be undetected.

Distances and dimensions. The distances to globular clusters in the Milky Way Galaxy range from 2 kiloparsecs for the nearest to more than 100 kiloparsecs. They are too great for geometric methods of distance determination, but photometric methods involving color-magnitude diagrams and RR Lyrae stars can be used. The apparent diameters range from 36′ to 1′, and the linear diameters from 100 to 10 parsecs.

Structure. Globular clusters differ markedly in their degree of central concentration. A few are noticeably elliptical. Three observable characteristics are the radius of the central core, its brightness, and the tidal radius (from tidal forces imposed by the Galaxy). In some globulars the frequency of stars falls off as the cube of the distance from the center. Star counts in Messier 3, which has 2.45×10^5 solar masses, show that 95% of the visual light comes from stars intrinsically brighter than the Sun, while 90% of the mass is contributed by fainter stars. Density near the center is high, 50 stars per cubic parsec compared with 1 star per 10 cubic parsecs near the Sun.

Color-magnitude diagrams differ appreciably from those of open clusters. Stars of absolute magnitude brighter than −3 are absent. The brightest stars are yellow-red; the main sequence is represented by large numbers of stars from type F down, with a horizontal branch near absolute magnitude 0. The spectra show low metal content (ratio of iron to hydrogen), but the ratio varies from cluster to cluster.

More than 2000 variable stars have been found in globular clusters with Messier 3 the richest, containing more than 200. Short-period RR Lyrae stars make up nearly 90% of the variables. *See* VARIABLE STAR.

In the 1970s the discovery of x-ray sources in eight globular clusters, some of them burst sources, produced a new interest in these clusters. Most of these sources are located close to the center of highly condensed clusters and are thought to be neutron stars in binary systems. *See* X-RAY ASTRONOMY.

The nature of apparent wisps of nebulosity in globular clusters is disputed. Radio sources have been detected in some globulars, but the radio emission does not indicate large masses of gaseous material.

Motions. The individual stars in a globular cluster are describing orbits about the cluster center, which was first proved observationally for Omega Centauri. The clusters themselves are describing large orbits about the center of the Galaxy. Radial velocities for 70 clusters with respect to the Sun range from 493 to −360 km/s.

Age, formation, and dissolution. Many clusters probably formed in various parts of the Galaxy during its first 10^9 years. Some globular cluster stars may be the oldest stars in the Galaxy. Their ages, based on low-metallicity evolutionary models, can be as great as 1.41×10^{11} years. The clusters are so stable dynamically that their individual stars will burn out before the clusters disintegrate. *See* GALAXY; STELLAR EVOLUTION.

Clusters in extragalactic systems. In the 1970s globular clusters were discovered by the thousands in distant external galaxies. Open clusters totaling hundreds have been detected only in the nearest galaxies. Besides normal globular clusters, the Magellanic Clouds have a second type whose bright stars are blue. In the Local Group in 12 low-mass irregular and spiral galaxies, some 500 globulars are known. From about 9000 globular clusters observed in 15 Virgo cluster elliptical galaxies, including Messier 87 with 6000, the largest number known, a total population of 45,000 is estimated. The total number of globular clusters belonging to a galaxy is considered proportional to the mass of the spheroidal system of the parent galaxy. From their average brightness, globular clusters can be used as distance indicators for very remote galaxies. *See* GALAXY, EXTERNAL; MAGELLANIC CLOUDS; STAR. [HELEN S. HOGG]

Bibliography: C. Payne Gaposchkin, *Stars and Clusters*, 1979; W. E. Harris and R. Racine, Globular clusters in galaxies, *Annu. Rev. Astron. Astrophys.*, 17:241–274, 1979.

Stellar evolution

The large-scale, systematic, and irreversible changes with time of the structure and composition of a star. Stars are born, age, and die much like living beings, but the time scales involved are incomparably longer: for stars like the Sun, the lifetimes are on the order of 10^{10} years, and even for the most massive stars, whose life expectancy is much shorter, the life span is still on the order of 10^7 or 10^6 years. As a consequence, stellar evolution cannot be observed directly, except for rare cases of objects "caught" at stages of unusually fast structural changes.

Hertzsprung-Russell diagram. Evidence on stellar evolution is therefore mainly indirect. Many stars are observed at various stages of evolution; the task is to identify these stages. It is useful to plot two of the most important stellar characteristics: luminosity (the total amount of energy emitted per second) and effective temperature (which represents the stars' surface conditions) against each other in a Hertzsprung-Russell diagram, as is done schematically in Fig. 1. Most stars will fall on the main sequence, which runs diagonally across the diagram from hot and luminous stars in the upper left corner to cool and faint stars in the lower right corner. (In precise diagrams, based only on directly observed quantities, visual or photograph-

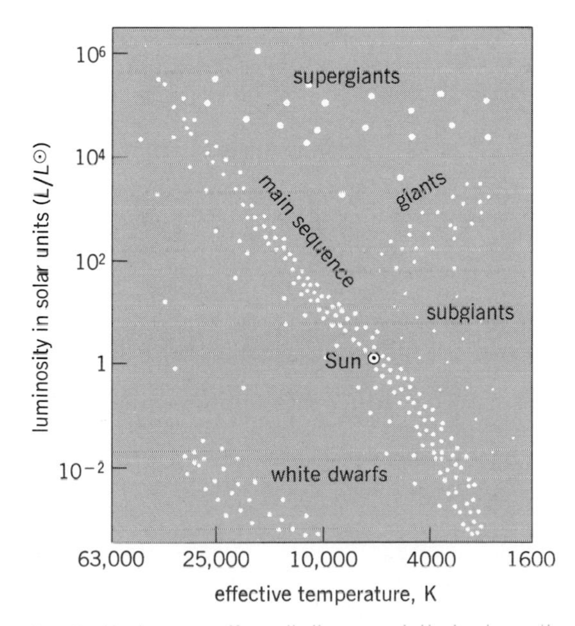

Fig. 1. Hertzsprung-Russell diagram plotted schematically for comparison with evolutionary tracks in Figs. 2–4.

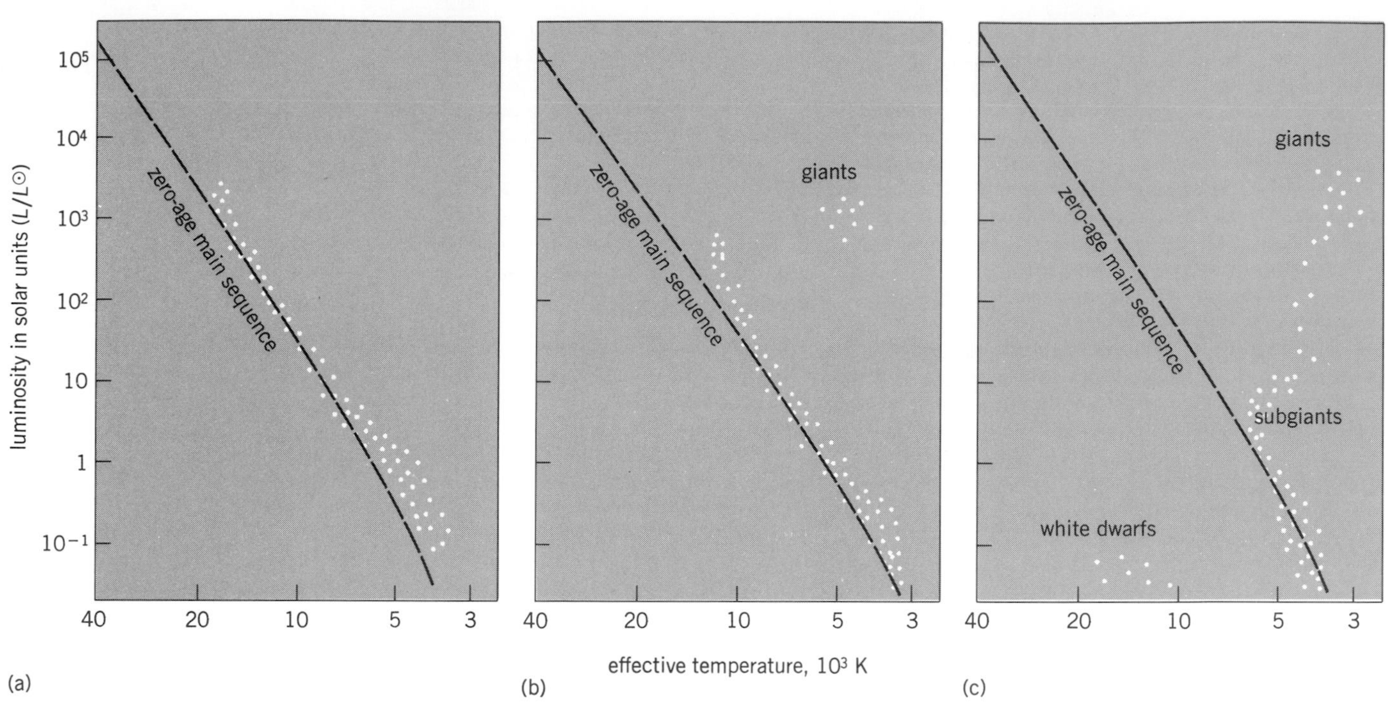

Fig. 2. Schematic Hertzsprung-Russell diagrams for open star clusters based on model calculations. (*a*) A young cluster (age about 10^7 years). Almost all stars lie along the main sequence; only those of very low mass are still contracting to it, while the most massive ones have already begun to move toward the giant region. (*b*) A middle-age cluster (age about 10^8 years). Low-mass stars are now on the main sequence, but most massive ones may have already exploded as supernovas or shed most of their mass and became white dwarfs. (*c*) An old open cluster (age about 4×10^9 years). All stars more massive than the Sun have evolved into subgiants or giants, there are some white dwarfs, and a few stars have disappeared in supernova explosions.

ic absolute magnitudes are usually given, and then the main sequence is not a straight line.) In addition, there are two fairly distinct groups: the rather cool but very luminous giants (and even more luminous supergiants) and the fairly hot but quite faint white dwarfs.

A star's luminosity depends on its effective temperature T (which characterizes the temperature at its surface) and on the total surface area as $L = 4 \pi\sigma R^2T^4$. The multiplicative constant $4 \pi\sigma$ disappears if the star's luminosity L is expressed in units of solar luminosity (which is 4×10^{33} ergs/S^{-1} or 4×10^{26} W), and the stellar radius R in terms of the solar radius (7×10^5 km): the $L = R^2 (T/T_\odot)^4$, where the solar effective temperature is $T_\odot =$ 5770 K. The effective temperatures of most giants are lower than that of the Sun, yet they are much more luminous; therefore their radii must be several orders of magnitude ($10^2 - 10^4$) larger than the solar radius. On the other hand, by a similar argument, white dwarfs must be much smaller, and indeed are of about the size of the Earth, that is, 0.01 solar radius. *See* HERTZSPRUNG-RUSSELL DIAGRAM.

Observational evidence. Although theoretical understanding of how stars are built and how they generate energy is necessary to determine what makes one star a giant and another one a white dwarf, observations alone contribute important information. Of particular value are studies of star clusters. These stellar systems consist of stars that were born at approximately the same time. Figure 2 shows schematic Hertzsprung-Russell diagrams of three star clusters. In Fig. 2*a* there are no giants; all stars lie along the main sequence band, and only the hottest and most luminous ones tend to be displaced to the right of it. In Fig. 2*b* this trend is much more pronounced; in fact, the top of the main sequence is missing, and a few cool giants (of reddish color) are observed instead. In Fig. 2*c* all the more luminous stars are now red giants, while the main sequence is populated only at its faint red end. From this evidence it can be inferred, and theoretical studies of stellar evolution confirm it, that stars evolve from the main-sequence to the giant stage. Thus the sequence of the three clusters in Fig. 2 is essentially a time sequence, from the youngest to the oldest cluster.

Stars are essentially thermal engines: they shine because they produce energy. The main-sequence stage is clearly one in which a typical star remains for a very long time, since so many stars are observed at this particular stage. Obviously, then, hot and luminous stars consume energy so fast that the main-sequence stage is shorter for them than for stars like the Sun. The mass M of certain stars (members of suitable binary systems) can be determined and it is found that along the main sequence there exists a clear mass-luminosity relation: more massive stars are more luminous. The dependence is very steep and can be written approximately as $L \propto M^4$. If the mass of the star is doubled, its total energy output increases 16 times; a star of 10 solar masses radiates as much energy as about 10,000 Suns. As discussed below, the source of stellar energy is mainly thermonuclear conversion of hydrogen into helium. Since the initial chemical composition of all stars in a star cluster is most likely the same, the amount of nuclear fuel initially available for any star is proportional to its mass. However, since the star consumes this energy at a rate proportional to M^4, its life expectancy t decreases rapidly with mass: $t \alpha M^{-3}$. *See* MASS-LUMINOSITY RELATION.

Stars do not start their lives on the main sequence. In the youngest clusters (Fig. 2*a*), the faint (that is, low-mass) stars tend to lie somewhat above the main sequence. Very young stars have been identified as rather cool objects, shining mainly in the infrared spectral region, and still surrounded by agglomerates of gas and dust, from which they obviously were born. As Fig. 2*a* shows, even the earliest stages of stellar life proceed more slowly for the less massive stars.

The Hertzsprung-Russell diagram in Fig. 1 has gaps between the main-sequence hot stars and the giants, and again between the main sequence and the white dwarfs: almost no stars have characteristics that would place them in these gaps. This is because the evolution across these gaps is relatively very fast. There is no doubt that the white dwarfs represent the final evolutionary stage for stars of about the solar mass. Since stars evolve first from the main sequence toward the region of the red giants, the subsequent evolution must bring them across the Hertzsprung-Russell diagram from the giants to the white-dwarf region. Such evolution must be rapid, since few transition objects are observed. One such kind of object is the planetary nebulae, of which the Ring Nebula is a well known example. In contradiction to their traditional but quite misleading name, planetary nebulae are actually old stars that have blown away their outer layers, which are still visible as rather symmetric nebulous shells surrounding the central star. The central star, a core of the parental star, is a small object evolving rapidly into the white-dwarf stage. *See* NEBULA; PLANETARY NEBULA.

Thus observations of various types of stars provide many clues about stellar evolution, but these clues are very incomplete, and at times ambiguous or confusing. The theory of stellar evolution must rely heavily on good understanding of the physical laws and principles on which the stars are built, and according to which they behave. Fortunately, these laws are in many respects simpler than the laws governing biological evolution, and there is confidence that many aspects of stellar evolution are reasonably well understood.

Role of gravitation. Stars are believed to be formed from condensations in the interstellar medium. Once a reasonably massive initial condensation forms (perhaps in a random process, or in local condensations due to traveling shock waves), it attracts surrounding particles by its force of gravitation. Although other forces may also assist, gravitation appears to be the force that builds stars. However, once they have formed, gravitation tends to destroy them. This paradox is inherent in the very nature of gravitation. According to Newton's law of universal gravitation, any two massive particles are mutually attracted by a force inversely proportional to the square of the distance between them. A particle located in the force field of many other particles, as inside or in the vicinity of a star, will be subject to a resultant net force due to the whole assembly; and if this assembly is spherically symmetric, it will act on any particle as if the total mass of all the others were located at the center of the star. Thus a particle of mass m is attracted by the total mass M of the star toward the star's center by a force $F = GMm/r^2$, where r is its distance from the center and G is the gravitational constant. If the particle can fall or move closer to the center, it does so, thereby increasing the force, since the distance r has decreased. This goes on all the time, and for all the particles forming the star. Gravitation tends to squeeze the star to smaller and smaller dimensions, but every contraction only strengthens the force, thereby compelling further contraction. *See* GRAVITATION.

A star's life is a perpetual struggle against self-destruction, namely collapse under the effect of self-gravitation. At the outset, the object (called a protostar) is little more than a huge contracting cloud of cold material. Its contraction accelerates all the time for the reasons just explained, and outright would collapse into a black hole if forces were not generated to counteract the gravitational contraction. Such a force is the thermal pressure of the gas. As the particles fall closer together, their potential energy is converted into kinetic energy of motion, and redistributed in numerous collisions; the resulting random motions are perceived as rising temperature of the gas, and associated with it is the pressure which eventually begins to balance gravitation. The thermal pressure is proportional to the density of the gas, and to its temperature. As the star contracts, its density increases, but its temperature is more efficient for producing a sufficiently high gas pressure, which can rise to millions of degrees if the star has contracted sufficiently. If a star has more mass, its self-gravitation is stronger; therefore a higher gas temperature is needed to halt the contraction. This results in higher luminosity. The deeper layers in the star must support a greater weight than the layers lying above them; therefore the gas pressure, as well as the temperature, must rise as one goes deeper into the star, and the temperature peaks at the center, where it must be very much higher than at the surface.

A temperature gradient is established in the star in the sense that temperature steadily decreases from the center to the surface. According to the second law of thermodynamics, energy must flow down the temperature gradient. A star must therefore shine simply because it is hotter inside, even if it does not possess adequate energy sources. Higher central temperature implies a steeper temperature gradient, hence also larger flow of heat. This is why more massive stars (which have a higher central temperature, as explained above) are more luminous. But loss of thermal energy means that the star would cool off, and hence the gas pressure would decrease because of decreasing temperature, gravitation would prevail, and the star would continue to contract. Indeed, this is what happens when the star does not have, or cannot tap its, nuclear energy resources. It is bound to contract, albeit rather slowly; each step in contraction releases a large amount of potential energy, of which about half is converted into internal thermal energy (temperature rises), while the other half escapes as radiation. Because of the stars' enormous masses, gravitational potential energy is a large source of radiative energy, and the Sun could shine at its present luminosity for about 10^8 years without nuclear sources. However, 10^8 years is much shorter than the known geological history of the Earth and of life on it, during which history the Sun must have been shining at about the same luminosity as it is now.

Thermonuclear reactions. When the central temperature of a contracting star reaches about 5×10^6 to 1×10^7 K, thermonuclear reactions are ignited in which hydrogen is converted into helium. For every gram of hydrogen involved in this reaction, 0.0071 g of material is converted into energy, yielding 6.4×10^{11} joules energy. Conversion of 6×10^8 metric tons of hydrogen per second is needed to cover the present energy output of the Sun; yet the Sun has such a large store of hydrogen that it can shine at this rate for about 10^{10} years, and has lived for only about half that time. The speed at which the conversion of hydrogen into helium proceeds is very sensitive to temperature. Practically no reactions occur at temperatures below 10^7 K; but with increasing temperature above this threshold, the rate increases so rapidly that a region at a temperature of 2×10^7 K generates about 130,000 times more energy than a region at a temperature of 10^7 K. Since more massive stars must be hotter at their centers, they also produce much more energy, and this explains the mass-luminosity relation mentioned above. *See* CARBON-NITROGEN-OXYGEN CYCLES; PROTON-PROTON CHAIN.

From main sequence to red giant. The onset of thermonuclear reactions provides sufficient heat to keep the pressure sufficiently high to halt the star's contraction. The star enters a long and quiet period of its life, and is now a main-sequence star. Eventually, however, all the available hydrogen in its interior will be consumed. This is only about 15% of all the hydrogen in the star, but the rest cannot be used since it lies in regions where the temperature is not sufficiently high. As the energy source at the star's center vanishes, gravitation prevails again, and contraction resumes. Contraction heats up the star everywhere, but most important is the rise in temperature in the layers immediately surrounding the helium-rich core in which hydrogen has been exhausted. In a relatively thin shell, which still contains hydrogen, the temperature will rise again high enough to ignite the thermonuclear conversion of hydrogen into helium. Instead of a nuclear-burning core, the star now has a nuclear-burning shell. This shell-burning stage is short, since the energetically inactive core cannot maintain a sufficiently high temperature to support the weight of the outer layers by gas pressure, and begins to contract again. However, this time the star is no longer a homogeneous body: the nuclear burning shell acts as a "watershed" for internal motions, which are roughly speaking, reflected on it. Thus, while the core is contracting, the envelope outside the nuclear-burning shell begins to expand. The size of the star begins to increase, but at the same time the outer layers cool off as they expand, there is a drop in the surface temperature, and the color changes from white to yellow to red: the star evolves rapidly into a red giant.

From red giant to supernova. If massive enough (above about 0.7 solar mass), the star reaches, in its contracting core, a temperature of or above 10^8 K. Then a new thermonuclear reaction starts in the core, namely, a transformation of three helium nuclei into one nucleus of carbon. This new energy source again temporarily stabilizes the star and halts core contraction and envelope expansion. A second quiet stage ensues, which explains why a fairly distinct and relatively populous category of red giants is observed. The high central temperature demands high energy production, which makes the giants very luminous. However, they are considerably less common objects than the main-sequence stars. This is because the quiet red giant stage is much shorter. As nuclear fuel, helium is much inferior to hydrogen, and is therefore consumed much faster. Another contraction of the core follows, accompanied eventually by an enormous expansion of the outer layers, and the star becomes a red supergiant, with a radius of 500 solar radii or so. If the star is sufficiently massive to become sufficiently hot at its center, further reactions may occur, building up heavier and heavier elements, all the way up to iron, which has the most compact nucleus. Heavier elements than iron cannot be built in thermonuclear reactions that release energy; on the contrary, energy must be supplied. In fact, the energy-releasing interactions of elements heavier than oxygen already yield so little that all the later stages of stellar evolution must be quite short. *See* ELEMENTS AND NUCLIDES, ORIGIN OF; SUPERGIANT STAR.

In addition, a number of factors contribute to instability of the star. Sooner or later the star runs out of energy, and it may happen so abruptly that the inevitable gravitational contraction changes to a real collapse. In a matter of hours, the cool material of the envelope falls quite close to the center and heats up enormously, and thermonuclear reactions ignite instantaneously. The implosion suddenly changes into an explosion, which releases a large amount of luminous energy but also shatters the star to pieces. A supernova is observed. *See* GRAVITATIONAL COLLAPSE.

A supernova explosion is a spectacular but very rare event. In a stellar system such as the Galaxy, containing about 10^{11} stars, one can expect only two supernovae per 100 years. This implies that only a small fraction of stars (probably only the very massive ones) end their lives in a supernova explosion. *See* SUPERNOVA.

Final stages of typical stars. An important property of stars of smaller mass, such as the Sun, is that the densities in their central cores are rather high. Near the center of the Sun, the density of material is on the order of 10^2 g/cm^3. In more massive stars, with higher central temperatures, the pressure of radiation becomes important and helps to support the weight of the outer layers; in Sun-like stars the radiation pressure is practically negligible, and the cores must become rather dense in order to develop sufficiently high gas pressure to balance the gravitational contraction. As discussed above, with the exhaustion of hydrogen in the core, further contraction must set in, raising the material density even higher. At a certain stage (at a lower density for less massive, cooler stars), the density becomes so high that electron degeneracy occurs. This phenomenon is a consequence of the uncertainty principle, one of the basic laws of quantum mechanics, which specifies, among other things, that the momentum of a particle known to be within a certain volume cannot be reduced below a value that depends on this volume, and that it increases as the volume decreases. Thus an increase in the density of the stellar material, diminishing as it does the volume per particle, introduces a new type of pressure, the pressure of the degenerate electrons. This pressure is independent of temperature and eventually stabilizes the star.

When degeneracy sets in at a certain stage of evolution (this happens during core contraction between two nuclear-burning stages), it halts the contraction. As a consequence, the core temperature no longer increases, and the next nuclear-burning stage is not reached. For example, a star with a mass of less than 0.7 solar mass, after it has exhausted all its core hydrogen, will never reach the temperature of 10^8 K needed to ignite the helium conversion into carbon. Such a core then produces no thermal energy through nuclear reactions, and no energy through contraction. However, the temperature gradient persists and the core continues to lose heat; therefore it begins to cool off.

The final evolutionary stage of a star of small mass is one of complete electron degeneracy, which stabilizes the mechanical structure of the star. The star does not generate energy, and shines only by gradually releasing the heat energy accumulated at earlier stages in the random motions of the particles. A thin nondegenerate surface layer slows down the escape of the heat, so that this evolutionary stage is unusually long ($10^9 - 10^{10}$ years). The star does not change in size; it only cools off gradually. At the early stages it is still hot enough to maintain a fairly high surface temperature, and it is observed as a white dwarf. As it cools off, it fades, and changes into something that ought to be called a dark dwarf. Because of their long lifetimes, many white dwarfs are known in the vicinity of the Sun, and only their extremely low luminosity (10^{-2} solar luminosities) makes it impossible to observe directly how numerous these objects are. *See* WHITE DWARF STAR.

The low luminosity, combined with high effective temperature, implies that white dwarfs are very small stars, about the size of the planets (a few percent of the solar radius). Yet their masses are comparable to that of the Sun; only when a solar mass is squeezed into the size of the Earth does the complete electron degeneracy set in. Again, greater weight of the outer layers demands higher counterpressure, and therefore greater density. Therefore the radius of a degenerate dwarf is smaller for a larger mass of the dwarf, and theoretical calculations show that the radius tends to zero when the mass approaches about 1.4 solar masses (this value is known as the Chandrasekhar limit). Beyond that limit no stable configuration is possible for a degenerate dwarf, since the pressure of the degenerate electrons is not strong enough to counteract self-gravitation. By the sheer weight of the material, electrons are squeezed into atomic nuclei and combine with protons to form neutrons. This permits further drastic contraction of the star, which continues until the neutrons are squeezed to such a small volume per particle that the uncertainty principle begins to operate again, this time on the degenerate neutron gas, and generates a new strong pressure opposing further contraction. A star can stabilize again, and exist for a long time as a neutron star. Neutron stars are incomparably smaller than even the white dwarf; a neutron star of 1 solar mass would have a radius of about 9 km. Objects so tiny are not easy to detect by usual optical observations. Fortunately, at least some neutron stars are observed as pulsars, since they produce flashes of radiation by spinning rapidly in a strong magnetic field. *See* NEUTRON STAR; PULSAR.

One way of producing neutron stars is the supernova explosion. The supernova observed in A.D. 1054 by Chinese astronomers in Taurus left behind a nebulosity (Crab Nebula) surrounding a pulsar. It is not clear if all supernovae leave a stellar remnant. Theoretical calculations show that if the mass of the remnant is larger than about 3 solar masses, not even the pressure of the degenerate neutrons can stop its collapse, and no other physical force is known that could counterbalance gravitation under such circumstances. Therefore it has been concluded that such an object would collapse without a limit and become a black hole. Actually, if a star of 3 solar masses were to contract to within a radius of 9 km, the object would appear as a black hole even if its further collapse were arrested by some unknown force inside that sphere, since the velocity necessary for escape from the surface would already surpass the speed of light, and therefore light would not be able to escape, a characteristic property of a black hole. Systematic searches for stellar black holes have so far revealed only one promising candidate, the x-ray source Cygnus X-1. Black holes, although of immense interest, do not seem to play an important role in stellar evolution. The vast majority of stars apparently do not collapse into black holes; some of them end as neutron stars, but for most the degenerate dwarf stage appears to be the end stage. *See* BLACK HOLE.

Mass loss from stars. The paucity of neutron stars and black holes, and the abundance of white dwarfs, pose an interesting problem considering that no star more massive than the Chandrasekhar limit (1.4 solar masses) can become a degenerate dwarf. It must therefore be assumed that many stars lose so much mass from their surfaces that their total masses eventually drop below the Chandrasekhar limit. Several ways are known by which a star can efficiently shed mass. A good percentage of all stars (30–50%) actually live in binary systems where the components are so close to each other that they cannot expand freely into the giant or supergiant size. At a certain distance from the center of the expanding star (depending on the mass and distance of the companion), the gravitational attraction of the companion prevails, and matter begins to flow either to that star or away from the system. A star can lose a large fraction of its mass in this type of evolution, and many interesting phenomena are associated with mass transfer in binary stars, for example, novae and dwarf novae. *See* BINARY STAR; NOVA.

Another way of shedding mass is the phenomenon of planetary nebula, described above, in which the star blows away its envelope and only the core is left; the core of a giant or supergiant is an almost readymade white dwarf. Possibly most of the stars with masses up to 4 solar masses pass through the planetary nebula stage and thus evolve into white dwarfs. Shedding of a planetary nebula envelope is a mild, rather slow process. An even less conspicuous way of losing mass is through stellar wind. For reasons not yet quite clear, luminous stars (both hot blue supergiants and cold red giants and supergiants) develop steady streaming from their surfaces, in which typically 10^{-6} solar mass per year of material leave the star. If this process continues for 10^6 years or longer, it can severely reduce the mass of the star, and thus alter completely its future evolution. *See* SOLAR WIND.

Evolutionary models of stars. Knowledge of certain evolutionary stages of stars, in particular of the end stages and of the very earliest stages, is still rather incomplete. On the contrary, evolution from the main-sequence to the red giant stage can be modeled successfully, and models of stellar structure and evolution match the observations very well. High-speed computers generate model sequences for different masses, such as those shown in Fig. 3. The models start on the main sequence at the moment when the star is supposed to start significant thermonuclear reactions (hydrogen burning). At that time, the star is assumed to be chemically homogeneous.

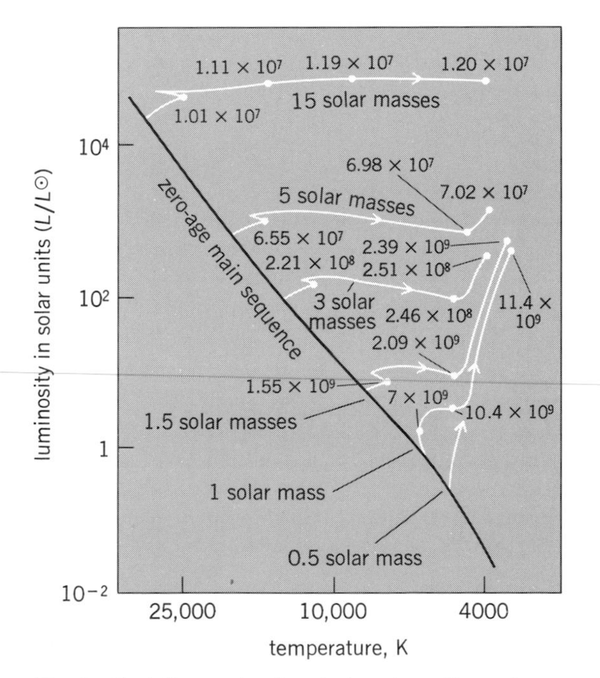

Fig. 3. Evolutionary tracks of stars from the main sequence to red giants, based on model calculations. Ages, in years, measured from the time when the star was chemically homogeneous (zero-age main sequence) are indicated along the track. (*From G. O. Abell, Realm of the Universe, 2d ed., Saunders College, 1980*)

It is assumed that the stars in the Sun's vicinity and in the spiral arms of the Galaxy (population I stars) contain about 70% hydrogen and 27% helium (by weight), the remaining 3% being heavier elements. The multielectron atoms of the heavier elements play an important role, since (contrary to hydrogen and helium) they are not completely stripped of all their electrons even deep inside the star, and contribute significantly to the opacity of the stellar material, that is, to the fact that photons originating deep in the stellar interior typically travel only a fraction of 1 cm before they are absorbed. Successive absorptions and reemissions slow down the progress of the photons to the surface, which then typically takes 10^6-10^7 years. This slow process significantly contributes to the stability of the star, and also provides important shielding for life on Earth from the pernicious radiation generated at the center of the Sun.

Thermonuclear reactions take place only in the innermost part of the Sun and stars; 94% of the entire solar energy output is generated in the innermost 33% of the mass occupying 0.8% of the volume of the Sun, according to one model. The radiation generated there consists of gamma rays and hard x-rays that would be completely lethal if actually emitted by the Sun. However, as these photons filter through the inert and cooler layers, they get degraded to much less energetic photons, until they eventually emerge as mostly visible light or infrared radiation, with some ultraviolet rays, corresponding to the effective surface temperature of 5800 K.

Since the evolution on the main sequence is so slow, conditions of equilibrium may be imposed, in computing models, which must be very nearly exactly fulfilled: (1) hydrostatic equilibrium, requiring that, at each level, gravitation must be exactly balanced by gas pressure; and (2) thermal equilibrium, meaning that exactly all the energy produced per second must be radiated away in a second. An additional simplification is that the gas behaves as a perfect gas, even under conditions prevailing at the center of the Sun. Although the density there is 148 g/cm³, compressing the atoms close together, the high temperature of 1.5×10^7 K significantly reduces the effective size of the atoms by stripping them of almost all their electrons (ionization is nearly complete), and enhanced thermal motions of particles more than compensate for the increased interactions due to overcrowding.

Energy generated in the core is transported through the star (and eventually to the surface) mainly by radiation; that is, the carriers are the photons. Only when the diffusion of photons becomes too slow because of high opacity does an additional mode of heat transport appear, namely, convection. Convective currents develop as hot masses of material rise upward, lose their surplus heat, and return back in a cooler descending current. Convection prevails in the cores of the hot and luminous stars, while the cores of the Sun-like and smaller stars are radiative. In cooler stars convection appears in the outer layers, where hydrogen and helium are only partially ionized. These extended convective envelopes are responsible for the outer appearance of red giants and supergiants, as well as for the pulsation instability of some of them, such as the cepheids. The third mode of heat transfer, conduction, is efficient only in degenerate dwarfs, where it is favored by the high density of material (10^5-10^8 g/cm^3) and by the large mobility of the free electrons. *See* CEPHEIDS.

Although convection may in some cases mix stellar material over large regions of the star, it does not reach far enough to provide effective mixing between the core and the surface. Thus the star (except in the case of very small and cool stars) cannot bring fresh supply of hydrogen to its core when it becomes depleted there. Also, the gradual differentiation of chemical composition in the deep interior, caused by successive nuclear reactions occuring always in smaller and smaller central regions (where the required high temperature exists), remains largely preserved and unmixed, so that an old massive star should have an onionlike chemical stratification, starting with an iron core and proceeding outward through layers of successively lighter elements, until the surface layers should preserve their initial, hydrogen-rich composition.

The figure of 70% hydrogen and 27% helium mentioned above for the initial composition is only approximate in the sense that the ratio of hydrogen to helium is not easy to determine. The relation between mass and luminosity, and mass and radius, is more sensitive to the abundance of the heavier elements, since they determine the opacity in the stellar interior. Population I stars have about 3% of their mass in heavier elements, while the halo-type stars (population II), observed, for example, in globular clusters, contain less than 1% of their mass in the form of the heavier elements.

Peculiar and variable stars. A small number of stars deviate from the norm in various respects. Some have anomalous chemical composition; others oscillate (regularly or less regularly) in size and light; some brighten or fade abruptly; others have unusually extended atmospheres and lose mass at an unusually high rate. In many cases these phenomena occur in interacting binary system. At other times they signal violation of the hydrostatic or thermal equilibrium, or local events which may be associated with magnetic fields, rapid rotation, and so forth. *See* STELLAR MAGNETIC FIELD; STELLAR ROTATION; VARIABLE STAR.

Star formation. Formation of new stars appears to take place in dense, dark gas clouds scattered along the spiral arms of the Galaxy. At the earliest stages, the newly born stars are shrouded by envelopes of in-falling dust, and can be observed only at infrared or radio wavelengths, to which the dust is more transparent. Theoretical models of early evolution agree that the collapsing cloud quickly develops a fairly dense core, which attracts more material from the surrounding cloud. When the

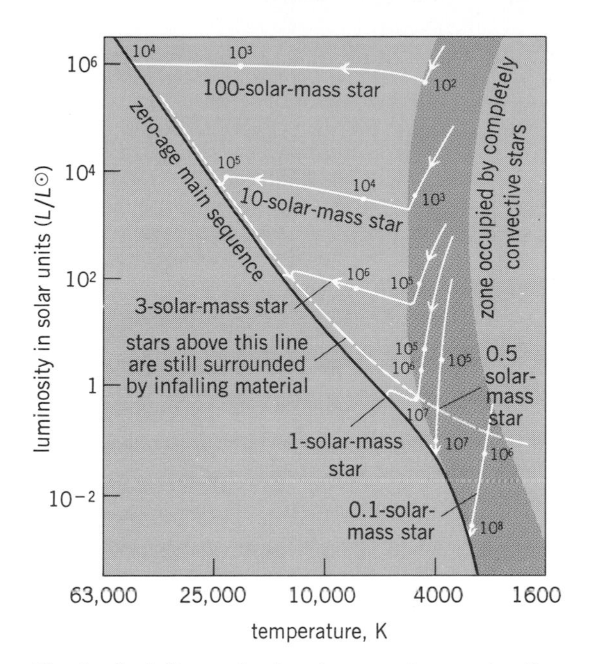

Fig. 4. Evolutionary tracks of young stars contracting toward the main sequence, according to model calculations. Numbers along tracks indicate ages in years. *(From G. O. Abell, Realm of the Universe, 2d ed., Saunders College, 1980)*

star establishes radiative transport of heat in its interior and begins to shine, its radiation pressure pushes away the nonaccreted remnant of the dust "cocoon." The contraction of the star slows down as it progresses toward the stable configuration of a main-sequence star, along a path shown in Fig. 4 according to model calculations. Stars lying above a line indicated on the figure are still completely surrounded by in-falling material, and are either completely hidden or shine only partially and to a variable degree. Stars of small mass descend at first vertically through the Hertzsprung-Russell diagram along the so-called Hayashi track, where convection prevails over almost the whole star and thoroughly mixes the material. When radiative energy transport prevails, the evolutionary tracks become predominantly horizontal. By that time the star becomes optically visible, but interaction with the remnants of the surrounding gas and dust cocoon, as well as possible internal instabilities, make it variable in luminosity and render its spectrum very peculiar. Such stars are known as the Herbig emission stars (when their masses are about 5 solar masses) and as T Tauri stars (with masses 0.5 to 2.5 solar masses). These objects typically appear in groups (associations) associated with nebulosities. Obviously stars are usually formed in groups, perhaps by fragmentation of a large protostar, and later on they disperse, although at times a strong gravitational bond holds them together in a star cluster. *See* STAR CLUSTER.

Solar neutrino puzzle. Although progress in understanding of stellar structure and evolution has been very impressive, there is one major puzzle that affects the understanding of the structure of the nearest star, the Sun. In the course of the thermonuclear reactions converting hydrogen into helium, a certain amount of neutrinos are generated, which should pass through the Sun without interaction. A tiny fraction of these may be captured in an ingeniously devised experiment conducted by R. Davis in a deep mineshaft. However, the observed flux of neutrinos is significantly smaller than the solar models predict. It is not clear where the fault is: in understanding of stellar structure, or in the nature of the neutrinos. *See* SOLAR NEUTRINOS; STAR; SUN.

[MIREK J. PLAVEC]

Bibliography: G. O. Abell, *Exploration of the Universe*, 4th ed., 1982; E. H. Avrett (ed.), *Frontiers of Astrophysics*, 1976; C. Payne-Gaposchkin, *Stars and Clusters*, 1979; V. C. Reddish, *Stellar Formation*, 1978; I. S. Shklovskii, *The Stars, Their Birth, Life, and Death*, 1978.

Stellar magnetic field

A magnetic field, far stronger than the Earth's magnetic field, which is possessed by many stars. Since 1945 it has become increasingly evident that gross electromagnetic phenomena are of universal importance in the physics of the stars and of the rarefied gases in the space between them.

Detection of magnetic fields of the Sun and stars and measurement of their intensities in gauss and their polarity as positive or negative or as north or south are accomplished by collecting the light with a telescope, analyzing it with a spectrograph, and studying the Zeeman effect in the spectral lines. Light-absorbing atoms in the atmosphere of a star spin about the lines of force of the stellar magnetic field so that the spectral lines are split into polarized components. The astronomer can photograph the spectrum and measure the Zeeman splitting effect, unless the lines are too much broadened by other effects such as the Doppler effect in a rapidly rotating star. *See* STAR.

The Earth has a magnetic field of about 0.6 gauss, and it was shown in 1953 that the Sun has a general or dipolar field, observable in high heliographic latitudes, some two or three times stronger than that of Earth. There are much stronger fields in transitory local magnetic regions, including sunspots, in lower latitudes. G. E. Hale in 1908 showed that sunspots have magnetic fields ranging up to 3000 gauss or more.

Observations with the 100- and 200-in. telescopes of the Hale Observatory resulted in the detection and measurement of magnetic fields in about 100 of the brighter stars that have sharp spectrum lines. Field intensity ranges up to 34,000 gauss in the A-type star HD 215441.

Stellar magnetic fields vary with time. Some of the variations are irregular, but many show periodicity, sometimes with superposed random fluctuations suggesting that intrinsic changes of a turbulent nature occur in the star's atmosphere. Modulation of the apparent field results from axial rotation. An outstanding example of a periodically varying magnetic star is 53 Camelopardalis, which has a magnetic field varying between the limits +3600 gauss and −5000 gauss in a period of 8.026 days. Several other large-amplitude magnetic variables have periods of about a week, and all of these show reversal of polarity. Magnetic variables of longer period are HD 188041 (226 days) and HD 187474 (2350 days).

Magnetic stars show anomalous abundances of the chemical elements that have not been satisfactorily explained. The rare-earth elements and elements of the iron group in particular seem to have a patchy distribution on the surfaces of magnetic stars. The field of a star controls the disposition of circumstellar ionized gases and so is of importance in cosmogony, in the evolution of stellar systems, and in acceleration of cosmic rays. *See* SOLAR MAGNETIC FIELD.

[HORACE W. BABCOCK]

Bibliography: T. G. Cowling, *Magnetohydrodynamics*, 2d ed., 1977; G. P. Kuiper (ed.), *Stars and Stellar Systems*, vol. 1: *Telescopes*, 1960, reprint 1977; E. N. Parker, *Cosmical Magnetic Fields: Their Origin and Activity*, 1979.

Stellar magnitude

The astronomical scale of brightness of stars. A difference of five magnitudes between two objects corresponds to a factor of 100 in their ratio brightness. The basis of this system is that what the eye notes as equal intervals of brightness are actually equal ratios. One magnitude corresponds to a light ratio of $(100)^{1/5} = 2.512$.

Color and intensity. Magnitudes should be defined monochromatically, or at least over a narrow and well-defined range of wavelengths; the subscript λ indicates the band used. Thus, in comparing the brightnesses I of objects A and B at a wavelength λ, the relations of magnitudes m would be given by Eqs. (1*a*) and (1*b*). The energy received

$$m_\lambda(A) = m_\lambda(B) - 2.5 \log_{10} \frac{I_\lambda(A)}{I_\lambda(B)} \qquad (1a)$$

$$\frac{I_\lambda(A)}{I_\lambda(B)} = 10^{-0.4[m_\lambda(A) - m_\lambda(B)]} \qquad (1b)$$

within two properly chosen filter bands can also be used to measure the colors of the stars. The zero point of the magnitude system is defined so that the brightest stars are near zero; the faintest photographed with the 200-in. (5-m) Hale reflector are near +24.7 magnitude. However, solid-state area detectors such as the charge-coupled device (CCD) have been used to reach beyond +26.5 magnitude with the 5-m reflector, and the ultimate limiting magnitude of the proposed Space Telescope using a CCD is expected to be near +28. The Sun is near −27 magnitude, so that the astronomical range covers 55 magnitudes, or 10^{22} in brightness.

The flux of radiation received outside the Earth's atmosphere from the Sun over all wavelengths is 1.4×10^6 ergs/(cm²)(s) or 1.4×10^3 W/m². The flux received from a star of apparent magnitude m is given by Eq. (2).

$$\begin{aligned} F(m) &= 2.4 \times 10^{-5}\, 10^{-0.4m} \text{ ergs/(cm}^2\text{)(s)} \\ &= 2.4 \times 10^{-8}\, 10^{-0.4m} \text{ W/(m}^2\text{)} \end{aligned} \qquad (2)$$

A calibration for scientific and engineering purposes can be derived from the *AB* magnitudes published by J. B. Oke for many stars. The units of F_ν in Eq. (3*a*) are ergs/(cm²)(s)(Hz), of F_ν in Eq. (3*b*)

$$AB = -2.5 \log F_\nu - 48.60 \quad (3a)$$
$$AB = -2.5 \log F_\nu - 56.10 \quad (3b)$$

$W/(m^2)$ (Hz), and of AB_ν magnitudes. There are 1050 photons arriving outside the Earth's atmosphere, per second per square centimeter within a 1-A bandwidth (or 1.05×10^8 photons per second per square meter within a 1-nm bandwidth), from a star of zero apparent magnitude, at a wavelength of 500 nm. For the absolute calibration of photoelectric and infrared magnitude scales, in units of flux density, $W/(m^2)(Hz)$, *see* INFRARED ASTRONOMY.

The energy received from a star depends on its distance r as well as on its intrinsic luminosity L. The absolute magnitude M is defined as the apparent magnitude a star would have if located at a standard distance of 10 parsecs (1 parsec = 3.0857×10^{16} m). *See* PARSEC.

From the inverse-square law, neglecting the possible interstellar absorption of light, Eq. (4) is

$$M = m + 5 - 5 \log r \quad (4)$$

obtained, where the units of r are parsecs. This permits one to deduce the absolute magnitude, given apparent magnitude and distance, or the distance, given M and m. The range of M is from about −10 to +20; the absolute visual magnitude of the Sun is about +5, dependent on the wavelength interval used. The bolometric magnitude, which measures the total radiation, is about +4.7, and the luminosity of the Sun is 4×10^{33} ergs/s or 4×10^{26} W. Thus for any star the bolometric (or total) luminosity L is given by Eq. (5), where $L = 2.9 \times 10^{35}\ 10^{-0.4M}$ ergs/s $= 2.9 \times 10^{28}\ 10^{-0.4M}$ W.

$$L/L_\odot = 10^{-0.4(M-4.7)} \quad (5)$$

The apparent or absolute magnitude of a star depends on the wavelength range over which the measurement is carried out.

Methods of measurement. Many combinations of photographic plates, photocells, or photomultipliers (with various sensitizations), and filters (isolating various wavelength regions) are used in astronomy. Such combinations give various effective wavelengths for the centers of the response curves. Table 1 contains some of the properties of the more frequently used response bands. The human eye and the visual magnitude system are used only for approximate work. Special narrow-band interference filters are also commonly used; broad-band filters are used with solid-state detectors in the infrared.

Multichannel devices with good spectral resolution prove useful. A simple, single-channel scanner traverses the spectrum by a moving slit and photocell. A dual-channel, sky-star, alternating radiometric type of device subtracts background sky instantaneously from the star. Finally, multichannel spectrophotometers, for high-resolution scanners, rival the spectrograph.

Color index. A wide variety of multicolor photoelectric systems are in use; correlations between such color and brightness measures are defined by the results for a group of standard stars. Although data on detector sensitivity and filter transmission partially define these systems, the actual relations found by experiment are often nonlinear.

The color of a star is given, in general, by a color index, on a particular base line, as in Eq. (6). By

$$\text{Color index} = m_{\lambda_1} - m_{\lambda_2} \quad (6)$$

convention, many color systems are adjusted so that the color of bright stars of spectral type AO is zero; such stars have color temperatures near 14,000 K. The International System is defined by the magnitudes, on these two scales, of certain stars near the north celestial pole and in other selected areas.

Many accurate photoelectric color systems exist, of which the most widely used is the Johnson-Morgan U,B,V three-color system. The accuracy of magnitudes on a photographic plate seldom exceeds ±0.1 magnitude; the precision and linearity of the photomultiplier makes ±0.01 magnitude attainable. Two color indices describe a star in the U,B,V system, that is, U-B and B-V. H. L. Johnson and W. W. Morgan (1953) have given standard colors and spectra of a large number of stars, together with the relation of the U,B,V system to the International System.

The data in Table 2 show the mean colors of main-sequence stars as a function of spectral type. The nonlinearity of the relation between U-B and

Table 1. Frequently employed magnitude systems

Name	Detector	Filter	Response band, nm
Bolometric	Theoretical, sensitive to all wavelengths		
Radiometric	Vacuum thermocouple	Earth's atmospheric transmission	UV – infrared
Infrared	Photographic infrared-sensitive plate (type IN)	Deep-red filter	700 – 900
Photo red	Red-sensitive plate (103aF)	Red filter	600 – 660
Photovisual	Yellow-sensitive plate (103aD)	Yellow filter	500 – 600
Visual	Human eye		350 – 500
Photographic	Blue-sensitive plate (103aO)	None, or minus UV	390 – 500
Photoelectric, V	1P21 Multiplier	Yellow	500 – 630
Photoelectric, B	1P21 Multiplier	Blue and minus UV	390 – 510
Photoelectric, U	1P21 Multiplier	Ultraviolet	300 – 390

Table 2. Colors of main-sequence stars in Johnson-Morgan system of standard photoelectric colors

Type	B-V	U-B	Type	B-V	U-B
B1	−0.28	−1.00	F5	+0.44	0.00
B3	−0.20	−0.71	G0	+0.60	+0.06
B5	−0.16	−0.56	G5	+0.68	+0.21
B8	−0.09	−0.29	K0	+0.82	+0.48
A0	0.00	0.00	K3	+1.01	+0.89
A3	+0.09	+0.07	K7	+1.37	+1.26
A7	+0.19	+0.08	M5	+1.69	+1.24
F0	+0.30	+0.02			

B-V colors is caused by the distortion of the energy curve of a star by continuous absorption or by lines.

Narrow-band filters with transmissions over only a few angstroms have been used to define nearly monochromatic photoelectric magnitudes. Combinations of these, as devised by B. Stromgren, are used to measure strengths of individual spectral lines as well as stellar temperatures.

Total radiation. Bolometric magnitudes m_b should measure the total energy of the star. The zero point of bolometric magnitudes is so defined that m_b should equal m_v (visual) for a star of surface temperature near 6000 K, like the Sun. A bolometric correction Δm_b must be applied to m_v, negative for all other temperatures, because either the infrared or ultraviolet will contain relatively more energy at other temperatures. For cool stars Δm_b can be evaluated by radiometric techniques, the excess energy being in the accessible infrared. For hot stars, the ultraviolet, not transmitted by the Earth's atmosphere, contains most of this energy, so that Δm_b values depend on observations from high-altitude balloons or rockets or from artificial satellites. [JESSE L. GREENSTEIN]

Stellar rotation

Surface rotational equatorial velocities, ranging from a few kilometers per second (km sec^{-1}) up to 500 km sec^{-1}, exhibited by stars. At 500 km sec^{-1} the stars become unstable. Observable directly in the Sun and in eclipsing stars, stellar rotation is detected through its broadening of the absorption lines by the Doppler effect. If the equatorial velocity is v_R and i is the inclination of the equator to the plane of the sky, the observed velocity at the limb is $\pm v_R \sin i$. A line is broadened, approximately, by $\Delta\lambda/\lambda = (v_R/c) \sin i$. The strength of the line is unchanged, so that the profiles vary with increasing velocity as shown in Fig. 1. The line in a nonrotating star is relatively narrow, absorbing about 45% of the light in the center and being about 1.0 nm wide. For rotations of 235, 430, 560 km sec^{-1} the line is broadened more and more, becoming shallower and wider. At the highest velocity shown, the star is near instability and the line is only 15% deep and 2.0 nm wide. Since rotational angular momentum tends to be conserved, stellar rotation is inherited from the gases out of which the star is formed. Galactic rotation ensures that the gas has a systematic angular velocity large enough to cause rapid stellar rotation. Since the interstellar gas is turbulent, individual eddies, as well as the stars born out of them, need not have parallel angular momentum vectors. Many stars exist in binary or multiple systems with orbital angular momentum of similar origin. If the rotation and orbital revolution are synchronous, the momentum is largely in the orbital motion; probably prestellar condensations with excessive momentum form binaries by fission. *See* ORBITAL MOTION.

Distribution of velocities. Studies of absorption line broadening by G. A. Shajin and O. Struve in 1929 showed that $\langle v_R \sin i \rangle$ was a steep function of spectral type, with a maximum in the hot stars of type B, and with largest velocities found in stars of type Be. The latter have emission lines indicating mass loss and surrounding gaseous shells. If the axes are randomly distributed, $\langle v_R \rangle = (4/\pi) \langle v_R \sin i \rangle$. Figure 2 shows, schematically, the distributions of $v_R \sin i$ for main-sequence stars of different spectral type, that is, different surface temperature, and for the B and Be stars. The F stars are slightly hotter than the Sun, and a few of them rotate with moderately high equatorial velocity. Most of them have very sharp lines near zero rotation. The hotter B and A stars have a relatively flat distribution of rotational velocities, reaching values of the order of 300 km sec^{-1}. The Be stars are peaked at over 300 km sec^{-1}. They are objects in which the rotation is so rapid that instability has occurred and gas envelopes surround the star, producing emission lines. The observed rapid de-

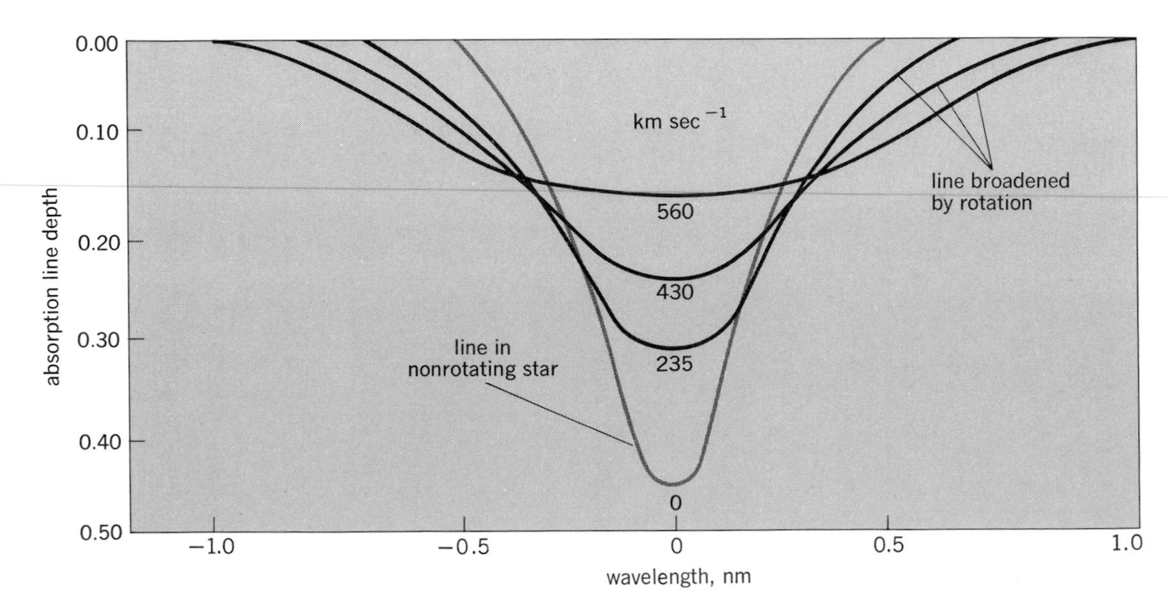

Fig. 1. The effect of rotation on the shape of a spectral absorption line.

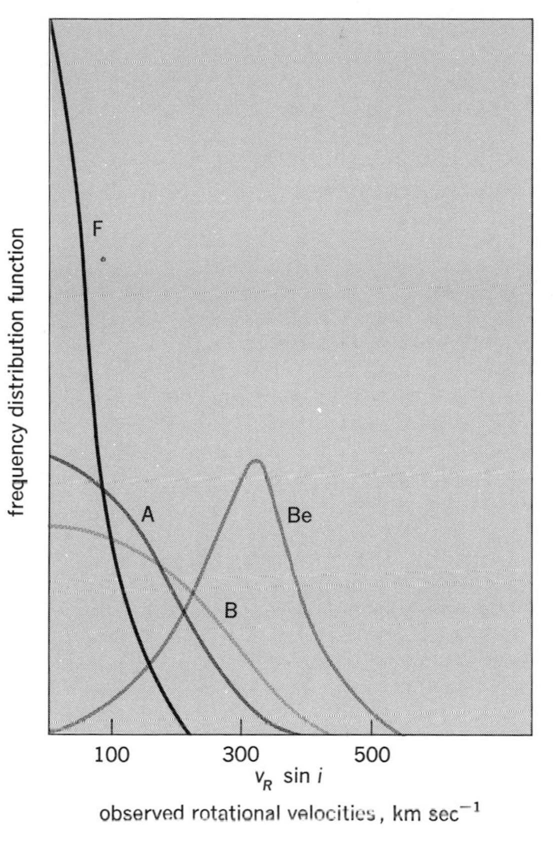

Fig. 2. Frequency distribution of observed rotational velocities for types of stars F, A, B, and Be.

crease of v_R with decreasing temperature is startling, and fundamental to much speculation on convection, stellar winds, and mass loss. *See* HERTZSPRUNG-RUSSELL DIAGRAM.

Observation of giants shows their $\langle v_R \rangle$ to be smaller at a given spectral type (with only a few exceptions) than for main-sequence stars. Clearly, their instability would begin at lower v_R, but the reduction of v_R has another cause. From stellar evolution theory, the envelope expands, resulting in a larger radius of gyration. The outcome is a reduction of the linear velocity of the surface layers. Depending on whether expanding stars rotate as rigid bodies, or whether angular momentum is conserved in shells, or is not conserved at all, one can expect different ratios of v_R for giants and main-sequence stars. Supergiants have turbulently broadened absorption lines, but in general show little or no rotation. Very old stars, which if unevolved must be cooler than the Sun, show no rotation. A few young stars in the process of formation show excessive rotation for their surface temperatures, and the v_R in young clusters is greater than in old. *See* GIANT STAR; STELLAR EVOLUTION; SUPERGIANT STAR.

Binary systems. Complex phenomena occur in close binary systems, where tidal drag ensures strong interaction between the stars. The theory of particle orbits in a rotating frame of reference was developed by J. L. Lagrange and has been extensively applied to the stellar case. Emission-line B stars often occur in double systems and, if an eclipse occurs, can be shown to be surrounded by a thin equatorial ring of hot gases. These gases, when traveling between the hot star and the observer, can also be seen by the absorption lines they produce. Complex, nonuniform streams exist and transfer matter from the larger to the smaller star. If two stars are separated by not more than their radii, a common gaseous envelope is formed (the so-called W Ursae Majoris stars). The brightness is continuously variable, in periods of a few hours, in these and other "contact" variables. This phenomenon has assumed importance with the discovery of binaries containing massive stars (or black holes), in which matter falls through an enormous potential gradient, emitting hard x-rays. This occurs either as the streams encounter an accretion ring or the collapsed star. *See* BINARY STAR; ECLIPSING VARIABLE STARS.

Slowing of rotation. The Sun's surface rotates with a shorter period at the equator than at the poles. Its internal rotation could be at a greater angular velocity, and meridional currents are required. The outer-convective zone, however, is the most important feature of the entire phenomenon of rotation. All stars with surface temperature below 7500 K, type F2, have their hydrogen partially ionized just below the surface layers; convection develops and transports much of the flux outward. In addition, above the region where hydrogen has become neutral, some energy is still transported to and through the surface. This energy is about 10^{-5} the radiated energy in the Sun; in more active stars the ratio may be 10^{-2}. One can observe the rotating, magnetized, solar wind by space probes. The ejection of matter, small in the Sun, may be as high as 10^{-6} $M_\odot$ year^{-1} in young stars. Since the matter is also magnetized, a convective star ejects rotating plasma which interacts with the star's environment. One obvious consequence is a mechanism which slows the rotation of the star by a drag, linking it to the stationary interstellar medium. Thus, one expects and finds that young stars with convective envelopes rotate more rapidly than old ones. In addition, the cutoff of rapid rotation for young stars, from observations in Fig. 2, lies near the temperature at which deep convection begins. The Sun cannot stop an initially rapid rotation with its present mass loss, but given an early, more active phase which young stars have, it may have done so.

Effect of rotation. The effect of rotation on the properties of stars is complex. The maximum possible angular velocity for a uniformly rotating star is $\omega_c^2/2\pi G\langle\rho\rangle = \varphi_c$, with ω_c the angular velocity, G the gravitational constant, and $\langle\rho\rangle$ the mean density, where $\varphi_c = 0.36075$. A normal hot star of 8 $M_\odot$, 4 $R_\odot$ gives $\omega_c = 1.6 \times 10^{-4}$ radian sec^{-1}, about 560 km sec^{-1}. At this breakup point, which corresponds to the observed Be stars in Fig. 2, the star is nonuniform in brightness, highly distorted in shape, and composite in spectrum and color. Heat must flow through the star from the equator to the hot poles. Thus, rotating stars differ from nonrotating stars, in an amount dependent on the inclination of the poles to the line of sight. Smaller-scale variations of luminosity, temperature, and spectrum are observed in stars with angular velocities less than ω_c, as deviations from the main sequence of nonrotating stars.

Neutron stars. The very condensed neutron stars rotate very rapidly—the Crab Nebula pulsar in 0.033 sec. The surface velocity of the Crab Nebula pulsar is 2000 km sec^{-1}. But if a magnetic field is fixed in its surface, its velocity of rotation reaches the velocity of light at 1600 km above the surface of the neutron star. The magnetic field is then seen as an electric field, and it is generally assumed that high-energy particles are produced by acceleration in that zone. Pulsars are observed to slow down; the longest known periods are near 3 sec. Presumably the rotational kinetic energy is converted into high-energy particles, thus producing the radio pulses (and the light and x-ray pulses in the central star of the Crab Nebula). Erratic period changes are also observed, which are presumably connected with either changes in the moment of inertia of the neutron star, expulsion of magnetic loops or superfluid effects. *See* CRAB NEBULA; NEUTRON STAR; PULSAR; STAR; SUN.

[JESSE L. GREENSTEIN]

Bibliography: K. J. Fricke and R. Kippenhahn, Evolution of rotating stars, *Annu. Rev. Astron. Astrophys.*, 10:45–72, 1972; S. Huang and O. Struve, Stellar rotation and atmospheric turbulence, in J. L. Greenstein (ed.), *Stellar Atmospheres*, 1960; L. Mestel, Meridian circulation in stars, in L. H. Aller and D. McLaughlin (eds.), *Stellar Structure*, 1965.

Subgiant star

A member of the family of stars intermediate between giants and the main sequence in the Hertzsprung-Russell (H-R) diagram. The mean luminosity of a subgiant is about 10 times the Sun; the surface temperature lies between 4000 and 7000 K. The masses are about 1.4 times that of the Sun. The subgiants often violate the mass-luminosity relation; that is, ζ Herculis A, a G subgiant, is four times as bright as its mass would predict. *See* HERTZSPRUNG-RUSSELL DIAGRAM; MASS-LUMINOSITY RELATION; STAR.

The subgiants are of particular importance in current theories of stellar evolution. If a main-sequence star has exhausted about 12% of its mass of hydrogen, the star begins to evolve, expanding, cooling at the surface, and brightening. Old stars of population II, of masses about 1.35 times the Sun, are now evolving into the subgiant region of the H-R diagram. The age of the oldest known stellar systems can be determined from the luminosities of the subgiants to be between 5,000,000,000 and 12,000,000,000 years. Similar stars, not in clusters, suggest an upper limit to the age of the galaxy, which, subject to some theoretical uncertainties, is of the same order. A different type of subgiant occurs also in close binary systems of the younger population I. Some have been found, unexpectedly, to emit bursts of radio-frequency energy, indicating mass transfer and the presence of some high-energy electrons. *See* STELLAR EVOLUTION.

[JESSE L. GREENSTEIN]

Sun

The star around which the Earth revolves, and the planet's source of light and heat (Fig. 1). The Sun is a globe of gas, 1.4 × 10^{6} km in diameter, held together by its own gravity. Because of the weight of the outer layers, the density and temperature increase inward, until a central temperature of over 15,000,000 K and density more than 90 times that of water is reached. At these great temperatures and densities, thermonuclear reactions converting hydrogen into helium take place, releasing the energy which streams outward.

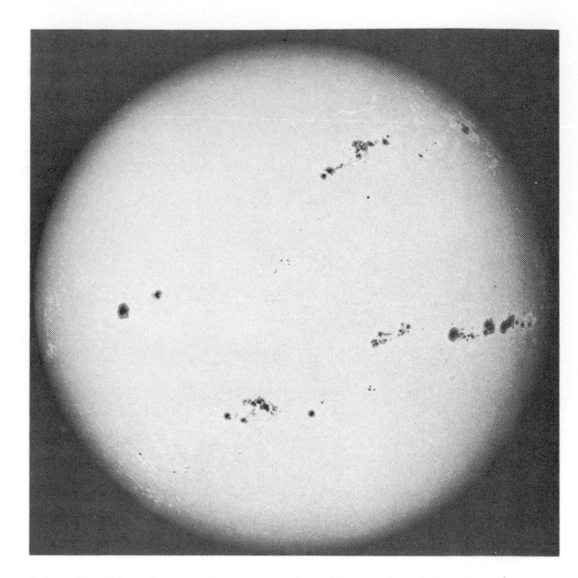

Fig. 1. The Sun, photographed in white light during the 1957 maximum of the sunspot cycle. (*Hale Observatories*)

The surface temperature of the Sun is about 6000 K; since solids and liquids do not exist at these temperatures, the Sun is entirely gaseous. Almost all the gas is in atomic form, although a few molecules exist in the coolest regions at the surface.

Table 1. Principal physical characteristics of Sun

Characteristic	Value
Mean distance from Earth (the astronomical unit)	1.4960 × 10^{8} km
Radius	(6.960 ± .001) × 10^{5} km
Mass	(1.991 ± .002) × 10^{33} g
Mean density	1.410 ± .002 g/cm^{3}
Surface gravity	(2.738 ± .003) × 10^{4} cm/s^{2} = 28 × terrestrial gravity
Total energy output	(3.86 ± .03) × 10^{33} erg/s = (3.86 ± 0.03) × 10^{26} W
Energy flux at surface	(6.34 ± .07) × 10^{10} erg/(cm^{2})(s) = (6.34 ± .07) × 10^{7} W/m^{2}
Effective surface temperature	5780 ± 50 K
Stellar magnitude (photovisual)	−26.73 ± .03
Absolute magnitude (photovisual)	+4.84 + .03
Inclination of axis of rotation to ecliptic	7°
Period of rotation	About 27 days; the Sun does not rotate as a solid body; it exhibits a systematic increase in period from 25 days at the equator to 31 days at the poles.

The Sun is a typical member of a numerous class of stars, the spectral type dG2 (d indicates dwarf). Other characteristics are given in Table 1.

Besides its great importance to human life, the Sun is of interest to all astronomers because it is the only star near enough for detailed study of its surface structure. Various surface and atmospheric phenomena, such as sunspot activity, and other behavior may be studied, and astronomers try to extrapolate these to the other stars which may be observed only as points of light.

The light and heat of the Sun make the Earth habitable. The Sun is, in fact, the ultimate source of nearly all the energy utilized by industrial civilizations in the form of water, power, fuels, and wind. Only atomic energy, radioactivity, and the lunar tides are examples of nonsolar energy. *See* EARTH.

SOLAR STRUCTURE

The interior of the Sun can be studied only by inference from the observed properties of the entire star. The mass, radius, surface temperature, and luminosity are known. Using the known properties of gases, it is possible to calculate that structure of the Sun which will produce the observed parameters at the surface (Fig. 2). The solution is complicated by uncertainties in the behavior of matter and radiation under the high temperature and density that are present in the solar interior. This is particularly true of the nature of the nuclear reactions. However, the general properties of the solution are quite reliable. A number of theoretical models using different assumptions have led to more or less similar results. A central density of near 90 g/cm^3 has been found, decreasing to 10^{-7} g/cm^3 at the surface. The central temperature is about 15,000,000 K, decreasing to 5000 K at the surface. Since this takes place over 700,000 km, the temperature gradient is only 20 K per kilometer. The radiation produced at the center by nuclear interactions flows outward rapidly.

The energy of the Sun is produced by the conversion of hydrogen into helium; because each hydrogen atom weighs 1.0078 atomic units and each helium atom is made from four hydrogen atoms, only 4.003 units, it follows that 0.0282 unit, or 0.7% of the mass m, is converted into energy E according to the Einstein formula $E = mc^2$, where c is the speed of light. Since the solar mass is 2×10^{33} g, conversion of 0.7% into helium yields 1.2×10^{52} ergs (1.2×10^{46} joules), enough to maintain the Sun for 10^{19} s, or over 10^{11} years. The rate of conversion required to produce the observed flux is 4×10^{38} atoms/s. For each hydrogen atom converted, one neutrino is produced, giving a flux of 1.3×10^{11} neutrinos/(cm$^2 \cdot$ s) at the Earth. These neutrinos cannot be detected; only higher-energy neutrinos produced by subordinate processes can be observed. These have been detected at the Earth, but in smaller quantities than expected. However, the neutrino emission theory has sufficient uncertainties so that the theory of nuclear burning is still generally accepted. *See* SOLAR NEUTRINOS.

Although the material at the center of the Sun is so dense that a few millimeters are opaque, the photons created by nuclear reactions are continually absorbed and reemitted and thus make their way to the surface. The atoms in the center of the Sun are entirely stripped of their electrons by the high temperatures, and most of the absorption is by continuous processes, such as scattering of light by electrons.

In the outer regions of the solar interior, the temperature is low enough for ions and even neutral atoms to form and, as a result, atomic absorption becomes very important. The high opacity makes it very difficult for the radiation to continue outward; steep temperature gradients are established which result in convective currents. Most of the outer envelope of the Sun is in such convective equilibrium. These large-scale mass motions produce many interesting phenomena at the surface, including sunspots and solar activity.

Since young stars rotate rapidly, astronomers have sought evidence that the central part of the Sun might be rotating rapidly, and only the outer surface rotating at the observed fairly slow rate. This would lead to a very small oblateness in the Sun's disk, about 1 part in 10^5. R. H. Dicke's detection in 1966 of such an oblateness was challenged by H. Hill, who found the surface to be oscillating with a 55-min period. Others have challenged these results, and both are still in doubt. *See* STELLAR ROTATION.

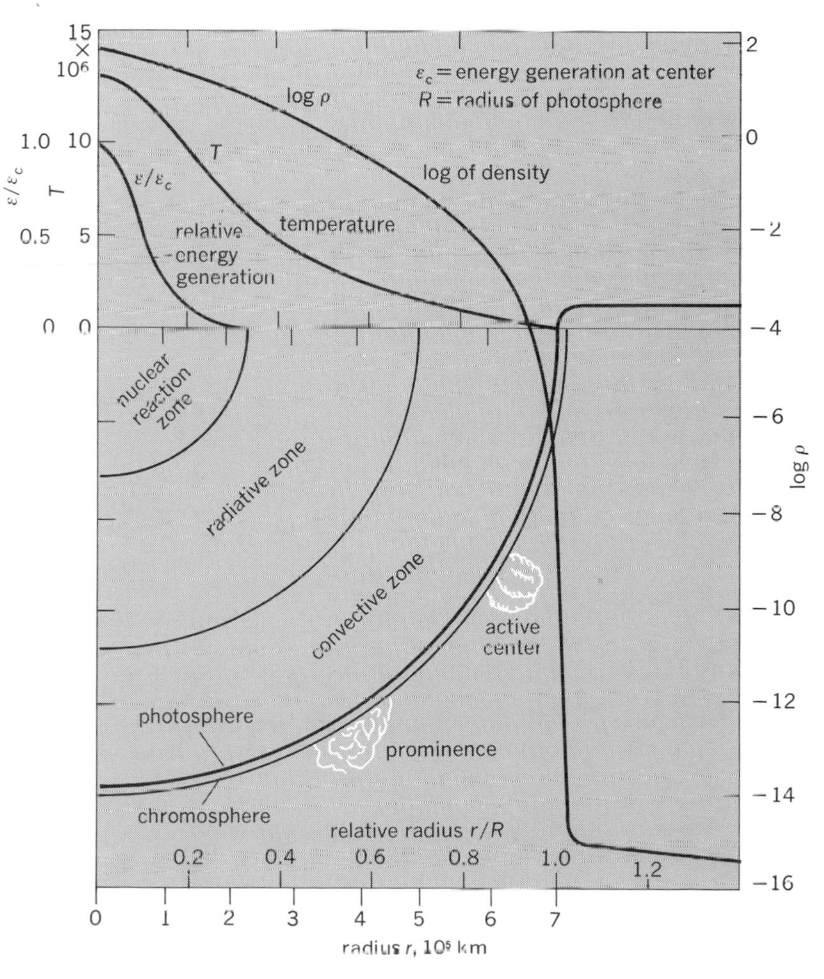

Fig. 2. Solar temperature, density, and energy generation in the interior and atmosphere of Sun. Curves in upper section show these as functions of radial distance from the center. Lower section shows the principal zones in the interior of Sun.

Radiation. Electromagnetic energy is produced by the Sun in essentially all wavelengths. Important radiation has been measured from long radio waves of 300 m down to x-rays of less than 0.1 nm (from rockets). In addition, considerable energy is emitted in the form of high-energy particles (cosmic rays). However, more than 95% of the energy is concentrated in the relatively narrow band between 290 and 2500 nm and is accessible to routine observation from ground stations on Earth. The maximum radiation is in the green region, and the eyes of human beings have naturally evolved to be sensitive to this range of the spectrum. The total radiation and its distribution in the spectrum are parameters of fundamental significance, because they measure the total energy output of the Sun and its effective surface temperature. The total radiation received from the Sun is termed the solar constant and has, in fact, been found within the limits of observation to be constant to about ±1%. The presently accepted value is 1.97 cal/(cm^2)(min). This is equivalent to 1.374×10^6 erg/(cm^2)(s) or 1.374×10^3 W/m^2. *See* SPACE PROBE.

The measurement of the solar constant has been greatly complicated by the absorption of solar radiation in all wavelengths by the Earth's atmosphere. Determinations have been made by observations at stations atop high mountains where the atmospheric perturbation is minimized, and by observing the variation during the day as the radiation passes through successively smaller distances in the terrestrial atmosphere. A great bulk of the studies was carried out by the Smithsonian Institution in a continuous program from the 1880s until 1955. The Smithsonian results showed that the solar constant was indeed constant to 1%, but there might be variation at a smaller scale. Spacecraft measurements suggest variations of 1 part in 2000, but cannot be confirmed because the constant is not measured from the ground.

The measurement of this immensely important quantity has been neglected, partly because many feel it is constant, and because the work is difficult, trying, and unexciting. A reduction of 7% in the solar constant would produce a completely frozen Earth, which only a 50% increase would thaw (because of the reflectivity of the ice). However, the fossil record shows that the solar constant has not changed much in the last billion years.

Atmosphere. Because the Sun is visible as a two-dimensional surface rather than as a point, the study of its atmosphere and surface phenomena is most interesting.

Although the Sun is gaseous, it is seen as a discrete surface from which practically all the heat and light are radiated. One sees through the gas to the point at which the density is so high that the material becomes opaque. This layer, the visible surface of the Sun, is termed the photosphere. Light from farther down reaches the Earth by repeated absorption and emission by the atoms, but the deepest layers cannot be seen directly. The surface is actually not sharp; the density drops off by about a factor of 3 each 150 km. However, the Sun is so far away that the smallest distance that can be resolved with the best telescope is about 300 km, and thus the edge appears sharp. *See* SOLAR CORONA; PHOTOSPHERE.

Looking at the Sun in isolated wavelengths absorbed by its atmospheric gases, one can no longer see down to the photosphere, but instead one sees the higher levels of the atmosphere, known as the chromosphere. This name results from the rosy color seen in this region at a solar eclipse. The chromosphere is a rapidly fluctuating region of jets and waves coming up from the surface. When all the convected energy coming up from below reaches the surface, it is concentrated in the thin material and produces considerable activity. As a result, this outer region is considerably hotter than the photosphere, with temperatures up to 30,000 K. *See* CHROMOSPHERE.

When the Moon obscures the Sun at a total solar eclipse, the vast extended atmosphere of the Sun called the corona can be seen. The corona is transparent and is visible only when seen against the dark sky of an eclipse. Its density is low, but its temperature is high (more than 1,000,000 K). The hot gas evaporating out from the corona flows steadily to the Earth and farther in what is called the solar wind. *See* SOLAR WIND.

Within the solar atmosphere many transient phenomena occur which may all be grouped under the heading of solar activity. They include sunspots and faculae in the photosphere, flares and plages in the chromosphere, and prominences and a variety of changing coronal structures in the corona. The existence and behavior of all these phenomena are connected with magnetic fields, and their frequency waxes and wanes in a great 22-year cycle called the sunspot cycle. *See* SOLAR MAGNETIC FIELD.

The sunspots and flares are sources of x-rays, cosmic rays, and radio emission which often have profound influence on interplanetary space and the upper atmosphere of the Earth.

The photospheric features of the Sun (sunspots and faculae) are readily observed with a small telescope by projection of the solar image through the eyepiece onto a shaded white card. This is the only safe method for observing the Sun without a specially designed solar eyepiece. Observations of chromospheric phenomena require special filters to exclude all light except that emitted by the atmospheric gases. Observations of the corona require either a total eclipse or a coronagraph.

Solar physics. Understanding of the Sun is derived from observations of the morphology of various phenomena and from physical analysis with the spectrograph. The former permits determination of what is happening, and the latter determination of the detailed physical parameters of the gas under observation. *See* ASTRONOMICAL SPECTROSCOPY.

When the light of the Sun is broken up into the different frequencies by the spectrograph, a bright continuum interrupted by thousands of dark absorption lines known as the Fraunhofer lines can be seen. Each line represents some discrete transition in a particular atom which occurs when a photon of light is absorbed and one of the atomic electrons jumps from one level to another. Of the 92 natural elements, 64 are represented in the Fraunhofer spectrum (Fig. 3), first observed by the German physicist J. Fraunhofer. The remaining atoms are undoubtedly present but remain undetected because they are rare or their lines are produced in spectral regions not accessible to present spectrographs. The relative abundances of the most numerous atoms have been estimated from the line intensities (Table 2). Many of these abundances have been confirmed by measurement of the relative abundances of different elements in the streams of particles coming from the Sun at the time of solar flares. The Sun can be described as a globe of chemically (but not spectroscopically) pure hydrogen and helium with traces of the other elements.

The nature of the formation of the spectrum lines can be understood in terms of a few simple rules which are an elaboration of those first put forth by G. R. Kirchhoff. A small amount of a hot gas will be transparent in all wavelengths except those characteristic of the atoms of the gas. In those wavelengths radiation is emitted by the jumping electrons, and an emission line spectrum is produced. If the volume of the gas is large enough and its density high enough, it will be opaque at all frequencies because of the existence of continuous processes which will absorb any wavelength of light. In this case the gas emits a continuous spectrum at all frequencies. However, at the outer edges the atoms of the gas absorb light coming from one direction (the inside) and reemit it in all directions, with a consequent reduction of intensity. This results in dark lines superposed against the bright background. Although the ultimate source of all energy is at the center of the Sun, at any one point the radiation is a result of either a scattering of light coming from within or the emission of radiation by an atom excited by collisions with the electrons of the gas.

By analyzing the nature of the spectrum lines observed, it can be determined which atoms are present, as well as their stages of ionization. This is so because, as electrons are successively removed from an atom by ionization (as a result of photoelectric effect or collisions with electrons), the spectrum changes completely. Obviously, the hotter the gas, the higher the stage of ionization. Thus when lines of iron ionized 14 times are observed in the solar corona, it may be assumed that the gas is very hot. On the other hand, by observing the spectra of molecules in the upper photosphere, it may be concluded that the temperature is near the lowest possible in the Sun.

In addition, the shape of the spectrum lines is a key to the physical conditions under which they are radiated. If the temperature is high, the atoms have a high velocity and the lines are broadened by the Doppler effect. If there are strong magnetic fields, the spectrum lines are split by the Zeeman effect. If there are strong electric fields, the lines are broadened by the Stark effect. The spectroscopist must disentangle these various effects. This task is aided by the fact that the spectrum lines can be observed at both the center of the Sun and near the limb, as well as from atoms which behave differently. For example, hydrogen atoms are very light and move extremely fast; hence the hydrogen lines show a strong broadening due to the Doppler effect. If the lines emitted by heavy atoms in the same region do not show this broadening, it may be concluded that this is purely a temperature effect, because the heavy atoms would not be expected to move so fast.

Table 2. Relative numbers of the most abundant atoms in Sun

Element	Number	Element	Number
Hydrogen, H	1,000,000	Silicon, Si	20
Helium, He	50,000–200,000	Iron, Fe	15
		Sulfur, S	8
Oxygen, O	500	Aluminum, Al	2
Nitrogen, N	400	Sodium, Na	2
Carbon, C	200	Calcium, Ca	1.5
Magnesium, Mg	33		

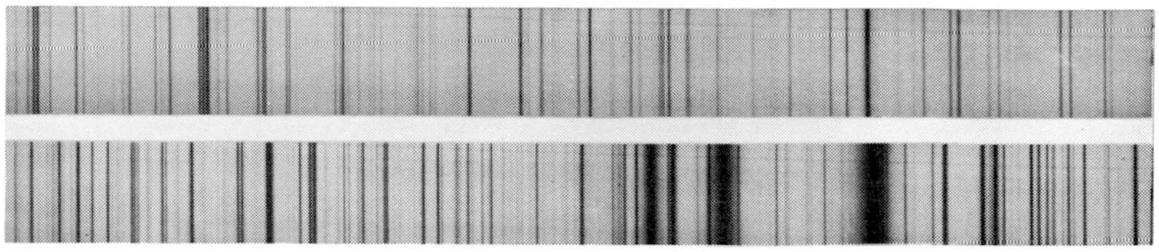

Fig. 3. Two sections of the Fraunhofer spectrum, showing bright continuum and dark absorption lines. The wavelength range covered by each strip is approximately 8.5 nm. The three strongest lines are produced by magnesium; the others, mostly iron. (*Sacramento Peak Observatory, operated by the Association of Universities for Research in Astronomy, Inc.*)

Helium, on the other hand, is inert and radiates only at high temperatures. Thus, when helium lines are seen, the gas is highly excited. In these ways the astrophysicist uses the spectrum lines to interpret the physical conditions in the atmosphere of the Sun.

Photosphere. As mentioned above, the photosphere is the visible surface of the Sun. In the visible wavelengths, its brightness decreases smoothly from the center of the solar disk to the limb. This limb darkening results from the fact that the line of sight to the observer passes through the atmosphere at an increasing angle to the normal as the point of observation approaches the limb. Hence the line of sight penetrates to a lower depth at the center than at the limb, where the path through the overlying material is much longer. Thus what one sees at the limb is a higher point in the atmosphere. The fact that the photosphere is darker at the limb indicates that the temperature decreases outward. Once the range of temperature is known, the variation of limb darkening with wavelength can be used to establish the absorbing properties of the materials at different wavelengths.

By combination of all the information on limb darkening at different wavelengths, it has been possible to construct reasonably reliable models of the temperature variation through the photosphere (Fig. 4).

Granulation. Except for sunspots and accompanying activity, the photosphere is quite uniform over the Sun. The only structure visible is the granulation, an irregular distribution having the shape of bright corn kernels with dark lanes in between (Fig. 5). The grains are quite small, of the order of 1000 km in diameter (about 1.3 seconds of arc as seen from the Earth), and have a life-span of about 8 min. The dark lanes between the granules are about 200 km across. Since the highest optical

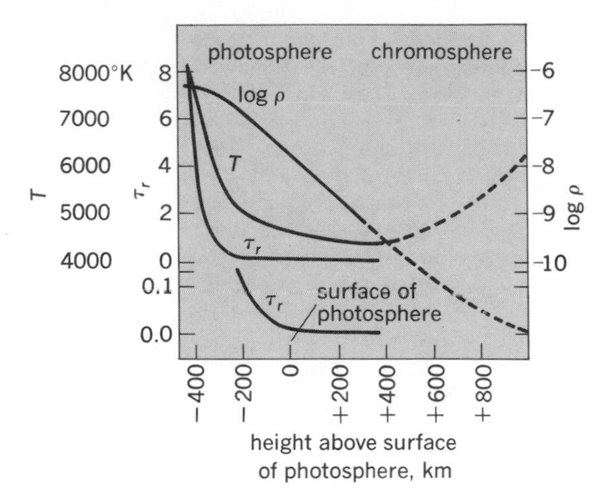

Fig. 4. The variation of temperature T, density ρ, and radial optical depth τ_r with height in the photosphere and (less certainly) in the chromosphere. Lower curve of optical depth shows variation with an expanded ordinate scale.

resolving power employed is of the order of 200–300 km, there may be many more fine lanes crossing the granules and dividing them into even smaller elements.

The granulation is visible evidence of convective activity below the surface. The bright grains are presumably the tops of hot, rising columns which bring energy up from the interior, while the dark intergranular may be the cool downward-moving material. High-resolution spectrograms (Fig. 6) show that each bright granule is marked by a violet displacement of the spectrum lines, indicating an upward velocity. The measured difference in brightness between a granule and an intergranule area is about 15%, indicating an effective temperature difference of the order of 200 K. Photographs of the granulation often show what appear to be distinct chains. However, no one has proved that these are anything other than a random association by the eye of the observer.

In Fig. 6 the brightness variations in the continuous spectrum are presumably the white-light granulation. The complex structure of the absorption lines is easy to see; there are some elements where the absorption lines are entirely missing, and others where tilted features indicate rotation. The brightest granules in the continuum correspond to arrow-shaped shifts to the blue.

Velocity fields. High-resolution Doppler effect measurements have shown the existence of important velocity fields in the photosphere.

These consist of (1) periodic oscillation with an amplitude of about 0.4 km/s and a period of about 250 s; and (2) a cellular flow in supergranules about 30,000 km across, in which material flows outward to the edge of the cells. Superposed on these velocity fields are substantially larger velocity fields in the overlying material of the chromosphere.

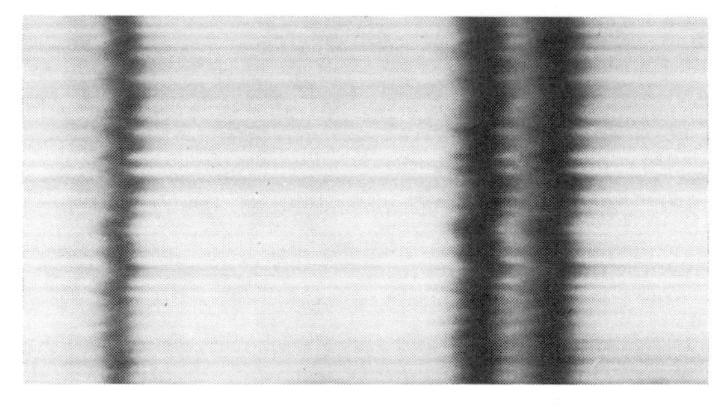

Fig. 6. High-resolution spectrogram at 518.85 nm made by J. W. Evans at Sacramento Peak Observatory, showing lines (left to right): Fe I 518.7922, Ti I 518.8700, and Ca I 518.8848. The total slit length is 120,000 km on the Sun. (*Sacramento Peak Observatory, Air Force Cambridge Research Laboratories*)

The 250-s oscillation has been shown to increase upward in amplitude and velocity, while the period decreases. It is obviously a resonant oscillation of the atmospheric material under the impetus of the convective motions below. Although the 5-min oscillation is the dominant pattern in the photosphere, careful measurement has revealed a whole pattern of weaker oscillations in which the length and period of oscillation are related in a definite way; that is, for a given length scale, only certain periods occur. These are interpreted as normal modes of oscillation of the Sun's envelope. The exact relation of period to scale is determined by the structure of the Sun, and this field of solar seismology makes it possible to probe the inner structure of the Sun for the first time.

The cellular pattern plays an important role in the structure of the upper atmosphere. The systematic outward flow of material in each cell tends to concentrate magnetic fields near the edges. The resulting chromospheric network shows strong magnetic fields distributed around the outside of

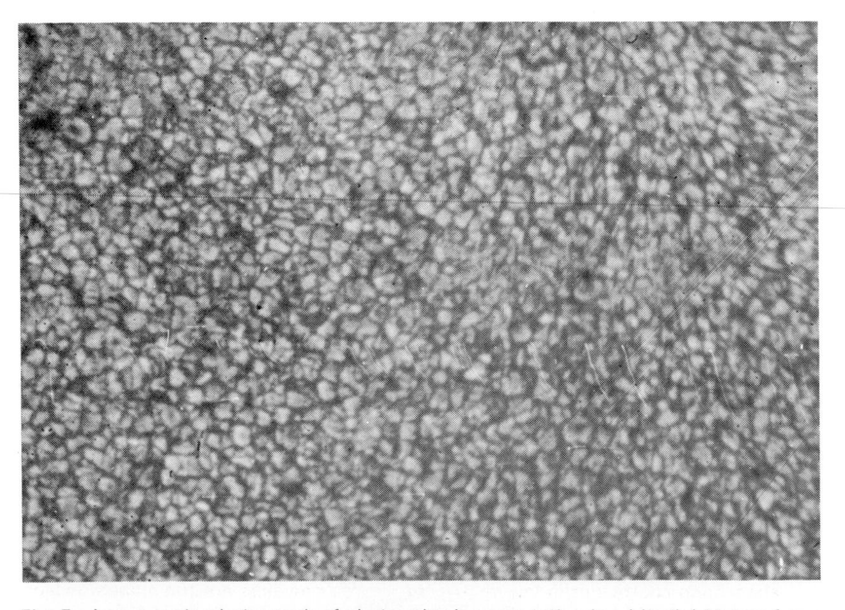

Fig. 5. Large-scale photograph of photospheric granulation in white light taken from an altitude of 80,000 ft (24 km) above sea level. The length of this section is about 55,000 km on the Sun. (*Princeton University Observatory*)

the cells, resulting in enhanced activity at those points.

Chromosphere. The chromosphere was first detected and named by early solar-eclipse observers. They saw it as a beautiful rosy arc which remained visible for a few seconds above the limb of the Moon when the photosphere had been covered. The red color is due to the dominating brightness of the H-α line of hydrogen at 656.28 nm in the chromospheric spectrum, which, except for a bare trace of continuum, is a pure emission spectrum of bright lines (Fig. 7). Because eclipses are rare and the chromosphere can be seen only edge-on at that time, astronomers have developed a special method of photographing the chromosphere in its characteristic lines, particularly H-α. In these spectrum lines the chromosphere is no longer transparent, and one looks at it instead of looking through it to the photosphere. The two principal devices for this observation are the birefringent filter, which utilizes combinations of calcite and quartz plates, and polarizers, which isolate all but a narrow band as small as 0.05 nm; and the spectroheliograph, which builds up a picture by moving the spectrograph across the image of the Sun. With these devices the overpowering background of photospheric light is removed, and only the chromosphere is seen. *See* SPECTROHELIOSCOPE.

The study of the morphology of the chromosphere is best carried out by time-lapse motion pictures. Photographs are made every 10 s or so, and then run through a projector at the normal rate of 16 frames per second. This gives a remarkable picture of the dynamic variations in the chromosphere. Such films of the center and limb of the Sun have provided remarkable pictures of this very complex zone just above the surface of the photosphere.

While the magnetic-field energy in the photosphere is considerably less than the energy of the material, the rapid falloff in density with height results in a situation where the magnetic field dominates the material motions, and the gas is ordered into large-scale patterns. This is not strikingly evident in the low chromosphere (0 – 1500 km), which seems an irregular extension of the underlying photosphere. In fact, the temperature continues to drop above the surface of the photosphere (as evidenced by limb darkening) to an apparent minimum of about 4000 K at a height of 1500 km. Even in the low chromosphere, however, the effect of the chromospheric network in the small bright regions or faculae that mark the edge of each network cell is seen. These faculae may be observed in white light near the limb of the Sun. In them the chromosphere shows a very fine structure of bright points, as though one were looking down at each line of force in the magnetic field (Fig. 8*a*).

If the edge of the Sun is examined in hydrogen light (H-α), the chromosphere is seen as an irregular band 3000 – 4000 km high, from which small jets, called spicules, protrude up to 7000 km. If the wavelength is tuned slightly off the line center, the uniform band disappears because it has only a narrow range of wavelength, and a forest of spicules down to the very surface replaces it. Since the limb is confusing, with many objects in the line of sight, the distribution is better seen on the disk.

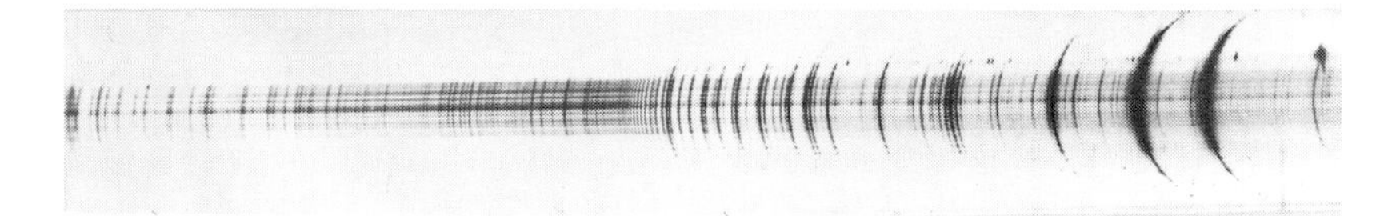

Fig. 7. The flash spectrum of the ultraviolet light from the chromosphere, photographed during an eclipse. This is a negative to enhance details. The strong H and K lines are at the right, the convergence of the Balmer series of hydrogen is near the center, and at the left the lines merge into the Balmer continuum toward shorter wavelengths. (*High Altitude Observatory*)

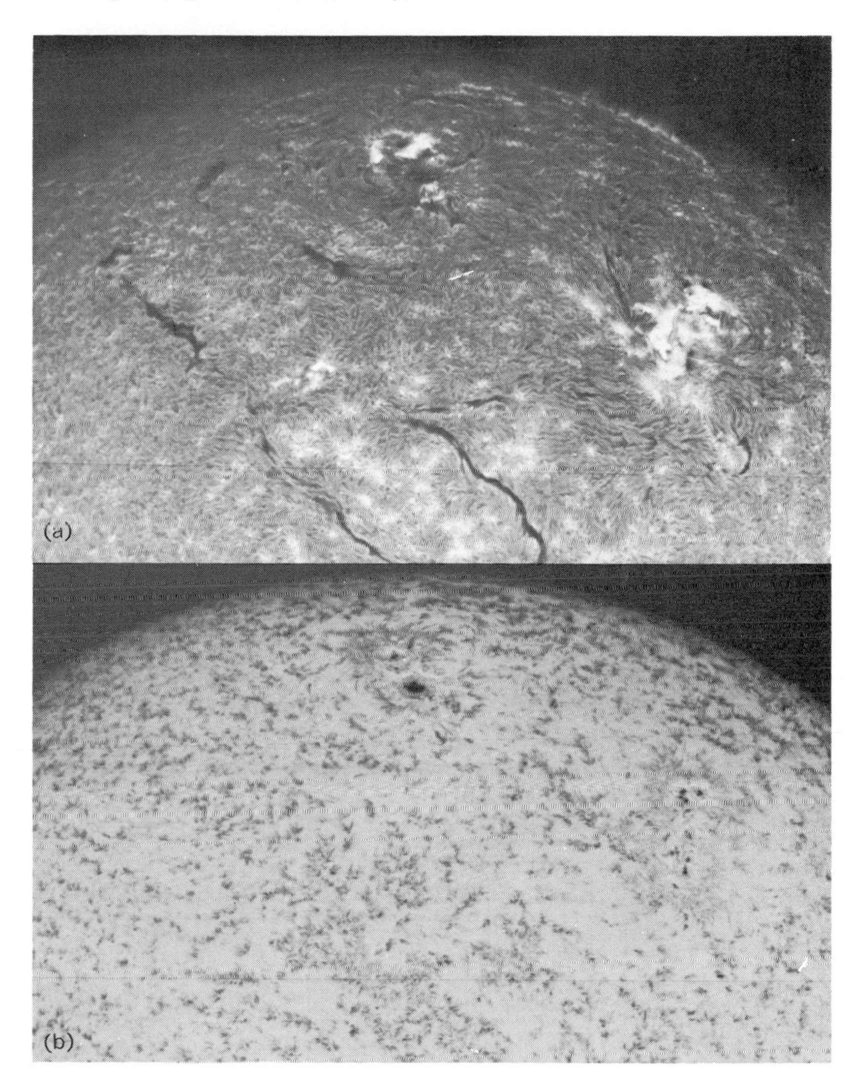

Fig. 8. Spectroheliograms. (*a*) Photocenter of H-α. Dark filaments form the boundary between regions of bright flocculi of different magnetic polarities. The flocculi themselves outline the cells of the chromospheric network and are brighter where the magnetic field is stronger. The fine dark features coming from each bright flocculus are spicules. At the bottom center two filaments converge in a region of weak magnetic activity. Two active regions around sunspots appear, surrounded by bright plages and "whirlpool structure." The dark filaments that curve into the sunspot regions also mark the boundary between fields of opposite polarity. These fields run horizontally along the filaments, as can be judged from the disk structure. The filament near the limb shows a typical bright rim structure underlying it. (*b*) At 0.07 nm showing structures with wide profiles. Only the spicules at the edges of the cells remain. These features are the spicules seen at the limb, and are seen to move with appropriate velocity on disk pictures. They must stand above the surface because they appear invariably as dark areas on the limbward side of flocculi near the limb. Although light from deeper layers in the Sun may reach the Earth off-band (because the absorption coefficient is lower), the structures one sees are the highest, which have the broadest profiles. The sunspot and penumbra appear clearly because one sees through overlying material. A bright rim is evident around the edges of the penumbrae. (*R. B. Leighton and coworkers, Hale Observatories*)

There the spicules are seen to be clumped and to come only from the small bright faculae that mark the chromospheric network. This is even more marked in the wing of the line (off-band), where the chromosphere is transparent, and only the spicules are seen (Fig. 8*b*). The spicules are visible off-band because they are moving rapidly and their spectrum lines are Doppler-broadened. Magnetic measurements show the faculae to have fairly strong magnetic fields, and velocity measurements show a gentle flow from the center to the edge of the chromospheric network cells and a downflow at the edges. Thus the chromosphere can be regarded as consisting of two components, the general chromosphere, which is evenly distributed (hence independent of magnetic effects) and dominates at lower levels, and the spicule component, which is connected with magnetic fields and dominates at greater heights.

Visibility through the Earth's atmosphere and instrumental inadequecies limit one's ability to resolve the structure of the spicules (Fig. 9). They shoot up to a height of about 6000 km above the photosphere with a velocity of about 20 or 30 km/s and then fade out. They may be very narrow, certainly less than 1000 km across. At the top of their trajectory some spicules fade out and others drop back into the chromosphere. It is believed that the spicule jets occur as a result of the focusing of mass motions in the low chromosphere by the strong magnetic fields at the edges of the chromospheric network. They form a channel by which energy travels from below into the solar corona.

Although the spicules were described above as jets, this structure is not certain. The spicules are difficult to observe, and it is possible that they are small flarelike eruptions or something more complicated. In pictures taken in the ultraviolet from Skylab, larger features called macrospicules were seen; H-α pictures from the ground showed small eruptions at their base.

When the solar surface is photographed in the ultraviolet, the hotter regions formed in the transition from chromosphere to corona are observed. These observations show the ultraviolet emission coming primarily from the edges of the network. This may be because the spicules are connected with the heating of the corona, or simply because the density there is higher, because of the spicule eruptions, so that the higher-temperature region is more apparent. It is not known why the chromosphere is hotter than the photosphere, and it may be that the temperature rises only in these network regions. The photosphere shows no variation at all with the chromospheric network, except for the velocity field discussed above.

Physical conditions in the chromosphere can be studied in the ultraviolet and radio ranges. If the emission in microwaves, say 3 cm, is measured, the temperature is found to steadily increase at longer wavelengths, which do not penetrate so deeply. This means the temperature increases outward, and the rate of increase can be estimated. At the edge of the Sun the observed emission should absorb greater (limb brightening) because higher layers should be observed, but the amount of limb brightening is, in fact, fairly small. This probably is because the temperature increase is irregular. Ultraviolet observations from Skylab have imaged the Sun and shown that the high-temperature areas are distributed with the chromospheric network, as well as the active regions around sunspots. Since all these places have strong magnetic fields, it can be inferred that the heating of the upper parts of the solar atmosphere depends on phenomena in regions of strong magnetic fields. Detailed observations of spicules show them to be flarelike, heating the atmosphere by input of magnetic and mechanical energy, and high-resolution observations in spectrum lines of C IV, which can exist only at temperatures above 60,000 K, show continual high-speed (up to 400 km/s) ejections.

Another unexpected discovery from the ultraviolet and radio observations is the extreme sharpness of the transition to temperatures of 10^6 K in the corona above; the whole process takes place in a few hundred kilometers, a small distance on the scale of the Sun. *See* RADIO ASTRONOMY; ULTRAVIOLET ASTRONOMY.

Inside the chromospheric network cells a confusing and interesting mass of small elements exists; the elements oscillate back and forth, almost like water in a bathtub, with velocities approaching 10 km/s and a period of about 180 s. This horizontal oscillation is very easy to see in motion pictures of the chromosphere. It is presumably coupled with the 250-s oscillation in the photo-

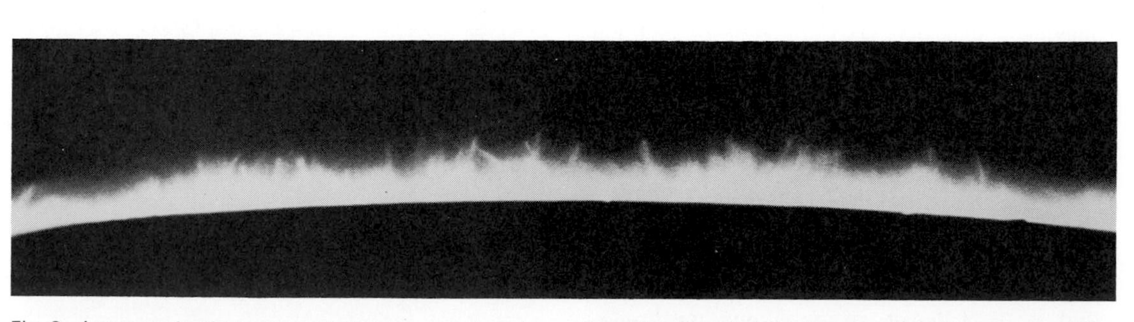

Fig. 9. Large-scale photograph of the chromosphere in H-α light, with the disk of the Sun artificially eclipsed and the hairy spicules projecting above the continuous chromosphere. The length of this section is about 140,000 km. (*Photograph by R. B. Dunn, through 15-in. telescope, Sacramento Peak Observatory, operated by the Association of Universities for Research in Astronomy, Inc.*)

sphere, and plays an important role in the transport of energy upward.

An important tool for the understanding of the transition region is the extreme ultraviolet spectrum as recorded by rockets and satellites (Fig. 10). Although the chromosphere is invisible against the disk in the visible region, in the ultraviolet region its high temperature makes it the dominant contributor, eventually superseded by the still hotter corona at even shorter wavelengths. In the extreme ultraviolet (xuv) spectrum, the lines of such ions as O II, O III, O IV, O V, O VI, C II, C III, C IV, and so forth, are seen. The roman numeral gives the number of electrons that have been removed, less one. From the relative intensity of these lines one can get some idea of the conditions at the successively greater heights in the atmosphere at which they are radiated. Furthermore, from the distribution of these lines on the disk it is possible to determine the temperature distribution. *See* ROCKET ASTRONOMY; SATELLITE ASTRONOMY.

In Fig. 10, the He II image shows the bright chromospheric network. This network disappears in regions at the left and right of the He II image; these regions are coronal holes at the poles of the Sun. Prominences are also visible. The brightest regions are active regions; since the coronal lines are active ones, they show up only in active regions. The brightest area is a flare.

Corona. The chromosphere can be seen only by blocking the bright photosphere, as in an eclipse, or observing in special wavelengths where the photosphere does not emit and the corona, which is even weaker and fainter, requires even more extreme measures. The halo of pearly light seen at solar eclipses (Fig. 11) is so weak that it is difficult to photograph even if the sky is clear. And while the chromosphere can be observed in the light of strong spectrum lines, the lines of the corona in the visible spectrum are so weak they can be detected only at the edge of the Sun. In the extreme ultraviolet and x-ray ranges this situation is different; only hot gases can emit at these wavelengths, so a picture of the Sun in x-rays is a picture of the corona. The x-ray pictures from Skylab (Fig. 12) have thus revealed the inner structure of the corona; unfortunately the radiation from outer regions is too weak to be detected that way. *See* X-RAY ASTRONOMY.

The light seen from the corona originates in three distinct processes which distinguish the F (Fraunhofer), K (Kontinuierlich-continuous), and E (emission) components.

The F corona is not directly associated with the Sun. It is a halo produced by the scattering of sunlight by interplanetary dust between the Sun and the Earth, and is properly regarded as the inner zone of the zodiacal light. Because the scattering is by solid particles, the Fraunhofer spectrum is seen in this light, and hence the name. The F component is negligible compared with the K component in the inner corona, but because of the rapid falloff of density in the corona, the F component dominates beyond 2.5 solar radii. *See* ZODIACAL LIGHT.

The K corona is photospheric light scattered by

Fig. 10. Photograph of the Sun in ionized helium (He II) at 30.4 nm made by Skylab. (*U.S. Navy photo*)

Fig. 11. The solar corona observed about 2 sec after second contact during the eclipse of Nov. 12, 1966, at Pulacayo, Bolivia (altitude 13,000 ft). This photograph, made with a radially symmetric, neutral density filter in the focal plane to compensate for the steep decline of coronal radiance with increasing distance, allows structural features to be traced from the chromosphere out to $4.5R_0$. The overexposed image of Venus appears in the NE quadrant. Typical "helmet" streamers overlie prominences at the SE and SW limbs, while another streamer at high latitude in the NW develops into a narrow ray at large distance. Arches in the corona and the absence of coronal material in domes immediately above prominences are particularly striking at the bases of the NW and SW streamers. A coronal condensation appears on the NW limb of a latitude of about 25°. (*G. A. Newkirk, National Center for Atmospheric Research*)

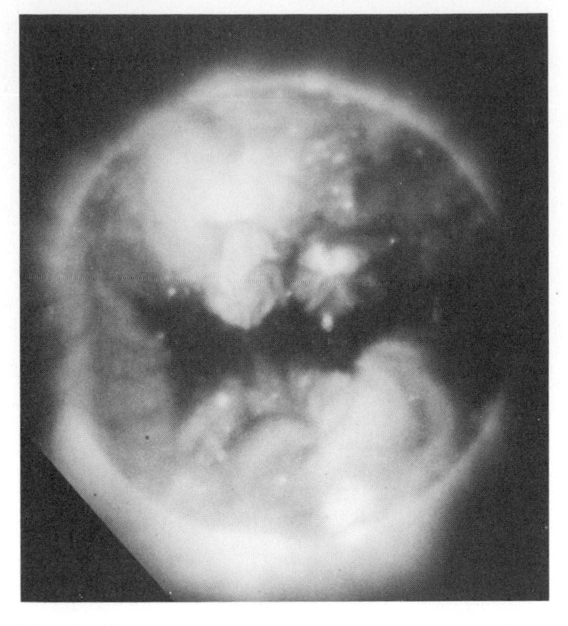

Fig. 12. X-ray photograph of solar corona taken from Skylab with a filter of passband 0.3–3.2 and 4.4–5.4 nm. Large coronal hole extends south from north polar cap. (*Solar Physics Group, American Science and Engineering, Inc.*)

the free electrons in the solar corona. Although this light is not truly emitted by the corona, it is an important tracer of the coronal material, particularly because its intensity is proportional to the electron density and does not fall off so greatly with height as the emission lines and x-rays, which fall off as density squared. Because it has no preferred wavelength, however, the K corona must be observed either during eclipse, or, if the time behavior is desired, from spacecraft, using the dark skies of space. The coronal electrons are moving so fast that the Fraunhofer spectrum lines are blurred and a continuous spectrum is seen. The light is also polarized by the scattering process, and this polarization is used to detect the K corona against the sky background. Because the K corona is produced by the simple process of electron scattering, it is possible to measure directly the electron density (and hence the total density) of the material in the corona. This is done by careful photometric measurements of the coronal brightness at eclipse. Since the material is predominately hydrogen and is entirely ionized, there is essentially one proton for each electron measured, and the density is equal to the proton mass times the number of electrons observed. The density at the base of the corona is found to be about 4×10^8 atoms/cm^3, falling off exponentially with a scale height of 50,000 km. The structure of the K corona varies markedly with the sunspot cycle. At sunspot maximum it presents a fairly symmetrical globular appearance with many domes and streamers in all directions. At sunspot minimum the globular shell shrinks toward the solar surface and individual dominant streamers are seen. At the poles there are brushes of small, symmetrically diverging streamers reminiscent of an iron-filings pattern. It is this appearance which suggested the presence of a general solar magnetic field.

The electron and temperature distribution in the corona may also be studied with radio telescopes. The propagation of radio waves through the corona is such that at each frequency one can see only down to a point characterized by the plasma frequency, which increases with density. At successively higher frequencies one sees deeper and deeper into the corona. Therefore the brightness at each frequency indicates the temperature at that height. It is possible, therefore, to map the temperature as a function of height. Further information may be obtained from the distribution of brightness across the solar disk in radio radiation. At high frequencies it is possible to see completely through the corona to the chromosphere, which appears as a uniform disk of about 6000 K. At longer wavelengths the corona is no longer transparent in the slanting direction of the limb, and therefore a peak of brightness near the edge of the Sun is seen. At the lower frequencies, where the radiation does not penetrate the corona, the Sun appears much larger than it does in visual wavelengths. *See* RADIO TELESCOPE.

The E, or emission-line, corona (Fig. 13) is the true emission of the ions in the corona. It is therefore the best source of information on the physical state of the corona. Originally the term E corona referred only to the emission lines in the visible, but it is now taken to include all the intrinsic radiation coming from processes in the corona: emission lines, continuous emission produced by interaction of free electrons and ions, and other minor sources. The emission lines are best observed from spacecraft in the XUV, where the photosphere is weak and the lines can be seen against the disk.

In spite of rather formidable technical difficulties, the coronagraph makes it possible to record the spectrum of the E corona in the visible, and even to photograph the structure directly through a birefringent filter whenever the sky is sufficiently clear. Since observations of the K corona outside of eclipse are limited to rather low resolution, the emission-line corona is an important tool in the observation of the day-to-day variation of the corona. The spectrograph dilutes the sky light by spreading it into a long, continuous spectrum. The coronal emission lines, on the other hand, are merely separated by the dispersion without dilution and stand out conspicuously against the sky

Fig. 13. Typical stable form of structure of the E corona in the light of the green line of Fe XIV photographed without an eclipse through a birefringent filter. (*Sacramento Peak Observatory, operated by the Association of Universities for Research in Astronomy, Inc.*)

continuum. Figure 14 shows a short length of spectrum with a green 530.3-nm line of the corona and lines of a bright prominence superposed on the Fraunhofer spectrum of a sky scatter. The K corona is so weak that it cannot be seen. The emission lines are proportional to the square of the density and fall off sharply with increasing height. *See* CORONAGRAPH.

The history of the identification of the coronal emission lines is a remarkable chapter in astronomy. The strongest, the Fe XIV green line, was independently discovered by W. Harkness and C. A. Young during the 1869 eclipse. This, and weaker coronal emission lines subsequently discovered, corresponded to no known spectrum lines, and were collectively attributed to the unknown element, coronium. More than 70 years later, they were identified by W. Grotrian and, more extensively, by B. Edlen as forbidden lines from highly ionized atoms which could be excited only under conditions of high temperature, low density, and enormous volumes quite beyond any conceivable laboratory resources.

The lines of the E corona in the visible are due to highly ionized Fe, Ni, Ca, or Ar. These complex ions have levels between which magnetic dipole transitions occur which have a relatively long wavelength for such highly ionized matter. In the ultraviolet the normal spectral lines of these and other ions may be observed. Typically the ions of a given element will be ionized down to the member of heliumlike sequence (C V, O VII, Ne IX) which requires great energy to ionize it further. The heliumlike ions are hard to excite; their lines are seen in the soft x-ray range. When the heliumlike ion captures an electron, even briefly, the lithiumlike ion is formed; the odd electron is very active spectroscopically, and the strongest lines in the ultraviolet are $2s$-$2p$ transitions in lithiumlike ions such as O VI and Ne VIII. The berylliumlike sequence is also strong. In solar flares, very highly ionized species up to Fe XXVI have been seen.

If the spectrum of the E corona is analyzed, the ions present are found to have ionization potentials of 250 to 350 eV, and are common at temperatures from 1.5 to 2.5×10^6 K. Measurement of the radio emission and the width of the spectrum lines confirms this.

The high temperature of the corona is a fascinating problem. Obviously the temperature of a star is highest at the center and decreases outward to the edge. One can only guess that the steep increase in temperature in the corona is due to boundary effects connected with the steeply decreasing density at the edge of the Sun and the convective currents beneath it. These, together with magnetic fields, give a "crack of the whip" effect in which the same energy is concentrated in progressively smaller numbers of atoms. The result is the production of the high temperature of the corona. It is now known by various indirect means that coronas are common in G and K stars like the Sun, and that many have enormously extensive coronas compared to the Sun. Many have been detected to be x-ray sources, similar to but much stronger than the solar corona. *See* STAR.

Solar wind. One of the most remarkable surprises of the age of rockets and satellites has been the discovery of the existence of the solar wind. This continual outflow of matter from the Sun was predicted by Eugene Parker on the basis that the high temperature in the corona must lead to a rapid outflow at great distances from the Sun. He predicted a velocity near 1000 km/s. Spacecraft such as *Mariner 2* and the IMP series have detected a continual flow of plasma from the Sun with a velocity ranging from 300 to 500 km/s and density about 1 atom/cm³ near the Earth. The flow occurs because the conductivity in the corona is so high that high temperatures exist at some distance from the Sun, where gravity can no longer hang on to the material. The particles flow along a spiral path dictated by magnetic fields from the Sun carried out into the interplanetary medium. The rotation of the Sun produces the spiral pattern. The magnetic field near the Earth is measured to be rather uniform over large sectors of the Sun, corresponding to one dominant polarity or another. The solar-wind flow has a continual effect on the upper atmosphere of the Earth.

Fig. 14. Green line of Fe XIV in E corona (long arc at left) with bright metallic lines of a short prominence, and yellow chromospheric helium line D_3 (arc at right), photographed through a small coronagraph. (*Sacramento Peak Observatory, operated by the Association of Universities for Research in Astronomy, Inc.*)

Coronal holes. Early coronal observations showed that the corona was occasionally not visible over certain regions. In particular, most of the time it was quite weak over the poles. The x-ray pictures, particularly the long sequences from Skylab, revealed great bands of the solar surface essentially devoid of corona for many months. These proved to be regions where the local magnetic fields were connected to quite distant places, so the fields actually reached out to heights from which the solar wind could sweep the gas outward. The poles, of course, were the extreme case because the field lines reached out to the other pole of the Sun, but of course were swept away by the solar wind. Analysis of solar wind data showed that equatorial coronal holes were associated with high-velocity streams in the solar wind, and recurrent geomagnetic storms were associated with the return of these holes. Thus the great intensity of the corona over sunspot regions is partly due to their strong, closed magnetic fields which trap the coronal gas. Coronal holes are particularly prominent in the late stages of the sunspot cycle.

SOLAR ACTIVITY

The foregoing section has been concerned with the structure of the normal steady state of the undisturbed Sun. There are, in addition, a number of transient phenomena known collectively as solar activity. These are all connected with sunspots or their remnants, and wax and wane in a remarkable

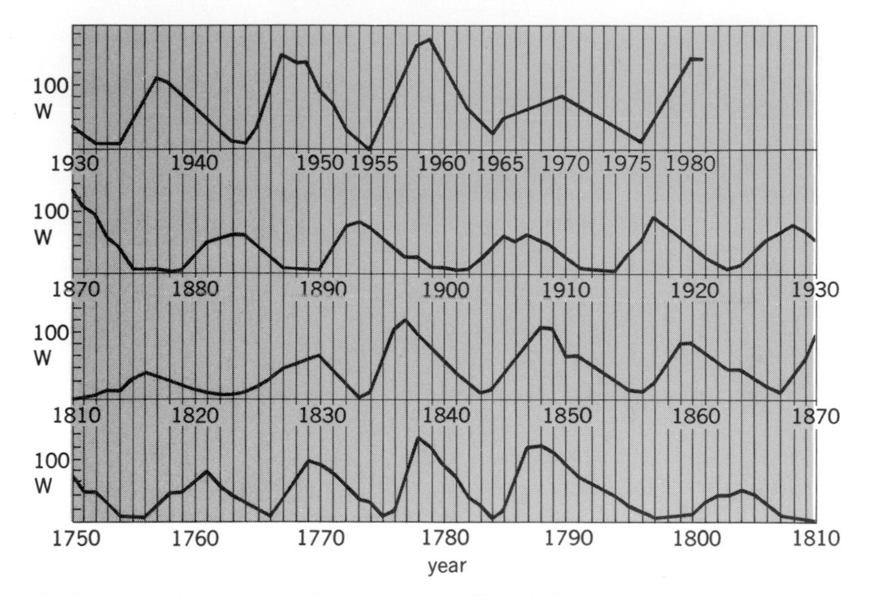

Fig. 15. Annual mean sunspot number from 1750 to 1980.

cycle of activity. The sunspot cycle consists of variations in the sizes, numbers, and positions of the sunspots, expressed quantitatively as the sunspot number (Fig. 15). The number of sunspots peaks soon after the beginning of each cycle and decays to a minimum in 11 years. The magnetic polarity of the sunspot groups reverses in each successive cycle so that the complete cycle lasts 22 years. The first spots of a cycle always occur at higher latitudes, between 20 and 35°, and as the spots increase in size they occur closer to the equator. Almost no spots are observed outside the latitude range of 5 – 35°. The average duration of the cycle from the first appearance of high-latitude spots to the disappearance of the last low-latitude spots is nearly 14 years. The difference between these two figures is the result of overlap of consecutive cycles. Thus there is hardly ever a time when there are no spots on the Sun at all. In the cycles peaking in 1958 and 1969, there was a great dominance of sunspots in the northern hemisphere, an asymmetry which has since disappeared. For periods of time there were few or no spots in the southern hemisphere. In addition, the two cycles with maximum activity in 1946 and 1957 were the two largest in history. Whether these facts are connected is not known.

Maunder minimum. Sunspots were discovered by C. Scheiner and by Galileo around 1610, and were extensively observed afterward. They can easily be seen with a small telescope or with the camera obscura, a dark room with a pinhole. But between 1650 and 1715 very few spots were seen, although experienced observers searched for them. After 1715 the spots returned. This period, first noted by E. W. Maunder, was associated with a long cold spell in Europe, known as the Little Ice Age. A number of secondary effects, particularly in carbon-14 in tree rings, confirm the low solar activity level at that time. Whether or not the Maunder minimum caused the Little Ice Age cannot be established, but there is little doubt that the sunspots dropped out during this period (although a weak 11-year period remained in the few spots present). So it is quite possible that the present era is a period of relatively high activity. Compared to stars of the Sun's class, it is about average.

Sunspots. Sunspots are the most conspicuous features of solar activity and are easily seen through a small telescope. They are dark areas on the Sun produced by the most intense magnetic fields. The magnetic fields produce a cooling of the surface, most likely by suppression of the normal convection which transports energy from the lower levels. Deprived of this supply of heat, the area cools by radiation and becomes a dark sunspot. *See* SUNSPOT.

Curiously enough, the region close to the cool sunspot is the scene of the hottest and most intense activity because of the great magnetic energy present in the sunspot.

The birth of sunspots is marked by the appearance of pairs of small spots about 20,000 km apart. In the light of the H-α line the chromosphere is seen to be particularly bright, with dark arches connecting the members of a pair. The arches mark lines of force connecting the spot fields. These fields are almost always in a particular ori-

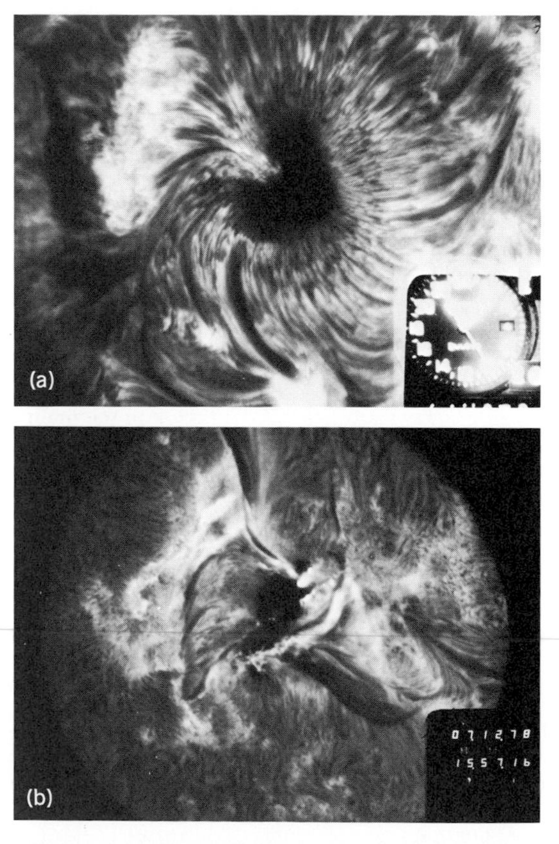

Fig. 16. Sunspots photographed in H-α light. In these photographs it is not possible to see down to the granulation. At this higher level the matter is dominated by field lines that delineate the magnetic configuration. (*a*) Large symmetric sunspot. Some lines of force terminate in the following plage (below) and some radiate outward. (*b*) Highly active sunspot with twisted field lines connected to satellite polarity all around it. This spot group produced huge flares. (*Big Bear Solar Observatory*)

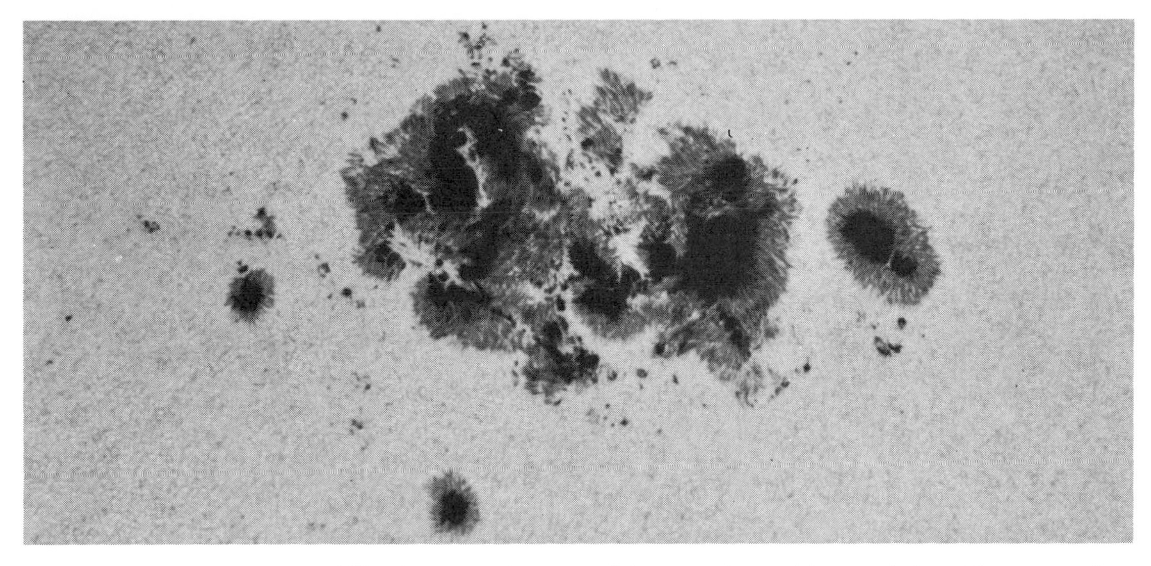

Fig. 17. Large sunspot group of May 17, 1951, photographed in white light showing the filamentary structure of the penumbra and the granulation of the surrounding surface. (*Hale Observatories*)

entation; in any given cycle, almost all the spot groups in a hemisphere will have the same magnetic polarity leading as the Sun rotates east to west in 27 days; in the other hemisphere the spot polarities will be exactly opposite. The fields are therefore designated p if they are the normal preceding polarity for that hemisphere, and f if they are the normal following magnetic polarity. In the next spot cycle the polarities reverse. If an emerging spot group has the wrong polarity, it usually dies out quickly but occasionally grows into a very great spot group.

As the sunspot group grows, remarkable things happen. The p spot rapidly moves westward while the f spot remains fixed. In most cases the f spot dies out, leaving a bright plage, while the p spot grows into a mature, round spot (Fig. 16). If there are little spots around, they merge into the main p spot. After a few days the p spot stops separating, and the sunspots no longer move. Often if the region is born with a tilt to the east-west line, the p spot will drag it into conformity with the rules. Doppler shifts in the dark arches show material moving upward at the center and flowing downward at the ends; this must result from magnetic tubes pushing up from below. When the sunspots form, the dark arches disappear, but they may remain at the center of the group to mark flux still emerging.

The typical spot group grows in a few days to the configuration of p spot and f plage, lasts a week or two, and disappears, leaving two puddles of p and f magnetic fields marked by weak plages. But occasionally a large spot group appears (Fig. 17), usually marked from the start by rapid growth of many small spot pairs, sometimes by new eruption of field in an older spot group. The new flux always emerges in pairs, because all magnetic fields must occur in pairs. As new flux pushes its way into old, great stresses and shears occur, which are relieved by flares.

A typical mature sunspot is seen in white light to contain a central dark area, the umbra, where the magnetic field is strong and vertical, surrounded by a less dark band called the penumbra, where the magnetic field spreads out radially, forming an aura of dark fibrils across the granulation (Fig. 18). The umbra appears dark because it is quite cool, only about 3000 K compared to 6000 K in the photosphere. The spot pressure, consisting of magnetic and gas pressure, must balance the outside pressure; hence the spot must somehow cool by magnetic effects until the inside gas pressure is

Fig. 18. Sunspot photographed in white light with extremely high resolution. Granulation at top is separated by fine dark lanes; the granules are about 1000 km across. At lower right is a sunspot umbra and in between is the penumbra, where fine fibrils mark the radial force lines. (*Big Bear Solar Observatory*)

considerably lower than the outside.

Observations of the chromosphere above the sunspot reveal the presence of strong oscillations, called umbral flashes, with a 150-s period. In the penumbra, running waves spread radially outward with a period of 300 s—roughly twice that of the umbral flashes—and a velocity of about 10 km/s.

There is a strong tendency for sunspot activity to break out all over the Sun at the same time. This enhances the idea that there is a general mechanism for their growth. There also may be preferred meridians for their occurrence, but the evidence is weak.

By means of the Babcock magnetograph, which permits the measurement of weak magnetic fields on the Sun, it has been possible to study the connection between the intense magnetic fields in the sunspot groups and the weak magnetic fields distributed over the surface (Fig. 19). The weak fields are concentrated on the edges of the chromospheric network, which consists of cells 30,000 km across. The fields vary considerably in strength from cell to cell. The origin of these magnetic fields appears to be the big sunspot groups. As the spots decay or are torn apart by motions in the solar atmosphere, their magnetic fields spread out and drift toward the poles. Because the poles rotate more slowly than the equator, the fields lag behind each sunspot group in a large region of one magnetic polarity shaped like the wing of a butterfly. Sometimes these regions may extend 90–120° in longitude. Their polarity is normally that of the following sunspot polarity. As time goes by, the weak fields reach the pole and establish there a dominant magnetic polarity which, because it is the same as the following spots, is opposite to the polarity of the preceding sunspots which dominate most spot groups.

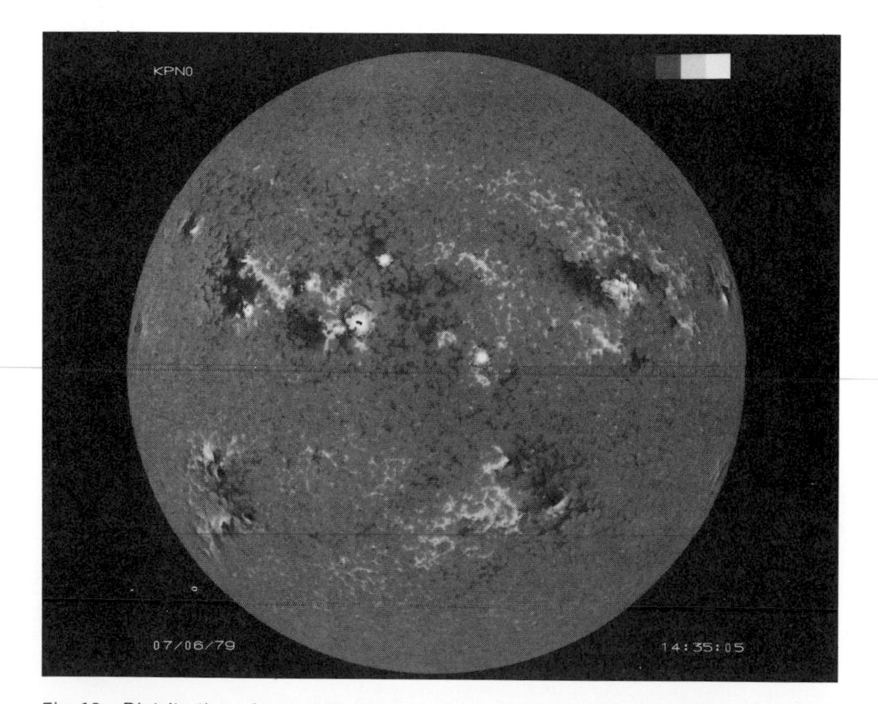

Fig. 19. Distribution of magnetic fields on the Sun at a time of high solar activity (July 6, 1979). Black areas are regions where the magnetic field points away from the Earth; white areas, toward the Earth. The strength of the field is indicated by the degree of blackness or whiteness. The north rotation pole of the Sun is at the top. (*Kitt Peak National Observatory*)

Thus, a few years after the outbreak of the sunspot cycle, the fields generated by the broken-up spots reach the poles, producing at each pole magnetic fields of polarity opposite to those of the preceding sunspots in each hemisphere. Some models of solar activity propose that, as the sunspots of the parent cycle die away, this large-scale dipole field is amplified by the differential rotation to produce a new cycle of sunspots, which naturally will be opposite in polarity to the preceding cycle.

The large unipolar magnetic regions spreading out from centers of activity are easily recognized on H-α or calcium monochromatic pictures of the Sun. This is because the boundaries between these regions, where the magnetic field becomes horizontal in order to change direction, are usually marked by large prominences, accumulations of material in the atmosphere supported by the horizontal field. These are seen as dark filaments, such as in Fig. 8. Because the unipolar regions are areas of enhanced magnetic field, they also correspond to enhanced emission in the chromospheric network, and thus can be distinguished in calcium spectroheliograms.

There are models for the evolution of solar magnetic fields and, in turn, the fields produce the sunspots and all their effects. However, these explanations mostly rest on differential rotation, since there is very little understanding of why the equator of the Sun should rotate so much more rapidly than the poles.

Ephemeral active regions. The picture of the sunspot cycle has been confused by the discovery of ephemeral active regions. These are tiny active regions which last only a day on the average and occur all over the Sun. The distribution of active regions in lifetime continues over to these small regions, until most of the flux erupting on the Sun at any one time is in the form of ephemeral regions; but they are so small and short-lived that they have no lasting effect on the field configuration. Ephemeral regions are best visible in x-ray pictures, where they appear as x-ray bright points (Fig. 18). At sunspot minimum they are quite prominent, but at maximum they disappear in x-ray photographs, although they still can be found in magnetic data. So it is possible that the sunspot cycle simply governs the maximum size to which ephemeral regions may grow.

Prominences. Although prominences appear dark against the disk, they appear bright against the dark sky. They occur only in regions of horizontal magnetic fields, because these fields support them against the solar gravity. Thus filaments on the disk are good markers of the transition from one magnetic polarity to the opposite. As can be seen from Figs. 20 and 21, prominences are among the most beautiful of solar phenomena. The spectra of prominences show a number of bright emission lines of various elements, mostly singly ionized (Fig. 22). Analysis of these lines shows that long-lived, stable prominences have a temperature of about 6000 K, while transient prominences connected with flares show many fewer lines and are over 30,000 K.

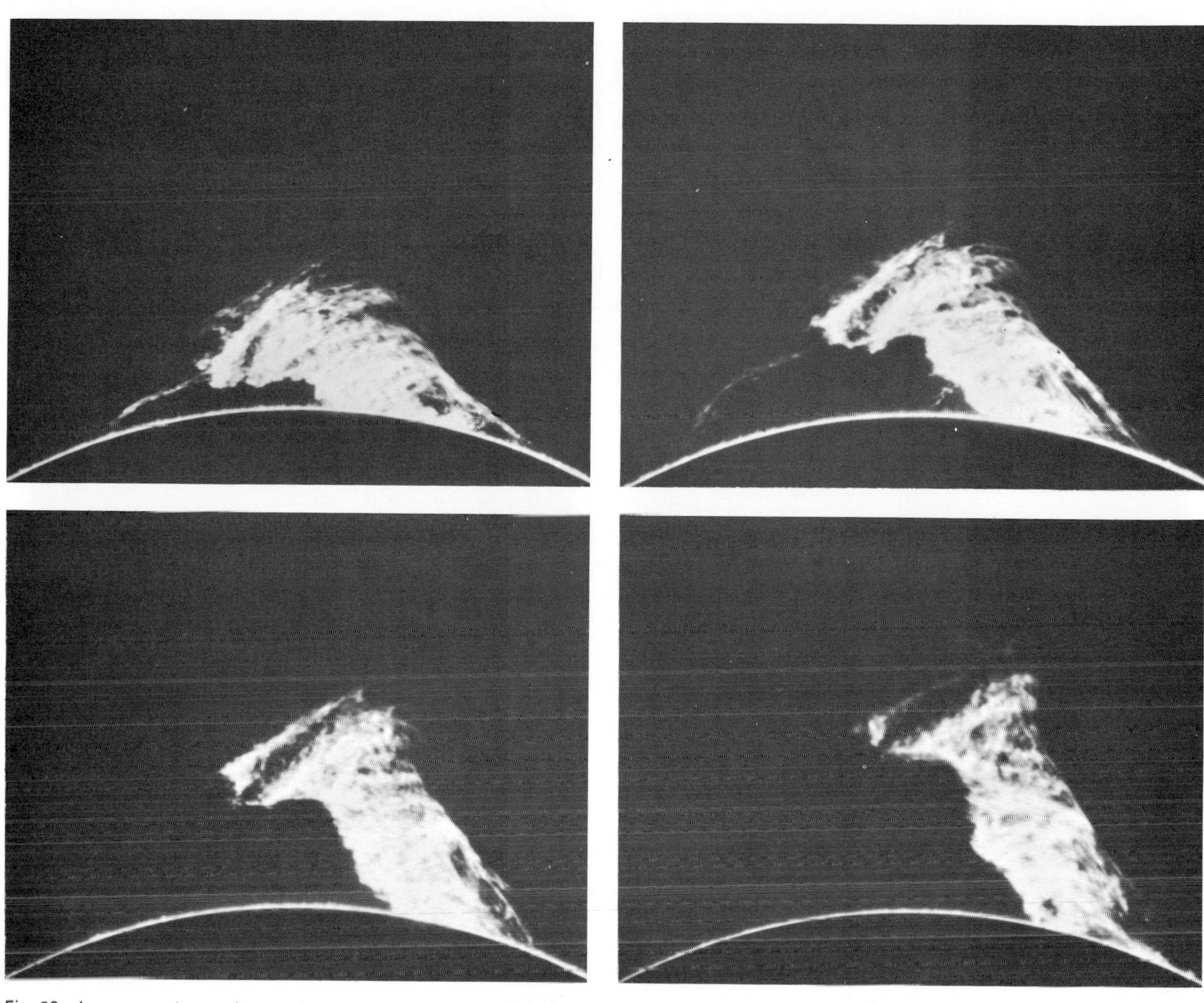

Fig. 20. Large prominence in eruptive phase after days or weeks of static inactivity. Four frames cover interval of 29 min. Horizontal width of single frame is about 670,000 km. H-α light. (*Sacramento Peak Observatory, operated by the Association of Universities for Research in Astronomy, Inc.*)

Plages. Just as prominences occur when the magnetic field changes from one sign to the other, plages occur whenever the magnetic field is vertical and relatively strong. They are bright regions visible in any strong spectrum line (Fig. 23). It is now believed that the field is, in fact, stable only when it exceeds 1000 gauss (0.1 tesla), and the difference between apparently weaker field regions, such as the chromospheric network, and plages is only the density of clumps of strong field. Plages are normally associated with sunspots; a typical active region will have the preceding magnetic field clumped in a sunspot and the following field spread out in a plage. In H-α light, the plage is seen to be connected to the sunspot by dark fibrils outlining the lines of force (Fig. 24); since the magnetic fluxes must be equal, the field in the plage is related to that of the spot by the ratio of their areas. The plage is bright in almost all wavelengths —x-ray, radio, ultraviolet—although in x-rays the peak brightness is in the connecting arches. The elements of the chromospheric network resemble small plages. If the field is just as strong as a plage, but horizontal, there is no plage brightening; therefore the vertical field plays a most important role in the heating of the plage and corona.

The appearance of plages in different wavelengths (Fig. 25) shows clearly the growing dominance of the magnetic fields with increasing height. If the energy of the magnetic field is compared with the kinetic energy of the gas in the low photosphere, only the magnetic fields inside sunspots are strong enough to dominate material motions. But in the chromosphere, the density drops so rapidly that the magnetic fields spreading out from the sunspot easily dominate. As a result little or no whirlpool structure is seen in white-light photospheric pictures, but it is easily seen in H-α, which reveals the chromospheric structure higher up. In white light the plages are visible only near

Fig. 21. Characteristic small prominence, showing fibrous structure. The generally vertical filaments are paths of downward moving material. Horizontal width of a single frame is 125,000 km. H-α light. (*Sacramento Peak Observatory, operated by the Association of Universities for Research in Astronomy, Inc.*)

the limb, where one sees higher in the photosphere. They are easily seen in H-α, but are even more widespread in calcium K-line spectroheliograms because of enhanced contrast in the violet. With rockets, spectroheliograms can be made in the high-ionization lines showing emission mainly from active regions and plages. These are the only regions in the atmosphere hot enough to produce considerable ultraviolet radiation.

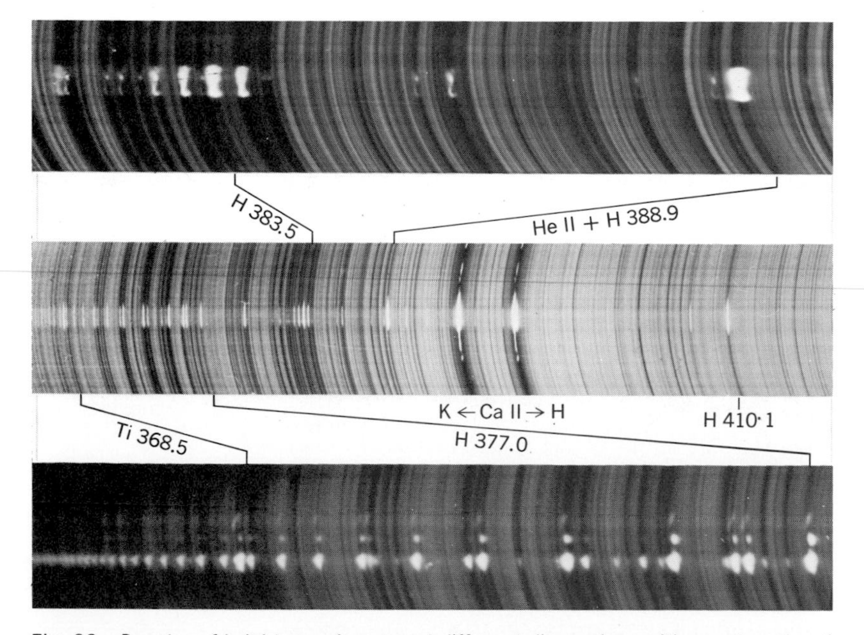

Fig. 22. Spectra of bright prominences at different dispersions with wavelengths of corresponding lines. From top to bottom the wavelength range covered by each spectrum is 8.8, 54.5 and 12.9 nm. The converging Balmer series merges into the Balmer continuum in bottom spectrum. (*Sacramento Peak Observatory, operated by the Association of Universities for Research in Astronomy, Inc.*)

Although sunspots are always accompanied by plages, the reverse is not true. Occasional fields of small plages develop in the sunspot zone and fade away without the appearance of any spots. In these cases the magnetic fields never reach sufficient strength to generate sunspots. Similarly, the plages usually remain for a few weeks to mark the location where a sunspot has died. Eventually they probably break up and spread out into the chromospheric network, but this has not been observed directly.

Flares. The most spectacular activity associated with sunspots is the solar flare (Fig. 26). A flare is defined as an abrupt increase in the H-α emission from the sunspot region. The brightness of the flare may be five times that of the associated plage; the rise time is seldom longer than a few minutes, sometimes only 10 s. Of course, the hydrogen brightening is only a symptom; observations by other means show that there is a tremendous energy release, of as much as 10^{33} ergs (10^{26} W) in a large flare. An active sunspot group will show many small flares the size of a large sunspot. Once or twice in the lifetime of a large group, a great flare will occur covering the entire region. The energy released in the three or four great flares each year equals that in all the small flares put together. The flares are ranked by area in three classes, with class 3 the greatest; there is also a class of subflares for smaller brightenings which are only marginally flares.

Solar flares occur only as the result of sunspot activity. Practically all sunspots produce some flares, but certain spots are prodigiously active. The source of the flare energy is the magnetic fields surrounding the sunspots. Every magnetic field has a certain minimum energy form, normally that where no currents flow, a so-called potential field. But if the fields have been strongly twisted, currents are set up which stabilize the twisted fields until they become unstable and jump to lower energy states, producing flares.

Complex magnetic fields occur only as the result of sunspot motion or emergence. But sunspots move only when they are emerging; hence all of the complexity results from flux emergence. Because the density of the solar atmosphere is low and its conductivity is high, magnetic field lines are very sticky. If a new spot emerges in old magnetic fields, it cannot connect to them immediately, but steep gradients are set up which eventually are released in the flare cataclysm. As mentioned above, when a new dipole emerges, the preceding sunspot runs forward, pulled by unseen forces below the surface. If the emergence is near an old spot and the p spot moves into like polarity, little will happen. If it hits a p spot, the two will merge. But if the field it runs into has opposite polarity, there will be a strong tendency for the new p spot to exchange its previous ties to connect to the local f spot. The boundary between these fields never takes a simple form, but usually is marked by a filament (a prominence seen against the disk), which marks a boundary with field lines running at right angles to the line between the spots. Of course, a lower-energy configuration would have the spots directly connected, but this never occurs. Eventually the filament becomes unstable and a flare occurs. All solar flares involve filaments, and one of the best flare warnings is a sudden lifting of a filament, marking a strong magnetic gradient.

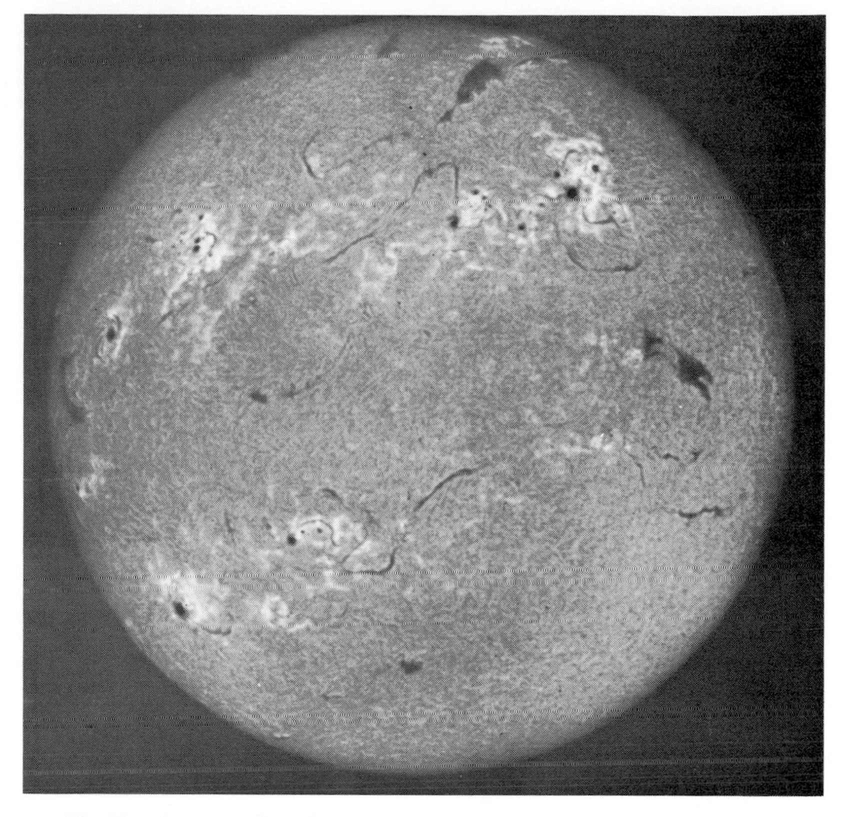

Fig. 23. The disk of the Sun photographed in H-α light, showing bright plages in centers of activity and dark filaments (prominences). (*Sacramento Peak Observatory, operated by the Association of Universities for Research in Astronomy, Inc.*)

Another source of flare energy is the decay of old sunspot fields. If the fields are not simply connected but are marked by large filaments and sheared magnetic fields, a filament will usually erupt in a large, slow flare as the fields decay.

Flares are rarely seen in white light because they are an atmospheric phenomenon and their density is so low that they are transparent. On the other hand, their temperature is so high that, in the

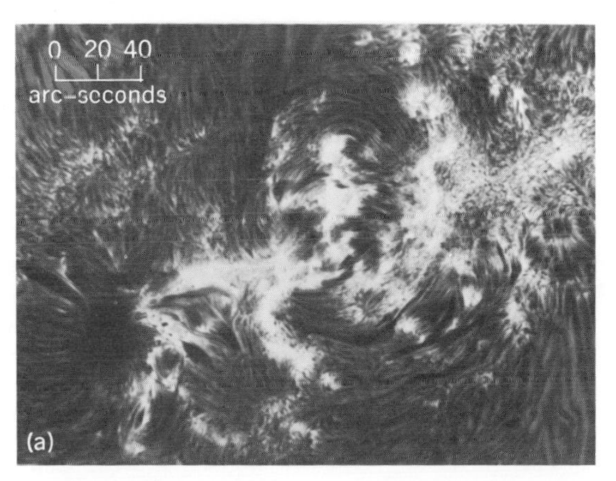

Fig. 24. Two photographs of the same active region matched to scale. The dominant p sunspot is on the left, with following plage on the right. (*a*) Photograph in H-α light. All areas of strong field are bright (except the sunspot), and all boundaries between opposite polarity are marked by dark arched fibrils. (*b*) Photograph of the magnetic field. White polarity is preceding, dark polarity following (the spot itself is saturated). The small inclusions of dark polarity near white are typically the loci of solar flares. (Large flares only occur with much more complex fields and sharper gradients.) (*Big Bear Solar Observatory*)

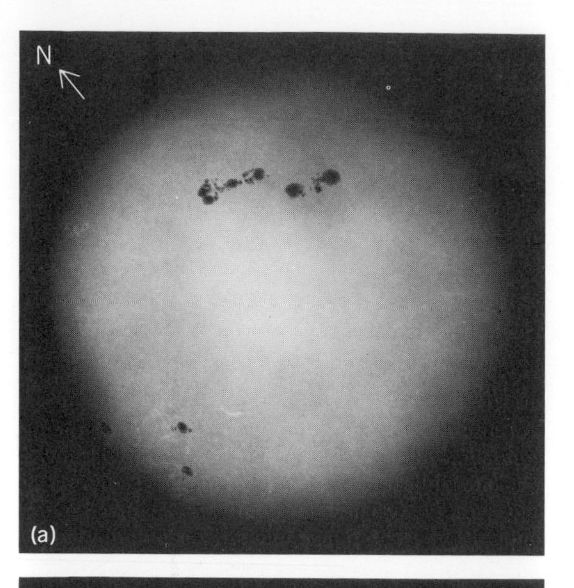

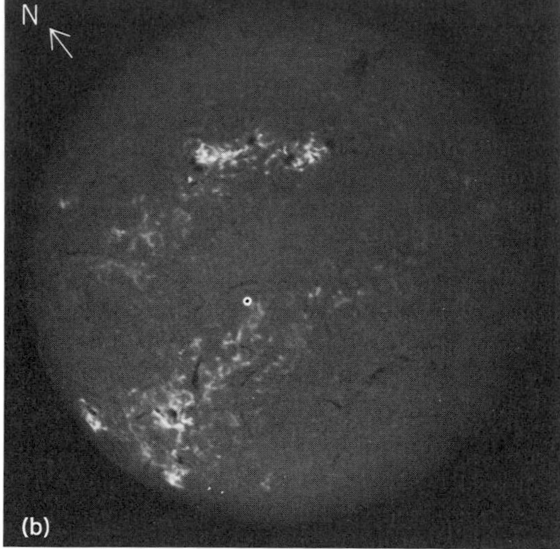

Fig. 25. The Sun photographed on April 8, 1980, in (*a*) white light and (*b*) red H-α light, showing the appearance of plages in different wavelengths. Plages can be seen in white light near the limb. The northern (upper) hemisphere shows two large bipolar groups which had limited flare activity because the magnetic configuration was simple. New groups are rotating onto the disk at lower left. (*Big Bear Solar Observatory*)

ultraviolet region, they may equal the intensity of the entire Sun. However, because the flares are most easily observed in H-α and because the H-α brightening is an extremely accurate indicator, they are mostly studied in this way.

It has become possible to observe the flare phenomenon over a great range of wavelengths, in which many diverse phenomena have been observed. In x-rays a sharp pulse of radiation is observed, with photon energies of 100,000 eV and higher. At lower x-ray energies, the time behavior of the pulse is not as steep. The x-ray pulse always occurs during the first sharp brightening. Simultaneously, a broad-band burst of radio waves is emitted in the microwave region, peaking around 3000 MHz. This is synchrotron emission from the same energetic electrons which cause the x-ray burst. At longer wavelengths, more complex phenomena (Fig. 27) are observed associated with the outward movement of disturbances from the flare through the corona. The corona at each point radiates a certain frequency depending on the density; thus as disturbances pass upward, radiation is excited at successively lower frequencies. The most intense of these is the type II burst, with a downward drift in frequency during a period of 2 to 5 min; this corresponds to an outward velocity of about 1000 km/s and is seen only in the largest flares. The type III bursts are much more frequent and show a very rapid frequency drift corresponding to disturbances propagating in the corona at 100,000 km/s and more. Some type III bursts are observed to turn downward, as though reflected by magnetic fields in the atmosphere. These are called U bursts. Following a large type II burst, one occasionally observes a type IV radio burst or storm which is a long-lasting wide-band emission from a great cloud in the corona. By analogy, the intense broad-band microwave burst accompanying the x-ray emission is called microwave type IV. The outward-moving type III bursts also produce an afteremission called type V emission.

Any large sunspot group is accompanied by a radio noise storm, and the detailed bursts of this storm are normally called type I bursts. They are strongly polarized by the coronal magnetic fields.

In addition to the x-ray and radio emissions from flares, which are all electromagnetic radiation, great numbers of energetic particles, both nuclei and electrons, are observed. Since energetic particles must follow the magnetic fields, the corpuscular radiation is not observed from all big flares, but only from those favorably situated in the western hemisphere of the Sun. Because of the solar rotation, the lines of force from the western side of the Sun (as seen from the Earth) lead back to the Earth and guide the flare particles to the Earth. The nuclei are mostly protons, because hydrogen is the

Fig. 26. The great "sea horse" flare of August 7, 1972, late in the flare, photographed in the blue wing of the H-α line. The neutral line between two bright strands is crossed by an arcade of bright loop prominences raining down from the corona. (*Big Bear Solar Observatory*)

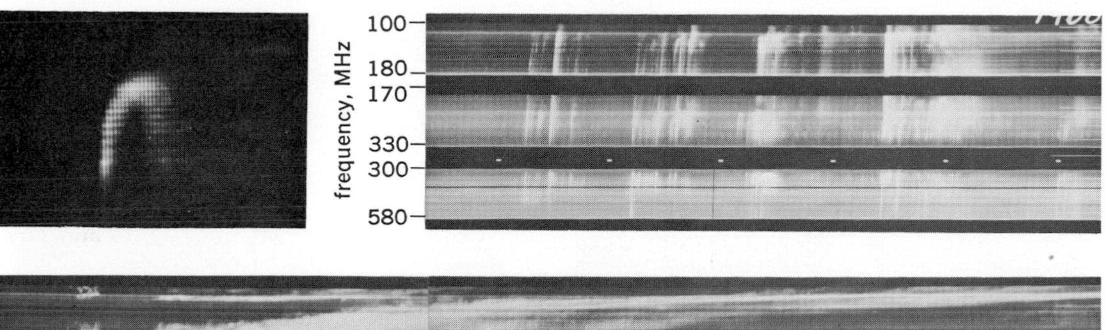

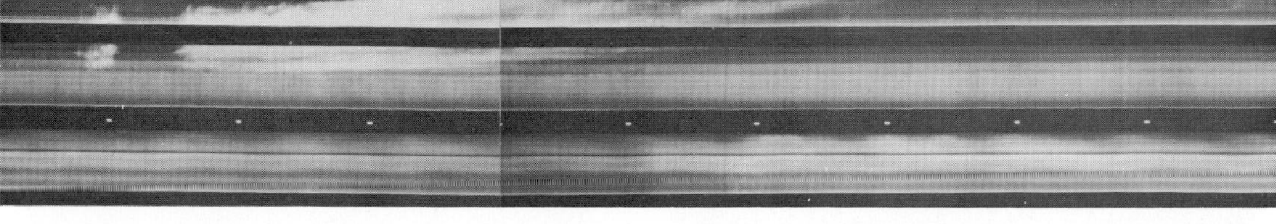

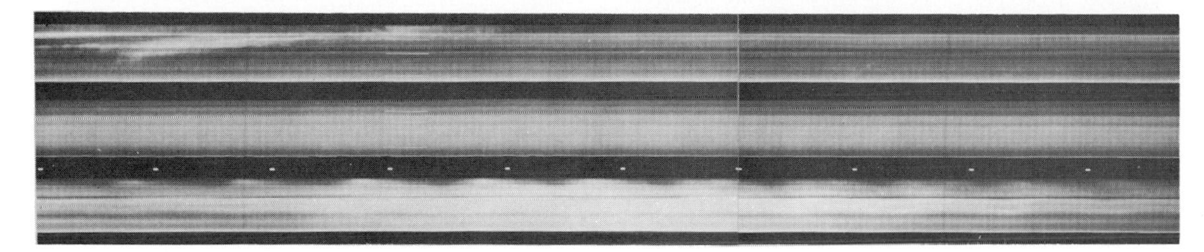

Fig. 27. Time variations in the solar radio spectrum. Frequency range from 100 to 580 MHz (the vertical coordinate) is divided into three overlapping bands with frequencies indicated in top row. Variations with time are shown in horizontal direction. Interval between the white pips between the middle and lower frequency bands represents 1 min. Top left, a highly magnified U burst in the 140-MHz band. The interval between successive vertical scans is 0.3 s. Top right, a series of type III bursts. The middle and lower rows are continuous, showing the development of a type II burst followed by a strong continuum in the 450-MHz band. *(Recordings from radio spectrometer, Fort Davis Station, Harvard College Observatory)*

dominant constituent of the Sun.

The flare-produced cosmic rays are most numerous at 3–5 MeV, but range upward in energy to hundreds of millions of electronvolts. Almost all flares produce cosmic rays, but the cosmic-ray production from the biggest flare of any year will probably equal that of all the other flares put together. The high-energy cosmic rays appear moments after the beginning of the H-α flare, but because of time of flight and the guiding by the interplanetary magnetic field, the lower-energy cosmic rays appear much later. A cosmic ray storm may last for days. The flux of low-energy particles in big flares is so intense that it endangers the lives of astronauts outside the terrestrial magnetic field. *See* COSMIC RAYS.

Relativistic electrons with energies of millions of electronvolts are also observed shortly after the beginning of optical flares. They are presumably accelerated by the same phenomenon that produced the cosmic rays.

A day or two after a very large flare, the pulse of arrival of a new group of particles of low energy (1–5 MeV) but of very large numbers may be observed. These particles, which produce what is called a geomagnetic storm, are a cloud ejected directly from the flare with a velocity of 1500 km/s or more.

It is known that any large sunspot group is continually producing large numbers of low-energy cosmic rays, presumably as the result of many small flares. The energy spectrum is very steep, about the inverse fifth power; hence, if the magnetic changes in a large flare increase the energy of all cosmic rays by a factor of 10, the number in some high-energy range will be increased by 10^5.

Until 1972 the MeV-energy cosmic rays could be studied only upon their arrival at the Earth, distorted by time of flight and magnetic field effects. Then studies of the high-energy gamma-ray spectrum revealed a series of gamma-ray lines due to nuclear reactions on the surface of the Sun produced by the energetic particles. The following reactions were observed:

1. Positrons created by collisions of protons with surface nuclei annihilate with ambient electrons, producing a line at 0.5 MeV.
2. Neutrons split off from ambient nuclei by protons recombine with ambient protons, forming deuterons, giving a strong line at 2.2 MeV.
3. Protons colliding with carbon and oxygen nuclei excite them to a higher state, giving lines at 4.4 and 6 MeV.

Many other lines are possible, but higher-resolution detectors adequate to detect them have not been orbited.

High-energy phenomena play a fundamental role in flare physics. All the evidence suggests that the primary energy input from the magnetic field to the flare gases is due to high-energy nonthermal

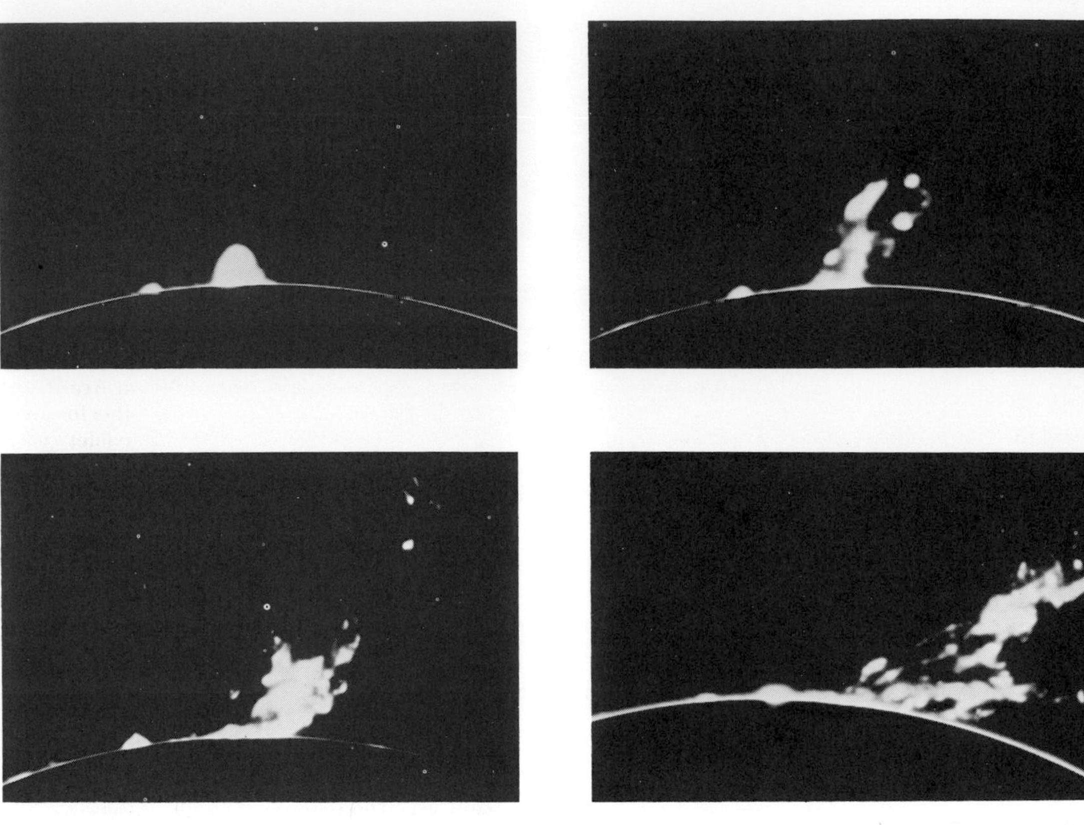

Fig. 28. Flare prominence, or spray, of Feb. 10, 1956. Universal time for successive frames: 211600, 211845, 212115, 213345 cover a total interval of 17 min 45 s. Some fragments exceeded a velocity of 1100 km/s. Each frame shows 670,000 km on the Sun. (*Sacramento Peak Observatory, operated by the Association of Universities for Research in Astronomy, Inc.*)

particles—electrons or protons—which transmit their energy to the ambient atoms by collisions. A few minutes later, thermal energy in the flare is released by compressing material in the corona; temperatures above 20×10^6 K are produced, and the resulting material condenses in the form of elegant loop prominences.

Flare emissions. Since the flare introduces a very energetic plasma into a relatively cool region, a whole range of particles and waves are produced. Each produces a characteristic emission.

As noted above, the energetic protons produce gamma-ray lines when they penetrate to the photosphere. If they are sufficiently numerous, they heat the photosphere faster than it can reemit and a white light flare is observed, usually in the form of bright transient flashes at the foot points of the flare loops.

Electrons radiate more efficiently. When they collide with protons, they produce hard x-rays by bremsstrahlung. For the first, impulsive burst of electrons the electron spectrum is hard and so are the x-rays. Later a thermal, softer spectrum is produced. The electrons also produce microwave emission as they spiral in the magnetic fields by a process known as synchrotron emission. From the spectrum of the microwaves and x-rays the number and spectrum of the electrons can be deduced. With huge radio antennas such as the Very Large Array in New Mexico, the source of microwaves can be mapped. It turns out to be a small kernel right at the top of the flare loops.

When the filament at the site of the flare heats up, a coronal cloud is formed with a temperature above 10^7 K (usually 30×10^6 K) and density 10^{10} atoms/cm^3, 10–100 times greater than the normal corona. This cloud is compressed by magnetic effects and produces strong, soft x-rays and ultraviolet emissions. This emission is full of spectrum lines, and these have been extensively studied by x-ray spectroscopists. The lines observed are due to highly ionized atoms all the way up to Fe XXVI (iron with 25 of 26 electrons stripped away). In the first stages of the thermal event the Fe XXVI and XXV lines are seen; then as the plasma cools down, lines of elements in lower ionization dominate, along with iron in lower ionization states. Permitted and forbidden transitions are observed, and the ratio of these is used to deduce the local density, the temperature being given by the ionization.

Flare ejecta. Almost every flare is accompanied by the ejection of material; the magnetic field acts like a rifle barrel to collimate the flare energy. Neutral hydrogen in the flare filament is usually ejected in a poorly collimated spray (Fig. 28) at speeds up to a few hundred kilometers per second (although 1500 km/s has been recorded). The sprays give rise to a magnetohydrodynamic shock wave in the corona which may be recognized in meter-wave radio emission by a so-called type II or slow-drift burst. The frequency of the emission drifts downward as the wave moves to lower densities and the radio emission excited decreases in frequency. The wave velocity is 1000 – 2000 km/s, and the waves have been observed all the way out to the Earth. The waves carry energetic trapped particles along, and a large increase in low-energy cosmic rays (1 – 5 MeV) is often observed when the wave reaches the Earth.

Flare waves are often observed on the solar surface in H-α light when their path is such as to include the surface. They usually appear as a bright front; sometimes filaments wink (that is, they drop out of the H-α bandpass by a down-and-up Doppler shift) as the front goes by.

There is a whole range of jetlike ejecta called surges which may be quite large. They occur when the flare is not too big and the magnetic field is strong enough to contain and collimate the material. Surge velocities are usually less than 200 km/s, and they often fall back along the same route.

Flares produce intense streams of electrons. These travel at about one-third the speed of light and produce type III or fast-drift radio bursts. In this case the frequency of the radio emission excited by the passage of the electron stream through the corona drifts downward rapidly as the stream reaches lower densities. The streams have been tracked all the way to the Earth, where pulses of electrons have been measured, typically at energies of 40 keV.

The development of spacecraft coronagraphs has permitted the observation of flare ejecta in the corona. Most of the eruptions produce large shock waves traveling outward through the corona.

Flare and corona. The normal processes of heating the corona cannot sustain all the material injected by the flare, which therefore cools down by radiation. When this material cools and condenses, it rains down on the surface. Because of the great strength of magnetic fields overlying the sunspots, the gas follows the magnetic lines of force in curved trajectories, forming loop prominences (Fig. 29). The cooler the material gets, the lower its degree of ionization and the faster it radiates and cools. In fact, a large fraction of the H-α radiation in flares is thought to be due to the falling material from the great loop prominences coming from the coronal cloud. Since the flare may last only minutes and the loop prominences will go on for many hours, they are often seen as a signal that a great center of activity is at the limb.

The birefringent filter enables one to see the appearance of the corona by isolating the coronal emission lines. The coronal material, as well as the prominence material, follows the magnetic lines of force, and many curved arches are seen. However, these do not change as rapidly as the condensing material in the prominences (Fig. 30). Often, series of five or six concentric arches are seen over an active center; they expand slowly and occasionally have been observed to break open at the top and whip into vertical streamers with apparent velocities of up to 600 km/s. This activity coincides with the eruption of material through the magnetic field at the time of a flare. The coronal cloud produced by the flare is responsible for low-energy x-rays in the 0.2 – 1.2 nm range, as well as for gradual increases of radiation in the microwave region. However, the hard x-rays and impulsive microwave bursts are associated with more transient groups of hard electrons which are accelerated by flares.

Solar terrestrial effects. Aside from heat, light, and the solar tides, the direct influences of the Sun on the Earth are generally too delicate for direct detection by the human senses. One exception is the aurora, which may be seen on occasions in temperate latitudes and more often in Arctic and Antarctic regions. The agents responsible for other solar influences on the Earth are ultraviolet and x-radiation and streams of charged particles emitted by the Sun. These are strongly variable with the level of solar activity, and the effects on the Earth are very large indeed when they are observed with sensitive radio equipment and magnetic com-

Fig. 29. Quiescent prominences in H-α light, seen (*a*) against the sky and (*b*) against the disk. (*Big Bear Solar Observatory*)

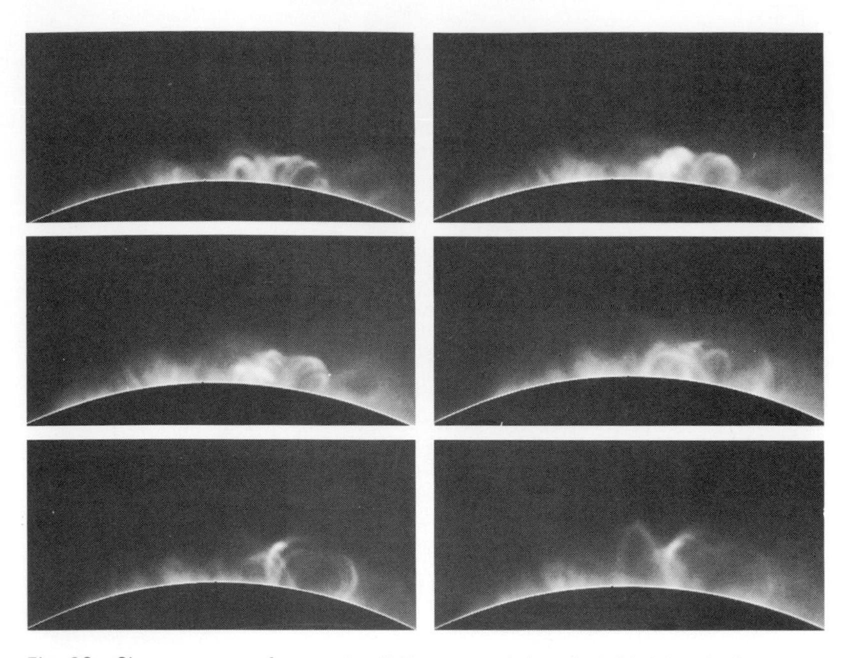

Fig. 30. Six exposures of coronal activity over an interval of 4 h taken in the green line (530.3 nm). Concentric arch structure in the first frame is a characteristic structure. (*Sacramento Peak Observatory, operated by the Association of Universities for Research in Astronomy, Inc.*)

passes, as well as with detectors on artificial Earth satellites.

The terrestrial ionosphere is a result of the steady flux of solar ultraviolet and x-radiation. Even the quiet corona and low-level chromospheric activity is sufficient to make possible an ionosphere with a low electron density. As the solar activity increases, the electron density in the ionosphere increases and higher frequencies are reflected. Most of the ionization is caused by solar ultraviolet rays between 20 and 100 nm, which ionize oxygen and nitrogen in the upper atmosphere. Only a small fraction of the molecules above 100 km are ionized.

The burst of hard x-rays emitted at the onset of a solar flare causes ionization much deeper in the atmosphere, around 60 km. At this level the density of the air is sufficiently high so that the electrons cannot oscillate freely but give up their energy by collisions with neutral atoms; thus, when they are excited by radio waves (which normally would penetrate and be reflected from the higher ionospheric layers), they absorb the radio waves and produce a shortwave fade-out (SWF). Long-distance radio communication, which depends on reflection from the ionosphere, deteriorates or is blacked out altogether for a few hours. The effect is almost instantaneous with the beginning of the flare, and serves as an indicator of its occurrence. It may also be observed by recording the radio emission from galactic sources which normally penetrates the ionosphere but is suddenly absorbed during an SWF. This phenomenon is called sudden cosmic noise absorption (SCNA). Temporary ionospheric currents in the beginning of the fade-out produce changes in the geomagnetic field strength, and if they are severe they may induce currents in long land lines sufficient to stop telephone communications. SWF is always associated with a flare, although occasionally flares will not produce SWF.

The ionizing radiation in SWF is hard x-rays in the 0.1 – 0.2 nm region. By a coincidence, there is also a small amount of ionization in this region (called the D layer) produced by the photoionization of nitric oxide by the Lyman-α line of hydrogen. However, this line does not change much during flares and does not produce a fade-out.

A whole series of other phenomena is associated with the creation of the D layer at the bottom of the ionosphere and its changing height. Among these phenomena are the sudden phase anomalies (SPA), produced by the lowering of the effective reflecting layer; and the sudden enhancement of atmospherics (SEA), produced by increased reflection of distant radio noise.

The particle effects on the Earth are extensive. The cosmic-ray storms produced by big flares do not reach the equatorial regions easily because of the Earth's magnetic field, but they spiral into the polar caps and produce what is called polar-cap absorption (PCA). This is the direct ionization of the ionosphere above the poles. Its effect is a polar blackout of radio communications across the polar regions. The very large number of particles associated with the low-energy pulse manages to penetrate the geomagnetic field to produce the geomagnetic storm. This is most intense near the poles, but may reach down into temperate and even tropic latitudes. The aurora is but one trace of the energetic particles precipitating in the upper atmosphere. The currents induced by these great numbers of particles produce sharp changes in the Earth's magnetic field, and the ionization produces considerable changes in the radio propagation. The magnetic field changes have also been known to produce severe effects in long power lines, creating surges and pulses which trip circuit breakers and produce power outages.

The relation between geomagnetic storms and flares is not completely determinate. Large flares on the western side of the Sun usually produce the most geomagnetic disturbances, but great magnetic storms often occur without a flare. The storms not associated with flares show a strong recurrence and, since for a long time their origin could not be identified, their source was termed the M region.

The M regions have now been identified with coronal holes. Spacecraft measurements have shown a close connection between coronal holes and high-velocity streams. The coronal holes are long-lived and produce a geomagnetic storm at the Earth each time they go by. The biggest geomagnetic storms, however, are produced by the shock waves from big flares.

Measurements of the plasma velocity in the solar wind show that there is a strong correlation between the velocity of the plasma and the geomagnetic activity. The passage of the M region and the beginning of the geomagnetic storm are usually connected with a shock wave in the interplanetary magnetic field which impinges on the Earth's field and produces a sharp lowering of the field, known as a sudden commencement, the signal for the beginning of a geomagnetic storm. The particles which closely follow produce the storm.

Observations of the aurora have identified a smaller phase of the geomagnetic field called the substorm. A substorm is a violent rearrangement in the outer fringes of the Earth's field, where it interacts with the solar wind and is drawn out into a long tail on the antisolar side. It has been found that substorms tend to occur when the north-south component of the magnetic field in the solar wind changes from parallel to antiparallel with the Earth's field.

Many astronomers believe there is a link between the sunspot cycle and long-term weather trends, but it has proved very difficult to detect. The evidence for a Little Ice Age during the Maunder minimum is persuasive, but there was no effect in Asia. There has been an apparent coincidence between great droughts in the western United States in 1910, 1932, 1954, and 1976 and alternate minima of the 11-year cycle; that is, the droughts appeared to agree with the 22-year magnetic cycle. Droughts are so sporadic, however, that direct statistical connections are not very obvious. For example, 1980 was a year of heat and drought too, but at the maximum of the sunspot cycle. However, some analyses show that the likelihood of severe local droughts does peak at the 22-year minima, but the locality of the drought will move about. Measurements of isotope ratios in fossil water, which indicate the temperature at which the water fell as rain, also indicate 22-year periods. So there are bits of evidence but little understanding.

SOLAR INSTRUMENTS

Although the observational instruments of the solar astronomer are the same in principle as those used by other astronomers, there are two points which determine the differences. First, the Sun is extremely bright, so that high concentration of light is not so important. Second, because of the disturbance of the Earth's atmosphere by solar heating during the day, the "seeing" is considerably worse in the daytime, and solar telescopes cannot hope to utilize as sharp an image as stellar telescopes. Furthermore, the telescopes must be designed so that the great heating produced by the Sun does not distort the images. This is usually done by liberal use of white paint and careful control of the atmospheric conditions inside the telescope.

Solar telescopes fall into two general classes: those designed for observations of the brilliant solar disk, and coronagraphs designed for the study of the much fainter prominences and the still fainter corona through the relatively bright, scattered light of the sky.

Disk telescope. Excellent definition in a solar image of reasonable size is usually the first requirement in a disk telescope. Definition is limited by the quality of the optical system, by the fundamental diffraction limit of resolution inherent in a given aperture, and by the quality of the seeing. Seeing is the term used to describe the blurring effect of the changing density gradients in the terrestrial atmosphere due to thermal convection. Seeing is never perfect, and there are very few locations where excellent definition is possible during perhaps 100 h in a year. These places are usually in lowland and maritime regions where stabilizing temperature inversions exist. Unfortunately the transparency is usually poor in such locations. Most solar telescopes have been built on top of mountains, because astronomers were used to building telescopes there and because the air is

Fig. 31. The 150-ft (46-m) solar-tower telescope of the Hale Observatories has a vertical underground spectrograph.

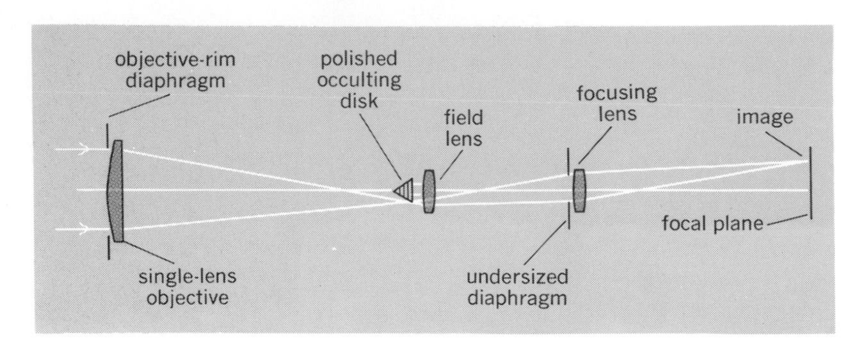

Fig. 32. Optical system of a coronagraph.

clear. Unfortunately, the air above such sites is quite turbulent and the seeing is very poor. In addition, clouds tend to form over mountain peaks. A few telescopes have been built in lakes, where daytime convection is reduced and much better results are obtained.

To maintain the definition of the telescope optical system, it must be guarded against the severe effects of solar heating. Solar heating produces air currents in the telescope and distorts the optics. This is not serious in refractors because little heat is absorbed in the lens and the lens acts as a cap against thermal convection. But reflectors, which are preferred for their achromatic properties, have severe difficulties, because the mirror absorbs heat and because the light travels up and down the tube. All modern reflecting solar telescopes are vacuum telescopes with large entrance windows. Special low-expansion materials are used to reduce changes in the mirror, and the gregorian design is preferred because heat load on secondary optics is reduced.

The most convenient type of disk telescope is the solar tower (Fig. 31). Two flat mirrors at the top of a tower, one of which is equatorially mounted to follow the diurnal motion of the Sun, reflect sunlight into a long-focus fixed vertical telescope. The telescope may be either a refractor or a compound reflector of 30 – 50 cm aperture. It produces a large image of the Sun near ground level, where the light can be conveniently reflected into any one of a number of large fixed accessory instruments.

A less expensive variant of the tower telescope is the horizontal fixed telescope, with the flat mirrors at ground level. This arrangement, when carefully designed, works well for small apertures, but convective disturbances in the long horizontal air path degrade the definition of large instruments.

For many purposes an equatorial telescope is preferred; in particular, it is easier to evacuate an equatorial system than a large tower, and for some purposes, such as polarimetry, a straight-through system is desirable. However, it is difficult to servo-guide an equatorial, because the whole telescope must be moved instead of a single mirror. *See* OPTICAL TELESCOPE; TELESCOPE.

Coronagraph. The coronagraph, invented by B. Lyot in 1931, is designed with one overriding consideration in mind: the elimination of instrumental scattered light. Its most delicate task is the observation of the corona immediately adjacent to the disk of the Sun, which is about a million times brighter. In an ordinary telescope, a little dust on the objective, diffraction at the edge of the aperture, and otherwise insignificant defects in the glass of the objective are all sources of diffuse scattered light which is usually some hundreds of times brighter than the corona.

The coronagraph is deceptively simple (Fig. 32). The critical component is the single-lens objective; it must be made of flawless glass, which is polished and cleaned to a perfection far beyond average standards. The rest of the system is more ordinary. The disk of the Sun is eclipsed by a polished conical disk, and photospheric light diffracted by the rim of the objective is intercepted in its image formed by the field lens, on an undersized diaphragm. The final lens then images the corona on the focal plane. The spectrum of the corona can be observed by placing the slit of the spectrograph at the focal plane, or direct photographs in the green line may be taken by inserting the appropriate birefringent filter behind the focusing lens and letting the image fall on a photographic film or plate.

The usual coronagraph has an aperture of 15 cm or less. The largest is a 40-cm instrument (Fig. 33). An internal mirror system reflects the light through the polar axis into an observing laboratory, where

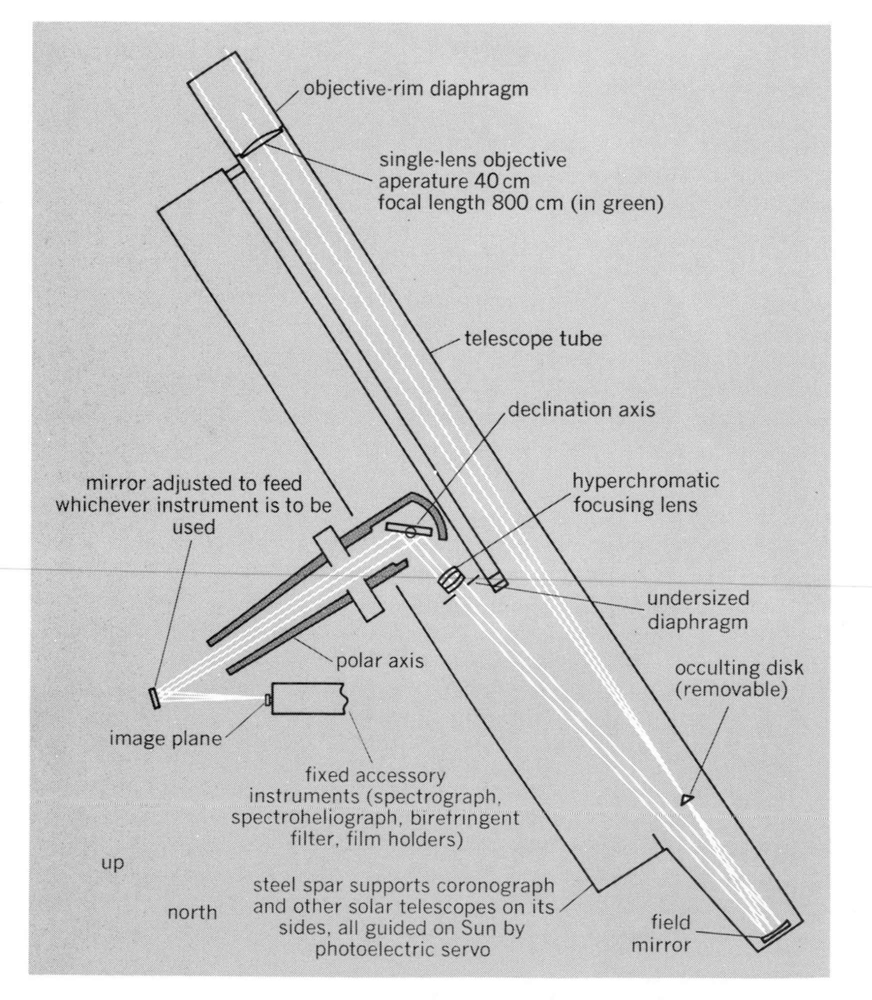

Fig. 33. Large 40-cm solar telescope of the Sacramento Peak Observatory.

the image can be directed to large analyzing instruments, as in a tower telescope. The focusing lens corrects the color aberration of the objective, and the telescope can be used equally well for disk and limb observations. Figure 21 was taken with it, in combination with a birefringent filter.

The coronagraph became outmoded when coronal observations from spacecraft became possible. The coronagraph can record the corona only at the limb, whereas soft x-ray telescopes reveal the whole corona on the disk. Also, a much greater range of spectral lines is available in the ultraviolet. Thus, only a few coronagraphs remain. However, these instruments still have capabilities that cannot be matched in space, in particular high-spatial resolution and flexibility. They are unexcelled for prominence observation. Since the coronagraphs must be located on mountains, they have limited good seeing.

Spectrograph. Modern solar spectrographs for use with disk telescopes utilize the great brightness of the solar image to achieve dispersion of the order of 100 mm/nm and spectroscopic resolution in excess of 500,000. These desirable characteristics require long focal lengths (10–25 m) and superlative diffraction gratings of the largest sizes that remain consistent with accuracy of ruling.

Large solar spectrographs are nearly always either of the Littrow autocollimating type or the reflecting type in which the collimator and camera element are long-focus concave mirrors (Fig. 34). They may be mounted either vertically in a well or horizontally on solid concrete piers. The Fourier transform spectrometer has proved extraordinarily powerful.

The spectrographs used with coronagraphs for observations of solar limb phenomena are generally much smaller and simpler, with medium dispersion of 5 mm/nm or less. They are usually simple Littrow grating spectrographs of about 200 cm focal length. The most important requirements are high light efficiency and stigmatic image. Most of them are small enough to be carried on the same mounting with the coronagraphs that feed them.

Spectroheliograph. The spectroheliograph and birefringent filter present an extended area of the solar image at one sharply defined wavelength. The wavelength chosen is usually at the center of the H-α- line of hydrogen or at the H or K line of ionized calcium. The chromosphere is opaque at these particular wavelengths, and the picture obtained is, therefore, that of the chromosphere. It is evident from Figs. 8, 23, 24*a*, and 25*b* that the structure is quite different from that of the photosphere, shown in Figs. 1, 5, 17.

The spectroheliograph is a stigmatic scanning monochromator made by inserting a slit in the focal plane of a spectrograph, accurately centered on the H-α line (or any other wavelength desired), so that only the light of this line is transmitted. Variations in image intensity along the first slit are faithfully reproduced in the exit slit. Several different scanning arrangements have been successfully used; one uses fixed optics. The monochromator is mounted on ways which permit smooth motion in the direction of dispersion. The solar image from a fixed telescope falls on the entrance slit of the monochromator. The light of the H-α line emerging from the second slit falls on a stationary photographic plate. As the monochromator moves along its ways, the first slit scans across the solar image, and the exit slit correspondingly scans the photographic plate. Thus an image of the Sun in the light of the H-α line is built up on the photographic plate continuously. Varying the second slit changes the output wavelength.

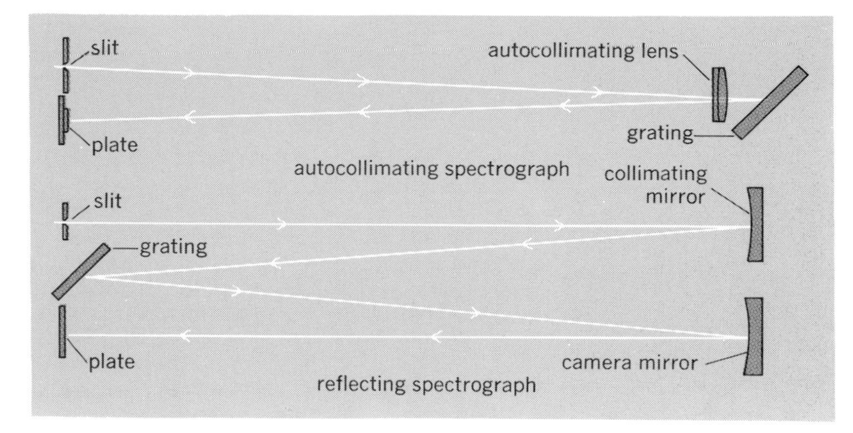

Fig. 34. Two forms of solar spectrograph. Horizontal scale is greatly compressed.

Birefringent filter. The birefringent filter consists of a multiple sandwich of alternate layers of polarizing films and plates cut from a birefringent crystal (usually quartz or calcite). The assembly transmits the light in a series of sharp, widely spaced wavelength bands. One or another of the polarizers absorbs the light of all intervening wavelengths. A multiplier filter is used to isolate the desired band and exclude the others. Filters made for observations on the disk of the Sun generally have transmission bands 0.025–0.075 nm wide, centered on the H-α line. For observations at the limb, bandwidths up to 1 nm are used for prominences in H-α, and bandwidths of about 0.2 nm for the green and red coronal lines. The birefringent filter is compact enough to be used with a conventional small telescope but is far less flexible than the bulkier spectroheliograph in choice of bandwidth and wavelength. Fabry-Perot interferometers have also been utilized. *See* ASTRONOMICAL OBSERVATORY; ASTRONOMICAL PHOTOGRAPHY.

[HAROLD ZIRIN]

Bibliography: D. E. Billings, *A Guide to the Solar Corona*, 1966; J. C. Brandt, *The Physics and Astronomy of the Sun and Stars*, 1966; E. Gibson, *The Quiet Sun*, NASA SP-303, 1972; R. Krüger, *Introduction to Solar Radio Astronomy and Radio Physics*, 1979; P. A. Sturrock (ed.), *Solar Flares*, 1980; Z. Svestka, *Solar Flares*, 1976; E. Tandberg-Hanssen, *Solar Activity*, 1967; O. R. White (ed.), *The Solar Output and Its Variation*, 1977; H. Zirin, *The Solar Atmosphere*, 1966; J. R. Zirker (ed.), *Coronal Holes and High Speed Windstreams*, 1977.

Sundial

An instrument for telling time by the Sun. It is composed of a style that casts a shadow and a dial plate, which is the surface upon which hour lines are marked and upon which the shadow falls. The style lies parallel to Earth's axis. The construction of the hour lines is based on the assumption that the apparent motion of the Sun is always on the celestial equator.

Sundials can be made in any form and on any surface. They may be large and stationary, or small and portable. They may be made for use in a particular place or anywhere. The most widely used form is the horizontal dial that indicates local apparent time (Sun time). Other forms of the sundial indicate local mean time, and standard time. *See* TIME.

The highest form of sundial construction is found in the heliochronometer, which tells standard time with great accuracy. Incorporated in its construction is the equation of time and the time difference in longitude between the place where it is to be used and the standard time meridian for that locality. This makes possible a sundial that can be read as a clock.

The sundial is said to be the oldest scientific instrument to come down to us unchanged. The underlying scientific principle of its construction makes it a useful device for educational purposes as well as for timekeeping. [ROBERT N. MAYALL]

Bibliography: R. N. Mayall and M. L. Mayall, *Skyshooting: Photography for Amateur Astronomers*, 1967; R. N. Mayall and M. L. Mayall, *Sundials: How to Know, Use, and Make Them*, 2d ed., 1973.

Sunspot

A dark area in the photosphere of the Sun caused by a lowered surface temperature. The temperature at the center of a spot is about 4000 K, and the surface brightness is one-fifth that of the normal photosphere. The sizes and numbers of sunspots vary in the celebrated 11-year sunspot cycle, which is shared by all other forms of solar activity. *See* SUN. [JOHN W. EVANS]

Superclusters

Associations of galaxy clusters and groups, typically composed of a few rich clusters and many poorer groups and isolated galaxies. Most investigators find superclusters to be arranged in broad sheets and chains, and some believe that they are connected, forming cells which enclose nearly empty volumes of space. If so, the large-scale structure of luminous matter in the universe is therefore apparently best visualized as spongelike, at least on scales up to about 10^9 light-years (9.46×10^{24} m), which is about as far as currently available data reliably extend.

Distribution of nearby galaxies. Until the mid-1970's, observational studies of superclusters—with one exception, the Local Supercluster—were primarily confined to the examination of the areal distribution of galaxies over the sky. In 1784 William Herschel noted that most of the "nebulae" he had discovered seemed to be concentrated in a wide belt across the northern sky, oriented roughly perpendicularly to the Milky Way. The Virgo Cluster, discovered previously by Charles Messier and Pierre Méchain, formed a dense concentration near the pole of the Milky Way in Herschel's "stratum" of nebulae. This great band of nebulae was further explored and commented on during the next 140 years, but not until Edwin Hubble conclusively demonstrated in 1924 that these nebulae were indeed extragalactic could a reasonable explanation be offered for this decidedly nonrandom distribution of the bright galaxies. However, it was 1953 before Gérard de Vaucouleurs realized that observers on Earth are simply seeing from the inside a great flattened system of groups and clusters of galaxies, with a few individual objects scattered between. De Vaucouleurs called this the Local Supercluster in 1958, though 5 years earlier he had called attention to neglected earlier work. Similar ideas had been strongly hinted at in studies by Cleveland Abbe in 1867 and by J. H. Reynolds in the early 1900s. The Swedish astronomers Knut Lundmark, Erik Holmberg, and Anders Reiz also visualized a similar nearby "metagalactic cloud" of galaxies in the 1920s and 1930s, as did Harlow Shapley at Harvard.

Local Supercluster. Figure 1 shows the apparent positions of all galaxies in the northern galactic hemisphere with known radial velocities less than 2000 km/s. These objects are the Galaxy's nearest neighbors in extragalactic space, and most are members of the Local Supercluster, which shows well as the band of objects running diagonally across the illustration. The Virgo Cluster is the concentration of points in this band to the upper left of center. The cross shows the point toward which de Vaucouleurs' analysis suggests the Galaxy is moving. If the Milky Way were shown here, it would be an irregular belt around the periphery of the illustration.

Rotation. The apparent flattening of the Local Supercluster naturally led de Vaucouleurs to test the idea that it might be rotating as well as expanding. Although not universally accepted, the resulting model fit, and continues to fit, the totality of the data on the velocities and positions of the nearer galaxies. It also explains an apparent anisotropy of

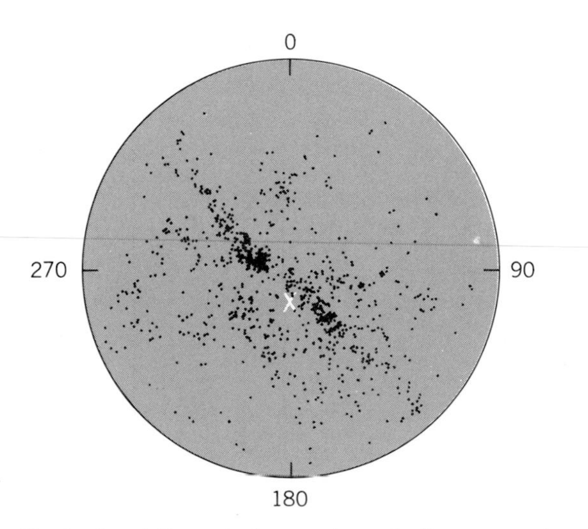

Fig. 1. Local Supercluster as shown by the apparent positions of all galaxies in the northern galactic hemisphere with known radial velocities less than 2000 km/s. Numbers are galactic longitudes. (*From R. B. Tully and J. R. Fisher, A picture of the Supercluster, Bull. Amer. Astron. Soc., 8:555, 1976*)

velocities of bright galaxies across the sky: at the tenth magnitude, galaxial velocities in the northern hemisphere are smaller by nearly a factor of 2 than velocities in the south. Furthermore, the very marked inequality of the numbers of galaxies brighter than the fourteenth magnitude between the two hemispheres (the southern sky has about half as many objects) arises because the Galaxy is not located in the center of the Local Supercluster.

One consequence of this model should be an observable anisotropy in the 3-K microwave background radiation due to the Earth's rotational motion around the center of the Local Supercluster. Such an anisotropy has been reported by radio astronomers since 1969. However, the speed and direction of motion of the Galaxy indicated by the 3-K anisotropy (545 km/s toward galactic longitude 268° and latitude +42°) is not in good agreement with de Vaucouleurs' model (350 km · s^{-1} toward galactic longitude 185°, galactic latitude +75°, a point in the sky near the supergalactic equator where it passes through southern Ursa Major). A possible interpretation is that the Local Supercluster itself is moving with respect to the microwave background radiation, though more observations at all wavelengths will be needed to clarify the situation. *See* COSMIC MICROWAVE RADIATION.

Nature and origin. The fact that the rotating-expanding model of the Local Supercluster fits the data so well has often been taken as evidence that it is a physical gravitationally bound system. This is almost certainly not the case. Since the Local Supercluster is about $1.5-2 \times 10^8$ light-years ($1.4-1.9 \times 10^{24}$ m) across, its period of rotation should be comparable to or greater than the age of the universe. Thus the Local Supercluster is probably better thought of as simply a relic of a vast inhomogeneity in the distribution of matter in the early history of the universe. Similarly, its "rotation" may be a reflection of internal motions that may have played an important role in its creation. Several theoretical studies have been directed toward this viewpoint. Among the more successful at explaining the observations are the gravitational instability concept and the idea of primeval turbulence.

Distribution of distant galaxies. Besides the Local Supercluster, there are perhaps 50 other superclusters that have been noted. These have been found primarily through examination of the various surveys of faint galaxies and of clusters of galaxies. Most useful among these surveys are those conducted by Shapley at Harvard and by C. D. Shane at Lick Observatory, both concentrating on faint galaxies, and the two catalogs of rich clusters of galaxies by George Abèll and by Fritz Zwicky, both working with the Palomar Sky Survey. Figure 2 shows the distribution of all galaxies brighter than magnitude 19 in the northern galactic hemisphere that were counted during the Lick survey. There are actually so many galaxies represented (over a million) that the numbers counted in 10 arc-minute by 10-arc-minute squares are represented by successively lighter shades of gray, from black for no galaxies to white for 10 or more.

This view of the cosmos is decidedly confused, however, since galaxies at all distances are seen projected onto the plane of the sky. Distant, intrinsically bright galaxies look little different at first glance from nearby faint ones. Thus, while all of the galaxies in Fig. 1 are actually also present in Fig. 2, they are completely lost in the maze of over a million fainter objects that are revealed by the deeper survey. This resultant smearing out of the actual spatial distribution of galaxies has led to general acceptance of the concept of the isotropy and homogeneity of the universe. However, the painstaking collection of more and better data for some thousands of the brightest and nearest galaxies, combined with new and more sophisticated methods of statistical data analysis applied to the surveys of the more distant objects, has given a drastically different view of the cosmos. Instead of a rather smooth distribution of galaxies where a variation of a factor of 2 or more in the counts could be passed over as a "random fluctuation," there now appear to exist superclusters separated by equally vast holes of amazingly empty space and extending over scales up to 3×10^8 light years. The filamentary ringlike structures so easily seen in Fig. 2 are being shown to be real associations of galaxies and clusters in space. Detailed observations of superclusters in Hercules, Coma Berenices, Perseus, Horologium, Indus, and other large areas of the sky have confirmed this general picture, as have studies in depth of smaller areas. Similarly, statistical investigations of the distribution of faint galaxies are best represented by models in which galaxies and clusters are arrayed in superclusters.

Observations at nonoptical wavelengths. Nearly all of the evidence set out above for the large-scale structure of the universe comes from optical observations of galaxies, though radio and infrared work on nearby galaxies has played an important role. For more distant objects, however, radio studies at various frequencies have given a considerably less clear picture. Distant radio sources are apparently not significantly clustered, whatever their nature (quasars, galaxies, and so forth), although inhomogeneities of up to 1.5×10^9 light-years are not ruled out; nor is there any evidence for small-scale (less than about 5°) anisotropy in the microwave background radiation in spite of the large-scale anisotropy noted above. Most rich clusters of galaxies emit x-rays, apparently thermal radiation from hot gas distributed throughout the clusters. But no extended x-ray sources have been unambiguously identified outside of clusters that would indicate the presence of similar gas clouds in intercluster space. Similarly, radio observations of extragalactic neutral hydrogen clouds have shown that these too are apparently closely associated with galaxies or with groups of galaxies. Thus it would seem that the galaxies themselves highlight the distribution of most matter in the universe, and that the great voids between the superclusters are indeed mostly or entirely empty. *See* RADIO ASTRONOMY; X-RAY ASTRONOMY.

However, since the sky surveys at wavelengths outside the narrow optical band are still mostly incomplete, and since many bright x-ray, infrared, and radio sources have no known optical counterparts, conclusions about the nature of the universe

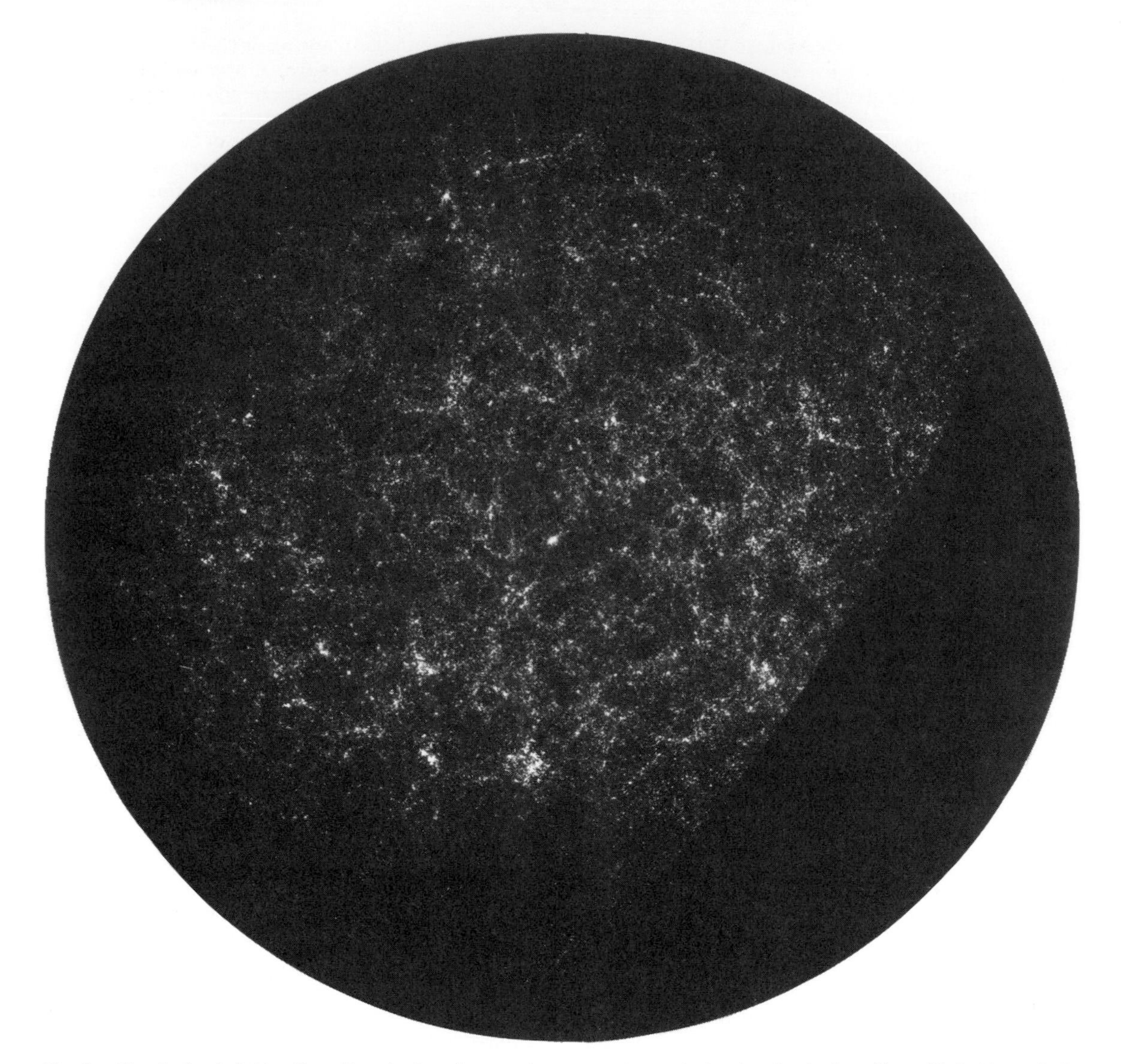

Fig. 2. All galaxies brighter than the nineteenth magnitude as counted on photographic plates taken at Lick Observatory by C. D. Shane and C. A. Wirtanen. Many clusters and superclusters are visible as chains, filaments, and clouds of galaxies. (*From M. Seldner et al., New reduction of the Lick catalogue of galaxies, Astron. J., 82:249–256, 1977*)

from optical observations alone must be treated with caution. Understanding of the physical processes taking place in stars, interstellar matter, and the galaxies themselves have changed radically as the result of collection of nonoptical data. Thus it is indeed possible that the spongelike picture of the distribution of matter in the universe may also need drastic revision when new data are collected and fully understood.

Extension of the clustering hierarchy. Finally, if associations of galaxies are thought of as a hierarchy—single galaxies forming groups and clusters which in turn form superclusters—there is marginal evidence in the distribution of the most distant galaxies and clusters that the next step in the hierarchy (clusters of superclusters) may exist. As noted above, however, studies of the most distant radio sources do not confirm these suggestions; nor do counts of faint galaxies of successively fainter magnitudes in different directions in the sky. Such counts, in fact, suggest upper limits to the scale of inhomogeneities of about 1.5×10^9 light-years. *See* COSMOLOGY; GALAXY, EXTERNAL.

[HAROLD G. CORWIN, JR.]

Bibliography: G. O. Abell, Superclustering of galaxies, in M. S. Longair (ed.), *Confrontation of Cosmological Theories with Observational Data*, IAU Symp. no. 63 (1973), pp. 79–92, 1974; M. S. Longair and J. Einasto (eds.), *The Large-Scale Structure of the Universe*, IAU Symp. no. 79 (1977), 1978; P. J. E. Peebles, *The Large-Scale Structure of the Universe*, 1980; G. de Vaucouleurs, The large-scale distribution of galaxies and clusters of galaxies, *Publ. Astron. Soc. Pac.*, 83:113–143, 1971.

Supergiant star

A member of the family containing the intrinsically brightest stars, populating the top of the Hertzsprung-Russell diagram. Supergiant stars occur at all temperatures, from 30,000 to 3000 K, and have luminosities ranging from 10^4 to 10^6 times that of the Sun. The hot supergiants have radii 20 times that of the Sun. The cool supergiants are the largest known stars, reaching several thousand solar radii. Among the bright supergiants are Deneb (α-Cygni), Rigel (β-Orionis), and Betelgeuse (α-Orionis). Because few have masses exceeding 20–40 Suns, the ratio of luminosity to mass is high. Thermonuclear energy sources will therefore be rapidly exhausted, hydrogen being consumed in a time scale of only 10^6 to 10^7 years. The supergiants are thus young but rapidly evolving massive stars. Nearly all are slightly variable in light and radial velocity, have highly turbulent and extended atmospheres, and show broadened spectral lines. *See* HERTZSPRUNG-RUSSELL DIAGRAM; STAR; STELLAR EVOLUTION.

The cool supergiants have such low mean density and surface gravity that they are clearly unstable and lose matter into space in an intense version of the solar wind. These stellar winds carry as much as 1 solar mass away from the supergiant in only 10^6 years. Rapid evolution by neutrino emission also shortens the life of supergiants.

An upper limit to the luminosity of the supergiants seems to be fairly established at well above 10^6 times that of the Sun. There is also a theoretical upper limit, set by A. S. Eddington, when the radiative flux carries momentum beyond that value which gravity can balance. Consequently, supergiants can be used to determine distances to a galaxy, with moderate accuracy, by assuming that they have the same limit in that galaxy as in the Milky Way. When observable, the spectra of supergiants in the Magellanic Clouds and nearby galaxies closely resemble those in the Milky Way. *See* GALAXY, EXTERNAL.

[JESSE L. GREENSTEIN]

Supermassive stars

Hypothetical objects with masses exceeding 60 solar masses, the mass of the largest known ordinary stars (1 solar mass = 2×10^{30} kg). The term is most often used in connection with objects larger than 10^4 solar masses that might be the energy source in quasars and active galaxies. Since these objects have not been observed and are hypothetical at present, very little is known with certainty about their nature and behavior. The description of their properties presented here is entirely based on theoretical calculations.

Dimensions. Just prior to the onset of instability, models indicate that a 10^4–solar mass supermassive star would have a radius of 3×10^6 km. The central temperature of such a nonrotating object would be roughly 10^9 K, and nuclear reactions would have occurred in a considerable part of the stellar interior. The corresponding numbers for a 10^8–solar mass object are a radius of 3×10^{12} km (500 times the size of Pluto's orbit) and a central temperature of 300,000 K, too low for fusion reactions to start.

Dynamics and evolution. About the only thing known for certain about supermassive stars is that they are unstable. Arthur Eddington, describing the properties of stars in general in the 1920s, showed that objects larger than ordinary stars would be supported mainly by radiation pressure rather than gas pressure. A star completely supported by radiation pressure, like a supermassive star, is unstable (because the ratio of specific heats $\gamma = 4/3$). However, supermassive stars might last a significant length of time before instability occurs. The lower-mass objects, in which nuclear reactions can provide some energy, could live 10^6 years. In the higher-mass objects, larger than 10^6 solar masses, for example, the lifetime is determined by the time taken for gravitational binding energy to dissipate, and is on the order of decades for 10^8 solar masses. Rotation affects these lifetimes. A rapidly rotating supermassive star, also called a relativistic disk, might have a considerably longer lifetime (and also a lower temperature). *See* STAR.

When instability occurs, a supermassive star can collapse catastrophically or explode. Models producing explosions have so far been confined to the lower-mass objects. The most reasonable scenario is that an unstable supermassive star will collapse extremely rapidly, possibly emit gravitational radiation, and become a giant black hole. *See* BLACK HOLE; GRAVITATION; GRAVITATIONAL COLLAPSE.

Observations. Supermassive stars, if they exist, would probably be formed in galactic nuclei. William Fowler and Fred Hoyle suggested in 1963 that they could provide the energy for the quasars, distant objects with luminosities as high as 10^{48} ergs/s (10^{41} W), 10^5 times the power emitted by the stars in the Milky Way Galaxy. This suggestion could be verified by the detection of periodicities in the light emission from quasars, dimly suggested but not confirmed by the observations of some objects. The final catastrophic collapse might generate long wavelength gravitational waves which could be detected by satellite facilities. Exploding supermassive stars may also be responsible for the rapid motions which occur in many galaxies, including the Milky Way Galaxy (as shown by the expanding spiral arms in the inner regions), and they may be the sites for some nucleosynthesis. *See* GALAXY; GALAXY, EXTERNAL; QUASARS.

[HARRY L. SHIPMAN]

Bibliography: F. Pacini and M. Rees, Rotation in high-energy astrophysics, *Sci. Amer.*, 228(2): 98–105, February 1973; S. Shapiro and S. Teukolsky, Gravitational collapse of supermassive stars to black holes: Numerical solution of the Einstein equations, *Astrophys. J. Lett.*, 234:L177–L182, December, 15, 1979; H. L. Shipman, *Black Holes, Quasars, and the Universe*, 2d ed., 1980; R. V. Wagoner, Physics of massive objects, *Annu. Rev. Astron. Astrophys.*, 7:553–576, 1969.

Supernova

The sudden brightening of a star, by a factor of up to 10^{10}, so that it may briefly outshine the entire galaxy in which it is located. Fading occurs slowly, over several years or more. The supernova phenomenon was first distinguished from that of ordinary novae by Walter Baade and Fritz Zwicky in 1934. They also suggested that the most likely source of the enormous energy required was the formation of a neutron star, and that a certain star in the Crab Nebula, remnant of the 1054 supernova, was probably such a neutron star. Recent observations and calculations have essentially confirmed their suggestions. *See* CRAB NEBULA; NEUTRON STAR; NOVA; PULSAR.

Frequency. A large galaxy like the Milky Way probably has about one supernova explosion every 30 years. Because dust blocks the view of much of the Milky Way, only six have been recorded in the last 1000 years (in 1006, 1054, 1572, 1604, and probably 1181 and 1680), by a wide range of astronomers and astrologers living in China, Japan, Europe, the Arab countries, and (possibly) among the North American Indians. More than 100 supernovae found in other galaxies since 1885 supplement knowledge gained from the local ones.

Supernova remnants. In all cases the spectra show large amounts of gas being thrown off at speeds of thousands of kilometers per second. The gas consists of the same chemical elements found in the Sun and other ordinary stars (though in one of the two major types of supernovae, hydrogen is grossly deficient, indicating a highly evolved parent star). The blown-off gas travels rapidly out into space, sweeping up more (interstellar) gas as it goes, and remains visible as a bright nebula for thousands of years. Such supernova remnants are also characteristically copious sources of radio and x-rays (indicating the presence of magnetic fields, relativistic electrons, and shock waves). Some dozens of them have thereby been identified in the galaxy. *See* ASTRONOMICAL SPECTROSCOPY; RADIO ASTRONOMY; X-RAY ASTRONOMY.

Energy release. The energy released by neutron star formation is believed to be responsible for the enormous luminosity increase and for blowing off the nebula. In addition, energy from supernova explosions probably also accelerates cosmic rays and produces many of the heavier chemical elements (those from iron on up to uranium) that can be made only in energy-absorbing reactions. Still more energy apparently comes off in neutrinos and gravitational radiation. *See* COSMIC RAYS; ELEMENTS AND NUCLIDES, ORIGIN OF.

Theory of formation. Theoretical work suggests that some sort of supernova should be the inevitable result of the death of every star more than about six (or four or eight) times as massive as the Sun. As such a star evolves, it exhausts first hydrogen and then helium as fuels to keep it shining. The next available fuel, carbon, may ignite explosively, blowing the star apart completely (one sort of supernova, though not the sort Baade and Zwicky had in mind, and not the most common sort, as it produces lots of iron and no pulsar). If carbon burns peacefully, other fuels (neon, oxygen, silicon) are used in turn until the star has an iron core, from which no further nuclear energy can be extracted. The iron core gradually increases in mass (as the assorted fuels burn in envelopes around it) to a critical value called the Chandrasekhar limit (after S. Chandrasekhar, who first calculated it with reference to white dwarfs). The core then collapses suddenly and catastrophically (because the electrons in its atoms combine with the protons to make neutrons, which can be packed much more tightly than the original particles), releasing some 10^{53} ergs (10^{46} J) of energy.

Details of how the energy is transferred to the envelope to blow it off are not very well understood, but may involve neutrinos, additional nuclear reactions, and bouncing of the core. The core becomes a neutron star or possibly sometimes a black hole, and the outer layers expand as a supernova remnant. A similar set of events would result from a white dwarf suddenly finding itself more massive than the Chandrasekhar limit (for example, because material has been dropped on it from a binary companion, or because it has been rotating and slows down), and may be responsible for the class of supernovae found among less massive stars. *See* BLACK HOLE; GRAVITATIONAL COLLAPSE; STELLAR EVOLUTION; WHITE DWARF STAR.

Observational evidence. The strongest confirmation of this scenario comes from the presence of a pulsar in the Crab Nebula. It is the star Baade and Zwicky pointed to, and is the most rapidly rotating pulsar known. Its rotation is slowing down at a rate that suggests its birth about the year 1054, and makes it possible to calculate the amount of energy being fed into the nebula (in the form of magnetic field and relativistic electrons) by the pulsar. This turns out to be just the amount (about 10^{37} ergs/s or 10^{30}W) needed to keep the nebula shining (in visible, radio, and x-ray wavelengths) at the observed rate, thereby solving the long-standing problem of what powers the Crab. One other supernova remnant (in the constellation Vela) contains a pulsar (period = 0.089 s). It is not known how many other remnants may contain pulsars that are not aimed at the Earth, or neutron stars that do not radiate as pulsars. Such neutron stars should be reasonably bright x-ray sources; one may have been seen with the HEAO series of x-ray satellite telescopes. *See* VARIABLE STAR.

[VIRGINIA TRIMBLE]

Bibliography: D. H. Clark and F. R. Stephenson, *The Historical Supernovae*, 1977; C. B. Cosmovici (ed.), *Supernovae and Their Remnants*, 1974; D. N. Schramm (ed.), *Supernovae*, 1977.

Taurus

The Bull, in astronomy, a winter constellation. Taurus is the second sign of the zodiac. The group contains two notable star clusters, the Hyades and the

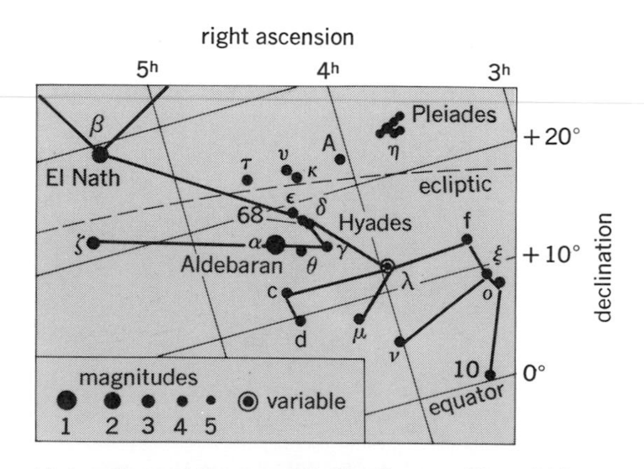

Line pattern of the constellation Taurus. The grid lines represent the coordinates of the sky. The apparent brightness, or magnitude, of the stars is shown by the sizes of the dots, which are graded by appropriate numbers as indicated.

Pleiades. The Hyades is a V-shaped cluster, the V forming the head of the charging bull, with the fiery bright star Aldebaran in the right eye (see illustration). This star has long been used in navigation. The long horns of the bull extend northeast to the constellation Auriga. Farther west lies the compact, beautiful cluster of six stars, the famous Pleiades, sometimes called the seven sisters, suggesting thereby that one of the stars has faded from naked-eye view. This group in the Bull's shoulder has the shape of a tiny dipper. *See* CONSTELLATION. [CHING-SUNG YU]

Tektite

A member of one of several groups of objects that are composed almost entirely of natural glass formed from the melting and rapid cooling of terrestrial rocks by the energy accompanying impacts of large extraterrestrial bodies. Tektites are dark brown to green, show laminar to highly contorted flow structure on weathered surfaces and in thin slices, are brittle with excellent conchoidal fracture, and occur in masses ranging to as much as tens of kilograms but are mostly much smaller to microscopic in size. The shapes of tektites are those of common fluid splash and rotational forms including drops, spheres, and dumbbells, unless they have been abraded together with surface gravels. A few tektites have shapes that are caused by two different heating events: the impact that melted the parent rock to form the glass, and a second event apparently due to reentry aerodynamic heating.

Occurrences, groups, and ages. With the discovery of tektites in the Soviet Union, there are now five major groups known: (1) North American, 34,000,000 years old, found in Texas (bediasites) and the Georgia Coastal Plains, with a single specimen reported from Martha's Vineyard, MA; (2) Czechoslovakian (moldavites), 15,000,000 years old, found both in Bohemia (green and transparent) and in Moravia (brown and turbid); (3) Ivory Coast, 1,300,000 years old; (4) Russian (irgizites; see illustration), 1,100,000 years old, found in the Northern Aral Region; and (5) Australasian, 700,000 years old, occurring notably in Australia, the Philippines, Belitung, Thailand, and numerous other localities. The North American, Ivory Coast, and Australasian tektites also occur as microtektites in oceanic sediment cores near the areas of their land occurrences. In the land occurrences, virtually all of the tektites are found mixed with surface gravels and recent sediments that are younger than their formation ages.

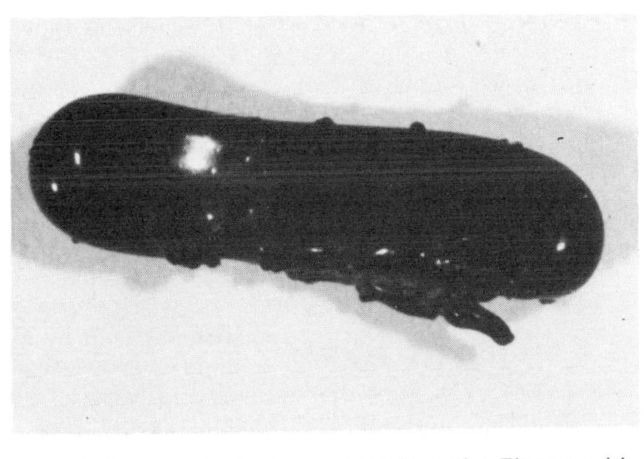

Dumbbell-shaped tektite (irgizite) from the Zhamanshin meteoritic impact crater, Northern Aral Region, Soviet Union. The glass is so dark brown that it appears black in reflected light.

Tektites, in general, are not rare objects. Millions of tektites have been recovered from the Australasian occurrence. However, some of the more pleasingly colored specimens from Bohemia and Georgia are prized by collectors and command substantial prices, either as specimens or potential gem material.

Three of the tektite groups are associated with known large impact craters. The irgizites occur in and immediately around the Zhamanshin meteoritic impact crater (10–15 km in diameter) in the Northern Aral Region of the Soviet Union. Both the Ivory Coast tektites and the moldavites occur close to probable source impact craters of identical age, the Bosumtwi Crater, Ghana (10.5 km in diameter) and the Ries Crater, Germany (24 km in diameter), respectively. The source craters for the North American and Australasian tektites have not yet been identified positively.

Composition. The chemical compositions of tektites differ from those of ordinary terrestrial rocks principally in that they contain less water and have a greater ratio of ferrous to ferric iron, both of which are almost certainly a result of their very-high-temperature history. The extreme ranges of major element compositions, expressed as oxides in weight percent, are as follows: SiO_2, 48–85 wt %; Al_2O_3, 8–18 wt %; FeO, 1.4–11 wt %; MgO, 0.4–28 wt %; CaO, 0.3–10 wt %; Na_2O, 0.3–3.9 wt %; K_2O, 1.3–3.8 wt %; TiO_2, 0.3–1.1 wt %. However, portions of the foregoing ranges are based on analyses of glass particles from oceanic sediment cores that may or may not be tektites, as this identification is difficult for these very small particles. If other glass occurrences from impact craters were included as tektites, such as splash form glass from the Wabar Craters in Saudi Arabia, and the Lonar Crater in India, the ranges would expand significantly. As indicated by the wide range of chemical compositions, there are corresponding wide ranges of properties such as specific gravity and refractive index.

Inclusions. Spherical vesicles ranging in size from microscopic to as much as several centimeters are common in most tektites, as are small lechatelierite (silica glass) particles. The presence of coesite, a high-pressure polymorph of silica, in some tektites from Southeast Asia is additional evidence of the impact genesis of tektites. Coesite is known to form at the Earth's surface only from the very high pressures of transient shock waves caused by large hypervelocity meteorite impacts and cratering events. Baddeleyite, monoclinic ZrO_2, is present in some tektites as a high-temperature decomposition product of the mineral zircon. Meteoritic nickel-iron also has been observed in a few tektites, which is further evidence of the origin of tektites as meteoritic impact melts. Such inclusions of coesite, baddeleyite, and nickel-iron are common in the glassy fusion products in and around a number of terrestrial impact craters. *See* METEORITE.

Historical perspective. Tektites have been recognized as unusual objects for more than a millennium, and they were the objects of intensive scientific research from the turn of the century until a few years before the return of lunar samples by the Apollo missions. This activity was generated by the possibility that tektites originated from the Moon as secondary ejecta from impact craters on the lunar surface. This point of view was effectively advocated by a number of scientists until the recognition of the associated impact craters with some tektite groups on the Earth. Later analyses of lunar samples demonstrated that they are not suitable parent rocks for tektites. Unfortunately, the discovery of the Russian tektites, in a clear relation to an impact crater on the Earth, came too late to influence these arguments significantly.

Unanswered questions. The source crater of the Australasian tektites has not yet been identified, although this is the youngest and most widespread tektite group. Also, present understanding of the mechanics and sequence of events in very large impact crater formation does not permit specifying unambiguously the mode by which some tektites escape the Earth's atmosphere to reenter and form the remelted layer as a result of aerodynamic heating.

[ELBERT A. KING]

Bibliography: P. W. Florenski, The meteoritic crater Zhamanshin (Northern Aral Region, USSR) and its tektites and impactites [in German], *Chem. Erde*, 36:83–95, 1977; E. A. King, The origin of tektites: A brief review, *Amer. Sci.*, 65:212–218, 1977; E. A. King, *Space Geology: An Introduction*, pp. 69–80, 1976; E. A. King and J. Arndt, Water content of Russian tektites, *Nature*, 269:48–49, 1977; S. R. Taylor, Tektites: A post-Apollo view, *Earth-Sci. Rev.*, 9:101–123, 1973.

Telescope

An instrument used to collect, measure, or analyze radiation from a distant object. Most commonly, telescope refers to an assemblage of lenses or mirrors, or both, that enhances the ability of the eye either to see objects more distinctly or to see fainter objects. In its most general meaning, telescope refers to a device that collects radiation, which may be in the form of electromagnetic radiation or particle radiation, from a limited direction in space.

Optical telescopes. Optical telescopes may be classified as refracting telescopes, which use only lenses; reflecting telescopes, which use mirrors to focus light; and catadioptric telescopes, which use both lenses and mirrors. Special types of reflecting telescopes which depart radically from conventional ground-based systems are the multimirror telescope and the space telescope.

Refracting telescopes. Small refracting telescopes are used in binoculars, cameras, gunsights, galvanometers, periscopes, surveying instruments, rangefinders, and a great variety of other devices. Parallel or nearly parallel light from the distant object enters from the left, and the objective lens forms an inverted image of it (illustration *a*). The inverted image is viewed with the aid of a second lens, called the eyepiece. The eyepiece is adjusted (focused) to form a parallel bundle of rays so that the image of the object may be viewed by the eye without strain. The angular separation A of two points on the object is magnified to the angle A' as seen by the eye. The magnification (A'/A) is numerically equal to the ratio of the focal lengths of the lenses (f_1/f_2). Both the objective lens and the eyepiece usually consist of several lenses combined in order to minimize deleterious effects (aberrations) that are caused by the wavelength dependence of the refraction of glass and by the use of lenses with spherical surfaces. Additional lenses are included in telescopes that are intended for terrestrial viewing to provide an upright image.

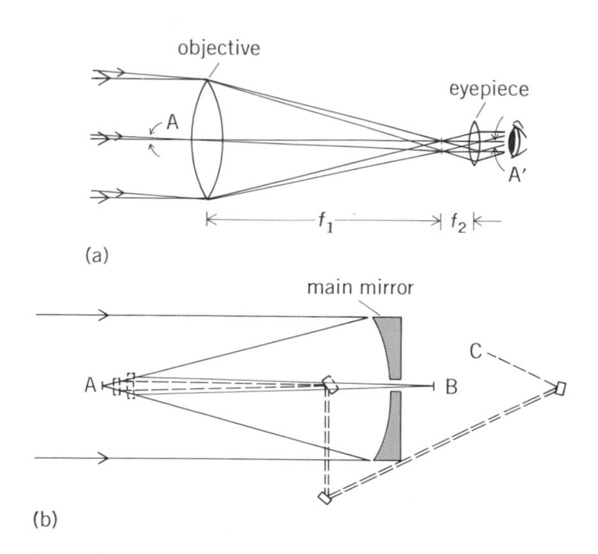

Simplified optical diagrams of (*a*) refracting and (*b*) large, versatile astronomical reflecting telescopes.

The telescope has been used in astronomy since its introduction by Galileo in the 17th century. Improvements in the ability to see faint objects have resulted largely from an increase in the diameter of the objective. The largest refracting telescope is the 40-in. (1.02-m) telescope at Yerkes Observatory. This size is about the limit for optical glass lenses.

Reflecting telescopes. Much larger telescopes have been made using mirrors instead of lenses (illustration *b*). The reflecting surface of the main mirror is a thin deposited layer of aluminum. Since the bulk of the mirror serves only to support the aluminum, the bulk does not have to be clear optical glass. However, it must be able to accept a high polish by an optician and also be mechanically stable. In the reflecting telescope shown in illustration *b*, auxiliary mirrors may be used near the prime focus at A to provide additional focus positions behind the main mirror at B or at a fixed position C where the collected light from a star may be analyzed in a laboratory. Although the eye is frequently used to verify the identification of the object to be studied, most scientific investigations are made by using photographic plates or electronic light detectors. The largest reflecting telescope is the 6-m instrument near Zelenchukskaya in the Soviet Union. *See* ASTRONOMICAL PHOTOGRAPHY; IMAGE TUBE.

Catadioptric telescopes. Catadioptric telescopes combine both mirrors and lenses. This combination is generally used to image a wide field of the sky. A notable example is the 48-in. (1.2-m) Schmidt telescope on Palomar Mountain in California. *See* SCHMIDT CAMERA.

Multimirror telescopes. The practical and economical limit to the size of mirrors has been nearly reached in the Soviet telescope. Therefore, newer designs and constructions have been directed toward utilizing a number of mirrors which are mounted so that the light collected by them may be brought to a common focus. A multimirror telescope (MMT) that is designed to combine six 1.8-m mirrors to yield the light-gathering power of a single 4.5-m mirror is located on Mount Hopkins in Arizona. Designs have been studied for multimirror telescopes with the equivalent light collection of a 25-m mirror.

Space telescope. The ability of large telescopes to resolve fine detail is limited by a number of factors. The ultimate limitation set by diffraction of light is rarely or never achieved by large telescopes. Instead, the finest resolved detail is limited by atmospheric inhomogeneities which set a limit of about 0.2 arc-second. Distortion due to the mirror's own weight causes additional problems in astronomical telescopes. These problems are eliminated in the Earth-orbiting space telescope, designed with an aperture of 2.3 m. The telescope is designed to be used in ultraviolet as well as visible light, resulting in a great improvement in resolution of small angles, not only by the elimination of the aforementioned terrestrial effects, but by the reduced blurring by diffraction in the ultraviolet. *See* SATELLITE ASTRONOMY.

Collectors of radiant energy. As collectors of radiation from a specific direction, telescopes may be classified as focusing and nonfocusing. Nonfocusing telescopes are used for radiation with energies of x-rays and above. Focusing telescopes, intended for nonvisible wavelengths, are similar to optical ones, but they differ in the details of construction.

Radio telescopes. Radio telescopes utilize mirrors of very large size which, because of the long wavelength for which they are used, may consist of only an open wire mesh. Because of the limitation by diffraction, arrays of large reflectors have been constructed. The output of each mirror of the array is combined in a process called aperture synthesis to yield a resolution roughly equivalent to that provided by a telescope the size of the array. An example of this type is the very large array (VLA) near Socorro, NM. *See* RADIO TELESCOPE.

Infrared telescopes. In infrared telescopes the secondary mirror (that near *A* in illustration *b*) is caused to oscillate rotationally about an axis through a diameter. This motion causes an infrared detector at *B* to see alternately the sky and the sky plus the desired object. The signals received at these two mirror positions are subtracted and, as a consequence, the large background radiation received both from the atmosphere and from the telescope is canceled. However, because of the random nature of thermal radiation, the fluctuations of the background emission are not canceled. Thus infrared telescopes are additionally designed to reduce the telescope background radiation and its fluctuations. Two large infrared telescopes (the 3.8-m United Kingdom Infrared Telescope and the 3-m NASA Infrared Telescope Facility) went into operation on Mauna Kea in Hawaii in 1979. *See* INFRARED ASTRONOMY.

Ultraviolet telescopes. Ultraviolet telescopes have special mirror coatings with high ultraviolet reflectivity. Since the atmosphere is not transparent below 300 nm, ultraviolet telescopes are usually flown above the atmosphere either in rockets or in orbiting spacecraft. An example is the International Ultraviolet Explorer. *See* ULTRAVIOLET ASTRONOMY.

X-ray telescopes. X-ray telescopes must be used above the atmosphere and are flown in rockets or satellites. The focusing type uses an unusual optical design in which the reflection from the surfaces occurs at nearly grazing incidence. This is the only way of achieving reflective optics for x-rays.

Nonfocusing x-ray telescopes use opaque heavy-metal (lead) channels or tubes in front of an x-ray detector to confine the directional sensitivity of the detector. The detectors may be proportional counters or scintillation detectors. *See* X-RAY TELESCOPE.

Gamma-ray telescopes. Gamma-ray telescopes use coincidence and anticoincidence circuits with scintillation or semiconductor detectors to obtain directional discrimination. With coincidence counting, two or more detectors in a line must give a simultaneous detection for a gamma ray to be counted. Other detectors are often used to surround the telescope to reduce the unwanted background arising from undesired particles. A simultaneous count received in one or more of these shielding detectors nullifies (anticoincidence) the detection otherwise registered in the coincidence detectors. Thus gamma rays that trigger only the coincidence circuits are detected. Since many gamma rays are produced within the atmosphere by other particles, the telescopes are usually flown in balloons, rockets, or satellites. *See* GAMMA-RAY ASTRONOMY.

Cosmic-ray telescopes. Cosmic-ray telescopes are used to detect primary protons or heavier-element nuclei, or to detect the products produced when these particles interact with the atmosphere. In its simplest form a cosmic-ray telescope may consist of nuclear track emulsions borne aloft in balloons or spacecraft. A very large cosmic-ray telescope deep in a mine in Utah detects penetrating mesons in the 1 to 100 TeV range. This telescope uses Cerenkov detectors in combination with plane parallel arrays of cylindrical spark-tube counters. The Cerenkov counter detects the presence of a high-energy particle and triggers the acoustic sensing of sparks in the counters. A computer is used to analyze the spark data and determine the direction of the incoming meson. The arrival of high-energy cosmic rays appears to be isotropic, most likely because of scattering by magnetic fields of the galaxy. *See* COSMIC RAYS.

Neutrino telescope. Planning for a neutrino telescope has been undertaken. Called the deep underwater muon and neutrino detector (DUMAND), it will consist of 1 km^3 of ocean water several kilometers beneath the ocean surface. An array of 22,000 Cerenkov counters, suspended in the water, will sense the shower of hadrons which results from the interaction of an extraterrestrial neutrino with the ocean water above the detector. Since the shower preserves the direction of the original neutrino, computer analysis of which detectors were excited will yield the neutrino's direction to within 1°. It is hoped that neutrinos from a number of potential astrophysical sources will be detected.

[WILLIAM M. SINTON]

Bibliography: J. G. Emming (ed.), *Electromagnetic Radiation in Space*, vol. 9, Astrophysics and Space Science Library, 1967; G. R. Miczaika and W. M. Sinton, *Tools of the Astronomer*, 1961; H. Ogelman and J. R. Wayland (eds.), *Introduction to Experimental Techniques of High Energy Astrophysics*, NASA, 1970.

Tide

Stresses exerted in a body by the gravitational action of another, and related phenomena resulting from these stresses. Every body in the universe raises tides, to some extent, on every other. This article deals only with tides on the Earth, since these are fundamentally the same as tides on all bodies. Sometimes variations of sea level, whatever their origin, are referred to as tides.

Introduction. The tide-generating forces arise from the gravitational action of Sun and Moon, the effect of the Moon being about twice as effective as that of the Sun in producing tides. The tidal effects of all other bodies on the Earth are negligible. The tidal forces act to generate stresses in all parts of the Earth and give rise to relative movements of the matter of the solid Earth, ocean, and atmosphere. The Earth's rotation gives these movements an alternating character having principal periodicities of 12.42 and 12.00 hr, corresponding to half the mean lunar and solar day, respectively.

In the ocean the tidal forces act to generate alternating tidal currents and displacements of the sea surface. These phenomena are important to shipping and have been studied extensively. The main object of tidal studies has been to predict the tidal elevation or current at a given seaport or other place in the ocean at any given time.

The prediction problem may be attacked in two ways. Since the relative motions of Earth, Moon, and Sun are known precisely, it is possible to specify the tidal forces over the Earth at any past or future time with great precision. It should be possible to relate tidal elevations and currents at any point in the oceans to these forces, making use of classical mechanics and hydrodynamics. Such a theoretical approach to tidal prediction has not yet yielded any great success, owing in great part to the complicated shape of the ocean basins. However, use of numerical-hydrodynamical models (such as the work of K. T. Bogdanov, N. Grijalva, W. Hansen, M. C. Henderschott, and C. L. Pekeris) has yielded some satisfactory results and undoubtedly will have practical importance.

The other approach, which consists of making use of past observations of the tide at a certain place to predict the tide for the same place, has yielded practical results. The method cannot be used for a location where there have been no previous observations. In the harmonic method the frequencies of the many tidal constituents are derived from knowledge of the movements of Earth, Moon, and Sun. The amplitude and epoch of each constituent are determined from the tidal observations. The actual tide can then by synthesized by summing up an adequate number of harmonic constituents. The method might loosely be thought of as extrapolation.

A "convolution" method of tidal analysis and prediction has been proposed by W. H. Munk and D. E. Cartwright. In this method past observations at a place are used to determine a numerical operator which, when applied to the known tide-producing forces, will calculate the resulting tide.

In the following discussion only the lunar effect is considered, and it is understood that analogous statements apply to the solar effect.

Tide-generating force. If the Moon attracted every point within the Earth with equal force, there would be no tide. It is the small difference in direction and magnitude of the lunar attractive force, from one point of the Earth's mass to another, which gives rise to the tidal stresses.

According to Newton's laws, the Moon attracts every particle of the Earth with a force directed toward the center of the Moon, with magnitude proportional to the inverse square of the distance between the Moon's center and the particle. At point A in Fig. 1, the Moon is in the zenith and at

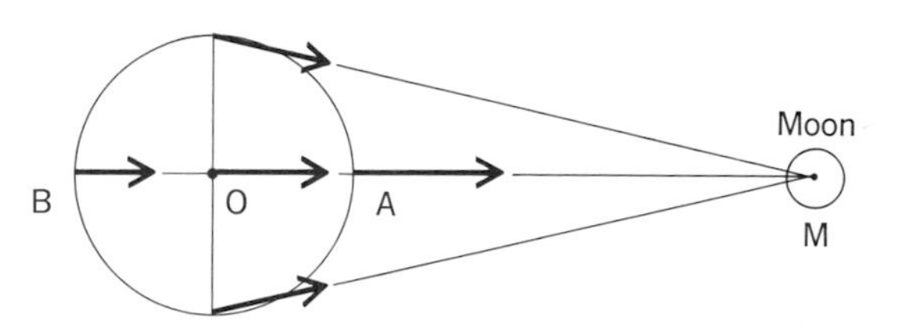

Fig. 1. Schematic diagram of the lunar gravitational force on different points in the Earth.

point B the Moon is at nadir. It is evident that the upward force of the Moon's attraction at A is greater than the downward force at B because of its closer proximity to the Moon. Such differential forces are responsible for stresses in all parts of the Earth. The Moon's gravitational pull on the Earth can be expressed as the vector sum of a constant force, equal to the Moon's attraction on the Earth's center, and a small deviation which varies from point to point in the Earth (Fig. 2). This small deviation is referred to as the tide-generating force. The larger constant force is balanced completely by acceleration (centrifugal force) of the Earth in its orbital motion around the center of mass of the Earth-Moon system, and plays no part in tidal phenomena. *See* GRAVITATION.

The tide-generating force is proportional to the mass of the disturbing body (Moon) and to the inverse cube of its distance. This inverse cube law accounts for the fact that the Moon is 2.17 times as important, insofar as tides are concerned, as the

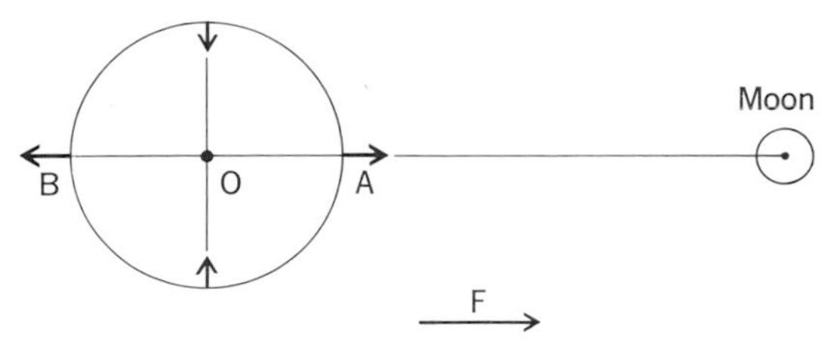

Fig. 2. Schematic diagram of the tide-generating force on different points in the Earth. The vector sum of this tide-generating force and the constant force F (which does not vary from point to point) produce the force field indicated in Fig. 1. Force F is compensated by the centrifugal force of the Earth in its orbital motion.

Sun, although the latter's direct gravitational pull on the Earth, which is governed by an inverse-square law, is about 180 times the Moon's pull.

The tide-generating force, as illustrated in Fig. 2, can be expressed as the gradient of the tide-generating potential, Eq. (1), where λ is the zenith

$$\psi = \frac{3}{2}\frac{\gamma M r^2}{c^3}\left(\tfrac{1}{3} - \cos^2\lambda\right) \qquad (1)$$

distance of the Moon, r is distance from the Earth's center, c is distance between the centers of Earth and Moon, γ is the gravitational constant, and M is the mass of the Moon. In this expression, terms containing higher powers of the smaller number r/c have been neglected. As ψ depends only on the space variables r and λ, it is symmetrical about the Earth-Moon axis.

It helps one visualize the form of the tide-generating potential to consider how a hypothetical "inertialess" ocean covering the whole Earth would respond to the tidal forces. In order to be in equilibrium with the tidal forces, the surface must assume the shape of an equipotential surface as determined by both the Earth's own gravity and the tide-generating force. The elevation of the surface is given approximately by Eq. (2), where ψ is evalu-

$$\bar{\zeta} = -\frac{\psi}{g} + \text{const} \qquad (2)$$

ated at the Earth's surface and g is the acceleration of the Earth's gravity. The elevation $\bar{\zeta}$ of this hypothetical ocean is known as the equilibrium tide. Knowledge of the equilibrium tide over the entire Earth determines completely the tide-generating potential (and hence the tidal forces) at all points within the Earth as well as on its surface. Therefore, when the equilibrium tide is mentioned, it shall be understood that reference to the tide-generating force is also being made.

Harmonic development of the tide. The equilibrium tide as determined from relations (1) and (2) has the form of a prolate spheroid (football-shaped) whose major axis coincides with the Earth-Moon axis. The Earth rotates relative to this equilibrium tidal form so that the nature of the (equilibrium) tidal variation with time at a particular point on the Earth's surface is not immediately obvious. To analyze the character of this variation, it is convenient to express the zenith angle of the Moon in terms of the geographical coordinates θ, ϕ of a point on the Earth's surface (θ is colatitude, ϕ is east longitude) and the declination D and west hour angle reckoned from Greenwich α of the Moon. When this is done, the equilibrium tide can be expressed as the sum of the three terms in Eq. (3), where a is the Earth's radius.

$$\bar{\zeta} = \frac{3}{4}\frac{\gamma M}{g}\frac{a^2}{c^3}\left[(3\sin^2 D - 1)\left(\cos^2\theta - \tfrac{1}{3}\right) + \sin 2D \sin 2\theta \cos(\alpha+\phi) + \cos^2 D \sin^2\theta \cos 2(\alpha+\phi)\right] \qquad (3)$$

The first term represents a partial tide which is symmetrical about the Earth's axis, as it is independent of longitude. The only time variation results from the slowly varying lunar declination and distance from Earth. This tide is called the long-period tide. Its actual geographical shape is that of a spheroid whose axis coincides with the Earth's axis and whose oblateness slowly but continuously varies.

The second term of Eq. (3) represents a partial tide having, at any instant, maximum elevations at 45°N and 45°S on opposite sides of the Earth, and two minimum elevations lying at similar, alternate positions on the same great circle passing through the poles. Because of the factor $\cos(\alpha + \phi)$ the tide rotates in a westerly direction relative to the Earth, and any geographical position experiences a complete oscillation in a lunar day, the time taken for α to increase by the amount 2π. Consequently, this partial tide is called the diurnal tide. Because of the factor $\sin 2D$, the diurnal equilibrium tide is zero at the instant the moon crosses the Equator; because of the factor $\sin 2\theta$, there is no diurnal equilibrium tidal fluctuation at the Equator or at the poles.

The third term of Eq. (3) is a partial tide having, at any instant, two maximum elevations on the Equator at opposite ends of the Earth, separated alternately by two minima also on the Equator. This whole form also rotates westward relative to the Earth, making a complete revolution in a lunar day. But any geographic position on the Earth will experience two cycles during this time because of the factor $\cos 2(\alpha + \phi)$. Consequently, this tide is called the semidiurnal tide. Because of the factor $\sin^2\theta$, there is no semidiurnal equilibrium tidal fluctuation at the poles, while the fluctuation is strongest at the Equator.

It has been found very convenient to consider the equilibrium tide as the sum of a number of terms, called constituents, which have a simple geographical shape and vary harmonically in time. This is the basis of the harmonic development of the tide. A great number of tidal phenomena can be adequately described by a linear law; that is, the effect of each harmonic constituent can be superimposed on the effects of the others. Herein is the great advantage of the harmonic method in dealing with tidal problems. The three terms of Eq. (3) do not vary with time in a purely harmonic manner. The parameters c and D themselves vary, and the rapidly increasing α does not do so at a constant rate owing to ellipticity and other irregularities of the Moon's orbit. Actually, each of the three partial tides can be separated into an entire species of harmonic constituents. The constituents of any one of the three species have the same geographical shape, but different periods, amplitudes, and epochs.

The solar tide is developed in the same way. As before, the three species of constituents arise: long-period, diurnal, and semidiurnal. The equilibrium tide at any place is the sum of both the lunar and solar tides. When the Sun and Moon are nearly in the same apparent position in the sky (new Moon) or are nearly at opposite positions (full Moon), the lunar and solar effects reinforce each other. This condition is called the spring tide. During the spring tide the principal lunar and solar constituents are in phase. At quadrature the solar effect cancels to some extent the lunar effect, the principal lunar and solar constituents being out of phase. This condition is known as the neap tide.

The entire equilibrium tide can now be expressed by Eq. (4), where $H=3\gamma Ma^2/g\bar{c}^3=54$ cm,

$$\zeta = H[\tfrac{1}{2}(1-3\cos^2\theta)\sum_L f_i C_i \cos A_i + \sin 2\theta \sum_D f_i C_i \cos(A_i+\phi) + \sin^2\theta \sum_S f_i C_i \cos(A_i + 2\phi)] \quad (4)$$

and $1/\bar{c}$ represents the mean (in time) value of $1/c$. Each term in the above series represents a constituent. Terms of higher powers of the Moon's parallax (a/c) are not included in Eq. (4) because of their different latitude dependence, but they are of relatively small importance. The subscripts L, D, and S indicate summation over the long-period, diurnal, and simidiurnal constituents, respectively. The Cs are the constituent coefficients and are constant for each constituent. They account for the relative strength of all lunar and solar constituents. In a purely harmonic development, such as carried out by A. T. Doodson in 1921, the A parts of the arguments increase linearly with time, and the node factors f are all unity. In George Darwin's "almost harmonic" development of 1882, the constituents undergo a slow change in amplitude and epoch with the 19-year nodal cycle of the Moon. The node factors f take this slow variation into account. The As increase almost linearly with time. Tables in U.S. Coast and Geodetic Survey Spec. Publ. no. 98 enable one to compute the phase of the argument of any of Darwin's constituents at any time, and values of the node factors for each year are given.

In spite of the many advantages of the purely harmonic development, Darwin's method is still used by most agencies engaged in tidal work. In Darwin's classification, each constituent is represented by a symbol with a numerical subscript, 0, 1, or 2, which designates whether the constituent is long-period, diurnal, or semidiurnal. Some of the most important of Darwin's constituents are listed in the table.

The periods of all the semidiurnal constituents are grouped about 12 hr, and the diurnal periods about 24 hr. This results from the fact that the Earth rotates much faster than the revolution of the Moon about the Earth or of the Earth about the Sun. The principal lunar semidiurnal constituent M_2 beats against the others giving rise to a modulated semidiurnal waveform whose amplitude varies with the Moon's phase (the spring-neap effect), distance, and so on. Similarly, the amplitude of the modulated diurnal wave varies with the varying lunar declination, solar declination, and lunar phase. For example, the spring tide at full Moon or new Moon is manifested by constituents M_2 and S_2 being in phase, thus reinforcing each other. During the neap tide when the Moon is at quadrature, the constituents M_2 and S_2 are out of phase, and tend to cancel each other. The other variations in the intensity of the tide are similarly reflected in the "beating" of other groups of constituents.

Darwin's constituents

Constituent	Speed, deg/hr	Coefficient
Long-period		
Mf, lunar fortnightly	1.098	0.157
Ssa, solar semiannual	0.082	0.073
Diurnal		
K_1, lunisolar	15.041	0.530
O_1, larger lunar	13.943	0.377
P_1, larger solar	14.959	0.176
Semidiurnal		
M_2, principal lunar	28.984	0.908
S_2, principal solar	30.000	0.423
N_2, larger lunar elliptic	28.440	0.176
K_2, lunisolar	30.082	0.115

Tides in the ocean. The tide in the ocean deviates markedly from the equilibrium tide, which is not surprising if one recalls that the equilibrium tide is based on neglect of the inertial forces. These forces are appreciable unless the periods of all free oscillations in the ocean are small compared with those of the tidal forces. Actually, there are free oscillations in the ocean (ordinary gravity seiches) having periods of the order of a large fraction of a day, and there may be others (planetary modes) having periods of the order of several days. For the long-period constituents the observed tide should behave like the equilibrium tide, but this is difficult to show because of their small amplitude in the presence of relatively large meteorological effects.

At most places in the ocean and along the coasts, sea level rises and falls in a regular manner. The highest level usually occurs twice in any lunar day, the times bearing a constant relationship with the Moon's meridional passage. The time between the Moon's meridional passage and the next high tide is called the lunitidal interval. The difference in level between successive high and low tides, called the range of the tide, is generally greatest near the time of full or new Moon, and smallest near the times of quadrature. This results from the spring-neap variation in the equilibrium tide. The range of the tide usually exhibits a secondary variation, being greater near the time of perigee (when the Moon is closest to the Earth) and smaller at apogee (when the Moon is farthest away).

The above situation is observed at places where the tide is predominantly semidiurnal. At many other places, it is observed that one of the two maxima in any lunar day is higher than the other. This effect is known as the diurnal inequality and represents the presence of an appreciable diurnal variation. At these places, the tide is said to be of the "mixed" type. At a few places, the diurnal tide actually predominates, there generally being only one high and low tide during the lunar day.

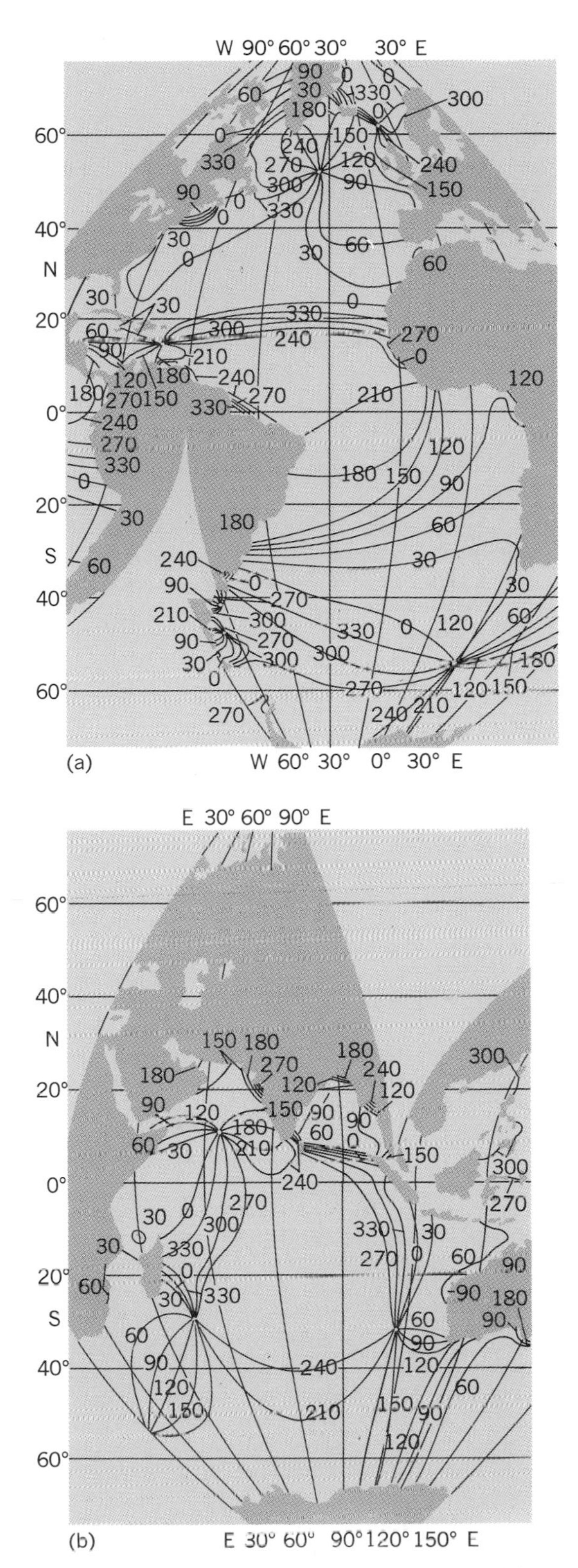

Fig. 3. Cotidal chart for M_2. (*a*) Atlantic Ocean. (*b*) Indian Ocean. (*G. Dietrich, Veroeff. Inst. Meeresk., n.s. A, Geogr.-naturwiss. Reihe, no. 41, 1944*)

Both observation and theory indicate that the ocean tide can generally be considered linear. As a result of this fact, the effect in the ocean of each constituent of series in Eq. (4) can be considered by itself. Each equilibrium constituent causes a reaction in the ocean. The tide in the ocean is the sum total of all the reactions of the individual constituents. Furthermore, each constituent of the ocean tide is harmonic (sinusoidal) in time. If the amplitude of an equilibrium constituent varies with the nodal cycle of the Moon, the amplitude of the oceanic constituent varies proportionately.

As a consequence of the above, the tidal elevation in the ocean can be expressed by Eq. (5),

$$\zeta = \Sigma f_i h_i \cos (A_i - G_i) \tag{5}$$

where $h_i(\theta,\phi)$ is called the amplitude and $G_i(\theta,\phi)$ the Greenwich epoch of each constituent. The summation in Eq. (5) extends over all constituents of all species. The fs and the As have the same meaning as in Eq. (4) for the equilibrium tide and are determined from astronomic data.

To specify completely the tidal elevation over the entire surface of the ocean for all time, one would need ocean-wide charts of $h(\theta,\phi)$, called corange charts, and of $G(\theta,\phi)$, called cotidal charts, for each important constituent. Construction of these charts would solve the ultimate problem in tidal prediction. Many attempts have been made to construct cotidal charts, the most notable those of W. Whewell, 1833; R. A. Harris, 1904; R. Sterneck, 1920; and G. Dietrich, 1944. These attempts have been based on a little theory and far too few observations.

Figures 3 and 4 show Dietrich's cotidal chart for M_2. Each curve passes through points having high water at the same time, time being indicated as phase of the M_2 equilibrium argument. A characteristic feature of cotidal charts is the occurrence of points through which all cotidal curves pass. These are called amphidromic points. Here the amplitude of the constituent under consideration must be zero. The existence of such amphidromic points has been borne out by theoretical studies of tides in ocean basins of simple geometric shape. The mechanism which gives rise to amphidromic points is intimately related to the rotation of the Earth and the Coriolis force.

The amplitude of a constituent, $h(\theta,\phi)$, is generally high in some large regions of the oceans and low in others, but in addition there are small-scale erratic variations, at least along the coastline. Perhaps this is partly an illusion caused by the placement of some tide gages near the open coast and the placement of others up rivers and estuaries. It is well known that the phase and amplitude of the tide change rapidly as the tidal wave progresses up a river.

The range of the ocean tide varies between wide limits. The highest range is encountered in the Bay of Fundy, where values exceeding 50 ft (15 m) have been observed. In some places in the Mediterranean, South Pacific, and Arctic, the tidal range never exceeds 2 ft (0.6 m).

The tide may be considerably different in small adjacent seas than in the nearby ocean, and here

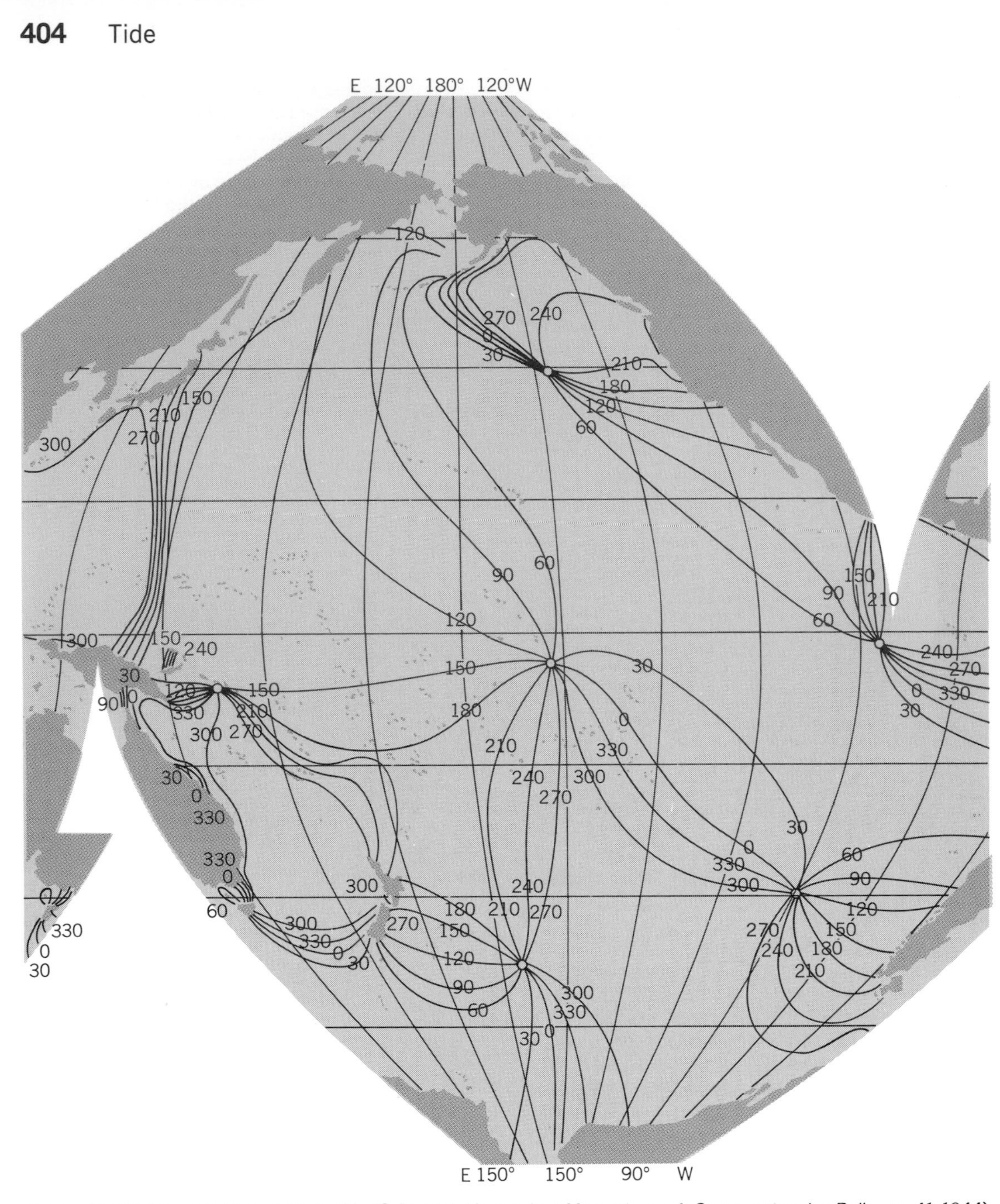

Fig. 4. Pacific Ocean cotidal chart for M_2. (*G. Dietrich, Veroeff. Inst. Meeresk., n.s. A, Geogr.-naturwiss. Reihe, no. 41, 1944*)

resonance phenomena frequently occur. The periods of free oscillation of a body of water are determined by their boundary and depth configurations. If one of these free periods is near that of a large tidal constituent, the latter may be amplified considerably in the small sea. The large tidal range in the Bay of Fundy is an example of this effect. Here the resonance period is nearly 12 hr, and it is the semidiurnal constituents that are large. The diurnal constituents are not extremely greater in the Bay of Fundy than in the nearby ocean.

In lakes and other completely enclosed bodies of water the periods of free oscillation are usually much smaller than those of the tidal constituents. Therefore the tide in these places obeys the principles of statics. Since there is no tidal variation in the total volume of water in lakes the mean surface elevation does not change with the tide. The surface slope is determined by the slope of the equilibrium tide, and the related changes in elevation are usually very small, of the order of a fraction of a millimeter for small lakes.

Tidal currents. The south and east components of the tidal current can be developed in the same way as the tidal elevation since they also depend linearly on the tidal forces. Consequently, the same analysis and prediction methods can be used. Expressions similar to Eq. (5) represent the current components, each constituent having its own amplitude and phase at each geographic point. It should be emphasized that the current speed or direction cannot be developed in this way since these are not linearly related to the tidal forces.

Only in special cases are the two tidal current components exactly in or out of phase, and so the tidal current in the ocean is generally rotatory. A drogue or other floating object describes a trajectory similar in form to a Lissajous figure. In a narrow channel only the component along its axis is of interest. Where shipping is important through such a channel or port entrance, current predictions, as well as tidal height predictions, are sometimes prepared.

Owing to the rotation of the Earth, there is a gyroscopic, or Coriolis, force acting perpendicularly to the motion of any water particle in motion. In the Northern Hemisphere this force is to the right of the current vector. The horizontal, or tractive, component of the tidal force generally rotates in the clockwise sense in the Northern Hemisphere. As a result of both these influences the tidal currents in the open ocean generally rotate in the clockwise sense in the Northern Hemisphere, and in the counterclockwise sense in the Southern Hemisphere. There are exceptions, however, and the complete dynamics should be taken into account.

The variation of the tidal current with depth is not well known. It is generally agreed that the current would be constant from top to bottom were it not for stratification of the water and bottom friction. The variation of velocity with depth due to the stratification of the water is associated with internal wave motion. Serial observations made from anchored or drifting ships have disclosed prominent tidal periodicities in the vertical thermal structure of the water.

Dynamics of ocean tide. The theoretical methods for studying tidal dynamics in the oceans were put forth by Laplace in the 18th century. The following assumptions are introduced: (1) The water is homogeneous; (2) vertical displacements and velocities of the water particles are small in comparison to the horizontal displacements and velocities; (3) the water pressure at any point in the water is given adequately by the hydrostatic law, that is, is equal to the head of water above the given point; (4) all dissipative forces are neglected; (5) the ocean basins are assumed rigid (as if there were no bodily tide), and the gravitational potential of the tidally displaced masses is neglected; and (6) the tidal elevation is small compared with the water depth.

If assumptions (1) and (3) are valid, it can readily be shown that the tidal currents are uniform with depth. This is a conclusion which is not in complete harmony with observations, and there are internal wave modes thus left out of Laplace's theory. Nevertheless the main features of the tide are probably contained in the equations.

The water motion in the oceans is, in theory, determined by knowledge of the shape of the ocean basins and the tide-generating force (or equilibrium tide) at every point in the oceans for all time. The theory makes use of two relations: (1) the equation of continuity, which states that the rate of change of water mass in any vertical column in the ocean is equal to the rate at which water is flowing into the column; and (2) the equations of motion, which state that the total acceleration of a water "particle" (relative to an inertial system, thus taking into account the rotation of the Earth) is equal to the total force per unit mass acting on that particle. Under the above assumptions, the equation of continuity takes the form of Eq. (6), where $d(\theta,\phi)$ is

$$\frac{\partial \zeta}{\partial t} = -\frac{1}{a \sin \theta}\left[\frac{\partial}{\partial \theta}(ud \sin \theta) + \frac{\partial}{\partial \phi}(vd)\right] \quad (6)$$

the water depth. The equations of motion in the southward and eastward directions, respectively, are given by Eq. (7), where ω designates the angu-

$$\begin{aligned} \frac{\partial u}{\partial t} - 2\omega v \cos \theta &= -\frac{g}{a}\frac{\partial}{\partial \theta}(\zeta - \bar{\zeta}) \\ \frac{\partial v}{\partial t} + 2\omega u \cos \theta &= -\frac{g}{a}\csc \theta \frac{\partial}{\partial \phi}(\zeta - \bar{\zeta}) \end{aligned} \quad (7)$$

lar rate of rotation of the Earth, and u and v the south and east components of the tidal current. All other quantities are as previously defined.

It is probable that exact mathematical solutions to Eqs. (6) and (7), taking even approximately into account the complicated shape of the ocean basins, will never be obtained. However, the equations have certain features which serve to give us some insight into the nature of ocean tides. For instance it is evident that if many equilibrium tides are acting simultaneously on the ocean, then the ocean tide will be the sum of the individual reactions. This linearity results directly from above assumption (6). In certain shallow regions of the ocean the tides are noticeably distorted, as would be expected if assumption (6) were violated. This distortion is usually considered as resulting from the presence of so-called shallow-water constituents having frequencies equal to harmonics and to beat frequencies of the equilibrium constituents. These must be considered, at some places, or there will be large discrepancies between prediction and observation. Certain mathematical solutions to Eqs. (6) and (7) have been obtained for hypothetical ocean basins of simple geometric shape. Laplace solved them for an ocean of constant depth covering the entire Earth. Several solutions have been obtained for an ocean of constant depth bounded by two meridians. The result of one of the solutions obtained by J. Proudman and A. Doodson is shown in Fig. 5, which represents a cotidal chart of the K_2 tide in an ocean of depth 14,520 ft bounded by meridians 70° apart. The K_2 tide was calculated because of mathematical simplifications, but the M_2 tide should be quite similar. Comparison of Fig. 5 with the Atlantic Ocean in Fig. 3 discloses no striking similarities except for the general occurrence of amphidromic systems.

The bodily tide. The solid part of the Earth suffers periodic deformation resulting from the tide-generating forces just as the oceans do.

The gravest known modes of free oscillation of the solid Earth have periods of the order of an hour, much shorter than those of the principal tidal constituents. Therefore, the principles of statics

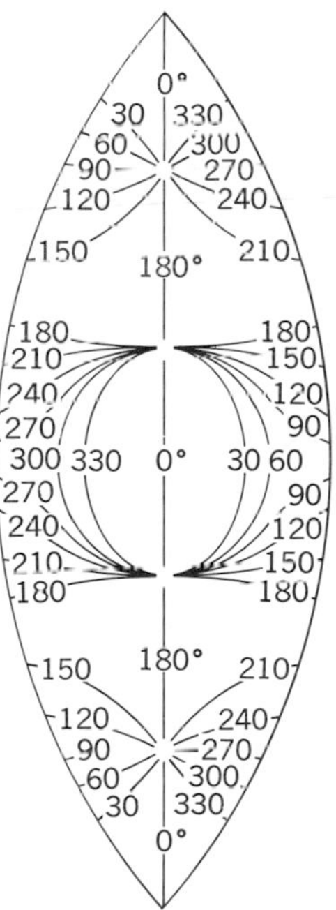

Fig. 5. Cotidal chart for K_2 in a hypothetical ocean of constant depth bounded by meridians 70° apart. (*J. Proudman and A. T. Doodson, from A. T. Doodson and H. D. Warburg, Admiralty Manual of Tides, London, 1941*)

can be used to describe the bodily tide, in contrast to tides in the oceans and atmosphere, where the inertial effect is important.

Associated with the bodily tide are periodic changes in gravity, manifesting themselves as (1) a variation of the vertical, or plumb line, with respect to any solid structure embedded in the Earth's crust; and (2) a variation in the magnitude of the acceleration of gravity at any point. These effects arise from the gravitational attraction of the tidally displaced matter of the Earth (solid, ocean, and atmosphere) as well as directly from the tide-generating forces. The magnitude of the former factor is of the order of several tens of microgals (1 gal = 1 cm/sec^2).

Atmospheric tides. Since air, as other matter, is subject to gravitational influence, there are tides in the atmosphere possessing many features of similarity with those in the ocean. One of the characteristics of these tides is a small oscillatory variation in the atmospheric pressure at any place. This fluctuation of pressure, as in the case of the ocean tide, may be considered as the sum of the usual tidal constituents, and standard tidal analysis and prediction methods may be used. The principal lunar semidiurnal constituent M_2 of the pressure variation has been determined for a number of places, and found to have an amplitude of the order of 0.03 millibars. The dynamical theory of these tides has been the subject of considerable study. The equations which have been considered have the same general form as those for ocean tides. The S_2 constituent shows a much larger oscillation with an amplitude of the order of 1 millibar, but here diurnal heating dominates the gravitational effects. If diurnal heating were the whole story one would expect an even larger S_1 effect, and the fact that S_2 is larger is attributed to an atmospheric resonance near 12 hr.

Tidal analysis and prediction. The distribution in space and time of the tidal forces within the Earth is precisely known from astronomic data. The effects of these forces on the oceans cannot, by present methods, be described in detail on a worldwide basis because of the difficult nature of the dynamical relationships and the complicated shape of the ocean basins. Practical prediction methods make use of past observations at the place under consideration.

The procedure is the same for prediction of any tidal variable—such as the atmospheric pressure, component displacements of the solid Earth, components of the tidal current, and so on—which depends linearly on the tidal forces. In the harmonic method the frequencies, or periods, of the tidal constituents are determined by the astronomic data, and the harmonic constants (amplitudes and epochs) are obtained from the observations. Equation (5) then represents the tide at all past and future times for the place under consideration, where the values of h are the amplitudes of whatever tidal variable is being predicted. In this discussion the sea-level elevation will be used as an example, since it is the variable for which predictions are most commonly made. The procedure is basically the same for each constituent, but is most easily described for the series of constituents, $S_1, S_2, S_3, \ldots$, whose periods are submultiples of 24 hr. Suppose that the tidal elevation at 1:00 A.M. is averaged for all the days of the tide record, and similarly for 2:00 A.M., 3:00 A.M., and for each hour of the day. The 24 values thus obtained represent the average diurnal variation during the entire record. Any constituent whose period is not a submultiple of 24 hr will contribute very little to the average of all the 1:00 A.M. values since its phase will be different from one day to the next, and its average value at 1:00 A.M. will be very close to zero for a long record. The same is true for each hour of the day, and so its average diurnal variation is small. The longer the record the freer will be the average diurnal oscillation from the effects of the other constituents. The diurnal oscillation is then analyzed by the well-known methods of harmonic analysis to determine the amplitudes and phases of all the harmonics of the 24-hr oscillation.

The same procedure is used for each other constituent; that is, the tide record is divided into consecutive constituent days, each equal to the period (or double the period in the case of the semidiurnal constituents) of the constituent. If the tide record is tabulated each solar hour, there is a slight complication due to the fact that the constituent hours do not coincide with the solar hours. This difficulty is overcome by substituting the tabulated value nearest the required time and later compensating the consistent error introduced by an augmenting factor.

Since the record length is always finite, the harmonic constants of a constituent determined by this method are somewhat contaminated by the effects of other constituents. A first-order correction of these effects can be made by an elimination procedure. In general it is more efficient to take the record length equal to the synodic (beat) period of two or more of the principal constituents. Of course, the longer the record the better. Standard analyses consist of 29 days, 58 days, 369 days, and so on.

It is not practical to determine the harmonic constants of the lesser constituents in this way if errors or uncertainties of the data are of the same order of magnitude as their amplitudes. If tidal oscillations in the oceans were far from resonance then the amplitude H of each constituent should be expected to be approximately proportional to its theoretical coefficient C, and the local epochs G all to be near the same value. In other words, for the semidiurnal constituent X, Eq. (8) should hold.

$$\frac{H(X)}{C(X)} = \frac{H(M_2)}{C(M_2)} \qquad G(X) = G(M_2) \tag{8}$$

Here X is referred to M_2 for the reason that the latter is one of the principal constituents whose harmonic constants can be determined with best accuracy. Any other important constituent could be used. Inferring the harmonic constants of the lesser constituents by means of Eq. (8) is sometimes preferable to direct means. It should be borne in mind that a constituent of one species cannot be inferred from one of another species because their equilibrium counterparts have different geographic shapes and no general relationship such as Eq. (8) exists.

Once the harmonic constants are determined, the tide is synthesized according to Eq. (5), usually with the help of a special "tide-predicting machine," although any means of computation could be used. Usually only the times and heights of high and low water are published in the predictions.

Tidal friction. The dissipation of energy by the tide is important in the study of planetary motion because it is a mechanism whereby angular momentum can be transferred from one type of motion to another. An appreciable amount of tidal dissipation takes place in the ocean, and possibly also in the solid Earth. In 1952 Sir Harold Jeffreys estimated that about half the tidal energy present in the ocean at any time is dissipated each day. A large part of this dissipation takes place by friction of tidal currents along the bottom of shallow seas and shelves and along the coasts. The rate of dissipation is so large that there should be a noticeable effect on the tide in the oceans.

If the planet's speed of rotation is greater than its satellite's speed of revolution about it, as is the case in the Earth-Moon system, then tidal dissipation always tends to decelerate the planet's rotation, with the satellite's speed of revolution changing to conserve angular momentum of the entire system. The Moon's attraction on the irregularly shaped tidal bulge on the Earth exerts on it a decelerating torque. Thus tidal friction tends to increase the length of day, to increase the distance between Earth and Moon, and to increase the lunar month, but these increases are infinitesimal. The day may have lengthened by 1 sec during the last 120,000 years because of tidal friction and other factors. [GORDON W. GROVES]

Bibliography: A. Defant, *Ebb and Flow: The Tides of Earth, Air, and Water*, 1958; G. Godin, *Analysis of Tides*, 1972; H. Lamb, *Hydrodynamics*, 6th ed., 1945; P. Schureman, *A Manual of Harmonic Analysis and Prediction of Tides*, U.S. Coast and Geodetic Survey Spec. Publ. no. 98, 1941.

Time

The dimension of the physical universe which, at a given place, orders the sequence of events; also, a designated instant in this sequence, as the time of day, technically known as an epoch.

Measurement. Time measurement consists in counting the repetitions of any recurring phenomenon and, if the interval between successive recurrences is sensible, in subdividing it. The phenomenon most used has been the rotation of Earth, where the counting is by days. Days are measured by observing the meridian passages of stars and are subdivided with the aid of precision clocks. The day is, however, subject to variations in duration; consequently, when the utmost precision is required, years instead of days are measured and subdivided. *See* DAY.

A determination of time is synonymous with the establishment of an epoch; it consists in ascertaining the clock correction, which is the correction that should be applied to the reading of a clock (positive if the clock is slow) at a specified epoch. A time interval may be measured in two ways: as the duration between two known epochs or simply by counting from an arbitrary starting point, as is done with a stopwatch.

Time units are the intervals between successive recurrences of phenomena, such as the period of rotation of Earth, also arbitrary multiples and subdivisions of these intervals, such as the hour, being one twenty-fourth of a day, and the minute, being one-sixtieth of an hour. *See* MONTH; YEAR.

Time bases. Several phenomena are used as the time base to be divided into hours. For astronomical purposes, sidereal time is used; for terrestrial purposes, solar time is used.

Sidereal time. The hour angle of the vernal equinox is the measure of sidereal time. It is reckoned from 0 to 24 hr; each hour is subdivided into 60 sidereal min and the minutes into 60 sidereal sec.

Sidereal clocks are used for convenience in most astronomical observatories, because a star, or other object outside the solar system, comes to the same place in the sky each night at virtually the same sidereal time.

Solar time. The hour angle of the Sun is the apparent solar time. The only true indicator of apparent solar time is a sundial.

Mean solar time has been devised to eliminate the irregularities in apparent solar time that arise from the obliquity of the ecliptic and the varying speed of Earth in its orbit around the Sun. It is the hour angle of a fictitious mean Sun, an imagined point moving uniformly along the celestial equator at the same rate as the average rate of the actual Sun along the ecliptic. In practice, it is intervals of sidereal time that are directly observed, and afterward converted into intervals of mean solar time by division by 1.00273790926.

Because sidereal and solar time are both defined as hour angles, at any instant they vary from place to place on Earth. When the mean Sun is on the meridian of Greenwich, the mean solar time is 12 noon at Greenwich. At that instant of absolute time, the mean solar time for all places west of Greenwich is earlier than noon and for all places east of Greenwich later than noon, the progression being at the rate of 1 hr for each 15° of longitude, or 12 hr for a semicircumference. Thus, at the same instant, at a short distance east of the 180th meridian, the mean solar time is 12:01 A.M., and at a short distance west of the 180th meridian it is 11:59 P.M. of the same day. Thus, persons going westward around the Earth must advance their time one day, and those going eastward must retard their time one day, in order to be in agreement with their neighbors when they return home. The International Date Line is the name given to a line following approximately the 180th meridian, but avoiding inhabited islands, where the change of date is made. Mean solar time at Greenwich is called Greenwich mean time.

Zone and standard times. To avoid the inconvenience of the continuous change of mean solar time with longitude, zone time or civil time is the time generally used. Earth is divided into 24 time zones, each approximately 15° wide and centered on standard longitudes 0, 15, 30°, and so on as illustrated. Within each of these zones, the time kept is the mean solar time of the standard meridian. Most developed nations use zone time.

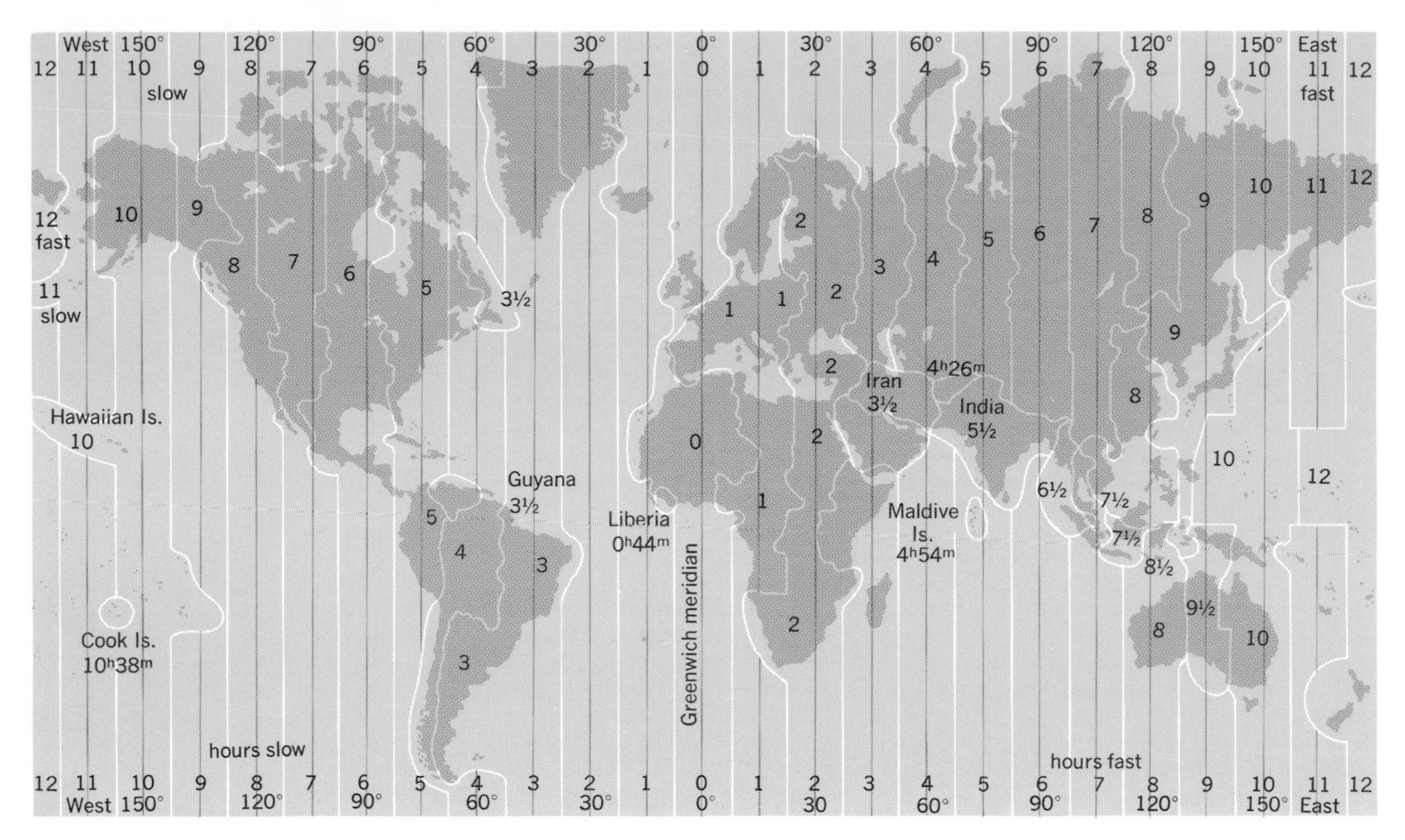

The world is divided into 24 standard time zones, progressively differing from Greenwich time by 1 hr. Some countries use half-hour intervals or fractional hours, Venezuela being 4½ hr slow and India being 5½ hr fast on Greenwich time.

Zone time is reckoned from 0 to 24 hr for most official purposes, the time in hours and minutes being expressed by a four-figure group followed by the zone designation, as 1009 zone plus 5, referring to the zone 75° west of Greenwich. The various zones are sometimes designated by letters, especially the Greenwich zone, which is Z, 1509 Z meaning 1509 Greenwich mean time. The zone centered on the 180th meridian is divided into two parts, the one east of the date line being designated plus 12 and the other minus 12. The time July 2, 2400 is identical with July 3, 0000.

In civil life the designations A.M. and P.M. are often used, usually with punctuation between hours and minutes; thus 1009 may be written as 10:09 A.M. and 1509 as 3:09 P.M. In this system, noon is correctly designated as 12:00 M. Sometimes July 2, 2400 is called July 2, 12:00 P.M. The designation July 3, 12:00 A.M. is not used, although it is logically the same as July 2, 12:00 P.M. The designations for noon and midnight are, however, often confused, and it is better to write 12:00 noon and July 2–3, 12:00 midnight in order to avoid ambiguity. In some occupations where time is of special importance, there is a rule against using 12:00 at all, 11:59 or 12:01 being substituted. The time 1 min after midnight is 12:01 A.M. and 1 min after noon is 12:01 P.M.

The figure shows the designations of the various time zones, the longitudes of the standard meridians, the letter designations, and the times in the various zones when it is noon at Greenwich.

In the United States the boundaries of the time zones are fixed by the Interstate Commerce Commission, and here as elsewhere the actual boundaries depart considerably from the meridians exactly midway between the standard meridians.

Ships at sea and transoceanic planes use Greenwich mean time for navigation and communications, but for regulating daily activities on board they use any convenient approximation to zone time, avoiding frequent changes during daylight hours. [GERALD M. CLEMENCE]

Daylight saving time. Many countries of the world, including the United States, advance their time 1 hr during the summer months, into daylight saving time. Thus 6 A.M. standard time becomes 7 A.M. daylight saving time. Such a practice effectively transfers an hour of little-used early morning light to the evening. It is particularly advantageous in urban areas where manufacturers and other industries can save on electric power and the residents can benefit from the daylight hour in the evening. This practice is of little value in areas far north, with naturally long days and short nights, or in the tropical areas where days and nights are more nearly equal.

The United States used daylight saving time, countrywide, in both world wars, and in 1966 adopted the Uniform Time Act, effective Apr. 1, 1967, which requires all states to use daylight saving time from the last Sunday in April to the last

Sunday in October. However, a state may, by legislation, exempt itself from daylight saving time provided that the entire state remains on standard time. [VAN H. ENGLISH]

Ephemeris time. The orbital motions of the Moon and planets are used for ephemeris time. It is free from the irregularities in mean solar time caused by variations in the rate of rotation of the Earth and is determined in practice as a correction to Greenwich mean time, which will bring observations of the right ascension and declination of the Moon into agreement with the theoretically calculated values. Clocks are not actually regulated to ephemeris time; it is sufficient to keep a record of the corrections necessary.

Time signals are pips emitted by radio stations in most civilized countries, enabling the listener to ascertain the zone time accurate to a small fraction of a second. In the United States time signals consist of coded seconds pulses emitted at frequent intervals by naval radio stations and by station WWV of the National Bureau of Standards, the precision being controlled by the U.S. Naval Observatory. *See* CALENDAR.

Atomic time. In 1967 the International Conference on Weights and Measures adopted a new definition of the second of time: 9,192,631,770 transitions between two specified hyperfine levels of the atom of cesium-133. As thus defined, the measurement of time becomes independent of astronomical observations. Navigators and surveyors still require mean solar time and astronomers still require ephemeris time for their special purposes. Comparisons of the three different ways of measuring time will eventually yield much information of scientific interest. *See* ATOMIC TIME.

[GERALD M. CLEMENCE]

Time-interval measurement

Time interval is measured with high precision, most conveniently, with an electronic, digital reading counter. This contains an oscillator, which has a frequency, say, of 10^7 hertz. A starting signal causes the counter to start counting cycles, and a second signal causes the counter to stop. The accumulated number is displayed by digital indicating lights and gives the time interval which has elapsed between the two signals in units of 0.1 microsecond.

Short time intervals may be measured with high precision, to 1 nanosecond (10^{-9} s), with a suitable oscilloscope. The time interval between two pulses is measured by the horizontal displacement between the pulses as displayed on the oscilloscope.

Ultra-short phenomena in the picosecond range (1 ps$=10^{-12}$ s) are studied by use of short pulses generated by lasers (about 0.5–10 ps). Time interval is measured in several ways—for example, by using an electrooptical shutter (Kerr cell) or by releasing electrons at a photocathode and sweeping them across a screen, as in an oscilloscope.

A stopwatch provides an accuracy of about 0.2 s.

Radioactive decay is used to measure long time intervals concerning human history, the Earth, and the solar system.

[WILLIAM MARKOWITZ]

Transit

The apparent passage of a planet across the surface of the Sun, of a satellite across the surface of the parent planet, or of a star, planet, or reference point across an adopted line of reference.

Only Mercury and Venus are seen in transit across the surface of the Sun, because they are the only planets orbiting inside the path of Earth.

Transits of Venus. These are very rare. The Earth must be essentially in a straight line with Venus and the Sun, therefore Venus must be in inferior conjunction at the same time that it passes one of the nodes of its orbit. Conjunction must occur within 2 days of June 7 and Dec. 9 to fulfill the conditions. The last transit of Venus took place in 1882. There will not be another one until 2004. *See* PLANET· VENUS.

Transits of Mercury. These are relatively more frequent and occur at the rate of about 13 per century. The same geometric conditions are required as for transits of Venus, but the limits are not as narrow. Conjunction must occur within 3 days of May 8 or within five days of Nov. 10 (see illustration). Because of this, November transits are about twice as frequent as May transits. Observations of November transits yield more accurate results than those of May transits, because in November the motion of the planet is more rapid, thus permitting better timing of the contacts. *See* MERCURY.

Four contacts are observed: exterior ingress, interior ingress, interior egress, and exterior egress, designating respectively the exterior and interior points of tangency between the planet and the Sun at the beginning of the transit (ingress) and the interior and exterior points of tangency at the end (egress).

Transits of Mercury are observed for the purpose of determining the exact position of the planet and to improve data on the elements of the orbit.

Transits of Jupiter's satellites. Transits of the Galilean satellites of Jupiter occur at each of their

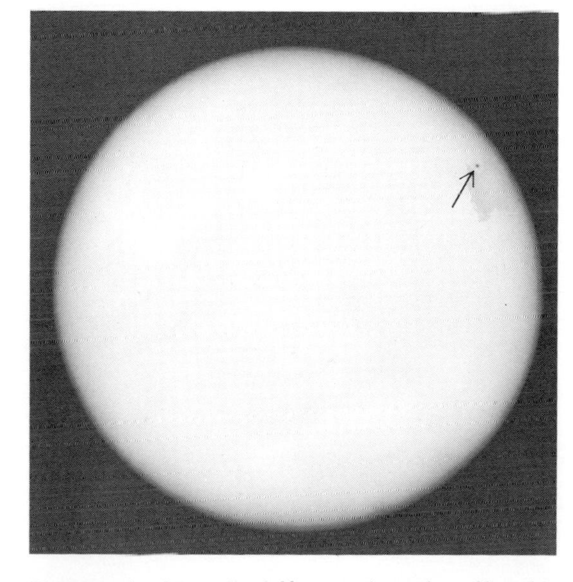

Photograph of transit of Mercury (arrow) on Nov. 14, 1953, taken at Washington, D.C. (*Official U.S. Navy Photograph*)

inferior conjunctions with the exception of satellite IV, which occasionally passes clear of the planet's disk. They are difficult to observe and are used mainly to estimate the albedo (reflectivity) of the satellites relative to that of Jupiter. As each satellite passes in front of the planet, it casts its shadow on the planet's disk and causes the phenomenon of shadow-transit. *See* JUPITER.

Transits of stars. Passages of stars across the local meridian are observed extensively for solving problems of fundamental positional astronomy, timekeeping, and navigation. At the precise instant of transit of a star across the local meridian, the local sidereal time is exactly equal to the star's right ascension, and the latitude of the observer is equal to the sum of the star's declination and its zenith distance.

Furthermore, the difference between the local sidereal time and the Greenwich sidereal time gives a measure of the longitude of the observer. Any of those quantities may be treated as the unknown to be determined by observations of meridian transits. *See* ASTRONOMICAL COORDINATE SYSTEMS; ECLIPSE.

[SIMONE D. GOSSNER]

Trojan asteroids

Asteroids located near the equilateral lagrangian stability points of the Sun-Jupiter system (see illustration). As shown by J. L. Lagrange in 1772, these are two of the five stable points in the circular, restricted, three-body system, the other three points being located along a line through the two most massive bodies in the system. In 1906 Max Wolf discovered an asteroid located near the lagrangian point following Jupiter in its orbit. Within a year, two more were found, one of which was located near the preceding lagrangian point. It was quickly decided to name these asteroids after participants in the Trojan War as given in Homer's *Iliad*. Hence the term Trojan asteroid refers to asteroids orbiting the Sun near one of Jupiter's equilateral lagrangian points. With the exception of (624) Hektor in the preceding "cloud" and (617) Patroclus in the following "cloud," asteroids in each of these clouds are named after Greek and Trojan warriors, respectively.

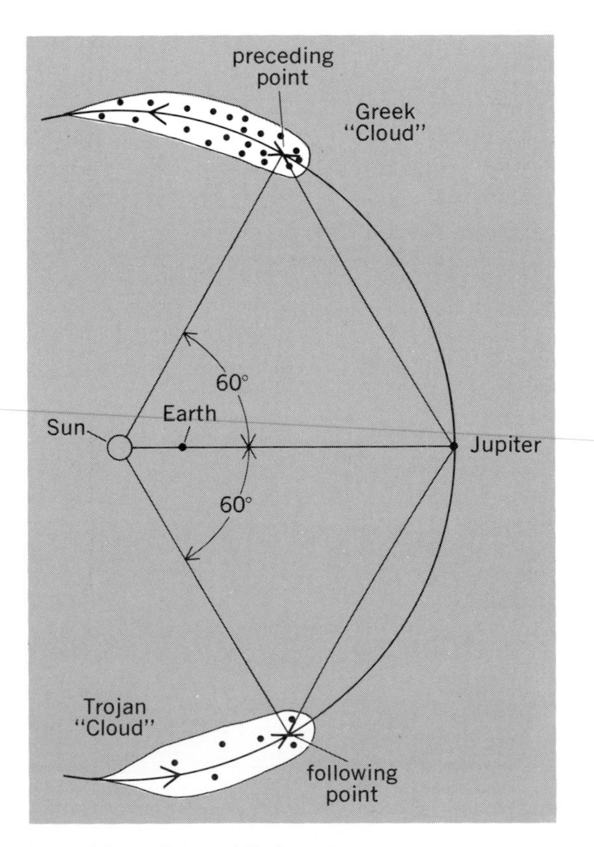

Lagrangian points and Trojan asteroids.

One sometimes uses the term Trojans in a generic sense to refer to hypothetical objects occupying the equilateral lagrangian points of other pairs of bodies. Unsuccessful searches have been made for Trojans of the Earth, Saturn, and Neptune, as well as for the Earth-Moon system. It is doubtful whether bodies near these lagrangian points would be in stable orbits, because of perturbations by the major planets. Indeed most of Jupiter's Trojans do not move in the plane of its orbit, but in orbits inclined by as much as 25° and at longitudes differing by up to 40° from the longitudes of the theoretical lagrangian points.

As of 1979, there were 25 numbered Trojan asteroids, about a dozen of which have diameters greater than 100 km. However, photographic surveys have shown that there are about 900 with diameters exceeding 15 km. Approximately 700 of these are located around the preceding point, while only about 200 are found near the following point.

Trojans are dark objects reflecting only between 2 and 5% of the visual light they receive. The majority are compositionally similar to the most common type of outer main-belt asteroid, but some, perhaps as many as one-third, have no known analog among the asteroids or meteorites. *See* ASTEROID; METEORITE.

[EDWARD F. TEDESCO]

Bibliography: J. Degewij and C. J. van Houten, Distant asteroids and outer Jovian satellites, in T. Gehrels (ed.), *Asteroids*, pp. 417–435, 1979; T. Gehrels, The asteroids: History, surveys, techniques and future work, in T. Gehrels (ed.), *Asteroids*, pp. 3–24, 1979.

Twilight

The period between sunset and darkness in the evening, and between darkness and sunrise in the morning. The following statements apply to evening twilight; the reverse would apply to morning twilight.

The characteristic light is caused by atmospheric scattering, which transmits sunlight to the observer for some time after the Sun has set. It depends geometrically on latitude, longitude, and elevation of the observer, and on the time of year. Physically it depends also on local conditions, particularly the weather.

Three degrees of twilight are conventionally distinguished. Civil twilight ends when the center of the Sun is 6° below the horizon; if the sky is clear, it is usually practicable to carry on ordinary outdoor occupations without artificial light during civil twilight. Nautical twilight ends when the depression of the Sun is 12°; at this time both the horizon and the brighter stars are visible. Astronomical twilight ends when the depression of the Sun is 18°; at this time no trace of illumination by the Sun appears in the sky. As thus defined, the times of ending of the three sorts of twilight can be precisely calculated.

[GERALD M. CLEMENCE]

Bibliography: G. V. Rozenberg, *Twilight*, 1966; U.S. Naval Observatory, *Tables of Sunrise, Sunset, and Twilight*, supplement to *American Ephemeris*, 1946, reprint 1962.

Twinkling stars

Stars exhibiting rapid fluctuations of the brightness and size of the images caused by the Earth's atmosphere; the phenomenon is also called astronomical seeing or scintillation. Irregularities of temperature, density, and water-vapor content, accompanied by changes of refractive index, are carried across the beam of light from the star and produce an irregularly corrugated wave front. Irregularities of the electron density in the ionosphere also change the refractive index at radio frequencies and cause twinkling of extraterrestrial sources of radio noise.

Small radio sources have been found to vary rapidly in brightness because of the flow of density fluctuations in the solar wind across the path between the radio source and the Earth. The method of interplanetary scintillations has proved a useful means of probing the interplanetary medium near the Sun and also of establishing the very small size of some of the radio sources. Radio sources whose angular diameters are large do not show interplanetary scintillation at all, just as planets do not share in the stellar twinkling. The pulsars show erratic changes of amplitude in their radio emission, caused both by radio scintillation near the neutron star or in interstellar space. See PULSAR; RADIO SOURCES; STAR.

The typical size of the optical irregularities is large compared to the eye but small compared to a telescope. If few are present and observed with a small telescope, they behave like independent lenses or wedges, moving the image bodily. If many are present, the image is blurred and large. Much seeing is caused by high-altitude turbulence and winds. However, the flow of air over local terrain obstacles and around buildings produces considerable microturbulence, which has been found to affect the performance of telescopes. Newer telescopes are placed at least 100–150 ft (30–45 m) above the ground level, where the microclimate becomes more stable. Prior to choice of a site, studies of the microturbulence are made from towers measuring wind and temperature fluctuation. Extraordinarily good conditions of astronomical stability have been found in the foothills of the Andes in Chile. The Inter-American Astronomical Observatory, funded by the National Science Foundation, has set up instruments at Cerro Tololo, and a 150-in. (4-m) reflector there takes advantage of the extraordinarily stable conditions. These conditions are similar to those in southern California but apparently are of even higher quality. A consortium of European countries has also built a large telescope in Chile, and another 100-in. (2.54-m) telescope has been completed by the Carnegie Institution of Washington.

The best astronomical seeing (at night) gives images about 0.3 second of arc; good seeing gives 1 second; poor seeing gives up to 10 seconds; daytime seeing is generally worse. Observations up to an altitude of 40,000 ft (12 km) indicate that seeing may improve with altitude, and balloon flights have given excellent solar photographs. Seeing is worst at large zenith distances, where chromatic effects appear. Seeing has less effect on photographs of the Earth taken from high-altitude satellites. The proposed large space telescope to be launched by the space shuttle will be the first to be immune to image degradation by seeing. See ASTRONOMICAL OBSERVATORY.

[JESSE L. GREENSTEIN]

Bibliography: H. G. Booker and W. E. Gordon, A theory of radio scattering in the troposphere, *Proc. Inst. Radio Eng.*, 38:401–412, 1950; S. Chandrasekhar, A statistical basis for the theory of stellar scintillation, *Mon. Not. R.A.S.*, 112: 475–483, 1952; G. Keller, The effect of electron screening on energy generation, *Astrophys. J.*, 118: 142–146, 1953; B. J. Rickett Interstellar scattering and scintillation of radio waves, *Annu. Rev. Astron. Astrophys.*, 15:479–504, 1977; J. Stock and G. Keller, Astronomical seeing, in G. P. Kuiper and B. Middlehurst (eds.), *Telescopes*, pp. 138–175, 1960.

Ultraviolet astronomy

Astronomical investigations utilizing observations carried out in the spectral region from approximately 350 to 90 nanometers. Astronomical observations in the ultraviolet and x-ray region of the spectrum must be carried out above the Earth's atmosphere, since the atmosphere ozone (O_3), oxygen (O_2), and nitrogen (N_2) are opaque at these short wavelengths. Such observations are made by using sounding rockets, satellites, or space probes.

Ultraviolet measurements have proved important for the studies of planets, stars, interstellar gas, and galaxies. Most of the light emitted from hot stars appears in the ultraviolet, and knowledge of the temperature and composition of these stars, usually quite young, depends upon such observations. The strongest spectral lines produced by the lighter and more abundant atoms occur in the ultraviolet. Studies of the ultraviolet line spectrum have proved important to understanding the chemical composition densities and temperatures of the interstellar gas and dust that exists in space between the stars. It is from this interstellar matter that stars are born. Ultraviolet observations of galaxies can provide vital information on the evolution of galaxies. See ASTRONOMICAL SPECTROSCOPY; GALAXY, EXTERNAL; INTERSTELLAR MATTER; STAR; STELLAR EVOLUTION.

Modern astrophysical knowledge depends upon the synthesis of observations from the entire electromagnetic spectrum. Ground-based optical and radio telescopes provide measurements in the near-ultraviolet, visual, and near-infrared spectrum, and the bulk of the radio region. Space observations yield data on infrared, ultraviolet, x-ray, and gamma-ray radiation. Each spectral region provides complementary insight on specific physical processes occurring in the cosmos. *See* GAMMA-RAY ASTRONOMY; INFRARED ASTRONOMY; OPTICAL TELESCOPE; RADIO ASTRONOMY; TELESCOPE; X-RAY ASTRONOMY.

Ultraviolet astronomy started with the advent of the space age. Early measurements in the 1950s were made with small telescopes which were carried aloft on sounding rockets and provided, at most, a few minutes of observations above the Earth's atmosphere. The first major astronomical observatory in space was the Orbiting Astronomical Observatory, which operated for several years in earth orbit. A number of scientific satellites have successfully continued these ultraviolet measurements. *See* ROCKET ASTRONOMY; SATELLITE ASTRONOMY.

The Space Telescope promises to be one of the most important advances in astronomical instrumentation in this century. It will carry out measurements never before possible throughout the ultraviolet optical and near infrared. With the advent of the space shuttle and Space Telescope, ultraviolet astronomy will become merely an extension of classical astronomy to somewhat shorter wavelengths.

[ARTHUR D. CODE]

Universe

The sum of all the matter and energy that exist. The true scale of the universe is even now only dimly perceived by astronomers, despite tremendous advances in both observing techniques and equipment. The planet Earth seems large to humans, yet it is small compared with the Sun, which is only one of a hundred billion stars in the Milky Way Galaxy. And as large as this "island universe" is, it too is small compared with the clusters of thousands of such galaxies which populate the universe out to and (undoubtedly) beyond the range of astronomical vision. *See* GALAXY.

Distance and time. The immense size of the universe and the great distances between objects introduce some fundamental limitations upon observations of remote galaxies. The basic restriction is caused by the finite speed of light in space. Light travels of 300,000 km/s; it therefore requires some 8.3 min to travel from the Sun to the Earth. Note that light travel time can therefore be used as a measure of distance (the Sun is 8.3 "light-minutes" from the Earth). A more important restriction is the following: the Sun is seen as it was when light left its surface 8.3 min ago. Therefore the past is always being seen, not the present; this "looking into the past" becomes more significant with increasing distance. The nearest large galaxy similar to the Milky Way is almost 2×10^6 light-years (1.89×10^{19} km) distant. A photograph made today of this galaxy records light which began its journey across space at a time when *Homo sapiens* was first beginning to appear on the Earth.

Most stars change very little in a hundred, million, or even a billion years, so the "time lag" caused by their distance is not a problem. Even the Andromeda Galaxy, 2×10^6 (1.89×10^{19} km) light-years away, has not changed significantly during the light travel time. Some stars have died, some new stars have formed, and the galaxy as a whole has rotated perhaps 1/100 of a turn. The overall structure and content of the galaxy have not changed, however, because light travel time is a small fraction of the evolutionary time scale for Andromeda. This is not true for very distant galaxies, however, whose light has been traveling across space for a large fraction of the age of the universe. A galaxy 10^{10} light-years (9.46×10^{22} km) away is seen as it was when the universe was half its present age; such a galaxy may have changed significantly during this interval. The universe and its galaxies can therefore be studied at earlier and earlier epochs merely by observing to greater and greater distances.

It is clear that the calculation of distances to remote galaxies is an important part of modern astronomy, because calculation of distance also fixes the relative age of the object being studied. The observation of spectral lines from distant galaxies shows them to be shifted from their normal laboratory wavelengths toward the red end of the spectrum. This so-called Doppler shift is due to the motion of these galaxies away from the Milky Way; the shift is proportional to the velocity of the galaxy. Most galaxies are moving away from the Milky Way, and fainter ones are moving more rapidly. E. P. Hubble showed that this is due to general expansion of the universe, and demonstrated that there is a relation between velocity v of a galaxy and its distance d, given by $v = H_o d$, where H_o, the constant of proportionality, is called the Hubble constant, and is a measure of the rate of expansion. It is possible to use this relation to determine the distance to a galaxy through observation of its velocity as shown in the shift of its spectral lines.

Cosmological principle. To be able to understand what is seen at very great distances from the Earth, the following assumption, called the cosmological principle, is required: the same laws of physics operate everywhere throughout the universe. Stated another way, that sector of the universe of which Earth is a part is assumed to be representative of all other sectors (on the scale of galaxies and clusters of galaxies). The force of gravity and the ways in which x-rays are produced are assumed to be the same throughout the universe. There is as yet no experimental basis for this assumption. *See* COSMOLOGY.

Organization. On the scale of atoms, the universe is over 90% hydrogen. Most of these atoms are collected into gaseous stars which, like the Sun, are transforming hydrogen into helium by means of nuclear fusion reactions in their extremely hot interiors. This hydrogen burning is therefore slowly changing the composition of the universe. Stars are collected into vast galaxies; the Milky Way Galaxy alone contains over 10^{11} stars. Galaxies in turn are grouped into clusters of galaxies; each cluster may contain thousands of galaxies, each having billions of stars. Galaxies, or perhaps clusters of galaxies, may be considered the structural blocks of the universe, but stars are the basic unit of these blocks.

Stars and planets. The Sun is a typical middle-aged star, with a mass of 2×10^{33} (that is, some 333,000 times the mass of the Earth). Larger and smaller stars are known. The smallest stars able to sustain hydrogen burning are roughly one-tenth the size of the Sun. The largest normal stars are

Model of the solar system*

Body	Diameter, km	Representation in model	Diameter in model	Distance from Sun (Earth = 1)	Distance from model "Sun"
Mercury	4,864	Apple seed	0.125 in. (3.2 mm)	0.387	124 ft (37.8 m)
Venus	12,100	Pea	0.311 in. (7.9 mm)	0.723	232 ft (70.8 m)
Earth	12,756	Pea	0.328 in. (8.3 mm)	1	321 ft (97.9 m)
Mars	6,788	Small ball bearing	0.175 in. (4.5 mm)	1.52	489 ft (149.1 m)
Asteroids	<1,000	Large grain of sand	0.026 in. (0.7 mm)	2.8	900 ft (274.5 m)
Jupiter	137,400	Tennis ball	3.533 in. (89.7 mm)	5.2	1,671 ft (509.7 m)
Saturn	115,100	Racquet ball	2.96 in. (75.2 mm)	9.5	3,054 ft (931.5 m)
Uranus	50,000	Cherry	1.29 in. (32.8 mm)	19.2	6,171 ft (1,882.2 m)
Neptune	49,400	Cherry	1.27 in. (32.3 mm)	30.1	9,675 ft (2,950.9 m)
Pluto	?	?	?	39.4	12,664 ft (3,862.5 m)
Comets	<100	Tiny grain of sand	0.03 in. (0.08 mm)	50,000	3,030 mi (4,878 km)
Nearest star		Beachball	3 ft (92 cm)	4.3 light-years	16,324 mi (26,282 km)

*In this model, the Sun is represented by a beachball 3 ft (92 cm) in diameter, and the scale is 3 ft = 1.4×10^6 km.

about 10 times the size of the Sun, although immense, swollen, evolved red giants do exist which may become so large as to swallow most of their inner planet system. Tiny white dwarfs the size of the Earth are also known; these cinders have been slowly cooling off after their nuclear fire died, and represent the end state for the Sun some 6 to 7×10^9 years from now. The lifetime of a star depends on its mass. The Sun will continue to burn hydrogen without significant change in its structure for about 5×10^9 years, thereby doubling its present lifetime. Heavier stars burn out faster and more violently, whereas the smaller, lighter stars will live 100 times longer than the Sun.

Stars are, in general, not isolated in the Galaxy. Most have companions of some kind, and most companions are other stars. Some companions are too small to sustain nuclear reactions; these are called planets. In addition to the nine planets which move about the Sun, there are 33 known satellites of the planets, more than 2000 well-observed asteroids (of 50,000 believed to exist), and an unknown number of comets and meteoroids. So tiny are planets, their satellites, and the comets compared with the Sun that these small companions are nearly undetectable from a distance of 10 light-years (9.46×10^{13} km). *See* SUN.

The table shows the solar system in terms of a model in which the Sun is represented by a beachball 3 ft (92 cm) across. On this scale the Earth is the size of a pea, 321 ft (97.9 m) from the Sun (roughly the distance across a football field)! The Moon is only a small grain of sand some 10 in. (25.4 cm) from the Earth. The largest planet, Jupiter, is represented as a tennis ball 3.5 in. (89.7 mm) across and almost 0.3 mi (509.7 m) from the Sun. Comets, which in reality are 10-km-diameter balls of frozen methane, water, and other gases that occasionally come into the inner solar system (but generally lie 50,000 times as far from the Sun as does the Earth), are represented by pinpoints more than 3000 mi (4878 km) from the Sun. This cloud of comets represents the edge of the solar system.

In terms of travel, the solar system seems large: signals from a spacecraft located at the distance of Pluto would take nearly 6 hr to reach the Earth. The radius of the cloud of comets is about 7.5×10^{12} km, or about 290 light-days; this effectively measures the size of the solar system. By comparison, the nearest star (another beachball in the model of the table) is 4.3 light-years from the Sun (about 16,300 mi or 26,282 km in the model). The entire solar system is only about two-tenths the distance to the nearest star.

Galaxies. The number of planetary systems in the Galaxy is not known, but estimates range into the hundreds of millions. But each of these systems is small compared with the galaxy's diameter of 100,000 light-years (9.46×10^{17} km). The 10^{11} stars of the Galaxy are spread out in a thin disk with a central bulge; only a small percentage of the stars lies above or below this plane. For the Galaxy as a whole, the average separation between stars is a few light-years. The Sun lies about 30,000 light-years (2.84×10^{17} km) from the dense galactic center, on the inner edge of one of the spiral arms (which contain hot young stars, gas, and dust) that characterize this type of galaxy. So great is this distance from the center (more than 36,000 times the radius of the solar system) that the Sun requires some 2.5×10^7 years to completely circle the Galaxy. That is, the solar system has orbited only 18 times since the Sun was born (compared with nearly 4.5×10^9 trips of the Earth around the Sun).

While spiral galaxies like the Milky Way Galaxy are composed mostly of stars, perhaps 15% of the total mass (of 10^{44} g) may be in the form of dust and gas (Fig. 1). This is the material from which new stars are formed, a process which continues today in spiral galaxies. No such star formation is observed in elliptical galaxies, however, which are spherical or oval systems of dim reddish stars with almost no gas or dust. Elliptical galaxies outnumber spirals; especially common are dwarf ellipticals of a few million stars spread over 6000 light-years (5.67×10^{16} km) or so. These small galaxies are seen often as satellites to larger systems such as the Andromeda Galaxy. Much rarer are the giant ellipticals with 10^{13} stars (or 100 times the number in the Milky Way). These giants are often found at the center of large clusters of galaxies, and may be three or four times larger than the large spirals.

One simple way to characterize these galaxies is by their mass-to-luminosity ratio (M/L).

Fig. 1. The great spiral galaxy in Andromeda (M31, NGC 224) and its two small elliptical companions (NGC 205 and 221), photographed with the 48-in. (1.2-m). Schmidt telescope. (*Hale Observatories*)

Both the mass and the brightness of a galaxy depend on the stars in that galaxy; if most of the stars are of the dim, red type, M/L will be a high number. Elliptical galaxies are of this type, and have M/L about 4–10 (M/L for the Sun is 1). Giant ellipticals, with their greater mass, may have $M/L = 30$. A spiral galaxy such as the Milky Way has a large number of hot and bright blue stars, so it is brighter compared with its size than an elliptical, and has $M/L < 1$. The M/L effectively measures the stellar content of a galaxy.

Clusters of galaxies. Essentially all the stars in the universe are found in galaxies (there are effectively none between the galaxies). Most galaxies themselves are not solitary, but are arranged into clusters, which may be composed of a number of subclusterings. The Milky Way Galaxy is part of a collection of about 24 galaxies called the Local Group. This cluster is some 3×10^6 light-years (2.838×10^{19} km) across and is dominated by two large spiral galaxies, the Milky Way and the Andromeda galaxies. A third, somewhat smaller spiral, four irregular, and four regular but small elliptical galaxies, are the major members of the Local Group. There are also an unknown number of dwarf ellipticals present (nine have been detected in a limited search). Observations suggest the presence of two or three wanderers that have penetrated the Local Group; one of them may be a giant elliptical, although observations of it are difficult due to obscuration by the dust of the Milky Way.

Small groups such as the Local Group are probably common, but much larger clusters also exist. The nearest of these, the Virgo Cluster, contains several thousand galaxies of all types within a diameter of about 6×10^6 to 7×10^6 light-years (5.67×10^{19} to 6.62×10^{19} km). The Virgo Cluster, which lies some 5×10^7 light-years (4.73×10^{20} km) from the Local Group, is a typical loose or irregular cluster of galaxies, in which spirals and normal ellipticals are roughly equal in number. In richer, denser regular clusters, spirals are absent, and the number of elliptical galaxies may be 10,000. For such a cluster, the density of galaxies near the center may be a thousand or more times the density of galaxies in the Local Group, making such concentrations the most tightly packed groupings in the universe, relative to the size of their members, as discussed below. The nearest rich regular cluster is called Coma; it lies some 3×10^8 light-years (2.84×10^{21} km) away and is at least 10^7 light-years (9.46×10^{19} km) across.

Thousands of clusters of galaxies have been observed, and some astronomers suggest that these are arranged into clusters of clusters, or superclusters. The Local Supercluster would then have some 100 clusters as members, including the large Virgo Cluster and the Local Group, and it would be some 10^8 to 1.5×10^8 light-years (9.46×10^{20} to 1.419×10^{21} km) across.

One way to compare these objects and how they are grouped is through the computation of a "packing factor," which may be defined as the separation of two objects divided by their mean diameter. This factor shows how closely spaced are objects relative to their size. For example, within the Milky Way Galaxy the average separation between two stars is about 2 light-years, or 1.9×10^{13} km. The diameter of a star such as the Sun is 1.4×10^6 km. Therefore stars are on the average 1.4×10^7 times farther apart than they are large. Compare this with the packing factor for galaxies in the Local Group: The Andromeda spiral galaxy is about 100,000 light-years (9.46×10^{17} km) across and is about 2×10^6 light-years (1.89×10^{19} km) distant. Therefore, within the Local Group spirals are about 20 times as distant as they are large, or 700,000 times more closely packed than the stars in the Milky Way! Inside dense clusters, such as the Coma Cluster, galaxies are spaced a thousand times more closely than this. The same type of computation for packing of clusters of galaxies suggests that clusters are 10 or less times more distant than they are large, at least within the Local Supercluster. *See* GALAXY, EXTERNAL; SUPERCLUSTERS.

Properties. The universe is dynamic. The early observation that all galaxies are receding led to the now well-accepted conclusion that the universe is expanding. This expansion implies that the universe is evolving. Observations of quasars support this view. These immensely powerful sources of radiation all seem to lie at great distances from Earth. The lack of proved nearby quasars suggests that quasars were a part of an earlier epoch of the universe. Also, very remote clusters of galaxies seem to be denser than nearby clusters, as would be expected if the universe had been smaller in the past, as implied by the expansion.

There are several ways to determine a rough value for the age of the universe. The universe must be older than the Milky Way Galaxy, whose age is estimated at something like 1.4×10^{10} years, based on studies of the oldest known stars. This represents a lower limit to the age of both the Galaxy and the universe. (The Galaxy may, in fact, be somewhat older than the oldest observed stars.) The expansion of the universe itself provides an upper limit to age. Extrapolating the expansion backwards (assuming that the observed rate has been constant since the expansion started), all of the observed galaxies would have started from the same point about 2×10^{10} years ago. The expansion rate was probably greater in the early history of the universe; the age of the universe is therefore somewhat less than the 2×10^{10} years given by the Hubble expansion time.

The universe is at least as large (and as old) as the most distant galaxy or cluster of galaxies observed thus far. But since more and more distant galaxies are still being found, it can only be said that the universe is larger than about 8×10^9 light-years (7.56×10^{22} km) across.

Evolution of the universe. One explanation for the observed expansion of the universe is the "big bang" theory, in which all of the matter and radiant energy of the universe were initially compressed into a single object. At the assumed temperatures of some trillions of degrees, matter could not have existed in its present form, but consisted of subatomic or even subnuclear particles. For reasons unknown, this "primordial egg" (which may have been as small as a planetary system) exploded, hurling matter and radiation outward in a slowly decreasing expansion that led to rapidly dropping temperatures. Ten seconds after the explosion, normal protons, neutrons and electrons could exist in the billion-degree environment. After 100 s of expansion, protons and neutrons began combining at temperatures of 100,000,000 K to form deuterium, a heavy isotope of hydrogen. Reaction with additional protons in the dense conditions produced helium. It has been estimated that some 20% of the original protons were converted into helium in this way within the first several minutes of the universe's existence. After this the expanding matter cooled below 10,000,000 K, making further nucleosynthesis (manufacturing of new elements) impossible. The other 90 natural elements which are part of the universe owe their existence to the nuclear fusion reactions which would occur much later inside stars. *See* BIG BANG THEORY.

Steadily dropping temperatures accompanied the continuing expansion for the next several million years. The density of the universe remained high enough to trap the radiant energy of the preexplosion time until temperatures fell to about 5000 K. At this point electrons combined with protons to produce neutral hydrogen atoms; as a result radiation was decoupled from matter and was released to flood the universe.

A small remnant of this fireball radiation is in fact observed to be uniformly filling the universe. As with any radiation from a very distant source, the original high-temperature energy has been degraded or red-shifted to much longer wavelengths. The radio observations of this 3 K blackbody radiation (as though the universe were at a temperature of 3 K) constitutes the most conclusive proof that some type of "big bang" did indeed occur. The remnant background of the original fireball cannot be explained by alternate theories of the universe which assume a steady state or nonevolutionary history.

Only after the universe had expanded to where the temperature dropped to about 300 K could galaxies begin to form out of gravitational condensations of the expanding gas. Within these large condensations, smaller centers of contraction became the stars which today constitute the major fraction of a galaxy's mass.

Future of the universe. The current expansion of the universe is a product of the "big bang" explosion, and the present rate of expansion should be less than the original rate because of the gravitational slowdown produced by the matter of the universe when it was still small and dense. How long the universe will continue to expand will depend on whether or not the explosion was stronger than the force of gravity due to all the matter of the universe. If the explosion was more powerful than the gravity of the universe, then the universe has "escape velocity" and will expand forever. Such an "open" universe is infinite in both time and space. Galaxies will continue to separate, and the average density of the universe will continue to decrease. The other extreme is given by the re-

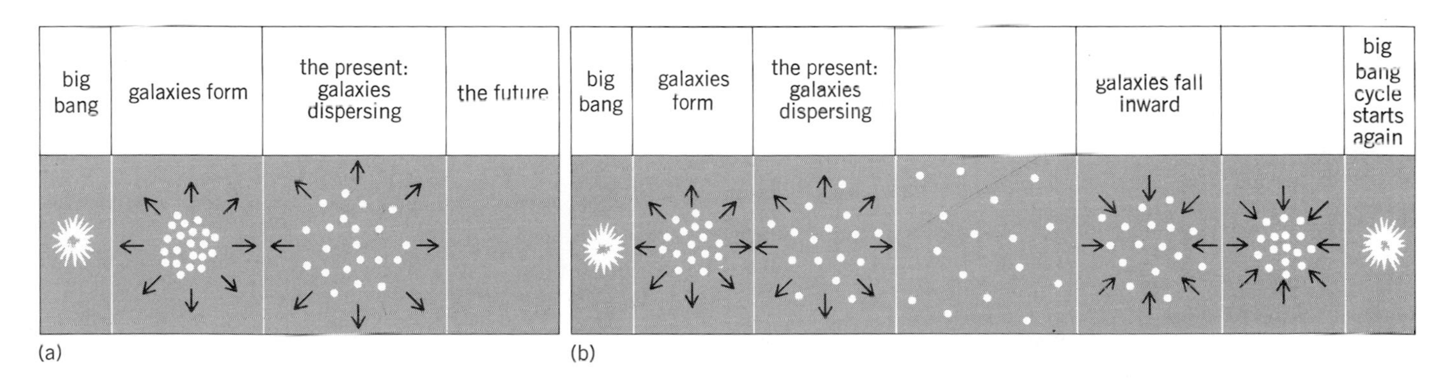

Fig. 2. Comparison of the open and oscillating universes. (*a*) In the open universe, matter continually thins out, and the galaxies continually move farther apart. (*b*) In an oscillating universe, expansion is followed by contraction until another "bang" causes renewed expansion. (*From R. Jastrow and M. H. Thompson, Astronomy: Fundamentals and Frontiers, pp. 269 and 277, 3d ed., copyright © 1977 by John Wiley & Sons; reprinted by permission*)

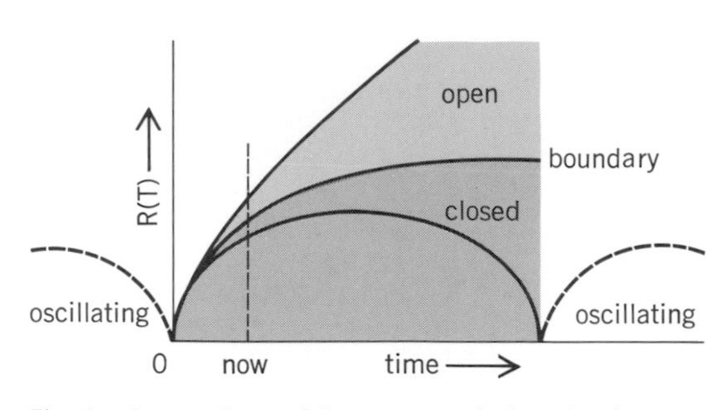

Fig. 3. Comparison of the open and closed universes. $R(T)$ effectively measures the radius of the universe, or the distance between two galaxies. *(From J. M. Pasachoff, Contemporary Astronomy, p. 555, W. B. Saunders, 1977)*

verse situation: the total density of matter in the universe is so great that its gravity is greater than the explosion-induced expansion. In this case the universe will eventually come to rest, then reverse and fall back on itself. Galaxies will approach one another, crowding together until they lose their separate identities. Matter will revert to basic atomic particles and then to more fundamental particles. Radiation will become trapped within the shrinking universe, temperatures will rise, and conditions will approach those of the previous "big bang." It seems reasonable that the universe will then "bounce" or explode in another "big bang" and start over. Such an "oscillating" universe is finite in extent but infinite in time; there is no way of knowing which oscillation is going on at present, as records of the previous oscillations are destroyed in each new "bang" (Figs. 2 and 3). The third but unlikely possibility is that the expansion exactly balances the force of gravity. The universe will then expand and stop, remaining motionless for all time.

To decide which of these three possible futures is most likely, it is necessary to determine the density of matter in the universe. Direct estimates from observed galaxies fall short of the mass needed to close the universe (halt the expansion) by a factor of 1000. Attempts to account for "invisible" mass in the form of nonluminous stars, gas, and black holes (which can be estimated from the motions of galaxies within a cluster) indicate that some clusters of galaxies contain 30 or more times as much matter as would be seen in the galaxies as stars. This still leaves the universe short by more than 10 times. Indirect clues are available from the observed amount of "big bang" deuterium, which depends on the density of matter at the time of formation. Again there seems to have been at least 10 times too little matter to halt the expansion. Finally there is the change in the expansion of the universe itself: the greater the gravity, the greater the slowdown of the expansion rate with time. Observations of the very distant clusters of galaxies show that the expansion was only slightly faster in the past, which implies that the total gravity of the universe is too small to halt the motion of the galaxies. The universe thus appears to be open, and the best evidence suggests that the expansion will continue. Refutation of this would necessitate the discovery of large amounts of previously unknown mass in the universe. [HERBERT FREY]

Bibliography: R. Jastrow and M. H. Thompson, *Astronomy: Fundamentals and Frontiers*, 3d ed., 1977; I. R. King, *The Universe Unfolding*, 1976; J. M. Pasachoff, *Contemporary Astronomy*, 1977.

Uranus

URANUS

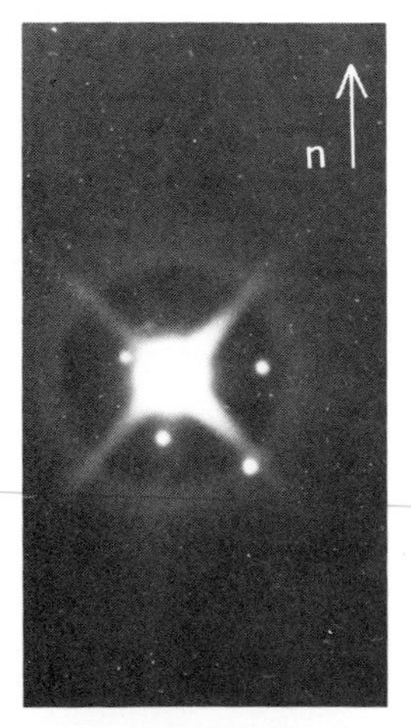

Fig. 1. Satellites seen nearly normal to the common orbital plane. In order of increasing distance: Miranda (16.8 magnitude), Ariel (14.8), Umbriel (15.4), Titania (13.9), and Oberon (14.3). Arrow indicates north direction. *(From G. P. Kuiper and B. M. Middlehurst, eds., Planets and Satellites, 1961)*

The first planet to be discovered with the telescope and the seventh in the order of distance from the Sun. It was found accidentally by W. Herschel in England on Mar. 13, 1781. Herschel's telescope was large enough to show that this object was not starlike in appearance; it appeared as a fuzzy patch of light. At first he thought that it was a comet, but subsequent calculations of the orbit demonstrated that Uranus was indeed a planet, about twice as far from the Sun as Saturn and therefore in an orbit that agreed almost exactly with the prediction of the Titius-Bode relation for planetary distances. *See* PLANET.

The planet and its orbit. The main orbital elements are the semimajor axis (mean distance to the Sun) of 1.79×10^9 mi (2.87×10^9 km); the eccentricity of 0.047; the sidereal period of 84.01 years; the orbital velocity of 4.25 mi/sec (6.8 km/sec); and the inclination of orbital plane to ecliptic of 0.8°.

These characteristics indicate a normal, well-behaved orbital motion. Therefore it is surprising to discover that the obliquity (inclination of rotational axis to orbit plane) of Uranus is the highest of any known planet: 97.9° This means that the axis is almost in the plane of the orbit, and thus the seasons on Uranus are very unusual. During summer in one hemisphere, the pole points almost directly toward the Sun while the other hemisphere is in total darkness. Forty-one years later, the situation is reversed. This means that "day" and "night" for an observer at the north or south pole of Uranus each last more than 40 years. Recent and future years of solstice (pole toward Sun) are 1944 and 1985, and of equinox (equator toward Sun) are 1966 and 2007. The actual period of rotation of Uranus is 12.3 hr as determined by spectroscopic observations.

The mean apparent equatorial diameter of the disk of this distant planet is about 3.6″ (the Moon has an apparent diameter of 31′). The corresponding linear equatorial diameter is 32,400 mi (51,800 km). The mass is 14.63 times the mass of the Earth, a value that can be well determined from the observed motions of the satellites. The corresponding mean density is 1.31 g/cm³, which is greater than that of Saturn even though Uranus is smaller than Saturn. This means that Uranus is richer than Saturn (or Jupiter) in elements heavier then hydrogen and helium but not nearly so rich in these elements as Earth (mean density 5.5). The same is true for Neptune; thus these two planets constitute a distinct subgroup in the outer solar system when compared with Jupiter and Saturn, which are composed of a nearly solar distribution of the elements. *See* JUPITER; NEPTUNE; SATURN.

The apparent visual magnitude of Uranus at mean opposition, that is, when closest to Earth, is +5.5. Therefore, this planet is just visible to the naked eye in a dark sky when its position among the stars is known. The corresponding value of the albedo, or reflectivity, is 0.5, a high value typical of the outer planets. The temperature Uranus would assume in simple equilibrium with the incident solar radiation is about 55 K, according to calculations. This is essentially identical with the value that is obtained by direct measurement. There is no evidence for the existence of an internal energy source or for a thermal inversion in the upper atmosphere, as observed for Jupiter, Saturn, Titan, and Neptune. Observations at radio frequencies have detected thermal radiation from the lower atmosphere of Uranus, indicating that the temperature increases with depth. At a wavelength of 21 cm, a temperature of 280±60 K has been measured.

Through the telescope, Uranus appears as a small, slightly elliptical blue-green disk. Some observers have reported the presence of faint, dusky bands on the disk when the equator is oriented toward the Earth, but these sightings are not fully confirmed. Spectroscopic observations have shown that the atmosphere of the planet contains large amounts of methane and hydrogen. The abundances of these gases are the subject of active research, but normalized path lengths as large as 6 km (CH_4) and 225 km (H_2) have been estimated. In comparison, the nitrogen and oxygen in the Earth's atmosphere give a path length of only 8 km. It is expected that helium and neon are also present, but they cannot be detected by ground-based observations. At the lower, warmer levels of the atmosphere, ammonia, hydrogen sulfide, water vapor, and other reduced compounds should be present. See PLANETARY PHYSICS.

Satellites and rings. Uranus has five known satellites that form a remarkably regular system, with low orbital eccentricities and inclinations (see table). Because the orbital planes are in the equatorial plane of the planet, the orbits are sometimes seen edgewise (at the equinoxes) and sometimes face-on (at the solstices; Fig. 1).

The satellites are too small to show measurable disks, and thus very little is known about them. The diameters given in the table are simply estimates based on the assumption that the satellites all have an albedo of 0.8, which is typical for the inner, icy moons of Saturn. Uranus is the only outer planet (besides Pluto) that has no known satellite in a retrograde orbit or in an orbit with high inclination. Such satellites are thought to be captured objects and tend to be at the outer fringes of a satellite system. If such an object is present in the vicinity of Uranus, it must be fainter than magnitude 20.

In 1977 a system of rings (Fig. 2) was discovered around Uranus during observations of a stellar occultation. There are at least nine rings, the largest having a width of only 50 km; they are designated by a mixture of Greek letters and Arabic numerals (Fig. 2). Since the rotational axis of Uranus is almost in the plane of the planet's orbit about the Sun, the rings are nearly perpendicular to this plane, which made it possible to discover them when they passed in front of a star. The mode of origin of these rings and the reasons for their stability are unknown. A promising hypothesis associates them with the presence of tiny, unseen satellites.

Satellites of Uranus

Satellite	Mean distance from Uranus, 10^3 mi (10^3 km)	Sidereal period, days	Diameter, mi (km)	Magnitude at mean opposition
I Ariel	119 (192)	2.520	380 (610)	(vis) 15.5
II Umbriel	166 (267)	4.144	250 (400)	(vis) 16
III Titania	273 (439)	8.706	620 (1000)	(vis) 14.0
IV Oberon	364 (586)	13.463	500 (800)	(vis) 14.2
V Miranda	81 (130)	1.414		(phot) 17

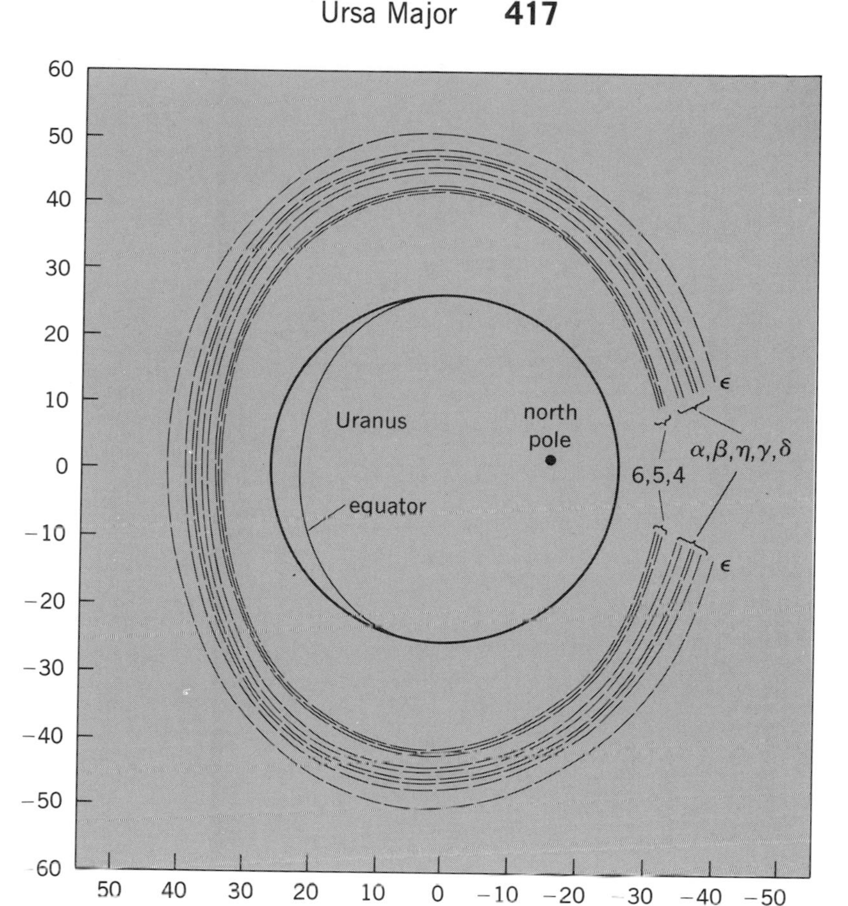

Fig. 2. Ring system of Uranus. Scales are distances from center of planet in units of 10^3 km.

Cosmogony. The fractionation process that led to the present diminished abundances of hydrogen and helium on Uranus is not well understood. The similarity in size and composition between this planet and Neptune is striking and suggests a common mode of origin. Yet Neptune does not exhibit the puzzling obliquity shown by Uranus, a characteristic that has been attributed to an oblique impact by a large fragment during the final stage of accretion. This aspect of the planet's origin also requires further study before an acceptable explanation can be expected.

[TOBIAS C. OWEN]

Bibliography: W. K. Hartmann, *Moons and Planets*, 1972; G. P. Kuiper and B. M. Middlehurst (eds.), *Planets and Satellites*, 1961.

Ursa Major

The most widely known and oldest of the astronomical constellations, Ursa Major, or the Great Bear, is a circumpolar group as viewed from the middle latitudes of the Northern Hemisphere. One part of the configuration, a group of seven bright

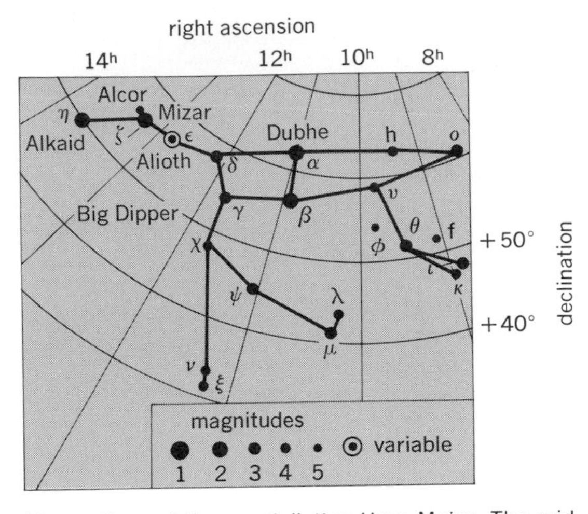

Line pattern of the constellation Ursa Major. The grid lines represent the coordinates of the sky. The apparent brightness, or magnitude, of the stars is shown by the sizes of the dots, which are graded by appropriate numbers as indicated.

stars, which is pictured as the tail of the Great Bear, is commonly known in the United States as the Big Dipper which it resembles (see illustration). This group of stars is also known in various lands as Charles' Wain (wagon) and the Plough. The Chinese call it the Northern Bushel. The two stars α and β at the front of the bowl of the dipper are called pointers, because a line joining them points to Polaris, the North Star. One of the pointers, the northern one, called Dubhe, is a navigational star. The star next to the end of the handle is Mizar, another navigational star, with its close companion Alcor. Next to Mizar is Alioth, the third navigational star in this group. *See* CONSTELLATION.

[CHING-SUNG YU]

Ursa Minor

The astronomical constellation Little Bear. Ursa Minor is a circumpolar constellation whose brightest star, Polaris, is almost at the north celestial

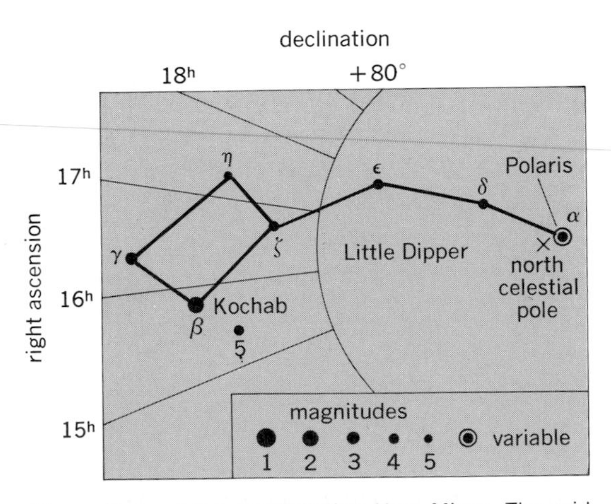

Line pattern of the constellation Ursa Minor. The grid lines represent the coordinates of the sky. The apparent brightness, or magnitude, of the stars is shown by the sizes of the dots, which are graded by appropriate numbers as indicated.

pole. Seven of the eight stars appear to form a dipper, hence the constellation is alternately known as the Little Dipper (see illustration). Polaris is at the end of the handle. Situated about 1 degree from the true celestial pole, Polaris is a variable star, pulsating in brightness periodically. The two bright stars β and γ at the front of the bowl are often called the Guardian of the Pole, because they circle about Polaris closer than other conspicuous stars. Kochab, the brighter of the two, was at one time closer to the true pole. It is an Arab word meaning polestar. The Little Bear is also known as Smaller Chariot by the Danes. *See* CONSTELLATION; URSA MAJOR.

[CHING-SUNG YU]

Variable star

A star that has a detectable change in its intensity, which is often accompanied by other physical changes. The changes in brightness of a variable star may be a few thousandths of a magnitude up to 20 magnitudes or even more. The cause of the variation may be either an eclipsing effect caused by the motions within a system of two or more bodies or an actual physical change within the star itself or in its surrounding atmosphere. The physical or intrinsic variables will be considered here. For a discussion of the extrinsic or eclipsing variables *see* BINARY STAR; ECLIPSING VARIABLE STARS.

Nomenclature. The nomenclature of variable stars is based on the Argelander system. In this system, the first variable star in a constellation is assigned the letter R, the next S, and so on until Z; then the letters are repeated, RR, RS, . . . , RZ; SS, ST, . . . , SZ, and so on to ZZ; then AA, AB, up to QZ, which is the 334th variable in a constellation. Later discoveries are assigned numbers V335, V336, . . . These letters are combined with the genitive of the Latin constellation names and often used with the convenient six-figure designation, the first four numbers of which give the hours and minutes of right ascension, and the last two the degrees of declination, for the epoch 1900. For example, 233815 R Aquarii (south declination), 213843 SS Cygni, 061115 CZ Orionis, and 184300 V603 Aquilae.

A committee appointed by the International Astronomical Union decides when a variable should be accepted as such and assigns its name. The third edition of the *General Catalogue of Variable Stars* lists 20,437 variable stars in the Milky Way Galaxy, and three supplements add 5401 more. Undoubtedly there are many thousands of undiscovered variable stars in this galaxy, bright enough to be observed, and as many thousands more too faint and too distant to be seen with present equipment. *See* GALAXY.

Importance. Studies of the variations of stars add to the general knowledge of the structure of the universe. Stars are the most useful means available for the determination of distances and dimensions of the Milky Way and other galaxies. Most of the variables show peculiarities in their variations when they are observed over long periods of time. This necessitates long-range programs for continuing observations over many years. *See* GALAXY, EXTERNAL.

Programs for the systematic visual observation of variables with relatively large ranges are carried out by groups of observers, such as the American Association of Variable Star Observers (AAVSO). These groups are composed mostly of amateur astronomers with their own equipment, who derive much satisfaction from the knowledge that their conscientious work is of great value to scientific research. Members of the AAVSO have contributed more than 4,000,000 observations since it was founded in 1911.

Professional astronomers at the great observatories usually concentrate on observations of specific variables. Their observations are made visually and photographically, or, in the case of variables with small or rapid changes, with photoelectric photometers. An example of a large photographic program is the Harvard Observatory photographic patrol, which covered the sky for more than 50 years. This collection of nearly 500,000 plates contains material for the study of variables brighter than twelfth magnitude.

Classification. Most intrinsic variables are either pulsating or eruptive (see table). They may be separated into three main divisions—those which vary with periodic regularity, those with cyclic or semiregular changes, and those which are completely erratic, or irregular. A few variables are unique and cannot be placed in any of these classes, and others seem to be combinations of several classes.

The light curve of a variable shows how the brightness changes with time. Observed magnitudes are plotted against a time scale, from which the times of maxima and minima may be determined. An inspection of the light curve may also indicate if the variable is periodic or cyclic and gives a first approximation of the period. Furthermore, the shape of the curve, that is, the steepness of the rise and fall, together with the time between maxima, may give a clue to the type of variability (Fig. 1). See LIGHT CURVES.

The spectra of variable stars cover the whole

Representative types of intrinsic variables

Class	Range of period or cycle	Color and luminosity	Amplitude (magnitude)	Number*
Pulsating				
RR Lyrae type	0.05–1.2 days	White	<2	4433
Classical cepheid	1–70 days	High-luminosity yellow	0.1–2	706
Long-period	80–1000 days	Red giants	2.5–6+	4566
Semiregular	30–1000 days	Red giants	<1.5–2	2227
Others				1856
				13,788
Eruptive				
U Gem and Z Cam	10–600 days	Subdwarfs	2–6	235
Flare stars	?	Red dwarfs	1–6	28
Recurrent novae	20 years–?	Dwarfs?	7–9	6
Novae	centuries?	Dwarfs?	8–16	160
Supernovae	?	?	20+	7
Nebular variables	?	Yellow dwarfs		869
Others				311
				1616

*From B. V. Kukarkin and P. P. Parenago, *General Catalogue of Variable Stars*, 3d ed., 1969. In addition there are listed 4062 eclipsing stars, 803 unstudied variables, and 183 others unique or constant.

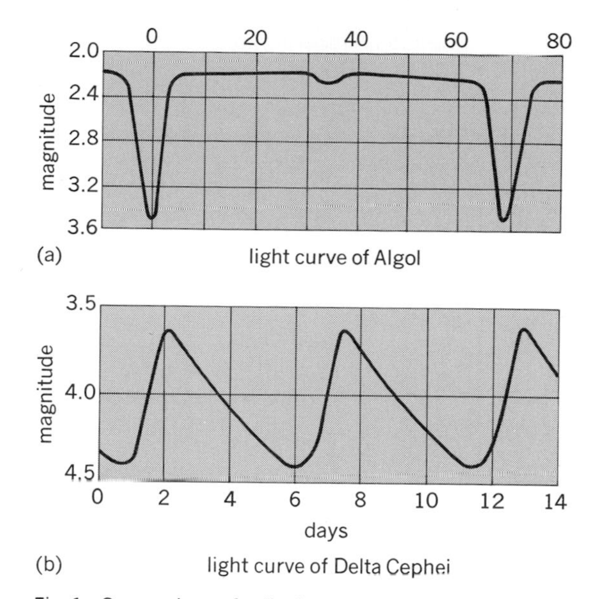

Fig. 1. Comparison of eclipsing variable stars and intrinsic variable stars. (*a*) An eclipsing variable star, with a period of 2.87 days. (*b*) A typical intrinsic variable star, with a period of 5.37 days.

sequence of spectral classes, from the hot, highly luminous blue stars at one end, to the cool, red stars at the other. Stars of some types of variability may be recognized as variables by means of their spectra. Specifically, a star with bands of metallic oxides or carbon, and with emission lines of hydrogen, is almost certainly a long-period variable or possibly a semiregular variable. Many fainter novae are discovered from their spectra, which at one stage of their development show many emission lines, including the so-called forbidden lines of oxygen, neon, iron and other elements. See ASTRONOMICAL SPECTROSCOPY; STAR.

Periodic variables are divided into two main classes: cepheids and long-period, or Mira-type, variables. There are many intermediate examples, so that no definite dividing lines may be drawn.

Short-period variables. Cepheid variables are characterized by several important relations. They are among the most luminous stars known, and thus can be seen at great distances. There is a definite correlation between period and luminosity. As a result of this relationship, they are used as a measuring tape for the universe. Other correlations exist between period and spectrum (temperature), period and radial velocity, and period and form of light curve.

Both spectrum and radial velocity of a cepheid change as the brightness changes. The spectrum at maximum is about a whole class earlier than at minimum. The greatest velocity of approach (negative velocity) occurs near maximum light and greatest recession velocity (positive) near minimum light. See CEPHEIDS.

Classical cepheids are population I stars and are situated within about 1 kiloparsec of the galactic plane.

The period-luminosity relation was first noticed in 1910 by Henrietta Leavitt at Harvard in the course of her work on the cepheids in the Magel-

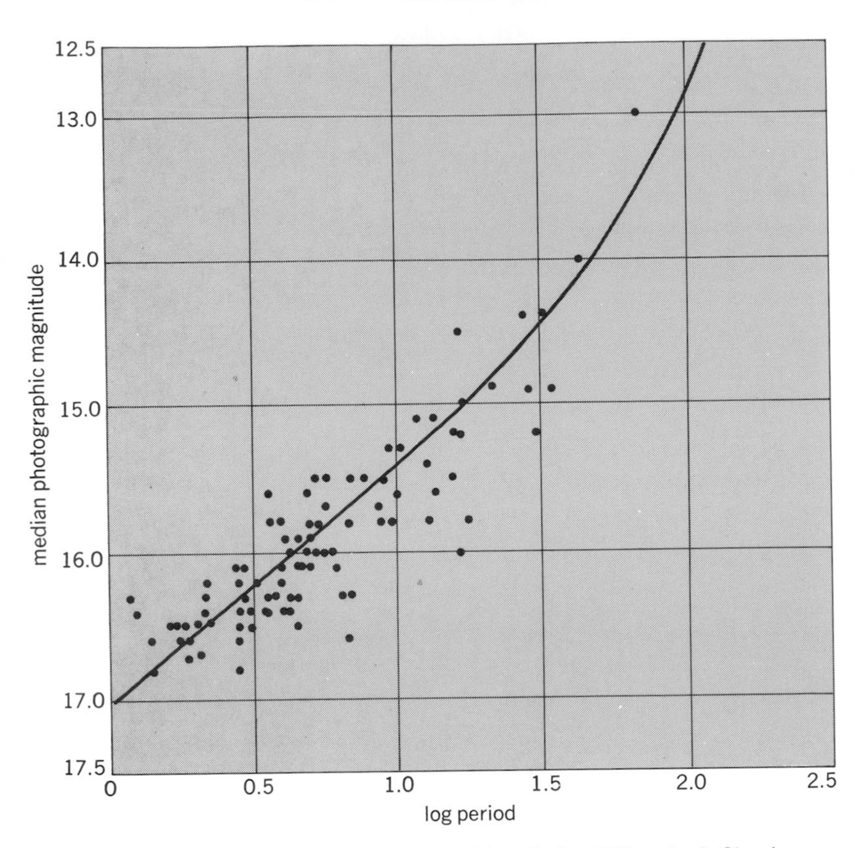

Fig. 2. Period-luminosity curve for cepheid variables in Small Magellanic Cloud.

lanic Clouds and was further developed by Harlow Shapley (Fig. 2).

If the zero point of absolute magnitude could be established for cepheids of any given period, the distances of any stellar system in the universe, in which cepheids are found, could be determined from the equation

$$M = m + 5 - 5 \log D$$

Here M is absolute magnitude, m is apparent magnitude, and D is distance in parsecs. In 1950 the zero point was finally established by Walter Baade at Mount Wilson and Palomar Observatories from his work on cepheids in the Andromeda Nebula.

Intrinsic variables with the shortest known period are the cluster-type cepheids or RR Lyrae stars. RR Lyrae itself is one of the brightest and best studied of the group, with a period of 13.5 hr and a range of brightness from magnitude 6.9 to 8.0. The shortest period known, 89 min, is that of CY Aquarii.

The RR Lyrae stars are fainter than the classical cepheids. They have an absolute photographic magnitude of 0.0 and show no progression of luminosity with period. Therefore, if an RR Lyrae star can be identified in a cluster or galaxy, and its apparent magnitude determined, its distance may be obtained from the relationship stated above.

RR Lyrae stars do not form an extension of the period-luminosity curve of the classical cepheids but lie in a branch about $1\frac{1}{2}$ magnitudes fainter. They are population II stars and form a spheroid in the Milky Way Galaxy, concentrated toward the galactic nucleus.

A group of peculiar cepheids, the W Virginis stars, are closely related to the RR Lyrae stars. They are also population II stars, have high velocities, and have periods greater than 10 days. A further continuation of this branch of the period-luminosity curve seems to be the RV Tauri stars, with periods between 40 and 150 days. Their spectra are more similar to those of RR Lyrae and W Virginis stars than to those of the classical cepheids.

The pulsation theory has been used to explain the variation of cepheids and related variables. The theory was first developed by Shapley (1914), by Arthur S. Eddington (1919), and later by many others. According to this theory, the star goes through regular cycles of expansion and contraction. As it expands, the pressure in its interior decreases, the temperature goes down, and it becomes redder and fainter. When the star reaches its maximum diameter, the process is reversed: It begins to contract, pressure increases, temperature rises, and it becomes bluer and brighter.

A simple process of expansion and contraction of the star as a whole fails to explain some of the observed peculiarities. It is assumed that when the interior of the star pulsates, the compression waves pass through the lower levels of the star's atmosphere and reach the higher levels at some later time. In this way, the observed lag of greatest velocity of approach relative to maximum brightness may be explained.

Long-period variables. Long-period, or Mira-type, variables are by far the most numerous of the known variables in the Milky Way Galaxy. They are red giants with spectra of either class M or one of the carbon types, and usually have hydrogen emission lines. The spectra are of later type (redder and cooler) at minimum rather than at maximum.

A study of the spectra of long-period variables provides information about the layering of the atmosphere of a star. The bright lines of hydrogen appear to be produced near the surface of the star, and the metallic vapors form a great cloud floating above the surface. The bright lines found in most stellar spectra originate in the distended atmospheres of high-temperature stars. *See* STELLAR EVOLUTION.

Several hundred long-period variables are brighter than tenth magnitude at maximum, and their large variations make them spectacular stars to observe with the unaided eye and small telescope. Many of them have been observed for more than 100 years. Mean light curves and mean periods have been determined for most of the brighter ones, and many of them show decided variations in the shape of light curve and length of period from epoch to epoch. The greatest irregularities occur in the heights and shapes of maxima. For a discussion of the most famous long-period variable *see* MIRA.

Many long-period variables are high-velocity stars and belong to population II, but some of the longer-period ones are more concentrated toward the galactic plane and have population I characteristics. Others seem to form a link between the two populations.

Long-period variability may be caused by the same action as with the cepheid variables. Their great size and lower density could account for their irregularities.

Irregular variables. Nonperiodic variables may be separated to some extent by their spectra. Semiregular red giants are the most numerous and are closely related to the long-period variables. Amplitudes are small, ranging from less than 2 magnitudes to infinitesimal amounts. Possibly all red stars with molecular bands in their spectra are variable.

Most red variables with periods less than 150 days are subject to great irregularities, and some lose all semblance of periodicity at times; others are completely irregular.

The most violent explosions or eruptions occur in the novae and supernovae. Increases of 10–12 magnitudes have been observed within a few hours. Previous to the explosion, novae are probably dense blue dwarfs. *See* NOVA; SUPERNOVA.

Other stars which have rapid increases of brightness are the U Geminorum and Z Camelopardalis stars (see eruptive stars in the table). These also are blue stars with dwarf characteristics and are often referred to as dwarf novae. Their eruptions appear at semiperiodic intervals. A relationship exists between the amplitude of the variation and the duration between eruptions; variables with greater ranges have longer periods.

A small group of novae, called recurrent novae, have been observed to have two or more explosions. Recurrent novae may be considered to be intermediate between true novae and the U Geminorum stars.

Flare stars are red dwarfs, with emission line spectra, which more than double their brightness in a few minutes. Little is known about the frequency of the eruptions, and most of them have been found by accident. Variable-star observers are making special studies of a group of red dwarfs to accumulate frequency statistics of flares.

The best known of the flare stars is UV Ceti, the typical star of its class. It is normally of the thirteenth magnitude and has been observed to increase more than 6 magnitudes in less than 1 min. The decrease is almost as rapid. Many lesser flares have been observed visually, spectroscopically, photographically, and photoelectrically.

Nebular variables, also known as T Tauri, RW Aurigae, or Orion variables, are subject to rapid changes at irregular and unpredictable intervals. Most of them are dwarf stars with absorption spectra of classes G to M; in addition, many of them have peculiar bright lines, especially the H and K lines of calcium. A few of them, such as R Monocerotis and R Coronae Australis, are definitely associated with funnel-shaped nebulae that also vary.

[MARGARET W. MAYALL]

Bibliography: L. Campbell and L. Jacchia, *Story of Variable Stars*, reprint, 1945; B. V. Kukarkin and P. P. Parenago, *General Catalogue of Variable Stars*, 3d ed., 1969, 3 suppl. 1971, 1974, 1976; C. Payne-Gaposchkin, *The Galactic Novae*, rev. ed., 1964; W. Strohmeier, *Variable Stars*, 1973.

Vega

One of the brightest stars in the sky, apparent magnitude 0.1. Vega, or α Lyrae, is a normal main sequence star of spectral type A0, having an effective temperature near 10,050 K. Its distance is 8 parsecs, its absolute magnitude is +0.5 (40 times as bright as the Sun), and its radius is three times that of the Sun. The angular diameter has been measured with the stellar interferometer in Australia as 0.00347 second of arc. The spectrum shows mainly the strong Balmer lines of hydrogen. *See* STAR.

[JESSE L. GREENSTEIN]

Venus

The second planet in distance from the Sun. This neighbor of the Earth is very similar to it in such gross characteristics as mass, radius, and density (see table). In other ways Venus is apparently different. Its atmospheric mass is almost a hundred times that of the Earth; its atmosphere is mostly carbon dioxide instead of nitrogen and oxygen; an extensive cloud layer of concentrated sulfuric acid is present; its surface temperature is an unbearable 750 K; and it rotates with a period of 243 days, and from east to west, in the opposite sense of most other planets. Some of these differences are due more to alternate evolutionary paths of the two planets than to totally different initial conditions.

Characteristics of Venus

Characteristics	Values
Mass	0.82 Earth's mass
Radius	0.95 Earth's radius or 6050 km
Mean density	0.94 Earth's value or 5.2 g/cm^3
Orbital distance from the Sun	0.72 Earth's distance or 1.1×10^8 km
Orbital period*	0.62 of an Earth year or 225 Earth days
Orbital eccentricity	0.4 Earth's value or 0.007
Orbital inclination to Earth's orbital plane	3.4°
Rotational period†	243 Earth days with respect to the stars, 117 Earth days with respect to the Sun

*Length of a Venus year. †Length of a Venus day.

Appearance. To the naked eye, Venus is the brightest starlike object in the sky. It is usually visible during the night either soon after sunset or close to sunrise. It can sometimes be seen during the daytime. As observed through a pair of binoculars or a telescope, Venus exhibits a crescentlike or gibbous appearance. Like the Moon, Venus ranges over a full set of phases from a "new moon" to a "full moon." This phase variation is caused by a changing fraction of the Sun-illuminated hemisphere facing toward the Earth. At new moon, only the dark nighttime side is seen, while at full moon all of the daytime hemisphere is seen. Venus undergoes a complete set of phase changes over its synodic period of 584 days.

Clouds. The light seen coming from Venus is almost entirely due to sunlight that is reflected from a dense cloud layer whose top is located about 70 km above the surface and whose bottom lies within 50 km of the surface. In contrast to the Earth's approximately 50% cloud cover, the clouds

VENUS

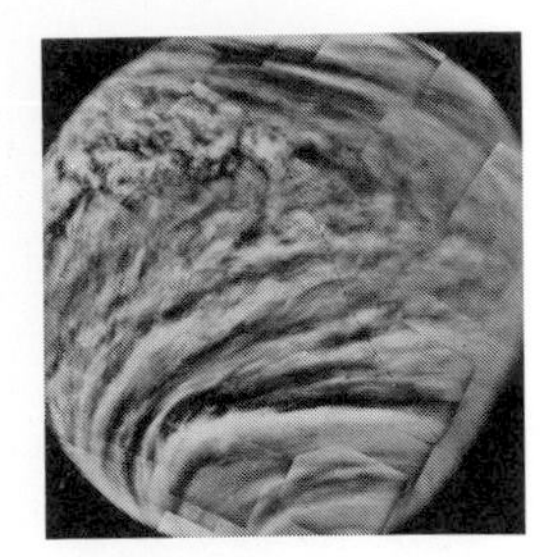

Fig. 1. Venus as seen in ultraviolet light. This figure was constructed from high-resolution photographs obtained from *Mariner 10*.

of Venus are present over the entire planet. In yellow or red light they present a uniform appearance. However, as illustrated in Fig. 1, the clouds show a banded and spotted pattern when viewed in ultraviolet light. These ultraviolet markings provide information about atmospheric motions.

The clouds of Venus consist of a large number of tiny particles, about 1 micrometer in size, that are made of a water solution of concentrated sulfuric acid. Such a composition may at first seem very surprising when compared with the water clouds of the Earth's lower atmosphere. However, sulfuric acid particles are the dominant type of particles in the Earth's upper atmosphere, althought the amount there is much less than the amount present in Venus's atmosphere. In the case of both planets, sulfuric acid is produced primarily from sulfur-containing gases that combine with water vapor and oxygen-containing gases. Compositional measurements made from the American Pioneer Venus Sounder probe, which descended through Venus's atmosphere on December 9, 1978, show that sulfur dioxide is the principal sulfur containing gas in Venus's atmosphere. Sulfur dioxide is also the major gas species injected into the Earth's stratosphere by volcanic explosions. Such injections cause a large, but temporary, increase in the amount of sulfuric acid there.

Atmospheric composition. Measurements conducted within Venus's atmosphere from American and Soviet probes and remotely from spacecraft and the Earth have provided a good definition of the gases that constitute Venus's atmosphere. By far, the chief gas species is carbon dioxide, which makes up 96% of the atmospheric molecules, while nitrogen accounts for almost all the remainder. Trace amounts of sulfur dioxide (~200 parts per million), water vapor (~100 ppm), carbon monoxide (20 ppm), argon (~70 ppm), helium (~10 ppm), neon (~5 ppm), hydrogen chloride (~0.4 ppm), and hydrogen flouride (~.01 ppm) are present in the lower atmosphere, with the concentration of the first two of these declining dramatically near the cloud tops due to the formation of new sulfuric acid there.

Carbon dioxide. In contrast to the dominance of carbon dioxide in Venus's atmosphere, the Earth's atmosphere consists mostly of nitrogen and oxygen, with carbon dioxide being present at a level of only 300 ppm. In part, this difference may stem more from temperature differences than from intrinsic differences. Over the lifetime of the Earth, an amount of carbon dioxide comparable to that in Venus's atmosphere was vented out of the Earth's hot interior. The outgassed carbon dioxide remained in the atmosphere for only a short time. Almost all of it was first dissolved in the oceans and subsequently converted into carbonate rocks, such as limestone. Venus's surface is much too hot for oceans of water to be present, and hence its atmosphere has been able to retain essentially all of the carbon dioxide vented from its interior.

Rare gases. There are two varieties of rare gases found in planetary atmospheres: those that were derived from the gas cloud (solar nebula) from which the planets formed (primitive rare gases), and those that were produced from the radioactive decay of certain elements, such as potassium, in the interior of the planets. A fundamental finding about the composition of Venus's atmosphere by spacecraft observations is the detection of much more primitive argon and neon than in the Earth's atmosphere. Furthermore, Mars's atmosphere has even less of these rare gases than does the Earth's. Because small stray bodies have about an equal chance of hitting the Earth and Venus, this difference in the abundance of primitive rare gases implies that they were not derived primarily from impacts of planets with gas-rich comets and asteroids over the planets' lifetime. Rather, it is consistent with an origin involving the sticking of rare gases to the surface of planet-forming dust grains in the solar nebula. After the inner planets formed, their interiors heated up and volcanoes released the rare gases into their atmospheres. Due to a higher pressure in the solar nebula at closer distances to the Sun, the grains that formed Venus had larger amounts of primitive rare gases. More generally, the other gases in the atmospheres of Venus, the Earth, and Mars may have been derived from chemical compounds that were present in the dust grains from which they formed and that were released into their atmospheres as portions of their interiors heated up.

Water vapor. The amount of water vapor in Venus's atmosphere is much less (about 10^5 times) than the amount of water in the Earth's oceans. Since water oceans would completely evaporate at the high temperature of Venus's surface and since Venus probably formed with an amount of water not much less than that of the Earth, Venus's atmosphere may have initially had much more water than it does today. Processes by which Venus's atmosphere could have lost water include chemical reactions between water and certain atmospheric gases, such as carbon monoxide; reactions between water and hot rocks at the surface and in the interior; and the breakdown of water into hydrogen and oxygen by ultraviolet sunlight. In each case, the oxygen in water was either freed or combined with some other chemical. Hydrogen was also freed and eventually escaped to space because Venus's gravity was unable to permanently retain this very light gas.

Temperature. By detecting long-wavelength heat radiation produced at the surface, radio telescopes first showed that the surface temperature was 730 K, that is, almost 400 K higher than the boiling point of water. This result has been confirmed by direct temperature measurements of the atmosphere made from Soviet and American spacecraft that have descended through Venus's atmosphere. The atmospheric temperature has a relatively cool value of 250 K (about 25 K below the freezing point of water) near the top of the cloud layer, which is at a pressure of about 1/20 that at the Earth's surface. The temperature gradually increases with decreasing altitude until it reaches 730 K at the surface, where the pressure is 90 times that at the Earth's surface.

The high value of Venus's surface temperature is not due to its being closer to the Sun than the Earth. Because its cloud layer reflects back to space about 75% of the incident sunlight, Venus actually absorbs less solar energy than does the Earth. Rather, the high temperature is the result of a very efficient greenhouse effect that allows a small but significant fraction of the incident sunlight to penetrate to the surface (about 2.5% according to Soviet and American spacecraft measurements), but prevents all except a negligible fraction of the heat generated by the surface from escaping directly to space. The thermal energy produced by the surface and hot lower atmosphere, which occurs at infrared wavelengths, is very effectively absorbed by the carbon dioxide, water vapor, sulfur dioxide, and sulfuric acid particles of the atmosphere. However, these materials are poor absorbers at visible wavelengths, where most of the solar energy lies. The greenhouse effect raises Venus's surface temperature by almost 500 K, but causes only a modest 35 K rise in the surface temperature of the Earth. This difference is due to the Earth's atmosphere being partially transparent at some infrared wavelengths and thus permitting some surface heat to escape to space.

Meteorology. Studies of the motion of the ultraviolet markings, the horizontal displacement experienced by atmospheric probes, and other data indicate that the atmosphere near the cloud tops is moving with a jet-stream-like velocity of about 100 m/s (~200 mi/h) from east to west in the direction of Venus's rotation and only about 5 m/s (10 mi/h) from the equator toward the pole. In contrast to the situation for the Earth where only a small portion of the atmosphere moves at jet stream speeds, the entire cloud top region on Venus moves at these large speeds. The wind speed for the most part gradually, but in a few places sharply, decreases with declining altitude and achieves values of a modest few m/s within 10 km of the surface. Like the winds on the Earth, the winds on Venus are produced ultimately by differences in the amount of solar energy absorbed by different areas of the planet, such as areas at different latitudes. The large wind speeds near the cloud tops may be the result of Venus's atmosphere being very deep. The transport of momentum by the mean circulation and the eddies from the deep dense portions near the surface to the much less dense regions near the cloud tops translates sluggish motions into fast ones.

Because Venus's atmosphere is massive, atmospheric motions are very effective in reducing the horizontal variations of temperature in its lower atmosphere. In the region of the clouds and at lower altitudes, temperature variations occur mostly in the north-south direction, with the equator being only a few degrees warmer than the poles close to the surface and some tens of degrees warmer within the cloud region. Above the clouds, the atmosphere is actually somewhat warmer at the poles than near the equator due to the effects of heat transported by atmospheric motions. As Venus's axis of rotation is almost exactly perpendicular to its orbital plane, its climate has little seasonal variability.

Interior. The very similar mean densities of Venus and the Earth imply that Venus is made of rocks similar to those that make up the Earth. However, because Venus formed closer to the Sun, in perhaps a somewhat warmer environment, it may have initially contained a smaller amount of sulfur and water-bearing compounds. Such an environmental difference would probably not have significantly affected Venus's content of the long-lived radioactive elements uranium, potassium, and thorium. Over the lifetime of the Earth, and presumably Venus, the decay of these elements may have generated enough heat to cause these planets to become chemically differentiated, as free iron melted and sank toward their centers. Also, both planets probably formed "hot" due to gravitational energy released in bringing small chunks of rock together to form them. In this case, they may have undergone substantial differentiation in their early histories, with an accompanying release of much of their atmospheric gases. Thus, Venus's interior may be qualitatively similar to that of the Earth in having a central iron core, a middle mantle made of rocks rich in silicon, oxygen, iron, and magnesium, and a thin outer crust containing rocks enriched in silicon in comparison with the rocks of the mantle. However, in contrast to the situation for the Earth, Venus's core may now be solid, which could account for the absence of a detectable magnetic field.

Surface. The ubiquitous cloud layer totally obscures Venus's surface at visible wavelengths. However, since the clouds and atmosphere are

Fig. 2. Photograph of Venus by a Soviet spacecraft (partly visible at bottom) on the planet's surface.

transparent to long-wavelength radio waves, observations of Venus's surface can be made with radar telescopes. Processing of radar signals bounced off Venus have yielded crude pictures of its surface which show the presence of large, circular features. These may be giant craters produced by impacting meteoroids, analogous to the ones that pockmark the Moon. In addition, there are features that resemble large volcanic constructs, enormous rift valleys, and jagged mountain ranges. Other radar observations conducted from the American Pioneer Venus orbiter show that there is little topographic relief over most of Venus's surface. However, in a few places, huge continental-sized plateaus rise about 10 km above the surrounding plains, while in other places isolated lowlands occur. The 20% of Venus's surface that consists of lowlands can be contrasted with the 70% of the Earth's surface that contains ocean basins.

Direct measurements of the properties of Venus's surface have been obtained from several of the Venera spacecraft that have functioned for a short time after landing on the surface. These observations show that some of the surface rocks have a high content of natural radioactive elements. The observed level of radioactivity is similar to that found in granitic and basaltic rocks on the Earth and is consistent with the type of surface rocks expected if Venus has been chemically differentiated. Photographs obtained from a landed spacecraft in the dim sunlight at the surface (Fig. 2) reveal the presence of numerous rocks. Some of these rocks are rounded, perhaps as a result of sandblasting by windblown sand.

Rotation. In contrast to the Earth and almost all other planets, Venus rotates in the opposite direction with respect to its orbital motion about the Sun. It rotates so slowly that there are only two sunrises and sunsets per Venus year. Tides raised in the body of the planet by the Sun may have greatly reduced the rate of rotation from an initially large value, similar to the Earth's, to its present low value. Because Venus's atmosphere is so massive, tides raised by the Sun in the atmosphere may have also been important, with its current rate of rotation being determined chiefly by a balance between the oppositely directed torques arising from these two types of solar tides. Venus apparently presents the same face to the Earth at times of closest approach. This suggests that tidal forces exerted by the Earth also played a role and helped lock Venus in its present rotational state.

Life. The current high surface temperature and acid constitution of the clouds preclude the existence of living organisms like those that inhabit the Earth.

Spacecraft. Observations of Venus have been carried out by the United States Mariner spacecraft that flew by the planet, by the Pioneer Venus spacecraft that consisted of an orbiter and five probes that descended into its atmosphere, and by a number of Soviet Venera spacecraft. Possible future spacecraft missions to Venus include a United States orbiter with a sophisticated radar that could obtain high-resolution "pictures" of the planet's cloud-shrouded surface and a Soviet balloon that could circle the planet at a constant altitude close to the cloud bottoms. *See* PLANET.

[JAMES B. POLLACK]

Bibliography: P. A. Moore, *The Planet Venus*, 3d ed., 1961; Pioneer Venus results, *Science*, 203: 743–808, February 23, 1979, and 205:41–121, July 6, 1979; C. Sagan, *The Planets*, 1965; A. Young and L. Young, Venus, *Sci. Amer.*, 223(3):70–78, September 1975.

Virgo

In astronomy, a constellation handed down from antiquity, visible throughout the summer months. This sixth sign of the zodiac represents a maiden in a half reclining position (see illustration). It is

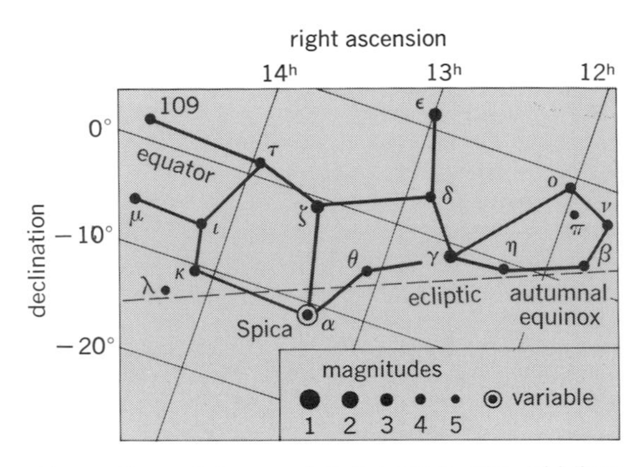

Line pattern of the constellation Virgo. The grid lines represent the coordinates of the sky. The apparent brightness, or magnitude, of the stars is shown by the sizes of the dots, which are graded by appropriate numbers as indicated.

identified with the goddess of justice, Astraea. The balance is seen by her side in the sky. The brightest star in the constellation, Spica, is a spectroscopic binary with a massive dark companion. It is a fine first-magnitude star of the purest white tint, and also a navigational star. *See* CONSTELLATION; LIBRA.

[CHING-SUNG YU]

White dwarf star

An intrinsically faint star of very small radius and high density. The mass of a typical white dwarf is about 0.7 that of the Sun, and the average radius is 8000 km. The mean density is 6×10^5 g/cm^3 (or about 10 tons/in.^{3}). Surface temperatures are between 85,000 and 4000 K, and brightness is from 10 to 10^{-4} that of the Sun. Such stars are composed largely of helium, carbon, or heavier elements.

Observations of the hotter white dwarfs of normal spectrum have confirmed the Einstein gravitational red shift. The average of 50 white dwarfs with hydrogen lines shows an apparent velocity or recession from the Earth between 40 and 50 km/s. The most probable mass deduced from these observations is about 0.7 times that of the Sun, and the radius 1% that of the Sun.

The theoretical upper limit of a nonrotating white dwarf mass is about 1.4 times greater than the Sun. Given the high density, at any plausible temperature the electrons can become degenerate.

All low-electron-energy states are occupied, with only electrons of exceptionally high energy being able to change their positions and momenta freely.

Atmosphere. Spectroscopic observations of white dwarfs demonstrate atmospheric compositions reflecting various stages of nuclear evolution of their red giant parents. Their cores must contain helium, carbon, and heavier elements. Their atmospheres, however, show the strangest dichotomy in composition of any stars. Either they are essentially pure hydrogen with no metals (H/He $> 10^3$), or they are pure helium, sometimes with some metals (He/H $> 10^4$). White dwarfs with hot hydrogen or helium atmospheres have been found from space observations of Apollo-Soyuz, Skylab, and the International Ultraviolet Explorer Satellite, confirming results from the ground, but also showing that hot coronas may exist. Results from space have, in general, confirmed those from the ground, and have given confidence in the higher surface temperatures, which reach 85,000 K in one star.

Thermal energy. A white dwarf is one final stage of stellar evolution, with thermonuclear energy sources extinct. The heavy nuclei are still mobile, and their thermal energy maintains the star's radiation. The star cools and becomes fainter and redder. The thermal energy, which is normally 3/2 kT (where T is the temperature and k is Boltzmann's constant) for a gas, per ion, becomes 3 kT for a solid. The transition between the state of free mobility and a lattice state occurs at about 2×10^7 K at these high densities. The total energy available is proportional to the total number of nucleons and is therefore inversely proportional to the atomic number; the lifetime at a given brightness depends on these considerations. About 3% of the stars are now white dwarfs, but in about 5×10^9 years all present stars of higher luminosity than the Sun will have burned out to the white dwarf stage. *See* STELLAR EVOLUTION.

An anomaly has been found in the number and relative frequency of cool, red white dwarfs. It has been expected that these would be very common, but, in fact, objects more than 10,000 times fainter than the Sun are rare. A further peculiarity of the available thermal energy may account for this. At surface temperature below 4000 K, the solid lattice of the core is below its Debye temperature. The specific heat is then very small, and the surface radiation cools the star at an unexpectedly rapid rate.

Binary systems. White dwarfs have been found to be one member of certain close, interacting binary stars of the type that may become classical novae (undergoing a giant explosion), recurrent novae, or dwarf novae. This class of stars, now called cataclysmic variables, provides fascinating insights into what must be happening on a larger scale near black holes. A main-sequence star near a white dwarf evolves and increases in size until the matter overflows its Roche lobe and finds itself in the gravitational grasp of the white dwarf. Since the velocity of infall at the surface of a white dwarf is 6000 km/s, a hydrogen atom would have an energy of 25 kV on impact. Furthermore, excessive hydrogen in a white dwarf would result eventually in a thermonuclear runaway. In fact, a dense accretion disk, like the rings of Saturn, is formed around the white dwarf. This provides far-ultraviolet and x-ray flux, rapid flickering in intensity, and occasionally violent outbursts. *See* BINARY STAR; BLACK HOLE; NOVA.

Magnetic fields. Some white dwarfs have intense magnetic fields, from a few megagauss (a few hundred teslas) to 300 megagauss (30 kiloteslas). In single white dwarfs, these produce so-far-unexplained features in the spectra. In binaries, the magnetic fields control the motion of the installing gases, focusing matter into an accretion funnel or spot.

Neutron stars. In a dense gas more massive than a white dwarf star, the central pressure is too high for the degenerate electron pressure to support the weight of the overlying layers. No normal, stable configuration exists until the pressure of the nucleons arrest the collapse. Electrons are squeezed back into protons (inverse beta decay), and the resulting neutrons are packed until "nuclear" densities (about 10^{14} g/cm^3) are reached. Such a neutron star is only about 10 km in radius, and the Einstein red shift at the surface is very large. Such neutron stars would not be visible by thermal radiation because of their small size. However, the discovery of pulsars has almost certainly established their existence. Detected at radio frequencies by periodic bursts with spacing from 0.02 to 3 s, the rotating neutron star beams radio waves like a lighthouse. Only one pulsar has been well studied at optical wavelengths. It is visible at the center of the Crab Nebula; its radio radiation, x-rays, and light are nonthermal in origin, generated by rotation of an intense magnetic field, approaching 3×10^{12} gauss (300 megateslas) at a speed approaching that of light. Thus, the two expected types of condensed star, white dwarfs and neutron stars, have now been found. *See* CRAB NEBULA; PULSAR.

[JESSE L. GREENSTEIN]

Wolf-Rayet star

A member of a class of very hot stars (100,000–35,000 K) which characteristically show broad, bright emission lines in their spectra. Wolf-Rayet stars are classified into two main groups, the WN (nitrogen) and the WC (carbon) stars. The emission lines show very high excitation (lines of He^+ and even up to O^{5+} are strong) and are broadened by an unexplained mechanism. The Australian stellar interferometer has shown that the angular size of the emission-line region is five times that of the region producing the continuous spectrum. Displaced absorption lines caused by fast-moving gas shells are sometimes seen, especially in the ultraviolet. The simple explanation of expansion of an outer layer with velocities up to 2000 km/sec is found insufficient from studies of Wolf-Rayet stars which happen to be members of a spectroscopic binary system. Scattering of light by fast-moving electrons has also been suggested. Luminosities must be very high, in the range of 10^4–10^5 times that of the Sun. These stars are probably very young and represent an unstable and short-lived early stage in stellar evolution. Many are close binaries with two stars nearly in contact. *See* STAR; STELLAR EVOLUTION.

[JESSE L. GREENSTEIN]

X-ray astronomy

The study of x-ray emission from extrasolar sources. It includes the study of virtually all types of astronomical objects, from stars to galaxies and quasars. X-ray astronomy is a recent addition to the ancient science of astronomy. The first nonsolar x-ray source was detected in 1962 during a rocket flight from White Sands, NM. Riccardo Giacconi and coworkers at American Science and Engineering built the experiment for this flight and observed the source, later identified with a star in the constellation Scorpius (Sco X-1), and a uniform background of x-rays which comes from beyond the Galaxy.

X-ray astronomy is a space science that requires rockets or space satellites to carry experiments above the Earth's atmosphere, which would otherwise absorb the radiation. The x-ray region of the electromagnetic spectrum extends from wavelengths of about 10 picometers (10^{-9} cm) to a few tens of nanometers (a few times 10^{-6} cm), with shorter wavelengths corresponding to higher-energy photons (1 nm corresponds to about 1000 eV). X-ray astronomy is subdivided into broad bands—soft and hard—depending upon the energy of the radiation being studied. Observations in the soft band (below about 10 keV) must be carried out above the atmosphere, while hard x-ray observations can be made at high altitudes achieved with balloons. There is a limit to the observable spectrum at low energies due to absorption by neutral interstellar hydrogen gas. The low-energy cutoff depends on direction, but can be as low as 250 eV. In the direction of the galactic center the cutoff is as high as 2000 eV. At high energies (short wavelength) space remains transparent through the entire x-ray spectrum. *See* ROCKET ASTRONOMY; SATELLITE ASTRONOMY.

Historical development. The short history of x-ray astronomy has seen rapid advances in the number of objects detected and their variety, and in understanding the nature of these objects. This can be compared to the surge in radio astronomy, following its early history, and is typical of the era of discovery in most fields. The first exploratory phase in x-ray astronomy was carried out with sounding rockets and balloon flights which spanned the late 1960s. In 1970 the first satellite devoted to x-ray astronomy was launched (*SAS-A* or *Uhuru*).

During the early 1970s this, and other satellites which followed, conducted the first full sky surveys for x-ray emissions. While rather insensitive compared to the current level of development, these early instruments provided rich returns, including the discovery of new types of objects previously unsuspected. Examples are x-ray emission from binary systems containing a collapsed star (white dwarf, neutron star, and possible black holes) and, in the extragalactic realm, the discovery of intergalactic high-temperature gas associated with clusters of galaxies which was previously unobservable.

Einstein Observatory. In 1978 a major advance came to the field with the launch of the Einstein Observatory (*HEAO 2*). This satellite introduced the use of focusing high-resolution optics to x-ray astronomy. The ensuing increase in sensitivity and the new ability to obtain images have resulted in a qualitative change in the scope of the field. *See* X-RAY TELESCOPE.

Identification of sources. The first nonsolar x-ray source to be identified with a specific celestial object was the Crab Nebula. This was done in 1964 with a rocket-borne experiment by Herbert Friedman and coworkers at the Naval Research Laboratories. Taking advantage of the lunar occultation of the Crab Nebula, the experiment detected the decrease in x-ray counting rate as the source was obscured by the Moon. This allowed a positive identification with the nebula, and also demonstrated that this particular x-ray source was itself extended in size (about 1 minute of arc in diameter). This first successful identification was quickly followed by several more as the experimental techniques were refined and better rocket control systems became available. The identification of Sco X-1 (first detected in the 1962 rocket flight) came in 1966 as a result of an accurate x-ray position and optical studies of objects within the x-ray position. The object is a faint (thirteenth-magnitude) ultraviolet star which is emitting about 1000 times as much power in x-rays as in visible light. This makes Sco X-1 a type of object which is entirely different from previously known objects, and its existence could not have been foreseen on the basis of either optical or radio observations alone. *See* CRAB NEBULA.

Types of sources. X-ray astronomy is traditionally subdivided into subspecialties depending upon the origin of sources—galactic or extragalactic—and the energy range of observations—soft (about 0.25–10 keV) and hard (greater than 10 keV). Presently, most observations have been in the soft band and have been carried out with a series of scientific satellites. These satellite observatories have progressed through the initial discovery phase of the field to systematic surveys of the entire sky, and now to detailed studies of specific objects and classes of objects. Sky surveys have detected and located hundreds of sources at a sensitivity of about 1/10,000 the strength of the brightest source Sco X-1. This is comparable to a survey of the sky in visible light which extends from the brightest stars to about ninth-magnitude stars. There are about 250,000 stars in this range of brightness and only 500 x-ray sources. Figure 1 shows how these x-ray objects are distributed about the sky. The concentration of sources along the equator of this plot corresponds to objects which lie in the Milky Way Galaxy, particularly in the disk, which contains most of the galactic stars and the spiral arms. The other x-ray sources which are plotted in Fig. 1 are associated mainly with extragalactic objects such as individual galaxies, clusters of galaxies, and quasars. *See* GALAXY.

Galactic sources. The first two identified nonsolar x-ray sources (Sco X-1 and the Crab Nebula) are galactic objects which are unusual and distinct from one another.

Supernova remnants. The Crab is a supernova

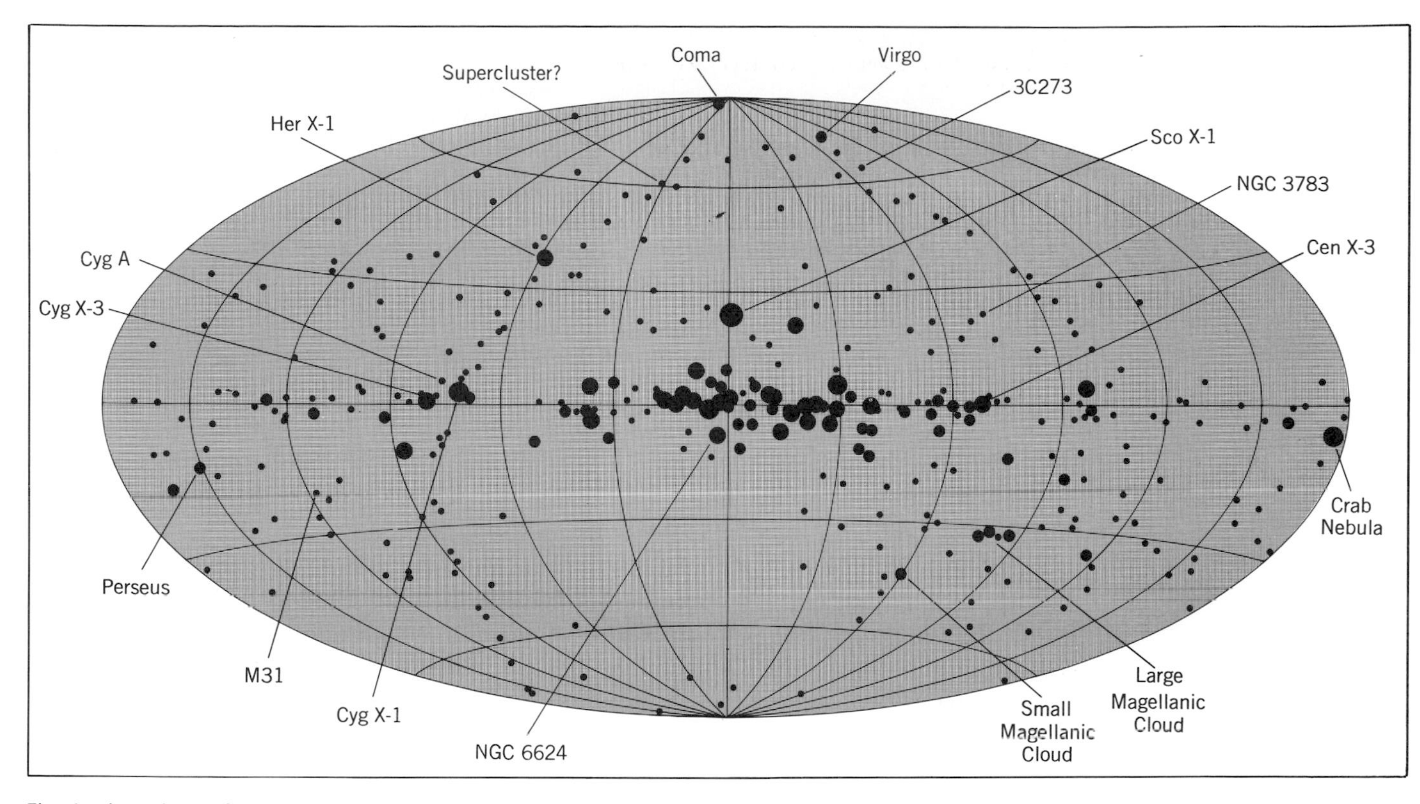

Fig. 1. Locations of about 400 x-ray sources in the Fourth Uhuru Catalog, plotted on an equal-area projection of the celestial sphere. The plane of the Galaxy is the equator of this plot. (*W. Forman et al., The Fourth Uhuru Catalog of x-ray sources, Astrophys. J. Suppl., vol. 38, 1978*)

remnant which is left over from the explosive death of a star. This particular supernova remnant is about 900 years old and contains a rapidly rotating neutron star at its center as well as a nebula consisting of hot gas and energetic particles. The neutron star is the remaining core of the star which exploded, and its rotation causes it to radiate pulses of electromagnetic energy in radio waves, visible light, and even x-rays. Other supernova remnants have been observed and found to emit x-rays. The youngest known supernova remnant in the Galaxy is Cas-A, in the constellation Cassiopeia, and is about 350 years old. Its appearance in x-rays is shown in Fig. 2, where the expanding shell of the explosion can easily be seen. More detailed analysis from the spectrum of x-ray emission shows that heavy elements, formed by nucleosynthesis within the preeruptive star, have been blown off during the explosion and are mixing into the interstellar medium. *See* ELEMENTS AND NUCLIDES, ORIGIN OF; SUPERNOVA.

Compact sources. Studies of galactic objects in x-ray astronomy have led to new insights concerning the later stage of a star's life—old age, death, and in some cases metamorphosis into collapsed objects, such as a white dwarf, neutron star, or black hole. Several types of galactic x-ray sources have been identified. In addition to the supernova remnants described above, there are compact sources which form the majority of galactic emitters. These sources are further classified by their variability, which ranges from time scales as long as years to as short as milliseconds. There are transient sources which suddenly appear and reach high levels of emission rapidly, then slowly fade over periods of weeks to months. This is similar to the behavior of optical novae. Other sources have periodic variations which result from binary systems where an x-ray emitting star orbits around a normal star and is periodically blocked from view. The x-ray star is often a collapsed object—a white dwarf, a neutron star, or even a black hole. In some instances the x-rays are not emitted as a steady flux, but appear pulsed with a period of a few seconds or less. This is similar to the observations of radio pulsars, and is believed to be due to the rotation of a neutron star beaming x-rays toward the Earth. Another type of time behavior is represented by the "bursters." These sources usually emit at some constant level, with occasional short bursts or flares of increased brightness. The bursts are minutes long and occur at random times. Sometimes groups of bursts occur, and often there are long quiescent periods followed by active intervals. These outbursts may well be instabilities in the x-ray emission processes associated with this type of source. *See* BLACK HOLE; NEUTRON STAR; NOVA; PULSAR; STELLAR EVOLUTION; WHITE DWARF STAR.

Galactic sources are also classified according to their locations within the Milky Way. Several sources are associated with globular clusters—collections of many thousands of stars—and are also burster-type sources. Other associations are those clustered toward the galactic center and those mainly associated with the spiral arms. It is widely accepted that many of the galactic x-ray sources are different types of collapsed objects

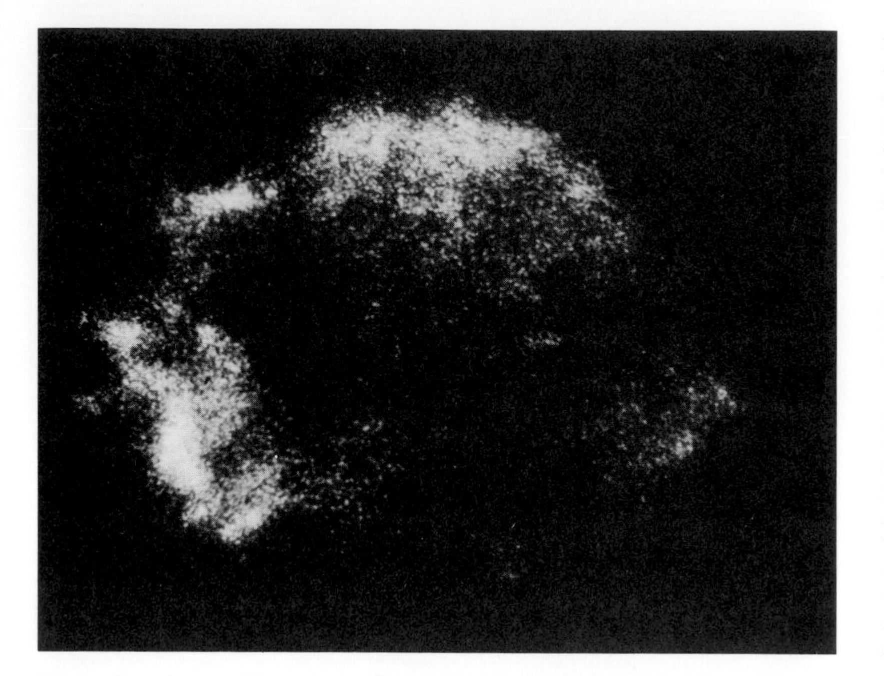

Fig. 2. X-ray image of the supernova remnant Cassiopeia A from the Einstein Observatory. (*S. S. Murray, Harvard-Smithsonian Center for Astrophysics*)

which are accreting material from their surroundings, and that the in-falling matter heats up and radiates in x-rays as it comes close to the surface of the compact stars. The variety in behavior is due to differing local conditions, so that the amount of in-falling material and the path taken vary from class to class. *See* X-RAY STAR.

Normal stars. With the increase in sensitivity obtained with the x-ray telescope of the Einstein Observatory, low levels of x-ray emission have been detected from all types of normal stars. This surprising result has revived interest in models of stellar structure and evolution, which must now be reevaluated. In these sources the x-ray emission is most likely coming from the corona of the stars, with energy being transported from the stellar interior, through the surface, and into the extended atmosphere.

Extragalactic sources. As with galactic x-ray astronomy, virtually all types of objects have been detected as x-ray sources as a result of the increased sensitivity available. These distant objects are radiating enormous quantities of energy in the x-ray band, in some cases more than visible light. Among the more interesting types of objects detected have been apparently normal galaxies, galaxies with active nuclei, radio galaxies, clusters of galaxies, and quasars.

Galaxies and quasars. The nearest spiral galaxy is M31—the Andromeda Galaxy. This was first detected as an x-ray source with the survey satellites of the early 1970s. It has now been observed with the Einstein Observatory, which makes it possible to see a normal galaxy like the Milky Way resolved into individual stellar sources. The distribution and association with galactic features is similar to the Milky Way. There appears to be no emission associated with M31 as a whole, but just the summed emission of the stars. For active nuclei galaxies, this is not the case. X-ray emission from these objects is orders of magnitude in excess of that from normal galaxies, and is clearly associated with the galaxy itself. In most cases the emission comes from the galactic nucleus, and must be confined to a relatively small region on the basis of variability and lack of structure at the current observational limit of a few seconds of arc. Many astrophysicists believe that galactic nuclei are the sites of massive black holes, at least 10^6 to 10^9 times the mass of the Sun, and that the radiation from these objects is due to gravitational energy released by in-falling material. Some scientists speculate that this is a common feature of all galaxies, and that the broad range of properties such as x-ray luminosity reflects the size of the black hole and the availability of in-falling matter. Thus, normal galaxies have dormant nuclei where there is little or no material available, while Seyfert galaxies are quite active, and the quasars represent the extreme case of this mechanism. X-ray observations are particularly relevant in understanding the nature of galactic nuclei, since the high-energy emission is a more direct probe of the basic energy source than either optical or radio. *See* ANDROMEDA GALAXY; QUASARS.

Clusters of galaxies. Clusters of galaxies are another class of extragalactic x-ray sources. These are collections of hundreds to thousands of individual galaxies which form gravitationally bound systems. They are among the largest aggregates of matter in the universe and can be detected at very large distances. This allows them to be studied at earlier epochs in the development of the universe, and they may yield information on the order of formation of galaxies and their evolution. The space between galaxies in such clusters has been found to contain hot (approximately 10^8 K), tenuous gas which glows in x-rays. The mass of the matter is equal to the mass of the visible galaxies; it contains heavy elements which were not present in the primordial mixture from which stars and galaxies are believed to have formed. Observations of some clusters which have multiple subclustering are interpreted as examples where formation is still in progress, and give further support to the concept of a hierarchy of small to large structures. *See* GALAXY, EXTERNAL.

X-ray background. In addition to observing a bright source which later was identified as Sco X-1, the 1962 rocket experiment also discovered a uniform background of x-ray emission around the sky. This was confirmed by following experiments and found to contain at least two components. At low energies the background is dominated by galactic emission, while above about 1 keV the radiation comes from beyond the Galaxy. For this harder component, there are two possible sources. Either there is a truly diffuse source of emission such as hot gas or very-high-energy particles, or the background is the superposition of many discrete sources associated with various extragalactic objects. Measurements of the spectrum of the background indicate a smooth shape. This can be well approximated by the combined spectra of several types of extragalactic sources or by the spectrum expected from a very hot gas. Studies of source counts have been carried out since the first all-sky survey found a large number of extragalactic objects. Depending upon the extrapolation models used, individual sources account for at least 30% and perhaps all of the background observed. This result limits the amount of hot gas that can be present in the universe to less than the critical amount necessary for closure. That is, there is not enough matter presently known to exist in the universe to allow gravitational forces to overcome the initial expansion of the big bang and cause a later contraction. If such matter exists, it must be in some unobservable form such as dark stars or isolated black holes. *See* COSMOLOGY; INTERSTELLAR MATTER. [STEPHEN S. MURRAY]

X-ray star

A source of x-rays from outside the solar system. Although the Sun emits some x-radiation, it cannot be called an x-ray star because its energy output in the optical region of the spectrum vastly exceeds that in the x-ray region. Observations of the 100 or so known galactic x-ray stars show that these objects have just the reverse characteristic: the overwhelming amount of their energy is given off in the x-ray band, while in visible light they are inconspicuous. *See* STAR; SUN; X-RAY ASTRONOMY.

Location. The majority of galactic x-ray stars are located in a narrow layer coincident with the plane of the Milky Way Galaxy. Because this coincides with the spatial distribution of bright young stars in the Galaxy, there is at least circumstantial evidence that many x-ray stars may be associated with such young stars, or at least may have had similar stages of early evolution. However, a few x-ray stars have also been observed in the spherical halo of the Galaxy, far from the plane, and in globular clusters; both of these regions are known to contain the oldest stars in the Galaxy. It is uncertain whether the x-ray stars in the plane and the halo are closely related, or a different phenomenon. *See* GALAXY; STELLAR EVOLUTION.

Binary star systems. The study of both x-ray and optical data from the region of x-ray stars has led to the conclusion that most, or perhaps even all, such stars are members of binary star systems, consisting of one relatively normal visible star and one subluminous compact star, in a gravitationally bound orbit about each other. In the case of the x-ray data, in several sources eclipses are observed; that is, the x-ray intensity regularly vanishes or is drastically reduced, later to reappear, in a strictly periodic cycle. These eclipses are interpreted as the passage of the normal star through the line of sight between the Earth and the x-ray star.

In cases where the position of the x-ray source has been very accurately measured, it is sometimes possible to locate a visible star associated with the system. About 10 such optical counterparts are known, although all are too faint to be seen with the unaided eye. In most cases, studies of the visible component show evidence independent of the x-ray data for the binary nature of these systems. The spectral lines of the visible star indicate a periodic shift in wavelength from the Doppler effect, as the two stars move about their common center of mass. Often the x-ray star heats or gravitationally distorts the normal star sufficiently that the side facing the x-rays and the side facing away differ substantially in brightness. The effect observed at Earth is then a cyclical change in the visible star's brightness, as the x-ray star moves in its orbit, with the period of the brightness variation equal to the orbital period of the star. *See* BINARY STAR.

Nature of x-ray stars. Although the geometry of the orbits of several x-ray binary systems is now quite accurately measured, and the visible components studied in some detail, the exact nature of the x-ray stars themselves is a matter of controversy. Most theories assume these objects to be highly evolved and quite compact. The compact star accretes gaseous matter from its nearby visible companion. This infalling matter reaches a very high temperature before it is stopped at or near the surface of the x-ray star, and the very hot gas then emits x-rays. In effect, a machine is at work to convert gravitational energy into x-radiation.

Support is lent to this theoretical picture by much of the observational data. In particular, many x-ray stars are seen to rapidly fluctuate in intensity on a time scale of a fraction of a second, implying that the emitting region is quite compact. However, these data cannot help to distinguish the precise evolutionary state of the compact star. One known type of highly evolved compact object which may be responsible for the x-ray emission is the white dwarf. Such stars have radii about equal to that of the Earth, and are numerous and well studied in the solar neighborhood. However, no x-ray system is close enough to Earth that the faint visible radiation from a white dwarf would be directly detectable. *See* WHITE DWARF STAR.

Some x-ray stars are known to emit extremely regular and rapid bursts of x-rays, in an analogous fashion to the radio pulsars. The pulsars are generally believed to be neutron stars: stellar remnants collapsed even more than white dwarfs, until they have radii of only several miles, but still possess a mass equivalent to that of the Sun. It seems likely that most of the x-ray pulsars are also neutron stars. *See* PULSAR.

Finally, the possibility exists that some of the x-ray stars may be black holes, the ultimate stage of stellar collapse. In these as yet hypothetical stars, matter is compressed into such a dense configuration that the intense gravitational field prevents even light from escaping. They could evidence themselves only indirectly, for example, in a binary system where the hot accreting matter could radiate x-rays just prior to falling into the black hole. *See* BLACK HOLE; GRAVITATIONAL COLLAPSE.

It is difficult to devise observational tests that will make it possible to choose unambiguously among these possible candidates for x-ray stars. One possibility may lie in measuring the mass of the compact object, by carefully noting its effects on the visible star. There are theoretical limits on the maximum masses of stable white dwarfs and neutron stars, of about 1.2 and 3 times the mass of the Sun, respectively. X-ray stars which can definitely be proven more massive than either of these limits may well provide the first observational evidence for the elusive black holes. It may emerge that all three types of compact stars—white dwarfs, neutron stars, and black holes—are candidates for different x-ray stars.

[JOSEPH SILK; BRUCE MARGON]

X-ray telescope

An instrument designed to collect and detect x-rays emitted from a source outside the Earth's atmosphere and to resolve the x-rays into an image. Absorption by the atmosphere requires that x-ray telescopes be carried to high altitudes. Balloons are used for detection systems designed for higher-energy (harder) x-ray observations, whereas rockets and satellites are required for softer x-ray detectors. *See* X-RAY ASTRONOMY.

Image formation. An image-forming telescopic lens for x-ray wavelengths can be based on the phenomenon of total external reflection at a surface where the index of refraction changes. In the case of x-rays, the index of refraction in matter is slightly less than unity. By application of Snell's laws, the condition for total external reflection is that the radiation be incident at small grazing angles, less than a critical angle of about 1°, to the reflecting surface. The value of the critical angle depends on the wavelength of the radiation and the material used. As the wavelength decreases (higher energy), the grazing angle required is smaller; as the atomic number (Z) of the material increases, the grazing angle for a given wavelength increases. The detailed reflectivity as a function of energy is complicated by x-ray absorption edges in the material.

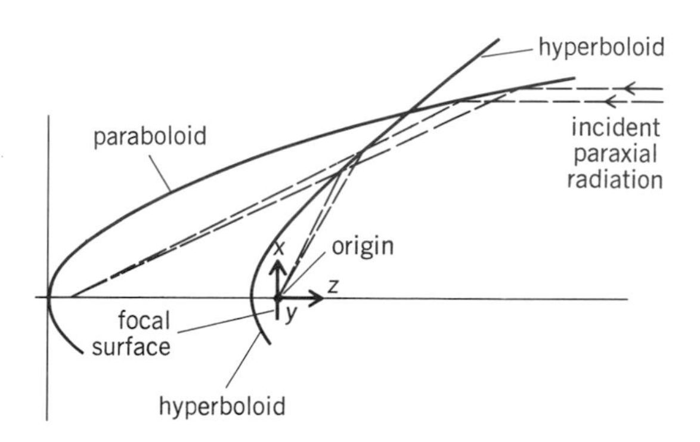

Fig. 1. Wolter type I x-ray telescope. Two surfaces of revolution (paraboloid and hyperboloid) reflect the incident x-rays to a common focus. The horizontal scale is greatly compressed.

Based on these properties, x-ray mirrors have been constructed which focus an image in two dimensions. Various configurations of surfaces are possible; Fig. 1 shows one which has been built and used successfully for x-ray astronomy on the Einstein Observatory (*HEAO 2*). Two reflective surfaces of revolution are used to produce a high-quality image. These mirrors are manufactured from fused quartz, which is coated with nickel after being shaped and polished. The surfaces are extremely smooth in order to focus well, and the shape of the mirror must also be within narrow tolerances. The Einstein mirrors are the largest ever built; they are about 0.6 m in diameter and 1.2 m long, and the total area of polished surface is about the same as that of the 5-m telescope at Palomar Mountain; yet the effective area of the telescope is only about 1000 cm², because of the grazing-angle geometry. *See* TELESCOPE.

Image detection. The telescope mirrors focus x-rays, producing an image in two dimensions in the same manner as the lenses of optical telescopes result in images of the sky. Suitable devices are required to detect and record these images, completing the functional requirements of an astronomical telescope. Various types of x-ray detectors have been developed for this purpose. These are position-sensitive devices which in effect are electronic cameras suitable for x-ray wavelengths.

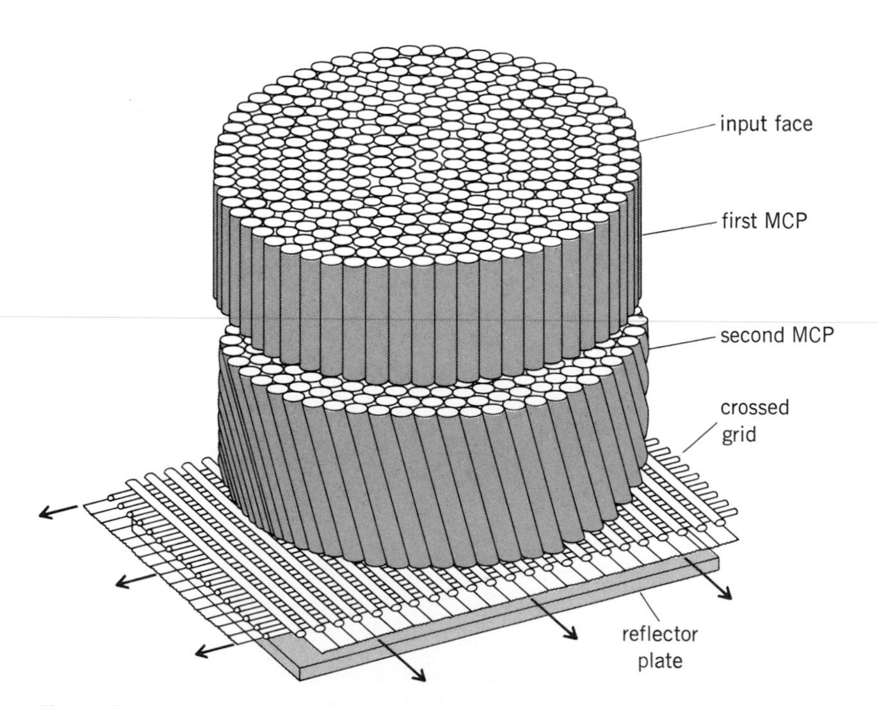

Fig. 2. Blow-up (not to scale) of the major components of a high-resolution imaging detector for an x-ray telescope. Microchannel plates (MCP) detect incident radiation and amplify the signal for position determination by the grid of crossed wires below.

The high angular resolution of the grazing-incidence telescope requires a camera which has correspondingly good spatial resolution. One type of detector uses microchannel plates (MCP; Fig. 2) and yields about 20 μm resolution for x-rays in the soft energy band (about 200–4000 eV). The MCP is an array of small hollow tubes (about 15 μm in diameter) or channels which are processed to have high secondary electron yield from their inner walls. A single x-ray photon which strikes the surface of a channel produces a free electron. The electric field produced by placing a high voltage across the MCP accelerates this electron, which collides with the wall of the tube to produce more electrons. This results in a cascade of electrons through the channel, multiplying in number until a sufficient signal is produced to be recorded electronically, giving the location of the event.

Another detector used with x-ray telescopes is a form of gas counter in which x-rays are photoelectrically absorbed, yielding an electron which is detected by the ionization it produces in the gas. By operation of the counter in the proportional mode and use of planes of wires to localize the electrical signals, the position and amplitude of each event are recorded. These detectors generally have lower spatial resolution (about 1 mm) than do MCP detectors, but they can be made larger in size and provide some energy measurement which is not possible in the MCP device.

[STEPHEN S. MURRAY]

Year

Any of several units of time based on the revolution of Earth around the Sun. The tropical year, to which the calendar is adjusted, is the period that is required for the mean longitude of the Sun to increase 360°. Its duration is approximately 365.24220 mean solar days. It is also the period after which the seasons repeat themselves.

The sidereal year, 365.25636 mean solar days in duration, is the average period of revolution of Earth with respect to a fixed direction in space.

The anomalistic year, 365.25964 mean solar days in duration, is the average interval between successive closest approaches of Earth to the Sun. *See* TIME. [GERALD M. CLEMENCE]

Zenith

The point directly overhead in the sky. The astronomical zenith, which is that usually meant, is the upper intersection of a plumb line with the celestial sphere. The zenith distance of a celestial object is its angular distance from the astronomical zenith and is identical with the complement of its altitude. The geocentric zenith is the upper intersection with the celestial sphere of an imagined line through the center of Earth and the observer. The point diametrically opposite the zenith is called the nadir. *See* ASTRONOMICAL COORDINATE SYSTEMS. [GERALD M. CLEMENCE]

Zodiac

A band of the sky extending 8° on each side of the ecliptic, within which the Moon and principal planets remain. It is divided into 12 conventional signs, each containing 30° of celestial longitude. These signs are named Aries, Taurus, Gemini, Cancer, Leo, Virgo, Libra, Scorpio, Sagittarius, Capricornus, Aquarius, and Pisces. The names are identical with the names of the constellations through which the Sun moves, but the actual constellations are not each 30° in extent, nor does the Sun actually enter the constellation of Aries at the vernal equinox, although the vernal equinox is sometimes conventionally called the first point of Aries. The signs of the zodiac are without practical astronomical significance.

[GERALD M. CLEMENCE]

Zodiacal light

A diffuse band of luminosity occasionally visible on the ecliptic. It is sunlight diffracted and reflected by dust particles in the solar system within and beyond the orbit of Earth. Zodiacal light is best seen after evening twilight or before morning twilight when the ecliptic is at right angles to the horizon. In the Northern Hemisphere this is in the evening during springtime, and in the morning during autumn.

The light steadily increases from about 170–30° away from the Sun.

Fraunhofer corona. Sunlight scattered within Earth's atmosphere normally prevents observation of zodiacal light at elongations (angular distance from the Sun) of less than 30°. During a solar eclipse, however, the scattered light is reduced and, at positions near the Sun, the zodiacal light may be intense enough to observe.

Part of solar coronal light is due to diffraction by zodiacal particles. Superposed on the diffracted light is light scattered by electrons in the solar corona itself. The Doppler effect of the rapidly moving electrons blurs the solar Fraunhofer absorption lines, leaving a continuum that dilutes the Fraunhofer lines in the diffracted light. This dilution allows the two components of the corona to be separated. The intensity due to the zodiacal particles is well represented by an extrapolation of the zodiacal light observed in elongations between 170 and 30°. A constant space density of particles, at distances greater than 0.1 astronomical units (AU) is sufficient to explain the observations. (Presumably such matter is vaporized by the Sun at lesser distances.) *See* SOLAR CORONA.

Gegenschein. At elongations of 180°, zodiacal light shows a marked increase in brightness. This patch of light, ordinarily about 10° in diameter, is called the Gegenschein or counterglow. It is attributed to the increased efficiency of reflections at large phase angles by diffusely reflecting particles.

Poynting-Robertson effect. A relativistic study of the equations of motion for a body moving in a radiation field was made by H. P. Robertson following an approximate solution of this motion problem by J. H. Poynting. If absorption and isotropic reemission of radiation occur, the body experiences a resisting force proportional to its velocity. For a body in orbit about the Sun, the semimajor axis must decrease for the orbit to be commensurate with orbital energy. Eventually, then, the body spirals into the Sun. Because the radiant energy absorbed is proportional to the area of the body and because the resisting force is inversely proportional to the mass, small bodies are influenced more than large. H. C. van der Hulst showed that the zodiacal particles are generally smaller than 0.35 mm in radius. Particles in this size range and with relatively small initial orbits (1–10 AU) are thrown into the Sun within a few million years and at a rate of about 1 ton per second. F. L. Whipple estimated that meteoric material is introduced into the solar system by cometary disintegration at a rate of about 30 tons per second. This probably represents the source of supply for the zodiacal particles. *See* SUN.

[RICHARD E. MC CROSKY]

List of Contributors

List of Contributors

A

Abell, Prof. George O. *Department of Astronomy, University of California, Los Angeles.* COSMOLOGY; LOCAL GROUP.

Adler, Dr. Ronald J. *Lockheed Palo Alto Research Laboratories, Palo Alto, CA.* RELATIVITY—in part.

Akasofu, Prof. S.-I. *University of Alaska.* MAGNETOSPHERE.

Aller, Prof. Lawrence H. *Department of Astronomy, University of California, Los Angeles.* ASTRONOMICAL SPECTROSCOPY; ASTROPHYSICS; MAGELLANIC CLOUDS; MESSIER CATALOG; NEBULA; other articles.

Anderson, Dr. John D. *Jet Propulsion Laboratory, California Institute of Technology.* ASTRONOMICAL UNIT

B

Babcock, Dr. Horace W. *Director, Hale Observatories, Pasadena, CA.* SOLAR MAGNETIC FIELD; STELLAR MAGNETIC FIELD.

Bahcall, Dr. John N. *Institute for Advanced Studies, Princeton, NJ.* SOLAR NEUTRINOS.

Beatty, J. Kelly. *"Sky and Telescope," Sky Publishing Corporation, Cambridge, MA.* MARS; PLANET; SATELLITE.

Bok, Dr. Bart J. *Professor Emeritus, Department of Astronomy, University of Arizona.* GALAXY; GLOBULE.

Bolz, Dr. Ray E. *Leonard S. Case Professor and Dean of Engineering, Case Western Reserve University.* NUTATION.

Brandt, Dr. John C. *Chief, Laboratory for Astronomy and Solar Physics, NASA Goddard Space Flight Center, Greenbelt, MD.* COMET; HALLEY'S COMET.

Brown, Dr. Robert L. *National Radio Astronomy Observatory, Charlottesville, VA.* RADIO ASTRONOMY; RADIO TELESCOPE.

Burke, Dr. James D. *Manager, Advanced Technical Studies, Jet Propulsion Laboratory, California Institute of Technology.* MOON.

C

Cameron, Dr. A. G. W. *Associate Director for Planetary Sciences, Center for Astrophysics, Harvard College Observatory.* PLANETARY PHYSICS; SOLAR SYSTEM.

Caughlan, Prof. Georgeanne R. *Department of Physics, Montana State University.* CARBON-NITROGEN-OXYGEN CYCLES; ELEMENTS AND NUCLIDES, ORIGIN OF; PROTON-PROTON CHAIN.

Chapman, Dr. Clark R. *Planetary Science Institute, Tucson, AZ.* ASTEROID.

Chapman, Dr. Robert D. *Section Head, Laboratory for Solar Physics and Astrophysics, NASA Goddard Space Flight Center, Beltsville, MD.* OPTICAL TELESCOPE—in part.

Clemence, Dr. Gerald M. *Deceased; formerly, Observatory, Yale University.* ASTRONOMICAL COORDINATE SYSTEMS; CHRONOMETER; EPHEMERIS; TIME—in part; ZENITH; other articles.

Cleminshaw, Dr. Clarence H. *Director Emeritus, Griffith Observatory, Los Angeles, CA.* ECLIPSE.

Code, Dr. Arthur D. *Department of Astronomy, University of Wisconsin.* ULTRAVIOLET ASTRONOMY.

Corwin, Harold G., Jr. *Department of Astronomy, Royal Observatory, Edinburgh, Scotland.* SUPERCLUSTERS.

Cox, Dr. Arthur N. *Los Alamos Scientific Laboratory, Los Alamos, NM.* CEPHEIDS.

Cromwell, Dr. Richard H. *Steward Observatory, University of Arizona.* IMAGE TUBE.

Cutler, Dr. Leonard S. *Director, Physical Research Laboratory, Hewlett-Packard Company, Palo Alto, CA.* ATOMIC CLOCK.

D

Danby, Dr. J. M. A. *Department of Mathematics, North Carolina State University.* GRAVITATION.

Dent, Dr. William. *Department of Physics and Astronomy, University of Massachusetts.* QUASARS.

Dickel, Dr. John R. *Department of Astronomy, University of Illinois.* RADIO SOURCES.

Duncombe, Dr. Raynor L. *Department of Aerospace Engineering, University of Texas, Austin.* APHELION; AREAL VELOCITY; KEPLER'S EQUATION; PERTURBATION; RETROGRADE MOTION; other articles.

E

English, Prof. Van H. *Department of Geography, Dartmouth College.* EQUINOX; TIME—in part.

Evans, Dr. John W. *Director, Sacramento Peak Observatory, Air Force Cambridge Research Laboratories, Sunspot, NM.* CHROMOSPHERE; CORONAGRAPH; SOLAR CONSTANT; SPECTROHELIOSCOPE—validated; SUNSPOT; other articles.

Evenson, Dr. Paul. *Fermi Institute, University of Chicago*. COSMIC RAYS.

F

Foster, Dr. Robert J. *Department of Physics, San Jose State University*. EARTH.

Frey, Herbert. *Geophysics Branch, NASA Goddard Space Flight Center, Greenbelt, MD*. UNIVERSE.

G

Gossner, Simone Daro. *U.S. Naval Observatory, Washington, DC*. OCCULTATION; TRANSIT.

Green, Dr. Louis C. *Department of Astronomy, Haverford College*. HUBBLE CONSTANT.

Greenstein, Prof. Jesse L. *Hale Observatories, Pasadena, CA*. ALDEBARAN; BLACK DWARF; CARBON STAR; FLUORESCENCE ASTRONOMY; STELLAR MAGNITUDE; other articles.

Grindlay, Dr. Jonathan E. *Center for Astrophysics, Cambridge, MA*. SATELLITE ASTRONOMY—in part.

Groth, Dr. Edward J. *Department of Physics, Princeton University*. SATELLITE ASTRONOMY—in part.

Groves, Dr. Gordon W. *Instituto de Geofisica, Torre de Ciencias, Ciudad Universitaria, Mexico*. TIDE.

H

Hayes, Dr. Donald S. *Kitt Peak National Observatory, Tucson, AZ*. HERTZSPRUNG-RUSSELL DIAGRAM.

Hendry, Dr. Elaine M. *Department of Physics and Astronomy, Georgia State University*. JUPITER.

Herget, Dr. Paul. *Cincinnati Observatory, University of Cincinnati*. EROS.

Herzberger, Dr. Max. *Consulting Professor, Department of Physics, Louisiana State University*. SCHMIDT CAMERA.

Hogg, Dr. Helen S. *David Dunlap Observatory, University of Toronto, Canada*. HYADES; PLEIADES; STAR CLOUDS; STAR CLUSTERS.

Hundhausen, Dr. A. J. *High Altitude Observatory, National Center for Atmospheric Research, Boulder, CO*. SOLAR WIND.

Hynek, Dr. J. Allen. *Department of Astronomy, Northwestern University*. ASTRONOMY; PRECESSION OF EQUINOXES.

K

King, Dr. Elbert A. *Department of Geology, University of Houston*. TEKTITE.

Klock, Dr. Benny L. *Director, Six Inch Transit Circle Division, U.S. Naval Observatory*. ASTROMETRY; OPTICAL TELESCOPE—both in part.

Kohman, Dr. Truman P. *Department of Chemistry, Carnegie-Mellon Institute*. COSMOCHEMISTRY.

Kuiper, Prof. Gerard P. *Lunar and Planetary Laboratory, University of Arizona*. APOGEE; PERIGEE; PHASE.

L

Lamb, Dr. D. Q. *Department of Physics, University of Illinois, Urbana*. GAMMA-RAY ASTRONOMY—in part.

Landshoff, Dr. Rolf K. M. *Consulting Scientist, Lockheed Missiles and Space Company, Palo Alto, CA*. COSMIC ELECTRODYNAMICS.

Latham, Dr. David W. *Center for Astrophysics, Cambridge, MA*. ASTRONOMICAL PHOTOGRAPHY.

Layzer, Prof. David. *Harvard College Observatory*. EINSTEIN SHIFT; RED SHIFT.

Liller, Dr. William. *Department of Astronomy, Harvard University*. COAL SACK.

Lunetta, Donald M. *Curator, Charles C. Gates Planetarium, Denver Museum of Natural History*. PLANETARIUM.

M

McCarthy, Dr. Dennis D. *Time Service Division, U.S. Naval Observatory, Washington, DC*. EARTH ROTATION AND ORBITAL MOTION; LATITUDE AND LONGITUDE.

McCrosky, Dr. Richard E. *Smithsonian Astrophysical Observatory, Cambridge, MA*. ENCKE'S COMET; METEOR; ZODIACAL LIGHT.

McMath, Prof. Robert R. *Deceased; formerly, Director, McMath-Hulbert Observatory, University of Michigan*. SPECTROHELIOSCOPE.

Manuel, Dr. Oliver K. *Department of Chemistry, University of Missouri*. METEORITE.

Margon, Dr. Bruce. *Space Sciences Laboratory, University of California, Berkeley*. SS 433; X-RAY STAR—coauthored.

Markowitz, Dr. William. *Department of Physics, Nova University; and Editor, "Geophysical Surveys."* ATOMIC TIME; CHRONOGRAPH; DYNAMICAL TIME; EQUATION OF TIME; TIME-INTERVAL MEASUREMENT.

Martin, Dr. P. G. *Department of Astronomy, University of Toronto, Canada*. INTERPLANETARY MATTER; INTERSTELLAR MATTER.

Mayall, Margaret W. *American Association of Variable Star Observers, Cambridge, MA*. MIRA; VARIABLE STAR.

Mayall, Robert N. *Director, Planning and Research Associates, Boston, MA*. SUNDIAL.

Montmerle, Dr. T. *Commissariat à l'Énergie Atomique, Centre d'Études Nucléaires de Saclay, Gif-sur-Yvette, France*. GAMMA-RAY ASTRONOMY—in part.

Moody, Capt. Alton B. *Navigation Consultant, La Jolla, CA*. CELESTIAL NAVIGATION.

Morrison, Dr. David. *Institute for Astronomy, University of Hawaii*. MERCURY.

Mulholland, Dr. J. Derral. *Department of Astronomy, University of Texas, Austin*. LUNAR LASER RANGING.

Muller, Dr. Richard A. *Lawrence Berkeley Laboratory, University of California, Berkeley*. COSMIC MICROWAVE RADIATION.

Murray, Stephen S. *Harvard College Observatory, Cambridge, MA*. X-RAY ASTRONOMY; X-RAY TELESCOPE.

O

Owen, Dr. Tobias C. *Department of Earth and Space Sciences, State University of New York, Stony Brook*. NEPTUNE; PLUTO; SATURN; URANUS.

P

Page, Dr. Thornton. *National Aeronautics and Space Administration, Houston, TX*. ASTRONOMICAL OBSERVATORY.

Pasachoff, Dr. Jay M. *Hopkins Observatory, Williams College*. BIG BANG THEORY.

Peters, Prof. Philip C. *Department of Physics, University of Washington*. BLACK HOLE.

Plavec, Dr. Mirek. *Department of Astronomy, University of California, Los Angeles*. BINARY STAR; STELLAR EVOLUTION.

Pollack, Dr. James B. *Ames Research Center, Moffett Field, CA*. VENUS.

Popper, Dr. Daniel M. *Department of Astronomy, University of California, Los Angeles*. MASS-LUMINOSITY RELATION.

R

Rockett, Frank H. *Engineering Consultant, Charlottesville, VA*. CARDINAL POINTS; CELESTIAL SPHERE; FOUCAULT PENDULUM; HOROLOGY.

S

Sandage, Dr. Allan. *Hale Observatories, Pasadena, CA.* ANDROMEDA GALAXY.

Schwartz, Dr. H. M. *Department of Physics, University of Arkansas.* RELATIVITY—in part.

Seidelmann, Dr. P. K. *Director, Nautical Almanac Office, U.S. Naval Observatory, Washington, DC.* ALMANAC; ROCHE LIMIT.

Shipman, Dr. Harry L. *Department of Physics, University of Delaware.* ACTIVE GALACTIC NUCLEI; GRAVITATIONAL LENS; NEUTRON STAR; SUPERMASSIVE STARS.

Silk, Dr. Joseph. *Department of Astronomy, University of California, Berkeley.* GALAXY, EXTERNAL; X-RAY STAR—coauthored.

Sinton, Dr. William M. *Department of Astronomy, University of Hawaii.* TELESCOPE.

Smith, Prof. Harlan J. *Department of Astronomy, University of Texas, Austin.* ABERRATION.

Soberman, Dr. Robert K. *Manager, Meteor and Planetary Physics, General Electric Company, Valley Forge Space Center, Philadelphia, PA.* MICROMETEORITE.

Sonett, Charles P. *Lunar and Planetary Laboratory, Department of Planetary Sciences, University of Arizona.* SPACE PROBE.

Stroke, Dr. George W. *Department of Electrical Sciences, and Head, Electro-Optical Sciences Center, State University of New York, Stony Brook.* DOPPLER EFFECT.

Suess, Dr. Hans E. *Department of Chemistry, University of California, San Diego.* ELEMENTS, COSMIC ABUNDANCE OF.

T

Taylor, Dr. Joseph H., Jr. *Department of Physics and Astronomy, University of Massachusetts.* PULSAR.

Tedesco, Dr. Edward F. *Lunar and Planetary Laboratory, University of Arizona.* CERES; TROJAN ASTEROIDS.

Thorne, Dr. Kip S. *Kellogg Radiation Laboratory, California Institute of Technology.* GRAVITATIONAL COLLAPSE.

Tousey, Dr. Richard. *U.S. Naval Research Laboratory, Washington, DC.* ROCKET ASTRONOMY.

Trimble, Dr. Virginia. *Department of Physics, University of California, Irvine.* NOVA; SUPERNOVA.

Turner, Dr. Edwin L. *Princeton University Observatory.* GALAXY, EXTERNAL—coauthored.

Turner, Joyce B. *Institute for Advanced Study, Princeton, NJ.* GALAXY, EXTERNAL—coauthored.

V

van de Kamp, Dr. Peter. *Sproul Observatory, Swarthmore College.* ASTROMETRY—in part; CELESTIAL MECHANICS; PARALLAX.

Vaucouleurs, Prof. Gerard de. *Department of Astronomy, University of Texas, Austin.* MIDNIGHT SUN.

Veverka, Prof. Joseph. *Laboratory for Planetary Studies, Cornell University.* ALBEDO.

W

Wood, Dr. Frank Bradshaw. *Department of Physics and Astronomy, University of Florida.* ECLIPSING VARIABLE STARS; LIGHT CURVES.

Y

Yu, Dr. Ching-Sung. *Professor Emeritus of Astronomy, Hood College.* AQUARIUS; CAPRICORNUS; PERSEUS; SCORPIUS; URSA MAJOR; other articles.

Z

Zirin, Dr. Harold. *Department of Physics, California Institute of Technology.* SUN.

Index

Index

Asterisks indicate page references to article titles.

H

I

T

U

V

W